A TEXTBOOK OF

ENGINEERING DRAWING

A TEXTBOOK OF
ENGINEERING DRAWING

Prof. P.J. SHAH
***M.E.* (*Mech.*)**
Formerly Professor of Mechanical Engineering
L.D. College of Engineering
Ahmedabad

Assisted by

Parag P. Shah
[B.E. Mechanical]

Keyur P. Shah
[B.E. Mechanical]

Umang P. Shah
[Graphic Designer]

With more than 1158 Diagrams, 606 Solved Problems and 549 Exercises

S Chand And Company Limited
(ISO 9001 Certified Company)

S Chand And Company Limited
(ISO 9001 Certified Company)
Head Office: Block B-1, House No. D-1, Ground Floor, Mohan Co-operative Industrial Estate, New Delhi – 110 044 | Phone: 011-66672000
Registered Office: A-27, 2nd Floor, Mohan Co-operative Industrial Estate, New Delhi – 110 044 Phone: 011-49731800
www.**schandpublishing.com**; e-mail: **info@schandpublishing.com**

Branches

Chennai : Ph: 23632120; chennai@schandpublishing.com
Guwahati : Ph: 2738811, 2735640; guwahati@schandpublishing.com
Hyderabad : Ph: 40186018; hyderabad@schandpublishing.com
Jalandhar : Ph: 4645630; jalandhar@schandpublishing.com
Kolkata : Ph: 23357458, 23353914; kolkata@schandpublishing.com
Lucknow : Ph: 4003633; lucknow@schandpublishing.com
Mumbai : Ph: 25000297; mumbai@schandpublishing.com
Patna : Ph: 2260011; patna@schandpublishing.com

First Edition 2008
Reprints 2009, 2010, 2012
Revised Multicolour Edition 2013 Reprint 2021 (Twice)
Reprint 2022 (Twice)

ISBN : 978-81-219-4182-2 **Product Code:** H3EDR62EDES10ENAB13R

PRINTED IN INDIA

By Vikas Publishing House Private Limited, Plot 20/4, Site-IV, Industrial Area Sahibabad, Ghaziabad – 201 010 and Published by S Chand And Company Limited, A-27, 2nd Floor, Mohan Co-operative Industrial Estate, New Delhi – 110 044.

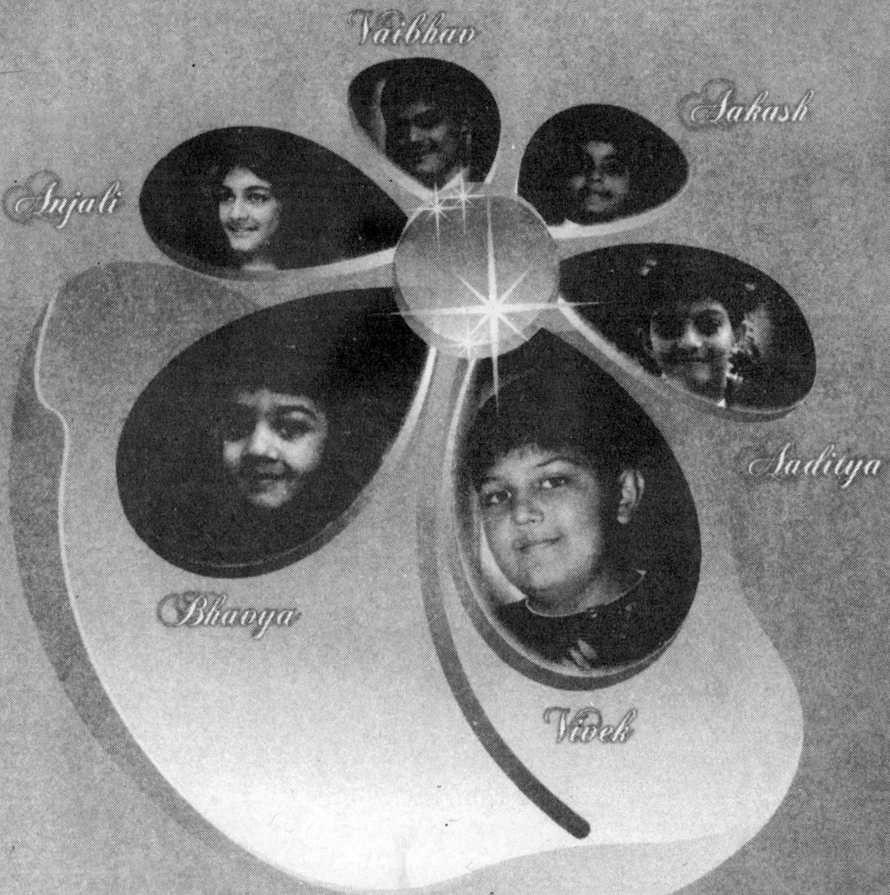

Vaibhav

Aakash

Anjali

Aaditya

Bhavya

Vivek

Dedicated To
My Grand Children

FOREWORD

There are several books on Engineering Drawing with variety of treatment and presentation. Each book naturally bears the imprint of the style of its author. Prof. P. J. Shah, having fairly long experience of teaching this subject, has tried to develop the subject-matter based on his comprehension of the genuine difficulties of the students and perception that here is a subject for which a live interest needs to be cultivated to synthesise the art and craft, grammar and composition, of the Engineer's language.

One may fear that with the advent of computer graphics as well as computer aided design and drawing, the study of Engineering Drawing in this manner might become irrelevant and unnecessary. But it need not be so. This is particularly so because software development in that area will also require a thorough understanding of all aspects of Engineering Drawing. Maybe, future editions of this book will have to incorporate these aspects also.

With standards defining the practice of art and presentation, the author has tried his best to deal with various topics skilfully. I am sure that the book will find ready acceptance in Engineering Colleges and Polytechnics at the level of the basic course in Engineering Drawing. I congratulate the author for this commendable effort and remarkable success.

Prof. N.R. Dave
(M. Tech.)
Director of Technical Education
Government of Gujarat
GANDHINAGAR

PREFACE

This publication deals with language of engineers, i.e., Engineering Drawing. It is based on the syllabi of degree and diploma courses of Indian Universities and the Technical Examination Boards of various states.

In this textbook, I have tried to lay emphasis on the fundamental principles. The test and numerous illustrative pictorial diagrams are presented in an easy-to-digest form. In fact, I have tried to cultivate the three dimensional imagination of the reader. Difficult topics, like projections of straight lines, planes, solids, sections of solids, development of surfaces of solids, penetration, orthographic reading and isometric drawings are made simpler by three dimensional illustrations. A unique notation system is framed and adopted by me to make understanding easier.

Many exercises are given at the end of each chapter with solutions given at the back of the book. Each chapter contains an adequate number of solved and unsolved problems to imbue self-confidence in the reader.

Readers may use this book as a supplement to classroom teaching or for self study. The structure of this textbook is framed to serve both the average and brilliant students.

I express my thanks to various universities of India for using questions, knowingly or unknowingly, from their question papers.

Extracts from I.S. SP-46-1988, which are to be followed by schools, polytechnics and colleges in the teaching of subject "Engineering Drawing" are given in this textbook with the permission of ISI authority. I hereby sincerely thank the ISI authorities.

Diagrams are prepared in accordance with I.S.I. For dimensioning the latest method is followed and Metric units are used.

Important chapter on "Computer Aided Drafing (CAD)" is added, since it is the need of today.

My thanks are also due to the **'ROTRING COMPANY OF INDIA'** for allowing me to use their literature on drafting equipments.

I must express a deep snese of gratitude to Prof. N.R. Dave, Director of Technical Education, Gujarat, for writing the foreword to this book.

The author is greatly indebted to Management and the Editorial department of S. Chand & Company Ltd. for all help and support in the publication of this book.

Corrections, suggestions, modifications, etc. to improve the book will be gratefully received by me. Remember there is always a scope for improvement.

Prof. P.J. Shah

CONTENTS

CHAPTER - 1

Drafting Equipments

Engineering drawing is a language of all persons involved in engineering activity. Engineering ideas are recorded by preparing drawings and execution of work is also carried out on the basis of drawings. Communication in engineering field is done by drawings. Like music drawing is a universal language.

To prepare engineering drawing, special drawing instruments are required. It is advisable to purchase simple and good quality instruments rather than many ordinary instruments. All those who are connected with engineering activity should know the use of drawing instruments, handling of instruments and proper maintenance of drawing instruments. A good draftsman is able to prepare correct, accurate and decent drawings in the least possible time with the proper use of instruments.

The following drawing equipments and materials are required in the preparation of drawing. (See Fig. 1.1)

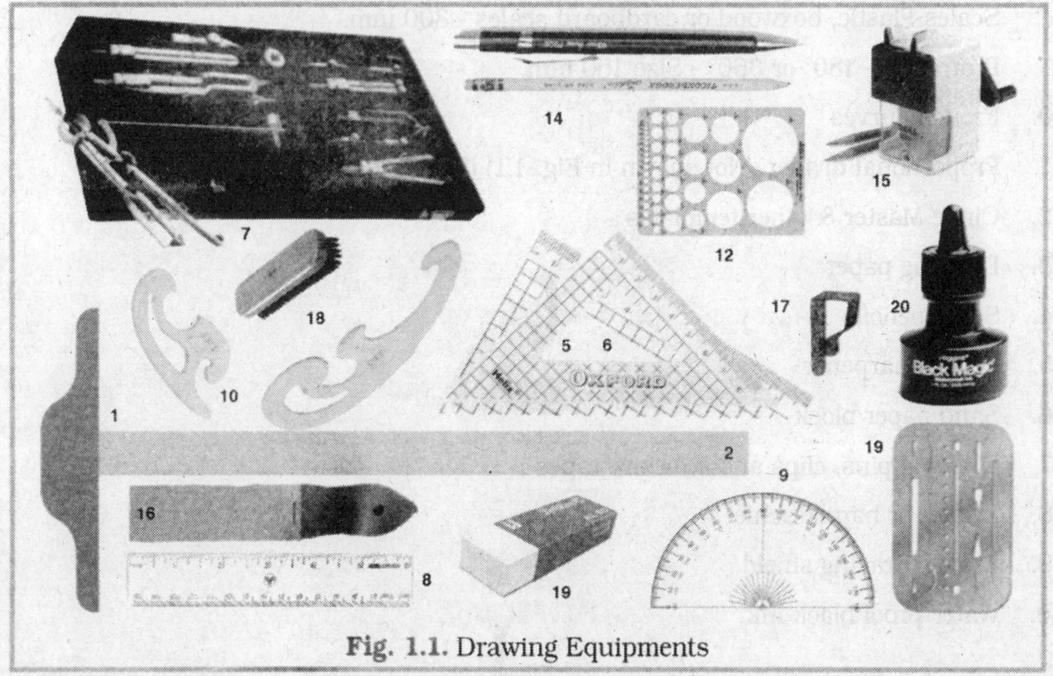

Fig. 1.1. Drawing Equipments

1. Drawing Board
2. Tee-Square
3. Mini-drafter (Not shown in Fig. 1.1) (See Fig. 1.3.)
4. Parallel Straightedge (Not shown in Fig. 1.1) (See Fig. 1.4)
5. Set-Square 45°
6. Set-Square 30° - 60°
7. Instrument Box
   (a) Large compass with interchangeable pencil and pen legs having knee joint
   (b) Large dividers
   (c) Small bow compass
   (d) Small bow divider
   (e) Small bow pen
   (f) Ruling pen
   (g) Interchangeable ruling pen used in compass leg
   (h) Container with spare parts
   (i) Case for lead
   (j) Screw driver
   (k) Drop compass - Rotor
8. Scales-Plastic, boxwood or cardboard scales - 300 mm.
9. Protractor - 180° or 360° - Size 150 mm.
10. French curves
11. Proportional divider. (Not shown in Fig. 1.1) (See Fig. 1.17)
12. Circle Master & other templates
13. Drawing paper
14. Set of pencils
15. Pencil sharpener
16. Sand paper block
17. Drawing pins, clips and adhesive tapes
18. Duster or handkerchief
19. Eraser-Erasing shield
20. Water-proof black ink.

1. **Drawing Board :** (See Fig. 1.2)

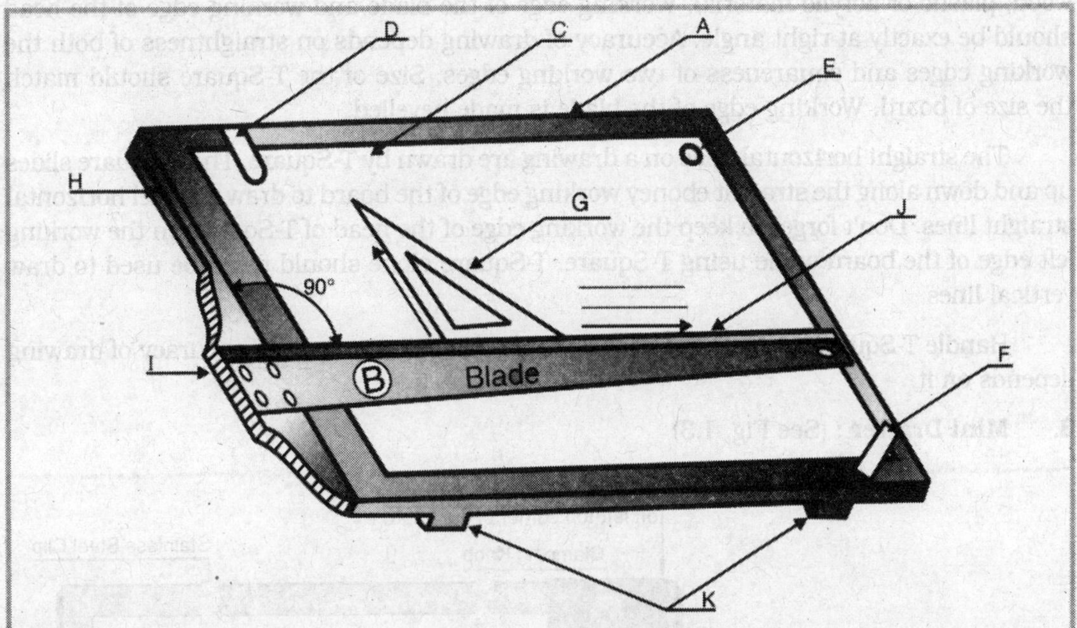

Fig. 1.2. (A) Board, (B) Tee, (C) Paper, (D) S.S. Clip, (E) Pin, (F) Adhesive Tape, (G) Set Square, (H) Eboney Edge of Board, (I) Stock or Head, (J) Working Edge of Tee (K) Battens

In order to prevent warping, the drawing board is made of narrow strips glued edge-to-edge of well seasoned white pine or bass soft wood with minimum of one straight eboney working edge, as a base for the T-Square. Surface of the board should be free from cracks and it should be absolutely flat. On the otherside of board two battens are fixed by screws fitted in the slots to allow for seasonal contraction and expansion.

The standard sizes of drawing boards available are as follows. These sizes are fixed by I.S. 1444.

Sr. No.	Designation	Dimension, mm	Name
1.	B_0	1500 × 1000	Antiquarian
2.	B_1	1000 × 700	Double Elephant
3.	B_2	700 × 500	Imperial
4.	B_3	500 × 350	Half Imperial
5.	B_4	350 × 250	Quarter Imperial

For convenience of working, the board is kept tilted at nearly 15° to 20°.

2. **Tee - Square :** (See Fig. 1.2)

Mostly T - Square is made of two parts

(i) The blade (ii) The stock or The head.

The blade and the head are rigidly fastened together. Tee-Square is made out of hard-wood, plastic or acrylic material. Working edge of the blade and working edge of the head should be exactly at right angle. Accuracy of drawing depends on straightness of both the working edges and squareness of two working edges. Size of the T-Square should match the size of board. Working edge of the blade is made bevelled.

The straight horizontal lines on a drawing are drawn by T-Square. The T-Square slides up and down along the straight eboney working edge of the board to draw parallel horizontal straight lines. Don't forget to keep the working edge of the head of T-Square on the working left edge of the board while using T-Square. T-Square alone should never be used to draw vertical lines.

Handle T-Square with atmost care because it is delicate and the accuracy of drawing depends on it.

3. Mini Drafter : (See Fig. 1.3)

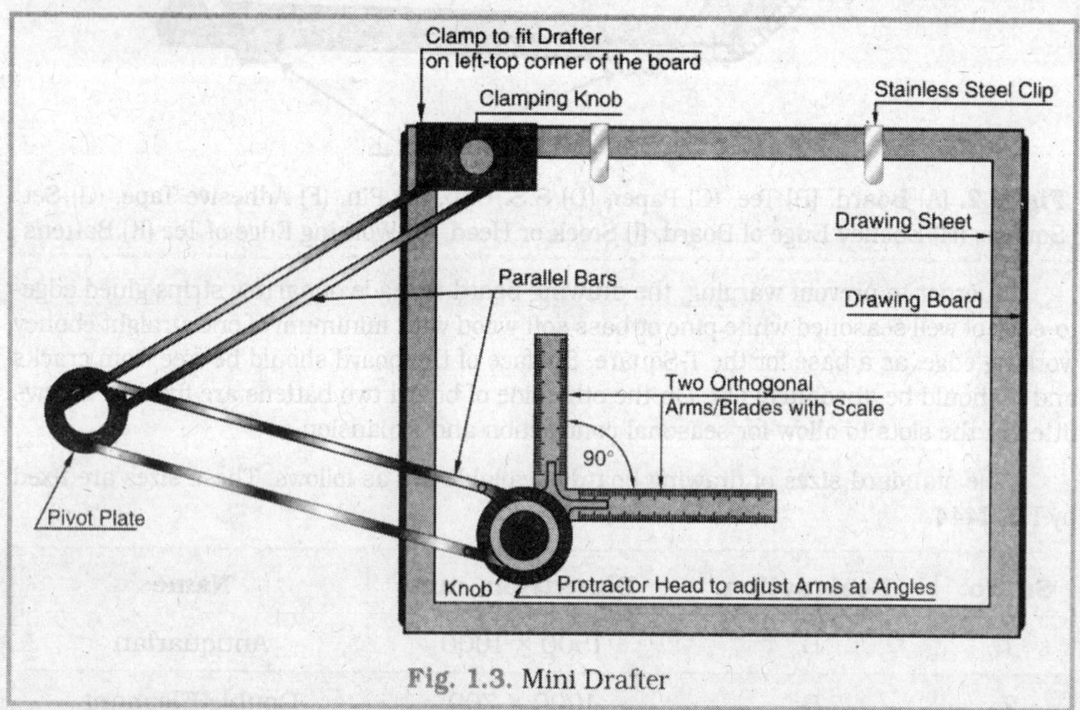

Fig. 1.3. Mini Drafter

Mini drafter is used in many drawing and design offices to do drawing work faster. It serves the purpose of T - Square, Set - Square, Protractor and Scales. During operation the two blades of the drafter always remain parallel to their set position no matter where they are moved on the sheet. Two blades of the drafter are accurately set at right angles to each other.

The blades are detachable and hence blades with different scales can be used. To set and clamp two blades at desired angle there is a provision of the adjusting head with vernier protractor.

4. Parallel Straight Edge : (See Fig. 1.4)

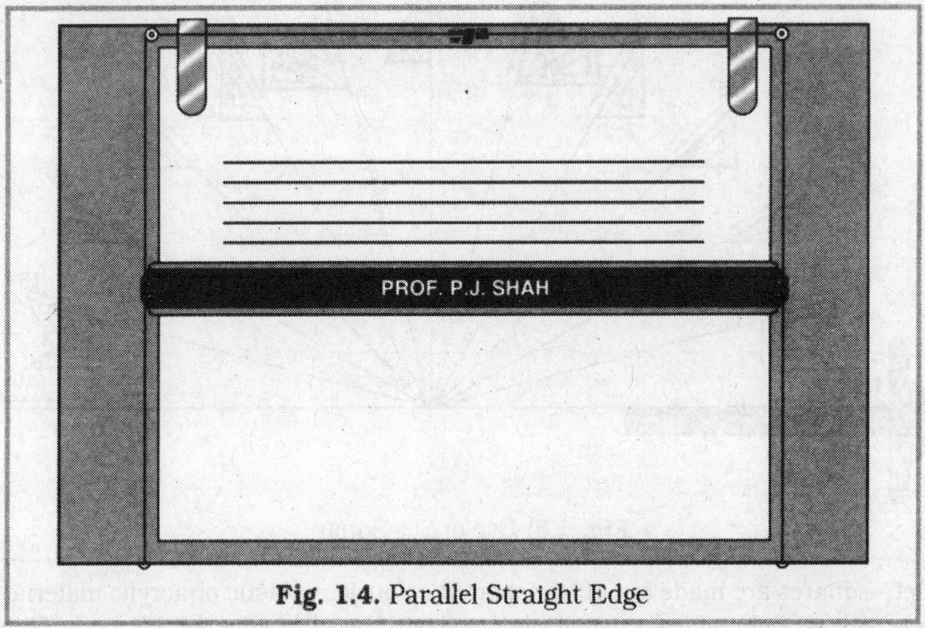

Fig. 1.4. Parallel Straight Edge

It is used in place of T - Square. Mostly it is preferred on large drawing boards. It is simply a straight edge of board size moved up and down on the board by inextensible string - pulley arrangement. It always remain parallel to previously set position.

5. Set-Square 45° : (See Fig. 1.5)

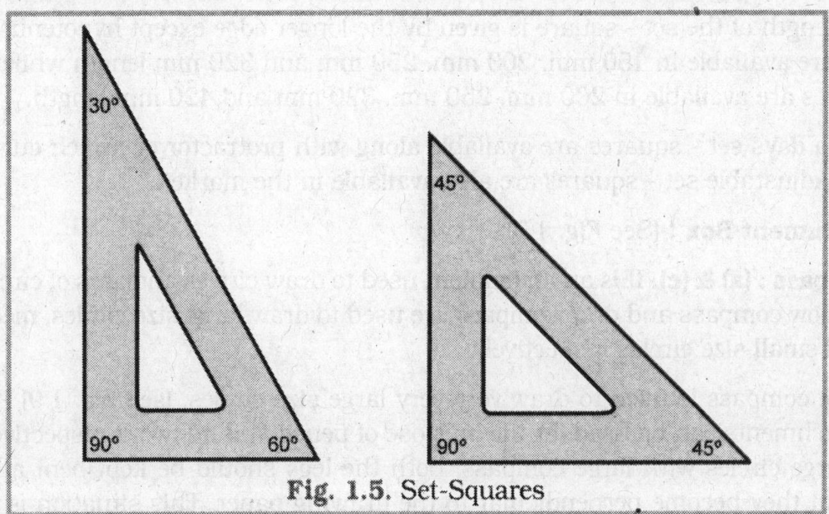

Fig. 1.5. Set-Squares

6. Set-Square 30° - 60° : (See Fig. 1.5)

Set - Squares are used with T - Square to draw lines at various angles with the horizontal. The two most common set - squares used by draftsman are 45° Set - Square and 30° - 60° Set - Square. With the combination of T - Square and two Set - Squares angles of 15°, 30°, 45°, 60°, 75°, 90°, 105°, 120°, 135°, 150°, 165°, and 180° are drawn as shown in Fig. 1.6 The above combination can divide the circle into 24 equal divisions.

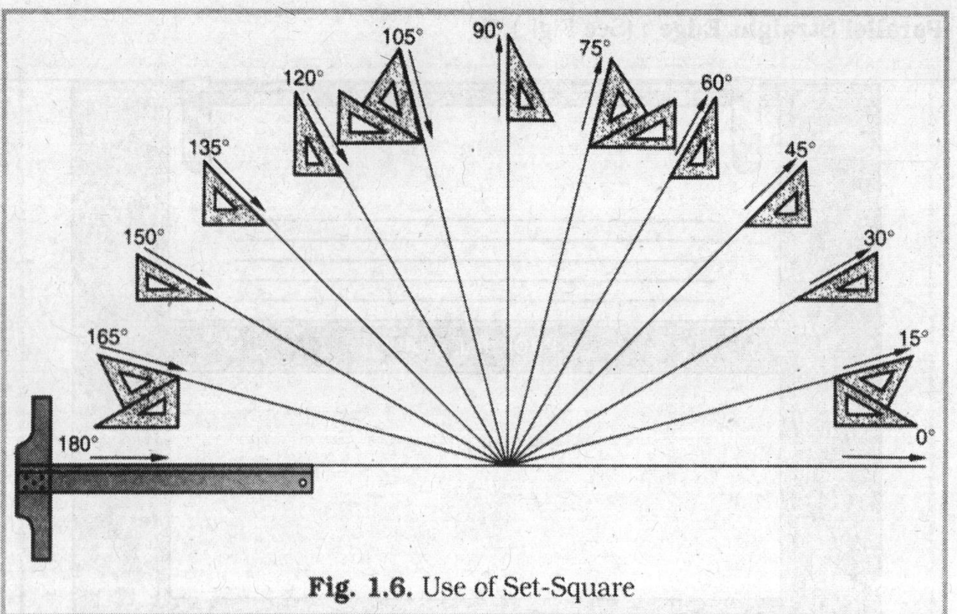

Fig. 1.6. Use of Set-Square

Set - squares are made out of transparent colourless plastic or acrylic material. They are made with straight edge for pencil work and with bevelled edge for ink work. Thickness of set squares varies from 1.5 mm to 2.5 mm.

When laying out lines, set - squares are placed firmly against the upper working edge of the T - Square. Parallel angular lines are drawn by moving the set - square or combination of two set - squares on the working edge of T - Square.

The length of the set - square is given by the longer edge except hypotenuse. 45° set - squares are available in 150 mm, 200 mm, 250 mm and 320 mm length while 30° - 60° set - squares are available in 200 mm, 250 mm, 320 mm and 420 mm length.

Now a days set - squares are available along with protractor or french curve carved with in it. Adjustable set - squares are also available in the market.

7. **Instrument Box** : (See Fig. 1.7)

Compass : (a) & (c). It is an instrument used to draw circles and arcs of circles; Large compass, bow compass and drop compass are used to draw large size circles, medium size circles and small size circles respectively.

Beam compass is used to draw very very large size circles. (see Fig. 1.9) Pencil and inking attachments can be used for the purpose of pencil and ink work respectively. While drawing large circles with large compass, both the legs should be kept bent at the knee joint so that they become perpendicular to the drawing paper. This situation is advisable for accurate circle.

Compass leads should extend approximately 9 mm. However, the metal needle point of the compass extended about 1 mm more to compensate for the distance it enters the paper, see Fig. 1.10 The flat side of the lead is kept facing outward. The compass is revolved between thumb and index finger, a little downward pressure on the metal needle side is necessary to keep it at the centre.

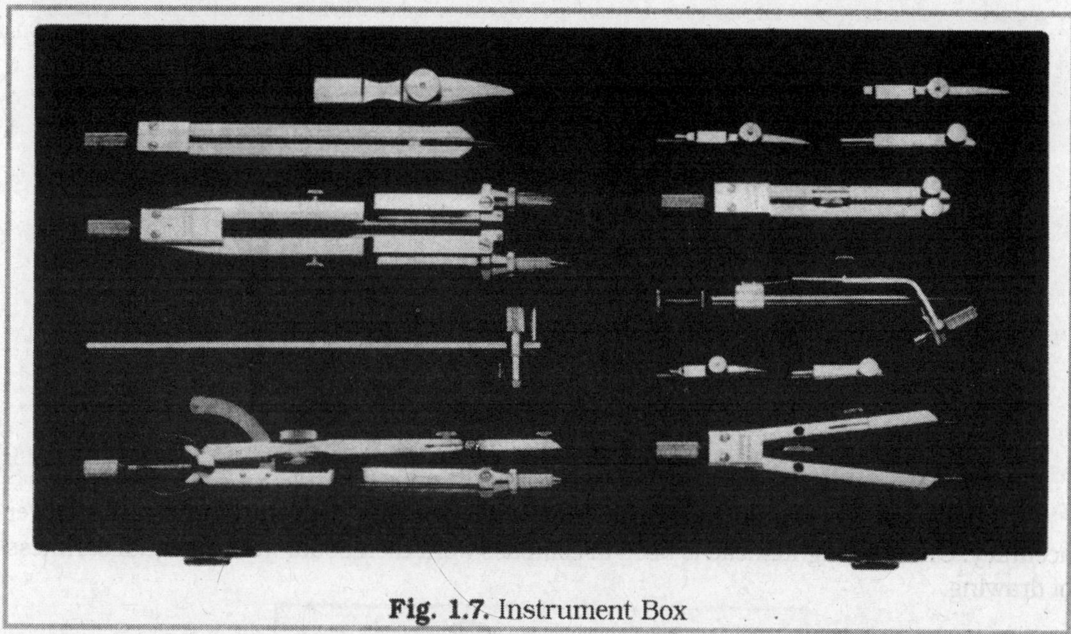

Fig. 1.7. Instrument Box

Lengthening bar can be used to draw very large size circle. (See Fig. 1.8)

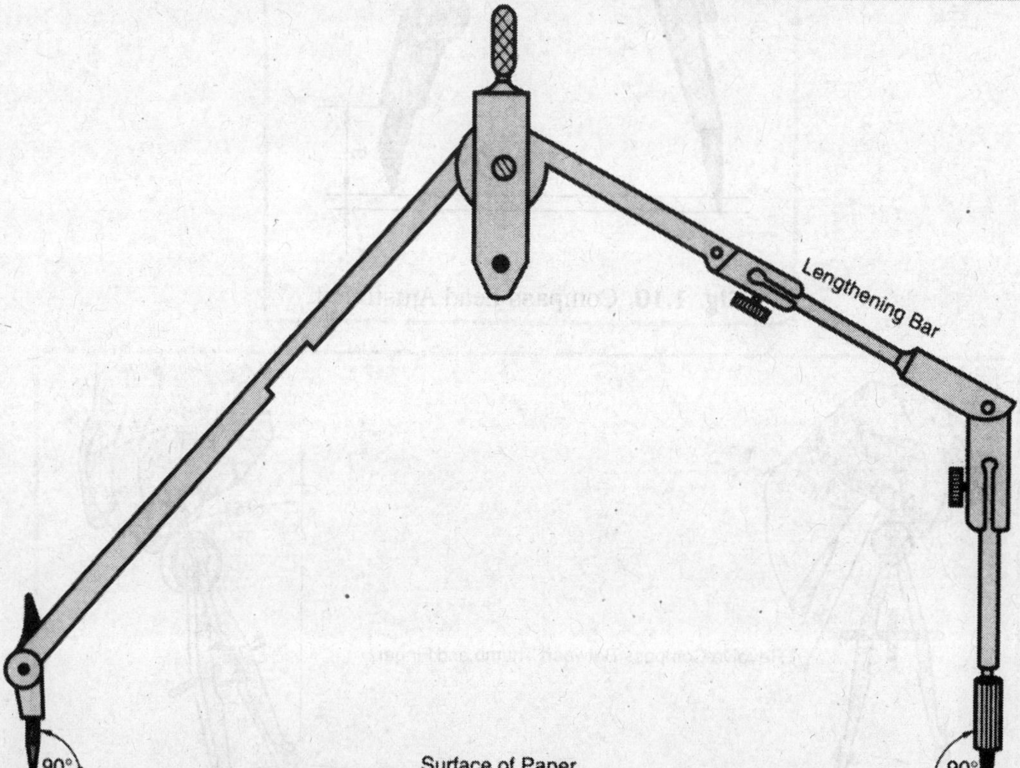

Lengthening Bar

90°

Surface of Paper

90°

Fig. 1.8. Use of Lengthening Bar

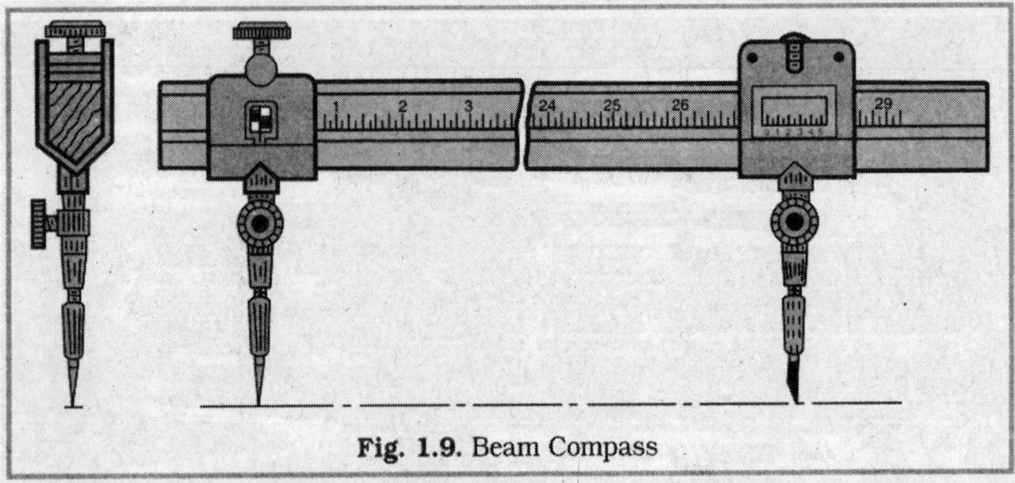

Fig. 1.9. Beam Compass

Smaller circles should be drawn first while drawing concentric circles for better accuracy. One grade lighter lead is used in compass work then usual work for equal darkness of drawing.

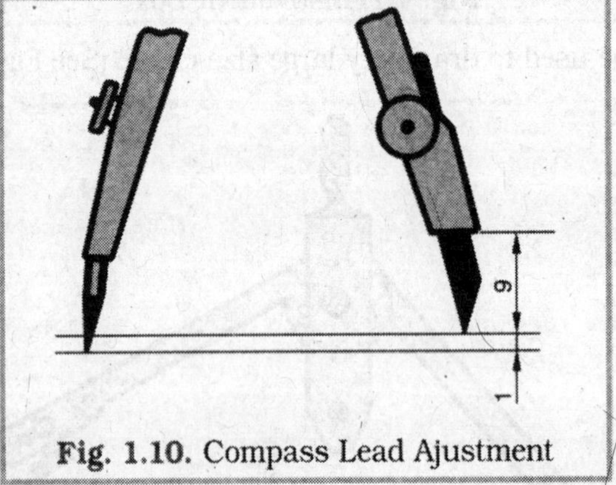

Fig. 1.10. Compass Lead Ajustment

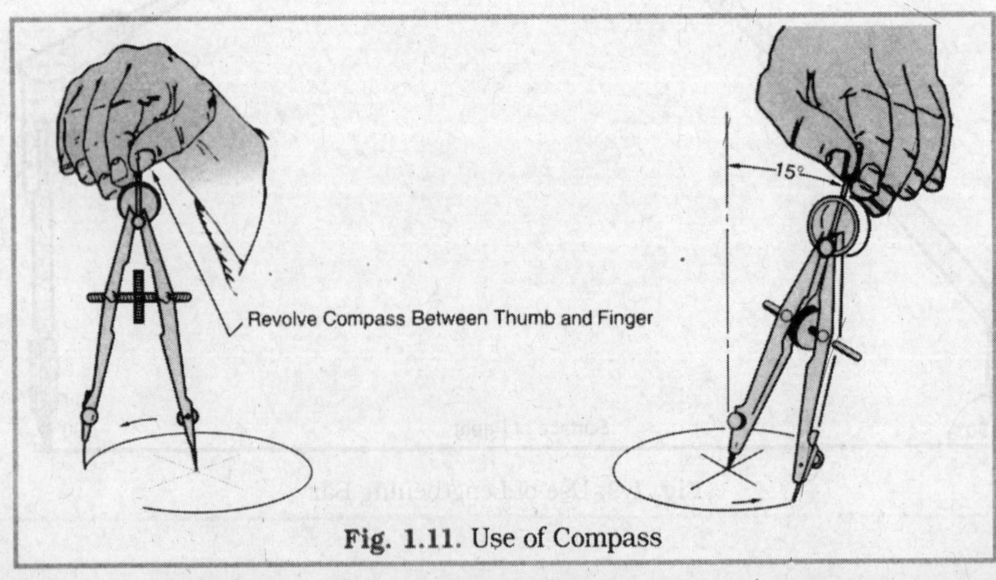

Fig. 1.11. Use of Compass

Divider : (b) and (d) (See Fig. 1.12)

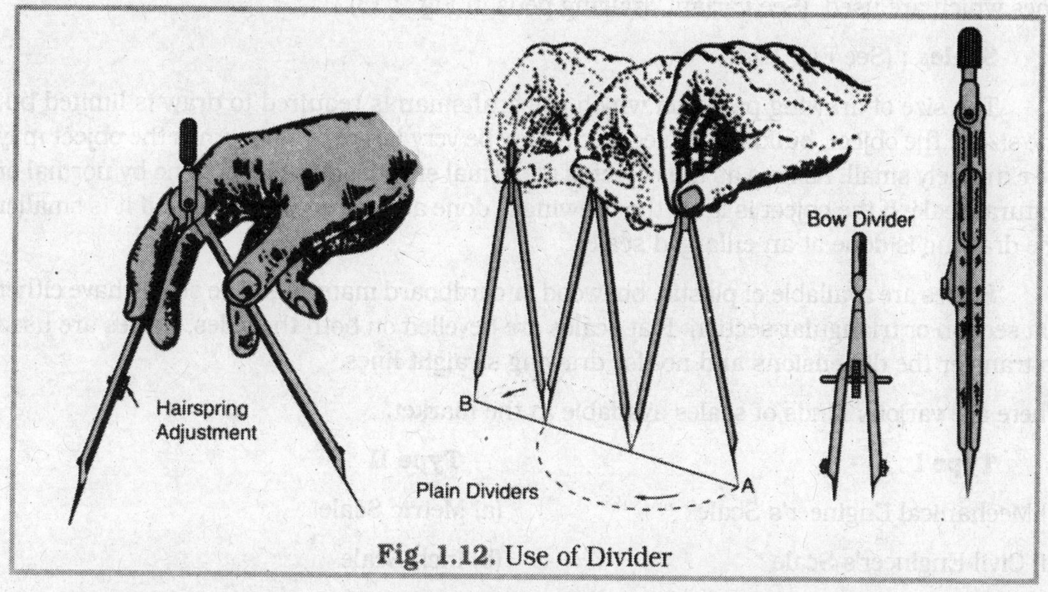

Fig. 1.12. Use of Divider

It is like a compass except it has metal points at legs. Dividers are used to transfer distances or dimensions and sometimes for dividing spaces into equal divisions. The bow divider is particularly useful for exact work because the wheel adjustment effectively holds the setting.

(f) **Ruling Pen and Variant Drawing Pens** : (See Fig. 1.13)

This pen is used for inking work. It consists of a pair of steel nibs fitted to a bone or metal holder. The gap distance between the nibs is adjusted to adjust thickness of line. For clear, neat and uniform inking work, it is advisable to clean the pen frequently during its use. Pen should not be dipped inside the ink bottle. Under no circumstances ink must not be allowed on the outside surface of the nibs.

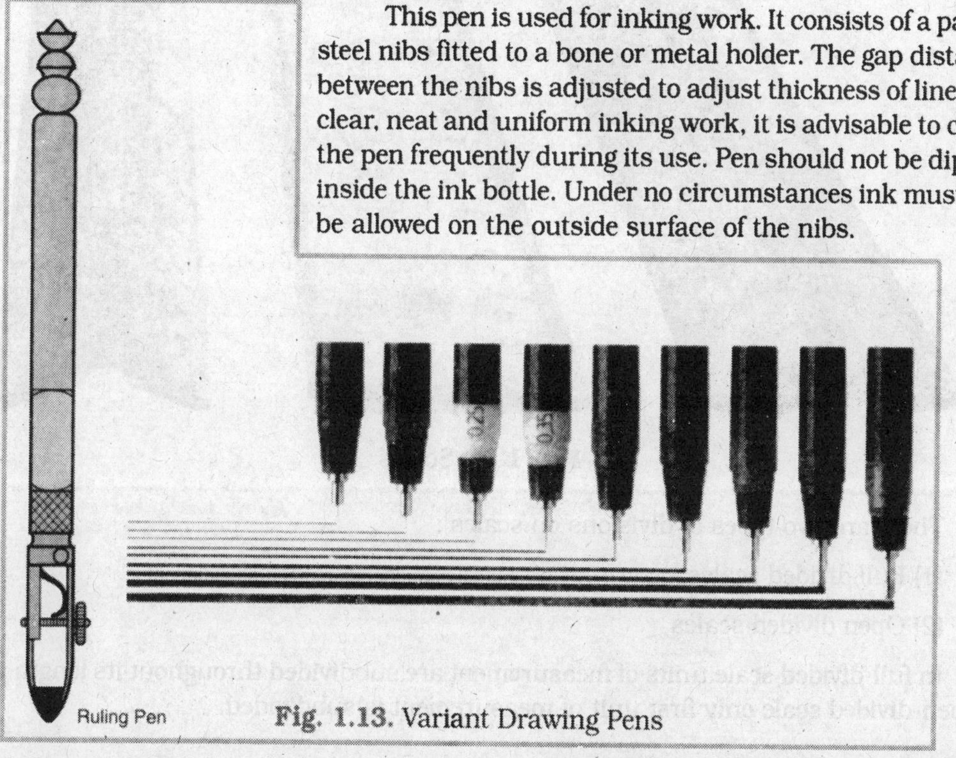

Fig. 1.13. Variant Drawing Pens

Now a days inking pens with different thickness heads are available and they are the only ones which are used. (See variant drawing pens in Fig. 1.13)

8. Scales : (See Fig. 1.14)

The size of drawing paper on which the draftsman is required to draw is limited but the size of the object, building, project etc.., may be very large, or sometimes the object may be extremely small. As long as the object is of normal size the drawing is done by normal or natural scale, If the object is large the drawing is done at a reduced scale and if it is smaller the drawing is done at an enlarged scale.

Scales are available of plastic, boxwood or cardboard materials. The scales have either flat section or triangular section. Flat scales are bevelled on both the sides. Scales are used to transfer the dimensions and not for drawing straight lines.

There are various kinds of scales available in the market.

Type I	**Type II**
(i) Mechanical Engineer's Scale	(a) Metric Scale
(ii) Civil Engineer's Scale	(b) Inch Scale
(iii) Architect's Scale	

Fig. 1.14. Scales

There are two types of divisions on scales :

(1) Full divided scales

(2) Open divided scales

In full-divided scale units of measurement are subdivided throughout its length while in open-divided scale only first unit of measurement is subdivided.

9. Protractor : (See Fig. 1.15)

Protractors are used for the measurement of angles. They are available in semi-circular and circular shapes. Protractors are made of paper, wood, plastic, brass metal or celluloid materials. Transparent protractors with bevelled edge are more often used. Normally it is graduated to ½° or 1° and numbered at an intervals of 10°. For very accurate work protractor with vernier attachment is used.

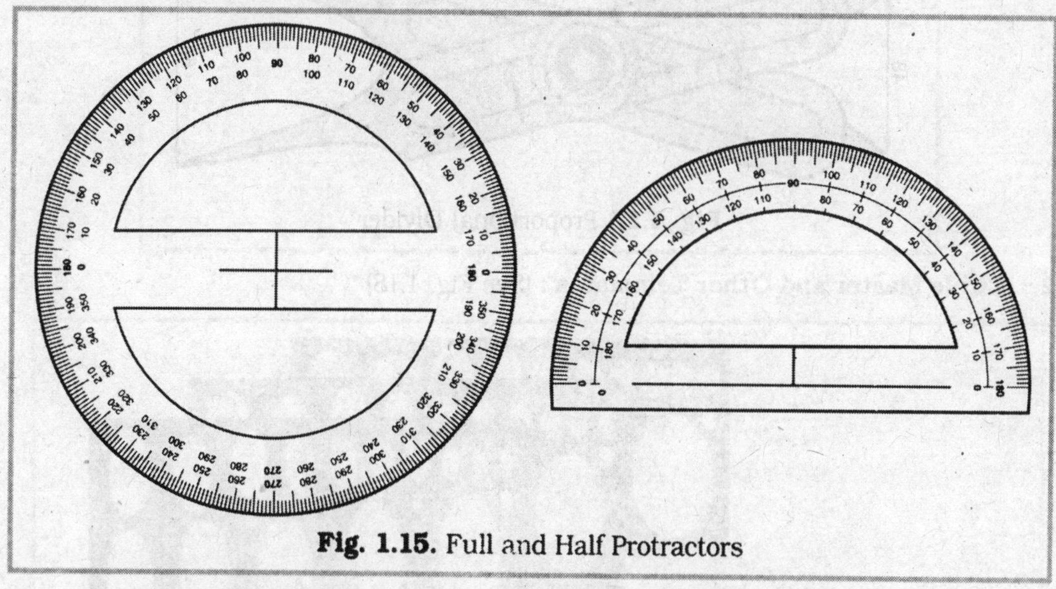

Fig. 1.15. Full and Half Protractors

10. French Curves : (See Fig. 1.16)

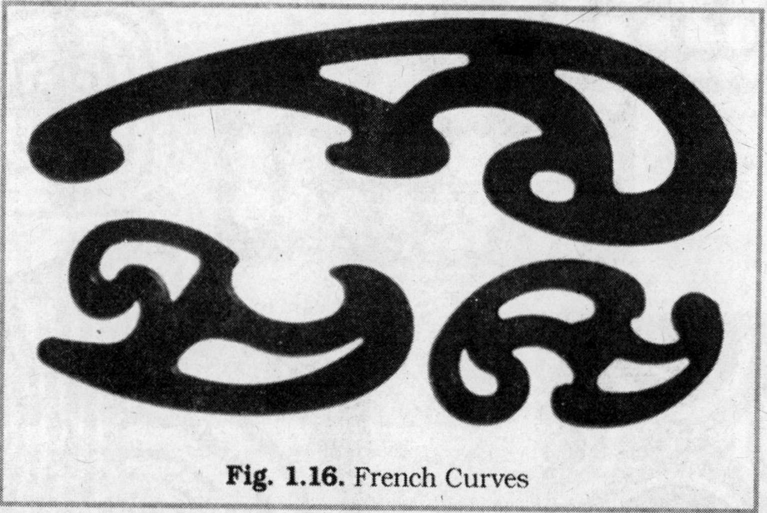

Fig. 1.16. French Curves

Curves other than arc of circles are drawn accurately and uniformly with the help of french curves. French curves are available in a set of assorted curves. They are made of transparent material. Use of french curves requires sufficient experience, especially in inking work.

11. Proportional Divider : (See Fig. 1.17)

Proportional divider is used to draw drawing at an enlarged or at a reduce scale from the available drawing. It has a sliding adjustable pivot. Pivot position decides the conversion of scale. Scale marks are marked on the proportional divider.

Fig. 1.17. Proportional Divider

12. Circle Master and Other Templates : (See Fig. 1.18)

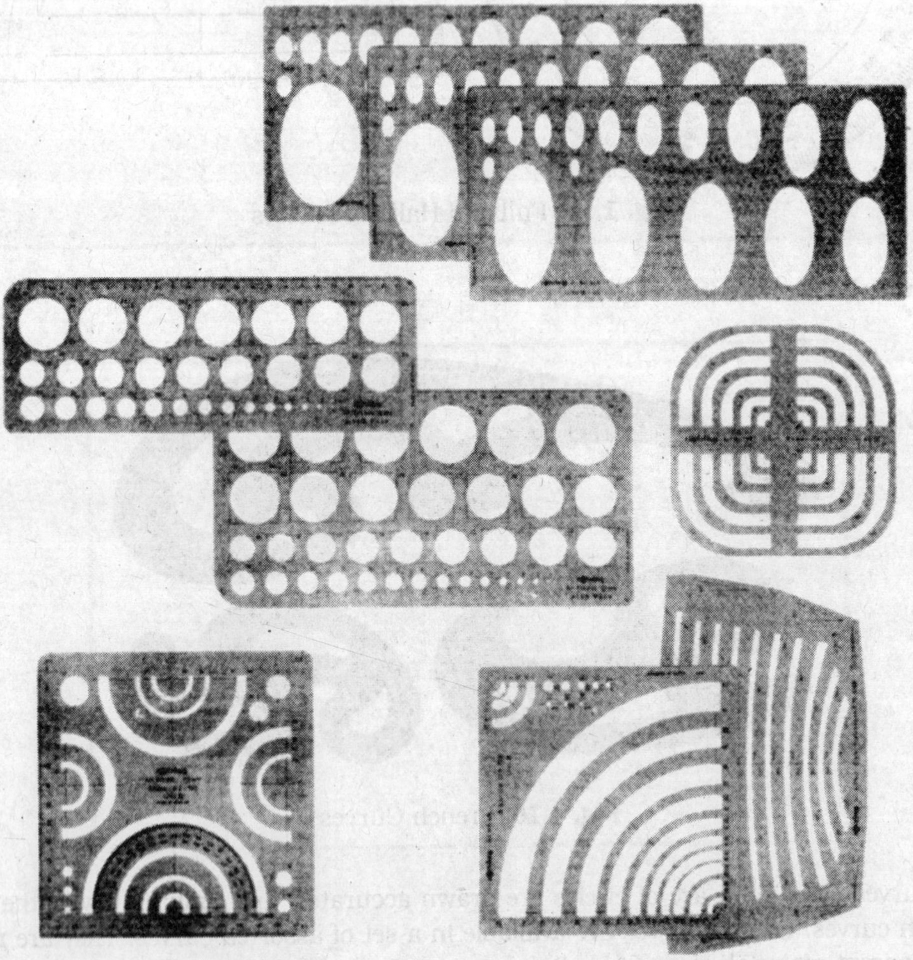

Fig. 1.18. Circle Master and other Templates

It is a thin, flat piece of plastic with different sizes of circles cut in it. It is used to speed up the drafting work. Every circle has 4 black line marks on the circumference along two right angle axes. These line marks are used for adjusting circle master on two axes of circle. It is one kind of template. Many such templates are available for ellipses, squares, triangles and other polygons.

13. Drafting Paper :

Hand made paper with smooth surface is used for the purpose of drawing. Drawing paper should be uniform and as white as possible. It should have good erasing quality. A paper with sufficient grains to produce sharp and clean pencil lines is desirable. It is available in rolls or sheets. Weight of drawing paper per rim is about 25 to 30 Kilograms.

The preferred sizes of drawing sheets according to I. S. 696-1972 are given in the table below :

Sr. No.		Trimmed		Untrimmed	
	Designation	Width	Length	Width	Length
1.	A_0	841	1189	880	1230
2.	A_1	594	841	625	880
3.	A_2	420	594	450	625
4.	A_3	297	420	330	450
5.	A_4	210	297	240	330
6.	A_5	148	210	165	240

14. Set of Pencils :

The main two ingredients in the pencil leads are graphite and clay. Pencils are available in 20 grades, 8B softest to 10H hardest. Hardness and softness of the pencils depend upon the proportions of graphite and clay. Soft pencils produce dark black lines while hard pencils produce thin gray lines. Softer pencils have larger size leads while harder pencils have smaller size leads. Even though the hardness number of the pencils may be same, hardness differ with different manufacturer and hence it is advisable to use pencils of the same manufacturer.

Now a days refill pencils or push pencils are used. Lead size in these pencils is generally 0.5 mm. Leads for these pencils are available in different grades. It is more convenient compared to usual wooden pencils. See Fig. 1.19 Grades of pencils and their uses are shown in the Fig. 1.20.

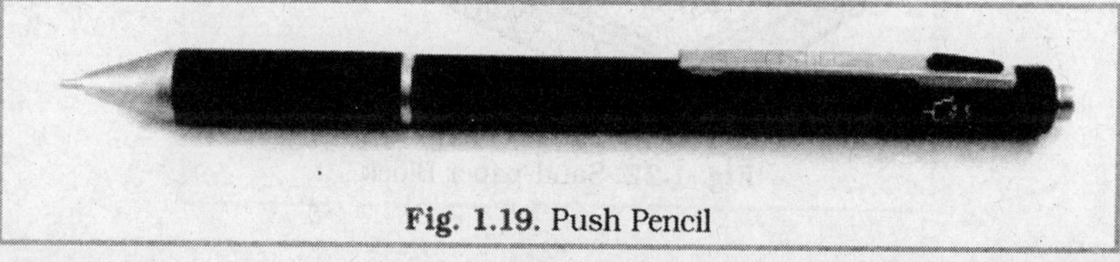

Fig. 1.19. Push Pencil

Different shapes of the pencil points are shown in Fig. 1.21. The conical point is used for general drafting and lettering. The wedge point is used for drawing uniform long straight lines. The bevel shape point is used in compass work.

15. Pencil Sharpener :

Knife edge can be used as a pencil sharpener. Pencil sharpening machines are also available and they are used in drawing design offices.

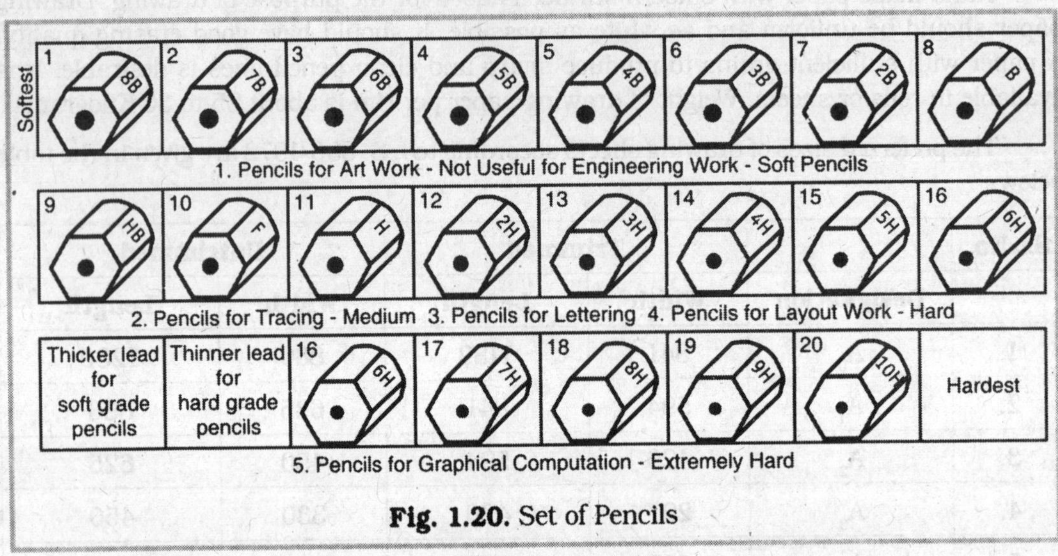

Fig. 1.20. Set of Pencils

Fig. 1.21. Pencil Points

16. Sand Paper Block : (See Fig. 1.22)

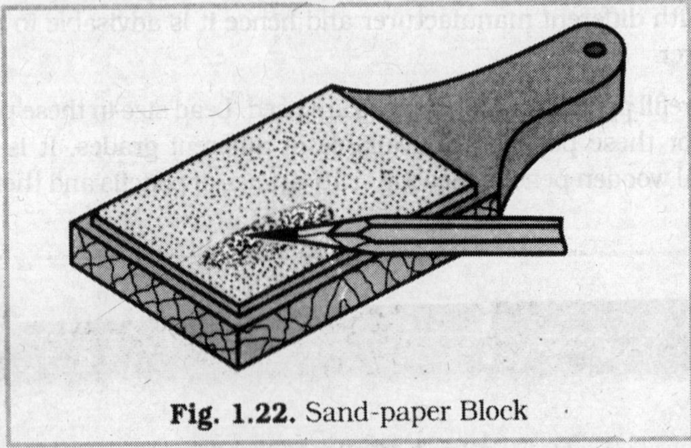

Fig. 1.22. Sand-paper Block

Sand paper block is used for sharpening lead points of desired shapes. It is a wooden block 50 × 50 × 12 mm size with sand paper pasted or nailed over it. Sand paper can be replaced when it warns out. Don't sharpen leads near or over drawing as the graphite powder will smear the drawing surface.

17. Drawing Pins, Clips and Adhesive Tapes : (See Fig. 1.23)

Drawing paper is fixed on the drawing board by pins or clips or adhesive tapes. Steel pins with brass tops are available in the market. Disadvantage of these pins is that they make holes in the drawing paper and spoil the surface of the board.

Adhesive tapes are available in rolls. Adhesive tapes do not make any hole in the paper as well as on the board.

Stainless steel elastic clips are available in the market. It is used extensively because it holds the paper without sticking to it or without making hole on it.

18. Duster or Handkerchief :

It is nothing but a piece of clean cloth used to clean drawing paper and instruments. Don't try to clean a drawing surface by hand as it tends to smudge the drawing. It is used to remove loose graphite and eraser crumbs from the drawing surface.

19. Eraser-Erasing Shield : (See Fig. 1.24) .

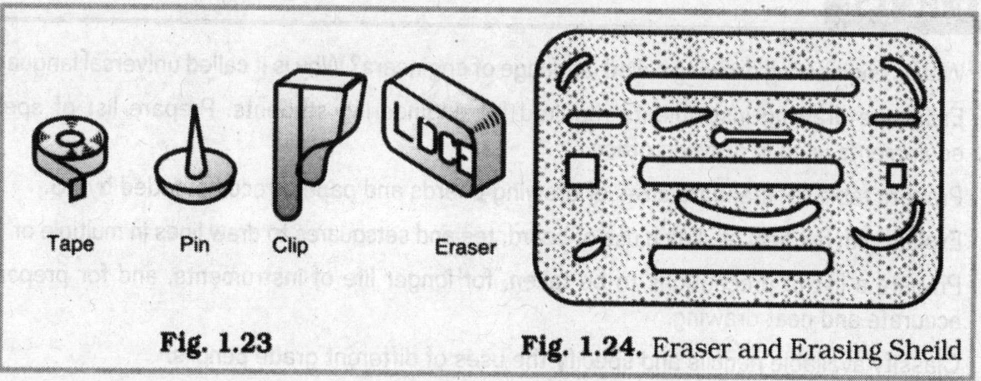

Fig. 1.23 **Fig. 1.24.** Eraser and Erasing Sheild

Pencil eraser, what we call it as rubber, are available in many varieties. Always use good quality soft rubber so that due to its use the surface of drawing paper does not get damaged. To restrict the erasing area pencil rubbers are used.

The erasing shield is a very convenient device for erasing unwanted line while protecting others. It also prevents paper from possibility of damage.

The electric eraser machine is commonly used in large drawing offices. The machines are available with interchangeable pencil and ink eraser points.

20. Roll-N-Draw Ruler : (See Fig. 1.25)

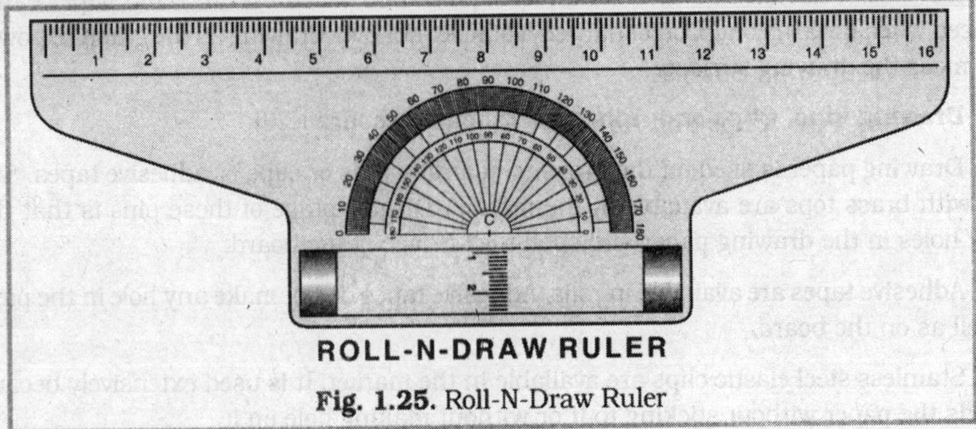

Fig. 1.25. Roll-N-Draw Ruler

With the help of this ruler parallel lines are drawn quickly and very easily. Straight edge of the ruler is kept on a given / known line and then it is rolled to new required location and then parallel line to the prior line is drawn. This way parallel horizontal, vertical or inclined lines are drawn faster with ease. Used for graphical solutions of problems in different subjects.

EXERCISE

1. Why is engineering drawing called language of engineers? Why is it called universal language?
2. Enlist the drawing instruments required for engineering students. Prepare list of special equipments used in drawing office.
3. Prepare table for standard sizes of drawing boards and papers recommended by I.S.
4. Explain with the help of sketch use of board, tee and setsquares to draw lines in multiple of 15°.
5. Prepare a list of precautions to be taken, for longer life of instruments, and for preparing accurate and neat drawing.
6. Classify available pencils and specify the uses of different grade pencils.
7. Name different compasses, specifying uses of each one.
8. Draw different types of pencil points. Specify uses of each one.
9. What are french curves and how are they used?
10. What are templates and how they help in preparation of drawing?
11. Why are scales needed?
12. Explain the use of proportional divider.
13. Compare pair of Tee-Set square with parallel straight-edge and drafting machine.
14. Explain the uses of a divider.
15. What is an eraser and what is an eraser shield?
16. Redraw the figures given below to practice proper use of drawing equipments. Use appropriate equipment to do the work accurately, quickly and correctly. Assume dimension if missing.

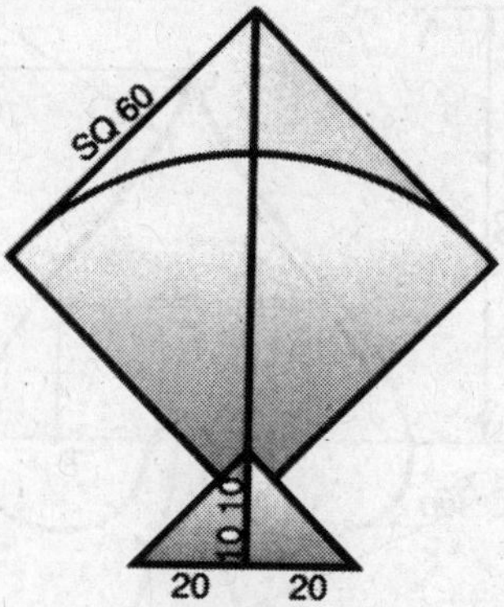

SQ 60
10 10
20 20

KITE

MARBEL DESIGN

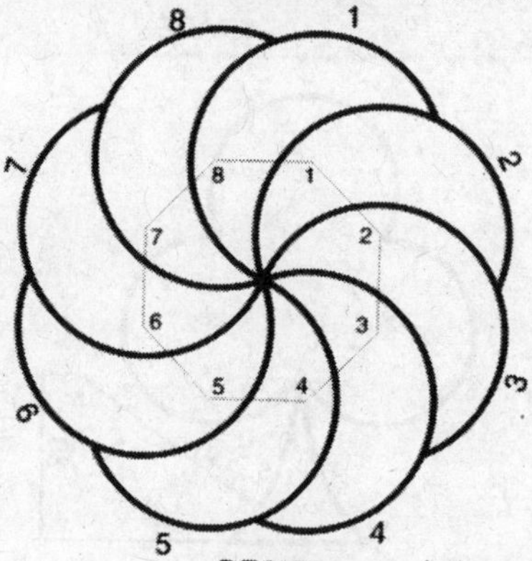

8 1 2 3 4 5 6 7
8 1 2 3 4 5 6 7

DESIGN

RANGOLI

45°
200 mm

DESIGN

BORDER DESIGN

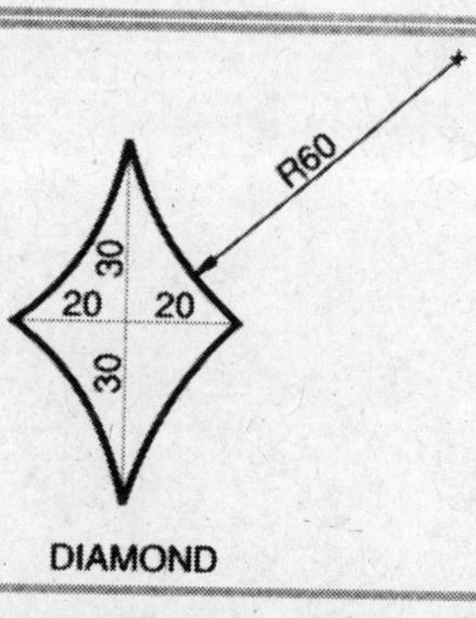

R60
30
20 20
30

DIAMOND

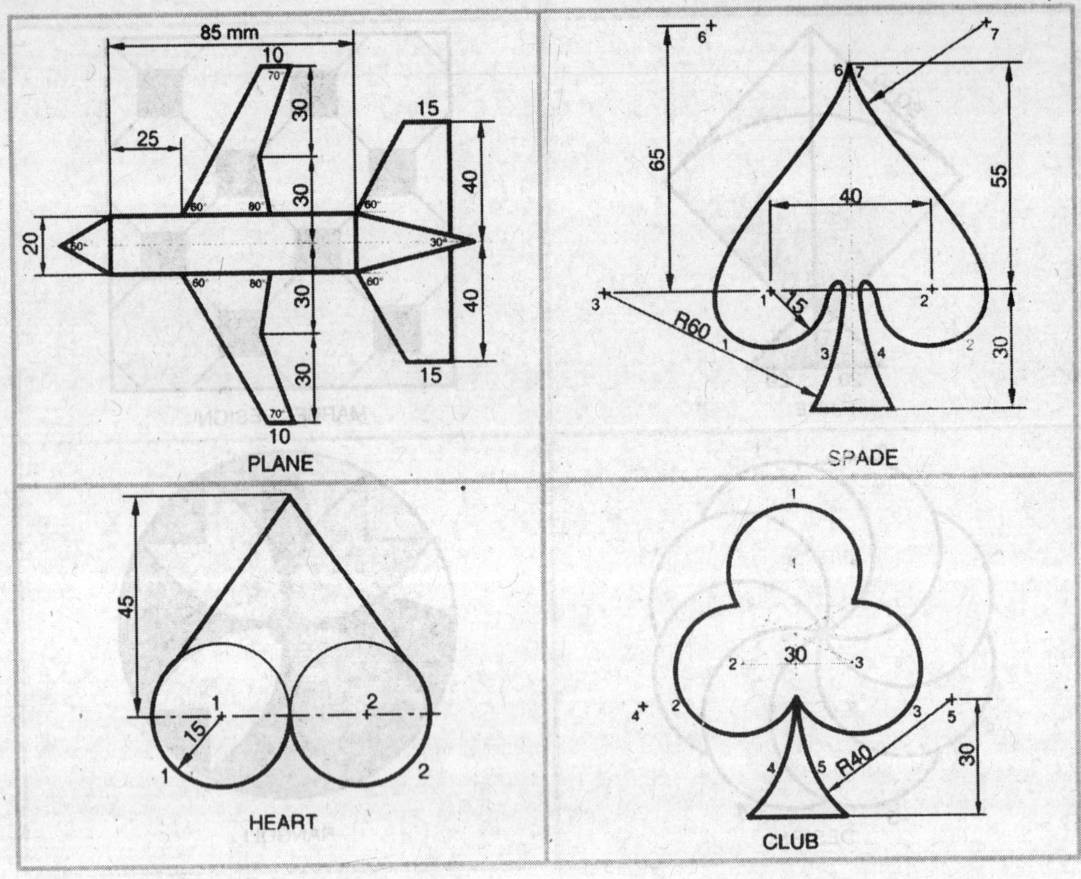

PLANE

SPADE

HEART

CLUB

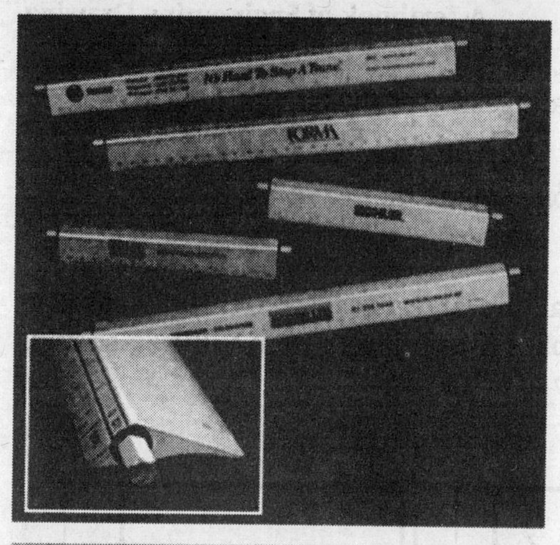

CHAPTER - 2

Sheet Sizes, Scales, Lines and Lettering

1. Sheet Sizes :

The preferred sizes of drawing sheets are given in Table-1 below. The second choice of drawing sheet sizes, both trimmed and untrimmed are given in Table-2 below. However, it is recommended that only preferred sizes should be used as far as possible.

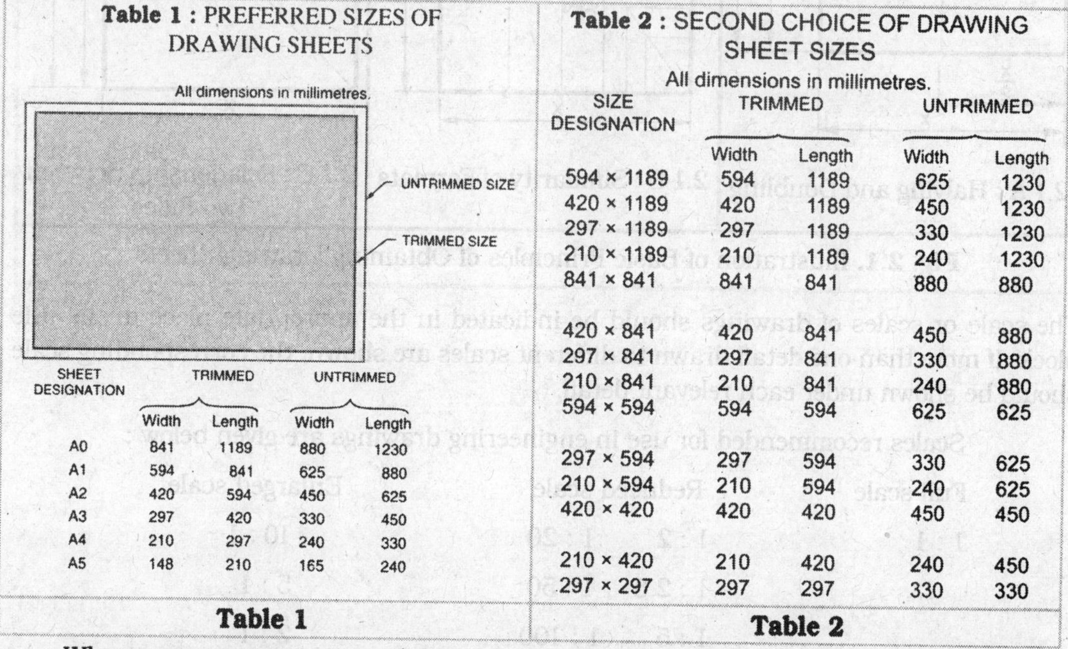

Table 1 : PREFERRED SIZES OF DRAWING SHEETS

All dimensions in millimetres.

SHEET DESIGNATION	TRIMMED		UNTRIMMED	
	Width	Length	Width	Length
A0	841	1189	880	1230
A1	594	841	625	880
A2	420	594	450	625
A3	297	420	330	450
A4	210	297	240	330
A5	148	210	165	240

Table 1

Table 2 : SECOND CHOICE OF DRAWING SHEET SIZES

All dimensions in millimetres.

SIZE DESIGNATION	TRIMMED		UNTRIMMED	
	Width	Length	Width	Length
594 × 1189	594	1189	625	1230
420 × 1189	420	1189	450	1230
297 × 1189	297	1189	330	1230
210 × 1189	210	1189	240	1230
841 × 841	841	841	880	880
420 × 841	420	841	450	880
297 × 841	297	841	330	880
210 × 841	210	841	240	880
594 × 594	594	594	625	625
297 × 594	297	594	330	625
210 × 594	210	594	240	625
420 × 420	420	420	450	450
210 × 420	210	420	240	450
297 × 297	297	297	330	330

Table 2

Wherever necessary, sizes of sheet with length 1189 m.m., may be further extended by steps of 210 m.m. on the length only.

In arriving at the trimmed size of drawing sheets, the following basic principles, which have been dealt with in detail in IS : 1064-1961 'Specification for paper sizes' have been taken into consideration.

(a) Two successive preferred sizes of drawing sheets are obtained either by halving or doubling. (See Fig. 2.1A) Consequently the surface areas of two successive preferred sizes are in ratio of 1 : 2.

(b) The formats or forms of preferred sizes are geometrically similar to one another, the sides of each size being in ratio of 1 : √2. (See Fig. 2.1B and Fig. 2.1C)

(c) The surface area of the basic size AO is one square metre.

2. Scales :

All drawings should be drawn to a scale. In general, the largest scale conveniently possible should be adopted. Drawings of parts or assemblies drawn larger than full size should, where practicable, include an undimensional view to actual size.

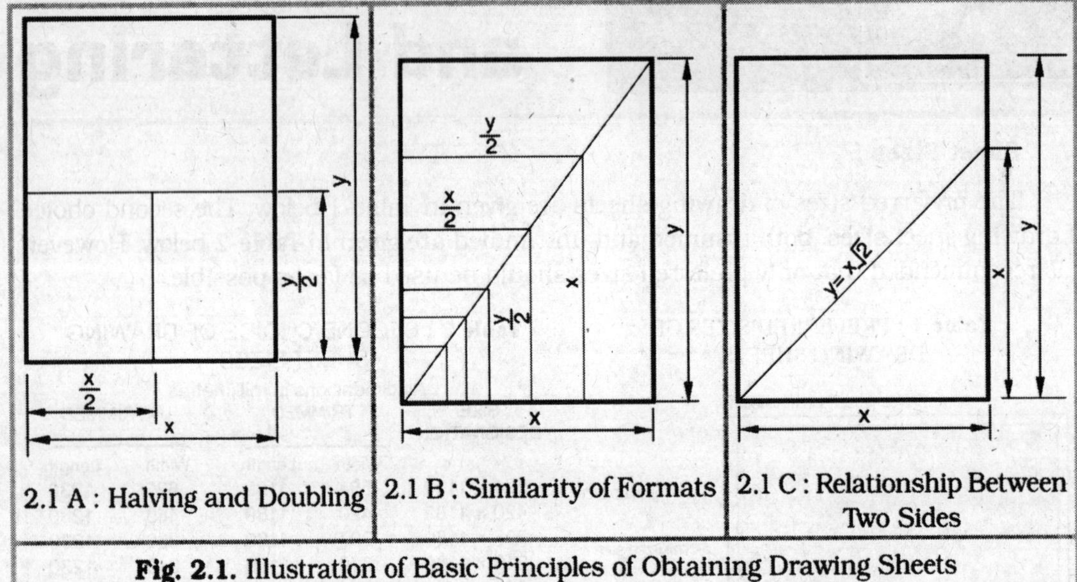

Fig. 2.1. Illustration of Basic Principles of Obtaining Drawing Sheets

The scale or scales of drawings should be indicated in the appropriate place in the title block. If more than one detail drawn to different scales are shown, the corresponding scale should be shown under each relevant detail.

Scales recommended for use in engineering drawings are given below :

Full scale	Reduced scale		Enlarged scale
1 : 1	1 : 2	1 : 20	10 : 1
	1 : 2.5	1 : 50	5 : 1
	1 : 5	1 : 100	2 : 1
	1 : 10	1 : 200	

3. Lines :

For general engineering drawings, the types of lines shown in Fig. 2.2 should be used. All lines should be sharp and dense to obtain good reproduction.

The thickness of lines should be chosen according to the type and size of drawing. The line group is identified by the thickest line. For a given view or section, the lines employed should be chosen from one of the lines of the group. (See Fig. 2.3)

Centre lines should project for a short distance beyond the outline to which they refer but, where necessary, to aid dimensioning or to correlate views, they may be extended.

Hidden lines to show interior or hidden surfaces should be included only where their use definitely assists in the interpretation of the drawing. An example of the use of various types of lines is shown in Fig. 2.4

Where drawings are to be reproduced to a smaller size by photographic process, the thickness of lines in the originals should be suitably accentuated to ensure sufficient legibility and clarity after reduction.

	Type	Illustration	Application
A	Continuous thick		Visible outlines
B	Continuous thin		Dimension lines, leader lines, extension lines, construction lines, outlines of adjacent parts, hatching and revolved section
C	Continuous thin-wavy		Irregular boundary lines, short break lines
D	Short dashes medium		Hidden outlines and edges
E	Long chain thin		Centre lines, locus lines, extreme positions of the moveable parts, parts situated in front of the cutting planes and pitch circles
F	Long chain thick at ends and thin elsewhere		Cutting plane lines
G	Long chain thick		To indicate surfaces which are to receive additional treatment
H	Ruled line and short zigzag thin		Long break lines

Fig. 2.2. Types of lines

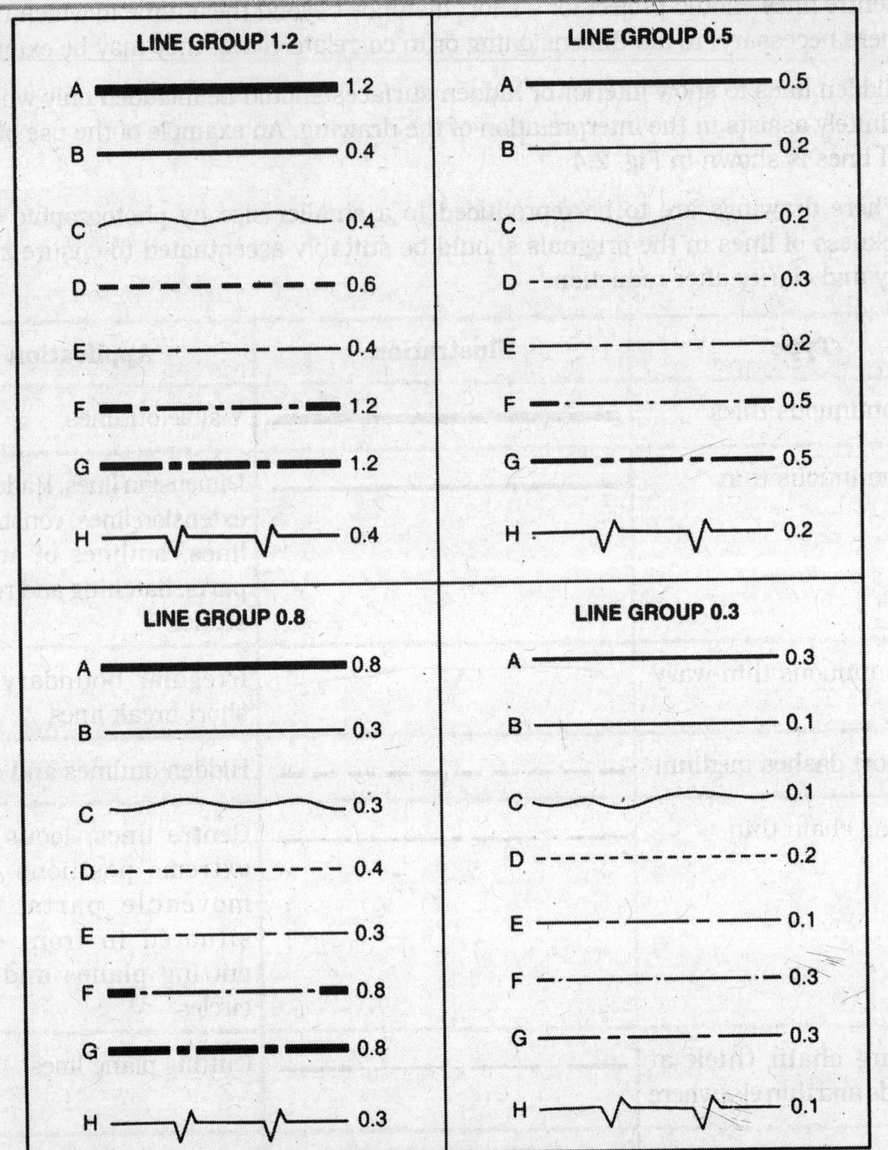

Fig. 2.3. Comparative Thickness in mm of Various Types of Line Groups

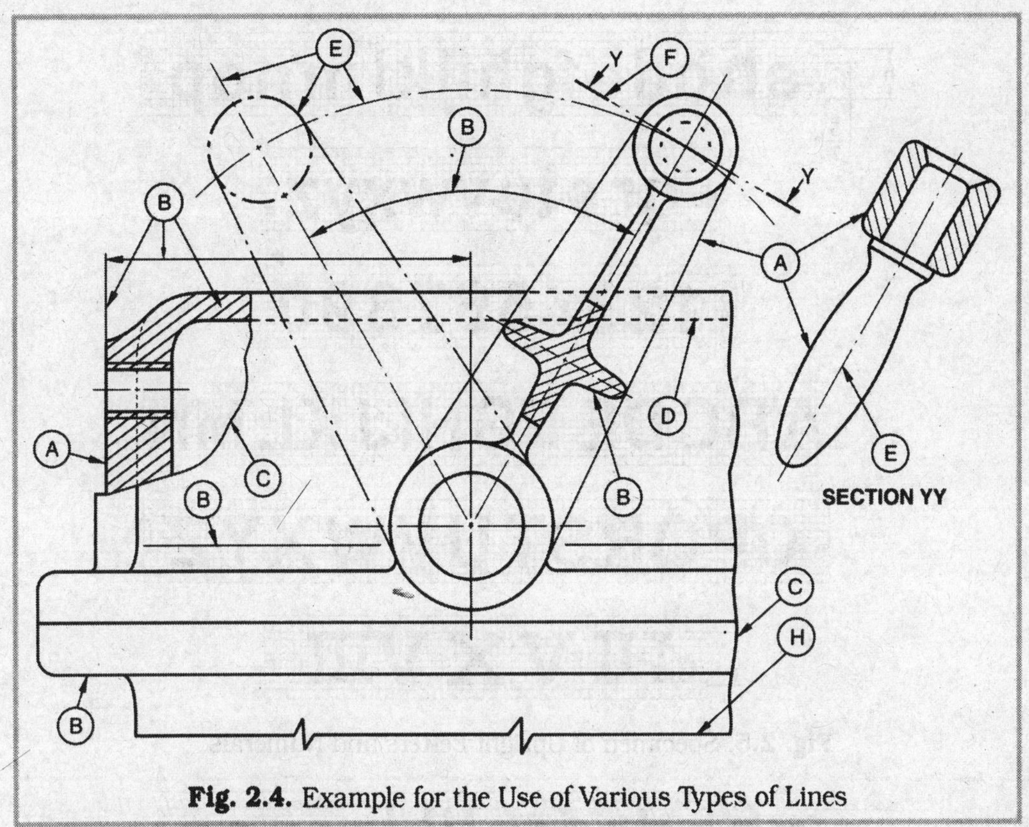

Fig. 2.4. Example for the Use of Various Types of Lines

4. Lettering :

The main requirements for 'lettering' on engineering drawings are legibility, uniformity, ease and rapidity in execution.

Both the upright and inclined letters and numerals are suitable for general use. All letters should be capitals, except where lower case letters are accepted internationally for abbreviations. If inclined letters are used, an inclination of 75° to the horizontal is recommended. Recommended specimens of letters and numerals for upright and inclined types are given in Figs.2.5 to 2.10 Fig.2.10 represents Hindi letters specified and recommended by IS : 69 6 - 19 72.

Letters and numerals are designated by their heights. Recommended sizes of letters and numerals to suit different purposes are given in Table - 3 below. However, actual sizes used depend upon the size of the drawing and purpose for which it is intended.

Lettering should be done on the drawing in such a manner that it may be read when the drawing is viewed from the bottom edge except where it is used, for dimensioning purposes.

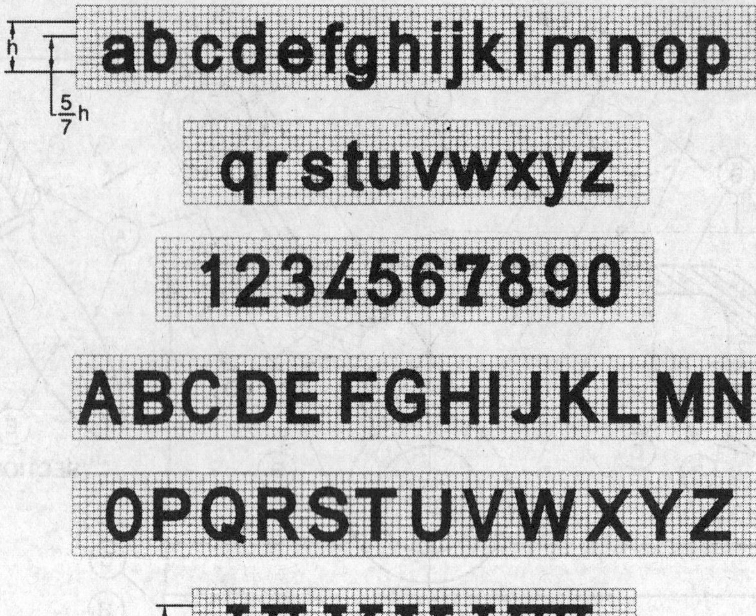

Fig. 2.5. Specimen of Upright Letters and Numerals

Fig. 2.6. Specimen of Inclined Letters and Numerals

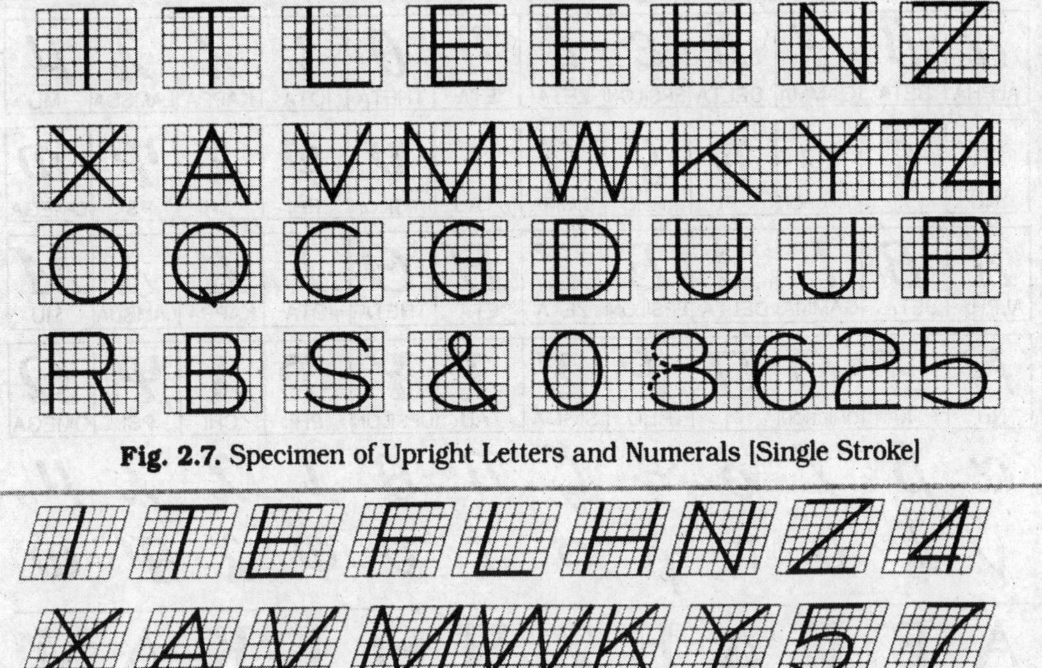

Fig. 2.7. Specimen of Upright Letters and Numerals [Single Stroke]

Fig. 2.8. Specimen of Inclined Letters and Numerals [Single Stroke]

Table 3 : Recommended Sizes of Letters and Numerals

Item	Size h, m.m.
Drawing number in title block and letters denoting cutting plane section	10, 12
Title of drawing	6, 8
Sub - Titles and headings	3, 4, 5, 6
Notes, such as legends, schedules, material list, dimensioning	3, 4, 5
Alteration entries and tolerances	2, 3

When drawings are to be reproduced to a smaller scale by photographic process, the size of letters in original drawing should be suitably accentuated to ensure sufficient legibility and clarity after reduction.

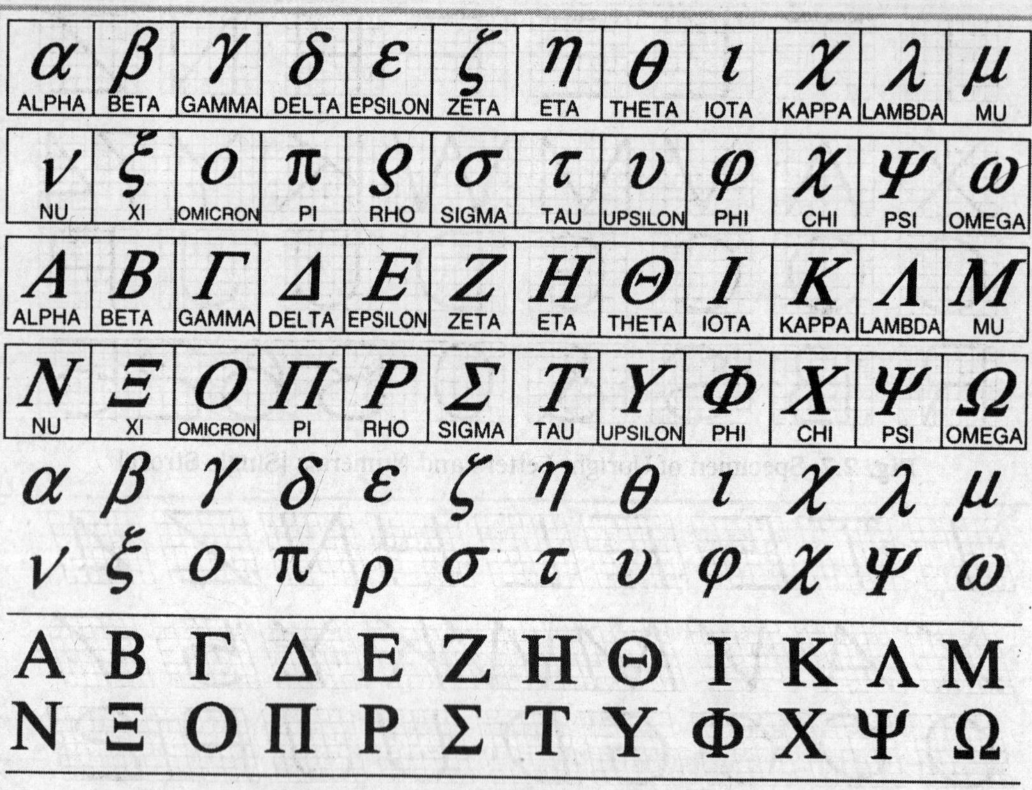

α β γ δ ε ζ η θ ι χ λ μ
ALPHA BETA GAMMA DELTA EPSILON ZETA ETA THETA IOTA KAPPA LAMBDA MU
ν ξ ο π ϱ σ τ υ φ χ Ψ ω
NU XI OMICRON PI RHO SIGMA TAU UPSILON PHI CHI PSI OMEGA
Α Β Γ Δ Ε Ζ Η Θ Ι Κ Λ Μ
ALPHA BETA GAMMA DELTA EPSILON ZETA ETA THETA IOTA KAPPA LAMBDA MU
Ν Ξ Ο Π Ρ Σ Τ Υ Φ Χ Ψ Ω
NU XI OMICRON PI RHO SIGMA TAU UPSILON PHI CHI PSI OMEGA
α β γ δ ε ζ η θ ι χ λ μ
ν ξ ο π ρ σ τ υ φ χ Ψ ω
Α Β Γ Δ Ε Ζ Η Θ Ι Κ Λ Μ
Ν Ξ Ο Π Ρ Σ Τ Υ Φ Χ Ψ Ω

Fig. 2.9. Specimen of Inclined and Upright Greek Letters

b a c c c b
d h e
अ आ इ ई उ ऊ
c c a a b b
d h e
ए ऐ ओ औ अं अः
h b c h c
h
क ख ग घ ङ
h h h b h
h
च छ ज झ ञ

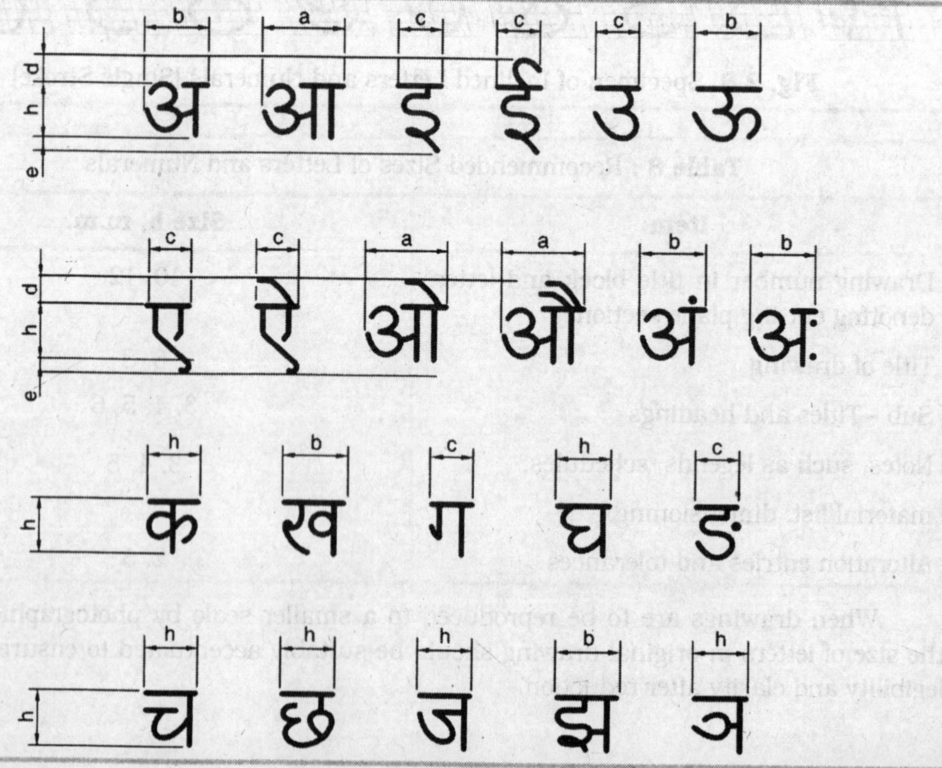

Fig. 2.10. Specimen of Hindi Letters

EXERCISE

1. What do you understand by trimmed and untrimmed drawing sheet sizes ? Prepare a list of recommended sizes of drawing sheets.
2. State the considerations used in preparing 'Specification for paper sizes'.
3. Why is a scale necessary in preparation of a drawing ? State the recommended scales.
4. Enlist various types of lines used for machine drawings. State their relative thicknesses.
5. How will you decide about height of lettering ? Prepare a drawing of all English and all Hindi alphabets in vertical as well as in inclined style.
6. What are single stroke letters ? Where are they used ?

TITLE BLOCK FOR STUDENTS OF EDUCATIONAL INSTITUTIONS

	20	15	15	135	
9		DATE	SIGN.	L. D. COLLEGE OF ENGINEERING	11
8	6 STD.				
8	FAIR			AHMEDABAD- 15.	11
8	COMP.			ORTHOGRAPHIC	13
8	SHAH U. P.				
8	B. E. SEM.- IA			READING	13
8	ROLL NO - 101			10 DRG. NO - 5.95-96	17
8	SCALE 1:2				
	50			65 / 70	

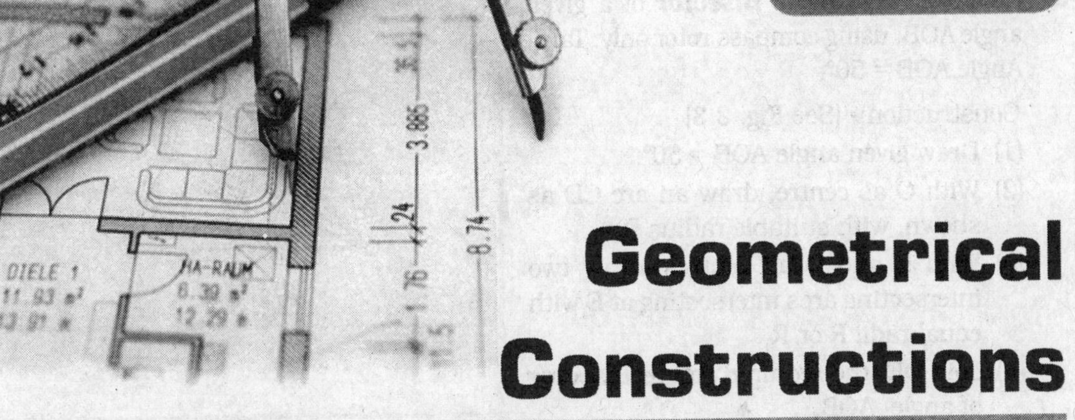

CHAPTER - 3

Geometrical Constructions

1. To draw ***A Line Parallel*** to a given line ***AB*** at a given distance R from it.

Construction:- [See Fig. 3.1]

(1) Draw a given line AB.

(2) With any two O_1 and O_2 points, on line AB, as centres and radius equal to R draw arcs 1 and 2 on the same side of AB as shown.

(3) Draw a line CD, common tangent to two arcs 1 and 2 already drawn.

(4) CD is the required ***Parallel Line to AB*** at a distance R.

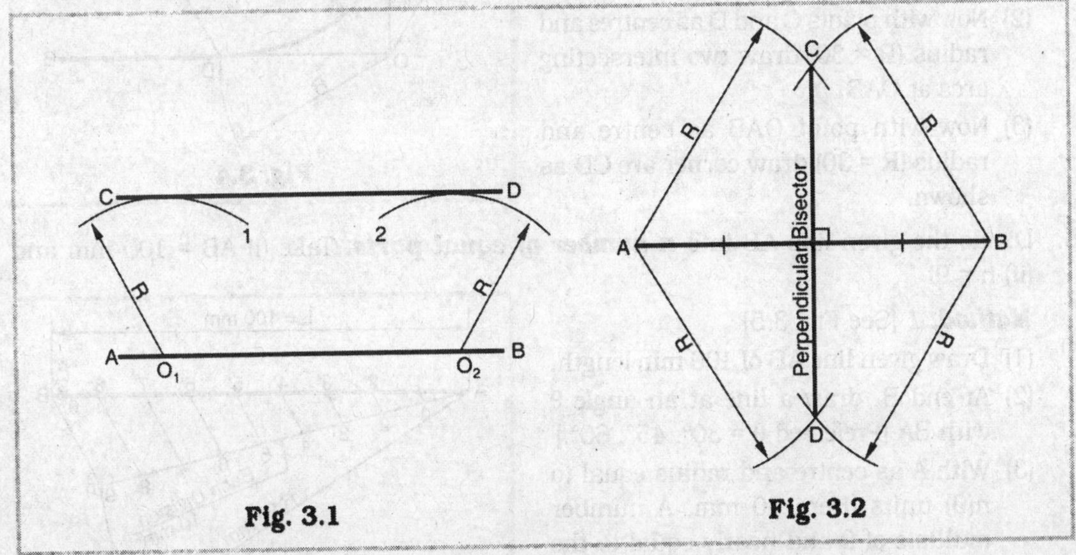

Fig. 3.1 **Fig. 3.2**

2. Construct ***A Perpendicular Bisector*** of a given line AB. Take AB = 80 mm. Construction:- [See Fig. 3.2]

(1) Draw a given line AB = 80 mm.

(2) With ends A and B as centres and radius R > AB/2, draw pair of two intersecting arcs intersecting at points C and D on either side of AB as shown.

(3) Join CD, ***The Perpendicular Bisector*** of a given line AB.

3. Construct ***An Angle Bisector*** of a given angle AOB, using compass rotor only. Take, Angle AOB = 50°.

Construction:- [See Fig. 3.3]

(1) Draw given angle AOB = 50°.

(2) With O as centre, draw an arc CD as shown, with suitable radius R.

(3) With C and D as centres, draw two intersecting arcs intersecting at E with equal radii R or R_1.

(4) Join OE, the required ***Angle Bisector*** of angle AOB.

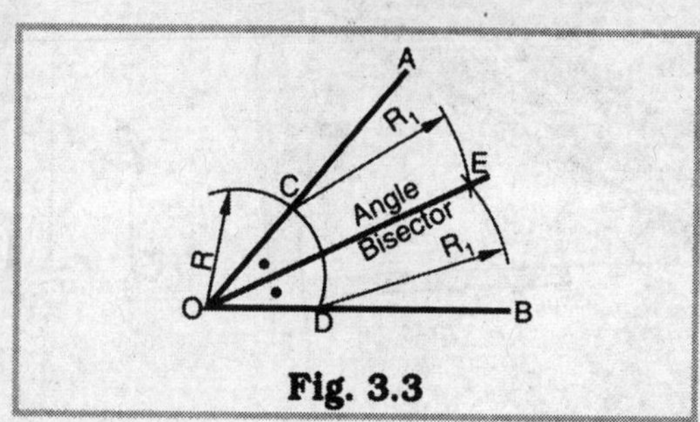

Fig. 3.3

4. To draw ***A Right Angled Corner Arc*** of radius R. Given AOB right angled corner. Given radius R = 30 mm.

Construction:- [See Fig. 3.4]

(1) With corner O as centre and radius equal to (R = 30) draw two arcs to get points C and D on lines OA and OB respectively.

(2) Now with points C and D as centres and radius (R = 30) draw two intersecting arcs at OAB.

(3) Now with point OAB as centre and radius (R = 30) draw corner arc CD as shown.

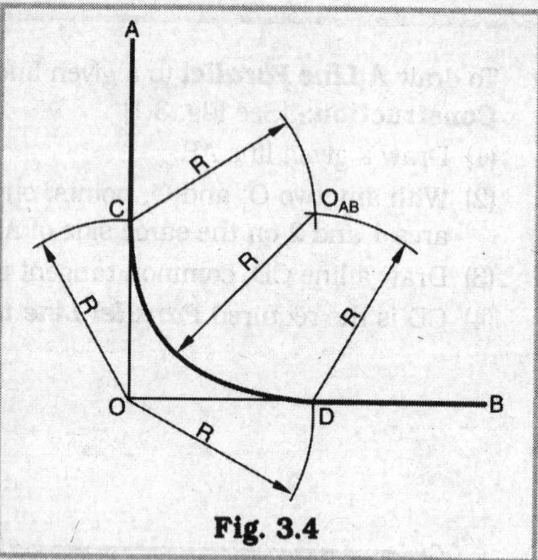

Fig. 3.4

5. Divide the given line AB into ***n number of equal parts.*** Take (i) AB = 100 mm and (ii) n = 9.

Method: 1 [See Fig. 3.5]

(1) Draw given line AB of 100 mm length.

(2) At end B, draw a line at an angle θ with BA [Preferred θ = 30°, 45°, 60°.]

(3) With A as centre and radius equal to n(9) units [Here 90 mm, A number multiple of 9 and nearly equal to the length of a given line AB = 100 mm] draw arc to cut previously drawn line at point n(9) as shown.

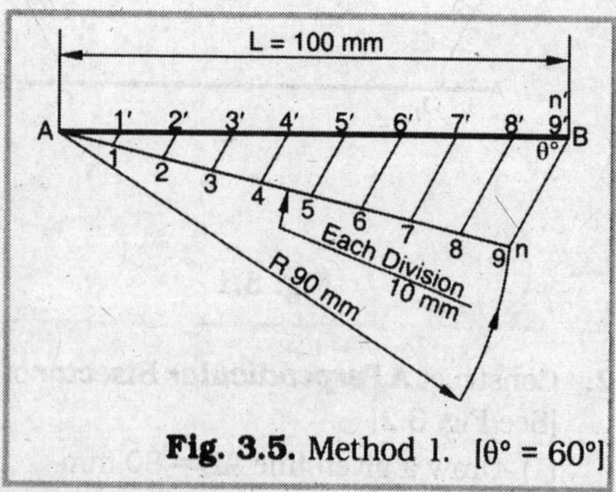

Fig. 3.5. Method 1. [θ° = 60°]

(4) With scale divide the line An into n(9) equal parts [10 mm each]. Mark them 1, 2, 3, ...n.

(5) Draw parallel lines to Bn line at θ angle from points 1, 2, 3, ...n. Mark corresponding points 1', 2', 3', ...n' on AB as shown.

(6) Points 1', 2', 3', ...n' divide the line AB into n(9) number of equal parts.

5A. Method: 2 [See Fig. 3.5A]

All other construction is same. Only A1 = 12 = 23 = ... = 89 are marked on a line by a suitable radius on compass rotor or divider.

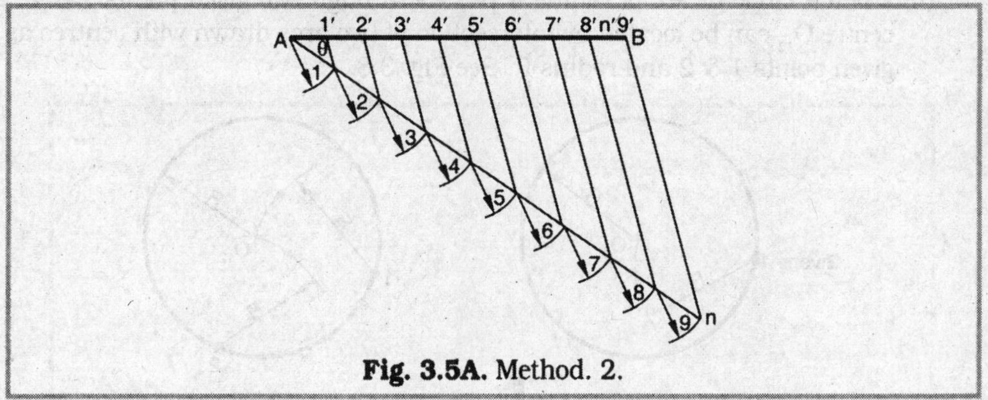

Fig. 3.5A. Method. 2.

6. To construct ***A Continuous Smooth Curve,*** consisting of arcs of circle, passing through given points. Given points (n = 6) with co-ordinates are O_1(5, 2), O_2(3, 4), O_3(2, 6), O_4 (4, 10), O_5(6, 14) and O_6(9, 12).

Construction:- [See Fig. 3.6]

(1) First plot 6 points O_1, O_2,...On using their given co-ordinates. Join O_{102}, O_2O_3, ... On–1On.

(2) Draw perpendicular bisectors of O_1 O_2, O_2O_3,...$O_{n-1}O_n$.

(3) Perpendicular bisectors of O_1O_2 and O_2O_3 intersects at centre C_{123}. Now with C_{123} as centre and radius equal to [$C_{123}O_1$] draw an arc of a circle $O_1O_2O_3$.

(4) Continue this method as shown in the figure to get centres C_{34}, C_{45}, C_{56}. Draw arcs similarly to get continuous curve $O_1O_2O_3O_4O_5O_6$ as shown.

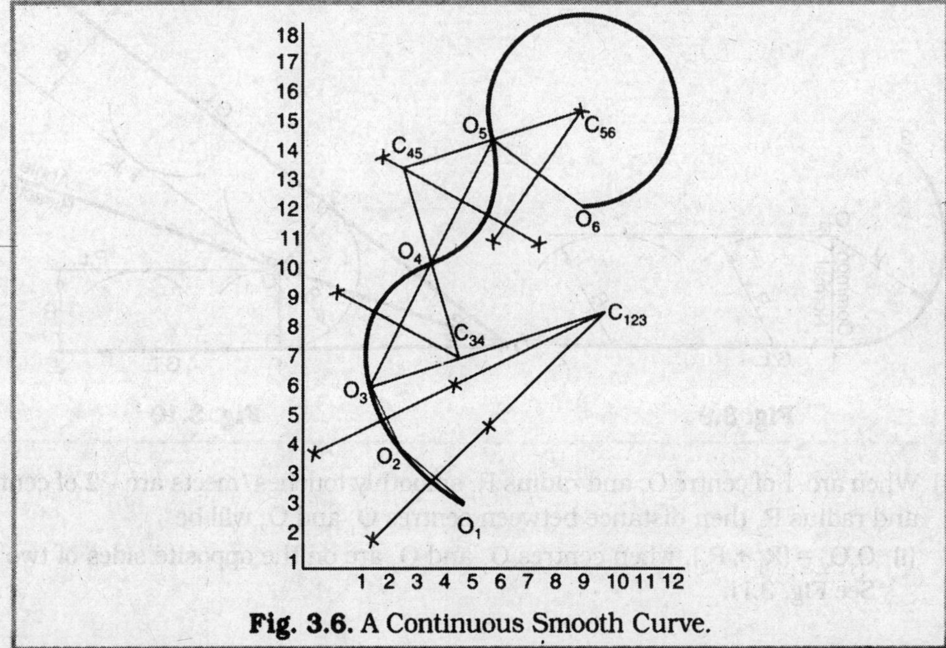

Fig. 3.6. A Continuous Smooth Curve.

7. *Important Rules For Figure Redrawing:-*

(1) (i) Centre (O_1) of a circle or arc, passing through a given point-1 will be at a radial distance R from the given point-1 See Fig. 3.7.

(ii) When a circle or arc of radius R passes through two given points 1 & 2, then centre O_{12} can be located by intersection of two arcs drawn with centres as two given points 1 & 2 and radius R. See Fig. 3.8.

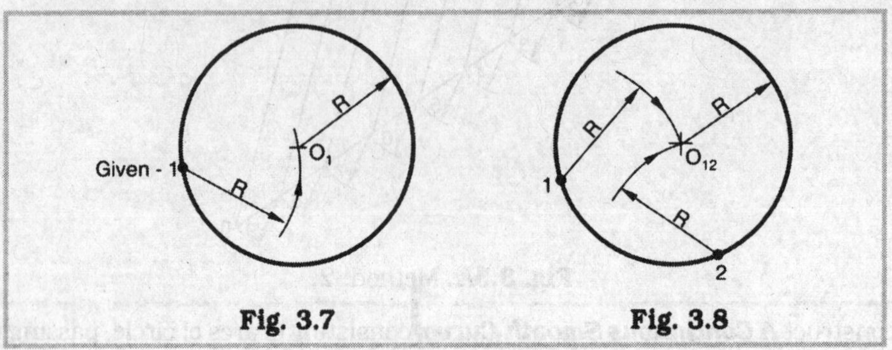

Fig. 3.7 **Fig. 3.8**

(2) Centre O_{13} of the arc of radius R[of the circle] touching/meeting a given st-line 12 will be on another st-line parallel to the given st-line 12 at a distance R. See Fig. 3.9.

When arc smoothly touches two given st-lines 12 and 34, then centre can be located by intersection of two such parallel st-lines. See Fig. 3.10. Centre will be on an angle bisector of angle 43012.

(3) When an arc drawn with centre O_{13} smoothly meets a st-line 12, then the joining common meeting point 1 and centre O_{13} will be a common normal to the arc 13 and to the st-line 12. See Fig. 3.9.

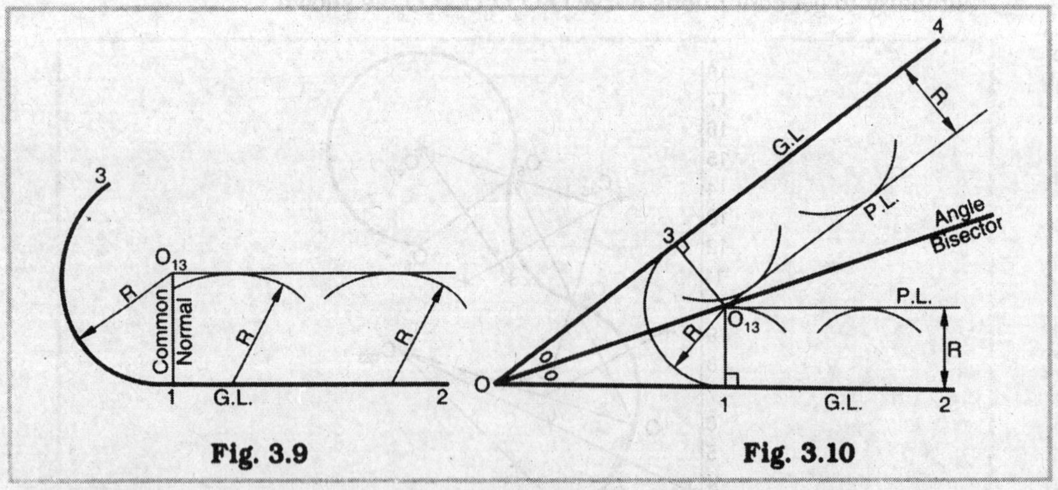

Fig. 3.9 **Fig. 3.10**

(4) When arc-1 of centre O_1 and radius R_1 smoothly touches/meets arc – 2 of centre O_2 and radius R_2 then distance between centres O_1 and O_2 will be

(i) $O_1O_2 = (R_1 + R_2)$, when centres O_1 and O_2 are on the opposite sides of two arcs. See Fig. 3.11.

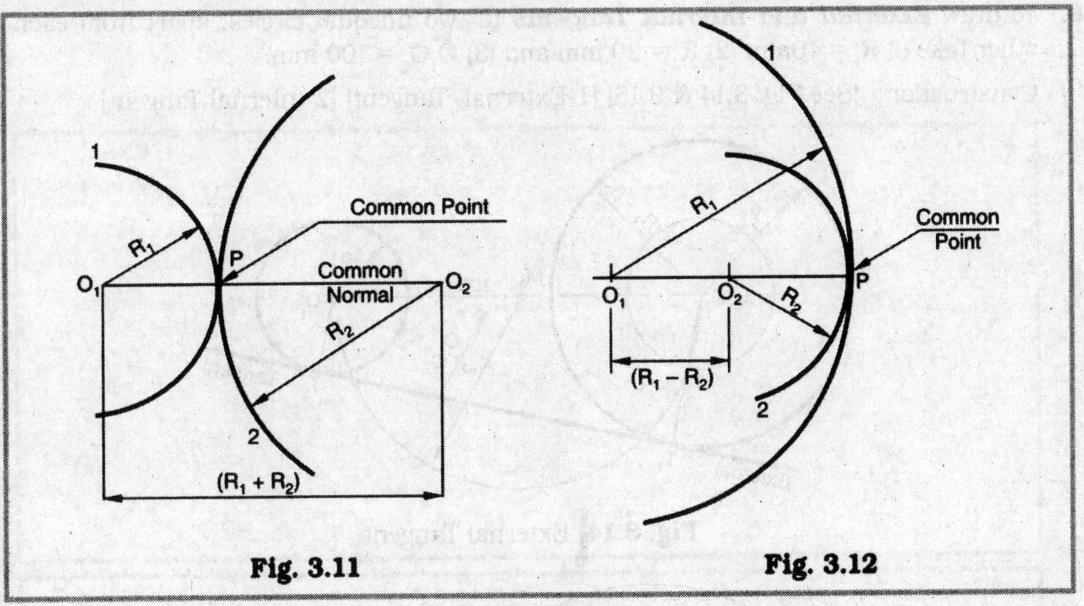

Fig. 3.11 **Fig. 3.12**

(ii) $O_1O_2 = \pm(R_1 - R_2)$, when centres O_1 and O_2 are on the same side of two arcs. See Fig. 3.12.

Common meeting point of two meeting arcs will be on a line joining centres of two arcs. This line joining centres is a common normal to two arcs.

(5) Centre O of arc smoothly touching two arcs 1 & 2 will be at a distance.

(i) $(R_1 + R)$ from centre O_1 and $(R_2 + R)$ from centre O_2, if centre O is on opposite sides of O_1 & O_2. See Fig. 3.13B.

(ii) $\pm(R - R_1)$ from centre O_1 and $\pm(R - R_2)$ from centre O_2, if centre O is on the same side of O_1 & O_2. See Fig. 3.13A.

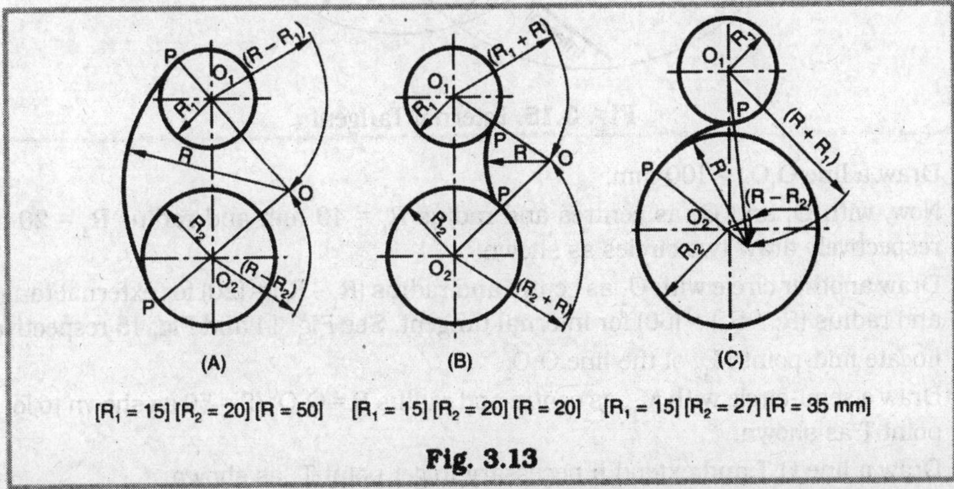

Fig. 3.13

In short if centres are on opposite sides summation of radii is to be done and if centres are on the same side subtraction of radii is to be done. Common point to two meeting arcs will on a line joining centres. This line is a common normal to two arcs.

8. To draw ***External and Internal Tangents*** to two unequal circles, apart from each other. Take (i) R_1 = 40mm (2) R_2 = 20 mm and (3) O_1O_2 = 100 mm.

Construction:- [See Fig. 3.14 & 3.15] [1-External Tangent] [2-Internal Tangent]

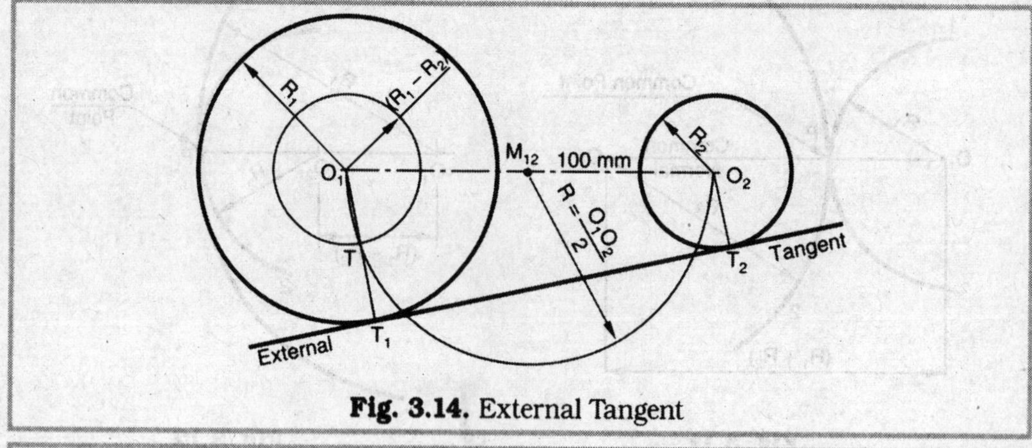

Fig. 3.14. External Tangent

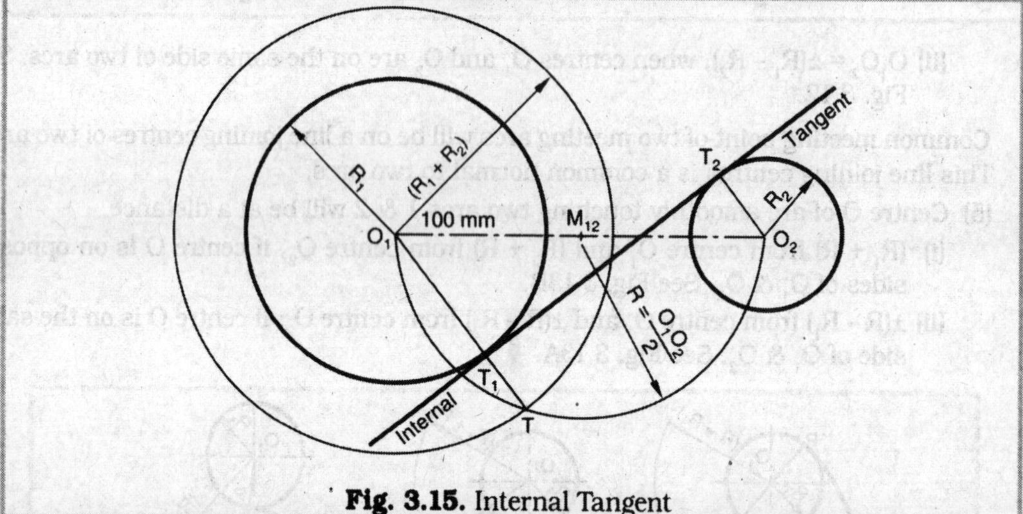

Fig. 3.15. Internal Tangent

(1) Draw a line O_1O_2 = 100 mm.

(2) Now, with O_1 and O_2 as centres and radius R_1 = 40 mm and radius R_2 = 20 mm respectively draw two circles as shown.

(3) Draw another circle with O_1 as centre and radius $(R_1 - R_2)$ = (20) for external tangent and radius $(R_1 + R_2)$ = (60) for internal tangent, See Fig. 14 and Fig. 15 respectively.

(4) Locate mid-point M_{12} of the line O_1O_2.

(5) Draw a semi-circle with M_{12} as centre and radius $R = O_1O_2/2 = 50$ as shown to locate point T as shown.

(6) Draw a line O_1T and extend if necessary to get point T_1 as shown.

(7) Draw a line O_2T_2 parallel to O_1T_1 as shown.

(8) Join T_1T_2 line, the required external/internal tangent.

(9) In this manner open and cross belt drive pulley systems can be drawn.

9. *Practice Problems of Figure Redrawing. (1 to 6)*

1. ***Redraw the Fig. 3.16 (a)***, given below. ***[M.S.B.T.E. November/December 1998] [8]***

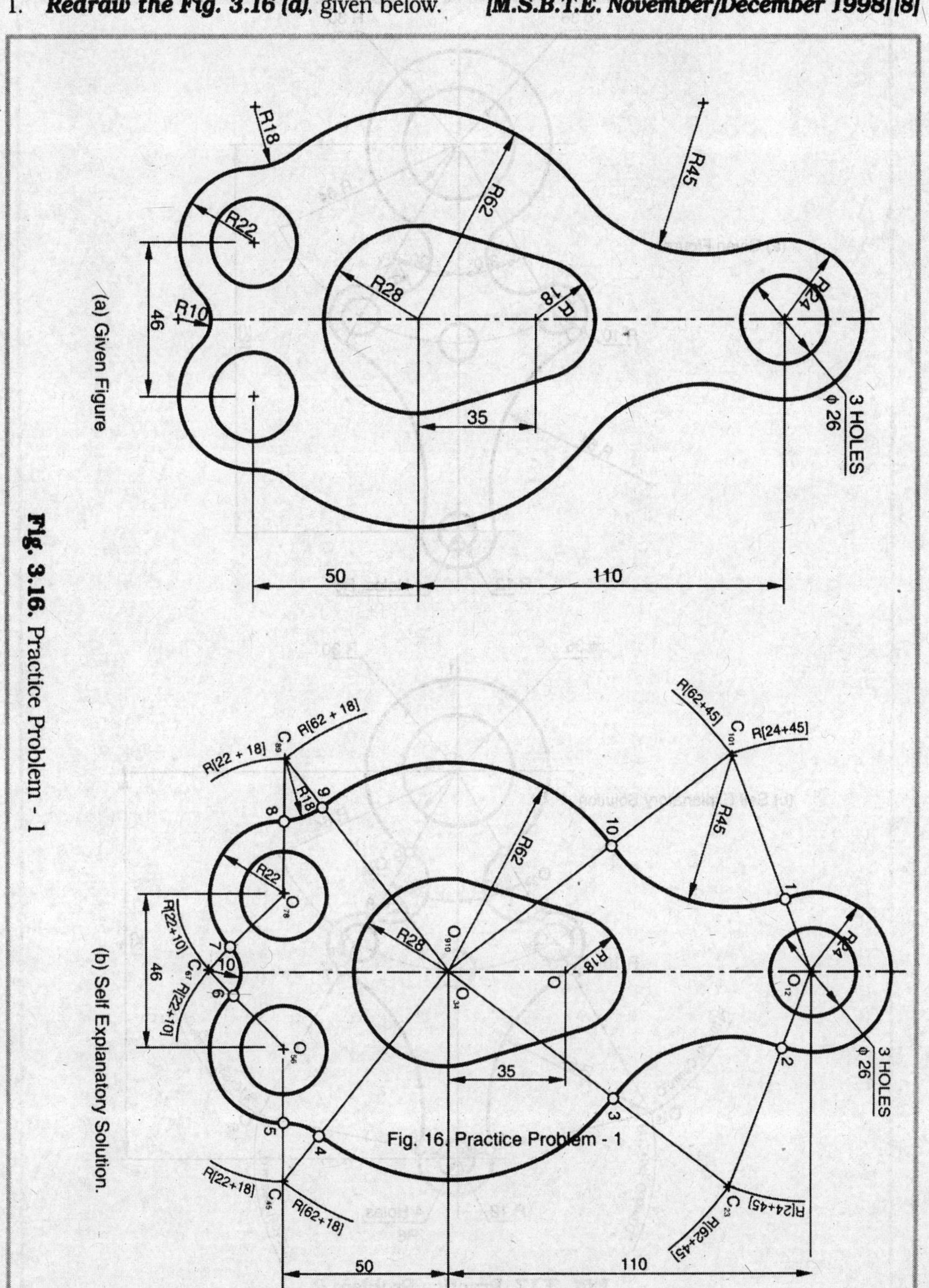

Fig. 3.16. Practice Problem - 1

2. ***Redraw the Fig. 3.17 (a),*** given below.

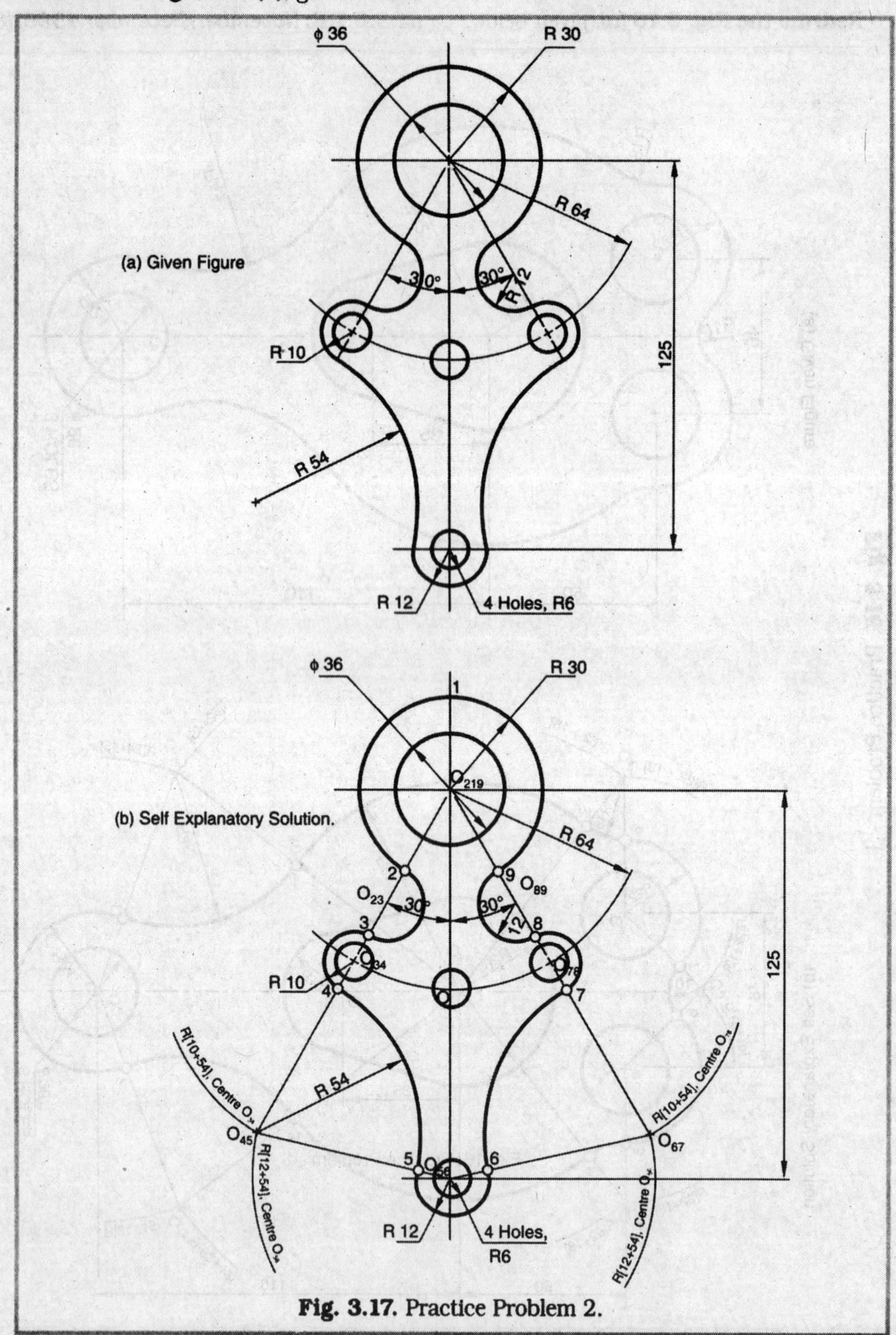

Fig. 3.17. Practice Problem 2.

3. ***Redraw the Fig. 3.18 (a),*** given below. [M.S.B.T.E. April/May 1998] [8]

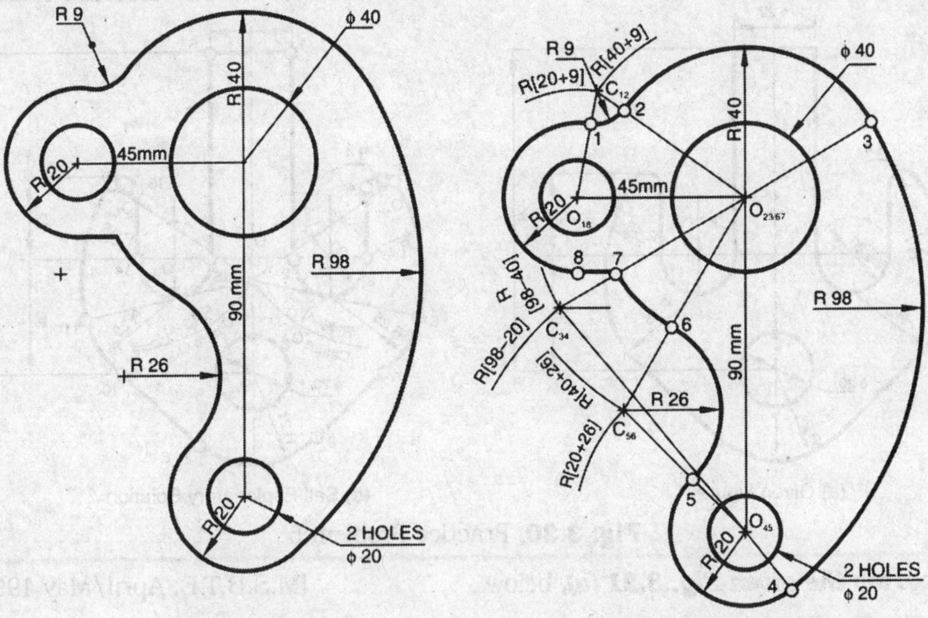

(a) Given Figure

(b) Self Explanatory Solution

Fig. 3.18. Practice Problem 3

4. **Redraw the Fig. 3.19(a)**, given below. [M.S.B.T.E. April 2000] [8]

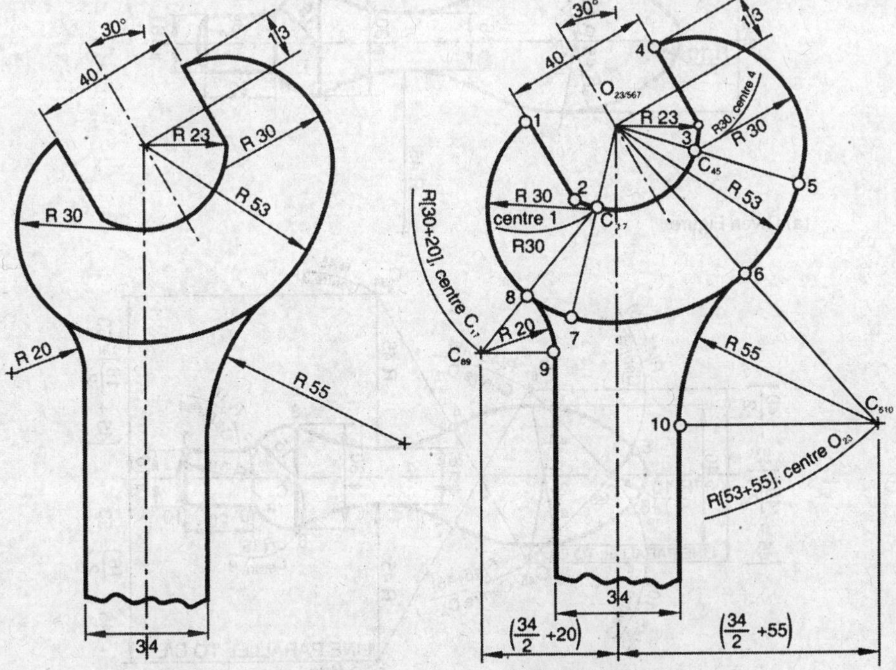

(a) Given Figure

(b) Self Explanatory Solution

Fig. 3.19. Practice Problem 4

5. ***Redraw the Fig. 3.20 (a),*** given below

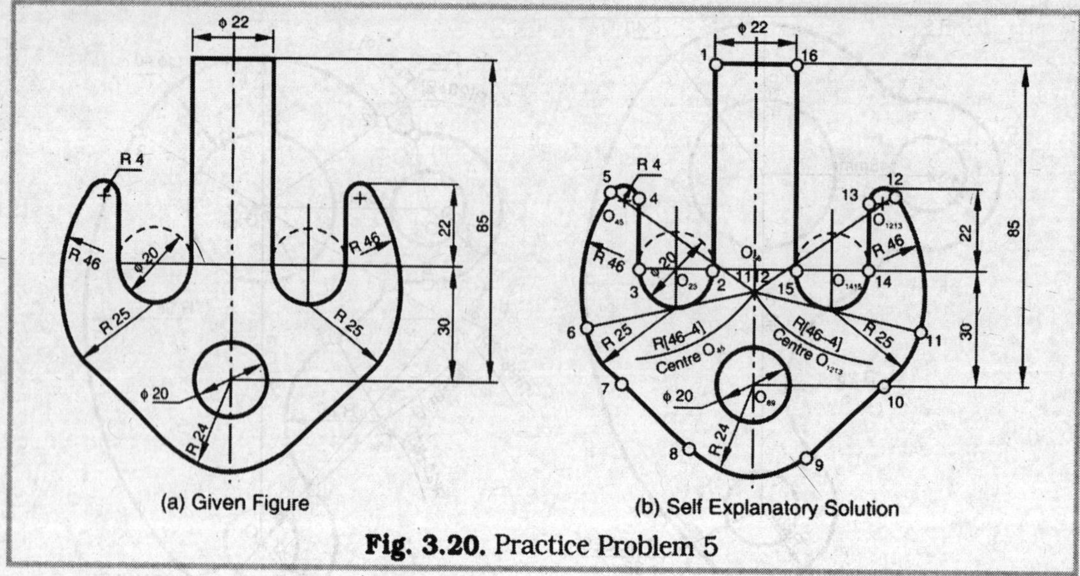

(a) Given Figure

(b) Self Explanatory Solution

Fig. 3.20. Practice Problem 5

6. ***Redraw the given Fig. 3.21 (a),*** below. [M.S.B.T.E. April/May 1996] [4]

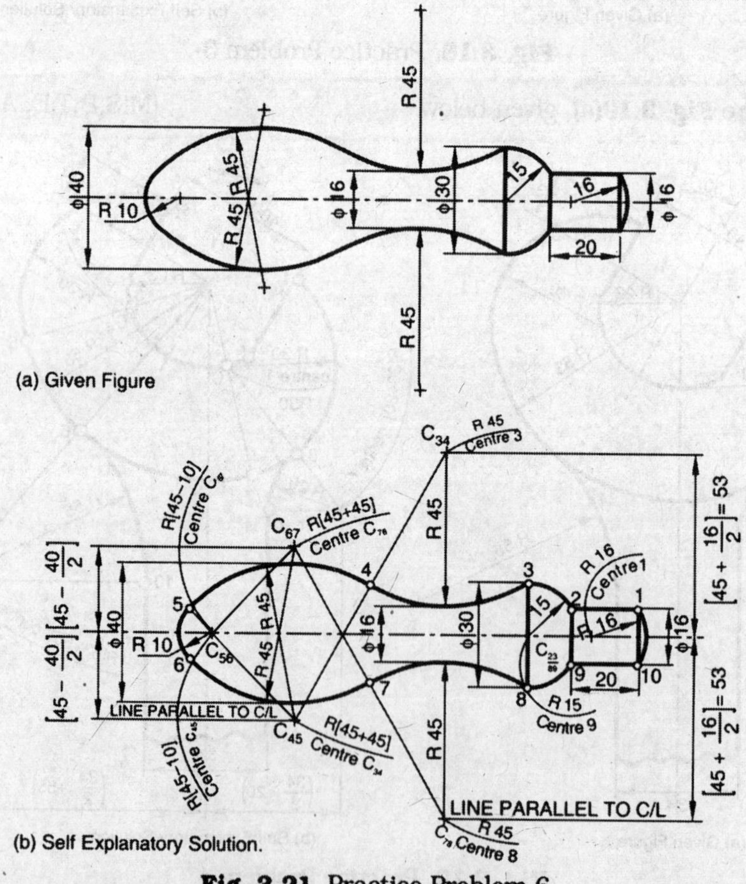

(a) Given Figure

(b) Self Explanatory Solution.

Fig. 3.21. Practice Problem 6

10. Construction of ***Regular Polygons***

Students should learn the geometry and construction of regular polygons like

(1) Equilateral Triangle. [$\theta = 60°$]

(2) Square. [$\theta = 90°$]

(3) Pentagon. [$\theta = 108°$]

(4) Hexagon. [$\theta = 120°$]

(5) Heptagon. [$\theta = 180 - \frac{360}{7}$] [$n = 7$]

(6) Octagon etc. [$\theta = 135°$]

Sum of the external angles of a regular polygon is 360°. Hence each external angle is $\left[\frac{360}{n}\right]^{\circ}$.

$$\therefore \begin{bmatrix}\text{Internal Angle of} \\ \text{Regular Polygon}\end{bmatrix}\theta = \left[180° - \frac{360}{n}\right]^{\circ}$$

Equilateral triangle and square are simple and common. Hence not discussed in this chapter.

(1) To construct ***A Hexagon by using only compass rotor.*** See Fig. 3.22.

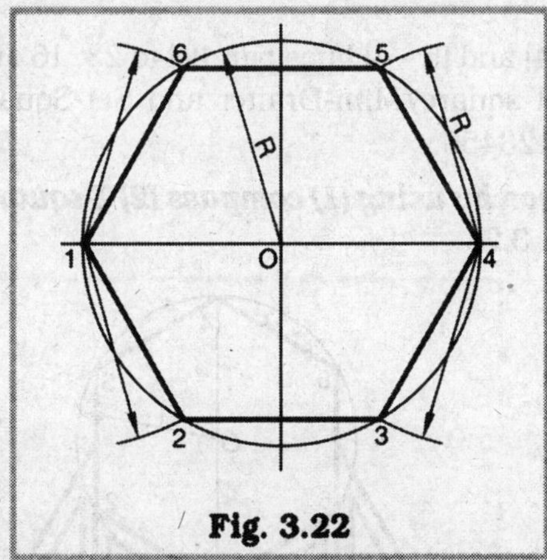

Fig. 3.22

(i) Draw a circle of radius R, equal to the side of hexagon.

(ii) Mark any point 1 on the circumference of a circle. Point 1 can be taken on extreme left side or on extreme top side depending upon orientation required of the hexagon.

(iii) Draw diameter 1-4 from point 1.

(iv) Now with points 1 and 4 as the centres and radius R draw set of two arcs on two sides cutting the circle at points [2 – 6] and [3 – 5] respectively as shown.

(v) Join points [1 – 2], [2 – 3], [3 – 4], [4 – 5], [5 – 6], [6 – 1] to get hexagon.

(2) To construct ***A Hexagon by using (1) T-square [or] Mini-Drafter alongwith (2) 30°-60° set-square.*** See Fig. 3.23.

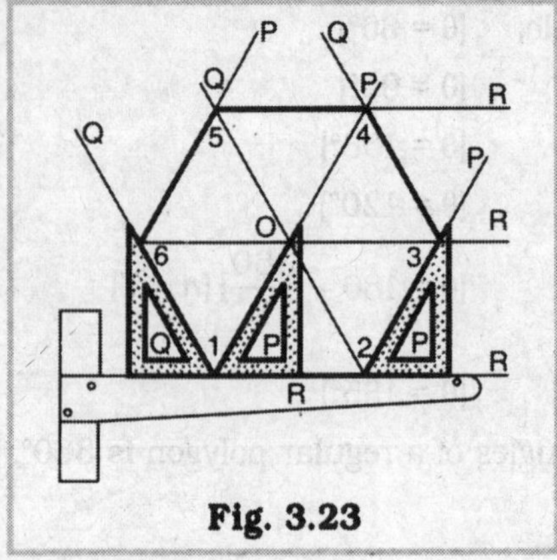

Fig. 3.23

(i) First draw horizontal line 1 – 2, of the length equal to the side of regular hexagon.

(ii) Draw two lines (P) 1 – 4 and (Q) 1 – 6 by the combination of T-square/Mini-Drafter and 30° – 60° set square as shown.

Similarly draw two lines (Q) 2 – 5 and (P) 2 – 3.

Mark the centre point O, where line 25 intersects line 14. Point O will be the centre of hexagon.

(iii) Draw horizontal line through the centre point O. Mark points of intersection 6 and 3 as shown.

(iv) Draw [6 – 5], [3 – 4] and [5 – 4] lines parallel to 23, 16 and 12 respectively by similar combinations of T-square/Mini-Drafter and Set-Square. This way completes the regular hexagon 123456.

(3) To construct ***A Hexagon by using (1) compass (2) T-square/Drafter and (3) 30° – 60° sets square.*** See Fig. 3.24.

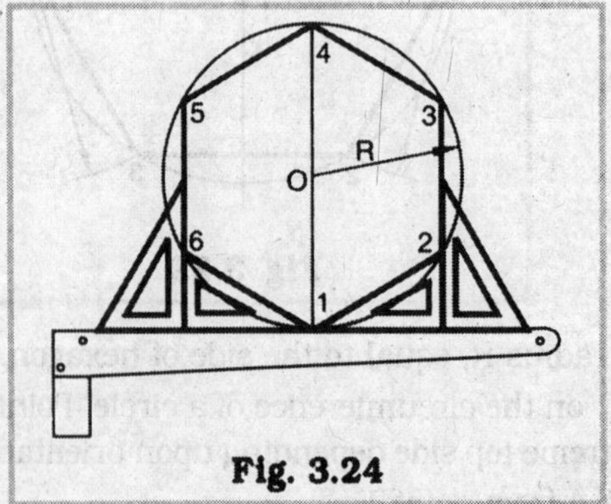

Fig. 3.24

(i) First draw a circle of radius R, equal to the given side of hexagon.

(ii) Mark lowest point 1 on the circumference of a circle, vertically below the centre of the circle, by using T-Square/Drafter.

(iii) Draw two chord lines 12 and 16 by combination of T-square/Drafter and 30° set square as shown.

(iv) Draw two vertical chord lines 65 and 23 by combination of T-square/Drafter and 90° set squares as shown. It can be drawn only with drafter.

(v) Draw two chord lines 54 and 34 in the similar manner as (iii) and complete the hexagon.

(4) ***To construct a regular polygon of n sides.*** Length L of side is given. In this illustrative problem ***A Heptagon (n = 7)*** is drawn. See Fig. 3.25.

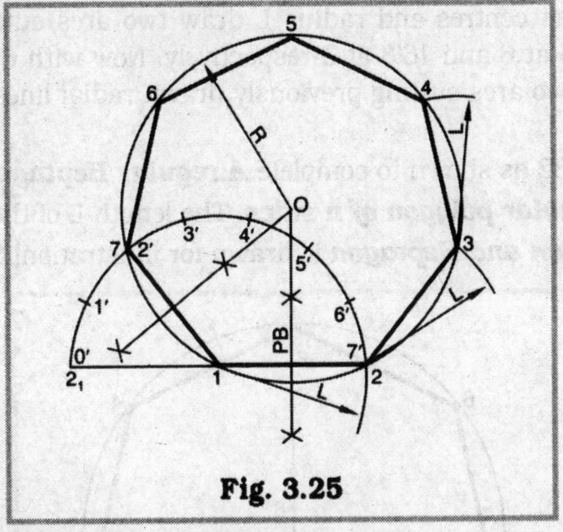

Fig. 3.25

(i) First draw a horizontal line 12 of given length L.

(ii) Draw a semicircle with centre 1 and radius R = 12 = L as shown.

Divide the circumference of the semicircle into n = 7 equal parts either by divider or by protractor. Mark points 0', 1', 2'...n' (7') as shown.

(iii) Join 12'.

(iv) Draw two perpendicular bisectors of lines 12 and 12' by using compass rotor as shown. Mark point of intersection O as shown.

(v) Draw a full circle with O as centre and radius equal to R = 01 = 02' = as shown.

(vi) By compass, of adjusted length L, mark points 3, 4, 5 and 6 on the circumference of full circle. Join them to complete ***A Heptagon.***

5. ***To construct a regular polygon of n sides.*** Length L of the side is given. In this illustrative problem ***A Heptagon*** (n = 7) is drawn. See Fig. 3.26.

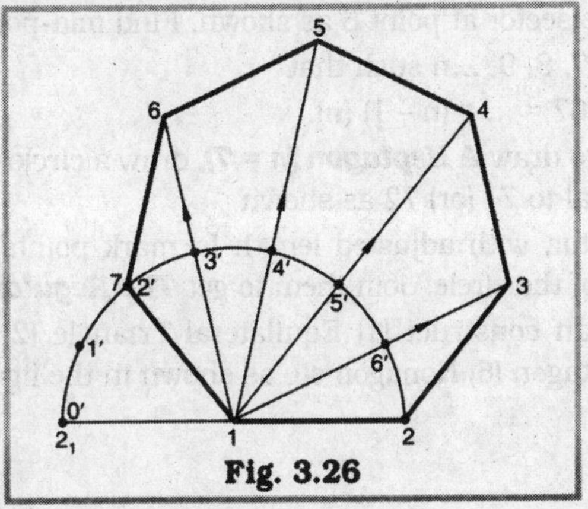

Fig. 3.26

(i) First draw a horizontal line 12 of given length L.

(ii) Draw a semicircle with centre 1 and radius R = 12 = L as shown.

Divide the circumferences of the semicircle into n = 7 equal parts either by divider or by protractor. Mark points 0′, 1′, 2′...n′ (7′) as shown.

(iii) Join 12′.

(iv) Draw radial lines 13′ 6, 14′ 5′ 15′ 4, 16′ 3and so on.

(v) With 2′ and 2 as centres and radius L draw two arcs cutting previously drawn radial lines 13′6 at 6 and 16′3 at 3 respectively. Now with 6 and 3 as centres and radius L draw two arcs cutting previously drawn radial lines 14′5 at 5 and 15′4 at 4 respectively.

(vi) Join lines and 32 as shown to complete ***A regular Heptagon.***

6. ***To construct a regular polygon of n sides.*** The length L of the side is given. [Note:- ***n = 7 is taken and Heptagon*** is drawn for illustration] See Fig. 3.27.

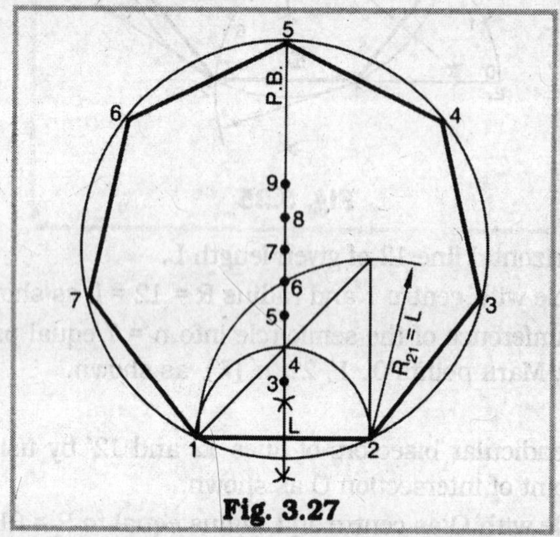

Fig. 3.27

(i) First draw a line 12 of length L.

(ii) Draw perpendicular bisector (P.B.) of line 12 as shown.

(iii) Draw a semicircle, with line 12 as diameter, cutting perpendicular bisector at point 4.

(iv) Draw a quarter circle, with point 2 as the centre and R21 = L as the radius, cutting perpendicular bisector at point 6 as shown. Find mid-point 5 of length 46.

(v) Mark points 3, 7, 8, 9, ...n such that

34 = 45 = 56 = 67 = ... = (n – 1) (n).

(vi) Since we want to draw ***A Heptagon (n = 7),*** draw a circle with point 7 as the centre and radius equal to 71 (or) 72 as shown

(vii) By compass rotor, with adjusted length L, mark points 3, 4, 5, 6 and 7 on the circumference of the circle. Join them to get ***The Regular Heptagon.***

(viii) Similarly we can construct (1) Equilateral Triangle (2) Square (3) Pentagon (4) Hexagon (5) Octagon (6) Nonagon etc as shown in the figure. See Fig. 3.28.

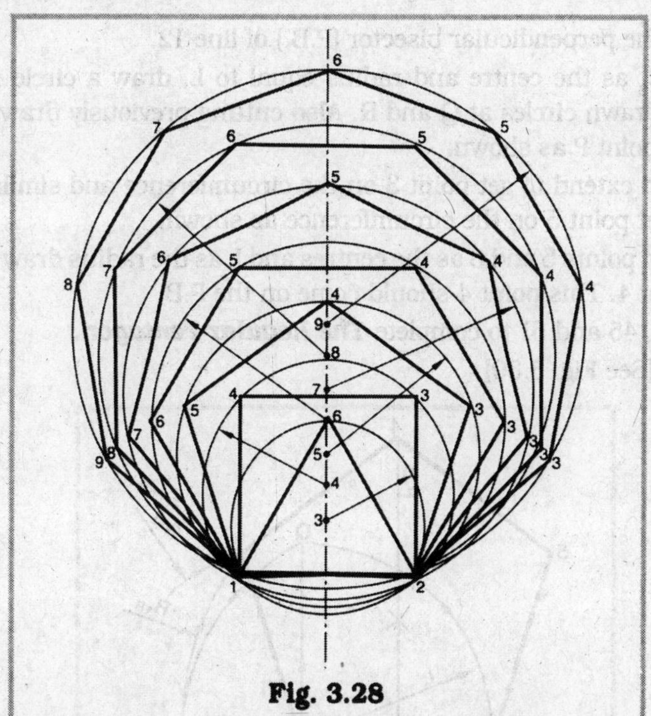

Fig. 3.28

Special Methods for Construction of A Regular Pentagon.

7. ***To construct A Regular Pentagon*** of given length of side L by special method-1 and method-2.

Method: 1 [See Fig. 3.29]

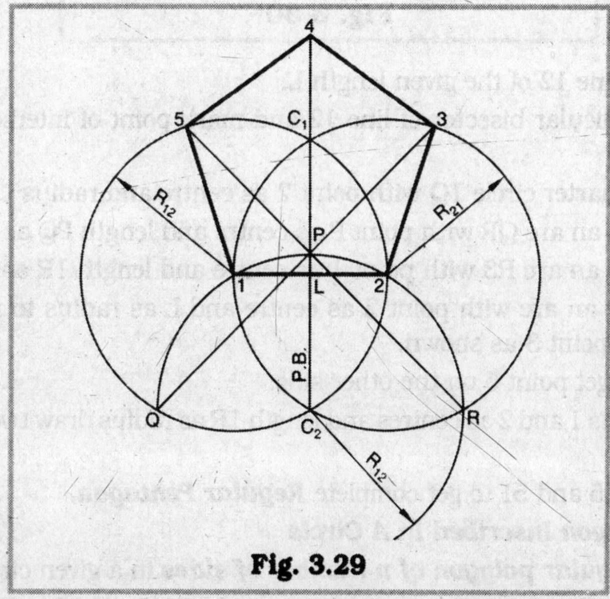

Fig. 3.29

(i) First draw a horizontal line 12 of given length L.

(ii) Draw two circles ⊙[1, R_{12} = L] and ⊙[2, R_{21} = L] intersecting at points C_1 and C_2 as shown.

Join C_1C_2, the perpendicular bisector (P.B.) of line 12.

(iii) Now with C_2 as the centre and radius equal to L, draw a circle intersecting two previously drawn circles at Q and R. Also cutting previously drawn perpendicular bisector at point P as shown.

(iv) Join QP and extend to get point 3 on the circumference and similarly join RP and extend to get point 5 on the circumference as shown.

(v) Finally, with points 5 and 3 as the centres and L as the radius draw two intersecting arcs at point 4. This point 4 should come on the P.B.

(vi) Join 23, 34, 45 and 51 to complete ***The Regular Pentagon.***

Method: 2 [See Fig. 3.30]

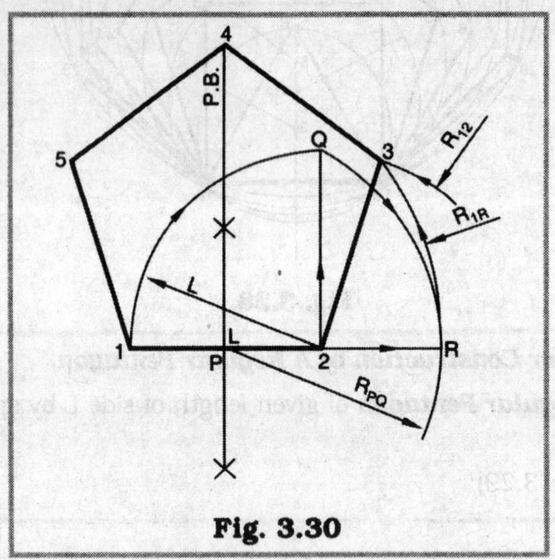

Fig. 3.30

(i) First draw a line 12 of the given length L.

(ii) Draw perpendicular bisector of line 12 and mark point of intersection P on 12 as shown.

(iii) (1) Draw a quarter circle 1Q with point 2 as centre and radius L as shown.

(2) Now draw an arc QR with point P as centre and length PQ as radius as shown.

(3) Now draw an arc R3 with point 1 as centre and length 1R as radius.

(4) Now draw an arc with point 2 as centre and L as radius to intersect with arc (1, 1R) at point 3 as shown.

(5) Similarly get point 5 on the other side.

(iv) Now with points 1 and 2 as centres and length 1R as radius draw two arcs intersecting at point 4.

(v) Join 23, 34, 45 and 51 to get complete ***Regular Pentagon.***

Regular Polygon Inscribed In A Circle

8. ***To inscribe a regular polygon of n number of sides*** in a given circle of radius R. [For illustration ***A Pentagon*** is inscribed in a circle **⊙(O, R).**] See Fig. 3.31

General Method:- For Polygon of n sides.

(i) First draw a circle with point O as centre and given radius R.

(ii) Draw diameter 15′ and divide it into n = 5 equal parts. Mark them as 1′, 2′, 3′,...n′ = 5′ as shown.

(iii) Now draw two arcs, with centres 1 and 5′ and radius $R1_{5'}$ and $R_{5'1}$ respectively and both equal to 2R as shown. Get the point of intersection C as shown.

(iv) Join C2′ and extend to get point 2 on the circumference of the circle ⊙(O,R).

(v) 12 is going to be the length of each side of the pentagon. Hence by compass, adjusted for length 12, draw arcs to get points 3, 4 and 5 on the circumference of the circle ⊙(O,R).

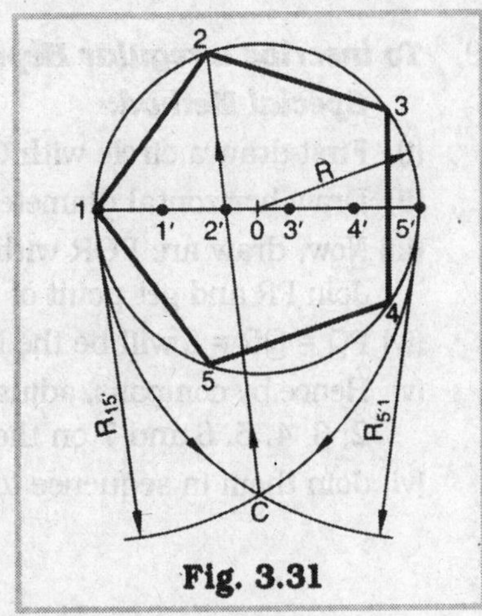

Fig. 3.31

(vi) Join lines 12, 23,...51 to get the required regular pentagon 123451 as shown.

Special Method For A Regular Pentagon:-

(i) First draw a circle with centre O and given radius R.

(ii) Draw horizontal diameter POQ and vertical diameter 101′.

(iii) Draw perpendicular bisector C_1CC_2 of OP as shown.

(iv) Now, with C as centre and radius C1, draw an arc 1S as shown. Further draw an arc S2 with centre 1 and radius 1S to get point 2 on the circumference of a circle ⊙(O,R). 12 is going to be the length of each side of a regular pentagon.

(v) Now by compass, adjusted for length 12, draw arcs one after the other to get points 3, 4 and 5 on the circumference of a circle ⊙(O,R). Join them in sequence to get the required ***Regular Pentagon.***

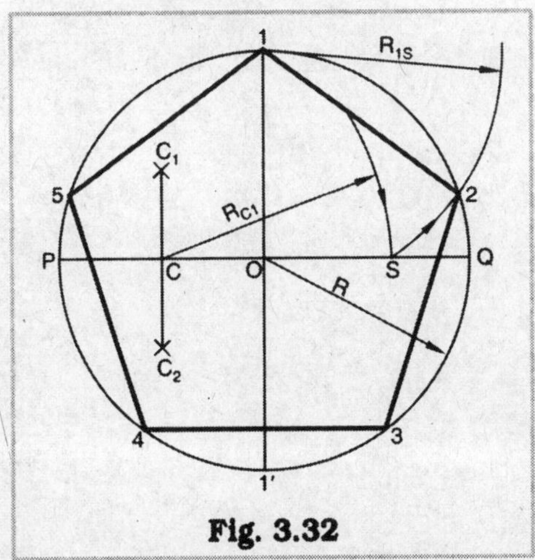

Fig. 3.32

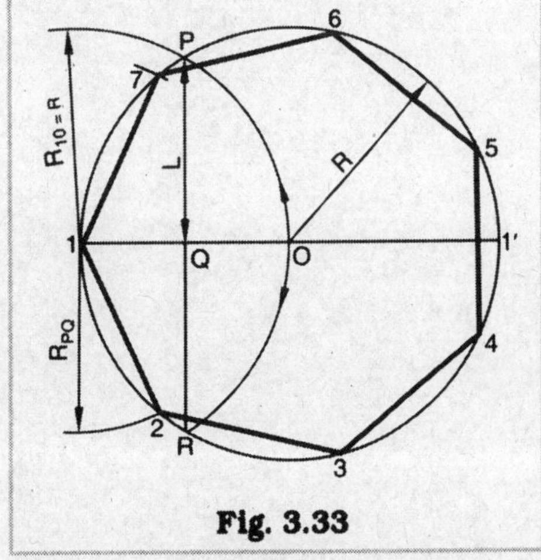

Fig. 3.33

9. ***To inscribe a regular Heptagon (n = 7) in a given circle ⊙(O,R)*** See Fig. 3.33

Special Method:-

(i) First draw a circle with O as centre and given radius R.

(ii) Draw horizontal diameter 101′ a shown.

(iii) Now, draw arc POR with point 1 as centre and radius equal to 10 = R as shown. Join PR and get point of intersection Q on diameter 101′ as shown.

(iv) PQ = QR = L will be the length of each side of a regular heptagon.

(v) Hence by compass, adjusted for length L, draw arcs one after the other to get points 2, 3, 4, 5, 6 and 7 on the circumference of a circle ⊙(O,R).

(vi) Join them in sequence to get ***A Regular Heptagon*** 12345671 as shown.

CHAPTER - 4

Scales

1. Introduction :

Drawings of objects are prepared on drawing paper. Size of objects may be extremely small, medium or extremely large. Drawing papers are available in standard sizes. Medium size objects are drawn on medium size drawing papers with their true sizes i.e. drawn to full size scale. Large objects are drawn with reduced scale to accomodate their drawings on small and medium size papers, e.g. The Map of India drawn on imperial size paper.

Sometimes extremely small objects, like watch mechanism, are drawn to an enlarged scale. Here the problem is not of accomodating the drawing on the drawing paper but of difficulty in drawing small size drawings accurately on the drawing paper.

I.S. have recommended the following standard scales.

(1) Full Scale (R.F.)	(2) Reduced Scale (R.F.)		(3) Enlarged Scale (R.F.)
1 : 1	1 : 2	1 : 20	10 : 1
	1 : 2.5	1 : 50	5 : 1
	1 : 5	1 : 100	2 : 1
	1 : 10	1 : 200	

Normally, ready available standard scales are adopted and objects are drawn either to full scale, reduced scale or to an enlarged scale. But sometimes it becomes necessary to prepare specific scale which suit the size of objects and corresponding to the size of the paper.

We shall now study the methods of preparing such non-standard specific scales.

2. Representative Fraction :

The ratio of the size of the element in the drawing to the size of the same element in the object is called the REPRESENTATIVE FRACTION (R.F.). Suppose a line of 10 mm length in drawing represents 1 metre length of the object than the R.F. is equal to

$$\frac{10 \text{ mm}}{1 \text{ metre}} = \frac{10 \text{ mm}}{1000 \text{ mm}} = \frac{1}{100}$$

Therefore, the scale of the drawing will be 1 : 100. Here size of the drawing is smaller than the actual size of the object and hence the scale is known as reduced scale (1 : 100).

Now, suppose a line of 10 mm length in drawing represents 1 mm length of the object than the R.F. is equal to

$$\frac{10 \text{ mm}}{1 \text{ mm}} = \frac{10}{1}$$

Therefore, the scale of the drawing will be 10 : 1. Here scale 10 : 1 is an enlarged scale. Enlarged scales are used to draw extremely small objects.

3. Types of Scales :

Scales are classified in two different manner as under :

1. (a) Mechanical Engineers' Scale
   (b) Architects' Scale
   (c) Civil Engineers' Scale

2. (a) Plain Scale
   (b) Diagonal Scale
   (c) Comparative Scale
   (d) Vernier Scale
   (e) Scale of Chords
   (f) Isometric Scale.

3.1 (a) Mechanical Engineers' Scale :

These scales are 300 mm long and each unit is sub-divided. Mechanical engineers generally use following scales.

1:1, 1:2, 1 : 2.5, 1 : 5, 2 : 1 and 5 : 1 only.

(b) Architects' Scale :

Architects are required to take very small R.F., since buildings are comparatively very big as compare to drawing paper size. Only the first main division of the architects' scale is sub-divided.

(c) Civil Engineers' Scale :

Civil engineers dealing with road maps and survey maps are required to take very very small R.F. These scales are sub-divided on their entire lengths.

According to I.S. 962-1967 metric scales used for architectural and building drawings are as below :

(a) Topographical Maps

1 cm = 1 km $\left[\frac{1}{100000}\right]$

1 cm = 0.5 km $\left[\frac{1}{50000}\right]$

(c) Large scale surveys and layouts

1 cm = 20 m $\left[\frac{1}{2000}\right]$

1 cm = 10 m $\left[\frac{1}{1000}\right]$

1 cm = 5 m $\left[\frac{1}{500}\right]$

(b) Town Surveys

2 cm = 1 km $\left[\frac{1}{50000}\right]$ 4 cm = 1 km $\left[\frac{1}{25000}\right]$

10 cm = 1 km $\left[\frac{1}{10000}\right]$ 1 cm = 50 m $\left[\frac{1}{5000}\right]$

(d) Preliminary or sketch drawing

1 cm = 5 m $\left[\frac{1}{500}\right]$ 1 cm = 2 m $\left[\frac{1}{200}\right]$

1 cm = 1 m $\left[\frac{1}{100}\right]$

(e) Working drawings, plans, elevations and sections

1 cm = 2 m $\left[\frac{1}{200}\right]$ 1 cm = 1 m $\left[\frac{1}{100}\right]$

1 cm = 0.5 m $\left[\frac{1}{50}\right]$

(f) Large scale drawings-General details

1 cm = 20 cm $\left[\frac{1}{20}\right]$ 1 cm = 10 cm $\left[\frac{1}{10}\right]$

(g) Enlarged details

1 cm = 10 cm $\left[\frac{1}{10}\right]$ 1 cm = 5 cm $\left[\frac{1}{5}\right]$

1 cm = 2 cm $\left[\frac{1}{2}\right]$ 1 cm = 1 cm or Full size

3.2 (a) Plain Scale : See Fig. 4.1.

A plain scale is nothing but a straight line divided into suitable number of equal parts or units, the first unit of which is further subdivided. Plain scale represents a unit and its fraction or two interconnected units.

The features of the plain-scale are as under :

1. The zero is placed at the end of the first sub-divided main division.
2. From the zero mark, the units are numbered towards right side and its sub-divisions are numbered towards left side.
3. Units of the sub-divisions and of the divisions are stated either below or at the respective ends.
4. R.F. must be mentioned just below the scale.

Problem 1 : Construct a plain scale of R.F. 1 : 100 to show metres and decimetres. Maximum measurement required is 10 metres. Indicate 8 m 7 dm on the scale.

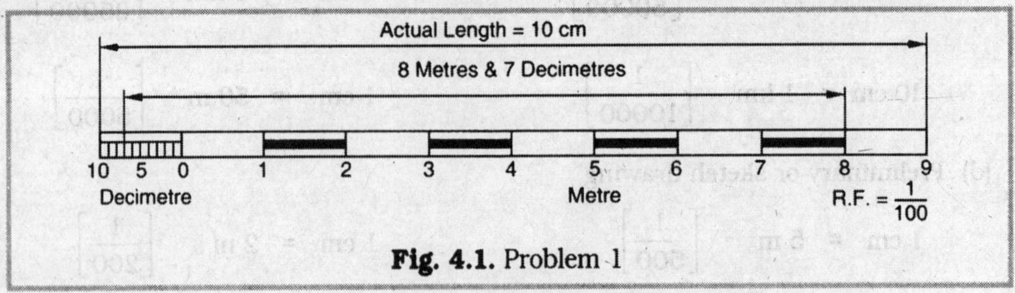

Fig. 4.1. Problem 1

For solution see Fig. 4.1 and follow the procedure as given below :

(i) First find the actual length of the scale by the equation as given below :

$$\text{Actual length of the scale} = [\text{R.F.}] \times \begin{bmatrix}\text{Units of maximum} \\ \text{length required to} \\ \text{be measured by the} \\ \text{scale.}\end{bmatrix}$$

$$= \frac{1}{100} \times 10 = 0.1 \text{ m} = 10 \text{ cm}$$

(ii) Draw a line of 10 cm length and divide it into a number of divisions (10) equal to number of units of required maximum measurement, each representing 1 m.

(iii) Sub-divide first main division into 10 sub-divisions, each representing 1 dm, since 1 metre = 10 decimetres.

(iv) Mark zero (0) at the end of the first division and towards right side mark 1, 2, ... 9 as shown.

(v) Towards left of zero (0), mark sub-divisions 5 and 10 only, since the space is limited.

(vi) Just below the sub-divided divisions write decimetre and just below 1 to 9 main divisions write metre. On the extreme right and at the bottom of the scale write R.F. as $\frac{1}{100}$

Problem 2 : Construct a plain scale of 1 cm = 1 kilometre, to show hectometres and kilometres. Scale should be long enough to measure distance between I.I.T Delhi and Rashtrapatibhavan, which is 15 kms. Indicate on the scale the distance between Caunought place and Rashtrapatibhavan, which is 3 km 7 hm or say 3.7 km.

For solution see Fig. 4.2 and follow the procedure as given below :

(i) First find actual length of the scale by the equation

$$\text{Actual length of the scale} = [\text{R.F.}] \times \begin{bmatrix}\text{Units of maximum} \\ \text{length required to} \\ \text{be measured by the} \\ \text{scale.}\end{bmatrix}$$

$$= \frac{1 \text{ cm}}{100000} \times 1500000 \text{ cm} = 15 \text{ cm}$$

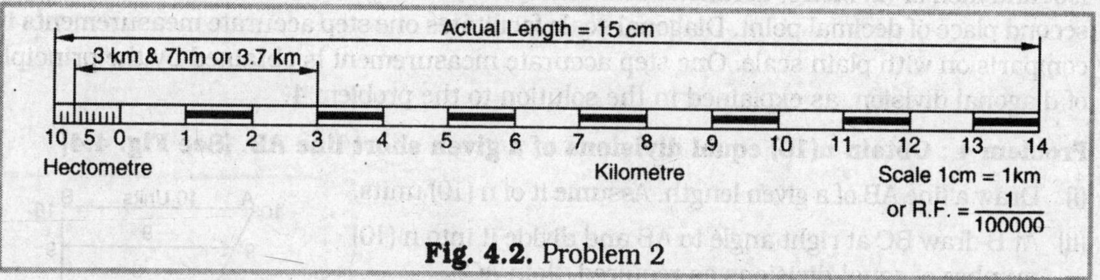

Fig. 4.2. Problem 2

(ii) Draw a line of 15 cm length and divide it into 15 equal parts, each representing 1 km.

(iii) Sub-divide first main division into 10 sub-divisions, each representing 0.1 km or 1 hectometre, since 1 km = 10 hm.

(iv) Mark zero (0) at the end of first division and towards right side mark 1, 2, ... 14 as shown. Towards left side mark sub-divisions 5 and 10 only.

(v) Just below the sub-divisions write hectometre and just below main-divisions write kilometre. On the right and at the bottom of the scale. write scale 1 cm = 1km or

$$\text{R.F.} = \frac{1}{100000}$$

(vi) Indicate 3 kms 7 hms or 3.7 kms on the scale as shown.

Problem 3 : On a map of Gujarat, 1 cm represents 5 kms. Construct a plain scale long enough to measure a distance between Ahmedabad and Baroda. Indicate on it a distance between Ahmedabad and Anand.

Distance : **(1) Ahmedabad - Baroda - 100 kms.**

(2) Ahmedabad - Anand - 65 kms.

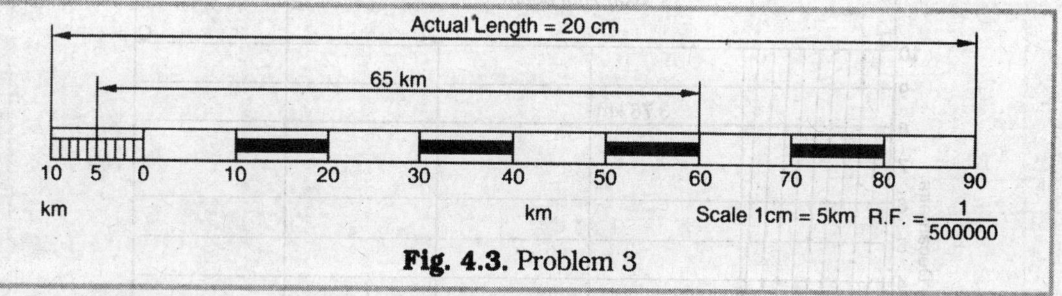

Fig. 4.3. Problem 3

(i) First find the actual length of the scale, as given below :

$$\text{Actual length of the scale} = \left[\frac{1 \text{ cm}}{5 \text{ kms}}\right] \times [100 \text{ kms}]$$

$$= 20 \text{ cms}$$

(ii) Draw a line of 20 cms length and divide it into 10 equal divisions, each representing 10 kms. Sub-divide first main division into 10 sub-divisions, each representing 1 km.

3.2 (b) Diagonal Scale :

A diagonal scale is used to represent either three interconnected units [like (1) Yard, foot and inch or (2) Metre, decimetre and centimetre] or only one unit and its fractions upto second place of decimal point. Diagonal scale facilitates one step accurate measurements in comparision with plain scale. One step accurate measurement is obtained by the principle of diagonal division, as explained in the solution to the problem 4.

Problem 4 : Obtain n(10) equal divisions of a given short line AB. [See Fig. 4.4]

(i) Draw a line AB of a given length. Assume it of n (10) units.

(ii) At B draw BC at right angle to AB and divide it into n (10) number of equal divisions as required. Join AC.

(iii) Through these n (10) divisions 1, 2, 3, ... etc. draw n (10) lines 11', 22', 33', ... etc. parallel to AB and get divisions 11', 22', 33',....... etc. as shown in fig.

(iv) From the geometry of the triangle, it is obvious that each horizontal line below AB is progressively shorter in length by $\frac{1}{(n)10}$ AB AB i. e., 99' = 0.9 AB, 88' = 0.8 AB, ... 11' = 0.1 AB.

Now, if length of AB is assumed as 10 units and if n = 10 then 99' = 9 units, 88' = 8 units, ... 11' = 1 unit.

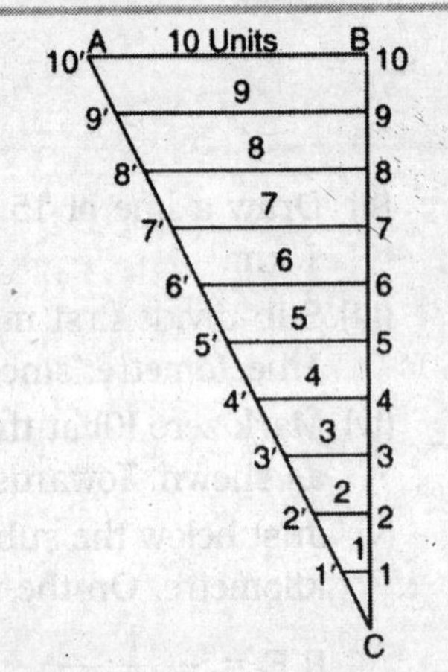

Fig. 4.4. Diagonal Scale

Problem 5 : Construct a diagonal scale with the scale 1 cm = 0.5 km. Showing kilometre, hectometre and decametre. Scale should be long enough to measure upto 5 kms. Indicate on the scale the distance of 3 kms 7 hm 8 dam.

For solution see Fig. 4.5 and follow the procedure as given below :

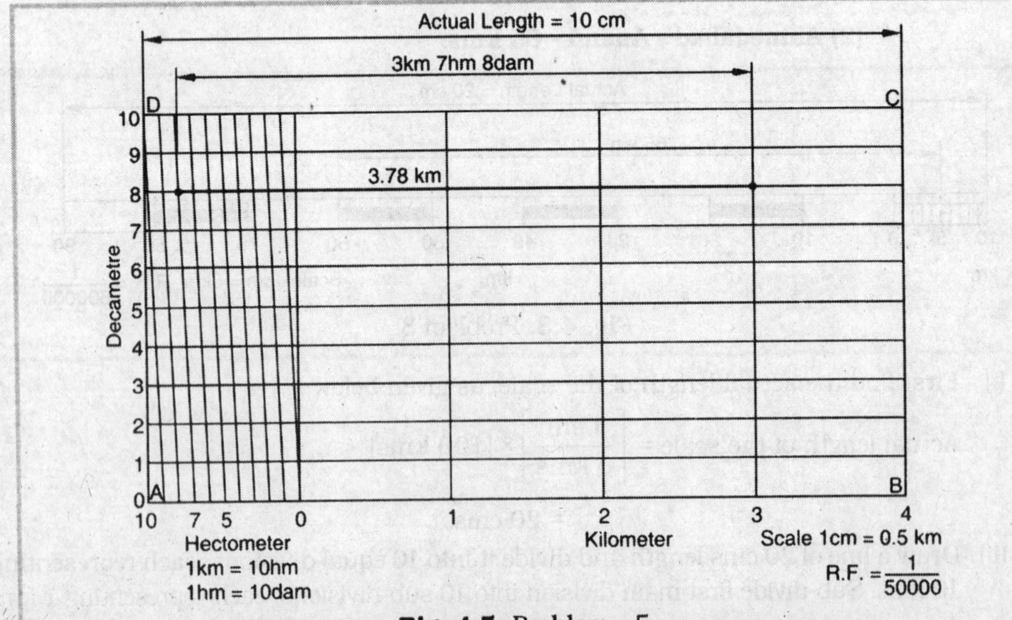

Fig. 4.5. Problem - 5

(i) Calculate actual length of the scale as usual.

$$\text{Actual length of scale} = [\text{R.F.}] \times \begin{bmatrix}\text{Maximum length of} \\ \text{Measurement}\end{bmatrix}$$

$$= \frac{1 \text{ cm}}{0.5 \text{ km}} \times 5 \text{ km} = 10 \text{ cm}$$

(ii) Draw a line AB of 10 cm length and divide it into 5 equal main divisions, each main division representing 1 kilometre.

(iii) Sub-divide 1st main division AO into 10 equal sub-divisions, since 1 km = 10 hm. Each sub-division will represent 1 hectometre.

(iv) On the line AB erect a rectangle ABCD with AD = BC = 10 unit as shown in the figure.

(v) Divide BC and AD into 10 equal divisions and join them by lines parallel to AB. Here 10 divisions are made, since the relation is 1 Hectometre = 10 Decametres.

(vi) Erect perpendiculars at 0, 1, 2 and 3. Mark E on CD, as shown and sub-divide DE into 10 parts. Join sub-divisions of AO with sub-divisions of DE diagonally, as shown in the figure. Thus by principle of diagonal division each of the hectometre is further divided into 10 equal parts and it will represent decametre. Now the diagonal scale is ready for use.

(vii) Observe and understand the representation of 3 km 7 hm 8 dam, shown in the figure.

Problem 6 : Construct a diagonal scale of R.F. = $\frac{1}{36}$ showing yard, foot and inch. Scale should be long enough to measure 5 yards. Indicate on it 3 yards 2 feet 9 inches.

For the solution see Fig. 4.6 and follow the procedure, as given below :

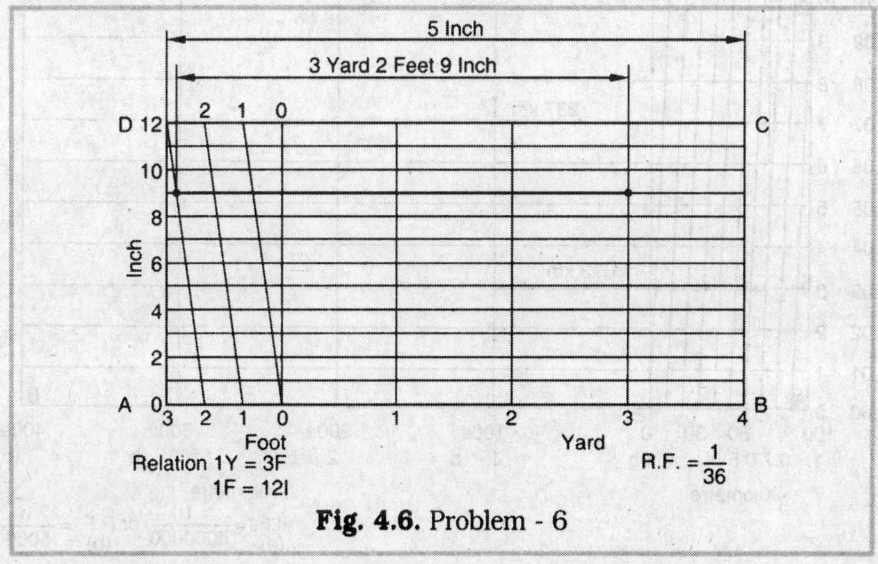

Fig. 4.6. Problem - 6

(i) Calculate the actual length of scale as under.

Actual length of scale = [R.F.] x [Max. length of Measurement]

$$= \frac{1}{36} \times 5 \text{ yards} = \frac{5}{36} \text{ yards}$$

$$= \frac{5}{36} \times \frac{3}{1} \times \frac{12}{1} = 5 \text{ inches.}$$

(ii) Draw a line AB of 5 inches length and divide it into 5 equal main divisions, each main division representing 1 yard.

(iii) Sub-divide 1st main division AO into 3 equal sub-divisions, since 1 yard = 3 feet. Each sub-division will represent 1 foot.

(iv) On the line AB erect a rectangle ABCD with AD = BC = 12 units, since 1 foot = 12 inches. Proceed further as per problem 5.

(v) Observe and understand the indicated distance 3 yards 2 feet and 9 inches.

Problem 7 : (a) The distance between Ahmedabad and Bombay is 500 kms. It is represented on a railway map by 10 cms. Construct a diagonal scale to measure kilometre. Show on scale the distance between Ahmedabad and Surat which is 237 kms.

(b) The distance between Rashtrapati Bhavan and Red Fort is 5.00 kms. It is represented on a Delhi city map by 10 cms. Construct a diagonal scale to measure a kilometre and its fraction upto two decimal. Show on the scale the distance between Rashtrapati Bhavan and India Gate which is 1.73 km.

For solution see fig. 4.7 and follow the procedure, as given below :

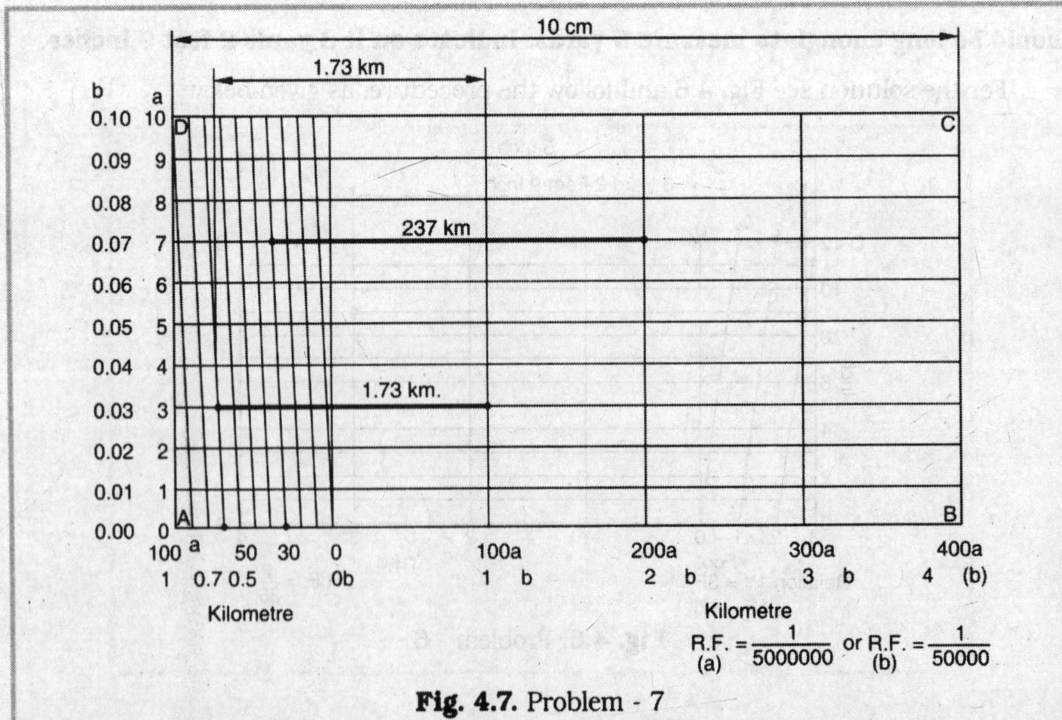

Fig. 4.7. Problem - 7

(i) Calculate the actual length of the scale as under :

Actual length of scale = [R.F.] x [Max. length of Measurement]

$$= \frac{10}{500} \times 500 = 10 \text{ cms.}$$

[OR] $$= \frac{10}{5} \times 5 = 10 \text{ cms.}$$

Proceed further as per the previous problem.

3.2 (c) Comparative Scale :

Scales having same representative fraction but graduated to read different units are known as comparative scales. Comparative scales are either plain scales or diagonal scales They are drawn separately in two different units (e.g. British and Metric) and superimposed one above the other.

Problem 8 : Old road map of Bombay city was drawn with 10 cm. on the map representing 25 miles. Construct a plain scale to read miles on this map and long enough to measure distance between Gateway of India and Borivalli which is 40 miles. Construct a comparative scale, attached to the above scale to read kilometres upto 70 kilometres.

Relation 1 mile = 1.609 km.

For solution see Fig. 4.8 and follow the procedure, as given below :

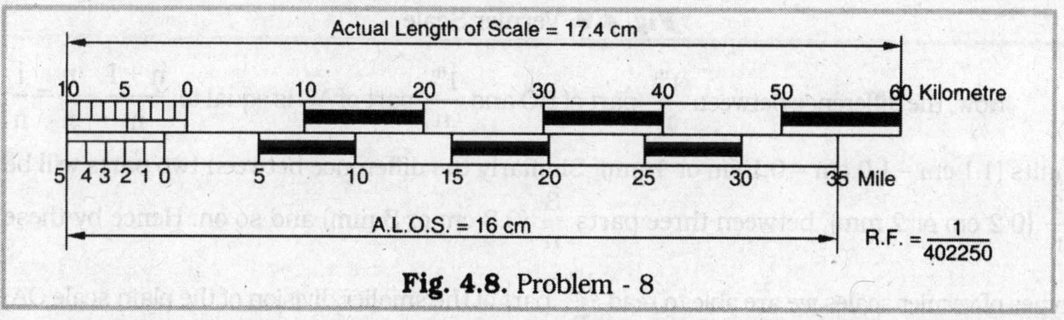

Fig. 4.8. Problem - 8

(i) First find the actual length of the two separate plain scales (1) to measure 40 miles and (2) to measure 70 kilometres with the same R.F. as given below :

$$\text{R.F.} = \frac{10}{25 \times 1.609 \times 1000 \times 100} = \frac{1}{402250}$$

$$\text{The actual length of the scale for 40 M} = \frac{10}{25} \times 40 = 16 \text{ cms.}$$

$$\text{The actual length of the scale for 70 kms} = \frac{10}{25 \times 1.609} \times \frac{70}{1} = 17.4 \text{ cms}$$

(ii) Draw two plain scales (1) 16 cm long with 8 main divisions with first main division sub-divided into 5 divisions and (2) 17.4 cm long with 7 main divisions with 1st main division sub-divided into 10 divisions one above the other and attached to it, as shown in Fig. 4.8.

As these two plain scales are side by side we get a comparative idea of scales for miles and kilometres and hence the scale is called the comparative scale.

3.2 (d) Vernier Scale : Vernier scale is used to read very small unit with greater accuracy. It consists of two scales (1) Main or Primary scale (2) Vernier scale. Vernier scale is attached above the primary scale, as shown in Fig. 4.9. It requires smaller space in comparision with the diagonal scale and it serves the purpose of diagonal scale.

To understand the principle of vernier scale see Fig. 4.9 and understand the following mathematics behind it.

In Fig. 4.9 line AO representing n units (10 cm) is divided into n(10) number of equal parts. Each part will represent $\frac{n}{n} = 1$ unit $\left[\frac{10 \text{ cms}}{10} = 1 \text{ cm}\right]$ Now line BO representing (n + 1) units (11 cm) is divided into n(10) number of equal parts. Each part will represent $\frac{n+1}{n} = 1 + \frac{1}{n}$ units $\left[\frac{11 \text{ cms}}{10} = 1.1 \text{ cm or } 11 \text{ mm.}\right]$

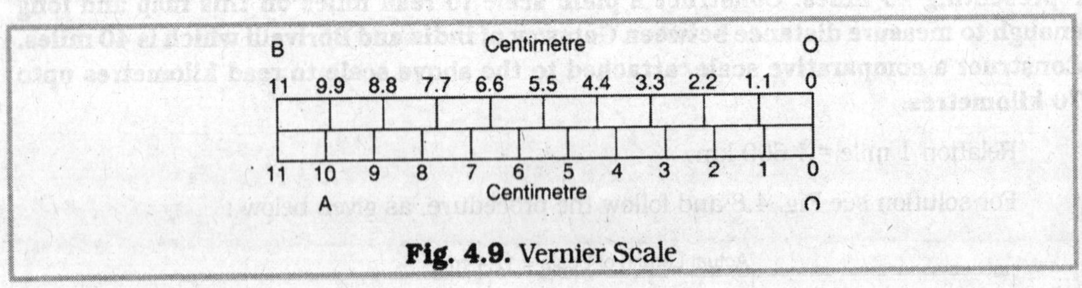

Fig. 4.9. Vernier Scale

Now, the difference between $\frac{1^{th}}{n}$ part of BO and $\frac{1^{th}}{n}$ part of AO is equal to $\frac{n+1}{n} - \frac{n}{n} = \frac{1}{n}$ units (1.1 cm – 1.0 cm = 0.1 cm or 1mm). Similarly the difference between two parts will be $\frac{2}{n}$ (0.2 cm or 2 mm), between three parts $\frac{3}{n}$ (0.3 cm or 3 mm) and so on. Hence by these types of vernier scales we are able to read $\frac{1^{th}}{n}$ part of the smaller division of the plain scale OA.

Now, we shall study vernier scale by illustration in the form of problem 9.

Problem 9 : Construct a vernier scale of R.F. = $\frac{1}{25}$ to read metre, decimetre and centimetre It should be long enough to measure 4 metres. Show on it length 2.47m, 1.62m and 0.91 m.

For solution see Fig. 4.10 and follow the procedure, as under :

(i) First find the actual length of the scale, as under :

The actual length of the scale = [R.F.] × [Max length]

$$= \frac{1}{25} \times 4 \text{ m} = \frac{4}{25} \text{ m} = 16 \text{ cm.}$$

(ii) Now draw a plain scale of length 16 cm. Divide it into 4 main divisions and sub-divide 1st main division into 10 sub-divisions. Each main division will represent one metre and each sub-division will represent one decimetre or 0.1 metre.

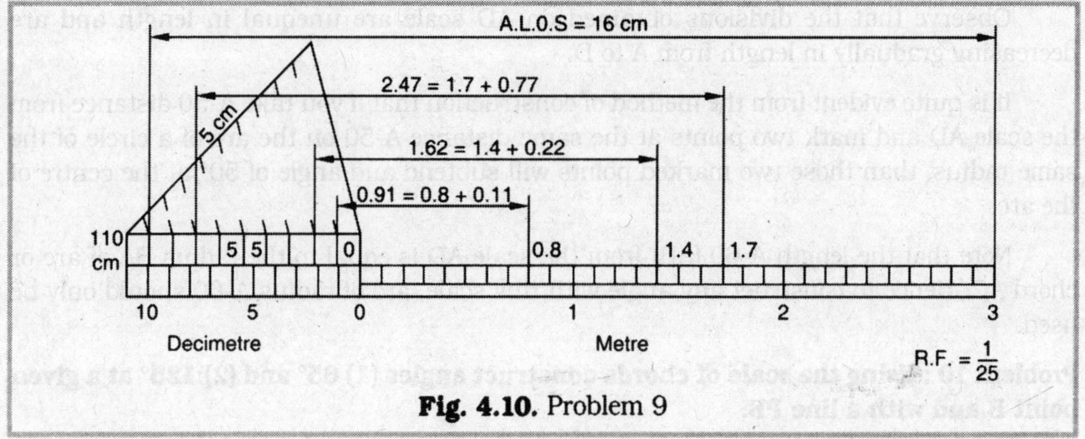

Fig. 4.10. Problem 9

(iii) Now prepare a vernier scale of length 11 decimetres just above decimetre scale and divide it into 10 equal divisions. Each division of vernier scale will represent 11 centimetres or 1.1 decimetre or 0.11 metre. Now the scale is ready for use.

(iv) Now on the vernier scale, lengths 11 cm (0.11 m), 22 cm (0.22 m), 33 cm (0.33 m), ...99 cm (0.99 m) and 110 cm (1.1 m) are available. To get length 2.47 metre we must use available length 0.77 m of vernier scale to adjust second decimal 0.07. Subtract 0.77 m , from 2.47 m to get 1.7 m. The length 1.7 m we should take from the plain scale or the main scale.

(v) The procedure for representing (1) 1.62 m = (1.4 + 0.22) and (2) 0.91 m = (0.8 + 0.11) is understood from the figure.

3.2 (e) Scale of Chords :

Scale of chords is used to measure and set out angles without the use of protractor. It is based on the principle of length of chords of different angles measured on the same arc. Method of construction is explained in Fig. 4.11 as under.

Draw a line AB of suitable length. At B erect a perpendicular to AB. With B as the centre and BA as the radius draw an arc AC. Here, the chord AC subtends an angle of 90° at the centre B.

Now divide arc/chord AC into 18 equal parts. First divide it into 3 equal parts by compass with radius equal to arc radius and centres A and C. Each part is further divided into 6 divisions by the divider and by trial and error method. Each small part on chord will subtend an angle of 5° at the centre B.

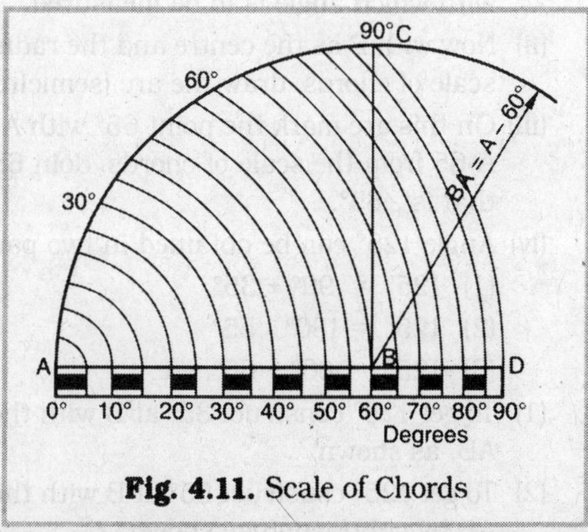

Fig. 4.11. Scale of Chords

Now transfer all division points from the arc on to the straight line AB produced with A as the centre and radii equal to A-5, A-10, A-15,.......A-90, as shown in the figure. The straight line AD, with divisions marked, is a scale of chords.

Observe that the divisions obtained on AD scale are unequal in length and are decreasing gradually in length from A to D.

It is quite evident from the method of construction that if you take A-50 distance from the scale AD and mark two points at the same distance A-50 on the arc of a circle of the same radius, than those two marked points will subtend and angle of 50° at the centre of the arc.

Note that the length A-60 (AB) from the scale AD is equal to the radius BA of arc or chord AC. Hence to construct any angle with this scale, arc of radius A-60 should only be used.

Problem 10 : Using the scale of chords construct angles (1) 65° and (2) 125° at a given point B and with a line PB.

For solution see Fig. 4.12 and follow the procedure, as given below :

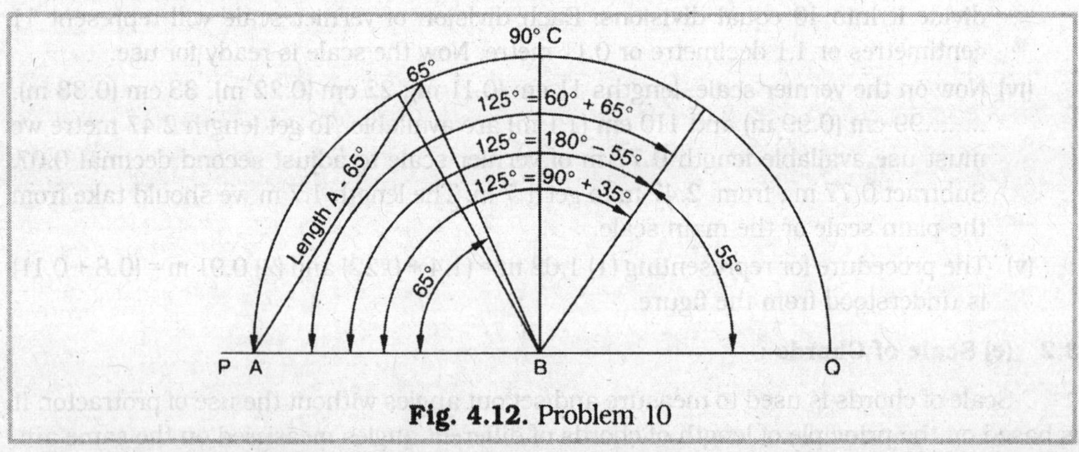

Fig. 4.12. Problem 10

(i) Draw a line PBQ. B is the point where angle is to be constructed and PB is the line with which angle is to be measured.

(ii) Now with B as the centre and the radius equal to the length A-60 or 0-60, from the scale of chords, draw the arc (semicircle) of a circle ACQ, as shown in the figure.

(iii) On this arc mark the point 65° with A as the centre and radius equal to the length A-65 from the scale of chords. Join 65 with B. Angle AB 65 is the required angle, that is ∠65°.

(iv) Angle 125° can be obtained in two parts, as given below :

(1) 125° = 90° + 35°

(2) 125° = 180° - 55°

(3) 125° = 60° + 65°, etc.

(1) To get 125° construct 35° at B with the line BC, and you will get 125° with the line AB, as shown.

(2) To get 125° construct 55° at B with the line BQ, and you will get 125° with the line AB by subtraction as shown.

(3) To get 125°, construct 65° with a line of 60°, and you will get 125° with the line AB by summation.

Problem 11 : Measure the given angle ∠POQ by means of the scale of chords.

For solution see Fig. 4.13 and follow the procedure, as given below :

(i) First draw the given angle ∠POQ.

(ii) Now with O as the centre and radius equal to the length A-60 or O-60, from the scale of chords, draw an arc cutting PO at A and QO at B, as shown. Put scale of chords on AB with 0° coinciding with A, as shown. Read angle on scale coinciding with the point B. Angle of scale coinciding with the point B is the measurement or value of angle ∠POQ.

Divide each division of 5° into 5 equal parts to measure and set out angle of any full number say 37°, 69°, etc. These divisions will show little approximate values.

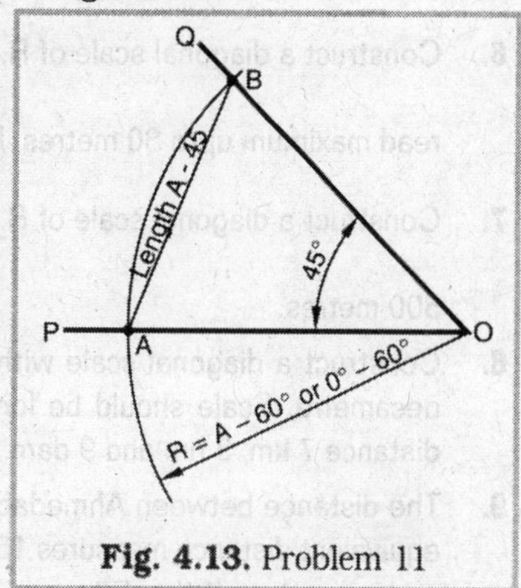

Fig. 4.13. Problem 11

3.2 (f) Isometric Scale :

See chapter of Isometric Projection for this scale.

4. Some of the important relations for length measurement are as shown :

1000 millimetres (mm)	= 1 metre (m)	1.609 kilometre (km)	= 1 mile
100 centimetres (cm)	= 1 metre (m)	1 yard	= 3 feet
10 decimetres (dm)	= 1 metre (m)	1 foot	= 12 inches
1 decametre (dam)	= 10 metres (m)	1 inch	= 2.54 cms
1 hectometre (hm)	= 100 metres(m)	1 nautical mile	= 1852 metres
1 kilometre (km)	= 1000 metres (m)	1 metre	= 1.0936 yards

EXERCISE

1. Construct a plain scale with R.F. = $\frac{1}{50}$ to read metre and decimetre. It should be long enough to read 5 metre. Show 3.6 metre on it.

2. Construct a plain scale 1 cm = 1 dm to read decimetre and centimetre. It should be long enough to read 20 dm. Show the length 157 cm on it.

3. Construct a plain scale of R.F. = $\frac{1}{5000}$ to read hectometre and decametre. It should be able to measure 10 hectometre. Measure a distance of 7 hm 6 dm on the scale.

4. On the road map of Ahmedabad - Gandhinagar 2 kms is represented by 1 cm. Ahmedabad Gandhinagar distance is 25 kms. Gandhi-Ashram is 18 kms from Gandhinagar. Represent it after preparing the scale for the above.

5. Construct a diagonal scale of R.F. = $\frac{1}{40000}$ showing kilometre, hectometre and decametre. It should be long enough to measure 5 kms. Measure a distance of 4 kms 7 hm 6 dam on the scale.

6. Construct a diagonal scale of R.F. $\frac{1}{500}$ to read metres and decimetres. It should be able to read maximum upto 80 metres. Measure a distance of 47.8 metres.

7. Construct a diagonal scale of R.F. $\frac{1}{2500}$ to show metres and long enough to measure upto 500 metres.

8. Construct a diagonal scale with scale 1 cm = 1.0 km showing kilometre, hectometre and decametre. Scale should be long enough to measure upto 15 kms. Indicate on the scale distance 7 km, 3 hm and 9 dam.

9. The distance between Ahmedabad and Surat is 250 km. On measurement on road map its equivalent distance measures 15 cm. What is R.F. ? Draw a diagonal scale showing kilometre and indicate on it the distances between following stations.

1.	Ahmedabad	-	Baroda	-	123 km.
2.	Baroda	-	Bharuch	-	73 km.
3.	Bharuch	-	Surat	-	54 km.

10. A rectangular map 200 cm x 125 cm wide represents an area of 62500 square metres. What is R.F. of the map ? Construct a diagonal scale to measure kilometre, hectometre and decametre. Show on scale a distance of 5 kilometres, 4 hectometres and 8 decametres.

11. An old road map of Rajkot - Morvi drawn with a scale in miles is available. On this map 1.0 cm long line represents 2.5 miles. Construct a scale to read miles and long enough to measure distance between Rajkot and Morvi which is 40 miles. Now attach to this scale a comparative scale in kilometres and long enough to measure 70 kilometres. Take 1 mile = 1.609 km.

12. Construct a vernier scale of Representative Fraction $\frac{1}{25}$ to read centimetres upto 6 metres. Show on it a length representing 2.07 metres.

13. Maximum distance of 80 kms in a map measures 6 inches. What is R.F. ? Construct comparative scales in kms (80) and in miles (50) with the same R.F. Take 1 km = 1093.6 yards.

14. Construct a diagonal scale of R.F. = $\frac{1}{48}$ showing yards, feet and inches to represent a maximum length of 8 yards. Mark on the scale a distance 5 yards 2 feet 8 inches.

15. Construct a vernier scale to read centimetres and long enough to measure upto 5 m. Take R.F. = $\frac{1}{25}$. Mark on it a distance 2.42 m.

16. Construct comparative diagonal scales one in metres and other in yards having R.F. = $\frac{1}{2700}$.

It should be long enough to show 400 metres. Take 1 metre = 1.0936 yards.

17. Construct a scale of chords having divisions of 5°. Using this scale set off angle of 25°, 35°, 50°, 65°, 125° and 140°.

18. Construct scale of chords and with the help of this scale measure angles B,E and Q of following three tiangles.

1. ΔABC having AB = 6 cm; BC = 8 cm and CA = 10 cm.
2. ΔDEF having DE = 5 cm; EF = 7 cm and FD = 10 cm.
3. ΔPQR having PQ = 5 cm; QR = 7 cm and RP = 8 cm.

CHAPTER - 5

Loci of Points

We are living in the modern age where we often come across machines and mechanisms. How the mechanism runs and what is going to be the path of various points of the mechanism will be of our interest. This information is also useful in design, force analysis and motion utilisation of the mechanism.

The path of a point keeping its distance constant from a fixed point, from a fixed straight line, from a fixed circle and various such combinations are also of our interest. In this chapter we shall study the path followed by different points of the various mechanisms along with the path followed by various points moving with given constraints or set laws.

Mechanism is the combination of various links so paired that the motion is completely constrained. Mechanisms are used to transform the motion, e.g. Rotary to reciprocating, reciprocating to rotary, rotary to oscillating, etc. In other words, by available mechanisms motion is converted into useful motion.

In this chapter we shall study following mechanisms :

(1) Slider Crank Chain.

  (a) Simple (See Fig. 5.11)

  (b) Off Set [See Fig. 5.12]

(2) Four Bar Chain.

  (a) Both the driver and driven cranks revolving (See Fig. 5.14)

  (b) Driver crank revolving and driven crank oscillating. (See Fig. 5.15)

  (c) Both driver and driven cranks oscillating. (See Fig. 5.16)

(3) Crank and connecting rod mechanism with connecting rod constrained to pass through guide named as trunnion (See Fig. 5.13)

(4) Combinations of above mechanisms (See Fig. 5.18)

Now we shall study first the path of various points moving in a plane with given conditions or set laws and then the path of various points of the mechanisms. Study will be carried out by taking concrete examples. The path followed by a point is called locus (plural loci).

Part - I

Problem 1 : Find the locus of a point P, moving in a plane, keeping its distance from the fixed point O as constant. Take constant distance equal to 20 mm (See Fig. 5.1)

Here the locus of the point P is going to be a circle of 20 mm radius and the fixed point O as the centre.

Problem 2 : Find the locus of a point P, moving in a plane, keeping its distance from the fixed straight line O_1O_2 as constant. Here the constant distance is taken as 20 mm (See Fig. 5.2)

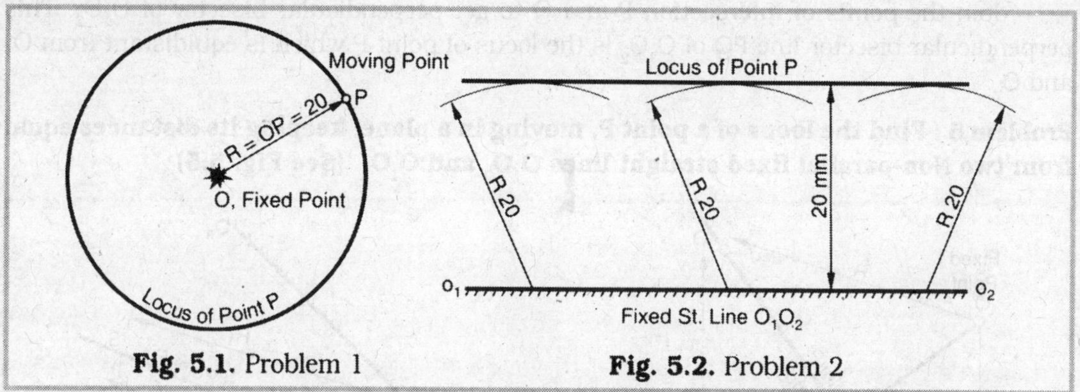

Fig. 5.1. Problem 1 **Fig. 5.2.** Problem 2

Take two or three points arbitrarily on the fixed straight line O_1O_2 and with these points as centres and radius equal to 20 mm draw arcs of circles on any one side of the straight line. Now draw a line in such a way that the line becomes tangent to those arcs of circles. Obviously this locus of P is going to be a straight line parallel to the straight line O_1O_2 at 20 mm distance.;

Problem 3 : Find the locus of a point P, moving in a plane, keeping its distance from the fixed circle ⊙ (0, 20) as constant and outside it. Here constant distance is taken as 10 mm (See Fig 5.3)

Take many points arbitrarily on the periphery of the circle ⊙(0. 20) as centres and with radius equal to 10 mm draw arcs in the plane of a circle, and outside it.

Now draw a curve in such way that this curve becomes tangent to all the arcs that you have drawn. Obviously this locus of P is going to be a circle of 30 mm radius with O as the centre.

If it is required to draw the locus inside then it will be a circle of 10 mm radius with the same point O as the centre.

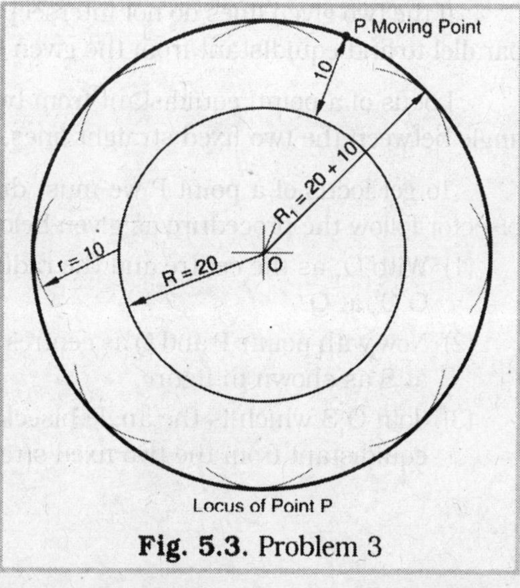

Fig. 5.3. Problem 3

Problem 4 : Find the locus of point P, moving in a plane, keeping its distances equal from the two fixed points O_1 and O_2. (See Fig. 5.4)

The locus of a point equidistant from the two fixed points is a perpendicular bisector of the line joining the two fixed points.

So to get the locus of point P equidistant from the fixed points O_1 and O_2, first join O_1 and O_2. Now with O_1 and O_2 as centres and R as radius draw two intersecting arcs on each side of O_1O_2 at P and Q as shown in figure. Take $R > \frac{O_1O_2}{2}$

Join the points of intersection P and Q to get perpendicular bisector of O_1O_2. This perpendicular bisector line PQ of O_1O_2 is the locus of point P which is equidistant from O_1 and O_2.

Problem 5 : Find the locus of a point P, moving in a plane, keeping its distances equal from two Non-parallel fixed straight lines O_1O_2 and O_1O_3. (See Fig. 5.5)

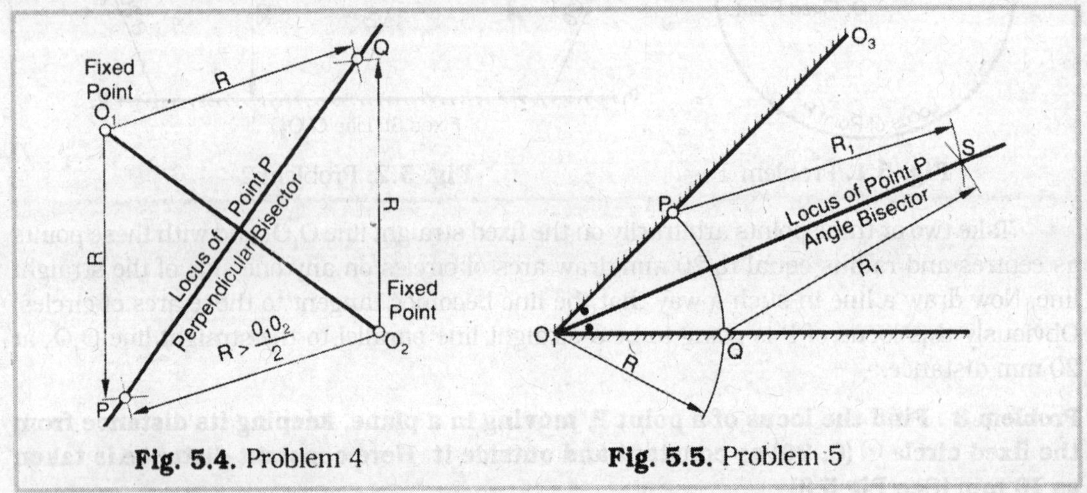

Fig. 5.4. Problem 4 **Fig. 5.5.** Problem 5

If the two given lines do not intersect in a given space then draw two intersecting lines parallel to and equidistant from the given two straight lines and then proceed for solution.

Locus of a point, equidistant from two fixed straight lines, is an angle bisector of the angle between the two fixed straight lines.

To get locus of a point P we must draw angle bisector of angle $O_3O_1O_2$. To get angle bisector follow the procedure as given below :

(1) With O_1 as the centre and the radius equal to R draw an arc cutting O_1O_3 at P and O_1O_2 at Q.

(2) Now with points P and Q as centres and radius equal to R_1 draw two arcs intersecting at S as shown in figure.

(3) Join O_1S which is the angle bisector of $O_3O_1O_2$ and so it is the locus of the point P equidistant from the two fixed straight lines O_1O_3 and O_1O_2.

Problem 6 : Find the locus of the point P, moving in a plane, keeping its distances equal from a fixed straight line O_1O_2 and a fixed point O. (See Fig. 5.6)

To get locus of the point P equidistant from the line O_1O_2 and the point O, follow the procedure as given below :

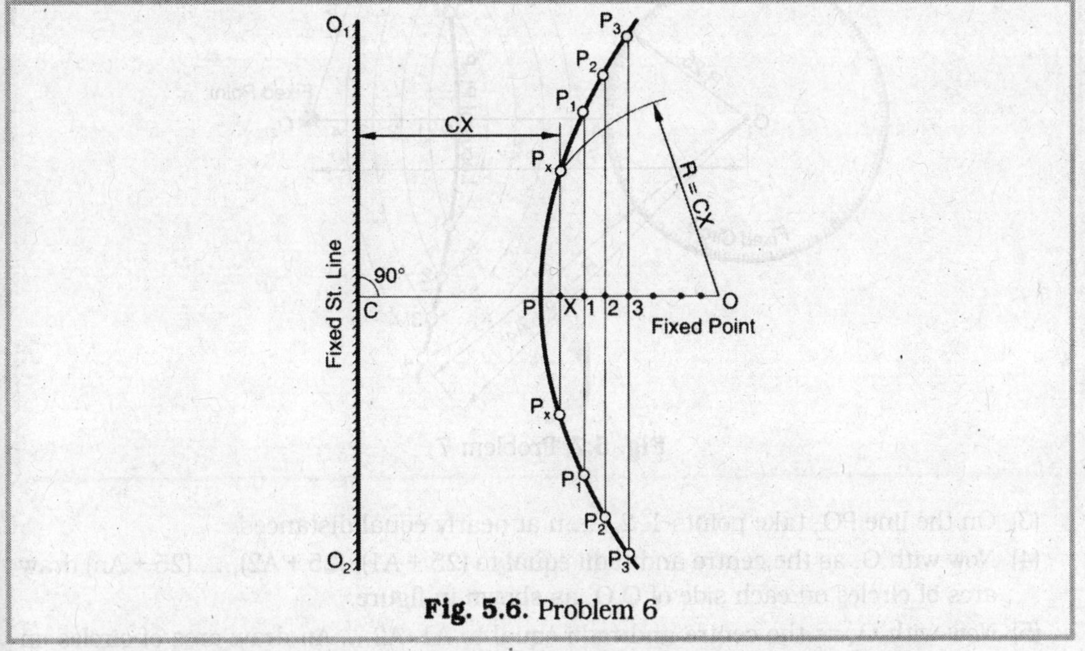

Fig. 5.6. Problem 6

(1) From the point O draw the perpendicular O C to the line O_1O_2.

(2) Find the mid point of OC and mark it as P.

(3) On the line PO take points X, 1, 2,...n at nearly equal distances, as shown in the figure.

(4) Through these points, X, 1, 2, 3,.....n draw lines parallel to O_1O_2.

(5) Now with O as the centre and the radius equal to CX, C1, C2.....Cn, draw two arcs one on each side of CO to intersect the parallel lines through X, 1, 2,......n respectively, as shown in the figure.

(6) On the parallel lines through X, 1, 2,....n, we get the points of intersections as (P_x, P_x), $(P_1, P_1)(P_2, P_2)$.....(P_n, P_n), as shown in figure.

(7) Join points P_n.......P_2, P_1, P_x, P, P_x, P_1, P_2.......P_n by a smooth curve. This smooth curve is the locus of the point P which is equidistant from the fixed line O_1 O_2 and the fixed point O. This curve is a Parabola.

Problem 7 : Find the locus of a point P, moving in a plane, keeping its distances equal from a fixed point O_2 and a fixed circle ⊙ $(O_1, 25)$. O_2 is 70 mm away from the centre O_1. (See Fig. 5.7)

To get the locus of the point P, equidistant from the fixed point O_2 and fixed circle $(O_1, 25)$, follow the procedure as given below:

(1) Join O_1 and O_2 and mark the point A where O_1O_2 intersects circle $(O_1, 25)$.

(2) Find mid point P of the line AO_2.

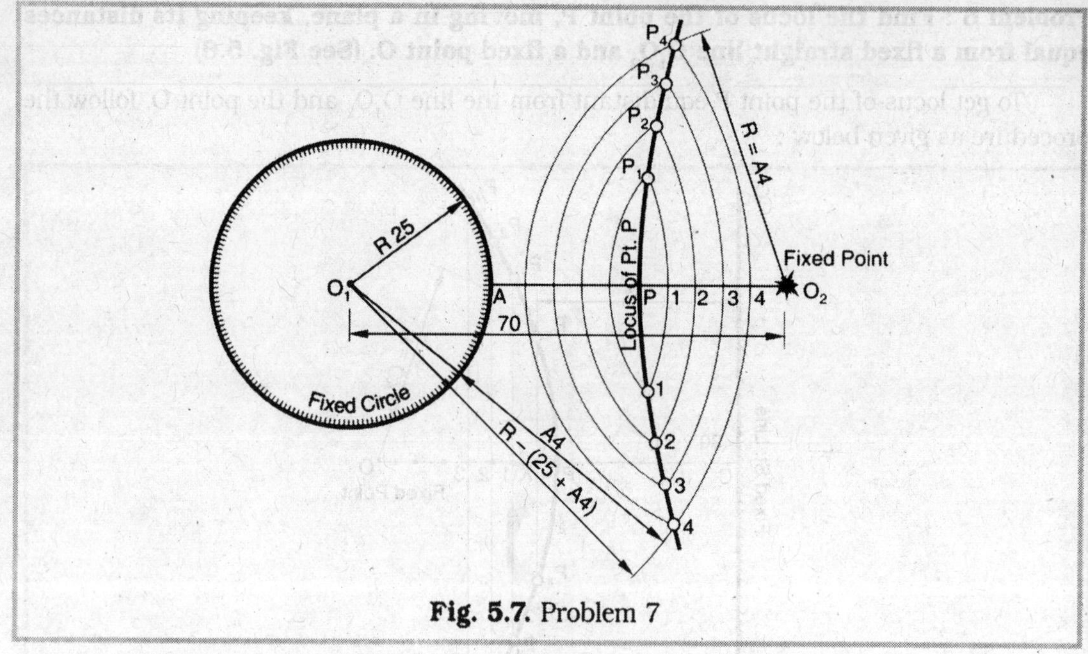

Fig. 5.7. Problem 7

(3) On the line PO_2 take points 1, 2,......n at nearly equal distances.

(4) Now with O_1 as the centre and radii equal to (25 + A1), (25 + A2),......(25 + An) draw arcs of circles on each side of O_1O_2 as shown in figure.

(5) Now with O_2 as the centre and radii equal to A1, A2,.....An draw arcs of circles on each side of O_1O_2 to intersect previously drawn corresponding arcs of the circles.

(6) By intersections we get points (P_1, P_1'), (P_2, P'_2)....(P_n, P_n'). Join all these points by a smooth curve. The smooth curve is the locus of the point P equidistant from a fixed point O_2 and a fixed circle (O_1, 25).

Problem 8 : Find the locus of a point P, moving in a plane, keeping its distances equal from a fixed straight line O_2O_3 and a fixed circle (O_1, 25). (See Fig. 5.8).

To get locus of a point P equidistant from a fixed straight line O_2O_3 and fixed circle (O_1, 25), follow the procedure as given below :

(1) From the point O_1 draw perpendicular O_1C to the fixed straight line O_2O_3.

(2) Mark the point A where O_1C intersects the fixed circle. Find the midpoint of AC and mark it as the point P.

(3) On the line PA take points 1, 2, 3,.....n at nearly equal distances, as shown in figure.

(4) From points 1, 2,.....n draw straight lines parallel to the fixed line O_2O_3, as shown in figure.

(5) Now with O_1 as the centre and radii equal to (25 + C1), (25 + C2), ...(25 + Cn) draw arcs of circles on each side of O_1C to intersect parallel lines from 1, 2,.....n at the points (P_1, P_1') (P_2, P_2').......(P_n, P_n') respectively.

(6) Join points P_n',....P_2',P_1', P, P_1, P_2,.....P_n by smooth curve. This curve is the locus of the point 'P' equidistant from the fixed straight line O_2O_3 and the fixed circle (O_1, 25).

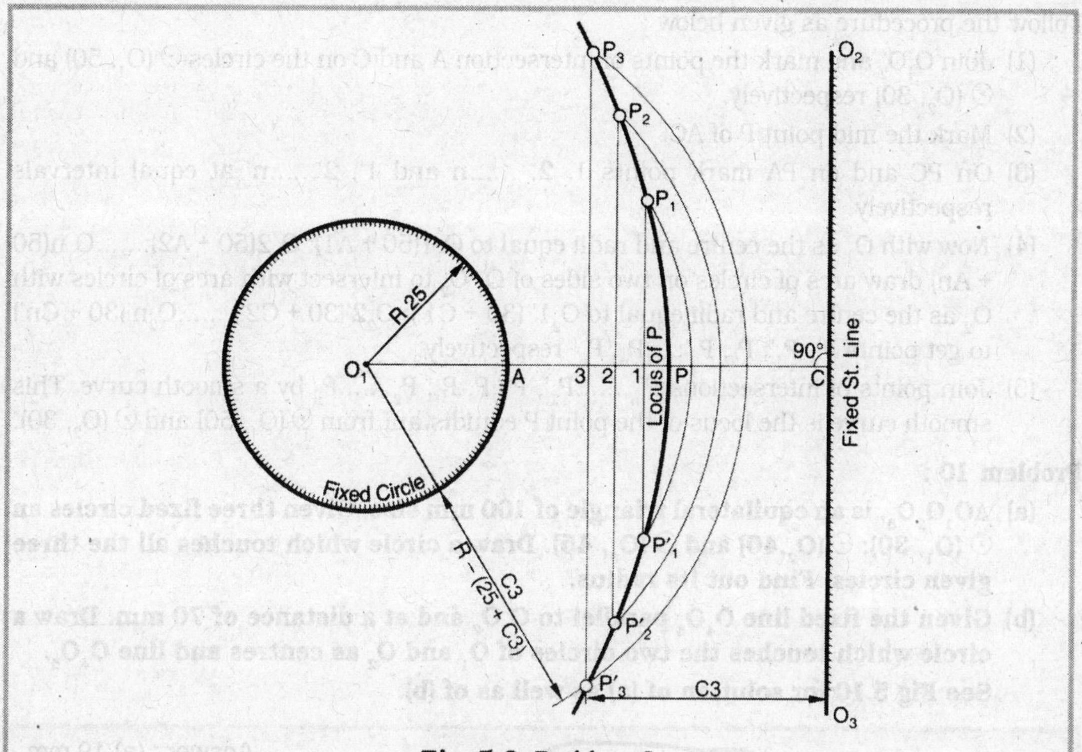

Fig. 5.8. Problem 8

Problem 9 : Find the locus of a point P, moving in a plane, keeping its distances equal from two fixed circles. Here the two fixed circles are (O_1, 50) and (O_2, 30). Take distance between O_1 and O_2 as 110 mm (See Fig. 5.9)

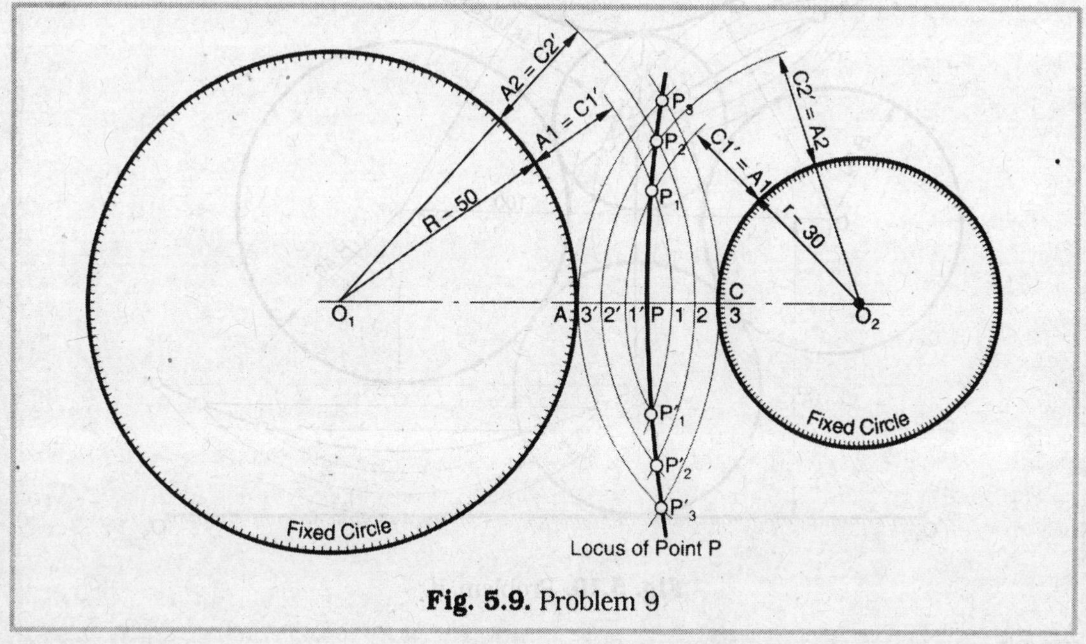

Fig. 5.9. Problem 9

Follow the procedure as given below :

(1) Join O_1O_2 and mark the points of intersection A and C on the circles ⊙ (O_1, 50) and ⊙ (O_2, 30) respectively.

(2) Mark the mid point P of AC.

(3) On PC and on PA mark points 1, 2, ...n and 1', 2'......n' at equal intervals respectively.

(4) Now with O_1 as the centre and radii equal to $O_1$1(50 + A1); $O_1$2(50 + A2);......O_1n(50 + An) draw arcs of circles on two sides of O_1 O_2 to intersect with arcs of circles with O_2 as the centre and radii equal to $O_2$1' (30 + C1'); $O_2$2'(30 + C2'),......O_2n'(30 + Cn') to get points P_1, P_1'; P_2, P_2';....P_n, P_n' respectively.

(5) Join points of intersections P'_n,......P_2', P'_1,P, P_1, P_2,......P_n by a smooth curve. This smooth curve is the locus of the point P equidistant from ⊙ (O_1, 50) and ⊙ (O_2, 30).

Problem 10 :

(a) $\Delta O_1O_2O_3$, is an equilateral triangle of 100 mm size. Given three fixed circles an ⊙ (O_1, 30); ⊙ (O_2,40) and ⊙ (O_3, 45). Draw a circle which touches all the three given circles. Find out its radius.

(b) Given the fixed line O_4O_5 parallel to O_1O_2 and at a distance of 70 mm. Draw a circle which touches the two circles of O_1 and O_2 as centres and line O_4O_5. See Fig 5.10 for solution of (a) as well as of (b).

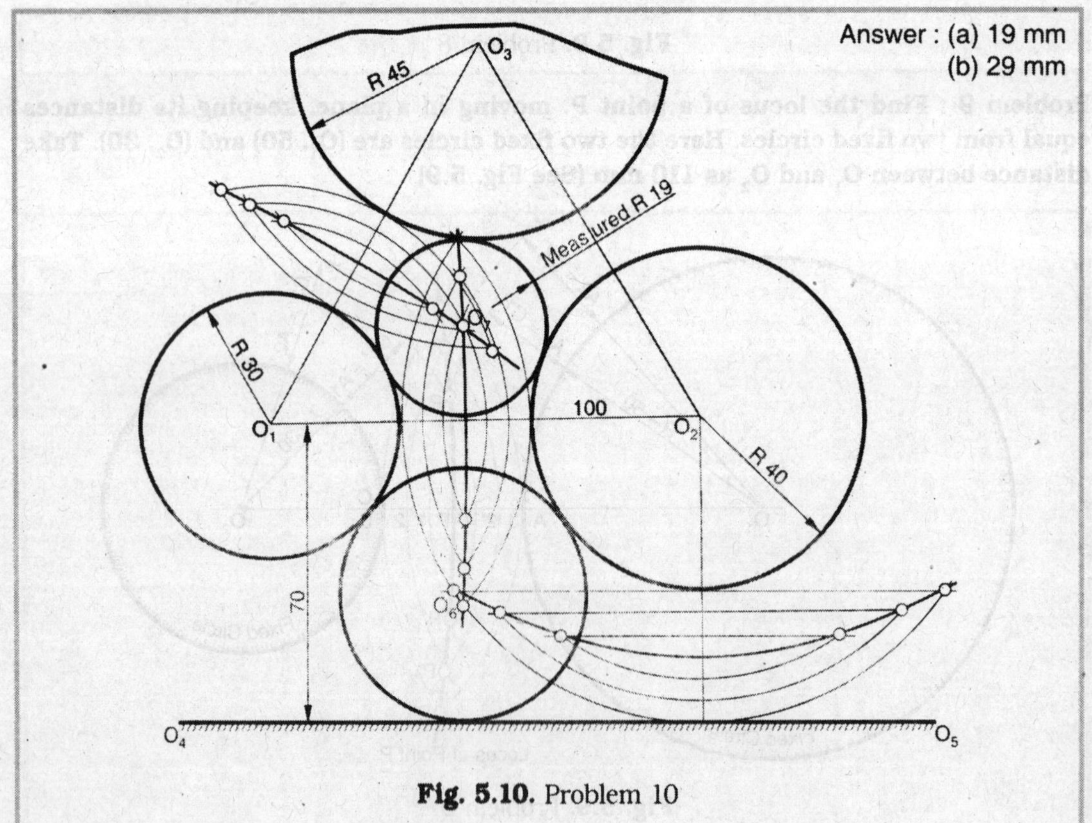

Fig. 5.10. Problem 10

To get the solution, follow the procedure as given below :

(a) Part (i) following the procedure of problem 9, find a curve equidistant from circles ⊙ (O_1, 30) and ⊙ (O_2, 40) and similarly, a curve equidistant from circles ⊙ (O_1,30) and ⊙ (O_3, 45). These two curves intersect at the point O_7. (ii) Draw a circle, with O_7 as the centre and of suitable radius, which touches the given three fixed circles. Measure the radius. Ans. 19 m.m.

(b) (i) Following the procedure of problem 9, find a curve equidistant from the two circles ⊙ (O_1, 30) and ⊙ (O_2, 40).

(ii) Following the procedure of problem 8, find a curve equidistant from one fixed circle, ⊙ (O_2, 40) and one fixed straight line O_4O_5.

(iii) Intersection of above two curves gives the point O_6 as shown in the figure. Therefore, the point O_6 is equidistant from ⊙ (O_1, 30); ⊙ (O_2, 40) and the line O_4O_5.

(iv) Now draw a circle with O_6 as the centre and with a suitable radius so that the circle touches the two given circles and the given straight line. Measure the radius. Ans:-29 m.m.

Part - II

Problem 11 : OBA is a simple slider crank chain. OB is a crank of 30 mm length. BA is a connecting rod of 90 mm length. Slider A is sliding on a straight path passing through point O. Draw the locus of the mid-point of the connecting rod AB for one complete revolution of the crank OB.

For the solution see Fig. 5.11.

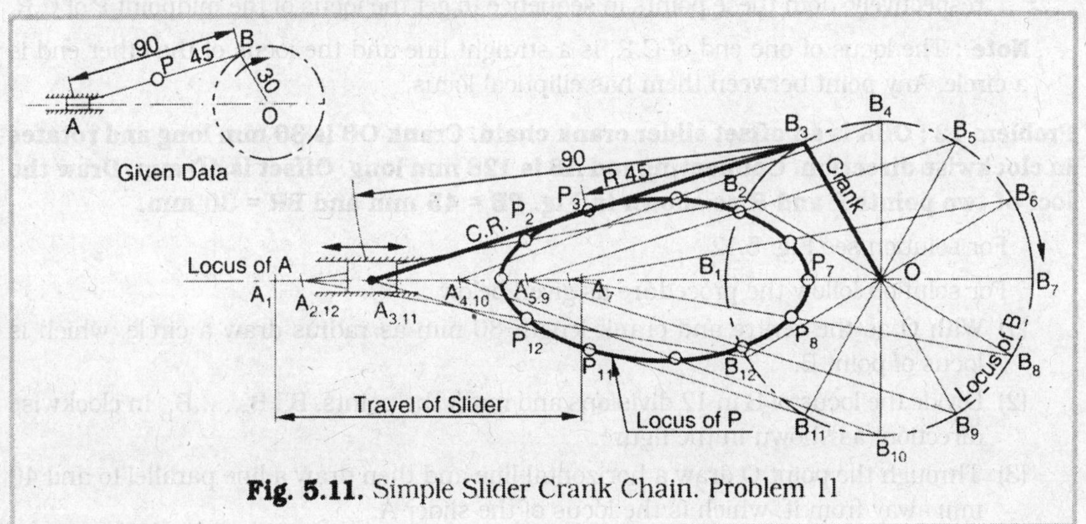

Fig. 5.11. Simple Slider Crank Chain. Problem 11

MECHANISM :

First of all we shall study about the mechanism. This mechanism is used in reciprocating engines, pumps, compressors, etc. This mechanism converts rotary motion into reciprocating or reciprocating motion into rotary. In oil engines, the piston reciprocates and thereby the crank revolves. In compressors the crank is given rotation by electric motor and thereby the piston reciprocates and compresses air.

In this mechanism there are four links : (i) crank (ii) connecting rod (iii) slider and (iv) fixed body having bearings for the crank and guide for the slider.

The crank is able to revolve about bearings provided at O. The big end of the connecting rod is connected to the crank by the turning pair (B) so that they can have relative angular velocity between them i.e. angle between them can change. The small end of C.R. is connected to the slider, also by the turning pair (A). Slider is able to reciprocate inside the guide provided in the main body which is fixed.

If the line of reciprocation of slider passes through the bearing of the crank i.e. here the point O, then it is called a simple slider crank chain. If it is offset then it is said to be an offset slider crank chain. (See Fig. 5.12)

For solution of the problem follow the procedure as given below. (See Fig. 5.11)

(1) Draw a circle ⊙ (0, 30). This circle is the locus of the point B.

(2) Divide this circle into 12 equal parts and mark them as B_1, B_2,...B_{12} as shown in the figure in the direction of rotation. OB_1, OB_2......OB_{12} represent the positions of the crank during the rotation.

(3) Draw a line through O preferably horizontal on which the slider A is to travel.

(4) Draw arcs of circles with B_1, B_2,......B_{12} as centres and length of C.R. as the radius to cut the straight line, locus of slider A, at various points. Mark those points as A_1, A_2......A_{12} respectively.

(5) Join A_1B_1, A_2B_2.....$A_{12}B_{12}$ to get various positions of C.R. during motion.

(6) Mark mid points P_1, P_2,......P_{12} on various positions of C.R. A_1B_1, A_2B_2.....$A_{12}B_{12}$ respectively. Join these points in sequence to get the locus of the midpoint P of C.R.

Note : The locus of one end of C.R. is a straight line and the locus of the other end is a circle. Any point between them has elliptical locus.

Problem 12 : OBA is an offset slider crank chain. Crank OB is 30 mm long and rotates in clockwise direction. Connecting rod AB is 128 mm long. Offset is 40 mm. Draw the loci of two points P and R as shown in Fig. PB = 45 mm and BR = 30 mm.

For solution see Fig. 5.12.

For solution follow the procedure as given below :

(1) With O as the centre and crank length 30 mm as radius draw a circle, which is locus of point B.

(2) Divide the locus of B in 12 divisions and mark the points, B_1, B_2......B_{12} in clockwise direction, as shown in the figure.

(3) Through the point O draw a horizontal line and then draw a line parallel to and 40 mm away from it, which is the locus of the slider A.

(4) Now draw arcs of circles with B_1, B_2......B_{12} as centres and length of C.R. i.e. 128 mm as the radius to cut the locus of A at the points A_1, A_2......A_{12} respectively.

(5) Join A_1B_1, A_2B_2......$A_{12}B_{12}$ and mark on them points P_1,P_2......P_{12} and R_1, R_2.....R_{12} taking 45 mm and 30 mm distances from B and in the opposite direction. of B respectively, as shown in the figure.

(6) Join P_1, P_2........P_{12} to get the locus of P and join R_1, R_2......R_{12} to get the locus of R.

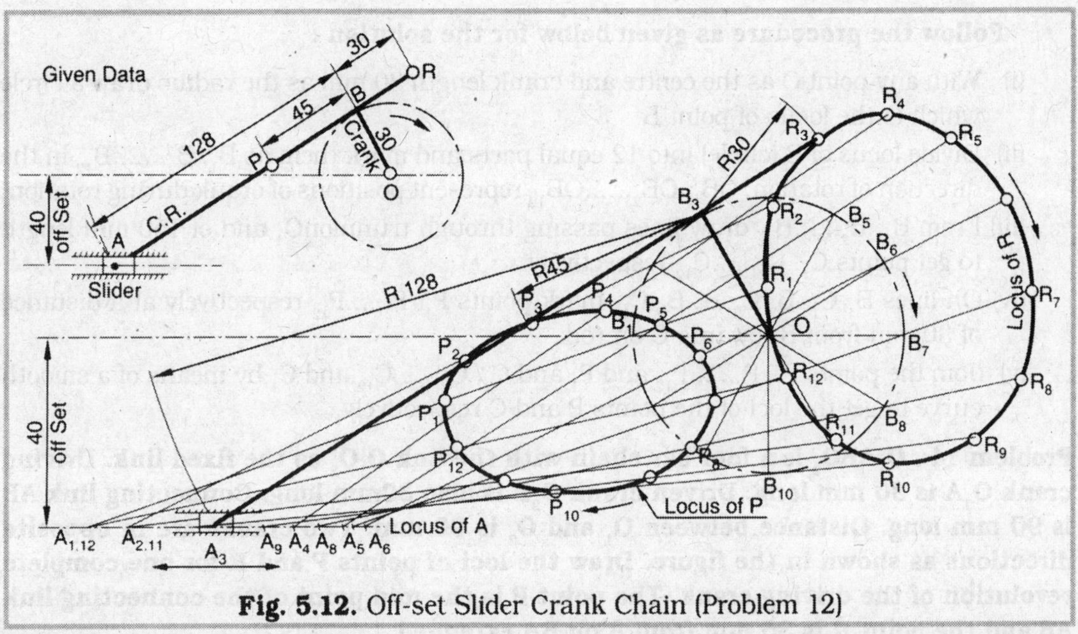

Fig. 5.12. Off-set Slider Crank Chain (Problem 12)

Problem 13 : Fig. 5.13 shows a mechanism in which OB is a crank of 30 mm length revolving in clockwise direction. BC is a rod connected to the crank at the point B by turning pair and rod BC is constrained to pass through the guide at O_1 called trunnion. Draw the loci of points P and C for one revolution of the crank. The point P is 30 mm from B on the rod BC. Length of BC is 150 mm (See Fig. 5.13). Point O_1 is 80 mm on the right and 15 mm below the point O.

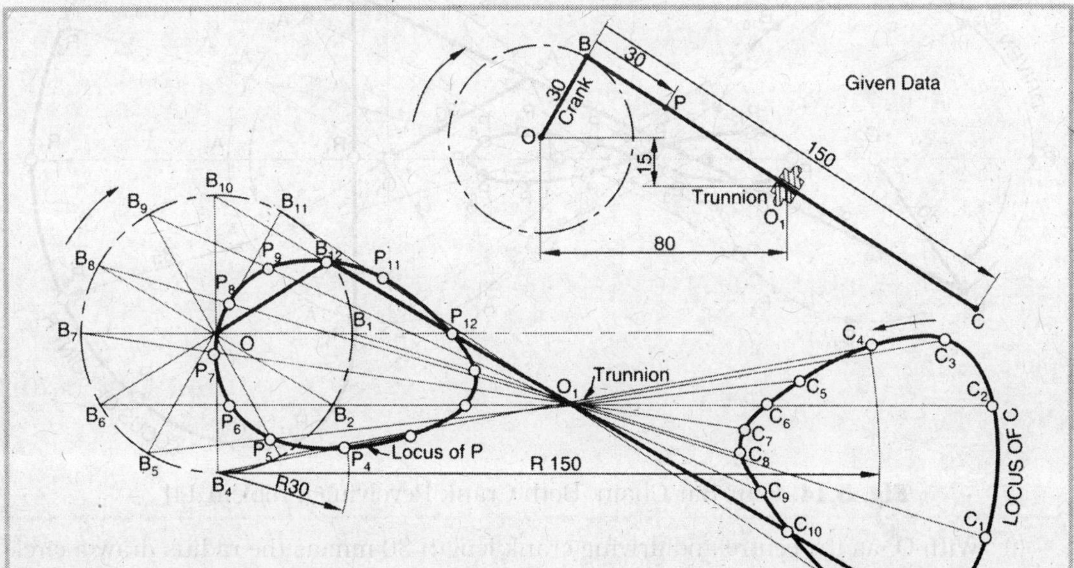

Fig. 5.13. Crank-Con-Rod-Trunnion Mechanism (Problem 13)

Follow the procedure as given below for the solution :

(i) With any point O as the centre and crank length 30 mm as the radius draw a circle which is the locus of point B.

(ii) Divide locus of B (circle) into 12 equal parts and mark them as B_1, B_2,......B_{12} in the direction of rotation. OB_1, OB_2,......OB_{12} represent positions of crank during rotation.

(iii) From B_1, B_2,.....B_{12} draw lines passing through trunnionO_1 and of 150 mm length to get points C_1, C_2,......C_{12} respectively.

(iv) On lines B_1 C_1, B_2 C_2,......$B_{12}C_{12}$ mark points P_1, P_2,.....P_{12} respectively at a distance of 30 mm from B towards C on rod.

(v) Join the points P_1, P_2,......P_{12} and P_1 and C_1, C_2,......C_{12} and C_1 by means of a smooth curve to get the loci of the points P and C respectively.

Problem 14 : O_1ABO_2 is a four bar chain with the link O_1O_2 as the fixed link. Driving crank O_1A is 30 mm long. Driven crank O_2B is also 30mm long. Connecting link AB is 90 mm long. Distance between O_1 and O_2 is 90 mm. Two cranks are in opposite directions as shown in the figure. Draw the loci of points P and R for one complete revolution of the driving crank. The point P is the mid point of the connecting link AB and the point R is 35 mm from A on BA extended.

See the Fig. 5.14 for solution and follow the procedure as given below :

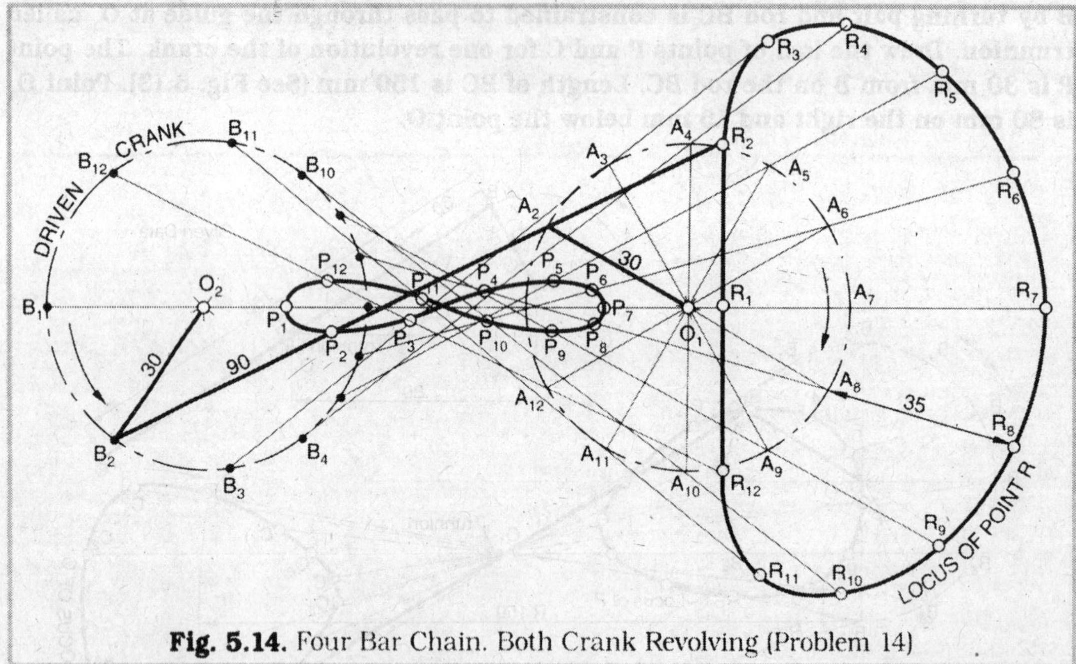

Fig. 5.14. Four Bar Chain. Both Crank Revolving (Problem 14)

(i) With O_1 as the centre and driving crank length 30 mm as the radius draw a circle which is the locus of the point A. Divide that circle into 12 equal parts and mark on them A_1, A_2......A_{12} in the direction of rotation.

(ii) Draw a horizontal line from O_1 and on it mark O_2 at 90 mm distance from O_1. With O_2 as centre and driven crank length 30 mm as the radius, draw a circle which is the locus of the point B.

(iii) With $A_1, A_2, \ldots A_{12}$ as the centres and the radius equal to the connecting link length 90 mm draw arcs to cut the locus of the point B at points $B_1, B_2, \ldots B_{12}$ as shown in the figure respectively. This cutting should be done on opposite side of O_1O_2 on the circle.

(iv) Join $A_1B_1, A_2B_2, \ldots A_{12}B_{12}$ and mark on them midpoints $P_1, P_2, \ldots P_{12}$ respectively. Join $P_1, P_2, \ldots P_{12}$ & P_1 to get the locus of the point P.

(v) Extend $B_1A_1, B_2A_2, \ldots B_{12}A_{12}$ by 35 mm to get points $R_1, R_2, \ldots R_{12}$ respectively. Join points $R_1, R_2, \ldots R_{12}$ & R_1 to get the locus of the point R.

Problem 15 : ABCD is a four bar chain with the link AD fixed and of 100 mm length. AB is a driving crank of length 30 mm. CD is a driven crank of 60 mm length. BC is 80 mm long connecting link. TM is a rod of 20 mm length which is attached to the connecting link BC at right angles to it and at the point M 40 mm from B on BC. S is a point on extension of BC 40 mm from C. Draw the loci of points T and S for one complete revolution of driving crank.

For solution see Fig. 5.15 and follow the procedure as given below :

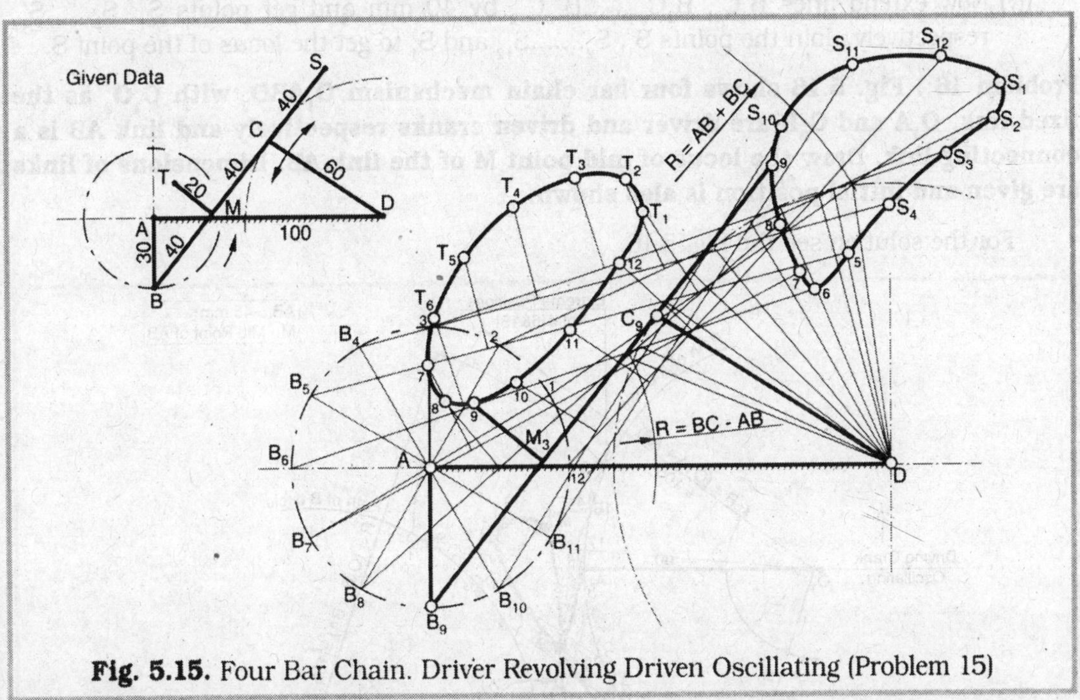

Fig. 5.15. Four Bar Chain. Driver Revolving Driven Oscillating (Problem 15)

First of all we should find whether the driven crank will rotate fully or oscillate. To find that follow the procedure as given below :

(i) Draw a circle ⊙ (D, 60) which is a locus of the point C.

(ii) With A as the centre and (AB + BC) i.e. 30 + 80 = 110 mm as the radius draw an arc of the circle and observe, whether the arc is going beyond the circle ⊙ (D, 60) or cutting it at two points. This is done because the maximum distance that the point C can go from A is [30 + 80 = 110] that is when both the links AB and BC are in one line. If that arc is beyond the circle ⊙ (D, 60), point C can travel anywhere on the circle ⊙ (D, 60) i.e. driven crank can rotate fully. If not, then the two cutting points of the arc with the circle ⊙ (D, 60) are the extreme positions of the point C. Here in this problem other extreme position of the point C is achieved by drawing an arc with A as centre and (BC - AB) i.e. (80 - 30 = 50) as the radius. In this condition BC and AB will be along one line but overlapping.

Now after finding the two extreme positions of the point C on the circle ⊙ (D, 60) follow the procedure as given below :

(i) With A as centre and the length of the driving crank 30 mm as the radius draw a circle. Divide it in 12 equal parts B_1, B_2, ...B_{12}.

(ii) With B_1, B_2,.....B_{12} as the centres and the radius equal to the length of the connecting link BC = 80 mm draw arcs cutting the circle ⊙ (D, 60), between two extreme positions of point C, at points C_1, C_2,......C_{12}. Join B_1C_1, B_2C_2,....$B_{12}C_{12}$ and mark on them points M_1, M_2,....M_{12} at distance 40 mm from points B_1, B_2,......B_{12} respectively.

(iii) At the points M_1, M_2,......M_{12} draw right angles of 20 mm length to the lines B_1C_1, B_2C_2,......$B_{12}C_{12}$ and mark the points T_1, T_2,......T_{12}. Join T_1, T_2,.....T_{12} and T_1 to get the locus of the point T.

(iv) Now extend lines B_1C_1, B_2C_2,.......$B_{12}C_{12}$ by 40 mm and get points S_1, S_2,.....S_{12} respectively. Join the points S_1, S_2,.......S_{12} and S_1 to get the locus of the point S.

Problem 16 : Fig. 5.16 shows four bar chain mechanism O_1ABO_2 with O_1O_2 as the fixed link. O_1A and O_2B are driver and driven cranks respectively and link AB is a connecting link. Draw the locus of mid point M of the link AB. Dimensions of links are given and initial position is also shown.

For the solution see the Fig. 5.16.

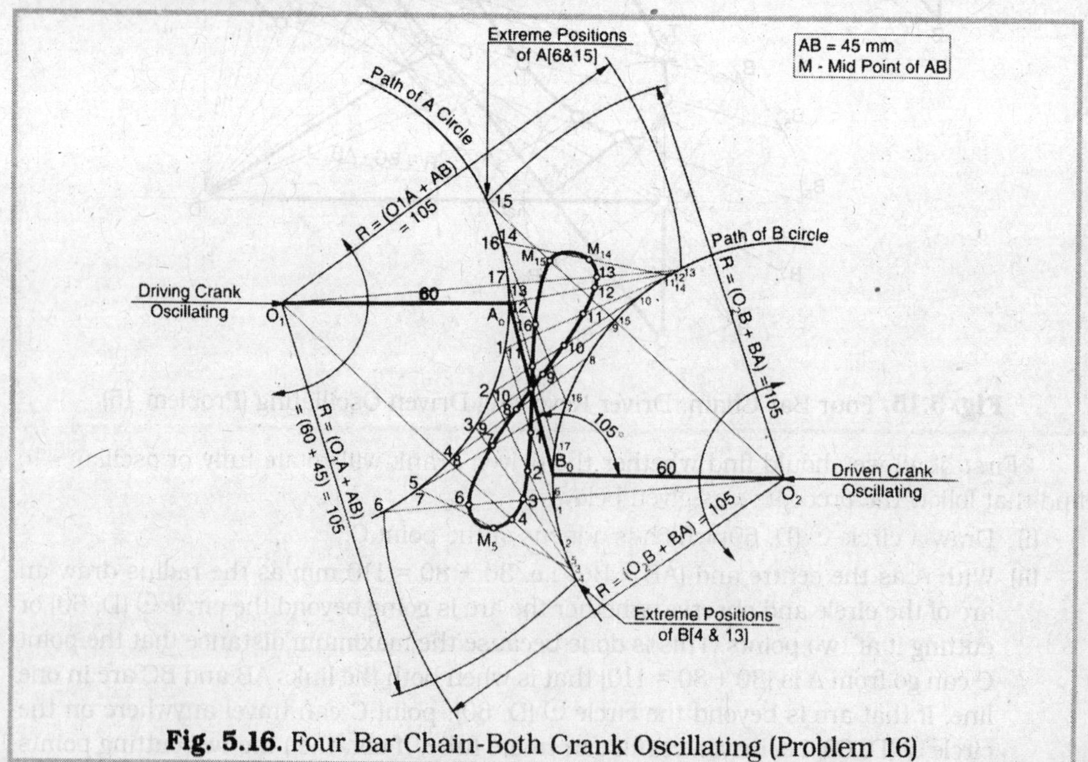

Fig. 5.16. Four Bar Chain-Both Crank Oscillating (Problem 16)

First of all find the extreme positions of the point A on the circle ⊙ (O_1, 60) with the help of intersection of arc of the circle ⊙ [O_2, (60 + 45)]. Similarly find extreme positions of the point B on circle ⊙ [O_2, 60] with the help of intersection of the arc of the circle ⊙ [O_1, (60 + 45]].

As shown in the figure, between the two extreme positions take different positions of the driver crank at nearly equal intervals, from the given configuration, in the direction of rotation and then back. Mark points A_1, A_2.....A_6.....A_{15}.......A_{17}.

Now with these points A_1, A_2.....A_6......A_{15}......A_{17} as centres and the radius equal to the length of connecting link AB, draw arcs of circles to intersect the circle ⊙ (O_2, 60) at points B_1, B_2.......B_4.......B_{13}.......B_{17}. Join A_1B_1, A_2B_2.......$A_{17}B_{17}$ and find their mid points M_1, M_2.......M_{17}. Now join M_1, M_2.....M_{17} and M_1 to get the locus of the point M.

[Note : Points are marked only by suffix due to less available space]

Problem 17 : The crank O_1A is 35 mm long and rotates about the point O_1 in the clock wise direction. The link AB is connected to the crank by turning pair at the point A. The link AB glides/slides over a fixed cylinder for which the circle ⊙ (O_2, 25) is shown in the figure. O_1O_2 = 100 mm, AB = 140 mm, AC = 15 mm ; BC = 155 mm. Draw the loci of the points B and C for one revolution of the crank.

For the solution see Fig. 5.17 and follow the procedure as given below :

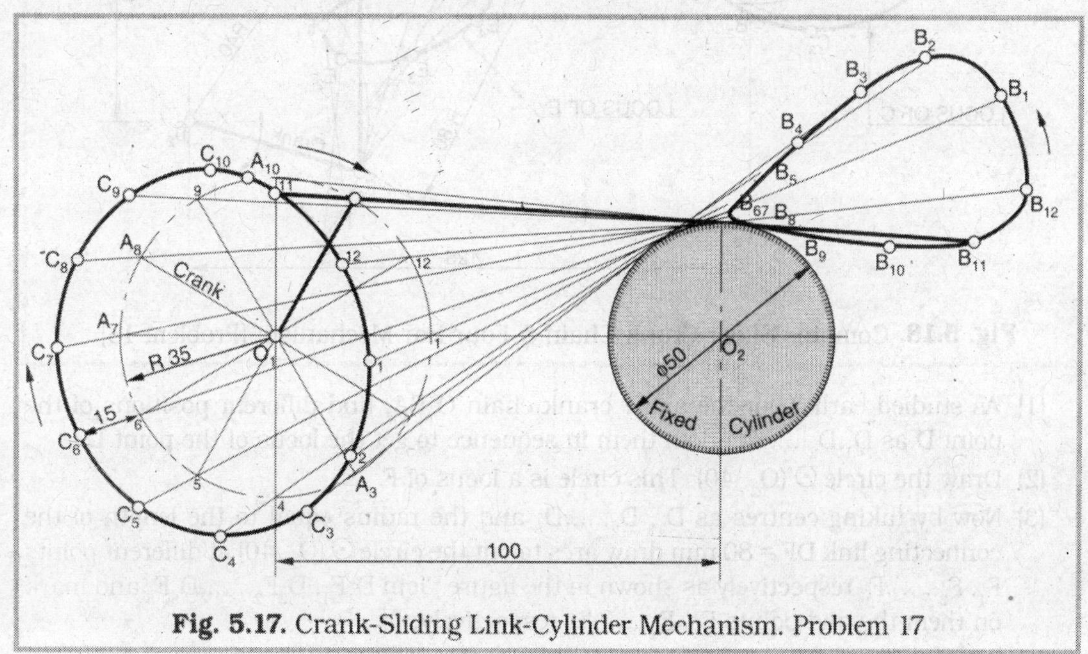

Fig. 5.17. Crank-Sliding Link-Cylinder Mechanism. Problem 17

(1) First draw the ⊙ (O_1, 35) and divide it into 12 equal divisions. Mark divisions in the direction of rotation as A_1, A_2.......A_{12}.

(2) Now draw lines tangent to the circle ⊙ (O_2, 25) on the upperside from A_1, A_2......A_{12} and mark on them B_1, B_2, ...B_{12} respectively by taking AB = 140 mm.

(3) Extend B_1A_1, B_2A_2.......$B_{12}A_{12}$ upto C_1,C_2.....C_{12} respectively by taking AC = 15 mm or BC = 155 mm.

(4) Join the points B_1, B_2, ...B_{12} and B_1 and C_1, C_2,.......C_{12} and C_1 to get the loci of the points B and C respectively.

Problem 18 : In the mechanism shown in Fig. 5.18, one end of the connecting link DF is connected to the point D of the connecting rod BA by a pin joint/turning pair and other end is connected to the oscillating crank O_2F also by a pin joint or a turning pair. Draw the loci of the point D and the mid point E of the connecting link DF for one revolution of the driving crank O_1B.

For the solution see the Fig. 5.18 and follow the procedure as given below :

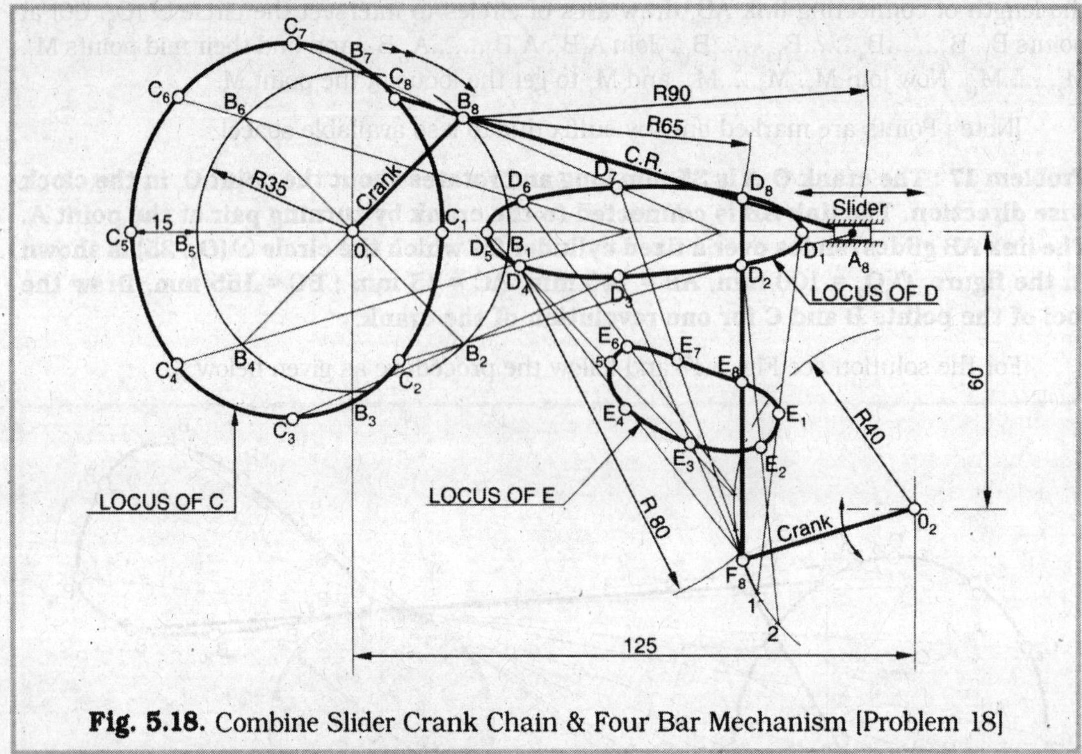

Fig. 5.18. Combine Slider Crank Chain & Four Bar Mechanism [Problem 18]

(1) As studied earlier, for the slider crank chain O_1BA, find different positions of the point D as D_1,D_2,.....D_8. Join them in sequence to get the locus of the point D.

(2) Draw the circle ⊙ (O_2, 40). This circle is a locus of F.

(3) Now by taking centres as D_1, D_2,.....D_8 and the radius equal to the length of the connecting link DF = 80 mm draw arcs to cut the circle ⊙ (O_2, 40) at different points F_1, F_2,.......F_8 respectively as shown in the figure. Join D_1F_1, D_2F_2,.......D_8F_8 and mark on them the mid points E_1, E_2,......E_8 respectively.

(4) Join E_1, E_2,.......E_8 and E_1 to get the locus of the point E.

Problem 19 : A circular disc of 72 mm diameter rotates about its centre in the clockwise direction. While the disc completes one revolution, an insect walks across the diameter of the disc. Plot the locus of the insect, assuming both the rotation of disc and movement of the insect as uniform.

[OR]

A link AB of 72 mm length rotates about its centre in the clockwise direction. While the link completes one revolution, the insect walks across the length from one end to the other. Plot the locus of the insect assuming the rotation of the link and the motion of the insect as uniform.

For the solution see Fig. 5.19 and follow the procedure as given below :

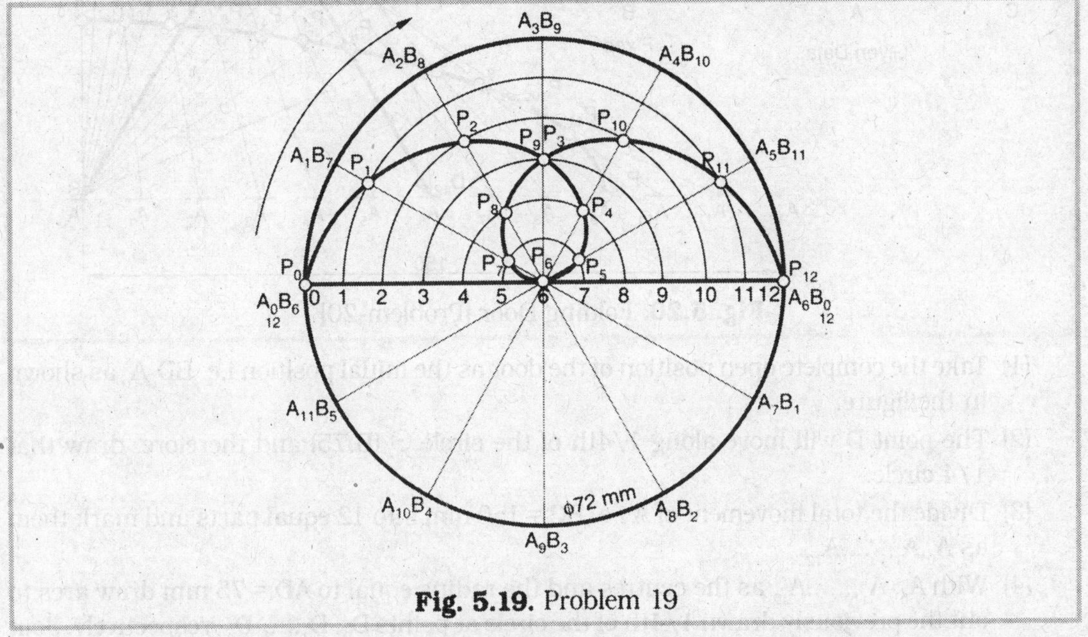

Fig. 5.19. Problem 19

(1) Draw the circle of 72 mm diameter and divide it into 12 equal parts. Mark them as A_0B_0, A_1B_1.......$A_{12}B_{12}$ in the clock-wise sense as shown in the figure.

(2) Divide the diameter AB also into 12 equal parts and mark them as 0,1,2.......12 as shown in the figure.

(3) Now when the disc/link has turned through 1/12 of the revolution and the diameter A_0B_0 must have achieved the position A_1B_1 and by that time the insect initially at A_0 must have walked 1/12 of the length of AB from A towards B.
Therefore, draw an arc of the circle passing through 1 to cut A_1B_1 at the point P_1, as shown in figure. Here the link/disc has turned from A_0B_0 to A_1B_1 and by that time the insect has moved by the distance A_1P_1(1/12 of diameter of disc or length of link).

(4) In the similar manner draw arcs of circles through the points 2,3......12 to cut the positions A_2B_2, A_3B_3.......$A_{12}B_{12}$ at points P_2, P_3.......P_{12} respectively. Join P_0, P_1....... P_{12} in a sequence to get the path of the insect as shown in the figure.

Problem 20 : As seen in the plan, AD and DB are two equal size portions of a folding door hinged joint or pinned joint at D. Span CB of the door is 150 mm. The end B is fixed and the end A is constrained to move along the line BC. Draw the locus of the mid point P of AD for a complete movement of the folding door.

For the solution see Fig. 5.20 and follow the procedure as given below :

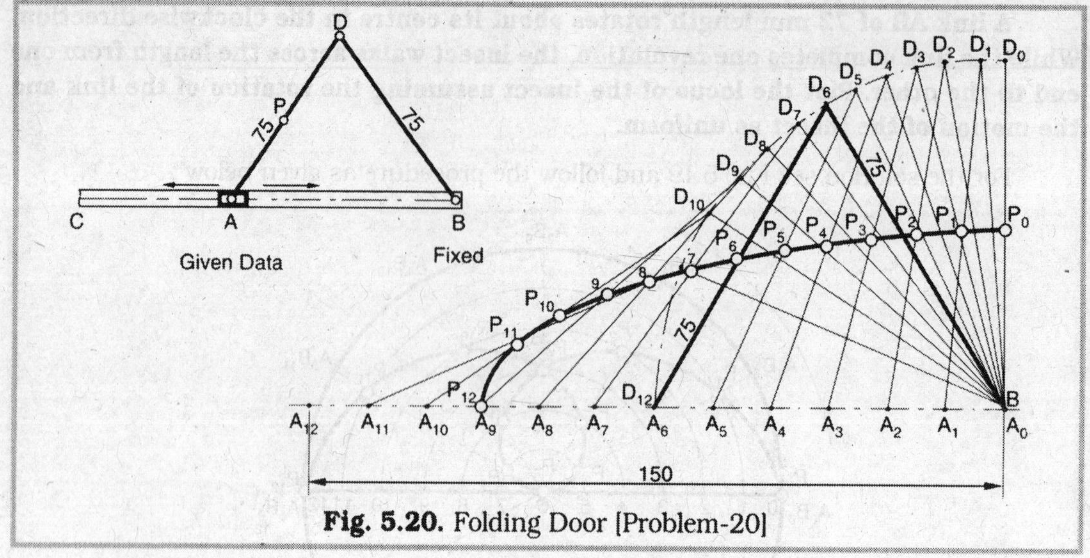

Fig. 5.20. Folding Door [Problem-20]

(1) Take the complete open position of the door as the initial position i.e. BD_0A_0 as shown in the figure.

(2). The point D will move along 1/4th of the circle ⊙ (B,75) and therefore, draw that 1/4 circle.

(3) Divide the total movement of A i.e. BC = 150 mm into 12 equal parts and mark them as $A_0, A_1, \ldots\ldots A_{12}$.

(4) With A_0, $A_1, \ldots\ldots A_{12}$ as the centres and the radius equal to AD = 75 mm draw arcs to cut the previously drawn 1/4th of the circle at points D_0, $D_1 \ldots\ldots D_{12}$ respectively. Join A_0D_0, $A_1D_1, \ldots\ldots A_{12}D_{12}$ and mark on them their mid points P_0, $P_1, \ldots\ldots P_{12}$ respectively.

(5) Join P_0, $P_1, \ldots\ldots P_{12}$ in sequence to get the locus of the mid point P of AD.

Problem 21 : Link OC, hinged at O, is 100 mm long. It carries a circular disc at C of radius 25 mm capable of rotating about the centre point C. Link, OC initially vertical, turns uniformly towards the right side by an angle of 45° and then towards the left side by the total angle 90° and then to the initial vertical position and during the same time the disc revolves uniformly in the clockwise sense through one complete revolution. Draw the locus of the point P on the disc, initially at the lowest position.

For the solution see Fig. 5.21 and follow the procedure as given below :

(1) Draw vertical line OC_0 of 100 mm length and draw the circle with C_0 as the centre and 25 mm as the radius to represent the initial position. Mark P_0 on the disc at the lowest position.

(2) At O, draw 45° angle to the right as well as to the left of the vertical OC_0 and divide each of them into three equal parts of 15°. In this way the total movement of the link OC is divided into 12 equal parts. Mark the positions of the centre C as C_0, C_1, $C_2, \ldots\ldots C_{12}$ as shown in the figure.

(3) Divide the circle of the disc into 12 equal parts 1,2,.....12 as shown.

(4) When the link has turned through 1/12th of its total angular movement(OC_0 OC_1) the line CP of the disc must have turned by 30° which is 1/12th of one revolution. So draw arc of a circle $1P_1$ with o as centre as shown in the figure and mark point

P_1. Similarly draw arcs $(2P_2)$, $(3P_3)$, ...$(11P_{11})$ as shown to get points P_2, P_3, ...P_{11}.

(5) Join points P_0, P_1,.......P_{12} in sequence to get the locus of the point P.

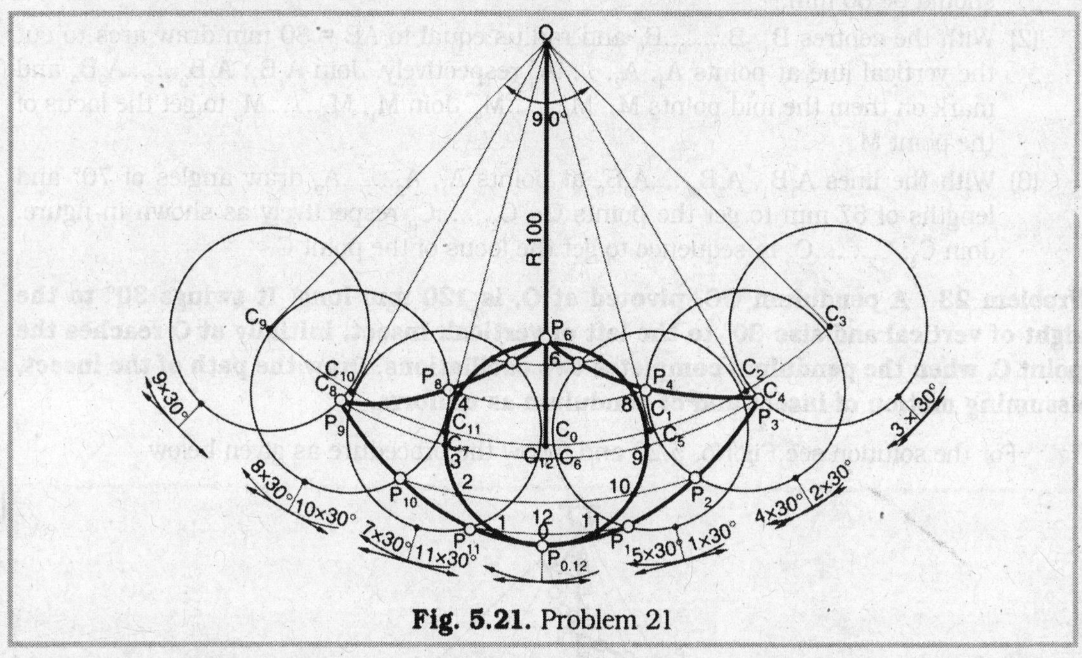

Fig. 5.21. Problem 21

Problem 22 : AB and AC are two links welded together at the point A at an angle of 70° to each other. The ends A and B of the link AB are constrained to slide in the vertical and the horizontal guides respectively. Draw the loci of the point C and the mid point M of the link AB as the link AB moves from the vertical to the horizontal position. AB = 80 mm and AC = 67 mm.

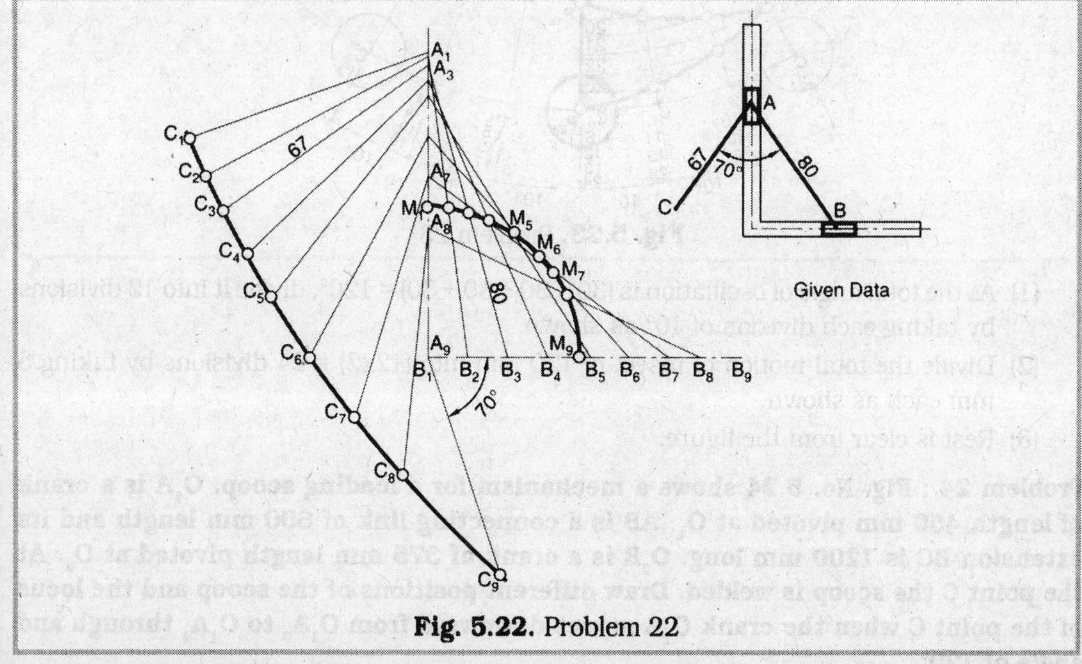

Fig. 5.22. Problem 22

For the solution see Fig. 5.22 and follow the procedure as given below :

(1) On the horizontal line take points B_1, B_2, B_3,......B_9 at nearly equal distances. B_1B_9 should be 80 mm.

(2) With the centres B_1, B_2,......B_9 and radius equal to AB = 80 mm draw arcs to cut the vertical line at points A_1, A_2, ...A_9 respectively. Join A_1B_1, A_2B_2,.......A_9B_9 and mark on them the mid points M_1, M_2,......M_9. Join M_1, M_2,......M_9 to get the locus of the point M.

(3) With the lines A_1B_1, A_2B_2,.....A_9B_9 at points A_1, A_2,.......A_9 draw angles of 70° and lengths of 67 mm to get the points C_1, C_2,......C_9 respectively as shown in figure. Join C_1, C_2,......C_9 in sequence to get the locus of the point C.

Problem 23 : A pendulum OC, pivoted at O, is 120 mm long. It swings 30° to the right of vertical and also 30° to the left of vertical. Insect, initially at O reaches the point C, when the pendulum completes two oscillations. Draw the path of the insect, assuming motion of insect and of pendulum as uniform.

For the solution see Fig.No. 5.23 and follow the procedure as given below.

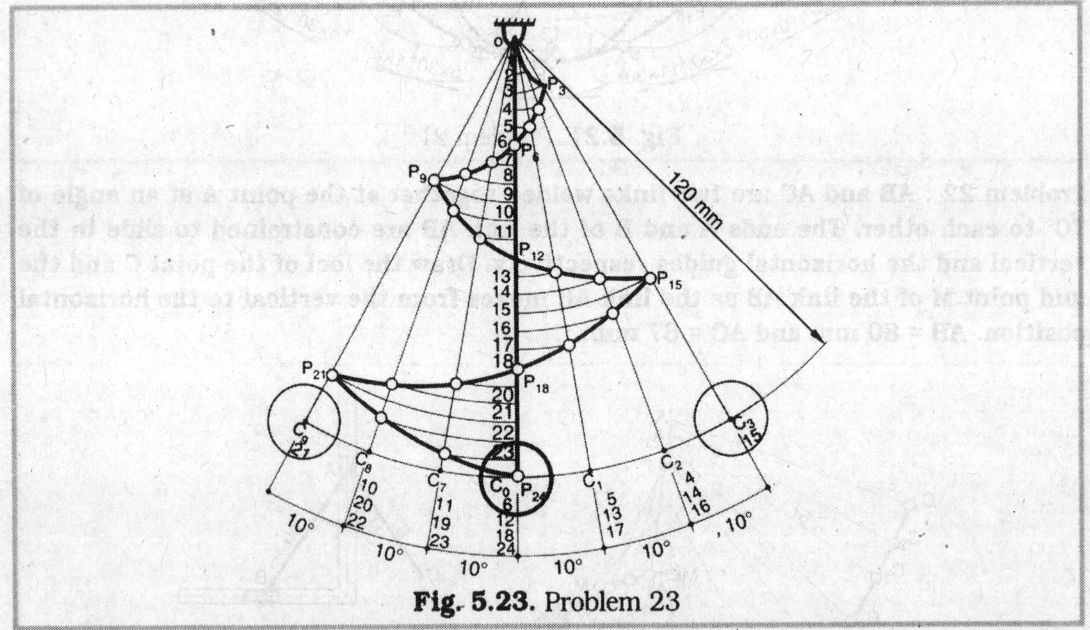

Fig. 5.23. Problem 23

(1) As the total angle of oscillation is (30 + 30 + 30 + 30) = 120°, divide it into 12 divisions by taking each division of 10° as shown.

(2) Divide the total motion of insect of 120 mm into (12x2) = 24 divisions by taking 5 mm each as shown.

(3) Rest is clear from the figure.

Problem 24 : Fig. No. 5.24 shows a mechanism for a loading scoop. O_1A is a crank of length 450 mm pivoted at O_1. AB is a connecting link of 600 mm length and its extension BC is 1200 mm long. O_2B is a crank of 375 mm length pivoted at O_2. At the point C the scoop is welded. Draw different positions of the scoop and the locus of the point C when the crank O_1A moves downward from O_1A_0 to O_1A_4 through and angle of 120°.

For the solution see Fig. 5.24 and follow the procedure as given below :

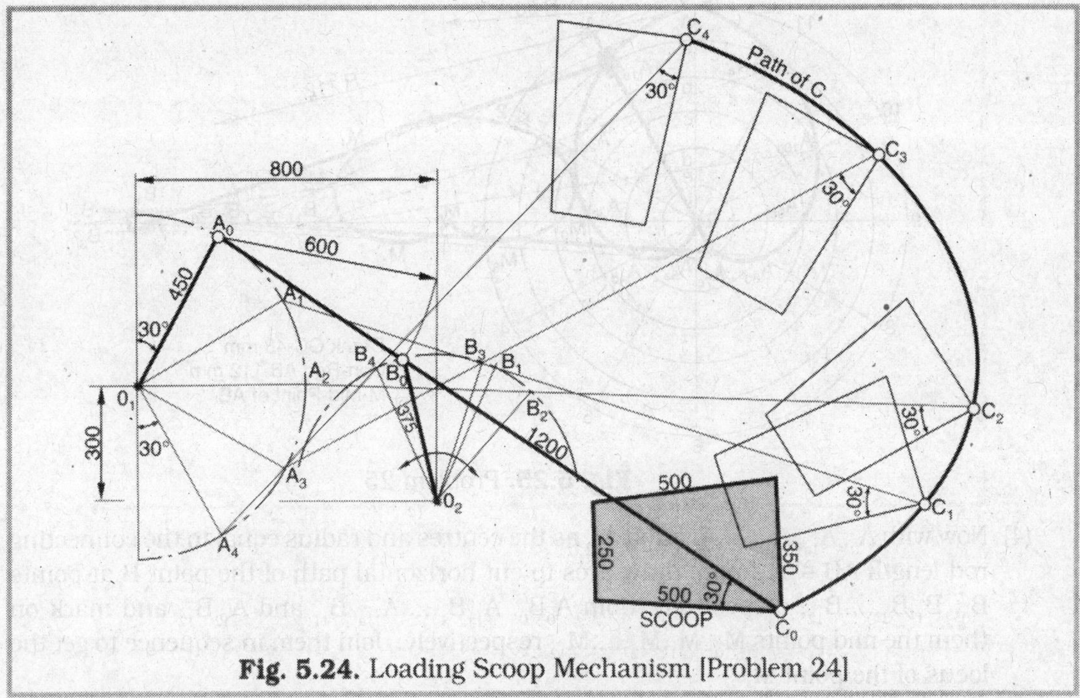

Fig. 5.24. Loading Scoop Mechanism [Problem 24]

(1) Divide 120° angle between O_1A_0 and O_1A_4 into four equal parts and mark points A_0, A_1,.....A_4.

(2) With centres A_0, A_1,........A_4 and radius equal to 600 mm draw arcs to cut the circle (O_2, 375) at points B_0, B_1, ...B_4 respectively.

(3) Join A_0B_0, A_1B_1,.......A_4B_4 and extend them by 1200 mm to get points C_0, C_1,.......C_4. Join C_0, C_1,.......C_4 to get the locus of the point C and draw various positions of the scoop keeping its position same with reference to the line A B C, by drawing 30° angles.

The different successive positions of the scoop give us the idea about how it is loaded.

Problem 25 : The crank OC rotates clockwise about O with a constant angular velocity. In one revolution of the crank the end A of the connecting rod AB moves from the outer most position C to the innermost position O and back to C at a constant speed, while the end B moves along a horizontal line through O. Draw the loci of the point A and the mid point M of AB for one revolution of the crank OC.

For the solution see Fig. 5.25 and follow the procedure as given below :

(1) With O as the centre and radius equal to the crank length OC = 48 mm, draw a circle and divide the circle into 12 equal parts.

(2) As the end A moves on the crank from the outermost to the innermost position and back, divide the crank length into 12/2 = 6 equal parts and mark them as a, b,....e, f as shown in the figure.

(3) When the crank has turned through 1/12 of a revolution the point A must have moved inside by 1/6. OA and so mark A_{1a} on O_1 by drawing an arc through a. Similarly, mark points A_{2b}, A_{3c},....A_{11a} and A_{12}. Join A_0, A_{1a}, A_{2b},....A_{11a} and A_{12} in sequence to get the locus of the point A.

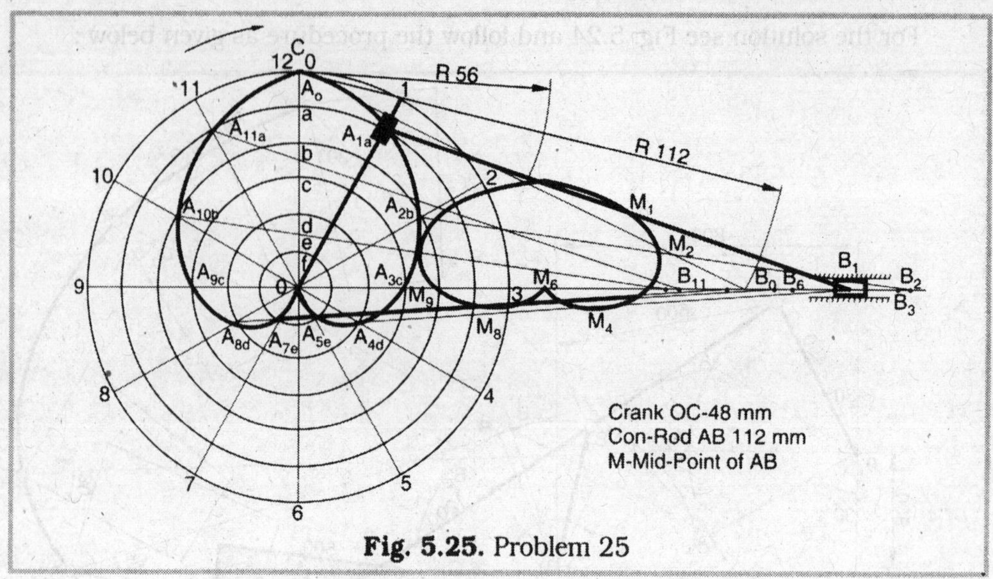

Fig. 5.25. Problem 25

(4) Now with A_0, A_{1a}, A_{2b},.....A_{11a} and A_{12} as the centres and radius equal to the connecting rod length AB = 112 mm, draw arcs to cut horizontal path of the point B at points B_0, B_1,B_2,.....B_{12}, respectively. Join A_0B_0, $A_{1a}B_1$,.....A_{11a} B_{11} and $A_{12}B_{12}$ and mark on them the mid points M_0, M_1,M_2,.....M_{12} respectively. Join them in sequence to get the locus of the point M.

EXERCISE

1. ΔPQR has sides PQ = 100 mm, QR = 75 mm and RP = 50 mm. Three circles ⊙ (P,30), ⊙ (Q, 40) and ⊙ (R, 15) are given. Draw two circles touching the three given circles.
2. The end P of the staircase PQ, 3000 mm long, slides vertically on a wall and its end Q slides horizontally away from the wall. Find the locus of the mid point of the staircase PQ.
3. In a slider crank chain OBA, the crank OB is 350 mm long and the connecting rod BA is 1050 mm long. Plot the loci of points P, Q and R where (i) Point P is on the connecting rod 350 mm from B, (ii) Point R is on extension of C.R. BA and 250mm from A (iii) Point Q is on extension of C.R. AB and 500 mm from B, see Fig. 5.26.

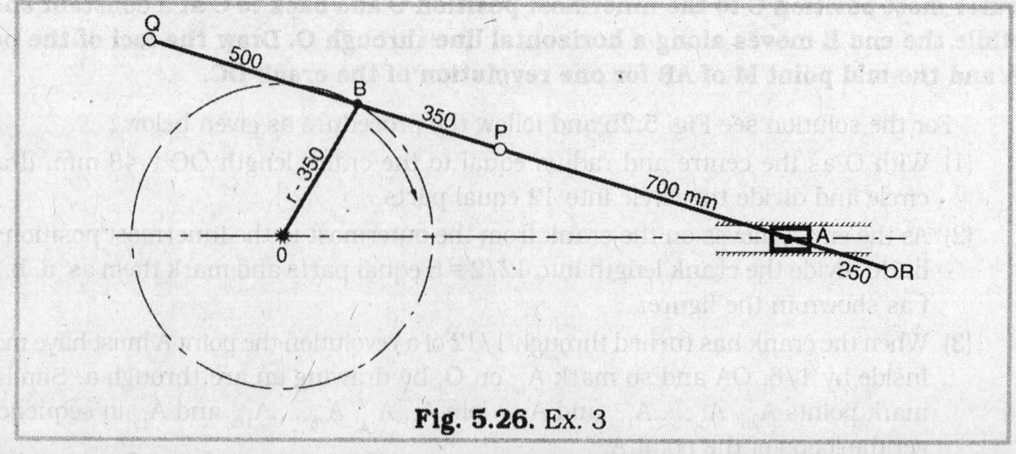

Fig. 5.26. Ex. 3

4. In an offset slider crank chain OBA, the crank OB is 300 mm long and the connecting rod BA is 1000 mm long. Slider 'A' slides in a horizontal guide 150 mm below the horizontal from O. Draw the loci of points P and Q where the point P is a point on the con-rod BA, 250 mm from B and the point Q is the end point of PQ, a rod attached at right angle to con-rod AB at P. (See Fig. 5.27)

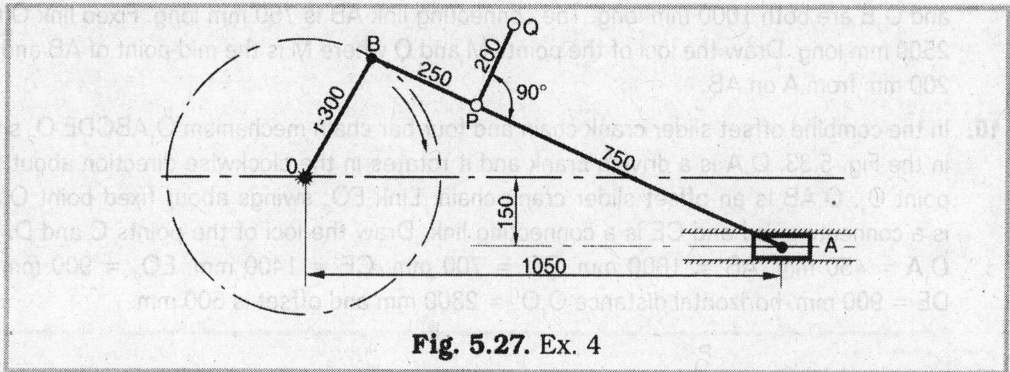

Fig. 5.27. Ex. 4

5. In the crank-connecting rod-trunnion mechanism shown in Fig. 5.28; crank OB is 400 mm long. Connecting rod BA is 1800 mm long and trunnion C is located 1250 mm on the right of O and 150 mm below O. Draw the loci of the points A, P and M where the point P is on the extension of AB and 300 mm from B and the point M is the mid point of AB.

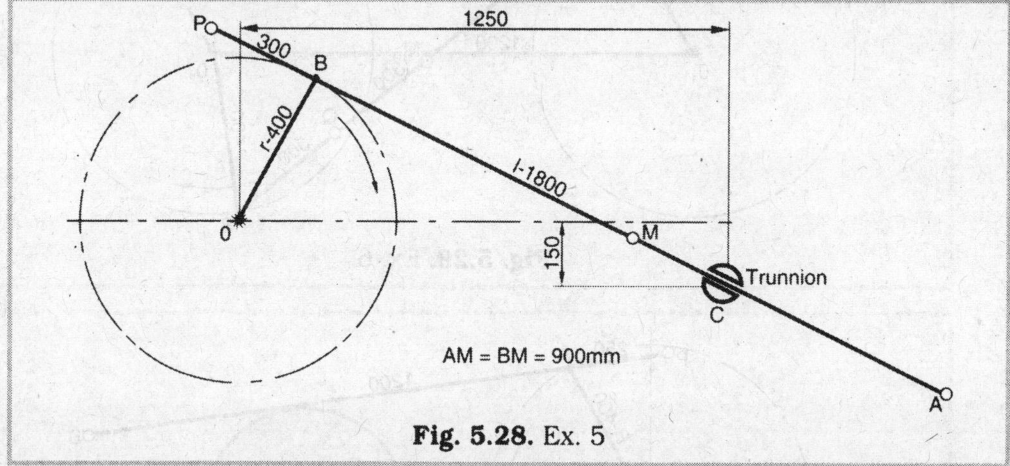

Fig. 5.28. Ex. 5

6. In the four bar chain mechanism O_1ABO_2 the horizontal fixed link O_1O_2 and the connecting link AB both are 1200 mm long. O_1A is a driving crank and is 400 mm long. O_2B is a driven crank and is also 400 mm long. Draw the loci of the points P, M and Q where the point P is on the extension of BA and 300 mm from A, the point M is the mid point of con-link AB and the point Q is on the con-link 300 mm from B. When (i) both the cranks rotate in the same direction (mechanism shown dotted) and (ii) both the cranks rotate in the opposite direction (mechanism shown full). (See Fig. 5.29)

7. In the mechanism shown in Fig. 5.30, the crank O_1 A rotates about O_1 and link PAB slides (in the same plane) on the curved surface of the cylinder (O_2, 250). Draw the loci of points P and B for one revolution of O_1A. Take O_1A= 400 mm, PA = 250 mm, AB - 1200 mm and horizontal distance O_1O_2 = 850 mm.

8. In the four bar chain mechanism O_1ABO_2 shown in Fig. 5.31, two cranks O_1A (driver) and O_2B (driven) are 400 mm and 1000 mm long respectively and the connecting link AB is 1300 mm long. Fixed link O_1O_2 is 1600 mm long. Draw the locus of the point P, on AB and 400 mm from A, for one revolution of the driving crank O_1A.

9. In the four bar chain mechanism O_1ABO_2 shown in Fig. 5.32 driving and driven cranks O_1A and O_2B are both 1000 mm long. The connecting link AB is 750 mm long. Fixed link O_1O_2 is 2500 mm long. Draw the loci of the points M and Q where M is the mid point of AB and Q is 200 mm from A on AB.

10. In the combine offset slider crank chain and four bar chain mechanism $O_1ABCDE\ O_2$ shown in the Fig. 5.33, O_1A is a driving crank and it rotates in the clockwise direction about fixed point O_1. O_1AB is an offset slider crank chain. Link EO_2 swings about fixed point O_2. AB is a connecting rod and CE is a connecting link. Draw the loci of the points C and D. Take O_1A = 450 mm, AB = 1800 mm, BC = 700 mm, CE = 1400 mm, EO_2 = 900 mm and DE = 900 mm, horizontal distance O_1O_2 = 2300 mm and offset is 600 mm.

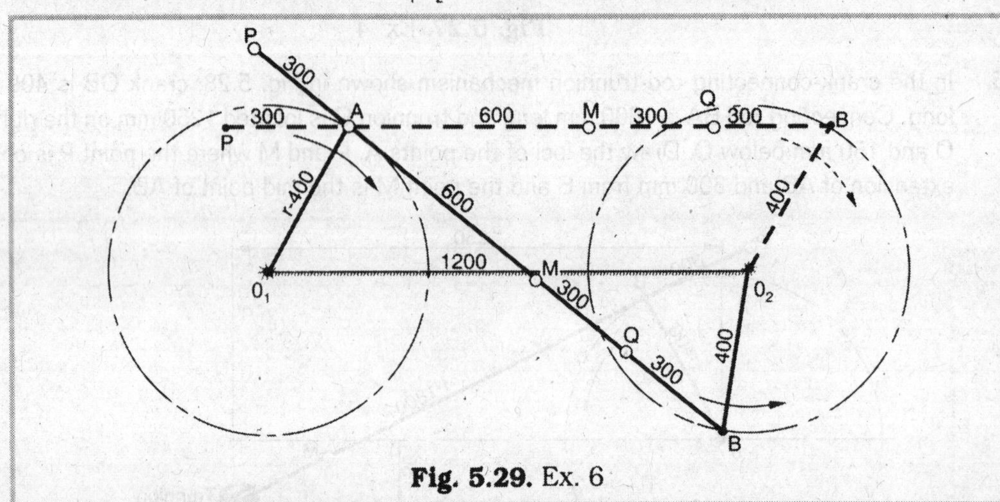

Fig. 5.29. Ex. 6

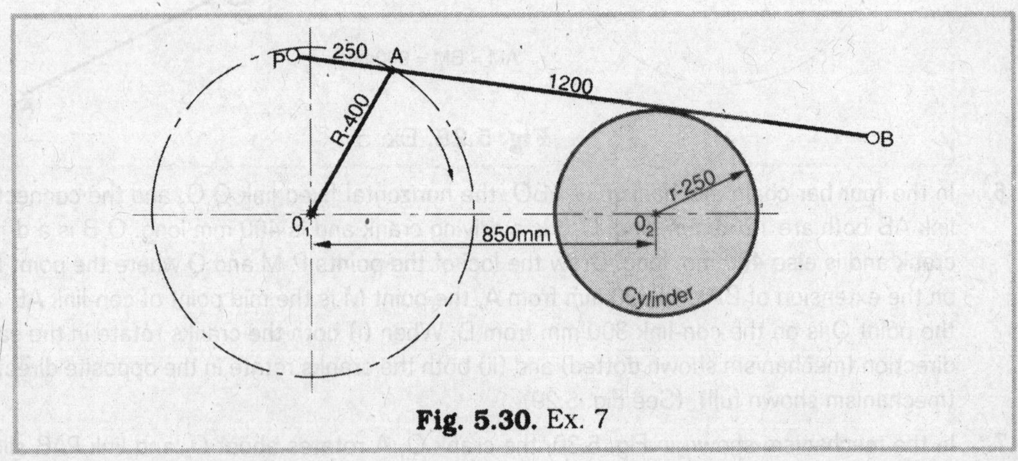

Fig. 5.30. Ex. 7

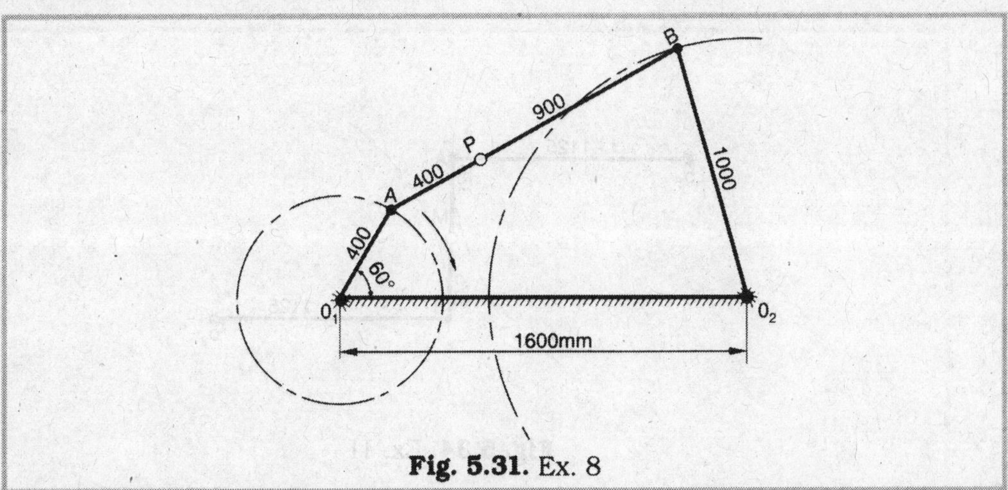

Fig. 5.31. Ex. 8

11. A modified Watt's straight line motion mechanism is shown in Fig. 5.34. Driving and driven cranks O_1A and O_2B are both 1125 mm long and swing about O_1 and O_2 respectively. The connecting link AB is 750 mm long. Draw the locus of the mid point M of AB.

12. The crank OA, of the mechanism shown in the Fig. 5.35, rotates clockwise at a uniform speed. During one revolution of the crank OA, the slider E at C slides to D and back to C at a uniform speed. Draw the locus of the point B for one revolution of the crank OA about O.

Take OA = 250 mm, AB = BC = 750 mm, CD = 500 mm.

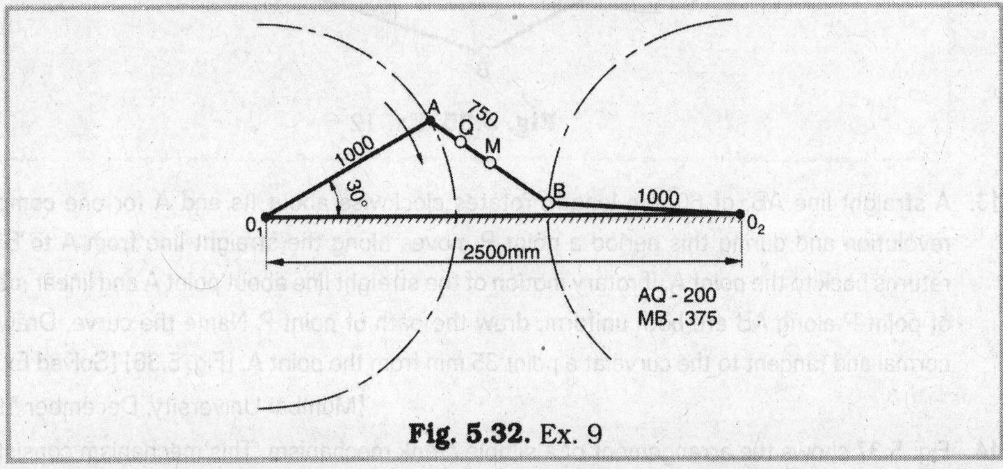

Fig. 5.32. Ex. 9

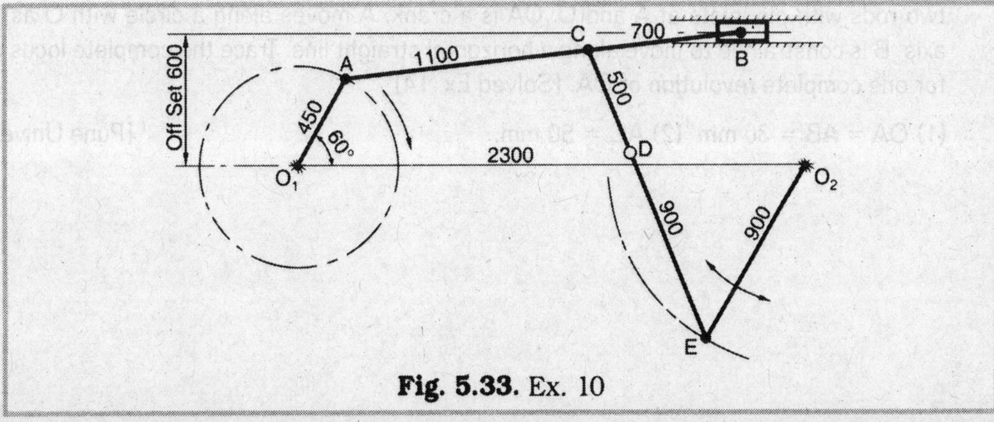

Fig. 5.33. Ex. 10

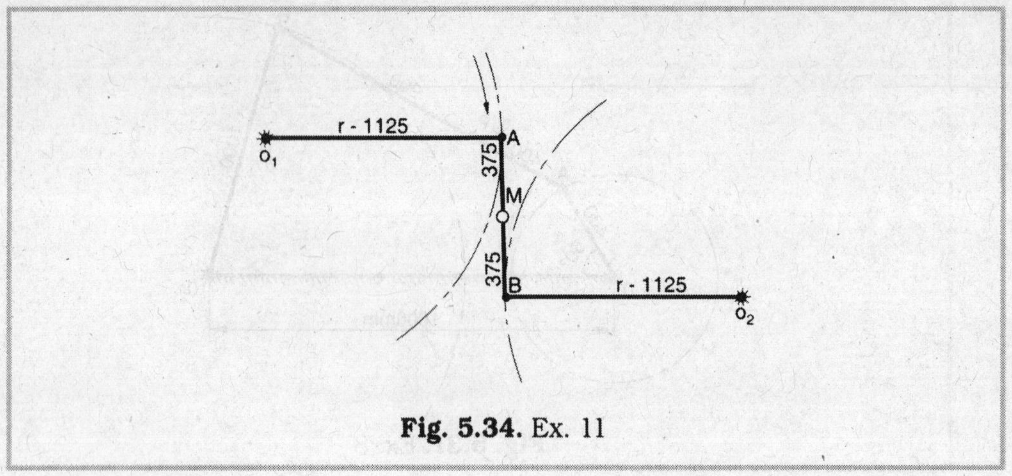

Fig. 5.34. Ex. 11

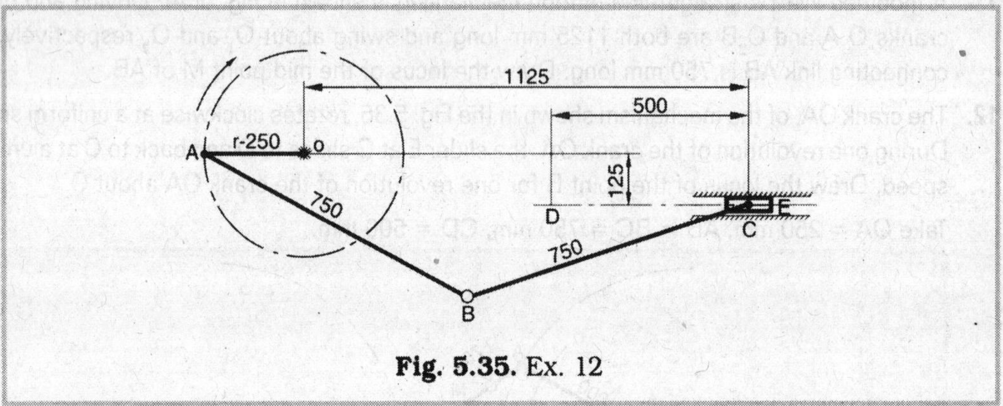

Fig. 5.35. Ex. 12

13. A straight line AB, of 60 mm length, rotates clockwise about its end A for one complete revolution and during this period a point P moves along the straight line from A to B and returns back to the point A. If rotary motion of the straight line about point A and linear motion of point P along AB are both uniform, draw the path of point P. Name the curve. Draw the normal and tangent to the curve at a point 35 mm from the point A. [Fig. 5.36] [Solved Ex. 13]

[Mumbai University, December 1994]

14. Fig. 5.37 shows the arrangement of a simple crank mechanism. This mechanism consists of two rods with pin joints at A and O. OA is a crank. A moves along a circle with O as fixed axis. B is constrained to move along a horizontal straight line. Trace the complete locus of C, for one complete revolution of OA. [Solved Ex. 14]

(1) OA = AB = 30 mm (2) AC = 50 mm. [Pune University]

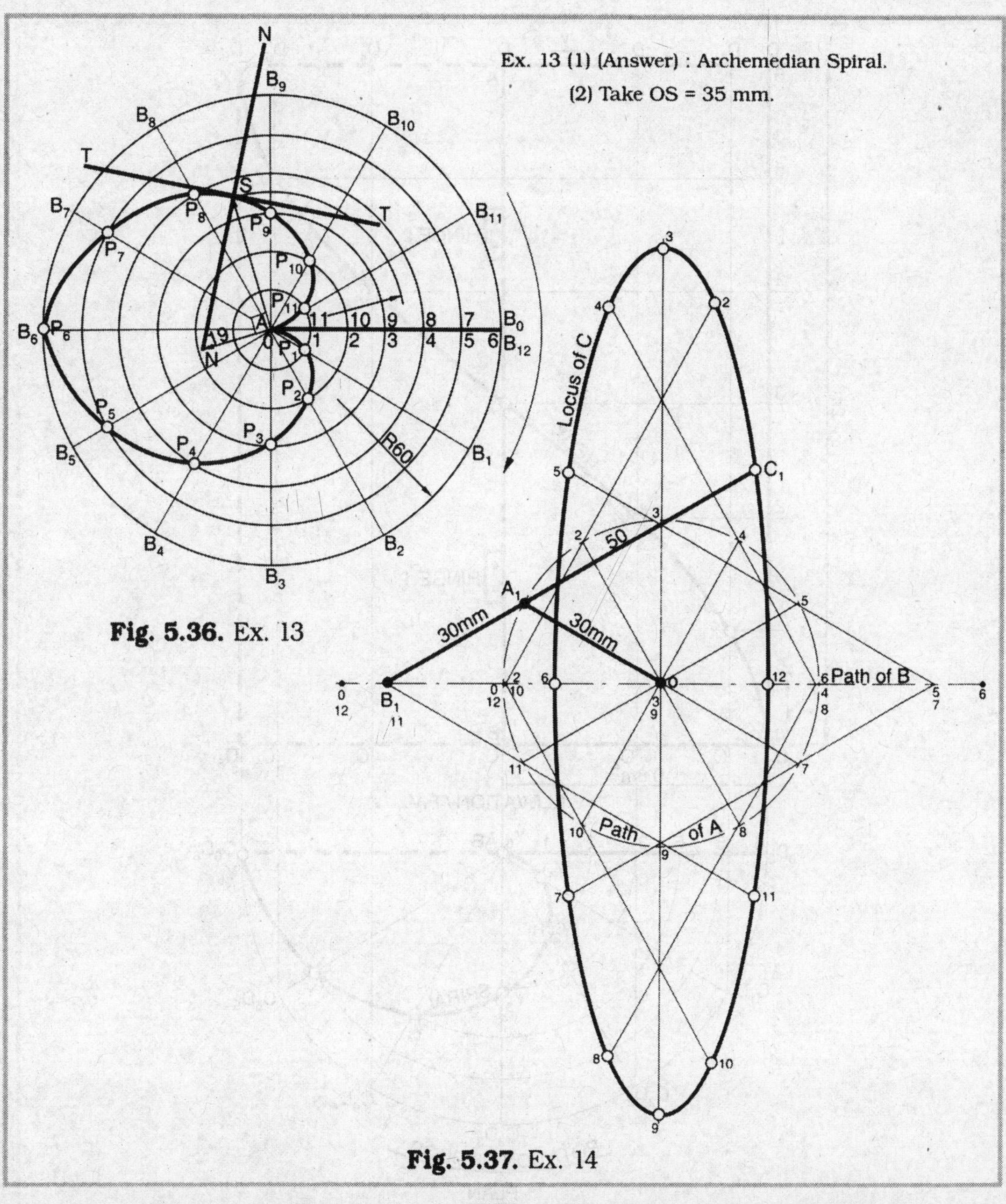

Fig. 5.36. Ex. 13

Fig. 5.37. Ex. 14

15. A rectangular door ABCD has its vertical edge AB 2 m long and a horizontal edge BC 0.8 m long. It is rotated about the hinged vertical edge AB as the axis and at the same time, a fly x moves from point C towards D and another fly Y moves from A towards D. By the time, the door rotates through 180°, both the flies reach point D. Using suitable scale, trace the paths of the flies in elevation and plan if the motions of the flies and the door are uniform. Name the curve traced out by the flies. Assume the door to be parallel to V. P. in initial position and the thickness of the door equal to that of your line.

[Fig. 5.38] [Solved Ex. 15] [Mumbai University, December 1995]

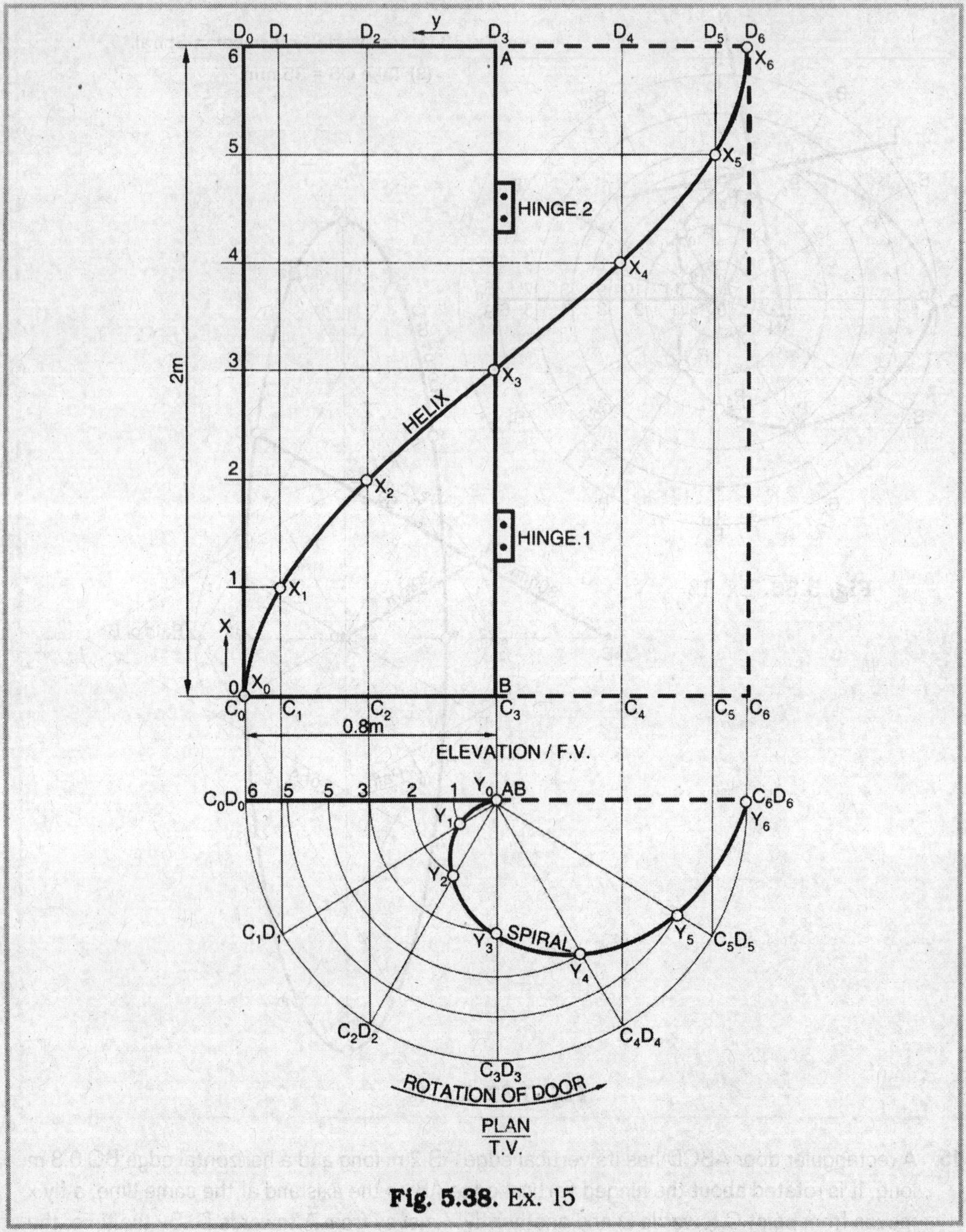

D0 D1 D2 y D3 D4 D5 D6
6 A X6
5 X5
HINGE.2
4 X4
2m
3 X3
HELIX
2 X2
HINGE.1
1 X1
X
0 X0 B
C0 C1 C2 C3 C4 C5 C6
0.8m
ELEVATION / F.V.
C0D0 6 5 5 3 2 1 Y0 AB C6D6
Y1 Y6
Y2
C1D1 Y3 SPIRAL Y5 C5D5
Y4
C2D2 C4D4
C3D3
ROTATION OF DOOR
PLAN
T.V.

Fig. 5.38. Ex. 15

CHAPTER - 6

Engineering Curves

In engineering we often come across various curves. These curves are useful by their nature and characteristics. Some of the laws of nature when represented on graph give us these curves. In space technology also these curves are used for deciding the path of space vehicles. In general, these curves are very very useful in engineering in understanding laws, in manufacturing of various items, in designing the mechanisms, in analysis of forces, in construction of bridges, dams, water tanks, in understanding the electrical power, etc.

Now first of all we shall study the classification of various types of curves used in engineering. There are six main curves of our interest. They are further divided as shown in chart form. Some of the methods of drawing the same is also mentioned. For chart see next page.

Having studied the classification of engineering curves, we shall study its definitions, uses and various methods of constructing the same one by one.

I. CONICS

The intersection of a cone by cutting plane gives us various curves of intersections known as conics. Different shapes by intersections are achieved by taking different positions of cutting planes, with respect to cone axis or base or generator.

This is shown in Fig. 6.1

(A) When the cutting plane contains the apex, we get a triangle as the section.

(B) When the cutting plane is perpendicular to the axis or parallel to the base in a right cone we get circle as the section.

(C) When the cutting plane is inclined to the axis but not parallel to generator we get an ellipse as the section.

(D) When the cutting plane is inclined to the axis and parallel to one of the generators of the cone we get a parabola as the section.

(E) When the cutting plane is parallel to the axis we get a hyperbola as the section.

[OR]

(E) When the inclined cutting plane to the base of cone cuts both the portions of double cone, we get a hyperbola as the section.

1.A and 1.B triangle and circle respectively are very common and hence not discussed.

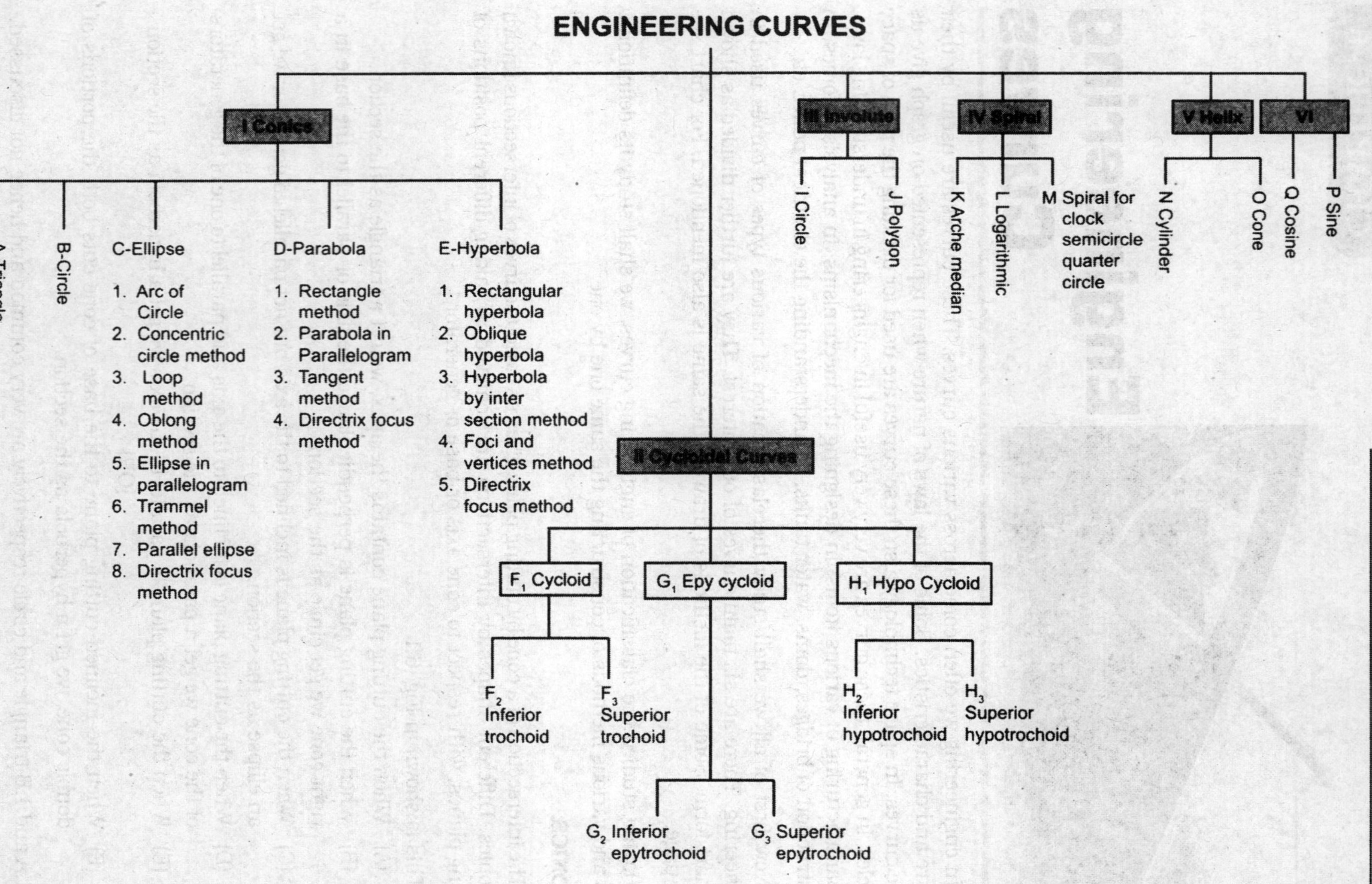
ENGINEERING CURVES
I Conics
A-Triangle
B-Circle
C-Ellipse
1. Arc of Circle
2. Concentric circle method
3. Loop method
4. Oblong method
5. Ellipse in parallelogram
6. Trammel method
7. Parallel ellipse
8. Directrix focus method
D-Parabola
1. Rectangle method
2. Parabola in Parallelogram
3. Tangent method
4. Directrix focus method
E-Hyperbola
1. Rectangular hyperbola
2. Oblique hyperbola
3. Hyperbola by inter section method
4. Foci and vertices method
5. Directrix focus method
II Cycloidal Curves
F_1 Cycloid
F_2 Inferior trochoid
F_3 Superior trochoid
G_1 Epy cycloid
G_2 Inferior epytrochoid
G_3 Superior epytrochoid
H_1 Hypo Cycloid
H_2 Inferior hypotrochoid
H_3 Superior hypotrochoid
III Involute
I Circle
J Polygon
IV Spiral
K Arche median
L Logarithmic
M Spiral for clock semicircle quarter circle
V Helix
N Cylinder
O Cone
VI
P Sine
Q Cosine

I. (C) ELLIPSE

Definition : (1) Ellipse is the locus of a point which moves in a plane so that the ratio of its distances from a fixed point (Focus) and a fixed straight line (Directrix) is a constant and less than one (1). (See Fig. 6.2)

[OR]

(2) Ellipse is the locus of a point, which moves so that the sum of its distances from two fixed points, called focal points or foci, is a constant. The sum of distances is equal to the major axis of the ellipse.

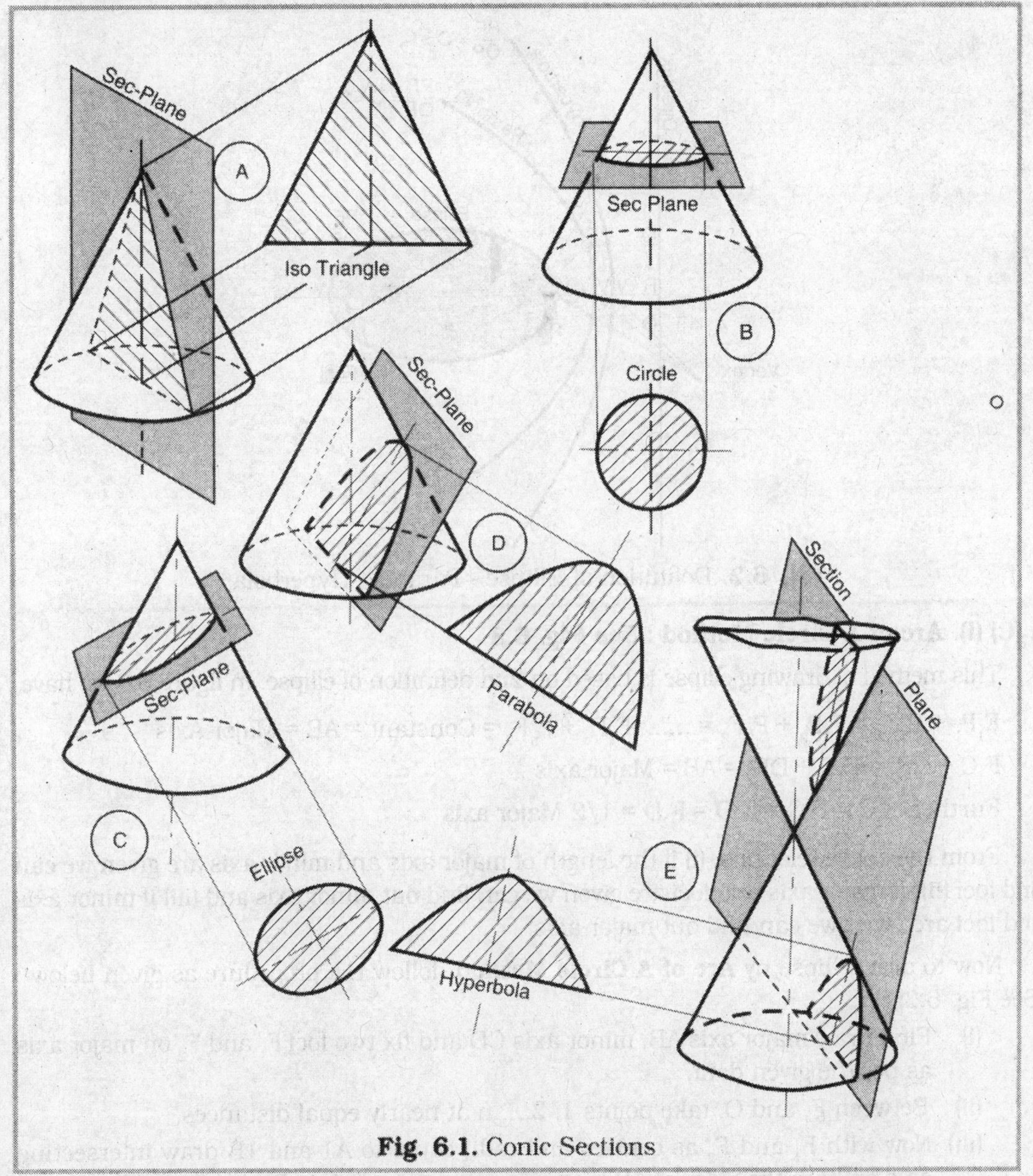

Fig. 6.1. Conic Sections

Uses : (1) Shape of a man-hole
(2) Shape of tank in a tanker
(3) Flanges of pipes, glands and stuffing boxes
(4) Shape used in bridges and arches
(5) Monuments
(6) Path of earth around the sun
(7) Shape of trays etc.

Now we shall study 8 methods of drawing ellipse one by one.

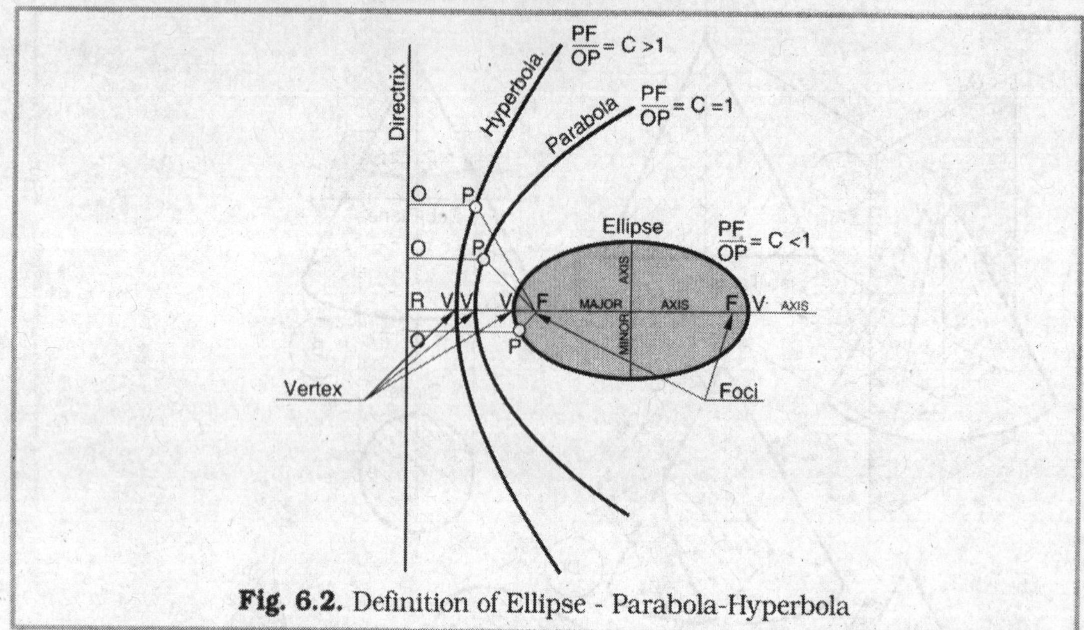

Fig. 6.2. Definition of Ellipse - Parabola-Hyperbola

1 (C) (i). Arc of A Circle Method : See Fig. 6.3

This method of drawing ellipse is based on 2nd definition of ellipse. In figure 6.3 we have,

$F_1P_1 + P_1F_2 = F_1P_2 + P_2F_2 = ...= F_1P_n + P_nF_2$ = Constant = AB = Major Axis

$F_1C + CF_2 = F_1D + DF_2$ = AB = Major axis

Further, $F_1C = F_2C = F_1D = F_2D$ = 1/2 Major axis

From the above relations, (i) if the length of major axis and minor axis are given we can find foci (ii). If major axis and foci are given we can find out minor axis and (iii) if minor axis and foci are given we can find out major axis.

Now to draw ellipse by **Arc of A Circle Method,** follow the procedure as given below : (See Fig. 6.3)

(i) First draw major axis AB, minor axis CD and fix two foci F_1 and F_2 on major axis as per the given data.

(ii) Between F_1 and O, take points 1, 2,....n at nearly equal distances.

(iii) Now with F_1 and F_2 as centres and radii equal to A1 and 1B draw intersecting arcs on both the sides of AB to get four P_1 points. Points P_3 and P_5 are illustrated in the figure. Similarly, get points P_2, P_3,......P_n.

(iv) Join above points of intersections by a smooth curve, including points A, B, C and D to get an ellipse.

I (C) (ii). Concentric Circle Method : See Fig. 6.4.

Follow the procedure as given below :

(i) Draw two concentric circles with major axis and minor axis as diameters and divide them into 12 equal parts as shown in Fig. 6.4 and mark 1,.......12 on both the circles.

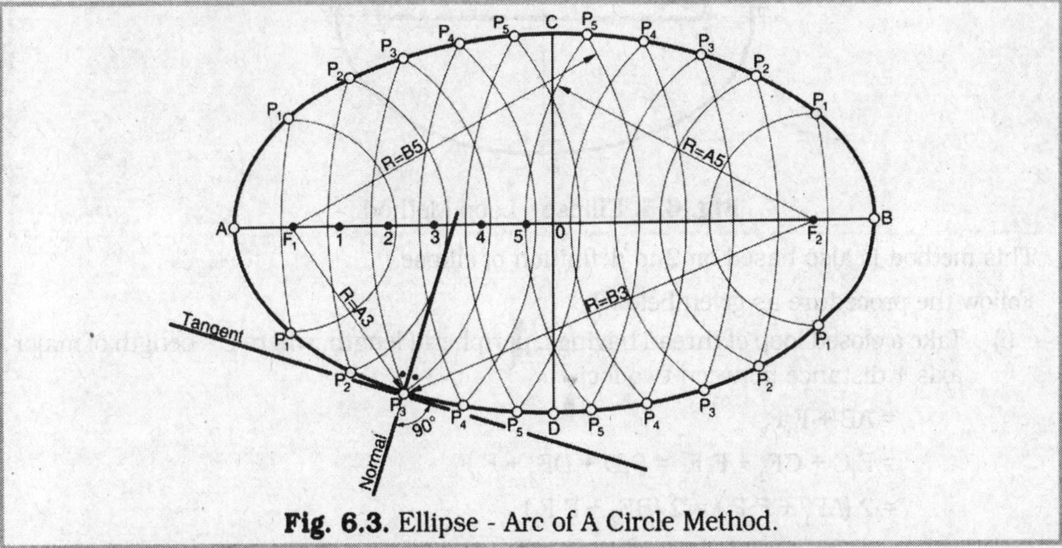

Fig. 6.3. Ellipse - Arc of A Circle Method.

(ii) From points 1,2.....12 of the larger circle draw parallel lines to minor axis (vertical lines) and from 1,2....12 of the smaller circle draw parallel lines to major axis (horizontal lines) intersecting at points P_1, P_2......P_{12} respectively.

(iii) Join all points of intersection P_1, P_2,......P_{12} in sequence by a smooth curve to get an ellipse.

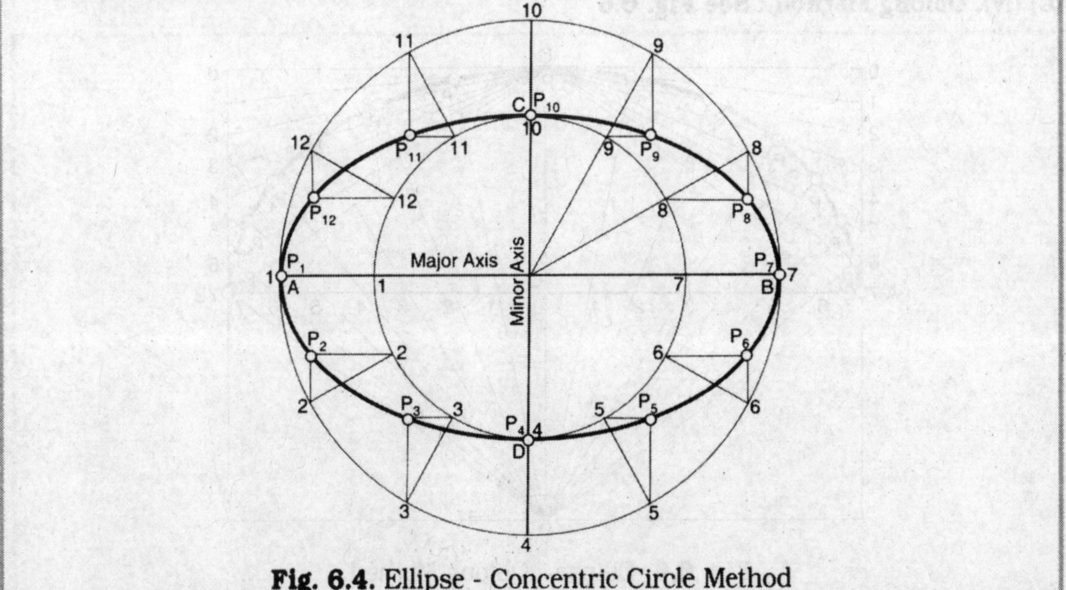

Fig. 6.4. Ellipse - Concentric Circle Method

I (C) (iii). Loop Method : See Fig. 6.5.

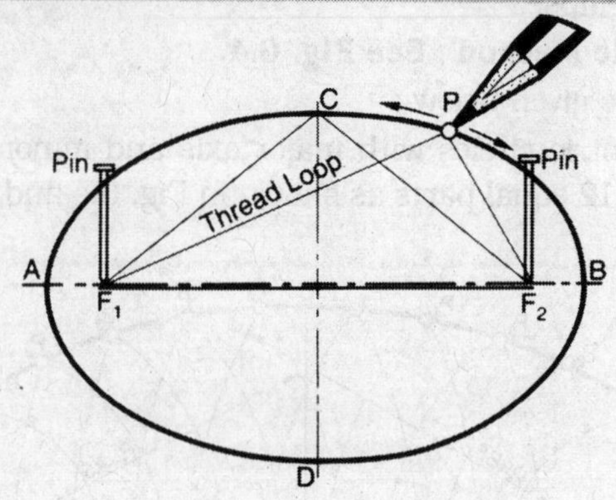

Fig. 6.5. Ellipse - Loop Method.

This method is also based on 2nd definition of ellipse.

Follow the procedure as given below :

(i) Take a closed loop of thread having L peripheral length, where L = Length of major axis + distance between two foci

$$= AB + F_1F_2$$

$$= F_1C + CF_2 + F_1F_2 = F_1D + DF_2 + F_1F_2$$

$$= 2\,(AF_1 + F_1F_2) = 2\,(BF_2 + F_1F_2)$$

$$= 2\,AF_2 = 2BF_1$$

(ii) Fix two pins at foci F_1 and F_2, as shown and arrange the loop of thread around pins as shown.

(iii) Insert pencil point inside the loop and move it, keeping the thread of loop always tight, then the pencil will draw an ellipse.

I (C) (iv). Oblong Method : See Fig. 6.6

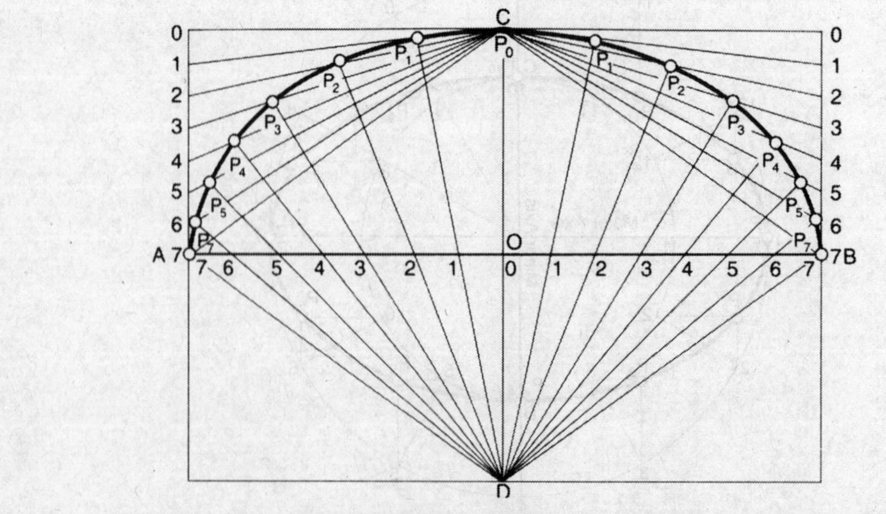

Fig. 6.6. Ellipse - Oblong Method.

Follow the procedure as given below :

(i) Draw rectangle with one side as major axis and another side as minor axis. Draw AB and CD centre lines of the rectangle, intersecting at O, as major axis and minor axis respectively.

(ii) Divide 1/2 of major axis OA into 7 equal parts and mark them as 0, 1, 2,......7 from O towards A. Also divide 1/2 of smaller side of rectangle into 7 equal parts and mark them as 0, 1,.....7, as shown in the figure.

(iii) Join DO, D1,....D7 and extend to intersect with lines joining CO, C1,......C7 at points P_0, P_1,.....P_7 respectively.

(iv) Do similar constructions for all four parts of rectangle and get points P_0, P_1,....... P_7. In all you will get 7 x 4 = 28 points. Join all these points by a smooth curve to get ellipse.

I (C) (v). Ellipse in Parallelogram : See Fig. 6.7

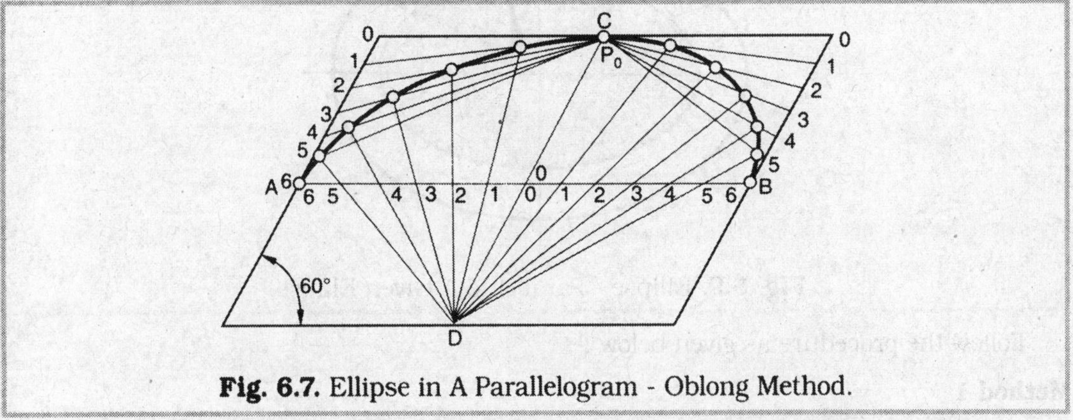

Fig. 6.7. Ellipse in A Parallelogram - Oblong Method.

(i) Draw A parallelogram first of given two sides (84 x 60, here) at given angle (60°,here) and follow the same procedure as followed in the oblong method.

I (C) (vi). Trammel Method : See Fig. 6.8.

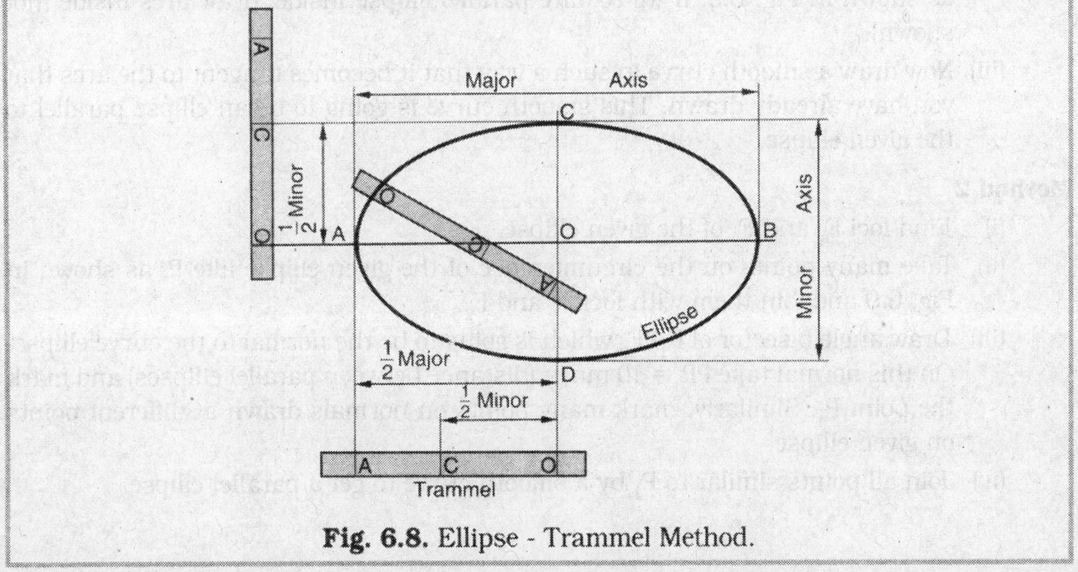

Fig. 6.8. Ellipse - Trammel Method.

(i) First draw two axes, major AB and minor CD, bisecting at right angle at point O.

(ii) Take trammel of hard board or hard paper and mark on it AO equal to (1/2) major axis and OC equal to (1/2) minor axis, as shown in Fig.6.8. Trammel is now ready for use.

(iii) Now arrange the trammel in all possible ways keeping the point A of the trammel on minor axis CD and the point C of the trammel on the major axis AB and mark points against O of the trammel on the paper.

(iv) Join the points, marked against the point O in different positions of the trammel, by a smooth curve to get an ellipse.

I (C) (vii). Parallel Ellipse to A Given Ellipse (Outside) : See Fig. 6.9.

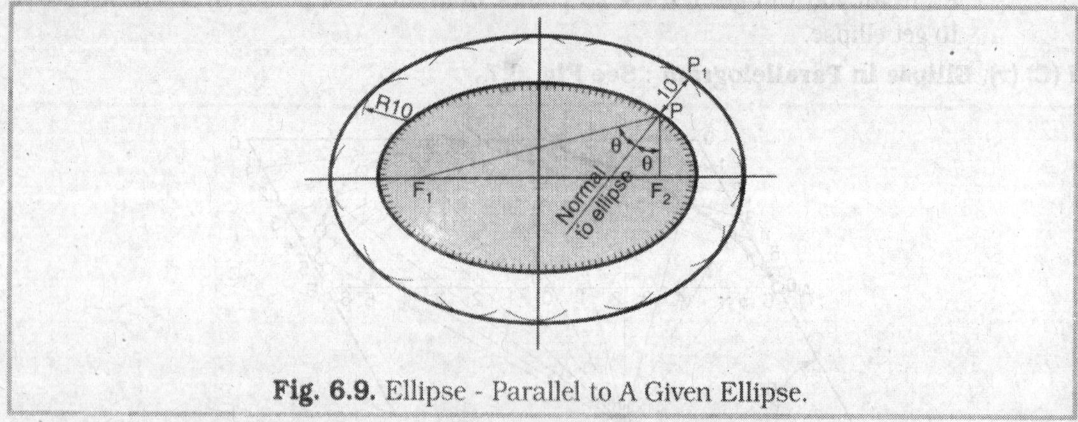

Fig. 6.9. Ellipse - Parallel to A Given Ellipse.

Follow the procedure as given below :

Method 1

(i) First draw the given ellipse.

(ii) Take many points, on the circumference of the given ellipse, as the centres and the radius equal to the distance between parallel ellipses, draw many arcs of circles, as shown in Fig. 6.9. If we require parallel ellipse inside, draw arcs inside (not shown).

(iii) Now draw a smooth curve in such a way that it becomes tangent to the arcs that you have already drawn. This smooth curve is going to be an ellipse parallel to the given ellipse.

Method 2

(i) Find foci F_1 and F_2 of the given ellipse.

(ii) Take many points on the circumference of the given ellipse like P, as shown in Fig. 6.9 and join them with foci F_1 and F_2.

(iii) Draw angle bisector of F_1PF_2 which is going to be the normal to the curve ellipse. On this normal take PP_1 = 10 mm = (distance between parallel ellipses) and mark the point P_1. Similarly, mark many points on normals drawn at different points on given ellipse.

(iv) Join all points similar to P_1 by a smooth curve to get a parallel ellipse.

I (C) (viii). Directrix - Focus Method : See Fig. 6.10

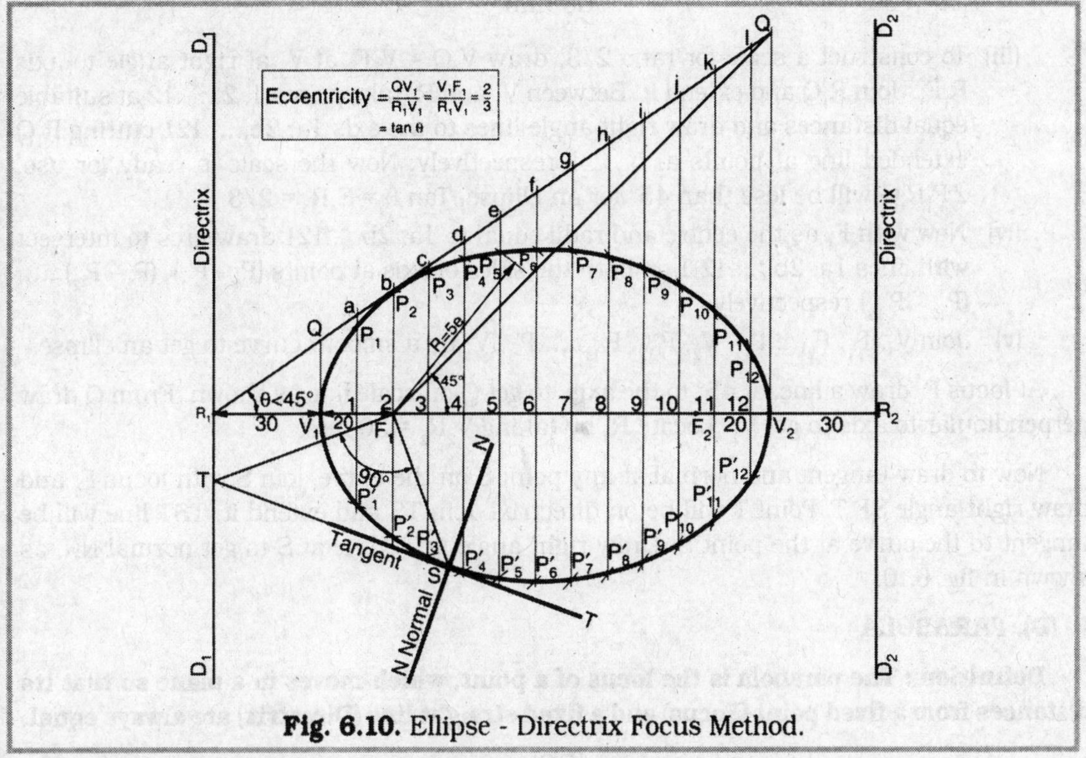

Fig. 6.10. Ellipse - Directrix Focus Method.

This method is based on the 1st definition of an ellipse. (See Fig. 6.2)

When a point moves in a plane keeping the ratio (known as eccentricity ratio) of its distances from focus to that of directrix equal to constant and less than 1 (one) then locus of the point is an **Ellipse.**

$$\text{Eccentricity Ratio} = \frac{PF}{OP} = \text{Constant} < 1. \text{ (See Fig. 6.2)}$$

If the constant ratio is one then its locus is a **Parabola.**

$$\text{Eccentricity Ratio} = \frac{PF}{OP} = \text{Constant} = 1. \text{ (See Fig. 6.2)}$$

If the constant ratio is more than one, then its locus is a **Hyperbola.**

$$\text{Eccentricity Ratio} = \frac{PF}{OP} = \text{Constant} > 1.$$

To construct an ellipse by the above method, follow the procedure as given below and see Fig. 6.10. Given the distance between the focus F_1 and directrix D_1D_1 as 50 mm and eccentricity ratio as 2/3.

(i) First draw directrix D_1D_1 and mark the focus F_1 at a distance 50 mm from R_1 on an axis line R_1R_2 drawn perpendicular to D_1D_1.

(ii) Divide F_1R_1 into ratio 2/3 by the point V_1 (Vertex) such that

$$\frac{F_1V_1}{V_1R_1} = \frac{2}{3} = \text{Eccentricity Ratio} = \frac{20 \text{ mm}}{30 \text{ mm}}$$

(iii) To construct a scale for ratio 2/3, draw $V_1Q = V_1F_1$ at V_1 at right angle to axis R_1F_1. Join R_1Q and extend it. Between V_1 and R_2 take points 1, 2,.....12 at suitable equal distances and draw right angle lines to the axis 1a, 2b,.....121 cutting R_1Q extended line at points a, b, ...l respectively. Now the scale is ready for use. $\angle F_1R_1Q$ will be less than 45° for an ellipse. Tan θ = E.R. = 2/3

(iv) Now with F_1 as the centre and radii equal to 1a, 2b, ...121 draw arcs to intersect with lines 1a, 2b,.....12 1 on both the sides of axis at points (P_1 - P'_1), (P_2 - P'_2)...... (P_{12} - P'_{12}) respectively.

(v) Join V_1, P_1, P_2,....P_{12}, V_2, P'_{12}, P'_{11}......P'_1, V_1 by a smooth curve to get an ellipse.

At focus F_1 draw a line, at 45° to the axis, to get Q on scale line as shown. From Q draw perpendicular to axis to get V_2. Locate R_2 by taking V_2R_2 = 30 mm.

Now to draw tangent and normal at any point S on the curve, join S with focus F_1 and draw right angle SF_1T. Point T will be on directrix. Join TS and extend it. TST line will be tangent to the curve at the point S. Draw right angle to TT line at S to get normal NN, as shown in fig. 6.10.

I - (D). PARABOLA

Definition : The parabola is the locus of a point, which moves in a plane so that its distances from a fixed point (Focus) and a fixed straight line (Directrix) are always equal.

[OR]

Ratio (known as Eccentricity) of its distances from focus to that of directrix is constant and equal to one (1). (See Fig. 6.2)

Uses : (1) Motor car head lamp reflector. (See Fig. 6.11)

(2) Sound reflector and detector

(3) Bridges and arches construction

(4) Shape of cooling towers

(5) Path of particle thrown at any angle with earth, etc.

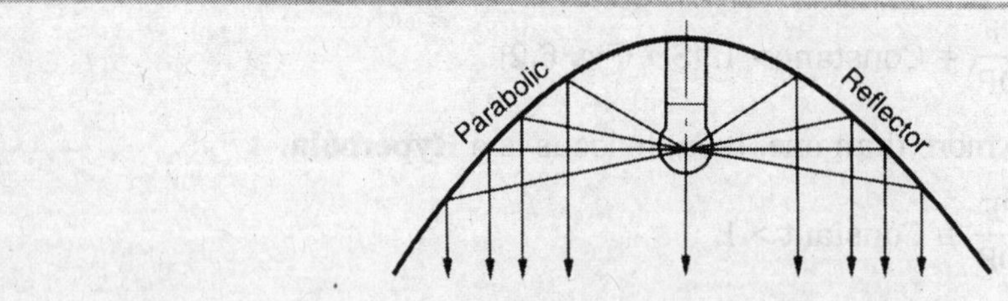

Fig. 6.11. Parabolic Reflector

Now we shall study 4 methods of constructing parabola one by one.

I (D) (i). Rectangle Method : See Fig. 6.12

Given the base length AB and the axis length OV. Axis OV is perpendicular to the base AB. Follow the procedure as given below and see Fig. 6.12.

1. Draw base AB and draw axis OV at right angle to AB at the mid point O of AB.
2. Complete the rectangle ABCD, as shown in the figure.

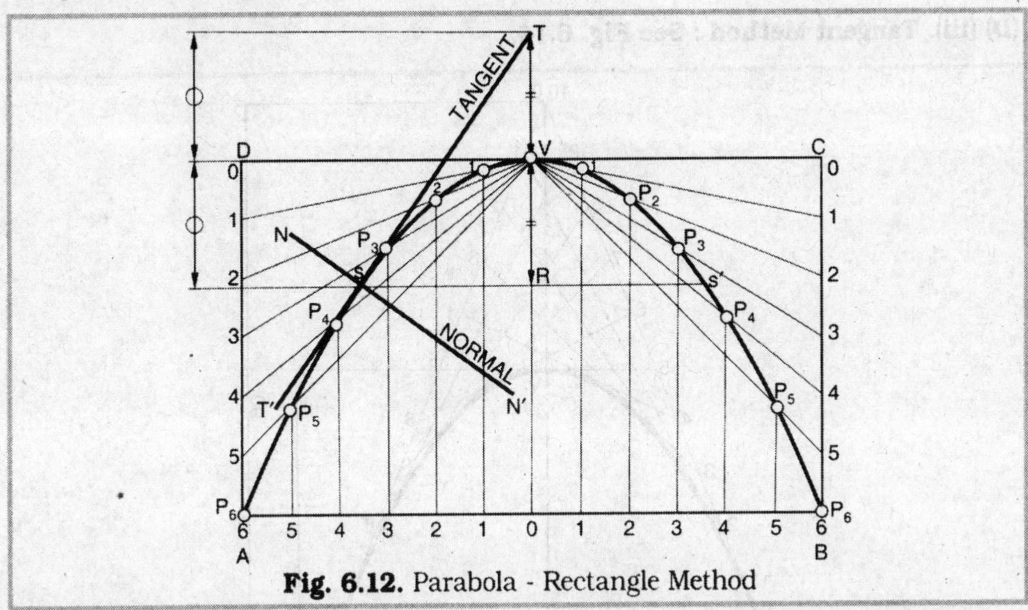

Fig. 6.12. Parabola - Rectangle Method

3. Divide OA and OB into 6 equal parts and mark them as 0, 1,....6. Similarly, divide CB and DA also into six equal parts and mark them as 0, 1,....6, as shown in the figure.
4. Draw parallel lines to the axis OV from 1, 2,....6 of AB to intersect with lines V1, V2,.... V6 at points P_1, P_2,.....P_6 respectively.
5. Join all points by a smooth curve to get a parabola.

I (D) (ii). Parabola in Parallelogram : See Fig. 6.13.

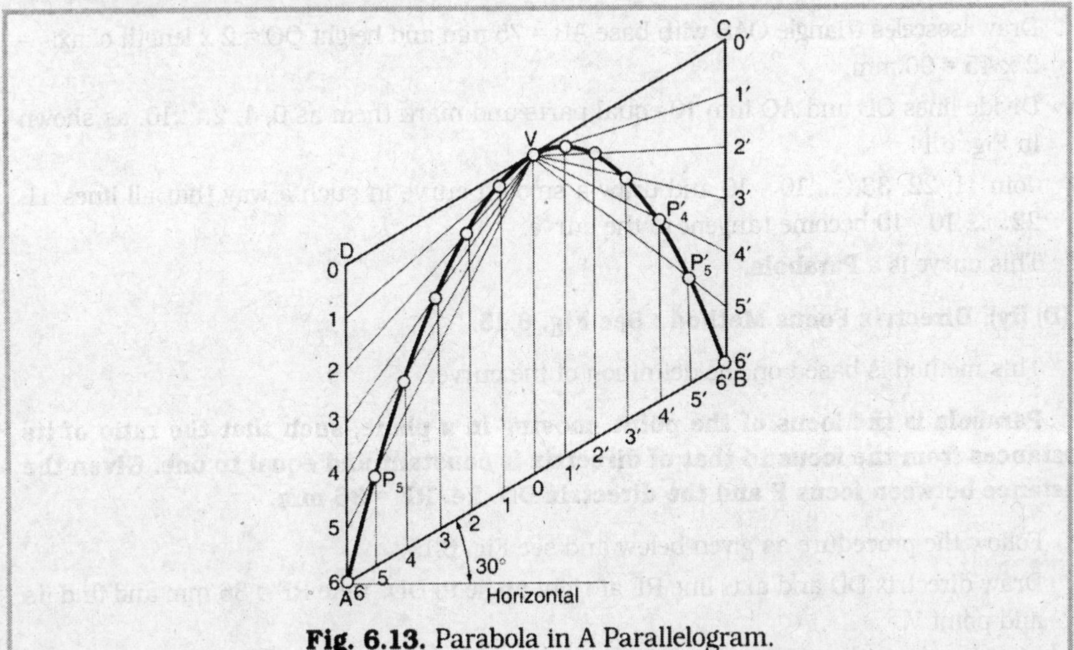

Fig. 6.13. Parabola in A Parallelogram.

This method is similar to the rectangle method. Here, instead of drawing a rectangle we have to draw a parallelogram.

I (D) (iii). Tangent Method : See Fig. 6.14.

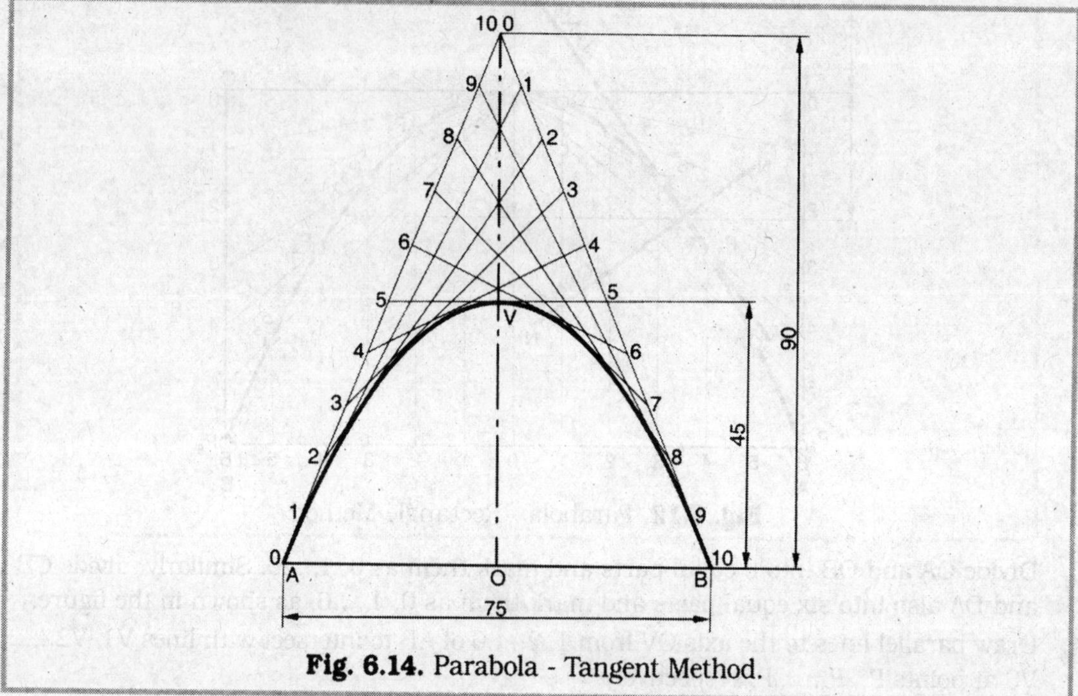

Fig. 6.14. Parabola - Tangent Method.

Given the base AB, 75 mm and axis OV, 45 mm, follow the procedure as given below and see Fig. 6.14.

1. Draw isosceles triangle OAB with base AB = 75 mm and height OO = 2 x length of axis = 2 x 45 = 90 mm.
2. Divide lines OB and AO into 10 equal parts and mark them as 0, 1, 2,.....10, as shown in Fig. 6.14.
3. Join 11, 22, 33,.......10 - 10 and draw a smooth curve in such a way that all lines 11, 22, ...10 - 10 become tangent to the curve.

   This curve is a **Parabola.**

I (D) (iv). Directrix Focus Method : See Fig. 6.15.

This method is based on the definition of the curve.

Parabola is the locus of the point, moving in a plane, such that the ratio of its distances from the focus to that of directrix is constant and equal to one. Given the distance between focus F and the directrix DD. i.e. RF = 36 mm.

Follow the procedure as given below and see Fig. 6.15.

1. Draw directrix DD and axis line RF at right angle to DD. Take RF = 36 mm and find its mid point V.

2. On the axis, take points 1, 2,....n and draw right angle lines to the axis.
3. Now with focus F as the centre and radii equal to R1, R2,.... Rn draw arcs on both the sides of axis to intersect with lines through 1, 2,.....n at points $P_1 - P'_1$, $P_2 - P'_2$,......$P_n - P'_n$.
4. Join points P'_n, P'_{n-1} ....P'_1, V, P_1, P_2.....P_n by means of a smooth curve to get **Parabola.**
5. Now to draw tangent and normal to the curve at any given point S, join the point S with focus F and draw ∠SFT = 90°. The point T is on the directrix. Join TS and extend it upto T. TT represents a tangent. At S draw right angle line NN to TT. Line NN represents normal to the curve at the point S.

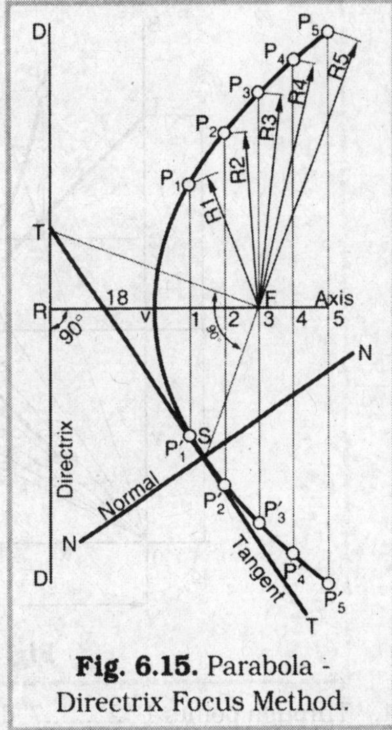

Fig. 6.15. Parabola - Directrix Focus Method.

I (E). HYPERBOLA

Definition : It is the locus of a point which moves in a plane so that the ratio of its distances from a fixed point (Focus) and a fixed straight line (Directrix) is constant and greater than one (1).

[OR]

It is the curve generated by a point, moving in a plane, so that the difference of its distances from the two fixed points called Foci is a constant and is equal to the distance between two vertices.

Uses : (1) Nature of graph of Boyle's law
(2) Shape of overhead water tanks
(3) Shape of cooling towers etc.

I (E) (i). Rectangular Hyperbola : See Fig. 6.16.

Draw a rectangular hyperbola passing through the point P0. Co-ordinates x and y of the point P_0 are given.

Follow the procedure as given below and see Fig.6.16.

1. First draw two axes OA and OB and mark point P_0 with the given co-ordinates x and y.
2. Through P_0 draw CD and EF lines parallel to OA and OB respectively.
3. Draw lines 01'1, 02'2, 03'3.....07'7 from 0 intersecting EF and CD lines at points 1', 2',.....7' and 1, 2,.....7 respectively.

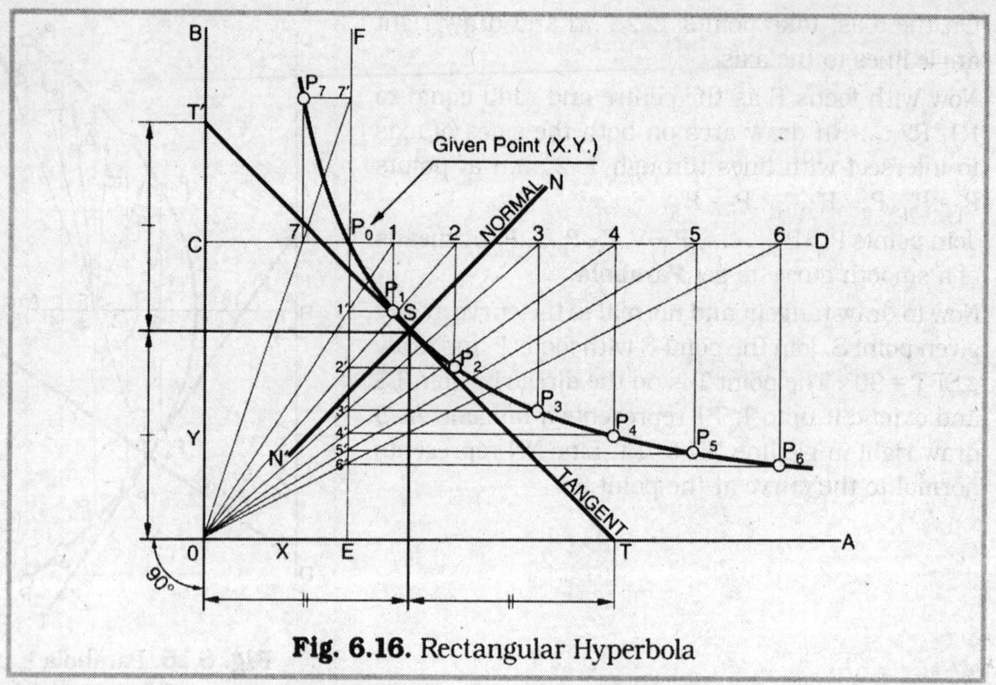

Fig. 6.16. Rectangular Hyperbola

4. Through points 1', 2',.......7' draw lines parallel to OA (horizontal) and through points 1, 2,.....7 draw lines parallel to OB (vertical) intersecting at points P_1,P_2,.....P_7 respectively.
5. Draw a smooth curve passing through P_7, P_0, P_1, P_2,....P_6 to get a **Hyperbola.**

I (E) (ii). Oblique Hyperbola : See Fig. 6.17.

Draw oblique hyperbola passing through point P_0, x and y co-ordinates of point P_0 are given. X and Y axes are at 75°.

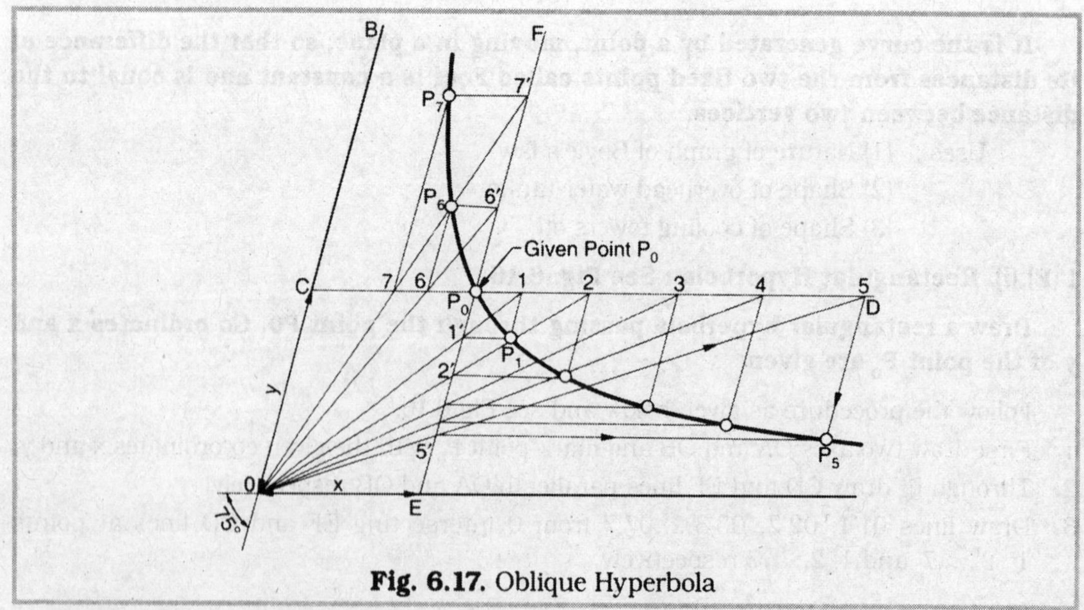

Fig. 6.17. Oblique Hyperbola

Procedure of construction is same as the rectangular hyperbola. (See Fig. 6.17)

I (E) (iii). Foci and Vertices Method : See Fig. 6.18.

Distance between two foci F and F1 are given on the transverse axis and distance 2a between vertices V and V1 are also given, draw the hyperbola.

This method is based on 2nd definition of hyperbola. Follow the procedure as given below. (See Fig. 6.18)

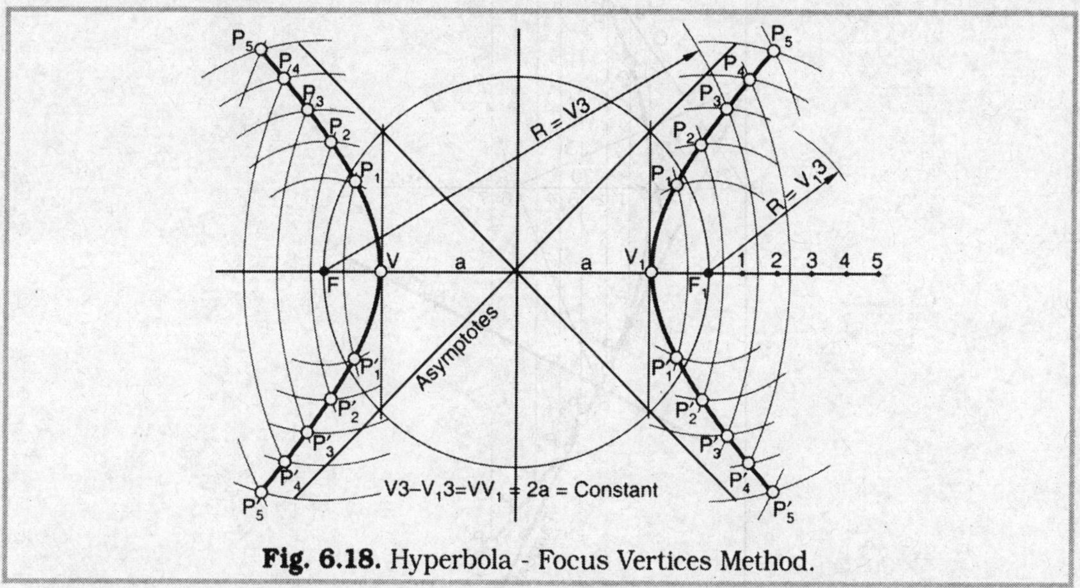

Fig. 6.18. Hyperbola - Focus Vertices Method.

1. First draw transverse axis and mark on it two foci F and F_1 and two vertices V and V_1, as shown.
2. On the axis mark points 1, 2,.....5 beyond F_1 at nearly equal distances.
3. Now with F and F_1 as centres and radii equal to (V1, $V_1$1), (V2, $V_1$2), ...(V5, $V_1$5) draw arcs on both sides to intersect at points (P_1, P'_1), (P_2, P'_2),.....(P_5, P'_5) respectively, as shown in Fig. 6.18.
4. Join P_5, P_4,.....V....P'_1, P'_2,.......P'_5 on two sides by a smooth curve to get two hyperbolas.
5. Now to get 2 Asymptotes, draw a circle with FF_1 as diameter. Draw right angle to the axis at vertices V and V_1 to intersect the circle drawn at two points. Draw two asymptotes passing through these points of intersections already obtained, as shown in Fig. 6.18.
6. To draw tangent and normal to the hyperbola at any point S, join the point S with F and F_1. Draw bisector of angle FSF_1. This bisector is the tangent to the hyperbola at the point S. Draw right angle to this tangent at S to get the normal at S. (Not shown).

I (E) (iv). Directrix - Focus Method : See Fig. 6.19.

Given the distance between focus F and the directrix DD. Eccentricity ratio is also given.

Here the distance between DD and focus F is taken = 50 mm and eccentricity ratio = 3/2. Method of drawing this curve is similar to the directrix focus method, for ellipse as well as for parabola. Scale angle θ will be more than 45°. Tanθ = E.R. = 3/2.

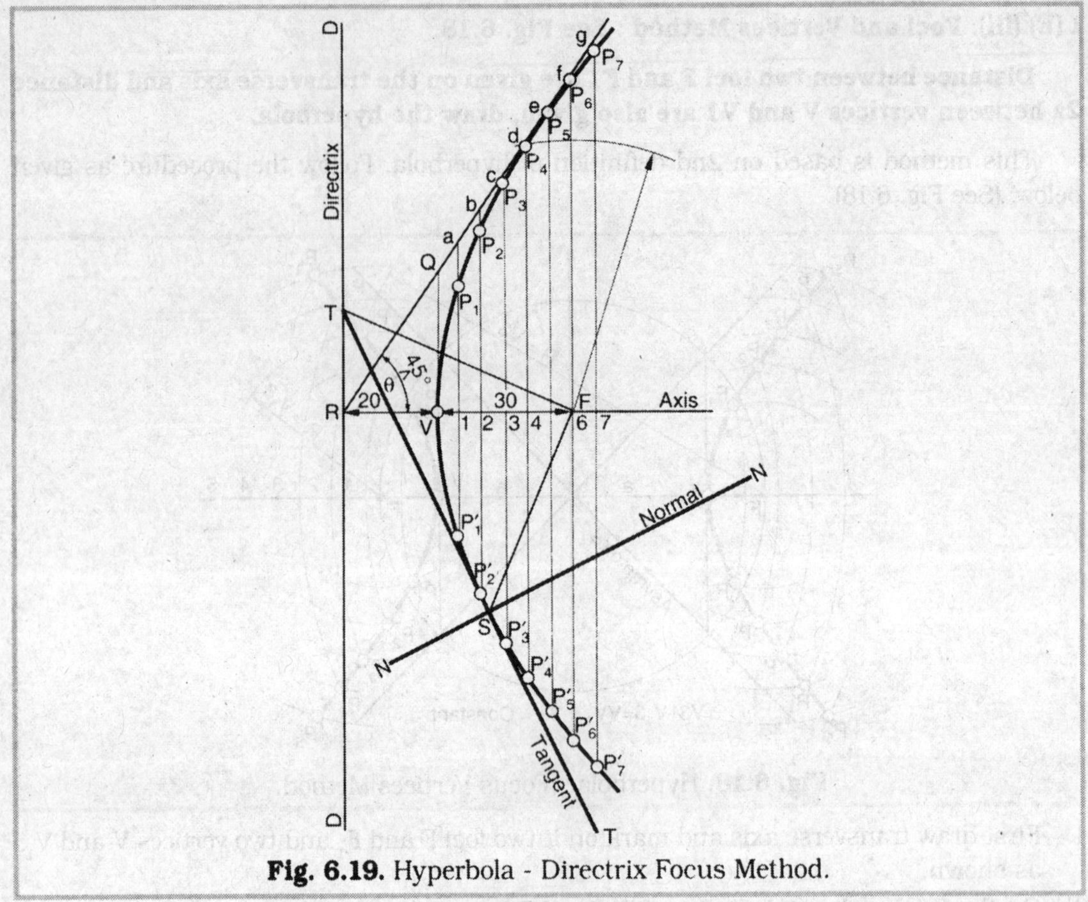

Fig. 6.19. Hyperbola - Directrix Focus Method.

Ellipse - Parabola - Hyperbola by Method of Intersection of A Plane and A Right Cone :-

Here we shall only study, how to locate the position of (1) Directrix (2) Vertex and (3) Focus, when a right cone is cut by a cutting plane in different manner. Method of drawing entire curve is discussed in the chapter of **Section of Solid.**

We know that, when a right cone is cut by a cutting plane inclined to the base and cutting all the generators of the cone, we get Ellipse as the section. See Fig. 6.20 (a).

Similarly when a right cone is cut by a cutting plane parallel to one of the generators of the cone, we get Parabola as the section. See Fig. 6.20(b).

Similarly when a right cone is cut by a cutting plane having inclination more with the base than for parabola, we get Hyperbola as the section. See Fig. 6.20(c).

Fig. 6.20(a), (b) and (c) are self explanatory, hence the explanation is not given.

II CYCLOIDAL GROUP OF CURVES

When one curve rolls over another curve without sliding or slipping, the path of any point of the rolling curve is called a **Roulette.**

When rolling curve is a circle and the curve on which it rolls is a straight line or a circle, we get cycloidal group of curves. There are nine different curves in this group. They are -

(1) F1 Cycloid (4) G1 Epycycloid (7) H1 Hypocycloid

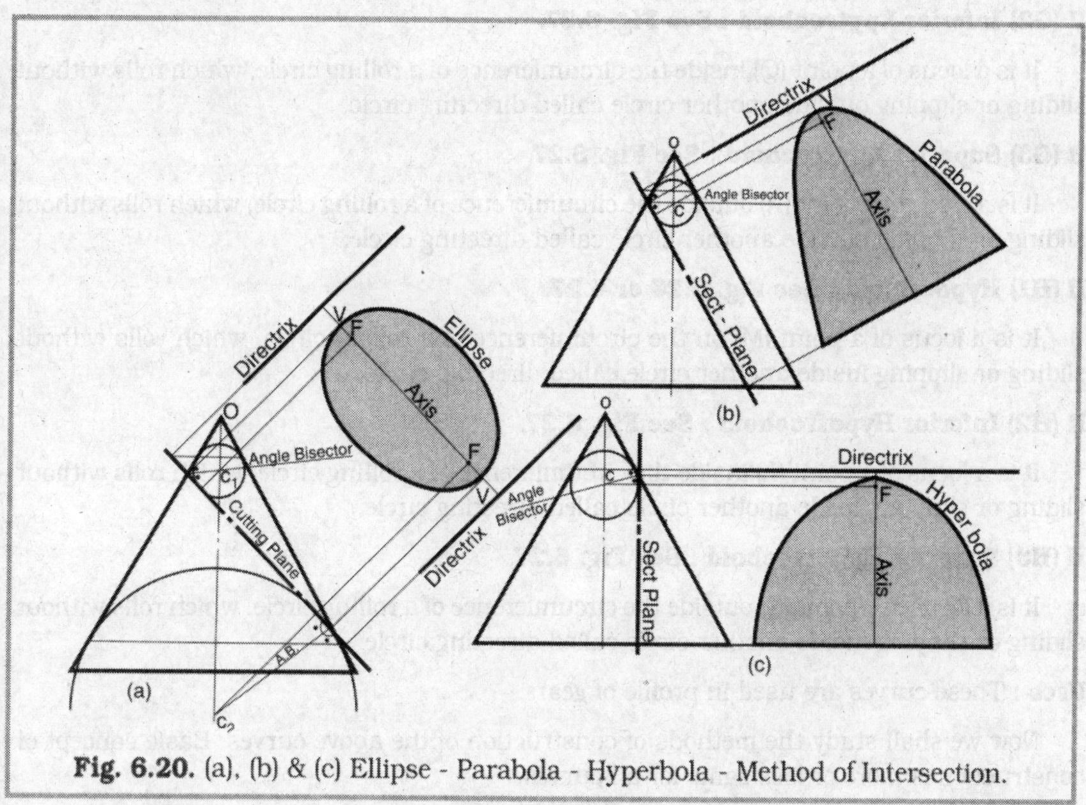

Fig. 6.20. (a), (b) & (c) Ellipse - Parabola - Hyperbola - Method of Intersection.

(2) F2 Inferior Trochoid (5) G2 Inferior Epy trochoid (8) H2 Inferior Hypotrochoid

(3) F3 Superior Trochoid (6) G3 Superior Epy trochoid (9) H3 Superior Hypotrochoid

Now we shall study the definition of each curve one by one.

II (F1) Cycloid : See Fig. 6.22 to understand definition.

Cycloid is a locus of a point (P) on the circumference of a rolling circle (generator), which rolls without slipping or sliding along a fixed straight line or a directing line or a director.

II (F2) Inferior Trochoid : See Fig. 6.23 or 6.24.

Inferior trochoid is a locus of a point (Q) inside the circumference of a rolling circle, which rolls without slipping or sliding along a fixed straight line.

II (F3) Superior Trochoid : See Fig. 6.23 or 6.25.

It is a locus of a point (R) outside the circumference of a rolling circle, which rolls without slipping or sliding along a fixed straight line.

II (G1) Epycycloid : See Fig. 6.26 or 6.27

It is a locus of a point (P) on the circumference of a rolling circle, which rolls without sliding or slipping outside another circle called directing circle.

II (G2) Inferior Epytrochoid : See Fig. 6.27.

It is a locus of a point (Q) inside the circumference of a rolling circle, which rolls without sliding or slipping outside another circle called directing circle.

II (G3) Superior Epytrochoid : See Fig. 6.27.

It is a locus of a point (R) outside the circumference of a rolling circle, which rolls without sliding or slipping outside another circle called directing circle.

II (H1) Hypocycloid : See Fig. 6.26 or 6.27.

It is a locus of a point (M) on the circumference of a rolling circle, which rolls without sliding or slipping inside another circle called directing circle.

II (H2) Inferior Hypotrochoid : See Fig. 6.27.

It is a locus of a point (L) inside the circumference of a rolling circle, which rolls without sliding or slipping inside another circle called directing circle.

II (H3) Superior Hypotrochoid : See Fig. 6.27.

It is a locus of a point (N) outside the circumference of a rolling circle, which rolls without sliding or slipping inside another circle called directing circle.

Uses : These curves are used in profile of gears.

Now we shall study the methods of construction of the above curves. Basic concept of construction is more or less same for all curves.

To understand one of the important basic concepts, see Fig. 6.21.

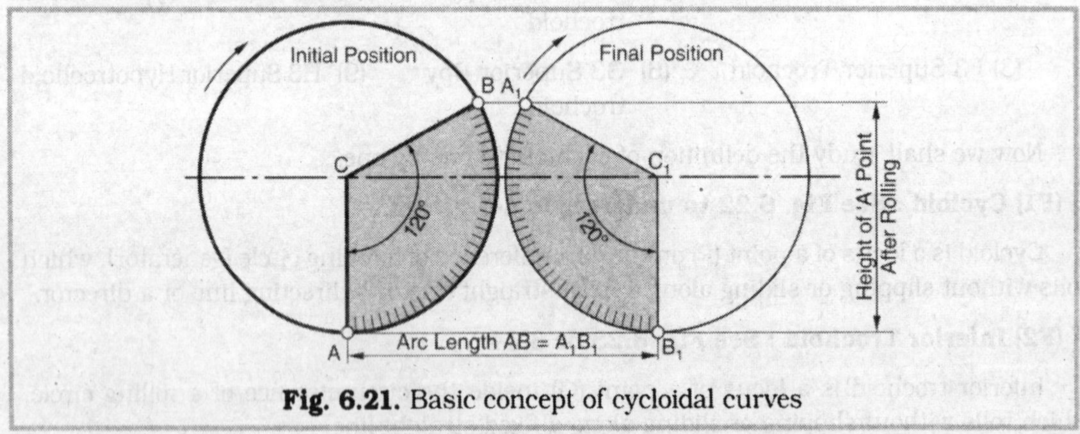

Fig. 6.21. Basic concept of cycloidal curves

Circle which rolls without slipping or sliding on a straight line (AB_1) is shown in initial (C) and final positions (C_1) after 120° rotation of the circle. When a circle rolls from position C to C_1, a distance equal to the arc AB = AB_1, point B moves down to B_1, while the point A moves upto A_1. The horizontal or parallel to the directing line from B of the initial position of the circle decides the height of the point (A) in the new position of a circle. This concept will be utilised in finding the positions of a point (P) after rotation of circle by 1/8th or 1/12th or multiple of its rotation either on straight lines or on arcs of circles.

II (F1) Cycloid : See Fig. 6.22

Problem 1 : Diameter of the rolling or the generating circle is D. Construct the cycloid. Draw tangent and normal at any point of curve. Initial position of point P is at the point of contact between generating circle and the directing line.

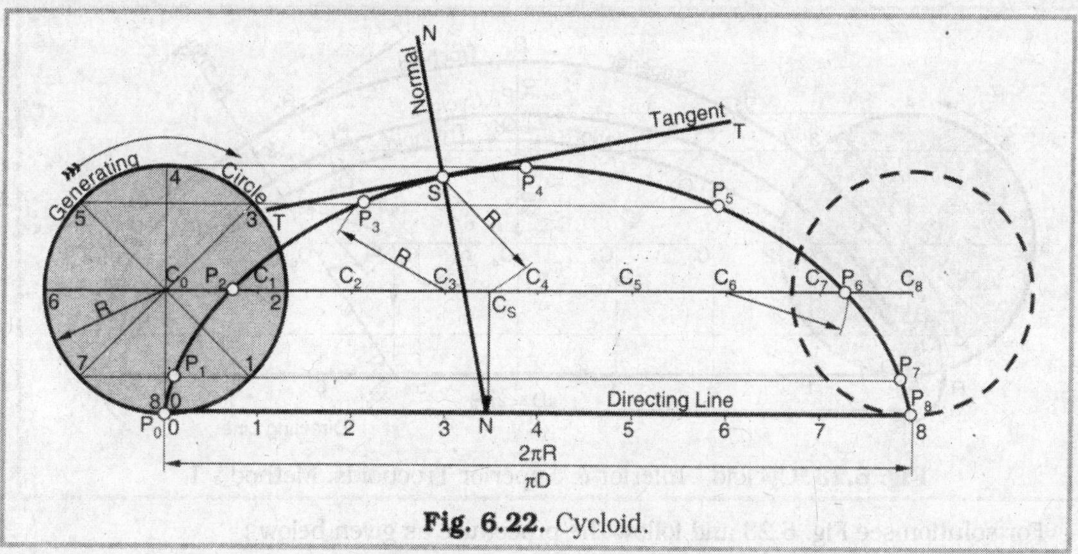

Fig. 6.22. Cycloid.

Follow the procedure as given below :

1. Draw a circle ⊙ (c_0, R) and mark P_0 on the circumference, the initial position of P, as shown in figure. Divide this circle into 8 equal parts and mark on the circle 0, 1, ...8 in opposite direction to rotation. Draw horizontal lines or parallel lines to directing lines through 0, 1, 2,....8.
2. Draw tangential directing line on it at P_0 and mark on it distance equal to πD from P_0. Divide this distance into 8 equal parts and mark them by 0, 1, 2,...8.
3. From C_0 draw parallel line to directing line and mark on it C_0, C_1, C_2,.....C_8 corresponding to 0, 1, 2,....8 of the directing line.
4. Now when the circle has rolled without slip by 1/8th of a revolution, the centre point C_0 must have moved to C_1 and the point P_0 must have moved and achieved the height of point 1(one) of the circle, according to the concept studied earlier.
5. Accordingly with C_1 as the centre and radius equal to R draw an arc to cut the horizontal line, through 1 of circle, at P_1. In a similar way get arc-line intersection points P_2, P_3,.... P_8 by drawing arcs with C_2, C_3,.....C_8 as centres and radius equal to R to intersect with lines through 2, 3,....8 of circle respectively.
6. Join P_0, P_1, P_2,....P_8 by means of a smooth curve to get a cycloid.
7. Take any point S on the curve. Now with S as the centre and radius equal to R draw an arc to cut centre line C_0C_s at some point C_s. Find the point N on the directing line corresponding to C_s. Join NS and draw perpendicular to it at S. The line NSN is normal and the line TST is tangent to the curve.

II (F2) and II (F3) Inferior and Superior Trochoids : See Fig. 6.23.

There are two methods of drawing trochoids.

Method 1 : ***Given the diameter of the rolling or generating circle as 50 mm. Given points P, Q and R are on the circumference, 5 mm inside the circumference, and 10 mm outside the circumference respectively on rolling circle. This circle is rolling without sliping/sliding on a fixed straight line. Draw the loci of points P, Q and R for one revolution of rolling circle. Draw tangent and normal to it at any point.***

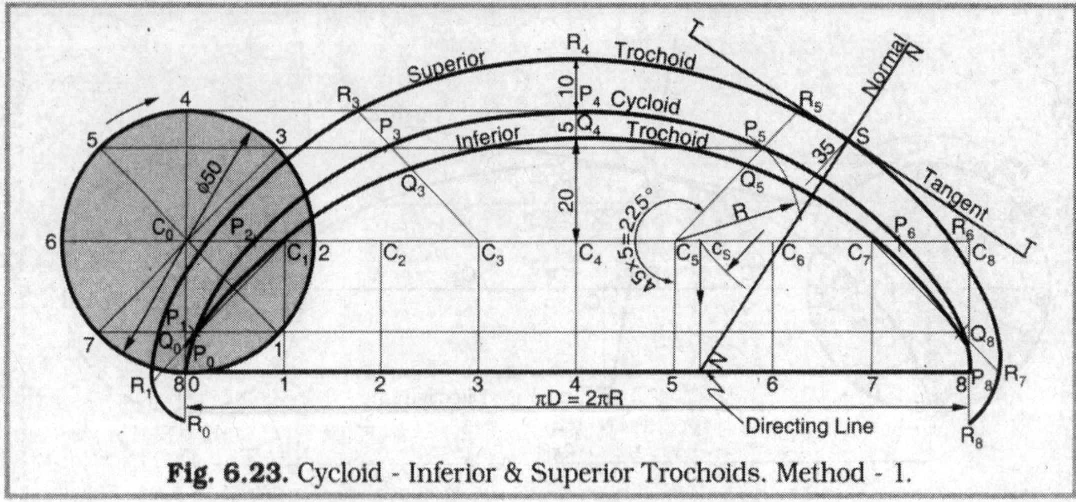

Fig. 6.23. Cycloid - Inferior & Superior Trochoids. Method - 1.

For solution see Fig. 6.23 and follow the procedure as given below :

1. Do the same construction as done for cycloid and obtain points C_0, C_1, C_2,....C_8 an P_0, P_1, P_2,......P_8, as shown in Fig. 6.23.
2. Join C_0P_0 and mark Q_0 on it and R_0 on the extension of it by taking P_0Q_0 = 5 mm and P_0R_0 = 10mm. Take Q_0 inside and R_0 outside. Similarly, find points (Q_1, R_1), (Q_2, R_2),...... (Q_8, R_8) on lines C_1P_1, C_2P_2,....C_8P_8 respectively.
3. Join points Q_0, Q_1, Q_2,......Q_8 in sequence by a smooth curve to get inferior trochoid. Similarly join points R_0, R_1, R_2,....R_8 in sequence by a smooth curve to get superior trochoid.
4. Take any point S on superior trochoid. With S as the centre and radius equal to 35 mm (10 + 25) draw an arc to cut the centre line at point C_s. From C_s draw right angle on the directing line to get the point N. Join NS and draw at S the line TST right angle to NS. TST and NS are tangent and normal respectively.
5. Similarly draw tangent and normal to inferior trochoid. Take radius 20 mm (25-5), instead of 35.

Method 2 : Inferior Trochoid. See Fig. 6.24.

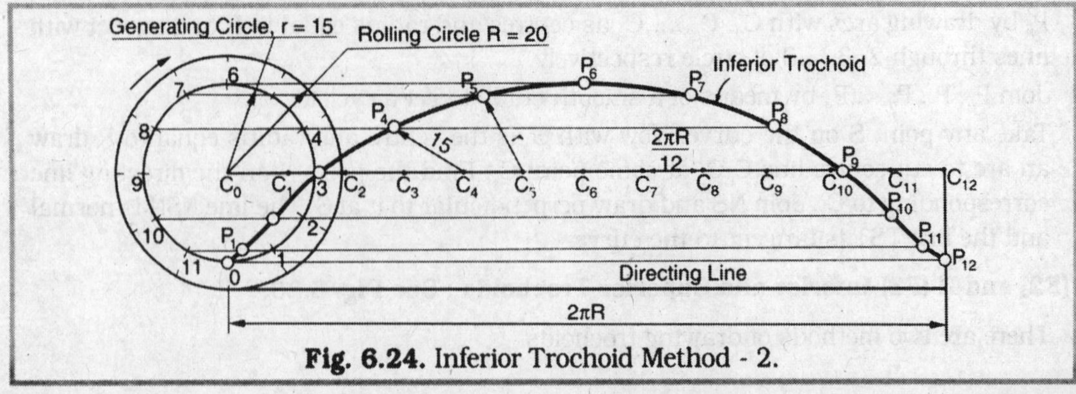

Fig. 6.24. Inferior Trochoid Method - 2.

A circle, of 20 mm radius, is rolling on a straight line without slip. A point P is inside the circle, 15 mm from the centre is initially at the bottom most position. Draw the locus of point P for one revolution and name the curve.

For solution follow the procedure as given below and see Fig. 6.24.

1. Draw two circles ⊙ (C_0, 20) and ⊙ (C_0, 15). Draw tangential directing line at bottom most point to circle ⊙ (C_0, 20). Take length of directing line equal to circumference of circle $2\pi R$. From centre C_0 draw path line of centre parallel to directing line also of $2\pi R$ length. Divide this centre line into 12 equal parts and mark them by points C_0, C_1, C_2,.....C_{12}. These points are going to be the positions of centres of rolling circle during its rotation by 1/12 of a revolution and multiples of it.
2. Divide ⊙ (C_0, 15) into 12 equal parts and mark on it points 0, 1, 2,.....12 in opposite direction to that of rotation. Draw parallel lines to the directing line from 0, 1, 2,.....12. On these lines point P will be located after 1/12 of a revolution and multiples of it.
3. Now with C_0, C_1, C_2,.....C_{12} as centres and radius equal to r = 15 mm draw arcs to cut lines through 0, 1, 2,....12 at points P_0, P_1, P_2,......P_{12} respectively. Join these points by a smooth curve to get inferior trochoid, as shown in the figure.

Method 2 : Superior Trochoid : See Fig. 6.25.

A circle of 15 mm radius is rolling on a straight line without slip. A point P is outside the circle 20 mm from the centre and is initially at the bottom most position. Draw the locus of the point P for one revolution and name the curve.

For solution follow the procedure as given below and see Fig. 6.25.

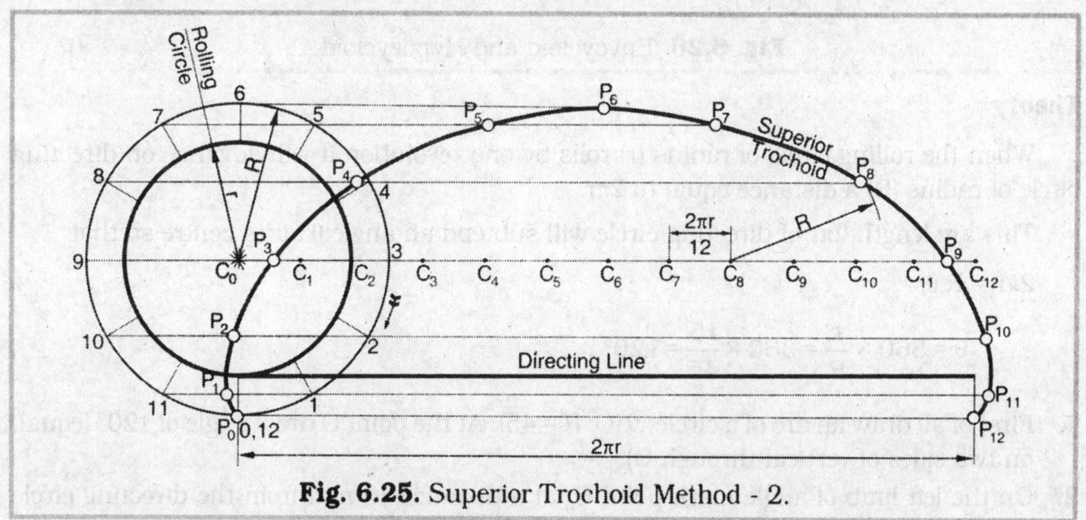

Fig. 6.25. Superior Trochoid Method - 2.

1. Here length of directing and centre lines should be same and equal to $2\pi r = 2\pi 15$ and circle ⊙ (C_0, 20) is to be divided into 12 equal parts.
2. Rest of the procedure is similar to inferior trochoid. The curve achieved is going to be superior trochoid.

II (G1) Epycycloid : See Fig. 6.26.

A circle, of 15 mm radius, is rolling on a circle of 45 mm radius without slip, on outside of it. Initially a point P is at the contact point of two circles. Draw the locus of

the point P for one revolution of the rolling circle. Name the curve and draw tangent and normal to the curve at any point S.

For solution see Fig. 6.26 and follow the procedure as given below :

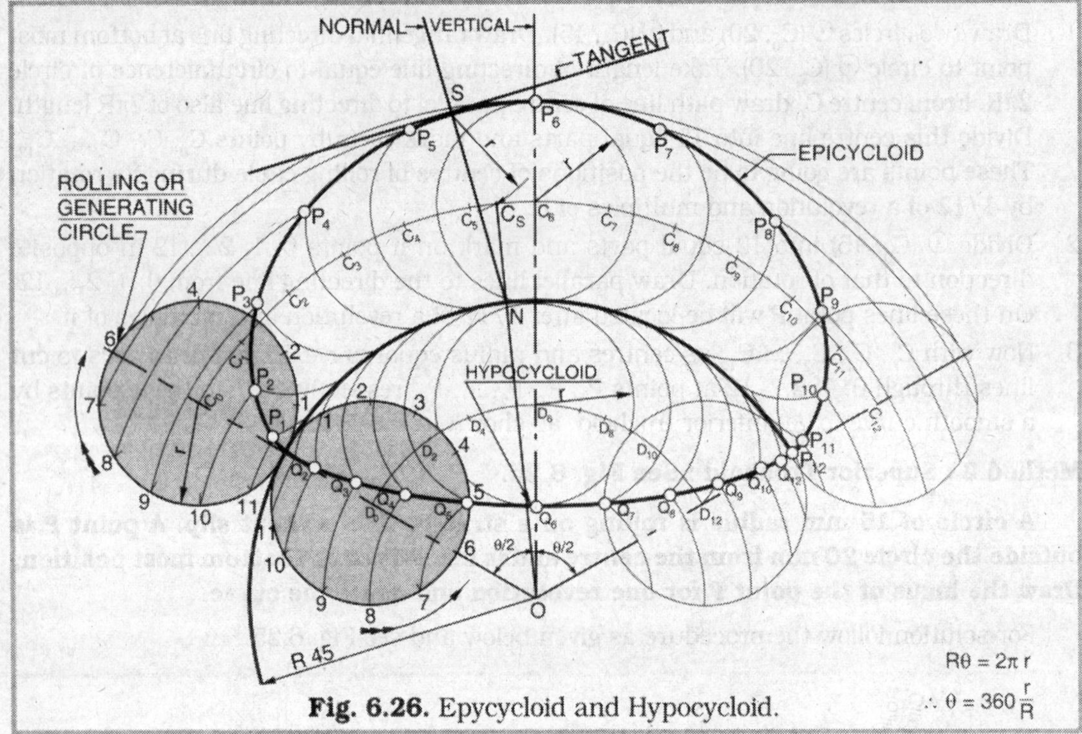

Fig. 6.26. Epycycloid and Hypocycloid.

Theory :

When the rolling circle of radius (r) rolls by one revolution it will advance on directing circle of radius (R) a distance equal to 2πr.

This arc length 2πr of directing circle will subtend an angle θ at its centre so that

$2\pi r = R.\theta.$

$$\therefore \quad \theta = 360 \times \frac{r}{R} = 360 \times \frac{15}{45} = 120°$$

1. First of all draw an arc of a circle ⊙ (O, R = 45). At the point O draw angle of 120° (equal, on two sides of vertical through O).
2. On the left limb of angle mark point C_0 at a distance 15 mm from the directing circle on outside of it. Draw rolling circle ⊙ (C_0, 15) touching directing circle at point P_0, the starting position of P.
3. Divide rolling circle ⊙ (C_0, 15) into 12 equal parts and mark them as 0, 1, 2,.....12, in opposite direction to that of rotation.
4. Now with O as the centre and radii equal to O1, O2,....O-12 draw arcs of required sufficient length to represent the position of the point P after rotation of rolling circle by 1/12 of a revolution and multiples of it. Also draw arc with O as centre and OC_0 (45+15) as the radius to get path line of the centre.

5. Divide angle of 120° into 12 equal parts and draw small radial lines through O, as shown in the figure, to intersect with the centre line at C_0, C_1, C_2,......C_{12}.
6. Now with C_0, C_1, C_2,......C_{12} as centres and radius equal to r = 15 mm draw arcs to intersect with arc lines through 0, 1, 2,.......12 at points P_0, P_1, P_2,.....P_{12} respectively as shown in Fig. 6.26.
7. Join P_0, P_1, P_2,.....P_{12} in sequence by a smooth curve to get Epycycloid.
8. Now to draw tangent and normal to the curve at any point S draw arc with S as the centre and radius equal to r = 15 to the intersect with the centre line at C_s. Draw C_sN radial line passing through O. Join NS which will be normal to the curve and now draw right angle to it at point S to get tangent.

II (G_2) Inferior Epytrochoid

(G_3) Superior Epytrochoid See Fig. 6.27.

Method 1 :

A rolling circle, of r = 27 mm radius, is rolling outside a directing circle of R_d = 81 mm radius without slip. A point P is on the circumference, a point Q is 7 mm inside the circumference and a point R is 13 mm outside the circumference of the rolling circle. Find out the loci of points P, Q and R for one revolution of the rolling circle.

For solution see Fig. 6.27 and follow the procedure as given below :

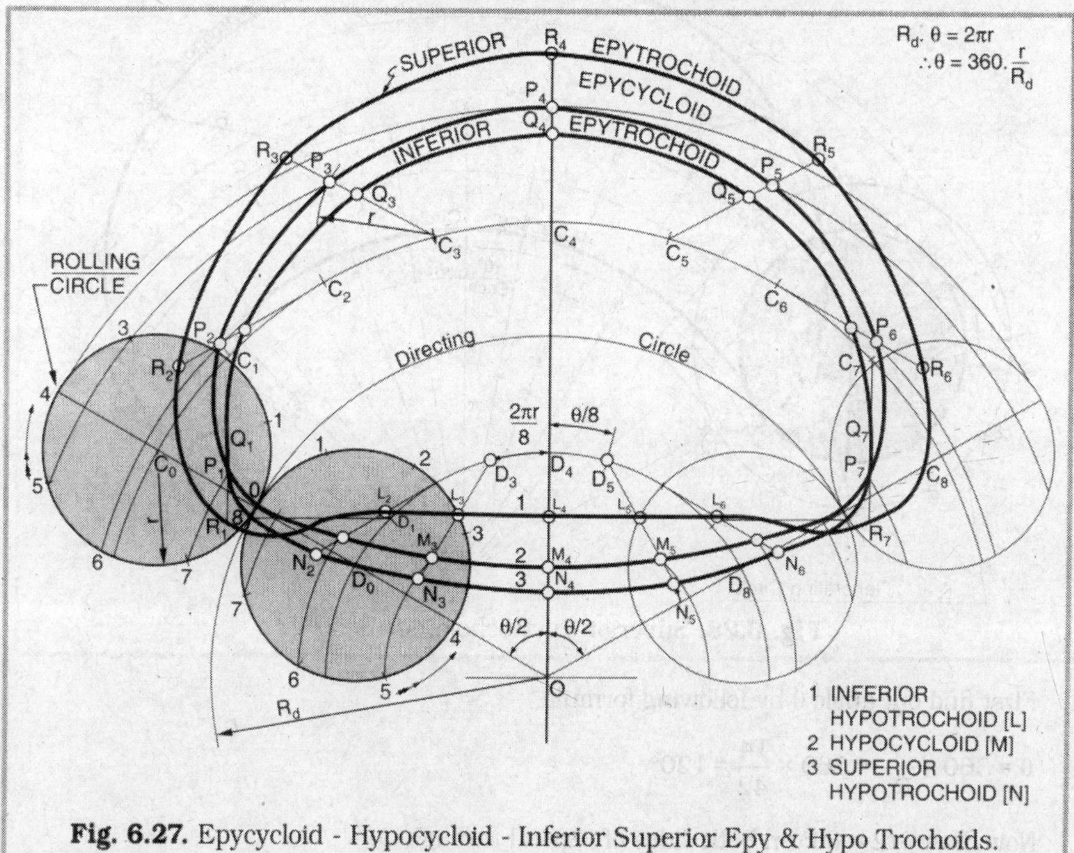

Fig. 6.27. Epycycloid - Hypocycloid - Inferior Superior Epy & Hypo Trochoids.

1. First follow the same procedure as followed for epycycloid and get points P_0 - C_0, P_1 - C_1,....P_8 - C_8.
2. Now join C_0P_0 and mark on it Q_0 and R_0 taking C_0Q_0 = 20 (27 - 7) and C_0R_0 = 40 (27 + 13). Similarly on C_1P_1, C_2P_2,......C_8P_8 mark points (Q_1,R_1), (Q_2, R_2),....(Q_8,R_8) respectively.
3. Join (P_0,P_1,.....P_8), (Q_0, Q_1, Q_2,....Q_8) and (R_0, R_1, R_2,.....R_8) by a smooth curve to get Epycycloid, Inferior Epytrochoid and Superior Epytrochoid respectively.

Method 2 : II (G_3) Superior Epytrochoid. See Fig. 6.28.

A circle, of 14 mm radius, is rolling outside another circle of 42 mm radius. A point P is outside the rolling circle 20 mm from the centre and is initially nearest to the directing circle centre. Draw the locus of the point P for one revolution and name the curve.

For solution follow the procedure as given below and see Fig. 6.28.

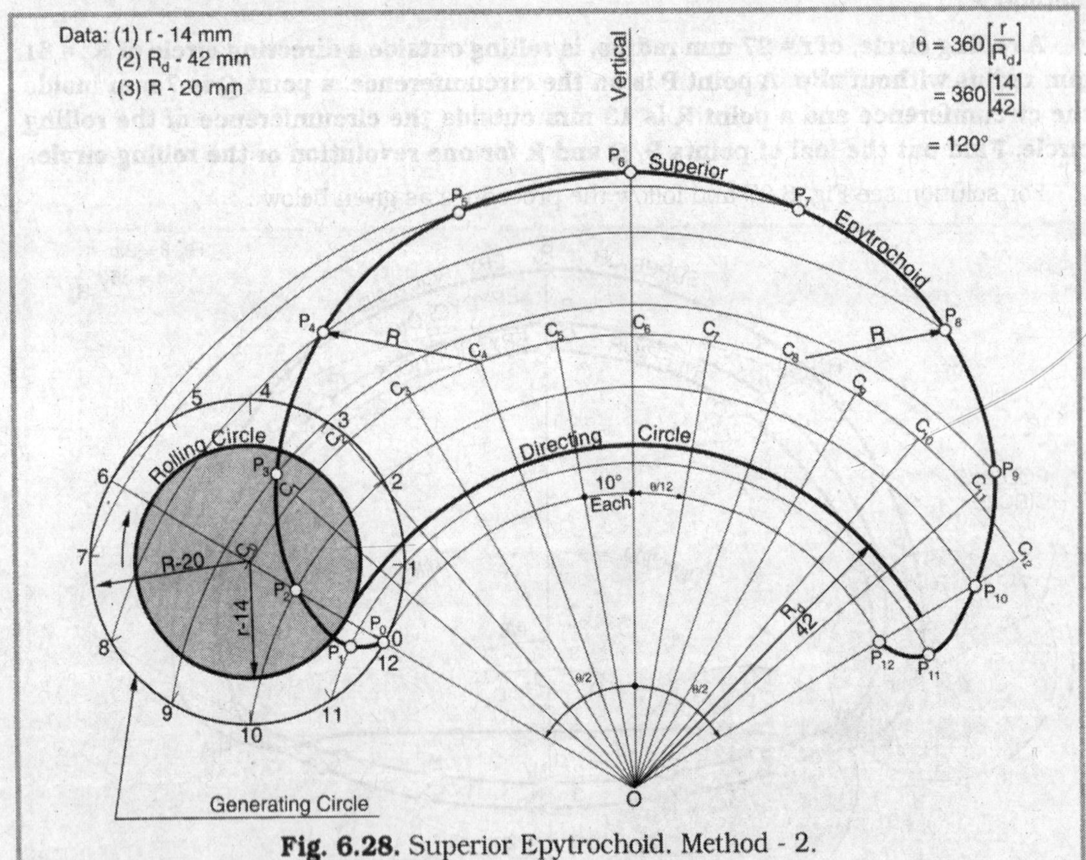

Fig. 6.28. Superior Epytrochoid. Method - 2.

1. First find out angle θ by following formula

$$\theta = 360 \times \frac{r}{R_d} = 360 \times \frac{14}{42} = 120°$$

Now draw θ/2 = 60° on both sides of vertical through O.

2. On the left limb of angle θ mark point C_0 at a distance 56(42 + 14)mm from O and draw

circle ⊙ (C_0, 14). Draw also circle ⊙ (C_0, 20) and divide it into 12 equal parts and mark points 0, 1, 2,....12 in the opposite direction to that of rotation.

3. Now draw arcs of circles with O as centre and radii equal to O-O, 01, 02,....012. On these arcs point P will be located after 1/12 revolution and multiple of it.
4. Draw arc with O as centre and radius equal to O-C_0. Divide angle θ into 12 equal parts and get 12 centres C_0, C_1, C_2,....C_{12} on the centre line.
5. Now with C_1, C_2,....C_{12} as centres and radius equal to R = 20 mm draw arcs to cut arcs of circle through 1, 2,....12 at points P_1, P_2,....P_{12} respectively. Arcs from centres C_4 and C_8 are shown in Fig. 6.28 to get P_4 and P_8 respectively.
6. Join P_0, P_1, P_2,....P_{12} in sequence by a smooth curve to get Superior Epytrochoid.

II (H1) Hypocycloid : See Fig. 6.26.

A circle, of r = 15 mm radius, is rolling inside a circle of R = 45 mm radius without slip. A point Q is on the circumference of the rolling circle. Initially point Q is at the contact point of two circles. Draw the locus of point Q for one revolution of rolling circle. Name the curve.

For solution see Fig. 6.26 and follow the procedure as given below :

1. Procedure is exactly similar to Epycycloid. Only difference is circle is rolling inside instead of outside. Name of the curve is Hypocycloid.

II (H_2) Inferior

(H_3) Superior Hypotrochoid : See Fig. 6.27.

Follow the same procedure as followed for Inferior and Superior Epytrochoid. Keep rolling circle inside instead of outside.

III INVOLUTE

Definition : (1) If a straight line is rolled round a circle without slipping or sliding, points on line will trace out involutes

[OR]

(2) Involute of a circle is a curve traced out by a point on a taut string unwound or wound from or on the surface of the circle.

If a polygon is taken instead of circle, then it is named as involute of a polygon e.g. Involute of a square, pentagon, hexagon, etc.

Uses : 1. Gear's profile is having involute shape.

III (I) Involute of A Circle : See Fig. 6.29.

A string is unwound from a circle of 20 mm diameter. Draw the locus of end of string P for unwounding the string's one turn. String is kept tight during unwounding. Draw tangent and normal to the curve at any point.

For solution see Fig. 6.29 and follow the procedure as given below :

Principle :

When the string is unwound, the length of the string goes on increasing by the amount equal to the arc length of the circle from which the string is unwound. Similarly, the length goes on decreasing during wounding operation. Further as the string is kept tight during unwounding or wounding operation it will remain tangential to the circle.

1. Draw a circle of 20 mm diameter and divide it into 8 equal parts and mark them as 0, 1, 2,....8.
2. Draw tangents to the circle at points 1, 2, ...8 in the direction of position of string during unwounding operation.
3. On tangents at points 1, 2,....8 take length equal to arc length 01, 02, 03,....08 to mark points P_1, P_2, P_3,....P_8 respectively.

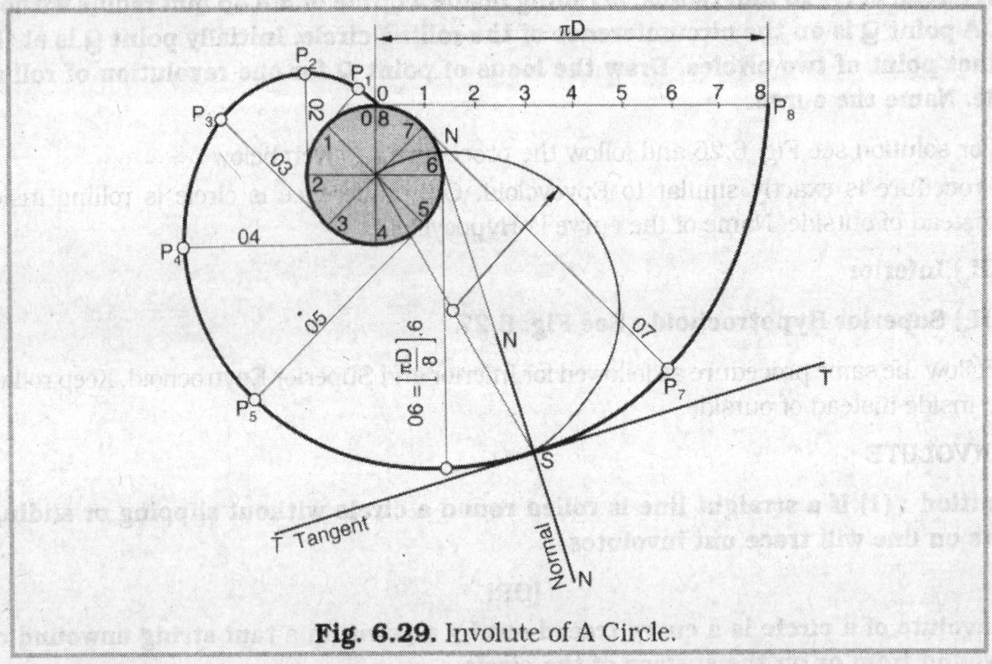

Fig. 6.29. Involute of A Circle.

For ease of taking arc length 01, 02,....08 etc. draw on tangent at point 8 distance equal to πD and divide it into 8 equal parts. Here length of each part is equal to 1/8 of the circumference (πD) of the circle.

4. Join points P_0, P_1,....P_8 by smooth curve to get involute of a circle as shown in Fig. 6.29.
5. To draw tangent and normal to the curve at any point S of curve, join point S with the centre of the circle. With this line as diameter draw a semicircle cutting the circle of involute at point N. Join N with S to get normal and draw right angle to this normal at point S to get tangent TT.

III (J) Involute of A Polygon : See Fig. 6.30.

A string is unwound from a square of 25 mm side. Draw the locus of end P of string for unwounding the string's one turn. String is kept tight during unwounding operation. Draw tangent and normal to the curve at any point.

Principle :

When the string is unwound from the surface of a square or any polygon, it turns on the corner of the square or polygon till it comes in line with the next surface. In Fig. 6.30 from P_0 to P_1 position the string turns on corner 1. At P_1 the string is in line with surface 12. With this principle follow the procedure as given below and see Fig. 6.30.

1. Draw a square of 25 mm side and mark corners 0, 1, 2,....4. Extend lines 21, 32, 43 and 14 by suitable amount.
2. Draw quarter circles with centres 1, 2, 3 and 4 and radii (1 × 01), (2 × 01), (3 × 01) and (4 × 01) to cut previously drawn lines at points P_1, P_2, P_3 and P_4 as shown in figure.

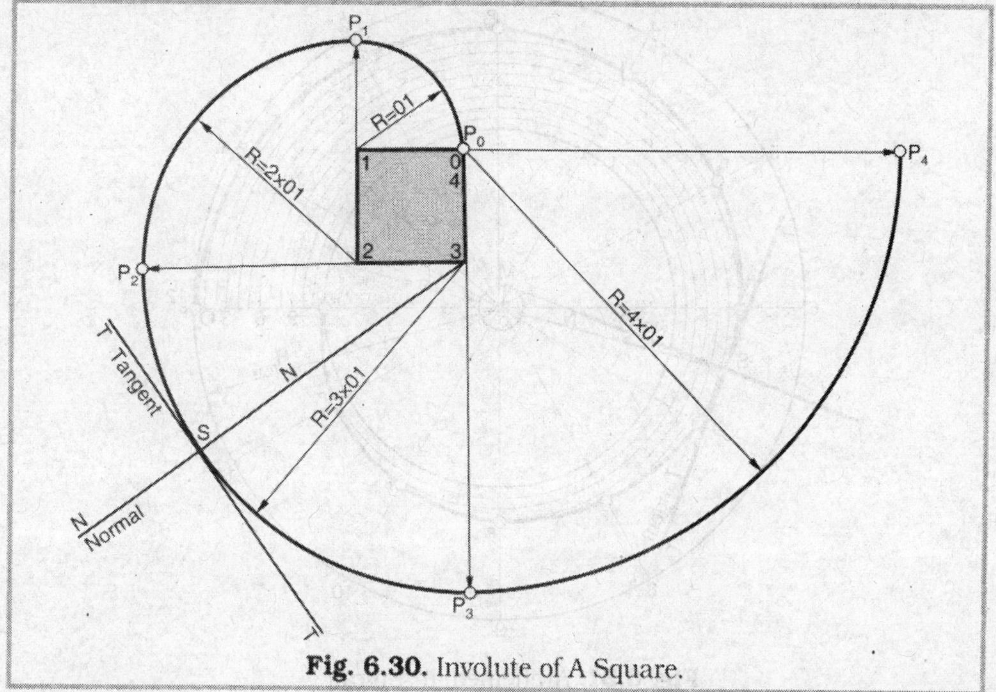

Fig. 6.30. Involute of A Square.

3. Combination of these four quarter circles P_0 P_1, P_1P_2, P_2P_3 and P_3P_4 is nothing but an involute of a square.
4. To draw tangent and normal to the curve at any point S, join S with centre of the corresponding circle, which will be normal to the curve. Draw right angle to it at S to get tangent.

IV SPIRAL

Definition :-

Archemedian Spiral : It is a curve generated by a point moving uniformly along a straight line, while the line swings or rotates about a fixed point with uniform angular velocity.

[OR]

It is a curve traced out by a point which moves uniformly both about the centre and at the same time away or towards the centre.

Logarithmic Spiral : It is the curve generated by the end of radius vector rotating about the centre so that the ratio of the lengths of the consecutive radius vectors for equal angular movements is constant.

Uses : 1. Shape of springs of watch mechanism, toys etc

2. Scroll plate of lathe chuck

3. Clamping devices of jigs and fixtures

4. Profile of cams for automation

IV (K) Archemedian Spiral : See Fig. 6.31.

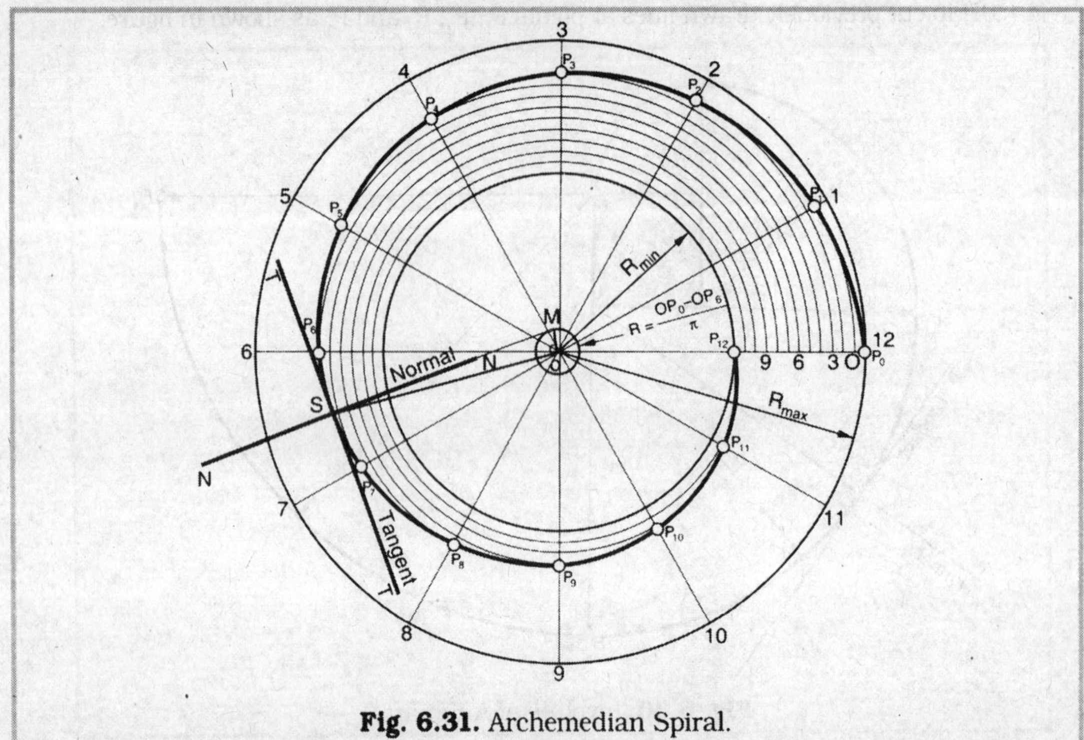

Fig. 6.31. Archemedian Spiral.

Problem 2 : Construct an Archemedian Spiral of one convolution, given the maximum and minimum radii as 55 mm and 31 mm respectively. Draw tangent and normal to the curve at any point.

See Fig. 6.31 and follow the procedure as given below :

1. With pole O as centre and radii equal to R_{max} = 55 mm and R_{min} = 31 mm, draw two circles.
2. Divide 360° at pole O into 12 equal parts and draw radial lines 00, 01, 02,....012.
3. Divide 24 mm [(55 - 31) or (R_{max} – R_{min})] length also into same 12 equal parts as shown in Fig. 6.31.
4. On lines 01, 02,....0.12 mark points P_1, P_2,.....P_{12} by successively decreasing the length of radius vector by one division èach time. For decreasing radius vector by one division, draw arcs of circles with O as centre and radii equal to 01, 02,....012 respectively.
5. Join points P_0, P_1, P_2.....P_{12} by smooth curve to get **Archemedian Spiral.**

6. Now to draw tangent and normal to the curve at any point S draw a circle with pole O as centre and radius equal to constant of the curve.

$$\text{Constant of the curve} = \frac{\text{Difference in lengths of any two radius vectors}}{\text{Angle between the corresponding radius vectors in radian}}$$

$$= \frac{OP_0 - OP_{12}}{2\pi} = \frac{55 - 31}{2\pi}$$

$$= 3.82 \text{ mm}$$

Join point S with pole O. Draw right angle at O with this SO line to get a point M on the circle drawn with radius, constant of the curve. Join SM. This is normal. Draw right angle to this normal at S to get tangent to the curve.

IV (L) Logarithmic Spiral :

Construct Logarithmic Spiral for one convolution. Given the length of the shortest radius vector equal to 11 mm and the ratio of the lengths of the successive radius vectors equal to 6/5 for vectorial angle of 30°.

Draw also tangent and normal to the curve at any point on the curve.

For solution see Fig. 6.32 and follow the procedure as given below :

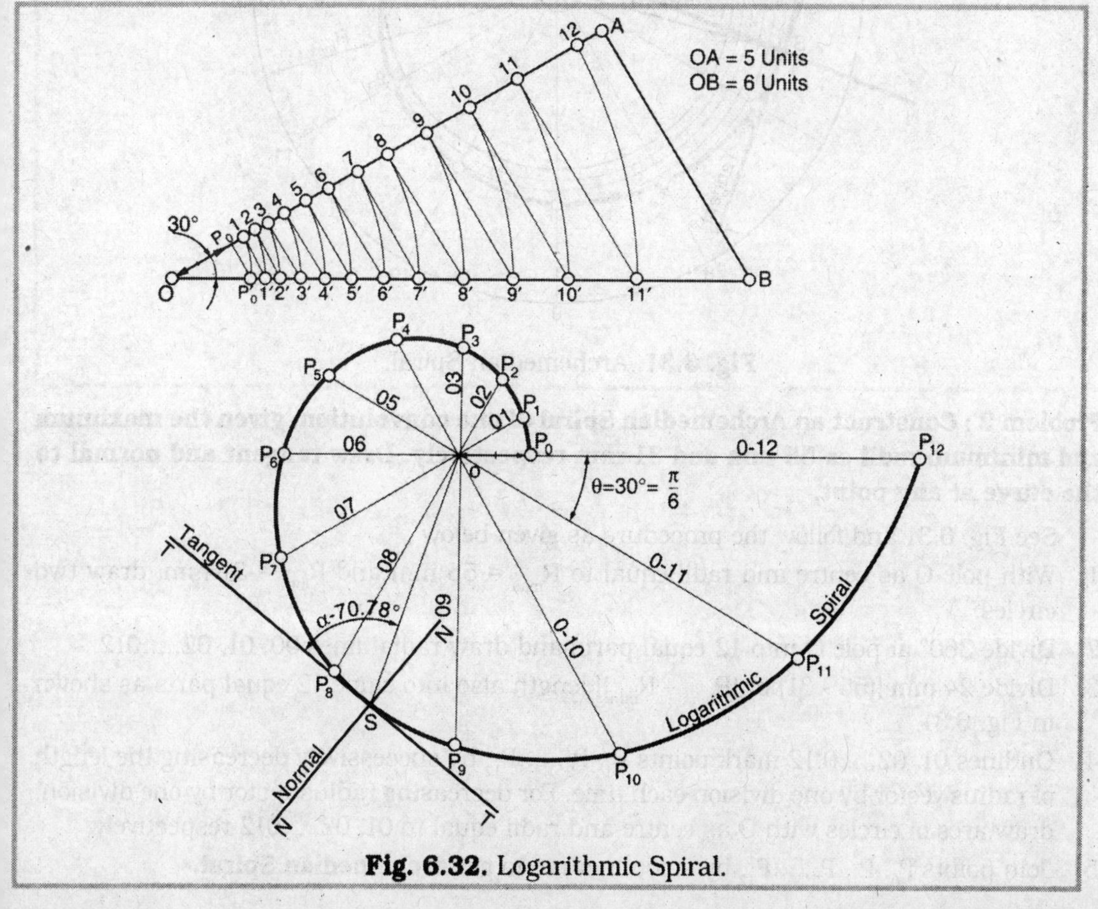

Fig. 6.32. Logarithmic Spiral.

Logarithmic Spiral : It is the curve generated by the end of radius vector rotating about the centre so that the ratio of the lengths of the consecutive radius vectors for equal angular movements is constant.

Uses :
1. Shape of springs of watch mechanism, toys etc
2. Scroll plate of lathe chuck
3. Clamping devices of jigs and fixtures
4. Profile of cams for automation

IV (K) Archemedian Spiral : See Fig. 6.31.

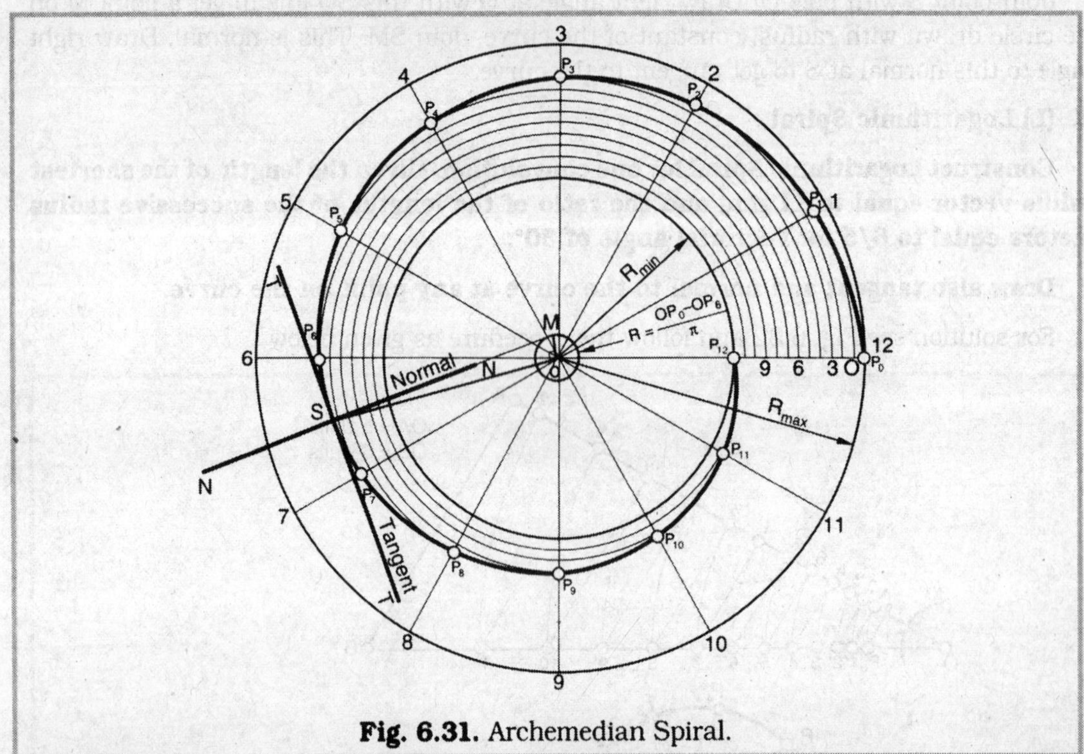

Fig. 6.31. Archemedian Spiral.

Problem 2 : Construct an Archemedian Spiral of one convolution, given the maximum and minimum radii as 55 mm and 31 mm respectively. Draw tangent and normal to the curve at any point.

See Fig. 6.31 and follow the procedure as given below :

1. With pole O as centre and radii equal to R_{max} = 55 mm and R_{min} = 31 mm, draw two circles.
2. Divide 360° at pole O into 12 equal parts and draw radial lines 00, 01, 02,....012.
3. Divide 24 mm [(55 - 31) or ($R_{max} - R_{min}$)] length also into same 12 equal parts as shown in Fig. 6.31.
4. On lines 01, 02,....0.12 mark points P_1, P_2,.....P_{12} by successively decreasing the length of radius vector by one division each time. For decreasing radius vector by one division, draw arcs of circles with O as centre and radii equal to 01, 02,....012 respectively.
5. Join points P_0, P_1, P_2,....P_{12} by smooth curve to get **Archemedian Spiral.**

V - HELIX

There are two types of helix. (i) Helix on cylinder and (ii) Helix on cone.

Definition :

Helix of Cylinder : It is the curve generated by a point which moves around the surface of a right circular cylinder, and at the same time advances in the axial direction at a speed which bears a constant ratio to the speed of rotation.

Helix of A Cone : It is the curve generated by a point which moves around the surface of a right circular-cone, and at the same time advances in the direction of axis at a speed which bears a constant ratio to the speed of rotation.

Uses of Helix :

1. Shape of Springs
2. Thread
3. Worm
4. Helical Stair cases
5. Helical grooved cam for automation.

GEOMETRY OF HELIX AND ITS CONSTRUCTION : See Fig. 6.35.

A point is moving around the surface of the cylinder of 25 mm diameter at uniform speed and at the same time it advances in the direction of axis by 35 mm also with uniform speed. Draw the path of moving point. Name the curve.

So for in this chapter we have studied curves of plane but path of moving point here does not remain in a plane but it is in a space. Helix, the path of moving point is drawn in the elevation or F.V. form to avoid three dimensional drawing.

The amount of advance (35)mm of moving point in the direction of axis during one revolution movement around the surface of the cylinder is termed as **Lead.** If the surface of cylinder is developed, the helix will become the hypotenuse of a right angle triangle whose base is the circumference of the cylinder (πD) and whose altitude is the lead (35) of helix. The angle between the hypotenuse and the base of the triangle is known as the **Helix Angle.**

Construction : See Fig. 6.35 and follow the procedure as given below :

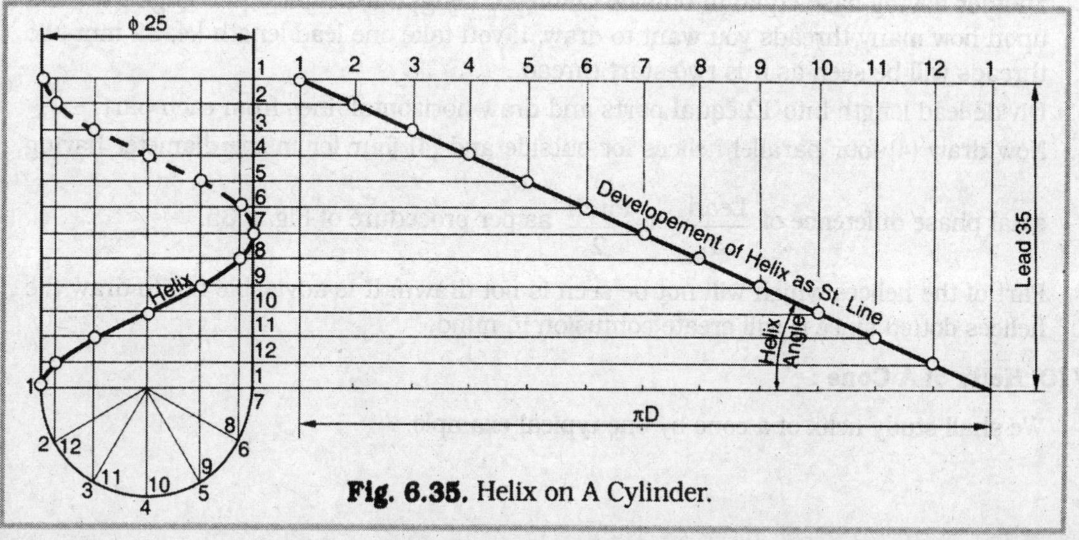

Fig. 6.35. Helix on A Cylinder.

1. Draw rectangle of 25 mm (Dia.) by 35 mm (Lead) size and draw semicircle attached to the rectangle at the bottom as shown in Fig. 6.35. Rectangle is the F.V. of cylinder.
2. Divide semicircle into 6 equal parts (virtually full circle into 12 equal parts) and mark them 1,2,....12. Divide 35 mm height of rectangle also into 12 equal parts and also mark them as 1, 2,....12 and 1.
3. Draw horizontal lines (11[rl] to the base of cylinder) and vertical lines (11[rl] to the axis of cylinder) from 1, 2,....12 of rectangle and 1, 2,....12 of circle respectively to intersect at corresponding points. Join all these points by smooth curve to get F.V. of helix. Half of the helix will be full and half will be dotted.
4. Development of the cylinder along with helix is also drawn in Fig. 6.35 to understand geometry of helix, helix angle and lead.

If n number of helices are drawn parallel then,

Lead = n x pitch

where pitch is the axial distance between corresponding points of adjacent helices. So far discussed method of construction of helix can be used to draw any types of threads, circular, square and rectangular helical springs, worm, etc. As an example we shall draw double start square threads having;

Inside diameter = 34 mm

Outside diameter = 50 mm

Lead of helix = 32 mm

Pitch of helix $= \frac{32}{2} = 16$ mm

See Fig. 6.36 (B) (Right) and follow the procedure as given below :

1. First draw two concentric circles of 34 mm and 50 mm diameters and divide them into 12 equal parts. From each division of smaller and bigger circles draw vertical lines.
2. Draw two rectangles having same axis with one having base equal to inside diameter and another having base equal to outside diameter. Length or height of rectangles depend upon how many threads you want to draw. If you take one lead length i.e. 32 mm two threads will be seen as it is two start thread.
3. Divide lead length into 12 equal parts and draw horizontal lines from each part.
4. Now draw (4) four parallel helices for outside and (4) four for inside diameter having axial phase difference of $\frac{\text{Lead}}{4}$ or $\frac{\text{Pitch}}{2}$ as per procedure of Fig. 6.35.

Part of the helices which will not be seen is not drawn. It is advisable not to draw the helices dotted since it will create confusion in mind.

V (O) Helix of A Cone :

We shall study helix of a cone by one typical example.

Problem 3 : A right angle triangle of two sides 33 mm and 53 mm and hypotenuse 62.5 mm is rotating around side of 53 mm as shown in Fig. 6.37. A point P is moving up on the hypotenuse during the rotation of the triangle. Point P moves from the bottom most position to highest position on hypotenuse while the triangle completes one revolution. Both the movements are uniform. Draw the path of the point P in elevation and plan. Name the curve of plan and curve of elevation.

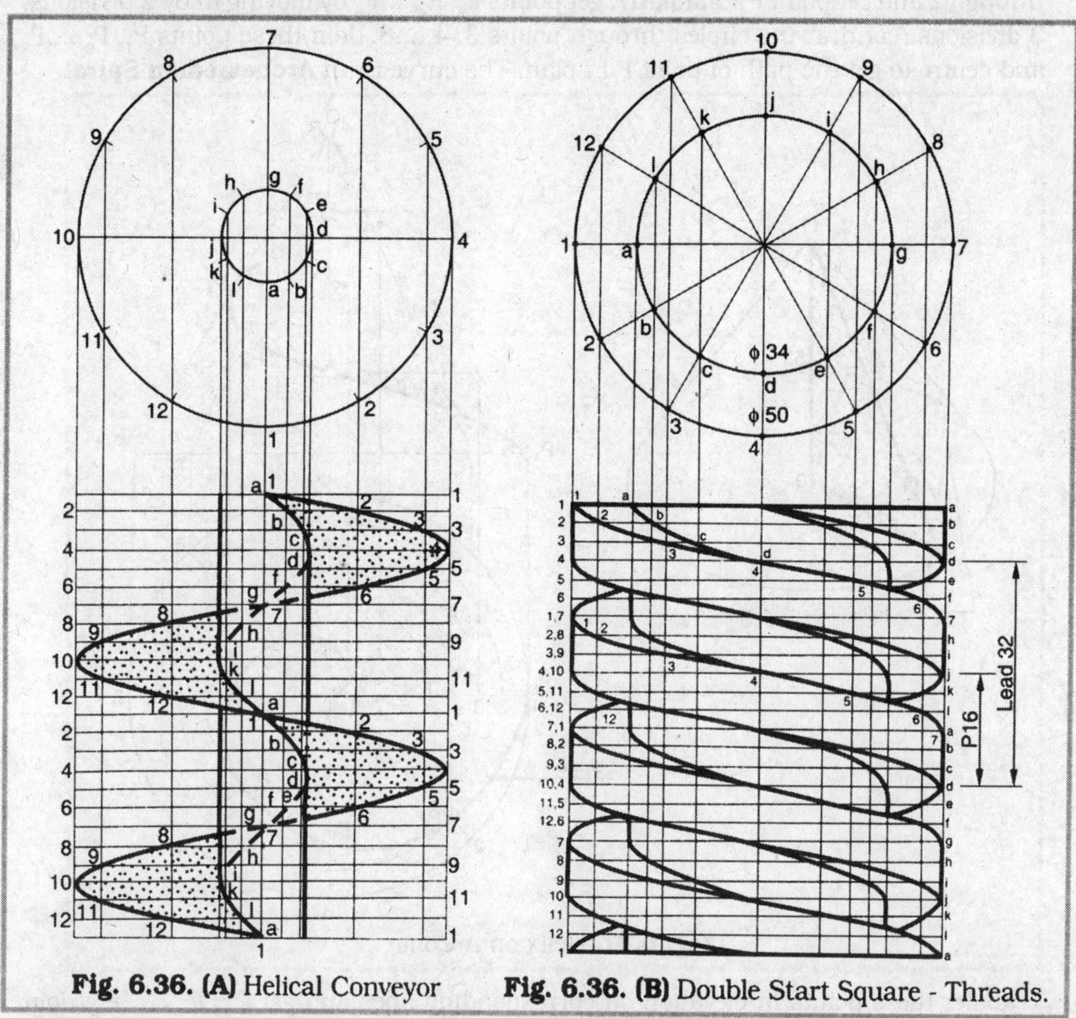

Fig. 6.36. (A) Helical Conveyor **Fig. 6.36. (B)** Double Start Square - Threads.

Here rotation of a triangle generates a cone of base diameter 66 mm and height 53 mm. Point P will move on the surface of this cone from lowest position to highest position for one revolution around the axis and so it will generate **Helix of A Cone.**

Now for solution see Fig. 6.37 and follow the procedure as given below :

1. Draw elevation (Δof 66 mm base and 53 mm height) and plan (circle of 66 mm diameter) of the cone, which is generated by revolution of the triangle.
2. Divide the circle of plan into 8 divisions and mark them as 1, 2,....8. Join plans of generators 01, 02,....08. Transfer these 8 points in elevation and draw elevations of generators.

3. Divide the radius of circle of plan into 8 divisions. Now when the triangle has turned through 1/8th of the revolution the point P must have moved up by 1/8th of the hypotenuse length and so in plan it must have moved in by 1/8th of the radius or one division of it.
4. Now according to above explanation on 02 line move in by one division or draw a circle through 2 and get point P_2. Similarly, get points P_3, P_4,.....P_8 by moving in by 2 divisions, 3 divisions....or drawing circles through points 3, 4,....8. Join these points P_1, P_2,....P_8 and centre to get the path of point P in plan. The curve is an **Archemedian Spiral.**

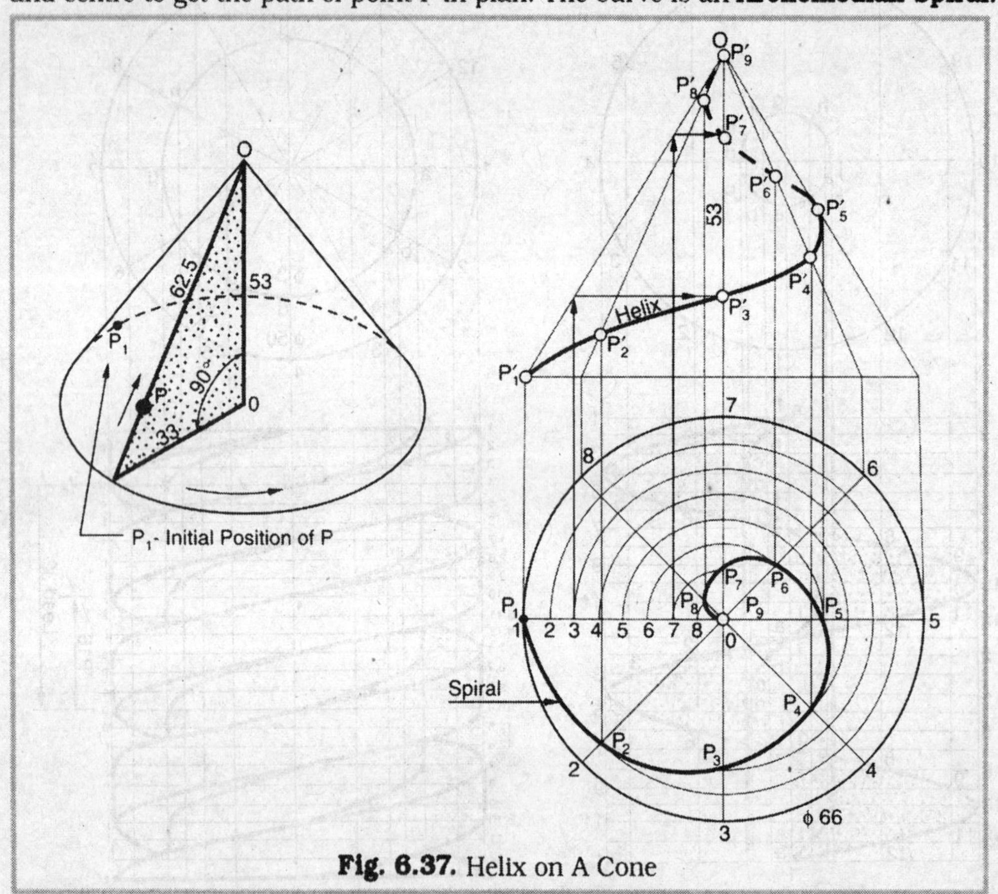

Fig. 6.37. Helix on A Cone

5. Transfer these points in elevation on corresponding lines and get P'_1, P'_2,......P'_8. Join these points in sequence to get path of the point P in elevation which is a **Helix of A cone.**

VI (P) Sine Curve (Q) Cosine Curve : See Fig. 6.38.

Sine and cosine curves are similar in appearance to helix but helix is a three dimensional curve whereas sine curve is a two dimensional curve. Both sine and cosine curves are similar. They have only phase difference between them. Methods of drawing two curves is also the same.

Definition :

Sine/Cosine curve is a graphical representation of projection of a point on diameters versus the angle turned by that point when moving on a circle with constant speed.

Use : 1. These two curves are very much useful in representing Alternating Current in electrical engineering.

2. It is also useful in understanding S.H.M.
3. Vibration characteristics.

Now to construct sine and cosine curves follow the procedure as given below and see Fig. 6.38.

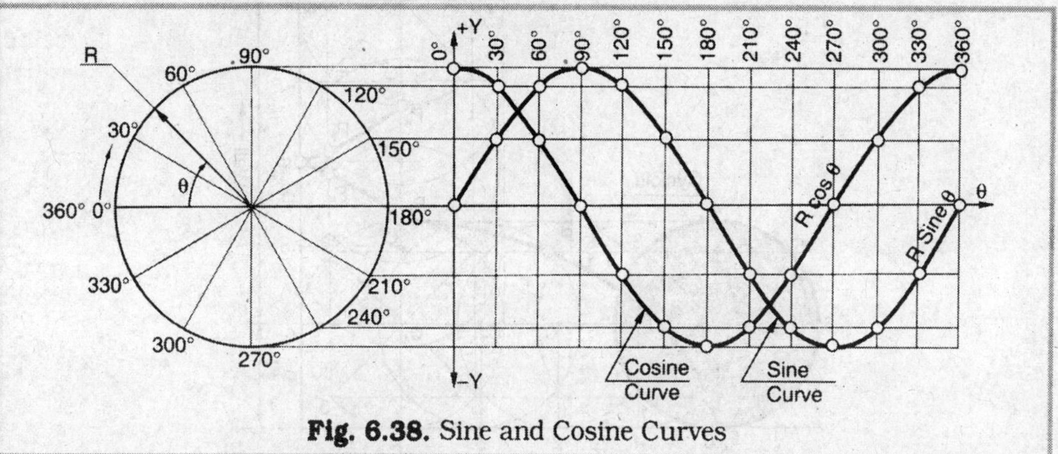

Fig. 6.38. Sine and Cosine Curves

1. First draw a circle of radius R and divide it into 12 equal parts and mark on it 0°, 30°, 60°,.....360°, as shown in Fig. 6.38.
2. Draw horizontal lines through 0°, 30°, 60°,.....360° and divide horizontal line of suitable length into 12 equal parts and mark on them θ = 0°, 30°,60°,....360°. Through these points draw vertical lines.
3. Join points of intersections of vertical and horizontal lines from respective angles to get sine curve.
4. Now join points of intersections of vertical lines through θ = 0°, 30°,60°,.....360° with horizontal lines through θ = (90 + 0°),(90 + 30°),.....(90 + 360°) to get the cosine curve.

In fact the cosine curve is in leading position to that of the sine curve.

Miscellaneous Problems

Problem 1 : A circle, of diameter D, rolls without slip on a horizontal surface (floor) by 1/2 revolution and then it rolls up a vertical surface (wall) by another 1/2 revolution. Initially the point P is at the bottom of circle touching the floor. Draw the path of the point P.

Take diameter of circle = 40 mm

Initially distance of centre of circle from the wall Ω 83 mm [Half circumference + D/2]

For solution see Fig. 6.39 and follow the procedure as given below :

1. Draw two lines one horizontal and another vertical to represent floor and wall respectively.
2. Draw ⊙ (C_0, 20) keeping distance of C_0 from the wall equal to 83 mm and just touching the floor. Mark position P_0 of point P at the bottom of the circle.
3. For first 1/2 revolution just do the procedure of drawing cycloid till the point P, reaches the top position (P_4), as shown in Fig. 6.39.

4. Now the circle will roll on the wall and for that initial position of Point P is P_4. Continue the same procedure as cycloid and get points P_5, P_6, P_7 and P_8. The curves P_4 to P_6 and P_6 to P_8 are nothing but last 1/4 part of a cycloid and first 1/4 part of a cycloid respectively. Join P_0, P_1,....P_4....P_6....P_8 by smooth curve to get the path of the point P.

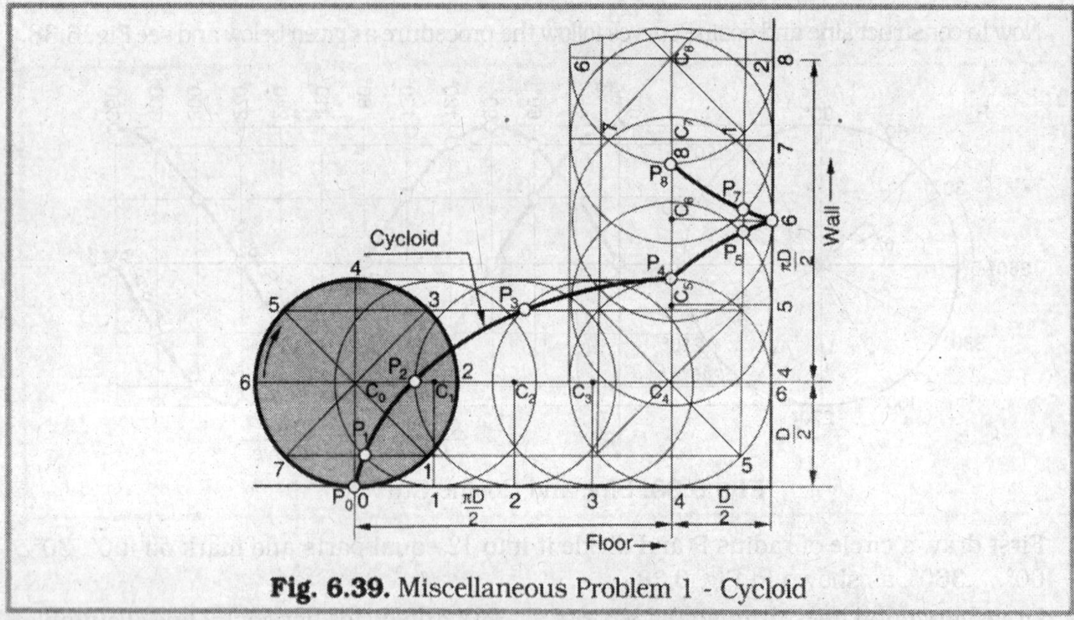

Fig. 6.39. Miscellaneous Problem 1 - Cycloid

Problem 2 : Prove by means of drawing that, if the diameter of the rolling circle is half the diameter of a directing circle, the hypocycloid is a straight line.

Take diameter of rolling circle = 70 mm and diameter of directing circle = 140 mm. For solution see Fig. 6.40.

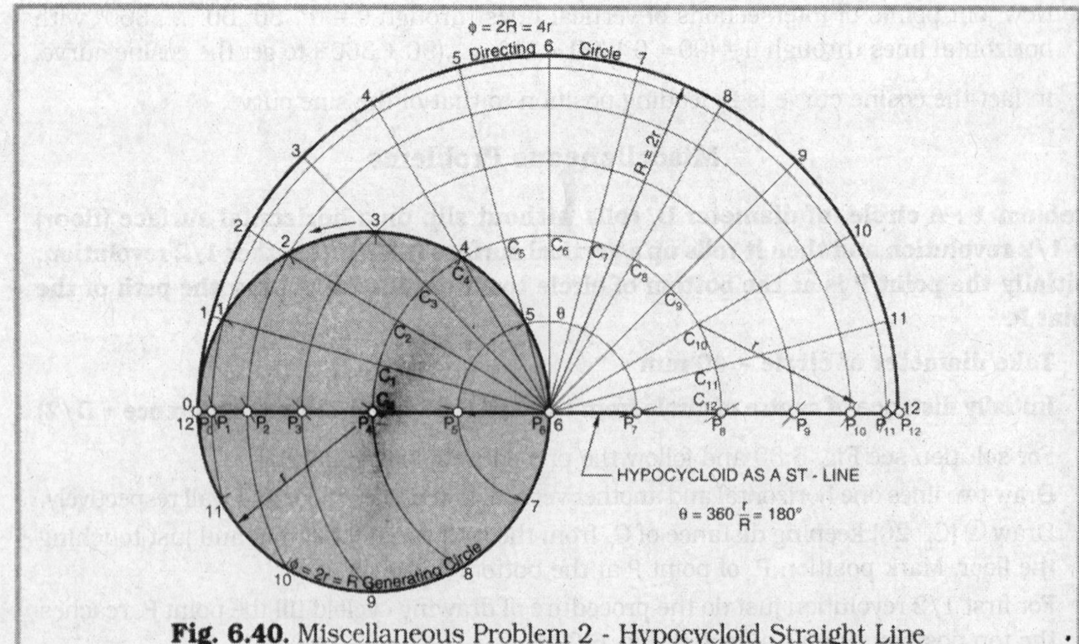

Fig. 6.40. Miscellaneous Problem 2 - Hypocycloid Straight Line

Do the same procedure as done for drawing hypocycloid. Do the drawing precisely, otherwise the points will deviate a little bit from the straight line and you will be confused while joining. Fig. 6.40 is self explanatory.

Problem 3 : A stick, of length equal to the circumference of a semicircle, is initially tangent to the semicircle on the right side of it. This stick now rolls over the circumference of a semicircle without sliding till it becomes tangent on the left side of the semicircle. Draw the loci of two end points of this stick. Name the curve. Take R = 42 mm.

For solution see Fig. 6.41 and follow the procedure as given below:

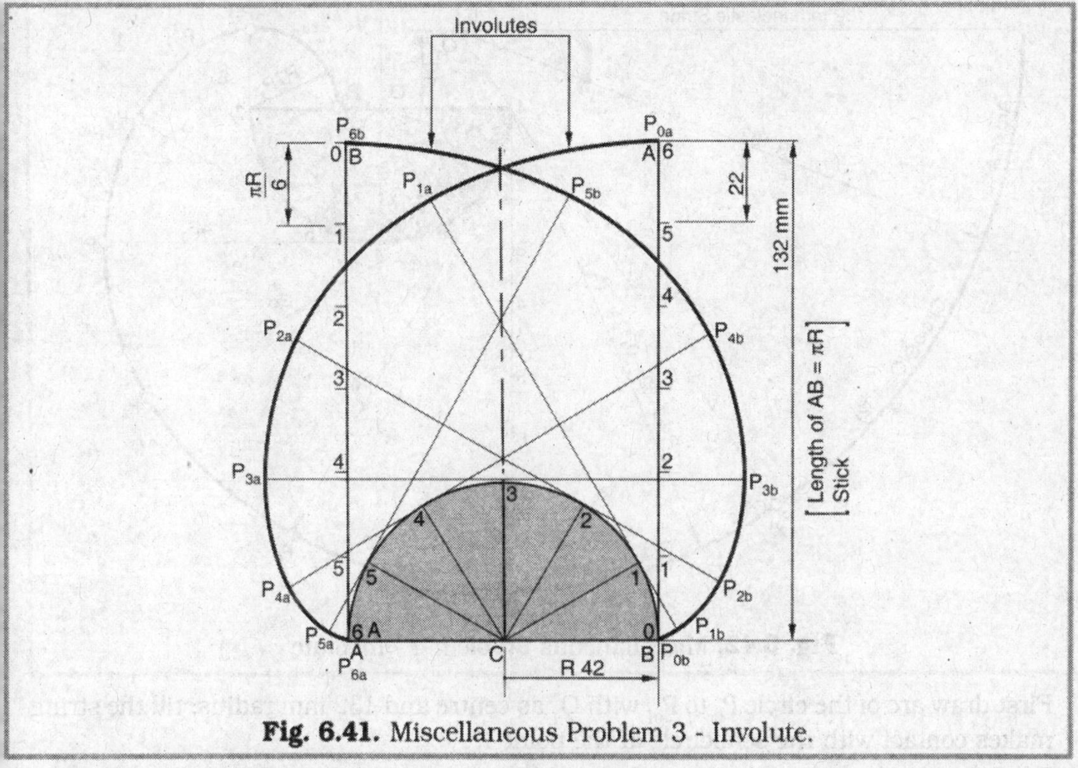

Fig. 6.41. Miscellaneous Problem 3 - Involute.

1. First draw a semicircle with C as the centre and R = 42 mm as its radius. AB is a diameter of semicircle. At point B draw tangent to this semicircle of length equal to πR = 132 mm.
2. Divide semicircle into 6 equal parts and mark them 0, 1, 2,....6 and also divide tangent at B into 6 equal parts. Each part = $\frac{132}{6}$ = 22 mm.
3. Now draw tangent at point 1 to the semicircle and mark on it one division on right side and 5 divisions on left side to get points P_{1b} and P_{1a} respectively. Similarly, on tangent at point 2 take two divisions on right and four divisions on left to get points P_{2b} and P_{2a} respectively. In the same manner get points (P_{3b}, P_{3a}), (P_{4b}, P_{4a}),....(P_{6b}, P_{6a}).
4. Join P_{0b}, P_{1b}, P_{2b},....P_{6b} and P_{0a}, P_{1a}, P_{2a},....P_{6a} by means of a smooth curve to get the loci of two end points of the stick. Both these curves are involute of a circle.

Problem 4 : A semicircle with O_2 as the centre and radius equal to 30 mm is fixed, as shown in Fig. 6.42. O_1P_0 is the inelastic string of 132 mm length. End O_1 of the string is fixed. Point O_1 is 18 mm on upper side and 18 mm on left side of O_2. String is turned in anticlockwise direction and simultaneously wound round the surface of the semicircle. Draw the locus of the point P, the free end of the string.

For solution see Fig. 6.42 and follow the procedure as given below :

1. Draw semicircle and divide it into 6 equal parts. Mark points 1, 2,.....7. as shown.

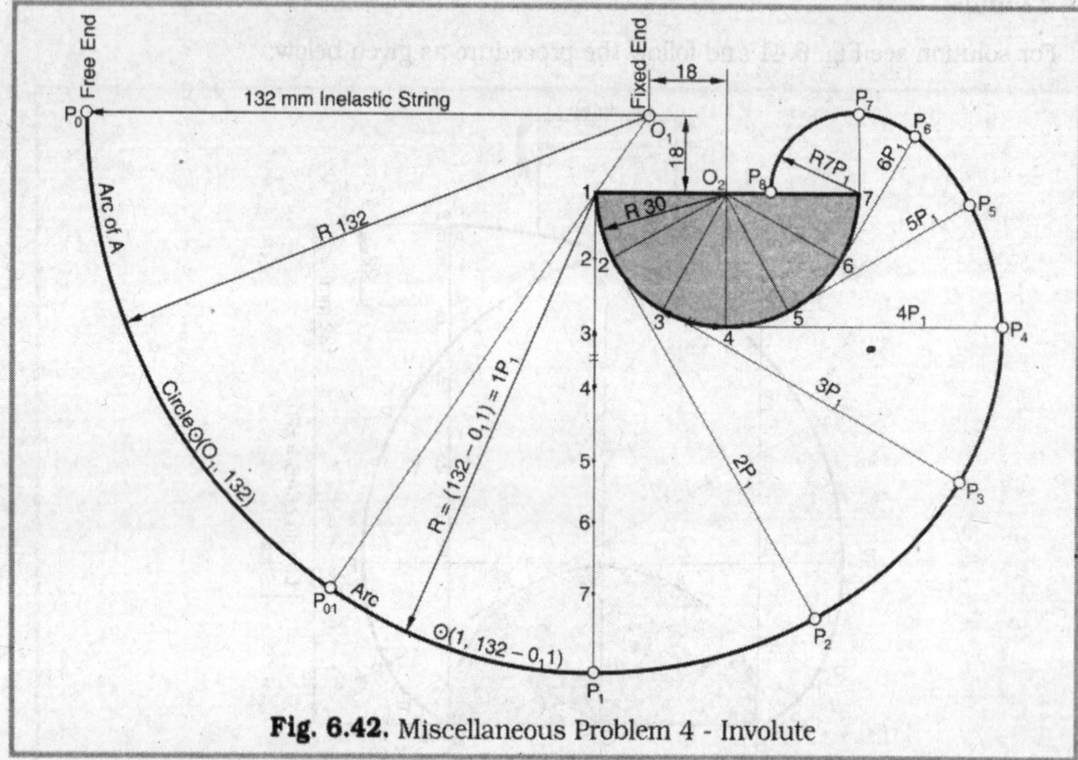

Fig. 6.42. Miscellaneous Problem 4 - Involute

2. First draw arc of the circle P_0 to P_{01} with O_1 as centre and 132 mm radius, till the string makes contact with the semicircle at the point 1.
3. Till the string becomes tangent to the circle at point 1, draw arc of circle P_{01} to P_1 with 1 as the centre and (132 - $O_1$1) mm as radius. Now when the string is further turned in anticlockwise direction it will be wound round the semicircle and the length of the string will go on decreasing. Therefore, in the direction 1 to P_1 divide, the length of the string equal to the circumference of semicircle, into 6 equal parts and mark them as points 2, 3,..7 as shown.
4. Now draw tangents to the semicircle at points 2, 3,....7 and mark on it length equal to $2P_1$, $3P_1$,....$7P_1$ to get points P_2, P_3,....P_7 respectively.
5. Now with 7 as the centre and radius equal to remaining length $7P_1$ draw quarter circle, as shown in the figure.
6. Join points P_1, P_2,....P_7 by means of a smooth curve. This part of the curve is an involute of a circle.
7. Total path of point P is from P_0, P_{01}, P_1, P_2,....P_7 and quarter circle P_7P_8.

Problem 5 : The foci of an ellipse are 110 mm apart. The minor axis is 70 mm long. Determine the length of the major axis and draw half ellipse by rectangle method and other half by concentric circles method. [8]

[Ans : Major Axis 130 mm] [B. T. E. M. S. April 2000 (Civil)]

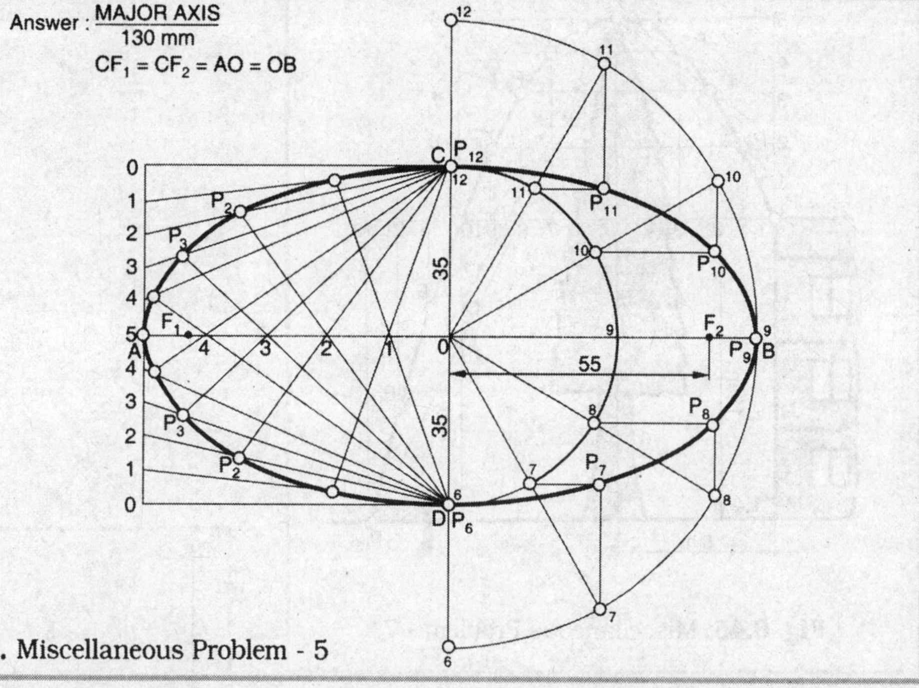

Answer : MAJOR AXIS 130 mm

$CF_1 = CF_2 = AO = OB$

Fig. 6.43. Miscellaneous Problem - 5

Problem 6 : Inscribe an ellipse in a parallelogram having sides 100 mm and 65 mm long and included angle of 120°. Determine the major and minor axis of an ellipse. Draw the tangent and normal to the ellipse at a point 20 mm above the horizontal axis and to the left of the centre. [Mumbai University, December 1997]

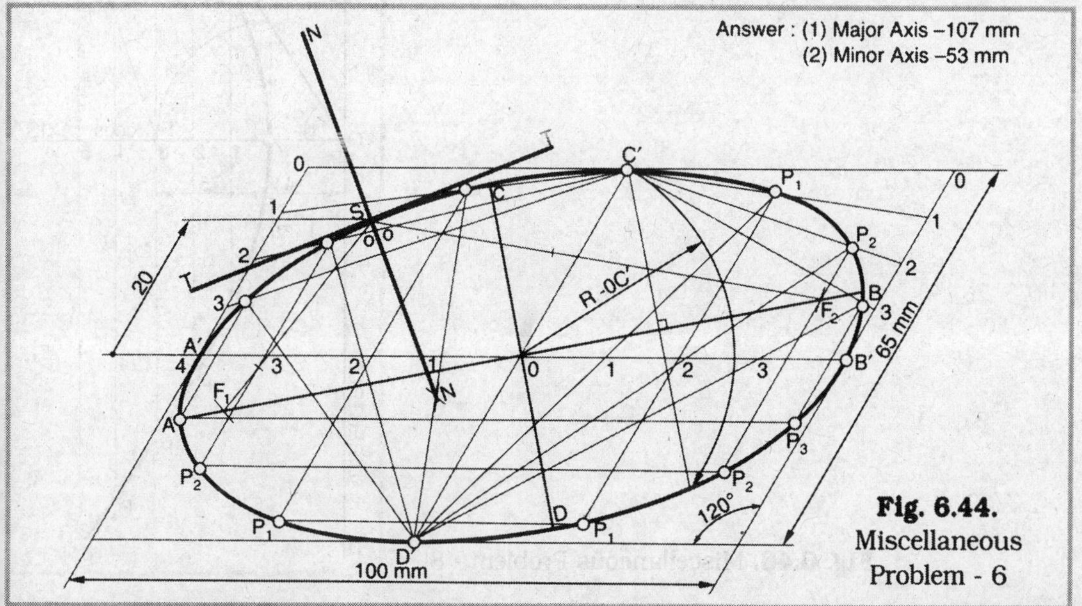

Fig. 6.44. Miscellaneous Problem - 6

Problem 7 : A boy, standing on the terrace of a building of 15 m height, throws a ball, which has its highest flight and just crosses a tree of 25 m height. Trace the path of the ball, if the distance between the building and the tree is 8 m.

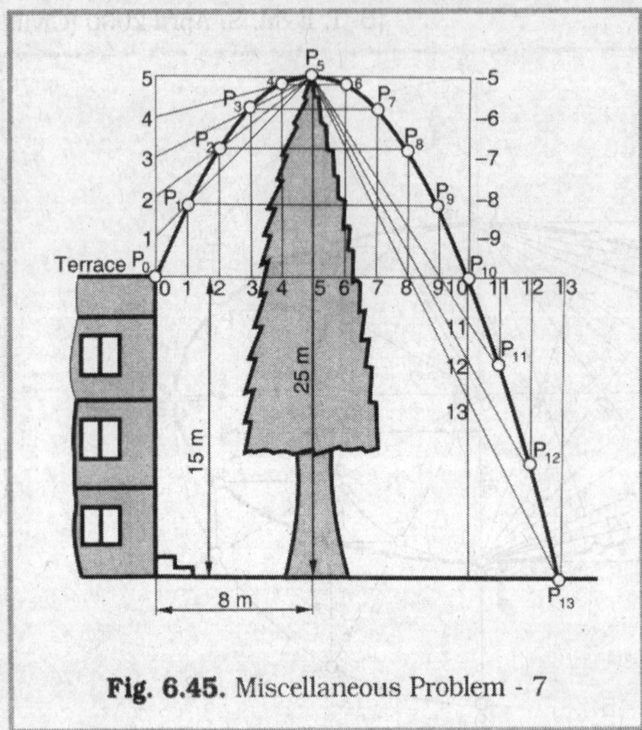

Fig. 6.45. Miscellaneous Problem - 7

Problem 8 : The vertex of hyperbola is 75 mm from its focus. Draw the curve if eccentricity is 3/2.[8]

[B. T. E. M. S. April/May 1997]

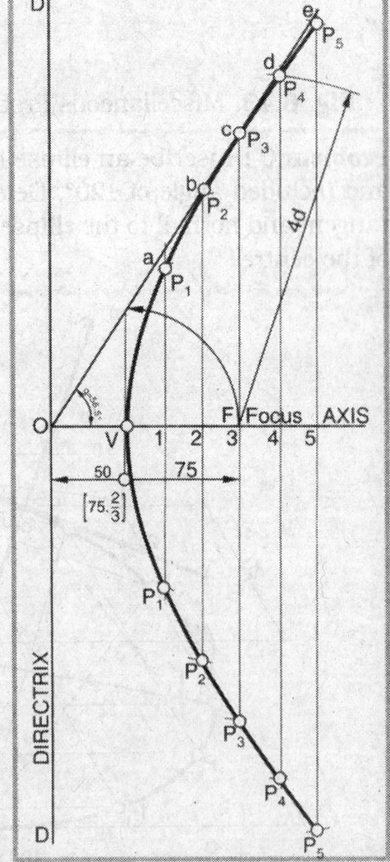

Fig. 6.46. Miscellaneous Problem - 8

Problem 9 : Draw the locus of a point P, which moves in such a way that the ratio of its distance from a fixed straight line to its distance from a fixed point is always constant and equal to 5/3. A fixed point is 60 mm away from the fixed straight line. Draw the tangent and normal to the curve at a point 70 mm from the fixed straight line. Name the curve.

[Ans : Ellipse] [Mumbai University, June 1996]

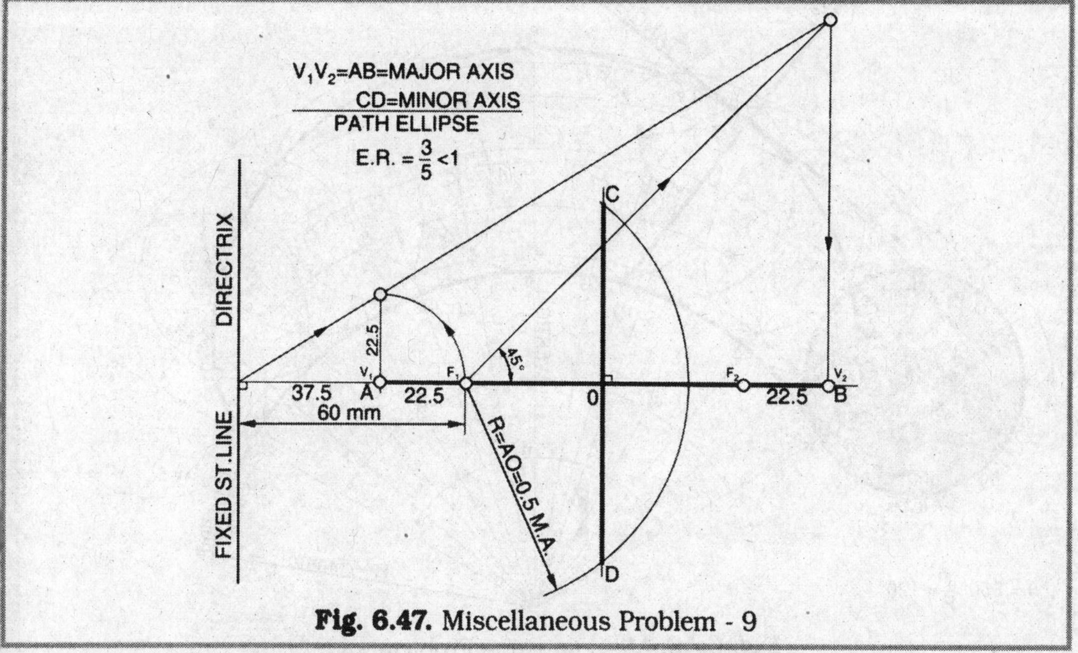

Fig. 6.47. Miscellaneous Problem - 9

Problem 10 : A circle of 50 mm diameter rolls along a straight line without slipping. Trace the path of a point on the circumference of the rolling circle for one complete revolution. Name the curve. [8]

[Ans : Cycloid] [B. T. E. M. S. April/May 1998]

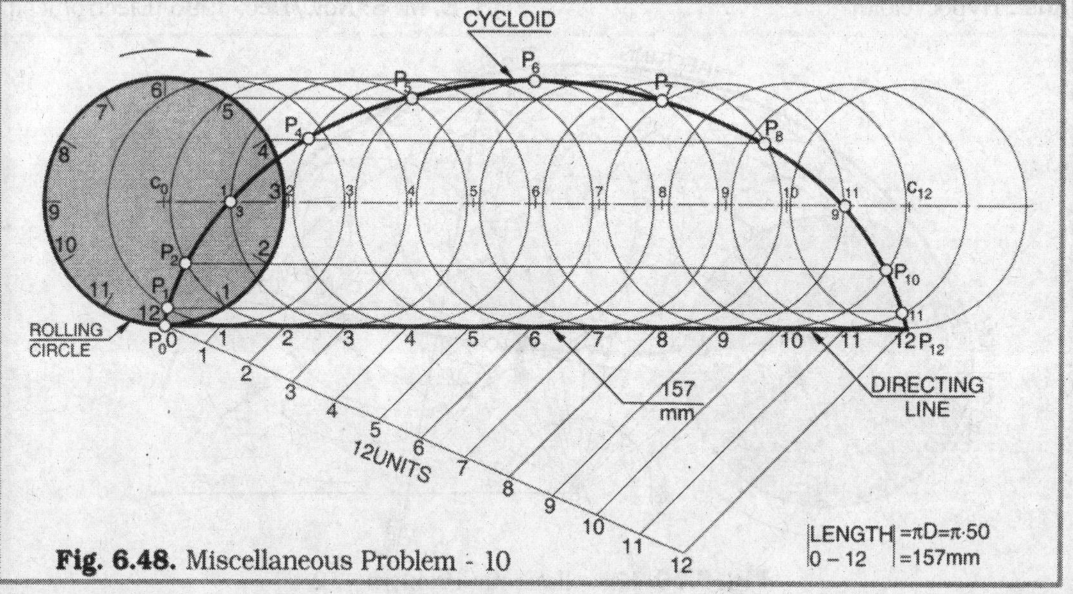

Fig. 6.48. Miscellaneous Problem - 10

Problem 11 : A circle of 50 mm diameter rolls along the circumference of another circle of 150 mm diameter from outside. Trace the path of a point P on the circumference of the rolling circle for one complete revolution. Name the curve. [7 + 1]

[Ans : Epycycloid] [B. T. E. M. S. November 1999 (Electronics)]

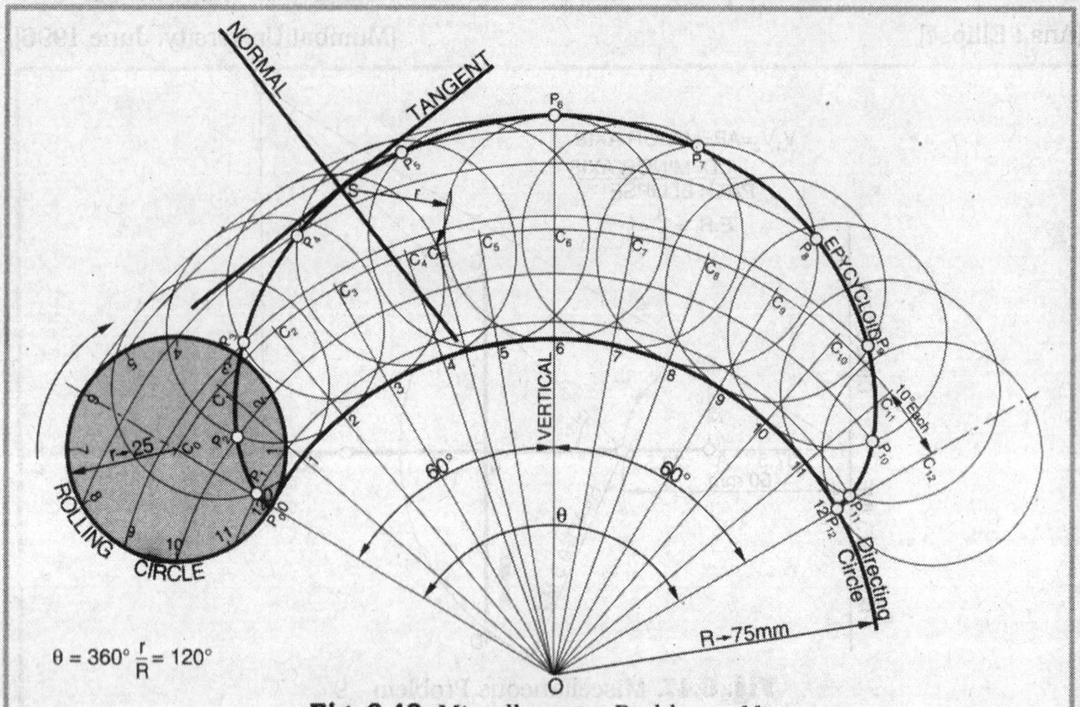

Fig. 6.49. Miscellaneous Problem - 11

Problem 12 : A circle, of 50 mm diameter, rolls along the circumference of another circle of 150 mm diameter from inside. Draw the path of a point P on the circumference of the rolling circle for one complete revolution. Name the curve. [7 +1]

[Ans : Hypocycloid] [B. E. M. S. Nov./Dec. 1996 (Electronics)]

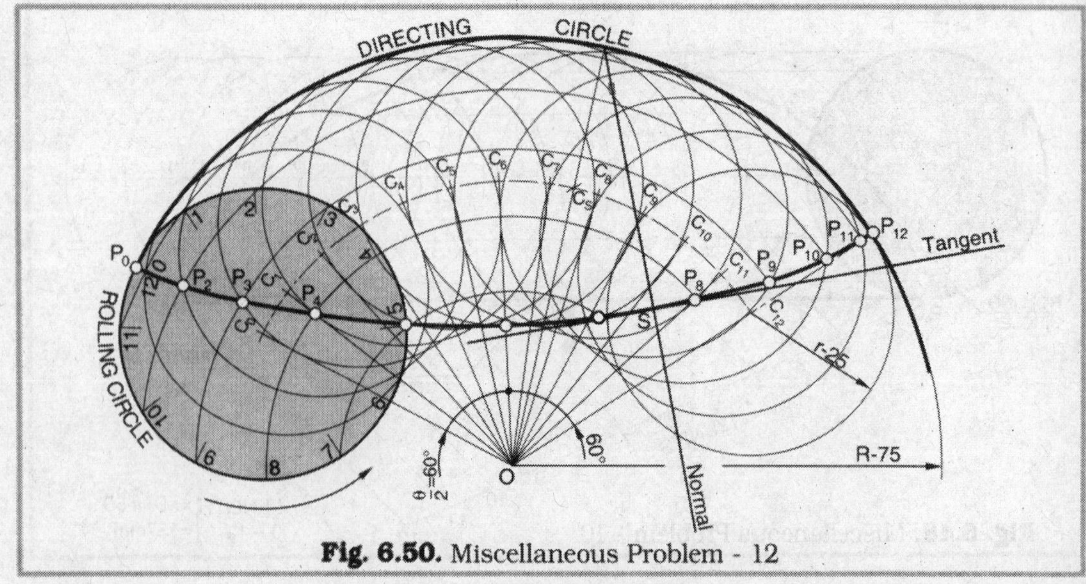

Fig. 6.50. Miscellaneous Problem - 12

Problem 13 : Show graphically that the hypocycloid is a straight line, when the diameter of the rolling circle is half that of the directing circle. Take radius of the rolling circle as 40 mm. [8] [B. T. E. *M. S. April/May 1998 and April 2000]

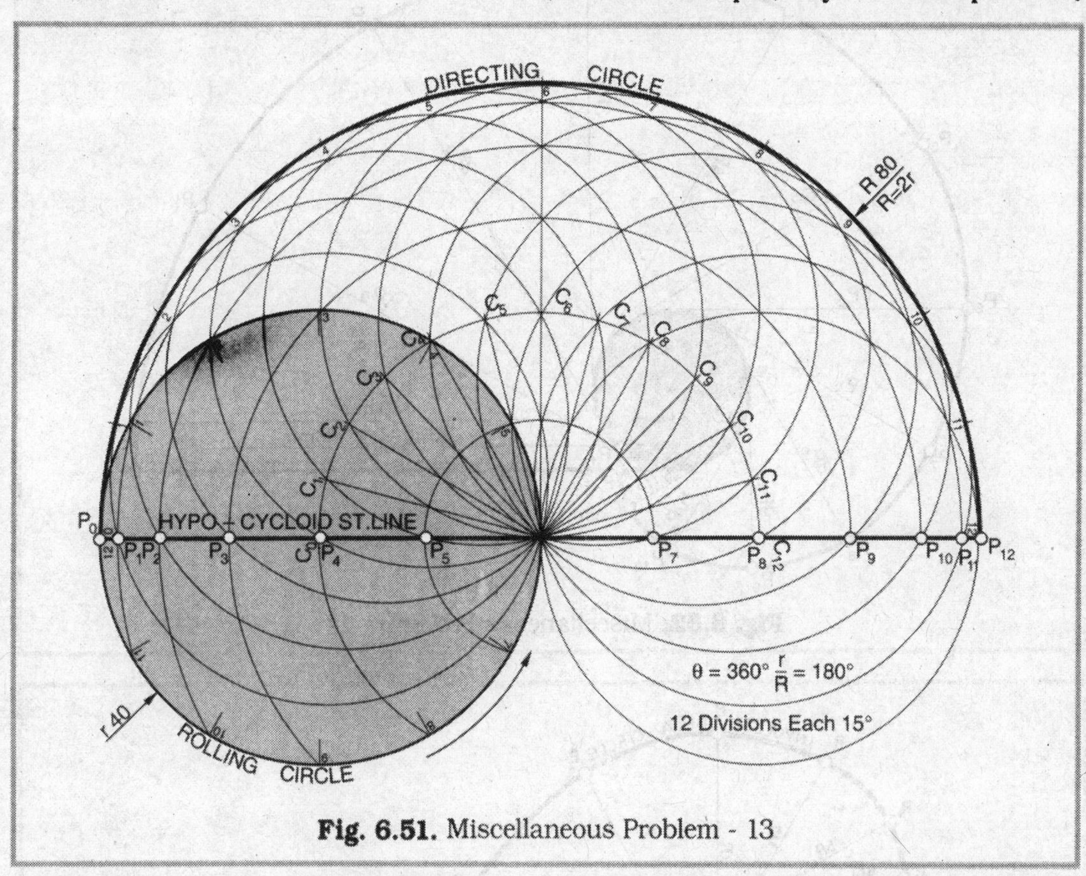

Fig. 6.51. Miscellaneous Problem - 13

Problem 14 : One end of an inelastic thread, 140 mm long, is attached to the circumference of a circular disc of 40 mm diameter. Draw the path of the free end of the thread, if it is completely wound round the disc, keeping the thread always tight. Name the curve. [8]

[Ans : Involute of A Circle] [B. T. E. M. S. April 2000 (Electronics)]

Problem 15 : An in-elastic string, of 100 mm length, is wound round a cylinder of 50 mm diameter, keeping the string always tight. Draw the curve generated by end point of string. Name the curve. [B. T. E. M. S. Nov. / Dec. 1997]

Problem 16 : A disc in the form of a semicircle and a semi-regular hexagon of thickness 10 mm is shown in the figure below.

A disc is firmly fixed at point O. An inelastic string of length 160 mm is fixed at point A and the free end B of a string is turned round the disc in anticlockwise direction. Draw the locus of B. Draw the tangent and normal to the curve at a distance of 110 mm away from pole O. Name the curve. [Mumbai University, May 1992]

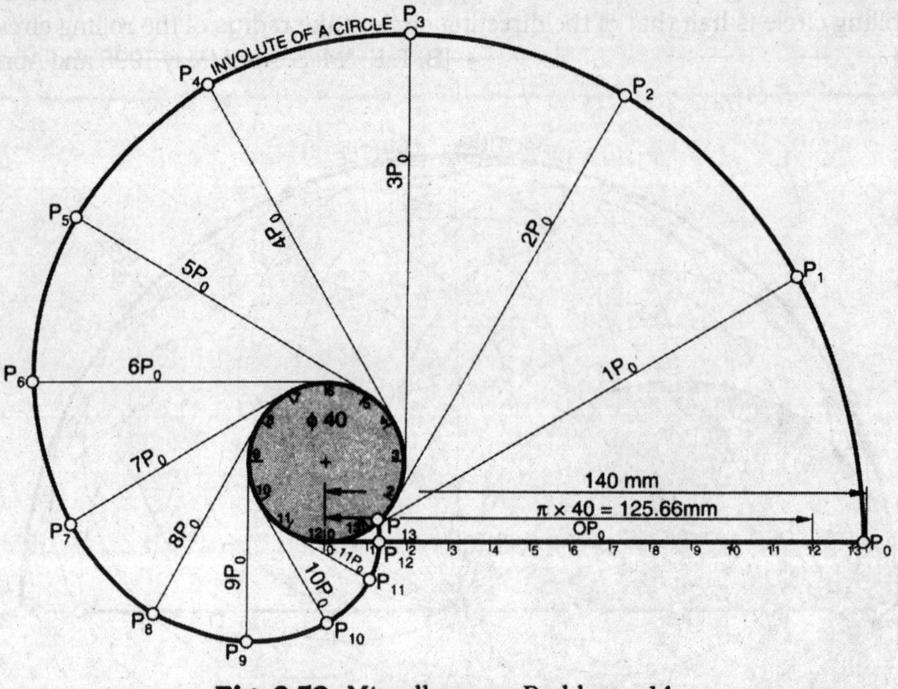

Fig. 6.52. Miscellaneous Problem - 14

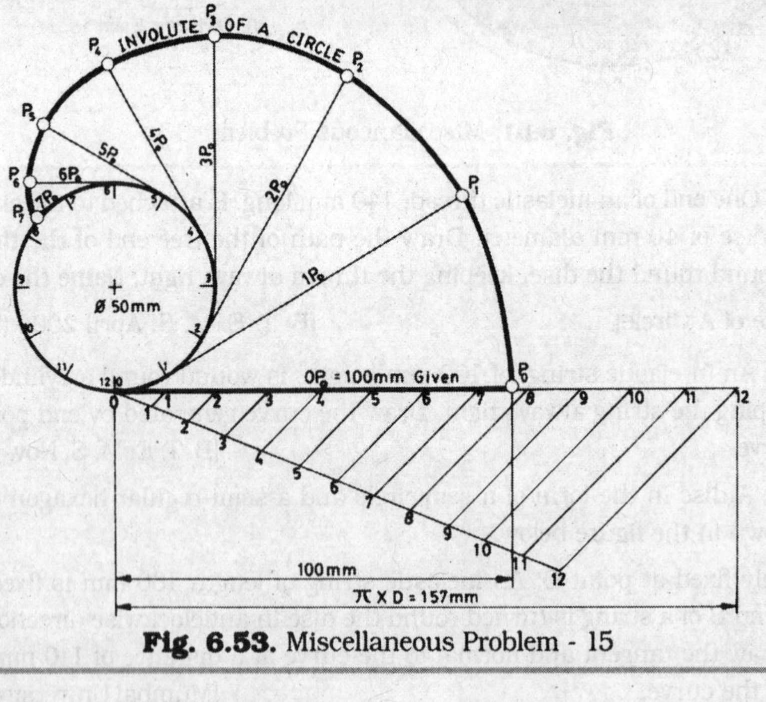

Fig. 6.53. Miscellaneous Problem - 15

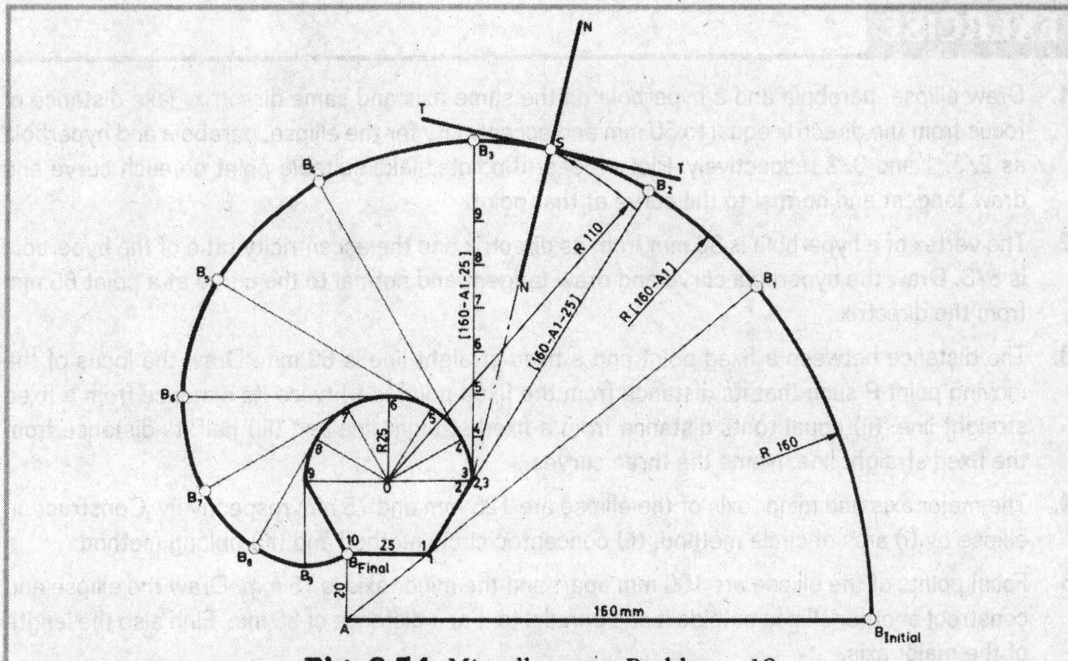

Fig. 6.54. Miscellaneous Problem - 16

Problem 17 : Draw one convolution of the Archemedian spiral represented by the polar equation r = [32 + 7.θ] where 'r' is in mm and θ is in radians. Draw the tangent and normal to the curve at point 55 mm from the pole. [Mumbai University, June 1996]

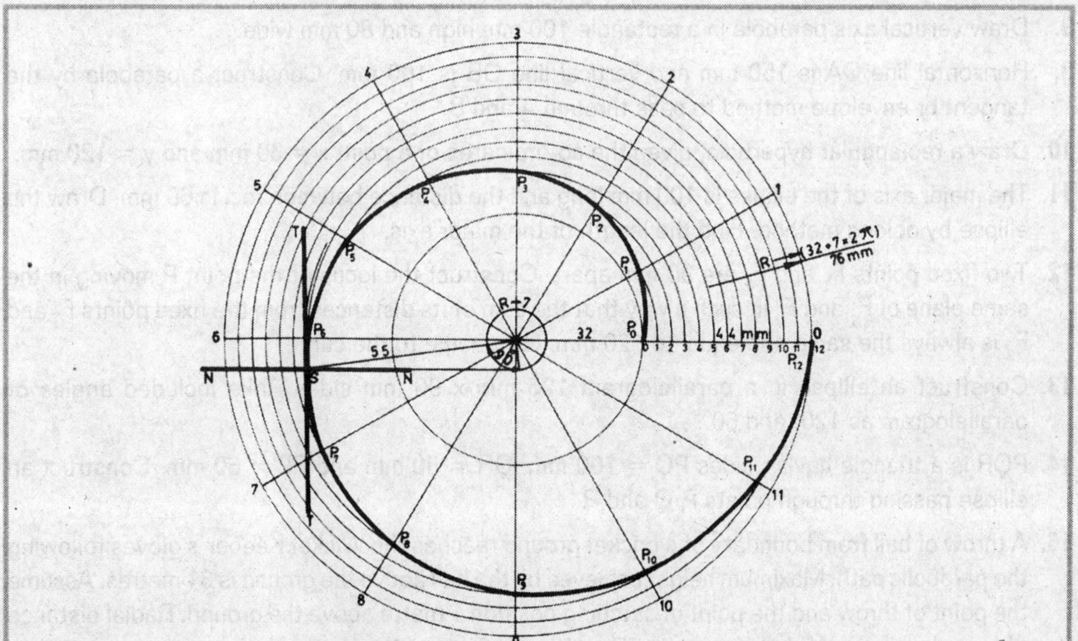

Fig. 6.55. Miscellaneous Problem - 17

EXERCISE

1. Draw ellipse, parabola and a hyperbola on the same axis and same directrix. Take distance of focus from the directrix equal to 50 mm and eccentricity for the ellipse, parabola and hyperbola as 2/3, 1 and 3/2 respectively. Plot at least 8 points. Take suitable point on each curve and draw tangent and normal to the curve at that point.
2. The vertex of a hyperbola is 30 mm from its directrix and the eccentricity ratio of the hyperbola is 5/3. Draw the hyperbola curve and draw tangent and normal to the curve at a point 65 mm from the directrix.
3. The distance between a fixed point and a fixed straight line is 60 mm. Draw the locus of the moving point P such that its distance from the fixed point is (i) twice its distance from a fixed straight line, (ii) equal to its distance from a fixed straight line and (iii) half its distance from the fixed straight line. Name the three curves.
4. The major axis and minor axis of the ellipse are 125 mm and 75 mm respectively. Construct an ellipse by (i) arcs of circle method, (ii) concentric circle method and (iii) oblong method.
5. Focal points of the ellipse are 100 mm apart and the minor axis is 75 mm. Draw the ellipse and construct another ellipse outside it and parallel to it at a distance of 25 mm. Find also the length of the major axis.
6. An elliptical hand hole flange of an air receiver tank has outside dimensions, major axis 100 and minor axis 70 mm. The inside hole is also elliptical and parallel to and 25 mm away from outer peryphery of the flange. Construct the drawing of gasket required for closing the hole.
7. Using locus method, construct a parabola having its vertext 30 mm from the focus.
8. Draw vertical axis parabola in a rectangle 100 mm high and 80 mm wide.
9. Horizontal line OA is 150 mm and vertical line OB is 100 mm. Construct a parabola by the tangent or envelope method to pass through A and B.
10. Draw a rectangular hyperbola, given the co-ordinates of a point x = 30 mm and y = 120 mm.
11. The major axis of the ellipse is 100 mm long and the distance between foci is 60 mm. Draw the ellipse by oblong method. Find the length of the minor axis.
12. Two fixed points F_1 and F_2 are 90 mm apart. Construct the locus of the point P moving in the same plane of F_1 and F_2 in such a way that the sum of its distances from the fixed points F_1 and F_2 is always the same and equal to 120 mm. Give name to the curve.
13. Construct an ellipse in a parallelogram 125 mm x 90 mm sides. Take included angles of parallelogram as 120° and 60°.
14. PQR is a triangle having sides PQ = 100 mm, QR = 80 mm and RP = 60 mm. Construct an ellipse passing through points P, Q and R.
15. A throw of ball from boundary of a cricket ground reaches the Wicket Keeper's gloves following the parabolic path. Maximum height achieved by the ball above the ground is 31 metres. Assume the point of throw and the point of catching position 1 metre above the ground. Radial distance of boundary from Wicket Keeper is 75 metres. Construct the path of ball.
16. Draw a rectangle of 125 mm x 100 mm sides and construct two parabolas in it with their axes parallel to two sides of a rectangle.

17. Two asymptotes OX and OY are at 75° angle. Point P is 35 mm from OX and 50 mm from OY. Draw hyperbola passing through the point P.

18. Two fixed points F_1 and F_2 are 60 mm apart. Trace two curves, described out by points P_1 and P_2 moving in the same plane of F_1 and F_2 such that the difference between distances from F_1 and F_2 is always constant and equal to 30 mm.

19. Draw a cycloid for a rolling circle, of 60 mm diameter rolling along a straight line without slipping. Take initial position of the tracing point at the bottom of the vertical centre line of the rolling circle. Draw tangent and normal to the curve at a point 35 mm above the directing line.

20. A circle of 49 mm diameter rolls along a straight line without slipping. Draw the curves traced by points Q, P and R located 15 mm inside the circle, on the circle and 25 mm outside the circle respectively. Take initial positions of points Q, P and R at the bottom on vertical centre line of circle. Name the curves traced.

21. A circle of 50 mm diameter rolls on the circumference of another circle of 150 mm diameter and outside it. Draw the locus of the point P on the circumference of the rolling circle for one complete revolution of it. Take initial position of point P at the contact point between two circle. Name the curve and draw tangent and normal to the curve at a point 115 mm from the centre of the bigger circle.

22. In problem 21 take rolling circle inside the directing circle and draw locus of point P. Name the curve and draw tangent and normal to the curve at point 60 mm from the centre of the directing circle.

23. Show by means of drawing that when the diameter of a rolling circle is half the diameter of directing circle, the hypocyloid is a straight line.

24. A circular ring of 100 mm diameter is rolling on the fixed cylinder of 50 mm diameter keeping contact from inside. Draw the locus of point P on the circumference of circular ring initially at the top contact point.

25. A wheel of 49 mm diameter rolls downward on the vertical wall by 1/2 a revolution and then on the floor by 1/2 a revolution without slipping. Draw the locus of point P on the circumference of the wheel. Take initial position of the point P at the contact point of the wheel with the wall.

26. A wheel of 50 mm diameter rolls on (i) inside and (ii) outside on another circle of 150 mm diameter. Draw and name the curves traced out by points Q and S 15 mm and 35 mm from the centre on the wheel along a straight line passing through the centre of the wheel.

27. Construct one complete turn of an involute of a
    (i) circle of 30 mm diameter.
    (ii) square of 30 mm side
    (iii) hexagon of 25 mm side

28. One end Q of inelastic string PQ, 150.5 mm long, is attached to the circumference of a half circular half hexagonal disc of 49 mm diameter. Draw the curve traced out by the other end of the string P when it is completely wound round the circumference of the disc, keeping the string always tight. Take initial position of string tangent at the mid point Q of circular portion.

29. A rod PQ 99 mm long is resting horizontally on the circumference of a circular disc of 63 mm diameter touching at the mid point. Rod PQ rolls without slipping on the circumference of a circular disc to the full extent on both the sides. Draw the loci of the points P and Q and name the curves.

30. Draw one complete turn of right hand and left hand helices given the diameter 50 mm and lead 75 mm.

31. Draw three complete turns of a right hand square helical spring given outside diameter 100 mm, inside diameter 70 mm and lead 40 mm.

32. Draw left hand, two start square threads of 160 mm length given the outside diameter of thread 100 mm and lead 80 mm.

33. Draw an Archemedian Spiral of 1.5 convolutions, the greatest and the least radii being 125 mm and 35 mm respectively. Draw tangent and normal to the spiral at a point 85 mm from the pole.

34. A point P moves radially outward from the centre of the circular disc to the peryphery when disc completes 2 revolutions. Radial movement of a point P and circular motion of disc is assumed uniform. Take diameter of the disc 120 mm Draw the locus of the point P and name the curve.

35. Construct logarithmic spiral for one convolution, given the length of the shortest radius equal to 15 mm and the ratio of the lengths of the successive radius vectors equal to 6/5 for vectorial angle of 30°.

36. Draw a square of 5 mm sides and then draw 4 turns of a spiral of quarter circles on it using four corners of the square as centres.

37. Taking 2 centres 5 m.m. apart on vertical line draw 4 turns of a spiral of half circles.

38. Draw curves for the following :

(i) $y = R_1 \sin\theta$ where $R_1 = 35$ mm.

(ii) $y = R_2 \cos\theta$ where $R_2 = 70$ mm.

Problem 39 :

A plot of ground in the shape of a parallelogram 15000 mm x 10000 mm, the angle between the sides being 60°. Inscribe an elliptical flower bed in it. Select a suitable scale.

Problem 40 :

Three points A, B and P while lying along a horizontal line in order, have AB = 60 mm and AP = 80 mm. While A and B are fixed and P starts moving such that AP + BP remain always constant and when they form an isosceles triangle, AP = BP = 50 mm. Draw the path traced out by the point P from the commencement of its motion back to its initial position.

Problem 41 :

The concrete arch for a water channel of a railway bridge is semi elliptical in shape with major axis 2.5 metres and minor axis 1.5 metres. Draw the boundary of concrete arc to suitable scale.

Problem 42 :

A cricket ball is thrown and reaches a maximum height of 10 metres and falls on the ground at a distance of 30 metres from the point of projection. Determine the angle of projection. Draw the path of the cricket ball and name the curve. Assume that the point of projection is on the ground level.

Problem 43 :

An autohead light reflector, for parallel rays of light, is parabolic in shape with its bulb at the focus point. The distance of the focus from where the parabola cuts the axis is 100 mm. Draw the shape of the reflector.

Problem 44 :

Motor car head lamp parabolic reflector is having an aperture (opening) of 175 mm and a depth of 135 mm Draw the shape of the reflector.

Problem 45 :

A stone is thrown from a building 7 metres high and its highest point of flight just crosses a palm tree 14 metres high. Trace the path of the projectile, if the distance between the building and the palm tree is 3.5 metres. Take a suitable scale.

Problem 46 :

For a perfect gas, the relation between the pressure P and the volume V in isothermal expansion is given by PV = constant. Draw the curve of isothermal expansion of an enclosed volume of gas if 0.056634 cubic metres of the gas correspond to a pressure of 0.3515 N. per sq.cm.

Problem 47 :

A wheel 1.5 metres in diameter has seven spokes connecting the rim and the hub. The wheel is rotating in anticlockwise direction at 80 rpm. A particle of dust starts from the centre and travels along a spoke with uniform velocity and reaches the rim after two seconds. Trace the path of the particle. Select a scale 1 : 25

Problem 48 :

Draw a logarithmic spiral for one convolution, the successive radii are of the ratio 9:8, final radius vector is 90 mm and the angle between the successive radii being 30°. Draw the tangent at any point of the curve.

Problem 49 :

A vehicle has 60 cm diameter wheels. For one complete revolution of the wheel, draw the locus of the point on its circumference when it passes over a segmental arched culvert of radius 3 metres. Take a scale of 1 : 20.

Problem 50 :

A circus man rides a motorcycle inside a globe of diameter 4 metres. The motor cycle wheel is 0.8 meter in diameter. Draw the locus of a point spot on the circumference of the motor cycle wheel for its one complete turn.

Problem 51 :

An ant is moving on a vertical cylinder in a circumferential direction at a constant angular speed in the clockwise direction as seen axially from the top and with simultaneous uniform rate of advance in the axial direction. If the diameter of the cylinder is 50 mm and the ant moves 75 mm in the axial direction during one turn, draw the path of the ant.

Problem 52 :

Construct two complete coils of a spring of round cross section, external diameter 90 mm, internal diameter 70 mm and pitch 40 mm.

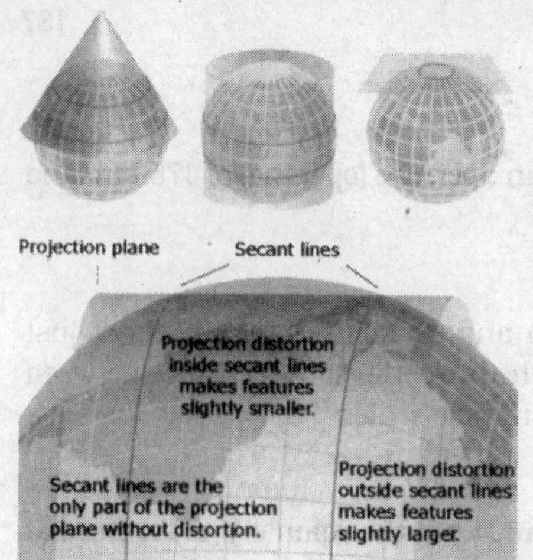

CHAPTER - 7

Projections, Planes of Projections and Systems of Projections

1. GENERAL

Engineers, Architects, Technicians and Craftsmen make use of various types of projections frequently. All connected with engineering must learn the methods of drawing projections on various planes of projections and they must also learn how to read all such projections. In this chapter we shall study different types of projections, different types of planes of projections and systems of projections.

Initilly students should not bother about systems of projections. Once the clear understanding is developed about projections and planes of projections, study of systems of projections will be carried out.

2. PROJECTION

There are three types of projections :-

(i) Orthographic Projections, (ii) Oblique Projections, (iii) Perspective Projections.

(i) Orthographic Projections :

See Fig. 7.1 (a)

Ortho means right angle or perpendicular. When from an object (Point, ends of a straight line or corners of solid) parallel rays perpendicular to the plane of projection (Picture Plane) are drawn to get projection on picture plane then the projection is known as Orthographic Projection. In this projection, an observer is assumed to be at infinite distance from the object. Parallel rays meet in the eyes of observer orthogonally positioned at infinity.

(ii) Oblique Projections :

See Fig. 7.1 (b)

Oblique means other than perpendicular i.e. inclined. When from the corners of an object parallel rays inclined to the plane of projection are drawn to get projection on the plane of projection (Picture Plane) then the projection is known as Oblique Projection. Oblique projections are rarely used in practice. This is a three dimensional projection. In this projection, an observer is assumed to be at infinite distance from the object. Parallel rays meet in the eyes of observer, obliquely positioned at infinity.

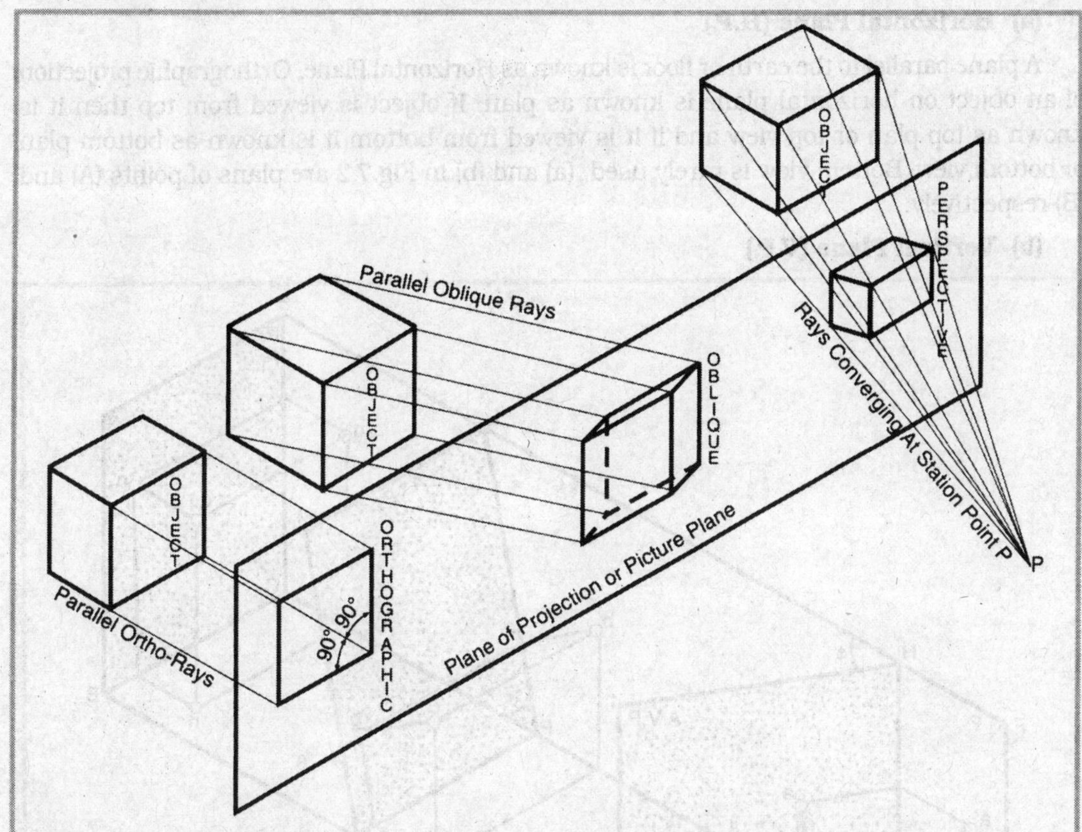

Fig. 7.1. (a) (b) & (c) Principles of Orthographic, Oblique and Perspective Projections

(iii) Perspective Projections :

See Fig. 7.1 (c)

In perspective projection rays from corners of an object converge to a finite point, where the eye of the observer is assumed to be located. If the plane of projection is between the eye and the object the projection will be smaller. If the plane of projection is on another side of the object than the projection will be larger than the object. Perspective projection gives real three dimensional picture of the object, which our eye is observing.

3. PLANES OF PROJECTIONS AND CORRESPONDING ORTHOGRAPHIC PROJECTIONS

Planes of projections or picture planes used in Engineering Drawing are :-

(i) Principle Planes. (H.P., V.P. & P.P. or A.V.P.)

(ii) Auxiliary Planes. (A.I.P. and A.V.P.) (Discussed in chapter no. 11).

(iii) Axonometric Planes (Discussed in chapter on Isometric)

(i) Principle Planes : See Fig. 7.2

(a) Horizontal Plane. (H.P.) (ABCD in Fig 7.2)

(b) Vertical Plane. (V.P.) (CDFE in Fig. 7.2)

(c) Profile Plane (P.P.) (BCEG in Fig. 7.2)

(a) Horizontal Plane (H.P.)

A plane parallel to the earth or floor is known as Horizontal Plane. Orthographic projection of an object on horizontal plane is known as plan. If object is viewed from top then it is known as top plan or top view and if it is viewed from bottom it is known as bottom plan or bottom view. Bottom view is rarely used. (a) and (b) in Fig.7.2 are plans of points (A) and (B) respectively.

(b) Vertical Plane (V.P.)

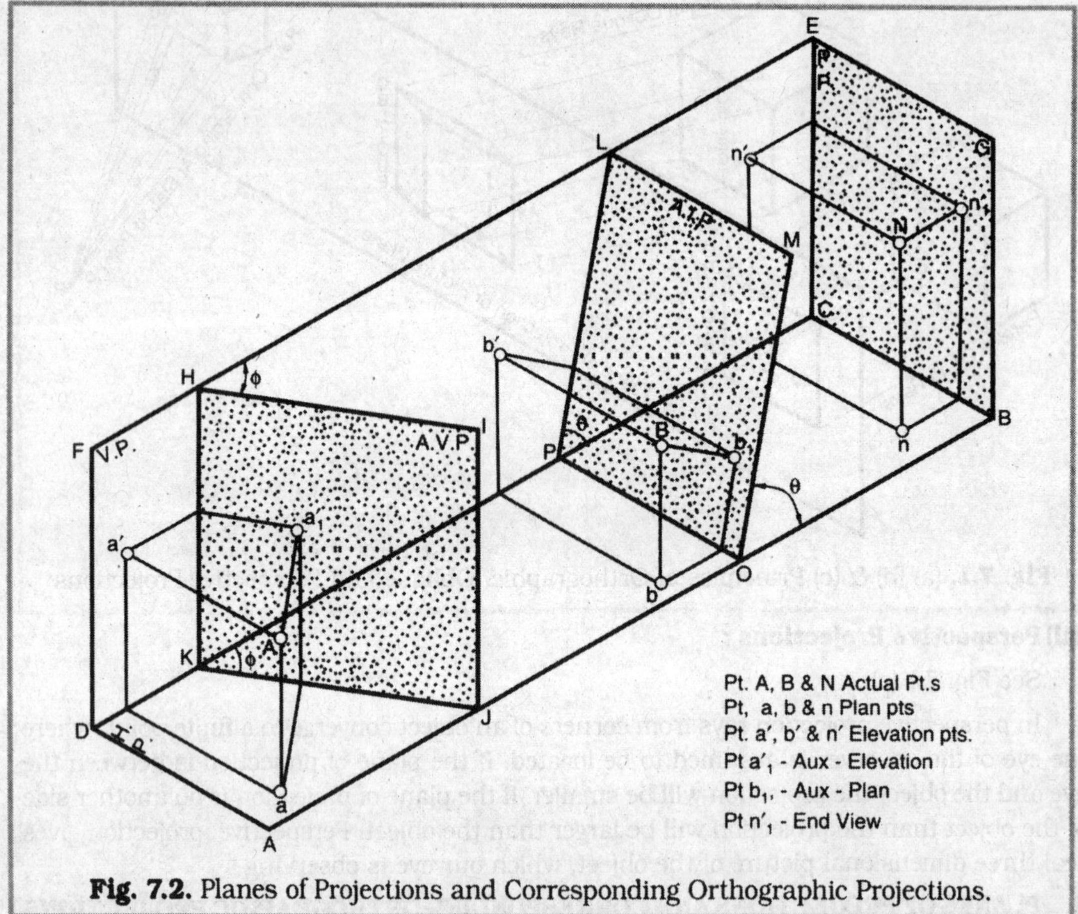

Fig. 7.2. Planes of Projections and Corresponding Orthographic Projections.

Any one plane conveniently selected out of the planes perpendicular to H.P. is known as vertical plane. Orthographic projection on V.P. is known as elevation. If an object is viewed from the front then it is known as front elevation or front view and if it is viewed from rear then it is known as rear elevation or rear view. Rear view is rarely used. (a') and (b') in Fig. 7.2 are elevations of points (A) and (B) respectively.

(c) Profile Plane or Auxiliary Vertical Plane (P.P. or A.V.P.)

A plane perpendicular to H.P. and V.P. both is known as profile plane. Orthographic projection on P.P. is known as end view, side view or end elevation. If the object is viewed from left then the view is known as left hand end view and similarly if it is viewed from right then it is known as right hand end view.

(ii) Auxiliary Planes :

There are two types of auxiliary planes. They are perpendicular to one of the principal planes and inclined to other principal plane. Projections on auxiliary planes are known as auxiliary views. Types of auxiliary planes are : (a) Auxiliary Vertical Plane (A.V.P.) (HIJK in Fig. 7.2) and (b) Auxiliary Inclined Plane (A.I.P.) (LMOP in Fig. 7.2).

(a) Auxiliary Vertical Plane (A.V.P.)

A plane perpendicular to H.P. and inclined to V.P. by an angle Ø is known as auxiliary vertical plane (A.V.P.) and projection on it (a_1') in Fig 7.2 is known as auxiliary elevation.

(b) Auxiliary Inclined Plane (A.I.P.)

A plane perpendicular to V.P. and inclined to H.P. by an angle θ is known as auxiliary inclined plane (A.I.P.) and projection (b_1) on it in Fig. 7.2 is known as auxiliary plan.

(iii) Axonometric Planes :

Planes inclined to all the three principal planes are known as Axonometric Planes. There are three types of Axonometric Planes.

(a) Isometric Plane.

(b) Diametric Plane.

(c) Trimetric Plane.

Projections on axonometric planes are known as axonometric projections. Plane QRG in Fig. 7.3 is an axonometric plane. C_2 is known as axonometric projection of point C.

(a) Isometric Plane :

It is an axonometric plane which is equally inclined to all the three principal planes. Projection on it is known as Isometric Projection. It is an important three dimensional projection. We shall study this projection in detail in chapter on Isometric.

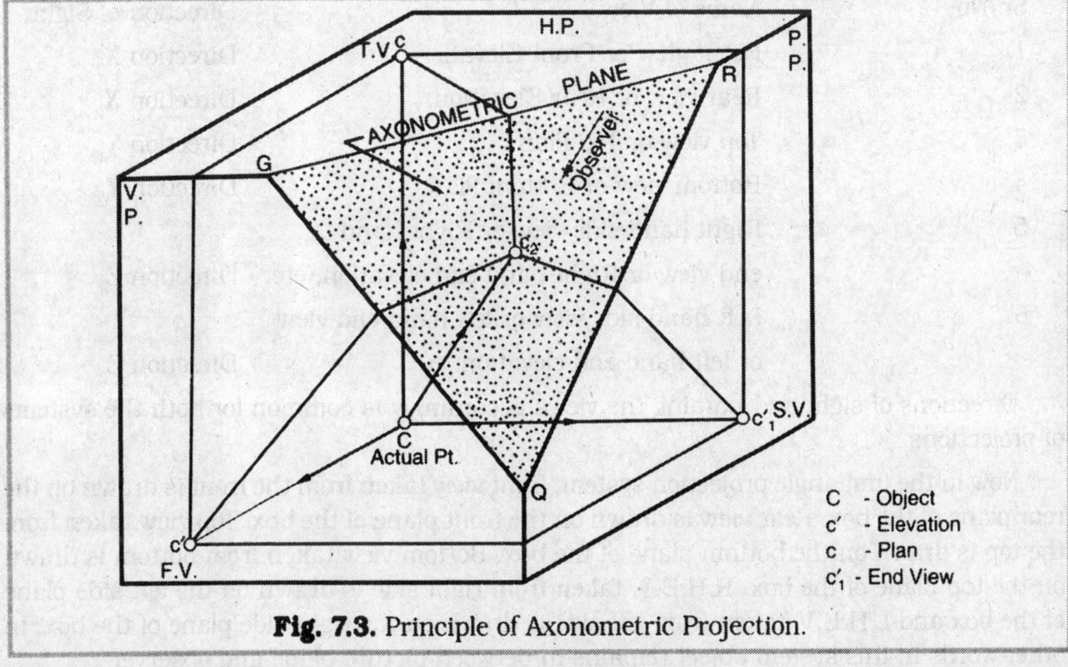

Fig. 7.3. Principle of Axonometric Projection.

(b) Diametric Plane :

It is an axonometric plane which is equally inclined to two of the three principal planes. Projection on it is known as diametric projection. It is little different than isometric projection.

(c) Trimetric Plane :

It is an axonometric plane which has different inclinations with all the three principal planes. Projection on it is known as trimetric projection. It is little more different than isometric projection.

4. SYSTEMS OF ORTHOGRAPHIC PROJECTIONS

There are two systems of orthographic projections.

(i) First angle projection system.

(ii) Third angle projection system.

At present first angle projection system is recommended for practice in engineering field as per I.S. Before few years third angle projection system was used in India. So we shall study both the systems of projections.

Although principle involved is the same, treatment appears to be different for problems of (a) Orthographic Projections of objects and (b) Problems of Solid Geometry for the same system of projections. So we shall study the same system for both the type of problems.

(a) First Angle Projection System for Orthographic Projections of Objects :

See Fig. 7.4 (i) (ii) (iii).

In Fig. 7.4 (i) object, whose orthographic projections are required to be drawn, is kept inside a glass box and is viewed from all six directions perpendicular to the 6 planes of the box and six views are achieved.

These six views of the object are named as under according to the direction of sight.

Sr. No	Name of View	Direction of Sight
1.	Front view or Front Elevation.	Direction X_1
2.	Rear view or Rear Elevation.	Direction X_2
3.	Top view or Top Plan.	Direction Y_1
4.	Bottom view or Bottom Plan.	Direction Y_2
5.	Right hand side view or Right hand end view or Right hand end elevation, etc.	Direction Z_1
6.	Left hand side view or left hand end view or left hand end elevation, etc.	Direction Z_2

Directions of sight and naming the views accordingly is common for both the systems of projections.

Now in the first angle projection system, front view taken from the front is drawn on the rear plane of the box. Rear view is drawn on the front plane of the box. Top view taken from the top is drawn on the bottom plane of the box. Bottom view taken from bottom is drawn on the top plane of the box. R.H.E.V. taken from right side is drawn on the left side plane of the box and L.H.E.V. taken from left side is drawn on the right side plane of the box. In other words in this system object remains in between picture plane and observer.

After drawing six views on six planes of the box, all planes of the box are opened about the plane which contains the front view. Front view is considered as the main and important view. F.V. in this system is on rear plane of the box and so all planes are opened around edges of rear plane. See Fig. 7.4 (ii).

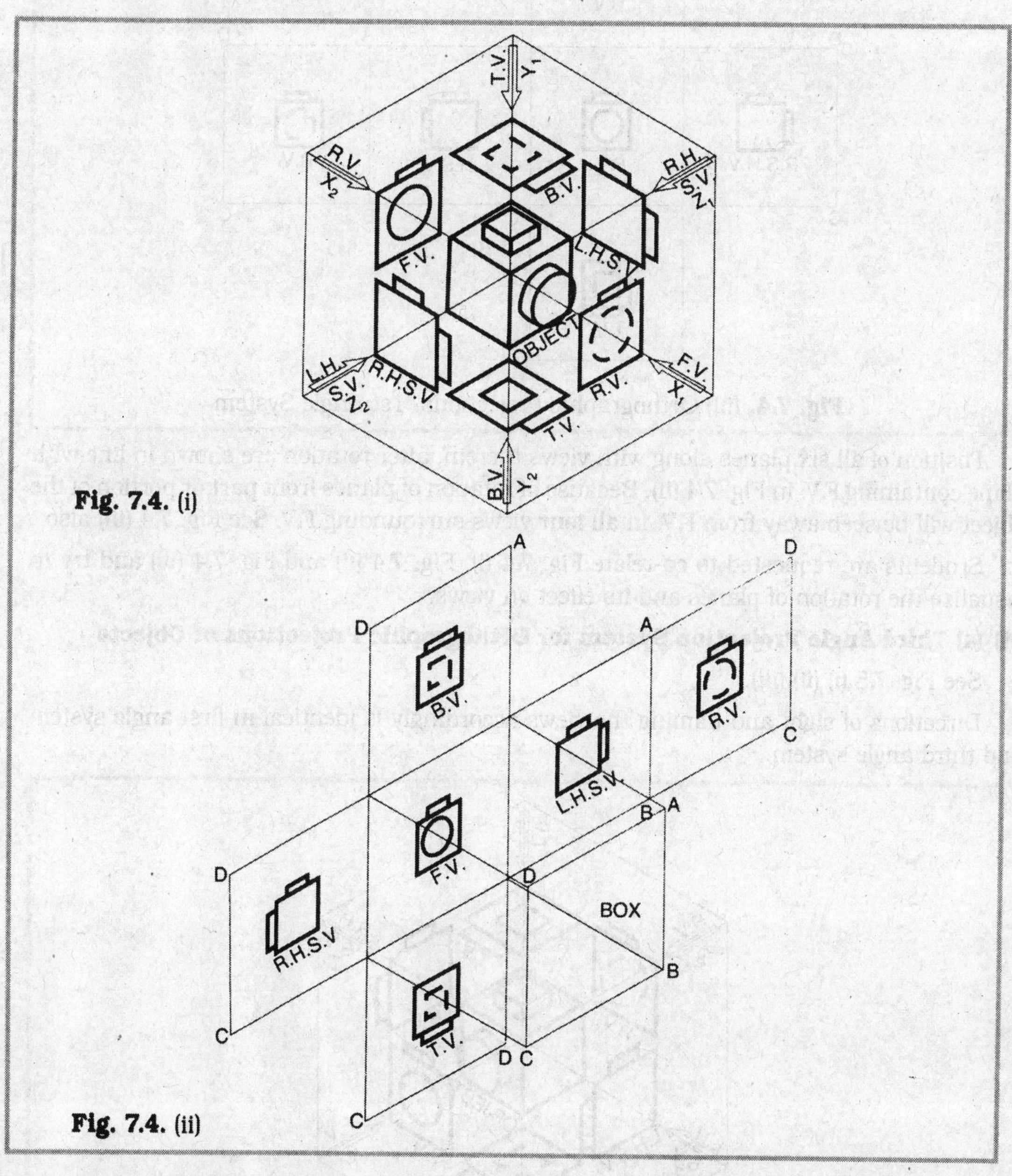

Fig. 7.4. (i)

Fig. 7.4. (ii)

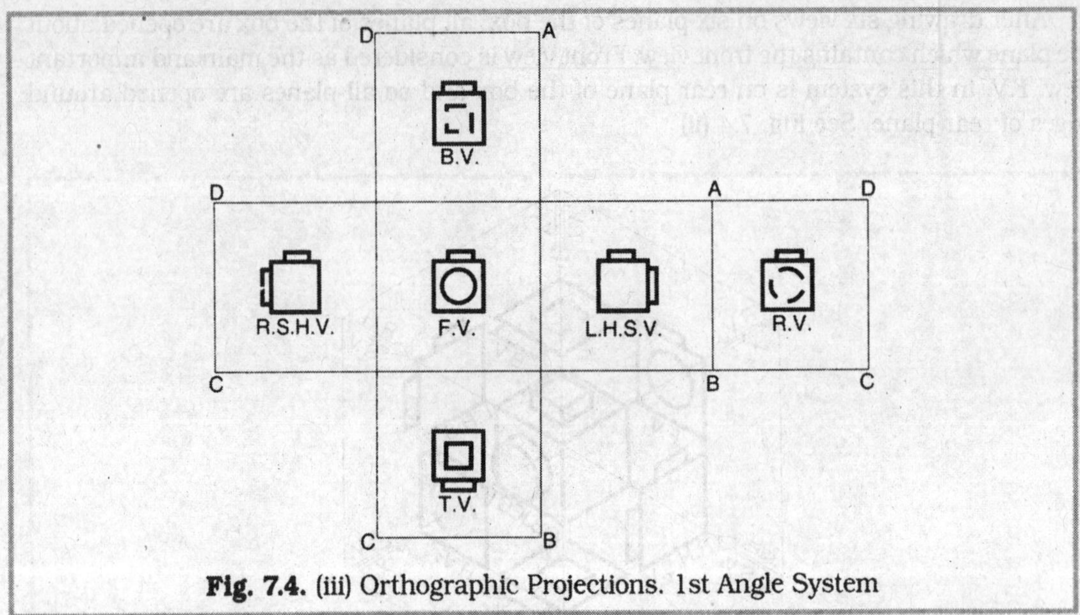

Fig. 7.4. (iii) Orthographic Projections. 1st Angle System

Position of all six planes along with views therein, after rotation are shown in line with plane containing F.V. in Fig. 7.4 (ii). Because of rotation of planes front part or portion of the object will be seen away from F.V. in all four views surrounding F.V. See Fig. 7.4 (iii) also.

Students are requested to co-relate Fig. 7.4 (i), Fig. 7.4 (ii) and Fig. 7.4 (iii) and try to visualize the rotation of planes and its effect on views.

(ii) (a) Third Angle Projection System for Orthographic Projections of Objects

See Fig. 7.5 (i) (ii) (iii).

Directions of sight and naming the views accordingly is identical in first angle system and third angle system.

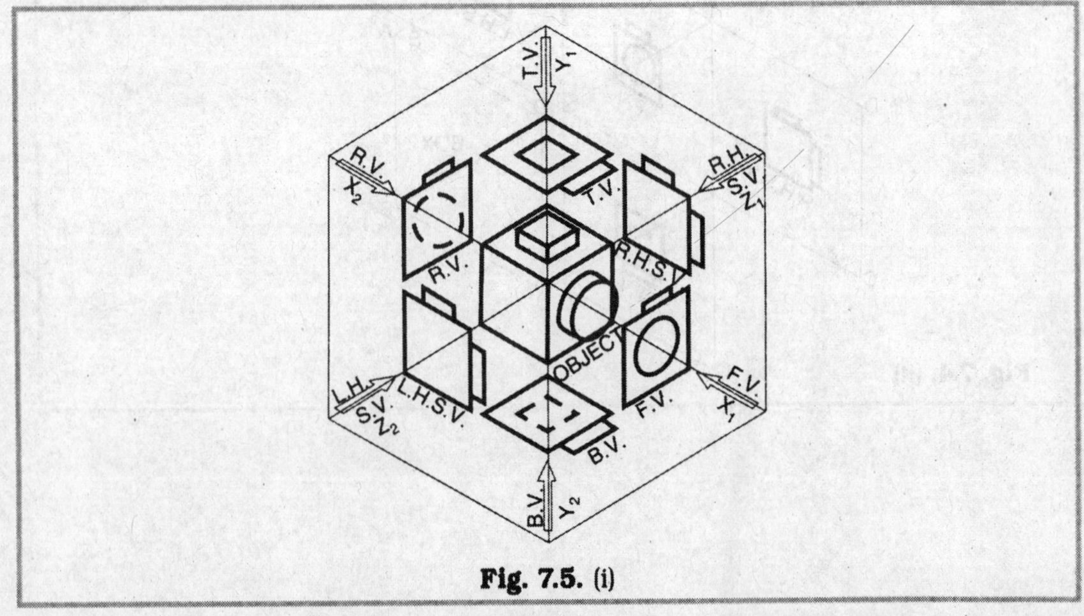

Fig. 7.5. (i)

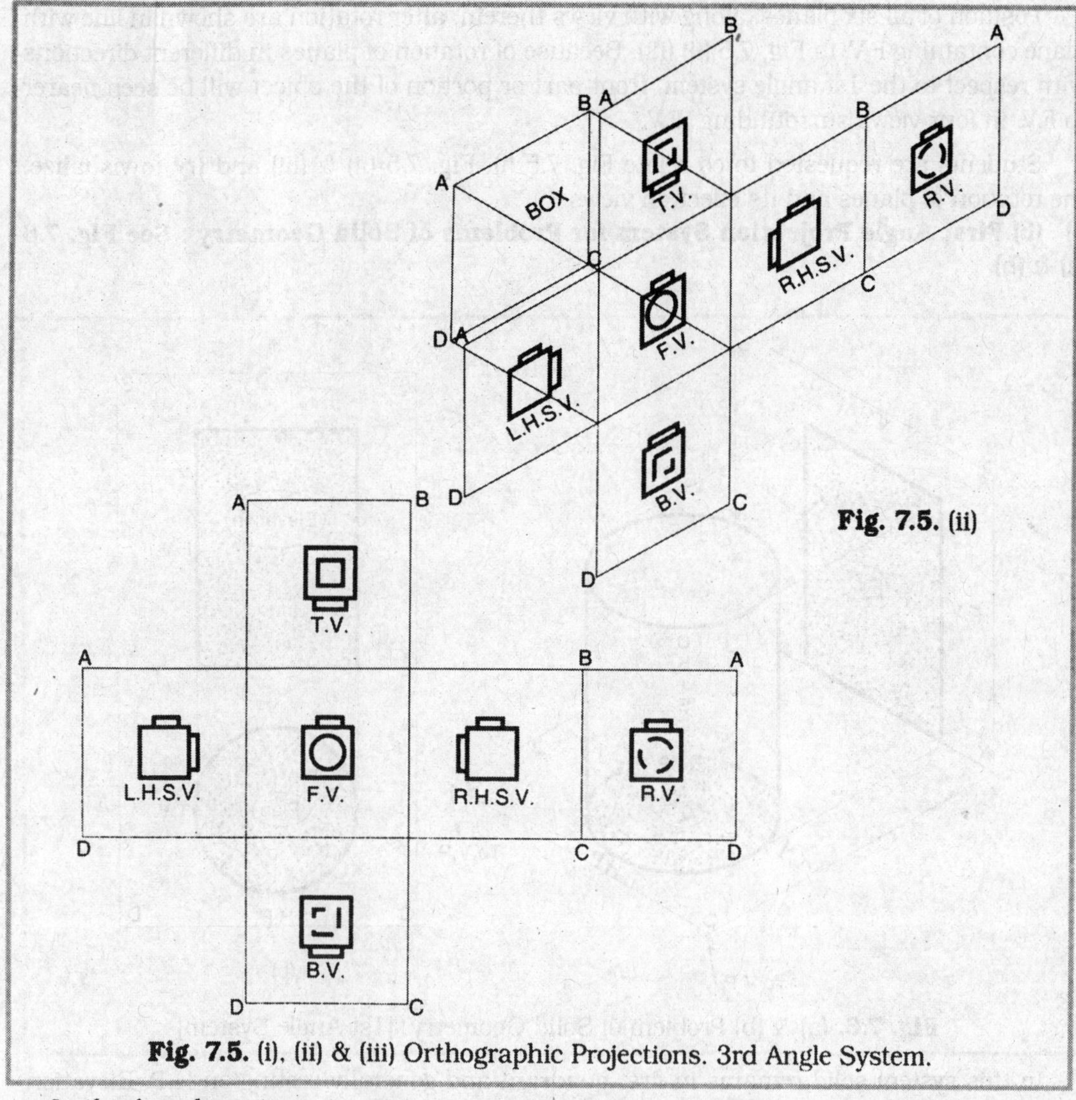

Fig. 7.5. (ii)

Fig. 7.5. (i), (ii) & (iii) Orthographic Projections. 3rd Angle System.

In third angle system :

1. F.V. taken from front is drawn on front plane of the box.
2. R.V. taken from rear is drawn on rear plane of the box.
3. T.V. taken from top is drawn on top plane of the box.
4. B.V. taken from bottom is drawn on bottom plane of the box.
5. R.H.S.V. taken from right side is drawn on right side plane of the box and
6. L.H.S.V. taken from left side is drawn on left side plane of the box.

In other words picture plane remains in between the object and observer.

After drawing six views on six planes of the box, all planes of the box are opened about the plane which contains front view. F.V. in this system is on front plane of the box and so all planes are opened around edges of front plane. See Fig. 7.5 (ii).

Position of all six planes, along with views therein, after rotation are shown in line with plane containing F.V. in Fig. 7.5 (ii) (iii). Because of rotation of planes in different directions with respect to the 1st angle system, front part or portion of the object will be seen nearer to F.V. in four views surrounding F.V.

Students are requested to co-relate Fig. 7.5 (i), Fig. 7.5 (ii) & (iii) and try to visualize, the rotation of planes and its effect on views.

(i) (b) First Angle Projection System for Problems of Solid Geometry : See Fig. 7.6 (a) & (b).

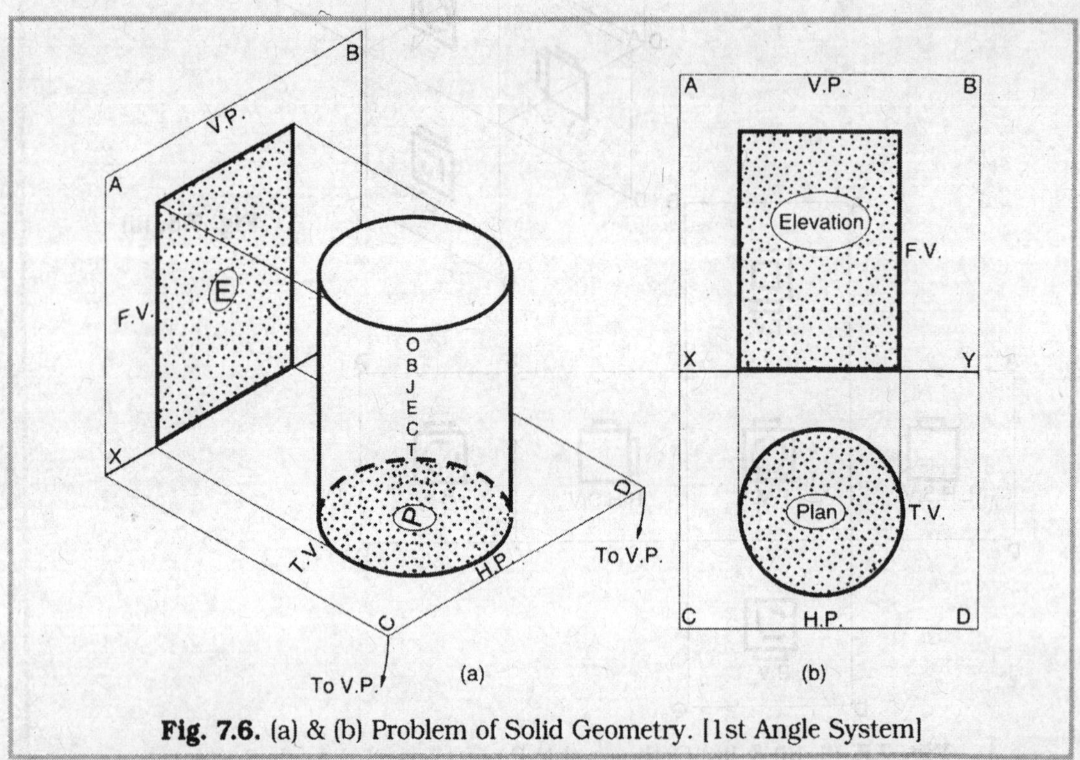

Fig. 7.6. (a) & (b) Problem of Solid Geometry. [1st Angle System]

In this system solid remains in first quadrant and generally resting on H.P. Elevation and plan of the solid are drawn on V.P.(Upper) and H.P. (Front) respectively. Afterwards H.P. (F) plane is given rotation of 90° as shown in Fig. 7.6 (a) to bring it in line with V.P.

In Fig. 7.6 (b) position of two planes V.P. and H.P. after rotation are shown along with projections, elevation and plan drawn therein respectively, XY line is generated by intersection of H.P. and V.P.

In short in this system plan is drawn below xy line and elevation is drawn above xy line. Ground line G.L. will be absent.

(ii) (b) Third Angle Projection System for Problems of Solid Geometry :

In this system solid remains in third quadrant and generally resting on ground. Elevation and plan of the solid are drawn on V.P. (lower) and H.P. (rear) respectively. Afterwards H.P. (R) plane is given rotation of 90°, as shown in Fig. 7.7 (a) to bring it in line with V.P.(L).

In Fig. 7.7 (b) position of two planes V.P.(L) and H.P.(R) after rotation are shown along with projections elevation and plan drawn therein respectively. XY line is generated by intersection of H.P. and V.P. and G.L. line is generated by intersection of ground and V.P.

In short in this system elevation is drawn between parallel G.L. line and xy line and plan is drawn above xy line as shown in Fig. 7.7 (b).

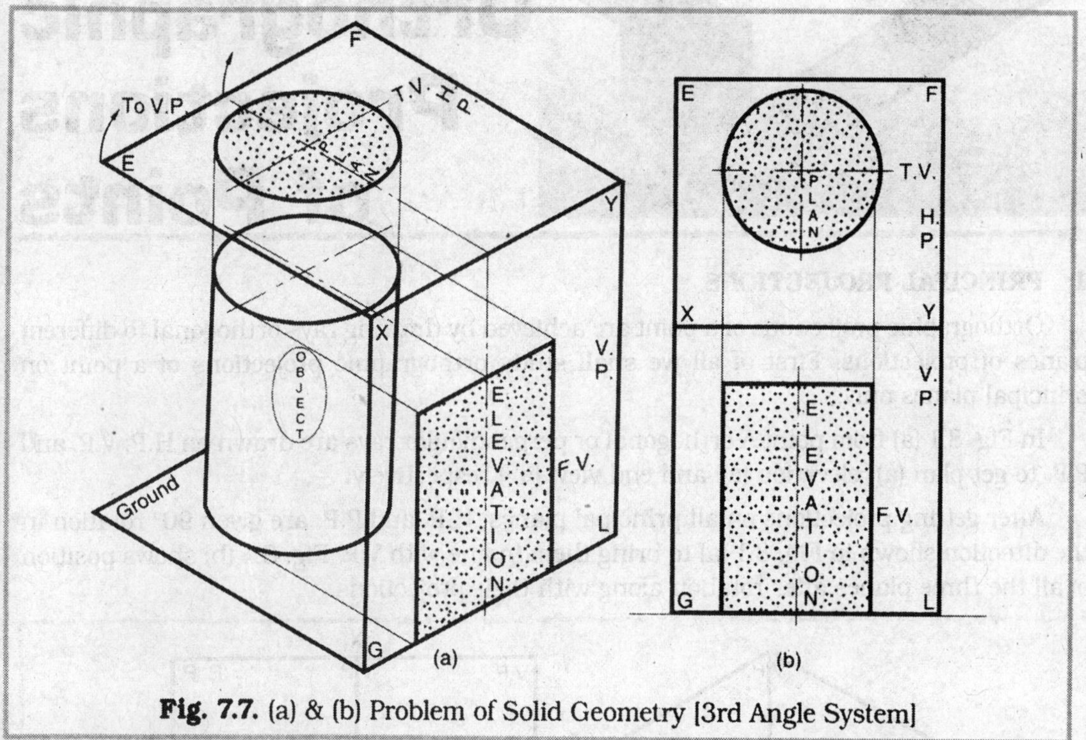

Fig. 7.7. (a) & (b) Problem of Solid Geometry [3rd Angle System]

EXERCISE

1. Name different types of projections you know. Explain the difference between them.
2. What is orthographic projection ? How it is named ?
3. Enlist clearly two systems of projections. Show clearly by sketch each one.
4. Explain clearly two systems of projections. How will you find out that particular projections given are in 1st angle system or in 3rd angle system ?
5. Explain with sketch how two systems of projections are applied in the solution of problems of solid geometry.
6. How the different projections are named ?
7. Explain box method used for orthographic projection.
8. Why orthographic projections are extensively used even though they are not three dimensional ?
9. Name three dimensional projections you know.

CHAPTER - 8

Orthographic Projections of Points

1. PRINCIPAL PROJECTIONS

Orthographic projections of a point are achieved by drawing rays orthogonal to different planes of projections. First of all we shall study orthographic projections of a point on principal planes only.

In Fig. 8.1 (a) from point A orthogonal or perpendicular rays are drawn on H.P; V.P. and P.P. to get plan (a); elevation (a') and end view (a'_1) respectively.

After getting projections on all principal planes, H.P. and P.P. are given 90° rotation in the direction shown in Fig. 8.1 (a) to bring them in line with V.P. Fig. 8.1 (b) shows position of all the three planes after rotation along with their projections.

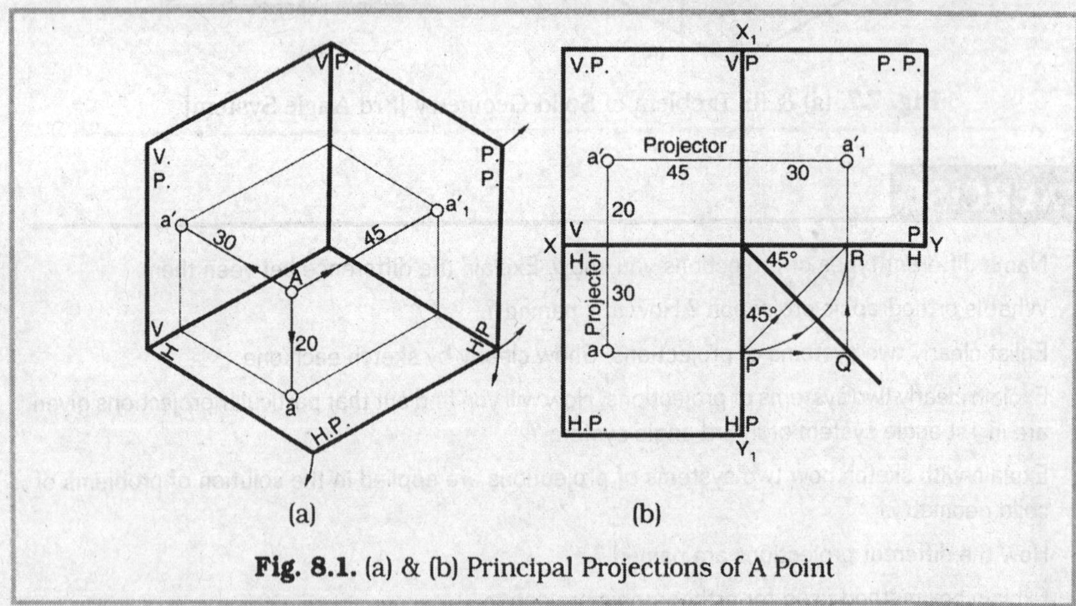

Fig. 8.1. (a) & (b) Principal Projections of A Point

It will be seen from Fig. 8.1 (a) and (b) that distance of point A from H.P.(20) is seen in elevation and end view from XY line or $\left[\frac{V}{H}\right]\left[\frac{P}{H}\right]$ line. Distance of the point A from V.P. (30)

is seen in plan from XY line or $\left[\frac{V}{H}\right]\left[\frac{P}{H}\right]$ line. Distance of the point A from P.P. (45) is seen in elevation and plan from X_1Y_1 line or $\left[\frac{V}{P}\right]\left[\frac{H}{P}\right]$ line.

Fig. 8.1 (b) indicates three methods of getting end view (a'_1) when elevation (a') and plan (a) are given. To get end view (a'_1) (1) draw parallel line to XY line from elevation a'. (2) Draw X_1Y_1 line perpendicular to XY line at suitable distance, here it is 45 mm (3) From plan (a) draw parallel line to XY line upto P on X_1Y_1 line. (4) Transfer P to R by 45° line PR or by quarter circle PR or by square PQR. (5) Draw parallel to X_1Y_1 from R to get a'_1 by intersection on line from a'. a'_1 is the required end view or end elevation.

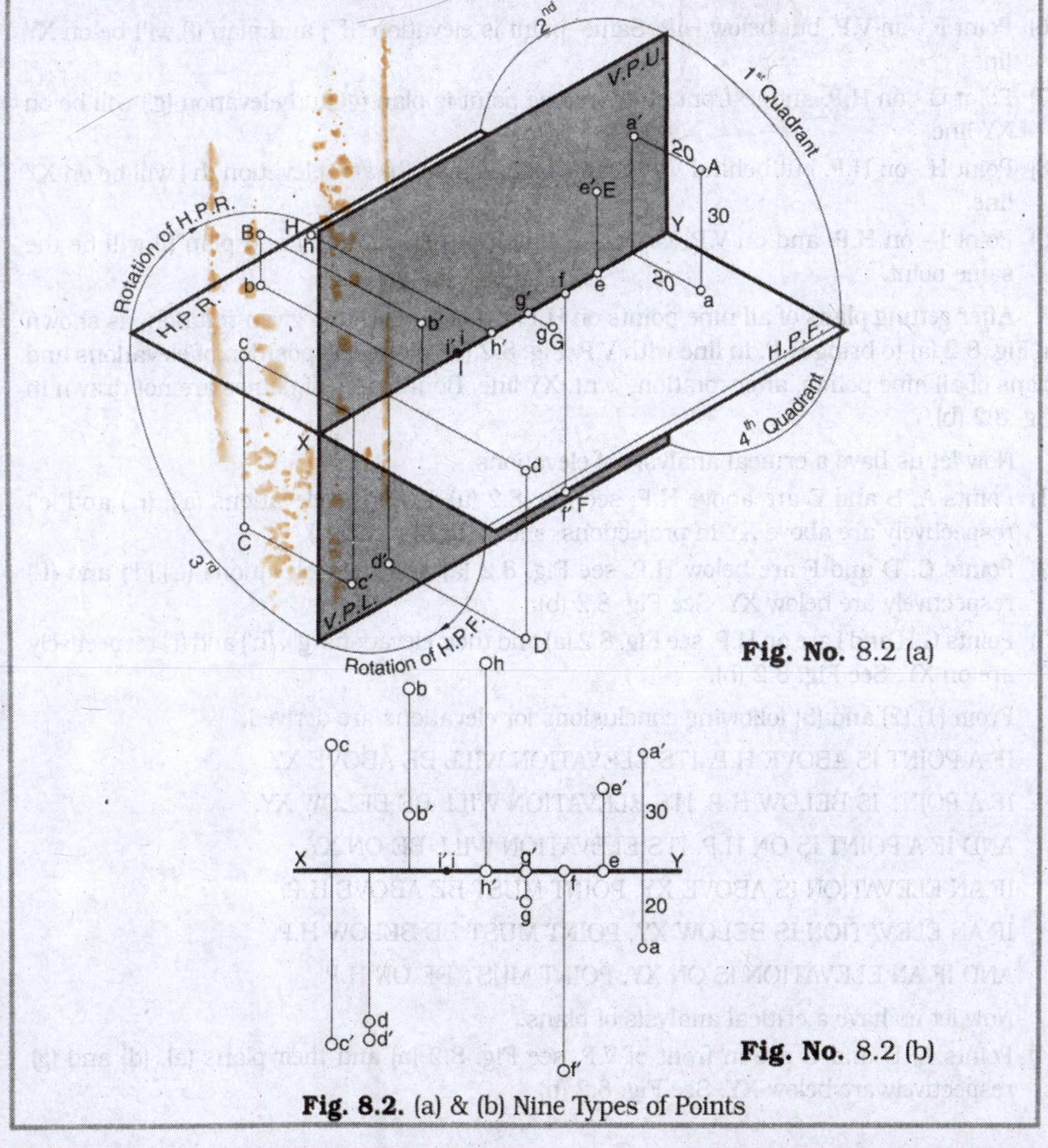

Fig. 8.2. (a) & (b) Nine Types of Points

If elevation and end view are given, plan can be drawn by following reverse method to above.

There are nine different types of points in space w.r.t H.P. and V.P. as shown in Fig. 8.2.

(1) Point A - 30 mm above H.P. and 20 mm in front of V.P. It is in 1st quadrant. (a') is elevation and (a) is plan.

(2) Point B - above H.P. and behind V.P. It is in the 2nd quadrant. (b') is elevation and (b) is plan.

(3) Point C - below H.P. and behind V.P. It is in the 3rd quadrant. (c') is elevation and (c) is plan.

(4) Point D - below H.P. and in front of V.P. It is in the 4th quadrant. (d') is elevation and (d) is plan.

(5) Point E - On V.P. and above H.P. Same point is elevation (e') and plan (e) will be on XY line.

(6) Point F - on V.P. but below H.P. Same point is elevation (f') and plan (f) will be on XY line.

(7) Point G - on H.P. and in front of V.P. Same point is plan (g) and elevation (g') will be on XY line.

(8) Point H - on H.P. but behind V.P. Same point is plan (h) and elevation (h') will be on XY line.

(9) Point I - on H.P. and on V.P. i.e. on XY line. Both elevation (i') and plan (i) will be the same point.

After getting plans of all nine points on H.P. (front & rear) it is given rotation, as shown in Fig. 8.2 (a) to bring H.P. in line with V.P. Fig. 8.2 (b) shows the position of elevations and plans of all nine points, after rotation, w.r.t. XY line. Boundaries of planes are not drawn in Fig. 8.2 (b).

Now let us have a critical analysis of elevations.

(1) Points A, B and E are above H.P. see Fig. 8.2 (a) and their elevations (a'), (b') and (e') respectively are above XY in projections, shown in Fig. 8.2 (b).

(2) Points C, D and F are below H.P. see Fig. 8.2 (a) and their elevations (c'),(d') and (f') respectively are below XY. See Fig. 8.2 (b).

(3) Points G,H and I are on H.P. see Fig. 8.2 (a) and their elevations (g'), (h') and (i') respectively are on XY. See Fig. 8.2 (b).

From (1),(2) and (3) following conclusions for elevations are derived;

IF A POINT IS ABOVE H.P. ITS ELEVATION WILL BE ABOVE XY.

IF A POINT IS BELOW H.P. ITS ELEVATION WILL BE BELOW XY.

AND IF A POINT IS ON H.P. ITS ELEVATION WILL BE ON XY.

IF AN ELEVATION IS ABOVE XY, POINT MUST BE ABOVE H.P.

IF AN ELEVATION IS BELOW XY, POINT MUST BE BELOW H.P.

AND IF AN ELEVATION IS ON XY, POINT MUST BE ON H.P.

Now let us have a critical analysis of plans.

(4) Points A, D and G are in front of V.P. see Fig. 8.2 (a) and their plans (a), (d) and (g) respectively are below XY. See Fig. 8.2 (b).

(5) Points B, C and H are behind V.P. see Fig. 8.2 (a) and their plans (b), (c) and (h) respectively are above XY. See Fig. 8.2 (b).

(6) Points E, F and I are on V.P. see Fig. 8.2 (a) and their plans (e), (f) and (i) respectively are on XY. See Fig. 8.2 (b).

From (4), (5) and (6) following conclusions for plans are derived;

IF A POINT IS IN FRONT OF V.P. ITS PLAN WILL BE BELOW XY.

IF A POINT IS BEHIND V.P. ITS PLAN WILL BE ABOVE XY, AND

IF A POINT IS ON V.P. ITS PLAN WILL BE ON XY.

IF A PLAN IS BELOW XY, POINT MUST BE IN FRONT OF V.P.

IF A PLAN IS ABOVE XY, POINT MUST BE BEHIND V.P. AND

IF A PLAN IS ON XY, POINT MUST BE ON V.P.

From the projections of point A, it is concluded that -

DISTANCE OF ELEVATION FROM XY IS EQUAL TO THE DISTANCE OF POINT FROM H.P. AND DISTANCE OF PLAN FROM XY IS EQUAL TO THE DISTANCE OF POINT FROM V.P.

See point A in Fig. 8.2 (a) and (b) for understanding above conclusion.

2. AUXILIARY PROJECTIONS ON A.V.P.

Now we shall study orthographic projections of a point on auxiliary planes i.e. on A.V.P. and A.I.P.

In Fig. 8.3 (a) three planes H.P; V.P; and A.V.P. are shown. A.V.P. is perpendicular to H.P and inclined to V.P. by an angle Ø.

Line along which A.V.P. meets the H.P. is known as horizontal trace (H.T.). It is shown in Fig. 8.3 (a) by H.T. Similarly line along which A.V.P. meets the V.P. is known as vertical trace (V.T.) and is shown in Fig. 8.3 (a) by V.T.

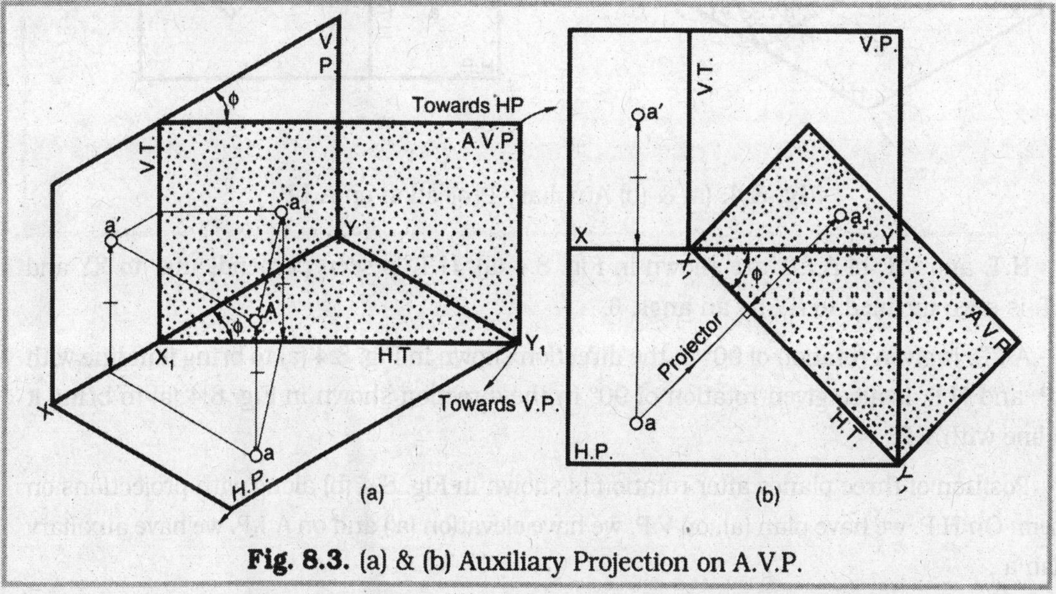

Fig. 8.3. (a) & (b) Auxiliary Projection on A.V.P.

V.T. is seen perpendicular to XY and H.T. is seen inclined to XY by an angle Ø.

H.P. is given rotation of 90° in the direction shown in Fig. 8.3 (a) to bring it in line with V.P. and A.V.P. is also given rotation of 90° in the direction shown in Fig. 8.3 (a) to bring it in line with H.P. and V.P. both.

Position of three planes after rotation is shown in Fig. 8.3 (b) along with projections on them. On H.P. we have plan (a), on V.P. we have elevation a' and on A.V.P. we have auxiliary elevation a'_1.

It is clear from Fig. 8.3 (a) and (b) that to get auxiliary elevation, plan is required to be projected on new X_1Y_1 line, which is nothing but H.T. of A.V.P. drawn at an angle Ø with XY. On projector, from plan (a), on X_1Y_1 take distance of elevation from XY and mark from X_1Y_1 to get a'_1, as shown in Fig. 8.3 (b).

3. AUXILIARY PROJECTIONS ON A.I.P.

Similarly, in Fig. 8.4 (a) three planes H.P; V.P. and A.I.P. are shown, A.I.P. is perpendicular to V.P. and inclined to H.P. by an angle θ.

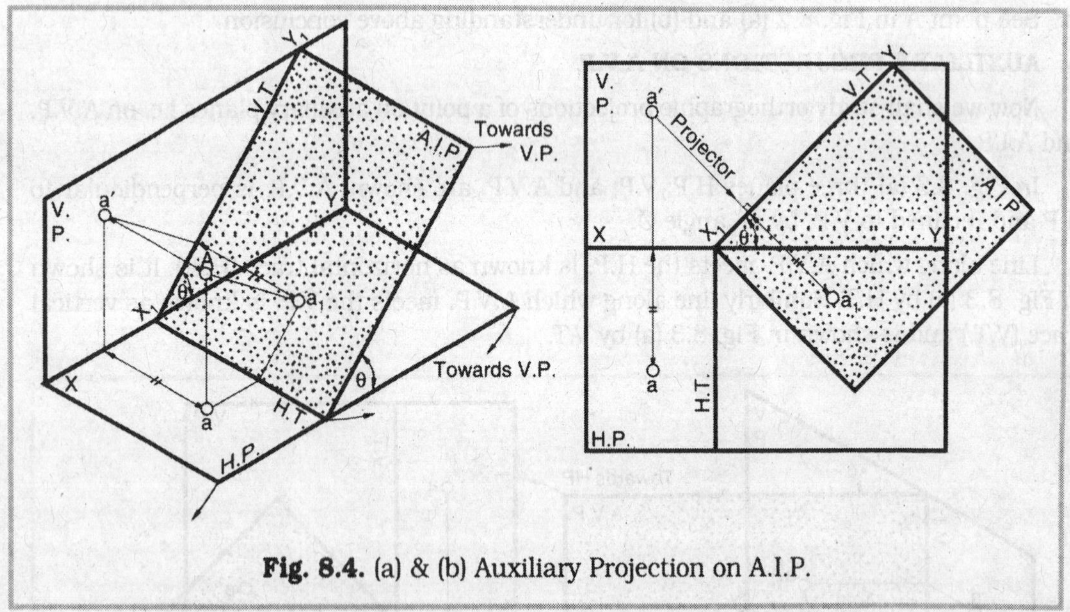

Fig. 8.4. (a) & (b) Auxiliary Projection on A.I.P.

H.T. and V.T. of A.I.P. are shown in Fig. 8.4 (a). H.T. is seen perpendicular to XY and V.T. is seen inclined to XY by an angle θ.

A.I.P. is given rotation of 90° in the direction shown in Fig. 8.4 (a) to bring it in line with V.P. and H.P. is also given rotation of 90° in the direction shown in Fig. 8.4 (a) to bring it in line with V.P.

Position of three planes after rotation is shown in Fig. 8.4 (b) along with projections on them. On H.P. we have plan (a), on V.P. we have elevation (a') and on A.I.P. we have auxiliary plan a_1.

It is clear from Fig. 8.4 (a) and (b) that to get auxiliary plan a_1, elevation a' is required to be projected on new X_1Y_1 line, which is nothing but V.T. of A.I.P. drawn at an angle θ with XY. On projector from elevation (a') on X_1Y_1 take distance of plan from XY and mark from X_1Y_1 to get a_1, as shown in Fig. 8.4 (b).

Boundaries of planes are shown just for understanding but in actual practice boundaries are not drawn.

4. PRACTICE PROBLEMS

Students are requested to keep conclusions ready on hand while solving problems.

Problem 1 : Draw the projections of the following points on the same x - y line.

(i) Point A in V.P. 30 mm below H.P.

(ii) Point B in H.P. 20 mm in front of V.P.

(iii) Point C 20 mm above H.P. and 20 mm behind V.P.

(iv) Point D 25 mm below H.P. and 40 mm behind V.P.

(v) Point E on H.P. and on V.P.

(vi) Point F 40 mm above H.P. and 10 mm in front of V.P.

(vii) Point G on V.P. 35 mm above H.P.

For solution see Fig. 8.5.

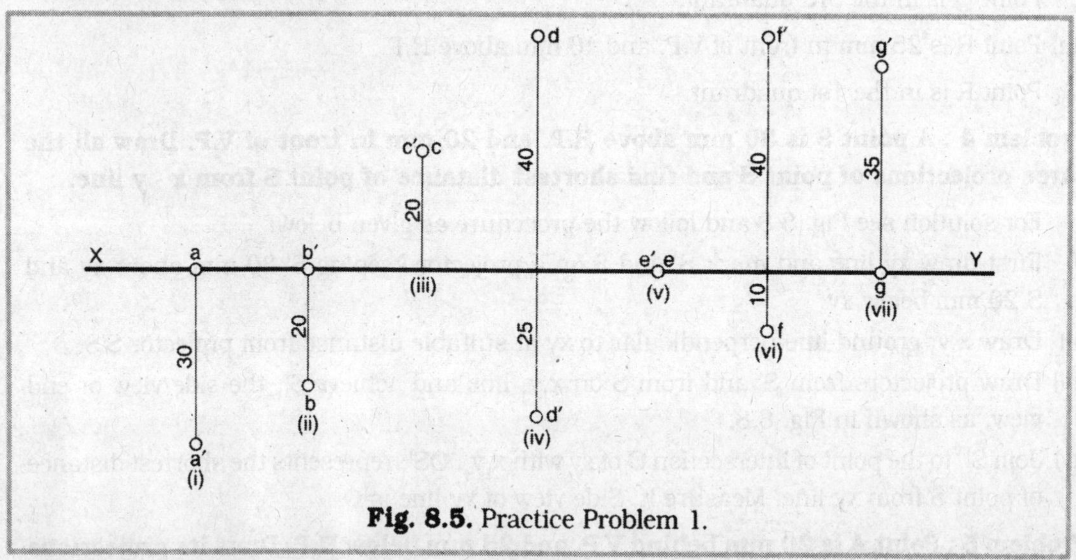

Fig. 8.5. Practice Problem 1.

Problem 2 : A point R is 25 m.m. from both the reference planes in all possible positions. Draw projections for all positions.

For solution see Fig. 8.6.

There are four possible positions of point R in 4 different quadrants and solution corresponding to each quadrant is given in Fig. 8.6 as R_1, R_2, R_3 and R_4.

Problem 3 : Projections of the points P,Q and R are given in Fig 8.7. State their positions w.r.t H.P. and V.P. and state their quadrants.

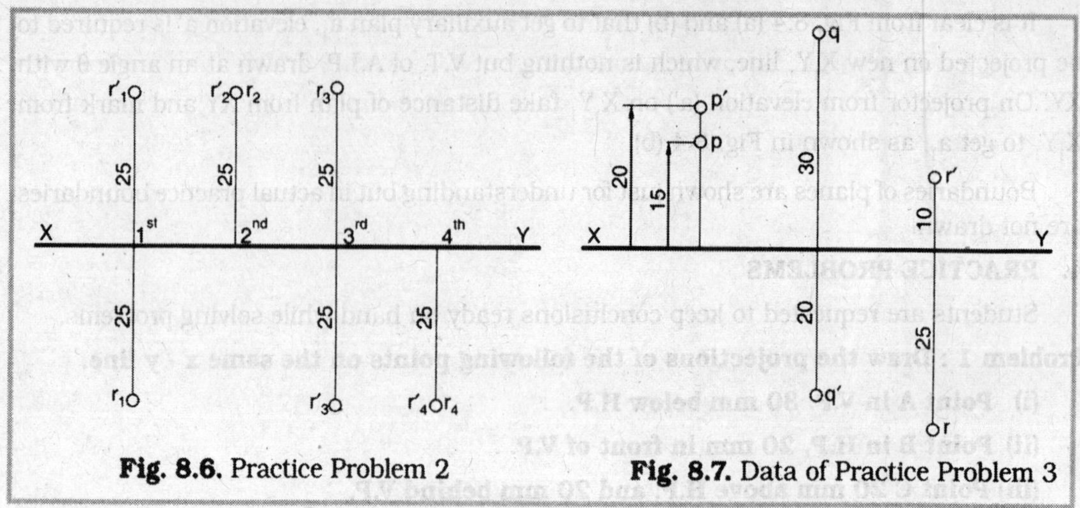

Fig. 8.6. Practice Problem 2 **Fig. 8.7.** Data of Practice Problem 3

Answer

(i) Point P is 20 m.m. above H.P. and 15 mm behind V.P.

Point P is in the 2nd quadrant.

(ii) Point Q is 30 mm behind V.P. and 20 mm below H.P.

Point Q is in the 3rd quadrant.

(iii) Point R is 25 mm in front of V.P. and 10 mm above H.P.

Point R is in the 1st quadrant.

Problem 4 : A point S is 30 mm above H.P. and 20 mm in front of V.P. Draw all the three projections of point S and find shortest distance of point S from x - y line.

For solution see Fig. 8.8 and follow the procedure as given below:

(i) First draw xy line and mark S' and S on a projector keeping S' 30 mm above xy and S 20 mm below xy.

(ii) Draw x_1y_1 ground line perpendicular to xy at suitable distance from projector S'S.

(iii) Draw projectors from S' and from S on x_1y_1 line and achieve S'_1 the side view or end view, as shown in Fig. 8.8.

(iv) Join S'_1 to the point of intersection O of xy with x_1y_1. OS'_1 represents the shortest distance of point S from xy line. Measure it. Side view of xy line is O.

Problem 5 : Point A is 20 mm behind V.P. and 25 mm below H.P. Draw its projections and project its auxiliary elevation on A.V.P. which makes 60° with V.P.

For solution see Fig. 8.9 and follow the procedure as given below :

(i) First draw plan (a) and elevation (a') of point A.

(ii) Draw x_1y_1 ground line inclined to xy at 60° angle and draw projector on it from plan point (a) as shown in Fig. On this projector mark distance of (a') from xy (25) beyond x_1y_1 and mark (a'_1). Then (a'_1) is an auxiliary elevation of point A.

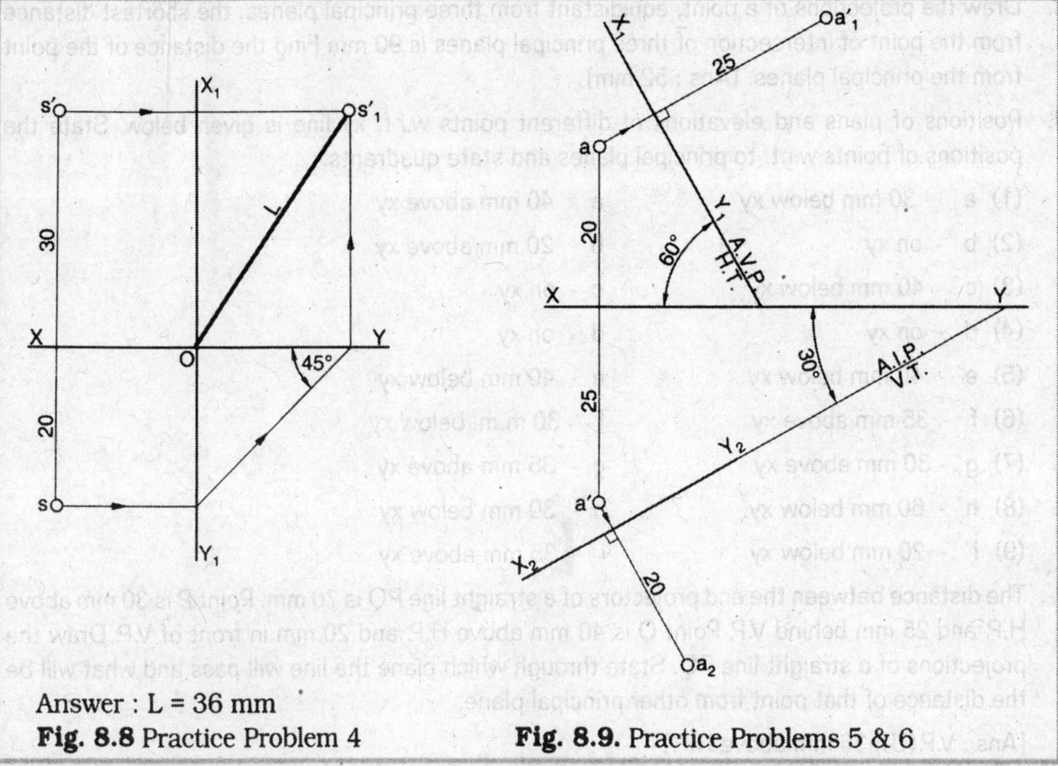

Answer : L = 36 mm

Fig. 8.8 Practice Problem 4

Fig. 8.9. Practice Problems 5 & 6

Problem 6 : Draw auxiliary plan of point A of problem 5, on an auxiliary inclined plane (A.I.P) inclined to H.P. by 30°.

For solution see Fig. 8.9 and follow the procedure as given below:

(i) Draw x_2y_2 ground line inclined to xy at 30° angle and draw projector on it from elevation point (a'), as shown in figure. On this projector mark distance of plan (a) from xy (20) beyond x_2y_2 and mark (a_2). (a_2) is an auxiliary plan of point A.

EXERCISE

1. Draw the projections of points, positions of which are given below :

(1) A point 'A' 30 mm above H.P. and 30 mm behind V.P.

(2) A point 'B' 35 mm below H.P; 25 mm behind V.P. and 20 mm behind the right side profile plane (P.P).

(3) A point 'C' on H.P.; on V.P. and on P.P.

(4) A point 'D' 40 mm below H.P. and 40 mm in front of V.P.

(5) A point 'E' on H.P. and 40 mm in front of V.P.

(6) A point 'F' on V.P; 25 mm above H.P. and 40 mm in front of right side P.P.

(7) A point 'G' is 50 mm from H.P.; V.P. Draw its projections in all possible positions.

2. Draw the projections of a point 'P' 20 mm above H.P. and in 1st quadrant if its shortest distance from xy line is 40 mm Find the distance of point P from V.P. [Ans : 34.5 m.m.]

3. Draw the projections of a point, equidistant from three principal planes, the shortest distance from the point of intersection of three principal planes is 90 mm Find the distance of the point from the principal planes. [Ans : 52 mm]

4. Positions of plans and elevations of different points w.r.t. xy line is given below. State the positions of points w.r.t. to principal planes and state quadrants.

(1) a′ - 30 mm below xy	a - 40 mm above xy
(2) b′ - on xy	b - 20 mm above xy
(3) c′ - 40 mm below xy	c - on xy
(4) d′ - on xy	d - on xy
(5) e′ - 40 mm below xy	e - 40 mm below xy
(6) f′ - 35 mm above xy	f - 30 m.m. below xy
(7) g′ - 30 mm above xy	g - 35 mm above xy
(8) h′ - 60 mm below xy	h - 30 mm below xy
(9) i′ - 20 mm below xy	i - 35 mm above xy

5. The distance between the end projectors of a straight line PQ is 70 mm. Point P is 30 mm above H.P. and 25 mm behind V.P. Point Q is 40 mm above H.P. and 20 mm in front of V.P. Draw the projections of a straight line PQ. State through which plane the line will pass and what will be the distance of that point from other principal plane.

[Ans : V.P.(U); 36 mm above H.P.]

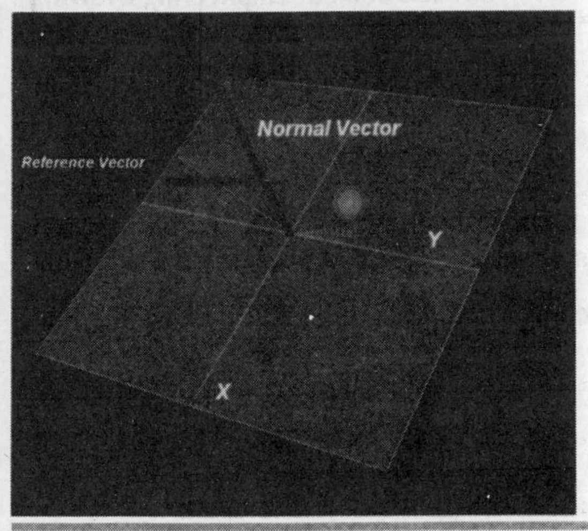

CHAPTER - 9

Projections of Straight Lines

1. INTRODUCTION

Lines in space are of three types considering their position with respect to principal planes :

1. Line parallel to two and perpendicular to one of the principal planes.
2. Line parallel to one and inclined to two principal planes.
3. Line inclined to all the three principal planes.

To study the projections of straight lines one must study five important theories dealt with in this chapter and understand the conclusions/discussions, etc. given at the end of theories. Theory No. 3 and Theory No. 5 are most important theories. Most of the problems on straight lines are based on the above two theories.

Projections of straight lines have nothing to do with 1st angle or 3rd angle system of projection. Projections of straight lines are drawn according to the positions of the end points of the straight lines w.r.t. H.P. and V.P. Therefore, three dimensional drawings are given in each theory with straight line drawn in 1st quadrant. As it is easy to understand the drawing of the first quadrant, three dimensional drawings are drawn with lines taken in 1st quadrant. Conclusion/discussion given at the end of each theory is also true for a line in any quadrant.

2. IMPORTANT THEORY - 1

Line parallel to two P.P.s and perpendicular to one principal plane :-

See Fig. 9.1 (a) and (b).

In Fig. 9.1 (a) we have lines AB, CD and EF perpendicular to H.P; V.P. and P.P. respectively. It is clear from Fig. 9.1(a) and (b) (i) that when line (AB) is perpendicular to H.P., plan (ab) is a point view and two other views i.e. elevation (a'b') and end view (a_1' b_1') are having true length and parallel to ground line generated by V.P. and P.P. i.e. X_1Y_1 or $\left[\frac{V}{P}\right]$ line.

Similarly, when line (CD), as shown in Fig. 9.1 (a), (b) (ii), is perpendicular to V.P., elevation (c'd') is a point view and two other views i.e. plan (cd) and end view (c_1' d_1') are having true length and parallel to ground line generated by H.P. and P.P. i.e. X_1Y_1/XY or $\left[\frac{H}{P}\right]$ line.

Similarly, when line (EF), as shown in Fig. 9.1 (a) (b) (iii), is perpendicular to P.P., end view (e_1' f_1') is a point view and other two views i.e. elevation (e'f') and plan (ef) are having true length and parallel to ground line generated by V.P. and H.P. i.e. XY line or $\left[\frac{V}{H}\right]$ line.

Conclusion

(1) WHEN A LINE IS PERPENDICULAR TO ONE OF THE PRINCIPAL PLANES IT IS PARALLEL TO TWO OTHER PRINCIPAL PLANES.

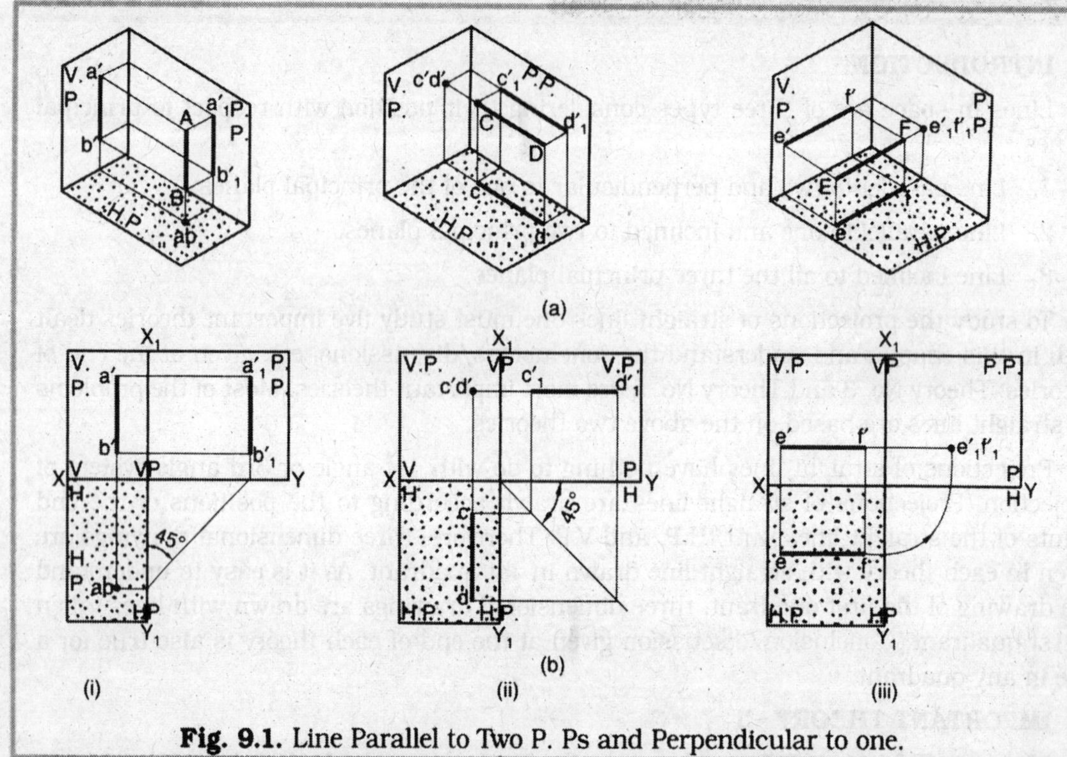

Fig. 9.1. Line Parallel to Two P. Ps and Perpendicular to one.

(2) PROJECTION OF A LINE ON A PLANE TO WHICH IT IS PERPENDICULAR IS A POINT VIEW AND PROJECTIONS ON OTHER TWO PRINCIPAL PLANES ARE OF TRUE LENGTH AND PARALLEL TO RESPECTIVE GROUND LINE. ALL CONDITIONS GO TOGETHER.

(3) ABOVE TWO CONDITIONS GO TOGETHER.

Having done the study of theory of straight line perpendicular to one of the principal planes and having gone through the conclusions and discussions we shall now solve few problems based on this theory.

Problem 1 : A line PJ, 50 mm long, is perpendicular to H.P. and it is below H.P. Point P is on H.P. and 30 mm behind V.P. Draw the projections of the line PJ.

For solution see Fig. 9.2 and follow the procedure as given below :

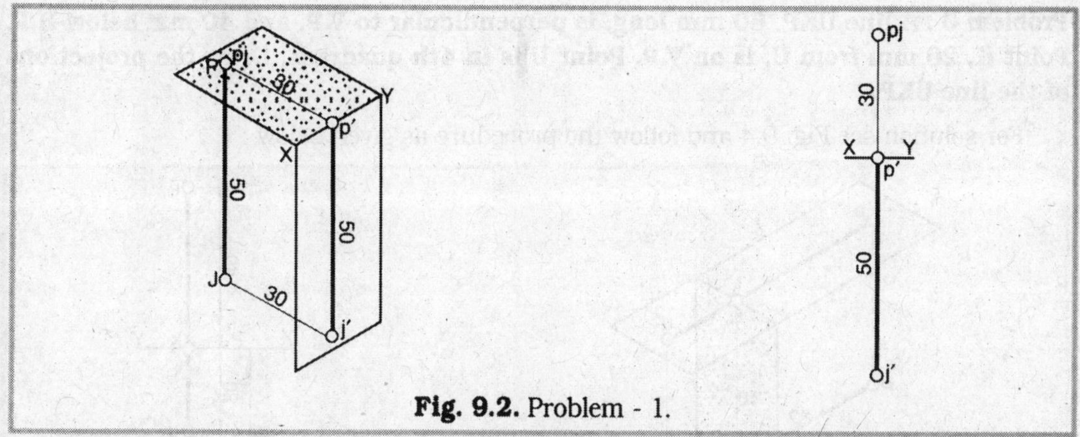

Fig. 9.2. Problem - 1.

Three dimensional drawing is given for understanding the location of the straight line w.r.t. planes. It is éxpected that it will also develop three dimensional thinking of the students.

(i) First of all mark the position of p and p'. As the point P is on H.P., p' will be on xy line and as it is 30 mm behind V.P., its plan p will be 30 mm above xy line.

(ii) As the line is perpendicular to H.P., its plan will be point view and it's elevation will be true length (50 mm) and perpendicular to xy line. So draw elevation p'j' of 50 mm length perpendicular to xy and below it and mark j at p only.

Problem 2 : A line MJ, 35 mm long, is perpendicular to the profile plane. The end M is 20 mm below H.P., 30 mm behind V.P. and 10 mm to the left of P.P. Draw all the three principal projections of the straight line MJ.

For solution see Fig. 9.3 and follow the procedure as given below:

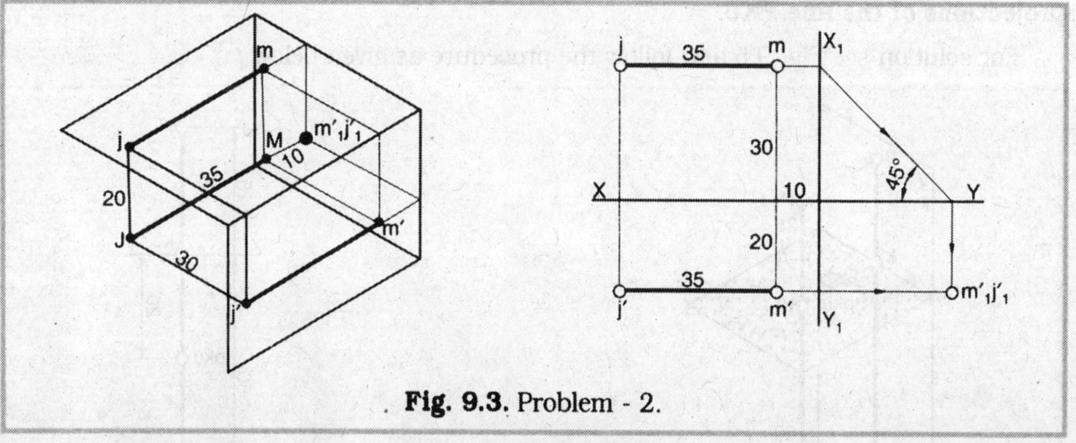

Fig. 9.3. Problem - 2.

(i) First draw xy and x_1y_1 lines and mark positions of projections m, m' and m'_1 of point M as per the given data. Follow the procedure of drawing end view, as studied earlier.

(ii) As the line is perpendicular to P.P., its end view will be point view and its elevation and plan will be true length (35) and parallel to xy line or perpendicular to x_1y_1 line. So draw jm and j'm' parallel to xy and of 35 mm length, as shown in Fig.9.3.

(iii) Point m'_1 and j'_1 will be the same point since end view is going to be point view.

Problem 3 : A line UKP, 60 mm long, is perpendicular to V.P. and 40 mm below H.P. Point K, 20 mm from U, is on V.P. Point U is in 4th quadrant. Draw the projections of the line UKP.

For solution see Fig. 9.4 and follow the procedure as given below :

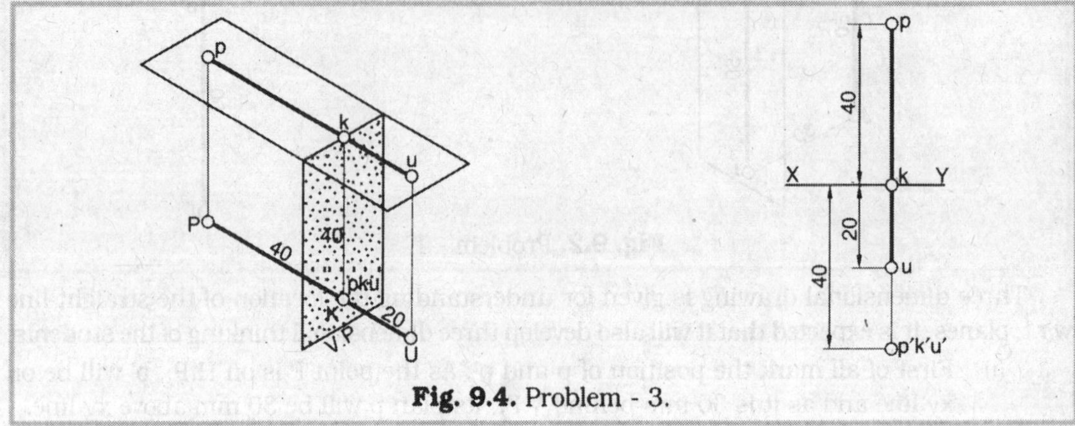

Fig. 9.4. Problem - 3.

(i) Elevation will be point view and plan will be perpendicular to xy line and will be of true length (60). So mark elevation p'k'u' 40 mm below xy line, since the line is given 40 mm below H.P.

(ii) As the point K is on V.P. mark plan point k on xy line. Mark u 20 mm below xy line and point p 40 mm above xy line on a perpendicular line to xy through k.

Problem 4 : A line PKU, 75 mm long, is perpendicular to H.P. Point P 20 mm away from V.P., is in the first quadrant, and point K is on H.P. PK is 25 mm long. Draw the projections of the line PKU.

For solution see Fig. 9.5 and follow the procedure as given below :

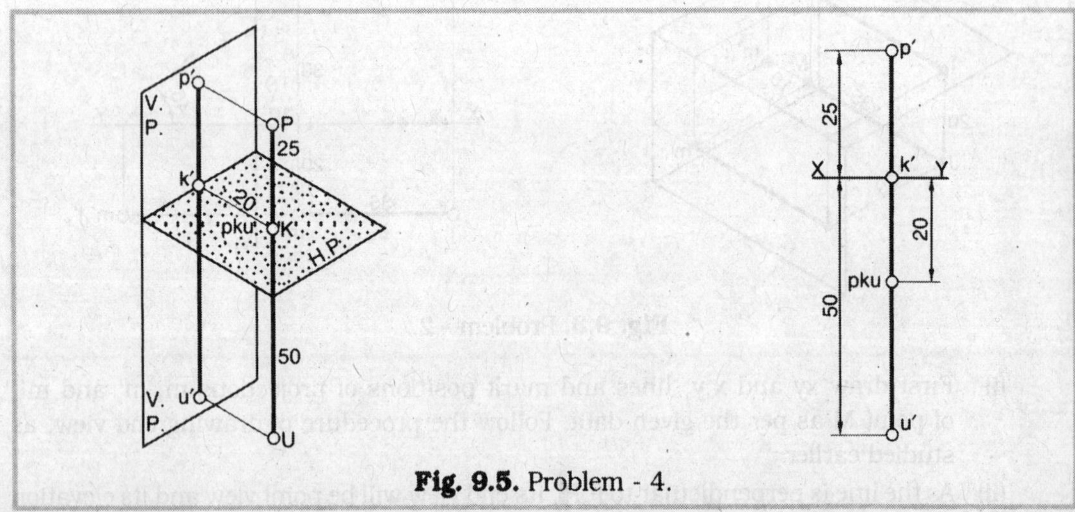

Fig. 9.5. Problem - 4.

Here plan will be point view and elevation will be true length (75) and perpendicular to xy line since the line is perpendicular to H.P. Follow the procedure similar to Problem 3.

3. IMPORTANT THEORY - 2

Line parallel to one and inclined to two other principal planes.

Important Notations :-

(1) θ - Inclination of line with H.P.

(2) Ø - Inclination of line with V.P.

(3) ψ - Inclination of line with P.P. or A.V.P.

In Fig. 9.6 (a) we have lines AB, CD and EF parallel to H.P., V.P. and P.P. respectively.

(i) Line AB is parallel to H.P. and inclined to V.P. and P.P. by an angle Ø and ψ respectively. It is seen from Fig. 9.6 (a) & (b)(i) that plan of AB i.e. ab is going to show true length and true inclination Ø and ψ with corresponding ground line. Its elevation a'b' and end view $a'_1b'_1$ will be less than true length and will be parallel to corresponding ground line Further it is clear from the figure that (Ø+ψ) = 90°.

(ii) Similarly, line CD is parallel to V.P. and inclined to H.P. and P.P. by an angle θ and ψ respectively. It is seen from Fig. 9.6 (a) and (b) (ii) that elevation of CD i.e. c'd' is going to show true length and true inclination θ and ψ with corresponding ground lines. Its plan cd and end view c'_1 d'_1 will be less than true length and will be parallel to corresponding ground line. Here (θ+ψ) = 90°.

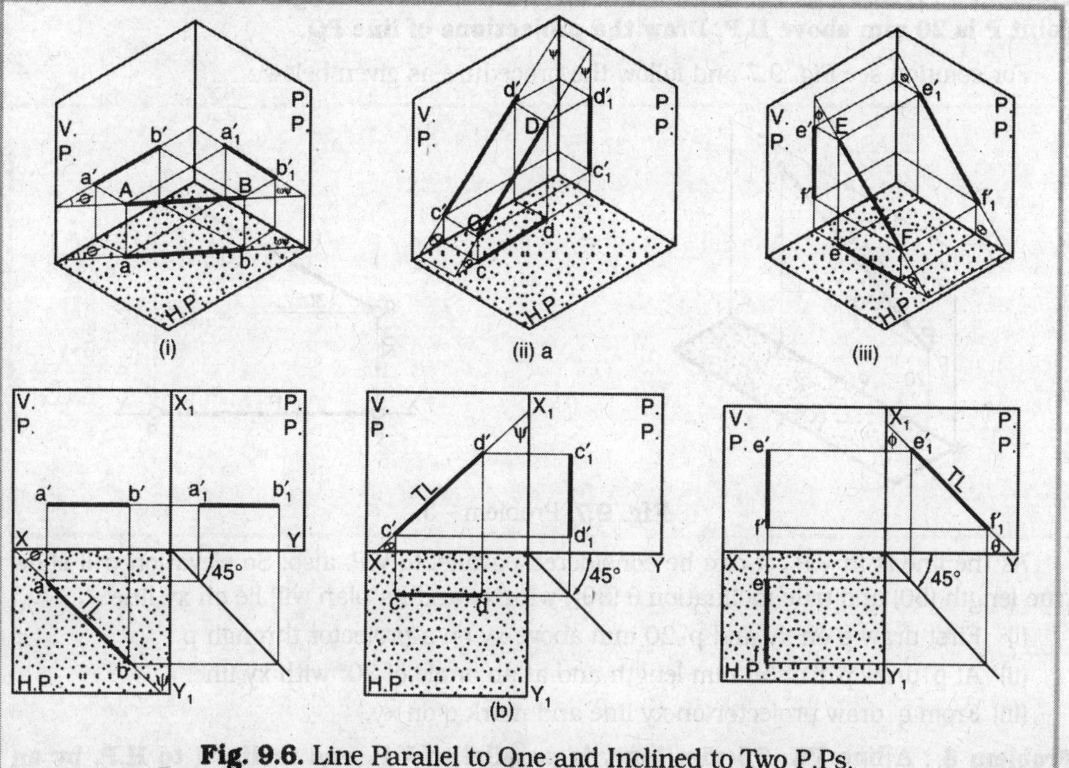

Fig. 9.6. Line Parallel to One and Inclined to Two P.Ps.

(iii) Similarly, line EF is parallel to P.P. and inclined to H.P. and V.P. by an angle θ and Ø respectively. It is seen from Fig. 9.6 (a) and (b) (iii) that end view $e'_1f'_1$ is going to show true length and true inclination θ and Ø with corresponding ground line. Its plan ef and elevation e'f' will be less than true length and will be parallel to corresponding ground line. Here (θ+Ø) = 90°. Further it is seen that, elevation and plan will be on a single projector.

CONCLUSION :

(1) WHEN A LINE IS PARALLEL TO ONE AND INCLINED TO TWO OTHER PRINCIPAL PLANES, ITS PROJECTION ON PLANE TO WHICH IT IS PARALLEL WILL SHOW TRUE LENGTH AND WILL ALSO SHOW TRUE INCLINATIONS WITH OTHER TWO PLANES WITH CORRESPONDING GROUND LINES.

(2) PROJECTIONS ON PLANES TO WHICH IT IS INCLINED AND NOT PARALLEL ARE SHORTER THAN THE TRUE LENGTH AND THEY WILL BE PARALLEL TO CORRESPONDING GROUND LINES.

(3) OUT OF 3 ANGLES θ, Ø and ψ, ONE WILL BE ZERO AND SUMMATION OF TWO OTHER WILL BE 90°.

(4) ALL THINGS GO TOGETHER AND CANNOT BE SEPARATED.

Having studied second theory of a straight line parallel to one of the principal planes and inclined to two other, we shall solve few problems based on it. Students are requested to go through conclusion/discussion etc. based on theory No.2 before solving problems based on it.

Problem 5 : A line PQ, 50 mm long, is in V.P. It makes an angle of 30° with the H.P. Point P is 20 mm above H.P. Draw the projections of line PQ.

For solution see Fig. 9.7 and follow the procedure as given below :

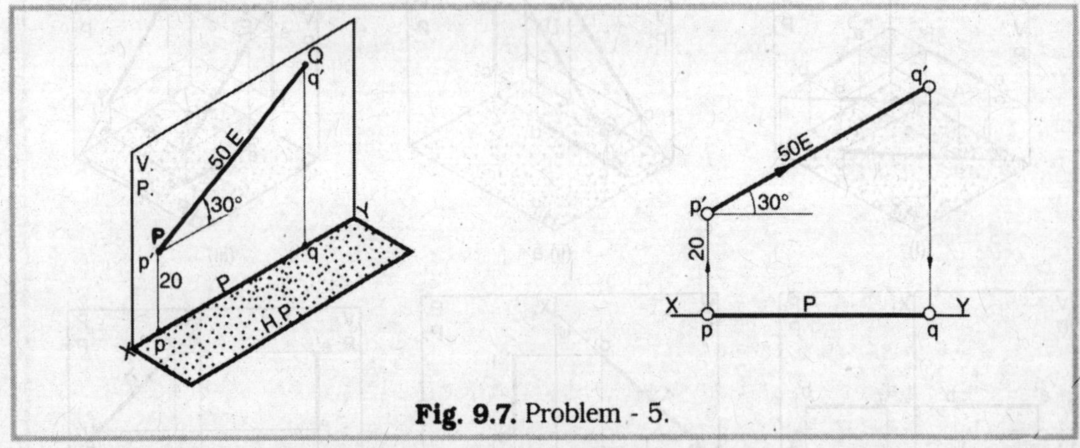

Fig. 9.7. Problem - 5.

As the line is in V.P., it can be considered parallel to V.P. also. So elevation will show true length (50) and true inclination θ (30°) with xy and its plan will lie on xy line.

(i) First draw p on xy and p' 20 mm above xy on a projector through p.

(ii) At p' draw p'q' of 50 mm length and at an angle of 30° with xy line.

(iii) From q' draw projector on xy line and mark q on xy.

Problem 6 : A line VK, 75 mm long, is parallel to V.P. and inclined to H.P. by an angle 45°. Point V is 30 mm below H.P. and 20 mm in front of V.P. Point K is in first quadrant. Draw the projections of the straight line VK.

For solution see Fig. 9.8 and follow the procedure as given below :

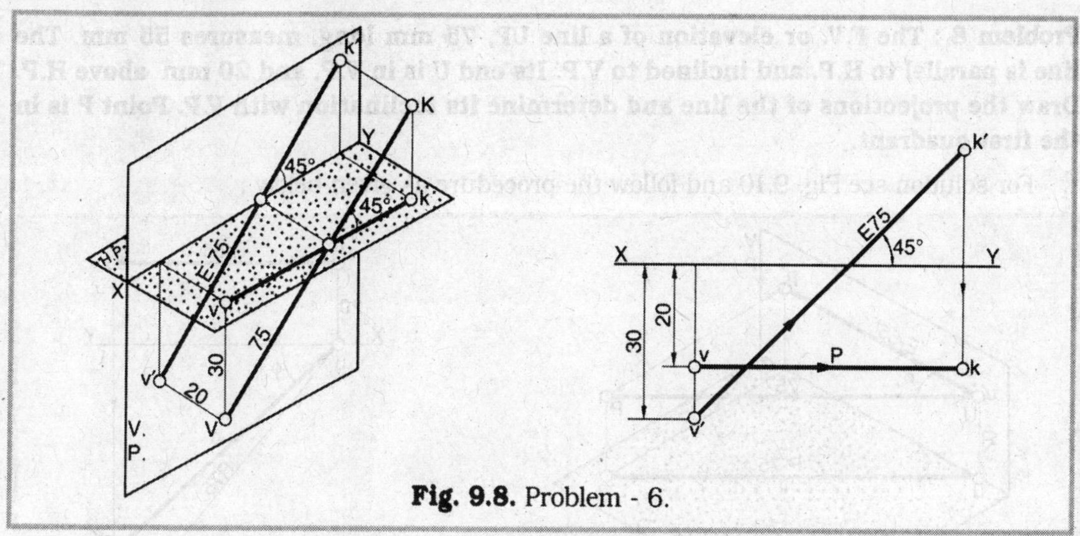

Fig. 9.8. Problem - 6.

Here, as the line is parallel to V.P. its elevation v'k' must show T.L. 75 mm and must make 45° with xy and its plan vk must be parallel to xy line.

(i) First plot v and v' on a projector 20 mm and 30 mm below xy respectively.

(ii) At v' draw v'k' of 75 mm length on upperside and it should make 45° angle with xy line. Since K is in 1st quadrant k' must be above xy line.

(iii) Draw projector through k' and parallel to xy from v to get k by intersection.

Problem 7 : A line KP, 70 mm long, is parallel to H.P. and inclined to V.P. by 60°. Point P is 20 mm above H.P. and 30 mm in front of V.P. Point K is behind V.P.Draw the projections of line KP.

For solution see Fig. 9.9 and follow the procedure as given below :

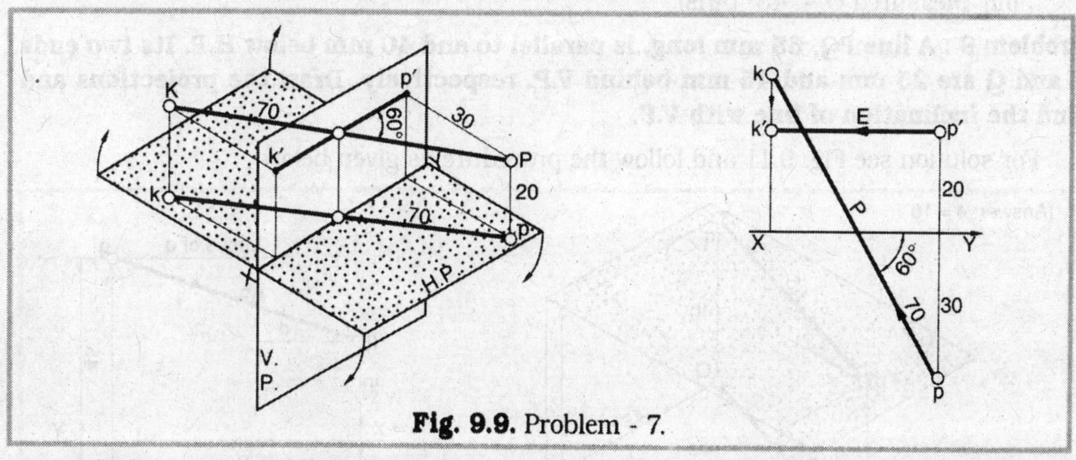

Fig. 9.9. Problem - 7.

As the line is parallel to H.P., plan kp will show T.L.(70) and will make true angle Ø(60°) with xy line and its elevation k'p' will be shorter and parallel to xy line.

(i) First plot p and p'.

(ii) From p draw pk line of T.L. 70 mm length at an angle of 60° with xy. Plan k should be above xy, since K is behind V.P.

(iii) Draw projector through k and parallel to xy from p' to get k' by intersection.

Problem 8 : The F.V. or elevation of a line UP, 75 mm long, measures 55 mm. The line is parallel to H.P. and inclined to V.P. Its end U is in V.P. and 20 mm above H.P. Draw the projections of the line and determine its inclination with V.P. Point P is in the first quadrant.

For solution see Fig. 9.10 and follow the procedure as given below :

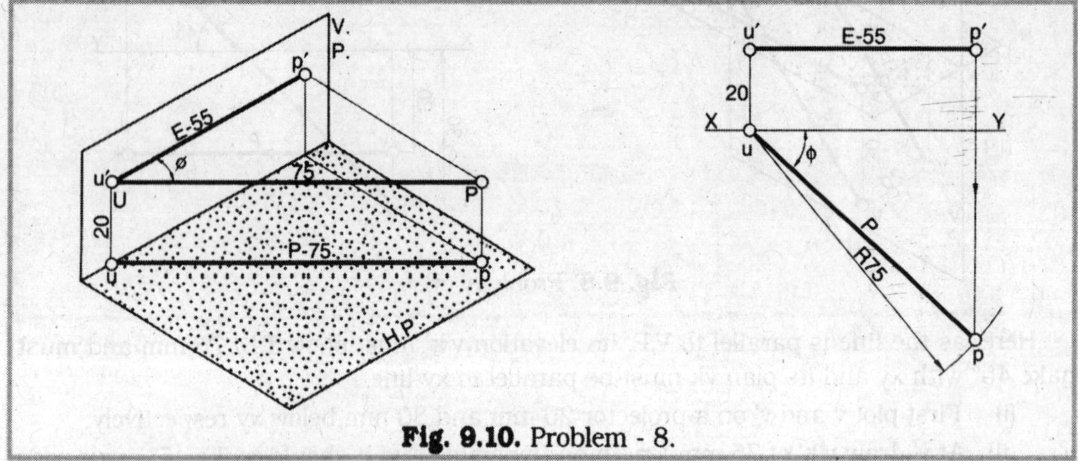

Fig. 9.10. Problem - 8.

As the line is parallel to H.P., elevation (55) will be parallel to xy and plan will be of T.L. (75), and will show true angle Ø with xy line.

(i) First draw u and u'. From u' draw line u'p' of 55 mm length and parallel to xy.

(ii) Draw projector through p' and draw arc with u as the centre and radius equal to T.L. (75) to get p, as shown in figure. Join up and measure angle of up with xy i.e. Ø .

(iii) Measured Ø = 43° (Ans).

Problem 9 : A line PQ, 65 mm long, is parallel to and 40 mm below H.P. Its two ends P and Q are 25 mm and 45 mm behind V.P. respectively. Draw the projections and find the inclination of line with V.P.

For solution see Fig. 9.11 and follow the procedure as given below:

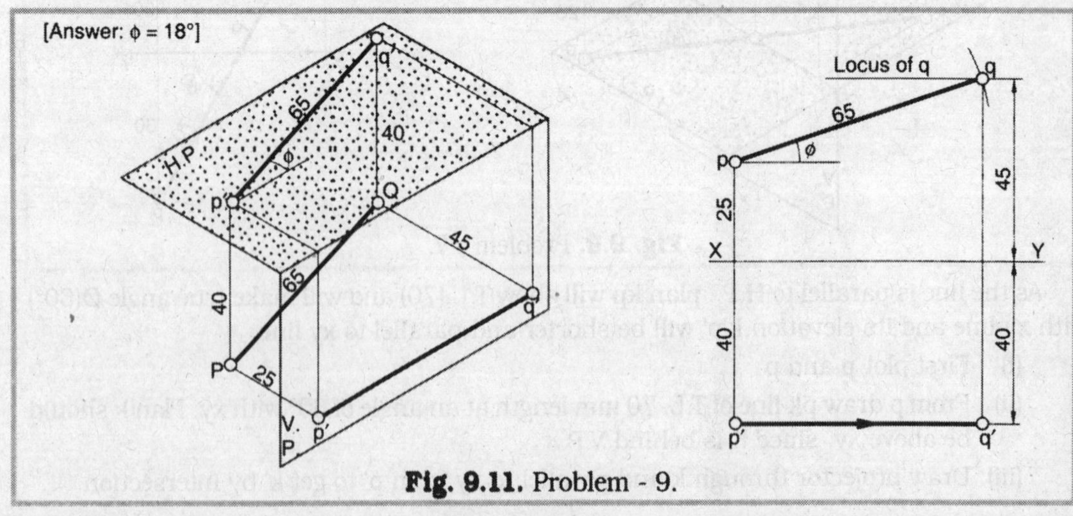

Fig. 9.11. Problem - 9.

(i) First draw p and p', on a projector, 25 mm above xy and 40 mm below xy respectively.

(ii) Draw locus of q, 45 mm above xy, since Q is 45 mm behind V.P.

(iii) Now with p as the centre and T.L. 65 mm as radius draw an arc to cut the locus of q at the point q.

(iv) Draw projector through q and parallel to xy from p' to get q' by intersection.

Problem 10 : Two pegs P and Q, fixed on the same wall, are 3000 mm (3m) apart. The distance between the pegs measured parallel to the floor is 2400 mm (2.4 m). If the peg P is 1000 mm (1m) above the floor, draw the projections of line joining two pegs. Find the height of the second peg Q. Also find the inclination of the line joining two pegs with the floor. The pegs are protuding out of the wall by 150 mm (0.15 m).

For solution see Fig. 9.12 and follow the procedure as given below :

As the two pegs are protuding out by the same length (150 mm) from a wall, line joining two pegs will be parallel to wall. Consider the wall as V.P. and floor as H.P. So line will be parallel to V.P. and inclined to H.P. Visualise this from the three dimensional drawing given in Fig. 9.12 Distance (2400) between two pegs measured parallel to the floor is going to be plan length.

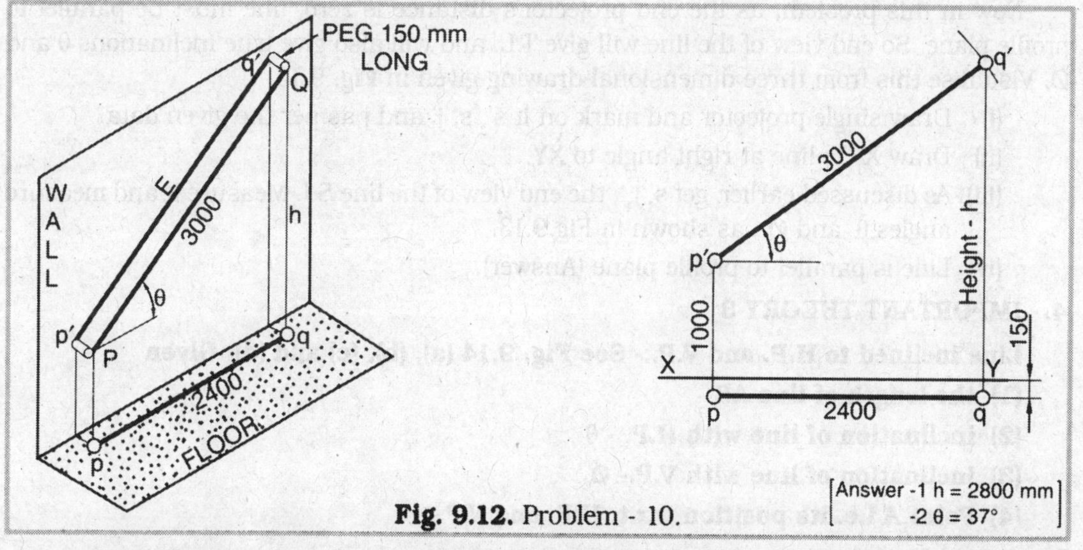

Fig. 9.12. Problem - 10.

(i) First draw p' and p, on a projector, 1000 mm above and 150 mm below xy respectively.

(ii) Draw plan pq of 2400 mm length and parallel to xy.

(iii) Draw projector through q and draw arc with p' as the centre and radius equal to T.L. 3000 mm (3m) to get by intersection the point q'. Join p'q' and measure the angle θ and height h of point q' from xy line. θ = 37° is the inclination of the line with the floor and h = 2800 mm (2.8 m) height of the peg Q above the floor.

Problem 11 : The distance between the end projectors of a line SJ is zero. Point S is 50 mm above H.P. and 30 mm in front of V.P. Point J is 20 mm above H.P. and 60 mm in front of V.P. Draw the projections and find the angle of the line with H.P. and V.P. and also find the true length of the line. State the position of the line w.r.t. P.P.

For solution see Fig. 9.13 and follow the procedure as given below :

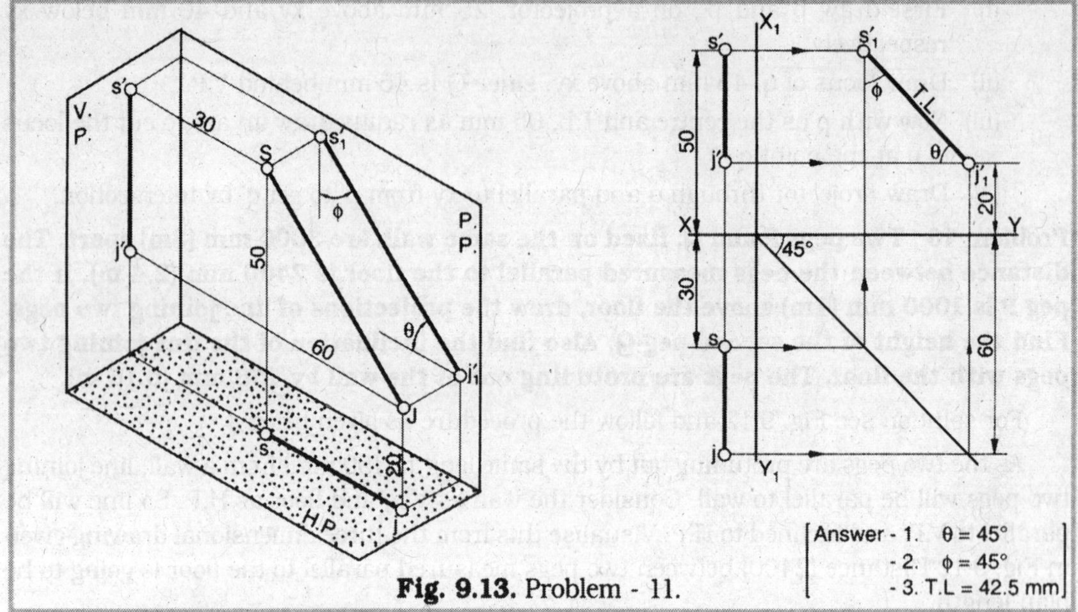

Fig. 9.13. Problem - 11.

Now in this problem, as the end projector's distance is zero, line must be parallel to profile plane. So end view of the line will give T.L. and will also give true inclinations θ and Ø. Visualise this from three dimensional drawing given in Fig. 9.13.

(i) Draw single projector and mark on it s', s, j' and j as per the given data.

(ii) Draw X_1Y_1 line at right angle to XY.

(iii) As discussed earlier, get $s_1'j_1'$, the end view of the line SJ. Measure it and measure angles θ and Ø, as shown in Fig.9.13.

(iv) Line is parallel to profile plane (Answer).

4. IMPORTANT THEORY-3

Line inclined to H.P. and V.P.:- See Fig. 9.14 (a), (b), (c) and (d). Given

(1) the length of line AB

(2) inclination of line with H.P. - θ

(3) inclination of line with V.P.- Ø

(4) Point A i.e. its position w.r.t. H.P. and V.P.

Important Notations :

(1) Actual line - AB

(2) Plan of line - ab

(3) Elevation of line - a'b'

(4) Line AB_1 parallel to V.P. - position of the line AB during its rotation about the point A keeping inclination θ with H.P. the same.

(5) Line AB_2 parallel to H.P. - position of the line AB during its rotation about the point A keeping inclination with V.P. the same.

(6) Inclination of elevation with XY line is α and inclination of plan with XY line is β.

Our line is AB, see Fig. 9.14 (a). Line AB is rotated about the point A, keeping the inclination of the line AB with H.P. constant and equal to θ.

During rotation the point B will rotate around the periphery of a horizontal circle and the line AB will generate a cone. Plan of this horizontal circle is also a circle of same size with plan(a) of point A as the centre. This circle is a circle on which plan of actual point B is going to lie and so we call it as the locus of b. Elevation of horizontal circle is a straight line parallel to XY. The elevation of the actual point B is going to lie on this straight line and so we will call it as the locus of b'.

Now during this rotation, the line AB will have its position as AB_1 when it will become parallel to V.P. See Fig. 9.14 (a).

As AB_1 is parallel to V.P., as per conclusion of line parallel to V.P. (See Fig. 9.14 (b) also),

(1) elevation $a'b'_1$ will be true length;

(2) elevation $a'b'_1$ will make an angle θ with XY line ; and

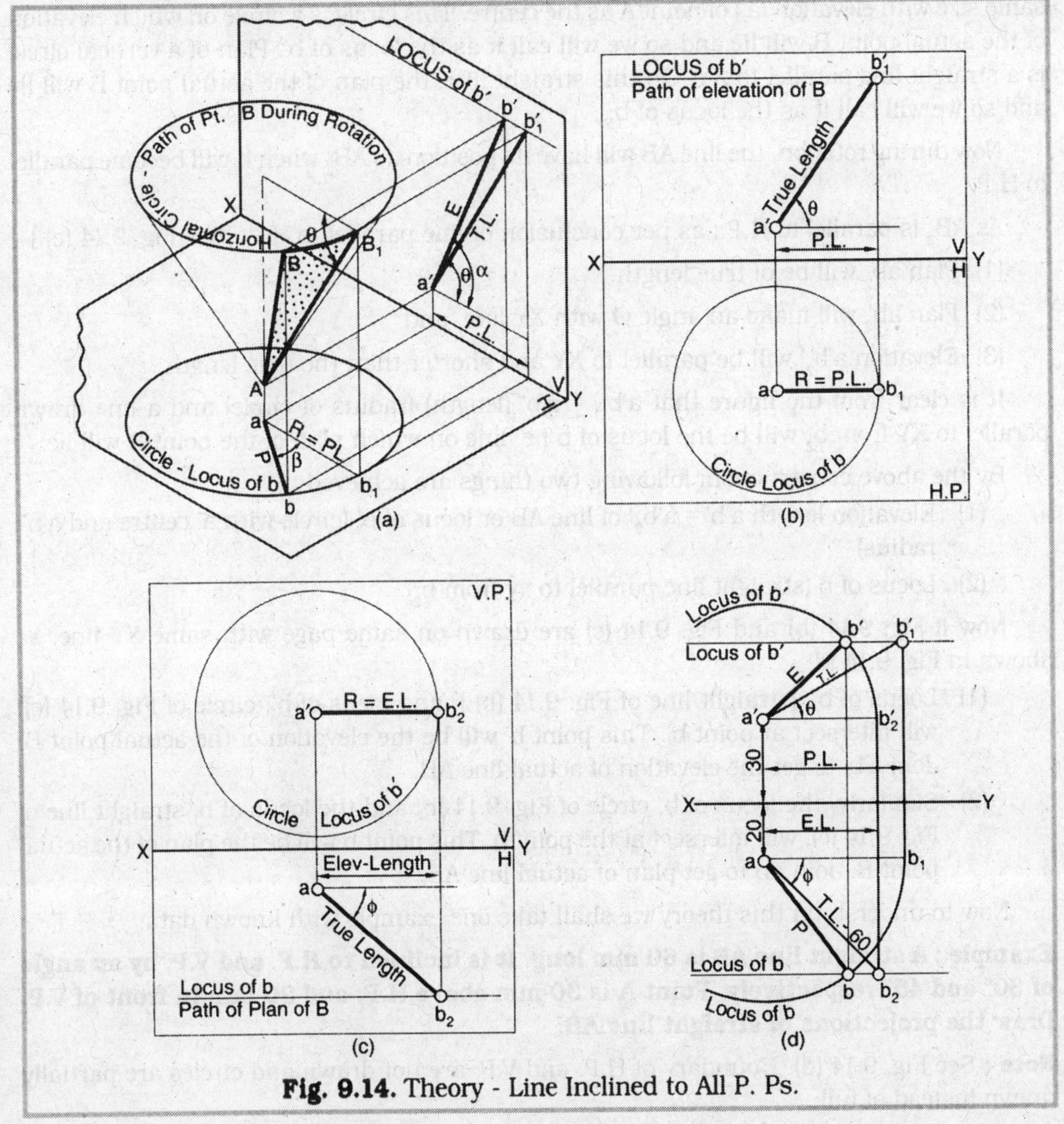

Fig. 9.14. Theory - Line Inclined to All P. Ps.

(3) plan ab_1 will be parallel to XY and shorter than the true length.

It is clear from the figure that ab_1 = ab (length) (radius of circle) and a line drawn parallel to XY from b'_1 is going to be locus of b' i.e. line on which elevation of point B is going to lie.

By this construction, the following two things are achieved. See Fig. 9.14 (b) also.

(1) Plan length ab = ab_1 of line AB or locus of b (circle with a centre and ab_1 radius)

(2) Locus of b'. (straight line parallel to XY from b'_1)

Similarly, a line AB is rotated about the point A keeping the inclination of line AB with V.P. constant and equal to Ø. As drawing is not given, imagine a similar drawing to Fig. 9.14 (a) with the path of B as vertical circle.

During rotation, the point B will rotate around the periphery of vertical circle (parallel to V.P.) and line AB will generate a cone. Elevation of this vertical circle will be a circle of the same size with elevation (a') of point A as the centre. This circle is a circle on which elevation of the actual point B will lie and so we will call it as the locus of b'. Plan of a vertical circle is a straight line parallel to XY. On this straight line, the plan of the actual point B will lie and so we will call it as the locus of b.

Now during rotation, the line AB will have its position as AB_2 when it will become parallel to H.P.

As AB_2 is parallel to H.P.; as per conclusion of line parallel to H.P. [See Fig. 9.14 (c).]

(1) Plan ab_2 will be of true length.

(2) Plan ab_2 will make an angle Ø with XY line, and

(3) Elevation $a'b_2'$ will be parallel to XY and shorter than the true length.

It is clear from the figure that $a'b'_2$ = a'b' (length) (radius of circle) and a line drawn parallel to XY from b_2 will be the locus of b i.e. line on which plan of the point B will lie.

By the above construction, following two things are achieved :

(1) Elevation length a'b' = $a'b_2'$ of line AB or locus of b' (circle with a' centre and $a'b'_2$ radius)

(2) Locus of b (straight line parallel to xy from b_2)

Now if Fig. 9.14 (b) and Fig. 9.14 (c) are drawn on same page with same XY line, as shown in Fig. 9.14 (d) -

(1) Locus of b', (straight line of Fig. 9.14 (b)), and locus of b', circle of Fig. 9.14 (c), will intersect at point b'. This point b' will be the elevation of the actual point B. Join a'b' to get the elevation of actual line AB.

(2) Similarly, the locus of b, circle of Fig. 9.14 (b) and the locus of b, straight line of Fig. 9.14 (c), will intersect at the point b. This point b will be the plan of the actual point B. Join ab to get plan of actual line AB.

Now to understand this theory we shall take one example with known data.

Example : A straight line AB is 60 mm long. It is inclined to H.P. and V.P. by an angle of 30° and 45° respectively. Point A is 30 mm above H.P. and 20 mm in front of V.P. Draw the projections of straight line AB.

Note : See Fig. 9.14 (d). Boundary of H.P. and V.P. are not drawn and circles are partially drawn instead of full.

Solution : First draw XY line and draw (a') and (a) 30 mm above and 20 mm below XY line respectively. At point (a') draw $a'b'_1$ line of true length i.e. 60 mm at an angle $\theta = 30°$ with horizontal line. Draw perpendicular line to XY from (b'_1) and horizontal line from point (a) to get by intersection point (b_1). This is done because line AB_1 is assumed parallel to V.P. and inclined to H.P. by $\theta = 30°$.

As AB_1 is parallel to V.P. we have -

(1) $a'b'_1$ = T.L, (2) $a'b'_1$ makes θ angle with XY and (3) plan ab_1, is parallel to XY, ab_1 = Plan length ab.

Now from (b_1') draw a line parallel to XY which is the locus of (b') (straight line) and draw a circle with plan point (a) as the centre and plan length ab_1 as the radius which is locus of b (circle). Full circle is not required and hence not drawn.

Similarly, at plan point (a) draw ab_2 line of true length i.e. 60 mm at an angle $Ø = 45°$ with horizontal line. Draw perpendicular line to XY from (b_2) and horizontal line from the point (a') to get by intersection point (b'_2). This is done because line AB_2 is assumed parallel to H.P. and inclined to V.P. by $Ø = 45°$.

As AB_2 is parallel to H.P, we have -

(1) ab_2 = T.L, (2) ab_2 makes Ø angle with XY and (3) Elevation $a'b'_2$ is parallel to XY. $a'b'_2$ = Elev-length a'b'.

Now from (b_2) draw a line parallel to XY to intersect with the circle, locus of (b), at the point b. Join ab to get plan.

Draw a circle with elevation point (a') as the centre and rad = Elev. length = $a'b'_2$ to intersect locus of (b') (straight line) at the point (b'). Join a'b' to get elevation.

Important Discussion on above theory, line inclined to both the planes H.P. and V.P. :-

(1) (a) In Fig. 9.14 (b) we have the following terms :-

(1) True length

(2) θ - Inclination of line with H.P.

(3) Plan length OR locus of b (circle)

(4) Locus of b' (straight line parallel to xy) or position of B w.r.t. H.P.

(b) Similarly, in Fig. 9.14 (c) we have the following terms :-

(1) True length

(2) Ø - Inclination of line with V.P.

(3) Elevation length or locus of b' (circle)

(4) Locus of b (straight line parallel to XY) or position of B w.r.t. V.P.

In (a) and (b) above, if any two terms out of four are given, same drawings can be constructed with available data and from there other two terms can be found out.

(2) Inclination of a line with H.P. i.e. θ is drawn at the point a' (or on elevation of one point). θ can be drawn in the upper or lower direction to horizontal line depending upon whether locus of b' (straight line) or locus of second point (straight line) is above a' or below a' respectively. If locus of b' is required above a', θ should be drawn on the upper side of straight line at a' and if locus of b' is required below a', θ should be drawn on the lower side of horizontal straight line at a'.

Similarly, inclination of line with V.P. i.e. Ø is drawn at the point (a) (or on plan of one point). Ø can be drawn in the upper or lower direction to horizontal line at (a) depending upon whether locus of (b) (straight line) is required above (a) or required below (a) respectively.

(3) Out of many points (e.g. two end points, V.T., H.T., mid point, etc) given of any straight line, only one point can be marked positively and for rest of the points only their loci can be drawn parallel to XY for elevations as well as for plans.

5. IMPORTANT THEORY-4

Traces of a line - see Fig. 9.15 (a) and (b).

Line has three traces : (1) Horizontal Trace (H.T.)

(ii) Vertical Trace (V.T.)

(iii) Profile Trace (not discussed). (P.T.)

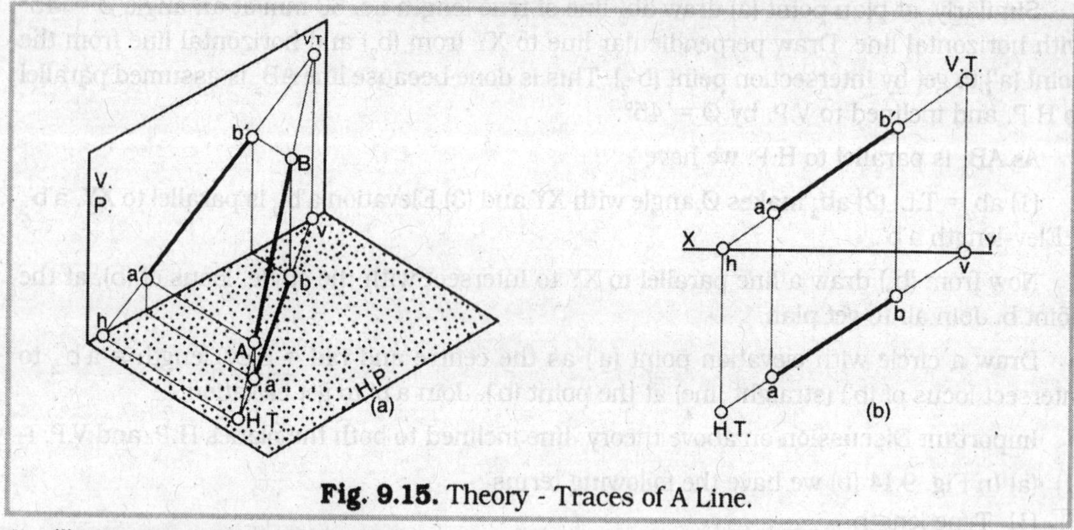

Fig. 9.15. Theory - Traces of A Line.

4. (i) Horizontal Trace (H.T.)

1. Point where the line meets (extended, if required) the horizontal plane is known as Horizontal Trace.
2. If a line is parallel to H.P., it will not meet H.P. and so will not have horizontal trace.
3. If two end points of a line are on same side of H.P., i.e. either above or below, H.T. will not be on actual line but it will be on the extension of the line.
4. To get horizontal trace extend your elevation, if necessary, upto the ground line XY, and mark the point h where elevation meets XY line. At h draw perpendicular to XY line and extend your plan, if necessary, to get by intersection of above two lines, the horizontal trace (H.T.) see Fig. 9.15 (a) & (b).
5. H.T. point will be on H.P. and it will be also on a plan line or on extension of it. Projection of H.T. point on V.P. will be on ground line XY and is denoted by h. Elevation of point H.T. (h) will be on elevation or on extension of elevation.

4 (ii) Vertical Trace (V.T.):

1. Point where the line meets (extend, if required) the V.P. is known as vertical trace (V.T.)

2. If a line is parallel to V.P. it will not meet V.P. and so it will not have V.T.
3. If two end points of a line are on same side of V.P. i.e.either in front or behind V.P., V.T. will not be on actual line but it will be on extension of the line.
4. To get V.T. extend your plan if necessary upto ground line XY, and mark point v where plan meets the XY line. At v draw perpendicular to XY line and extend your elevation if necessary to get by intersection of above two lines the V.T. (See Fig. 9.15 (a) and (b)).
5. V.T. point will be on V.P. and it will be on the elevation line or on extension of it. Projection of V.T. point on H.P. will be on ground line XY and is denoted by v. Plan of point V.T. (v) will be on plan or on extension of plan.

Important Discussion :

(1) WHEN ELEVATIONS OF TWO END POINTS OF A STRAIGHT LINE ARE ON OPPOSITE SIDES OF XY LINE THEN H.T. WILL BE ON THE LINE AND NOT ON EXTENSION OF LINE. LINE MUST BE PASSING THROUGH H.P.

(2) SIMILARLY, WHEN PLANS OF TWO END POINTS OF A STRAIGHT LINE ARE ON OPPOSITE SIDES OF XY LINE THEN V.T. WILL BE ON THE LINE AND NOT ON EXTENSION OF LINE. LINE MUST BE PASSING THROUGH V.P.

(3) H.T. POINT IS THE PLAN OF H.T. AND (h) IS THE ELEVATION OF H.T. SIMILARLY, V.T. POINT IS THE ELEVATION OF V.T. AND (v) IS THE PLAN OF V.T. (h) AND (v) ARE ALWAYS ON GROUND LINE.

(4) ELEVATIONS OF END POINTS (e.g. a', b') OF STRAIGHT LINE, (h) AND V.T. WILL BE ALWAYS ON ONE STRAIGHT LINE. SIMILARLY, PLANS OF END POINTS (i.e. a,b) OF STRAIGHT LINE, (v) AND H.T. WILL BE ALWAYS ON ONE STRAIGHT LINE.

6. IMPORTANT THEORY-5

Line inclined to both the planes : See Fig. 9.16 (a) & (b).

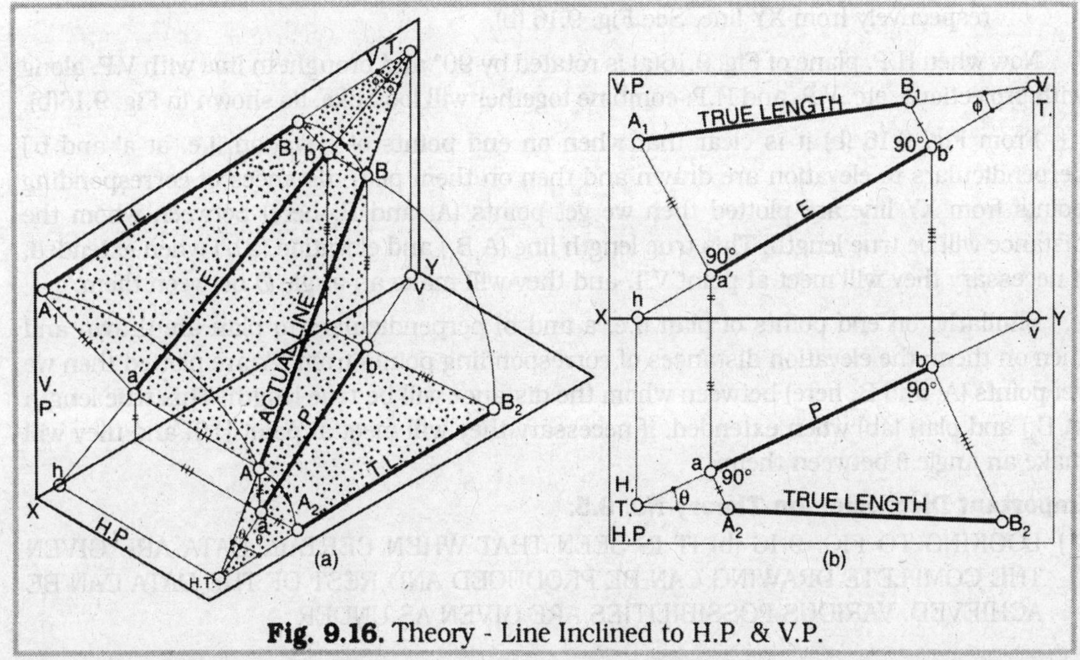

Fig. 9.16. Theory - Line Inclined to H.P. & V.P.

In Fig. 9.16 (a) line AB is shown inclined to both the principal planes H.P. and V.P. by an angle θ and Ø respectively.

AB is our actual line. a'b' and ab are the elevation and plan respectively. It is seen from Fig. 9.16 (a) that when AB is extended it meets V.P. at a point denoted as V.T. It is seen that a'b' (elevation) and actual line AB meet at point V.T. and they make an angle Ø between them.

Similarly, it is seen from Fig. 9.16 (a) that when BA is extended it meets H.P. at a point denoted as H.T. It is seen that ab (plan) and actual line AB meet at point H.T. and they make an angle θ between them.

Now if a ΔVT.a'.A is given rotation about line VT.b'.a' and positioned on V.P., as shown in Fig. 9.16 (a), we get new position of ΔVT.a'.A as ΔVTa'A_1.

In ΔV.Ta.'A_1 we have -

(1) A_1B_1 = true length of line. Since it is the new position of line AB after rotation.

(2) Angle a'.VT.A_1, i.e. angle between a'b' and true length A_1B_1, is Ø, and

(3) a'A_1 and b'B_1 are perpendicular to elevation a'b' at points (a') and (b') respectively, These perpendicular distances are equal to distances of a plan point a and plan point b respectively from XY line. See Fig. 9.16 (b) also for it.

Similarly, now if a ΔHT.b.B of Fig. 9.16 (a) is given rotation about the line HT.a.b and brought it to H.P., as shown in Fig. 9.16 (a). We get new position of ΔHT.b.B as ΔHT.b.B_2.

In ΔHT.b.B_2, we have-

(1) A_2B_2 = true length of line. Since it is the new position of line AB after rotation.

(2) Angle b.HT.B_2, i.e. angle between ab and true length A_2B_2, is θ, and

(3) aA_2 and bB_2 are perpendiculars to plan ab at points a and b respectively. These perpendicular distances are equal to distances of an elevation points a' and b' respectively from XY line. See Fig. 9.16 (b).

Now when H.P. plane of Fig. 9.16(a) is rotated by 90° and brought in line with V.P. along with projections, etc. V.P. and H.P. combine together will look like, as shown in Fig. 9.16(b).

From Fig. 9.16 (b) it is clear that when on end points of elevation (i.e. at a' and b') perpendiculars to elevation are drawn and then on them plan distances of corresponding points from XY line are plotted then we get points (A_1 and B_1 here) between whom the distance will be true length. This true length line (A_1B_1) and elevation (a'b') when extended, if necessary they will meet at point V.T. and they will make an angle Ø between them.

Similarly, on end points of plan (i.e. a and b) perpendiculars to plan are drawn and then on them, the elevation distances of corresponding points from XY are plotted then we get points (A_2 and B_2 here) between whom the distance will be true length. This true length (A_2B_2) and plan (ab) when extended, if necessary they will meet at point H.T. and they will make an angle θ between them.

Important Discussion on Theory No. 9.5:

(1) LOOKING TO FIG. 9.16 (b) IT IS SEEN THAT WHEN CERTAIN DATA ARE GIVEN THE COMPLETE DRAWING CAN BE PRODUCED AND REST OF THE DATA CAN BE ACHIEVED. VARIOUS POSSIBILITIES ARE GIVEN AS UNDER :-

DATA GIVEN	DATA ACHIEVED	DATA GIVEN	DATA ACHIEVED	DATA GIVEN	DATA ACHIEVED
(i) Elevation (ii) V.T. (iii) Ø	(i) TRUE LENGTH (ii) PLAN (iii) θ (iv) H.T.	(i) PLAN (ii) H.T. (iii) θ	(i) TRUE LENGTH (ii) ELEVATION (iii) Ø (iv) V.T.	(I) PLAN (ii) ELEVATION	(i) T. L. (ii) H.T. & V.T. (iii) θ and Ø

(2) WHEN PLANS OF TWO END POINTS OF A STRAIGHT LINE ARE ON OPPOSITE SIDES OF XY LINE THEN PERPENDICULARS TO THE ELEVATION ARE DRAWN IN OPPOSITE DIRECTIONS, AND IN THIS CASE V.T. WILL BE BETWEEN TWO END POINTS OF A STRAIGHT LINE. REVERSE IS ALSO TRUE.

SIMILARLY, WHEN ELEVATIONS OF TWO END POINTS OF A STRAIGHT LINE ARE ON OPPOSITE SIDES OF XY LINE THEN PERPENDICULARS TO THE PLAN ARE DRAWN IN OPPOSITE DIRECTIONS AND IN THIS CASE H.T. WILL BE ON THE LINE AND NOT ON EXTENSION OF LINE. REVERSE IS ALSO TRUE.

After studied 3rd, 4th and 5th theories of straight line and its conclusion/discussion, etc., we shall now solve different types of problems on it. Three dimensional figures are given to visualise the position of straight line. Arrows on lines in the solution represent the direction in which the problem is approached in arriving at the solution. Whatever values which are given in data are mentioned in the drawing and answers are written separately.

In the begining, simple examples are solved, then some applied problems are solved and at the end most typical problems from university papers are solved.

Problem 12 : The distance between the end projectors of a straight line AB is 60 mm. Point A is 5 mm above H.P. and 30 mm in front of V.P. Point B is 40 mm above H.P. and 50 mm behind V.P. Draw the projections and find the inclination of straight line AB with H.P. and V.P. and the true length of the line. Find also the traces.

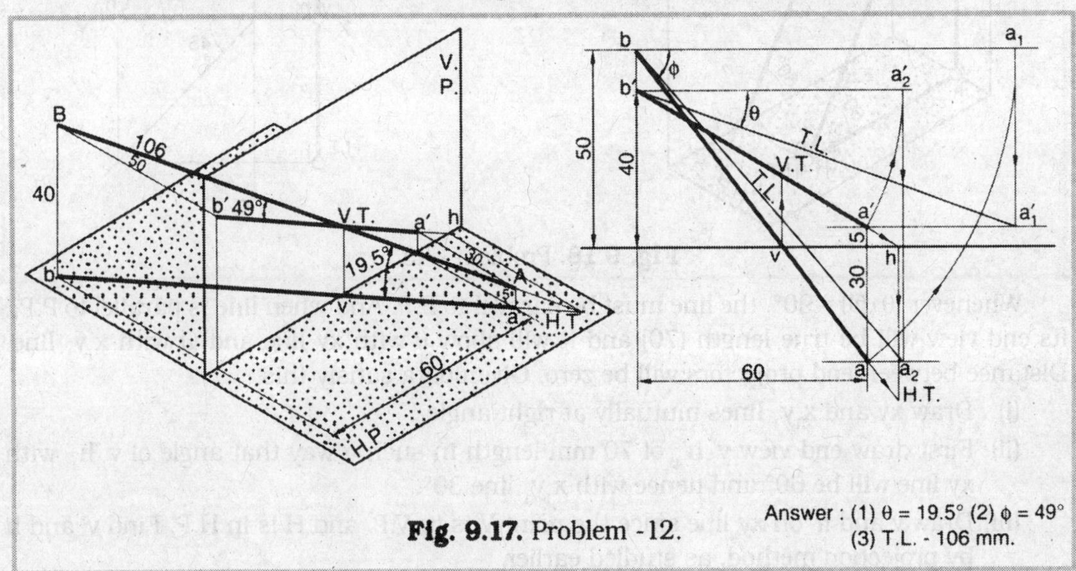

Fig. 9.17. Problem -12.

This problem can be solved by two methods. One by theory No.3 and another by theory No.5. Solution is done with the help of theory No.3. See Fig. 9.17 for solution and follow the procedure as given below :

(i) First draw two end projectors 60 mm apart and plot b, b' and a, a' as per given data. Join plan ab and elevation a'b'.

(ii) Rotate the plan ba about b to position ba_1 to make it parallel to xy so that its corresponding elevation $b'a'_1$ gives us true length and true inclination θ, as shown in Fig. 9.17, Measured value of θ = 19.5° and T.L. = 106 mm.

(iii) Similarly, rotate the elevation b'a' about b' to new position $b'a'_2$ (parallel to xy) so that its corresponding plan ba_2 gives us true length and true inclination Ø, as shown in Fig. 9.17. Measured value of T.L. = 106 mm and measured Ø= 49°.

(iv) To get H.T. extend elevation b'a' upto xy line and mark the point as h. Now draw projector through h and extend plan ba. They will intersect at H.T.

(v) Similarly, mark point v where plan ba intersects with xy line and draw projector through v to intersect with elevation at V.T.

Problem 13: A line VH, 70 mm long, has its end V in V.P. and end H in H.P. Line is inclined to H.P. by 60° and to V.P. by 30°. Draw the projections and find traces of a line VH.

For solution see Fig. 9.18 and follow the procedure as given below:

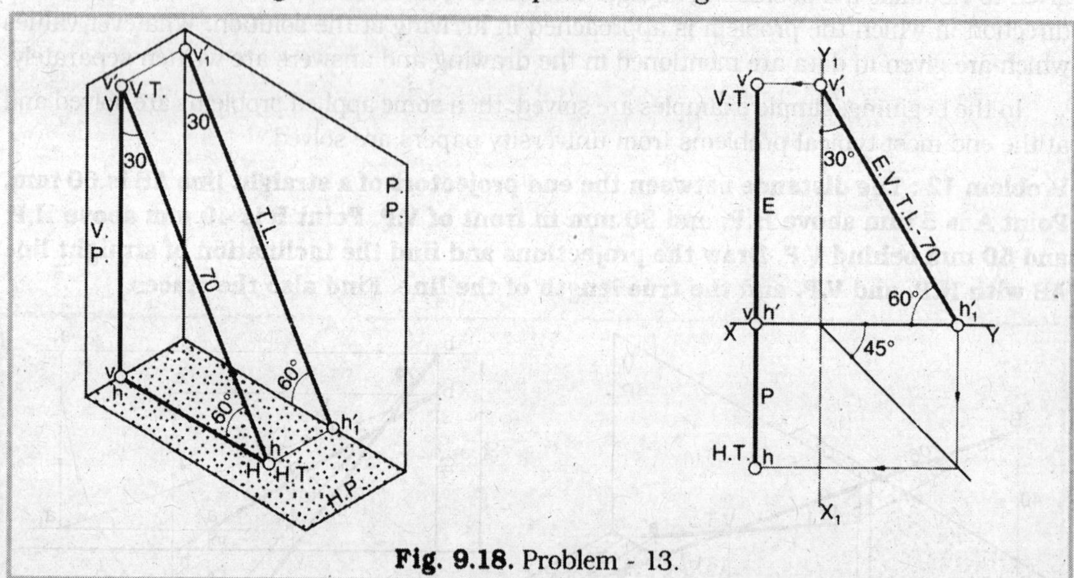

Fig. 9.18. Problem - 13.

Whenever (θ+Ø) = 90°, the line must be parallel to P.P. And when line is parallel to P.P., its end view will be true length (70) and it will show θ with xy line and Ø with x_1y_1 line. Distance between end projectors will be zero. Given data satisfy this.

(i) Draw xy and x_1y_1 lines mutually at right angle.

(ii) First draw end view $v'_1h'_1$ of 70 mm length in such a way that angle of $v_1'h_1'$ with xy line will be 60° and hence with x_1y_1 line 30°.

(iii) Draw v and h' on xy line since the point V is in V.P. and H is in H.P. Find v' and h by projection method, as studied earlier.

(iv) Point H (h) and point V (v') are H.T. and V.T. respectively.

Problem 14 : A line PQR, 100 mm long, is inclined to H.P. by 30° and V.P. by 45°. PQ : QR : 2 : 3. Point Q is in V.P. and 25 mm above H.P. Draw the projections of the line PQR when point R is in the 1st quadrant. Find the position of point P. Draw also the traces of line PQR.

For solution see Fig. 9.19 and follow the procedure as given below :

As position of point Q is given and as the quadrant of point R is given, do the solution first considering only the line QR. Length of QR will be 60 mm. So now for line QR, the point Q is completely given, T.L. = 60 mm, θ - 30° and Ø = 45°.

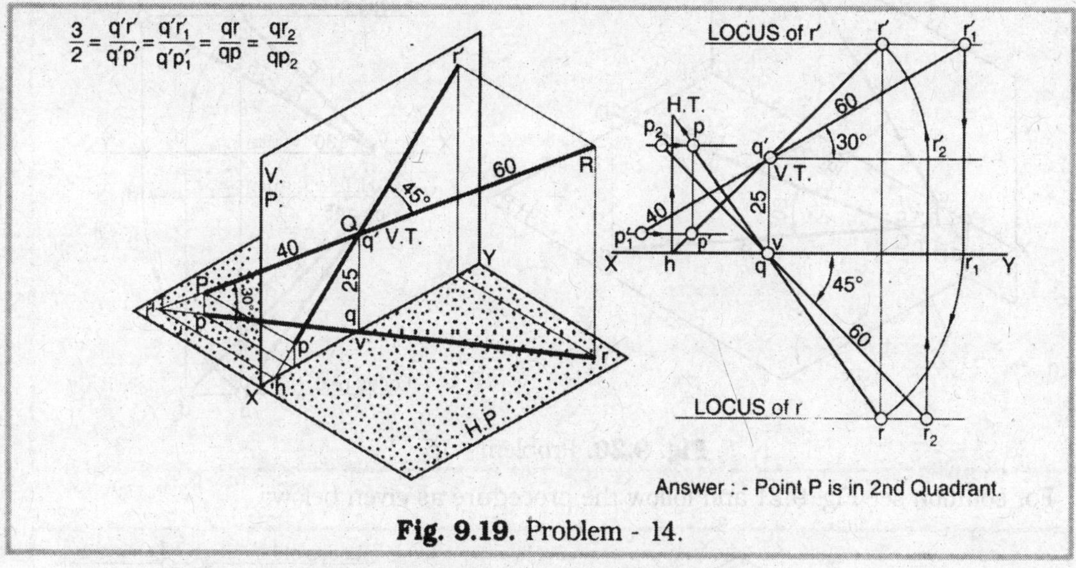

Fig. 9.19. Problem - 14.

(i) Draw the projections of line QR as per the theory No.3.

(ii) After getting the solution for line QR extend it by 3 : 2 proportion on other side of q and q' to get the solution of line QP, as shown in the figure.

(iii) Both p and p' are above xy line and so the point P is located in the 2nd quadrant. (Answer)

(iv) Get the traces as usual.

Problem 15 : The top view and the front view, of the line CD, measure 65 mm and 53 mm respectively. The line is inclined to H.P. and to V.P. by 30° and 45° respectively. The end C is on the H.P. and 12 mm in front of V.P. Other end D is in the 1st quadrant. Draw the projections of the line CD and find its true length and draw traces.

For solution see Fig. 9.20 and follow the procedure as given below:

(i) First draw c and c', 12 mm below xy line and on xy line respectively.

(ii) Draw $c'd'_2$ = 53 mm parallel to xy and draw cd_1 = 65 mm parallel to xy. Their corresponding plan cd_2 and elevation $c'd'_1$ are achieved by drawing 45° angle line at c and 30° angle line at c' respectively. $c'd'_1$ and cd_2 both will give T.L.

(iii) From d'_1 and d_2 draw parallel lines to xy line to get locus of d' and locus of d respectively.

(iv) Now with c' as the centre and radius equal to 53 mm draw an arc to get d' on locus of d'. Similarly, with c as the centre and radius equal to 65 mm draw an arc to get d on locus of d. Join c'd' and cd.

(v) Find the traces as usual.

Problem 16 : A line EF, 75 mm long, has its end E 20 mm below H.P. and 25 mm behind V.P. The end F is 50 mm below H.P. and 65 mm behind VP. Draw the projections of line EF and find its inclinations with H.P. and V.P.

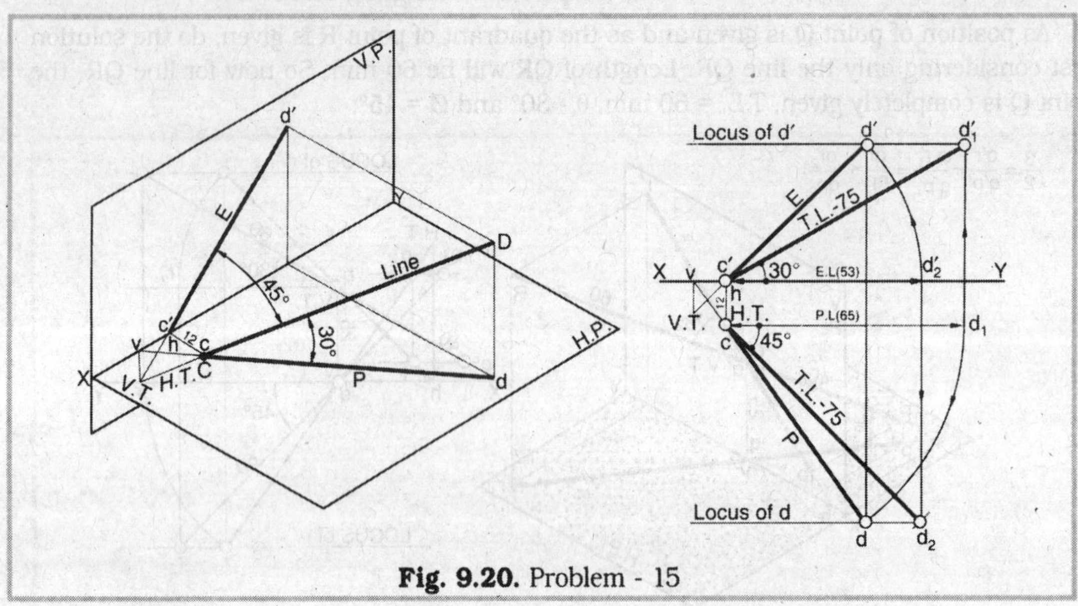

Fig. 9.20. Problem - 15

For solution see Fig. 9.21 and follow the procedure as given below:

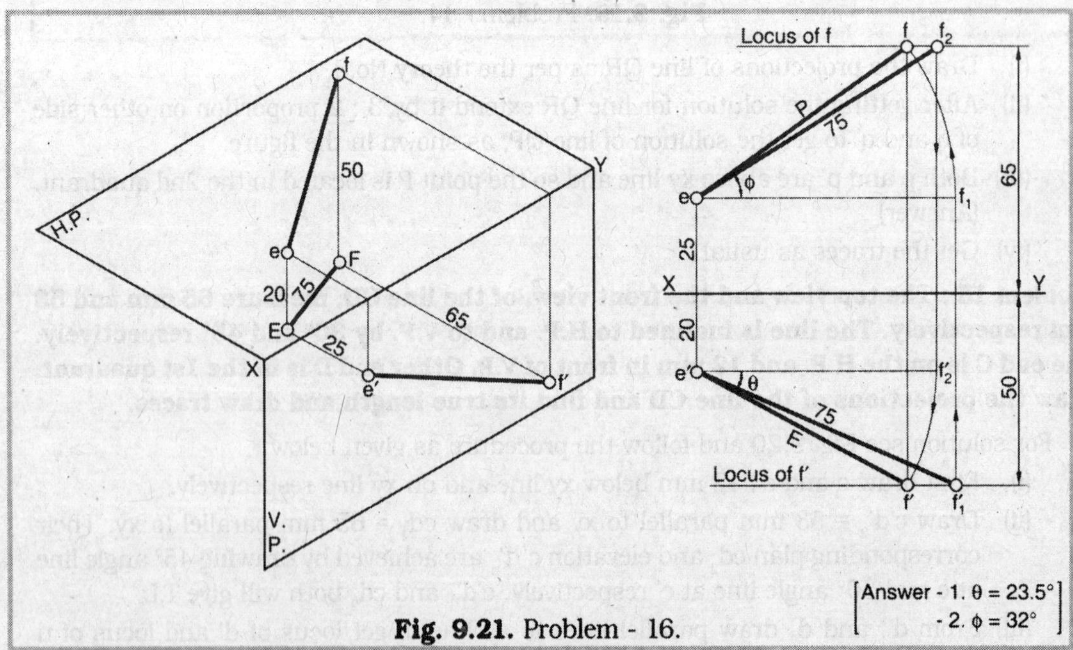

Fig. 9.21. Problem - 16.

(i) First of all mark e and e' and locus of f and locus of f'', as per given position of the point E and the point F from two reference planes.

(ii) Now with e as the centre and radius equal to 75 mm (T.L.) draw an arc to cut locus of f at f_2. Join ef_2 and measure angle Ø that ef_2 makes with xy. Similarly, with e' as the centre and radius equal to 75 mm draw an arc to cut locus of f' at f'_1. Join $e'f'_1$ and measure angle θ that $e'f'_1$ makes with xy.

(iii) Get plan ef and elevation e'f' by the procedure, as done earlier by proceeding in the directions of arrows, as shown in Fig. 9.21.

Problem 17 : The top view and the front view, of the line GH, measures 53 mm and 65 mm respectively. The line is 75 mm long. Point G is on the ground and 40 mm behind V.P. Draw the projections of the line GH and determine its inclinations with H.P. and V.P.

For solution see Fig. 9.22 and follow the procedure as given below :

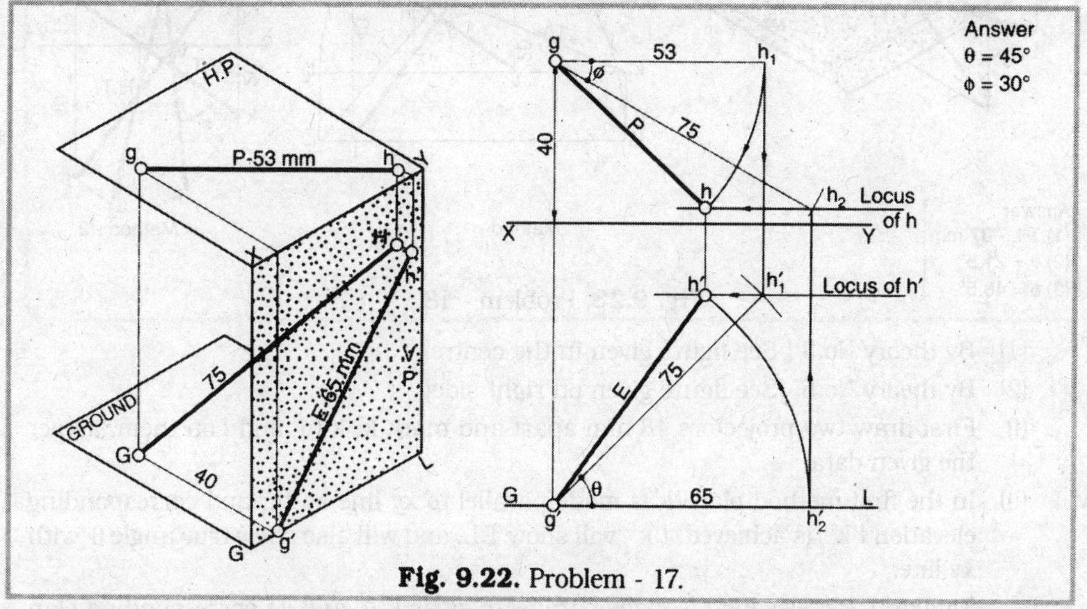

Fig. 9.22. Problem - 17.

Ground is a surface parallel to H.P. and taken below it at a suitable distance. In third angle projection system, instead of taking object resting on H.P. it is taken resting on ground. Intersection of ground with V.P. generates ground line (G.L.). Ground line is drawn parallel to xy and below it. Elevation is drawn between G.L. and xy line while plan is drawn above xy line, since H.P. plane after rotation takes its place above xy.

(i) First draw xy line and G.L. line. Mark (g') on G.L. line, since point G is on ground, and mark (g) 40 mm above xy, since point G is 40 mm behind V.P.

(ii) Draw gh_1 = 53 mm (T.V.L.) and $g'h'_2$ = 65 mm (F.V.L) parallel to xy and find their corresponding elevation $g'h'_1$ (T.L.) and plan gh_2 (T.L.) respectively by arc of 75 mm radius with g' and g as the centres respectively.

(iii) From h'_1 and h_2 draw lines parallel to xy to get locus of h' and locus of h respectively.

(iv) Now with g as the centre and radius equal to gh_1 = 53 mm (T.V.L.) draw arc of a circle to get h on locus of h. Similarly, with g' as the centre and radius equal to $g'h'_2$ = 65 mm (F.V.L.) draw arc of a circle to get h' on locus of h'. Join gh and g'h'. gh_2 (T.L.) will make an angle Ø and $g'h'_1$ (T.L.) will make θ angle with xy line. Measure them.

Problem 18 : The distance between the end projectors of the straight line KL is 48 mm. The end K is 20 mm below H.P. and 25 mm behind V.P. The end L is 12 mm above H.P. and 40 mm in front of V.P. Draw the projections and find the T.L. of the line, its inclinations with H.P. and V.P. and also its traces.

For solution see Fig. 9.23 and follow the procedure as given below:

The problem is solved by two methods :

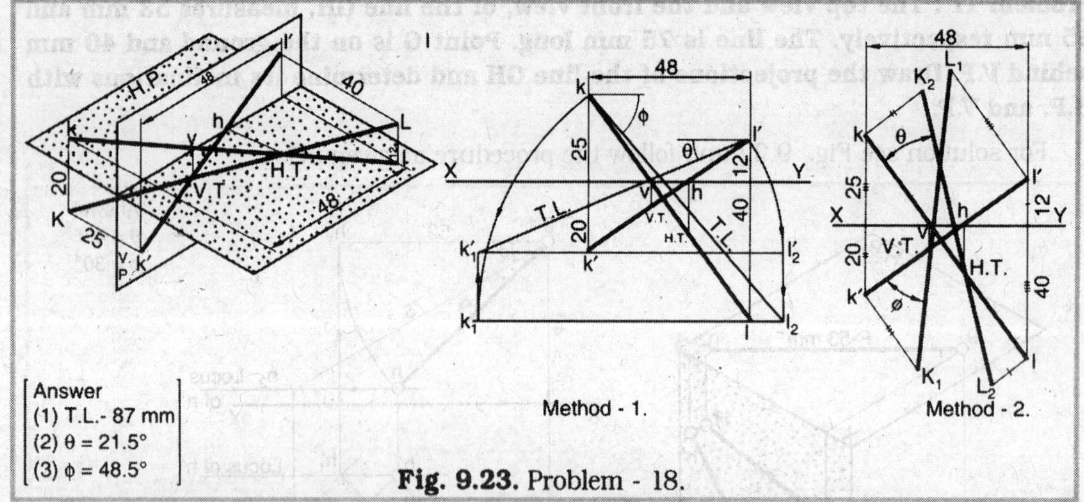

Fig. 9.23. Problem - 18.

(1) By theory No.3 [See figure given in the centre]

(2) By theory No.5. [See figure given on right side]

(i) First draw two projectors 48 mm apart and mark k, k′, *l*′ and *l* on them, as per the given data.

M.1 (ii) In the first method plan *l*k is made parallel to xy line as lk_1 and corresponding elevation $l'k'_1$ is achieved. $l'k'_1$ will show T.L. and will also show true angle θ with xy line.

Similarly, k′*l*′ elevation is made parallel to xy as $k'l'_2$ and its corresponding plan kl_2 is achieved. This plan kl_2 will show true length and will show true angle Ø with xy line.

Find out traces as usual.

M.2 (iii) Method-2 is based on theory No.5. On end points of plan k*l* i.e. at k and at *l* draw perpendiculars to plan k*l* and mark on them distance of k′ (20 mm) and of *l*′ (12 mm) from xy line to get points K_2 and L_2 respectively. As elevation points k′ and *l*′ are on opposite sides of xy, perpendiculars on plan are drawn in opposite directions.

Join K_2L_2. K_2L_2 will show true length of line KL. K_2L_2 and plan k*l* will intersect at H.T. and will make an angle θ between them.

Similarly, on end points of elevation k′*l*′ draw perpendiculars to elevation k′*l*′ and mark on them distance of k (25 mm) and of *l* (40 mm) from xy line to get points K_1 and L_1 respectively. Here also perpendiculars are drawn on opposite sides since plan points k and *l* are on opposite sides of xy line.

Join K_1L_1. K_1L_1 will show true length of line KL. K_1L_1 and elevation k′*l*′ will intersect at V.T. and will make an angle Ø between them.

This method - 2 based on theory No.5 is useful when,

(1) plan and elevation are given.

(2) plan, H.T. and θ are given.

(3) elevation, V.T. and Ø are given.

Problem 19 : The front view of the line AB, inclined at 30° to the V.P., measures 65 mm. Two ends of the line A and B are 50 mm and 20 mm below H.P. respectively. The V.T. of the line is 5 mm below H.P. Draw the projections and determine the T.L. of AB, its inclination with H.P. and its H.T. Assume line AB in 3rd quadrant.

For solution see Fig. 9.24 and follow the procedure as given below.

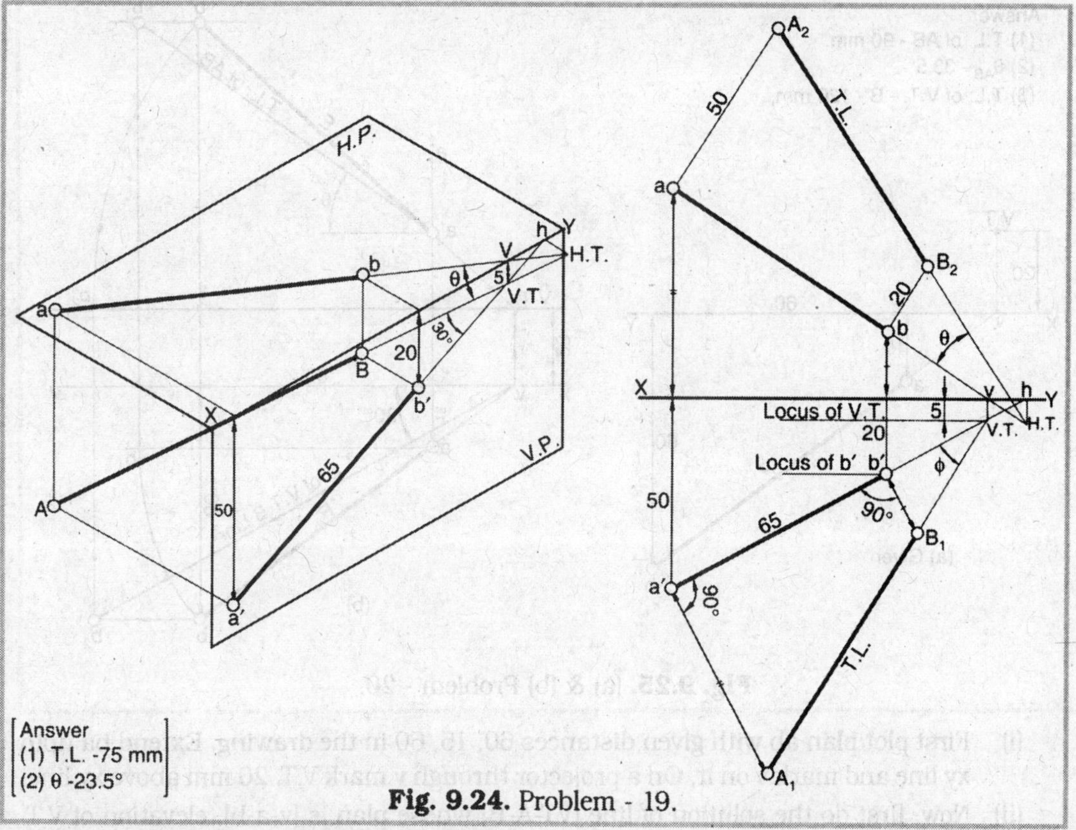

Fig. 9.24. Problem - 19.

This solution is based on theory No.5.

(i) Draw xy line and fix point a' 50 mm below xy.

(ii) Draw locus of b' and locus of V.T. 20 mm and 5 mm below xy respectively.

(iii) Now with a' as the centre and F.V. length 65 mm as the radius draw an arc to cut locus of b' at point b'. Join elevation a'b' and extend it upto locus of V.T. to get the point V.T.

(iv) At V.T. draw an angle Ø = 30° with a'b' V.T. line and draw right angle lines to a'b' at a' and b'. By intersections we will get points A_1 and B_1. A_1B_1 will show true length of line AB. a'A_1 and b'B_1 will give distances of plan points (a) and (b) respectively from xy line.

(v) Draw projectors through a' and b' and plot on them distances $a'A_1$ and $b'B_1$ respectively from xy line to get plan points (a) and (b). Join plan ab.

(vi) Now on plan plot distances of corresponding points of elevation to get A_2B_2. A_2B_2 will show true length of line AB.

A_2B_2 and plan ab extended will meet at H.T. and they will make an angle θ between them. Measure T.L. and θ.

Problem 20 : The plan ab and the V.T. of the line AB inclined at 30° to the V.P. are given in Fig. 9.25 (a). Draw the projections and determine (i) The T.L. (ii) inclination of line with H.P. and (iii) T.L. of line upto V.T.

For solution see Fig. 9.25 (b) and follow the procedure as given below :

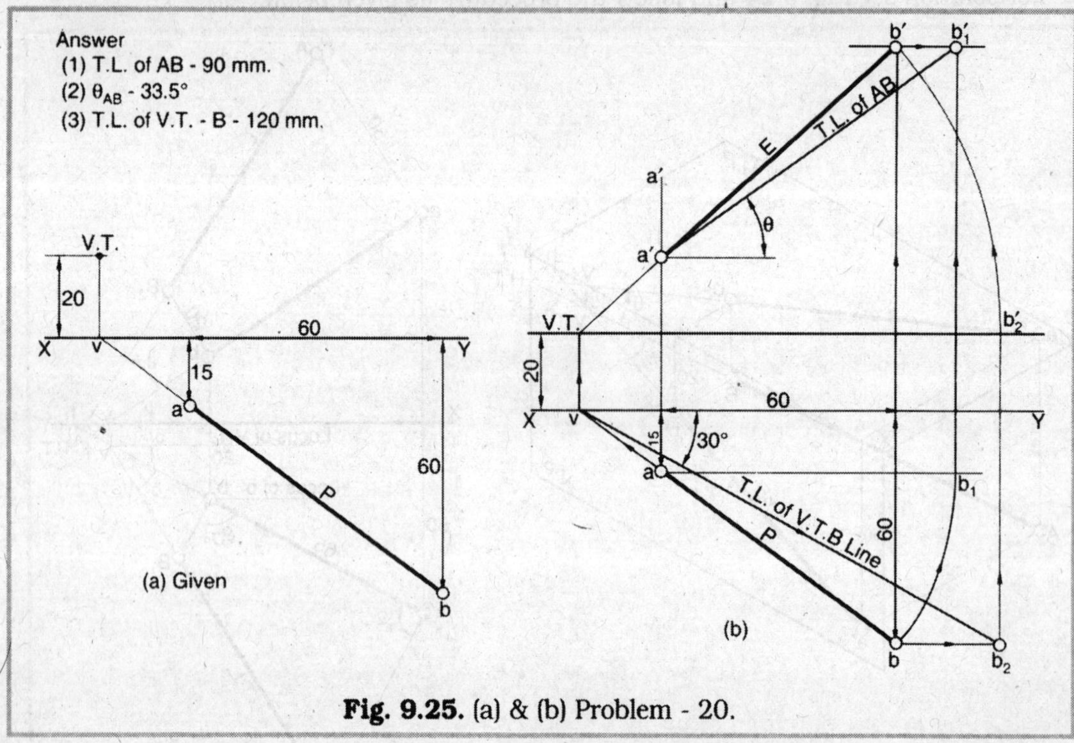

Fig. 9.25. (a) & (b) Problem - 20.

(i) First plot plan ab with given distances 60, 15, 60 in the drawing. Extend ba upto xy line and mark v on it. On a projector through v mark V.T. 20 mm above xy line.

(ii) Now, first do the solution of line [VT-A-B] whose plan is [v-a-b], elevation of V.T. as V.T. and line is inclined to V.P. by 30°.

At (v) draw 30° line with xy and parallel to xy from (b) to get by intersection point (b_2). vb_2 represents T.L. of VT-B line. [V.T.-b'_2] represents elevation length of line VT-B.

(iii) Now draw projector through b and draw arc of a circle with V.T. as the centre and elevation length [V.T.-b'_2] as the radius to get by intersection elevation point b'. Join VT-b' and mark on it a' by taking projector through a. a'b' represents elevation and ab represents plan of the line AB.

(iv) Find T.L. of AB and angle θ by usual method studied earlier.

Problem 21 : The distance between the end projectors of the straight line PJ is 55 mm and between projectors through H.T. and V.T. is 80 mm. The H.T. of the line is 35 mm in front of V.P. and V.T. of the line is 50 mm above the H.P. The end P is 10 mm above H.P. Draw the projections of the line PJ and find its inclinations with H.P. and V.P.

For solution see Fig. 9.26 and follow the procedure as given below:

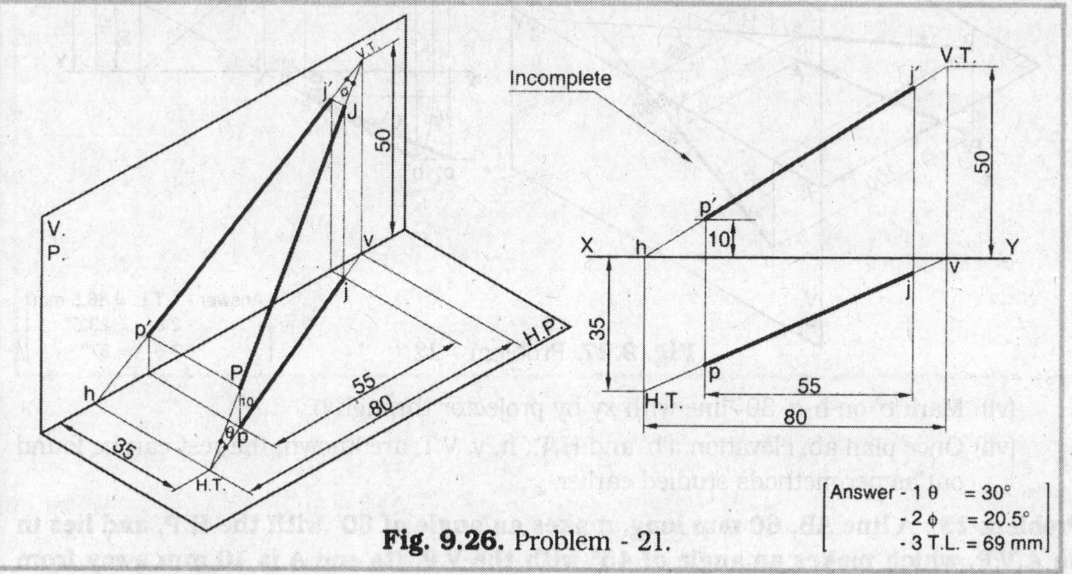

Fig. 9.26. Problem - 21.

(i) First plot two projectors 80 mm apart and mark on one h, H.T. and on other v and V.T. as per the given data.

(ii) Join h-VT and H.T.-v on which elevation points p', j' and plan points p, j are going to lie respectively. Draw locus of p' 10 mm above xy and mark p' by intersection with h-VT line. Projector through p' will fix p on HT-v line.

(iii) Now use end projectors distance 55 mm to get j' and j.

(iv) Once plan, elevation, H.T. and V.T. are known, the rest can be found easily (and so the problem is kept incomplete).

Problem 22 : The F.V. of a line AB makes an angle of 30° with xy. The H.T. of the line is 45 mm behind V.P., while its V.T. is 30 mm above H.P. One end of the line is 10 mm below H.P. and the other end is 100 mm behind V.P. Draw the projections of the line and determine (i) T.L. of the line AB and (ii) its inclinations with H.P. and V.P.

For solution see Fig. 9.27 and follow the procedure as given below:

(i) First plot h and H.T., on xy and 45 mm above xy respectively.

(ii) Now at h draw a line (b'a'h VT) at an angle of 30°, since elevation makes 30° with xy.

(iii) Mark a', on above 30° line, 10 mm below xy and mark V.T., on above 30° line, 30 mm above xy.

(iv) On a projector through V.T. on xy line mark v.

(v) Join v-H.T. and extend it. Plan points (a) and (b) will lie on this line. Mark (a) by projector through a'. Mark (b) on the same extended line v-H.T. at 100 mm above xy line.

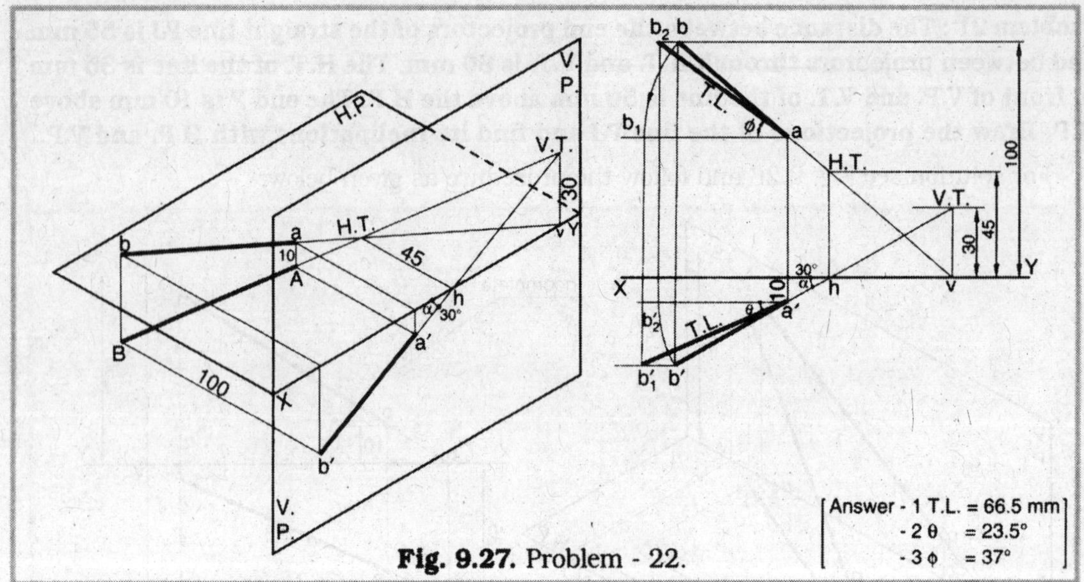

Fig. 9.27. Problem - 22.

(vi) Mark b' on h-a' 30° line with xy by projector through b.

(vii) Once plan ab, elevation a'b' and H.T., h, v, V.T. are known, the rest can be found out as per methods studied earlier.

Problem 23 : A line AB, 60 mm long, makes an angle of 60° with the H.P. and lies in an A.V.P. which makes an angle of 45° with the V.P. Its end A is 10 mm away from both the H.P. and V.P. Draw the projections of AB and determine (i) its inclination with V.P. and (ii) its traces. Assume the line in the first quadrant.

For solution see Fig. 9.28 and follow the procedure, as given below :

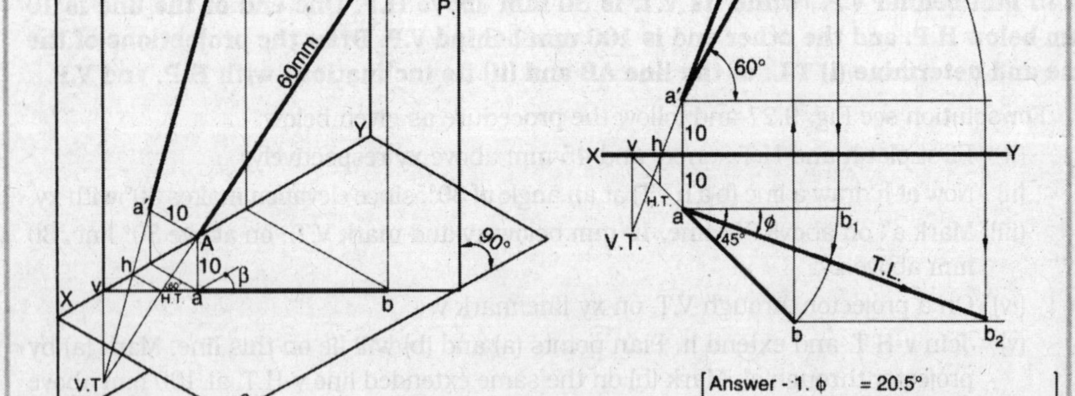

Answer - 1. ϕ = 20.5°
- 2. V.T. - 13 mm below H.P.
- 3. H.T. - 6 mm in front of V.P.

Fig. 9.28. Problem - 23.

Hints : As the line lies in an A.V.P. which makes an angle of 45° with the V.P., its plan will make an angle of 45° with xy. See 3-D drawing.

Proceed for solution in the direction of arrows given in the drawing.

Now we shall take few applied problems. In this type of problems generally earth is considered as H.P. or ground and according to situation prevailing in the problem a suitable vertical plane is selected.

Problem 24 : A fan is hanging in the centre of the room of 4m x 3m x 3m height. The centre of the fan is 0.75 m below the ceiling. The switch of this fan is on 3m x 3m size wall at the centre height and 0.5 m from the adjacent wall. Find the distance between the fan centre and the switch.

For solution see Fig. 9.29 and follow the procedure, as given below :

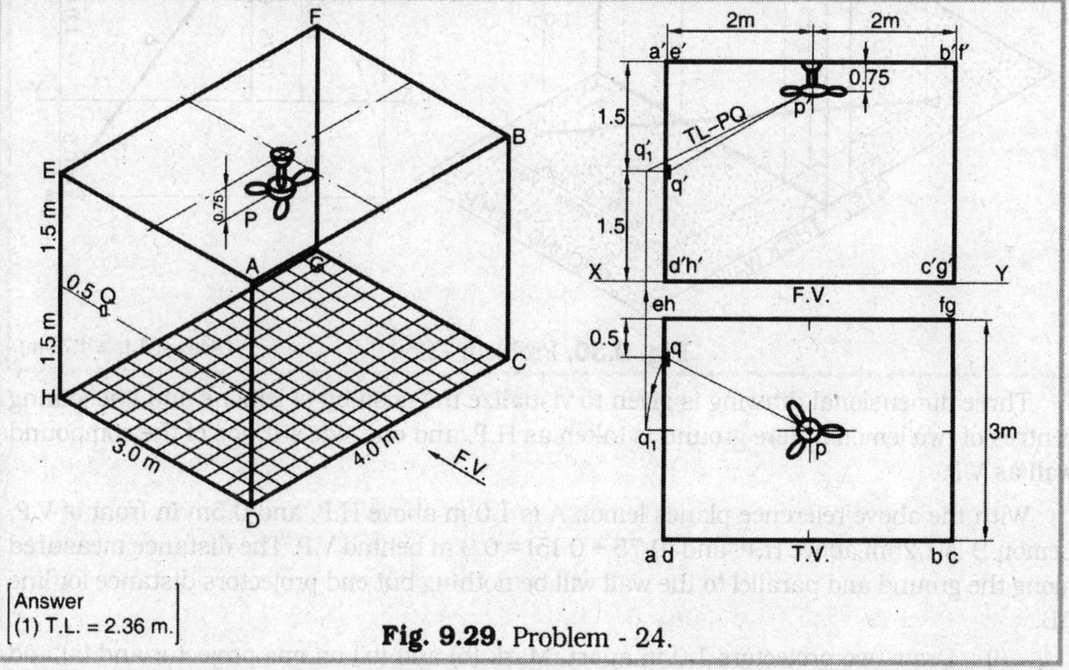

Fig. 9.29. Problem - 24.

Here, in this problem floor of the room is considered as H.P. and V.P. is taken parallel to 4m x 3 m size wall. After taking these suitable reference planes, plan and elevation of fan centre and switch are drawn along with the room.

Actual position of fan and switch is shown in three dimensional drawing. Elevation or F.V. is taken from front and plan is taken from top.

(i) First draw x-y line. Draw elevation above xy line and plan below xy line.

(ii) Elevation of the room will be 4m x 3m size rectangle and plan of the room will be also same size rectangle.

(iii) Mark plan points p and q and elevation points p' and q' in plan and elevation respectively. P is the centre of the fan and Q is the switch point.

(iv) Make plan pq parallel to xy line to position pq_1 and get its corresponding elevation $p'q'_1$. $p'q'_1$ will show true length of the line PQ i.e. distance between centre of the fan and switch point.

Problem 25 : Two lemons on a tree, planted near the compound wall of a bunglow, are 1.0 m and 1.25 m above the ground and 0.5 m and 0.75 m from a 15 cm thick compound wall but on the opposite sides of it. The distance between lemons measured along the ground and parallel to the wall is 1.0m. Determine the real distance between centres of two lemons.

For solution see Fig. 9.30 and follow the procedure, as given below :

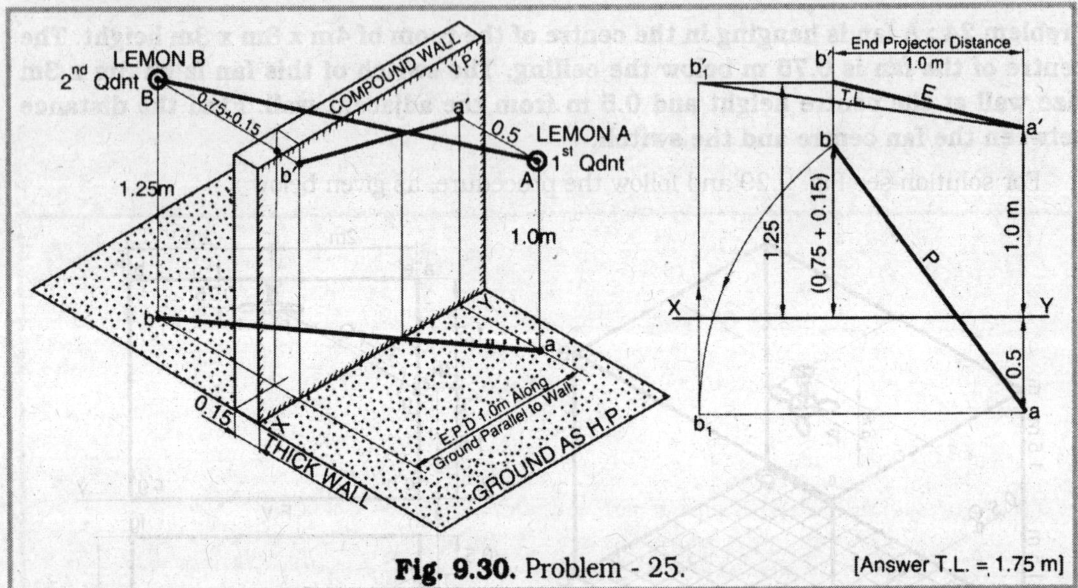

Fig. 9.30. Problem - 25. [Answer T.L. = 1.75 m]

Three dimensional drawing is given to visualize the position of lemons and line joining centres of two lemons. Here ground is taken as H.P. and one side surface of the compound wall as V.P.

With the above reference planes lemon A is 1.0 m above H.P. and 0.5m in front of V.P. Lemon B is 1.25m above H.P. and (0.75 + 0.15) = 0.9 m behind V.P. The distance measured along the ground and parallel to the wall will be nothing but end projectors distance for line AB.

(i) Draw two projectors 1.0 m apart. Mark (b) and (b') on one projector and (a) and (a') on another projector.

(ii) Make plan ab parallel to xy to position ab_1 and get its corresponding elevation $a'b'_1$, as shown in Fig. 9.30. $a'b'_1$ will show T.L. of line AB and will give real distance between the centres of two lemons.

Problem 26 : Two unequal legs AB and AC, hinged at A, make an angle of 135° between them in their elevation and plan. Leg AB is perpendicular to the P.P. Determine the real angle between them.

For solution see Fig. 9.31 and follow the procedure, as given below :

As the leg AB is perpendicular to P.P., its plan and elevation both will be parallel to xy and will show T.L.

(i) Draw b'a' and ba parallel to xy line of any length. At (a') and (a) draw line at 135° and mark on them c' and c respectively at suitable distance but on the same projector.

(ii) Find true lengths of AC and BC by usual method studied earlier.

(iii) a'b' is the true length of lines AB. So draw triangle a'b'C showing all three lines in true lengths. So in this triangle angle b'a'C will show the real angle (θ) between two legs AB and AC. Measure it. θ = 125° (Answer).

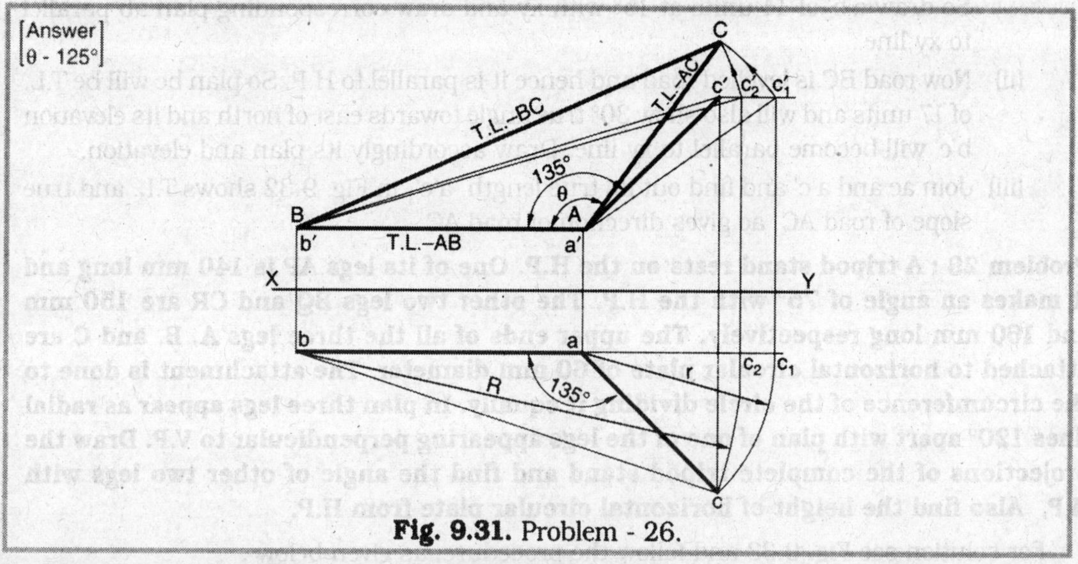

Fig. 9.31. Problem - 26.

Problem 27 : A straight road going uphill from station A, due east, to another station B and has a slope of 15°. Road distance between station A and station B is 14 kms. Another levelled (horizontal) road of 17 kms length, to join station B to station C, is in the direction 30° east of north when looked from B. Determine the length, direction and slope of the straight road joining station A to station C.

For solution see Fig. 9.32 and follow the procedure, as given below :

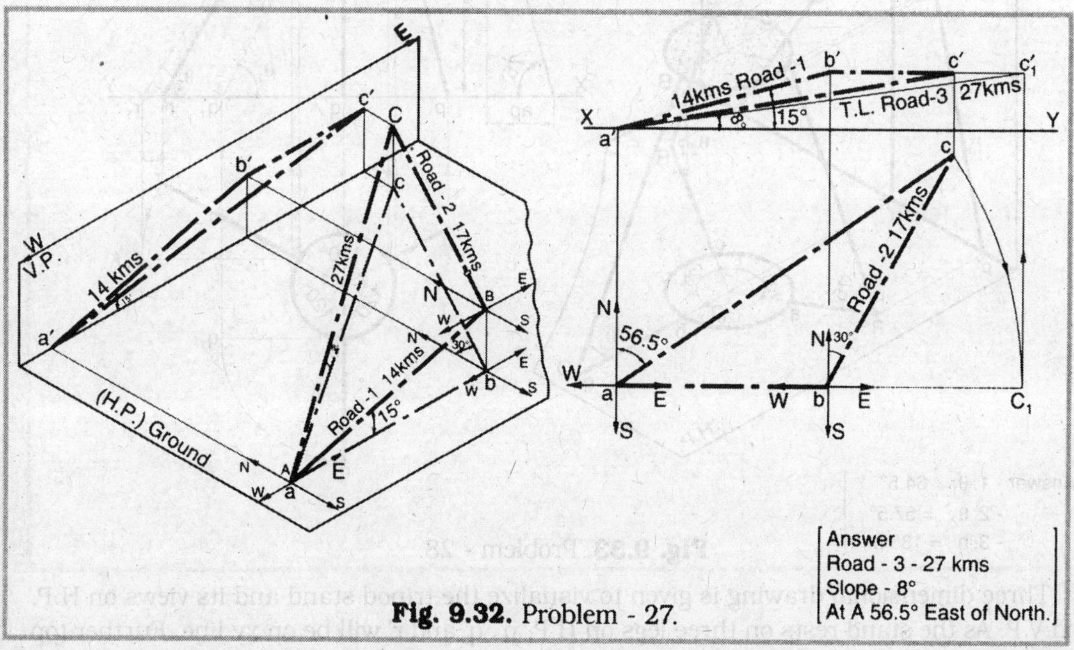

Fig. 9.32. Problem - 27.

(i) Here take earth as the H.P. and the vertical plane along east-west direction as V.P. So road AB, as it is going towards east, will become parallel to V.P. and hence a'b' will show T.L. 14 kms and will show true angle 15° with xy line and its plan ab will become parallel to xy line.

So draw a'b' of 14 units at 15° with xy and draw corresponding plan ab parallel to xy line.

(ii) Now road BC is levelled road and hence it is parallel to H.P. So plan bc will be T.L. of 17 units and will also show 30° true angle towards east of north and its elevation b'c' will become parallel to xy line. Draw accordingly its plan and elevation.

(iii) Join ac and a'c' and find out its true length. $a'c'_1$ in Fig. 9.32 shows T.L. and true slope of road AC. ac gives direction of road AC.

Problem 28 : A tripod stand rests on the H.P. One of its legs AP is 140 mm long and it makes an angle of 75° with the H.P. The other two legs BQ and CR are 150 mm and 160 mm long respectively. The upper ends of all the three legs A, B, and C are attached to horizontal circular plate of 60 mm diameter. The attachment is done to the circumference of the circle dividing it equally. In plan three legs appear as radial lines 120° apart with plan of one of the legs appearing perpendicular to V.P. Draw the projections of the complete tripod stand and find the angle of other two legs with H.P. Also find the height of horizontal circular plate from H.P.

For solution see Fig. 9.33 and follow the procedure, as given below :

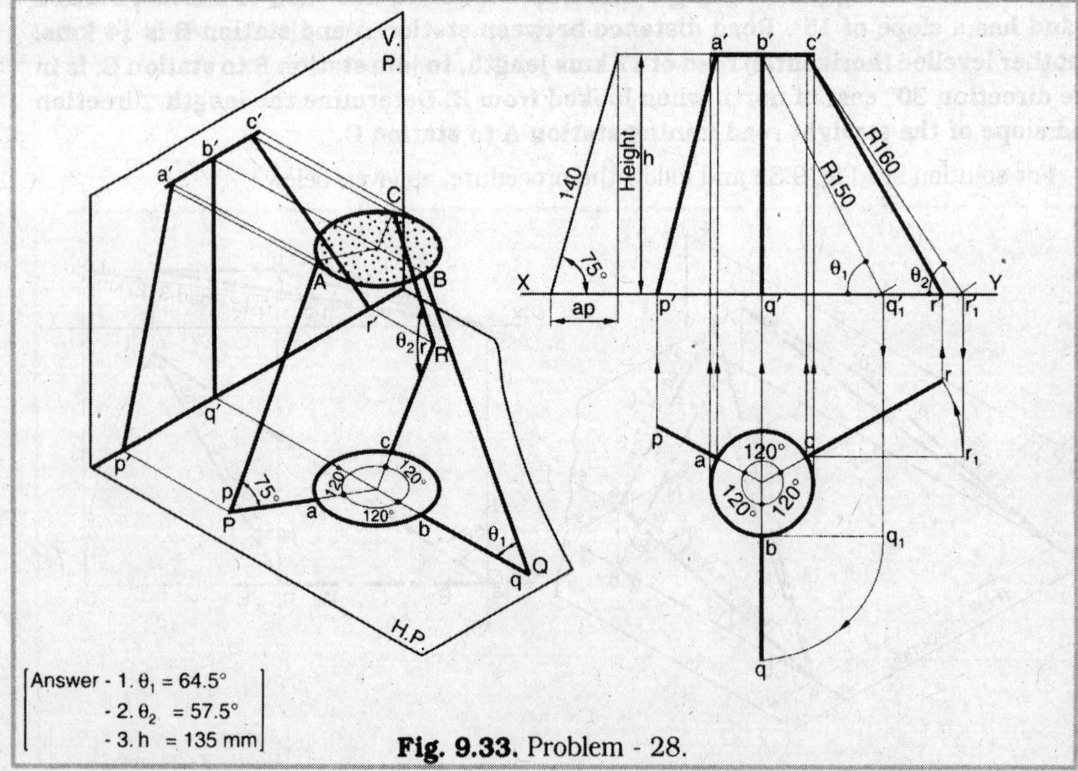

Fig. 9.33. Problem - 28.

Three dimensional drawing is given to visualize the tripod stand and its views on H.P. and V.P. As the stand rests on three legs on H.P. p', q' and r' will be on xy line. Further top surface is horizontal circular plate and so its plan will be circle (abc) and its elevation (a'b'c') will be a straight line parallel to xy line. In plan ap, bq and cr will be radial lines 120° apart, out of which bq will be perpendicular to xy line.

(i) Draw xy.

(ii) Draw 140 mm long line at 75° to xy line to get height h and plan length ap, as shown in Fig. 9.33.

(iii) In plan draw circle of 60 mm diameter and mark on it a, b and c 120° apart and draw radial lines. On one line mark ap by taking plan length ap.

(iv) Locate p' on xy line and a', b' and c' at height h from xy line by drawing projectors through p, a, b and c respectively.

(v) Now with b' and c' as centres and radii equal to 150 mm and 160 mm draw arcs on xy line to get points q'_1 and r'_1 respectively. $b'q'_1$ will make θ_1 angle and $c'r'_1$ will make θ_2 , with xy line. θ_1 and θ_2 represent angles of legs BQ and CR with H.P.

Further bq_1 and cr_1 represent their corresponding plan or in other words plan lengths of BQ and CR. So rotate plan lengths on radial lines to get bq and cr respectively.

Projectors through q and r on xy line will fix q' and r' respectively. Join b'q' and c'r'. Project also extreme two points of the circle.

TYPICAL PROBLEMS

Problem 29 : The elevation length of a straight line AB, inclined at 35° to the V.P., is 80 mm. The end A is 40 mm above H.P. while the end B is 80 mm in front of V.P. Draw the projections of the straight line when its V.T. is 15 mm above the H.P. and the line is completely in the first quadrant. Find the T.L. of the line and locate its H.T.

For solution, see Fig. 9.34 and follow the procedure, as given below :

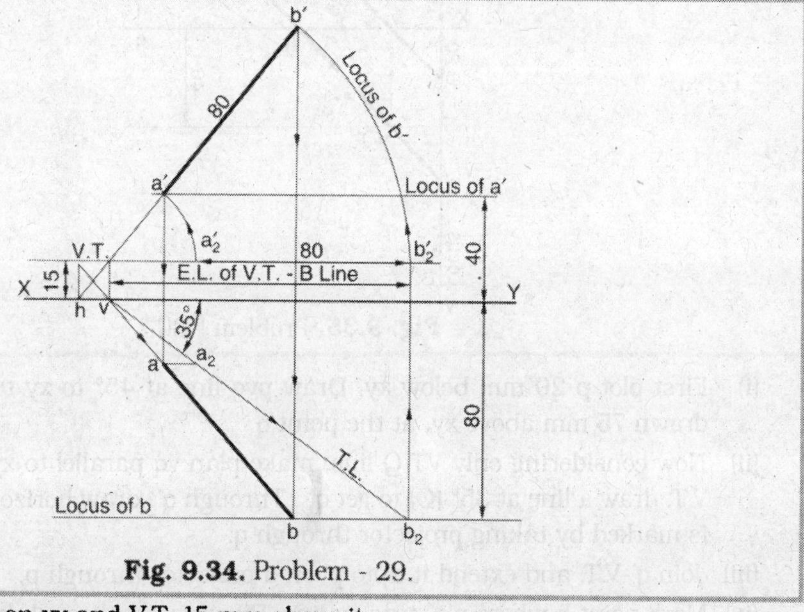

Fig. 9.34. Problem - 29.

(i) First plot (v) on xy and V.T. 15 mm above it.

(ii) Considering line VT-B, draw at (v) angle of 35° (Ø) below xy, since locus of (b) is going to be below xy.

(iii) Draw locus of (b) 80 mm below xy to cut 35° line at (b_2). vb_2 represents T.L. of VT.B line.

(iv) Get elevation length VT-b'_2 of line VT-B. From this length cut 80 mm from b'_2 side, since the length of elevation a'b' is given 80 mm.

(v) Now draw locus of (a') 40 mm above xy line.

(vi) Now with V.T. as the centre and radii equal to [VT-a'_2] and (VT-b'_2) draw two arcs of circles, the first arc will cut locus of (a') at point (a').

(vii) Join VT-a' and extend it upto (b') on second arc.

(viii) From (b') draw projector to get (b) on locus of (b). Join (v) with (b) and on it mark (a) by projector through (a').

(ix) From (a) draw horizontal to cut 35° line at point (a_2). a_2b_2 represents true length of line AB.

Problem 30 : The straight line PQ is inclined at 35° to H.P. while its plan is at 45° to the XY line. The end P is 20 mm in front of V.P. and it is below H.P. The end Q is 75 mm behind V.P. and it is above H.P. Draw the projections of the straight line when its V.T. is 40mm below H.P. Find the T.L. of the portion of the straight line which is in the second quadrant and locate its H.T.

For solution see Fig. 9.35 and follow the procedure, as given below :

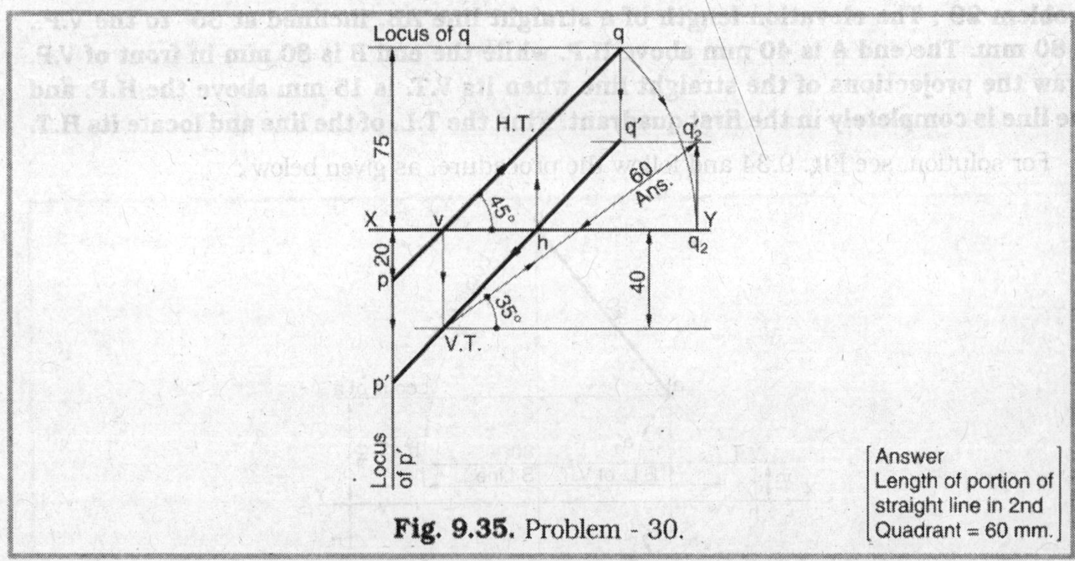

Fig. 9.35. Problem - 30.

(i) First plot p 20 mm below xy. Draw pvq line at 45° to xy to cut the locus of q, drawn 75 mm above xy, at the point q.

(ii) Now considering only VT-Q line, make plan vq parallel to xy and through point V.T. draw a line at 35° (Ø) to get q'_2. Through q'_2 draw horizontal line on which q' is marked by taking projector through q.

(iii) Join q'-V.T. and extend it upto p' on a projector through p.

(iv) Mark point h where p'q' cuts xy line. By projector through h on plan mark H.T.

(v) Measure on V.T.-q'_2 length above xy to get the length of the portion of the straight line which is in the second quadrant.

Problem 31 : The distance between the end projectors of a straight line AB is 60 mm. The end A is 25 mm above H.P. while the end B is in the second quadrant. The elevation and plan of the line are inclined at 50° and 40° to the xy line respectively. Draw the projections of the line when its V.T. is 50 mm above H.P. Find the T.L. of the portion of the line which is in the second quadrant and its inclination with H.P. and V.P. Locate the H.T.

For solution see Fig. 9.36 and follow the procedure, as given below :

(i) First draw two projectors 60 mm apart.

(ii) On one projector mark (a'), 25 mm above xy. Through (a') draw 50° line on upperside to get b' on second projector.

(iii) Draw locus of V.T., 50 mm above xy line, which will cut a'b' at V.T.

(iv) Mark v on xy by projector through V.T.

(v) At v draw 40° line ab, as shown in figure. Now ab and a'b' are plan and elevation of straight line AB respectively. The rest of the procedure can be easily done by students.

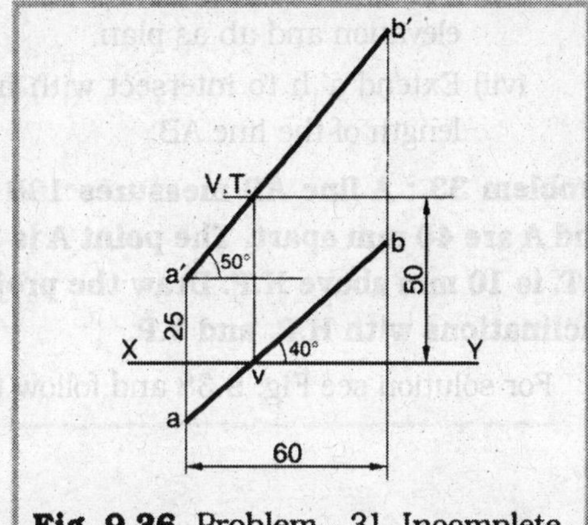

Fig. 9.36. Problem - 31. Incomplete.

Problem 32 : A straight line AB is inclined at 30° to the H.P. while its elevation is inclined at 45° to the xy line. The end A is 60 mm above H.P. and it is in front of V.P. The end B is 80 mm behind V.P. and it is in the third quadrant. Draw the projections of the straight line when its H.T. is 20 mm behind V.P. Find the T.L. of the line and locate its V.T.

For solution see Fig. 9.37 and follow the procedure, as given below :

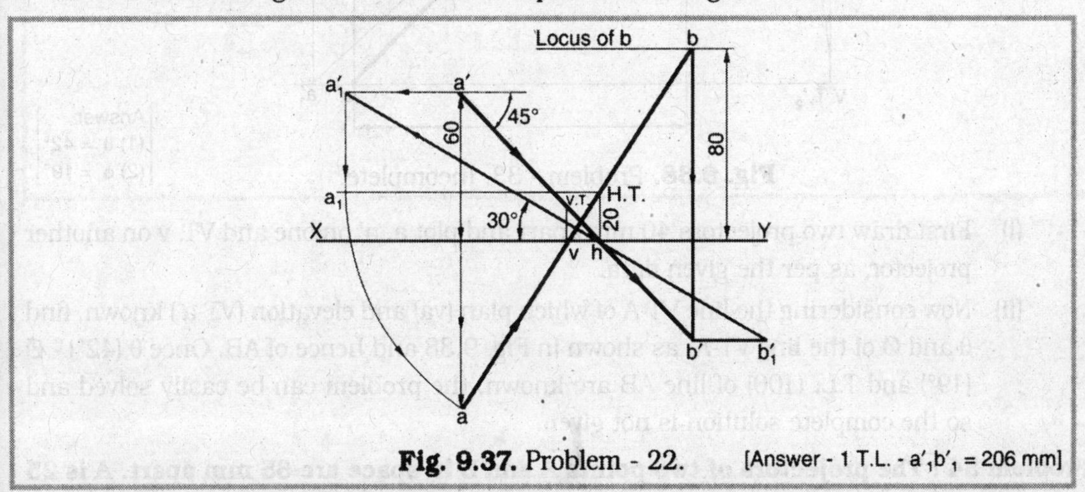

Fig. 9.37. Problem - 22. [Answer - 1 T.L. - $a'_1b'_1$ = 206 mm]

(i) First mark a' 60 mm above xy line. Draw 45° line at a' and get h on xy line.

(ii) Mark H.T. 20 mm above h.

(iii) Now considering only line H.T.-A, draw 30° (θ) at h on upperside and draw horizontal from a' to get a'_1 by intersection. a'_1h represents true length of line H.T.-A.

(iv) H.T.-a_1 represents plan length of H.T.-A. So with H.T. as the centre and radius equal to H.T.-a_1 draw an arc to cut projector through a' at a.

(v) Join a, H.T. and extend it upto b which will be on the locus of b, previously drawn 80 mm above xy.

(vi) From b draw projector to intersect with extended line a'h at b'. a'b' represents elevation and ab as plan.

(vii) Extend a'_1h to intersect with horizontal from b' at b'_1. $a'_1b'_1$ represents the true length of the line AB.

Problem 33 : A line AB measures 100 mm. The projectors through its V.T. and the end A are 40 mm apart. The point A is 30 mm below H.P. and 20 mm behind V.P. The V.T. is 10 mm above H.P. Draw the projections of the line and determine its H.T. and inclinations with H.P. and V.P.

For solution see Fig. 9.38 and follow the procedure, as given below :

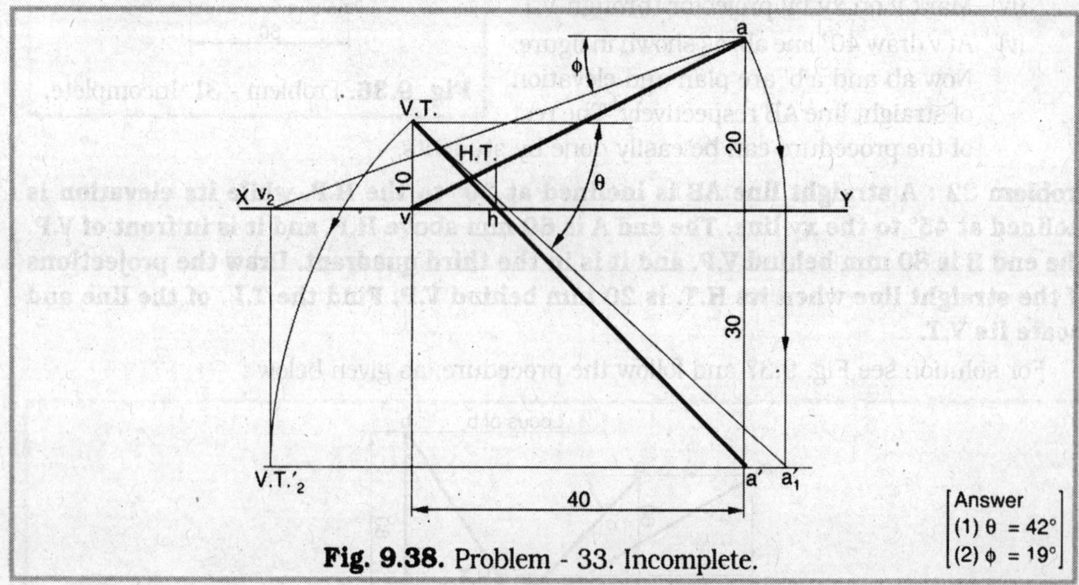

Fig. 9.38. Problem - 33. Incomplete.

(i) First draw two projectors 40 mm apart and plot a, a' on one and VT, v on another projector, as per the given data.

(ii) Now considering the line VT-A of which plan (va) and elevation (VT-a') known, find θ and Ø of the line VT-A, as shown in Fig. 9.38 and hence of AB. Once θ (42°) - Ø (19°) and T.L. (100) of line AB are known, the problem can be easily solved and so the complete solution is not given.

Problem 34 : The projectors of two points A and B in space are 65 mm apart. A is 25 mm behind the V.P. and 30 mm below H.P. The point B is 40 mm above H.P. and 15 mm in front of V.P. Find the third point C in the H.P. and in front of V.P. such that its distance from a point A is 90 mm and that from B is 60 mm.

For solution see Fig. 9.39 and follow the procedure, as given below :

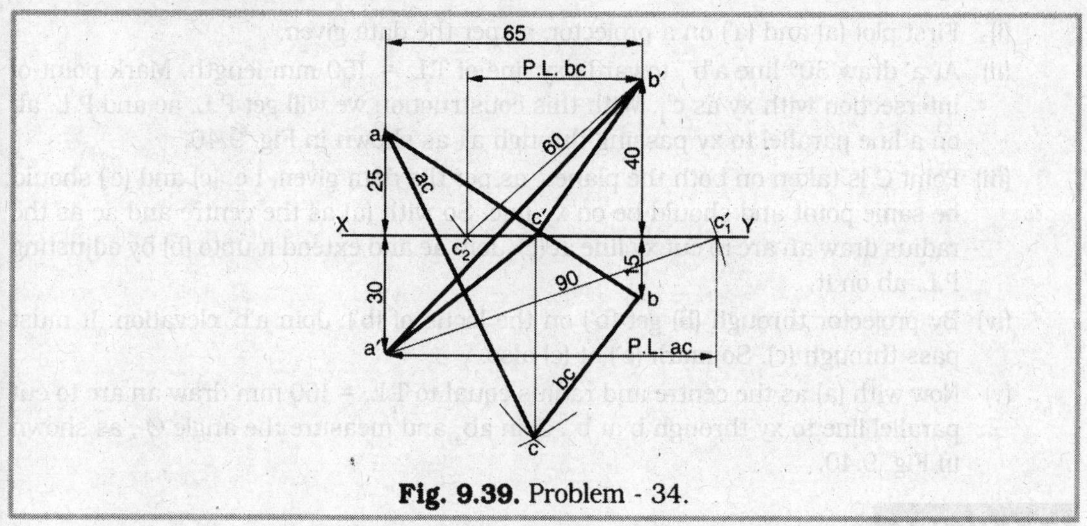

Fig. 9.39. Problem - 34.

(i) First draw two projectors 65 mm apart and plot a, a' on one and b, b' on another projector as per the given data. Join ab and a'b'.

(ii) xy line is going to be the locus of c' since point C is in H.P.

(iii) So with a' as the centre and R = 90 mm draw an arc to cut xy line at c'_1 and get plan length ac. Similarly, get plan length bc.

(iv) Now with (a) as the centre and R = ac draw an arc and draw another arc with (b) as the centre and R = bc to get the point (c) by intersection. These two arcs should be drawn below xy and not above xy, since point C is in front of V.P.

(v) From c draw projector to get c' on xy line.

Problem 35 : The end A of the straight line AB, 150 mm long, is 50 mm behind V.P. and 35 mm below H.P. The line is inclined at an angle of 30° with the H.P. and has a point on it in both the reference planes. Draw the projections of the line and find its inclination with V.P.

For solution see Fig. 9.40 and follow the procedure, as given below :

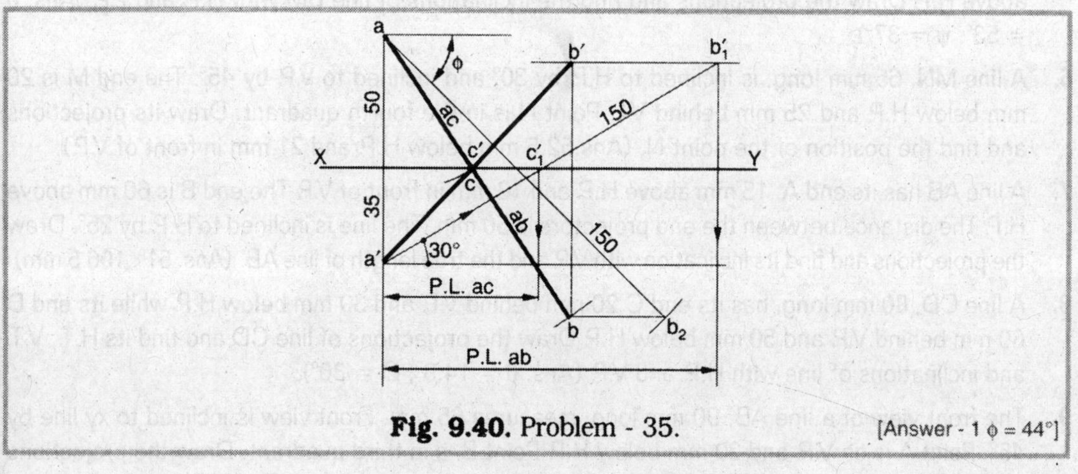

Fig. 9.40. Problem - 35.

[Answer - 1 ϕ = 44°]

(i) First plot (a) and (a') on a projector, as per the data given.

(ii) At a' draw 30° line $a'b'_1$ towards xy line of T.L. = 150 mm length. Mark point of intersection with xy as c'_1. With this construction we will get P.L. ac and P.L. ab on a line parallel to xy passing through a', as shown in Fig. 9.40.

(iii) Point C is taken on both the planes, as per the data given, i.e. (c) and (c') should be same point and should be on xy line. So with (a) as the centre and ac as the radius draw an arc to cut xy line at (c). Join ac and extend it upto (b) by adjusting P.L. ab on it.

(iv) By projector through (b) get (b') on the locus of (b'). Join a'b' elevation. It must pass through (c). So mark (c') at (c) also.

(v) Now with (a) as the centre and radius equal to T.L. = 150 mm draw an arc to cut parallel line to xy through b at b_2. Join ab_2 and measure the angle Ø , as shown in Fig. 9.40.

EXERCISE

1. A line AB is 75 mm long. It is perpendicular to profile plane. The point A is 30 mm above H.P. and 40 mm in front of V.P. Draw the projections.
2. A line CD, 70 mm long, is parallel to V.P. and P.P. Point C is nearer to H.P., 35 mm below H.P. and is behind V.P. Draw the projections.
3. A line EF, 80 mm long, is having its elevation point view. Point E is 25 mm above H.P. and 30 mm in front of V.P. and the point F is in the second quadrant. Draw its projections and find the ratio of length of line in first quadrant to the length in the second quadrant. (Ans. 3 : 5).
4. A line PQ, 75 mm long is parallel to V.P. and inclined to H.P. by 60°. The farthest point P from H.P. and P.P. is 70 mm from both the planes and 30 mm from V.P. Line PQ is in the third quadrant. Draw three projections and find inclination of line with P.P. State the position of the point Q. (Ans. 30°, 5 mm below H.P. and 30 mm behind V.P.)
5. A line BM, 100 mm long, is having its plan 60 mm long and parallel to xy and its side view is 80 mm long and parallel to x_1y_1. Point B is 30 mm in front of V.P., 80 mm from P.P. and 10 mm above H.P. Draw the projections and find the inclinations of line BM with H.P. and P.P. (Ans. $\theta = 53°$, $\psi = 37°$).
6. A line MN, 65 mm long, is inclined to H.P. by 30° and inclined to V.P. by 45°. The end M is 20 mm below H.P. and 25 mm behind V.P. Point N is in the fourth quadrant. Draw its projections and find the position of the point N. (Ans.52.5 mm below H.P. and 21 mm in front of V.P.).
7. A line AB has its end A, 15 mm above H.P. and 10 mm in front of V.P. The end B is 60 mm above H.P. The distance between the end projectors is 50 mm. The line is inclined to H.P. by 25°. Draw the projections and find its inclination with V.P. and the true length of line AB. (Ans. 51°; 106.5 mm).
8. A line CD, 80 mm long, has its end C 20 mm behind V.P. and 30 mm below H.P. while its end D 60 mm behind V.P. and 50 mm below H.P. Draw the projections of line CD and find its H.T., V.T. and inclinations of line with H.P. and V.P. (Ans. $\theta = 14.5°$; Ø = 30°).
9. The front view of a line AB, 90 mm long, measures 65 mm. Front view is inclined to xy line by 45°. Point A is on V.P. and 20 mm below H.P. Point B is in third quadrant. Draw the projections and find inclinations of line with V.P. and H.P. Also find traces. (Ans. Ø = 44°, $\theta = 30.5°$).

10. A line ABC, 120 mm long, has point B on xy line. AB:BC:::4:6. Its end A is in the first quadrant and B is in the third quadrant. The line is inclined to H.P. by 45° and to V.P. by 30°. Draw the projections of the line and measure elevation and plan length and state where is H.T. and V.T. (Ans : 104 mm ; 85 mm; H.T. and V.T. will be on xy line).

11. Distance between the end projectors of line CD is 75 mm. Points C and D are 25 mm and 75 mm below H.P. respectively. The line is inclined to V.P. by 30° and its V.T. is 50 mm below H.P. Draw the projections of CD and find the true length of line and its inclination with H.P. and H.T. Assume the point C in the third quadrant. (Ans T.L. = 104 mm ; θ = 28.5°)

12. A room is 5 m x 4.5m x 4 m high. Determine by method of projections of straight lines, distance between diagonally (solid) opposite corners of the room. (Ans. 7.83 metre).

13. The distance between end projectors of a straight line PQ is 130 mm. Point P is 40 mm below H.P. and 25 mm in front of V.P. Point Q is 75 mm above H.P. and 30 mm behind V.P. Draw the projections and find out its T.L., θ, Ø, H.T. and V.T. (Ans.T.L. =182 mm, θ = 39°, Ø = 17.5°).

14. Two mangoes A and B on a tree are 0.5 m and 1 m above the ground respectively. P and Q are two compound walls at right angle. Mango A is 0.30 m from the wall P and 0.6m from the wall Q. Mango B is 1.5m from the wall P and 2 m from the wall Q. Draw three projections of two mangoes and find the real distance between their centres. Assume both the mangoes inside the compound.

Also find distances of mangoes A and B from the corner, where two walls and the ground meet. (Ans.1.91 m; 0.837 m; 2.69 m).

15. Gurushikhar (P), Achalgadh (Q) and Delwada (R) are three hill stations 4000 m, 3000 m, 2000 m above the sea level respectively. They are connected by rope-way with each other. Hill stations P, Q and R are seen at an angle of elevation of 25°, 20° and 15° respectively from a station in the valley O at sea level. From the point O, P is due south-west, Q is due north and R is due north-west. Find the lengths of the rope for rope-way. Assume ropes tight and so consider them as straight lines.

(Ans. PQ-15.57 km; QR-6.7 km and RQ = 11.5 kms).

16. A line LMN, 120 mm long, is inclined to V.P. by 30° and inclined to H.P. by 45°. Its mid point M is the V.T. of a line and is 20 mm above H.P. Assuming the point L in the first quadrant, draw the projections of the straight line and find out the quadrant of point N. (Ans. Third quadrant).

17. End points A and B of a line AB, 90 mm long, are H.T. and V.T. respectively. The line AB is inclined at 30° and 60° with H.P. and V.P. respectively. Draw its projections and find positions of point A and B. Assume the line in first quadrant. (Ans. A on H.P. and 78 mm in front of V.P.; B is on V.P. and 45 mm above H.P.) (Hints : When θ+ Ø = 90°, line is parallel to P.P. and hence distance between end projectors will be zero).

18. Draw the projections of a straight line 80 mm long inclined at 60° to H.P. and 30° to V.P. with end A in the H.P. and the end B in V.P.

19. AB is straight line whose front view measures 60 mm and makes an angle of 45° with the XY line. The end A is in the H.P. and the vertical trace of the line is 10 mm above H.P. The straight line is inclined at 30° to V.P.

Draw the projections of AB and find its true length and true inclination with the H.P. Also locate its horizontal trace. (Ans. 69.3 mm ; θ = 38°).

20. A room measures 8 metres long, 5 metres wide and 4 metres high. An electric point hangs in the centre of the ceiling and 1 metre below it. A thin straight wire connecting the point to a switch kept in one of the corners of the room and is 2 metres above the floor. Draw the projections of the wire. Find the true length and the slope angle of the wire with the floor. (Ans. T.L. = 4.82 m, θ = 12°).

21. Three vertical poles AB, CD and EF are respectively 2.5m, 4m and 6m long. Their ends B, D and F are on the ground and form the corners of an equilateral triangle of 5 m long sides. Determine graphically the distances between top ends of the poles, namely AC, CE and AE also the inclinations of these with the ground. Scale 1 : 50. (Ans. AC = 5.22 m, CE = 5.385 m, AE = 6.1m; θ_{AC} = 16.7°, θ_{CE} = 21.8°, θ_{AE} = 35°)

22. Three points A, B and C are connected with each other by rods. A is 4 m above the gorund level, B is on the ground and C is 5 m below the ground level. The angles of depression of A, B and C when seen from a point O on a hillock 10 m above the ground are respectively 10°, 20° and 30°. A is 60° towards east of north, B is due north and C is due south east of O. Find the lengths of the connecting rods. Scale 1 : 100. (Ans. AB = 31.53 m ; BC = 49.64 m ; AC = 38.16 m).

23. Two fan motors hang from the ceiling of a hall, 12 m x 5m x 8m high at a height of 3 m from the floor. The motors are 3 m and 9 m from the end wall, 2 m and 3 m from the front wall. Determine graphically the distance of each motor from a corner of the hall of the floor. (Ans. 4.7 m; 5.2m ; 9.7 m and 9.95 m).

24. A line MN measures 120 mm. Its top and front views measure 80 mm and 96 mm respectively. A point P on the line, dividing it in the ratio of 1 : 2 i.e, MP : PN = 1 : 2, is contained by both the reference planes. Draw the projections of line, and determine its traces and inclinations with the reference planes. (Ans. θ = 48°; Ø = 37°).

25. A line PQ measuring 150 mm has its vertical trace 15 mm below horizontal plane. The end P is 40 mm below the horizontal plane and 30 mm behind the vertical plane. The projectors through its vertical trace and end P are 60 mm apart. Determine the projections and horizontal trace of the line. Also find its inclination to reference planes.

A point R is in the vertical plane 80 mm from the end P and 100 mm from the end Q of the above line PQ. Determine its distance from the horizontal plane.

If the distance of the point R from P remains constant and the point remains in the vertical plane, what is its least distance from Q ? (Ans. = 20.5°; Ø = 25°; H.T. is 18 mm in front of V.P.).

26. A chimney of a hostel kitchen is 15 m high and 0.5 m in diameter. This chimney is supported by three guy wires AP, BQ and CR which appear in plan at equal angles to each other. The ends P,Q and R are pegged to the ground at distances 2.5 m, 3.5 m and 4.5 m respectively from the centre of the chimney. The other ends A, B and C of wires are connected to a ring 3 m below the top end of the chimney. Draw the projections and find the lengths of the three guy wires. (Ans. 12.2 m; 12.43m ; and 12.73 m).

27. The projectors through the traces of a line PQ are 120 mm apart. Distance between end projectors of line PQ is 75 mm. End point P is 15 mm below H.P. The H.T. and V.T. are 50 mm behind V.P. and 80 mm below H.P. respectively. Draw the projections of PQ and determine its true length and inclination with H.P. and V.P. (Ans. 95.4 mm : θ = 31.5° ; Ø = 19°).

28. Two legs of the divider AB and AC are seen at 135° between them in their end elevation and plan. Leg AB is perpendicular to V.P. Draw the projections and find the real angle between two legs of the divider. (Ans. 125°).

CHAPTER - 10

Projections of Planes

1. INTRODUCTION :

A plane surface has only two dimensions. Third dimension thickness of a plane surface is negligible and assumed to be zero.

Planes are of three types. They are classified according to their three different types of position in space with respect to three principal planes, H.P., V.P. and P.P.

(1) Parallel to one and perpendicular to two other principal planes

(2) Perpendicular to one and inclined to two other principal planes

(3) Oblique or inclined to all P.P s.

2. TRACES OF A PLANE :

Just like traces of a line, planes also have traces. A plane, extended if necessary, will meet principal planes along straight lines unless it is parallel to any one of them. These straight lines are known as traces. Plane will meet, H.P. along horizontal trace (H.T.), V.P. along vertical trace (V.T.) and P.P. along profile trace (P.T.).

3. PLANE PARALLEL TO ONE AND (HENCE) PERPENDICULAR TO TWO OTHER P.Ps.

We shall study projections of these types of planes by taking few concrete examples, in 1st Angle System.

Problem 1 : A circular plate, 30 mm in diameter, is parallel to V.P. Its centre is 20 mm above H.P. and 22 mm in front of V.P. Draw its three projections and find its traces.

For solution see Fig. 10.1 and follow the procedure as given below. Solution is done in 1st angle system.

Looking at the three dimensional drawing of the circular plate in the given position, we can visualize that Elevation will be a circle of the same size. Plan will be straight line parallel to xy line and side view will be straight line parallel to x_1y_1 line. Plan and end view will have length equal to the diameter.

(i) First plot O' and O, 20 mm above xy and 22 mm below xy respectively.

(ii) With O' as the centre and radius equal to 15 mm draw a circle to get elevation.

(iii) Through O draw line parallel to xy and project extreme points a' and c' of the circle of elevation to get plan.

(iv) Get side view by the usual method.

(v) Since the plane is parallel to V.P. it will not have V.T. Its plan will be H.T. and its side view will be P.T.(Profile Trace).

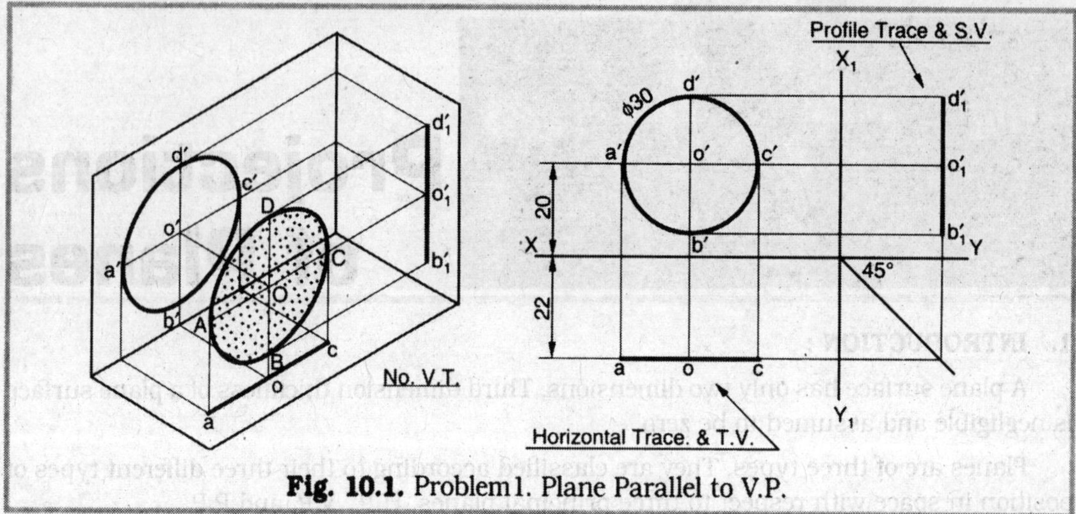

Fig. 10.1. Problem1. Plane Parallel to V.P.

Problem 2 : A square plate ABCD, side 25 mm is parallel to H.P. and is in the first quadrant. One of the edges of square plate is parallel to V.P. Draw its three projections and find out its traces.

For solution see Fig. 10.2 and follow the procedure, as given below

In three dimensional drawing, square plate is shown parallel to H.P. It is seen that plan will show true shape (square of 25 mm side) and its elevation and end view will give straight lines.

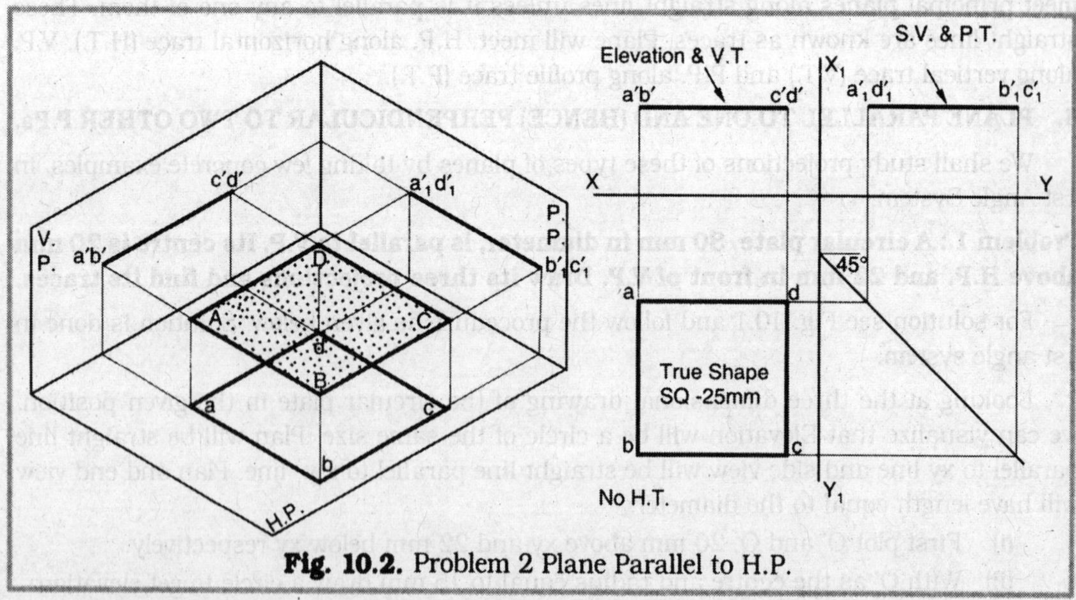

Fig. 10.2. Problem 2 Plane Parallel to H.P.

Further as the plate is in 1st quadrant, plan will be below xy and elevation and end view will be above xy line.

(i) Draw XY and X_1Y_1 lines mutually at right angle.

(ii) Draw plan (square of 25 mm side) below xy with edge parallel to xy at suitable distance and project its corners a,b,c and d for elevation above xy and end view above xy but through x_1y_1 line as per the usual method.

(iii) Draw elevation and end view at suitable distance parallel to xy line between projectors, as shown in figure.

(iv) Elevation will be V.T. of the plane and side view will be P.T. of the plane.

Problem 3 : An equilateral triangular plate UKP, side 40 mm, is parallel to V.P. and 40 mm from it with one of the edges of the plate inclined at 45° to the H.P. Draw its projections.

(Solve it by first angle projection system.)

For solution see Fig. 10.3 and follow the procedure, as given below:

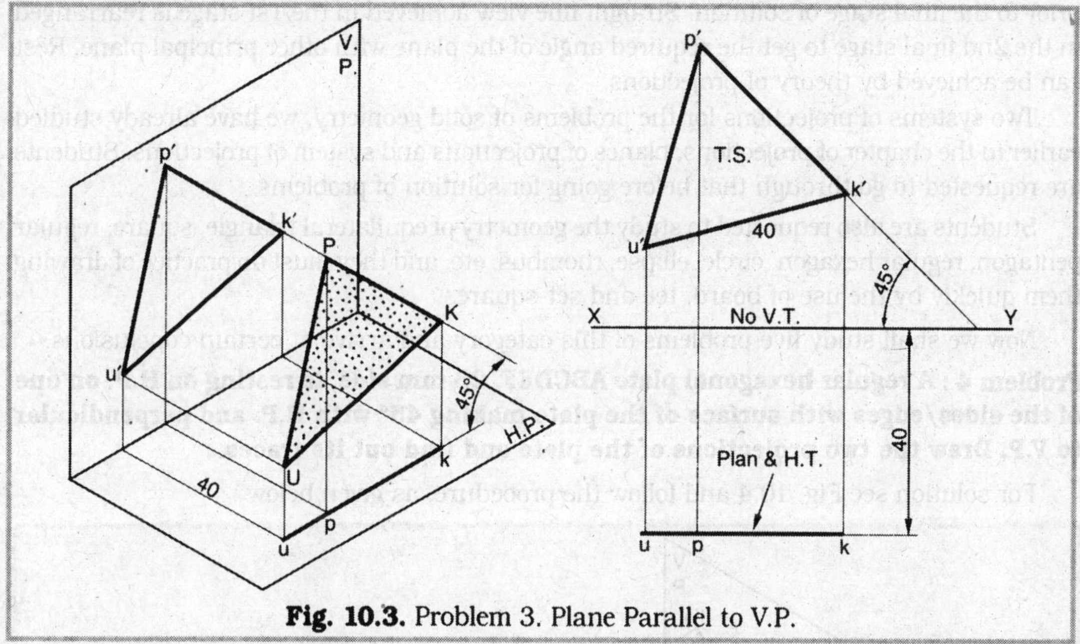

Fig. 10.3. Problem 3. Plane Parallel to V.P.

As the solution required is in first angle projection system, imagine equilateral ΔUKP in the first quadrant in the given position, as shown in Fig. 10.3. Observing this three dimensional drawing, we can say that its elevation will show the true shape and elevation k'p' of one edge KP of ΔUKP will make 45° with xy line and plan of ΔUKP will be parallel to xy and 40 mm below it.

(i) Draw equilateral triangle u'k'p' of 40 mm side above xy line at suitable distance with edge k'p' making 45° with xy.

(ii) Draw projectors from u', k' and p' below xy and draw parallel line to xy and 40 mm below it between projectors to get plan upk, as shown in Fig. 10.3.

Critically analysing the solutions of above three problems, we can conclude as under for plane parallel to one of the P.Ps.

(i) IF A PLANE IS PARALLEL TO ONE OF THE PRINCIPAL PLANES, IT WILL BE PERPENDICULAR TO TWO OTHER P.Ps.

(ii) PROJECTION OF PLANE ON A PRINCIPAL PLANE TO WHICH THE PLANE IS

PARALLEL WILL SHOW TRUE SHAPE AND TRUE ANGLES OF EDGES WITH OTHER PRINCIPAL PLANES.

(III) PROJECTIONS ON PRINCIPAL PLANES TO WHICH PLANE IS PERPENDICULAR WILL BE STRAIGHT LINES AND PARALLEL TO CORRESPONDING GROUND LINES e.g. xy or x_1y_1.

(iv) PLANE WILL NOT HAVE ANY TRACE ON P.P. TO WHICH IT IS PARALLEL AND OTHER TWO VIEWS WILL BE CORRESPONDING TRACES.

4. PLANE PERPENDICULAR TO ONE AND INCLINED TO TWO OTHER P.Ps.

It is obvious that as the plane is perpendicular to one of the principal planes the projection on that plane will be straight line. Projections on other two planes due to inclination will not show true shape. So to start the problem we have to draw one of the projections showing true shape and other projection showing straight line view. This is done by taking a stage prior to the final stage of solution. Straight line view achieved in the 1st stage is rearranged in the 2nd final stage to get the required angle of the plane with other principal plane. Rest can be achieved by theory of projections.

Two systems of projections for the problems of solid geometry, we have already studied earlier in the chapter of projections, planes of projections and system of projections. Students are requested to go through that before going for solution of problems.

Students are also requested to study the geometry of equilateral triangle, square, regular pentagon, regular hexagon, circle, ellipse, rhombus, etc. and they must do practice of drawing them quickly by the use of board, tee and set-squares.

Now we shall study five problems of this category and arrive at certain conclusions.

Problem 4 : A regular hexagonal plate ABCDEF, 25 mm side, is resting on H.P. on one of the sides/edges with surface of the plate making 45° with H.P. and perpendicular to V.P. Draw the two projections of the plate and find out its traces.

For solution see Fig. 10.4 and follow the procedure, as given below :

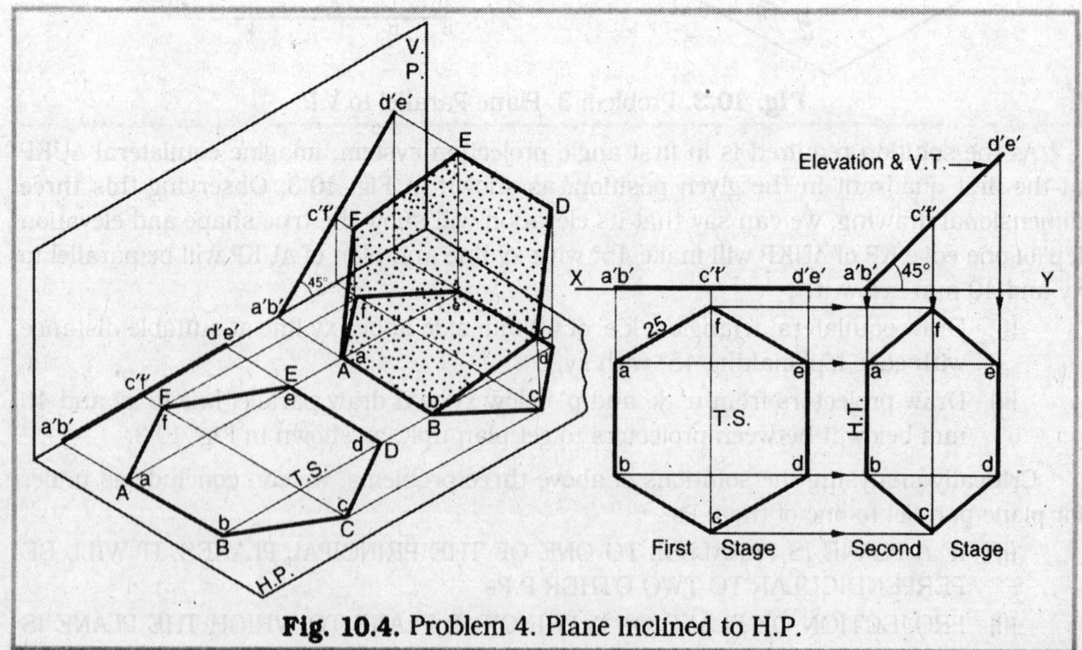

Fig. 10.4. Problem 4. Plane Inclined to H.P.

Whenever it is given that object, plane, solid, etc. rests on H.P. means the object, plane or solid is in the first quadrant, just as it is shown in Fig. 10.4 Whenever it rests on ground instead of H.P., it is in third quadrant.

In first angle system only xy line is drawn. Plan is drawn below xy and elevation is drawn above xy.

In third angle system two parallel ground lines are drawn (1) xy line (2) G.L. line below xy. Elevation is drawn between G.L. line and xy line and plan is drawn above xy line.

(i) First draw regular hexagon a b c d e f, which is the plan of regular hexagonal plate A B C D E F kept on H.P. with one edge out of six perpendicular to xy line, since plate is required to rest on H.P. on one of its edges. By keeping one edge a b perpendicular to xy line elevation of that edge a' b' gives us point view (a'b') is kept on xy, to keep two end points (A and B) of edge, i.e. edge, on H.P.

Get on xy line elevation [a'b' – c'f' – d'e'] by projecting this true shaped plan.

(ii) Rearrange this straight line elevation to new position (a'b' – c'f' – d'e') keeping a'b' on xy line and straight line elevation making 45° with xy line.

Project all points vertically downward and project all plan points of 1st stage horizontally to get by intersection plan points a, b, c, d, e and f, as shown in Fig.10.4 Join them to get plan.

(iii) Elevation and plan of 2nd stage is the answer. Projections of stage one is merely construction.

(iv) H.T. and V.T. are already shown in the figure.

Problem 5 : A circular plate, 50 mm diameter, is resting on H.P. on one of the points of its periphery with surface of the plate perpendicular to V.P. and inclined to H.P. by 30° . Draw the two projections of the circular plate.

For solution see Fig. 10.5 and follow the procedure, as given below:

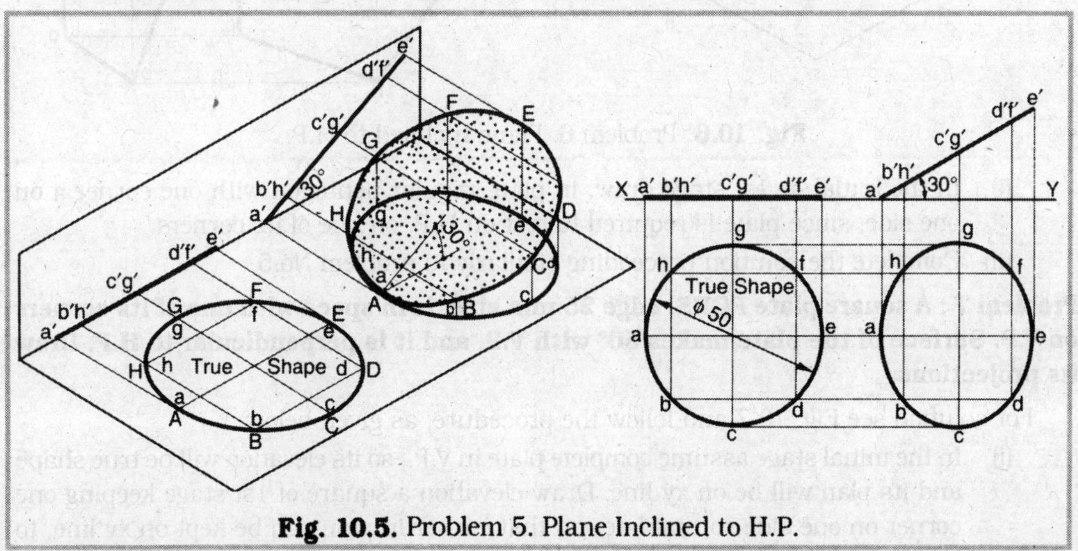

Fig. 10.5. Problem 5. Plane Inclined to H.P.

(i) Draw xy line.

(ii) In the first stage assume circular plate to be lying on the H.P. so that its plan will be a circle of the same size and its elevation will be straight line on xy line. First draw top view or plan (circle) and project elevation on xy line. Take help of 8 points equally spaced on its periphery.

(iii) Tilt circular plate about the point A by 30° or in other words rearrange straight line view of elevation to new position (a' - b'h' - c'g' - d'f' - e') so that it makes 30° with the xy line and point a' remains on xy line.

(iv) Draw projectors from elevation and horizontal lines from plan of previous stage to get by intersection new plan points a,b,c,.... h. Join them by smooth curve to get plan.

Problem 6 : A regular pentagonal plate ABCDE, 25 mm edge/side size, is resting on H.P. on one of its corners with surface of the plate perpendicular to V.P. and inclined to H.P. by 60°. Draw its two projections.

For solution see Fig. 10.6 and follow the procedure, as given below:

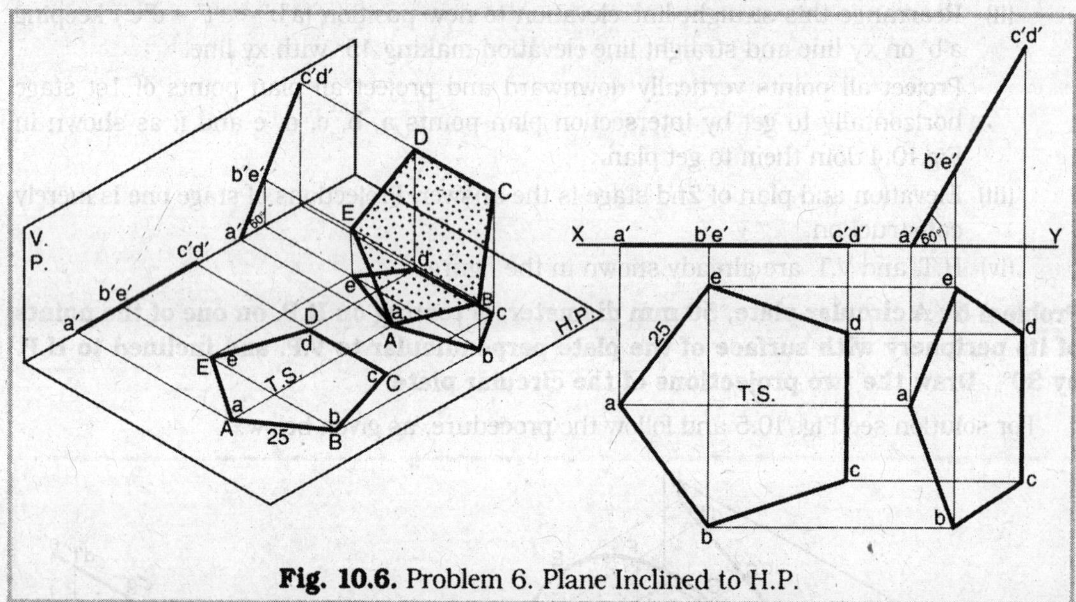

Fig. 10.6. Problem 6. Plane Inclined to H.P.

(i) In the initial or 1st stage draw, in plan, regular pentagon with one corner a on one side, since plate is required to rest on H.P. on one of its corners.

(ii) Complete the solution proceeding similarly to problem No.5.

Problem 7 : A square plate PQRS, edge 25 mm size, is in space with one of its corners on V.P. Surface of the plate makes 50° with V.P. and it is perpendicular to H.P. Draw its projections.

For solution see Fig. 10.7 and follow the procedure, as given below:

(i) In the initial stage assume complete plate in V.P., so its elevation will be true shape and its plan will be on xy line. Draw elevation a square of 1st stage keeping one corner on one side so that later on that corner in plan can be kept on xy line, to keep that point in V.P.

(ii) In second or final stage tilt plan of 1st stage so that it makes 50° with xy or in other words rearrange plan (p - qs - r) to new position (p - qs - r) keeping p on xy and line making 50° with xy.

(iii) Complete the solution as usual.

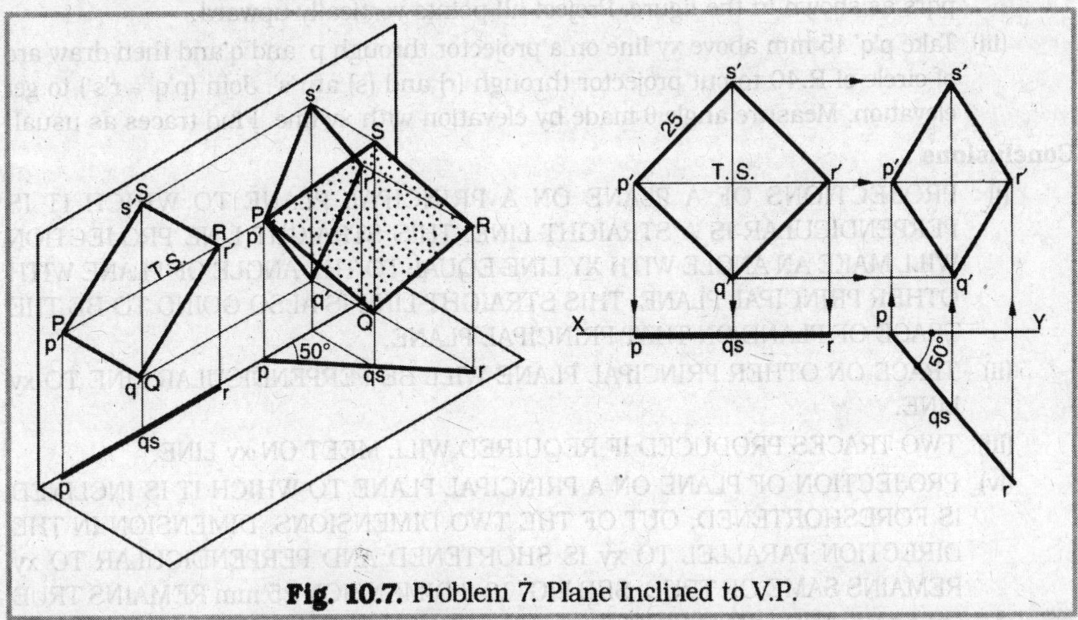

Fig. 10.7. Problem 7. Plane Inclined to V.P.

Problem 8 : A rectangular plate PQRS, 25 mm x 40 mm size, is in space with shorter edge parallel to H.P. and 15 mm above it. Plate PQRS is perpendicular to V.P. and inclined to H.P. by such an angle so that its plan becomes square. Draw the projections and find out its H.T; V.T. and inclination with H.P.

For solution see Fig. 10.8 and follow the procedure, as given below :

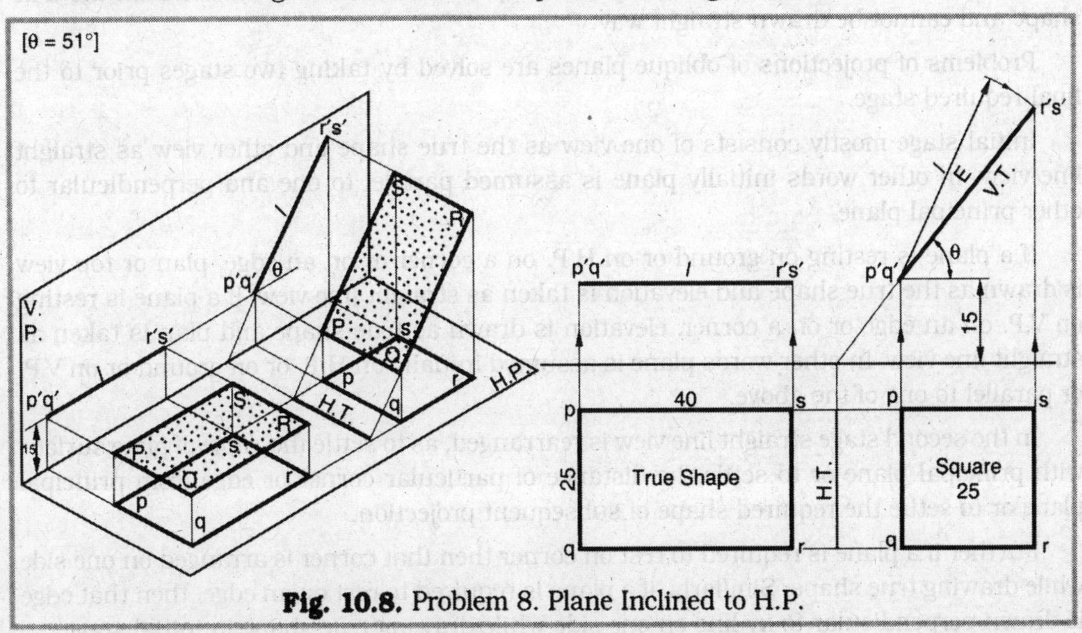

Fig. 10.8. Problem 8. Plane Inclined to H.P.

(i) In the initial stage take plate P Q R S parallel to H.P. and 15 mm above it as shown in figure. Its plan will be T.S. and its elevation will be straight line parallel to and 15 mm above xy line.

(ii) As the plan of 2nd and final stage is to be square draw plan (sq. of 25 mm size) pqrs as shown in the figure. Project all points vertically upward.

(iii) Take p'q' 15 mm above xy line on a projector through p and q and then draw arc of circle of R.40 to cut projector through (r) and (s) at r's'. Join (p'q' - r's') to get elevation. Measure angle θ made by elevation with xy line. Find traces as usual.

Conclusions :

(i) PROJECTIONS OF A PLANE ON A PRINCIPAL PLANE TO WHICH IT IS PERPENDICULAR IS A STRAIGHT LINE. THIS STRAIGHT LINE PROJECTION WILL MAKE AN ANGLE WITH XY LINE EQUAL TO THE ANGLE OF PLANE WITH OTHER PRINCIPAL PLANE. THIS STRAIGHT LINE IS ALSO GOING TO BE THE TRACE OF PLANE ON THAT PRINCIPAL PLANE.

(ii) TRACE ON OTHER PRINCIPAL PLANE WILL BE PERPENDICULAR LINE TO xy LINE.

(iii) TWO TRACES PRODUCED IF REQUIRED WILL MEET ON xy LINE.

(iv) PROJECTION OF PLANE ON A PRINCIPAL PLANE TO WHICH IT IS INCLINED IS FORESHORTENED. OUT OF THE TWO DIMENSIONS, DIMENSION IN THE DIRECTION PARALLEL TO xy IS SHORTENED AND PERPENDICULAR TO xy REMAINS SAME OR TRUE. SEE FIG. 10.8 DIMENSION 25 mm REMAINS TRUE OR SAME AND DIMENSION 40 mm IS REDUCED TO 25 mm.

5. OBLIQUE PLANES :

General :

When a plane is inclined to both the principal planes neither elevation nor plan will give straight line view nor will show the true shape i.e. views will be smaller than the true shape and cannot be drawn straight way.

Problems of projections of oblique planes are solved by taking two stages prior to the final required stage.

Initial stage mostly consists of one view as the true shape and other view as straight line view. In other words initially plane is assumed parallel to one and perpendicular to other principal plane.

If a plane is resting on ground or on H.P. on a corner or on an edge, plan or top view is drawn as the true shape and elevation is taken as straight line view. If a plane is resting on V.P. on an edge or on a corner, elevation is drawn as true shape and plan is taken as straight line view. In other words plane is assumed initially on H.P. or on ground or on V.P. or parallel to one of the above.

In the second stage straight line view is rearranged, as to settle the angle of plate surface with principal plane or to settle the distance of particular corner or edge from principal plane or to settle the required shape of subsequent projection.

Further if a plane is required to rest on corner then that corner is arranged on one side while drawing true shape. Similarly, if a plane is required to rest on an edge, then that edge is drawn perpendicular to xy line on one side while drawing true shape in initial stage.

Further solution of problems can be done by 1st angle system or by 3rd angle system. Shape of projection or views will not change due to change in system. Only their relative positions will change.

Problems can be solved also by taking auxiliary planes e.g. problem No 9, 11, 14, etc.

In auxiliary plane method required number of projections to be drawn are reduced. In fact repetition of same view is eliminated. This method is also known as rotation of ground line method.

Now we shall study the method of solution to this type of problems by solving a few concrete problems.

Problem 9 : A regular pentagonal plate, of 50 mm sides, has one of its corners on H.P. The plane of the pentagon is inclined at 30° to the H.P. The side of the pentagon, which is opposite to the corner, which is on H.P., is inclined at 45° to the V.P. Draw the projections of the plate.

For solution see Fig. 10.9 and follow the procedure as given below:

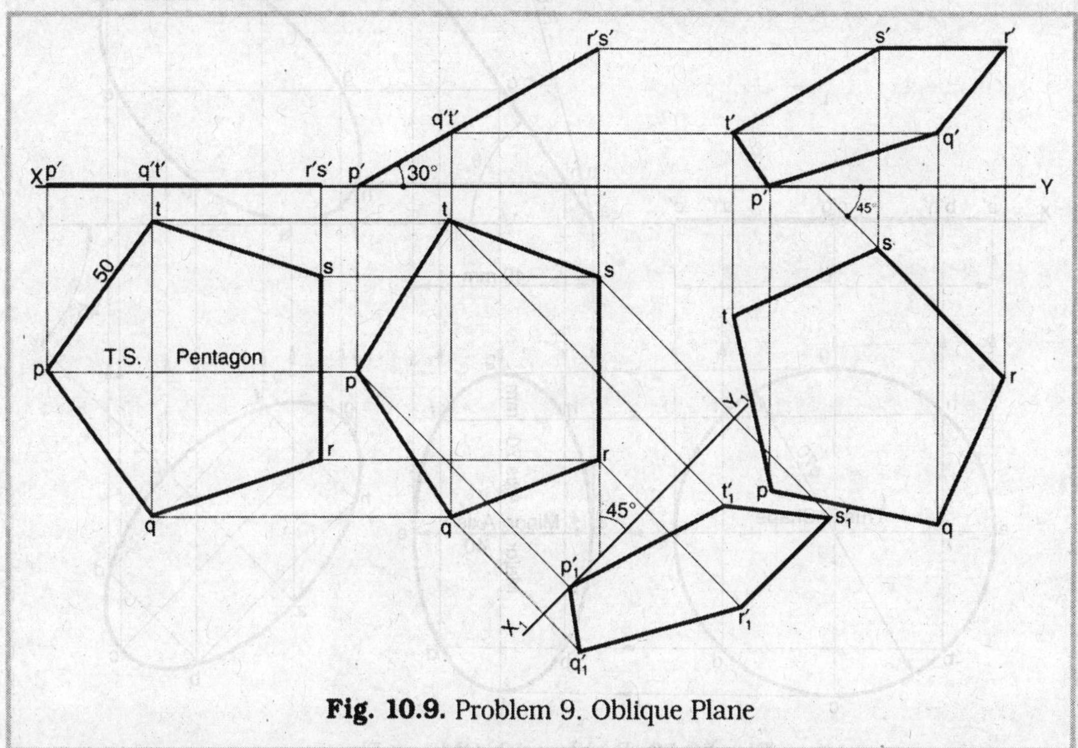

Fig. 10.9. Problem 9. Oblique Plane

The pentagonal plate must be in first quadrant, since it is given that corner is on H.P. And hence the solution is done by first angle projection system.

(i) In the initial stage pentagonal plate is assumed on H.P. and hence plan is drawn of true shape with one of the corners on one side. And elevation is drawn straight line on xy line.

(ii) As the plane of the plate is inclined to the H.P. by 30° and as one corner is on H.P., rearrange straight line view of 1st stage to the position given in Fig. 10.9 i.e. line view must make 30° with xy line and point p' should be xy line.

(iii) Draw projectors downward from the rearranged line view and draw horizontal lines from projection showing T.S. to get plan points by intersections.

(iv) Now line RS, opposite to corner P, on which it rests on H.P. is making 45° with

V.P. and hence draw new x_1y_1 line at 45° to rs line and project all points on x_1y_1 line. On these projectors mark points by taking distances of elevation points from xy line.

Problem 10 : Draw the projections of a circle, of 70 mm diameter, resting on the H.P. on a point A of the circumference. Plane is inclined to the H.P. such that the plan of it is an ellipse of minor axis 40 mm. The plan of the diameter, through the point A, is making an angle of 45° with the V.P. Measure the angle of the plane with the H.P.

For solution see Fig. 10.10 and follow the procedure, as given below :

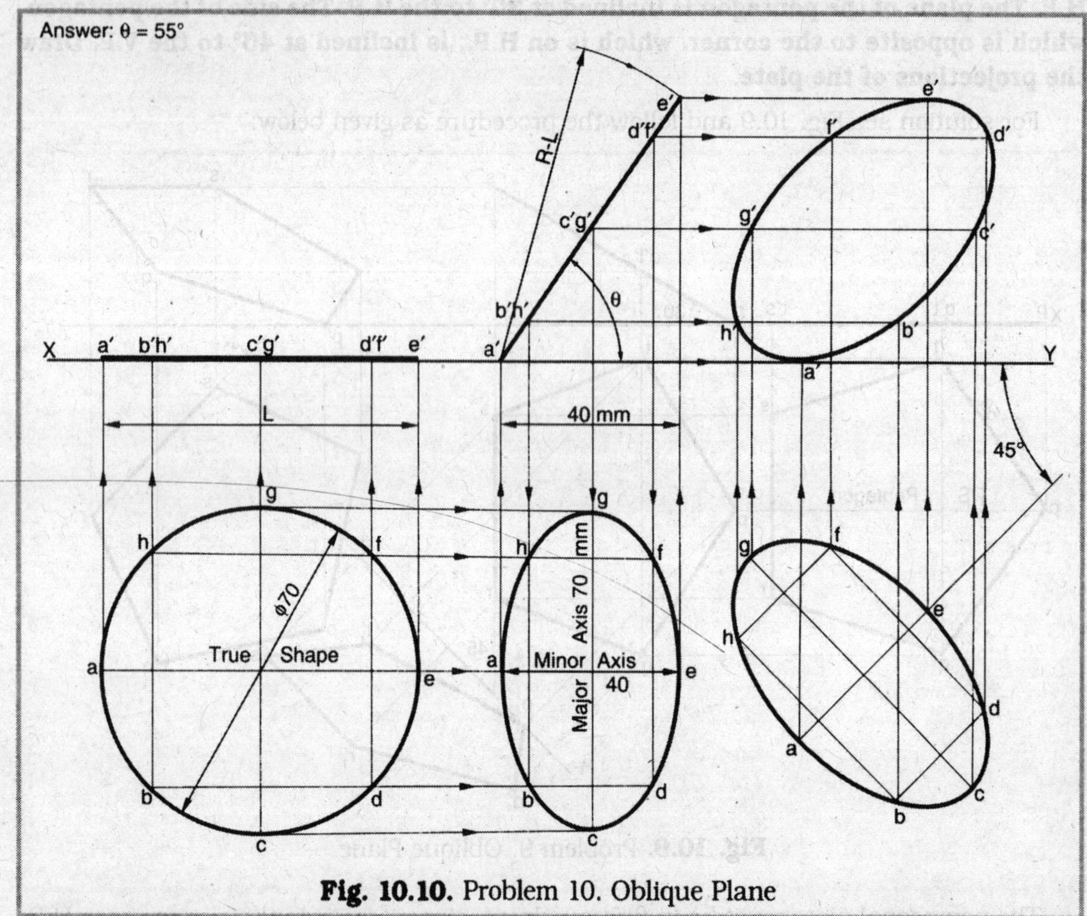

Fig. 10.10. Problem 10. Oblique Plane

Given that circle rests on H.P. and hence the solution is done according to 1st angle system. Further assume initially circle on H.P. and hence.

(i) Draw plan as circle of 70 mm diameter and draw its elevation, straight line on xy line.

(ii) Now rearrange this straight line view of elevation to such an inclination so that the distance between two extreme projectors becomes 40 mm. This is done to get minor axis of 40 mm of ellipse in subsequent projection. Measure this inclination θ. In fact in this position the plane of circle makes an angle θ with H.P.

(iii) By projection as usual, get ellipse in plan.

(iv) Since the plan of the diameter through A is making 45° with V.P., rearrange the ellipse to new position such that plan of the diameter AE i.e. ae makes 45° with xy line. Project all points and as usual get the final elevation.

Problem 11 : A square plate, of side 60 mm, is held on a corner on H.P. with a diagonal horizontal and inclined at 45° to V.P. (FRP). The plate is seen as a rhombus in plan with one of its diagonals measuring 30 mm. Draw the principal views of the plate and determine the angle it makes with H.P.

For solution see Fig. 10.11 and follow the procedure, as given below :

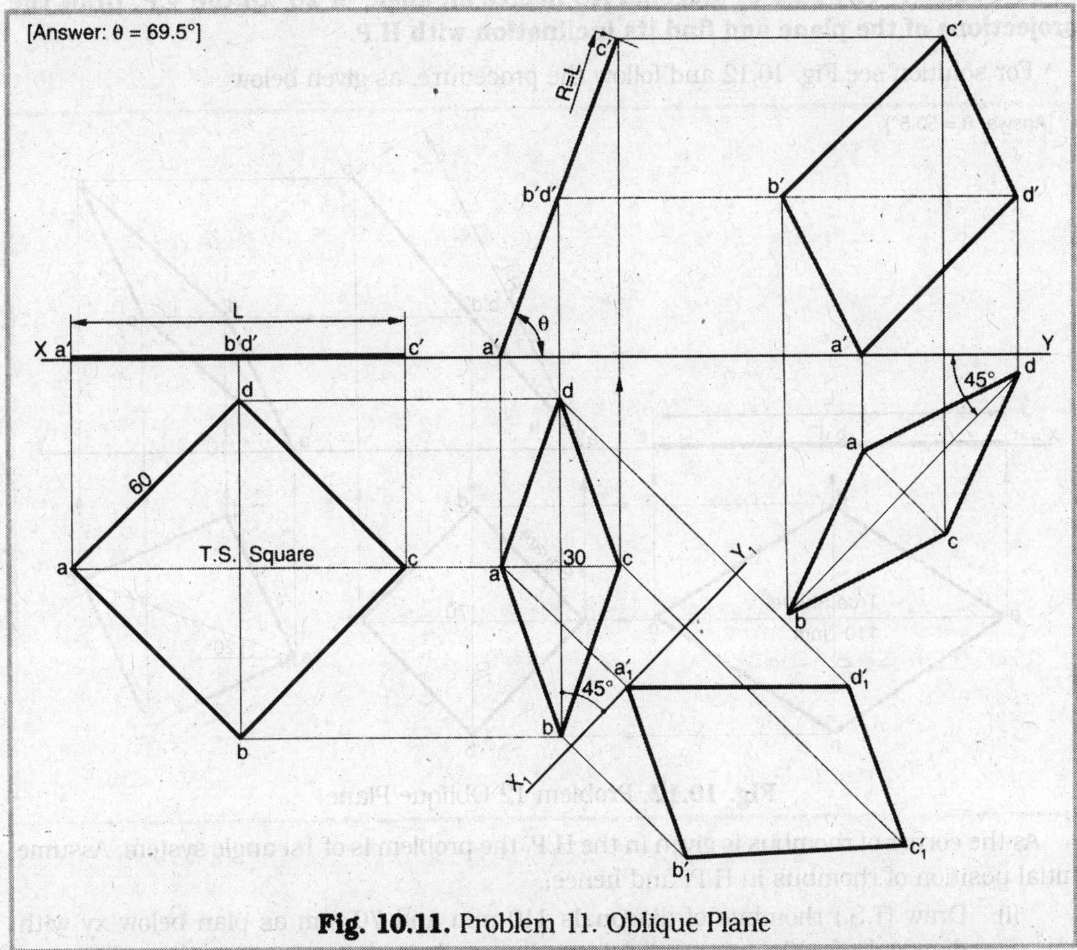

Fig. 10.11. Problem 11. Oblique Plane

As the corner is on H.P. use 1st angle system and keep the plate initially on H.P.

(i) Draw plan T.S. (square of 60 mm) below xy line keeping a corner on one side and draw elevation as straight line on xy line.

(ii) As the required plan is rhombus of 30 mm diagonal, rearrange this straight line elevation to such an inclination such that projectors through extreme points become 30 mm apart. Keep a' on xy line to keep corner A on H.P. Measure this inclination θ, which is the inclination of plate with H.P.

(iii) By projection as usual, get plan as rhombus with one diagonal as 30 mm and another diagonal as true length.

(iv) Since it is given that the horizontal diagonal is inclined at 45° to V.P., either draw new X_1Y_1 at 45° to bd and project elevation on X_1Y_1 or rearrange (rhombus) plan with bd line making 45° with xy line and project elevation from it on xy. Follow any one method. Here both methods are used to show that the shape of elevations achieved by different methods are the same. Further their positions w.r.t. corresponding XY lines is also the same.

Problem 12 : ABCD is a rhombus of diagonals AC = 110 mm and BD = 70 mm. Its corner A is in the H.P. and the plane is inclined to the H.P. such that the plan appears to be a square. The plan of diagonal AC makes an angle of 20° to the V.P. Draw the projections of the plane and find its inclination with H.P.

For solution see Fig. 10.12 and follow the procedure, as given below :

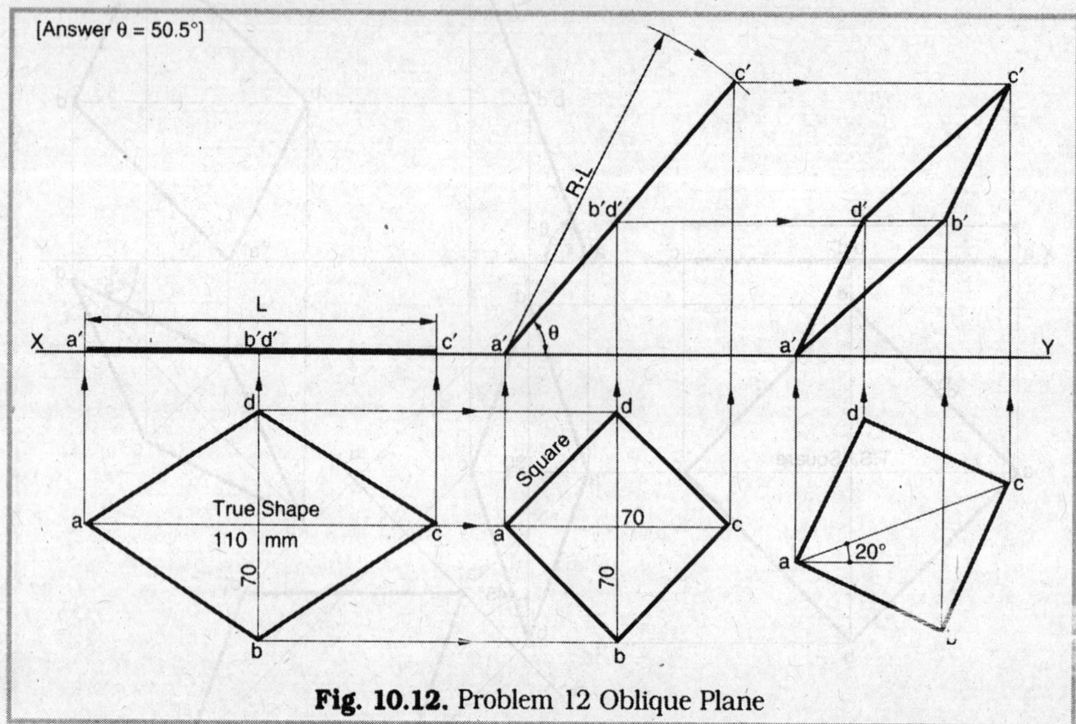

Fig. 10.12. Problem 12 Oblique Plane

As the corner of rhombus is given in the H.P. the problem is of 1st angle system, Assume initial position of rhombus in H.P. and hence.

(i) Draw (T.S.) rhombus of diagonals 110 mm and 70 mm as plan below xy with diagonal of 110 mm parallel to xy line and draw its corresponding elevation, straight line on xy line itself.

(ii) Rearrange this straight line elevation to new position as shown in the figure, keeping a' on xy line and projectors through end points 70 mm apart. Purpose of keeping 70 mm distance between projectors is to get a square in plan of diagonals 70 mm. Measure inclination θ, which is the inclination of rhombus with H.P.

(iii) Rearrange (square) plan, with plan of AC i.e ac making 20° with xy and project corresponding elevation.

Problem 13 : A regular hexagonal plate, 50 mm side, is resting on one of its corners in H.P. The diagonal through that corner is inclined at 40° to H.P. and (a) the plan of that diagonal inclined to V.P. by 30° and (b) diagonal is inclined at 30° to V.P.

For solution see Fig. 10.13 and follow the procedure, as given below :

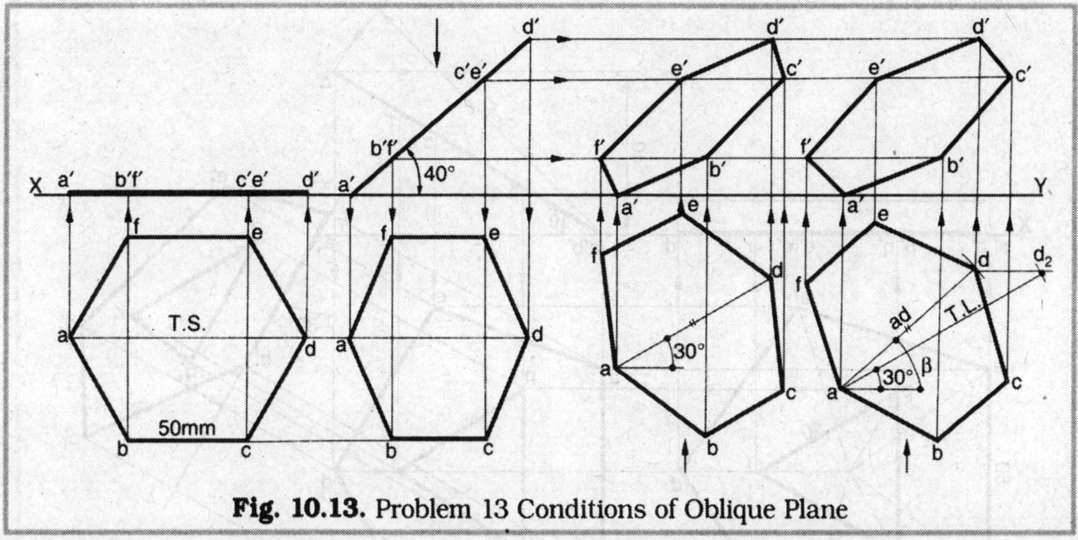

Fig. 10.13. Problem 13 Conditions of Oblique Plane

As one of the corners of the plate is given resting on H.P., the problem is of 1st quadrant and so the solution is given in 1st angle system. Moreover, initially the plate is assumed to be in H.P. and hence.

(i) Draw plan below xy and T.S., keeping one corner on one side, and draw its corresponding straight line elevation on xy line.

(ii) Rearrange straight line elevation at 40° to xy line keeping a' on xy, since the point A is on H.P. and diagonal AD makes 40° to H.P. Project its corresponding plan.

(iii) In part (a) as the plan of the diagonal is inclined to V.P. by 30°, redraw plan again with plan of AD i.e. (ad) making 30° with xy and project its corresponding elevation.

(iv) In part (b) as the diagonal itself, and not its plan, makes 30° with V.P., draw first T.L. (ad_2) of diagonal AD at 30° and adjust the plan (ad) length, as shown in the figure at β angle. Redraw the complete plan with ad making an angle β with xy and project its corresponding elevation.

Problem 14 : A pentagonal plate, of sides 50 mm, has a central equilateral triangular hole of 40 mm sides, with a side of plate and that of triangle parallel to each other. The plate is kept on H.P. on this side, the side being inclined at 30° to FRP (V.P.). If the highest point of the plate is 40 mm above the H.P., Determine the angle the plate makes with H.P. Project the triangular hole in all the views.

For solution see Fig. 10.14 and follow the procedure, as given below :

As the plate is to rest on H.P. on an edge, use 1st angle system and assume initial position of plate completely on H.P. with one of the edges of the plate perpendicular to xy and on one side.

(i) Draw regular pentagon abcde of 50 mm side as plan below xy with edge ab perpendicular to xy. Draw its corresponding elevation on xy line. Draw equilateral triangle of 40 mm side pqr in pentagon keeping the centre same and one edge of triangle pq parallel to ab. Project pqr also in elevation.

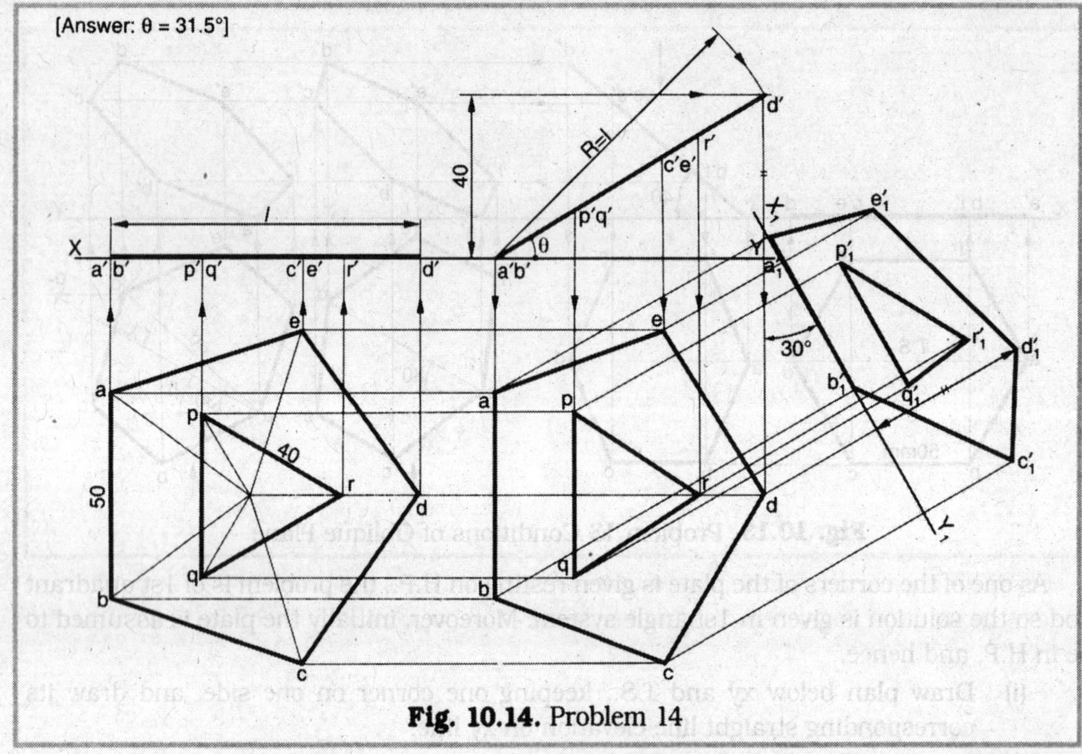

Fig. 10.14. Problem 14

(ii) The plate rests on H.P. on edge AB and point D, opposite to this edge, is going to be the highest point and 40 mm above H.P. And hence rearrange straight line elevation keeping a'b' on xy line and d' at 40 mm above xy line, as shown in the figure. Project all points and get its corresponding plan. Measure inclination θ, which is going to be the inclination of plate with H.P.

(iii) Now as the edge, on which it rests on H.P., makes 30°with V.P, draw new x_1y_1 line at 30° to ab line and project plan to get elevation $a'_1b'_1c'_1d'_1e'_1$ and $p'_1q'_1r'_1$, as shown in the figure.

Problem 15 : A regular pentagon ABCDE, of 30 mm sides, has its side AB in the V.P. and inclined at an angle of 30° to the H.P. The corner A is 15 mm above H.P. and the corner D is 20 mm in front of V.P. Draw the projections of the plane and find its inclination with the V.P.

For solution see Fig. 10.15 and follow the procedure, as given below :

As the side AB is required to be in V.P., assume initially pentagon ABCDE in V.P. with side AB perpendicular to xy or H.P. First angle system is used.

(i) Draw regular pentagon a'b'c'd'e' as elevation above xy line with edge or side a'b' perpendicular to xy line. Draw its corresponding plan on xy line.

(ii) It is given that the corner D is 20 mm in front of V.P. and the side AB is in V.P. and hence rearrange straight line plan view (plan) in such a way that ab remains on xy line and d remains 20 mm below xy. Measure angle Ø between xy line and rearranged straight line plan view. This angle Ø is the inclination of plane with V.P. Project corresponding elevation a'b'c'd'e'.

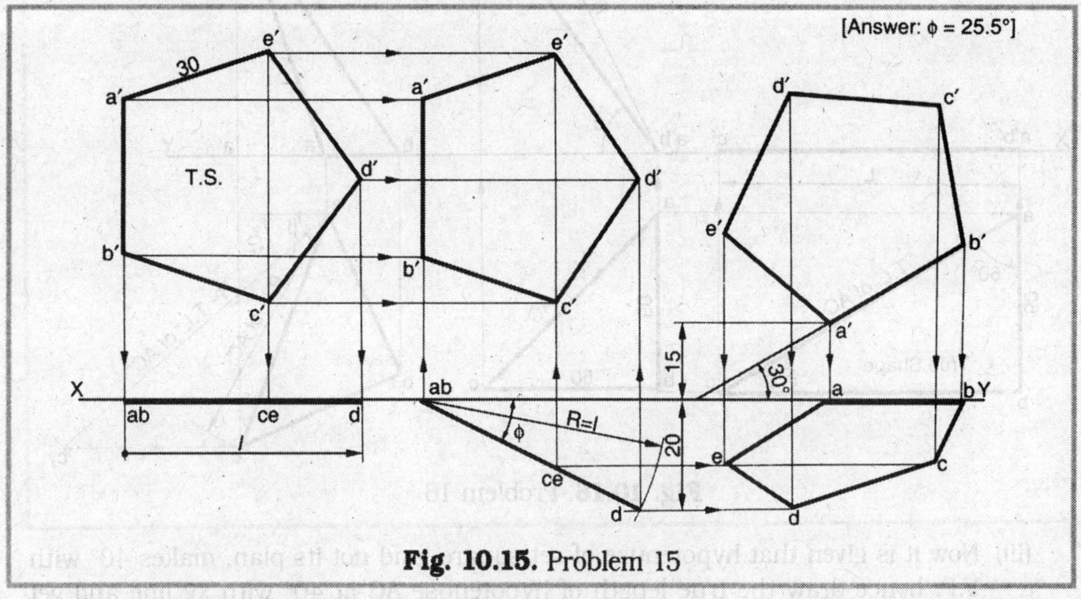

Fig. 10.15. Problem 15

(iii) It is given that side AB is inclined to H.P. by 30° and corner A is 15 mm above H.P. and hence rearrange elevation a'b'c'd'e' to new position a'b'c'd'e' with point a' 15 mm above xy and line a'b' making 30° with xy. Project its corresponding plan abcde.

Problem 16 : A 30° - 60° set square has its shortest side 50 mm long and is in the H.P. The top view of the set square is an isosceles triangle and the hypotenuse of the set square is inclined at an angle of 40° with the V.P. Draw the projections of the set square and find its inclination with the H.P.

For solution see Fig. 10.16 and follow the procedure, as given below :

It is given that shortest side of 50 mm length rests on H.P., assume initial position of set square in H.P. with the shortest side perpendicular to V.P. As it rests on H.P., use 1st angle system.

(i) First draw Δabc T.S. of set square. Project its corresponding straight line elevation a'b'c' on xy line.

(ii) It is given that the top view of the set square is an isosceles triangle and hence first draw isosceles triangle of 50 x 50 x hypotenuse, as shown in the figure and rearrange straight line view a'b'-c' such that it suits or fits to its corresponding plan already drawn. Measure the angle θ that rearranged straight line elevation a'b'-c' makes with xy line. This angle θ is the inclination of plane of set square with H.P.

[Answer: θ = 55°]

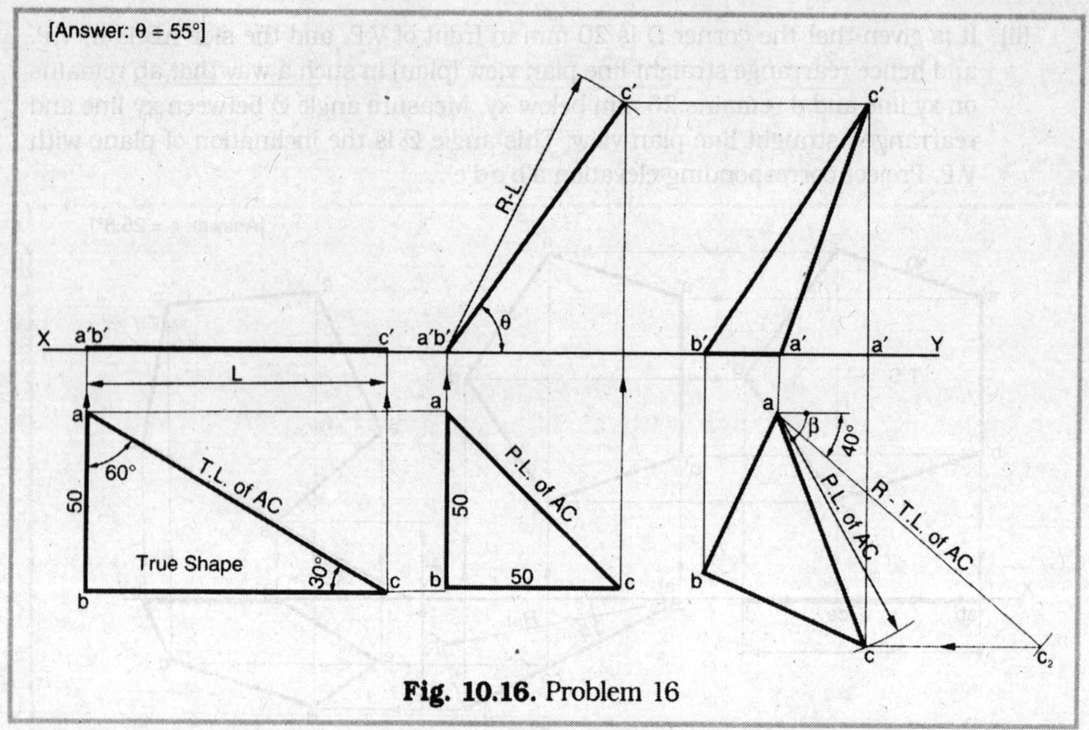

Fig. 10.16. Problem 16

(iii) Now it is given that hypotenuse of set square, and not its plan, makes 40° with V.P. hence draw the true length of hypotenuse AC at 40° with xy line and get point c_2. Draw parallel to xy line from c_2 and cut it at the point c by drawing an arc with (a) as the centre and radius equal to plan length (ac) of hypotenuse. Join ac and rearrange plan on this line ac. Project its corresponding elevation a'b'c', as shown in figure.

Problem 17 : ABCDE is a regular pentagonal plate, of 40 mm sides, has its corner A on the H.P. The plate is inclined to the H.P. such that the plan length of the edges AB and AE is each 35 mm. The side CD is parallel to both the reference planes. Draw the projections of the plate and find its inclination with the H.P.

For solution see Fig. 10.17 and follow the procedure, as given below :

As the corner A is given on H.P. the problem is required to be treated in 1st angle system.

Further assume initially the pentagonal plate on H.P. with corner A on one side.

(i) Draw regular pentagon abcde of 40 mm side as plan below xy line with corner a on one side and project its corresponding straight line elevation on xy line.

(ii) It is given that plan of AE and AB are of 35 mm length so first arrange 35 mm length lines ae and ab on parallel lines to xy line from a, e and b, as shown in figure. Draw projector through a to get a' on xy line.

(iii) Draw common projector through b and e and draw arc with a' as the centre and radius equal to r to intersect at b'e' as shown in the figure. Join a' and b'e' and extend it upto c'd'. This straight line a' - b'e' - c'd' is the rearranged position of straight line elevation a' - b'e' - c'd'. Project points downward to complete plan abcde.

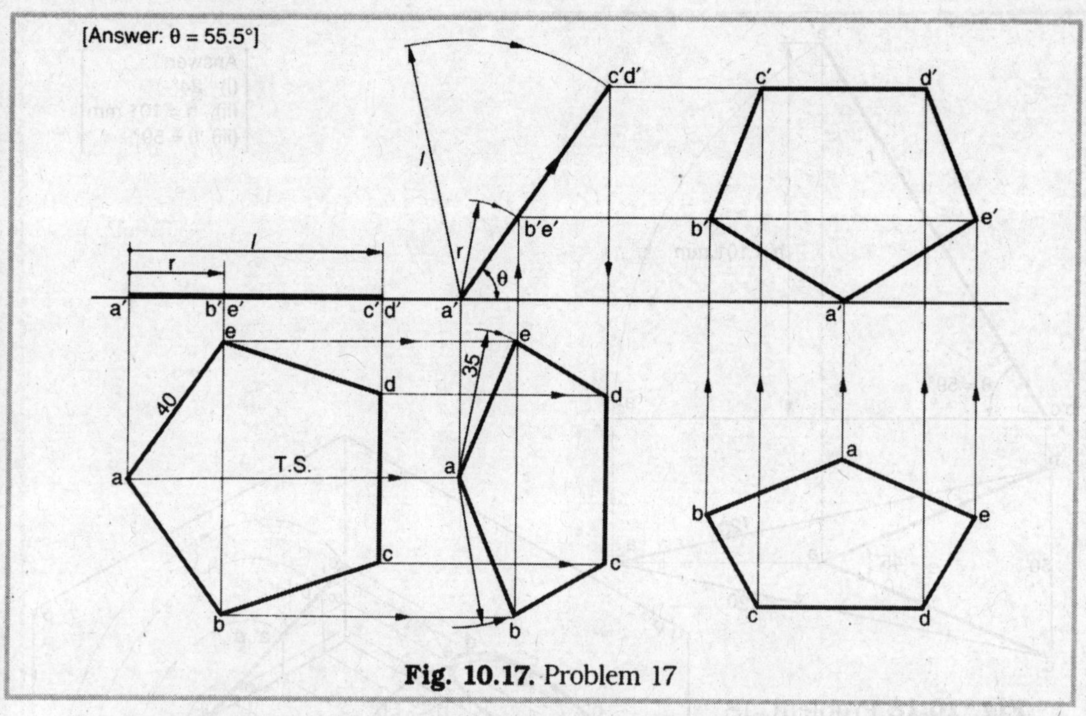

Fig. 10.17. Problem 17

(iv) Rearrange plan of CD i.e. cd parallel to xy and redraw the plan and project its corresponding elevation.

Angle θ is the inclination of the plate with H.P. Measure it.

Problem 18 : The top plan, of a pair of equal legs AB and AC of compass, appears as an isosceles triangle having base bc 50 mm and vertex angle at A 45°. Actual lengths of compass legs AB and AC are 120mm. Assume points B and C on H.P. and line connecting B and C perpendicular to V.P. Draw the projections and find

(1) The actual angle between two legs.

(2) The height of point A above the H.P. and

(3) Angle the plane, containing compass, makes with H.P.

For solution see Fig. 10.18. It is self explanatory.

Problem 19 : Draw the three projections of a circular lamina, of 50 mm diameter, having one end of the diameter resting on H.P. and the other end of the diameter on V.P. and the surface of the lamina inclined at 30° to the H.P. and at 60° to the V.P.

For solution see Fig. 10.19. It is self explanatory.

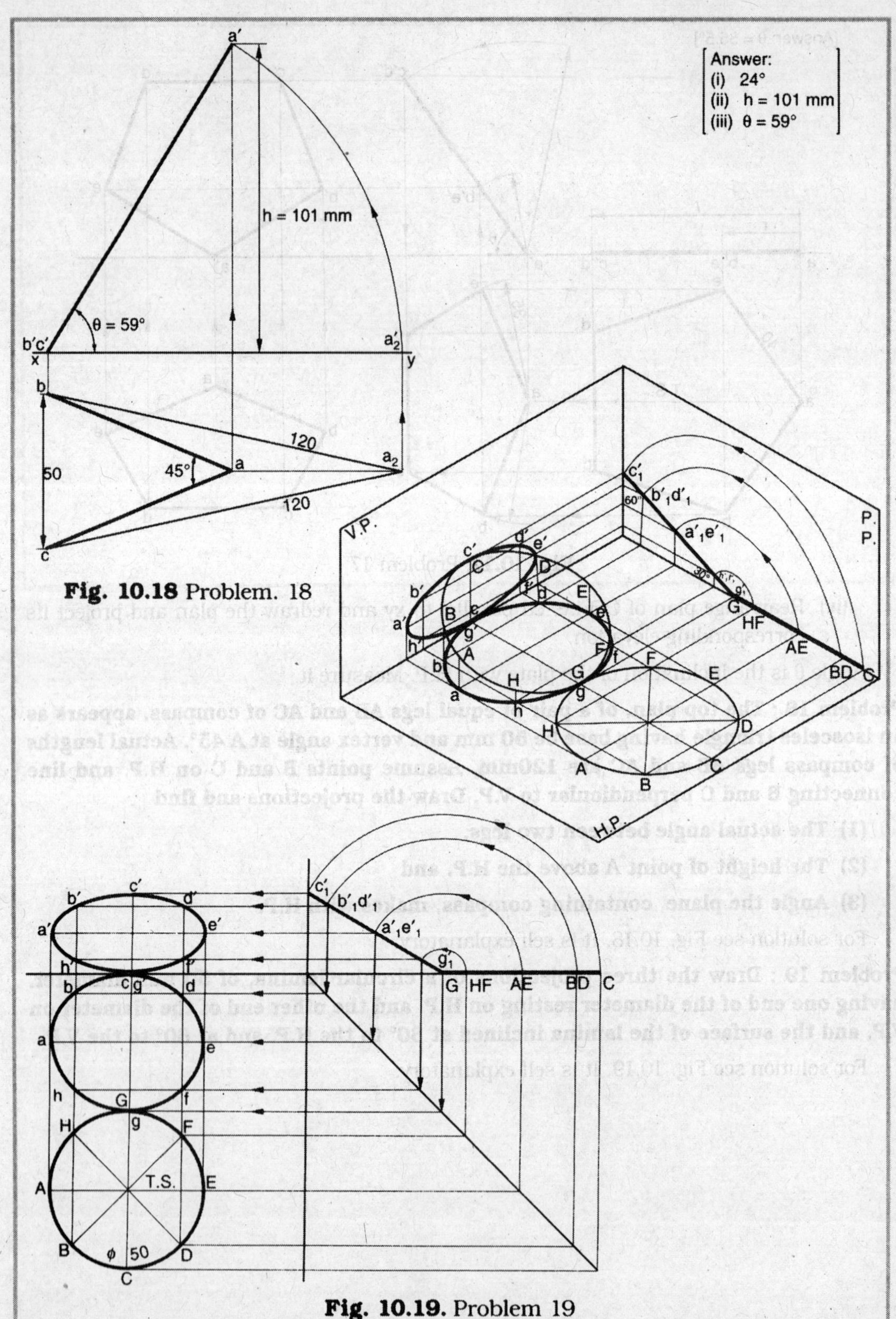

Fig. 10.18 Problem. 18

Fig. 10.19. Problem 19

(1) A semi-circular thin plate, of 60 mm diameter, rests on the H.P. on its diameter, which is inclined at 45° to the V.P. and the surface is inclined at 30° to the H.P. Draw the projections of the plate. [Fig. 10.20]

[Mumbai University, December 1994]

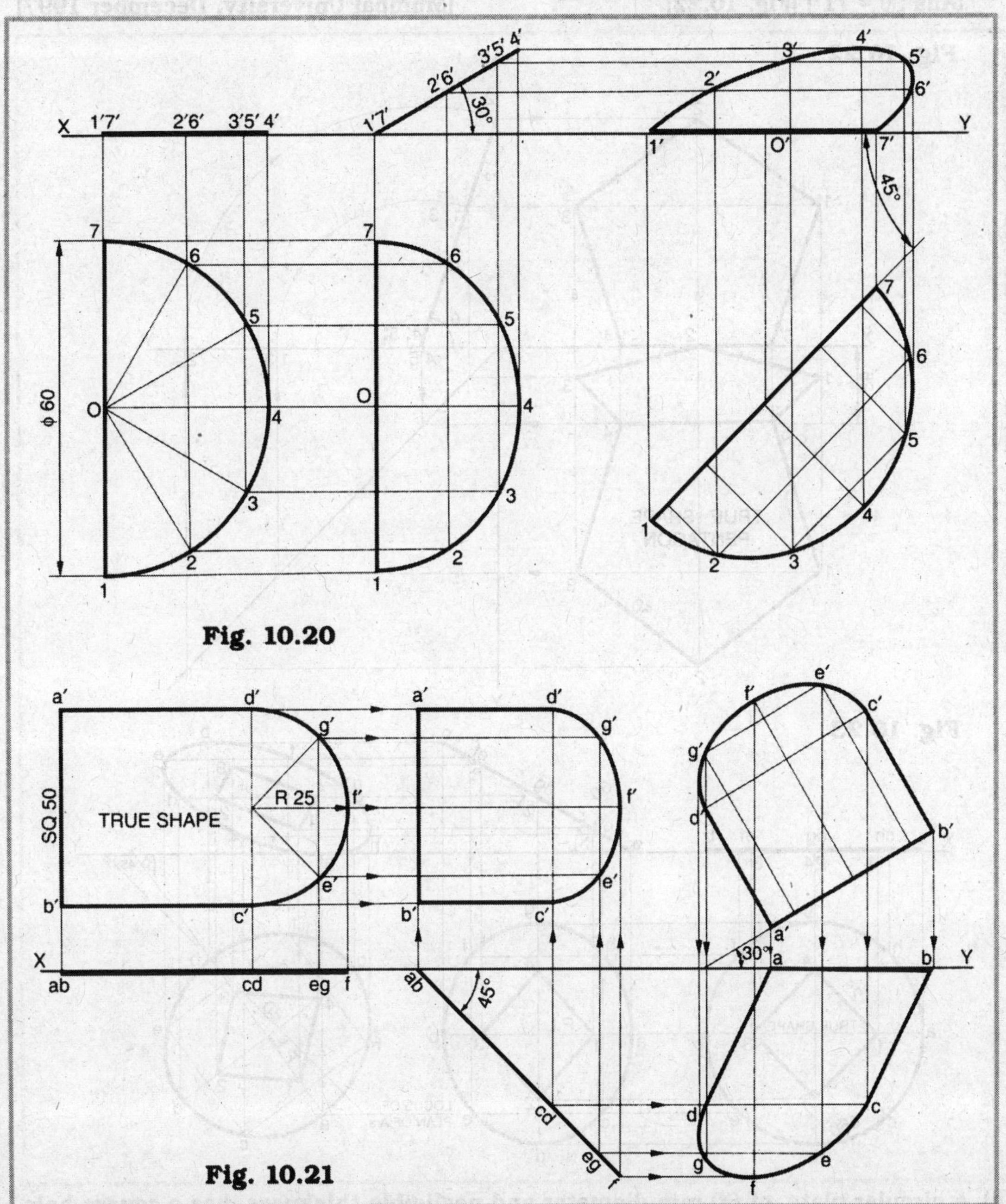

Fig. 10.20

Fig. 10.21

(2) A thin composite plate, consists of a square ABCD of 50 mm sides with an additional semi-circle constructed on CD as a diameter. The side AB is in the V.P. and makes 30° with the H.P. and the surface of a plate makes 45° with the V.P. Draw its projections. [Fig. 10.21] **[Mumbai University, June 1996]**

(3) A regular pentagon, of 50 mm sides, is resting on one of its sides on the H. P. such that it is parallel to and 25 mm infront of the V. P. If the highest corner of the pentagon rests in the V. P. Draw its projections and find the angle made by a plane with the H. P., the H. T. and the V. T. of the plane.

[Ans : θ = 71°] [Fig. 10.22] **[Mumbai University, December 1997]**

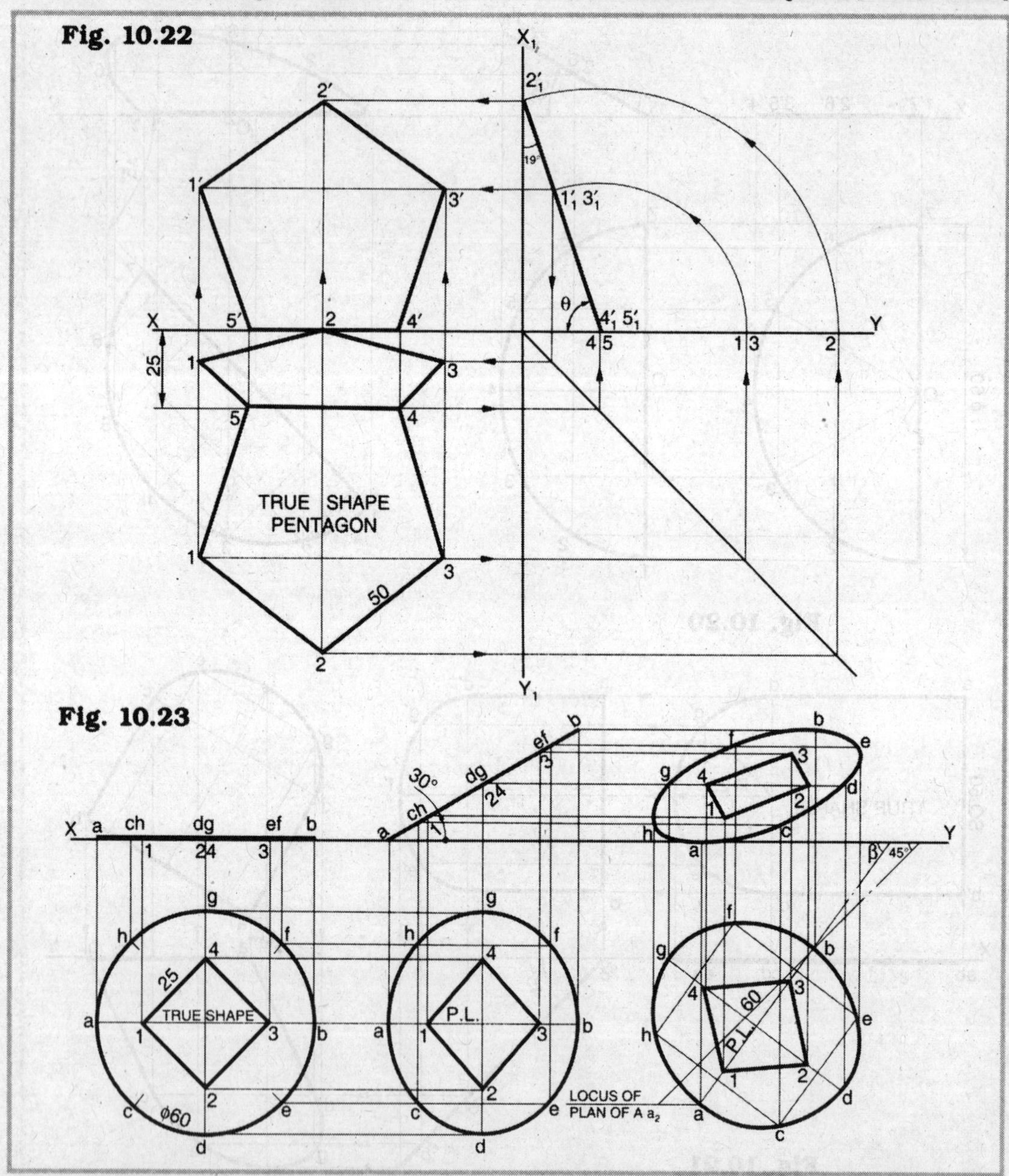

(4) A circular plate, of 60 mm diameter and negligible thickness, has a square hole side 25 mm, punched centrally. A plate is resting on the H. P. on point A of its rim with its surface inclined at 30° to the H. P. and the diameter AB, through A, is inclined at 45° to the V. P. Draw the projections of a plate with the hole. [Fig. 10.23]

[Mumbai University, December 1996]

(5) An isosceles triangular plate, of 50 mm base and 75 mm altitude, appears as an equilateral triangle of 50 mm in top view. Draw the projections of a plate if its 50 mm long edge is on the H. P. and inclined at 45° to the V. P. What are the inclinations of a plate with the H. P. and the V. P.? [Fig. 10.24]

[Ans : (1) θ = 55° (2) θ = 55°] **[Mumbai University, December 1995]**

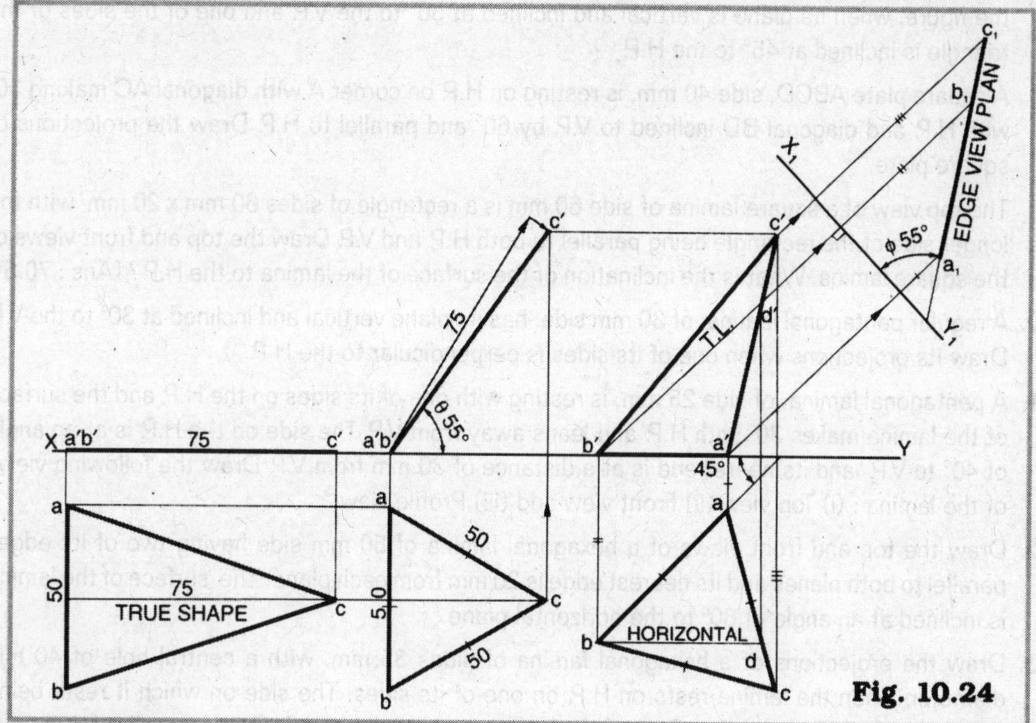

Fig. 10.24

EXERCISE

1. A regular hexagonal plate, side 30 mm size, is resting on H.P. on one of its corners with opposite corner in V.P. The plate is inclined to H.P. by 30° and to V.P. by 60°. Draw all the three projections of plate neglecting the thickness of plate.
2. A pentagonal plate, side 25 mm, is resting on H.P. on one of its corners with opposite edge to the corner making 30° with V.P. The plate is inclined to H.P. by 45°. Draw the projections.
3. An equilateral triangular plate, side 40 mm, is resting on H.P. on one of its sides such that the plan of it is an isosceles triangle with altitude 20 mm. The side, which is on H.P. is inclined to V.P. by 60°. Draw the projections and find out the inclination of plate with H.P. [Ans : 55°]
4. A hexagonal lamina of 30 mm sides is resting on the H.P. on one of its sides. The side which is on the H.P. is perpendicular to the V.P. and the surface of the lamina is inclined to the H.P. at 45°. The lamina is then rotated through 90° such that the side on the H.P. is parallel to the V.P., while the surface is still inclined to the H.P. at 45°. Draw the front view and the top view of the lamina in its final position.
5. ABC is a triangle of sides AB = 75 mm, BC = 60 mm and CA = 45 mm. Its longest side AB is in V.P. and inclined at 30° to H.P. Its surface makes an angle of 45° with the V.P. Draw its projections.

6. The top view of a 45° set square, with the side BC in H.P. and the side AB in the V.P., is a triangle abc. The side bc = 200 mm, being perpendicular to the XY line and angle bca = 25°. Draw the top and front views and measure the inclination of the set square to the H.P. Draw the auxiliary view on X_1Y_1 line perpendicular to xy line. [Ans : 62°]
7. Draw an equilateral triangle of 75 mm side and inscribe a circle in it. Draw the projections of the figure, when its plane is vertical and inclined at 30° to the V.P. and one of the sides of the triangle is inclined at 45° to the H.P.
8. A square plate ABCD, side 40 mm, is resting on H.P. on corner A with diagonal AC making 30° with H.P. and diagonal BD inclined to V.P. by 60° and parallel to H.P. Draw the projections of square plate.
9. The top view of a square lamina of side 60 mm is a rectangle of sides 60 mm x 20 mm with the longer side of the rectangle being parallel to both H.P. and V.P. Draw the top and front views of the square lamina. What is the inclination of the surface of the lamina to the H.P.? [Ans : 70.5°]
10. A regular pentagonal lamina, of 30 mm side, has its plane vertical and inclined at 30° to the V.P. Draw its projections when one of its sides is perpendicular to the H.P.
11. A pentagonal lamina, of side 25 mm, is resting with one of its sides on the H.P. and the surface of the lamina makes 30° with H.P. and leans away from V.P. The side on the H.P. is at an angle of 40° to V.P. and its nearer end is at a distance of 30 mm from V.P. Draw the following views of the lamina : (i) Top view (ii) Front view and (iii) Profile view.
12. Draw the top and front views of a hexagonal lamina of 50 mm side having two of its edges parallel to both planes and its nearest edge is 20 mm from each plane. The surface of the lamina is inclined at an angle of 60° to the horizontal plane.
13. Draw the projections of a hexagonal lamina of sides 35 mm, with a central hole of 40 mm diameter, when the lamina rests on H.P. on one of its sides. The side on which it rests being parallel to the V.P. and the surface of the lamina being inclined to H.P. at an angle of 40°.
14. A circular lamina, of 60 mm diameter, is resting on the H.P. on a point A of the circumference, with its plane inclined at 45° to H.P. and the top view of the diameter through A makes 30° with V.P. Draw the top and front views of the lamina.
15. Draw the projections of a rhombus, diagonals 125 mm and 50 mm size, having smaller diagonal parallel to H.P. and V.P. both and the bigger diagonal inclined to H.P. such that plan of rhombus becomes a square. Draw the projections and find the inclination of plane with H.P. Use 1st angle projection system. [Ans : 66.5°]
16. A regular hexagonal lamina is resting in V.P. on one of its sides with lamina making 45° with V.P. The side on which it rests on V.P. makes 60° with H.P. Draw the projections using rotation of ground line method.
17. The T.V. of a pair of equal legs AB and AC of compass, appears as two lines of 10 cms length meeting at 30° angle. If the actual length of compass legs is 15 cms, find the actual angle between two legs and height of point A above the H.P. For projections use 1st angle system of projections. Assume ends B and C on H.P. [Ans : 20°, 11.2cms.]

CHAPTER - 11

Projections of Point, Line and Plane on Auxiliary Planes

1. Introduction :

We have studied two types of auxiliary planes :

(1) Auxiliary Inclined Plane (A. I. P.)

(2) Auxiliary Vertical Plane (A. V. P.)

We also know that projections on A. I. P. are known as auxiliary plans and projections on A. V. P. are known as auxiliary elevations. When plan and elevation of a point, line or plane are given, we can get auxiliary plan and auxiliary elevation on A. I. P. and A. V. P. respectively by the method studied in the chapter of projection of a point. These auxiliary views play most important role in determining the following :

(a) The true length of a line

(b) Point view of a straight line

(c) Distance between two skew lines

(d) Edge view or line view of a plane

(e) True shape of a plane

(f) Angle between two intersecting planes

(g) Distance between two parallel lines

(h) Distance of A point from plane surface

(i) In orthographic projections of objects real or true shapes of inclined surfaces are seen in auxiliary views taken on auxiliary planes parallel to inclined surfaces of objects.

'Ve shall now take some illustrative practical problems and find the solutions of them by the use of auxiliary views.

2. Illustrative Problems :

Problem 1 : A point A is 45 mm above H. P. and 20 mm in front of V. P. Draw plan and elevation of the point A.

Project auxiliary plan of the point A on A. I. P. which is perpendicular to V. P. and inclined to H. P. by θ (60°).

For solution see Fig. 11.1 and follow the procedure, as given below :

(i) First draw xy line and locate a' and a.

(ii) Draw now new ground line either X_1Y_1 or X_2Y_2 or X_3Y_3 or X_4Y_4 at θ (60°) to xy line around elevation a'. Draw perpendicular line from a' to either X_1Y_1 or X_2Y_2 or X_3Y_3 or X_4Y_4 and mark on it a_1 or a_2 or a_3 or a_4 respectively by taking distance of plan a from xy line i.e. 20 mm beyond X_1Y_1 or X_2Y_2 or X_3Y_3 or X_4Y_4, as shown in the figure.

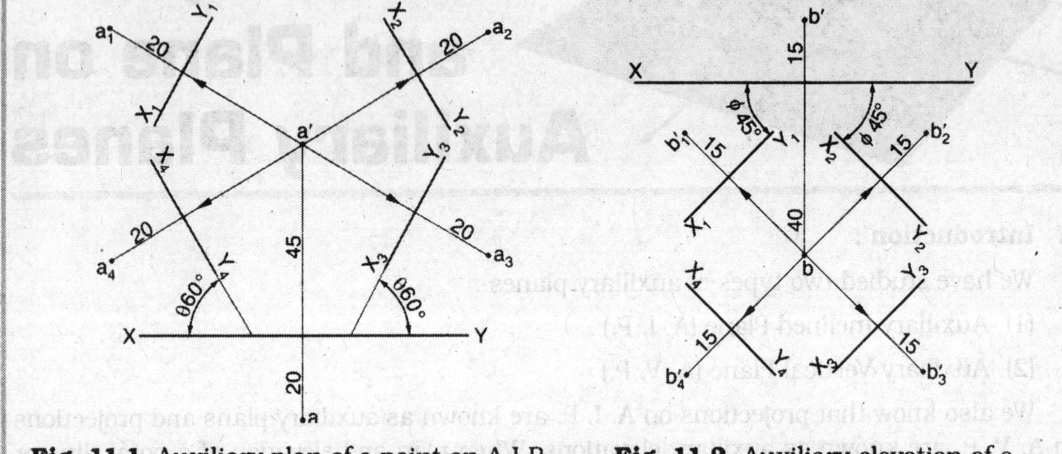

Fig. 11.1. Auxiliary plan of a point on A.I.P.

Fig. 11.2. Auxiliary elevation of a point on A.V.P.

Problem 2 : A point B is 15 mm above H.P. and 40 mm in front of V.P. Draw the projections and find its auxiliary elevation on A.V.P. perpendicular to H.P. and inclined to V.P. by an angle Ø (45°).

For solution see Fig. 11.2 and follow the procedure, as given below :

(i) First draw b' and b.

(ii) Draw now new G.L. either X_1Y_1 or X_2Y_2 or X_3Y_3 or X_4Y_4 at Ø (45°) to xyline around plan b. Draw perpendicular line from b to either X_1Y_1 or X_2Y_2 or X_3Y_3 or X_4Y_4 and mark on it b'_1 or b'_2 or b'_3 or b'_4 respectively by taking distance of elevation from xy line i.e. 15 mm beyond X_1Y_1 or X_2Y_2 or X_3Y_3 or X_4Y_4, as shown in the figure.

1. (a) True Length of A Straight Line :

We have studied in the chapter on straight lines that when a plan of a straight line is parallel to ground line, its corresponding elevation will show T.L. and true inclination with H.P.

Similarly, when elevation of a straight line is parallel to ground line, its corresponding plan will show T.L. and true inclination with V.P. This concept is used to get T.L., θ and Ø when plan and elevation are given.

Problem 3 : The distance between the end projectors of a straight line PJ is 45 mm. Point P is 45 mm above H.P. and 10 mm in front of V.P. Point J is 15 mm above H.P. and 25 mm in front of V.P. Draw the projections and find its T. L, θ and Ø by auxiliary plane method.

For solution see Fig. 11.3 (a) and (b) and follow the procedure, as given below :

(i) First draw elevation p'j' and plan pj.

(ii) Draw X_2Y_2 ground line parallel to elevation p'j' and project on it auxiliary plan p_2j_2, as shown in the figure. This view will be auxiliary plan. It will show T.L. of the line PJ and its inclination with X_2Y_2 will be Ø, the inclination of the line with V.P.

(iii) Similarly, draw X_1Y_1 ground line parallel to plan pj and project on it auxiliary elevation $p'_1j'_1$, as shown in the figure. This view will be auxiliary elevation. It will show T.L. of the line PJ and its inclination with x_1y_1 will be θ, the inclination of line with H.P. Three dimensional drawing is given here to visualize this concept.

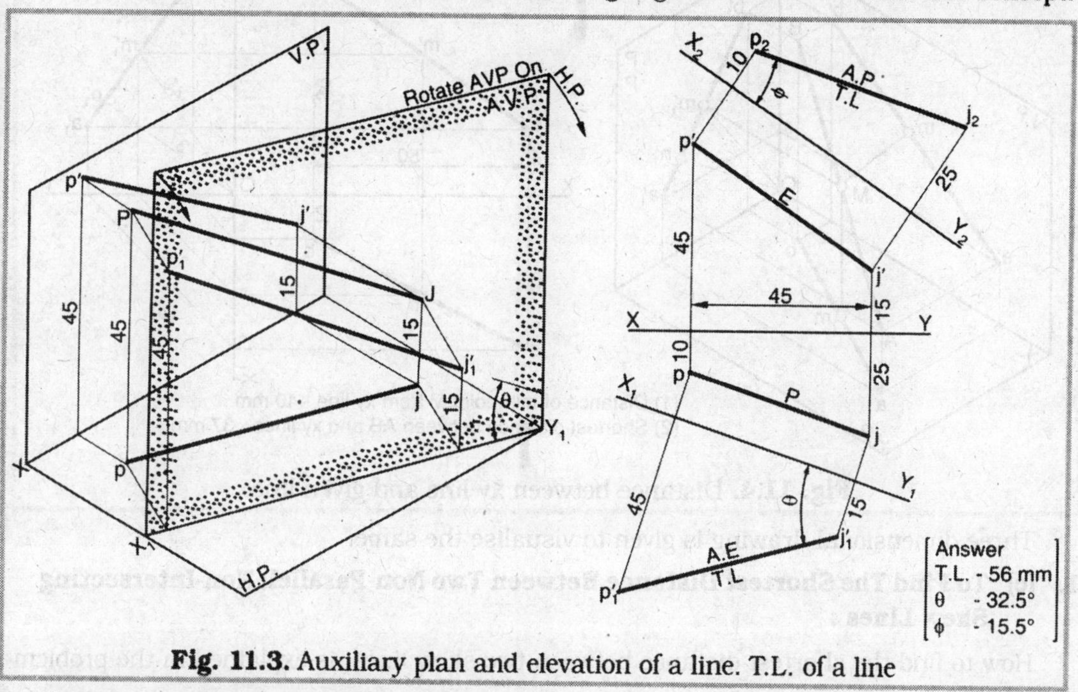

Fig. 11.3. Auxiliary plan and elevation of a line. T.L. of a line

1. (b) Point View of A Straight Line and

1. (c) Distance Between Two Skew Lines :

Whenever point view of a line is required from the given elevation and plan of a line, one must first find out T.L. view of a line by the procedure similar to problem 3, and then take new ground line perpendicular to T.L. view of a line already achieved.

xy line is always seen as a point view in an end view or side view. So whenever distance from xy line is required, it can be achieved from the side view.

Problems 4, 5 and 6 are solved by this concept. Whenever two skew pipelines are required to be connected by shortest possible pipeline, then problem can be solved by this concept.

Problem 4 : The distance between the end projectors of a line AB is 50 mm. Point A is 15 mm above H.P. and 35 mm in front of V.P. Point B is 50 mm above H.P. and 10 mm in front of V.P. Draw the projections of AB and find out (a) distance of mid point M of AB from xy and (b) the shortest distance between AB and xy lines.

For solution see Fig. 11.4 (a) and (b) and follow the procedure, as given below :

(i) Draw plan ab and elevation a'b', with the given data. Mark the mid points m and m' on them respectively.

(ii) Draw X_1Y_1 ground line perpendicular to xy and project on it end view $a'_1b'_1$ by the usual method. Project m'_1 also on it. xy line will be seen in end view as a point view O.

(iii) Join Om'_1 and measure it. This is the distance of mid point M from xy line.

(iv) Draw perpendicular On'_1 from O on $a'_1b'_1$ and measure it. It will be the shortest distance between AB and xy lines.

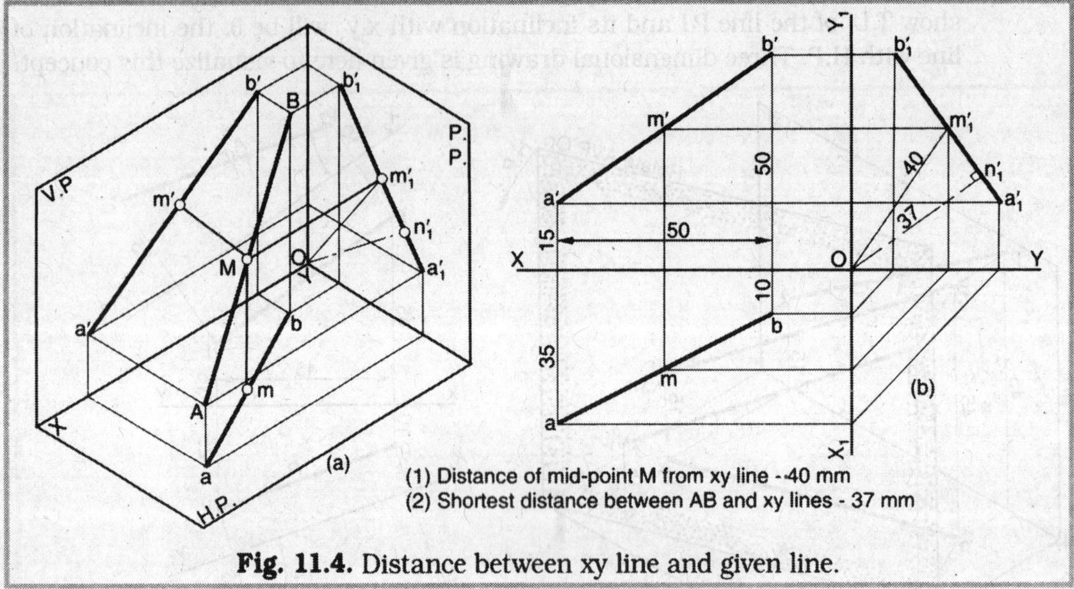

Fig. 11.4. Distance between xy line and given line.

Three dimensional drawing is given to visualise the same.

1. (c) To Find The Shortest Distance Between Two Non-Parallel, Non-Intersecting Skew Lines :

How to find the shortest distance between two skew lines, is explained in the problem given below :

Problem 5 : AB and PQ are two skew lines. Distance between end projectors of AB and PQ are 48 mm and 45 mm respectively. Projector of the point Q is 12 mm to the left of projector of B. Point A is 20 mm above H.P. and 1 mm in front of V.P. Point B is 40 mm above H.P. and 28 mm in front of V.P. Point P is 55 mm above H.P. and 45 mm in front of V.P. Point Q is 10 mm above H.P. and 15 mm in front of V.P. Draw the projections of the lines AB and PQ and find the shortest distance between the lines AB and PQ.

For solution see Fig. 11.5 and follow the procedure, as given below :

First of all draw projectors of points P, A, Q and B keeping distances between projectors of AB = 48 mm, of PQ = 45 mm and of QB = 12 mm. Knowing the distances of the points P, A, Q and B from V.P. and H.P., draw a' and a on projector of A, p' and p on projector of P, q' and q on projector of Q and b' and b on projector of B. Join a'b', ab, p'q' and pq.

Important discussion :

Out of the two lines AB and PQ, we must get point view of any one line. Further to get point view of a line, we must get the true length of the same line in the previous view. Here point view of line AB is obtained.

Now to get the true length of the line AB, select and draw X_1Y_1 ground line at any convenient distance parallel to plan of AB i.e. ab.

On this new ground line X_1Y_1, draw perpendicular projectors from a, b, p and q and on them, plot distances of a', b', p' and q' respectively from XY line on the other side of X_1Y_1, as shown in the figure for the point A.

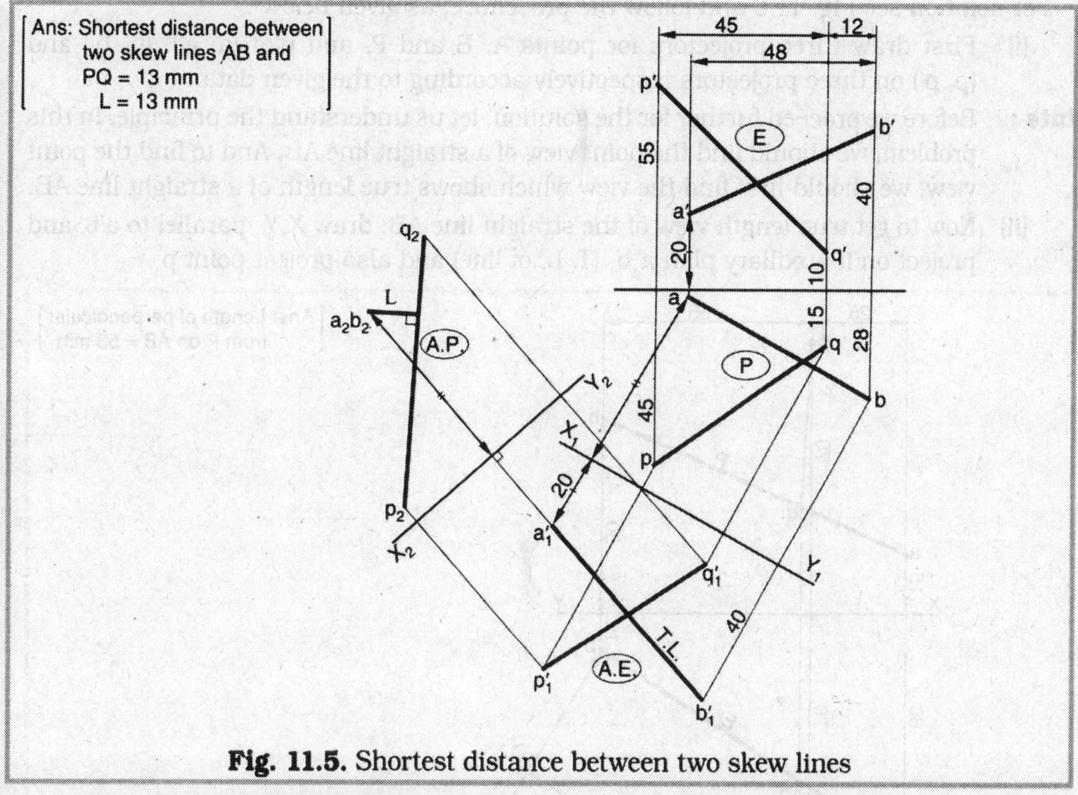

Fig. 11.5. Shortest distance between two skew lines

New projections of AB and PQ i.e. $a'_1b'_1$ and $p'_1q'_1$ are auxiliary elevations since the projectors are from plan points and distances are of elevation points. In this view, we have a'_1 b'_1 showing the true length since its plan ab is parallel to X_1Y_1. In fact X_1Y_1 is taken parallel to ab with this purpose only.

Now select and draw at any convenient distance, new ground line X_2Y_2 perpendicular to a'_1 b'_1 (or line showing T.L.). Draw perpendicular projectors from a'_1, b'_1, p'_1 and q'_1 on this new ground line X_2Y_2 and on them plot distances of a, b, p and q from X_1Y_1 line on the other side of X_2Y_2, as shown in the figure.

New projections of AB and PQ i.e. a_2b_2 and p_2q_2 are auxiliary plans since the projectors are from elevation (auxiliary) points and distances are of plan points. In these new projections, we have a_2b_2 auxiliary plan of AB as point view.

From this point view a_2b_2, draw perpendicular to p_2q_2 and measure it. This distance (L) as shown in the figure is the shortest distance between skew lines AB and PQ. It comes out as 13 mm.

Problem 6 : The distance between the end projectors of a straight line AB is 80 mm. The line lies in the third quadrant. End A is 15 mm from V.P. and 60 mm from H.P.,

while B is 50 mm from V.P. and 10 mm from H.P. A point P, situated on the projector at a distance of 25 mm from the projector through A, (the distance being measured towards the projector of B), is also in the third quadrant and is 70 mm from each plane. A perpendicular is drawn from P on AB. Locate it in plan and elevation and measure it.

For solution see Fig. 11.6 and follow the procedure, as given below :

(i) First draw three projectors for points A, B and P, and plot (a, a'), (b, b') and (p, p') on three projectors respectively according to the given data.

Hints : Before we proceed further for the solution, let us understand the principle. In this problem, we should find the point view of a straight line AB. And to find the point view, we should first find the view which shows true length of a straight line AB.

(ii) Now to get true length view of the straight line AB, draw X_1Y_1 parallel to a'b' and project on it auxiliary plan a_1b_1 (T. L. of line) and also project point p_1.

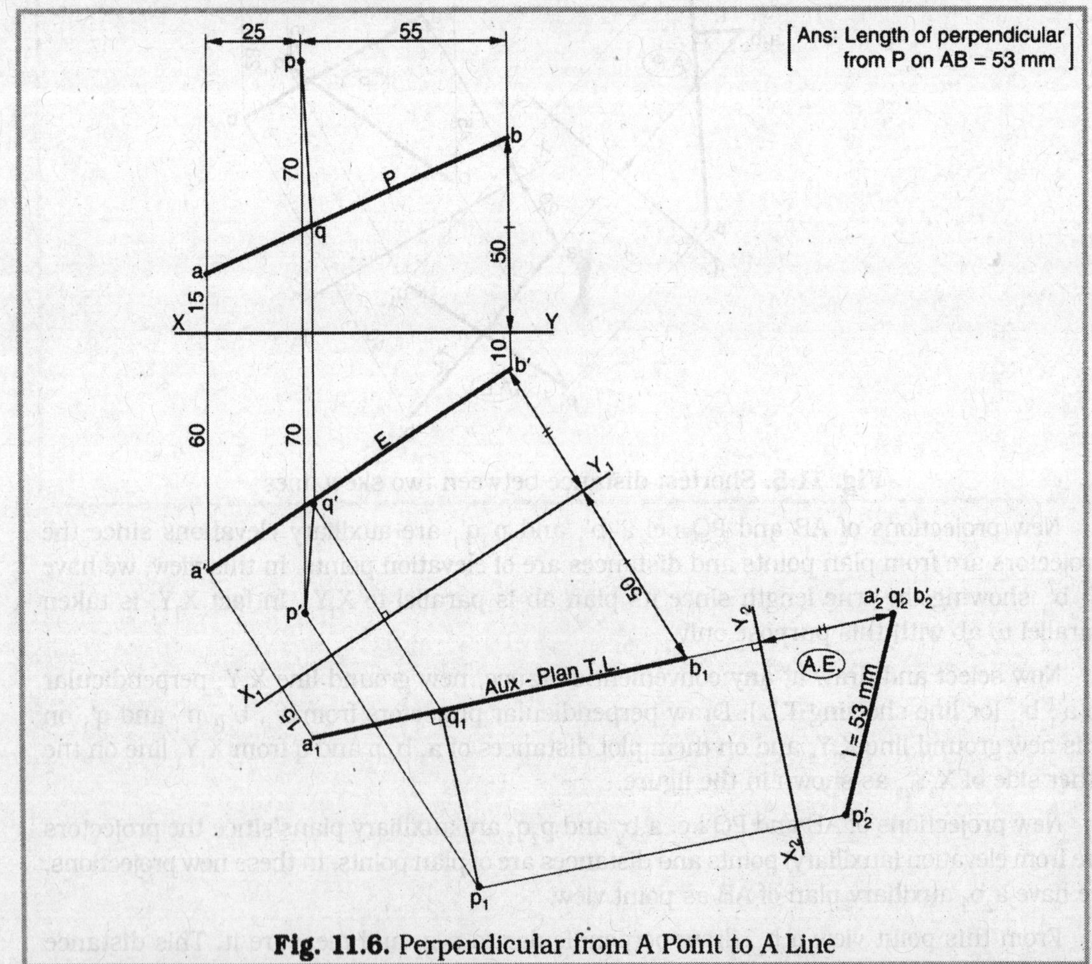

Fig. 11.6. Perpendicular from A Point to A Line

(iii) Now take new ground line X_2Y_2 perpendicular to a_1b_1 (perpendicular to T.L. of a line) and project on it straight line as well as p_1. We get a'_2, b'_2, a point view of the straight line AB, and view of point P as p'_2. Join these two points to get the true length of a perpendicular line from P on AB.

(iv) In the previous view i.e. auxiliary plan, draw perpendicular p_1q_1 from p_1 to a_1b_1 and transfer the point Q to all views to get views of perpendicular in all views, as shown in Fig. 11.6 Length of perpendicular measures 53 mm.

1. (d) & (e) Edge View of A Plane and True Shape of A Plane :

It is known to us that when a plan of a plane is a straight line and parallel to ground line, its corresponding elevation will show true shape. Similarly, when elevation of a plane is a straight line and parallel to ground line, its corresponding plan will show T.S.

Further, if a ground line is taken perpendicular to the true length of any straight line of plane, then subsequent view of the plane will give a straight line.

How to find the true shape of a plane when plan and elevation are given, is explained in problem 7, given below :

Problem 7 : a'b'c'd' and abcd are the elevation and plan of a plane ABCD, respectively. Find the true shape of a plane ABCD and its inclination with H.P.

For solution see Fig. 11.7 and follow the procedure, as given below :

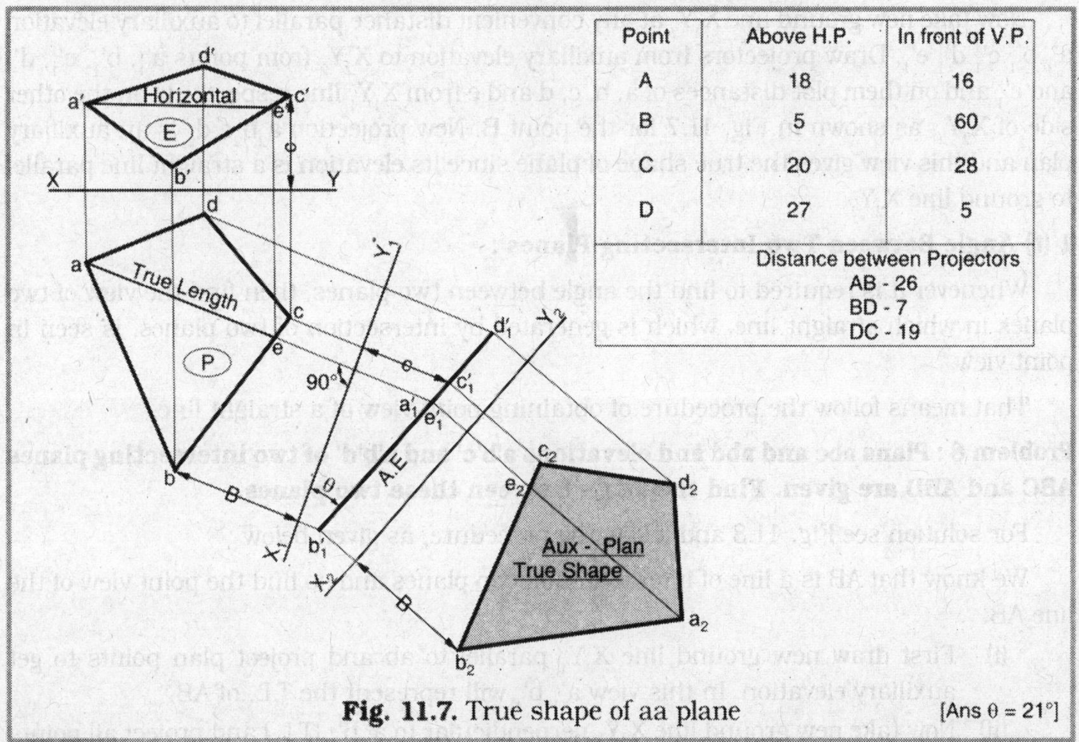

Point	Above H.P.	In front of V.P.
A	18	16
B	5	60
C	20	28
D	27	5

Distance between Projectors
AB - 26
BD - 7
DC - 19

Fig. 11.7. True shape of aa plane [Ans θ = 21°]

Solution :

First of all draw elevation a'b'c'd' and plan abcd, as per the data given.

Important discussion :

To get the true shape of a plane, we must get line view of a plane in the previous view or projection.

Again to get line view of a plane, we must get true length of any straight line of that plane in previous projection.

Again to get true length of any straight line in elevation or plan of a plane, the plan or elevation respectively of that straight line must be parallel to x-y line.

So as per the above discussion, we draw a'e' parallel to XY in elevation of a plane. Point e' falls on b'c' and so transfer it to plan on bc and get the plan of E as e. ae will give the true length.

Now to this true length line ae draw perpendicular line X_1Y_1 (new G.L.) at any convenient distance.

Draw perpendicular projectors to X_1Y_1 from points a, b, c, d and e and on them plot distances of a', b', c', d' and e' from XY line respectively on other side of X_1Y_1, as shown in Fig. 11.7 for the point C.

New projection a'_1 b'_1 c'_1 d'_1 e'_1 is an auxiliary elevation since the projectors are from plan points and distances are from elevation points. In this view, we have achieved straight line view of a plane. Whatever angle (θ) this straight line view (here auxiliary elevation) makes with X_1Y_1 is the inclination of the plane with H.P.

Now take new ground line X_2Y_2 at any convenient distance parallel to auxiliary elevation a'_1 b'_1 c'_1 d'_1 e'_1. Draw projectors from auxiliary elevation to X_2Y_2 from points a'_1, b'_1, c'_1, d'_1 and e'_1 and on them plot distances of a, b, c, d and e from X_1Y_1 line respectively on the other side of X_2Y_2, as shown in Fig. 11.7 for the point B. New projection $a_2b_2c_2d_2$ is an auxiliary plan and this view gives the true shape of plane since its elevation is a straight line parallel to ground line X_2Y_2.

1 (f) Angle Between Two Intersecting Planes :

Whenever it is required to find the angle between two planes, then find the view of two planes in which straight line, which is generated by intersection of two planes, is seen in point view.

That means follow the procedure of obtaining point view of a straight line.

Problem 8 : Plans abc and abd and elevations a'b'c' and a'b'd' of two intersecting planes ABC and ABD are given. Find the angle between these two planes.

For solution see Fig. 11.8 and follow the procedure, as given below :

We know that AB is a line of intersection of two planes and so find the point view of the line AB.

(i) First draw new ground line X_1Y_1 parallel to ab and project plan points to get auxiliary elevation. In this view a'_1 b'_1 will represent the T.L. of AB.

(ii) Now take new ground line X_2Y_2 perpendicular to $a'_1b'_1$ (T.L.) and project all points of A.E. to get A.P. In this A.P., the AB line is seen as a point view and hence planes are seen as edge views or line views. In this view, angle between two edge views of two planes i.e. θ is the real angle between two planes. Measure it.

(iii) If one wants to get θ in A. E., initially draw X_1Y_1 parallel to a'b' instead of ab and proceed in the same way.

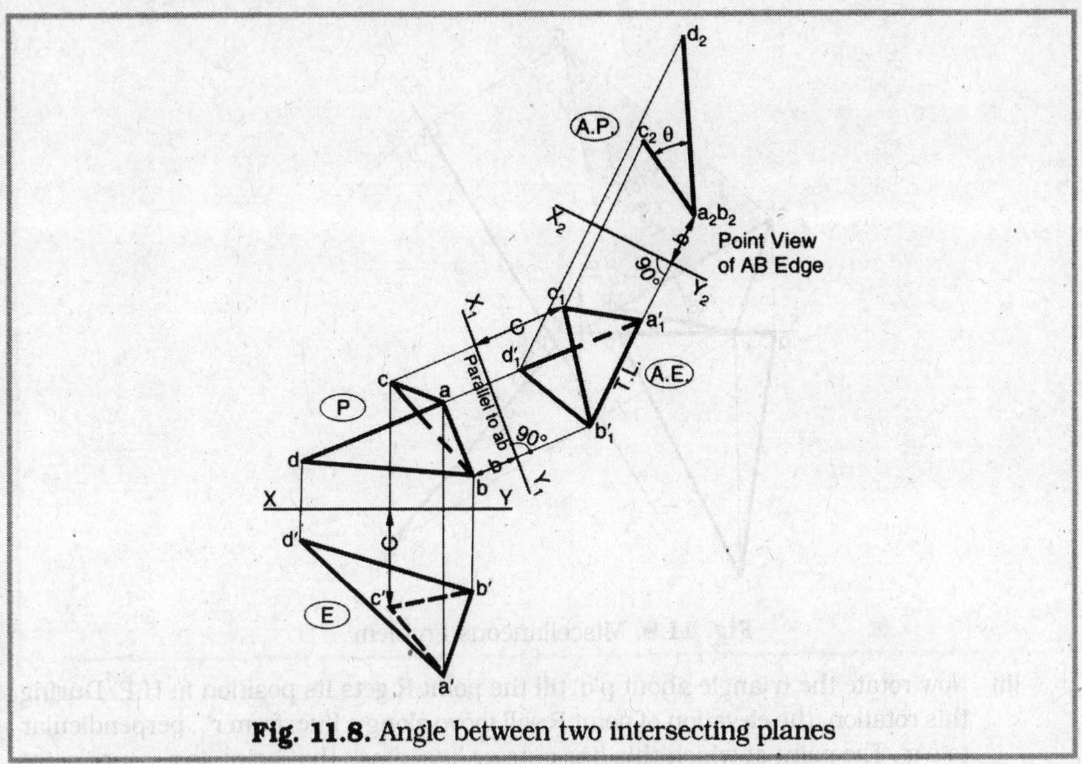

Fig. 11.8. Angle between two intersecting planes

1 (g) Distance Between Two Parallel Straight Lines :

The solution of finding the distance between two parallel straight lines, whose plan and elevation are given, can be done by following any one method given below :

(a) By finding the point views of both the straight lines and measuring distance between these point views,

[OR]

(b) By constructing a plane in each view out of two views of two given straight lines and then finding the true shape of that constructed plane. In the true shape of plane, distance between the two lines is the distance between two parallel straight lines.

3. Miscellaneous Problems :

Problem 9 : An isosceles triangle PQR having the base PQ 60 mm and altitude 85 mm has its base in the V.P. making an angle of 30° with the H.P. The corner P is 10 mm above the H.P. and the corner R is in the H.P. Draw the projections of the plane.

For solution see Fig. 11.9 and follow the procedure, as given below :

(i) Initially, assume the triangle to be lying completely in the V.P. with edge PQ making 30° with H.P. and the point P 10 mm above H.P., and hence draw $\Delta p'q'r'_1$ with base p'q' of 60 mm making 30° with xy and with point p' 10 mm above xy and the altitude of triangle 85 mm.

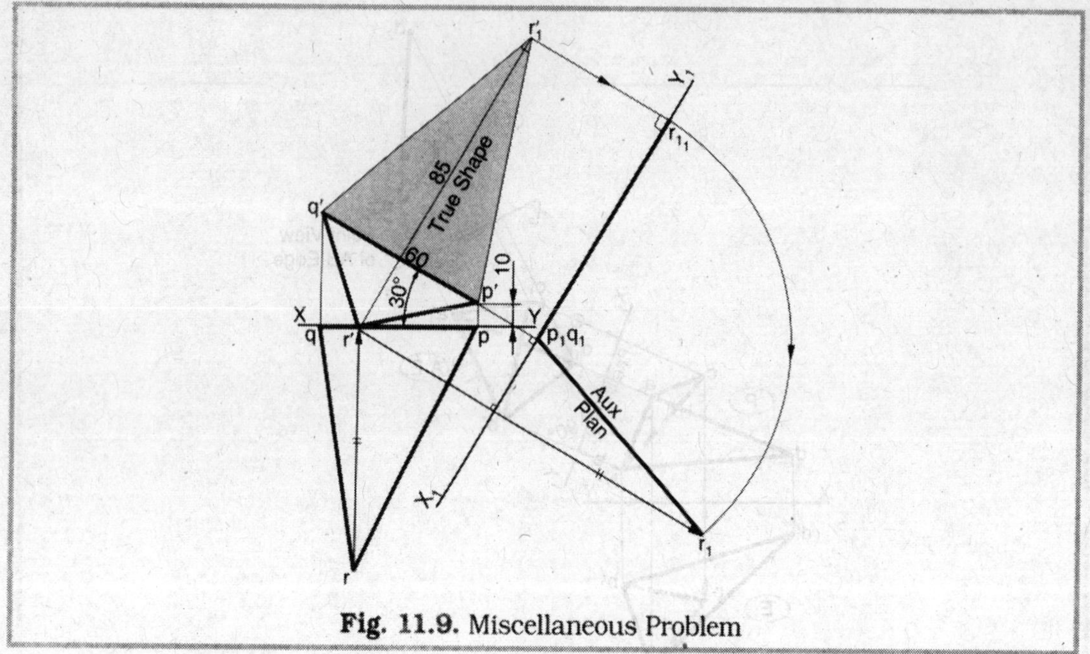

Fig. 11.9. Miscellaneous Problem

(ii) Now rotate the triangle about p'q' till the point R gets its position in H.P. During this rotation, the elevation of point R will move along a line, from r'_1, perpendicular to p'q'. The point at which this line cuts xy line, mark that point as r', since point R is in H.P. During this rotation, the auxiliary plan of point R on A.I.P, whose x_1y_1 line is perpendicular to p'q', is an arc of circle with point p_1q_1 as the centre and radius equal to p_1q_1 - r_1 (85 mm). From r', draw projector on x_1y_1 to cut the above arc at r_1. Measure the distance of r_1 from x_1y_1 and take the same distance to mark r on projector to xy from r'.

(iii) Join q'r' and p'r' to complete elevation and join qr and pr to complete plan.

Problem 10 : A picture frame, 2 m wide and 1 m high, is to be fixed on a horizontal wall railing by two straight wires attached to the top corners. The frame is to make an angle of 40° with the wall and the wires are to be fixed to a hook on the wall on the centre line of the frame and 1.5 m above the railing. Find the length of the wires and the angle between them.

For solution see Fig. 11.10 and follow the procedure, as given below :

Discussion :

To visualise the position of the frame and wire with respect to the wall, three dimensional drawing is given. The true width of the frame will be seen in elevation and plan and the true height of the frame will be seen in the side view or the end view. True angle (40°) of the frame plane with wall will be seen in the side view. So here we shall have to draw all views simultaneously.

(i) Draw plan, elevation and end view of the frame along with wires, as shown in Fig. 11.10

(ii) Find the true length of one wire. Second wire will be of the same length. Procedure for T.L. is shown in the figure for wire OA.

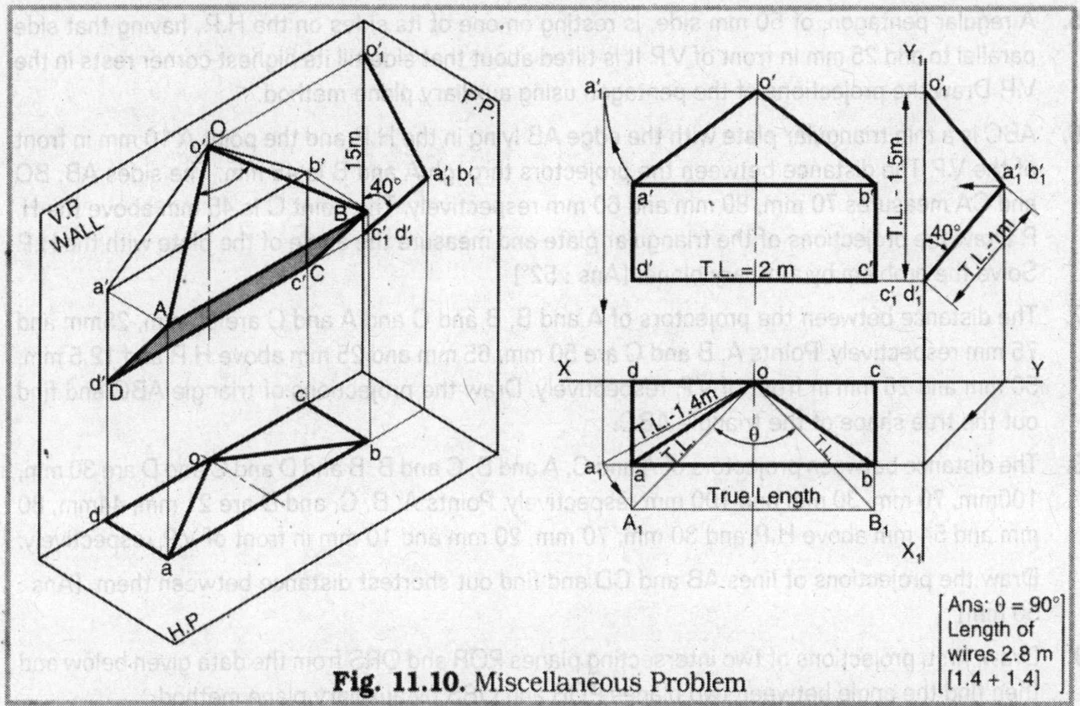

Fig. 11.10. Miscellaneous Problem

(iii) Draw triangle with two sides as T.L. of the wire and one side as 2 m width of the frame. In this triangle θ will be the true angle, so measure it. Measured true length of wire is 1.4 m and hence total length of two wires is 2.8 m and the measured true angle θ is 90° between wires.

EXERCISE

1. The distance between the end projectors of a straight line PQ is 70 mm. The end P is 45 mm above H.P. and 25 mm in front of V.P. The end Q is 35 mm above H.P. and 55 mm. in front of V.P. Find out the true length of a line and its inclination with H.P. and V.P. both, by auxiliary plane method. [Ans : 74 mm, θ =7.5° ø=23°]

2. A hexagonal plate, of 30 mm long sides, has an edge on the H.P. and inclined at 45° to the V.P. Hexagonal plate makes an angle of 30° with the H.P. Draw the projections using auxiliary plane method.

3. The distance between the end projectors of a line AB is 75 mm. The end A is 40 mm below H.P. and 25 mm behind V.P. The end B is 60 mm below H.P. and 10 mm behind V.P. Find (i) the shortest distance between two lines AB and xy and (ii) the distance of mid point M of AB from xy line. [Ans : 45 mm ; 53 mm]

4. A divider POQ has two equal legs OP and OQ of 100 mm length. It is resting on ground on two leg's ends P and Q. The distance between P and Q is 35 mm and the line PQ is inclined to the V.P. by 60°. The hinge point O is 50 mm above the ground.

   Draw the projections of the divider by auxiliary plane method and find out the inclination of the plane, containing the divider with ground or H.P. [Answer 30.5°]

5. A regular pentagon, of 50 mm side, is resting on one of its sides on the H.P., having that side parallel to and 25 mm in front of V.P. It is tilted about that side till its highest corner rests in the V.P. Draw the projections of the pentagon using auxiliary plane method.

6. ABC is a thin triangular plate with the edge AB lying in the H.P. and the point A 10 mm in front of the V.P. The distance between the projectors through A and B is 45 mm. The sides AB, BC and CA measures 70 mm, 80 mm and 60 mm respectively. The point C is 45 mm above the H. P. Draw the projections of the triangular plate and measure the angle of the plate with the H.P. Solve the problem by auxiliary plane. [Ans : 52°]

7. The distance between the projectors of A and B, B and C and A and C are 50 mm, 25mm and 75 mm respectively. Points A, B and C are 50 mm, 65 mm and 25 mm above H.P. and 12.5 mm, 50 mm and 25 mm in front of V.P. respectively. Draw the projections of triangle ABC and find out the true shape of the triangle ABC.

8. The distance between projectors of A and C, A and B, C and B, B and D and C and D are 30 mm, 100mm, 70 mm, 30 mm and 100 mm respectively. Points A, B, C, and D are 21 mm, 44mm, 80 mm and 54 mm above H.P. and 30 mm, 70 mm, 20 mm and 10 mm in front of V.P. respectively.

   Draw the projections of lines AB and CD and find out shortest distance between them. [Ans : 50 mm]

9. Draw, first, projections of two intersecting planes PQR and QRS from the data given below and then find the angle between two planes PQR and QRS by auxiliary plane method.

   The distance between projectors of R and P, P and Q, R and S, R and Q and S and P are 62 mm, 28 mm, 44 mm, 90 mm and 18 mm respectively. Points R, S, P and Q are 60 mm, 24 mm, 84 mm and 44 mm above H.P. and 44 mm, 80 mm, 70 mm and 20 mm in front of V.P. respectively. [Ans. 72°]

10. ABC is a triangular plate in the first quadrant, the projections of which appear as follows :

    Elevation triangle : a'b' = 80 mm, b'c' = 45 mm and c'a' = 80 mm a'b' making an angle of 45° with xy line and the corner A of the triangle is 5 mm above the H.P.

    Plan triangle : ab makes 30° with the xy line, bc = 95 mm, corner B is 5 mm in front of the V.P.

    Draw the projections of the plane and determine the true shape with the help of auxiliary views. Also find the inclinations of the plane with the H.P. and V.P. both. [Ans. : 42°, 68°]

11. In the triangular plate ABC of problem 10, draw a perpendicular from corner B to the opposite side AC of the triangle and measure its length. Also show the perpendicular line in the elevation and plan of the triangle. [Ans. : 80 mm]

CHAPTER - 12

Projections of Solids

1. INTRODUCTION :

Solids have three dimensions - length, breadth and depth/thickness. Solids are classified in five different ways and they are as below :

1. (i) Right (ii) Oblique
2. (i) Regular (ii) Irregular
3. (i) Polyhedra (ii) Solids of Revolution
4. (i) Full Solid (ii) Frustum of Solid
   (iii) Truncated Frustum of Solid
5. (i) Prism (ii) Pyramid
   (iii) Cylinder (iv) Cone
   (v) Sphere (vi) Octahedron
   (vii) Dodecahedron (viii) Icosahedron

1. (i) **Right Solids :** See Fig. 12.1 Axes of these solids are at right angle to their bases. Whenever not mentioned, assume right solid only.
   (ii) **Oblique Solids :** See Fig. 12.1 Axes of these solids are not at right angle to their bases. (Not treated in this book.)
2. (i) **Regular :** When the base of the solid is a regular polygon or a true circle, the solid is said to be regular. Wherever not mentioned, assume regular.
   (ii) **Irregular :** When the base of the solid is not a regular polygon or not a true circle, the solid is said to be irregular. (Not treated in this book.)
3. (i) **Polyhedra :** When the solid is bounded by planes, called faces then the solid is marked as polyhedron, e.g. tetrahedron, cube, prism, pyramid, etc. See Fig. 12.2, Nos. 1 to 10.
   (ii) **Solids of Revolution :** When a solid is generated by revolution of a plane figure about a line (axis) then this solid is named as a solid of revolution e.g. cylinder, cone, sphere, etc. See Fig. 12.2, Nos. 11 to 15.

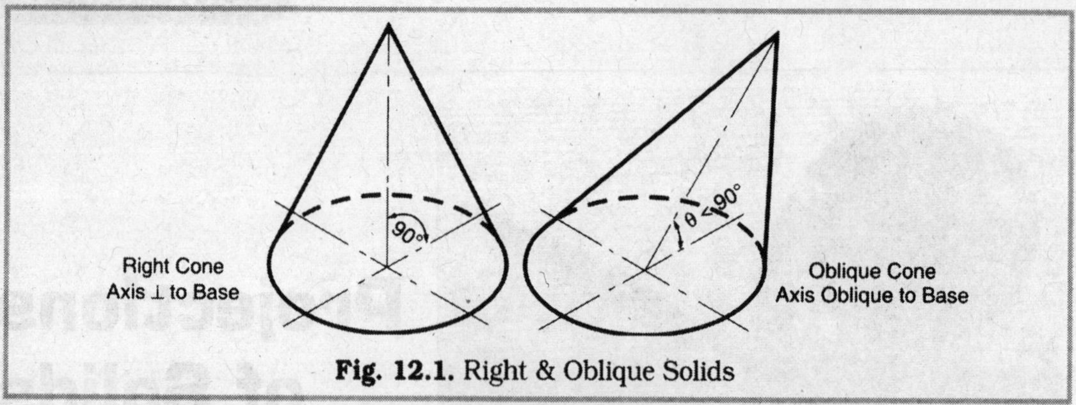

Fig. 12.1. Right & Oblique Solids

4. (i) **Full Solid** : When a solid is having its shape full, it is called a full solid. See Fig. 12.2, Nos. 1 to 13. Wherever not mentioned, take a full solid.

(ii) **Frustum of Solid** : When a solid is cut by a plane parallel to the base, then cut solid is called frustum. We can have frustum of a cone or a frustum of a pyramid. Frustum of a cone is shown in Fig. 12.2 as No.15.

(iii) **Truncated Frustum** : When a solid is cut by a plane inclined to the base, then the cut solid is named as truncated solid. A truncated cylinder is shown in Fig. 12.2 as No.14.

5. (i) Prisms : See Fig. 12.2, Nos.1 to 5.

Prism is a kind of polyhedron having two identical faces called its ends or bases. These two bases are parallel to each other and are joined by parallelograms or rectangular faces. Imaginary line passing through the centres of two bases is called axis.

Prisms are of following types :

(a) Triangular
(b) Square
(c) Rectangular
(d) Pentagonal
(e) Hexagonal
(f) Octagonal, etc.

When a prism is having infinite number of edges of base, then it becomes a Cylinder.

When a square prism is having its height equal to the length of edge of base, then it becomes a **Cube.** See Fig. 12.2, No.5. All prisms have totally (n x 3) number of edges, where n is the number of edges of base.

(ii) Pyramids : See Fig. 12.2, Nos. 6 to 10.

Pyramid is a kind of polyhedron having a base and an apex. Edges of base is connected to the apex by slant triangular faces. The imaginary line connecting the apex to the centre of the base is called the axis. Slant triangular faces of pyramids are isosceles triangles.

In case of a triangular pyramid, when all edges of base and slant edges are equal in length, then that pyramid is called a **Tetrahedron.** All pyramids have totally (n x 2) number of edges. Where n is the number of edges of base. Pyramids have similar types, as prisms have, i.e. triangular, square, etc.

By cutting a pyramid parallel to the base we get frustum of the pyramid and cutting it inclined to the base, we get truncated frustum of pyramid.

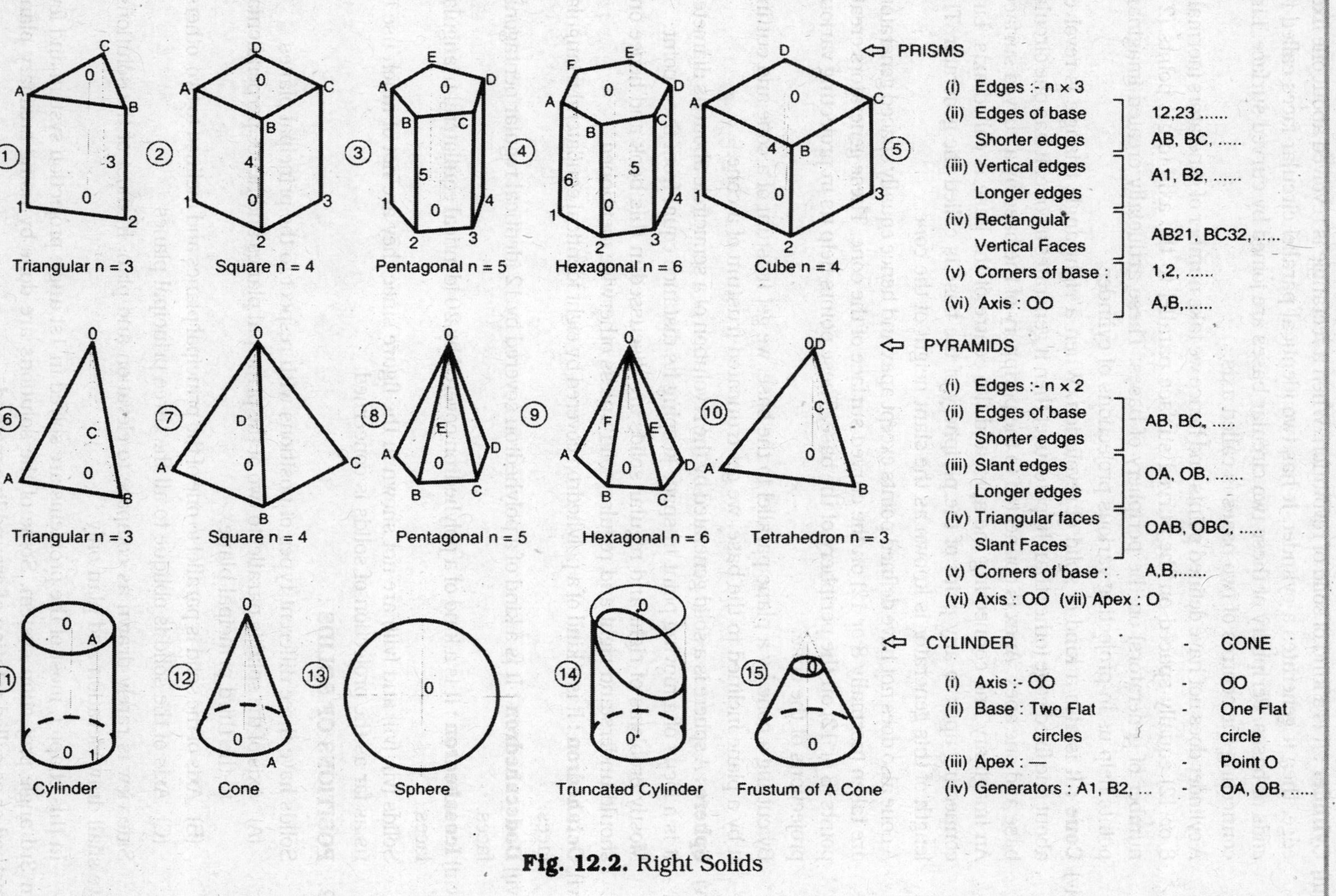

Fig. 12.2. Right Solids

(iii) **Cylinder :** It is a kind of solid of revolution. When a rectangle is revolved about one fixed edge, then it generates a cylinder. It has two identical parallel circular faces called its ends or bases. Periphery of these two circular bases are joined by curved surface. Line connecting the centres of two bases is called axis.

A cylinder does not have defined points and hence we take number of generators (normally 8 or 12) equally spaced on the curved surface parallel to the axis to get points (2 x number of generators) on the periphery of bases. These artificially created imaginary points help us in projecting various projections of cylinder.

(iv) **Cone :** It is also a kind of solid of revolution. When a right angle triangle is revolved about one fixed edge (other than hypotenuse) then it generates a cone. It has one circular base and one apex. Apex is connected to the periphery of the base by a curved surface. An imaginary line connecting the apex and the centre of the base is called axis. Line connecting apex to any point of the periphery of the base is called the generator. The length of this generator is known as the slant height of the cone.

A cone also does not have defined points except apex and hence equally spaced generators are taken (normally 8 or 12) on the curved surface of the cone. These generators create points (8 or 12) on the periphery of the base. These points help us in projecting various projections of the cone.

By cutting a cone by a plane parallel to the base, we get frustum of a cone and cutting it by a plane inclined to the base, we get truncated frustum of a cone.

(v) **Sphere :** A sphere is a solid generated by the revolution of a semicircle about its diameter. It is a locus of a moving point in space keeping its distance equal from fixed point.

Mostly problems of right and regular solids are discussed in this book and hence one should understand right and regular solid unless otherwise mentioned.

(vi) **Octahedron :** It is a kind of a polyhedron covered by eight identical equilateral triangular faces.

(vii) **Dodecahedron :** It is a kind of a polyhedron covered by 12 identical regular pentagonal faces.

(viii) **Icosahedron :** It is a kind of a polyhedron covered by 20 identical equilateral triangular faces.

Solids (vi), (vii) and (viii) are not shown in the figure since they are not of much use to us as far as the projection of solids is concerned.

2. POSITIONS OF SOLIDS :

Solids have three different types of positions with respect to the principal planes.

(A) Axis of the solid is parallel to two of the principal planes and hence perpendicular to the third principal plane.

(B) Axis of the solid is parallel to one of the principal planes and inclined to two others.

(C) Axis of the solid is oblique to all the three principal planes.

Side view is rarely drawn as compare to elevation and plan. In most of the solutions, we shall draw elevation and plan only.

In this chapter, most of the problems are solved in 1st angle projection system and few in 3rd angle projection system. Some of the solutions are done by using auxiliary plane method or so called rotation of ground line method.

(A) Axis of the solid parallel to two and perpendicular to one of the principal planes :

For right solids when axis is perpendicular to one of the principal planes, its base will be parallel to that principal plane. We have seen in the chapter of projections of planes that when a plane is parallel to principal plane, its projection on that principal plane will be true shape. And hence, the projection of the base on the plane, to which axis of the solid is perpendicular, will be true shape. Solution of problem is started by drawing this true shape of base.

If the axis is perpendicular to H.P. or ground in plan or in T.V., true shape of base is drawn first. Similarly, if the axis is perpendicular to V.P., in elevation, T.S. of base is drawn first and when perpendicular to P.P. in side view, T.S. of base is drawn first.

Further, as the axis of the solid is parallel to two principal planes and hence projections of the axis on these two planes will be T.L. So length of the axis or the height of the solid can be plotted true in these two views.

Now we shall study few examples of projections of solids of this type.

Problem 1 : A cylinder, diameter of base D mm and height h mm, is resting on H.P. on its base with axis of the cylinder *l* mm away from V.P. Draw the projections.

For solution see Fig. 12.3 (a) and (b)(i) and follow the procedure, as given below :

(i) First draw in plan T.S. of base i.e. a circle of D mm diameter. Take the centre of the circle *l* mm below xy, since axis is *l* mm in front of V.P.

(ii) Project the circle and take the height of the axis h mm to get a rectangle (D × h) as the elevation of the cylinder. Base of the rectangle should lie on xy line, since the cylinder is resting on H.P. on its base.

Problem 2 : A square prism, side of base 25 mm and height 40 mm, is having its axis perpendicular to V.P. and 30 mm above H.P. and parallel to it. One of the bases nearer to V.P. is 10 mm in front of V.P. Draw its projections.

For solution see Fig. 12.3 (a) and (b) (ii) and follow the procedure, as given below :

(i) Here, as the axis is perpendicular to V.P., first draw T.S. of base in elevation i.e. a square of 25 mm side. Keep the centre of the square 30 mm above xy, since the axis is 30 mm above H.P.

(ii) Draw projections from elevation and take length of the axis 40 mm in plan to get a rectangle of (25 × 40) as plan. Axis is parallel to H.P. and hence it will be seen of true length in plan. Further, draw one base of rectangle 10 mm below xy.

Problem 3 : A cone, diameter of base 35 mm and height 48 mm, is having its axis perpendicular to profile plane and parallel to H.P. and V.P. Draw its projections. Use 1st angle projection system.

For solution see Fig. 12.3 (a) and (b) (iii) and follow the procedure, as given below :

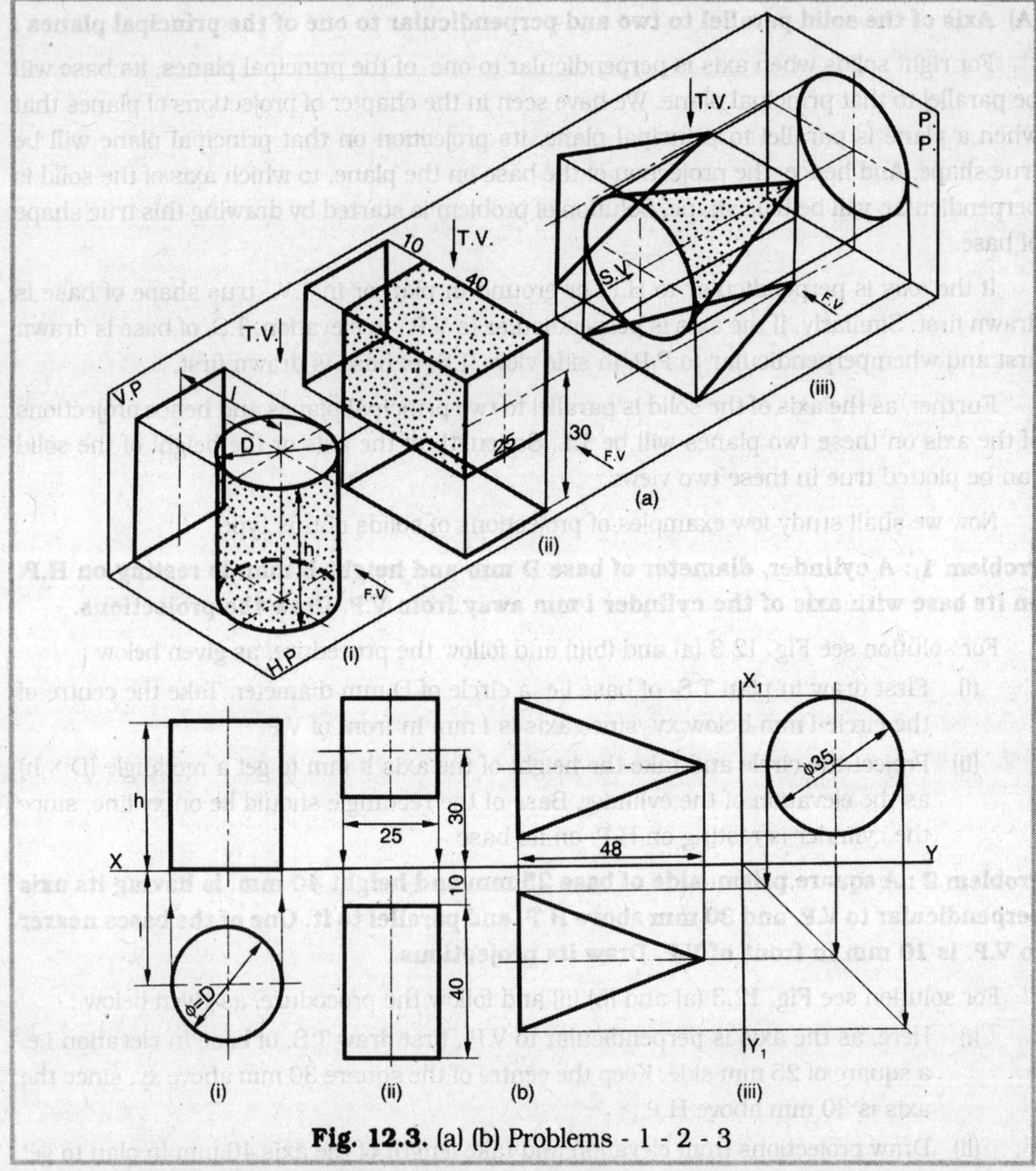

Fig. 12.3. (a) (b) Problems - 1 - 2 - 3

(i) As the axis of the cone is perpendicular to P.P., in end view, T.S. of the base i.e. circle of 35 mm diameter will be seen. First draw circle of 35 mm diameter in end view.

(ii) Elevation and plan of cone will be triangles of base 35 mm and height 48 mm, project it from end view or side view.

Three dimensional drawing is given for all the above problems to visualize the positions of objects and their projections.

Problem 4 : A hexagonal prism, side of base 25 mm and height 45 mm, is resting on H.P. on its base with one of the edges of the base parallel to V.P. The axis of the prism is parallel to and 35 mm away from V.P. Draw the projections of prism.

For solution see Fig. 12.4 (a) and (b) and follow the procedure, as given below :

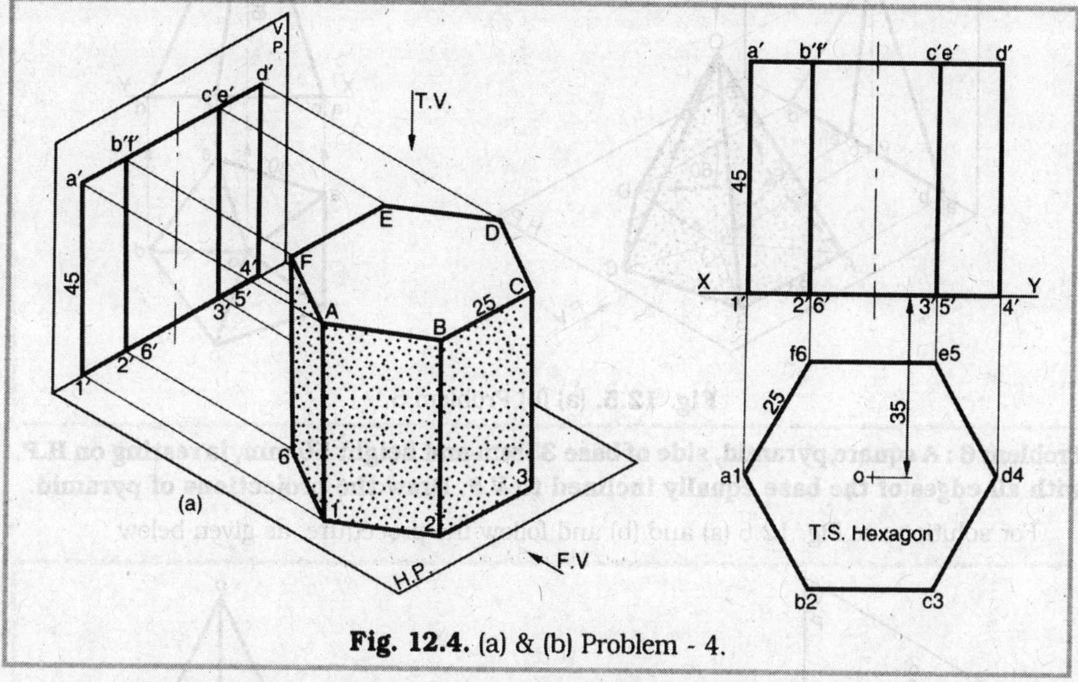

Fig. 12.4. (a) & (b) Problem - 4.

(i) First draw T.S., hexagon of 25 mm side with one side parallel to xy and the centre of the hexagon 35 mm below xy, as the plan.

(ii) Project all 6 points upwards from plan. Mark points 1', 2',....6' on xy line, since base of the prism is on H.P.

(iii) Now take height of 45 mm above xy line, since it will be seen in T.L. as its axis is parallel to V.P. and mark points a', b',....f' on corresponding projectors and complete the elevation as shown in the figure.

Problem 5 : A pentagonal pyramid, side of base 30 mm and height 60 mm, is resting on H.P. on its base with one of the edges of base perpendicular to V.P. Draw its projections.

For solution see Fig. 12.5 (a) and (b) and follow the procedure, as given below :

(i) As the pyramid is resting on H.P. on its base, base will be seen in T.S. in plan and hence first draw in plan regular pentagon abcde of 30 mm size with one edge ab perpendicular to xy line. Draw oa, ob,. .....oe by bisecting angles ∠ a, ∠ b, ... ∠ e respectively. oa, ob,.....oe represent plans of slant edges OA, OB...... OE respectively.

(ii) Project a, b,....e and o from plan and mark points a', b',....e' on corresponding projectors on xy line, since base ABCDE is on H.P. On projector through O, take 60 mm height from the base or xy line to mark point O'. Join o'a', o'b',....o'e' to complete the elevation.

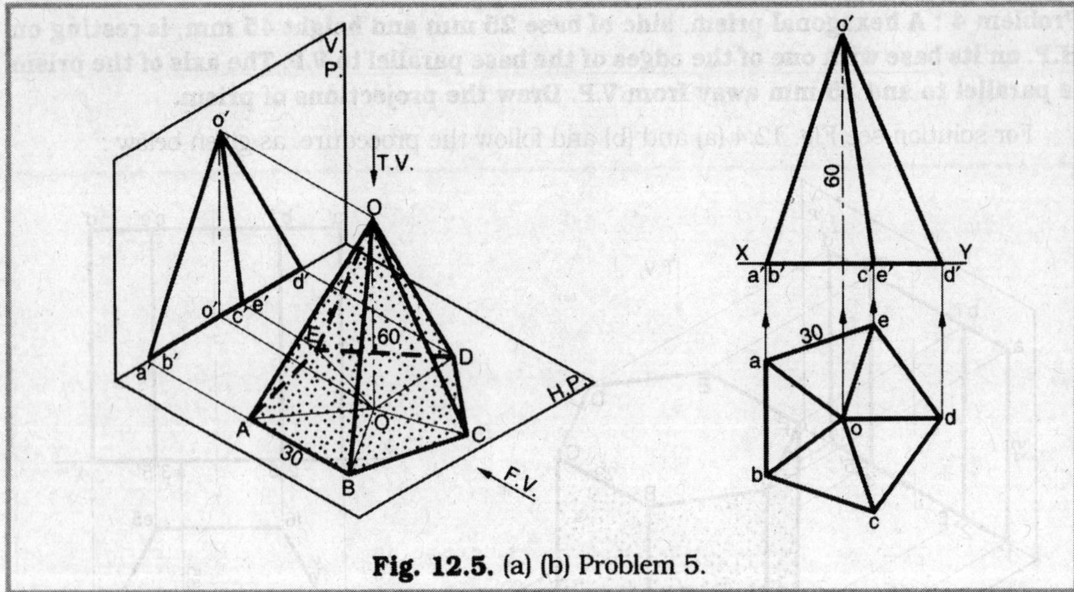

Fig. 12.5. (a) (b) Problem 5.

Problem 6 : A square pyramid, side of base 35mm and height 50 mm, is resting on H.P. with all edges of the base equally inclined to V.P. Draw the projections of pyramid.

For solution see Fig. 12.6 (a) and (b) and follow the procedure, as given below :

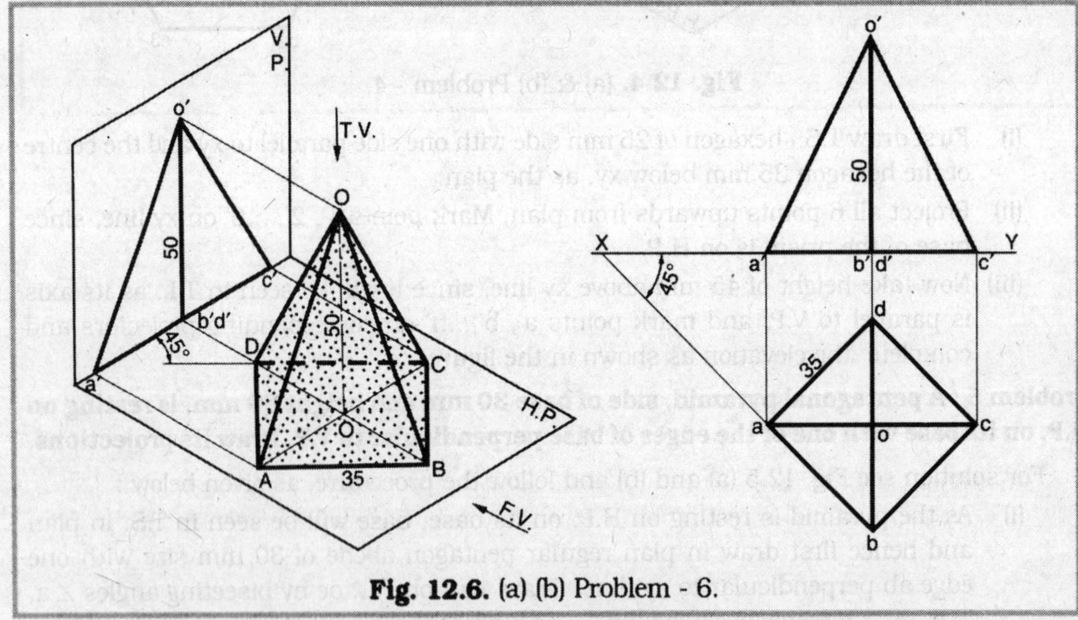

Fig. 12.6. (a) (b) Problem - 6.

(i) First draw T.S. of base in plan i.e. square of 35 mm. In drawing square, draw one edge ab at 45° to xy line, automatically all other edges will make 45° with xy also. Join ac and bd and get the centre o.

(ii) Project all points a, b, c, d and o upward and mark points a', b', c' and d' on xy line on corresponding projectors. On projector through o, take 50 mm from base or xy line to mark o'. Join o'a', o'b',....o'd' to complete the elevation.

Problem 7 : A square prism, side of base 35 mm and height 60 mm, is resting on its base in V.P. and axis perpendicular to V.P. and 30 mm above the H.P. One of the edges of the base is inclined at 30° to the H.P. Draw the projections of the square prism.

For solution see Fig. 12.7 (a) and (b) and follow the procedure, as given below :

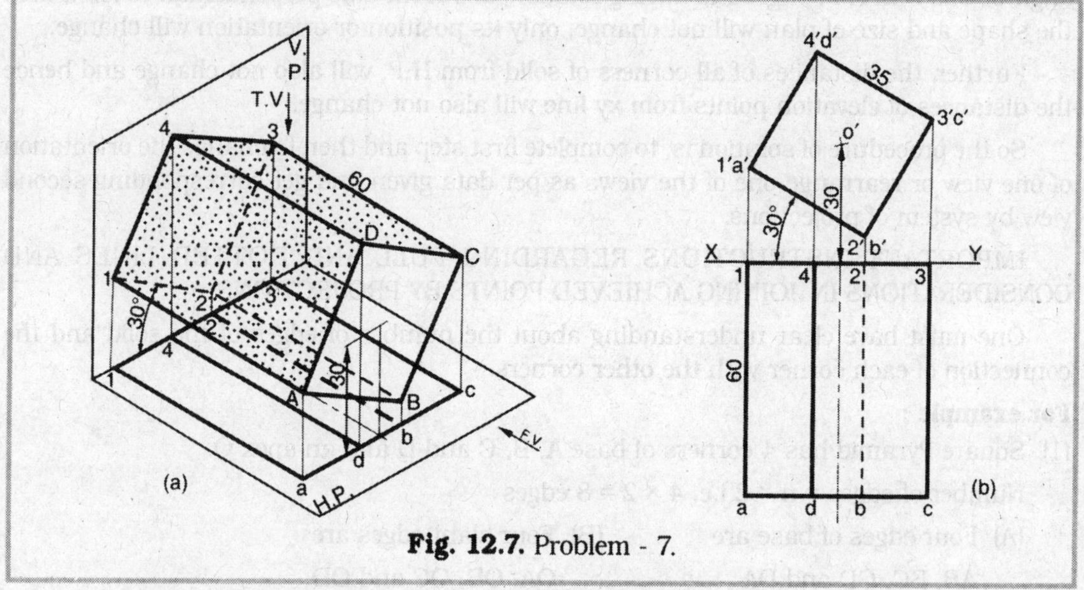

Fig. 12.7. Problem - 7.

(i) As the base is in V.P. elevation of base will be T.S. and its plan will be on xy line. So first draw elevation, a square of 35 mm side with one edge making 30° with xy line and the centre (o') 30 mm above xy line, as shown in the figure. Mark corners of square by 1', 2', 3' and 4'. a',b',c'and d' will also lie on respective corners.

(ii) Draw projectors, from elevation points, in the downward direction and take height of plan equal to 60 mm to complete the plan.

(B) Axis of the solid parallel to one of the principal planes and inclined to two others :

Projections of this type of solid are obtained in the second step. In the first step prior to the final or second step, assume the axis of solid parallel to two and perpendicular to one of the principal planes. Mostly assume axis of the solid perpendicular to the plane to which it is given inclined, e.g. if the axis of solid is parallel to V.P. and inclined to H.P. or ground, assume axis perpendicular to H.P. or ground and start the problem by drawing T.S. of base in plan. Get plan and elevation in the first step, as per study of part (A).

Important Aspects :

Now if you rotate the solid, changing angle of axis with H.P. or ground and not changing angle of axis with V.P. during rotation, about the axis perpendicular to V.P., then the shape and size of elevation will not change only its position or orientation will change.

Further, distances of all corners of solid from V.P. will also not change and hence the distances of plan points of these corners from xy line will also not change.

These two aspects are very very important in arriving at the final solution. Try to visualize these aspects from three dimensional drawing given for problems.

Similarly, if the axis of the solid is given parallel to H.P. or ground and inclined to V.P., assume axis perpendicular to V.P. and start the solution of problem by drawing T.S. of base in elevation. Get elevation and plan in first step, as per previous study.

Now if you rotate the solid, changing angle, the axis makes with V.P., and not changing angle of axis with H.P. or ground during rotation, about the axis perpendicular to H.P., then the shape and size of plan will not change, only its position or orientation will change.

Further, the distances of all corners of solid from H.P. will also not change and hence the distances of elevation points from xy line will also not change.

So the procedure of solution is, to complete first step and then to change the orientation of one view or rearrange one of the views as per data given and get corresponding second view by system of projections.

IMPORTANT INSTRUCTIONS REGARDING FULL AND DOTTED LINES AND CONSIDERATIONS IN JOINING ACHIEVED POINTS BY PROJECTIONS

One must have clear understanding about the number of edges of the solid and the connection of each corner with the other corners.

For example :

(1) Square Pyramid has 4 corners of base A, B, C and D and an apex O.

Number of edges = n × 2 i.e. 4 × 2 = 8 edges

(A) Four edges of base are AB, BC, CD and DA

(B) Four slant edges are OA, OB, OC and OD

i.e corner B is required to be joined with A, C and O and not with point D.

(2) Hexagonal Prism has 6 corners of bottom base 1, 2, 3, 4, 5 and 6, and 6 corners of top base A, B, C, D, E and F.

Number of edges = n × 3 i.e. 6 × 3 = 18 edges.

Twelve edges of two bases are	12,	23,	34,	45,	56	and	61.
	AB,	BC,	CD,	DE,	EF	and	FA
Six vertical edges are	1A,	2B,	3C,	4D,	5E	and	6F

Corner E is required to be connected to D, F and 5 to get edges DE, EF and 5E. E is not to be joined to A, B, 2, 4 etc. Remember each corner of solid has three (3) edges.

After getting projections of all points, join edges/corners at the boundary by full lines. Remember boundary of any projection can never be dotted. Out of two intersecting edges one will be full and other will be dotted. Now to decide about full and dotted edges following considerations are used.

(1) In first angle system, points which are away from xy line in elevation are likely to be full in plan than points which are nearer to xy.

Points which are away from xy line in plan are likely to be full in elevation than points which are nearer to xy.

(2) In third angle system, points which are nearer to xy line in elevation are likely to be full in plan than points which are away from xy.

Points which are nearer to xy line in plan are likely to be full in elevation than points which are away from xy.

Now we shall study a few examples of this type:

Problem 8 : A square prism, side of base 30 mm and height 50 mm, is resting on H.P. on one of its edges of base with (i) axis making an angle of (60°) with H.P. or (ii) rectangular face, containing the edge of base on which it rests on H.P., is making an angle (60°) with H.P. or (iii) base making an angle of (30°) with H.P. or (iv) opposite edge of other base or top base is having h mm height from H.P. Draw the projections of the prism when the axis of prism is parallel to V.P.

For solution see Fig. 12.8 (a & b) and follow the procedure, as given below :

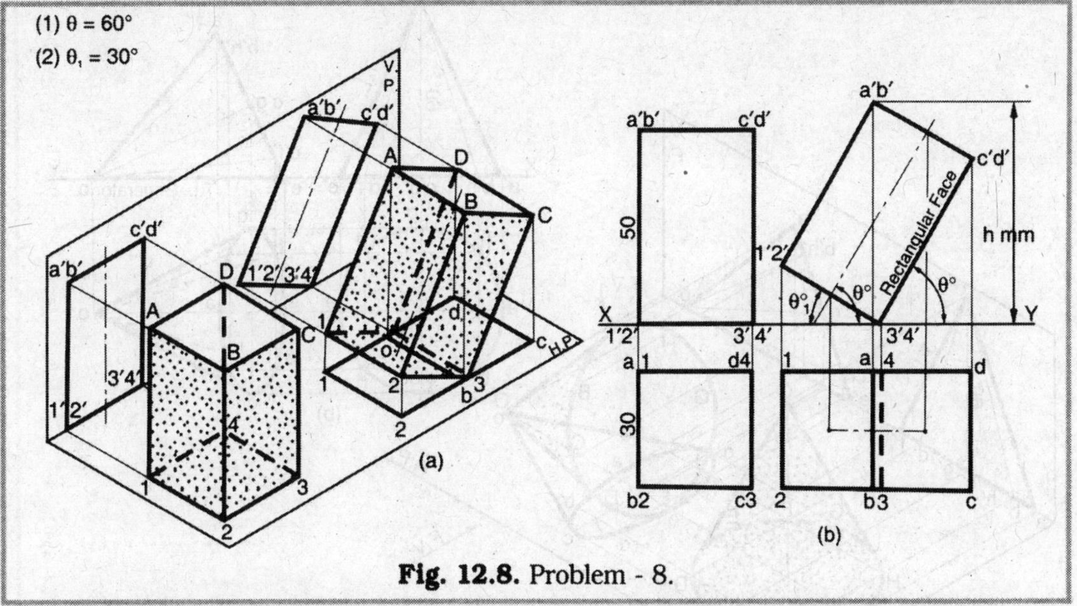

Fig. 12.8. Problem - 8.

As it is given that the prism is resting on H.P., the prism is in first quadrant, so solve the problem in 1st angle system.

First of all visualize the two stages of prism with the help of three dimensional drawing given. In three dimensional drawing projections are also shown. Students are not required to draw this three dimensional drawing. It is given here just to develop three dimensional thinking of students and to create correct understanding of problem.

(i) In first stage, assume axis of prism perpendicular to H.P. and hence base will be seen in T.S. in plan. So start the problem by drawing 30 mm size square in plan below xy line.

Further the prism is resting on edge of base on H.P. and hence draw one edge of square perpendicular to xy in plan. Edge 34 here.

Draw the plan and elevation, as studied in part A.

(ii) Now rearrange or reorient elevation already obtained in 1st stage to the new position considering following aspects.

(a) 3'4' must be on xy since edge 34 is in H.P.

(b) Axis should make $\theta°$ with xy or rectangular face 3'4'-d'c' must make $\theta°$ with xy or base 4'3'-1'2' must make $\theta_1°$ with xy or edge a'b' of top base which is opposite edge to 4'3' should have height h mm above xy line.

(iii) Draw projectors through all corners of rearranged elevation and parallel to xy from corners of plan of first stage to get plan points of 2nd stage by intersection of projectors, from corresponding points. Join them to get plan. Line 43 in plan will be dotted since it is near to xy line in elevation.

Problem 9 : A cone, diameter of base 55 mm and height 60 mm, is resting on H.P. on one of its generators with axis parallel to V.P. Draw the projections of cone.

For solution see Fig. 12.9 and follow the procedure, as given below :

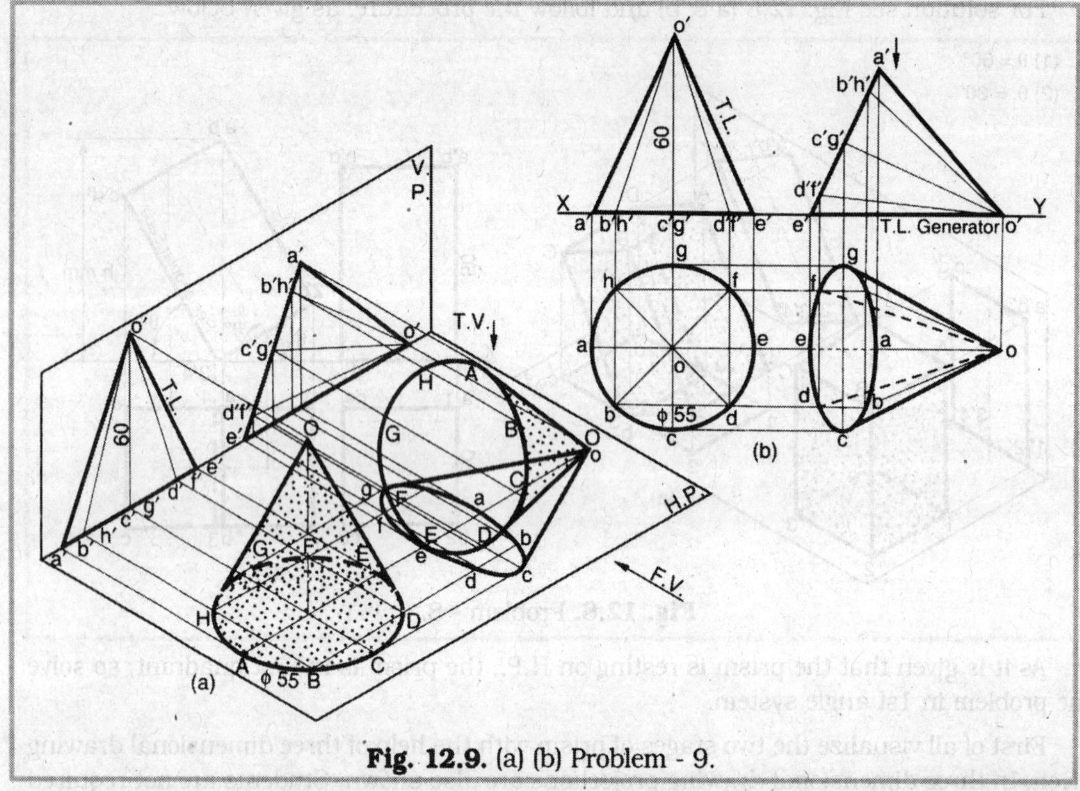

Fig. 12.9. (a) (b) Problem - 9.

As the cone is resting on H.P., the problem is required to be solved by 1st angle system. Further, a cone does not have defined lines or corners of its own and so we have to divide base into 8 or 12 equal parts and generate points and lines (generators).

When a cone is resting on H.P. on its generator, the axis will be inclined to H.P. by some angle and so assume initially the axis perpendicular to H.P.

(i) In the first stage, assume the cone resting on H.P. on its base with axis perpendicular to H.P. and draw plan and elevation, as shown in the figure.

(ii) Reorient elevation with generator showing T.L. on xy line and project plan as usual.

Base of the cone will be totally open on upper side and so will be seen completely from top and hence in plan it is drawn full.

Generators artificially created are shown in drawing just to give idea about projections. Actually they do not exist and so latter on in solutions of problems of cones and cylinders, these generators may or may not be shown in drawing. Only they will be used for the purpose of projections.

Problem 10 : A square pyramid, side of base 30 mm and height 45 mm, is resting on H.P. on one of the edges of the base with axis parallel to H.P. and V.P. both. Draw the projections of the pyramid.

For solution see Fig. 12.10 and follow the procedure, as given below :

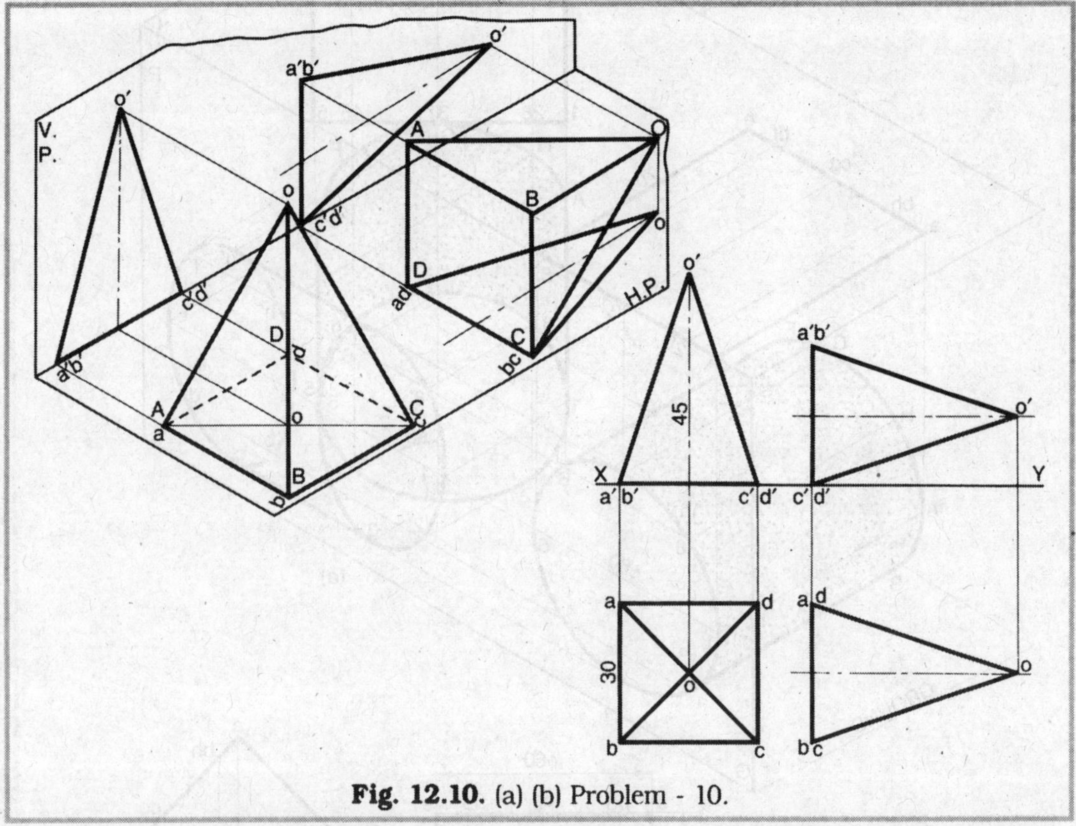

Fig. 12.10. (a) (b) Problem - 10.

Solution to this problem can be started by drawing T.S. of base in end view since the axis will be perpendicular to P.P. But here the problem is solved without drawing end view.

(i) First assume square pyramid resting on its base on H.P. with one edge of base perpendicular to V.P. and draw corresponding plan and elevation.

(ii) Reorient elevation with c'd' on xy line and axis parallel to xy line and project its corresponding plan as usual.

As the object is resting on H.P., 1st angle system of projection is used. Three dimensional drawing is given for visualization of the object in given position.

Problem 11 : A cylinder, diameter of base 60 mm and height 70 mm, is having a point of its periphery of base in V.P. with axis of cylinder inclined to V.P. by 45° and parallel to H.P. or ground. Draw the rojections of the cylinder.

For solution see Fig. 12.11(a) and (b) and follow the procedure, as given below :

This problem can be solved by either 1st angle or 3rd angle system of projection. Here, we have solved by 3rd angle system of projection.

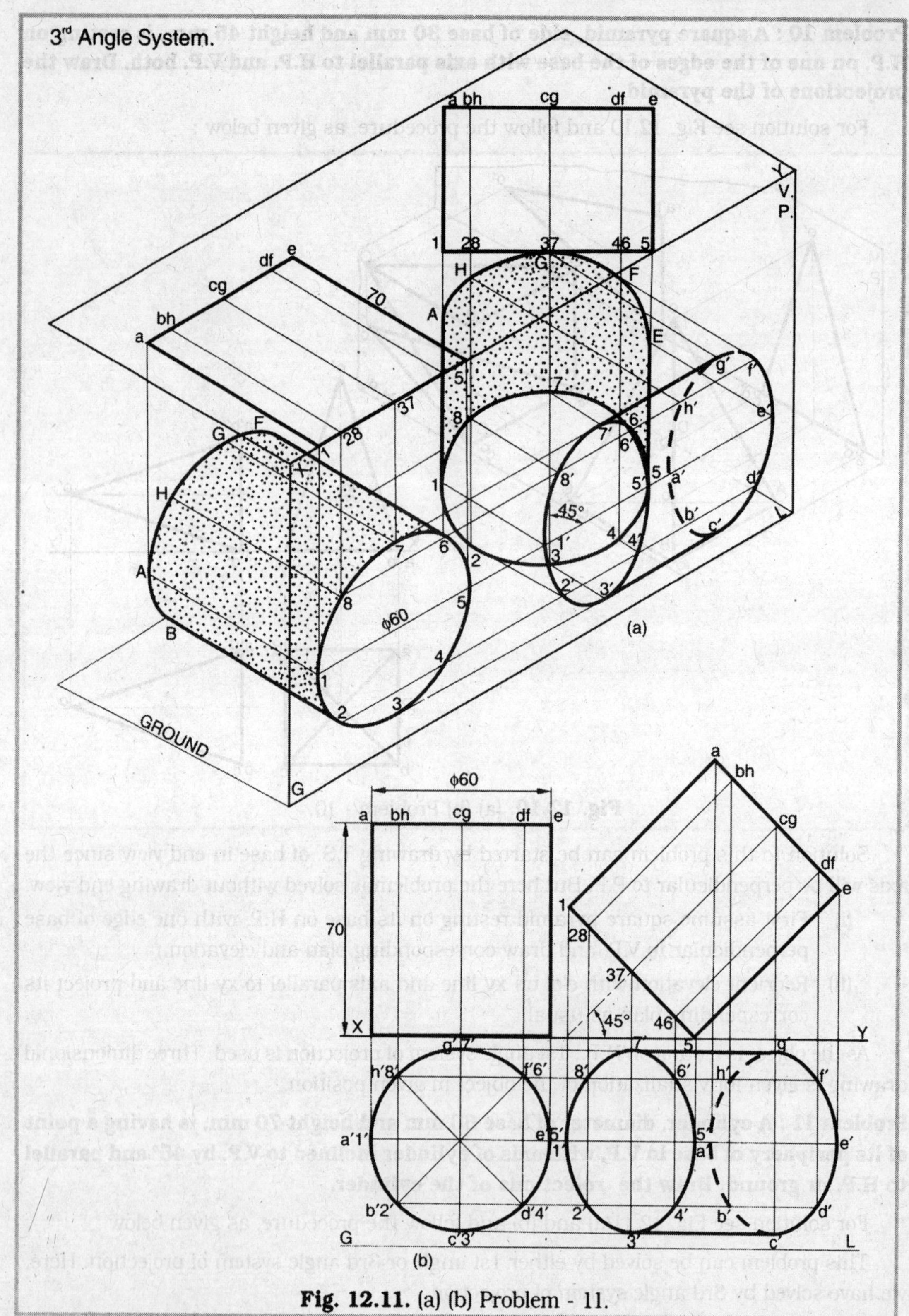

Fig. 12.11. (a) (b) Problem - 11.

(i) In first or initial stage assume axis of cylinder perpendicular to V.P. and assume base of cylinder in V.P. Draw plan and elevation of cylinder in this stage. Divide circle in elevation into 8 equal parts and draw corresponding generators in plan.

(ii) Reorient plan to new position such that point (5) of its periphery remains on xy line and axis makes 45° with xy line. Project all 16 points downward and draw horizontal lines from 16 points of elevation of first stage to get 16 points by intersections.

Join all points of two bases by smooth curve. Points 1, 2,....8 are nearer to xy line and open towards xy, so they are seen full and points a, b,....h are far away from xy and not seen from xy side. So they will be dotted. Boundary lines will be full.

Problem 12 : A cylinder, diameter of base 70 mm and height 20 mm, is resting on H.P. on its base. A hexagonal pyramid, side of base 25 mm and height 50 mm, is resting on the flat surface of the cylinder coaxially with one of the edges of the base parallel to V.P. Draw the projections of the cylinder and pyramid.

Project auxiliary plan on A.I.P. inclined to H.P. by 60° and project auxiliary elevation on A.V.P. inclined to V.P. by 45°.

For solution see Fig. 12.12 and follow the procedure as given below :

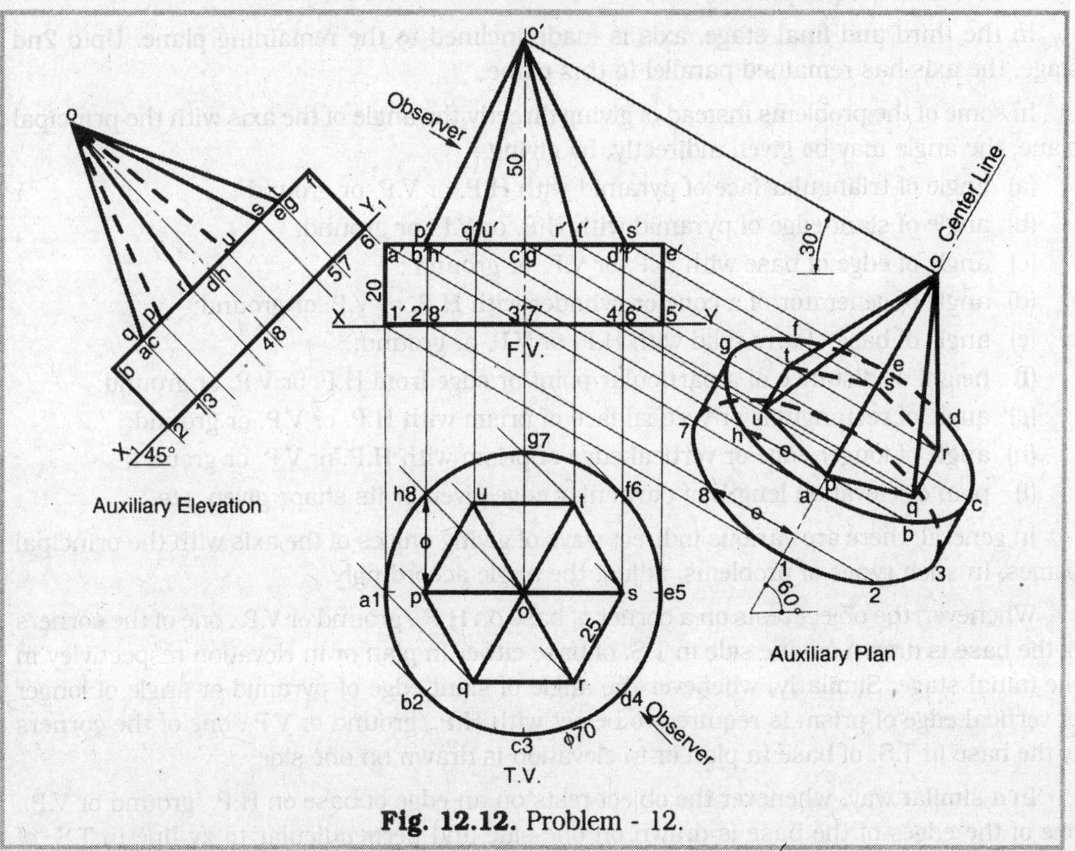

Fig. 12.12. Problem - 12.

(i) First draw plan and elevation of two objects in 1st angle system as the cylinder is given resting on H.P.

(ii) Draw new centre line at an angle of 60° with XY line. Project all points of elevation a', b',....h', 1', 2',.......8', p', q',.....u' and o' on new centre line and take distances of corresponding plan points a, b, ...h, 1, 2,....8, p, q,.....u and o from centre line and mark distances on two sides of centre line to get auxiliary plan points a, b, ...h, 1, 2,......8, p, q,.......u and o respectively.
Distance of points h and 8 are specifically shown in the figure.
Join all points to get auxiliary plan.

(iii) Draw new ground line X_1Y_1 at an angle of 45° with XY line. Project all points of plan and take on them distances of elevation points from xy to get auxiliary elevation. Join all points of auxiliary elevation to get auxiliary elevation.

(C) Axis of the solid inclined to all the three principal planes:

General Discussion : (Important)

To get the solution of this type of problems, we must assume initially axis parallel to two and hence perpendicular to one of the principal planes. Generally in initial stage, axis is taken perpendicular to the plane on which it is resting. (Solution is done as per part A).

In the second and intermediate stage axis is made inclined to the plane to which its axis is taken perpendicular in initial stage or to the plane on which the object is resting. (Solution is done as per part B.)

In the third and final stage, axis is made inclined to the remaining plane. Upto 2nd stage, the axis has remained parallel to this plane.

In some of the problems instead of giving directly the angle of the axis with the principal plane, the angle may be given indirectly, by giving :

(a) angle of triangular face of pyramid with H.P. or V.P. or ground ;
(b) angle of slant edge of pyramid with H.P. or V.P. or ground;
(c) angle of edge of base with H.P. or V.P. or ground ;
(d) angle of generator of a cone or cylinder with H.P. or V.P. or ground;
(e) angle of base of any solid with H.P. or V.P. or ground;
(f) height or distance of a particular point or edge from H.P. or V.P. or ground;
(g) angle of rectangular or vertical face of prism with H.P. or V.P. or ground;
(h) angle of longer edge or vertical edge of prism with H.P. or V.P. or ground;
(i) plan or elevation length of particular edge given or its shape given, etc.

In general, there are various indirect ways of giving angles of the axis with the principal planes. In such types of problems, adjust the angle accordingly.

Whenever, the object rests on a corner of base on H.P./ground or V.P., one of the corners of the base is drawn on one side in T.S. of base either in plan or in elevation respectivley in the initial stage. Similarly, whenever the angle of slant edge of pyramid or angle of longer or vertical edge of prism is required to be set with H.P./ground or V.P., one of the corners of the base in T.S. of base in plan or in elevation is drawn on one side.

In a similar way, whenever the object rests on an edge of base on H.P./ground or V.P., one of the edges of the base is drawn on one side and perpendicular to xy line in T.S. of base either in plan or in elevation respectively in the initial stage. Similarly, whenever the angle of slant or triangular face of pyramid or angle of vertical or rectangular face of prism is required to be set with H.P./ground or V.P. one of the edges of the base in T.S. of base in plan or in elevation respectively is drawn on one side and perpendicular to xy line in the initial stage. By taking edge of base perpendicular to xy, triangular face of pyramid and rectangular face of prism is seen in line view. Angle can be set between line view of face and xy line.

Now we shall take problems of different types, one by one and study the methods of solving it. In the starting, three dimensional drawings are given to develop understanding and later on students will have to imagine similar drawings for problems where three dimensional drawings are not given.

Methods of joining points remains same. Rules of dotted and full lines is also same. Most of the solutions are in 1st angle system and some are in 3rd angle system. Some problems are solved by auxiliary plane or rotation of ground line method.

Problem 13 : A square prism, edge/side of base 30 mm and height 45 mm, is resting on H.P. on one of the edges of the base. The edge on which it rests on H.P. makes 45° with V.P. The base of the prism makes 30° with H.P. or the axis of the prism makes 60° with H.P. or rectangular face, containing the edge on which it rests on H.P., makes 60° with H.P. Draw the projections of the prism, when (i) base is away from the observer or nearer to V.P, (ii) base is nearer to observer or away from V.P.

For solution see Fig. 12.13 (a) and (b) and follow the procedure, as given below :

As the object rests on H.P., solve the problem in 1st angle system.

(i) Assume initially the axis of the prism perpendicular to H.P. and the prism resting on H.P. on its base with one of the edges of the base perpendicular to V.P.

Therefore, draw square of 30 mm side, with one edge 34 perpendicular to xy and below xy, as plan, which is T.S. of the base. Project corresponding elevation.

(ii) Now rearrange or reorient elevation keeping 3'4' on xy line and either axis making 60° or base making 30° or vertical/rectangular face 4' 3' c' d' making 60° with xy line and project its corresponding plan. From three dimensional drawing, we can visualize that when object is tilted from axis perpendicular to H.P. to axis inclined to H.P. keeping axis parallel to V.P., the shape and size of elevation does not change.

(iii) Now without changing the angle of axis with H.P, turn the object from edge 43 perpendicular to V.P. to edge 43 inclined at 45° to V.P., as shown in Fig. 12.13 (a). As long as the angle of axis with H.P. is unchanged, the shape and size of plan will not change which we can visualize from Fig. 12.13 (a). This tilting of prism on edge 43 can be done in two ways out of which only one is shown in Fig. 12.13 (a) but in solution, Fig. 12.13 (b) both are drawn.

(iv) Now reorient plan of stage 2 such that edge 34 makes 45° with xy line in two ways and project its corresponding elevations. In Fig. 12.13 (b) (i) base is kept away from observer or nearer to V.P. compared to top base and in Fig.12.13(b) (ii) base is kept nearer to the observer or away from V.P.

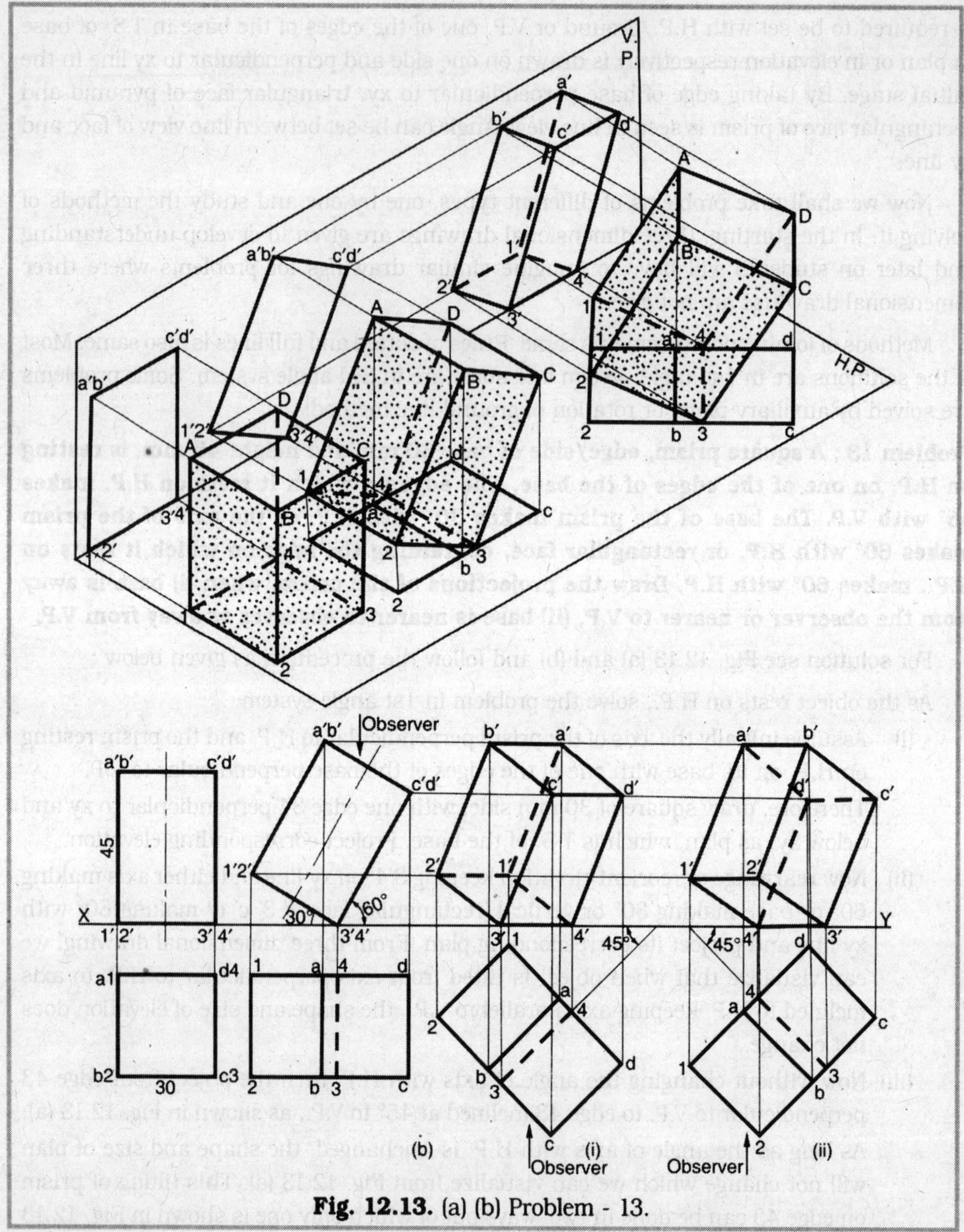

Fig. 12.13. (a) (b) Problem - 13.

Problem 14 : A hexagonal pyramid, of 30 mm side of base and 45 mm length of axis, is resting on one of its triangular faces on H.P. Draw the projections of the pyramid when its edge of base which is in H.P. is inclined at 60° to the V.P.

For solution see Fig. 12.14 (a) and (b) and follow the procedure, as given below :

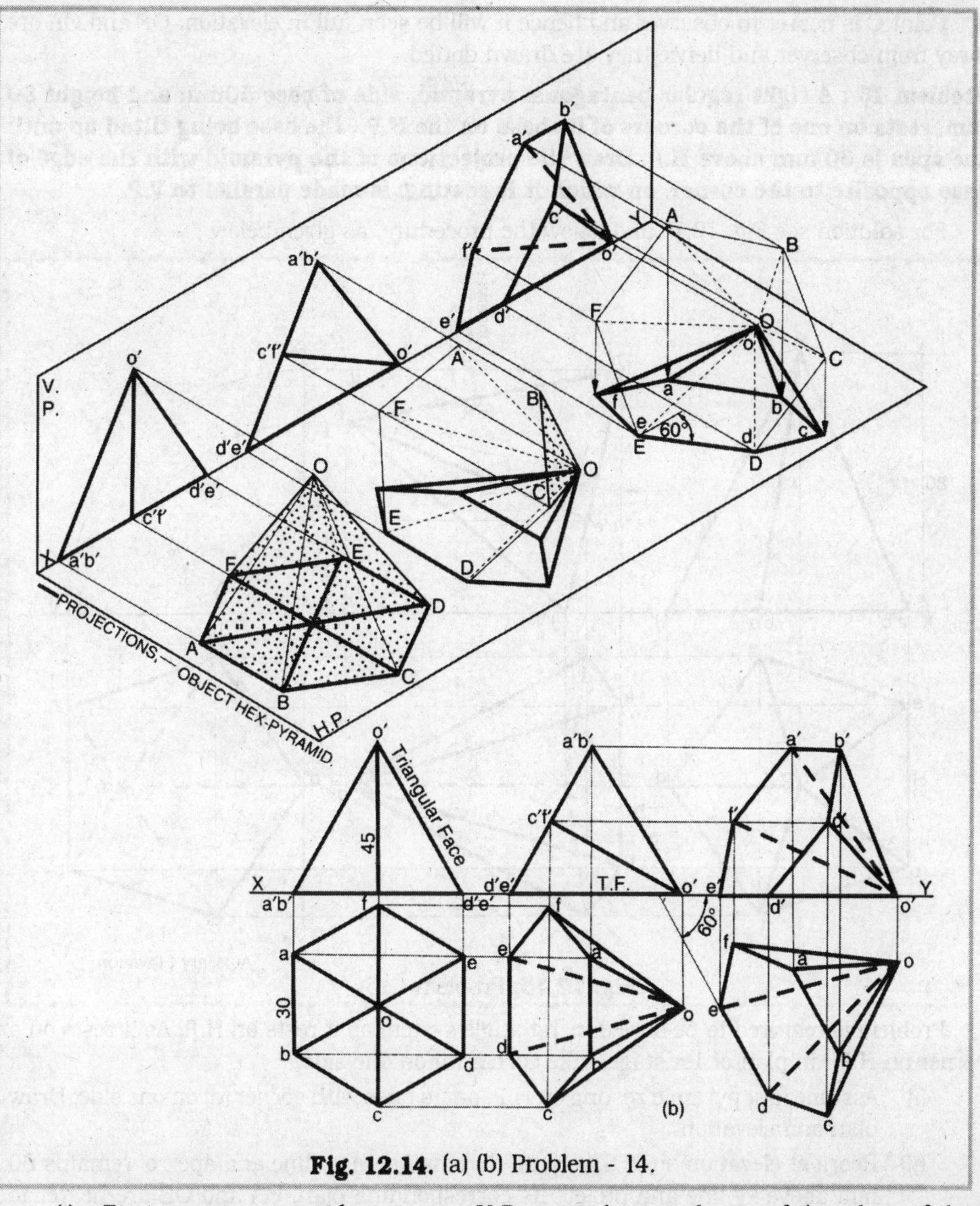

Fig. 12.14. (a) (b) Problem - 14.

(i) First assume pyramid resting on H.P. on its base with one of the edges of the base perpendicular to V.P. By drawing edge de perpendicular to xy, we can get elevation of triangular face ODE as straight line which can be later on kept on xy line. In this position draw plan and elevation.

(ii) Now reorient elevation such that triangular face i.e. line o'-d'e' gets its position on xy line and project its corresponding plan.

(iii) Now reorient the plan obtained, such that de of plan makes 60° with xy line and project its corresponding elevation.

Point C is nearer to observer and hence it will be seen full in elevation. OF and OA are away from observer and hence they are drawn dotted.

Problem 15 : A right regular pentagonal pyramid, side of base 50mm and height 80 mm, rests on one of the corners of its base on the H.P., the base being tilted up until the apex is 60 mm above H.P. Draw the projections of the pyramid with the edge of base opposite to the corner, on which it is resting, is made parallel to V.P.

For solution see Fig. 12.15 and follow the procedure, as given below :

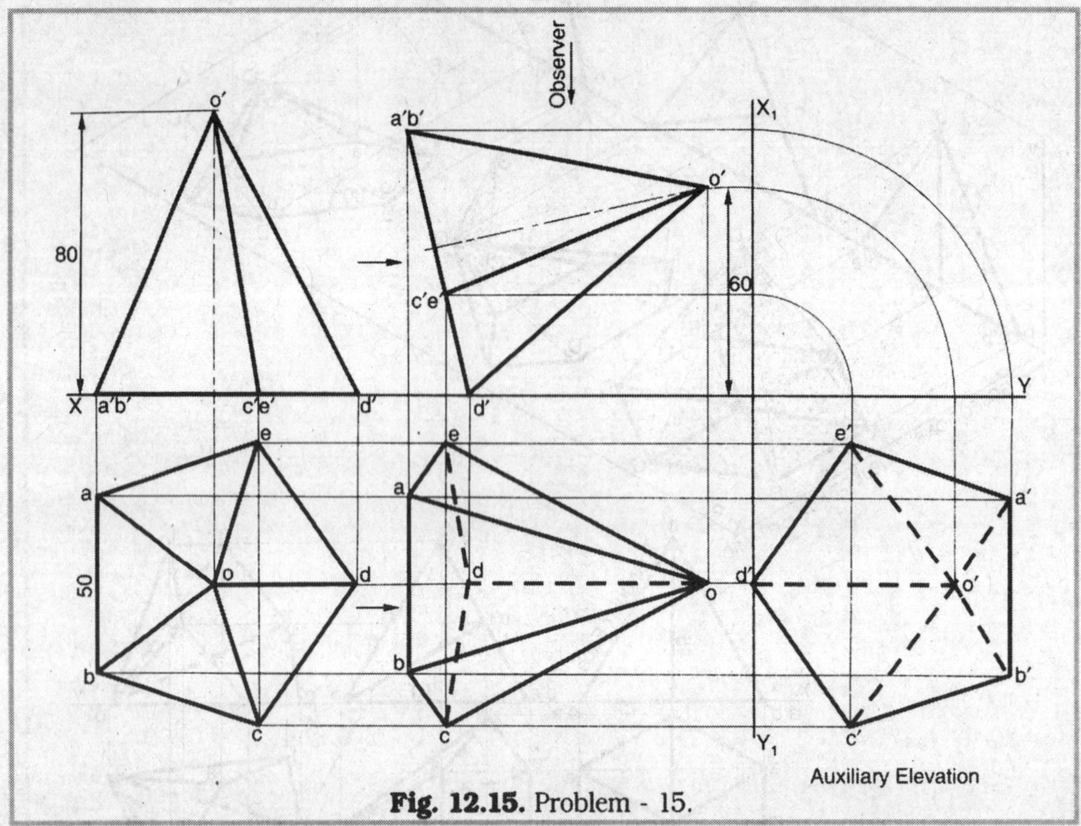

Fig. 12.15. Problem - 15.

Problem is required to be solved in 1st angle system as it rests on H.P. As it rests on a corner on H.P., in plan of 1st stage, take corner (d) on one side.

(i) Assume first pyramid resting on H.P. on its base with corner (d) on one side. Draw plan and elevation.

(ii) Reorient elevation such that point d' remains on xy line and apex o' remains 60 mm above xy line and project its corresponding plan. OA and OB are nearer to observer and so they are drawn full in plan.

(iii) Now either tilt the achieved plan such that line ab, which is opposite to corner d, is made parallel to xy or select new X_1Y_1 ground line parallel to ab. Then project elevation on XY or auxilliary elevation on X_1Y_1. We have done the latter.

Problem 16 : A pentagonal pyramid, edge of base 50 mm and height 75 mm, is resting on a corner of its base on H.P. in such a way that the slant edge containing that corner makes an angle of 60° with the H.P. and 30° with the V.P. Draw the projections. State which edge of base will be seen in true length in plan.

For solution see Fig. 12.16 and follow the procedure, as given below :

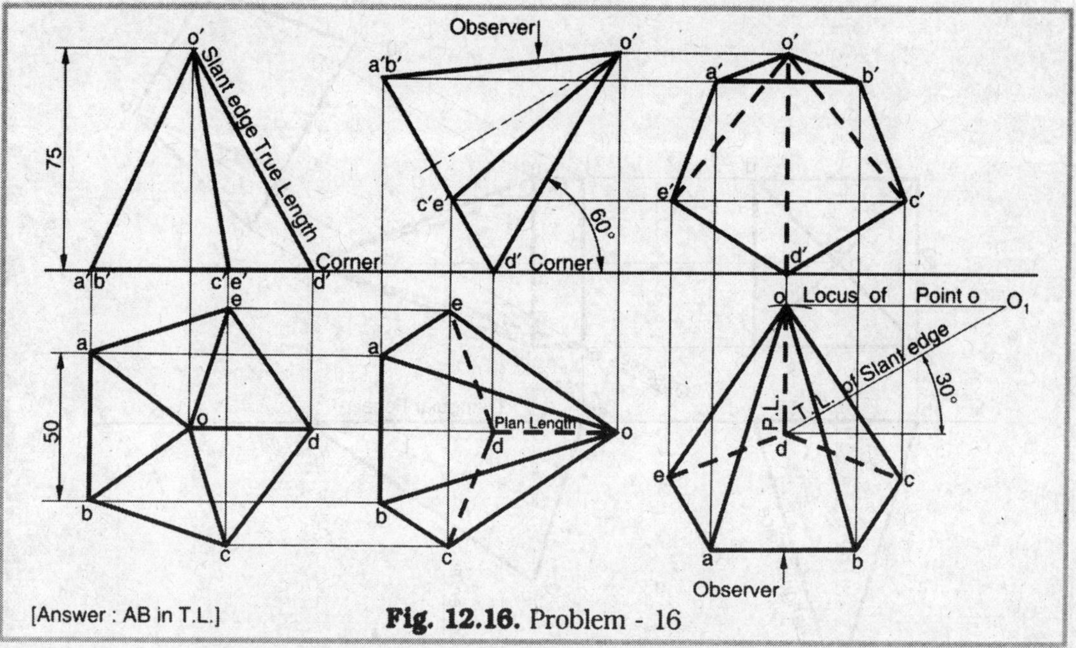

Fig. 12.16. Problem - 16

(i) As the pyramid is resting on a corner and as the angle is required to be set with slant edge, we have to draw T.S. of base of pyramid in plan with corner (d) on one side. In fact, pyramid is assumed resting on H.P. on its base. Draw plan and elevation in this position.

(ii) Rearrange elevation with (d') corner on xy line and slant edge d'o' making 60° with xy and project its corresponding plan. OA and OB are nearer to observer and so they will be full in plan.

(iii) Now slant edge is inclined to both the planes and hence adjust angle of 30° with true length and adjust plan length od with point o on locus of o, as shown in the figure. Now redraw the plan on od line already set and project corresponding elevation. OA and OB are nearer to the observer and hence o'a' and o'b' will be full. Points C, D, and E are away from the observer and hence o'c', o'd' and o'e' are dotted. Base of the pyramid is nearer to the observer and hence it is drawn full. Since $(\theta + \phi) = 90°$ for OD, o'd' and od will be vertical.

a'b' in final elevation is parallel to xy line and hence plan ab will be T.L.

Problem 17 : A square pyramid, side of base 50 mm and axis 80 mm long, has one of its triangular faces in the V.P. and edge of its base contained by that face makes an angle of 30° with the H.P. Draw its projections in 1st angle system.

For solution see Fig. 12.17 and follow the procedure, as given below :

(i) Assume pyramid having its base in V.P. and hence the axis perpendicular to V.P. with one edge of base perpendicular to H.P. initially. Draw its corresponding plan and elevation.

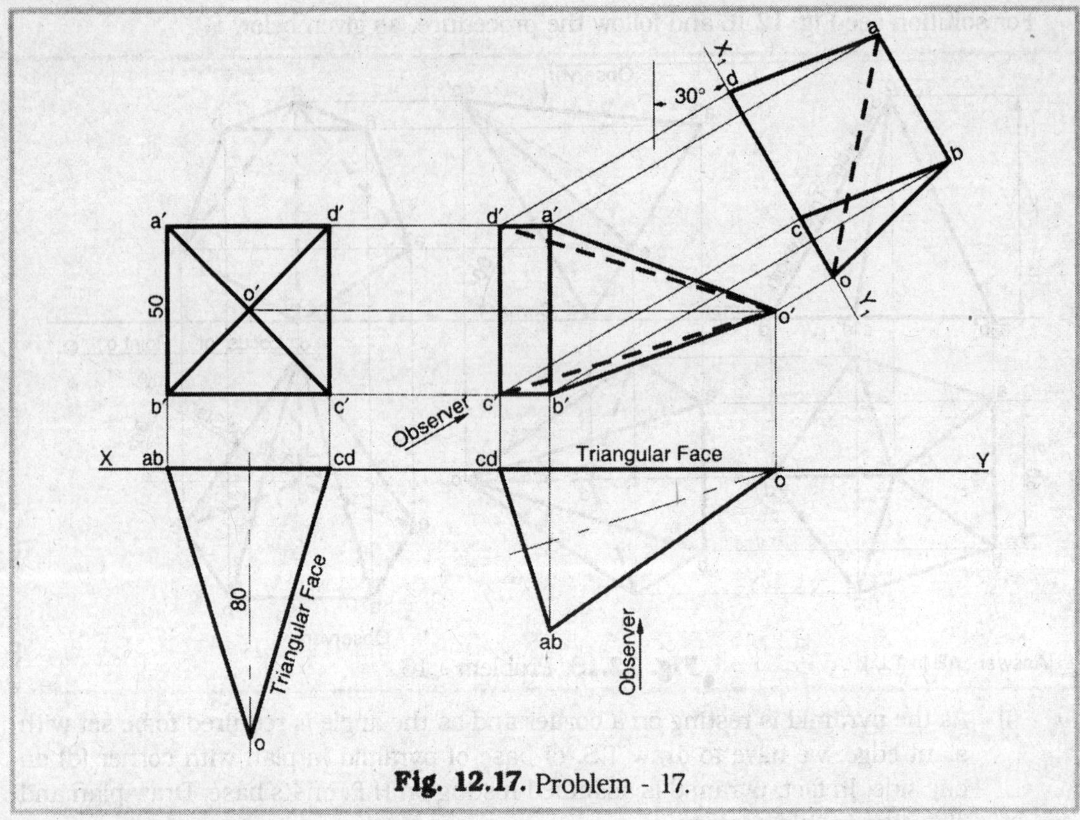

Fig. 12.17. Problem - 17.

(ii) Now rearrange plan with triangular face (i.e. line o - cd) on xy line and project its corresponding elevation.

(iii) Draw new X_1Y_1 ground line at 30° to c'd' and project auxiliary plan on it.

Problem 18 : A cone, diameter of base 60 mm and height 90 mm, is resting on H.P. on the point of peryphery of the base. Axis of the cone makes 60° with the H.P. and 30° with the V.P. Draw the projections of the cone, when the apex is nearer to V.P.

For solution see Fig. 12.18. Solution is self explanatory. Method of solution is similar to the Problem 16. For axis O – O, $(\theta + \phi) = 90°$.

Problem 19 : A cone, diameter of base 60 mm and height 70 mm, has one of its generators in H.P. and making an angle of 45° with the V.P. Draw the projections of the cone when (i) the apex is towards the observer and (ii) the apex is away from the observer.

For solution see Fig. 12.19 and follow the procedure, as given below :

As generator is in H.P., use 1st angle system.

(i) Initially assume the cone resting on H.P. on its base and draw plan and elevation accordingly.

(ii) Rearrange elevation with T.L. generator o'e' lying on xy line and project plan.

(iii) Rearrange plan such that plan of generator OE, i.e. oe makes 45° with xy line.

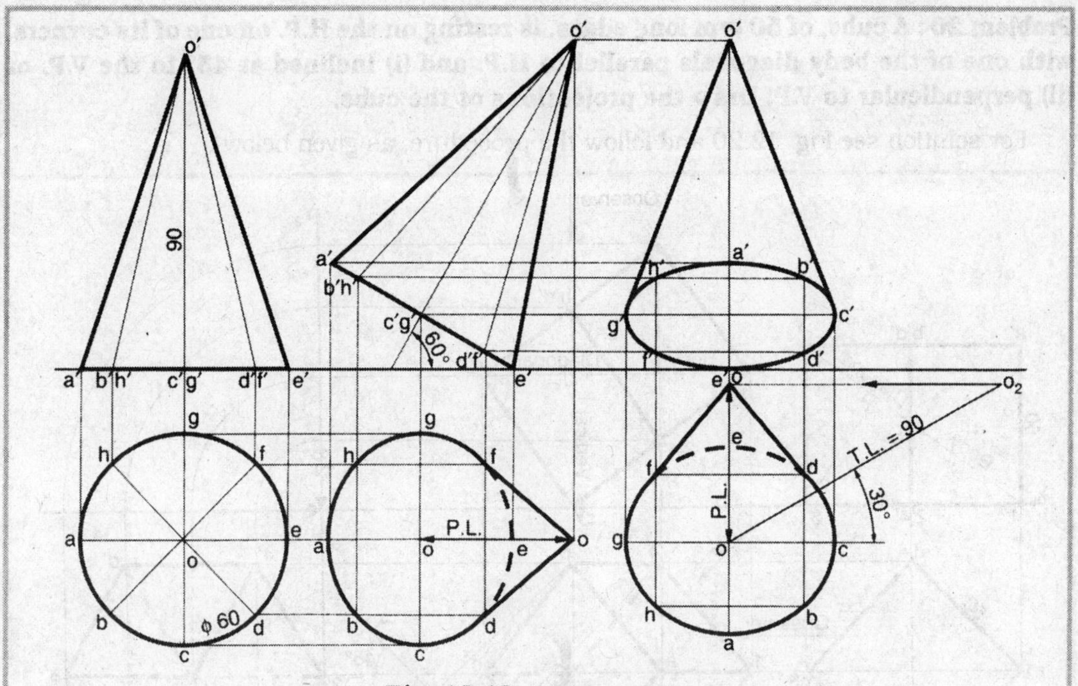

Fig. 12.18. Problem - 18.

This rearrangement can be done in two ways, (i) Apex nearer to observer and (ii) Apex away from observer, as shown in the figure and project corresponding elevations.

Points which are nearer to the observer will be seen full as compare to points which are away from the observer. Draw full and dotted accordingly. In case (ii) base is nearer to the observer and hence seen full.

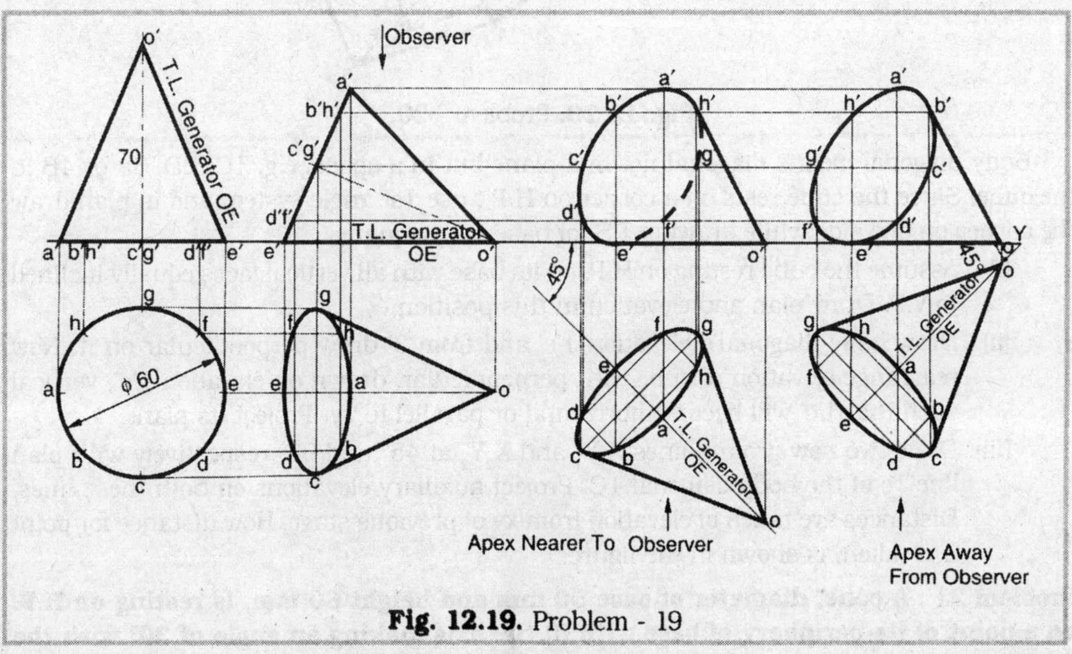

Fig. 12.19. Problem - 19

Problem 20 : A cube, of 50 mm long edges, is resting on the H.P. on one of its corners, with one of the body diagonals parallel to H.P. and (i) inclined at 45° to the V.P. or (ii) perpendicular to V.P. Draw the projections of the cube.

For solution see Fig. 12.20 and follow the procedure, as given below :

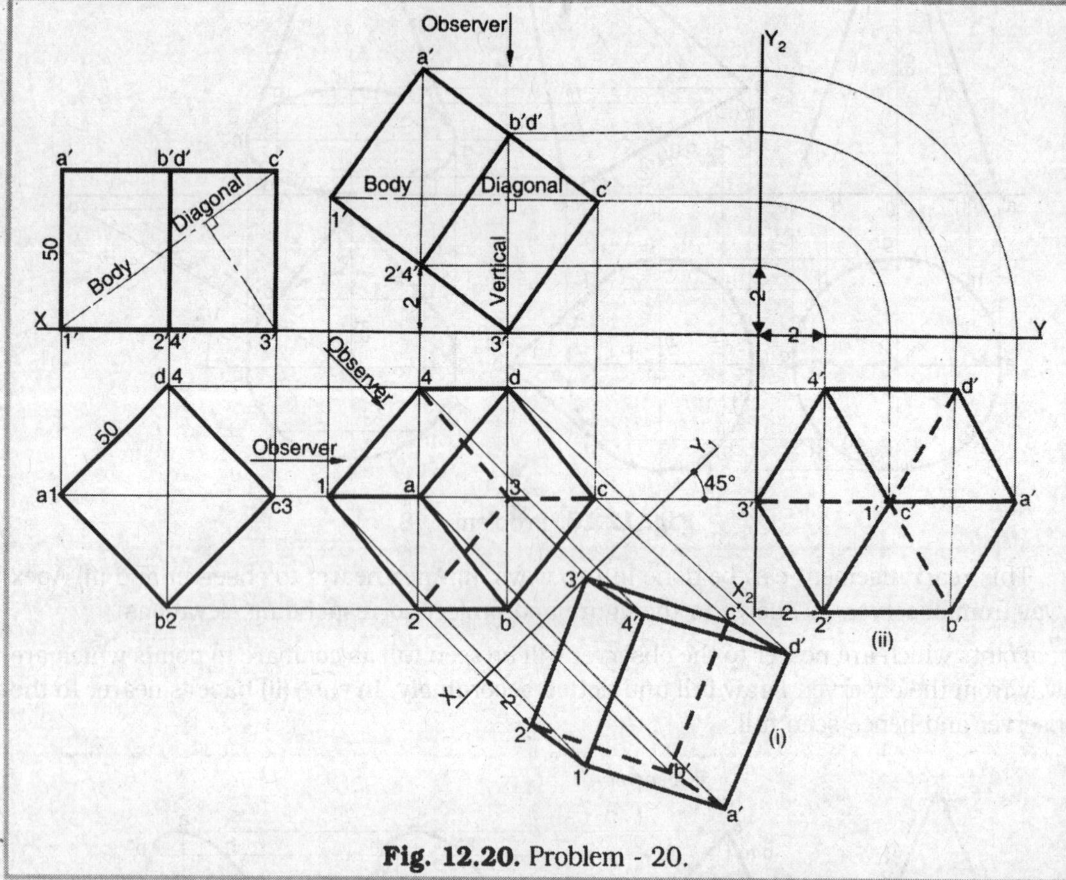

Fig. 12.20. Problem - 20.

Body diagonal means diagonal not in a plane but in a space, e.g. 1C, 2D, 3A or 4B in the cube. Since the cube rests on a corner on H.P., use 1st angle system and in plan draw the corner on one side while drawing T.S. of base i.e. a square.

(i) Assume the cube resting on H.P. on its base with all vertical faces equally inclined to V.P. Draw plan and elevation in this position.

(ii) Draw body diagonal's elevation 1'c' and from 3' draw perpendicular on it. Now rearrange elevation of cube with perpendicular, drawn on elevation 1'c', vertical such that 1'c' will become horizontal or parallel to xy. Project its plan.

(iii) Draw two new ground lines X_1Y_1 and X_2Y_2 at 45° and 90° respectively with plan line 1c of the body diagonal 1C. Project auxiliary elevations on both these lines. Distances are taken of elevation from xy of previous stage. How distance for point 2 is taken, is shown in the figure.

Problem 21 : A cone, diameter of base 50 mm and height 60 mm, is resting on H.P. on a point of its periphery of base with (i) the axis making an angle of 30° with the

H.P. and 45° with the V.P., (ii) the axis making an angle of 30° with the H.P. and the plan of the axis making 45° with the V.P. Draw the projections of the cone.

For solution see Fig. 12.21 and follow the procedure, as given below :

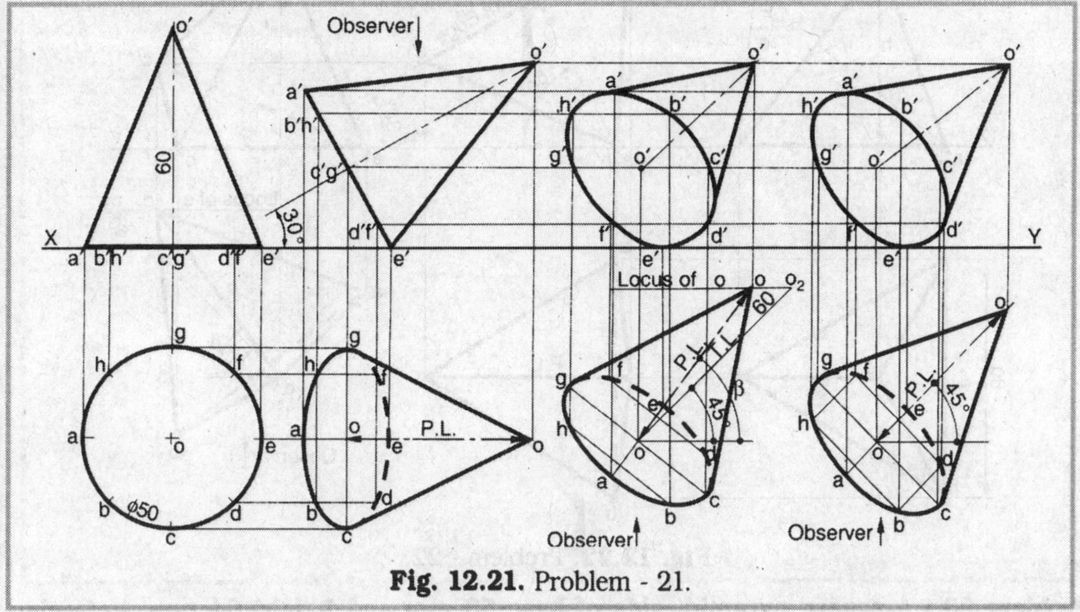

Fig. 12.21. Problem - 21.

(i) Assume initially cone resting on H.P. on its base and draw plan and elevation accordingly.

(ii) Rearrange elevation with its axis making 30° and project its corresponding plan.

(iii) In case (i), draw true length of axis at 45° and cut plan length by arc of a circle at some other angle such that plan of apex o lie on locus of o. Reproduce plan around this line and project its corresponding elevation.

(iv) In case (ii), draw P.L. itself at 45° and around this line reproduce plan and project its corresponding elevation.

Problem 22 : ABCD is a tetrahedron of 60 mm long edges. The edge AB is in the H.P. The edge CD is inclined at an angle of 30° to the H.P. and 45° to the V.P. Draw the projections of the tetrahedron.

For solution see Fig. 12.22 and follow the procedure, as given below :

Tetrahedron is a triangular pyramid with all six edges equal. Here height of the pyramid is not given. It can be found out by drawing true length of one of the slant edges in one view whose other view is parallel to xy. Here dc is parallel to xy and hence d'c' is T.L. = 60mm.

(i) Assume initially tetrahedron resting on H.P. on its face ABC with edge AB perpendicular to V.P. Draw plan and elevation in this position.

(ii) As the tetrahedron is required to rest on AB on H.P., draw ab in plan perpendicular to xy.

(iii) Reorient elevation with a'b' on xy and d'c' making 30° with xy line. Project its corresponding plan.

(iv) As edge CD is making 30° with H.P. and 45° with V.P., solve the problem of straight line CD of 60 mm true length and get the position of plan cd. Around this plan cd, reproduce plan of 2nd stage and project its elevation.

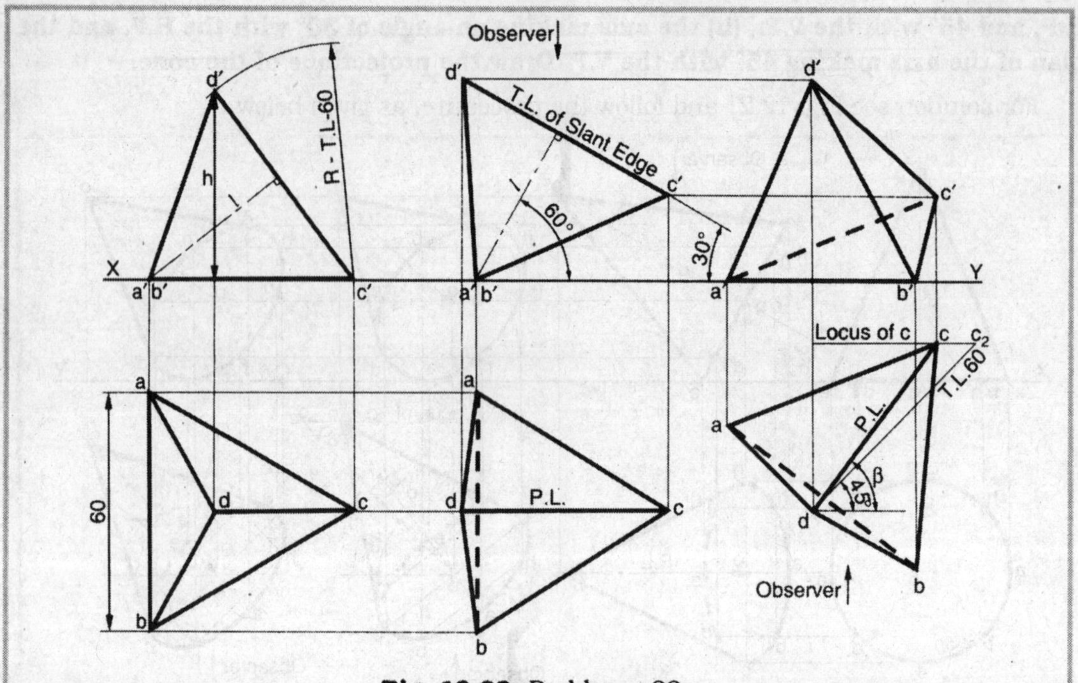

Fig. 12.22. Problem - 22.

Problem 23 : A square pyramid, side of base 50 mm and height 64 mm, is freely suspended from one of the corners of the base. Draw its projections when vertical plane containing axis makes an angle of 45° with the V.P.

For solution see Fig. 12.23 and follow the procedure, as given below :

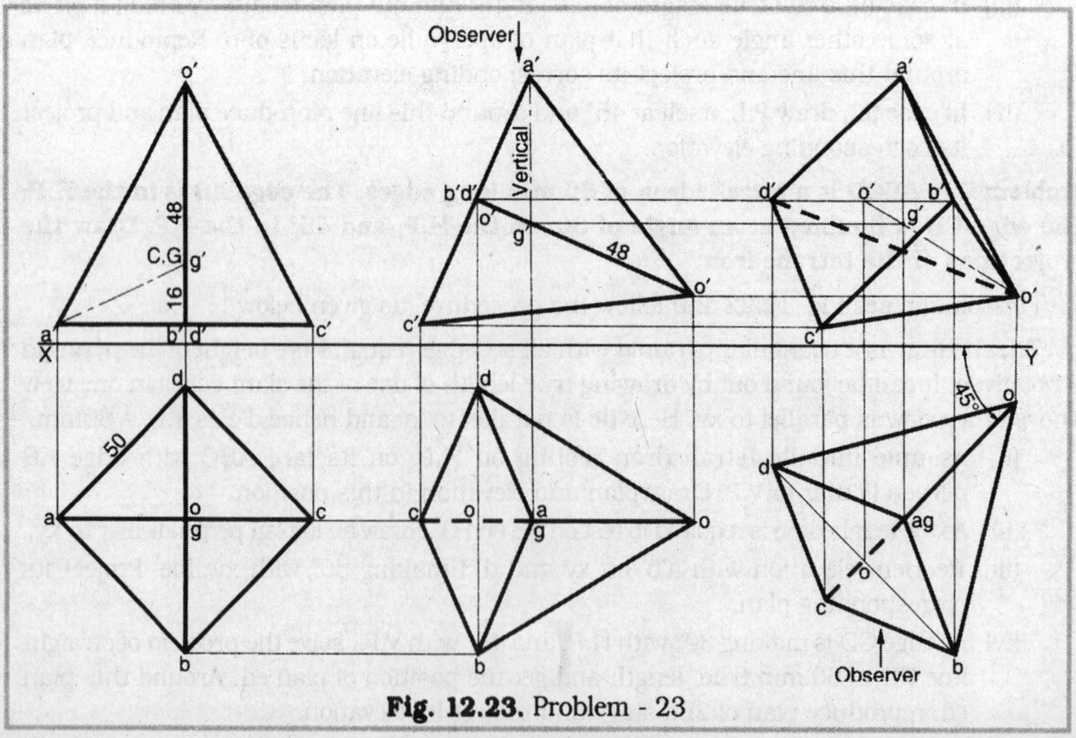

Fig. 12.23. Problem - 23

C.G. of the solid pyramid and solid cone is located at 1/4 height from the base. So here C.G. will be at 16 mm from the base. Further, whenever any thing is freely suspended, C.G. will remain vertically below the point of suspension.

(i) As it is required to suspend pyramid from a corner, draw corner of base on one side. Assuming initially axis of pyramid perpendicular to H.P., draw plan and elevation. Mark C.G. on axis in elevation. Join corner a' with C.G. in elevation.

(ii) Redraw or reproduce elevation in such a way that line a'g' joining corner A and C.G. becomes vertical with point of suspension A i.e. corner a' on upward side. Project its plan.

(iii) Now rearrange plan with the plan of axis making 45° with xy line. Vertical plane in plan always gives a line view and so vertical plane containing the axis will make 45°, as desired. Project its elevation. In plan point of suspension A and C.G. will be the same point and in elevation it will be on a vertical line.

Problem 24 : A pentagonal pyramid, side of base 30 mm and height 60 mm, stands on an edge of its base on the H.P. The edge makes an angle of 45° with the V.P. The slant face containing this edge of the base, makes an angle of 50° with the H.P. Draw the projections of the pyramid keeping (i) Apex nearer to the observer, (ii) Apex away from the observer.

Use auxiliary plane method or rotation of ground line method.

For solution see Fig. 12.24 and follow the procedure, as given below :

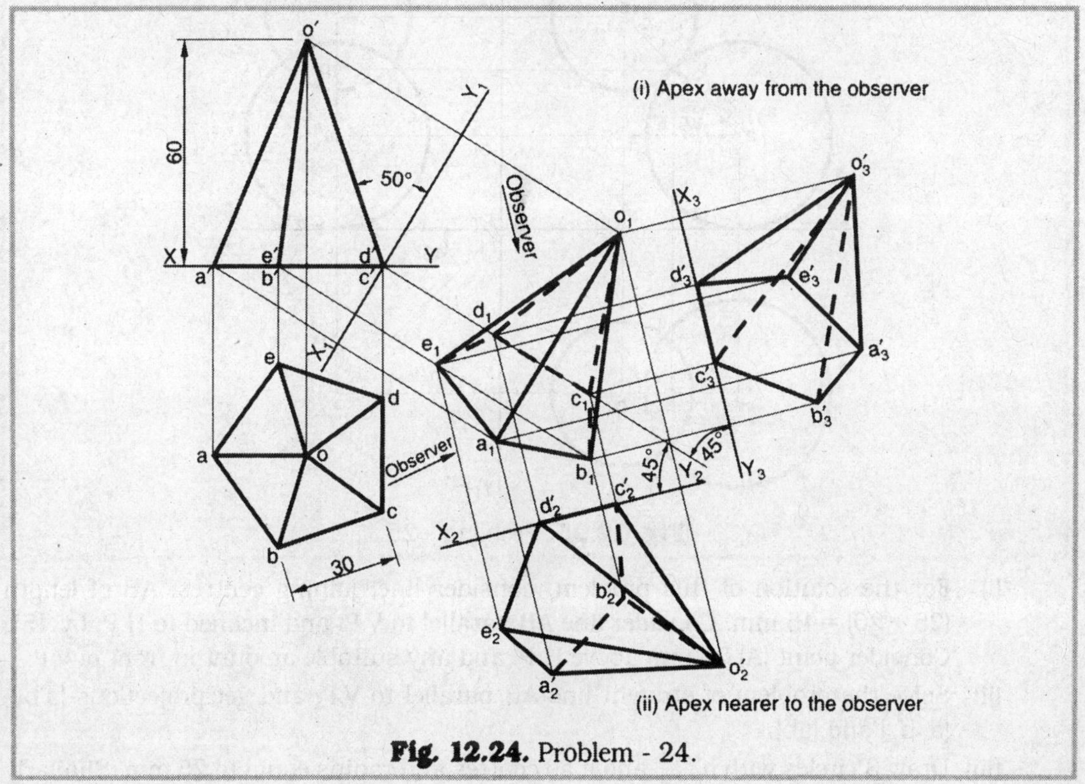

Fig. 12.24. Problem - 24.

As the object rests on H.P. on edge of its base and further angle is required to be set with slant triangular face, draw T.S. of base in plan with one edge (cd) perpendicular and on one side.

(i) Assuming initially pyramid resting on H.P. on its base, draw plan and elevation. Keep one edge (cd) perpendicular to V.P.

(ii) Draw X_1Y_1 ground line passing through c'd' and making 50° with straight line view o'c'd' of slant face OCD. Project on it auxiliary plan, as usual.

(iii) Now as edge CD, on which it rests on H.P., makes 45° with V.P., draw X_2Y_2 and X_3Y_3 lines at 45° with c_1d_1, plan of CD. In one case apex remains nearer and in other apex remains away from the observer. Project on these two lines two auxiliary elevations.

PROBLEMS OF PROJECTIONS OF SPHERES

Problem 25 : A sphere (A), radius 25 mm, is resting on H.P. Another sphere (B), radius 20 mm, is nailed to sphere (A). Line joining the centres of sphere (A) and sphere (B) is parallel to V.P. and inclined to H.P. by 45°. Draw three projections and show the contact point in all the three projections.

For solution see Fig. 12.25 and follow the procedure as given below.

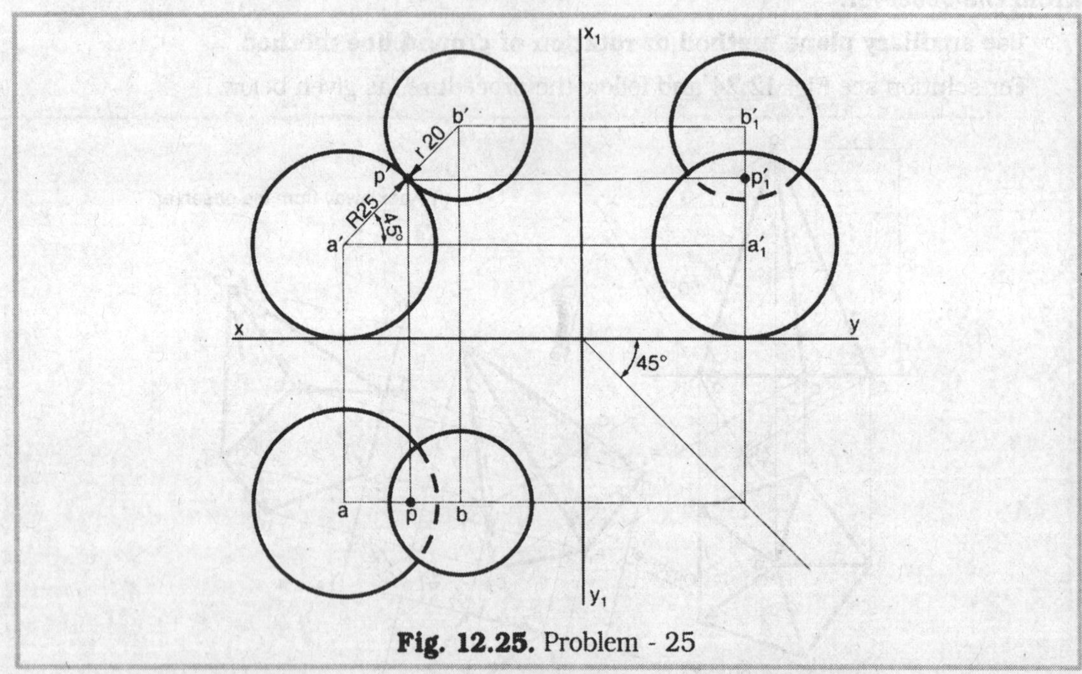

Fig. 12.25. Problem - 25

(i) For the solution of this problem, consider line, joining centres, AB of length (25 + 20) = 45 mm. Consider line AB parallel to V.P. and inclined to H.P. by 45°. Consider point (A) 25 mm above H.P. and any suitable amount in front of V.P.

(ii) Solve the problem of straight line AB parallel to V.P. and get projections [a'b'], [a_1'b_1'] and [ab].

(iii) Draw 3 circles with a', a_1' and a as centres and radius equal to 25 mm. Similarly draw 3 circles with b', b_1' and b as centres and radius equal to 20 mm. Draw full an dotted as required.

(iv) Contact point will be always on a line joining centres of two spheres. Project contact point from elevation to S.V. and to plan as shown.

Problem 26 : A sphere (A), radius R = 45 mm, is resting on H.P. Another sphere (B), radius r = 35 mm, is nailed to sphere - A. Line joining centres of sphere- A and sphere - B is inclined to H.P. by 30° and inclined to V.P. by 45°. Sphere (B) is on upperside of sphere (A) and is nearer to V.P. or away from the observer. Draw three projections along with contact point.

For solution see Fig. 12.26 and follow the procedure as given below.

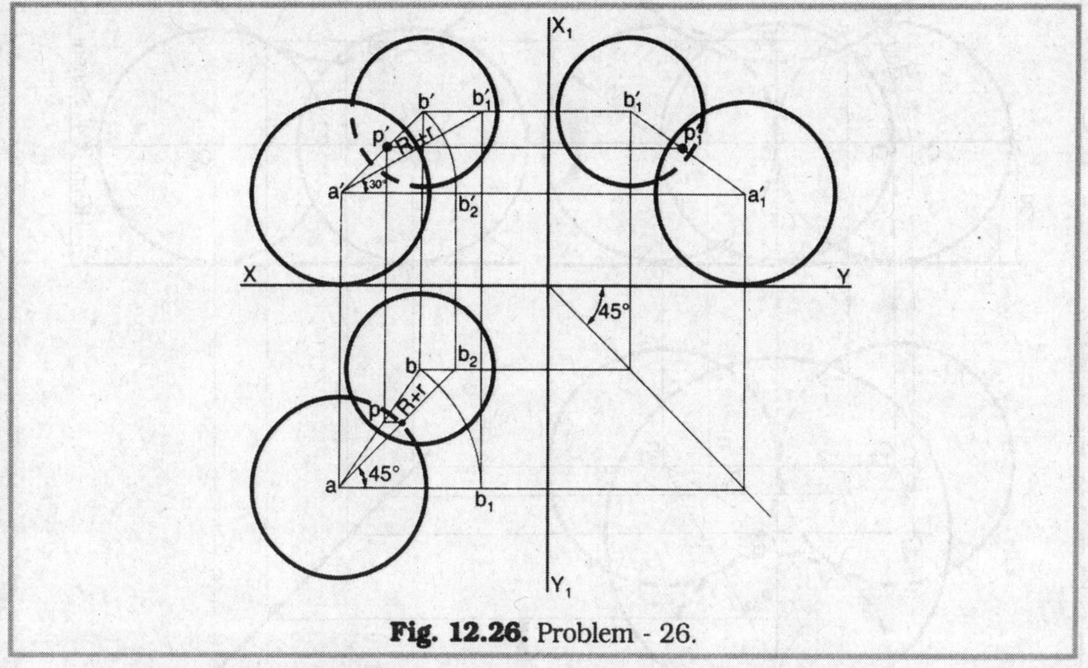

Fig. 12.26. Problem - 26.

(i) Draw projections of line AB. (1) Length = (R + r) = 80mm (2) θ = 30° (3) Ø = 45° and (4) Point A 45mm above H.P. and any suitable distance infront of V. P. Rest is clear from the figure.

(ii) First mark contact point on a true length line $a'b'_1$ and true length line ab_2 and than transfer on a'b', ab and $a'_1b'_1$ as shown in the figure.

Problem 27 : Three equal spheres, of 30 mm radius, are resting on the H.P. so that each touches the other two and line joining the centres of two of them is parallel to V.P. Third sphere is nearer to the observer. A fourth sphere, of 45 mm radius, is placed on the top of the three spheres. Draw three projections of 4 spheres and find the distance of the centre of the fourth sphere from the H.P. Show also contact points of fourth sphere with other spheres.

For solution see Fig. 12.27 and follow the procedure, as given below :

Before going for solutions of problems of spheres, let us understand following aspects useful for the solutions of problems of projections of spheres. Students are requested to visualize the same with the help of balls at home.

(a) Length of line, joining the centres of two spheres in contact, is equal to the sum of their radii.

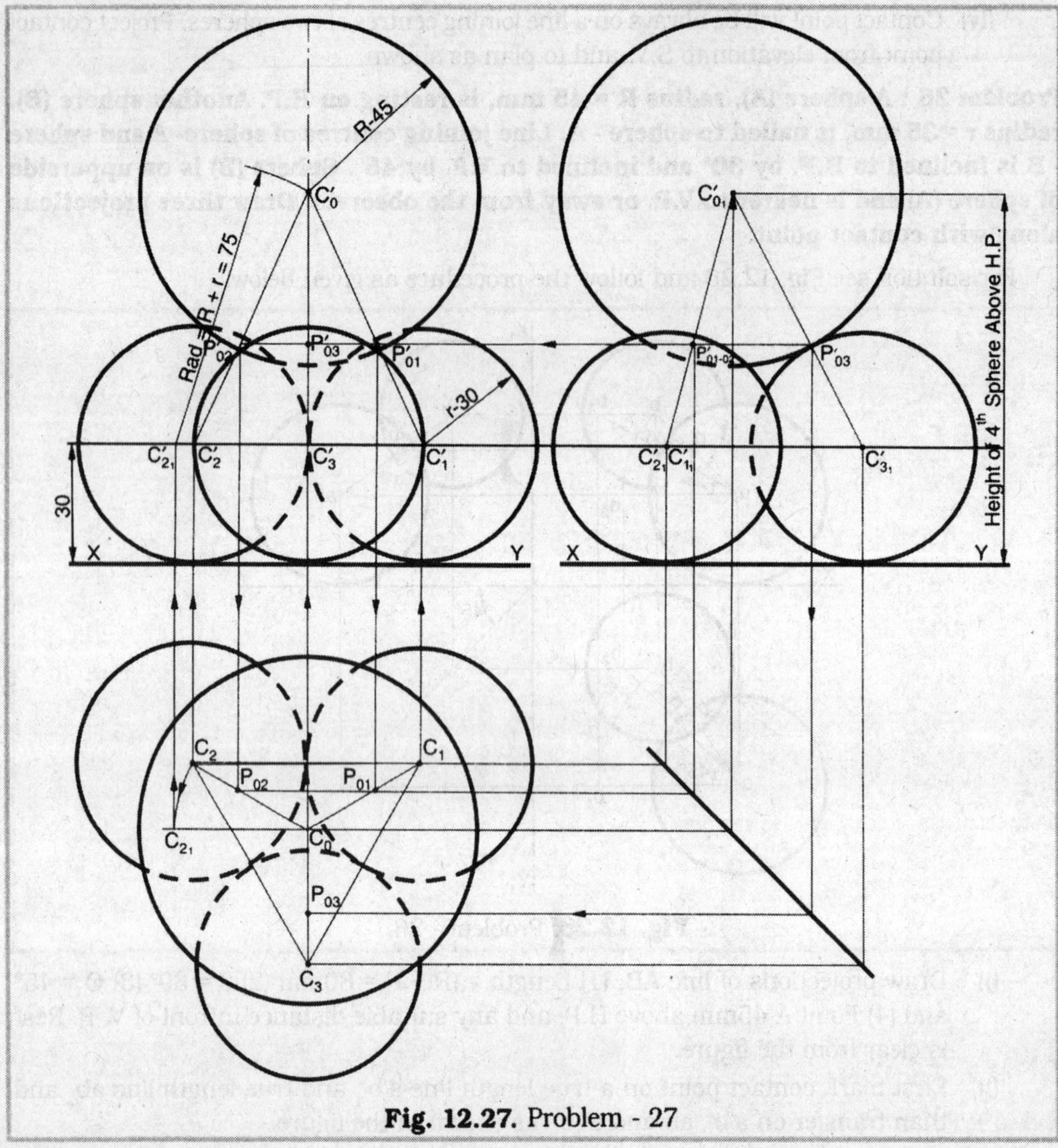

Fig. 12.27. Problem - 27

(b) Point of contact between two spheres always lies on the line joining the centres of the two spheres.

(c) When a sphere is resting on ground or on H.P., the distance of its centre from ground or H.P. will be equal to its radius.

(d) When one sphere, resting on ground or on H.P., is rotated, around another sphere, also resting on ground or on H.P., keeping contact between them, the loci of the centre of rotating sphere and the contact point between them, will be circles parallel to the ground or H.P. Hence the plans of these loci will be circles and their elevations will be straight lines parallel to G.L. line or xy line.

(e) Smaller sphere in contact with bigger sphere on ground or on H.P. will not be seen completely from the top.

(f) When spheres are equal in size and resting on H.P. or on ground in contact, the line joining centres will be parallel to H.P. or ground.

(g) When two spheres resting on ground or on H.P. in contact with line joining centres parallel to V.P., elevation length of line joining centres will be equal to the sum of the two radii and its plan will be parallel to XY line.

Now with above points in mind we shall solve few problems.

(i) Draw first equilateral triangle $C_1C_2C_3$ of side equal to (30 + 30) = 60 mm in plan with C_1C_2 parallel to xy and C_3 away from xy or nearer to the observer. With centres C_1, C_2 and C_3 and the radius equal to 30 mm, draw three circles in plan. Find out centre C_0 of $\Delta C_1C_2C_3$ and draw full circle with C_0 as the centre and 45 mm as the radius. Draw smaller circles dotted inside the bigger circle.

(ii) Draw projectors upward from C_1, C_2, C_3 and C_0. Mark C_1', C_2' and C_3' above xy line at height equal to the radius of the sphere. With C_1', C_2' and C_3' as the centres and the radius equal to 30 mm, draw three circles.

(iii) As in plan, lines joining the centre of the fourth sphere with other spheres are not parallel to xy, make any one parallel to xy so that its elevation will be seen in T.L. (30 + 45). Here C_0C_2 is made parallel to xy as C_0C_{21} and correspondingly in elevation T.L. of line joining the 2nd sphere with the fourth sphere is taken equal to 75 mm and arc of circle is drawn with C_{21}' as the centre and the radius equal to 75 mm to get C_0' on projector through C_0. Now with C_0' as the center and radius equal to 45 mm draw a circle.

(iv) Construction given in (iii) is not required, when S.V. is drawn. In S.V. C'_{01} C'_{31} will be true in length i.e.75mm.

In elevation, circle of C_3' as the centre will be seen full, since the third sphere is the nearest sphere to the observer, The second nearest sphere to the observer is the fourth bigger sphere and hence draw a circle dotted only in side elevation of third sphere. Other two circles in elevation of C_1' an C_2' as the centres should be drawn dotted inside the third and the fourth spheres since they are the farthest spheres to the observer.

Problem 28 : Three unequal spheres, of radii 35 mm, 20 mm and 10 mm, are resting on H.P. Sphere -2 is in contact with sphere-1 and sphere-3 and plane containing 3 centres is parallel to P.P. Smallest sphere is nearest to the observer and largest is nearest to V.P. Draw three projections alongwith contact points.

Fig. 12.28 is self-explanatory and hence discussion is not given.

Problem 29 : Three spheres 1, 2 and 3 of 70 mm, 40 mm and 15 mm radii respectively, rest on the H.P. each touching the other two. Draw their projections when the line joining the centres of the spheres 1 and 2 is parallel to the V.P. Show also the contact points in projections. Sphere 3 is nearer to the observer.

For solution see Fig. 12.29 and follow the procedure, as given below :

(i) First mark centre O'_1 above xy line at 70 mm height and draw circle with 70mm radius.

(ii) Mark centre O'_2 at 40 mm height above xy line and (70 + 40) = 110 mm distance from O'_1 centre. Draw a circle with O'_2 as the centre and the radius equal to 40 mm. Mark the touching point P'_{12}.

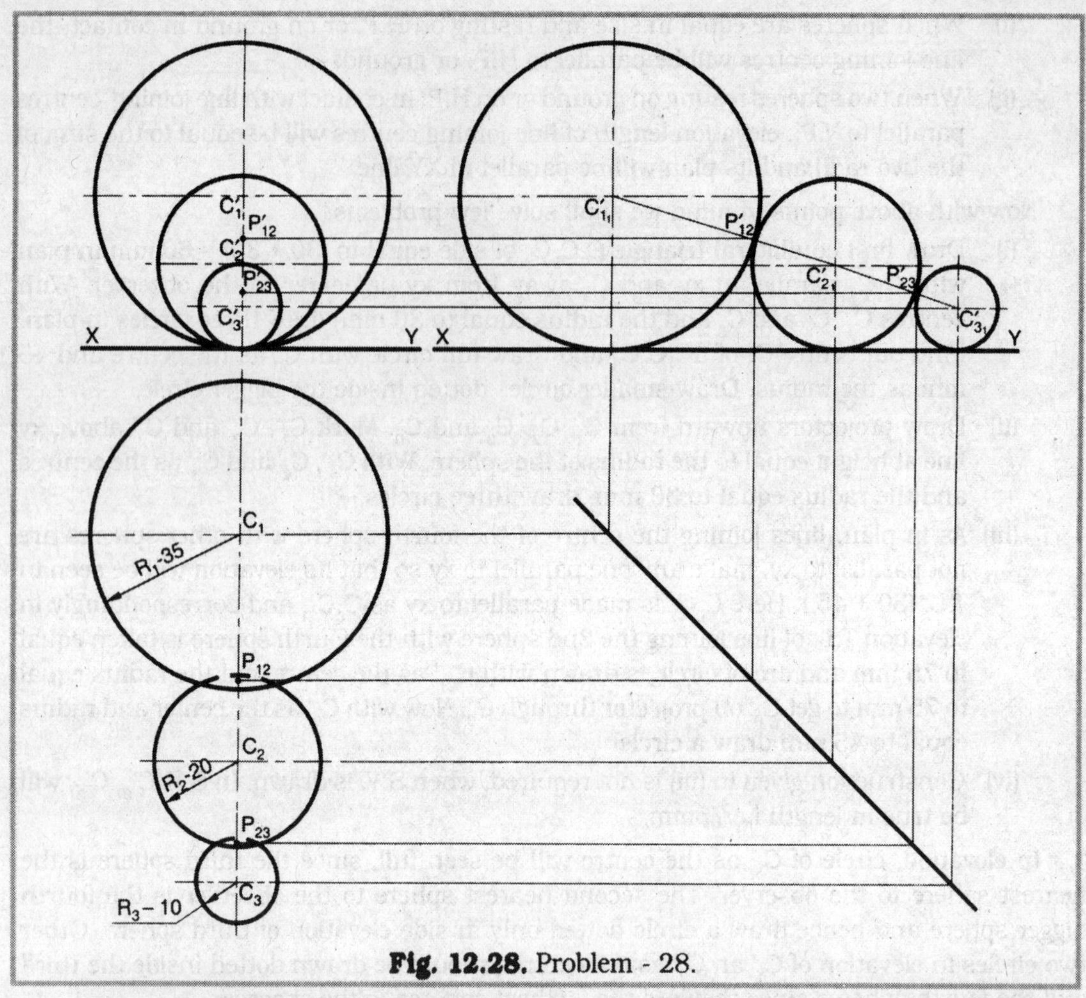

Fig. 12.28. Problem - 28.

(iii) Draw projectors from O'_1 and O'_2 and mark on it O_1 and O_2 respectively such that O_1O_2 line becomes parallel to the XY line. Draw two circles with O_1 and O_2 as the centres and radii equal to 70 mm and 40 mm respectively. Circle of 40 mm radius inside 70 mm radius circle should be drawn dotted. O_1O_2 is drawn parallel to XY and $O'_1O'_2$ is drawn of true length since the line joining centres O_1, O_2 is parallel to V.P. Draw projector from P'_{12} and mark P_{12} on O_1O_2 line in plan.

(iv) Now assume two positions of sphere three in contact with 1 as well as 2 such that lines joining centres of 1 and 3, and 2 and 3 become parallel to V.P. Afterwards, rotate sphere 3 around 1 and 2.

Draw O'_{32} and O'_{31} accordingly and project their plans O_{32} and O_{31} respectively. Mark in elevation contact points between 1 and 3 and between 2 and 3 and from these contact points, draw parallel lines to xy line to get locus of P_{13} and P_{23} in F.V. respectively.

Project above contact points in plan and draw circles passing through these points with O_1 and O_2 as the centres respectively to get locus of P_{13} in T.V. and locus of P_{23} in T.V.

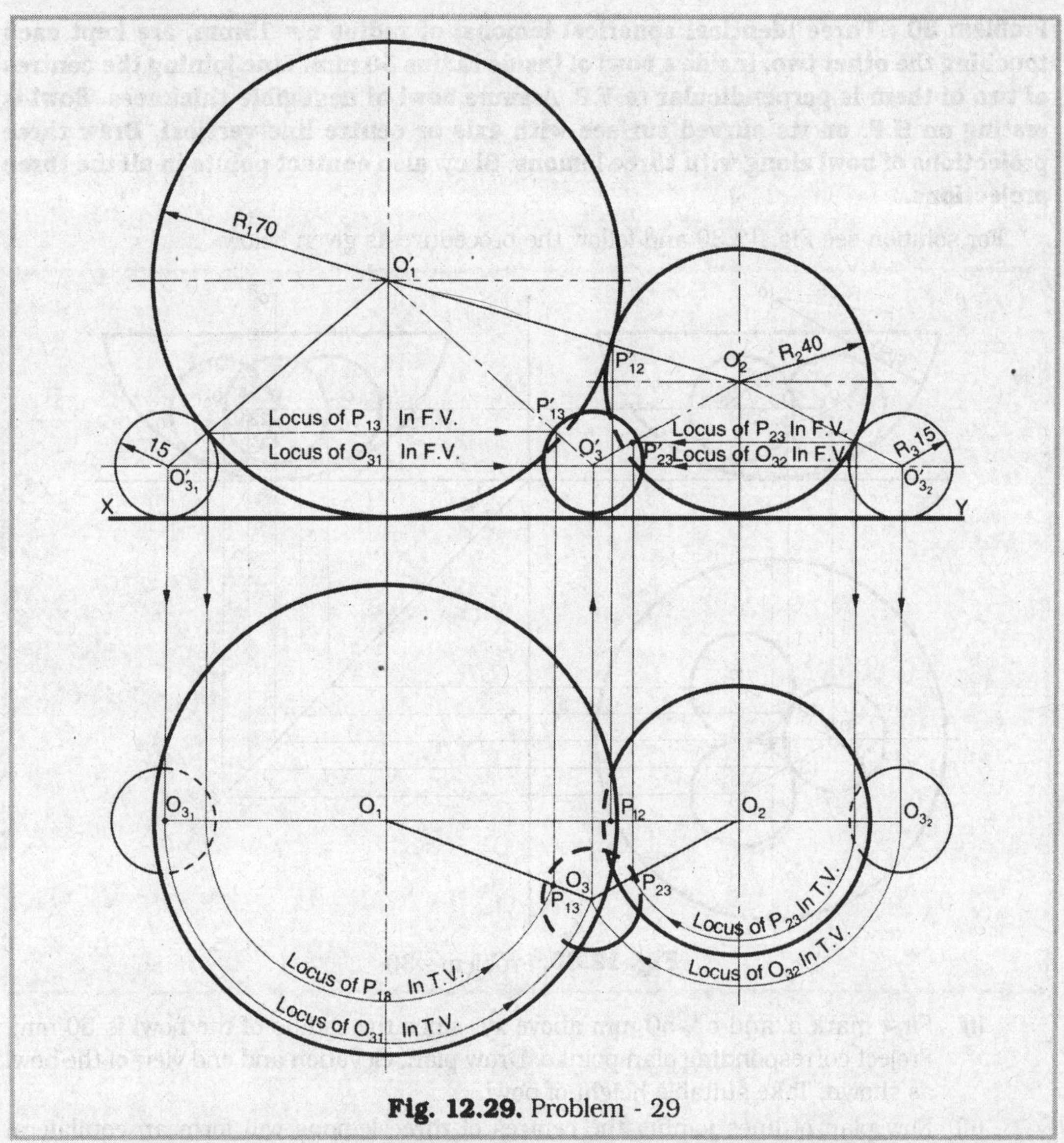

Fig. 12.29. Problem - 29

(v) Now draw two circles with O_1 and O_2 as their centres and radii equal to O_1O_{31} and O_2O_{32} respectively. These two circles represent the loci of plans of the centres of spheres 3 when sphere 3 is being rotated around spheres 1 and 2 respectively. Intersection of these circles will give rise to centre O_3. plan of sphere 3, when in contact with 1 as well as 2. Draw a circle with O_3 as the centre and radius equal to 15 mm. Circle No.3 inside 1 and 2 should be drawn dotted.

(vi) Project upward point O_3 and mark O'_3. 15 mm above xy line. Draw full circle with O'_3 as the centre and of 15 mm radius. Circle 1 and 2 inside circle 3 should be dotted.

(vii) Intersection of lines joining the centres in plan and elevation with loci of P_{13}, P_{23} in T.V. and F.V. will give us contact point positions in plan as well as in elevation.

Problem 30 : Three identical spherical lemons, of radius r = 15mm, are kept each touching the other two, inside a bowl of inside radius 50 mm. Line joining the centres of two of them is perpendicular to V.P. Assume bowl of negligible thickness. Bowl is resting on H.P. on its curved surface with axis or centre line vertical. Draw three projections of bowl along with three lemons. Show also contact points in all the three projections.

For solution see Fig. 12.30 and follow the procedure as given below.

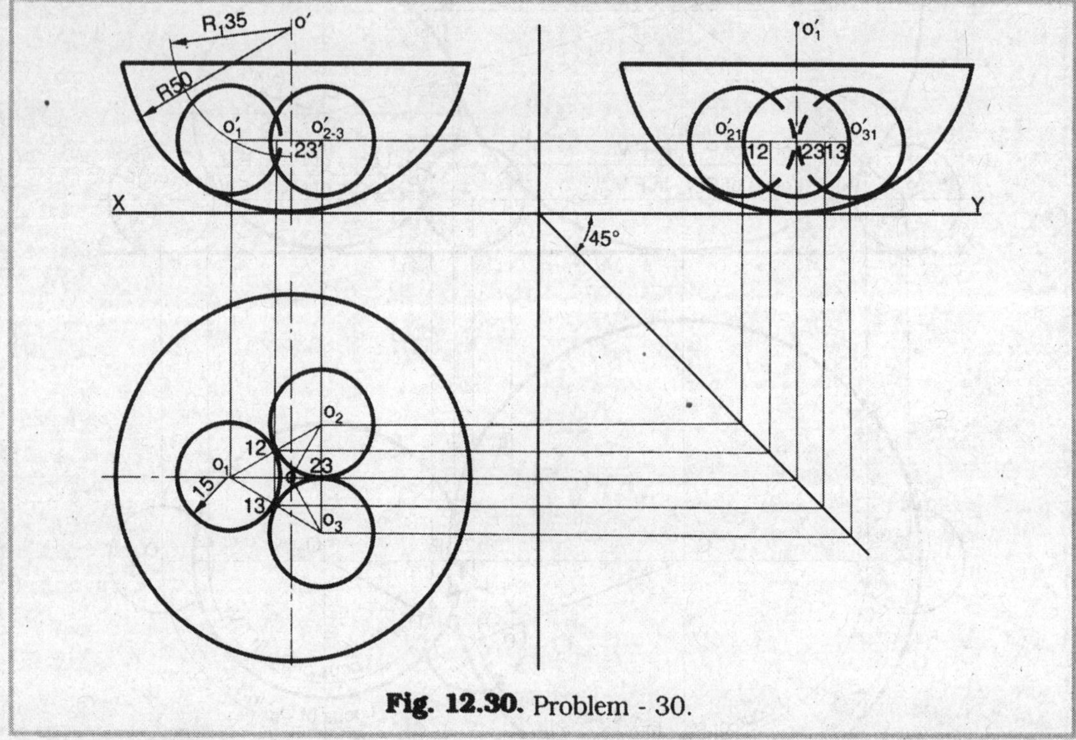

Fig. 12.30. Problem - 30.

(i) First mark o' and o_1', 50 mm above xy, since the radius of the bowl is 50 mm. Project corresponding plan point o. Draw plan, elevation and end view of the bowl as shown. Take suitable height of bowl.

(ii) Now plan of lines joining the centres of three lemons will form an equilateral triangle of (15 + 15) = 30 mm side. One side out of three will be perpendicular to xy and centre of the triangle will coincide with the plan of bowl centre. Draw as shown.

(iii) Now line O_1O will be parallel to V.P., since its plan o_1o is parallel to xy line. Hence elevation o'o_1' will be true in length i.e. (50 - 15) = 35 mm. Draw an arc with o' centre and radius 35 mm and draw a projector through o_1 to get o'$_1$ by intersection.

(iv) From elevation point o_1', draw a horizontal line upto side view and projectors from o_2 and o_3 to get centres in elevation and side view. Rest is clear from the figure.

Problem 31 ; A square prism, side of base 30 mm and height 80 mm, is resting on H.P. on its base with two sides of base parallel to V.P. Four equal spheres are resting on H.P., each sphere touching the other two spheres and touching the vertical rectangular face of the prism. Draw the projections of prism and spheres.

For solution see Fig. 12.31 and follow the procedure as given below.

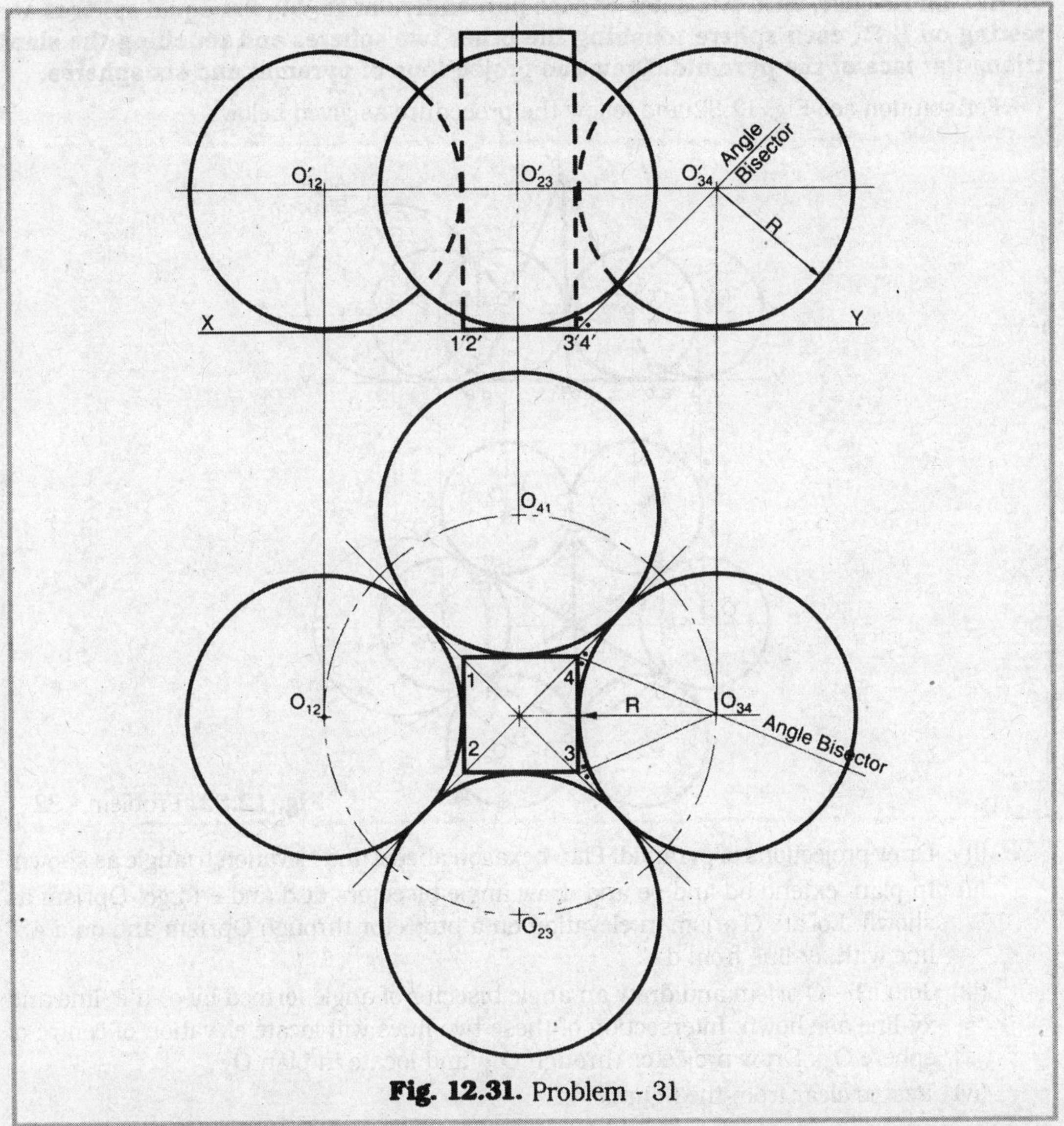

Fig. 12.31. Problem - 31

(i) First draw projections of square prism. Plan square 1234 and elevation rectangle.

(ii) In plan, draw lines through corners 1,2,3 and 4 meeting at the centre of the polygon square as shown.

(iii) Draw angle bisectors at corners 3 and 4 and get centre O_{34} by intersection. From O_{34} draw perpendicular to side 34 and get radius R as shown.

(iv) In elevation, at 3'4' draw an angle bisector (45° line) and projector through plan O_{34} to get centre O'_{34} by intersection.

(v) In elevation all 4 centres will be on a horizontal line through O'_{34} and in plan all 4 centres will be on a circle through O_{34} as shown. Locate centres and use radius R to draw circles.

Problem 32 : A hexagonal pyramid, side of base 30 mm and height 60 mm, is resting on H.P. on its base, with two sides of base perpendicular to V.P. Six equal spheres are resting on H.P., each sphere touching the other two spheres and touching the slant triangular face of the pyramid. Draw the projections of pyramid and six spheres.

For solution see Fig. 12.32 and follow the procedure as given below.

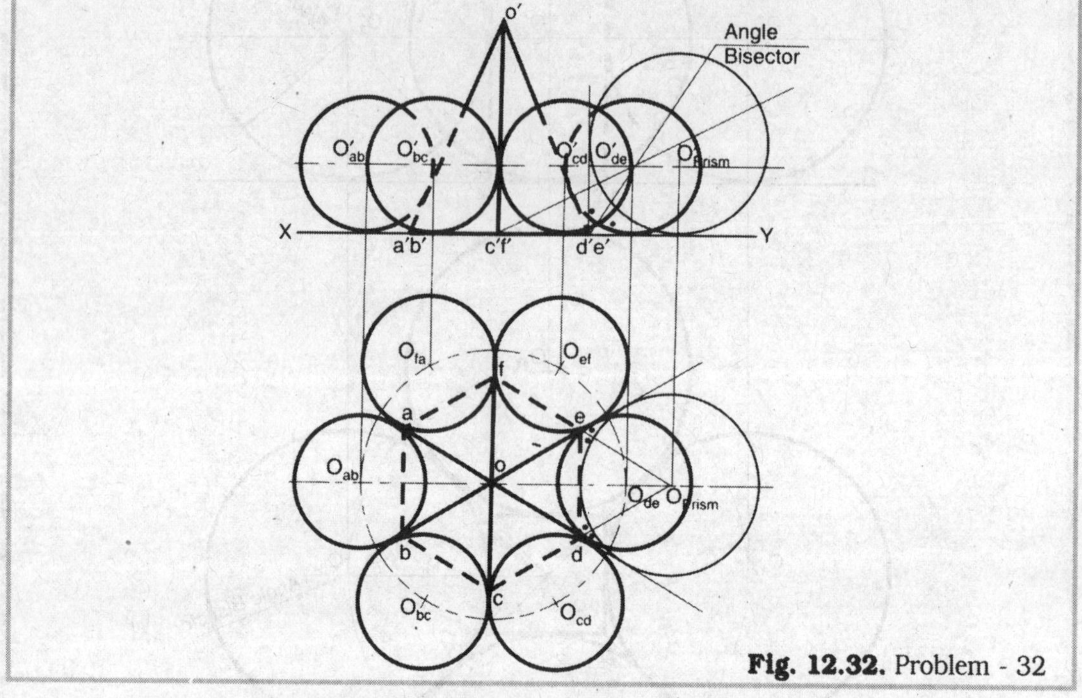

Fig. 12.32. Problem - 32

(i) Draw projections of pyramid. Plan hexagon abcdef and elevation triangle as shown.

(ii) In plan, extend od and oe and draw angle bisectors at d and e to get Oprism as shown. Locate O'prism in elevation on a projector through Oprism and on a 45° line with xy-line from d'e'.

(iii) Join c'f' - O'prism and draw an angle bisector of angle formed by o'-d'e' line and xy-line as shown. Intersection of these two lines will locate elevation of centre of sphere O_{dc}. Draw projector through O'_{de} and locate in plan O_{dc}.

(iv) Rest is clear from the figure.

Problem 33 : A pentagonal pyramid has a height of 60 mm and the side of a base 30 mm. The pyramid rests with one of the sides of a base on the H.P. such that the triangular face containing that side is perpendicular to the H.P. and makes an angle of 30° with the V.P. Draw its projections. [Fig. 12.33] [Mumbai University, December 1996]

Problem 34 : A pentagonal pyramid, side of base 35 mm and axis 70 mm long, is lying on one of its corners on the H.P. such that the two base edges passing through the corner on which it rests makes an equal inclination with the H.P. One of its triangular surface is parallel to the H.P. and perpendicular to the V.P. and the base edge containing that triangular surface is parallel to both H.P. and the V.P. Draw the projections of the solid, when the apex of the pyramid is nearer to the observer. [Fig. 12.34] [Mumbai University, May 1994]

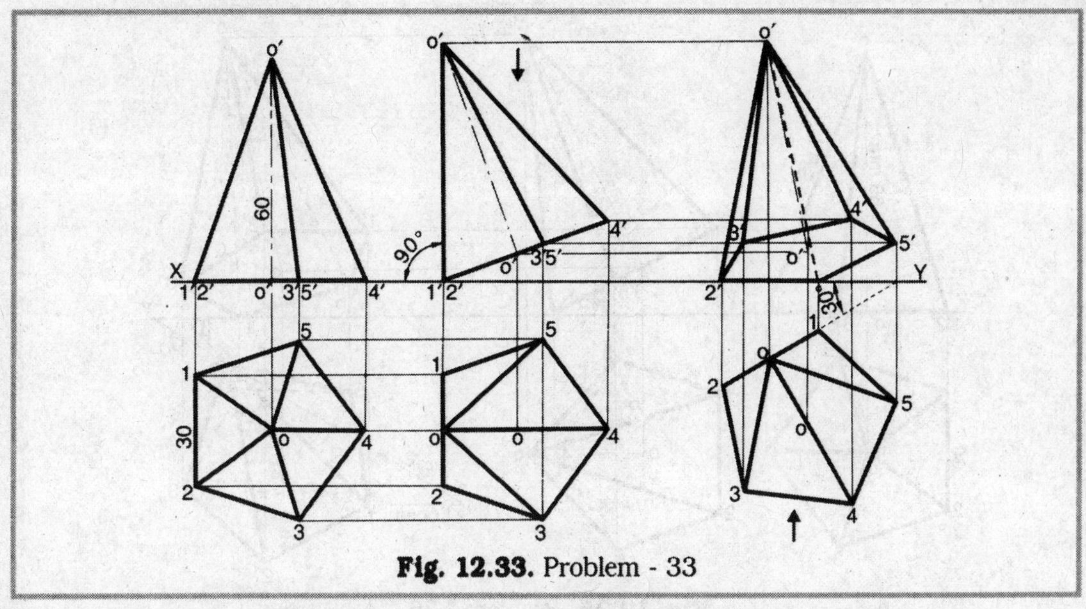

Fig. 12.33. Problem - 33

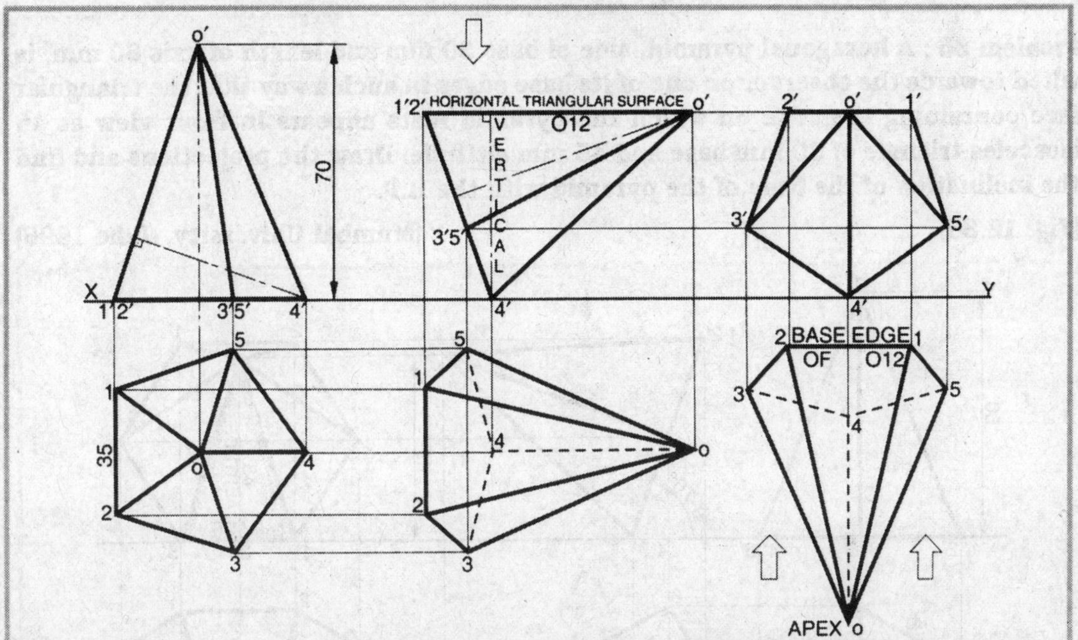

Fig. 12.34. Problem - 34

Problem 35 : A pentagonal pyramid, has a corner of its base on the H.P. with the triangular face opposite to it inclined at 45° to the H.P. and a slant edge within that triangular face inclined at 30° to the V.P. Draw the projections of a pyramid if an edge of its base is 30 mm and the axis is 65 mm long. [Fig. 12.35]

[Mumbai University, December 1995]

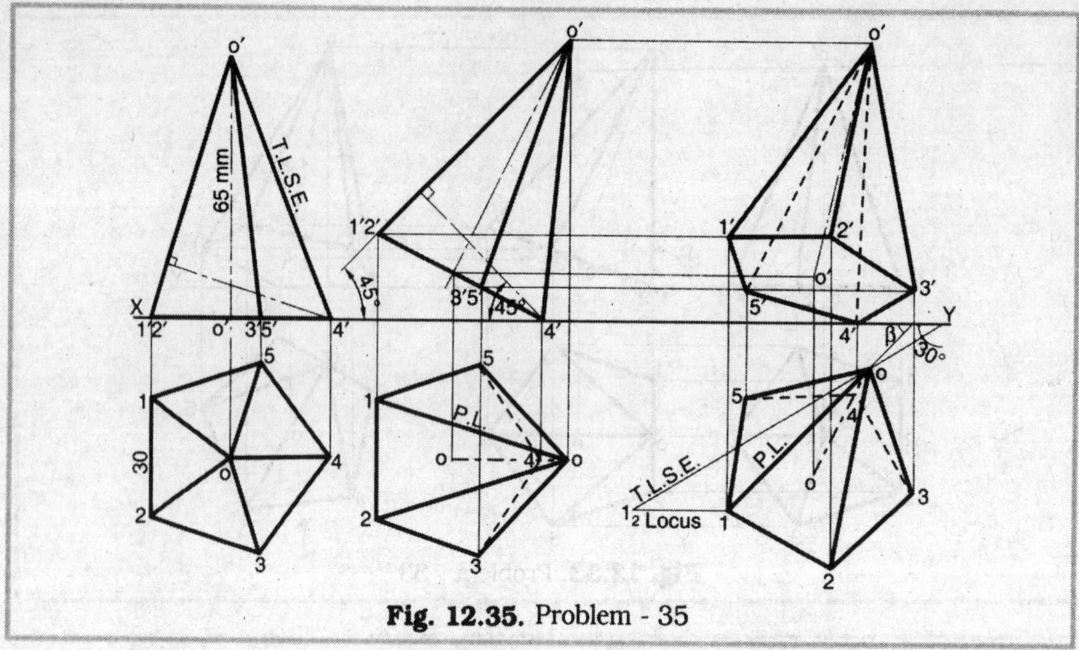

Fig. 12.35. Problem - 35

Problem 36 : A hexagonal pyramid, side of base 30 mm and length of axis 60 mm, is tilted towards the observer on one of its base edges in such a way that the triangular face containing the edge on which the pyramid rests appears in front view as an isosceles triangle of 30 mm base and 45 mm altitude. Draw the projections and find the inclination of the base of the pyramid with the H.P.

[Fig. 12.36] **[Mumbai University, June 1996]**

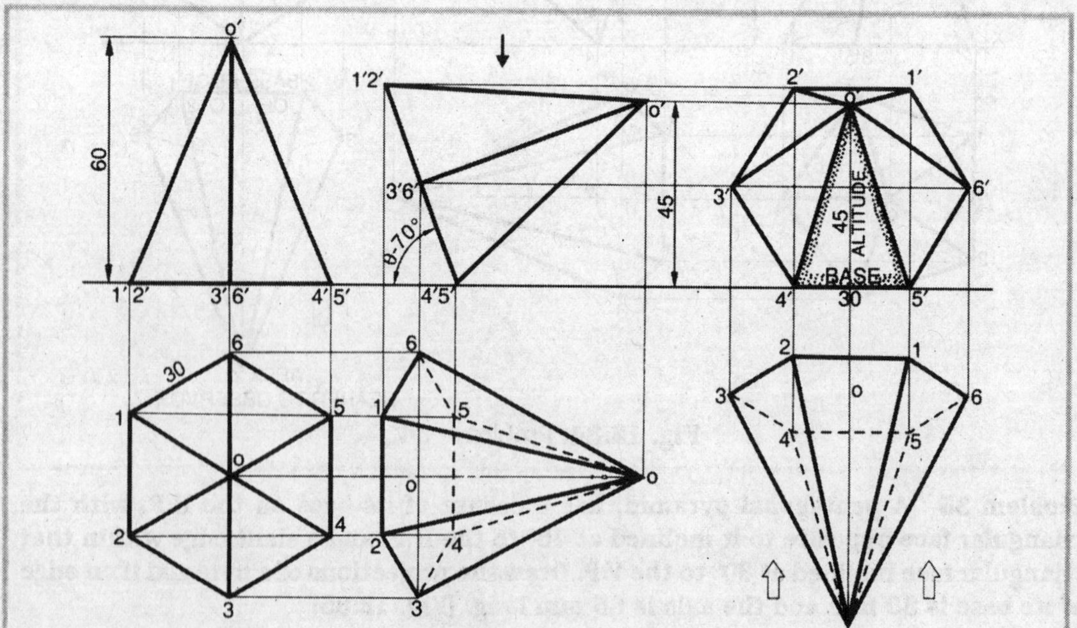

[Answer : $\theta_{Base} = 70°$]

Fig. 12.36. Problem - 36

Problem 37 : The frustum of a cone, having bottom base diameter 70 mm, top base diameter 30 mm and axis 50 mm, is resting on one of its generators on the H.P. Its T.V. of axis is inclined at 30° to the V.P. Draw its projections. [Fig. 12.37]

[Mumbai University, November 1990]

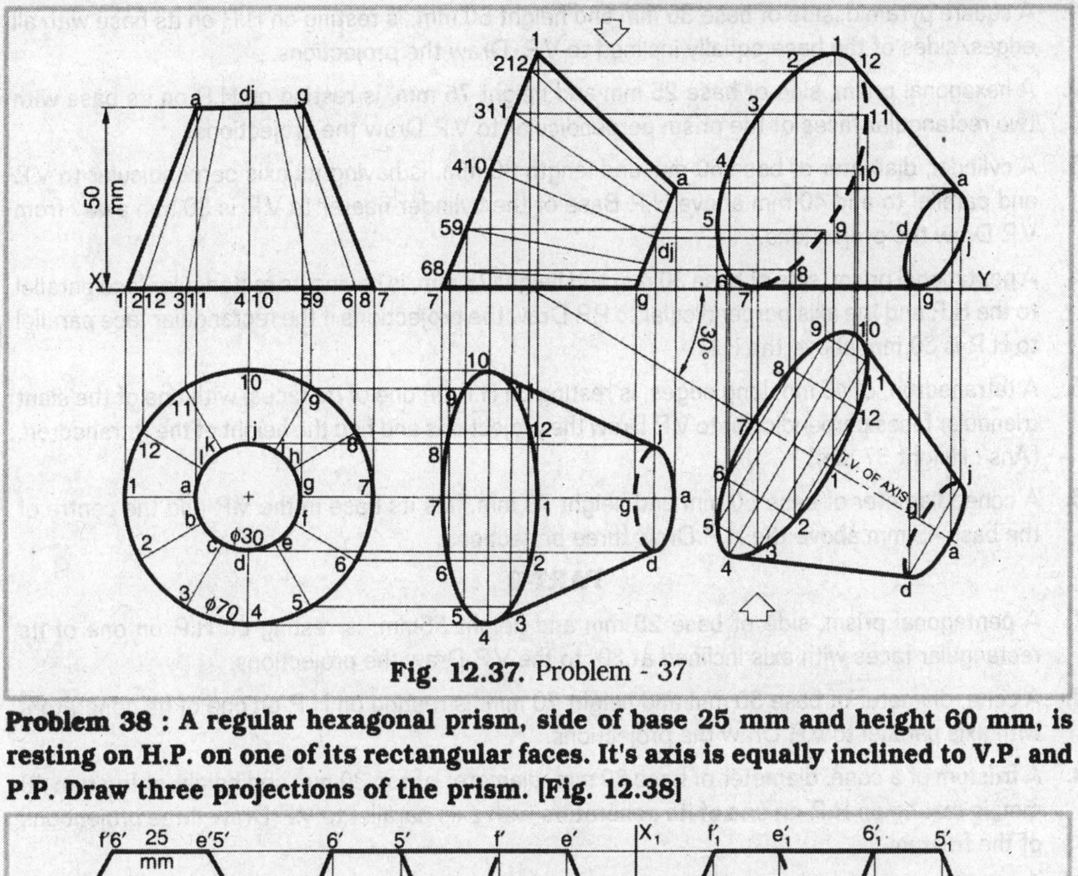

Fig. 12.37. Problem - 37

Problem 38 : A regular hexagonal prism, side of base 25 mm and height 60 mm, is resting on H.P. on one of its rectangular faces. It's axis is equally inclined to V.P. and P.P. Draw three projections of the prism. [Fig. 12.38]

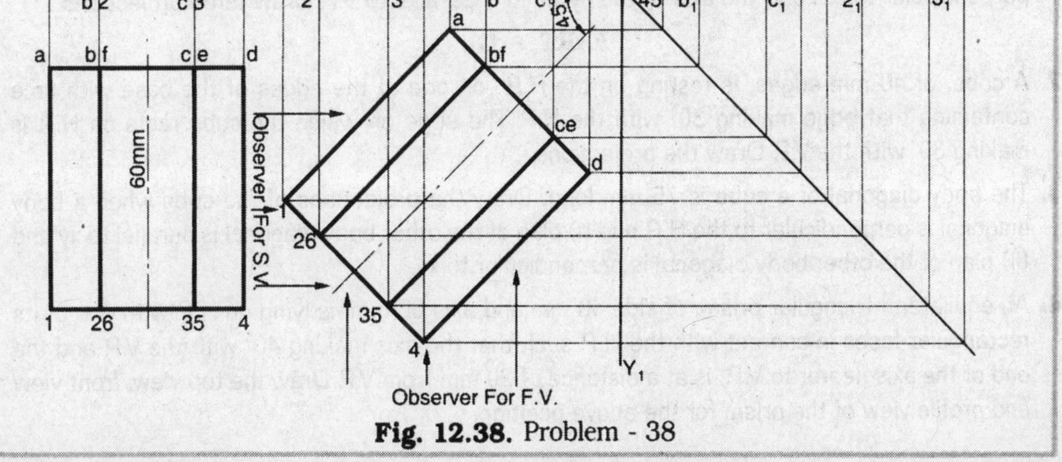

Fig. 12.38. Problem - 38

EXERCISE

PART - A

1. A square pyramid, side of base 30 mm and height 50 mm, is resting on H.P. on its base with all edges/sides of the base equally inclined to V.P. Draw the projections.
2. A hexagonal prism, side of base 25 mm and height 75 mm, is resting on H.P. on its base with two rectangular faces of the prism perpendicular to V.P. Draw the projections.
3. A cylinder, diameter of base 50 mm and length 60 mm, is having its axis perpendicular to V.P. and parallel to and 40 mm above H.P. Base of the cylinder nearer to V.P. is 30 mm away from V.P. Draw the projections.
4. A pentagonal prism, side of base 30 mm and height 75 mm, is having its rectangular face parallel to the H.P. and the axis perpendicular to P.P. Draw the projections if the rectangular face parallel to H.P. is 30 mm above the H.P.
5. A tetrahedron, of 45 mm long edges, is resting on H.P. on one of its faces, with one of the slant triangular faces perpendicular to V.P. Draw the projections and find the height of the tetrahedron. [Ans : Height 37 mm]
6. A cone, diameter of base 60 mm and height 70 mm, has its base in the V.P. and the centre of the base 45 mm above the H.P. Draw three projections.

PART-B

7. A pentagonal prism, side of base 25 mm and height 55mm, is resting on H.P. on one of its rectangular faces with axis inclined at 30° to the V.P. Draw the projections.
8. A cone, diameter of base 50 mm and height 70 mm, is resting on H.P. on one of its generators with axis parallel to V.P. Draw the projections.
9. A frustum of a cone, diameter of base 60 mm, diameter of top 30 mm and height of frustum 40 mm, is resting on H.P. on one of its generators with axis parallel to V.P. Draw three projections of the frustum.
10. A cylinder, diameter of base 50 mm and height 75 mm, is nailed on V.P. on one of its generators with axis making 45° with H.P. Draw three projections.
11. A hexagonal pyramid, side of base 25 mm and height 60 mm, is resting on the H.P. on one of the corners of the base such that the slant edge, containing the corner on which it rests on H.P., is perpendicular to P.P. and the axis of the pyramid is parallel to V.P. Draw three projections.

PART - C

12. A cube, of 40 mm edges, is resting on the H.P. on one of the edges of the base with face containing that edge making 30° with the H.P. The edge on which the cube rests on H.P. is making 30° with the V.P. Draw the projections.
13. The body diagonal of a cube is 75 mm long. Draw the projections of the cube when a body diagonal is perpendicular to the H.P. and (i) plan of the other body diagonal is parallel to xy and (ii) plan of the other body diagonal is perpendicular to xy.
14. An equilateral triangular prism, of side 40 mm and axis 60 mm, is lying on H.P. with one of its rectangular faces in contact with the H.P. such that the axis making 40° with the V.P. and the end of the axis nearer to V.P. is at a distance of 20 mm from V.P. Draw the top view, front view and profile view of the prism for the above position.

15. A square prism, side of base 40 mm and height 60 mm, is resting on H.P. on one of the corners of the base with vertical/longer edge containing the corner on which it rests on H.P. is inclined at 50° to the H.P. and 30° to the V.P. Draw the projections when (i) top end of the prism is nearer to V.P. and (ii) top end of the prism is away from V.P.

16. A pentagonal prism, of 25 mm side of base and height 50 mm, rests with one of its longer edges on the H.P. such that, one of the rectangular faces, containing the longer edge on which the prism rests, is inclined at 30° to the H.P. and is nearer to the observer. The axis of the prism is inclined at 40° to the V.P. Draw the top and front views.

17. A hexagonal prism, of 30 mm side of base and 70 mm height, has one of its rectangular faces in the V.P. and the edge of the base contained by that face makes an angle of 30° with the H.P. Draw the projections of the prism.

18. A square pyramid, of side 40 mm and axis 80 mm long, is situated in space below H.P. and behind V.P. such that the base of the pyramid makes 30° with H.P. and one pair of parallel sides of the base makes 20° with the V.P. and parallel to H.P. The apex is away from V.P. and the end of the axis in the base is at a distance of 50 mm from V.P. and 100 mm from H.P. Draw (i) top view, (ii) front view, (iii) profile view. The base of the pyramid is visible to the observer.

19. A square pyramid, whose faces are isosceles triangles of 50 mm base and 75 mm altitude, is lying on one of its faces on H.P. The projection of the axis of the pyramid on the H.P. makes an angle of 30° with the XY line. Draw the top, front and profile views of the pyramid.

20. A square pyramid, of side of base 30 mm and axis 55 mm, is resting with its apex on the H.P. such that this point of contact of the apex on the H.P. is at a distance of 50 mm from the V.P., the inclined edge connecting the apex to one of the corners of the base is vertical and that triangular face bounded by that vertical edge is perpendicular to the V.P. Draw the following views of the pyramid in the above position. (i) top view, (ii) front view.

21. A pentagonal pyramid, side of pentagon 30 mm and height 80 mm is resting on the H.P. on one of its base edges such that the triangular face containing that edge is perpendicular to H.P. and parallel to V.P. Draw the projections of the pyramid.

22. A cylinder, diameter of base 60 mm and height 90 mm, is resting on the H.P. on the point of its periphery of the base. The axis of cylinder is inclined to H.P. by 30° and (i) the top view of the axis is inclined at 45° to the V.P. and (ii) the axis is inclined at 45° to the V.P. Draw the projections. Keep top end of the cylinder nearer to V.P. in one case and away from V.P. in other.

23. A cone, diameter of base 50 mm and height 75 mm, is resting on the H.P. on the point of its periphery of the base. The axis of the cone is inclined to the H.P. by 30° and to the V.P. by 45°. Draw the projections when (i) apex is nearer to the observer and (ii) apex is away from the observer.

24. A tetrahedron, of 75 mm long edges, is resting on H.P. on one of the edges with slant face containing that edge is perpendicular to H.P. The edge, on which it rests on H.P. makes 45° with the V.P. Draw the projections.

25. A cone, diameter of base 50 mm and height 75 mm, is nailed on V.P. keeping one of its generators in V.P. The plane containing that generator and the axis of the cone makes an angle of 30° with the H.P. Draw the projections when (i) apex is nearer to H.P. and (ii) apex is away from H.P.

26. A waste paper basket has two cylindrical rings one at the top and another at the bottom end. Middle portion is a frustum of a cone. The diameter and height of the upper ring are 75 cm and

15 cm respectively. The diameter and height of the lower ring are 40 cm and 20 cm respectively. The height of the frustum of a cone is 50 cm. This waste paper basket is resting on the H.P. vertically on its bottom ring. Draw the plan and elevation and project its (i) auxiliary elevation on A.V.P. making 30° with V.P. and (ii) auxiliary plan on A.I.P. making 45° with H.P.

27. A square slab 75mm x 75 mm x 25 mm thick is resting on the H.P. with its vertical faces equally inclined to the V.P. A cylindrical slab of 75 mm diameter and 20 mm thickness is placed centrally on the top of the square slab. A hexagonal pyramid, side of base 30 mm and height 60 mm, is placed centrally on the top of the cylindrical slab. Axes of three solids coincides along a vertical straight line. Two edges of the base of the pyramid are perpendicular to the V.P. Draw the elevation and plan of the solids and project (i) auxiliary elevation on A.V.P. inclined at 30° to the V.P. and (ii) auxiliary plan on an A.I.P. making 45° with the H.P. Do not draw dotted lines in projections for clarity of projections.

28. A hexagonal nut, side of base 40 mm and height 40 mm, has a central hole of 40 mm diameter. Draw its projections when it is resting on H.P. on one of its edges of base with its axis making 45° with the H.P. and two of its faces remaining perpendicular to the V.P. Assume that the nut does not have chamfer.

29. A triangular prism, side of base 50 mm and length 75 mm, is resting on the H.P. on one of its rectangular faces with the axis perpendicular to the P.P. A cone, diameter of base 50 mm and height 75 mm, is resting on the H.P. and is leaning on one of the generators centrally on the rectangular face of the prism, with its axis parallel to P.P. The cone is nearer to the observer. Draw the projections.

30. Three unequal spheres, of 70 mm, 50 mm and 30 mm diameters, are resting on the H.P. so that each touches the other two. Line joining the centres of two smaller spheres is contained by a plane perpendicular to V.P. Draw three projections of spheres showing the contact points on them.

31. A pentagonal pyramid, side of base 35 mm and height 60 mm, is resting on the H.P., such that two of its slant edges are on the H.P. and one of which is normal to the V.P. Draw the projections, when the apex of the pyramid is nearer to the observer with respect to its base.

32. Three equal spheres, of 50 mm diameter, are resting on the H.P. so that each touches the other two. The line joining the centres of two of them is parallel to the V.P. and nearer to the observer. A fourth sphere, of 65 mm diameter, is resting on the top of the three spheres. Draw three projections showing the contact points. Find also the height of the centre of the 4th sphere. [Ans : Height 75 mm]

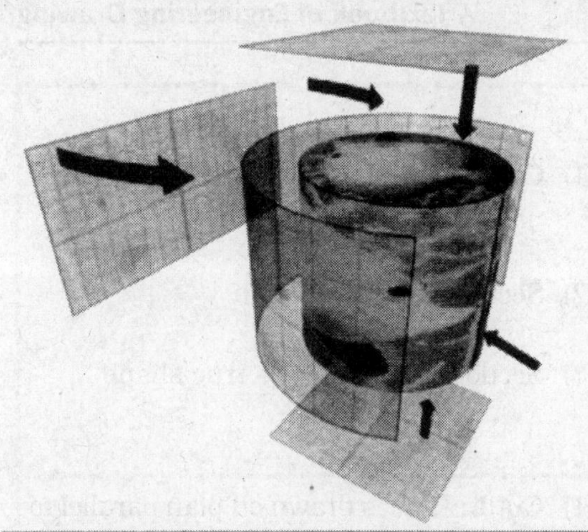

CHAPTER - 13

Sections of Solids

1. Introduction

Study of various sections of different solids taken at different angles with respect to axes of solids or bases of solids, is very important. Knowledge of how to cut a particular solid to get a desired shape and size of section, is necessary with design and fabrication engineers. In short knowledge of this topic is necessary for work of fabrication industries.

In this chapter, we shall imagine the object cut by cutting plane or section plane. The portion of the object between imaginary cutting plane and the observer is assumed to be removed and then the view is taken of the cut object. View of the cut object is known as sectional view. The surface produced by cutting the object is called section. This surface of section is indicated by hatching lines generally drawn at 45°.

2. Section Planes or Cutting Planes :

See Figs. 13.1, 13.2 and 13.3.

There are five different types of cutting planes or section planes.

They are :

(1) Horizontal cutting plane

(2) Vertical cutting plane parallel to V.P.

(3) Cutting plane parallel to Profile Plane

(4) Auxiliary vertical plane (A.V.P.)

(5) Auxiliary inclined plane. (A.I.P.)

2.1 Horizontal Cutting Plane :

This type of cutting plane is perpendicular to V.P. and parallel to H.P. Vertical trace of the cutting plane is used as cutting plane line. C.P. line is drawn on elevation parallel to xy. Plan will be sectional view along with section in true shape. (See Figs. 13.1, 13.2 and 13.3 Plane No.1)

2.2 Vertical Cutting Plane Parallel to V.P. :

This type of cutting plane is perpendicular to H.P. and parallel to V.P. Horizontal trace of the cutting plane is used as cutting plane line. C.P. line is drawn on plan parallel to xy line. Elevation will be sectional view along with the section in true shape. (See Figs. 13.1, 13.2 and 13.3. Plane No.2.)

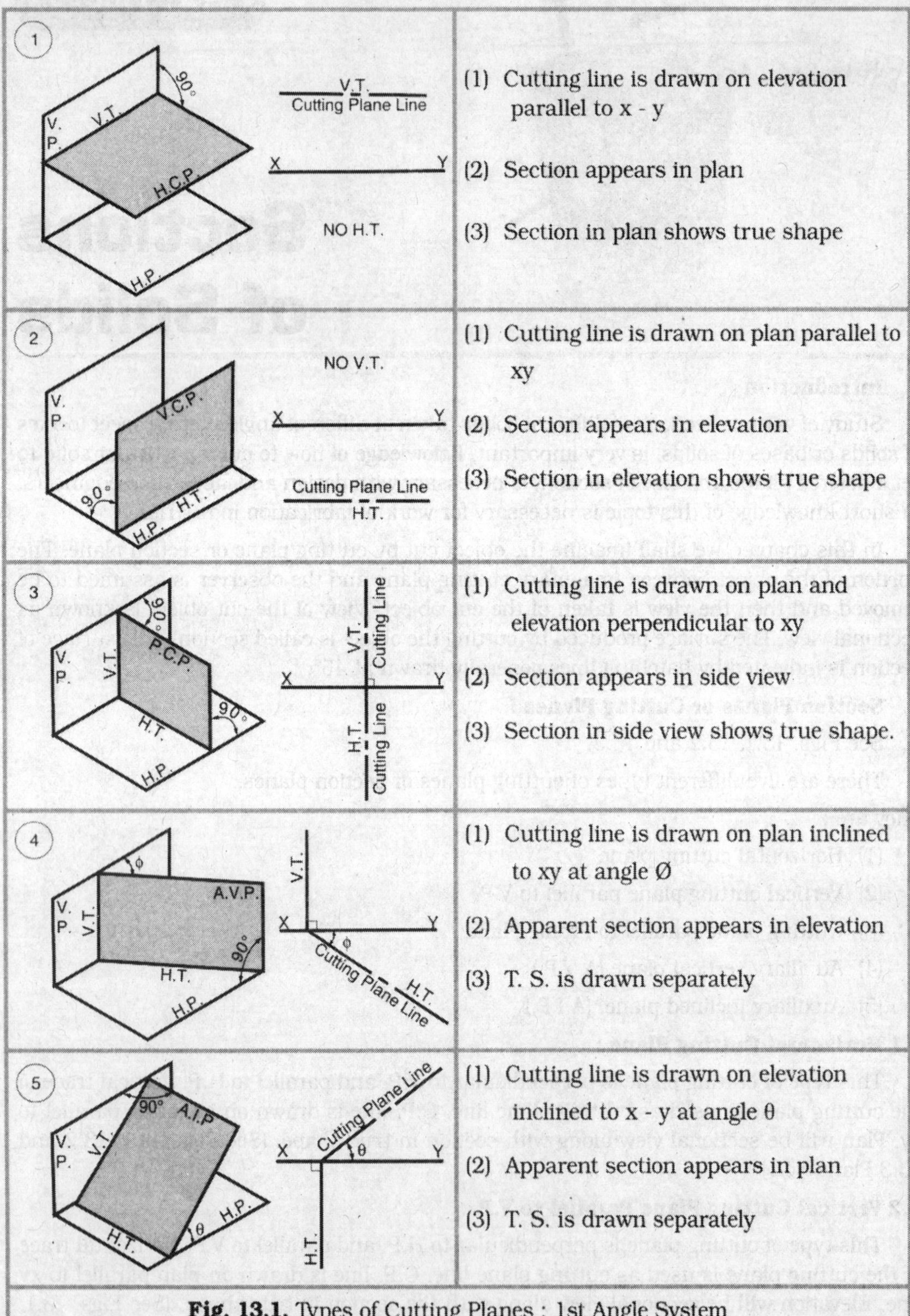

Fig. 13.1. Types of Cutting Planes : 1st Angle System

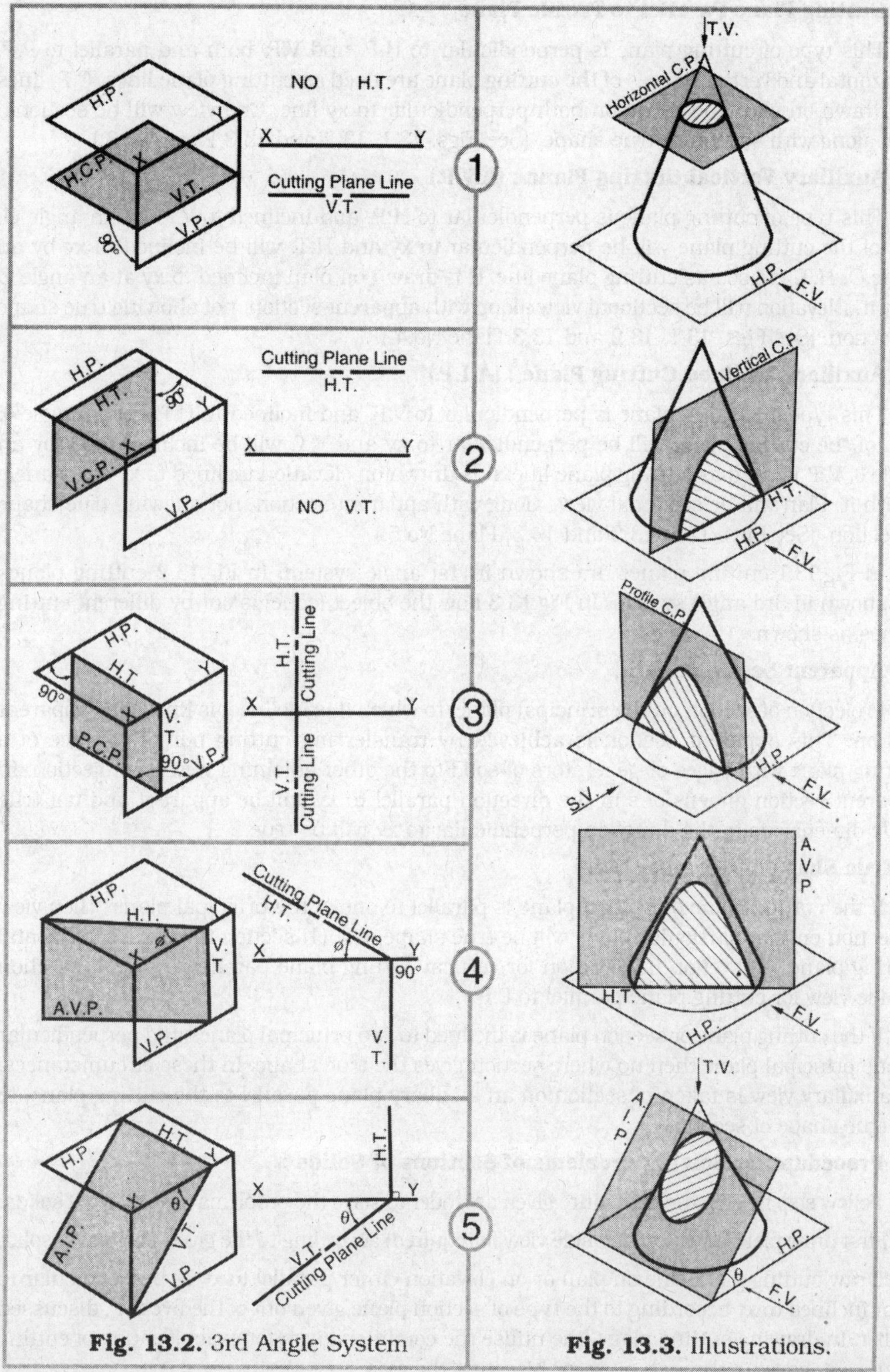

Fig. 13.2. 3rd Angle System

Fig. 13.3. Illustrations.

2.3 Cutting Plane Parallel to Profile Plane :

This type of cutting plane is perpendicular to H.P. and V.P. both and parallel to P.P. Horizontal and vertical traces of the cutting plane are used as cutting plane lines. C.P. lines are drawn on plan and elevation both perpendicular to xy line. End view will be sectional view along with section in true shape. (See Figs. 13.1, 13.2 and 13.3 Plane No.3.)

2.4 Auxiliary Vertical Cutting Plane : (A.V.P.)

This type of cutting plane is perpendicular to H.P. and inclined to V.P. by an angle Ø. V.T. of the cutting plane will be perpendicular to xy and H.T. will be inclined to xy by an angle Ø. H.T. is used as cutting plane line. It is drawn on plan inclined to xy at an angle Ø with it. Elevation will be sectional view along with apparent section, not showing true shape of section. (See Figs. 13.1, 13.2 and 13.3 Plane No.4.)

2.5 Auxiliary Inclined Cutting Plane : (A.I.P.)

This type of cutting plane is perpendicular to V.P. and inclined to H.P. by an angle θ. H.T. of the cutting plane will be perpendicular to xy and V.T. will be inclined to xy by an angle θ. V.T. is used as cutting plane line. It is drawn on elevation inclined to xy at an angle θ with it. Plan will be sectional view, along with apparent section, not showing true shape of section. (See Figs. 13.1, 13.2 and 13.3. Plane No.5.)

In Fig.13.1 cutting planes are shown in 1st angle system. In Fig.13.2 cutting planes are shown in 3rd angle system. In Fig.13.3 how the object (cone) is cut by different cutting planes, is shown.

3. Apparent Section : (A.S.)

Projection of section on the principal plane, to which it is inclined, is known as apparent section. This apparent section is achieved by transferring cutting points of trace of a cutting plane with edges or generators of solid to the other adjoining view or projection. In apparent section dimensions in the direction parallel to xy will be apparent and not true while dimensions in the direction perpendicular to xy will be true.

4. True Shape of Section : (T.S.)

If the cutting plane or section plane is parallel to one of the principal planes then view of section on that particular plane will be true shape, e.g. (1) section in plan for horizontal cutting plane, (2) section in elevation for vertical cutting plane parallel to V.P., (3) section in side view for cutting plane parallel to P.P.

If the cutting plane or section plane is inclined to two principal planes and perpendicular to one principal plane then no where section gives the true shape. In these circumstances, an auxiliary view is taken of section on an auxiliary plane parallel to the cutting plane, to get true shape of section.

5. Procedure for Solving Problems of Sections of Solids :

Follow step by step the procedure given as under to solve the problems of sections of solids.

5.1 First draw plan, elevation and side view (if required) according to the given position of solid.

5.2 Draw cutting plane line on plan or on elevation either parallel to xy or perpendicular to xy or inclined to xy according to the type of section plane given out of the five S.P. discussed earlier. In drawing cutting plane line utilise the condition given for exact location of cutting plane, e.g. passing through apex, or bisecting the axis, etc.

5.3 Find how many edges or generators are cut by cutting plane line. Two illustrations are given below :

5.3 (1) Refer Problem No.13, Fig.13.16.

Pentagonal Pyramid has 10 edges

AB BC CD DE EA - Edges of base
OA OB OC OD OE - Slant edges

Cutting plane line in elevation is cutting six edges, enrectangled above. Therefore, section will be polygon of six sides.

5.3 (2) Refer Problem No.11, Fig.13.14.

Cylinder has 12 generators and two bases.

A1 B2 C3 D4 E5 F6
G7 H8 I9 J10 K11 L12 } Generators

A B C D E F G H I J K L A
1 2 3 4 5 6 7 8 9 10 11 12 1 } Bases

Cutting plane line in plan is cutting six generators enrectangled above and it is cutting both the bases of cylinder at (2 × 2) = 4 points. Totally we have [6 + 4] = 10 cutting points.

5.4 Having found the cutting points on edges or generators of cutting plane line, transfer them by projectors perpendicular to x-y line on the other view or projection on the corresponding edges or generators.

5.5 Join these projected points by straight lines or curve depending upon the nature of solid to get the section.

5.6 If the section obtained does not show the true shape follow the procedure for getting the true shape of section as under :

Draw new ground line parallel to the cutting plane line. Project all cutting points from cutting plane line and take dimensions of respective points, from other view, from its ground line. Join the projected points obtained on new ground line by straight lines or curve to get the true shape of the section.

Draw hatching lines at 2 to 5 mm spacings inclined at 45° to the horizontal or boundary of section in apparent as well as the true section. Follow definite notation system.

5.7 Remember the following : Prisms have [n × 3] edges [n × 2 = 2n] edges of base (AB, BC....) (12, 23....) and [n] vertical or longer edges (A1,B2,......)]

Pyramids have [n × 2] edges [n edges of base (AB, BC....) or (12,23,....)] and [n] slant edges [(OA, OB,....) or (01, 02,....)] and have one apex o.

Cylinders have 12 generators A1, B2, C3, D4, E5, F6, G7, H8, I9, J10, K11 and L12 and 2 bases.

Cones have 12 generators either OA, OB, OC, OD, OE, OF, OG, OH, OI, OJ, OK and OL or O1, O2, O3, O4, O5, O6, O7, O8, O9, O10, O11 and O12 and have one base and one apex O.

To get polygon of [m] number of edges as section, cutting plane must cut [m] number of edges.

Now we shall study section of different solids one by one in the form of typical problems.

6. Sections of Prisms :

Problem 1 : A square prism, side of base 40 mm and height 50 mm, is resting on H.P. on its base with all vertical faces equally inclined to V.P. It is cut by a vertical section plane parallel to and away from V.P. and remaining 9 mm away from the axis of the prism, Draw plan and sectional elevation.

For solution see Fig. 13.4 and follow the procedure, as given below :

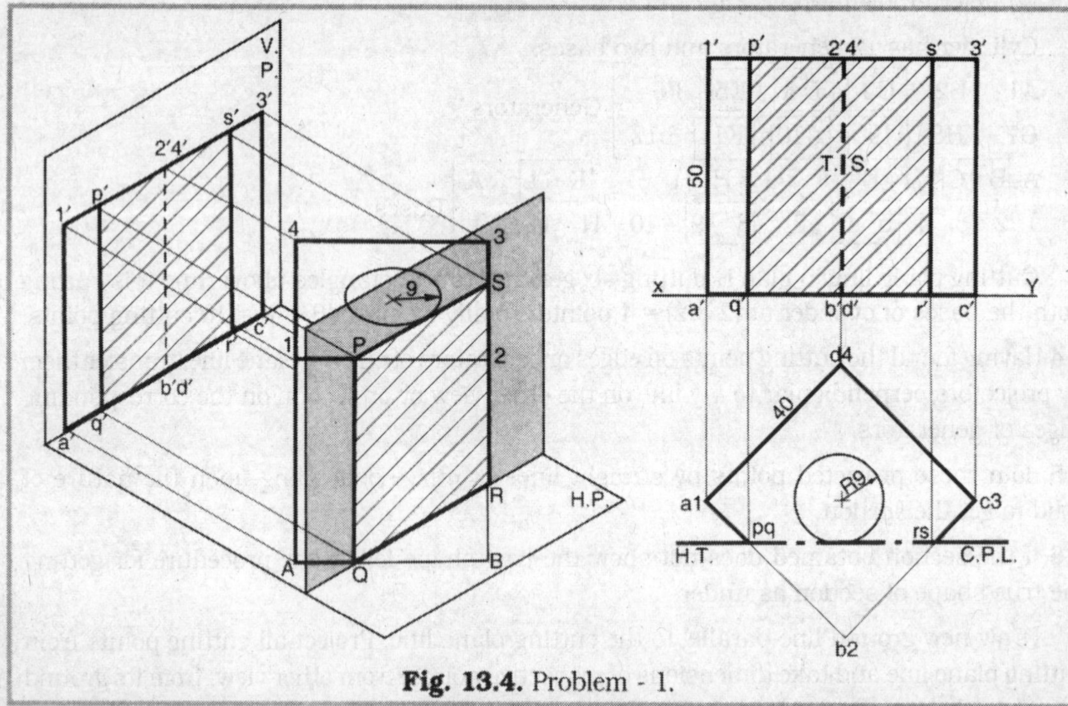

Fig. 13.4. Problem - 1.

Three dimensional drawing is given to visualise the solution of the problem.

(i) First draw plan and elevation of the prism.

(ii) With the point view plan of axis as the centre and radius equal to 9 mm draw a circle. Draw a cutting plane line (C.P.L.) parallel to xy, tangent to circle and away from xy.

(iii) C.P. line is cutting 4 edges of the prism 12, AB, BC and 23 at points P, Q, R and S respectively. Transfer cutting points p, q, r and s of plan to elevation on corresponding lines by projectors to get p', q', r' and s' respectively.

(iv) Join p', q',r'and s' to get the section. This section will show true shape since C.P.L. is parallel to xy line.

Problem 2 : A pentagonal prism is resting on H.P. on one of its rectangular faces with axis of the prism parallel to H.P. and inclined to V.P. by 30°. It is cut by a horizontal section plane 40 mm above the H.P. Draw elevation and sectional plan of the prism. Take side of base 40 mm and height 93 mm.

For solution see Fig. 13.5.

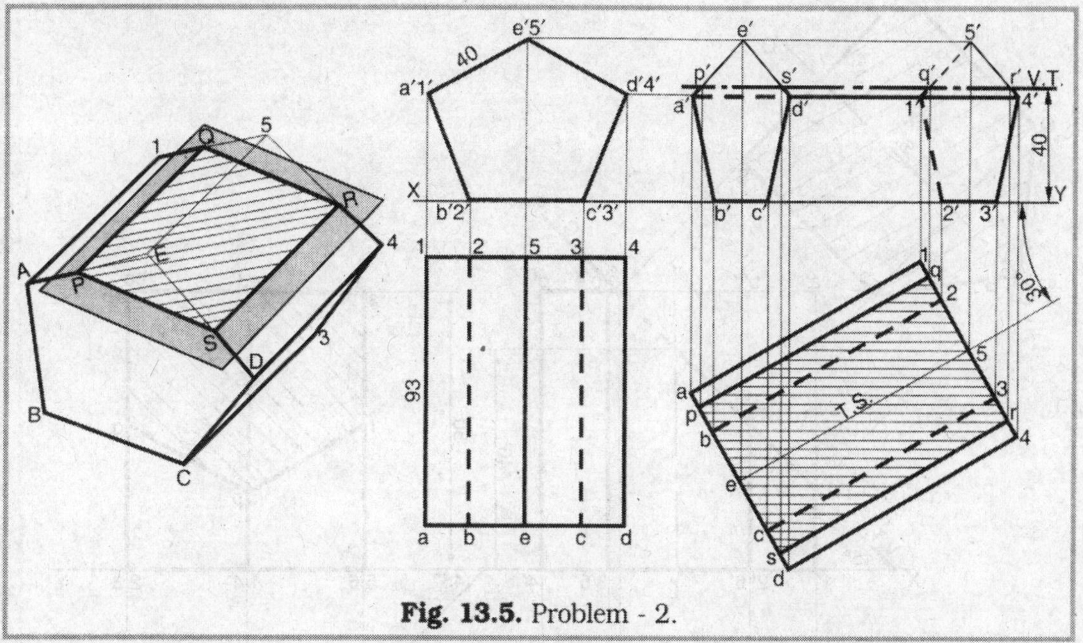

Fig. 13.5. Problem - 2.

Three dimensional drawing is given to visualise the solution of the problem. Solution is self explanatory.

Problem 3 : A hexagonal prism is resting on H.P. on its base with two edges/sides of base parallel to V.P. It is cut by A.I.P. perpendicular to V.P. and inclined to H.P. by 45° passing through a point of axis 40 mm above the base. Draw three principal projections and find the true shape of section. Take side of base 25 mm and height 50 mm.

For solution see Fig. 13.6 and follow the procedure, as given below :

(i) First draw three principal projections in the given position. Mark all corners by usual notations.

(ii) Draw cutting plane line (V.T. of C.P.) on elevation at 45° to xy, passing through a point 40 mm above xy and on the axis.

(iii) Cutting plane line is cutting vertical edges 1A, 2B, and 6F and sides of base BC and EF at points P, Q, T, R and S respectively.

(iv) Transfer these cutting points of elevation to plan and side view on respective edges and join them to get apparent sections.

(v) To get true shape of section draw perpendiculars to C.P.L. or V.T. at cutting points p', (q',t') and (r', s'). Take distances of respective points from plan from the xy line and mark the same on perpendiculars drawn at p', (q', t')and (r's') to mark points P, (Q,T) and (R, S) as shown in the figure. Join points P, Q, R, S and T by straight lines to get T.S. of section.

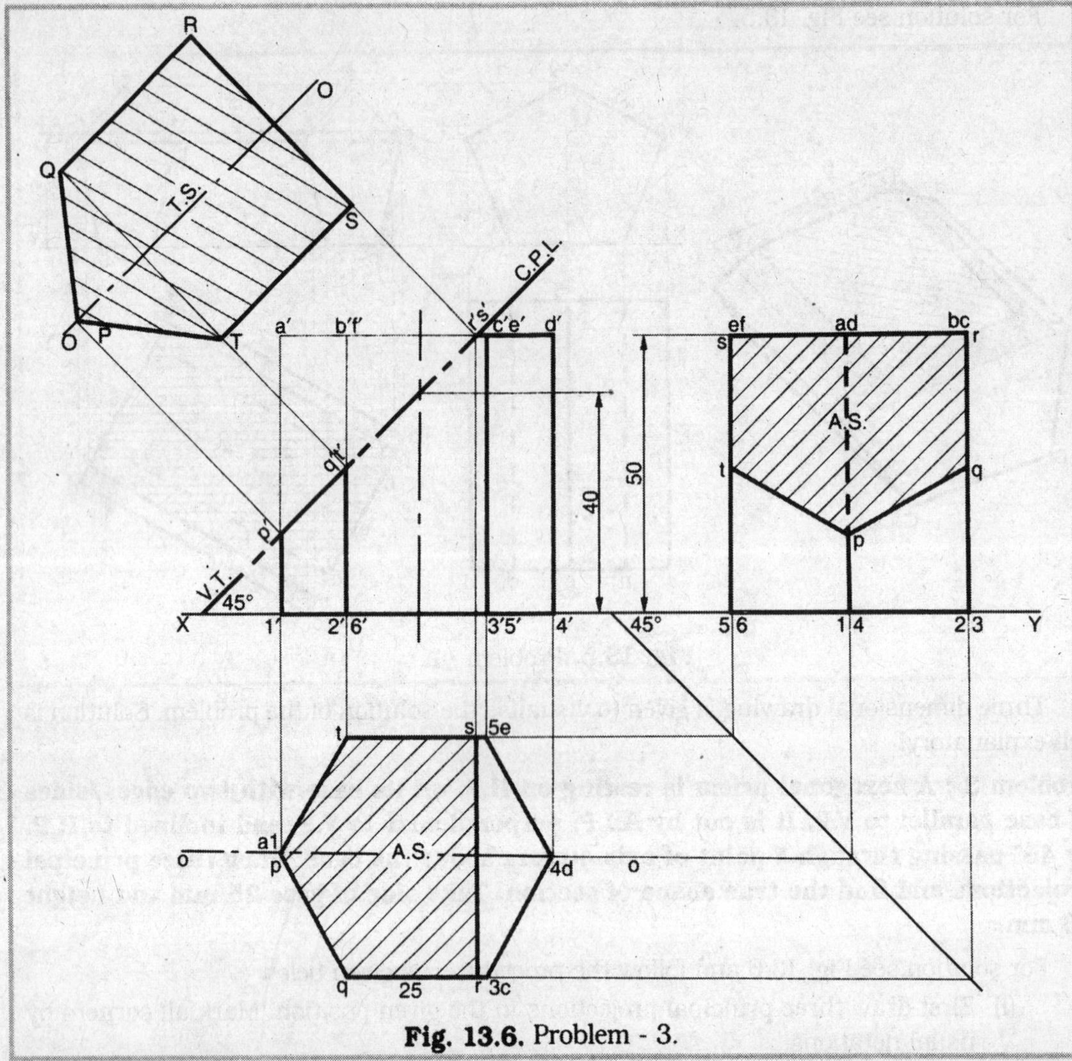

Fig. 13.6. Problem - 3.

Problem 4 : A square prism, side of base 50 mm and height 75 mm, is resting on H.P. on its base with all vertical faces equally inclined to V.P. It is cut by A.I.P. inclined at 45° to the H.P. passing through a point 55 mm above the base on the axis. Draw elevation, sectional plan and sectional auxiliary plan along with section on a plane parallel to cutting plane.

For solution see Fig. 13.7. Solution is self explanatory. For auxiliary projection dimensions are taken on two sides of centre line o - o.

Problem 5 : A pentagonal prism is resting on H.P. on one of its rectangular faces with axis of the prism parallel to H.P. and V.P. both. It is cut by Auxiliary Vertical Plane perpendicular to H.P. and inclined to V.P. by 45° passing through a point on the axis 15 mm from one end. Draw three principal projections with sections and also draw the true shape of section. Take side of base 45 mm and height 80 mm.

For solution see Fig. 13.8 and follow the procedure, as given below :

(i) First draw side view, elevation and plan of the prism for the given position.

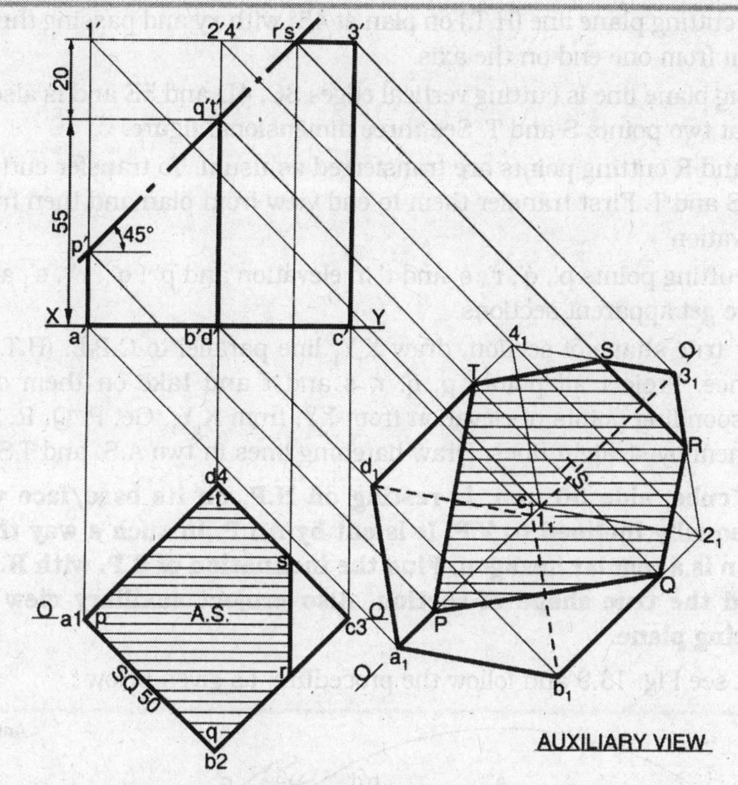

Fig. 13.7. Problem - 4.

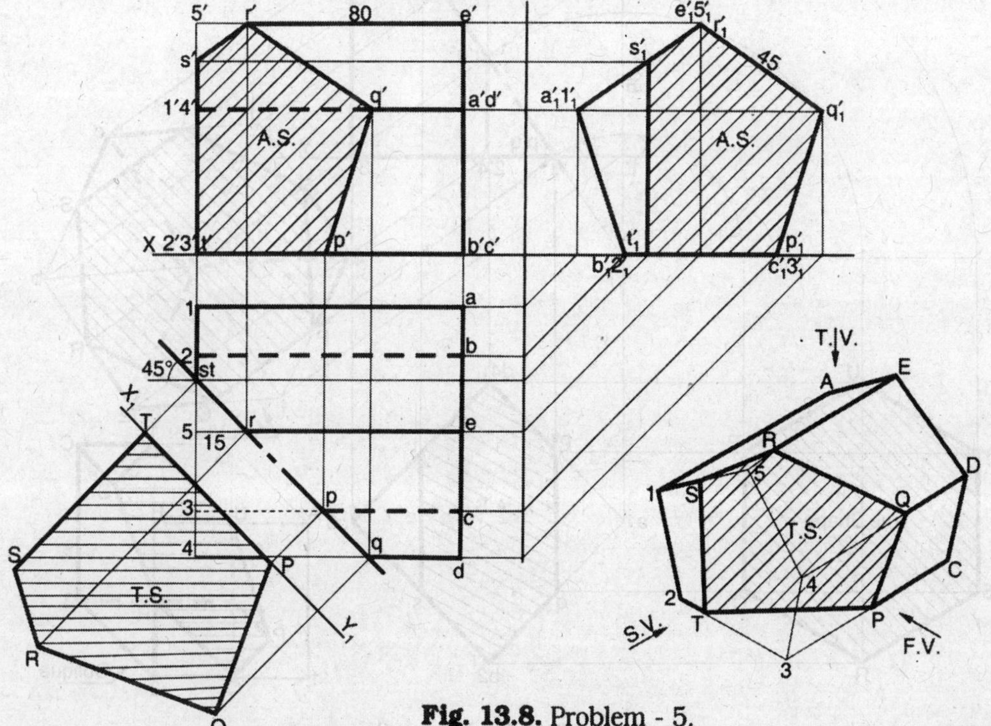

Fig. 13.8. Problem - 5.

(ii) Draw cutting plane line (H.T.) on plan at 45° with xy and passing through a point 15 mm from one end on the axis.

(iii) Cutting plane line is cutting vertical edges 3C, 4D and 5E and is also cutting the base at two points S and T. See three dimensional figure.

(iv) P, Q and R cutting points are transferred as usual. To transfer cutting points of base S and T. First transfer them to end view from plan and then from end view to elevation.

(v) Join cutting points p', q', r', s' and t' in elevation and p'_1, q'_1, r'_1, s'_1 and t'_1 in side view to get apparent sections.

(vi) To get true shape of section, draw X_1Y_1 line parallel to C.P.L. (H.T.) at suitable distance. Project all points p, q, r, s and t and take on them distances, of corresponding points of elevation from XY, from X_1Y_1. Get P, Q, R, S and T and join them by straight lines. Draw hatching lines in two A.S. and T.S.

Problem 6 : A cube, side 50 mm, is resting on H.P. on its base/face with all the vertical faces equally inclined to V.P. It is cut by A.I.P. in such a way that the true shape of section is a regular hexagon. Find the inclination of C.P. with H.P. and draw projections and the true shape of section. Also project auxiliary view on a plane parallel to cutting plane.

For solution see Fig. 13.9 and follow the procedure, as given below :

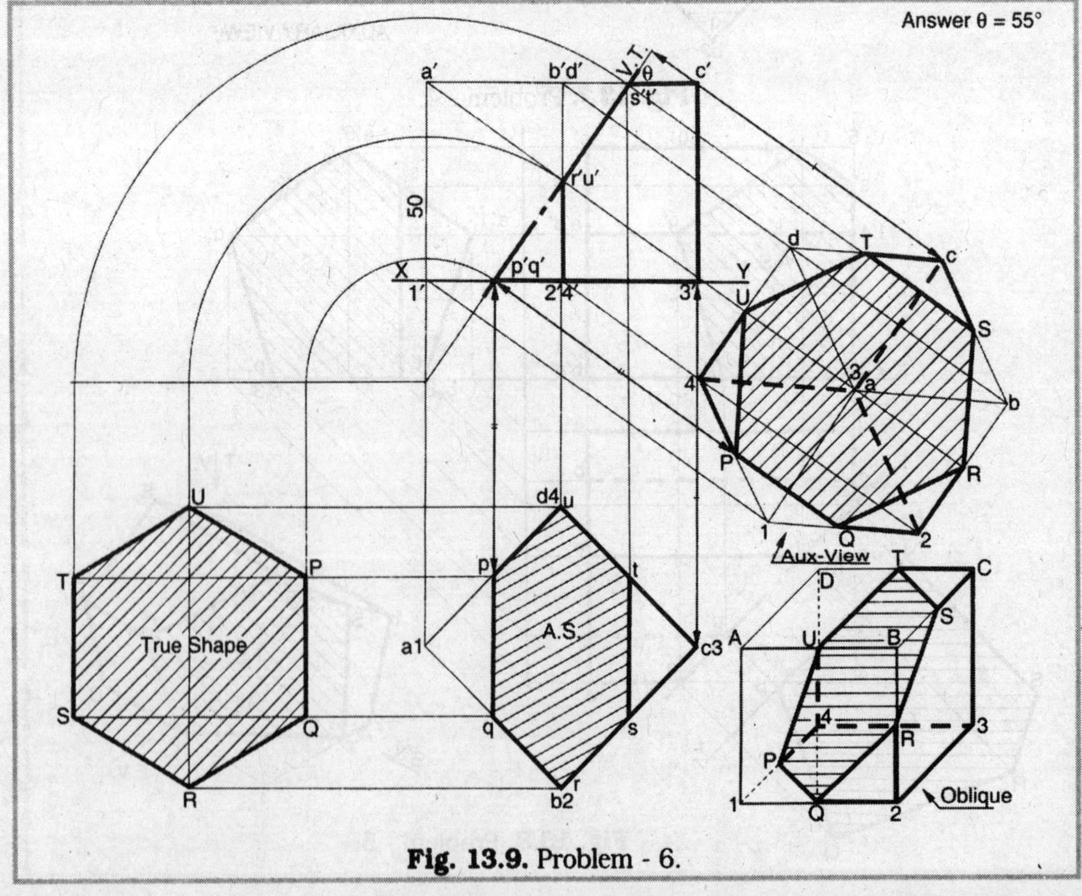

Fig. 13.9. Problem - 6.

(i) First draw two views, elevation and plan.

(ii) Draw cutting plane line on elevation in such a way that it must cut exactly 6 edges of the cube and in apparent section in plan pq and ts should have length 1/2 ru. This criterion is considered because the side of regular hexagon is equal to 1/2 the diagonal of it. Further dimensions of lines in apparent section in the direction perpendicular to xy is going to be true lengths.

(iii) The rest is usual. Figure itself is self explanatory. Note new method of drawing the true shape.

(iv) For auxiliary view project[(8 + 6) = 14] points. Draw cut part by light lines.

7. Sections of Cylinders :

Problem 7 : A cylinder is resting on H.P. on its base. It is cut by A.V.P. perpendicular to H.P. and inclined to V.P. by 45° and cutting it remaining 12 mm away from the axis. Draw the projections with section and draw also the true shape of section. Take diameter of cylinder 55 mm and height 60 mm.

[OR]

Problem 7 : A cylinder is resting on H.P. on its base. It is cut by A.V.P. inclined at 45° to V.P. in such a way that the true shape of section is a rectangle of 50 mm x 60 mm. Draw the projections and find the distance of A.V.P. from the axis.

For solution see Fig. 13.10 and follow the procedure, as given below :

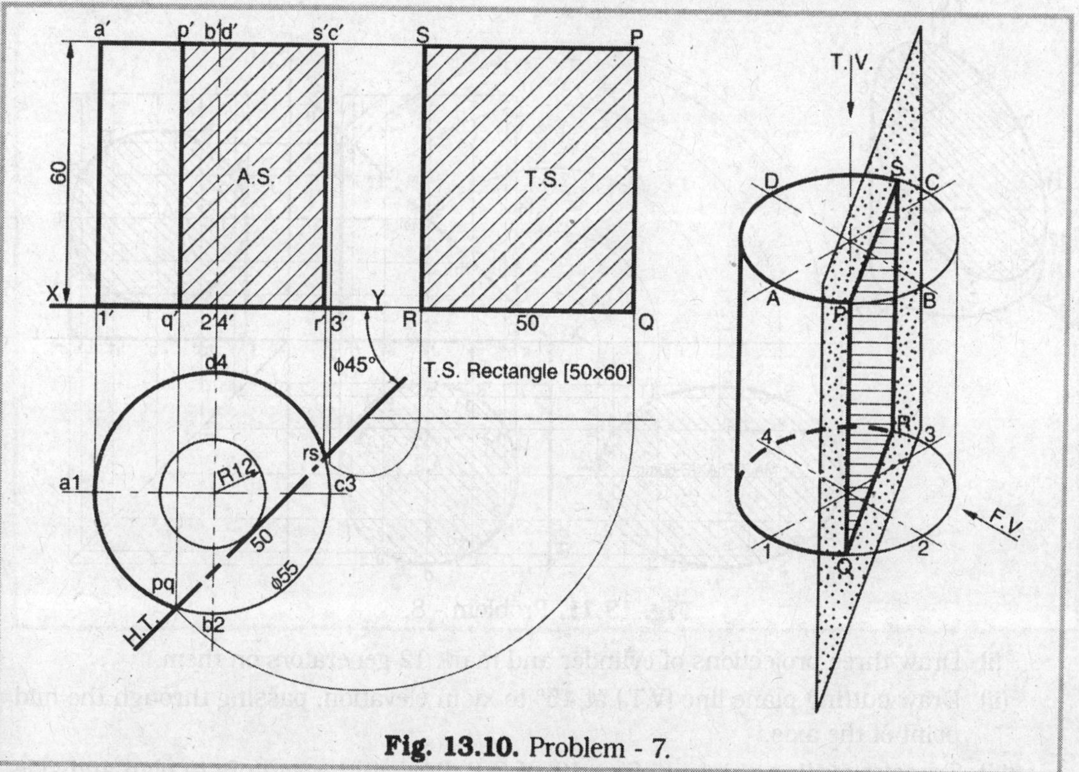

Fig. 13.10. Problem - 7.

(i) First draw plan and elevation of the cylinder.

(ii) Draw a circle of 12 mm radius, with point view plan of axis as the centre.

(iii) Now draw cutting plane line (H.T.) tangent to 12 mm radius circle and at 45° angle with xy line.

(iv) Cutting plane line is cutting two bases of cylinder at [2 × 2] totally 4 points p, q, r and s. Transfer these points on elevation on bases and get p', q', r' and s'.

(v) Draw also true shape of section, as usual.

[OR]

If the required true shape of section is a rectangle of 50 × 60 mm size then draw cutting plane line (H.T.) at 45° angle in plan such that distance between pq and rs is adjusted equal to 50 mm. Measure the distance of C.P. from the centre.

Problem 8 : A cylinder, diameter of base 43 mm and height 58 mm, is resting on H.P. on its base. It is cut by A.I.P. inclined at 45° to H.P. bisecting the axis. Draw three projections with section and find the true shape of section.

[OR]

Problem 8 : A cylinder, diameter of base 43 mm and height 58 mm is resting on H.P. on its base. It is cut by A.I.P. in such a way that the true shape of section is an ellipse with major axis 60 mm and minor axis 43 mm. Find the inclination of A.I.P. with H.P. and draw three projections.

For solution see Fig. 13.11 and follow the procedure, as given below :

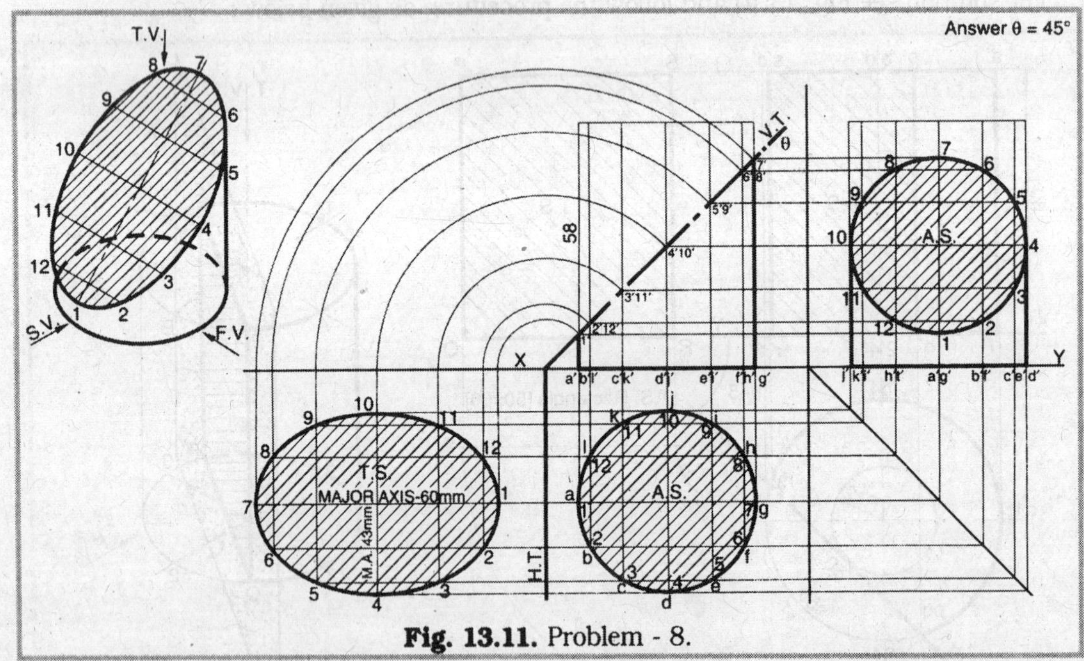

Fig. 13.11. Problem - 8.

(i) Draw three projections of cylinder and mark 12 generators on them.

(ii) Draw cutting plane line (V.T.) at 45° to xy in elevation, passing through the mid point of the axis.

(iii) Transfer cutting points 1', 2',....12' of C.P. line with generators to plan and side view. Join points 1, 2, 3....12 in plan and points 1, 2, 3,......12 in side view to get apparent sections. To get the true shape of section follow the method shown in the figure.

[OR]

If the required true shape of section is an ellipse with major axis 60 mm and minor axis 43 mm, draw cutting plane line (V.T.) such that length 1'7' becomes equal to 60 mm. Diameter of the cylinder will be the minor axis of ellipse.

Problem 9 : A hollow cylinder, outside diameter 60 mm, inside diameter 50 mm and height 80 mm, is resting on H.P. on its base. Determine the shape and size of the maximum possible size of the ellipse it can accomodate.

For solution see Fig. 13.12 and follow the procedure as given below.

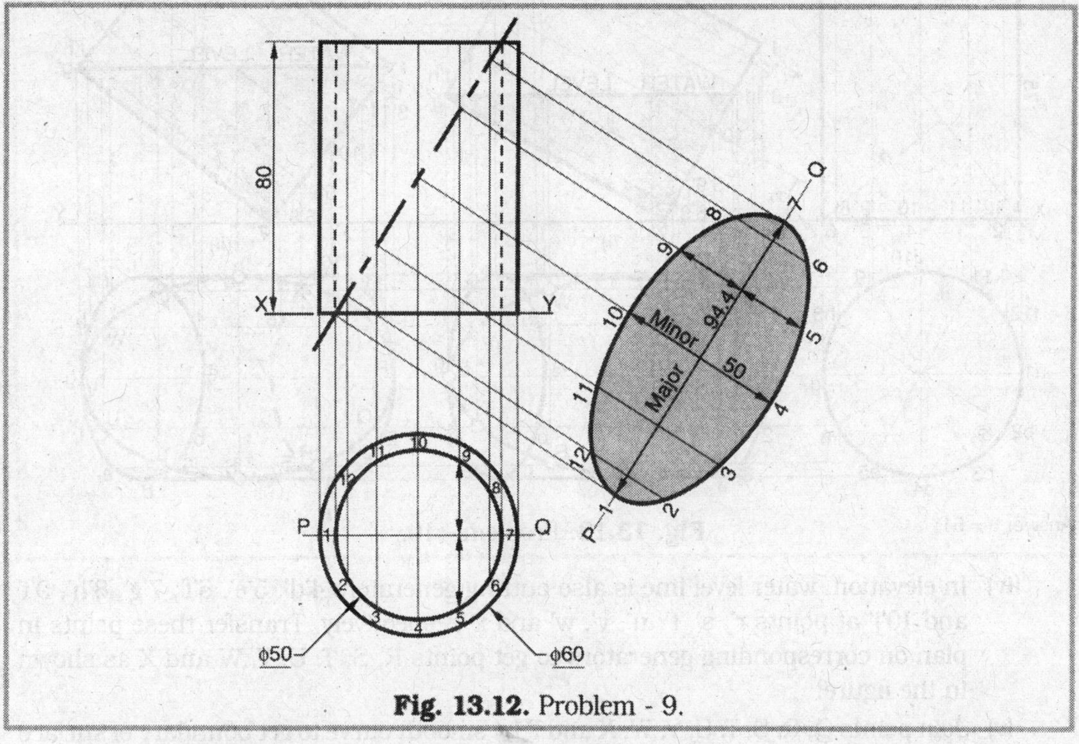

Fig. 13.12. Problem - 9.

(i) First draw plan and elevation of the hollow cylinder for the given position.

(ii) Draw 12 generators in plan and in elevation only on inside cylindrical surface.

(iii) Draw cutting plane line as shown in figure and get corresponding true shape of section, ellipse. Major axis 94.4mm, minor axis 50 mm.

Problem 10 : A transparent cylindrical container, diameter of base 60 mm and height 75 mm, is full of water. (1) It is tilted by 60° from vertical so that some water from it is drained out. (2) It is tilted by θ° from vertical so that half the water is drained out, find angle θ°.

Draw the surface of water in plan and in elevation.

(i) Draw first plan and elevation of cylinder, when it is resting on H.P. on point of its periphery of base with axis inclined at 30° with H.P. or 60° with the vertical and parallel to V.P.

(ii) Since surface of water always remains parallel to horizontal and as water is allowed to drain from the top of cylinder, in the given position water level line will be parallel to xy line and passing through the lowest point of top base i.e. g'. Draw horizontal line passing through g' in elevation. Consider this as cutting plane line.

(iii) Mark points p' and q' in elevation where water level line cuts the line for base. Transfer these two points in plan on elliptical base and get P and Q. Join PQ by straight line.

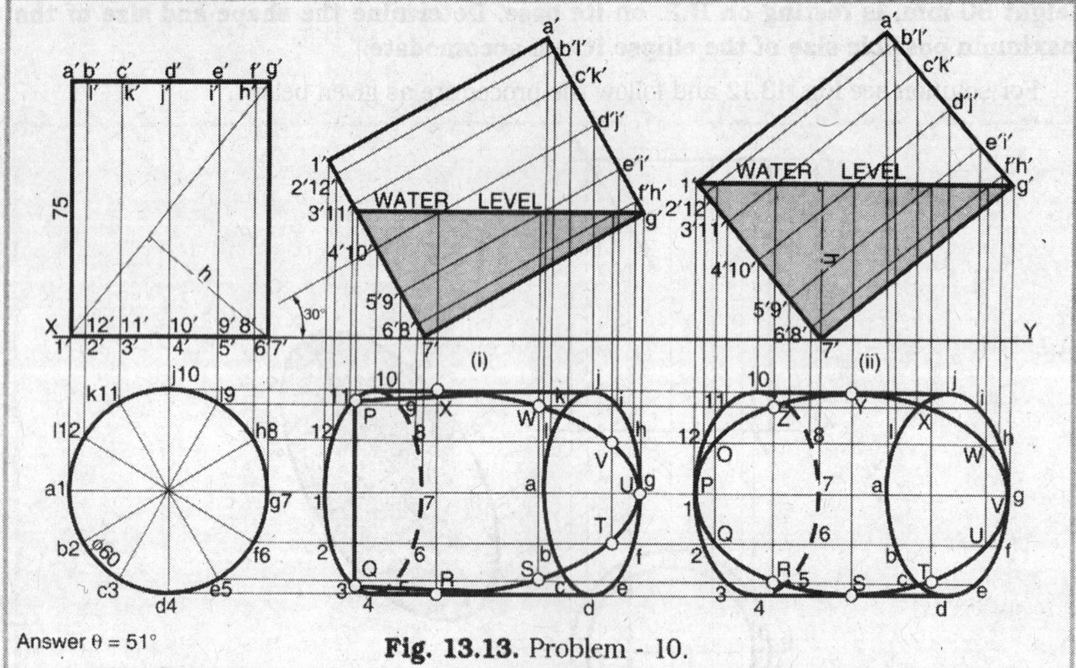

Fig. 13.13. Problem - 10.

(iv) In elevation, water level line is also cutting generators 4'd', 5'e', 6'f', 7'g', 8'h', 9'i' and 10'j' at points r', s', t', u', v', w' and x' respectively. Transfer these points in plan on corresponding generators to get points R, S, T, U, V, W and X as shown in the figure.

(v) Join points Q, R, S, T, U, V, W, X and P by smooth curve to get boundary of surface of water in plan. This will show the true shape.

(vi) For half the water to drain, tilt the cylinder to such an extent so that 1'g' becomes horizontal.

Problem 11 : A cylinder, diameter of base 70 mm and height 90 mm, is resting on H.P. on one of its generators with the axis remaining parallel to V.P. It is cut by an A.V.P. inclined to V.P. by 30° and passing through a point on the axis 50 mm from its one end. Draw the projections with the section.

For solution see Fig. 13.14 and follow the procedure, as given below :

(i) First draw plan, elevation and S.V. of cylinder in the given position.

(ii) Draw C.P. line (H.T.) on plan at 30° to xy and passing through a point of axis 50 mm from its one end.

(iii) C.P. line is cutting two bases at (2 + 2) = 4 points, p, y and u, t, as shown in the figure.

(iv) Transfer these points (p, y) and (u, t) from plan to S.V. (circle) and then to elevation (rectangle) by horizontal lines.

(v) Cutting plane line (H.T.) is also cutting generators 3c, 4d, 5e, 9i, 10j and 11k at points q, r, s, v, w and x respectively. Transfer these points on corresponding generators in elevation to get q', r', s', v', w' and x' as shown. Join these points by smooth curve and get apparent section. True shape can be drawn by the usual method.

8. Sections of Pyramids :

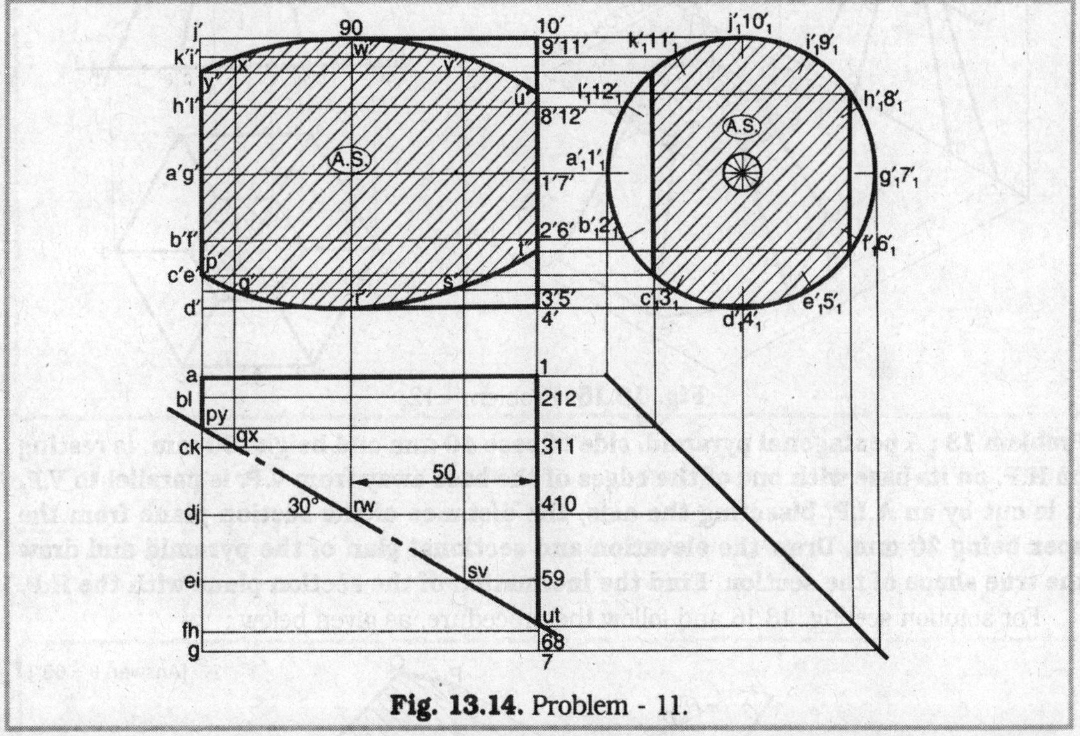

Fig. 13.14. Problem - 11.

Problem 12 : A hexagonal pyramid, side of base 30 mm and height 63 mm, is resting on H.P. on its base with two edges of base parallel to V.P. It is cut by horizontal section plane passing through a point on axis 27 mm from the apex. Draw projections with the section.

For solution see Fig. 13.15 and follow the procedure, as given below :

(i) First draw plan and elevation of the pyramid.

(ii) Draw C.P. line (V.T.) parallel to xy on elevation 27 mm below the apex.

(iii) C.P. line is cutting slant edges o'a', o'b', o'c', o'd', o'e' and o'f at points 1', 2', 3', 4', 5' and 6' respectively.

(iv) Transfer these points to plan by projectors on corresponding slant edges and get points 1, 2, 3, 4, 5, and 6. Join them by straight lines to get the section in plan which will be the true shape also.

Note that the section is also a regular hexagon. Three dimensional figure helps in understanding the problem.

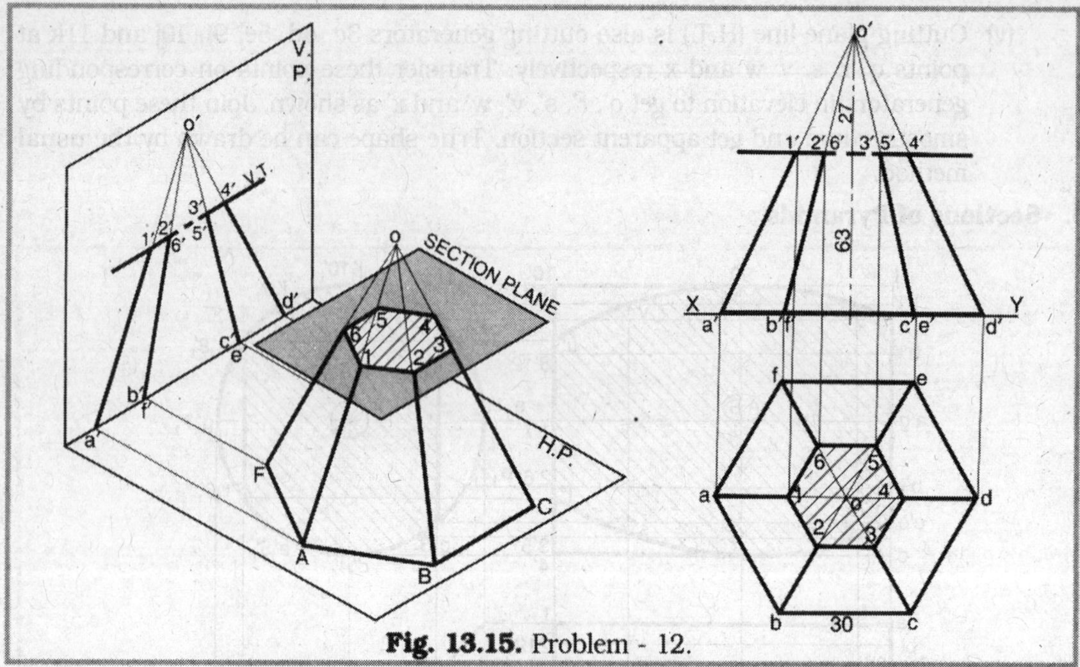

Fig. 13.15. Problem - 12.

Problem 13 : A pentagonal pyramid, side of base 40 mm and height 80 mm, is resting on H.P. on its base with one of the edges of the base away from V.P. is parallel to V.P. It is cut by an A.I.P. bisecting the axis, the distance of the section plane from the apex being 20 mm. Draw the elevation and sectional plan of the pyramid and draw the true shape of the section. Find the inclination of the section plane with the H.P.

For solution see Fig. 13.16 and follow the procedure, as given below :

[Answer: θ = 60°]

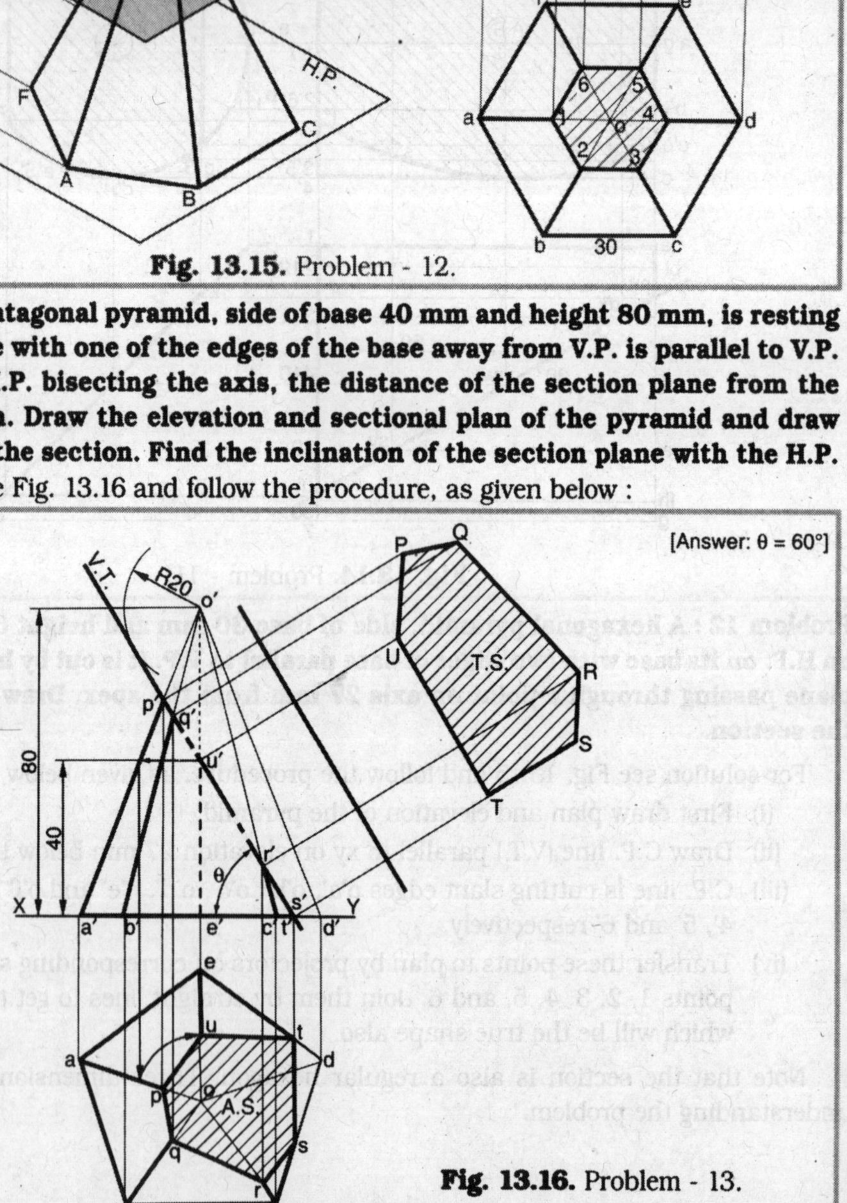

Fig. 13.16. Problem - 13.

(i) First draw plan and elevation of pyramid.

(ii) Draw and arc of 20 mm radius circle with elevation of apex O' as the centre. From the mid point of the axis draw tangent to arc of circle, already drawn. This tangent will represent the cutting plane line (V.T.).

(iii) C.P. line is cutting slant edges o'a', o'b', o'e', o'c' and is also cutting two sides of base c'd' and d'e'. Transfer 6 cutting points to plan on the corresponding edges by projectors. Join them to get apparent section in the plan.

(iv) Draw the true shape of section by the usual method.

Problem 14 : A hexagonal pyramid, side of base 33 mm and height 66 mm, is resting on H.P. on its base with two sides of base perpendicular to V.P. It is cut by an A.I.P., inclined to H.P. by 45°, passing through a point 25 mm from the apex on the axis. Draw three projections with sections and also draw the true shape of the section.

Solution is self explanatory. See Fig. 13.17.

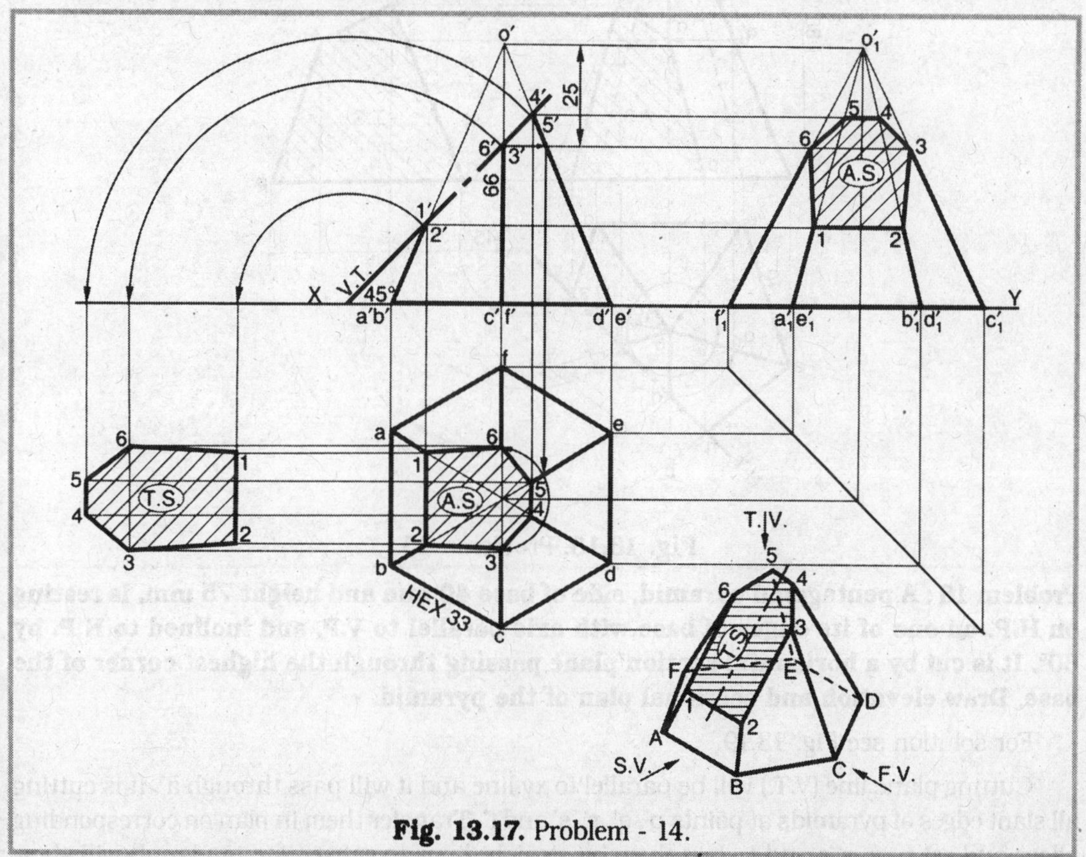

Fig. 13.17. Problem - 14.

Problem 15 : A pentagonal pyramid, side of base 40 mm and height 80 mm, is resting on H.P. on its base with one of the edges of the base, nearer to V.P., is parallel to V.P. It is cut by A.V.P. inclined to V.P. by 45°. Cutting plane remains 12 mm away from the axis. Draw sectional elevation, plan and the true shape of the section.

For solution see Fig. 13.18 and follow the procedure, as given below :

(i) Draw plan and elevation of pyramid in the given position.

(ii) Draw circle of 12 mm radius with plan of axis O as the centre. Draw cutting plane line in plan tangent to circle already drawn and at 45° to the xy line.

(iii) C.P. line is cutting 4 edges ab, ob, oc and cd at the points p,q,r and s respectively.

(iv) Transfer p,r and s to elevation by projectors on corresponding edges and get p', r' ans s'. To transfer q from ob (vertical) to ob' (also vertical), first transfer it to oa and get q_1 by arc of the circle of O centre. Transfer q_1 from oa to o'a' by projector and get q'_1. Draw horizontal line from q'_1 to o'b' to get q' on it. Join p', q', r' and s' by straight lines to get apparent section in elevation.

(v) Get the true shape by the usual method.

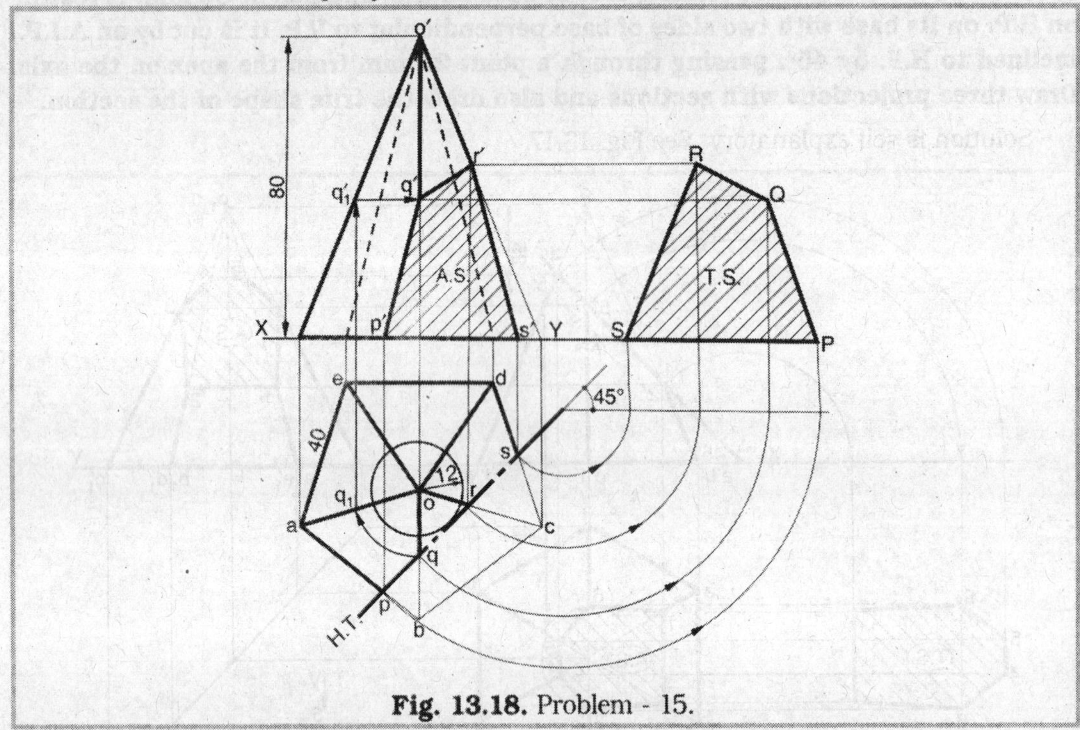

Fig. 13.18. Problem - 15.

Problem 16 : A pentagonal pyramid, side of base 40 mm and height 75 mm, is resting on H.P. on one of its edges of base with axis parallel to V.P. and inclined to H.P. by 60°. It is cut by a horizontal section plane passing through the highest corner of the base. Draw elevation and sectional plan of the pyramid.

For solution see Fig. 13.19.

Cutting plane line (V.T.) will be parallel to xy line and it will pass through a'. It is cutting all slant edges of pyramids at points p', q', r', s' and t'. Transfer them in plan on corresponding edges and get p, q, r, s and t. Join them by straight lines to get section in plan. It will show the true shape of section also, since cutting plane is horizontal.

Problem 17 : A tetrahedron of 70 mm side is resting on one of the faces on the H.P. having a side of that face perpendicular to the V.P. It is cut by an A.I.P. so that the true shape of the section is a square. Set the required cutting plane and draw the sectional F.V. and T.V. and project the true shape of section. Find out also the inclination of the cutting plane with H.P.

For solution see Fig. 13.20 and follow the procedure, as given below :

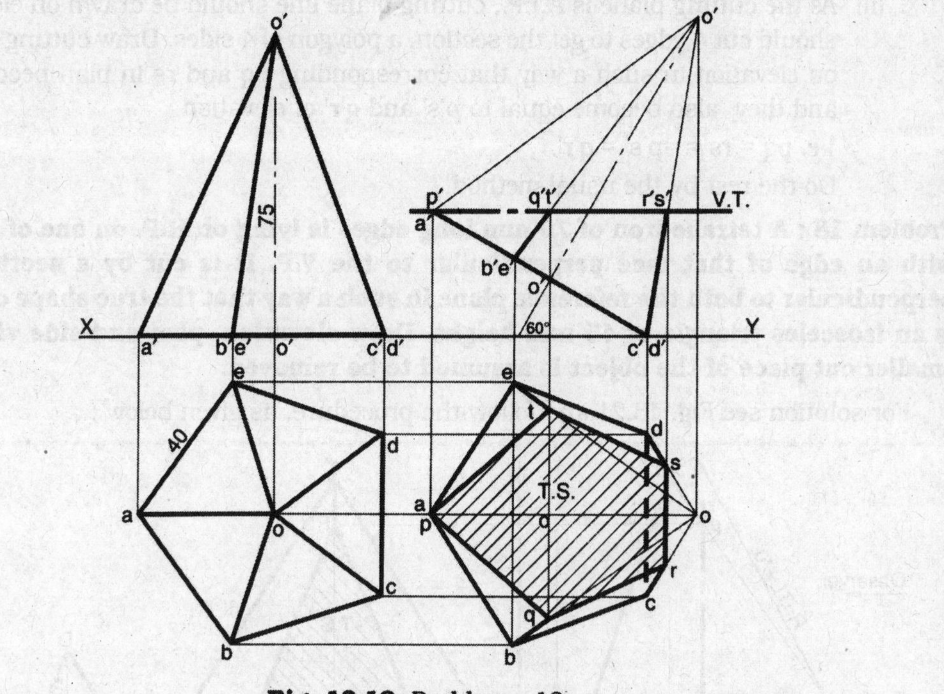

Fig. 13.19. Problem - 16.

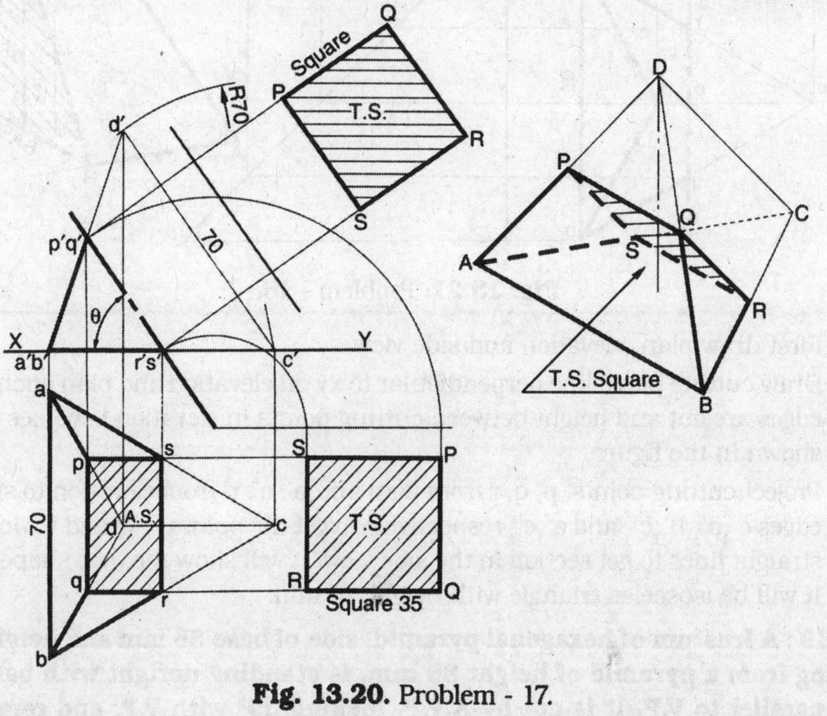

Fig. 13.20. Problem - 17.

(i) First draw plan and elevation of tetradron.

(ii) As the cutting plane is A.I.P., cutting plane line should be drawn on elevation. It should cut 4 edges to get the section, a polygon of 4 sides. Draw cutting plane line on elevation in such a way that corresponding pq and rs in plan become equal and they also become equal to p's' and q'r' of elevation.

i.e. pq = rs = =p's' = q'r'.

Do the rest by the usual method.

Problem 18 : A tetrahedron of 70 mm long edges is lying on H.P. on one of its faces, with an edge of that face perpendicular to the V.P. It is cut by a section plane perpendicular to both the reference plane in such a way that the true shape of section is an isosceles triangle of 45 mm height. Draw elevation, plan and side view when smaller cut piece of the object is assumed to be removed.

For solution see Fig. 13.21 and follow the procedure, as given below :

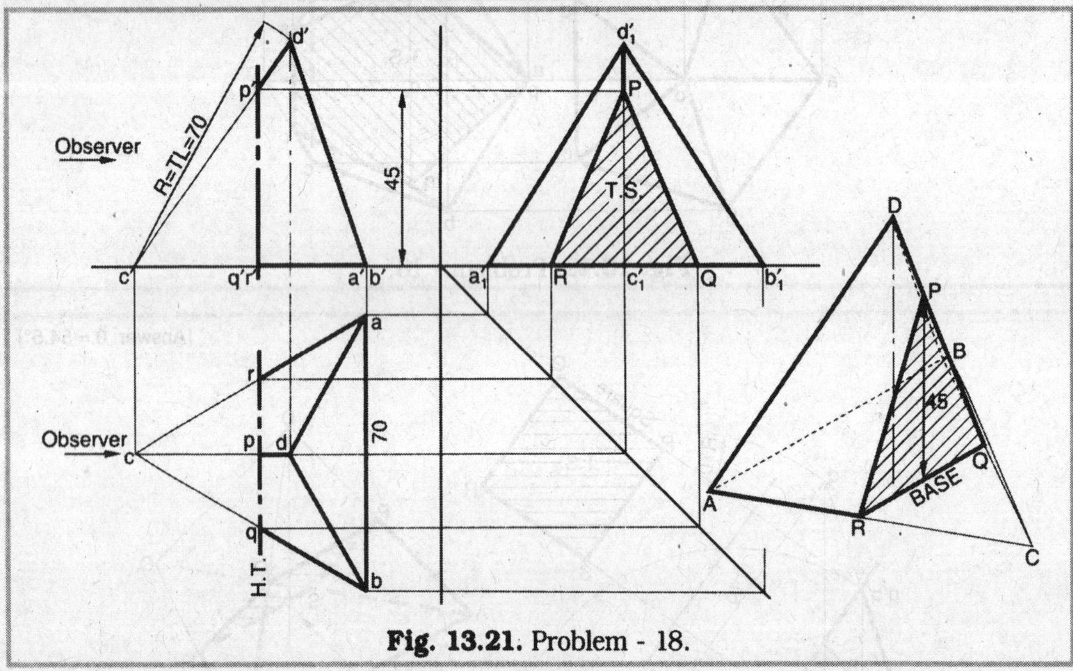

Fig. 13.21. Problem - 18.

(i) First draw plan, elevation and side view.

(ii) Draw cutting plane line perpendicular to xy on elevation and plan such that three edges are cut and height between cutting points in elevation becomes 45 mm, as shown in the figure.

(iii) Project cutting points, p, q, r from plan and p', q', r' from elevation to side view on edges $c'_1d'_1$, $b'_1c'_1$ and $a'_1c'_1$ respectively to get the points P, Q and R. Join them by straight lines to get section in the side view. It will show the true shape of section. It will be isosceles triangle with height 45 mm.

Problem 19 : A frustum of hexagonal pyramid, side of base 35 mm and height 50 mm, is a cutting from a pyramid of height 85 mm, is standing upright with base on H.P. and axis parallel to V.P. It is cut by A.V.P. making 55° with V.P. and remaining 20

mm away from the axis. Draw plan, sectional elevation, sectional side view and the true shape of section. Keep side of base parallel to V.P.

For solution see Fig. 13.22 and follow the procedure, as given below :

(i) First draw plan, elevation and side view of complete pyramid of base 35 mm and height 85 mm. Cut from it frustum of 50 mm height in elevation and make necessary changes in plan and in the side view.

(ii) Draw cutting plane line (H.T.) in plan at 55° to xy and 20 mm away from the axis.

(iii) It is cutting edges a1, b2, bc and af at points p, q, r and s respectively in plan.

(iv) Transfer p, q, r and s to elevation and side view on corresponding edges and get points p', q', r' and s' and p'_1, q'_1, r'_1 and s'_1 respectively. Join them by straight lines to get apparent sections.

(v) Follow the procedure shown in the figure to get true shape of section.

Three dimensional drawing gives an idea about how cutting takes place.

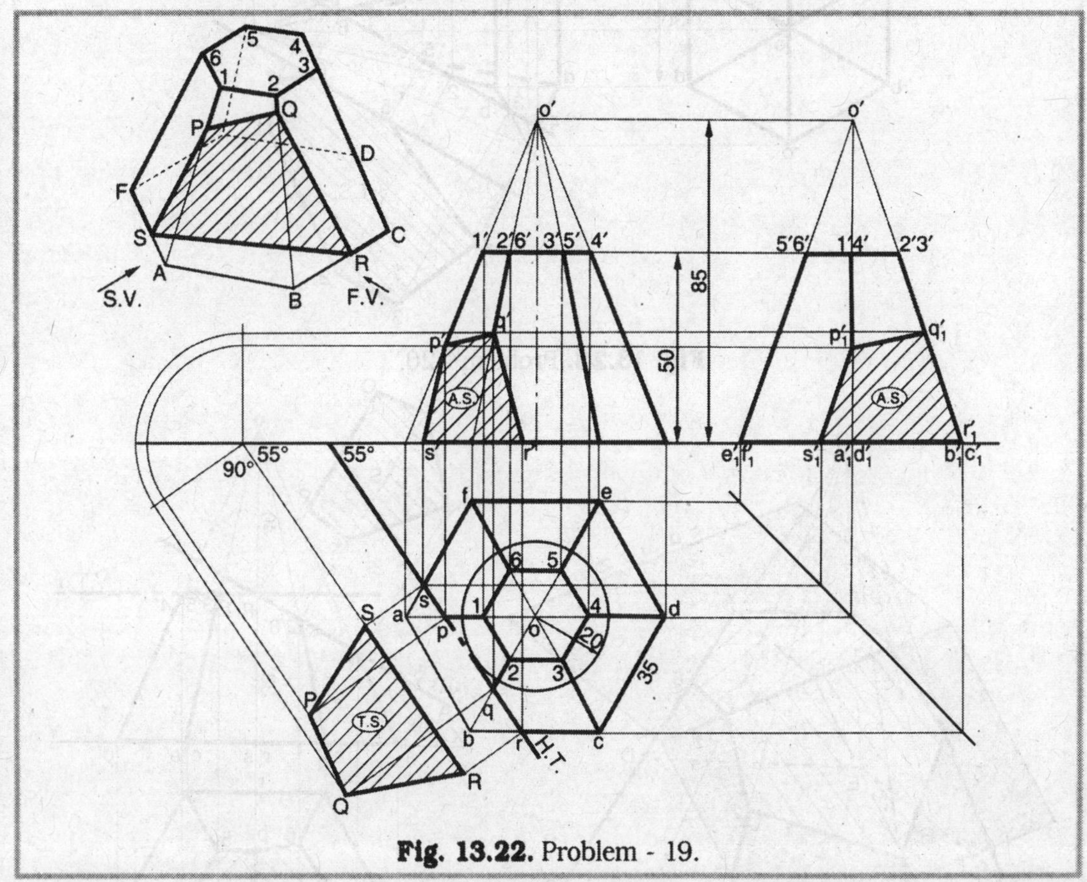

Fig. 13.22. Problem - 19.

Problem 20 : A hexagonal pyramid is resting on H.P. on one of its triangular faces with axis remaining parallel to V.P. It is cut by A.V.P. making 30° with V.P. passing through a point on the axis 33 mm from the apex. Draw plan, sectional elevation and the true shape of section. Take side of base 30 mm and height 75 mm.

For solution see Fig. 13.23. Solution is very clear from the figure. Take 33 mm dimension where axis is in true length.

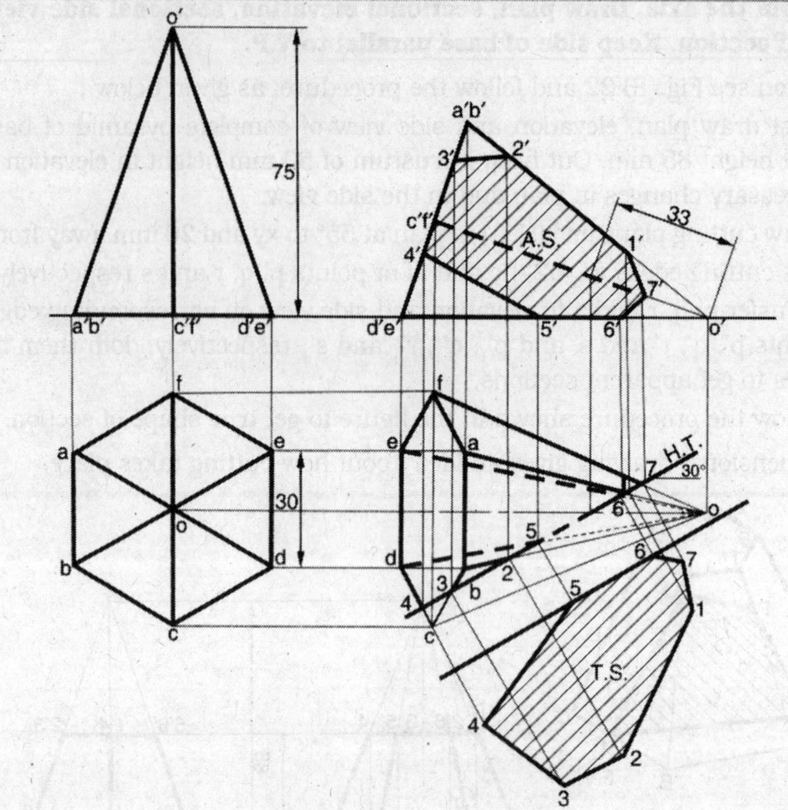

Fig. 13.23. Problem - 20.

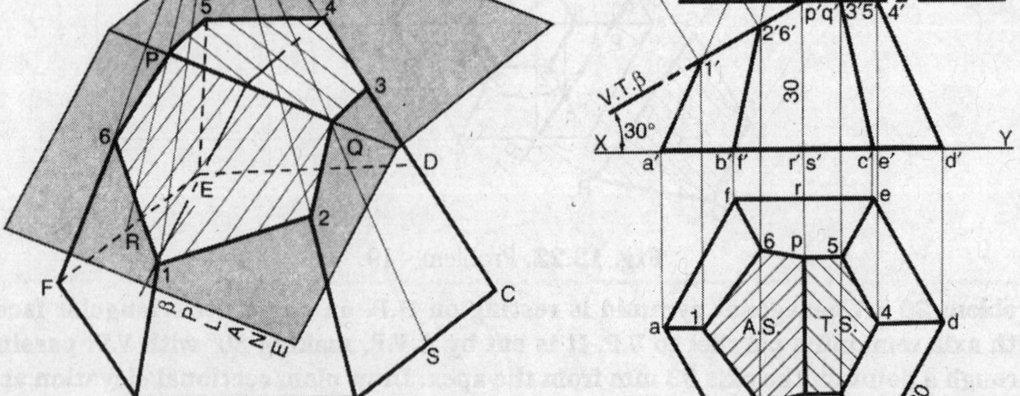

Fig. 13.24. Problem - 21.

Problem 21 : A hexagonal pyramid, side of base 30 mm and height 70 mm, is resting on H.P. on its base with two sides of base parallel to V.P. It is cut by A.I.P. inclined at 30° to H.P. and also by horizontal cutting plane, both passing through a point on the axis 40 mm from the apex. Draw elevation, sectional plan and the true shape of section.

For solution see Fig. 13.24 and follow the procedure, as given below :

(i) First draw plan and elevation of pyramid.

(ii) Draw two cutting planes alpha and beta parallel to xy and at 30° to xy respectively on elevation.

(iii) Take lines OS and OR on faces OBC and OEF such that its elevation o's' and o'r' pass through a point where two planes meet on the axis.

(iv) With alpha cutting plane parallel to xy, in plan we get section having sides parallel to plan of base i.e. pq345p is similar to rscder. This section will show T.S.

(v) For beta cutting plane get points 1,2 and 6 in plan, as usual. Join 12, 2q, p6 and 61 by straight lines to get apparent section.

(vi) Draw the true shape of section by the usual method.

Three dimensional drawing is given for clear understanding of the problem.

9. Sections of Cone :

Problems of section of cone can be solved by two methods -

(i) Generator Method

(ii) Circle Method or Section Plane Method.

9. (i) Generator Method :

In this method 8 or 12 generators are taken equally spaced on the curved surface of the cone. Cutting points of cutting planes with generators are transferred from one view to other views with the help of projectors. We shall solve different problems of cones by generator method.

Problem 22 : A cone, diameter of base 60 mm and height 60 mm, is resting on H.P,. on its base. It is cut by A.I.P. inclined to H.P. by 30°. The H.T. of the cutting plane remains 10 mm away from the periphery of the base. Draw elevation, sectional plan, sectional side view and the true shape of section.

For solution see Fig. 13.25 and follow the procedure as given below :

Visualise section from three dimensional drawing.

(i) First draw plan, elevation and side view and mark 12 generators on them.

(ii) Draw cutting plane line (V.T.) on elevation, 30° inclined to xy such that H.T. of cutting plane remains 10 mm away from the circle, the plan of base.

(iii) Cutting plane line (V.T.) is cutting all generators o'a', o'b', o'c',......o'l' at points 1', 2', 3',....12' respectively. Transfer these points on corresponding generators in plan and in side view by vertical projectors and horizontal projectors respectively.

(iv) Join 1, 2, 3,....12 and 1 and 1', 2', 3',.....12' and 1' by smooth curve to get apparent sections in plan and side view respectively.

(v) Project the true shape by usual method.

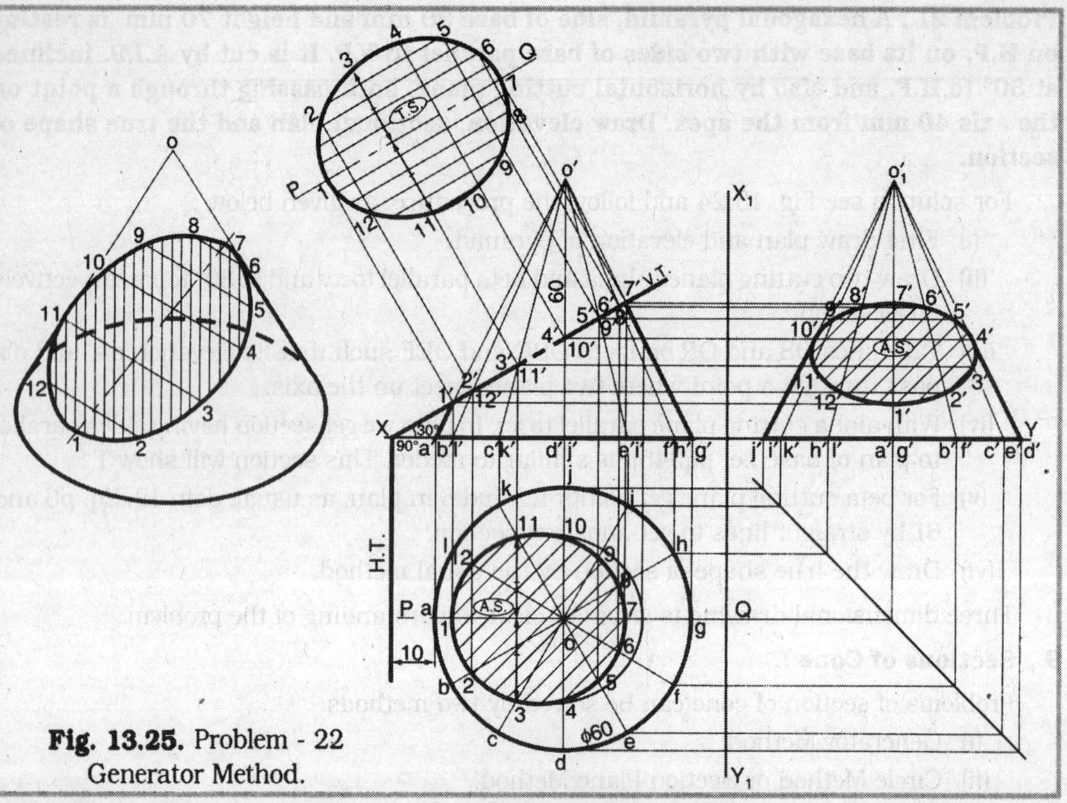

Fig. 13.25. Problem - 22
Generator Method.

Problem 23 : A cone, diameter of base 60 mm and height 75 mm, is resting on H.P. on one of its generators with axis parallel to V.P. It is cut by horizontal section plane passing through a point on the axis 50 mm away from the apex. Draw elevation and sectional plan.

For solution see Fig. 13.26 and follow the procedure, as given below :

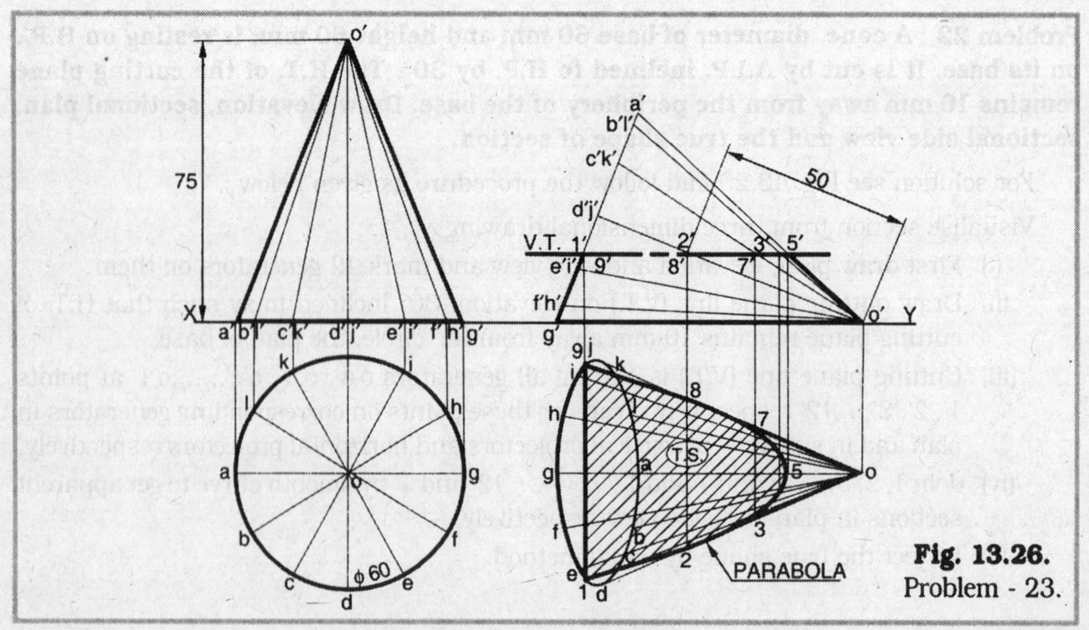

Fig. 13.26. Problem - 23.

(i) First draw plan and elevation of cone, with 12 generators marked on them, for the given condition.

(ii) Draw cutting plane line (V.T.) parallel to xy and passing through a point on axis 50 mm from apex.

(iii) Cutting plane line is cutting base at two points and is cutting generators o'a', o'b', o'c', o'd', o'j', o'k' and o'1' at 9 points. Transfer these nine points to plan on corresponding base and generators to get nine points. Join them, as shown in the figure to get section in plan. This section will show the true shape of section. It will be parabola.

Problem 24 : A cone, diameter of base 70 mm and height 80 mm, is resting on H.P. on its base. It is cut by a cutting plane perpendicular to V.P. and H.P. C.P. remains 15 mm away from the axis. Draw plan, elevation and sectional side view. State the nature of section.

For solution see Fig. 13.27 and follow the procedure, as given below :

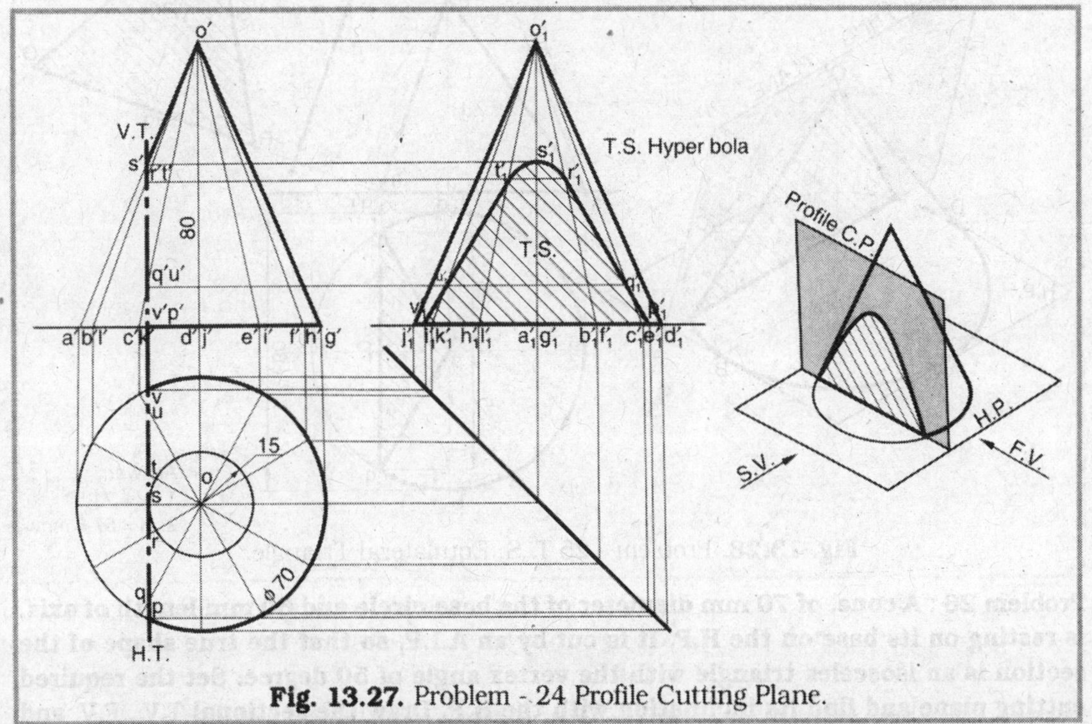

Fig. 13.27. Problem - 24 Profile Cutting Plane.

(i) Draw plan, elevation and side view with 12 generators marked on them.

(ii) Draw a circle of 15 mm radius in plan with point view plan of the axis as the centre. Now draw cutting plane line both in plan and elevation perpendicular to xy and tangent to the circle drawn.

(iii) Transfer cutting points of cutting plane with base and generators of cone from plan and elevation to side view on the corresponding lines to get section in side view. Section in side view will show the true shape. It will give a **Hyperbola** curve.

Problem 25 : A vertical cone, diameter of base 80 mm, is resting on its base on the H.P. It is cut by an A.I.P. so that the true shape of the section is an equilateral triangle

with 70 mm side. Determine the length of the axis of the cone and draw projections and the true shape of the section.

For solution see Fig. 13.28.

As the true shape of section of cone is an equilateral triangle of 70 mm side following things must exists.

1. Length of generator of cone should be 70 mm.
2. Cutting plane must pass through the apex and should cut the base of cone and the length between two cutting points of the base should be 70 mm.

Satisfy the above conditions and get projections and the true shape of section by the usual method. Three dimensional drawing helps in visualising and understanding the problem.

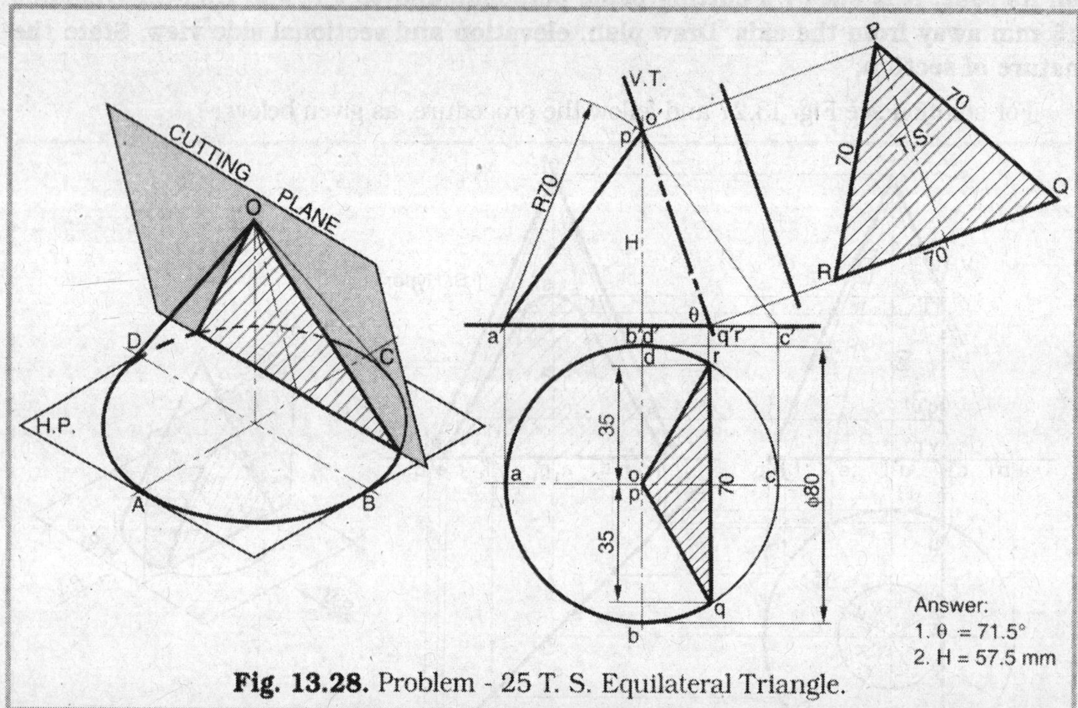

Fig. 13.28. Problem - 25 T. S. Equilateral Triangle.

Problem 26 : A cone, of 70 mm diameter of the base circle and 60 mm length of axis, is resting on its base on the H.P. It is cut by an A.I.P. so that the true shape of the section is an isosceles triangle with the vertex angle of 50 degree. Set the required cutting plane and find its inclination with the H.P. Draw the sectional T.V., F.V. and project the true shape of the section.

For solution see Fig. 13.29 and follow the procedure, as given below :

As the true shape of section of cone is an isosceles triangle with the vertex angle of 50 degree, following things must exist.

1. Section or cutting plane must pass through the apex.
2. If an isosceles triangle is drawn with two equal sides equal to generator length and angle between them equal to 50° then whatever H, the length of base of isosceles triangle, that you get must be the distance between cutting points of cutting plane with the base of cone.

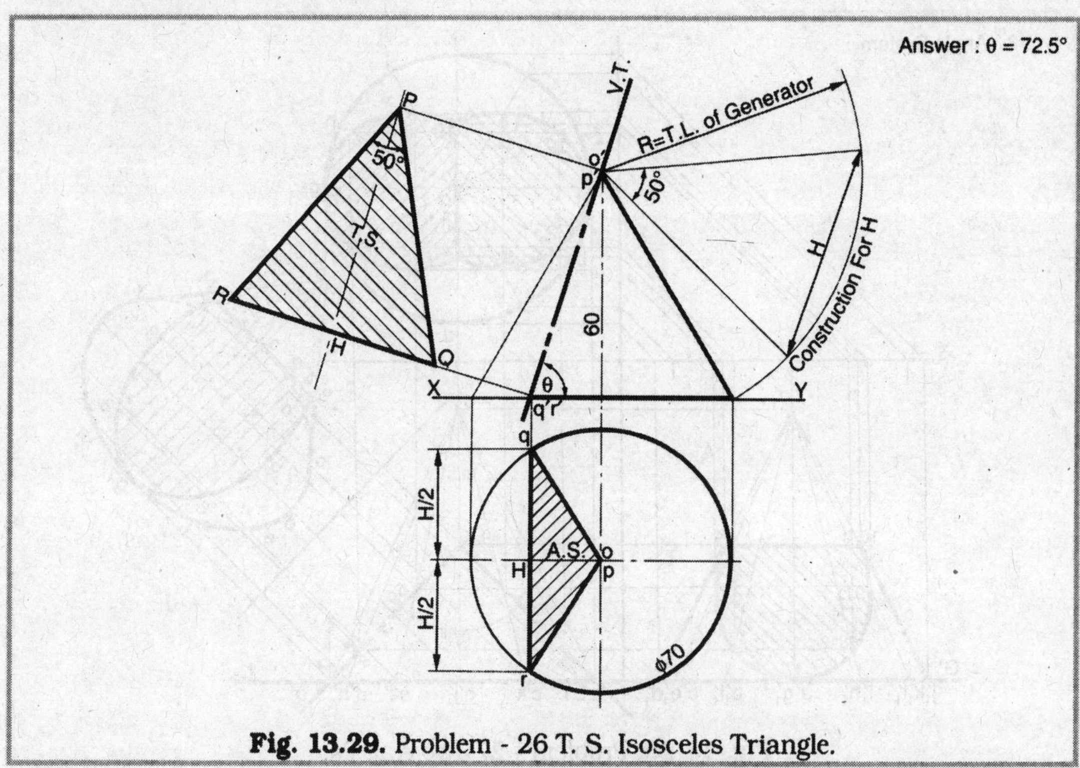

Fig. 13.29. Problem - 26 T. S. Isosceles Triangle.

Draw plan and elevation of cone. Draw cutting plane line on elevation satisfying above given conditions. The rest is very clear from the figure.

Problem 27 : A cone, of 70 mm diameter of the base circle and 70 mm length of axis, is resting on its base on the ground. It is cut by two A.I. cutting planes which are mutually perpendicular and intersect each other on the axis of the cone. One of the cutting planes passes through the extreme left point of the base circle while the other passes through the extreme right point of the base circle. Draw the sectional T.V., F.V. and project the L.H.S.V. of the remaining solid. Draw also auxiliary plan on a plane parallel to one of the cutting planes.

For solution see Fig. 13.30 and follow the procedure, as given below :

(i) Draw plan, elevation and L.H.S.V. of cone and mark on them 12 generators.

(ii) Draw 2 cutting plane lines, on elevation, at right angle passing through extreme points (a', g') of the base.

(iii) Two cutting planes combine together cut all the 12 generators . Transfer these 12 cutting points from elevation to plan and to L.H.S.V. on the corresponding generators.

(iv) Line 4-10 separates two sections. Draw hatching lines in two different directions in two sections.

(v) True shape along with cone is projected as auxiliary plan on auxiliary inclined plane parallel to cutting plane by the usual method.

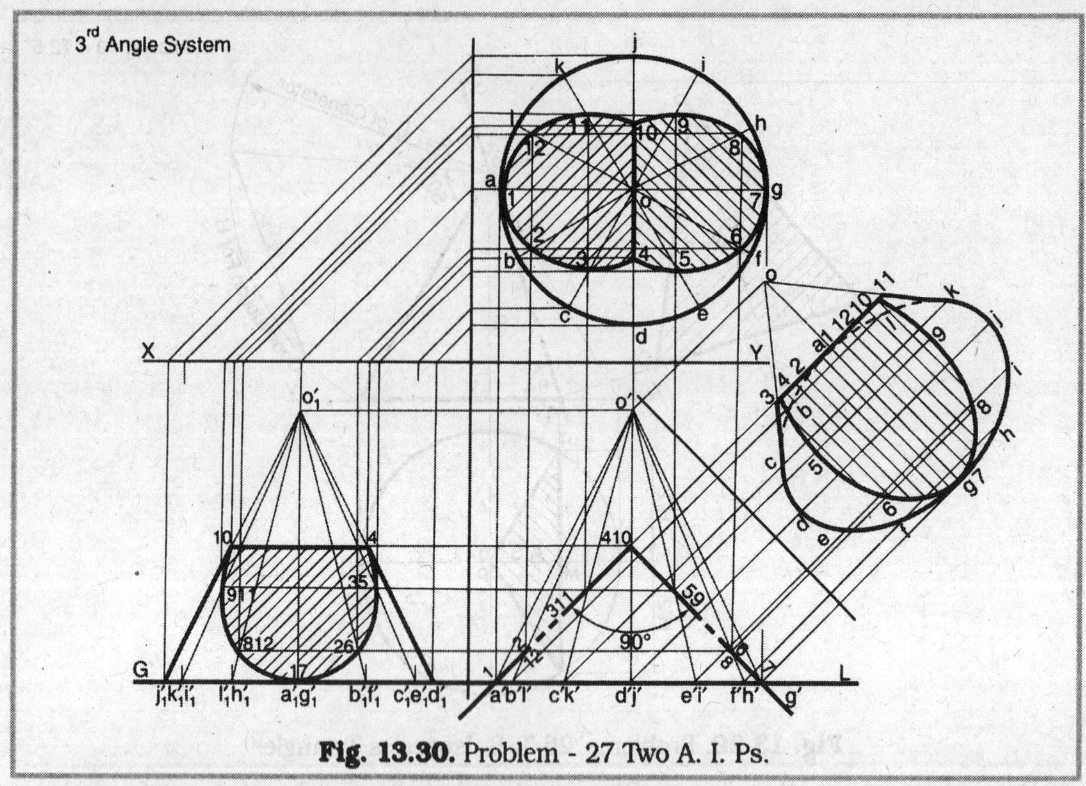

Fig. 13.30. Problem - 27 Two A. I. Ps.

9.2 Circle Method or Section Plane Method : In this method instead of using generators, section planes parallel to the base of cone are used. It is clear from the Fig.13.31 that when a cone is cut by a section plane (A) parallel to the base, we get a circle as section. Diameter of circle in plan will be equal to the length of cutting plane line between two extreme generators of elevation. Hence, if the actual cutting plane is on elevation, draw section plane lines on elevation at nearly equal distances and also draw corresponding circles in plan. Similarly, if the cutting plane is in plan then first draw many concentric circles in plan and also draw corresponding cutting plane lines in elevation. With the help of these circles of plan and corresponding cutting plane lines of elevation transfer cutting points by projectors from one view to another view.

Problem 28 : A cone, diameter of base 50 mm and height 57 mm, is resting on H.P., on its base. It is cut by a horizontal cutting plane A passing through a point of the axis 25 mm away from the apex. Draw elevation and sectional plan.

For solution see Fig. 13.31.

First draw plan and elevation. Draw horizontal C.P. line (A) (V.T.) 25 mm below the apex in elevation. Cutting plane line A is cutting extreme generators o'a' and o'c'. Transfer these two cutting points in plan on oa and oc. Draw a circle in plan with diameter equal to distance between two points on oa and oc, already obtained. This circle is a section and it shows the true shape of section.

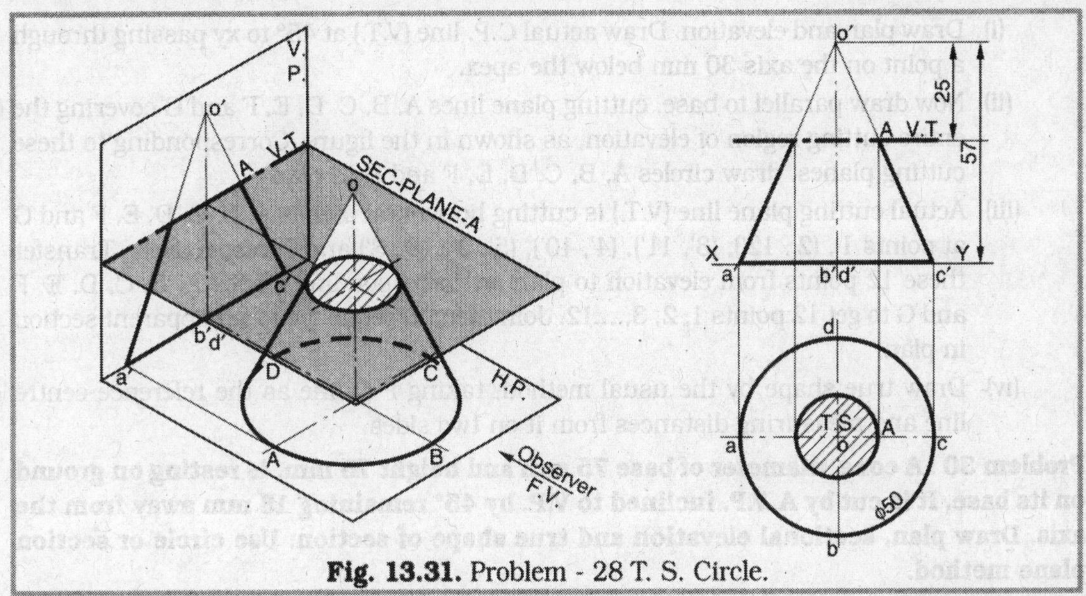

Fig. 13.31. Problem - 28 T. S. Circle.

Problem 29 : A cone, diameter of base 75 mm and height 75 mm, is resting on H.P. on its base. It is cut by an A.I.P. inclined to H.P. by 45° passing through a point on the axis 30 mm below the apex. Draw elevation, sectional plan and the true shape of section. Use circle or section plane method.

For solution see Fig. 13.32 and follow the procedure, as given below :

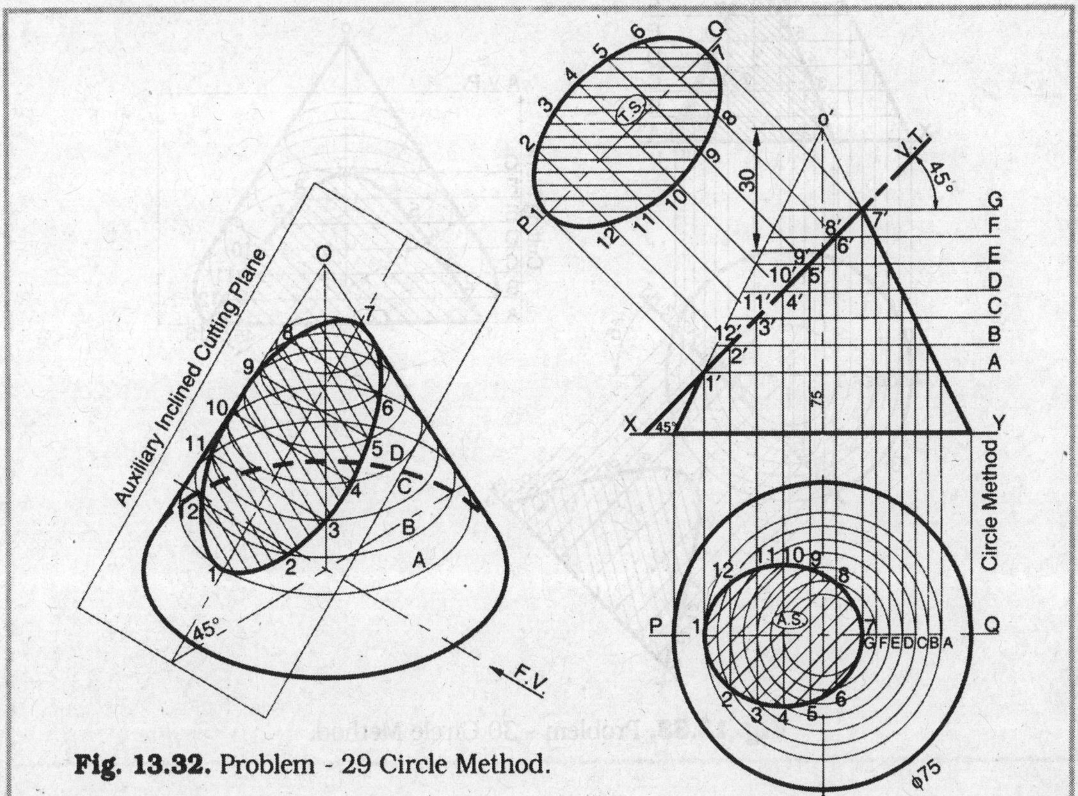

Fig. 13.32. Problem - 29 Circle Method.

(i) Draw plan and elevation. Draw actual C.P. line (V.T.) at 45° to xy passing through a point on the axis 30 mm below the apex.

(ii) Now draw parallel to base, cutting plane lines A, B, C, D, E, F and G covering the entire cutting region of elevation, as shown in the figure. Corresponding to these cutting planes, draw circles A, B, C, D, E, F and G in plan.

(iii) Actual cutting plane line (V.T.) is cutting horizontal planes A, B, C, D, E, F and G at points 1', (2', 12'), (3', 11'), (4', 10'), (5', 9'), (6', 8') and 7' respectively. Transfer these 12 points from elevation to plan on corresponding circles A, B, C, D, E, F and G to get 12 points 1, 2, 3,....12. Join them in sequence to get apparent section in plan.

(iv) Draw true shape by the usual method, taking P-Q line as the reference centre line and measuring distances from it on two sides.

Problem 30 : A cone, diameter of base 75 mm and height 75 mm, is resting on ground on its base. It is cut by A.V.P. inclined to V.P. by 45° remaining 15 mm away from the axis. Draw plan, sectional elevation and true shape of section. Use circle or section plane method.

For solution see Fig. 13.33 and follow the procedure, as given below :

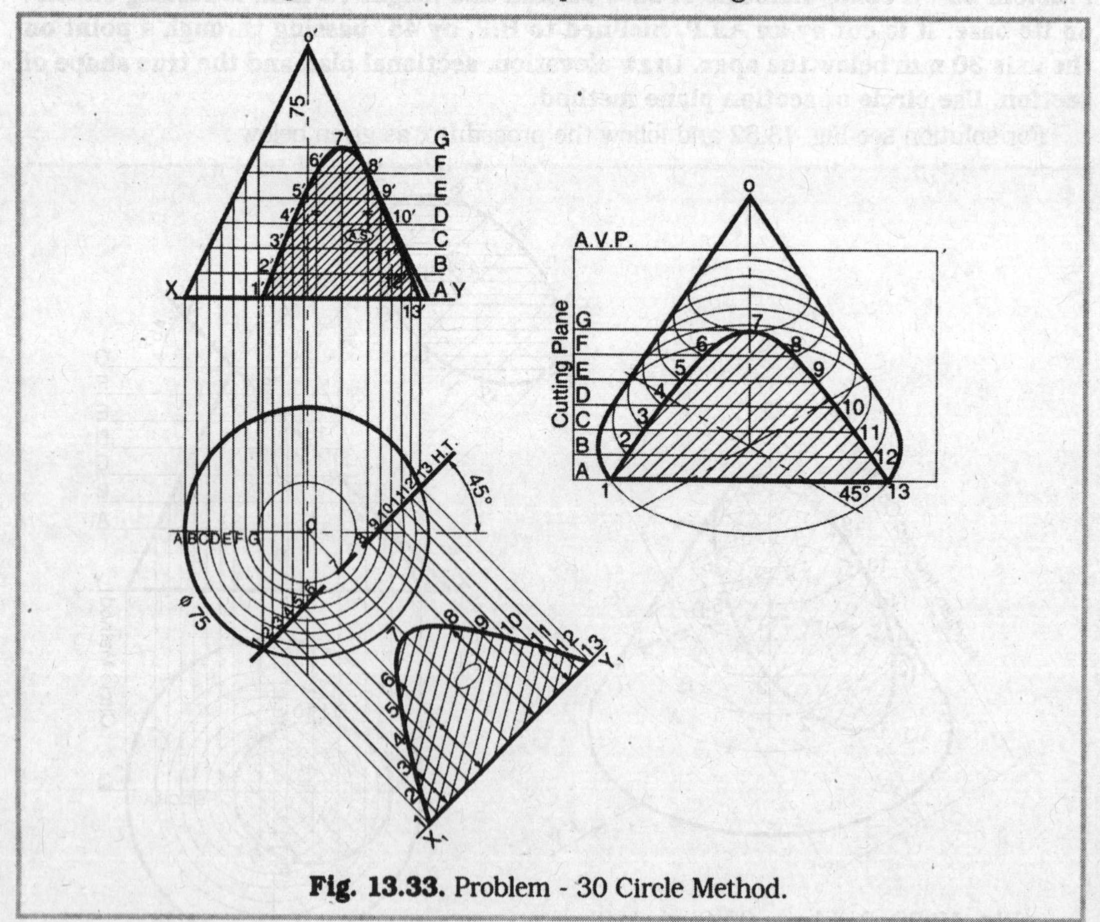

Fig. 13.33. Problem - 30 Circle Method.

(i) First draw plan and elevation of the cone. Draw C.P. line (H.T.) on plan at 45° to xy and tangent to the circle of 15 mm radius.

(ii) Between plan of base circle and 15 mm touching circle draw many circles (5 here) at nearly equal distances. Mark them as A, B, C, D, E, F and G. Draw, corresponding to these circles, horizontal cutting plane lines A, B, C, D, E, F and G in elevation.

(iii) Actual cutting plane line (H.T.) is cutting circles A, B, C, D, E, F and G at points (1, 13), (2, 12), (3, 11), (4, 10), (5, 9), (6, 8) and 7 respectively. Transfer these points from plan to elevation to corresponding horizontal lines A, B, C, D, E, F and G and get points 1', 2',.....13'. Join them in sequence to get apparent section in elevation.

(iv) Now to get true shape of section, draw X_1Y_1 ground line parallel to cutting plane line (H.T.) at suitable distance. Project perpendiculars to H.T. from plan points 1, 2,.....13 upto X_1Y_1 line and measure on them from X_1Y_1 distances of corresponding elevation points from G.L. line, e.g. see points 3, 6, 8 and 11.

Problem 31 : A cone, diameter of base 80 mm and height 80 mm, is resting on H.P. on its base. It is cut by an A.I.P. in such a way that the periphery of the true shape of the cut surface is a parabola of axis length 50 mm. (1) Draw the projections and true shape of section. (2) Draw the projections of the portion of the cone containing the apex when it is kept on the H.P. on its cut surface with the plan of the axis parallel to V.P.

For solution see Fig. 13.34 and follow the procedure, as given below :

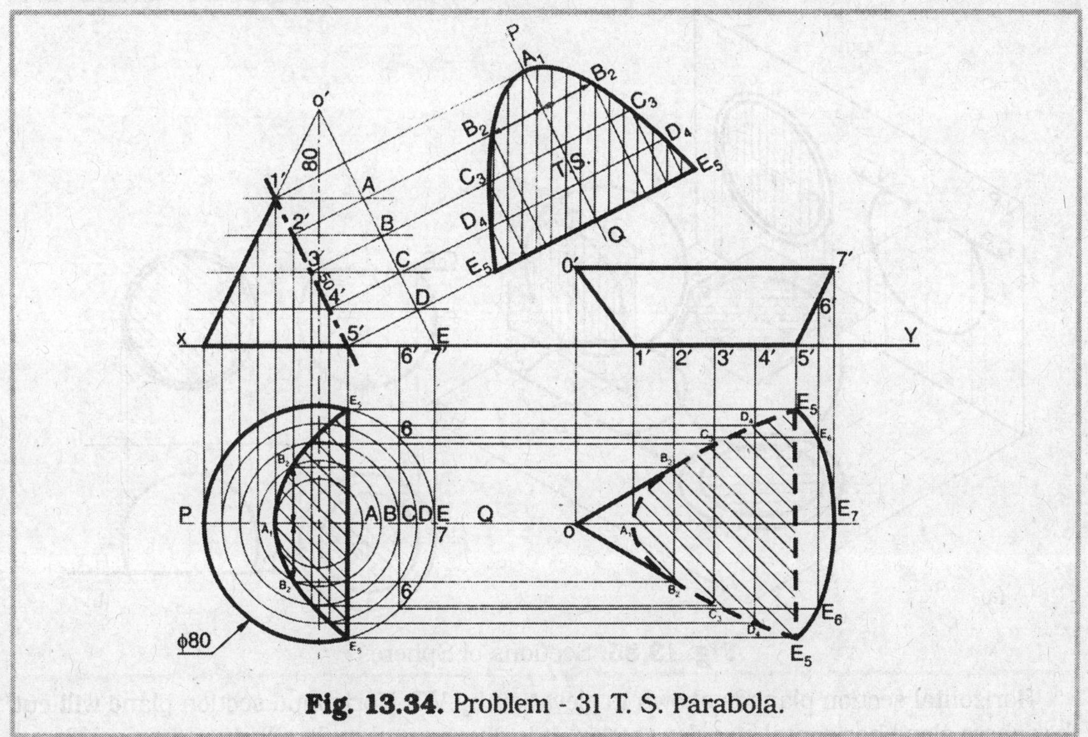

Fig. 13.34. Problem - 31 T. S. Parabola.

(i) First draw plan and elevation of the cone.

(ii) To get parabola in section cutting plane should be parallel to generator (Extreme right here). As the required axis length of parabola is 50 mm, draw cutting plane line parallel to generator in such a way that length 1'5' on cutting plane becomes 50 mm.

(iii) Draw horizontal lines for cutting planes A, B, C, D, and E, as shown in the figure. Draw circles corresponding to them in plan and mark them as A, B, C, D and E. Get apparent section and the true shape of section by the usual method of transferring cutting points. Here true shape is achieved by taking distances around the centre line PQ, drawn parallel to cutting plane line (V.T.) e.g. pt. B2.

(iv) Now redraw elevation of cut cone, containing apex, keeping its cut surface on the xy line. See the figure. Mark different points on it and draw vertical projectors from them to intersect with horizontal lines from corresponding points of previous plan. Join them to get plan along with true shape of section of the cut cone. Section will be dotted.

10 Sections of Spheres :

When a sphere is cut by any kind of cutting plane, the true shape of the section is always a circle. Diameter of the circle will be equal to the length of cutting plane line between two extreme cutting points, generated by cutting plane line with projection (circle) of sphere.

In Fig.13.35(a) two spheres are shown. Their plans and elevations are also shown on H.P. and V.P. respectively. The sphere of left hand side is cut by a horizontal section plane and the sphere of right hand side is cut by a vertical section plane parallel to V.P.

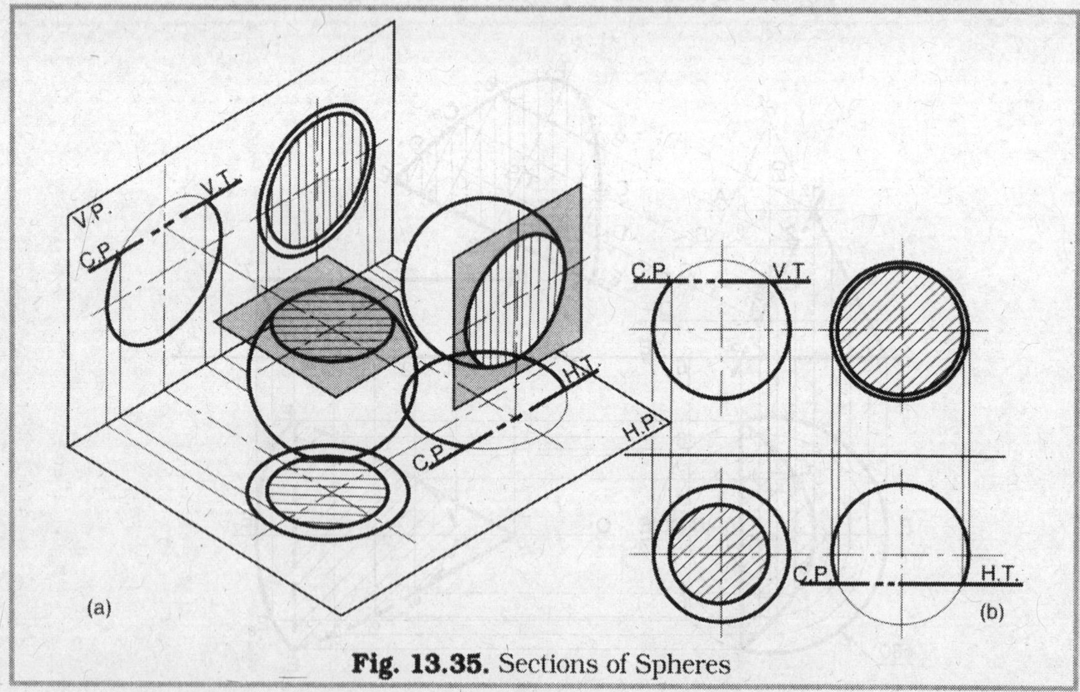

Fig. 13.35. Sections of Spheres

Horizontal section plane is shown in elevation by V.T. Horizontal section plane will cut the sphere along horizontal circular section. It is shown in plan. It will show the true shape of section.

Vertical section plane is shown in plan by H.T. It will cut the sphere along vertical circular section parallel to V.P. It is shown in elevation. It will show the true shape of section.

Projections of above two cut spheres are shown in Fig.13.35(b). It is clearly seen from the figure that diameter of circular section is equal to the length of cutting plane line (V.T. or H.T.) between two extreme cutting points (C.P.)

Problem 32 : A sphere, of diameter 80 mm, is resting on H.P. It is cut by A.I.P. in such a way that the true shape of section is a circle of 70 mm diameter and apparent section in plan is an ellipse with minor axis 35 mm. Draw elevation, sectional plan and true shape of section. Find also the inclination of A.I.P. cutting plane with H.P. What will be the length of major axis of ellipse ?

For solution see Fig.13.36 and follow the procedure, a given below :

(i) First draw plan and elevation of sphere. Both will be circle of 80 mm diameter. Elevation (circle) will be just touching x-y line.

(ii) Now draw inclined cutting plane line (V.T.) on elevation in such a way that its length between cutting points with elevation (circle) becomes 70 mm and its horizontal projection becomes 35 mm, as shown in the figure.

(iii) Now draw horizontal cutting plane lines A, B, C, D, E, F, E, D and C in cutting region, as shown in the figure. We have seen that if a sphere is cut by horizontal cutting plane we get circle in plan. Draw accordingly circles in plan corresponding to cutting planes A, B,.....F. Take C.Ps, C, D and E at equal distances from the central cutting plane F.

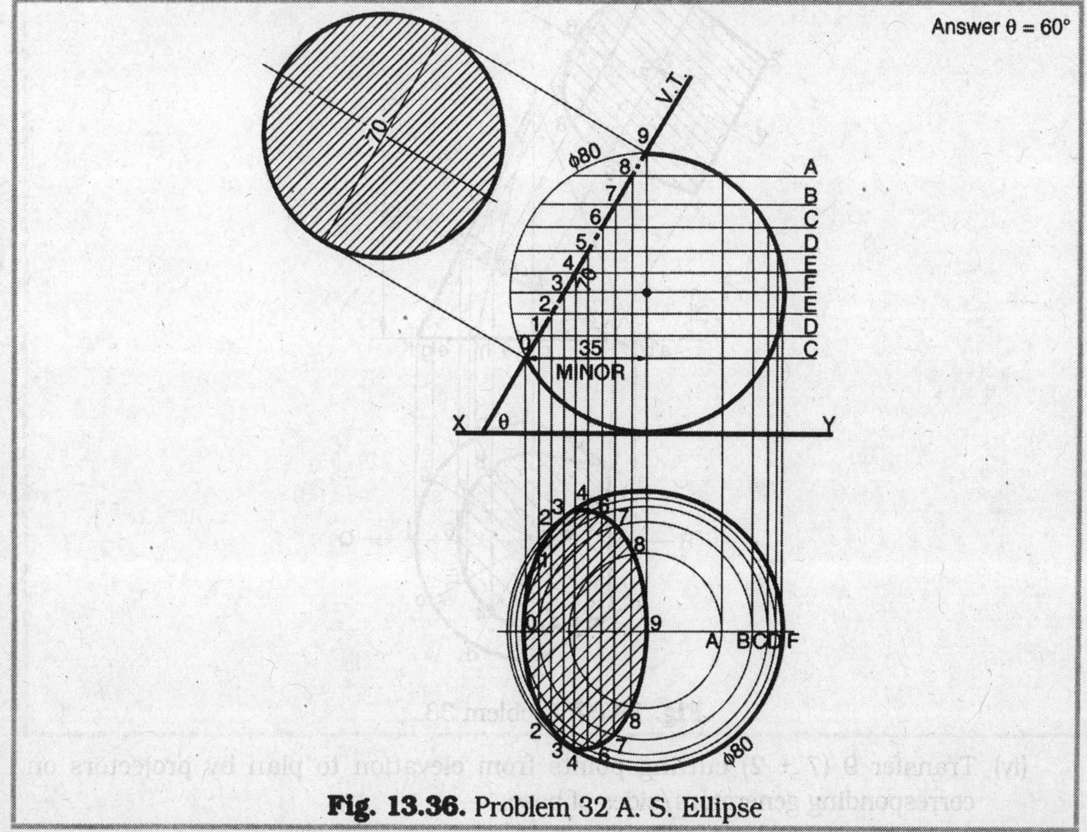

Fig. 13.36. Problem 32 A. S. Ellipse

(iv) Now transfer cutting points 0, 1, 2, ...9 from elevation to plan by projectors on corresponding circles, as shown in the figure.

(v) Join points achieved in plan by a smooth curve to get apparent section. This section will be ellipse with minor axis 35 mm and major axis 70 mm.

(vi) True shape will be a circle of 70 mm diameter. It is clear from the figure.

11. Miscellaneous Problems :

Problem 33 : A solid, made of half cone, diameter of base 60 mm, and half hexagonal pyramid, side of base 30 mm, is having 60 mm height. It is resting on H.P. on its base with middle edge of base perpendicular to V.P. It is cut by A.I.P. inclined at 30° to H.P., passing through a point on axis 12 mm above the base. Draw elevation, sectional plan and true shape of section.

For solution see Fig. 13.37 and follow the procedure, as given below :

(i) First draw plan (half hexagon and half circle, with middle edge of hexagon perpendicular to xy line) and elevation of combined solid. Draw seven generators on the surface of the cone.

(ii) Draw cutting plane line (V.T.) on elevation, inclined to xy at 30°, passing through a point on the axis 12 mm above the base.

(iii) Cutting plane line (V.T.) is cutting all seven generators of cone and two sides of the base BC and AI of the pyramid.

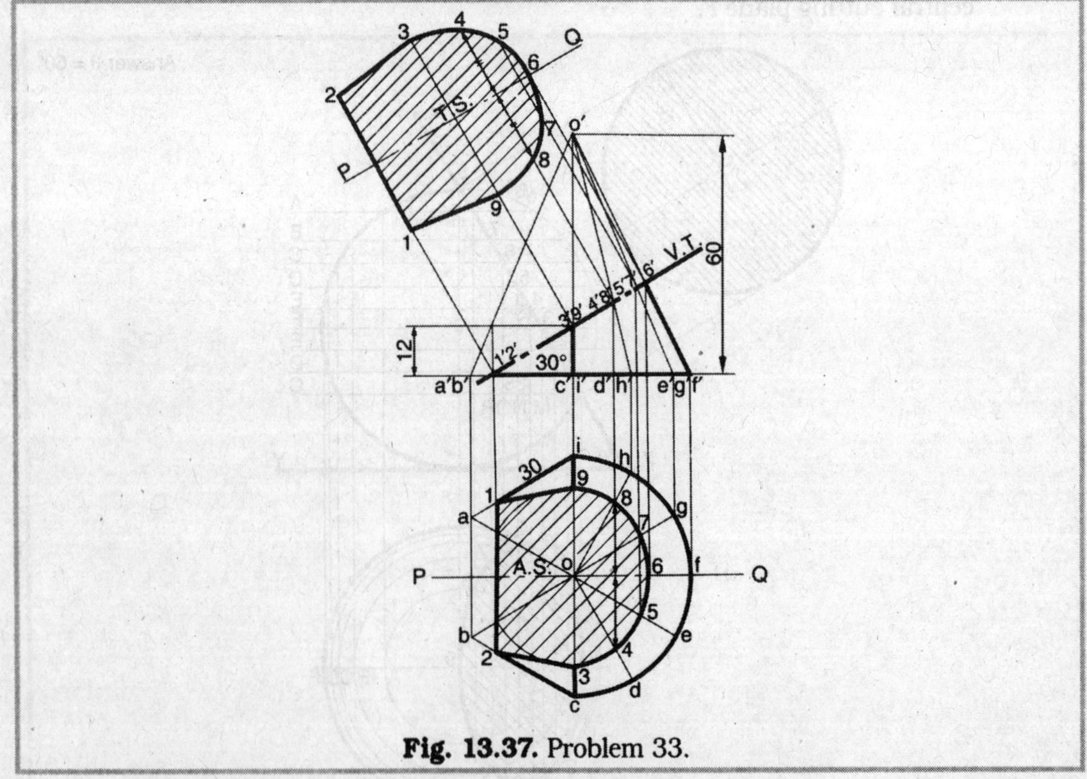

Fig. 13.37. Problem 33.

(iv) Transfer 9 (7 + 2) cutting points from elevation to plan by projectors on corresponding generators/sides of base.

(v) Join points of cone by a smooth curve and points of pyramid by straight lines to get apparent section in plan.

(vi) Project the true shape by the usual method taking reference line for true shape as the centre line PQ.

Problem 34 : A cylinder, diameter of base 80 mm and height 30 mm, is resting on H.P. on its base. A hemisphere of 30 mm radius is centrally placed on its top with flat face of hemisphere resting on cylinder. The formed composite solid is cut by A.I.P. inclined at 45° to H.P., passing through the highest point of hemisphere. Draw F.V., sectional T.V. and the true shape of section.

For solution see Fig. 13.38. Co-relate 3-D drawing and plan-elevation to understand the method of solution.

Circle or cutting plane method similar to Problem - 32 Fig.13.36 is used for getting section of hemisphere. True shape of section of hemisphere is a circle. The rest is clear from the figure.

Problem 35 : A cylinder, of 70 mm base diameter and 100 mm length of axis, is resting on its base on the H.P. It has a square hole of 35 mm side cut through its flat ends so that the axis of the hole coincides with that of the cylinder. The faces of the hole are equally inclined to the V.P. It is cut by an A.I.P. which passes through the extreme left point of the top end and the extreme right point of the bottom end of cylinder. (1) Draw elevation, sectional plan and true shape of the section (2) The cut portion of the solid is allowed to rest on its cut surface on the H.P. axis being parallel to the V.P. Draw the projections of that portion.

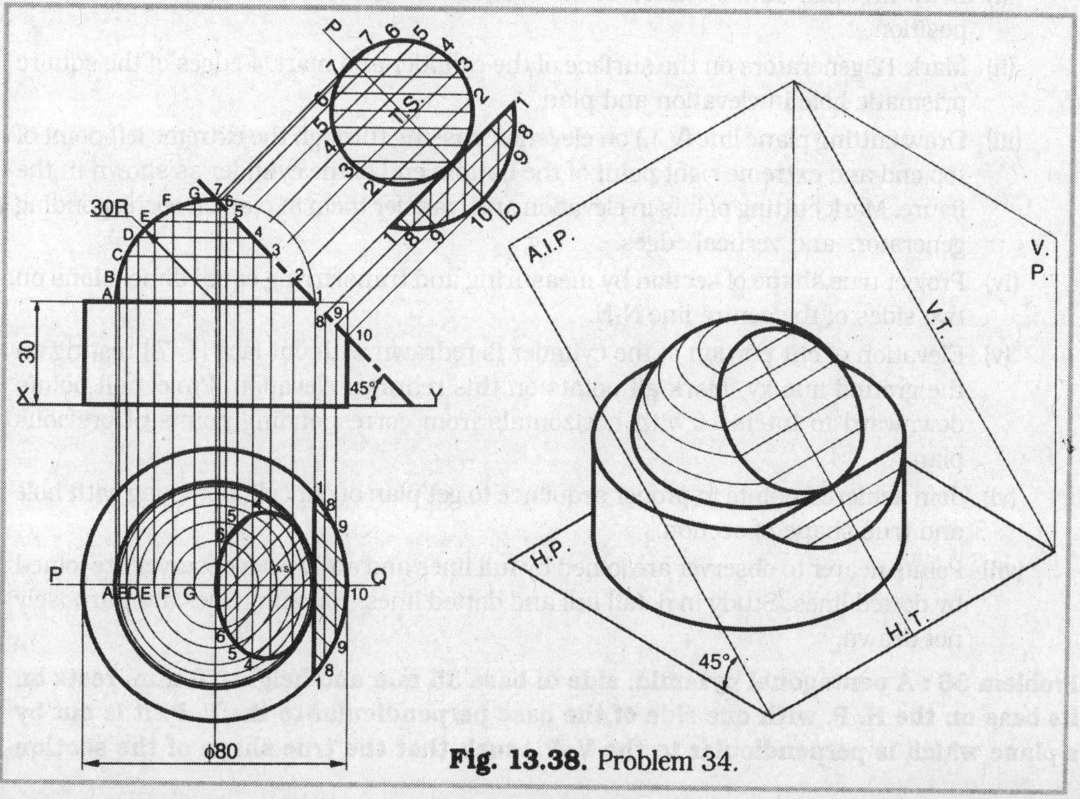

Fig. 13.38. Problem 34.

For solution see Fig. 13.39 and follow the procedure, as given below :

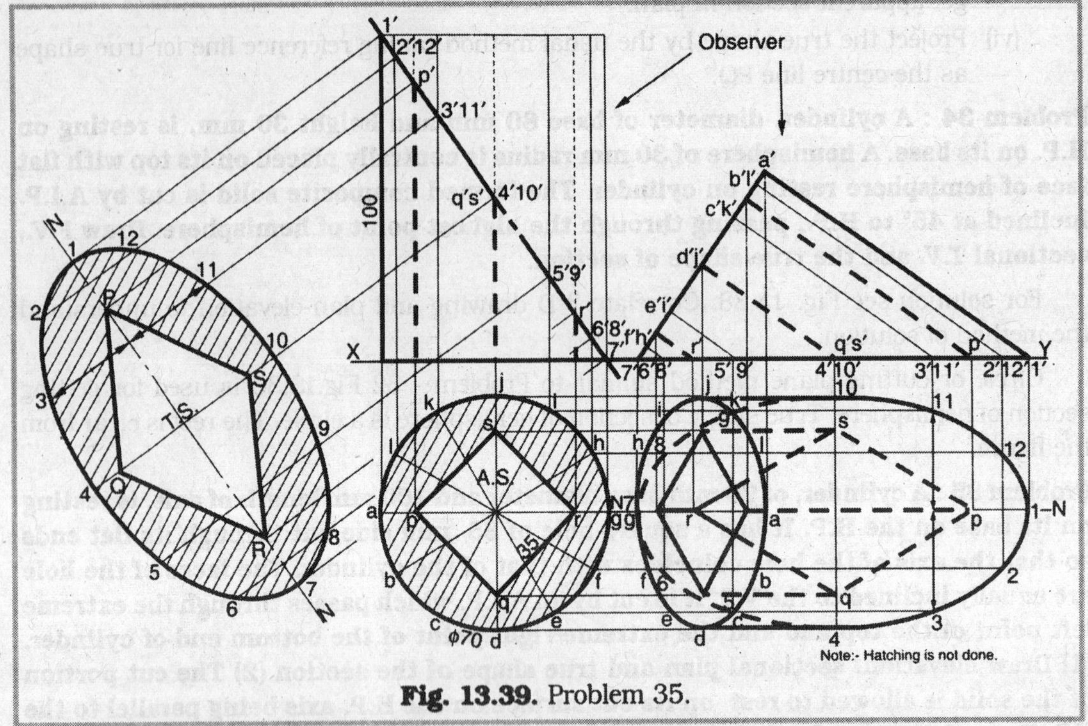

Fig. 13.39. Problem 35.

(i) Draw first plan and elevation of the cylinder along with square hole for the given position.

(ii) Mark 12 generators on the surface of the cylinder and mark 4 edges of the square prismatic hole in elevation and plan.

(iii) Draw cutting plane line (V.T.) on elevation passing through the extreme left point of top end and extreme right point of the bottom end of the cylinder as shown in the figure. Mark cutting points in elevation and transfer them in plan on corresponding generators and vertical edges.

(iv) Project true shape of section by measuring and transferring equal dimensions on two sides of the centre line N-N.

(v) Elevation of cut portion of the cylinder is redrawn with cut face (1'-7') resting on the ground line xy. Mark all points on this redrawn elevation. Project all points downward to intersect with horizontals from corresponding points of previous plan.

(vi) Join achieved points in proper sequence to get plan of cut cylinder along with hole and true shape of section.

(vii) Points nearer to observer are joined by full lines and away from observer are joined by dotted lines. Study in detail full and dotted lines. Hatching lines are purposely not drawn.

Problem 36 : A pentagonal pyramid, side of base 35 mm and height 70 mm, rests on its base on the H. P. with one side of the base perpendicular to the V. P. It is cut by a plane which is perpendicular to the V. P., such that the true shape of the section

is an isosceles triangle of maximum possible base and maximum height. Draw its F. V., sectional T. V. and true shape of the section. [Fig. 13.40]

[Mumbai University, May 1998]

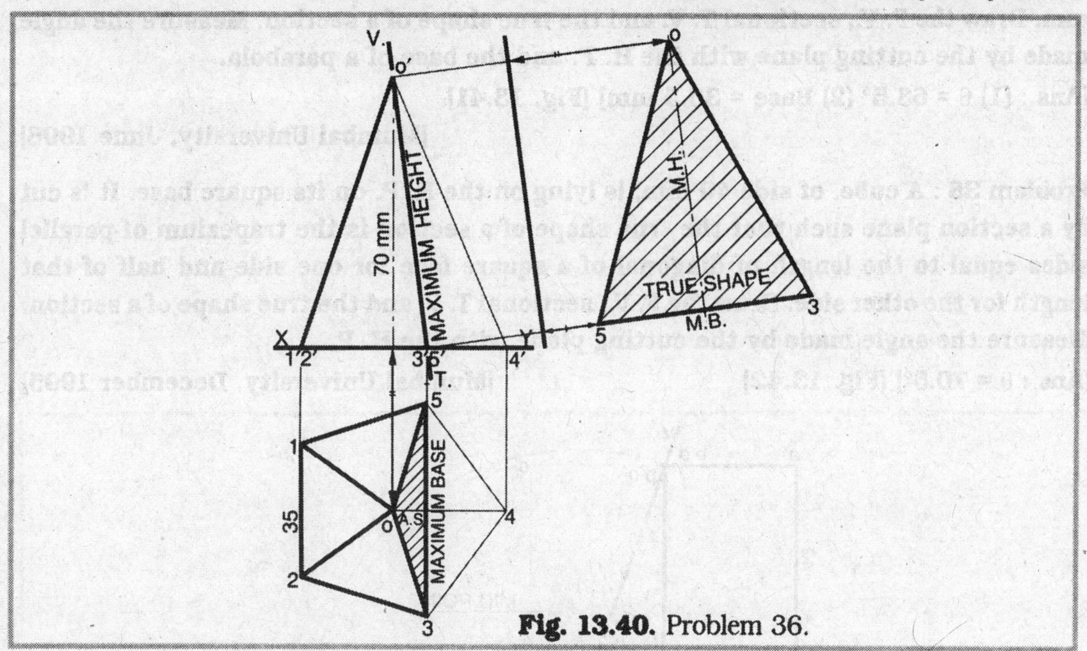

Fig. 13.40. Problem 36.

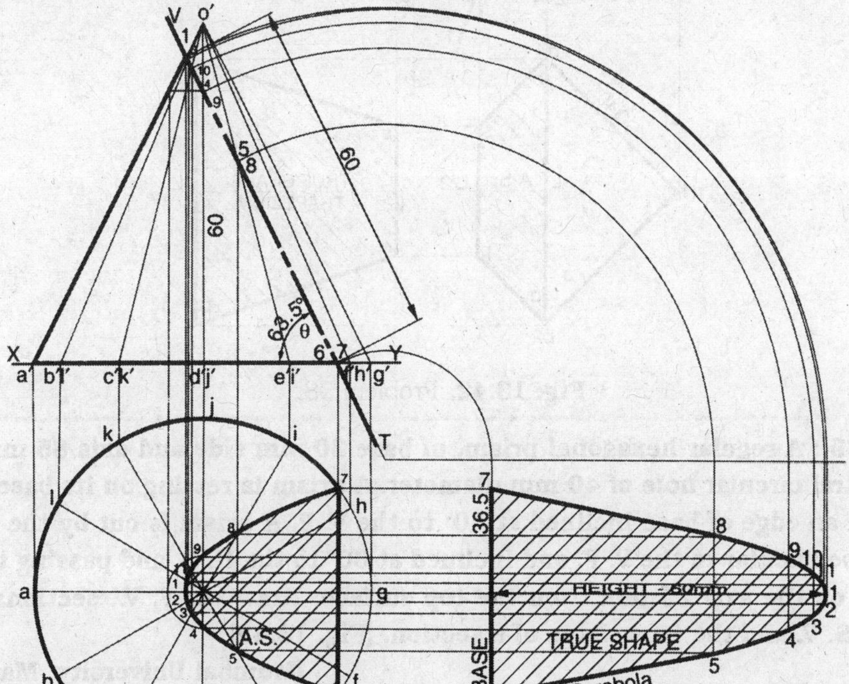

Fig. 13.41. Problem 37.

Problem 37 : A right circular cone, of diameter 60 mm and length of the axis 60 mm, is resting on the H. P. on its base. It is cut by a cutting plane perpendicular to the V. P. and inclined to the H. P. such that the true shape is a parabola of height 60 mm. Draw the F. V., sectional T. V. and the true shape of a section. Measure the angle made by the cutting plane with the H. P. and the base of a parabola.

[Ans : (1) θ = 63.5° (2) Base = 36.5 mm] [Fig. 13.41]

[Mumbai University, June 1996]

Problem 38 : A cube, of side 40 mm, is lying on the H. P. on its square base. It is cut by a section plane such that the true shape of a section is the trapezium of parallel sides equal to the length of diagonal of a square face for one side and half of that length for the other side. Draw the F. V., sectional T. V. and the true shape of a section. Measure the angle made by the cutting plane with the H. P.

[Ans : θ = 70.5°] [Fig. 13.42] **[Mumbai University, December 1995]**

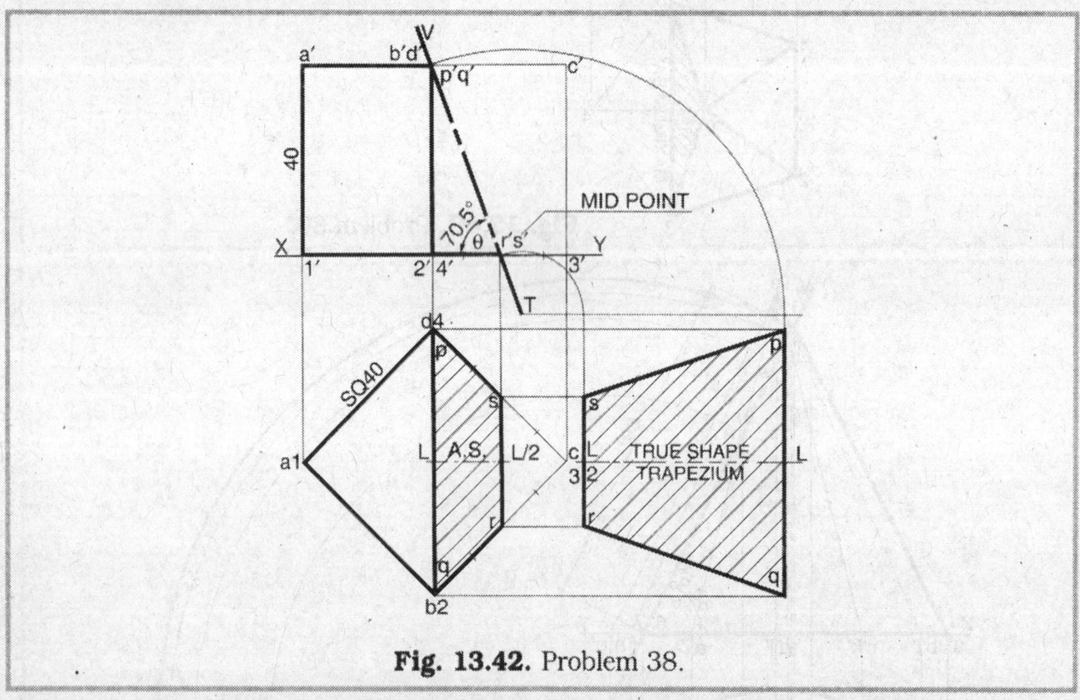

Fig. 13.42. Problem 38.

Problem 39 : A regular hexagonal prism, of base 30 mm side and axis 85 mm long, has a central circular hole of 40 mm diameter. A prism is resting on its base on the H. P. with an edge of base inclined at 40° to the V. P. A prism is cut by the section plane perpendicular to the V. P. and inclined at 60° to the H. P. and passing through the point on the axis 25 mm from the top surface. Draw the F. V., sectional T. V., sectional S. V. and the true shape of a section. [Fig. 13.43]

[Mumbai University, May 1997]

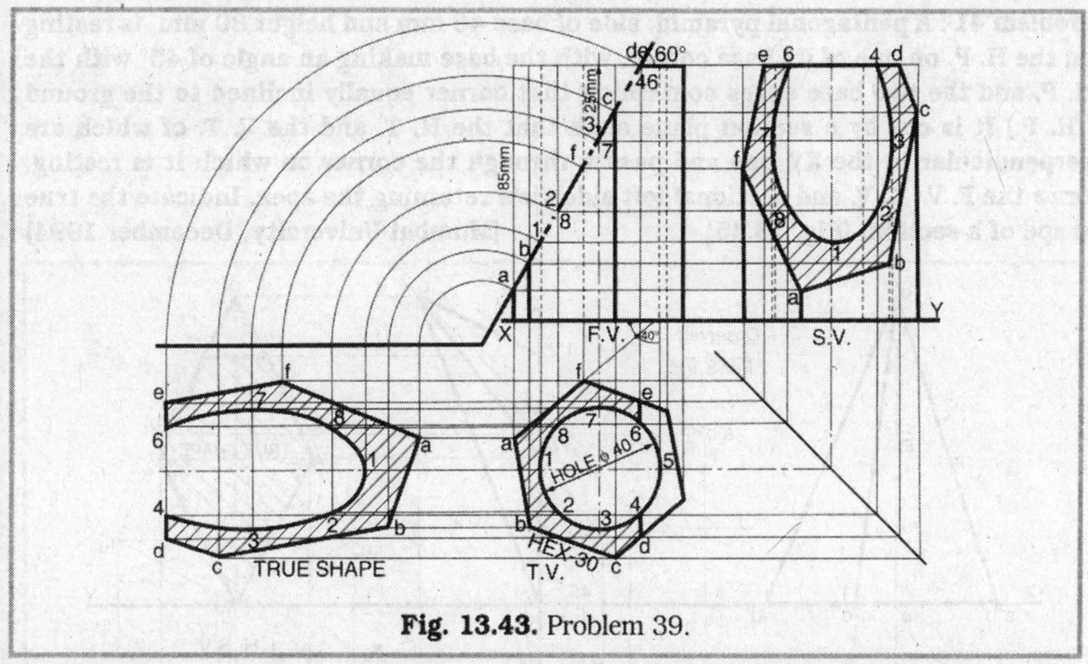

Fig. 13.43. Problem 39.

Problem 40 : A semicone, of diameter 80 mm and 90 mm axis length, is resting on its semicircular base on the H. P. such that the triangular face of a semicone is parallel to the V. P. and away from the observer. It is cut by a section plane perpendicular to the V. P. and inclined at 45° to the H. P. passing through the mid-point of an axis. Draw the sectional plan, elevation and the truc shape of a section. Also add the right hand side view, with given sectional details on it. [Fig. 13.44]

[Mumbai University, June 1995]

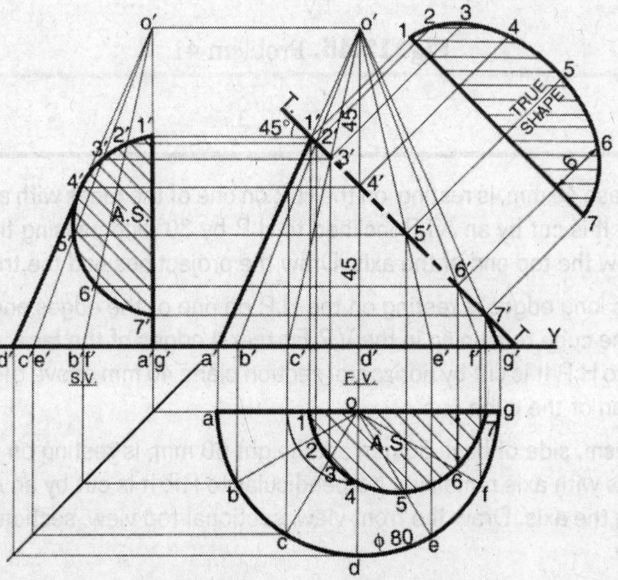

Fig. 13.44. Problem 40.

Problem 41 : A pentagonal pyramid, side of base 40 mm and height 80 mm, is resting on the H. P. on one of its base corner, with the base making an angle of 45° with the H. P. and the two base edges containing that corner equally inclined to the ground (H. P.) It is cut by a section plane such that the H. T. and the V. T. of which are perpendicular to the XY line and passes through the corner on which it is resting. Draw the F. V., T. V. and sectional left side view retaining the apex. Indicate the true shape of a section. [Fig. 13.45] [Mumbai University, December 1994]

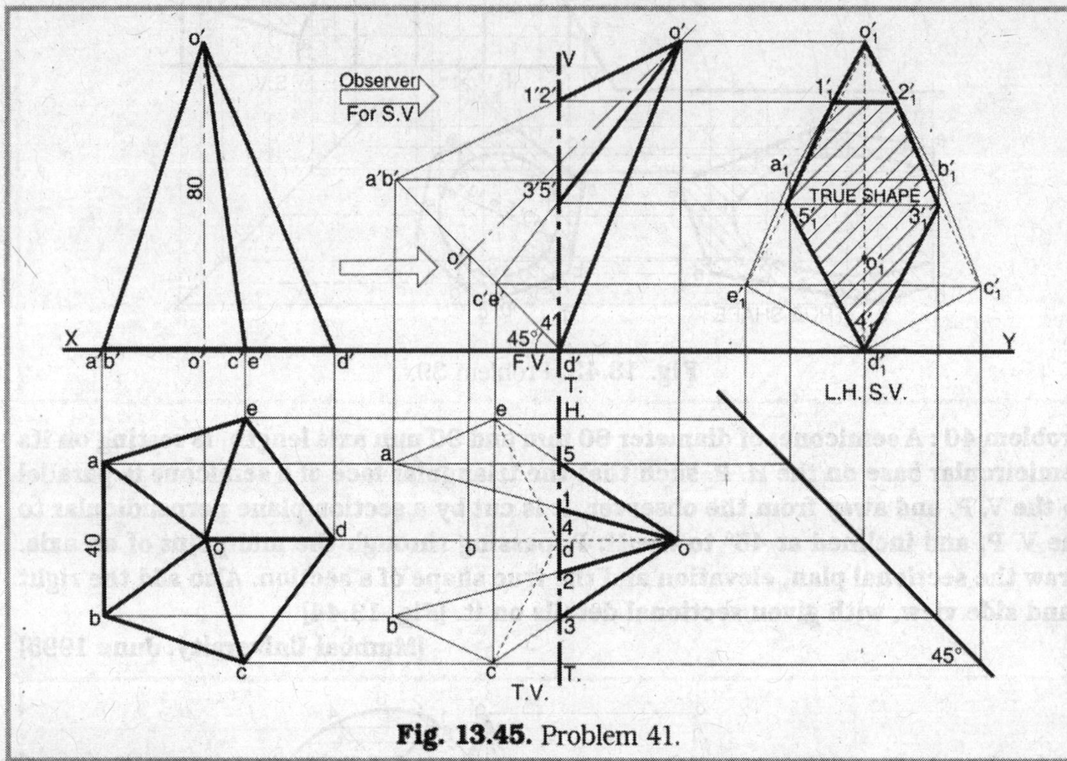

Fig. 13.45. Problem 41.

EXERCISE

1. A cube, side of base 40 mm, is resting on the H.P. on one of the faces with a vertical face inclined at 60° to the V.P. It is cut by an A.I.P. inclined to H.P. by 30° and passing through a point on the axis, 10 mm below the top end of the axis. Draw the projections and the true shape of section.

2. A cube of 50 mm long edges is resting on the H.P. on one of the edges and simultaneously one of the faces of the cube remaining in the V.P. Further 4 edges of the face, which is in the V.P., is equally inclined to H.P. It is cut by horizontal section plane 40 mm above the H.P. Draw elevation and sectional plan of the cube.

3. A pentagonal prism, side of base 30 mm and height 60 mm, is resting on the H.P. on one of its rectangular faces with axis remaining perpendicular to P.P. It is cut by an A.I.P., inclined to H.P. by 30°, bisecting the axis. Draw the front view, sectional top view, sectional side view and true shape of section.

4. A cube of 50 mm long edges is resting on the H.P. on one of the faces with vertical faces equally inclined to the V.P. It is cut by an A.I.P. in such a way that (a) the true shape of the section is

triangle with base 60 mm and the altitude 40 mm. Find the inclination of A.I.P. with H.P. (b) the true shape of the section is a trapezium with two parallel sides of 25 mm and 65 mm. Find the inclination of A.I.P. with H.P. (c) the true shape of the section is a regular hexagon. Find the inclination of an A.I.P. with the H.P. and length of the side of the hexagon. [Ans : $\theta_a = 42°$, $\theta_b = 69°$, $\theta_c = 55°$, L = 35.5 mm]

5. A pentagonal prism, side of base 35 mm and height 50 mm, rests on one of its rectangular faces on H.P. The axis of prism is parallel to both the H.P. and V.P. It is cut by A.I.P. inclined to H.P. by 30° bisecting the axis. Draw the true shape of section and its projections.

6. A hexagonal prism, side of base 40 mm and height 60 mm, is resting on H.P. on its base with two edges of base perpendicular to V.P. It has hexagonal hole of 20 mm side with sides of the hexagon parallel and their centres coinciding. It is cut by A.I.P. making 60° with H.P. and bisecting the axis. Draw three projections with sections and draw also the true shape of the section.

7. A hexagonal prism, side of base 25 mm and height 70 mm, rests on H.P. on one of its longer edges. The rectangular faces containing the longer edge, on which it rests, make equal angles with the H.P. The axis of the prism remains parallel to V.P. It is cut by A.I.P. making 30° with H.P. bisecting the axis of the prism. Draw projections and the true shape of section.

8. A square pyramid, of 50 mm edges of base and 70 mm height, is resting on its base on H.P. with one of the edges of the base perpendicular to V.P. It is cut by an inclined plane in such a way that the true shape of section is a trapezium whose parallel sides measures 40 mm and 20 mm. Draw the projections and the true shape of section and find the inclination of A.I.P. cutting plane with H.P. [Ans : $\theta = 39°$]

9. A hexagonal pyramid, side of base 25 mm and height 65 mm, is lying on the H.P. on one of its triangular faces with the axis parallel to V.P. It is cut by A.V.P. inclined to V.P. by 30° bisecting the axis. Draw the projections and draw also the true shape of section.

10. A tetrahedron, of 65 mm long edges, is lying on H.P. on one of its faces with one of its edges perpendicular to V.P. It is cut by a section plane which is perpendicular to V.P. so that the true shape of section is an isosceles triangle of base 50 mm and altitude 38 mm. Find the inclination of the section plane with H.P. Draw the front view, sectional top view and the true shape of section. [Ans: $\theta = 16°$ or 53°]

11. A cylinder, diameter of base 50 mm and height 75 mm is lying on H.P. on one of its generators with the axis parallel to V.P. It is cut by A.I.P., inclined to H.P. by 30°, passing through a point on the axis 30 mm from one end. Draw projections and the true shape of section.

12. Draw the top and front views of a cylinder cut by a section plane in such a manner that the true shape of section is an ellipse of 50 mm and 100 mm as its minor and major axes respectively. Find the slope angle of the cutting plane. Take the smallest generator to be 35 mm in length. [Ans : 60°]

13. A vertical cylinder, 50 mm diameter and 80 mm height, is cut by an A.V.P. making 30° to V.P. in such a way that the true shape of the section is a rectangle of 40 mm and 80 mm sides. Draw the projections and the true shape of section.

14. A cylinder, base 50 mm diameter and axis 75 mm long, has a square hole of 25 mm side cut through it so that the axis of the hole coincides with that of the cylinder. The cylinder is lying on the H.P. with the axis perpendicular to V.P. and the faces of the square hole equally inclined to the H.P. A vertical section plane inclined at 30° to the V.P. cuts the cylinder in two equal halves. Project the front view of the cylinder on an auxiliary vertical plane parallel to the section plane.

15. A right circular cylinder, of 60 mm diameter and 100 mm height, is resting on its rim with its axis parallel to V.P. and inclined at 30° to the H.P. It is cut by a plane inclined to V.P. such that the true shape of section is an ellipse of major axis 80 mm. Find the inclination of the section plane to the V.P. Draw the sectional front view, top view and true shape of section. [Ans : Ø = 48.5°]

16. A right cone, of 100 mm base circle diameter and 150 mm height, is resting on its base on the H.P. It is cut by a vertical plane inclined at 40° to the V.P., the plane being at a minimum distance of 15 mm from the axis of the cone. Draw the projections of this cone, clearly showing the section. Also project the true shape of the section.

17. A cone, diameter of base 50 mm and axis 60 mm long, is resting on its base on the H.P. It is cut by a section plane perpendicular to both the reference planes in such a way that the true shape of section is a hyperbola having 40 mm base. Draw the front view, top view and the sectional side view.

18. A cone, base 60 mm diameter and axis 70 mm long, has one of its generators on H.P. with its axis parallel to V.P. It is cut by an A.V.P. making 45° to V.P. and passing through the mid point of the axis. Draw the sectional F.V., T.V. and the true shape of the section.

19. A cone, diameter of base 60 mm and axis 70 mm long, is resting on its base on the H.P. It is cut by an A.I.P. so that the true shape of section is an isosceles triangle having 50 mm base. Draw the front view, sectional top view and sectional side view. Also find the true length of the sides of isosceles triangle in the true shape of section. [Ans : L = 76 mm, θ = 76.5°]

20. A plastic bucket, of height 400 mm and bottom and top diameter of 200 mm and 300 mm respectively, contains certain quantity of water. The water from the bucket is about to come out when the bucket is tilted about the base-rim through an angle of 45°. Draw the top view and front view of this inclined bucket and also the true shape of the free surface of the water, which is about to come out from it.

21. A square prism, axis 100 mm long, is resting on its base on H.P. The edges of the base are equally inclined to V.P. The prism is cut by an A.I.P. passing through the mid-point of the axis in such a way that the true shape of the section is rhombus having diagonals of 50 mm and 100 mm. Draw the projections and the true shape of the section. [Ans : Side of base = 35.5 mm, and θ = 60°]

22. A pentagonal prism, of side of base 40 mm and axis 80 mm long, is resting with its base on the H.P. such that one of the sides of the base is at an angle of 45° to the V.P. and the axis is at a distance of 50 mm from V.P. Draw the views of the solids in the above position if it is cut by two section planes, one is at 45° with H.P. sloping upwards and the other is horizontal. Both the section planes intersect each other at the mid-point of the axis.

23. A right regular pentagonal pyramid, of vertical height 62.5 mm and slant height 100 mm, stands on the H.P. on its base such that the axis is vertical and one of the sides of the base inclined at 30° to F.P. (V.P.). It is cut by a section plane perpendicular to F.P. but inclined to T.P. (H.P.) at 60° and passing through a point on the axis at a distance of 40 mm from the base. Draw the sectional top view and the true shape of section.

24. A pentagonal pyramid, side of base 40 mm, axis 75 mm long, rests vertically on H.P. with one of its inclined edges parallel to V.P. It is cut by a section plane perpendicular to V.P., inclined at 45° to H.P. and bisecting the axis. Draw the sectional top view, sectional side view and true shape of section.

25. A pentagonal pyramid, side of base 25 mm and height 60 mm, lies on one of its triangular faces on H.P. and its axis is parallel to V.P. It is cut by a plane perpendicular to H.P. and inclined to V.P. at 30° and passing through the middle point of the top view of axis. Draw the sectional front view and the true shape of section.

26. A sphere, of diameter 80 mm, resting on H.P. is cut by a vertical plane making an angle of 45° to the V.P. and at the shortest distance of 10 mm from the centre of the sphere. Draw the sectional front view of the sphere.

27. A sphere, 60 mm diameter, is resting on the H.P. It is cut by a plane inclined at 30° to H.P. and at right angle to the V.P., passing through a point 12 mm from the centre of the sphere on a line from the centre meeting the cutting plane at right angle. Draw the view from the front and the sectional view from above.

28. A cone, of 35 mm base diameter and 50 mm height, rests on a cylinder of 45 mm diameter and 40 mm height resting on the H.P., such that their axes coincide and are vertical. They are cut by a plane perpendicular to the horizontal plane and inclined to the vertical plane at 45° and passing through a point 10 mm away from the common axis. Draw the sectional front view and the true shape of section.

29. The frustum of a hexagonal pyramid, side of base 35 mm, side of top 25 mm and height 50 mm rests on ground on its base with two edges of base parallel to V.P. A hemisphere of 50 mm diameter is placed centrally on the top of a frustum such that the flat face of the hemisphere remains at the top. This combination of solids is cut by A.V.P. inclined to V.P. by 45° keeping the distance from axis of frustum 10 mm. Draw the projections and project auxiliary elevation on a plane parallel to the cutting plane.

30. The frustum of a cone, diameter of base 60 mm, diameter of top 30 mm and height 50 mm is resting on H.P. on its base. A hemisphere, of diameter 60 mm, is centrally placed on the top of a frustum such that the flat face of the hemisphere remains at the top. The combination is cut by an A.V.P. inclined to V.P. by 30° in such a way that the true shape of section of hemisphere is a half circle of 50 mm diameter. Draw the projections and project auxiliary elevation on A.V.P. parallel to cutting plane. Find the distance of cutting plane from the axis of frustum of the cone.

[Ans : Distance of cutting plane from the axis = 16.5 mm]

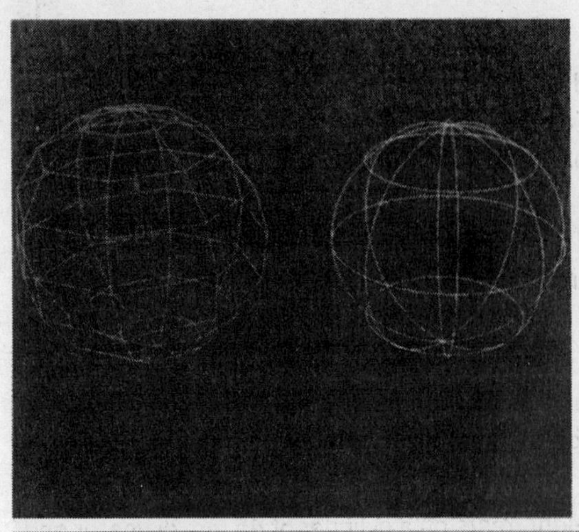

CHAPTER - 14

Development of Surfaces of Solids

DEVELOPMENT OF SURFACES OF SOLIDS

Important Note : In this chapter of development, plan or T.V. is not of much importance and hence sometimes not drawn. Draw plan or T.V. above F.V. in 3rd Angle System and below F.V. in 1st Angle System without any change in its geometry. Plan or T.V. does not contain effect of cutting in some of the problems.

1. Introduction

The study of development of surfaces of solids is very important from the point of view of its practical utility in fabrication industry. Any thing which is made of sheet of paper, raksin, plastic, cloth, card board, plywood or metal are usually cut first from flat sheet and then rolled or folded into required shape.

To make any article out of sheet metal, we have to first prepare its development or rolled out form and then we have to cut the sheet metal according to the shape and size of the development called blank and then roll it or fold it into the required shape of the article. Articles like ducts, containers, cup boards, metal furniture, lamp shades, pipe work in chemical and petrochemical industries, packing containers, radio and T.V. covers, car and ship body, etc. are made of sheets and hence in fabrication of above articles knowledge of development of surfaces of solids is necessary.

The term development refers to the complete surface of solid laid out on a single plane. As the shape and size of the development is used as pattern or template for cutting sheet blank, it should be drawn with strictest accuracy. Every thing in development must appear in true shape, true length and at the appropriate position.

In other words, the surfaces forming the solid body must be represented on one and same plane, because the sheet metal from which the blank is cut out is also plane. The procedure of unfolding the surfaces of an object is called developing and the representation so obtained is called development or develop view. The development should be so prepared that the material out of which it is cut can be utilised to the maximum with the minimum of loss.

We know that articles required for industrial use or for the use of society are having shapes made up of combinations of prisms, cylinders, pyramids, cones, spheres, their frustums and their truncated frustums.

We shall now study the procedure of development for different solids one by one by taking varieties of examples. Students should study and note the method of notations adopted for different solids. In the starting, the ends or bases of the solids are shown in development but later on they are omitted. They can be easily incorporated if required.

When solids are cut by sections their true shapes of sections also become the part of the development. However, they are omitted and not shown, in the development. If required they can be drawn as per the procedure studied in the chapter of sections of solids.

In the development (1) Draw outline of the development absolutely thick and dark. (2) Folding lines medium thick and medium dark. (3) All other construction lines, generators etc absolutely thin and light. (4) As far as possible keep some distance between elevation and development (5) In full development, draw the development of lateral surfaces, bases and true shape of section. (6) In the development of lateral surfaces don't consider base and true shape of section. (7) Write appropriate notations in good hand-writing.

2. Prisms

All right and regular prisms have at their two ends regular polygons of n sides and have n number of rectangular surfaces connecting these two polygons. So if surfaces of prism are unfolded in a single plane we get n number of rectangles having side of base as one side and length of vertical edge or height of prism as another side. Moreover, we get two polygons of n sides. (See Fig.14.1)

Problem 1 : Develop only the lateral surfaces of hexagonal prism resting on H.P. on its base with two edges of base parallel to V.P. Take the edge of base equal to [A] and height of prism equal to [B].

For solution see Fig. 14.1 and follow the procedure, as given below :

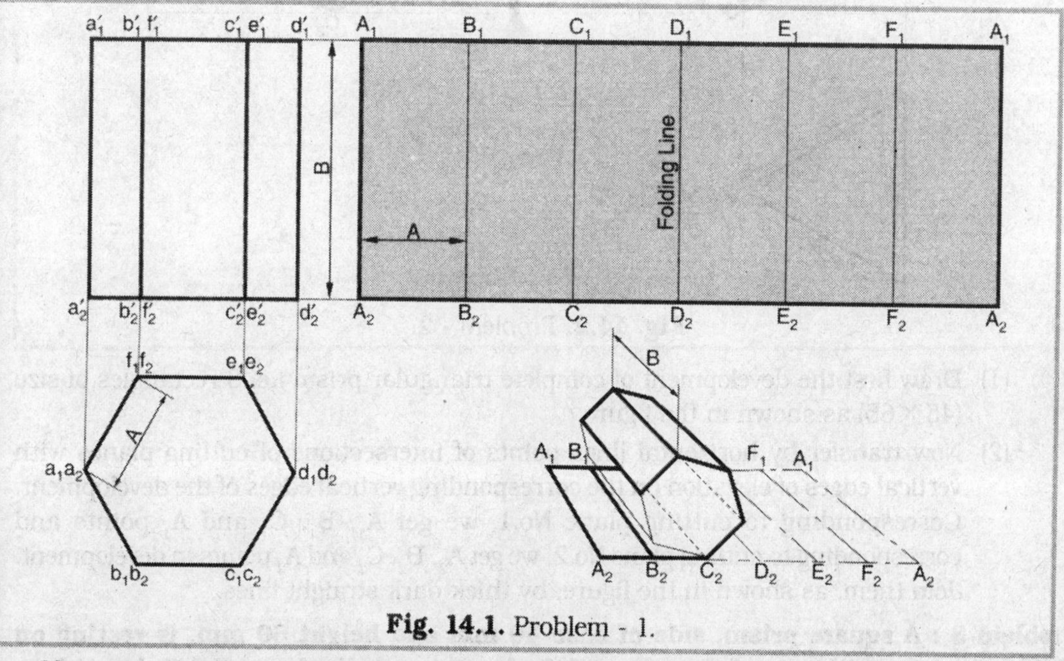

Fig. 14.1. Problem - 1.

Observe unfolding of the lateral surfaces of the prism shown in the three dimensional drawing. It is very clear from the figure that we shall get 6 rectangles.

(1) First draw plan and elevation of the prism according to the given condition.

(2) Stretch out two parallel horizontal straight lines, one from the top and another from the bottom surface of the elevation of the prism. Automatically the distance between two parallel lines will be maintained equal to the height of the prism (B).

(3) Keeping suitable gap between development and elevation, draw vertical line A_1A_2. Now draw vertical lines B_1B_2, C_1C_2,.....F_1F_2 and A_1A_2 at the intervals of dimension A to get 6 rectangles, as shown in the figure as the development of lateral surfaces of prism. All 6 rectangles are of (A × B) size.

Problem 2 : A triangular prism is resting on H.P. on its base with one of the rectangular faces perpendicular to V.P. Take edge of base 45 mm and height of prism 65 mm. Two parallel cutting planes perpendicular to V.P. and inclined to H.P. by 30° are cutting the prism, one passing through the edge of the top side base and another passing through the corner of the bottom base. Draw the development of lateral surfaces of the portion of prism between two cutting planes.

For solution see Fig. 14.2 and follow the procedure, as given below :

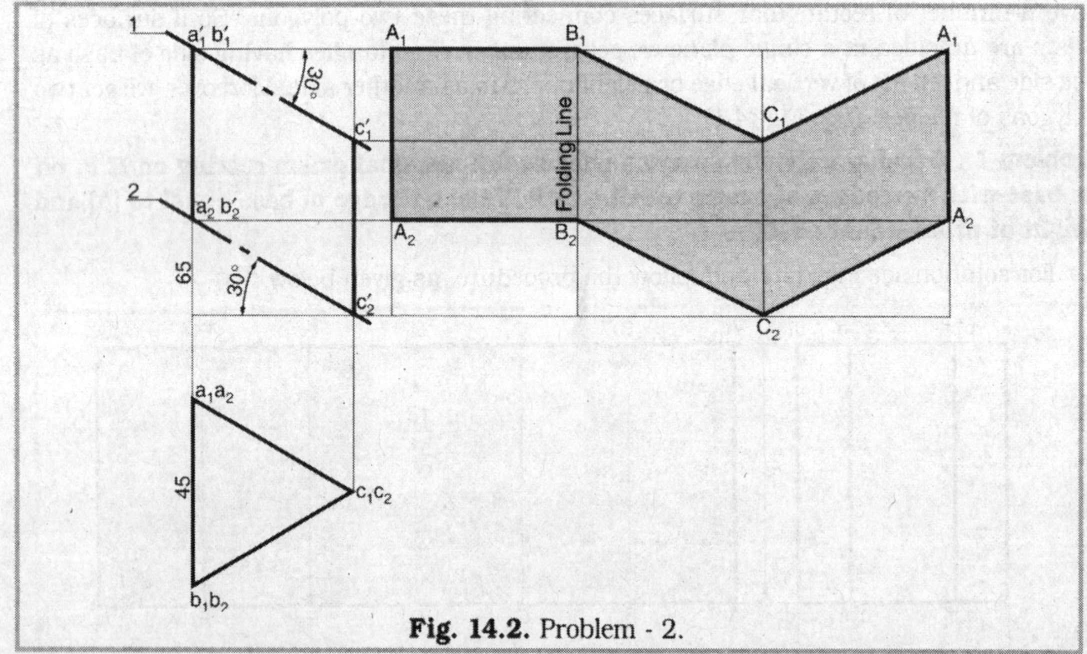

Fig. 14.2. Problem - 2.

(1) Draw first the development of complete triangular prism i.e. 3 rectangles of size (45 × 65) as shown in the figure.

(2) Now transfer by horizontal lines points of intersections of cutting planes with vertical edges of elevation on the corresponding vertical edges of the development. Corresponding to cutting plane No.1, we get A_1, B_1, C_1 and A_1 points and corresponding to cutting plane No.2, we get A_2, B_2, C_2 and A_2 points in development. Join them, as shown in the figure, by thick dark straight lines.

Problem 3 : A square prism, side of base 40 mm and height 60 mm, is resting on H.P. on its base with one of the edges of the base perpendicular to V.P. It is cut by a section plane perpendicular to V.P. and inclined to H.P. at 45° and passing through

a point 15 mm from the top end of the axis. Draw the complete development of the complete portion of the prism between section plane and the bottom of prism.

For solution see Fig. 14.3 and follow the procedure, as given below :

Three dimensional drawing of the cut prism is shown in the figure. This drawing will help in understanding the development of the prism.

(1) First of all draw the development of complete square prism. 4 rectangles and two squares for top and bottom bases.

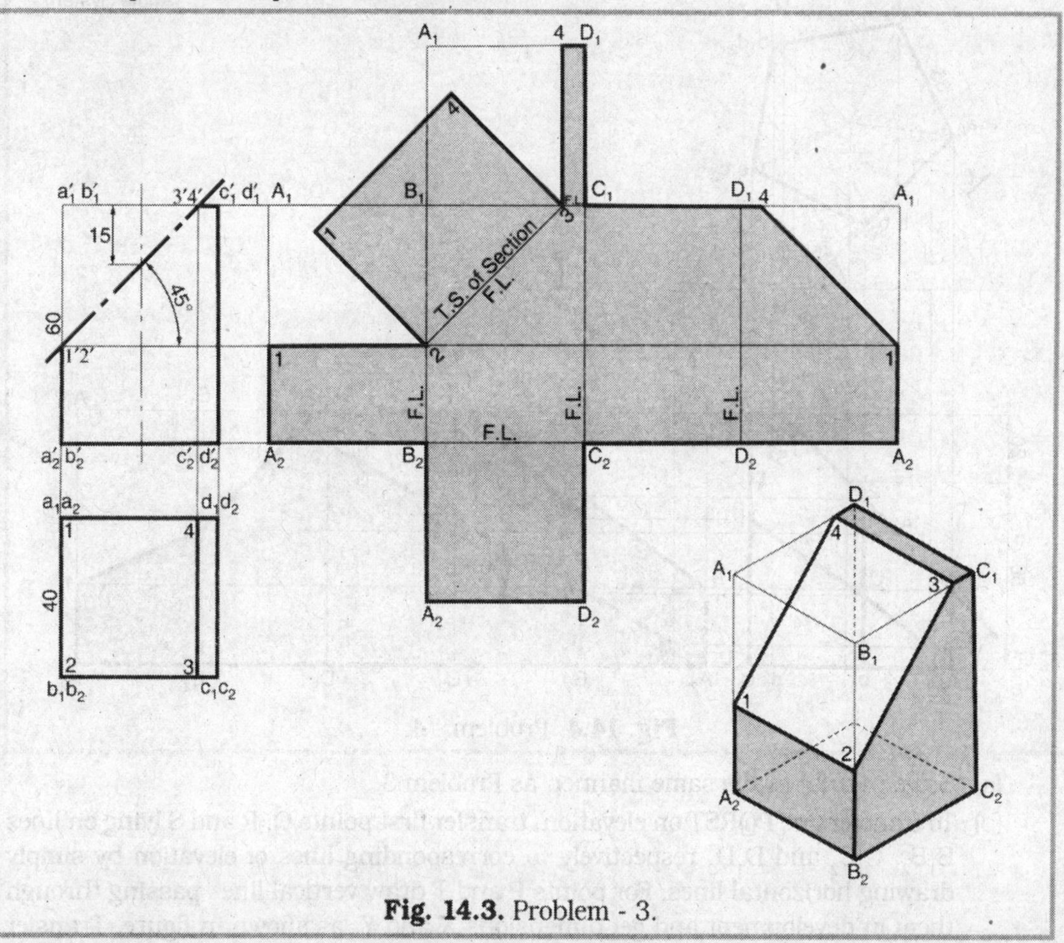

Fig. 14.3. Problem - 3.

(2) Cutting plane is cutting four edges of the prism A_1A_2, B_1B_2, B_1C_1 and D_1A_1 at four points 1, 2, 3 and 4 respectively.

(3) Transfer points 1 and 2 on lines A_1A_2 and B_1B_2 of development by horizontal lines from corresponding points 1' and 2' of elevation.

(4) Transfer 3' and 4' points of elevation to plan on corresponding lines b_1c_1 and d_1a_1 and mark points 3 and 4 respectively.

(5) Take dimensions $3c_1$ and $4d_1$ from plan on divider and transfer them on corresponding lines in development to get points 3 and 4.

Development is shown by thicker boundary lines. Try to visualise six faces of development by co-relating them with corresponding 6 faces of three dimensional drawing of cut prism. True shape of section 1234 is also shown in the development.

Problem 4 : (a) A pentagonal prism, side of base 35 mm and height 70 mm, is resting on ground on its base with one of the edges of the base inclined at 45° to the V.P. It is cut by A.I.P. inclined to H.P. by 45° passing through a point on the axis 20 mm from top end of axis. Draw the development of the cut prism. (b) Given the line P Q R S T in the development transfer it to the elevation of prism.

For solution see Fig. 14.4 and follow the procedure, as given below :

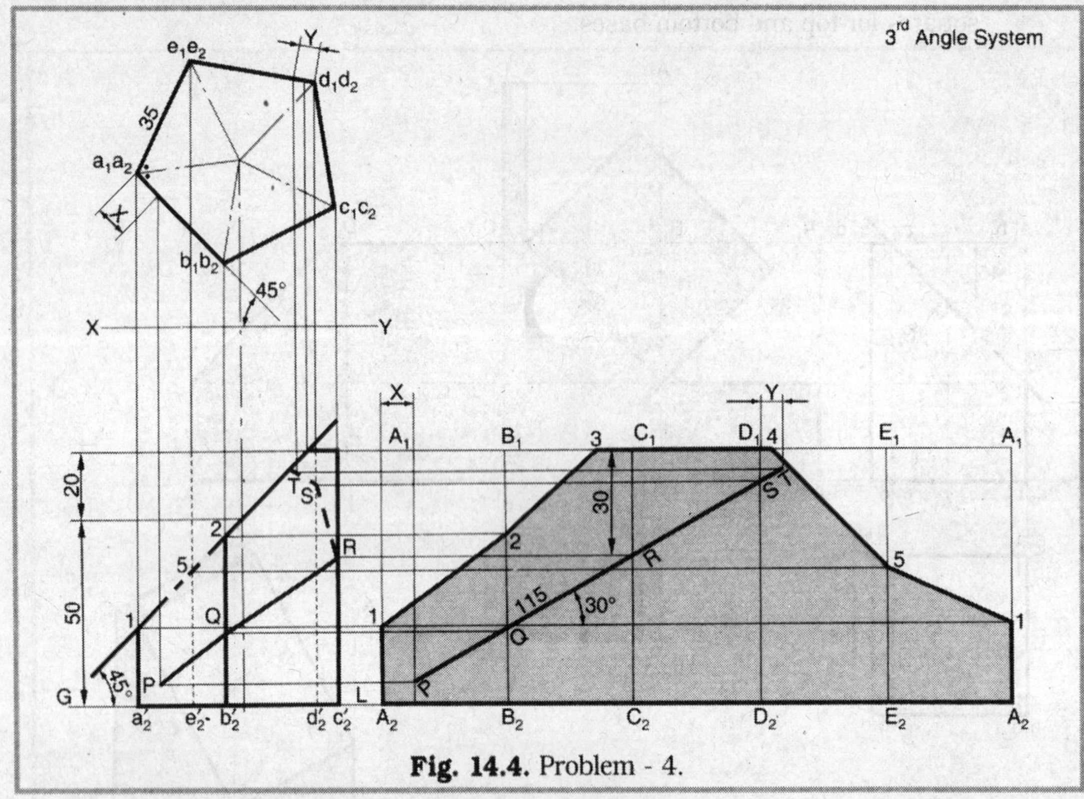

Fig. 14.4. Problem - 4.

(a) Solve part (a) in the same manner, as Problem 3.

(b) To transfer line PQRST on elevation, transfer first points Q, R and S lying on lines B_1B_2, C_1C_2 and D_1D_2 respectively to corresponding lines of elevation by simply drawing horizontal lines. For points P and T draw vertical lines passing through them in development and get dimensions X and Y, as shown in figure. Transfer these dimensions X and Y in plan on corresponding lines a_1b_1 and d_1e_1 respectively. And then draw vertical lines in elevation corresponding to those points. On these vertical lines transfer points P and T by horizontal lines from development.

Lines PQ and QR on AB and BC faces will be seen and hence draw full, while lines RS and ST on CD and DE faces will not be seen and hence draw dotted in elevation.

Problem 5 : A square prism, side of base 50 mm and height 75 mm, is resting on H.P. on its base with all vertical faces equally inclined to V.P. Hole of 60 mm diameter is drilled centrally through the prism. Axis of the hole is perpendicular to V.P. Draw the development of only the lateral surfaces of the prism.

For solution see Fig. 14.5 and follow the procedure, as given below :

(i) First draw the development of lateral surfaces of complete square prism.

(ii) Divide the circle of hole, in elevation, into 8 or 12 parts and draw vertical lines through those points. These vertical lines will generate points b, c, d, e, f, h, i, j, k and l in plan.

(iii) Transfer those points of plan to the development on the corresponding lines by transferring dimensions with the help of divider.

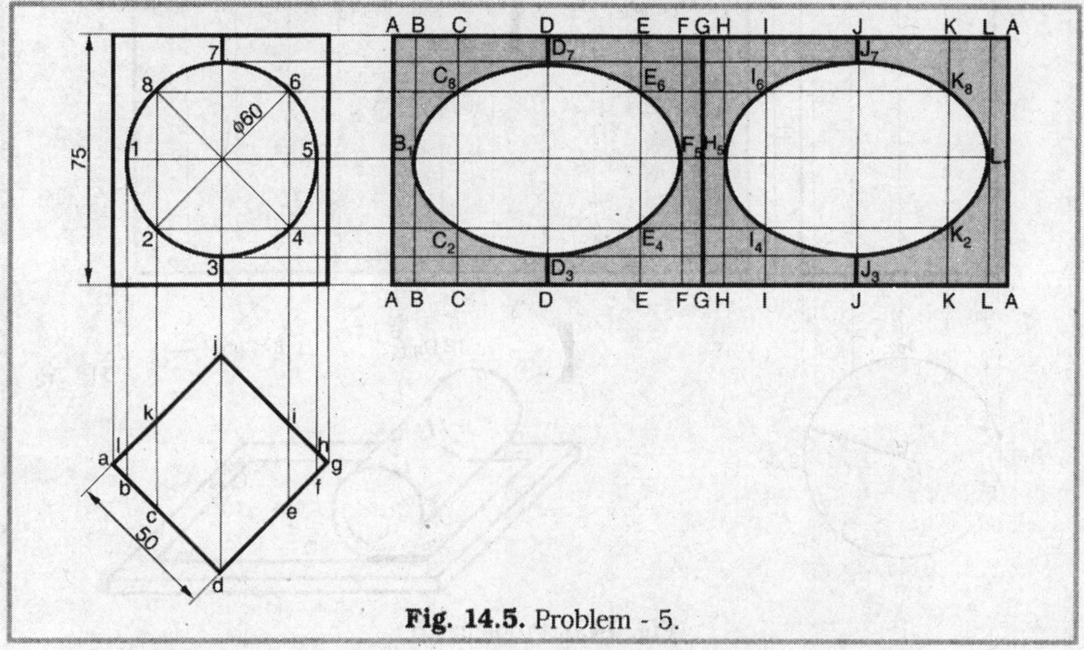

Fig. 14.5. Problem - 5.

(iv) Draw vertical lines in development through these points B, C, D, E, F, H, I, J, K and L already marked.

(v) From points 1, 2,.......8 of elevation draw horizontal lines upto (B, L), (C, K), (D, J), (E, I), (F, H), (E, I), (D, J) and (C, K) vertical lines of development to get points (B_1, L_1), (C_2, K_2) ....respectively.

(vi) Join B_1 C_2 D_3 E_4 F_5 E_6 D_7 C_8 and B_1 and similarly $L_1K_2J_3I_4$ H_5 I_6 J_7 K_8 and L_1 to complete the development.

3. Cylinders

Cylinders have at their two ends circles and have curved lateral surface connecting two circles. So if the surface of the cylinder is unfolded in a single plane we get a rectangle, having length equal to (πD) and height equal to height of the cylinder. See Fig. 14.6. Moreover, we get two circles of diameter D for top and bottom. These circles are generally not drawn.

Just as prisms, cylinders do not have defined corners and edges. Therefore, 12 generators are taken on the curved surface of a cylinder. These 12 generators and 24 end points of 12 generators are used for the purpose of projections.

Problem 6 : A cylinder, diameter of base D and height H, is resting on H.P. on its base. Draw the development of the curved surface of the cylinder. For solution see Fig. 14.6 and follow the procedure, as given below :

(1) First draw plan and elevation of the cylinder.

(2) Divide circle of plan into 12 equal divisions and mark points (a_1, a_2) (b_1, b_2) ....(l_1, l_2).

(3) Transferring above points to elevation, get generators $(a'_1 a'_2)$ $(b'_1 b'_2)$......$(l'_1 l'_2)$ in elevation.

(4) Draw rectangle in line with elevation of size ($\pi D \times H$), as shown in the figure, which represents development. Visualise this from three dimensional drawing.

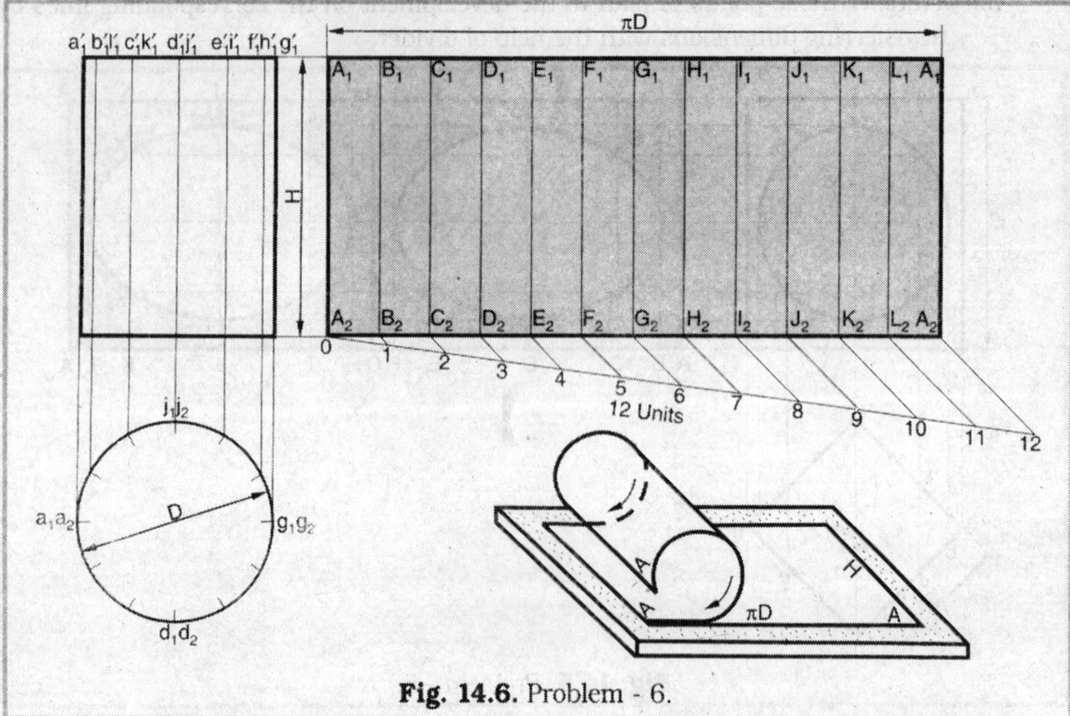

Fig. 14.6. Problem - 6.

(5) Divide the rectangle into 12 equal parts and draw generators (A_1A_2) (B_1B_2)...... (L_1L_2), (A_1A_2), as shown in the figure. Method to divide odd length (πD) into 12 equal divisions is shown in the figure

[OR]

(5) Transfer dimension a_1b_1 from plan to measure A_1B_1 in development. Get 12 parts in developments by taking a_1b_1 dimension 12 times. This is an approximate method, but convenient.

Chord a_1b_1 = arc a_1b_1 = A_1B_1 = B_1C_1 =......=L_1A_1

Approximate method is easier, convenient and hence preferable.

Problem 7 : Plan and elevation of a cylinder with section is given. Draw the development of the cut cylinder.

For solution see Fig. 14.7 and follow the procedure, as given below :

(1) First draw plan, elevation and development of the complete cylinder along with 12 generators marked on them.

(2) Draw cutting plane line in elevation and mark points 1', 2',......12' where cutting plane line is cutting generators $(a'_1 a'_2)$, $(b'_1 b'_2)$.......$(l'_1 l'_2)$ respectively.

(3) Now draw horizontal lines from 1', 2',....12' upto corresponding generators (A_1A_2), (B_1B_2)......(L_1L_2) in development to get points 1, 2,.....12, as shown in the figure.

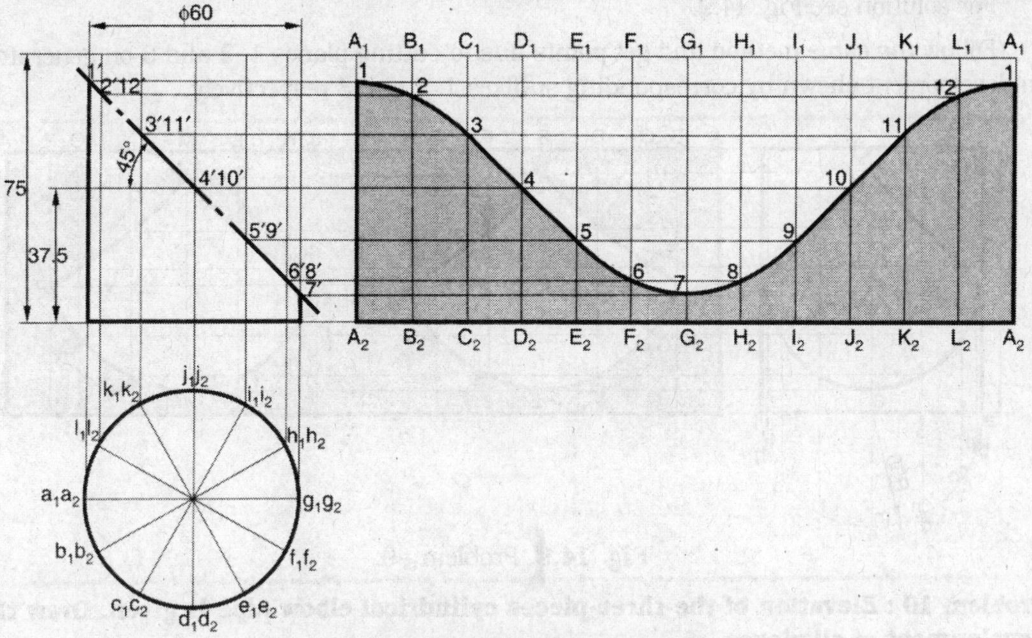

Fig. 14.7. Problem - 7.

Join them by a smooth curve to get the development.

(4) Development is shown by thicker boundary lines.

Problem 8 : Elevation of the cylinder, with three cuttings, is given. Draw the development.

For solution see Fig. 14.8 and follow the procedure, as given below :

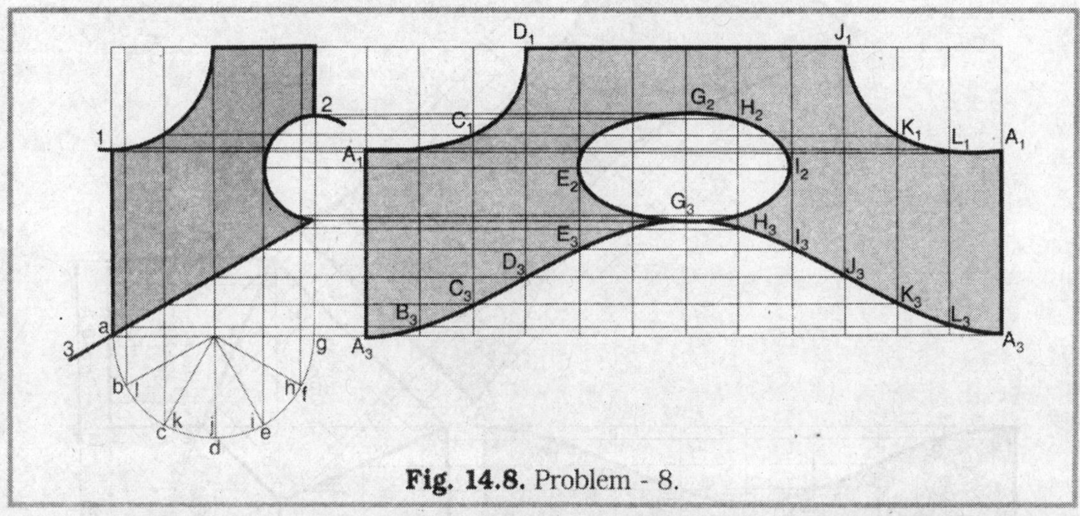

Fig. 14.8. Problem - 8.

(1) Follow the same method as followed in Problems 6 and 7. Mark three different cuttings by No.1, 2, and 3 and mark corresponding points in development by suffixes 1, 2 and 3.

Problem 9 : Elevation (with 3 cuttings) of a cylinder is given. Draw the development of the cut cylinder.

For solution see Fig. 14.9.

Follow the same method and get points due to cutting planes 1, 2 and 3 on generators in development shown by corresponding suffixes 1, 2 and 3 respectively.

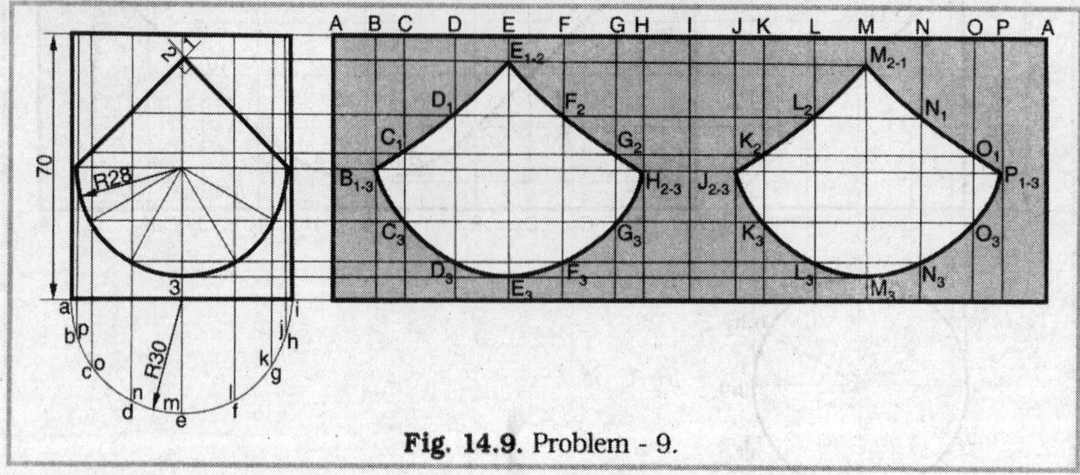

Fig. 14.9. Problem - 9.

Problem 10 : Elevation of the three-pieces cylindrical elbow pipe is given. Draw the development of all pieces.

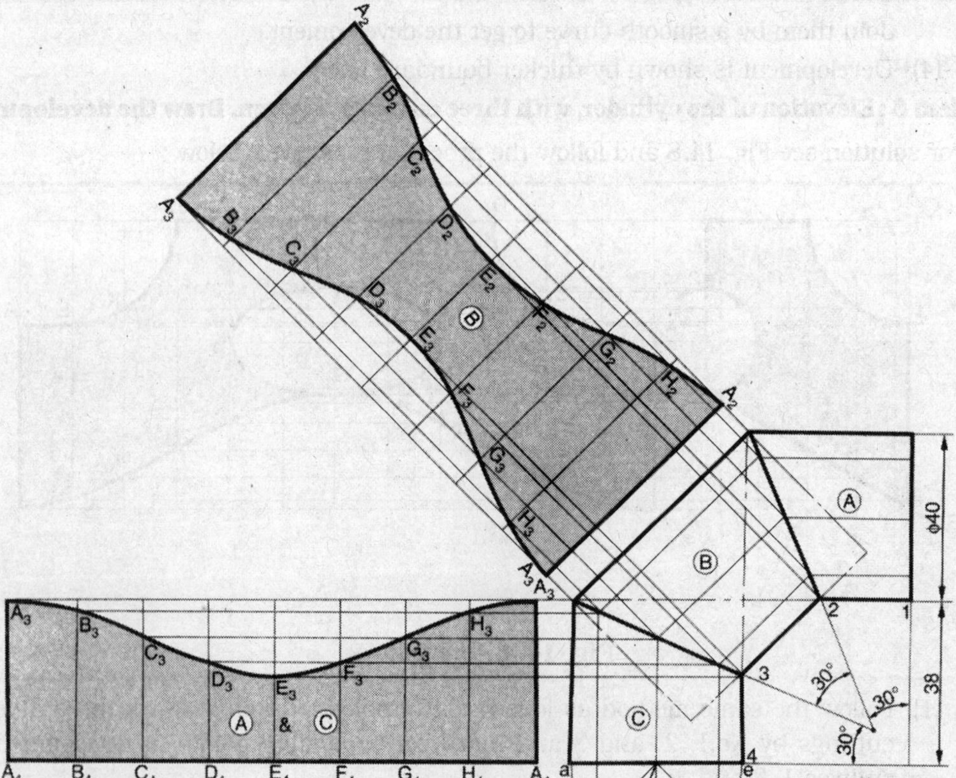

Fig. 14.10. Problem - 10.

For solution see Fig. 14.10 and follow the procedure, as given below :

(1) Draw three rectangles enclosing elevation of the piece A, piece B and piece C, as shown in the figure. These rectangles represent the elevations of the uncut (full) cylinders from which these three pieces are cut out by the application of cutting planes No.2 and 3.

(2) To mark 8 generators on all the 3 pieces, draw a semicircle attached to the piece C and divide it into 4 parts. Mark points a, b, c, ...h and draw from them lines parallel to the axis of three pieces successively.

(3) Now parts A and C are similar. Piece C is truncated by the section plane No.3 and piece A by S.P. No.2. Develop any one by following the method, studied earlier.

(4) Part B is truncated at both the ends by S.P. 3 and S.P. 2. Develop it.

Problem 11 : Three cut pieces of cylindrical pipes form a 'Y' joint as given in Fig.14.11. Draw the development of surfaces of all pieces.

For solution see Fig. 14.11.

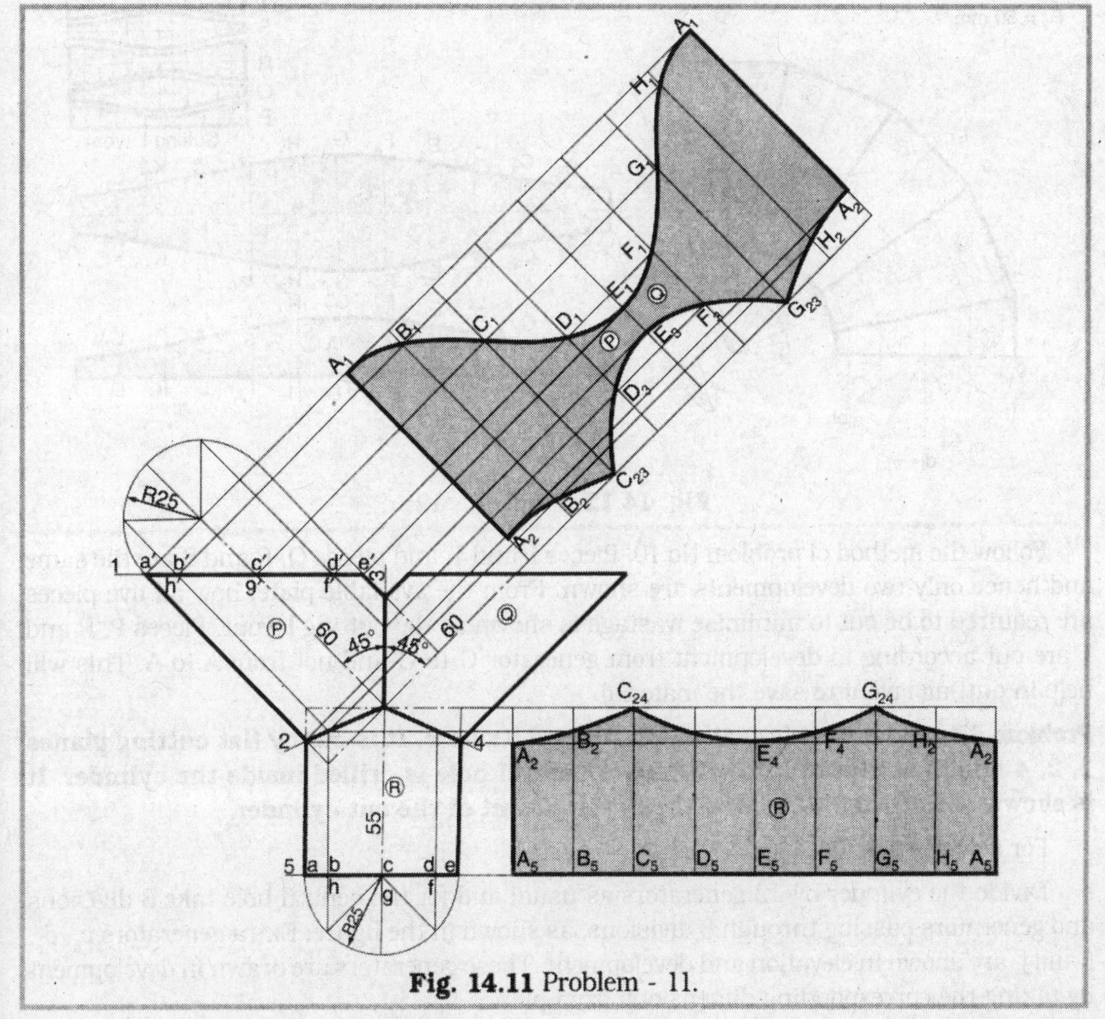

Fig. 14.11 Problem - 11.

(1) Draw 2 semi circles on the bases of pieces R and P. Divide them into four equal parts. Draw from them successively parallel lines to an axis in each piece and get 8 generators in all the three pieces.

(2) Enclose pieces P and R in two rectangles, as shown in the figure. These two rectangles represent the uncut (full) cylinders. Piece P has three cuttings : No 1, 2 and 3. While piece R has two cuttings No.2 and 4.

(3) As studied earlier develop pieces P and R.

(4) As pieces P and Q are the same, Q is not developed.

Suffix indicates the cutting plane number in the development.

Problem 12 : Elevation of the five pieces round 90° turn duct is given. Draw the development of all pieces.

For solution see Fig. 14.12.

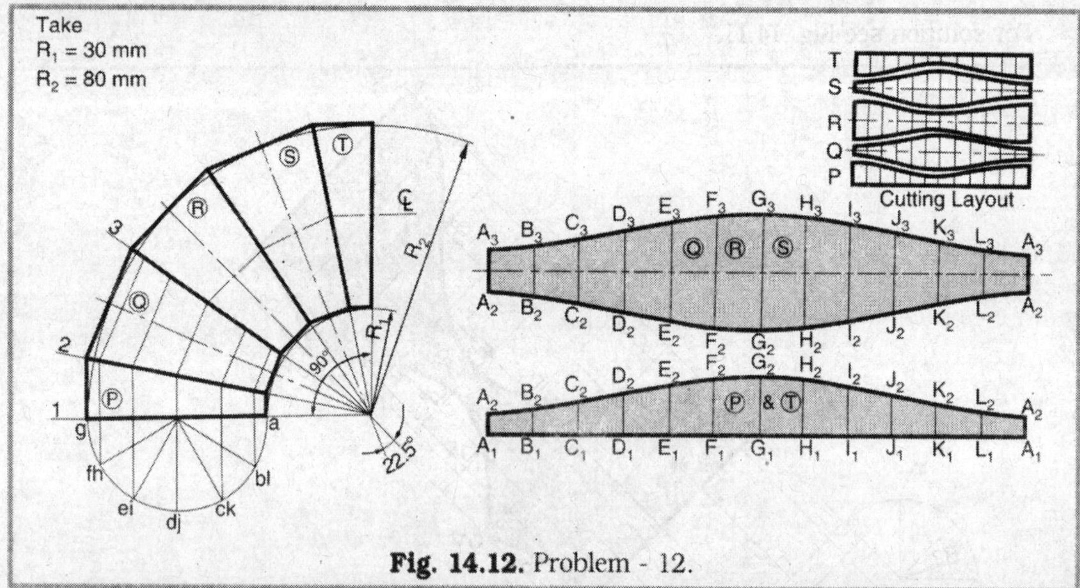

Fig. 14.12. Problem - 12.

Follow the method of problem No.10. Pieces P and T, and pieces Q, R and S are the same and hence only two developments are shown. From the available plate, how all five pieces are required to be cut to minimise wastage is shown as the cutting layout. Pieces P, R and T are cut according to development from generator G to G and not from A to A. This will help in cutting layout to save the material.

Problem 13 : A cylinder is resting on H.P. on its base. It is cut by flat cutting planes 1, 2, 4 and 5, as shown in the figure. A central hole is drilled inside the cylinder. It is shown as cutting No.3. Draw the development of the cut cylinder.

For solution see Fig. 14.13.

Divide the cylinder by 12 generators as usual and for the central hole take 8 divisions and generators passing through 8 divisions, as shown in the figure. Extra generators c_d, d_e, i_j and j_k are shown in elevation and development. These generators are drawn in development by taking the corresponding dimensions from plan.

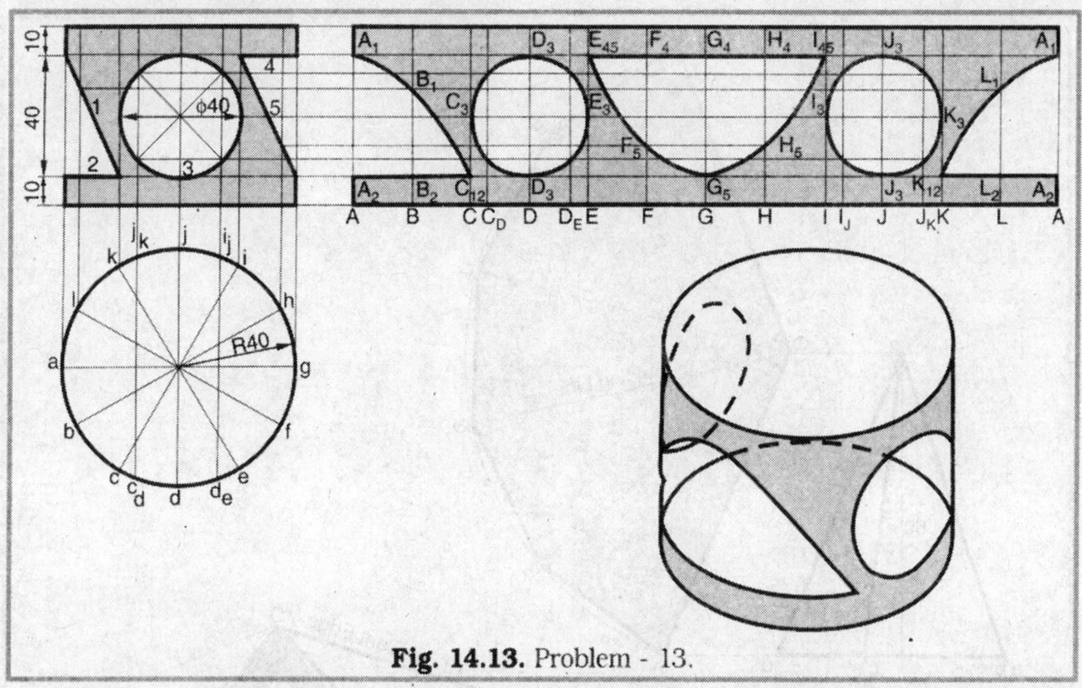

Fig. 14.13. Problem - 13.

Now points of intersections of generators with cuttings 1, 2, 3, 4 and 5 from elevation are transferred to development by simply drawing horizontal lines on corresponding generators. Join points obtained in development by a smooth curve to complete development.

4. Pyramids

All pyramids have at their one end (base) regular polygon of n sides and have n number of (Isosceles) triangular surfaces connecting regular polygon to the apex. So if the surfaces of the pyramid are unfolded in a single plane we get n number of equal isosceles triangles, having side of base of pyramid as one side base and slant edge of pyramid as other two sides. Moreover, we get a regular polygon of n sides. See Fig.14.14.

Regular polygon, for the base, is generally not drawn.

Problem 14 : A right regular pentagonal pyramid (30 × 60) is resting on H.P. on its base with one of the sides of base perpendicular to V.P. Draw its development.

For solution see Fig. 14.14 and follow the procedure, as given below :

(1) First draw plan and elevation.

(2) Plan (od) of slant edge OD is parallel to XY line and hence elevation o'd' will show T.L. In fact S.E. OD is parallel to V.P. in the given position.

(3) Draw OA parallel to o'd' as shown.

(4) Now draw an arc of a circle with O as the centre and the radius equal to (T.L.) true length of slant edge, as shown in Fig.14.14.

(5) On this arc cut AB = BC = CD = DE = EA = side of base of pyramid = ab = bc = cd = de = ea (Plan)

(6) Join AB, BC, CD, DE, EA, OA, OB, OC, OD and OE to complete the development.

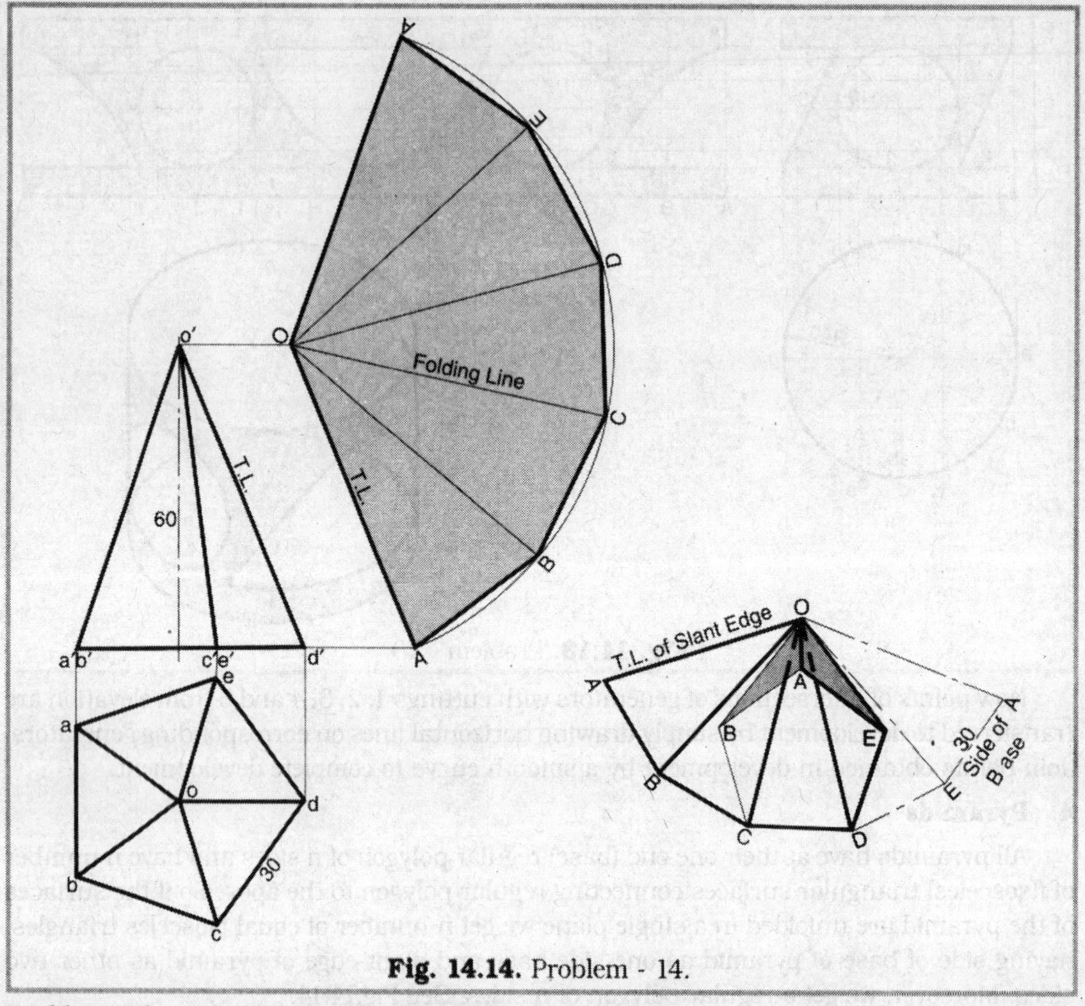

Fig. 14.14. Problem - 14.

Problem 15 : Pyramid in same position, as in Problem 14 is cut by a cutting plane No.1, as shown in Fig. 14.15. Draw the development of the cut pyramid.

For solution see Fig. 14.15 and follow the procedure, as shown below :

(1) First draw plan, elevation and development of the complete pyramid. Development is done from D to D instead of A to A, to get development in one piece.

(2) Draw cutting plane according to the given conditions.

(3) Cutting plane is cutting the following edges of the pyramid.

OA, OB, OC and OE ——— slant edges.

CD and DE ———— edges of base.

(4) For transferring cutting points of slant edges to the development draw first parallel lines to the base upto true length line (o' d') in elevation then to OD and then continue them as arcs of circles in development with O as the centre upto corresponding slant edges.

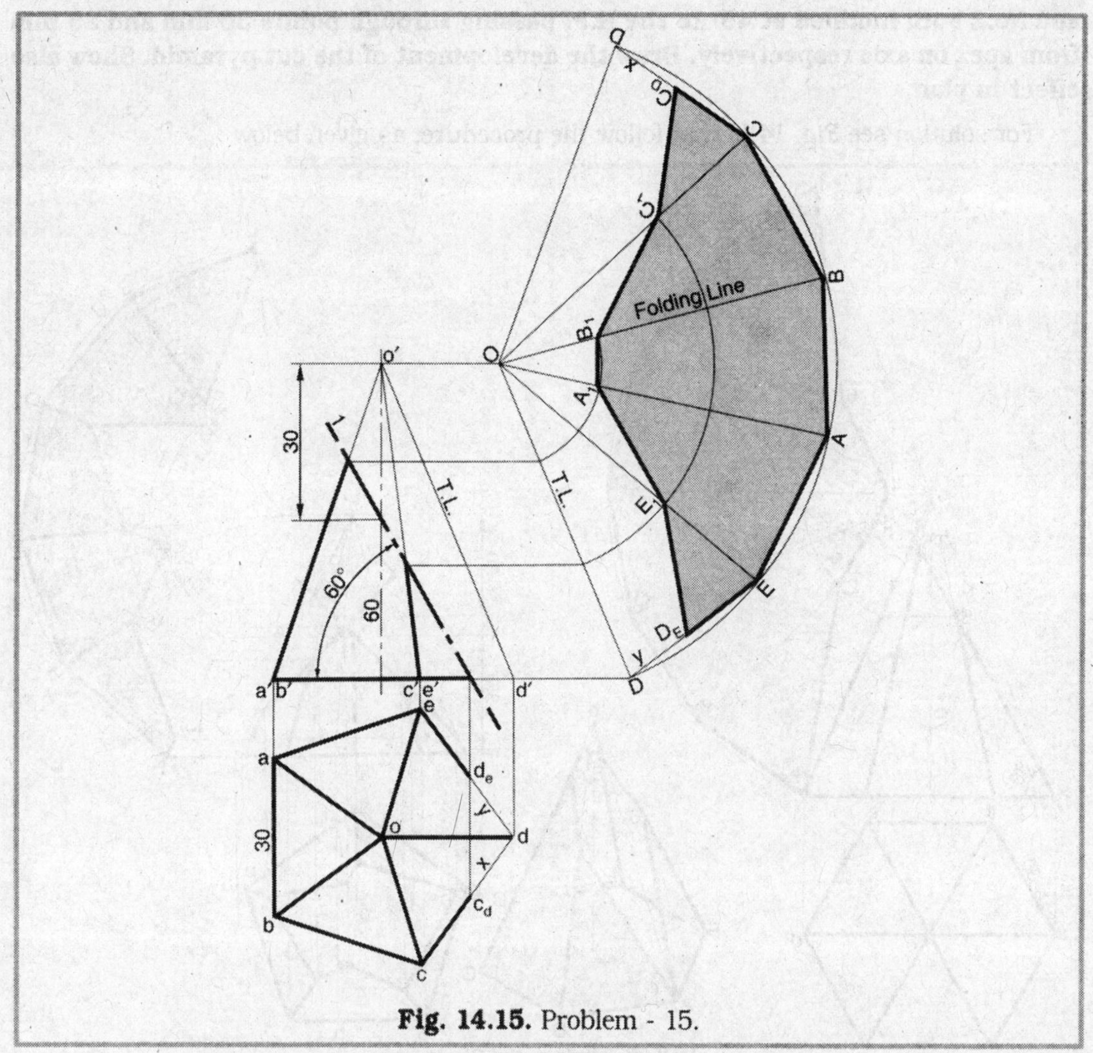

Fig. 14.15. Problem - 15.

(5) For transferring cutting points of the base to the development, transfer first cutting points to plan on corresponding sides of base. See points c_d and d_e. Then take dimensions x and y on the divider from plan and transfer them on CD and DE in development to get points C_D and D_E.

(6) Join these points by straight lines to complete development.

Problem 16 : A regular hexagonal pyramid (30 × 70) is resting on H.P. on its base with two edges of base parallel to V.P. It is cut by A.I.P. making 60° with H.P. and passing through one of the corners of the base. Draw the development of the truncated pyramid.

For solution see Fig.14.16. Method of solution is the same as studied earlier. Three dimensional view of the cut object will help in visualising the shape of the development. Compare lateral surfaces 2AB, 23CB, 34DC, 45ED, 56FE and 6AF of cut pyramid and the surfaces in the development.

Problem 17 : A pentagonal pyramid (40 × 70) is resting on H.P. on its base with one of the edges of base away from V.P. is parallel to V.P. It is cut by two A.I.P.s No.1

and No.2 both inclined at 45° to the H.P. passing through points 30 mm and 35 mm from apex on axis respectively. Draw the development of the cut pyramid. Show also effect in plan.

For solution see Fig. 14.17 and follow the procedure, as given below :

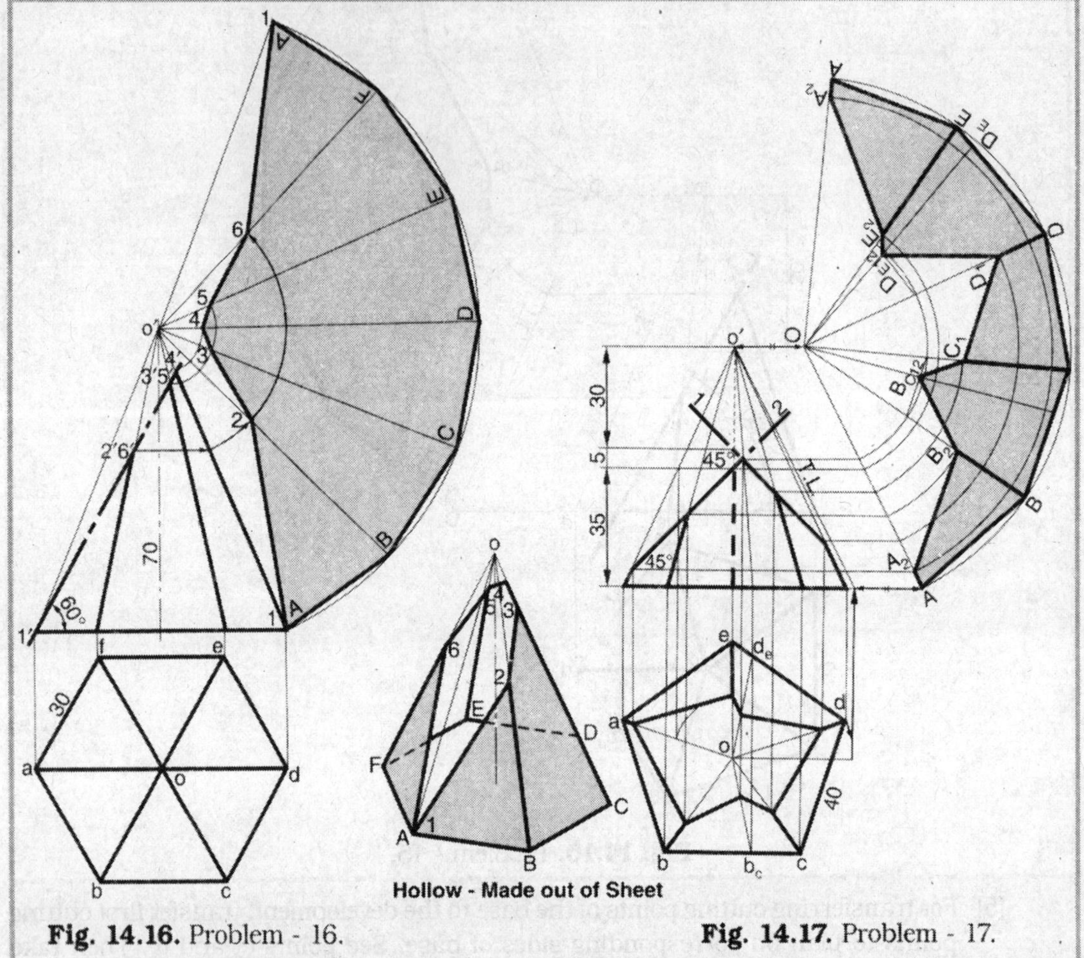

Fig. 14.16. Problem - 16.

Fig. 14.17. Problem - 17.

(1) In this problem first find out true length of slant edge by making plan of OD parallel to xy. Then as usual draw development of the complete pyramid, i.e. 5 Isosceles triangles.

(2) Take two extra lines $o'b'_c$ and $o'd'_e$ in elevation passing through a point of intersection of C.P.s 1 and 2. Transfer b'_c and d'_e in plan to get b_c and d_e. Mark points B_c and D_E in development by taking $BB_c = bb_c$ and $DD_E = dd_e$. Join OB_c and OD_E in development.

(3) Now transfer all cutting points of slant edges and extra lines with C.P.s 1 and 2 on corresponding lines of development by the usual method. Note the method of construction for locating points on extra lines in development.

(4) Join points to complete development.

(5) To get solution in third angle system simply draw T.V. above F.V.

Problem 18 : A hollow square pyramid, side of base 45 mm and height 65 mm, is resting on H.P. on its base with all sides of base equally inclined to V.P. A square hole of side 20 mm is drilled through the pyramid. Sides of the square hole are equally inclined to H.P. Axes of the pyramid and square hole intersect at right angle 20 mm above the base of the pyramid. Axis of the hole is perpendicular to V.P. Draw elevation, plan and development of the lateral surfaces of the pyramid.

For solution see Fig. 14.18. Solution is self explanatory.

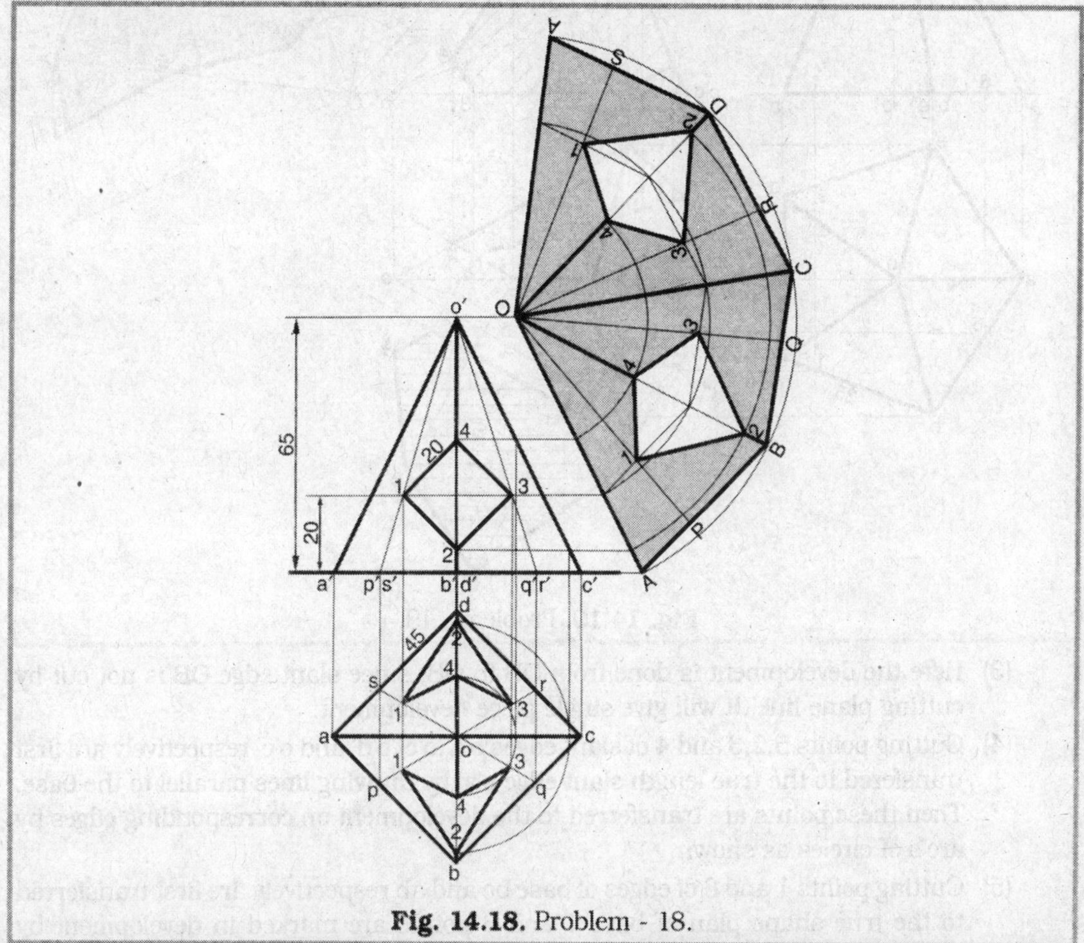

Fig. 14.18. Problem - 18.

Problem 19 : A pentagonal pyramid, side of base 35 mm and height 60 mm, is resting on H.P. on one of its triangular faces. It is cut by A.V.P. inclined to V.P. by 30° bisecting the axis. Draw sectional elevation, true shape of section and draw the development of lateral surfaces of the pentagonal pyramid. Assume axis of the pyramid parallel to V.P.

For solution see Fig. 14.19. and follow the procedure as given below.

(1) Draw the projections as per the given data. Draw cutting plane line on plan at 30° to the xy line, passing through the mid-point of the axis. Project apparent section and true shape of section as usual.

(2) For development o'a' in elevation represents the true length of the slant edge. Use

length o'a' as radius and draw arc of a circle with o as centre. Mark on this arc BC=CD=DE=EA=AB=side of base of pyramid = 35 mm.

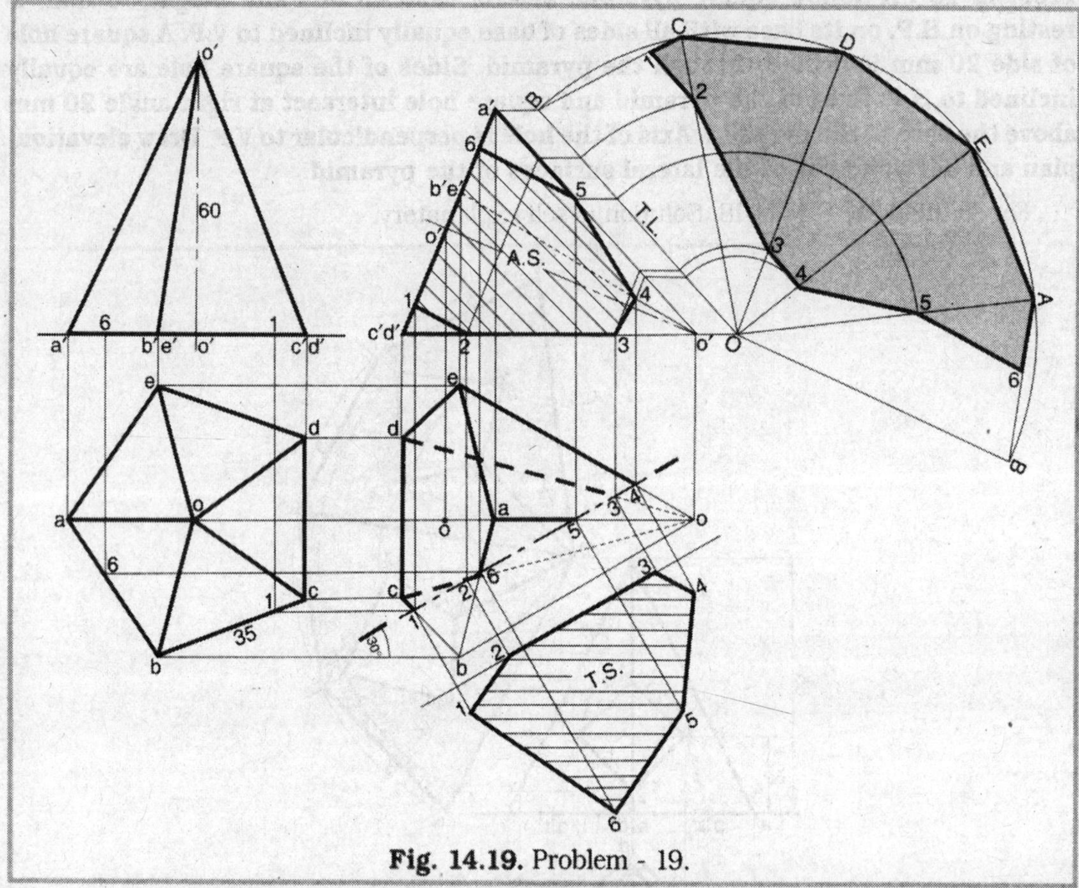

Fig. 14.19. Problem - 19.

(3) Here the development is done from OB to OB, since slant edge OB is not cut by cutting plane line. It will give single piece development.

(4) Cutting points 5,2,3 and 4 of slant edges o'a',o'c',o'd' and o'e' respectively are first transfered to the true length slant edge o'a' by drawing lines parallel to the base. Then these points are transferred to the development on corresponding edges by arc's of circles as shown.

(5) Cutting points 1 and 6 of edges of base bc and ab respectively are first transferred to the true shape plan of base 1 and 6 points are marked in development by measuring and transferring c1 = C1 and a6 = A6 by divider.

(6) Complete the development by joining points by straight lines as shown.

Problem 20 : A frustum of a square pyramid has its bottom of 60 mm side, top 30 mm side and height 70 mm. It is resting on H.P. on its bottom with two sides of bottom parallel to V.P. Draw the development of the frustum showing in it the string connecting the mid point of side of one face of top with the mid point of side of bottom of the opposite face by the shortest length. Show also string in plan and elevation.

For solution see Fig. 14.20 and follow the procedure, as given below :

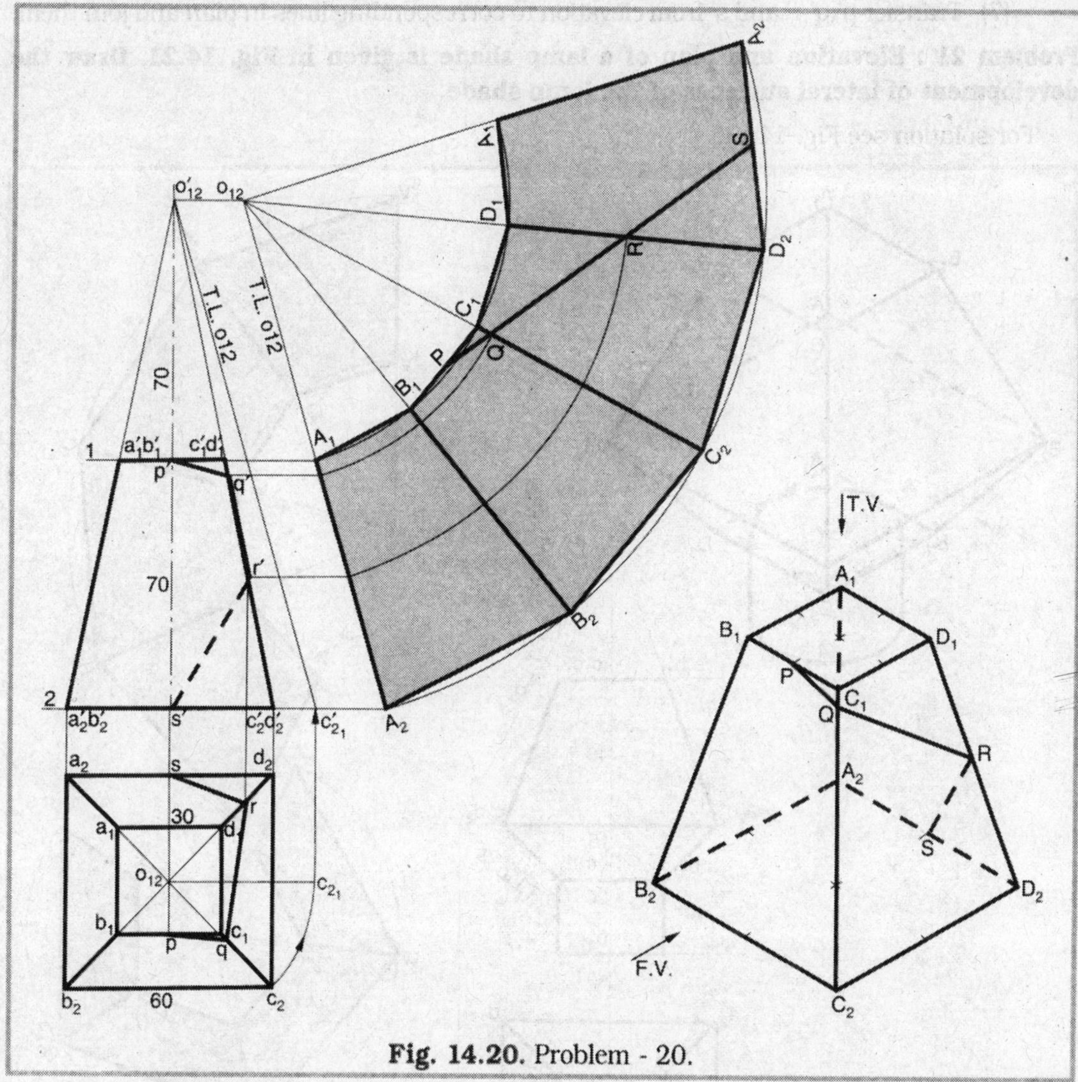

Fig. 14.20. Problem - 20.

(1) First draw plan and elevation of the frustum. Complete the pyramid and get the apex points O'_{12} and O_{12}, in elevation and plan respectively.

(2) Make $O_{12}C_2$ parallel to xy in plan to get true length in elevation.

(3) Draw the development of complete square pyramid as studied earlier and considering the top of frustum as horizontal cutting plane 1 get A_1, B_1, C_1, D_1 and A_1.

(4) In development take P, mid point of B_1C_1 and S as the mid point of D_2A_2 and join them by straight line.

(5) It is cutting slant edge C_1C_2 at Q and D_1D_2 at R.

(6) To transfer points Q and R from the development to the elevation first draw arcs of circle, from them, with O_{12} as the centre upto T.L. line and then horizontal lines upto corresponding lines of elevation $c'_1c'_2$ and $d'_1d'_2$ respectively. We get q' and r'. Now mark p' and s' mid points of $b'_1c'_1$ and $d'_2a'_2$ respectively. Join p'q', q'r' by full lines and r's' by dotted line.

(7) Transfer p',q',r' and s' from elevation to corresponding lines in plan and join them.

Problem 21 : Elevation and plan of a lamp shade is given in Fig. 14.21. Draw the development of lateral surfaces of the lamp shade.

For solution see Fig. 14.21.

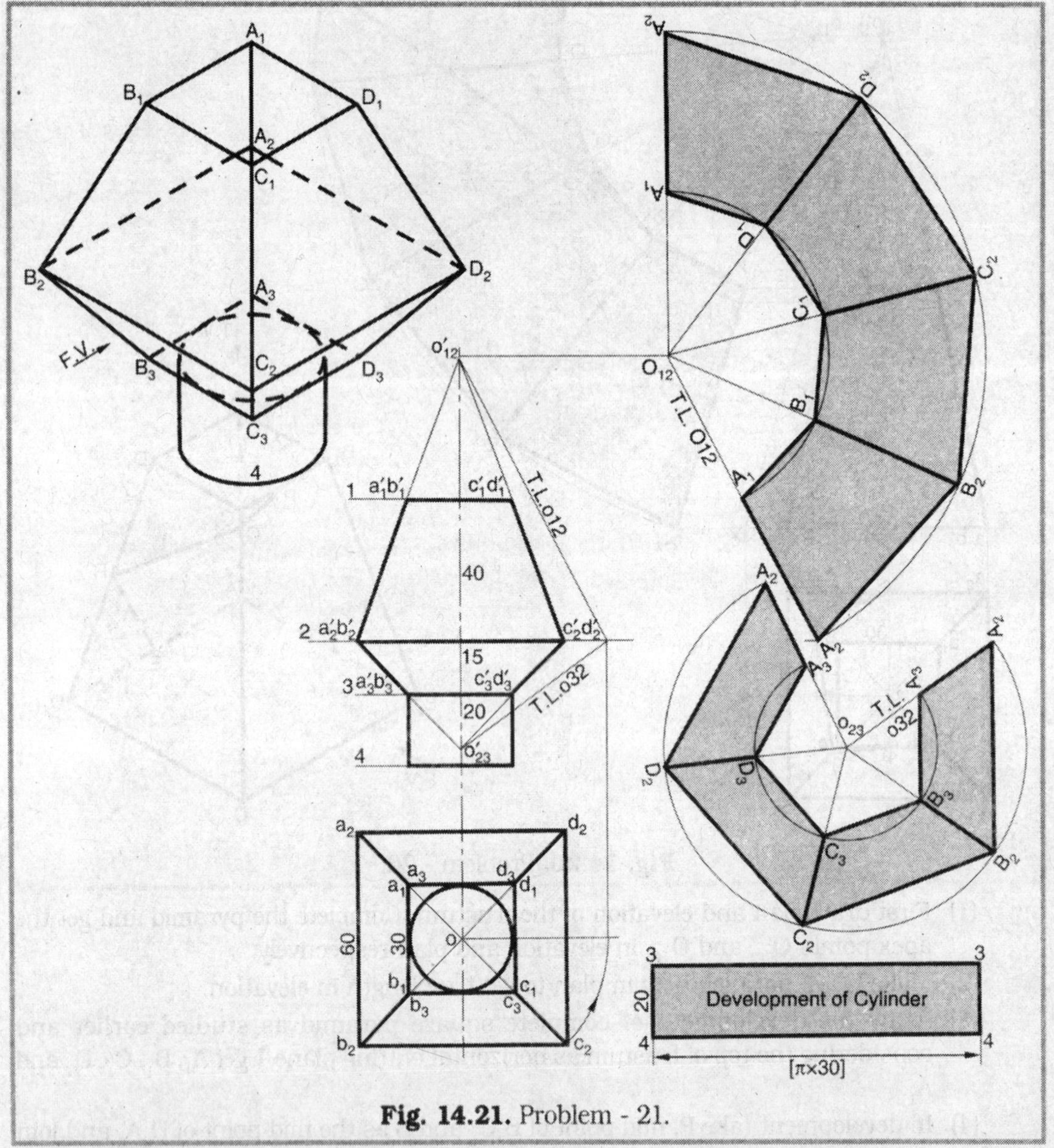

Fig. 14.21. Problem - 21.

Three dimensional drawing of the lamp shade is given in the figure for better understanding of problem. The lamp shade is made up of three hollow solids.

(1) Frustum of a square pyramid having cutting plane 1 and base 2.

(2) Inverted frustum of a square pyramid having base 2 and cutting plane 3, and

(3) The cylinder.

Develop three hollow solids separately, as usual.

Problem 22 : A square pyramid, side of base 50 mm and height 50 mm, is resting on H.P. on its base with all the edges of the base equally inclined to V.P. A rectangular slot, 40 mm wide and 15 mm high, is made in the centre at the bottom of the pyramid. Draw the development of the lateral surfaces of the pyramid. Show also the effect of the slot in the plan.

For solution see Fig. 14.22 and follow the procedure, as given below. Three dimensional drawing helps in understanding the solution of the problem.

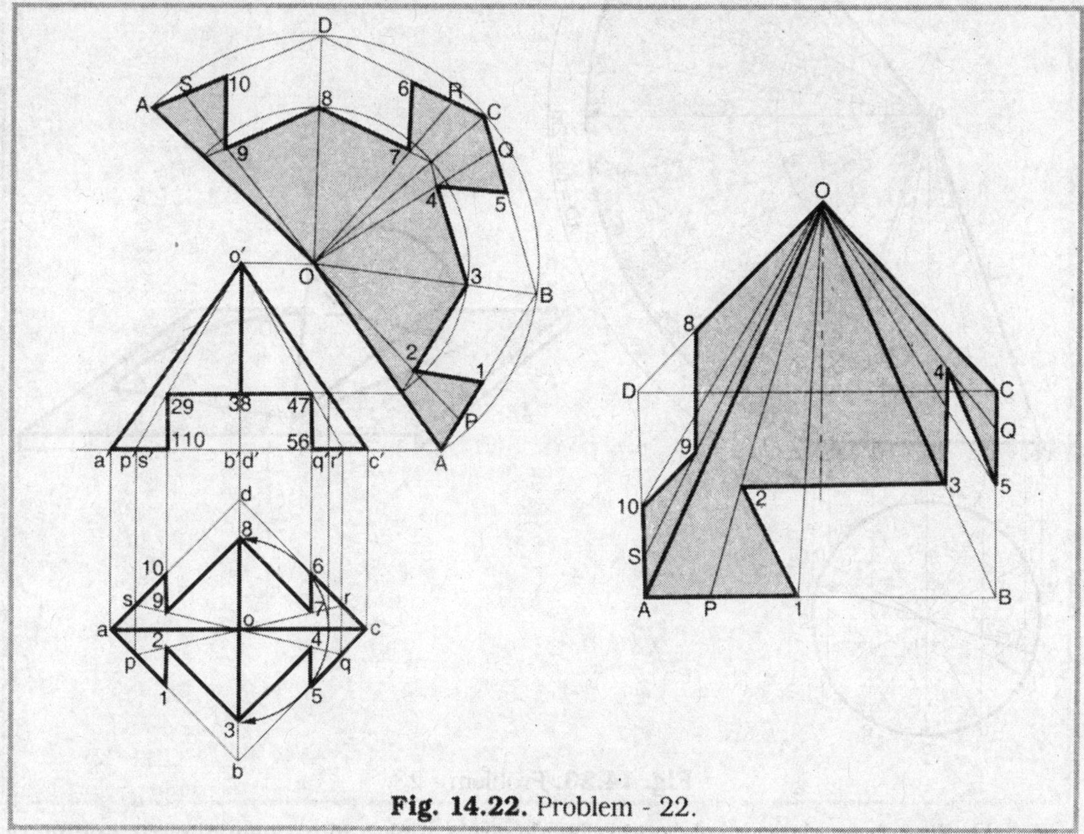

Fig. 14.22. Problem - 22.

Here in this problem extra lines are taken passing through corners of the rectangular slot in elevation. These lines are also shown in plan, in three dimensional drawing and in development. With the help of these lines many points are marked in development, other procedure is as usual. Note special construction adopted for locating points 2,4,7 and 9 in plan and in development.

5. Cones

Cone has at its one end (base) circle and has curved lateral surface connecting circle to the apex. So if the surfaces of the cone are unfolded as a single plane we get (1) one sector of a circle having its radius equal to the length of generator L, arc length of sector equal to πD and θ, angle of sector, equal to $(180° \times D/L)$ see Fig.14.23 and (2) circle of diameter D for base of cone. (Not shown in figure)

Problem 23 : A cone, diameter of base D and generator length L, is resting on H.P. on its base. Draw development of only lateral surface of cone.

For solution see Fig. 14.23 and follow the procedure, as shown below :

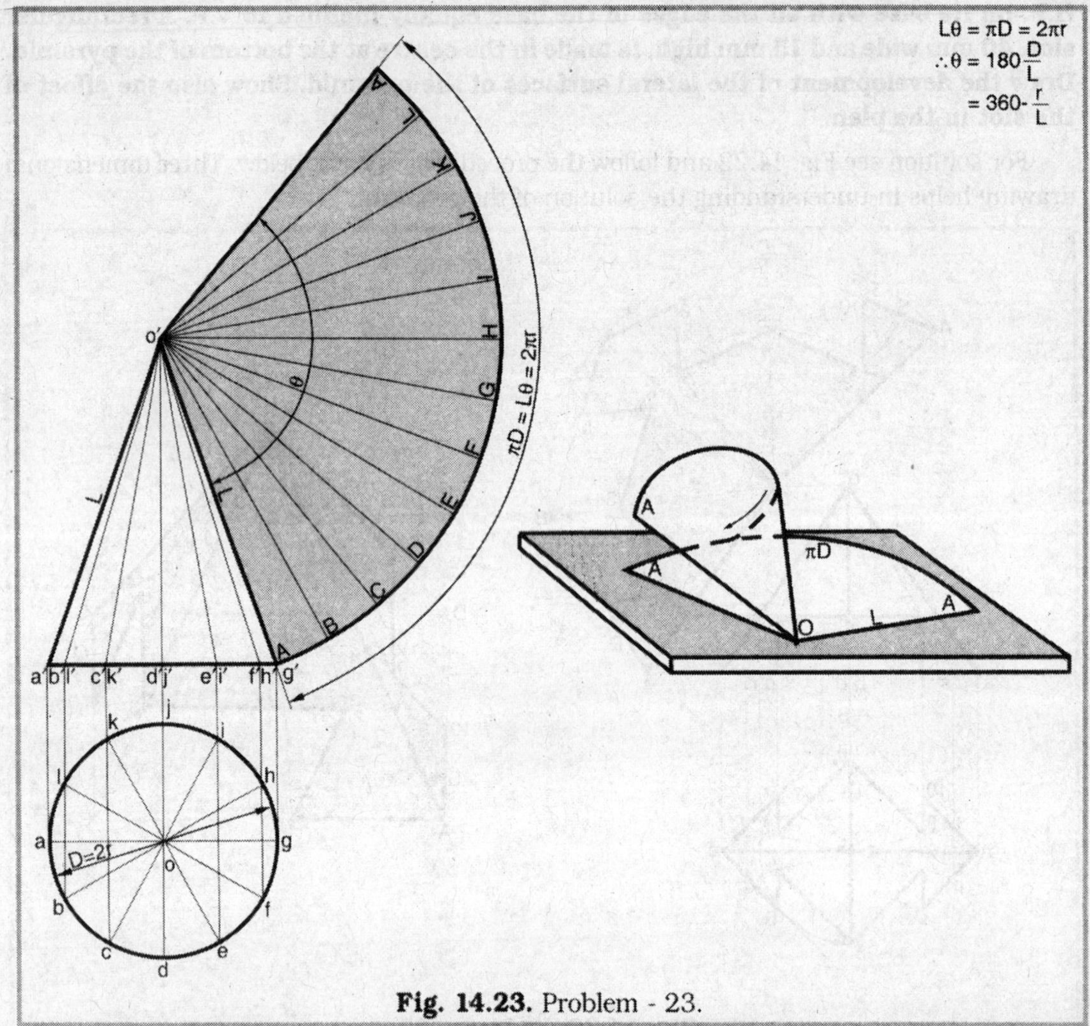

Fig. 14.23. Problem - 23.

(1) First draw plan and elevation. Divide circle of plan into 12 equal parts and mark 12 generators in plan and elevation.

(2) Now draw a circle with point O' as the centre and radius equal to L. On arc of the circle mark AB = BC = CD = ...= LA = ab = bc = cd = ...=la and get sector AO'AGA. This sector represents the development of the curved surface of the cone.

[OR]

(2) Draw angle AO'A = θ = 180 . D/L = 360°r/L and then divide angle into 12 equal parts. Where D is the diameter and r is the radius of the base circle of cone.

Problem 24 : A cone, radius of base 40 mm and slant height 120 mm, is resting on H.P. on its base. It is cut by three cutting planes perpendicular to V.P. and inclined to H.P. as shown. Draw the development of the portion of cone between cutting planes. Use exact method.

For solution see Fig. 14.24 and follow the procedure, as given below :

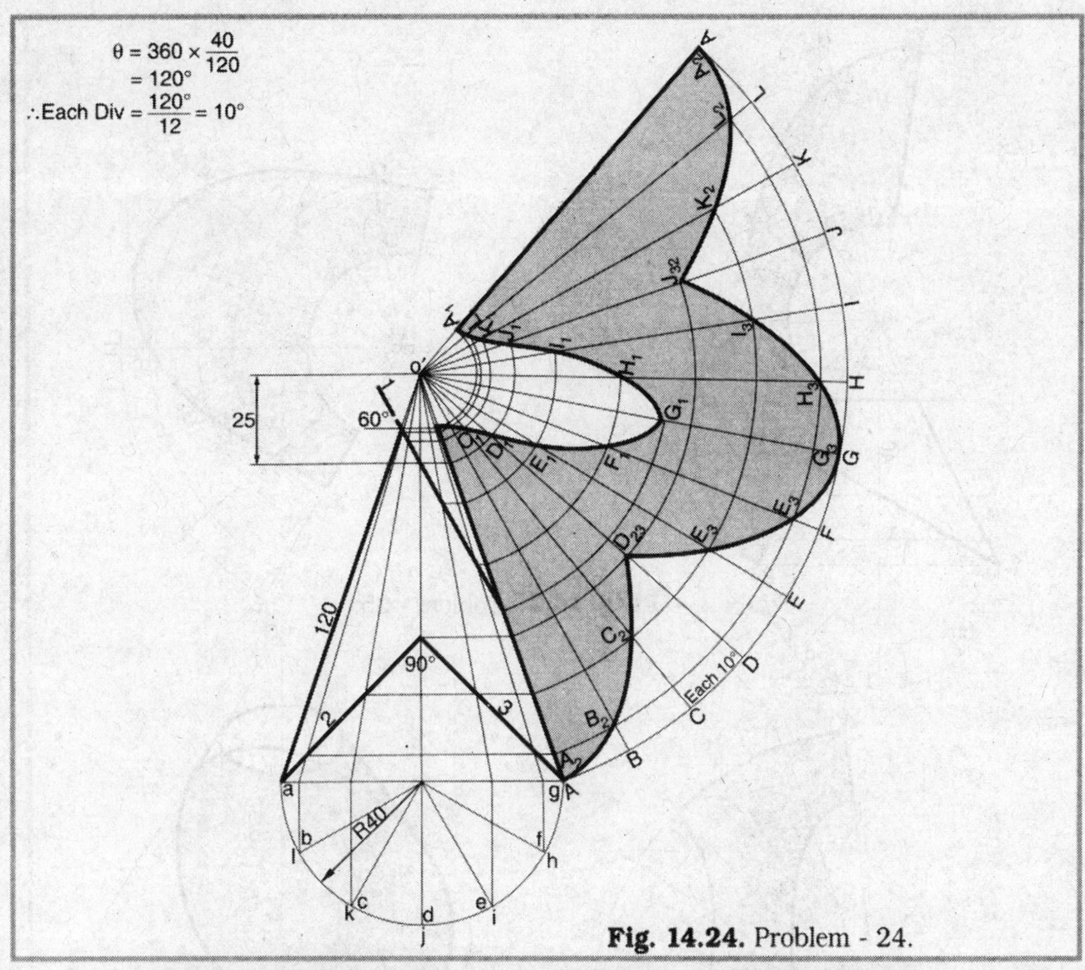

Fig. 14.24. Problem - 24.

(1) First draw elevation. Draw a semicircle adjoining elevation. Divide the semicircle into 6 parts to mark 12 generators on the curved surface of the cone. Measure true length of the generator o'g = 120 mm.

(2) Draw an arc of a circle with 120 mm radius and the centre o'. Take angle θ = 360 × 40/120 = 120° and divide θ = 120° into 12 equal parts and mark generators OA, OB, OC,......OL, OA in development, as shown in Fig. 14.24.

(3) Now transfer cutting points of 12 generators with C.P. 1,2, & 3 of elevation to development by lines parallel to base upto T.L. generator first and then by arcs of circles with o' as the centre upto the corresponding generators.

(4) Join points obtained by a smooth curve and complete the development.

Problem 25 : Elevation of a lamp shade in the form of frustum of a cone truncated at the bottom is given. Draw the development of the lamp shade.

For solution see Fig. 14.25.

Method of solution is similar to the previous problem. Three dimensional drawing is given to understand the shape of development. Cutting planes 1 and 2 generate points in development with suffix 1 and 2 respectively. Development is done in 2 ways as shown. One is for generator A - A and another is for E-E.

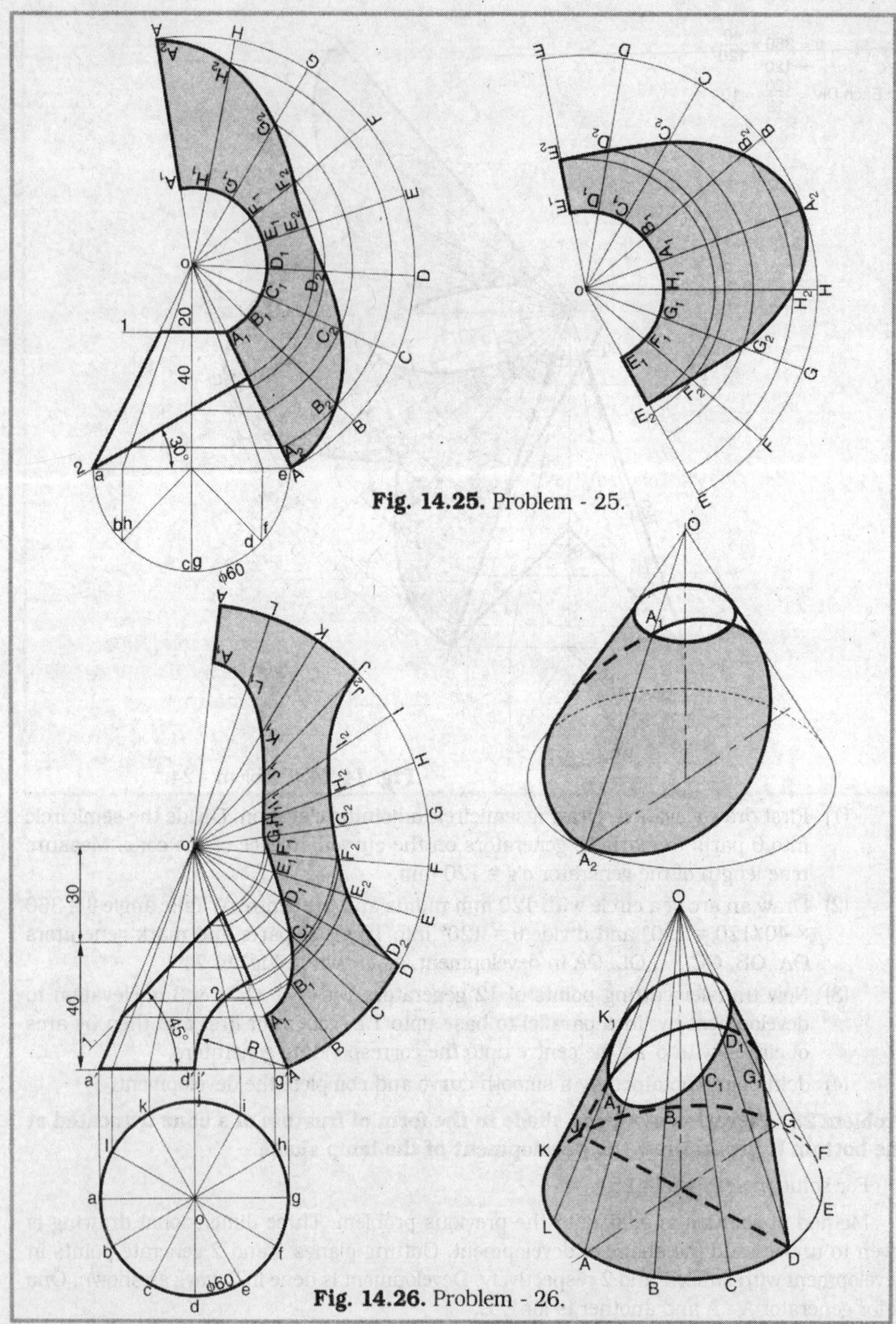

Fig. 14.25. Problem - 25.

Fig. 14.26. Problem - 26.

Problem 26 : Elevation of a cone, with cuttings 1 and 2, is given in Fig. 14.26. Draw the development of the cut portion of the cone.

For solution see Fig. 14.26. on previous page.

Observe three dimensional drawing to understand development. Solution is self explanatory. Draw plan above the F.V. if solution is required in 3rd Angle System.

Problem 27 : A cone is resting on H.P. on its base. A semicircular hole is drilled in the cone, as shown in Fig.14.27. Draw the development of the cone. Show also effect of cutting in plan.

For solution see Fig. 14.27 and follow the procedure, as given below :

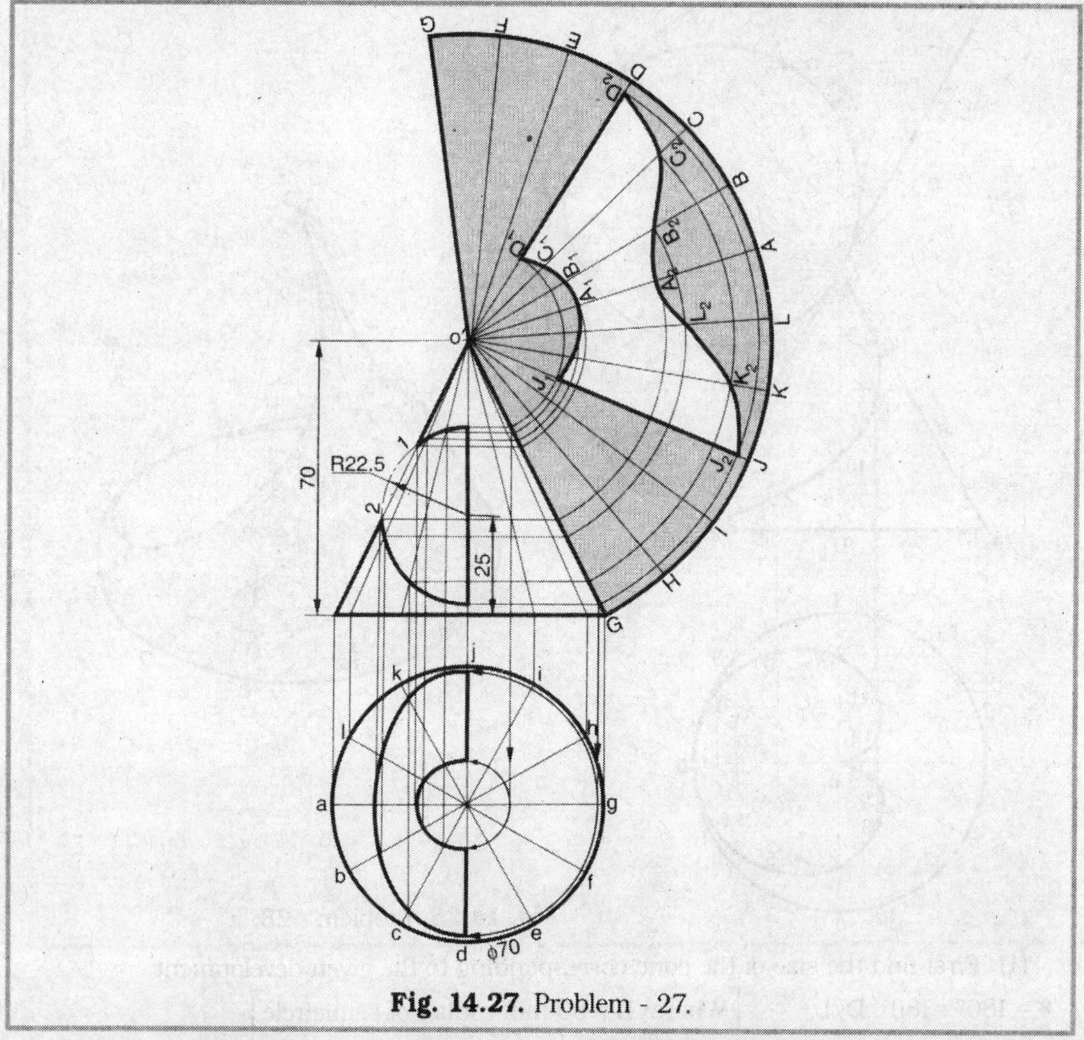

Fig. 14.27. Problem - 27.

(1) Draw plan and elevation of the full cone and draw also development of the full cone. Draw the semicircular cutting in elevation as shown. Mark 12 generators in plan, elevation and development.

(2) Transfer cutting points of 12 generators with semicircular hole of elevation to corresponding generators in plan and join them as shown.

(3) Transfer cutting points of generators with semicircular hole of elevation to development on corresponding generators by the usual method of horizontal lines and arcs of circles.

(4) Join the points obtained to complete the development.

Problem 28 : The development of a cone is a semicircle of 80 mm radius having a circular hole of 80 mm diameter. Draw the plan and elevation of the cone along with periphery of a circular hole shown on them.

For solution see Fig. 14.28 and follow the procedure, as given below :

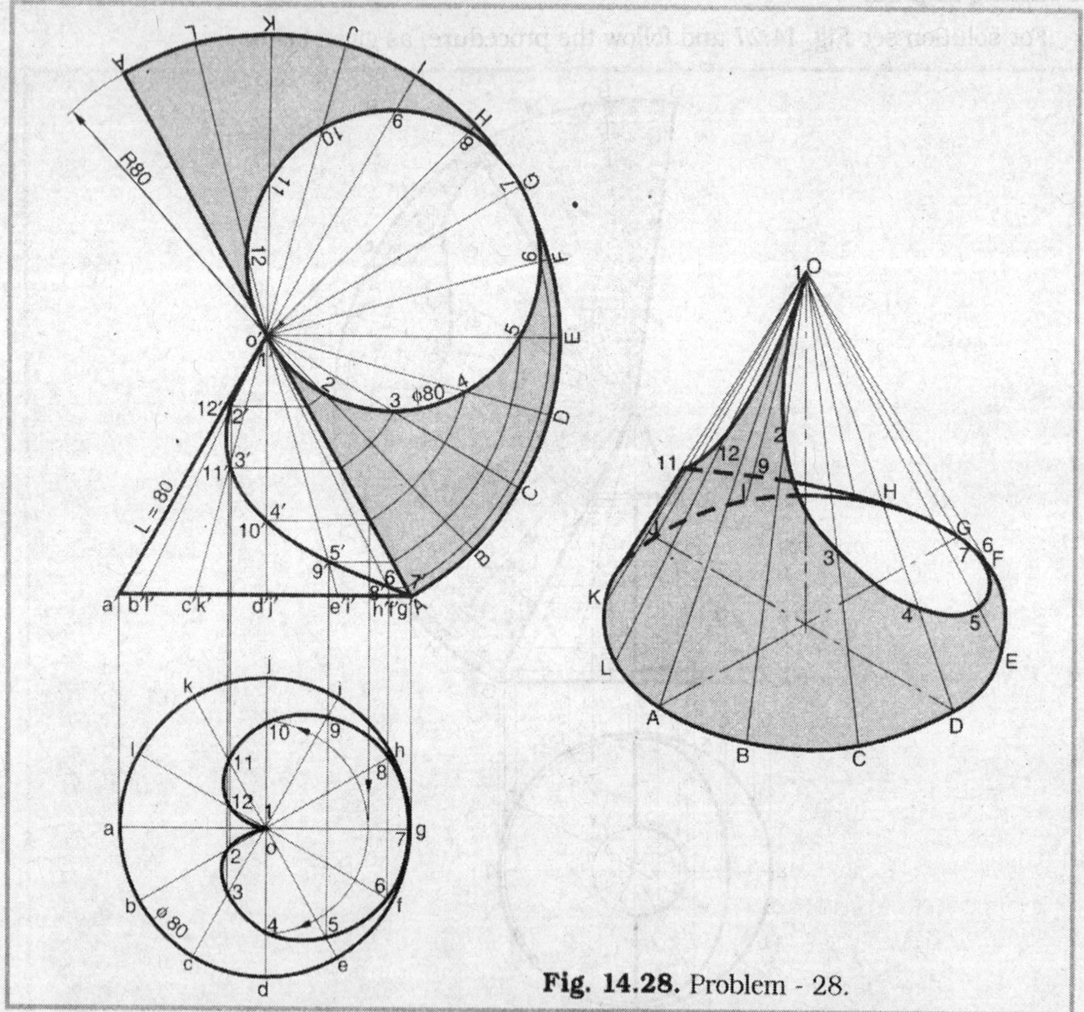

Fig. 14.28. Problem - 28.

(1) First find the size of the cone corresponding to the given development.

θ = 180° =180°. D/L [Where L = 80 mm radius of semicircle
L = length of generator of cone]

∴ D = 80 mm

(2) Draw plan and elevation of the cone, having diameter of base 80 mm and length of generator L = 80 mm. Draw adjoining to elevation a semicircular development, as shown in Fig. 14.28.

(3) Mark 12 generators in plan, elevation and development by the usual method. Draw a circular hole of 80 mm diameter in the development.

(4) Now transfer the cutting points 1, 2, 3,......12 from development to elevation first by drawing arcs of circles with o' as the centre upto T.L. generator and then by horizontal lines upto corresponding generators and get points 1', 2', 3',......12' respectively.

(5) Now draw projectors from 1', 2', 3',......12' to corresponding generators in plan to get points 1, 2, 3,....12.

(6) Join 1, 2, 3,.....12 and 1 and 1', 2', 3',......12' and 1' by a smooth curve to complete plan and elevation.

Problem 29 : Draw the plan and elevation of a cone resting on H.P. on its base. Show on them, the shortest path followed by a fly moving round the cone and returning to the same starting point. Fly starts from a point on the periphery of base. Take base diameter of cone 80 mm and height of axis 90 mm.

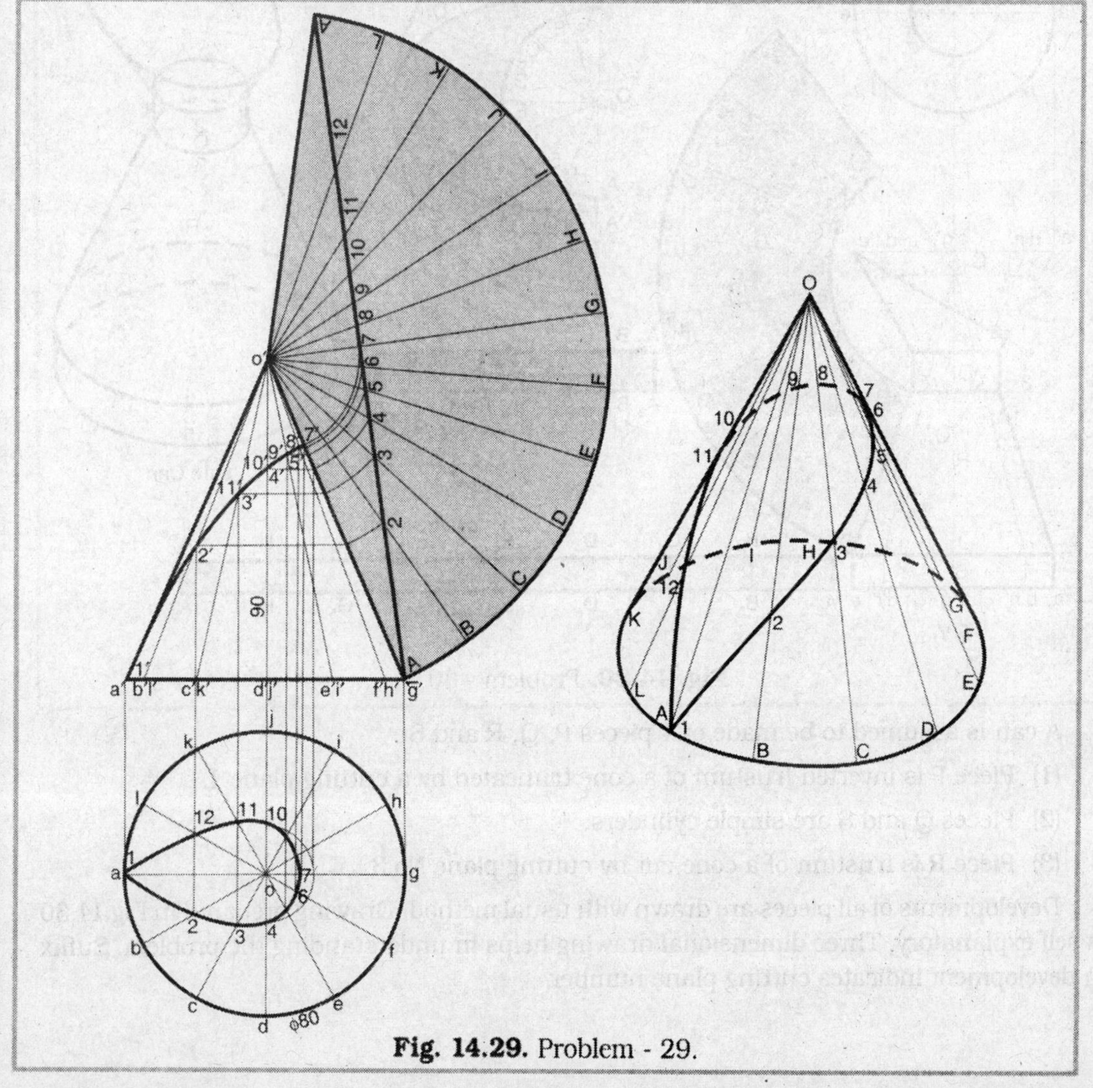

Fig. 14.29. Problem - 29.

For solution see Fig. 14.29 and follow the procedure, as given below :

(1) First draw plan, elevation and development of the cone. Mark on them 12 generators as usual.

(2) Join the point A with A in the development by a straight line. This straight line will represent the shortest path followed by the fly in the development.

(3) Transfer cutting points 1, 2, 3,.......12 and 1 from development to elevation and then to plan by the usual method on the corresponding generators. Join 1', 2', 3',.....12' and 1' and 1, 2, 3,.....12 and 1 in elevation and plan respectively to show path of the fly in them.

Problem 30 : Elevation of a 'Litre Can' is given in Fig.14.30. Prepare shape and size of tin sheets required to prepare the can.

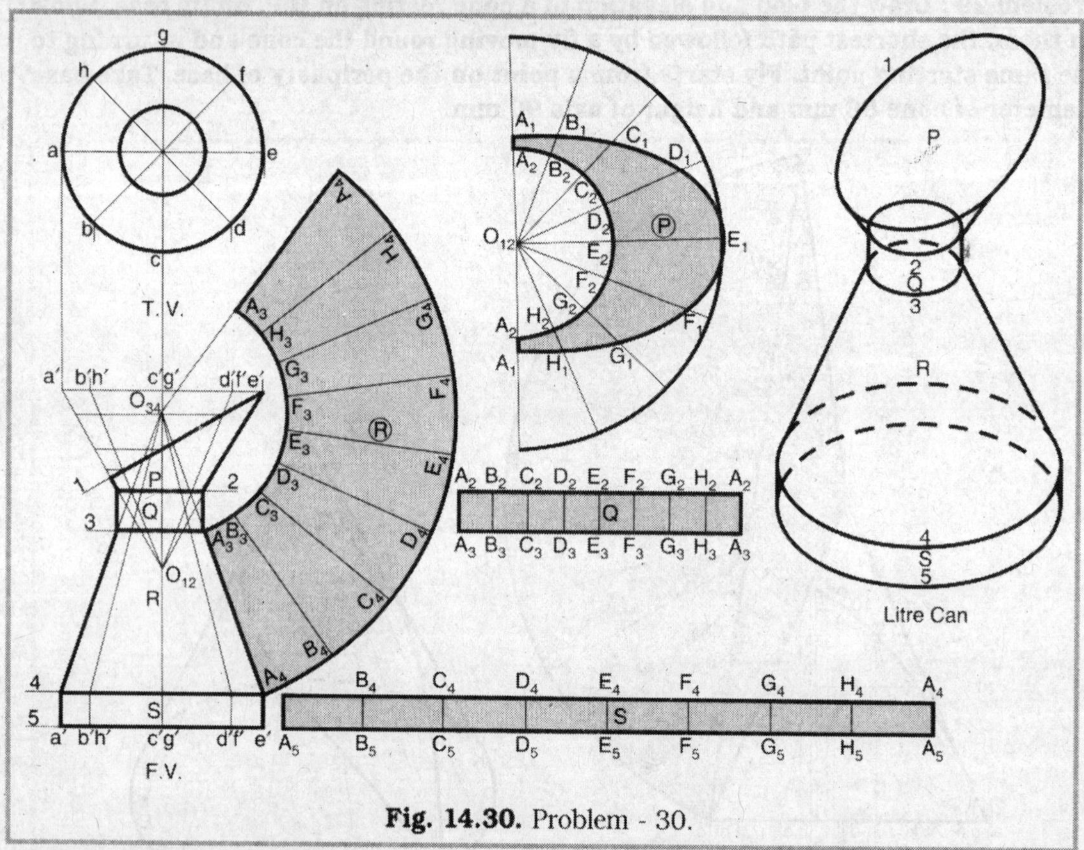

Fig. 14.30. Problem - 30.

A can is assumed to be made of 4 pieces P, Q, R and S..

(1) Piece P is inverted frustum of a cone truncated by a cutting plane 1.

(2) Pieces Q and S are simple cylinders.

(3) Piece R is frustum of a cone cut by cutting plane No.3.

Developments of all pieces are drawn with usual method. Drawing prepared in Fig.14.30 is self explanatory. Three dimensional drawing helps in understanding the problem. Suffix in development indicates cutting plane number.

6. Spheres

A sphere has totally curved surface. Development of the surface of a sphere can be done by any one of the two approximate methods. These two approximate methods are explained in the solutions of Problems 31 and 32.

Problem 31 : Draw the development of the surface of sphere by zone method.

For solution see Fig. 14.31 and follow the procedure, as given below :

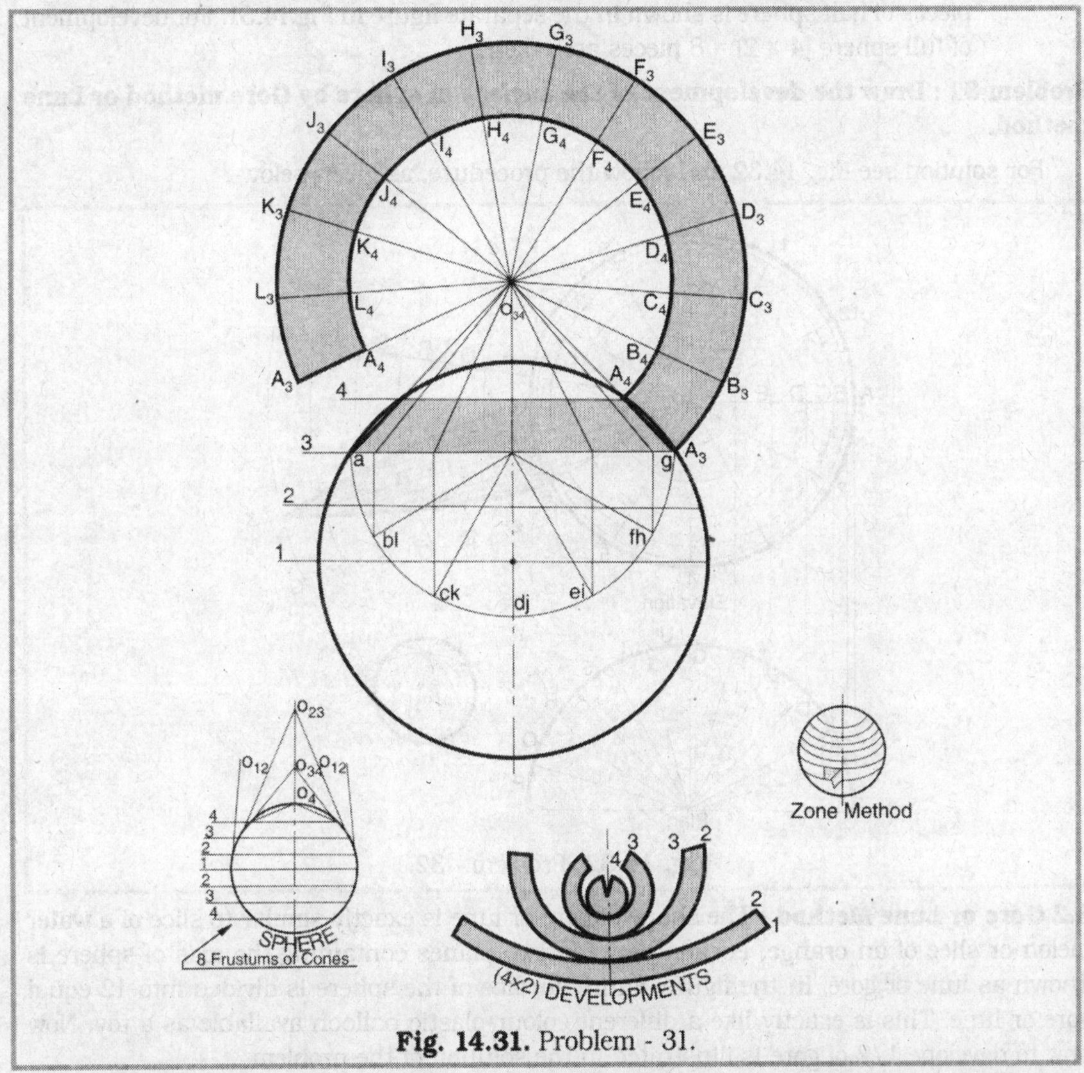

Fig. 14.31. Problem - 31.

6.1 Zone Method : Portion of the sphere between two parallel cutting planes perpendicular to the axis of the sphere is known as zone. In Fig.14.31 sphere is shown divided into many zones. For all practical purposes each zone can be considered as a frustum of a cone except the upper most and the lower most zone. They can be considered as cones of small height.

(1) Draw a circle to represent elevation of the sphere. Divide half circle into four zones by taking section or cutting planes 1, 2, 3 and 4 as shown in the figure.

(2) To illustrate the method of solution, zone between cutting planes 3 and 4 is taken.

(3) Zone between C.Ps. 3 and 4 is considered as a frustum of a cone. Complete the cone and get apex O_{34}. Draw a semicircle adjoining the base of the cone and divide it into 6 equal parts to mark 12 generators on the cone. Draw the development of the complete cone and mark points for cutting plane 4 by the method studied earlier in the problems of cone.

(4) Similarly, zones between 1 and 2, 2 and 3 and cone above 4 can be developed in the similar way and development of half sphere can be obtained. Development of 4 pieces of half sphere is shown in the separate figure in Fig.14.31. For development of full sphere (4 × 2) = 8 pieces are taken.

Problem 32 : Draw the development of the surface of sphere by Gore method or Lune method.

For solution see Fig. 14.32 and follow the procedure, as given below :

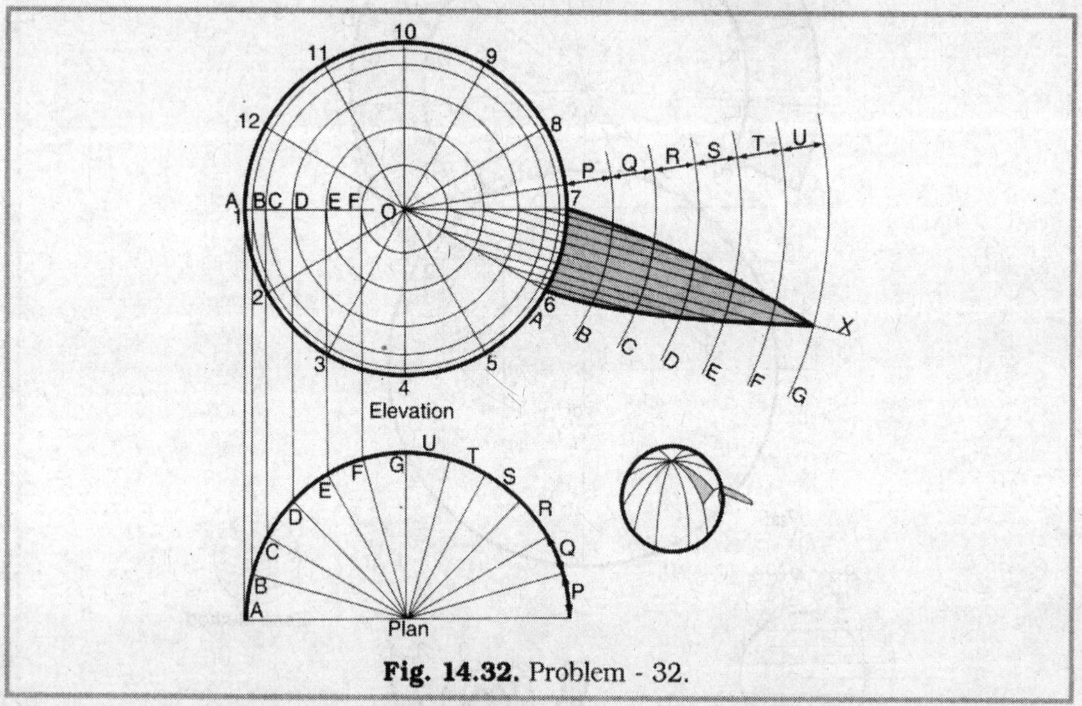

Fig. 14.32. Problem - 32.

6.2 Gore or Lune Method : The shape of gore or lune is exactly similar to slice of a water melon or slice of an orange. Portion between two planes containing the axis of sphere is known as lune or gore. In the figure, entire surface of the sphere is divided into 12 equal gore or lune. This is exactly like a different colour plastic balloon available as a toy. Now how to develope 1/2 of gore is illustrated in the solution of the problem.

(1) Draw elevation (circle) and plan (half circle) of half sphere. Divide elevation into 12 equal parts and circumference of plan into 12 equal parts and mark on it A, B, C,......Project A, B, C,....on the diameter of elevation and draw circles passing through them with the centre O.

(2) Draw arcs of circles with O as the centre and radii equal to (OA + P), (OA + P + Q), (OA + P + Q + R), and so on, as shown in the figure. Where P, Q, R,......etc. is the length of arc of small parts of plan.

(3) On these arcs project dimensions from arc of circles B, C, D,.... as shown in the figure. Join all points by a smooth curve to get development of 1/2 gore. Such 24, 1/2 gore is the development of a complete sphere.

7. Miscellaneous Problems :

Having studied the development of prisms, cylinders, pyramids, cones and spheres, let us study a few miscellaneous problems.

Problem 33 : Solid made up of a half cylinder, half hexagonal prism is resting on H.P. on its base with edge of prism perpendicular to V.P. It is cut by two flat cutting planes 1 and 2 and circular cutting plane 3, as shown in elevation in Fig.14.33. Draw the development of the surfaces of the cut solid, between 3 cutting planes.

For solution see Fig. 14.33 and follow the procedure, as given below :

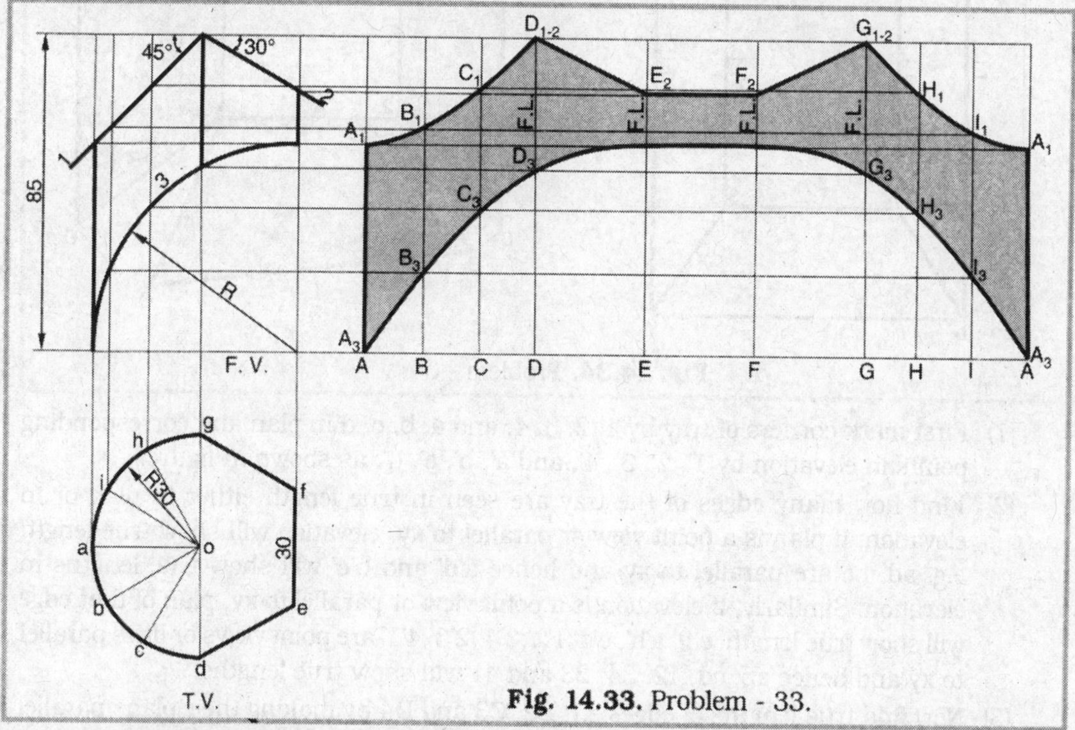

Fig. 14.33. Problem - 33.

(1) First draw plan and elevation of the composite solid. Divide cylindrical portion into 6 equal parts and mark 6 generators. Draw three cutting planes on elevation. Draw the development of full solid and mark on it 6 generators of a cylindrical portion and 4 vertical edges of prism.

(2) Now transfer cutting points of generators and vertical edges with cutting planes from elevation to development, on the corresponding lines, simply by drawing horizontal lines.

(3) Join points obtained by flat cutting with prism by straight lines and all other points by a curve to complete development Suffix indicates the C.P. number.

Problem 34 : Elevation and plan of an inverted truncated frustum of an oblique rectangular pyramid shape tray are given in Fig.14.34. Draw the development of the surfaces of the tray.

For solution see Fig. 14.34 and follow the procedure, as given below :

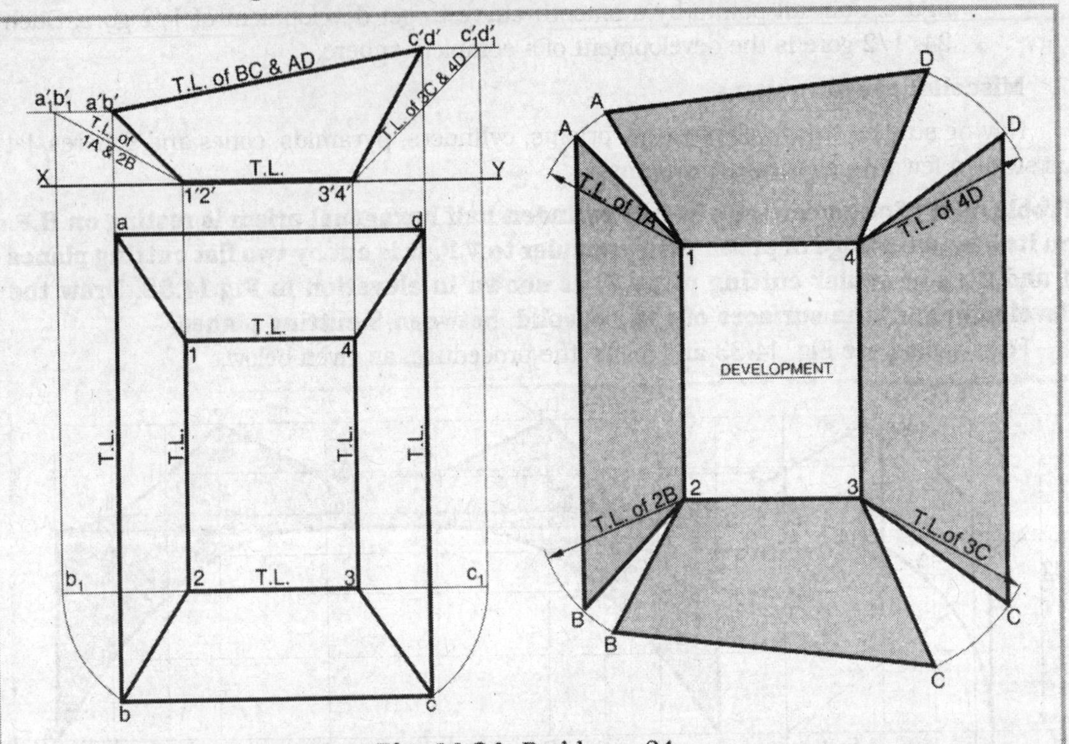

Fig. 14.34. Problem - 34.

(1) First mark corners of tray by 1, 2, 3, 4, and a, b, c, d in plan and corresponding points in elevation by 1', 2', 3', 4', and a', b', c', d', as shown in figure.

(2) Find how many edges of the tray are seen in true length either in plan or in elevation. If plan is a point view or parallel to xy, elevation will show true length e.g. ad, bc are parallel to xy and hence a'd' and b'c' will show true lengths in elevation. Similarly, if elevation is a point view or parallel to xy, plan of that edge will show true length. e.g. a'b', c'd',1'2',3'4',2'3',4'1' are point views or lines parallel to xy and hence ab, cd, 12, 34, 23 and 41 will show true lengths.

(3) Now find true lengths of edges A1, B2, C3 and D4 by making their plans parallel to xy and getting their elevations as true lengths.

(4) Now knowing the true lengths of all edges draw each face of tray with true lengths, as shown in the figure to get the development of the tray.

Problem 35 : Elevation and plan of a transition piece of a duct, circular at the top and rectangular at the bottom, is given in Fig.14.35. Draw the development of the surfaces of the duct piece.

Shape of the duct piece is shown in three dimensional drawing. Divide the circle into 12 parts and mark them by 1, 2, 3,....12. Mark rectangular base by A, B, C, D and E, as shown in the figure. Find true lengths of all possible edges and lines in elevation by making their plans parallel to xy as shown in the figure. Draw each face of duct piece by available true lengths of edges and lines to get development.

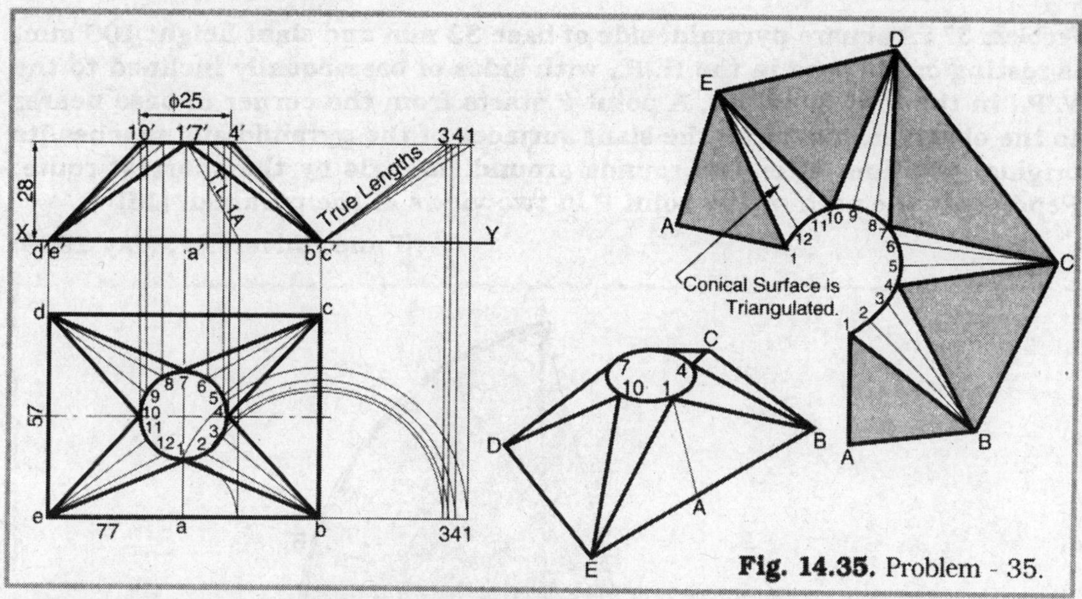

Fig. 14.35. Problem - 35.

Problem 36 : Draw a semi-circle of 150 mm diameter and inscribe in it the largest possible equilateral triangle. The semi-circle represents the lateral surface development of a cone and triangle as a line on its surface. Draw the projections of the cone resting on its base on the H. P. in first quadrant, showing the line in both the views. [10]

[Similar to Pune University, May 1995]

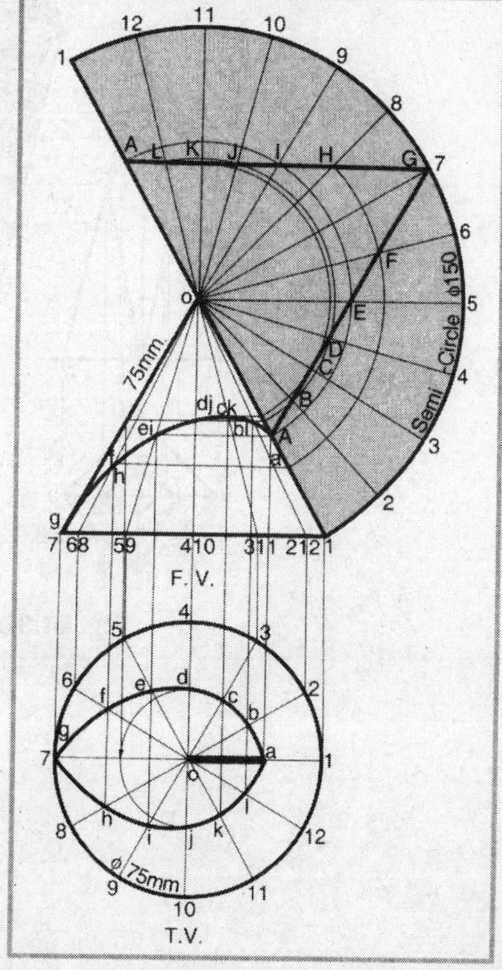

Fig. 14.36. Problem - 36.

Problem 37 : A square pyramid, side of base 32 mm and slant height 100 mm, is resting on its base in the H. P., with sides of base equally inclined to the V. P., in the first quadrant. A point P starts from the corner of base nearer to the observer, moves on the slant surfaces of the pyramid and reaches its original position, after two rounds around the axis by the shortest route. Represents the path of the point P in two views of the pyramid. [15]

[Pune University, May 1999]

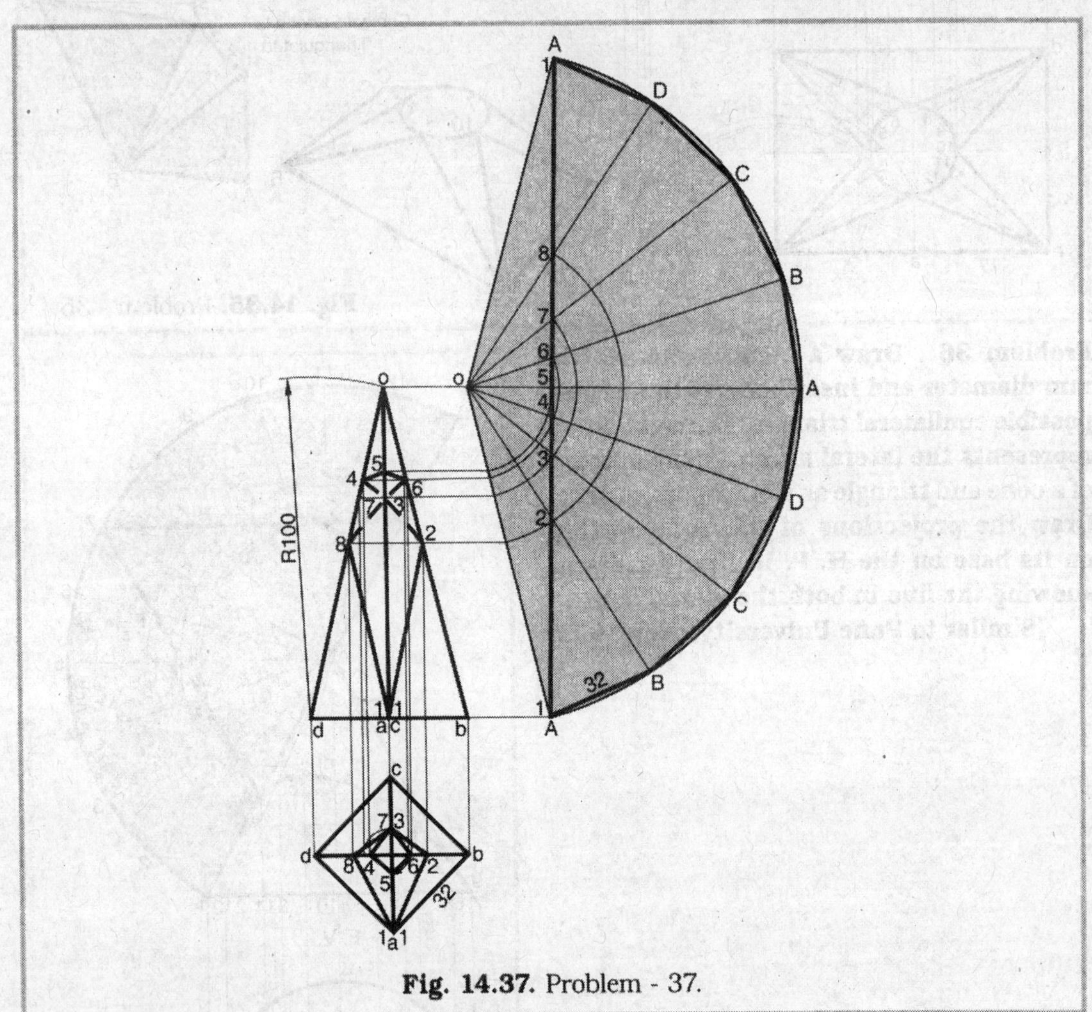

Fig. 14.37. Problem - 37.

Problem 38 : A square pyramid, base 60 mm side and axis 90 mm long, is resting on its base on the H. P. with the sides of the base equally inclined to the V. P. A vertical square hole is drilled through the pyramid such that their axes coincide, while a face of it is parallel to the V. P. The diagonal of the hole is 60 mm. Draw the projections showing lines of intersection. Draw the lateral surface development of the pyramid with hole. [16]

[Pune University, December 1995]

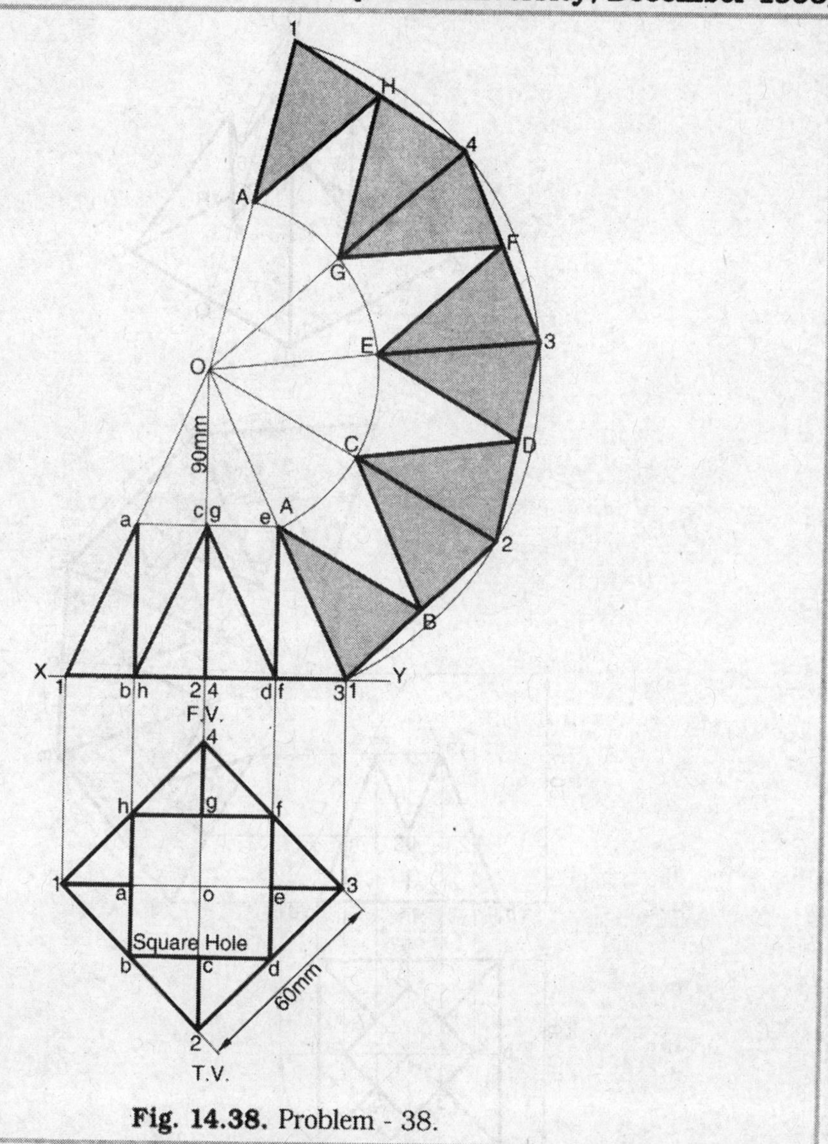

Fig. 14.38. Problem - 38.

Problem 39 : A square pyramid (side of base 50 mm and axis 60 mm] is kept on the H. P. on its base with two of its sides of base parallel to V. P. An axial square hole, of side 25 mm, is cut through the pyramid, so that the vertical faces of the hole are equally inclined to V. P. Draw projections of the solid and show the development of the lateral surface of the solid. [16]

[Pune University]

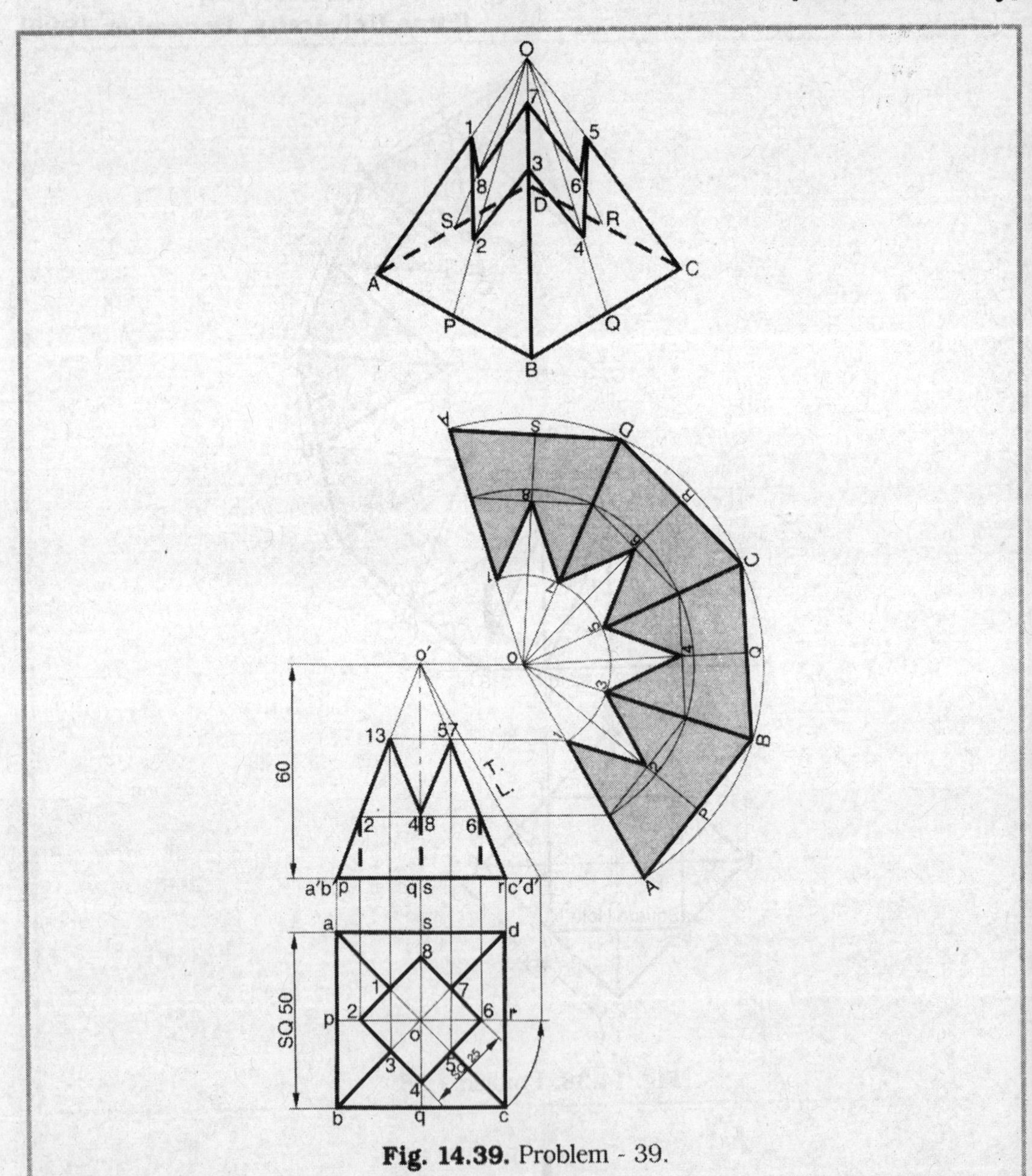

Fig. 14.39. Problem - 39.

EXERCISE

Note : In Figs.14.40 to 14.51 Plans or T.V.s are given in 1st Angle System without effect of cutting of object shown therein.

1. In Fig.14.40, plan and elevation of a cut square prism are given. Draw the complete development of the square prism.
2. In Fig.14.41, T.V. and F.V. of cut pentagonal prism are given. Develop only lateral surfaces of prism.
3. In Fig.14.42, T.V. and F.V. of cut hexagonal prism, cut by flat and curved cutting planes, are given. Draw the complete development of the prism.
4. In Fig.14.43, T.V. and F.V. of hexagonal prism with through circular hole are given. Develop only lateral surfaces of prism.
5. In Fig.14.44, plan and F.V. of cylinder with two flat cuttings and one semicircular cutting are shown. Draw the development of only lateral surfaces of the cylinder.
6. In Fig.14.45, plan and F.V. of a cylinder with cutting of arrow shape are given. Draw the development of the cut cylinder.
7. In Fig.14.46, plan and F.V. of cylinder with half circular and half square hole are given. Draw the development.
8. In Fig.14.47, plan and F.V. of cylinder with two flat cuttings are given. Draw the development.
9. In Fig.14.48, T.V. and F.V. of a square pyramid with flat cutting are given. Draw the complete development.
10. In Fig.14.49, plan and elevation of pentagonal pyramid with one cutting are given. Draw the development.
11. In Fig.14.50, T.V. and F.V. of hexagonal pyramid with half round cutting near the base are given. Draw the complete development.
12. In Fig.14.51, plan and elevation of composite solid (half cone and half hexagonal pyramid) are given with two cuttings. Draw the development.
13. In Fig.14.52, F.V. of a cone, with two flat cuttings, is shown. Draw the development.
14. In Fig.14.53, F.V. of cone, with two flat cuttings, is shown. Draw the development.
15. In Fig.14.54, F.V. of cone, with rectangular cutting near the base, is given. Draw the development.
16. In Fig.14.55, F.V. of cone, with round hole is given. Draw the development of surfaces of the cone.
17. Elevation of a cylindrical chimney, coming out from a roof, is given in Fig.14.56. Draw the development of surfaces of A and B portions of the chimney.
18. Elevation of a scoop (Quarter spherical and 2 pieces half cylindrical) is given in Fig.14.57. Draw the development.

19. F.V. of 'T'. pipe joint is given in Fig.14.58. Draw the development of both the pieces.

20. F.V. of round section 4 pieces 90° bend duct is given in Fig.14.59. Draw the development of all pieces.

21. F.V. of 'Y' pipe joint is given in Fig.14.60. Draw the development of all pieces.

22. F.V. of A Litre Can is given in Fig.14.61. Draw the development of surfaces of the can.

23. T.V. and F.V. of a square to square transition piece are given in 3rd Angle System in Fig.14.62. Draw the development of surfaces.

24. T.V. and F.V. of round to rectangular transition piece are given in 3rd Angle System in Fig.14.63. Draw the development of surfaces.

25. F.V. of off-set hopper is given in Fig.14.64. Draw the development of surfaces of all pieces.

26. Development of the cylinder is given in Fig.14.65. Draw the elevation of the cylinder showing line in it.

27. Development of the cone is given as a semicircle with a circular hole of maximum size in it. See Fig.14.66. Draw the plan and elevation of the cone showing circle in each view.

28. Development of the surfaces of cone with straight line on it is given in Fig.14.67. Draw the projections of a cone showing line in each projection.

29. Development of the surface of cylinder, with semicircular line on it, is given in Fig.14.68. Draw the projections of the cylinder showing semicircular line in F.V.

30. Development of the surface of a cone, with square cut in it, is given in Fig.14.69. Draw the projections of the cone showing cut in both the projections.

31. Development of the surface of a cone, with square and two semi-square cuts in it, is given in Fig.14.70. Draw the projections of the cone showing cuts in both the projections.

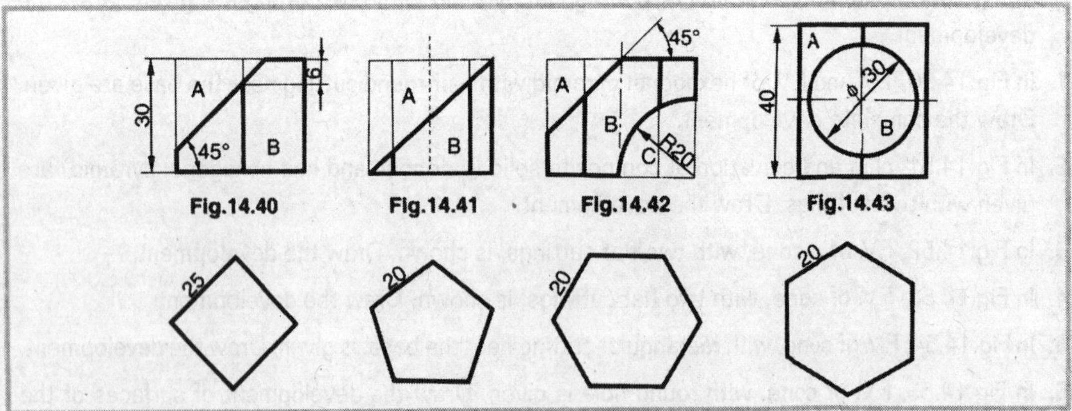

Fig.14.40 Fig.14.41 Fig.14.42 Fig.14.43

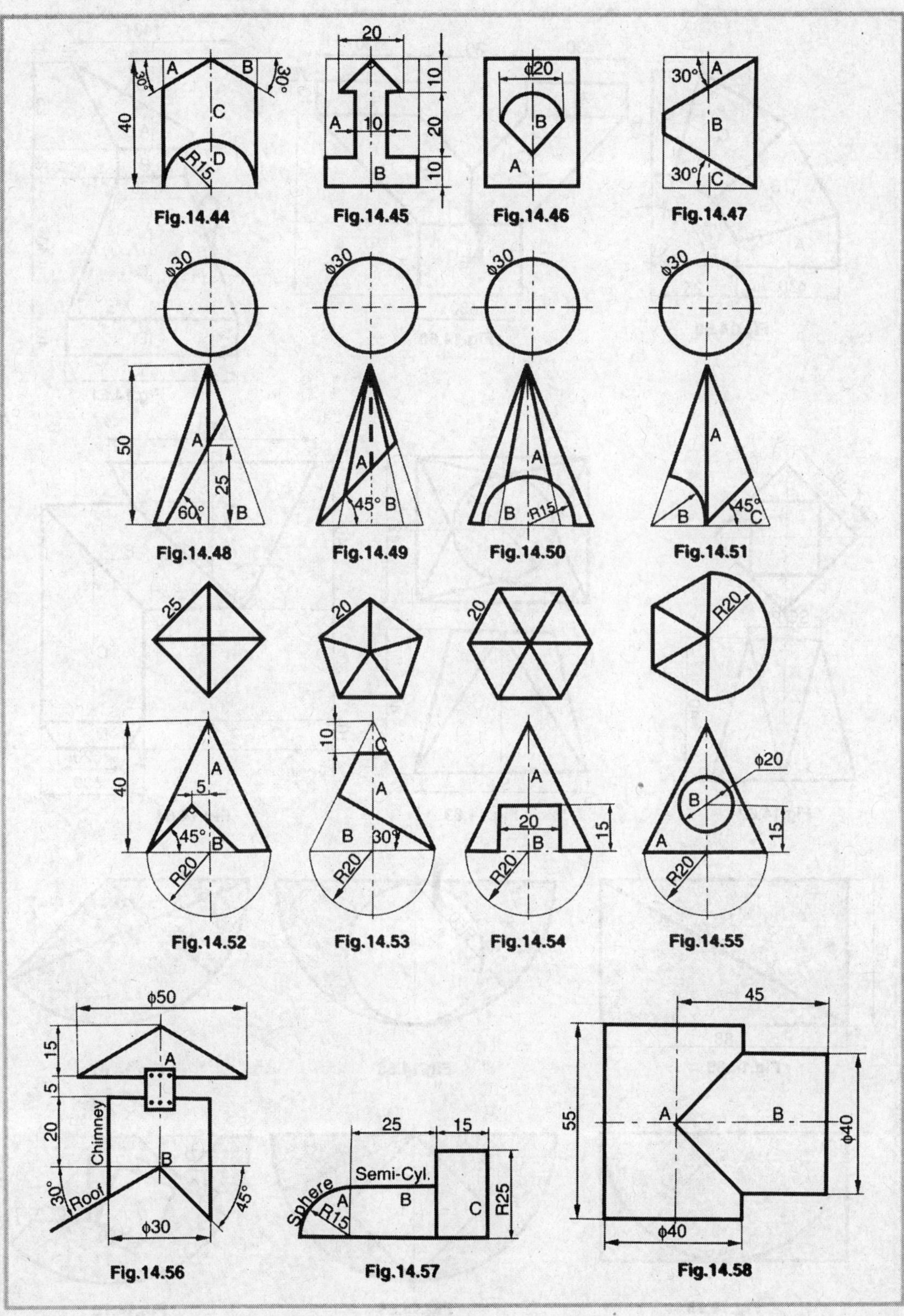

A
B
30°
30°
C
40
D
R15
Fig.14.44
20
10
A
10
20
B
10
Fig.14.45
ϕ20
B
A
Fig.14.46
30°
A
B
30°
C
Fig.14.47
ϕ30
50
A
25
60°
B
Fig.14.48
ϕ30
A
45°
B
Fig.14.49
ϕ30
A
B
R15
Fig.14.50
ϕ30
A
B
45°
C
Fig.14.51
25
40
A
5
45°
B
R20
Fig.14.52
20
10
C
A
B
30°
R20
Fig.14.53
20
A
20
B
15
R20
Fig.14.54
R20
ϕ20
B
A
15
R20
Fig.14.55
ϕ50
15
A
5
20
Chimney
B
30°
Roof
45°
ϕ30
Fig.14.56
25
15
Semi-Cyl.
Sphere
A
R15
B
C
R25
Fig.14.57
45
55
A
B
ϕ40
ϕ40
Fig.14.58

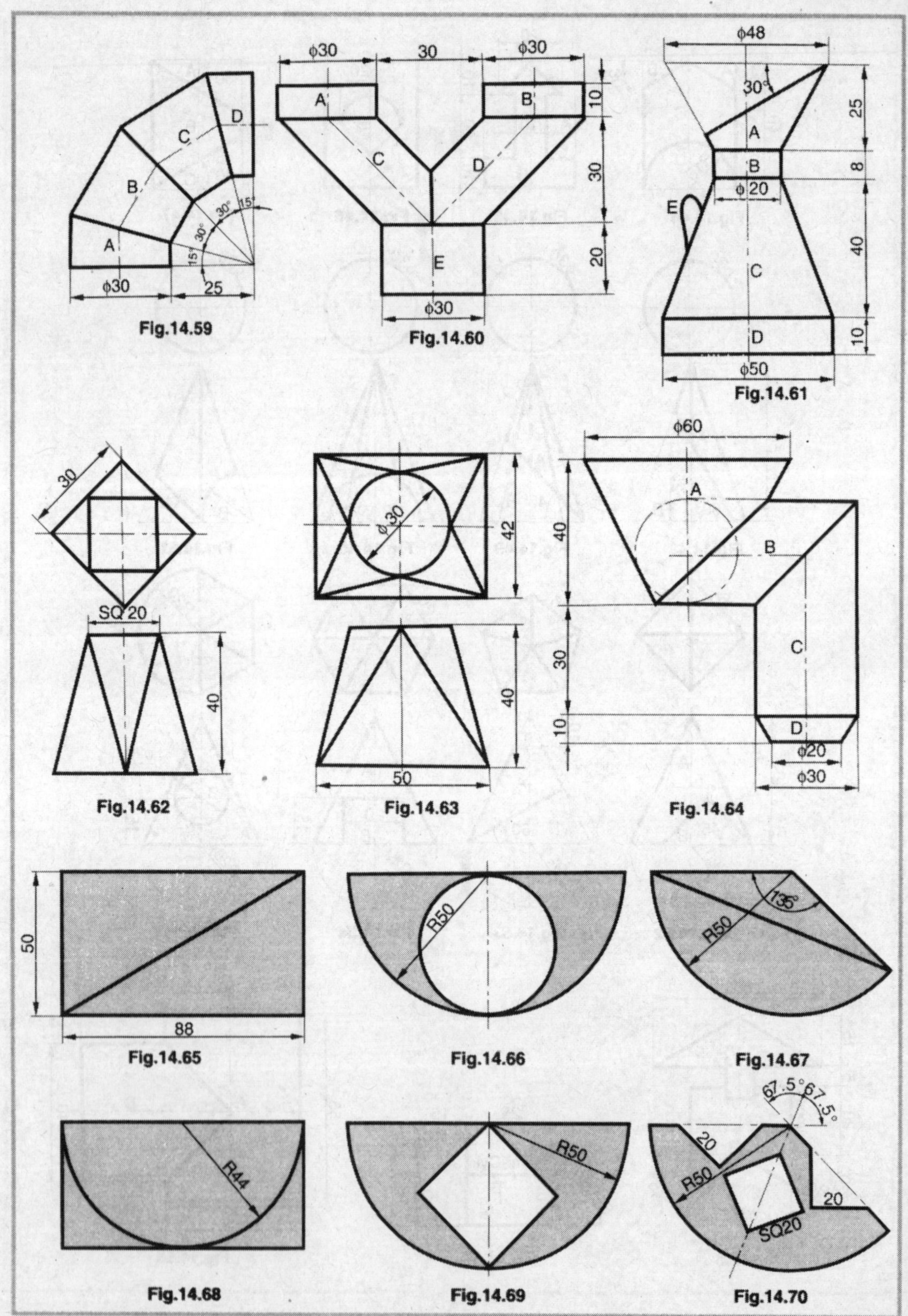

Fig.14.59

Fig.14.60

Fig.14.61

Fig.14.62

Fig.14.63

Fig.14.64

Fig.14.65

Fig.14.66

Fig.14.67

Fig.14.68

Fig.14.69

Fig.14.70

CHAPTER - 15

Interpenetration of Solids and Lines/Curves of Penetration

1. Introduction :

This topic of interpenetration is equally important as the topic of development of surfaces. A thorough knowledge of this topic will help the engineers in fabrication works.

Elements of equipments of chemical industry, petrochemical industry, cement industry, paper industry, fertilizers, etc. are made from sheet metal. Different shape ducts, vessels, tanks, etc. are required to be joined in various positions. Shape of the curve along which two parts meet, is of our interest, since two parts are welded along this curve of penetration. Further the development of surfaces of two parts can be done accurately if the curve of penetration is known.

Engineers in charge of fabrication work must first determine accurately the curve of penetration and then they should prepare the developments of two elements on the basis of curves of penetration. This procedure will save the time of fabrication and will also save the material.

2. Lines of Intersection or Curves of Intersection :

It is the line/curve along which surfaces of two parts or constituents meet. Imagine a vertical dome fitted on a horizontal boiler drum of railway engine (locomotive). The axis of dome is at right angles to the drum axis. The vertical round surface of dome extends only upto line or curve of intersection. This line of intersection is taken as cutting line for the purpose of accurate development of boiler drum is also done accurately by the use of same line of intersection.

Line of intersection or curve of penetration is a common line/curve to two surfaces of two meeting or penetrating elements or parts.

Lines of intersection may be straight lines or may be curved lines. It depends entirely upon the nature of surfaces of two meeting elements. If, out of the two surfaces, one is curved then line of intersection will be curved. Only if both meeting elements have flat surfaces, then the line of intersection will be a straight line.

When one solid penetrates the other solid or when the hole is drilled in a solid, the curve of penetration or line of intersection is marked only on one solid which is either penetrated or drilled. Penetrating solid remains intact. In both cases there will be two lines of intersection or curves of penetration.

3. Methods of Construction for Lines/ Curves of Intersection or Penetration :

There are two methods of construction ;

1. Line Method;
2. Cutting Plane Method.

3.1 Line Method :

Lines/edges/ generators are marked in three projections on the lateral surface/surfaces of one of the solids (A) in the region of intersection of two solids (A) and (B). These marked lines of (A) will intersect with edge-wise or line- wise view of lateral surface/surfaces of solid (B) in any one projection out of three principal projections. Locate and mark these points of intersection of lines/edges/generators of (A) with edge-wise view of the lateral surface of (B). Transfer these points of intersection to other views on corresponding lines simply by drawing projectors. Join the points thus obtained in sequence to get line of intersection.

Lines of intersection will appear in two views in case of pyramids and cones and will appear in one view in case of cylinders and prisms.

3.2 Cutting Plane Method :

Number of cutting planes are taken to cut both the solids in the region of intersection of two solids. These cutting planes may be taken either -

1. Perpendicular to H.P. and parallel to V.P.
2. Perpendicular to V.P. and parallel to H.P.

The cutting planes are selected in such a way that they will produce sections of regular shapes. Different methods of solution of problems are illustrated in the solved problems. There are five different solids; prisms, cylinders, cones, pyramids and spheres. There can be many combinations for intersection, as given below :

(1) Prism to prism
(2) Cylinder to cylinder
(3) Cone to cone
(4) Cylinder to prism
(5) Prism to cylinder
(6) Cone to cylinder
(7) Cylinder to cone
(8) Sphere to cylinder
(9) Cone to prism
(10) Sphere to prism
(11) Pyramid to prism, etc.

In this chapter problems of different varieties are solved and discussed. Knowledge of projections of solids is essential for drawing projections of given two solids, as per given data in the initial stage.

In most of the problems, three dimensional drawings of both the solids along with line/ lines of intersection are given. These drawings will help the students a lot to understand the problems and the following aspects.

(1) How two solids intersect?
(2) How the solution is done ?
(3) What is the nature of line of intersection ?
(4) Which lines of intersection are going to be full and which are going to be dotted ?
(5) How the joint will look like ? etc.

Students are not required to draw such drawings.

Students are requested to follow a particular system of notation. Do not just try to read problems, but always practice by actually solving them.

4. Cylinder to Cylinder :

Cylinders have curved lateral surfaces and hence the lines of intersection will be curved lines. Cylinders do not have edges or sides and hence generators are marked. Points where the curve of intersection is likely to change its direction, must be undertaken by drawing special generators passing through them.

Problem 1 : A vertical cylinder, diameter of base 42 mm and height 65 mm, is resting on H.P. on its base. It is penetrated by a horizontal cylinder, diameter of base 35 mm and height 73 mm. Axes of two cylinders bisect each other at right angles. Draw their projections showing on them curves of intersection assuming the axis of penetrating cylinder parallel to V.P.

For solution see Fig. 15.1 and follow the procedure, as given below :

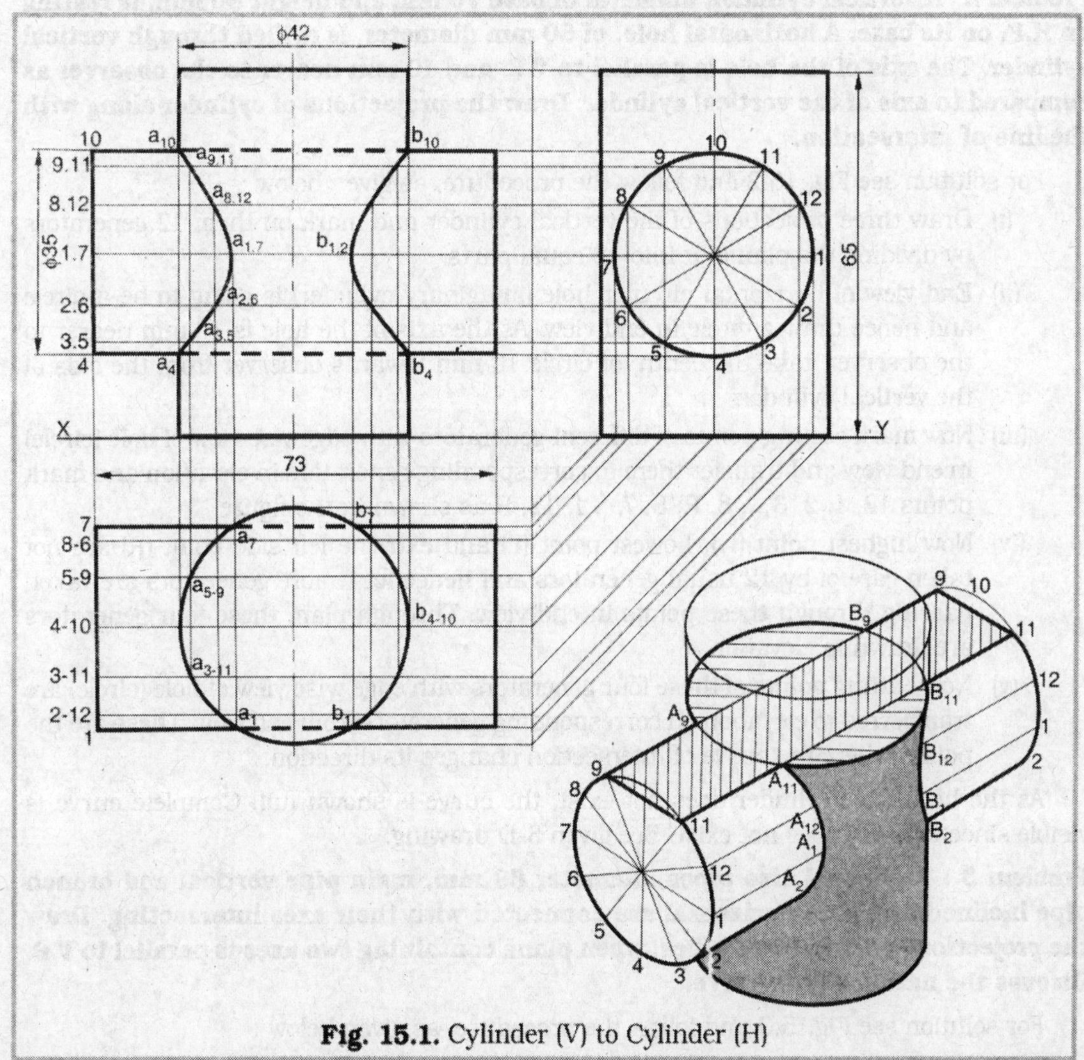

Fig. 15.1. Cylinder (V) to Cylinder (H)

(i) First draw three principal projections of both the solids in given position.

(ii) Mark 12 generators on three projections of only horizontal cylinder by dividing end view (circle) of horizontal cylinder into 12 equal parts.

(iii) Now entire lateral surface of the vertical cylinder is seen in line view or edge view (circle here) in plan. Mark points of intersection of generators 1, 2, 3,.....12 with line view (circle) as points a_1, a_2, a_3,....a_{12} and b_1, b_2, b_3,....b_{12} respectively, as shown in the figure. Try to understand these points of intersection from three dimensional drawing. Points A_9, B_9, A_{11} and B_{11} are shown.

(iv) Transfer these points by projectors to corresponding generators on elevation. Join located points of elevation in sequence to get the curve of penetration.

(v) From the three dimensional drawing, it is easy to understand that points A_4, A_3, A_2, A_1, A_{12}, A_{11} and A_{10} are seen in elevation and points A_9, A_8, A_7, A_6, A_5 are not seen. Although here hidden portion of the curve and visible portion of the curve coincides and hence hidden curve is not shown.

Problem 2 : A vertical cylinder, diameter of base 70 mm and height 80 mm, is resting on H.P. on its base. A horizontal hole, of 60 mm diameter, is drilled through vertical cylinder. The axis of the hole is parallel to V.P. and 10 mm nearer to the observer as compared to axis of the vertical cylinder. Draw the projections of cylinder along with the line of intersection.

For solution see Fig. 15.2 and follow the procedure, as given below :

(i) Draw three projections of the vertical cylinder and mark on them 12 generators by dividing the plan of it into 12 equal parts.

(ii) End view of horizontal circular hole (imaginary cylinder) is going to be a circle and hence draw a circle in end view. As the axis of the hole is 10 mm nearer to the observer, take the centre of circle 10 mm towards observer from the axis of the vertical cylinder.

(iii) Now mark cutting points of different generators with edgewise view of hole (circle) in end view and transfer them to corresponding generators in elevation and mark points 12, 1, 2, 3,....8, R, 8, 7,....1, 12, R as shown in the figure.

(iv) Now highest point (H), Lowest point (L) and extreme left side point (R) are not taken care of by 12 usual generators and hence four more generators are taken passing through these points in end view. Through plan, these four generators are drawn in elevation.

(v) Now cutting points of these four generators with edge wise view of hole (circle) are transferred to elevation on corresponding generators from end view. These are the points where the curve of intersection changes its direction.

As the horizontal cylinder does not exist, the curve is shown full. Complete curve is visible since cylinder does not exist. See given 3-D drawing.

Problem 3 : Two equal size pipes, diameter 60 mm, main pipe vertical and branch pipe inclined at 30° to horizontal are connected with their axes intersecting. Draw the projections of connected pipes, when plane containing two axes is parallel to V.P. Discuss the nature of the curve.

For solution see Fig.15.3 and follow the procedure, as given below :

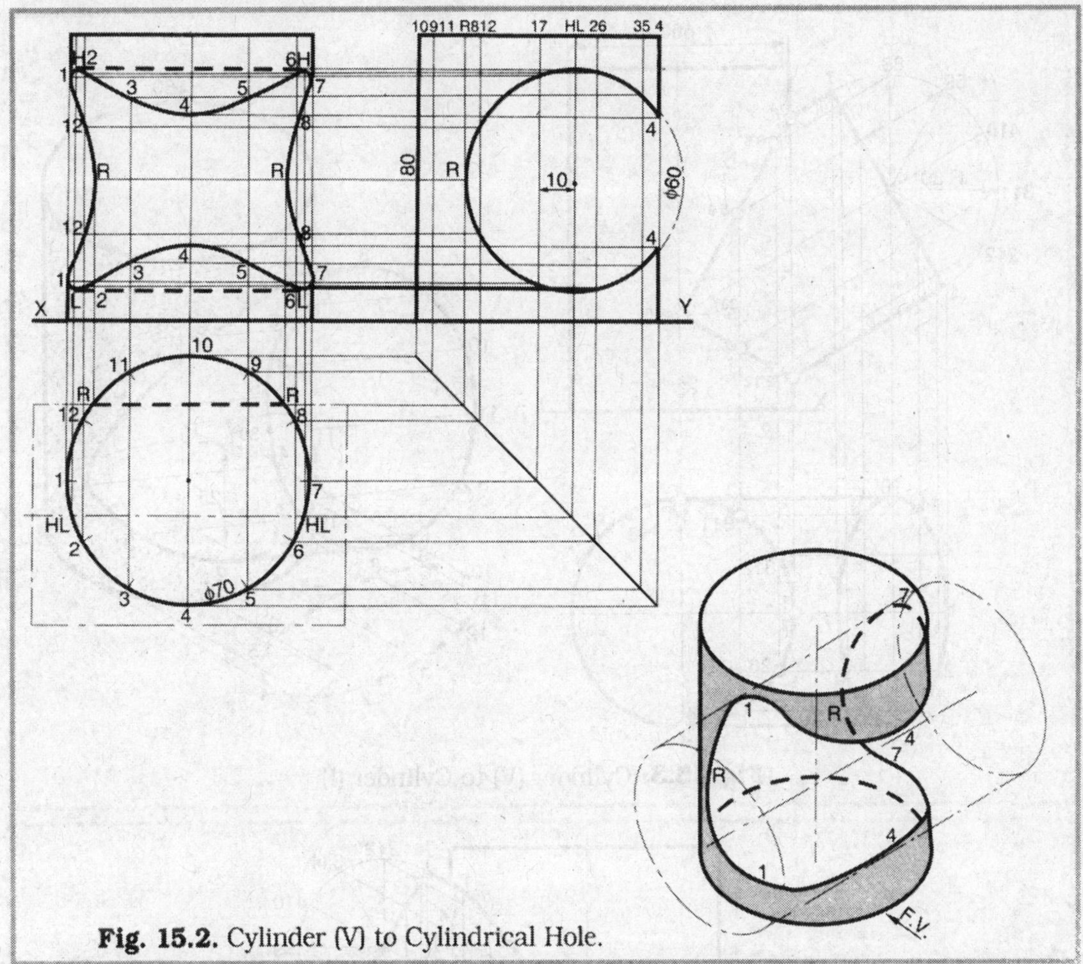

Fig. 15.2. Cylinder (V) to Cylindrical Hole.

(i) First draw plan of vertical cylinder and elevation of two cylinders in given position. Draw semicircle on the base of inclined cylinder and divide it into 6 divisions and thereby get 12 generators on the inclined cylinder. Plan of inclined cylinder along with 12 generators is drawn by projecting points from the base of elevation and taking distances from semicircle e.g. distances for points 5 and 9.

(ii) In plan, mark points where generators of inclined cylinder are cutting edgewise view of vertical cylinder and transfer them to elevation on corresponding generators. Join cutting points in elevation.

When diameters of two cylinders are same and when their axes intersect, the nature of penetration curve in elevation will be straight line.

Problem 4 : Two unequal size pipes, main pipe vertical and of 80 mm diameter and branch pipe inclined at 60° to vertical and of 50 mm diameter, are connected with their axes off-set and parallel to V.P. Axis of branch pipe is nearer to the observer by 15 mm as compared to the axis of the vertical main pipe. Draw the projections along with curves of intersection.

For solution see Fig. 15.4. Procedure of solution is the same as of the previous problem. Rest is clear from the figure.

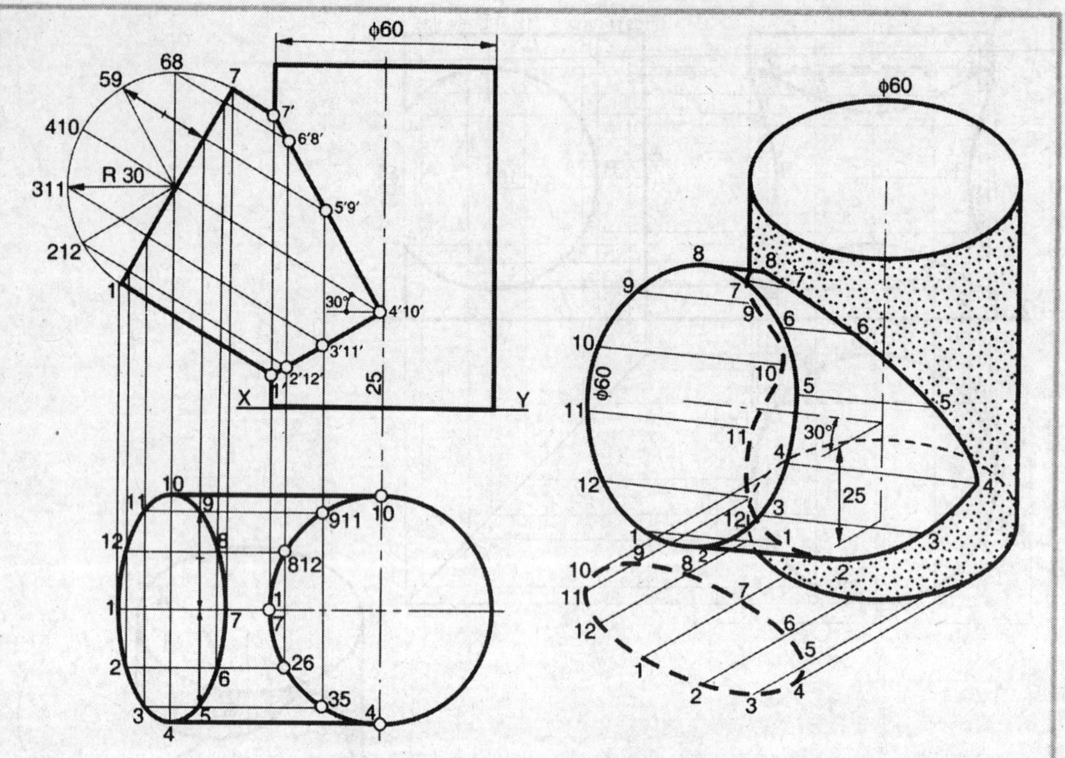

Fig. 15.3. Cylinder (V) to Cylinder (I)

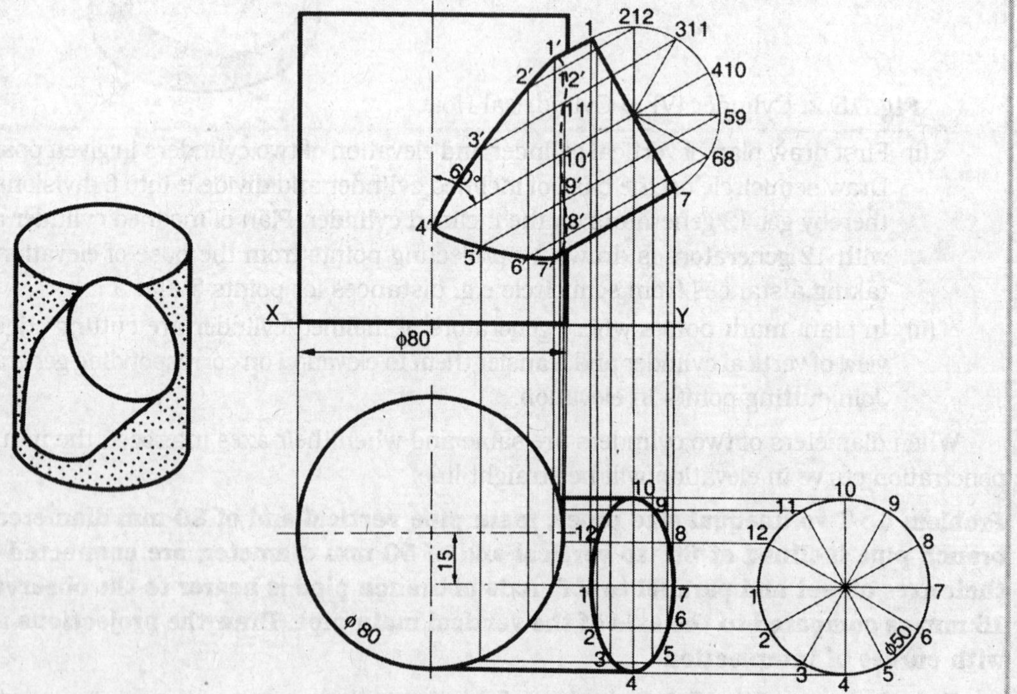

Fig. 15.4. Cylinder (V) to Cylinder (I)

Problem 5 : Vertical and horizontal cylinders, of 80 mm (equal) diameters, intersect at right angle. Draw the projections, in 1st Angle System, of two cylinders showing line of intersection when axes of both the cylinders are parallel to V.P.

For solution see Fig. 15.5 and follow the procedure, as given below :

(i) First draw plan, elevation and side view of both the cylinders. Take length of cylinders of any suitable amount.

(ii) Take 12 generators on horizontal cylinder and mark them in all the three views.

(iii) Now transfer cutting points of generators of horizontal cylinders with edge view of vertical cylinder (circle in plan) to elevation on corresponding generators and then join them. Lines of intersection in elevation are two cross lines, see Fig.15.5.

5. Prism to Prism :

Prisms have flat rectangular faces as their lateral surfaces. The line of intersection between two prisms will be closed figure composed of straight lines. Points, where edges of one of the prisms penetrate the surface of other prism, are marked and then joined by straight lines in sequence to get lines of intersection. Here notation A_{14} represents the point where the edge A is entering the surface 14.

Problem 6 : A square prism, side of base 50 mm and height 70 mm, is resting on H.P. on its base with one of the edges/sides of the base inclined at 30° to V.P. The above prism is penetrated by horizontal prism, side of base 32 mm and length 96 mm. Rectangular faces of the horizontal prism are equally inclined to H.P. and V.P. both. The axes of the two prisms bisect at right angle. Draw the projections and show the line of intersection.

For solution see Fig. 15.6 and follow the procedure, as given below :

(i) First draw three projections of both the prisms and mark vertical edges 11, 22, 33, 44 and AA, BB, CC and DD on vertical and horizontal prisms respectively.

(ii) Now observe that edge view of lateral surfaces of vertical prism is seen in plan and of horizontal prism is seen in side view.

(iii) Now edges aa, bb, cc and dd of the horizontal prism penetrate surfaces of vertical prism at points (a_{12}, a_{23}), (b_{41}, b_{23}), (c_{41}, c_{34}) and (d_{41}, d_{23}) respectively. Suffix indicates the rectangular face of vertical prism. Similarly in end view edges 1'1' and 3'3' of vertical prism penetrate surfaces of horizontal prism at points (1_{ab}, 1_{ad}) and (3_{bc}, 3_{cd}) respectively.

(iv) Transfer above points from plan and side view to elevation on corresponding edges and get points. Join them by straight lines to get lines of intersection.

(v) Faces 12 and 23 of the vertical prism and faces AB and AD of the horizontal prism are seen and hence points generated by them are drawn full.

Similarly, faces 34 and 41 of the vertical prism and faces BC and CD of the horizontal prism are not seen and hence they are drawn dotted. Visualise full, dotted, etc. from three dimensional drawing given.

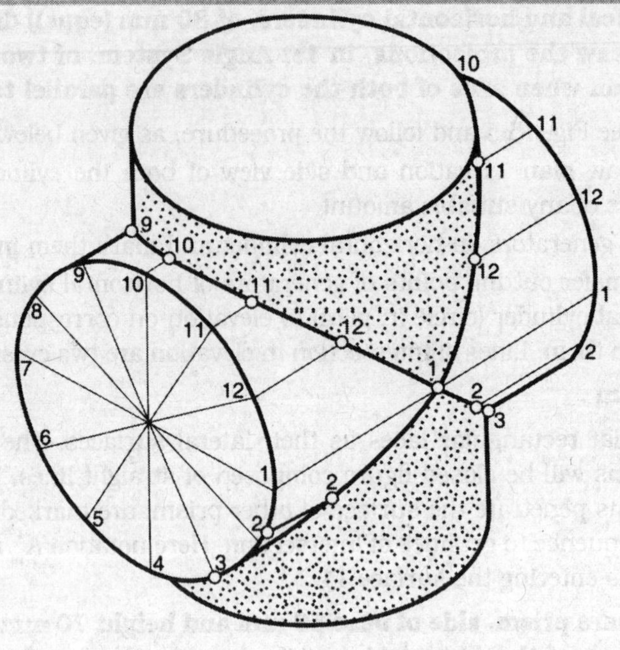

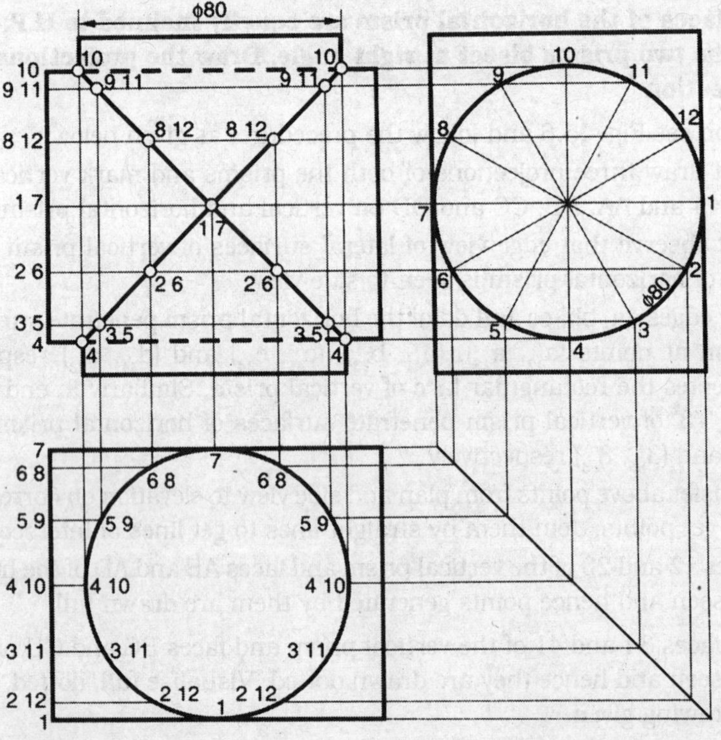

Fig. 15.5. Cylinder (V) to Cylinder (H) Equal and Intersecting

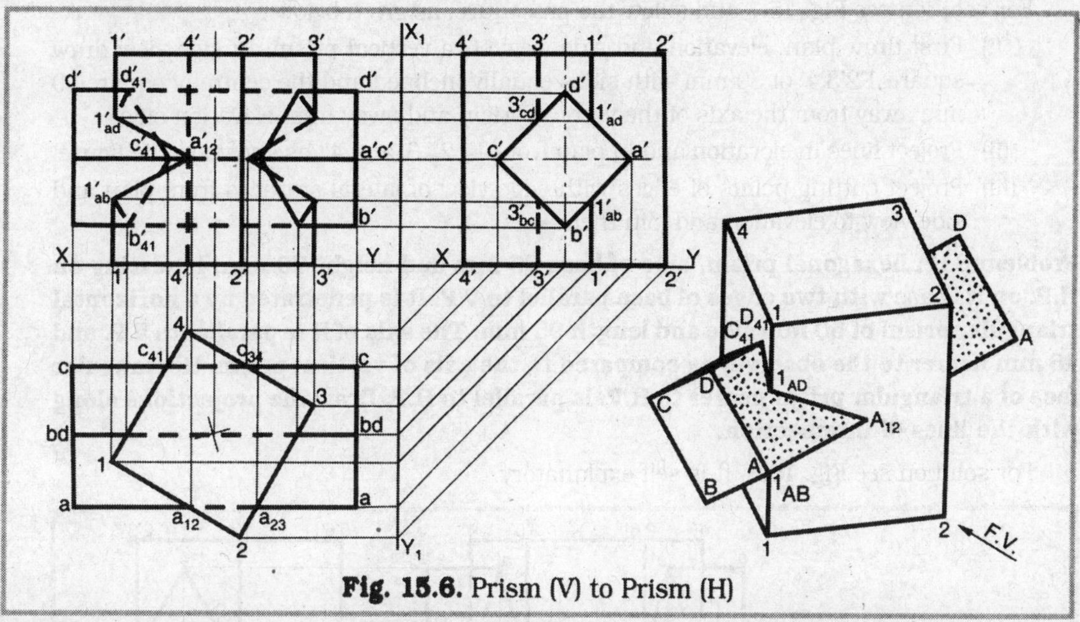

Fig. 15.6. Prism (V) to Prism (H)

Problem 7 : A square prism, side of base 50 mm and height 80 mm, is resting on H.P. on its base with all rectangular faces equally inclined to V.P. It has a horizontal square hole of 30 mm side with its axis parallel to V.P. and 10 mm in front of the axis of the vertical prism. Flat faces of the hole are equally inclined to H.P. and V.P. Draw the projections and show the lines of intersection.

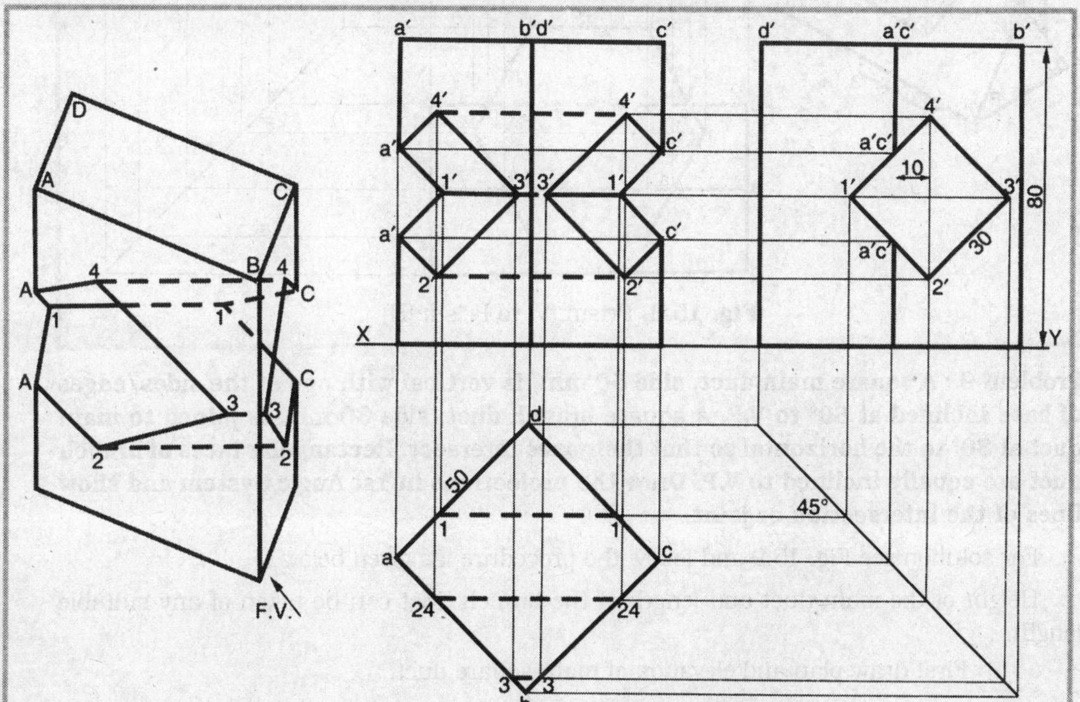

Fig. 15.7. Prism (V) to Prismatic Hole (H). Axes Offset

For solution see Fig. 15.7 and follow the procedure, as given below :

(i) First draw plan, elevation and side view of a vertical prism. In side view draw square 1'2'3'4' of 30 mm with sides equally inclined and the centre of square 10 mm away from the axis of the vertical prism and away from elevation side.

(ii) Project lines in elevation and in plan from 1', 2', 3' and 4', as shown in the figure.

(iii) Project cutting points of edges with edge view of lateral surfaces from plan and side view to elevation and join them.

Problem 8 : A hexagonal prism, side of base 35 mm and height 70 mm, is resting on H.P. on its base with two edges of base parallel to V.P. It is penetrated by a horizontal triangular prism of 50 mm side and length 95 mm. The axis of it is parallel to V.P. and 16 mm nearer to the observer as compared to the axis of vertical prism. Rectangular face of a triangular prism nearer to H.P. is parallel to H.P. Draw the projections along with the lines of intersection.

For solution see Fig. 15.8. It is self explanatory.

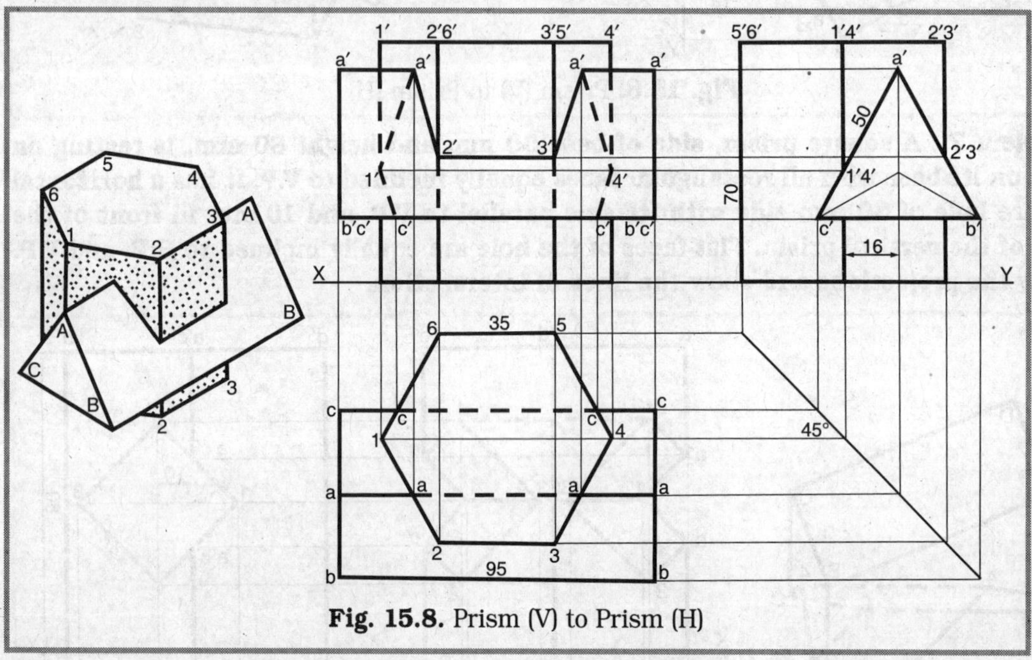

Fig. 15.8. Prism (V) to Prism (H)

Problem 9 : A square main duct, side 50 mm, is vertical with one of the sides/edges of base inclined at 60° to V.P. A square branch duct, side 30 mm, is joined to main duct at 30° to the horizontal so that their axes intersect. Rectangular faces of branch duct are equally inclined to V.P. Draw the projections in 1st Angle System and show lines of the intersection or joint.

For solution see Fig. 15.9 and follow the procedure, as given below :

Height of the main duct and length of the branch duct can be taken of any suitable length.

(i) First draw plan and elevation of main square duct.

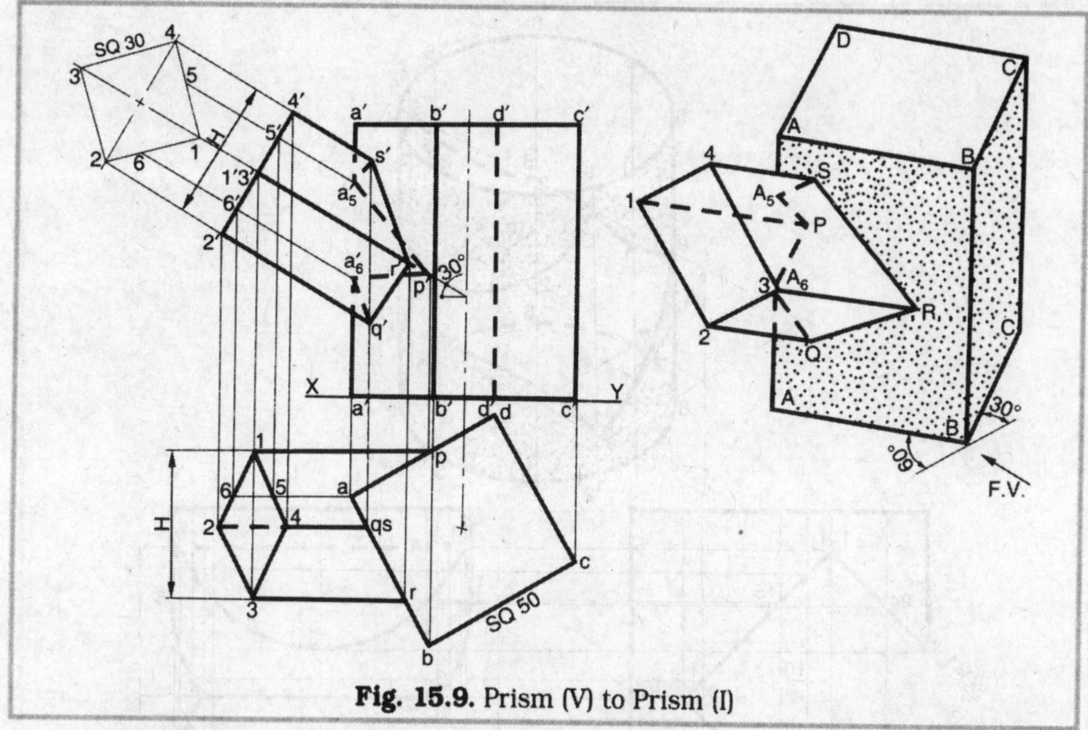

Fig. 15.9. Prism (V) to Prism (I)

(ii) In elevation draw a centre line of branch duct at 30° to the horizontal. At suitable distance draw 2 - 4 line at right angle to the centre line already drawn. Draw square 1234 of 30 mm side, as shown in the figure.

(iii) Draw lines, at 30° to horizontal, from 1, 2, 3 and 4 to get elevation of branch duct, as shown in the figure.

(iv) Draw plan of branch duct by transferring/projecting points 1', 2', 3' and 4' from elevation and taking height H/2 on both the sides of the centre line, as shown in the figure.

(v) Draw horizontal line from a, to get points 5 and 6 on plan of the square. Transfer them by projectors to get 5' and 6' in elevation. Draw lines at 30° to the horizontal from 5' and 6'.

(vi) Transfer cutting points p, q, r, s and a, from plan on corresponding lines of elevation and get points p', q', r', s' and (a'_5, a'_6) respectively. Join them by straight lines.

Faces AB and BC of the main duct and faces 23 and 34 of branch duct are seen and hence SR and QR are seen.

6. Cylinder to Prism :

Problem 10 : A cylinder, diameter of base 60 mm, is resting on H.P. on its base. A square hole, of 41 mm sides, is cut through it. The axis of the hole is parallel to H.P. and V.P. both and 10 mm nearer to the observer than the axis of the cylinder. The faces of the square hole are equally inclined to H.P. and V.P. Draw the projections of the cylinder showing the hole on it.

For solution see Fig. 15.10 and follow the procedure, as given below :

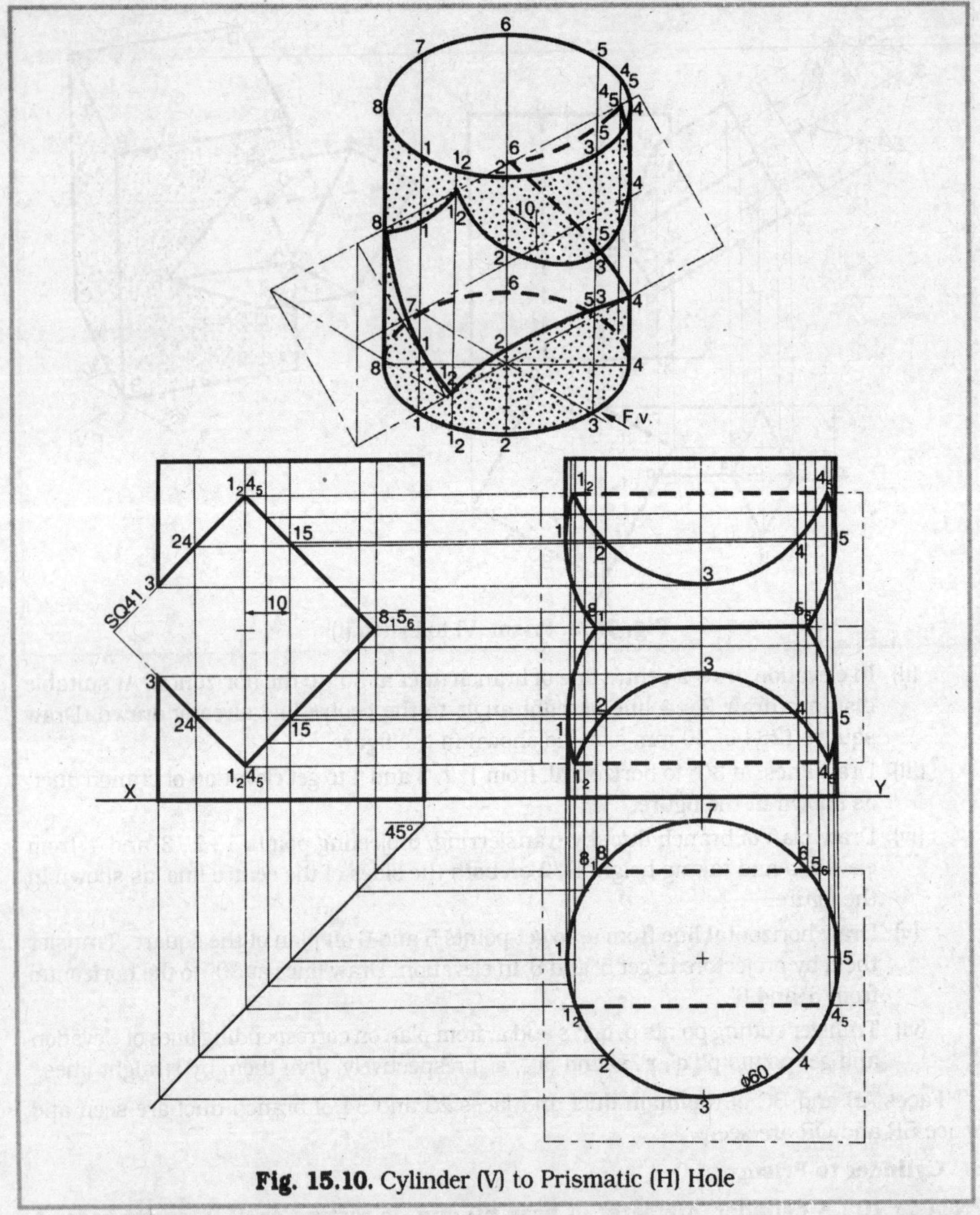

Fig. 15.10. Cylinder (V) to Prismatic (H) Hole

(i) First of all draw three projections of cylinder and mark on them 8 generators as usual.

(ii) Draw square of 41 mm side, with all four sides equally inclined (i.e. at 45°) to xy, in side view keeping the centre of square 10 mm away from the axis of the cylinder towards opposite side to elevation side or towards observer.

(iii) Now take 2 extra generators passing through top and bottom corners of square in side view and mark them in plan and then in elevation, as shown in the figure.

(iv) Now transfer the cutting points of generators 1, 1_2, 2, 3, 4, 4_5, 5, 5_6 and 8_1 with square in side view to elevation on the respective generators to get points 1, 1_2, 2, 3, 4, 4_5, 5, 5_6 and 8_1 respectively, as shown on elevation in the figure. Join them by a smooth curve. Points 5_6 and 8_1 are joined by a straight line.

Note :

Dashes and suffix for elevation and side view are purposely avoided for keeping drawings clearer.

Problem 11 : A cylinder, diameter of base 65 mm and height 110 mm, is resting on H.P. on its base. A square prism, side of base 36 mm, intersects the axis of cylinder at an angle of 60° (i.e. 30° with horizontal). The faces of the square prism are equally inclined to H.P. and V.P. both. Draw three projections of the cylinder and prism showing on them the line of intersection. Assume plane containing two axes parallel to V.P.

For solution see Fig. 15.11. Three dimensional drawing helps in understanding the procedure of solution and helps in visualising the joint. In the previous problem generators were taken on the surface of cylinder but here lines are taken on the surfaces of prism.

First of all 16 lines are drawn on all the three projections and then points of intersections of these lines with edge-view of lateral surface of cylinder are transferred from plan to elevation and to side view on the respective lines.

Rest is clear from the figure.

Problem 12 : A cylinder, of 80 mm diameter, is resting on H.P. on its base. A cone, diameter of base 90 mm and height 110 mm, penetrates the cylinder at right angle. The axis of the cone is parallel to H.P. and V.P. both. Apex of the cone reaches 60 mm beyond the axis of the cylinder. Draw projections showing the lines of intersection.

For solution see Fig. 15.12 and follow the procedure, as given below :

(i) First draw 2 projections of cylinder.

(ii) Draw 2 projections of cone, as shown in the figure and mark 12 generators on them.

(iii) Generators of cone 01, 02, 03,....012 enter and leave the edge view of cylinder (i.e. circle) at points a, b, c,....l respectively and similar points (not marked in the figure). Transfer them to elevation by projectors and get points a', b', c',....l' on one side and similar points on the other side (not marked in figure).

(iv) Join points by a smooth curve to get lines of intersection.

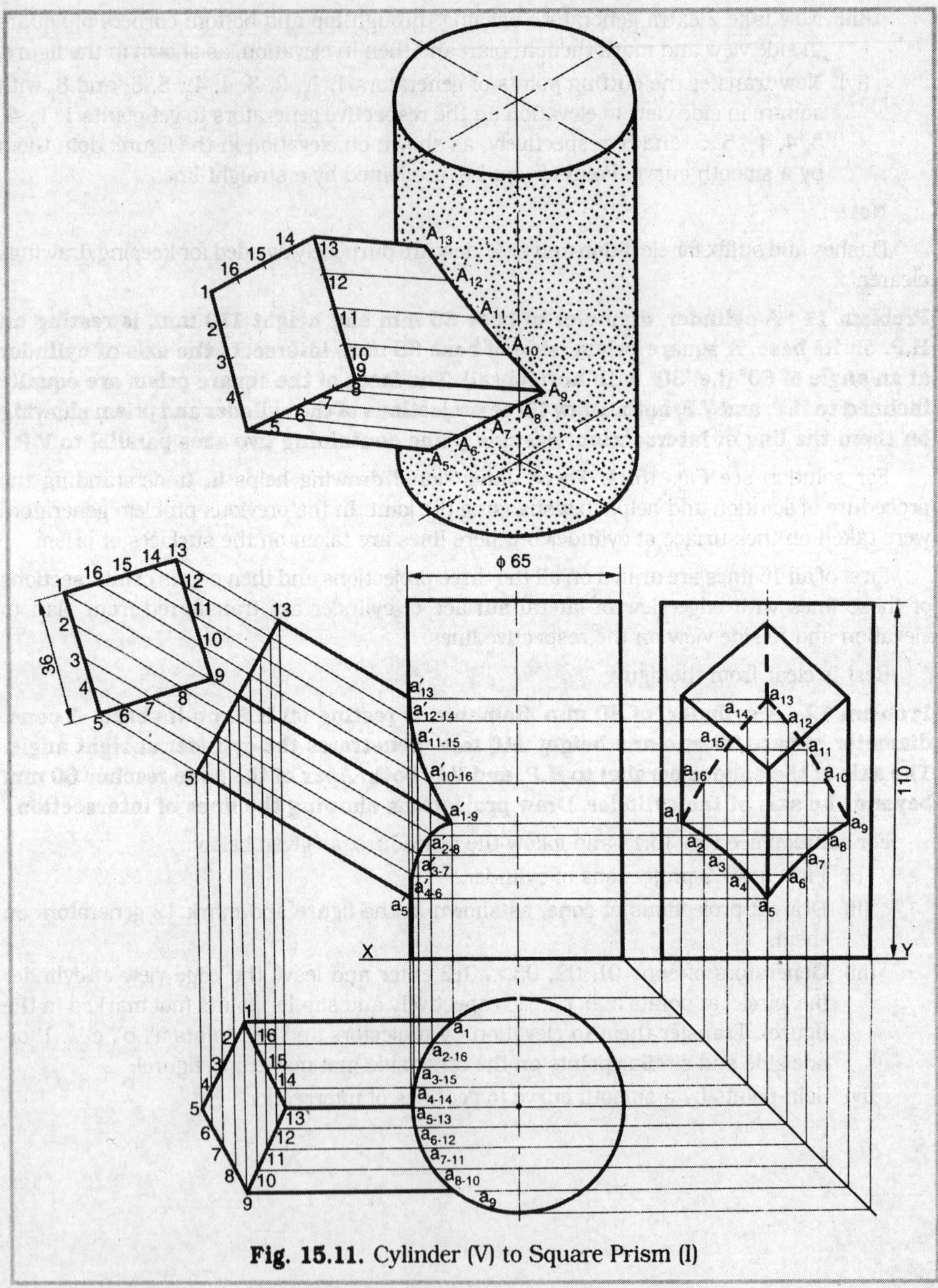

Fig. 15.11. Cylinder (V) to Square Prism (I)

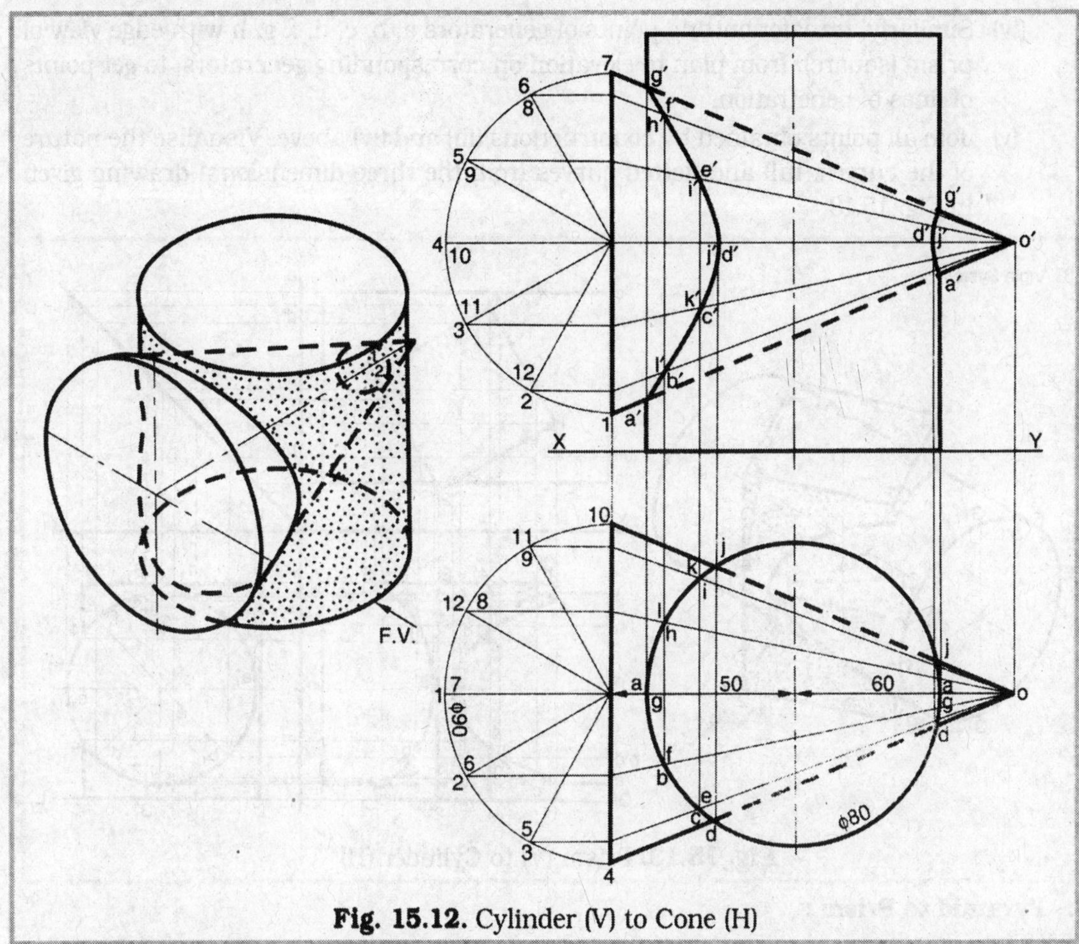

Fig. 15.12. Cylinder (V) to Cone (H)

7. Prism to Cylinder :

Problem 13 : A square prism, side of base 50 mm, is resting on ground on its base with one of the edges of the base inclined to V.P. by 30°. It is penetrated by a cylinder of 70 mm diameter. The axis of the cylinder is parallel to H.P. and V.P. both and is 5 mm away from the axis of the prism. Draw the projections of solids showing lines of penetration on them. Take height of the prism and length of the cylinder of any suitable value.

For solution see Fig. 15.13 and follow the procedure, as given below :

(i) First draw plan, elevation and side view of square prism in 3rd angle system, since it rests on ground.

(ii) In side view, take the centre of cylinder 5 mm away from the axis of prism on the other side then elevation side. Draw a circle with 70 mm diameter. Divide this circle into 8 equal parts and mark 8 generators on all projections.

(iii) Now transfer cutting points of edges $1'_1$, $3'_1$ and $4'_1$ with edge view of cylinder (circle) from side view to elevation, on corresponding edges to get points of lines of penetration.

(iv) Similarly, transfer cutting points of generators a, b, c, d, f, g, h with edge view of prism (square), from plan to elevation on corresponding generators, to get points of lines of penetration.

(v) Join all points obtained by constructions (iii) and (iv) above. Visualise the nature of the curves, full and dotted curves from the three dimensional drawing given in Fig. 15.13.

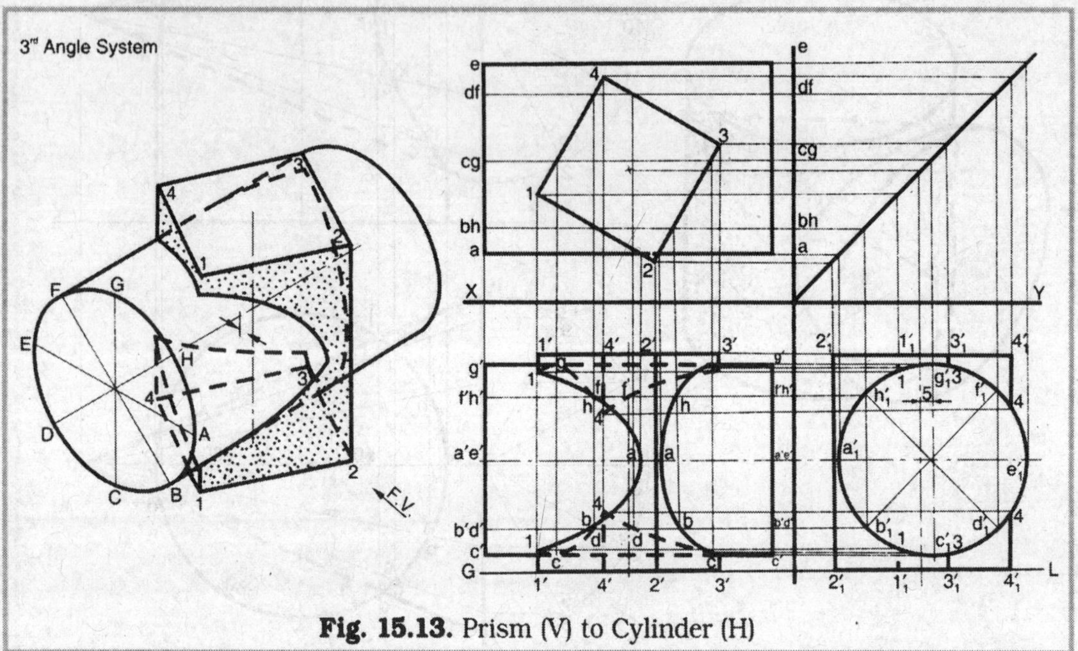

Fig. 15.13. Prism (V) to Cylinder (H)

8. Pyramid to Prism :

Problem 14 : A square pyramid, side of base 70 mm and height 125 mm, is resting on H.P. on its base with all edges of base equally inclined to V.P. A square prism, side of base 30 mm, intersects the axis of pyramid at right angle 50 mm above the base. The axis of the prism is parallel to H.P. and V.P. both and rectangular faces of the prism are equally inclined to H.P. and V.P. both. Draw projections showing on them lines of intersection.

For solution see Fig. 15.14 and follow the procedure, as given below :

(i) First draw plan and elevation of pyramid.

(ii) Draw a horizontal line 50 mm above x-y line/base and draw on it square ABCD of 30 mm side at suitable distance from elevation. Keep AB, BC, CD and DA equally inclined to xy line.

(iii) Draw elevation and plan of the square prism (rough).

(iv) Take horizontal section planes 1, 2 and 3 at heights of A, (B, D) and C respectively on elevation.

Now we have seen in the chapter of Sections of Solids that if a pyramid is cut by a plane parallel to the base, then its section will be similar in shape to the base. And hence draw squares 1,2 and 3 in plan corresponding to section planes 1, 2 and 3 respectively, as shown in the figure.

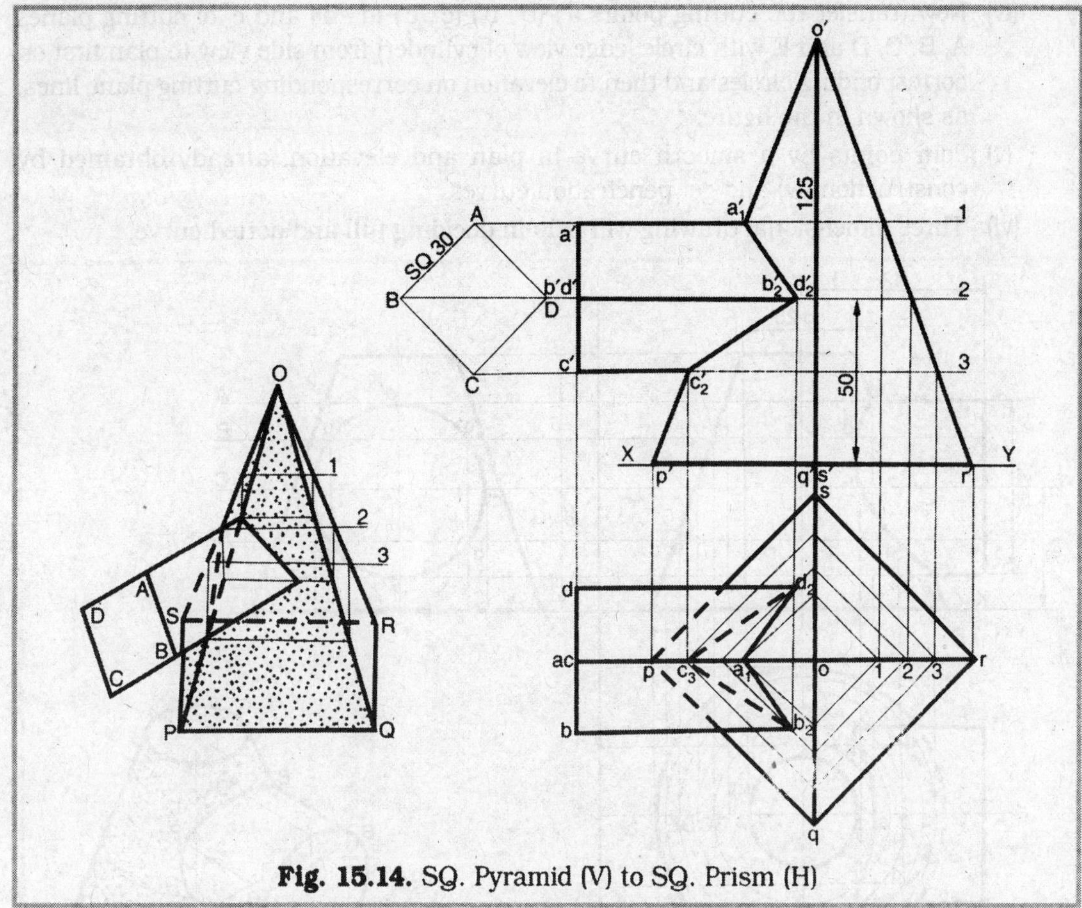

Fig. 15.14. SQ. Pyramid (V) to SQ. Prism (H)

(v) Now mark points a_1, b_2, c_3 and d_2 in plan where edges a, b, c and d meet squares 1, 2, 3 and 2 respectively. Transfer them to elevation on corresponding edges to get a'_1, b'_2, c'_3 and d'_2 respectively.

(vi) Join them by straight lines to get lines of intersection.

9. Cone with Cylinder :

Problem 15 : A frustum of a cone, diameter of base 65 mm, diameter of top 25 mm and height 56 mm, is resting on H.P. on its base. A cylinder, diameter of base 38 mm, intersects the axis of frustum of a cone at right angle, 27 mm above the base. The axis of the cylinder is parallel to H.P. and V.P. both. Draw three projections with lines of intersection.

For solution see Fig. 15.15 and follow the procedure, as given below :

(i) First draw plan, elevation and side view of the frustum of the cone.

(ii) In the side view, mark the centre of circle, on the axis of cone, 27 mm above the base and draw circle of 38 mm diameter. Project it in elevation and plan.

(iii) Take 5 horizontal cutting planes, A, B, C, D and E in the region where cylinder is meeting the cone both in elevation and side view at suitable distance between them and draw corresponding (sections) circles in plan.

(iv) Now transfer the cutting points a', (b', b') (c', c') (d', d') and e' of cutting planes A, B, C, D and E with circle (edge view of cylinder) from side view to plan first on corresponding circles and then to elevation on corresponding cutting plane lines, as shown in the figure.

(v) Join points by a smooth curve in plan and elevation, already obtained by construction (iv) and get penetration curve.

(vi) Three dimensional drawing will help in deciding full and dotted curve.

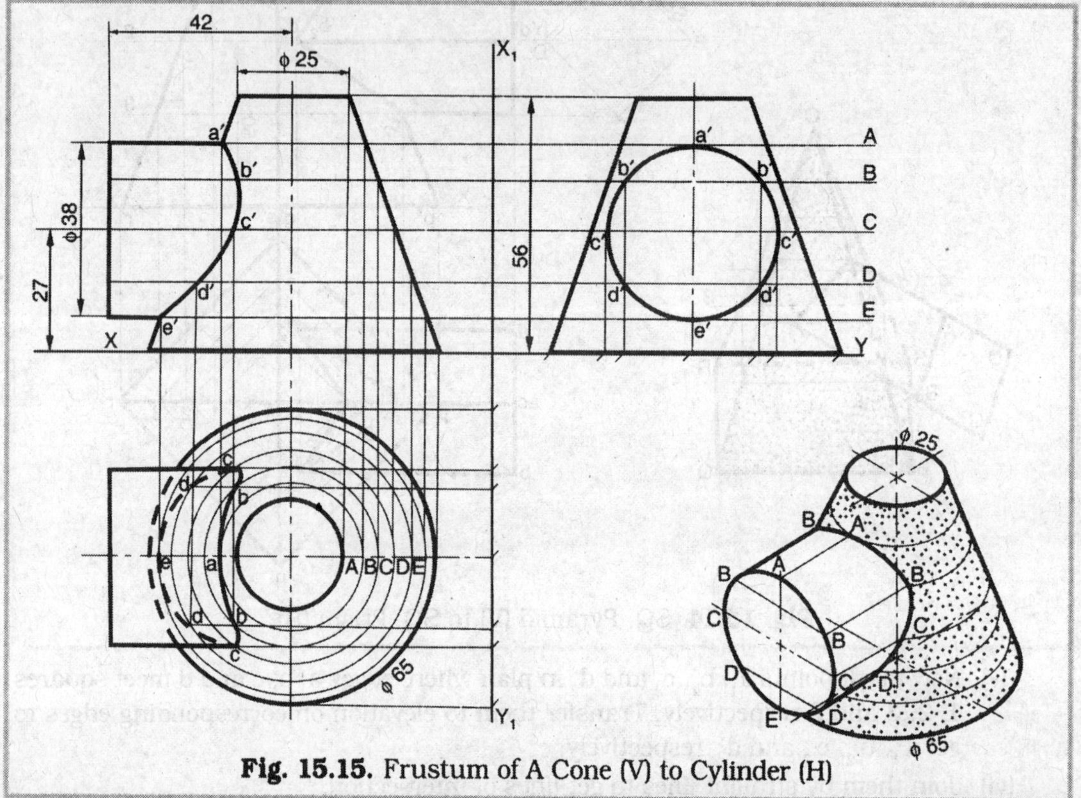

Fig. 15.15. Frustum of A Cone (V) to Cylinder (H)

Problem 16 : A cone, diameter of base 80 mm and height 90 mm, is resting on H.P. on its base. A hole, of 50 mm diameter, is drilled through the cone. The axis of the hole is 28 mm above and parallel to the base of cone. The axis of the hole is parallel to the V.P. also. Draw three projections of cone showing lines of intersections of hole with cone.

For solution see Fig. 15.16 and follow the procedure, as given below :

(i) First draw plan, elevation and side view of a cone and mark 12 generators on them.

(ii) In side view, mark a centre [c] on the axis 28 mm above the base. Draw a circle of 50 mm diameter with [c] as the centre. This circle represents the end view of the hole.

(iii) Now mark cutting points a, b, c, ...1 where generators 01, 02, 03,....012 of cone are cutting edge view of hole (circle) respectively. Transfer them to the corresponding generators in elevation first and then to plan, as shown in the figure. Join points, thus obtained, by a smooth curve in elevation and in plan to get lines of intersections. Lines of intersection are two cross straight lines in elevation and two circles in plan. This situation will arise when the circle of the cylindrical hole is touching two extreme generators of the cone in side view.

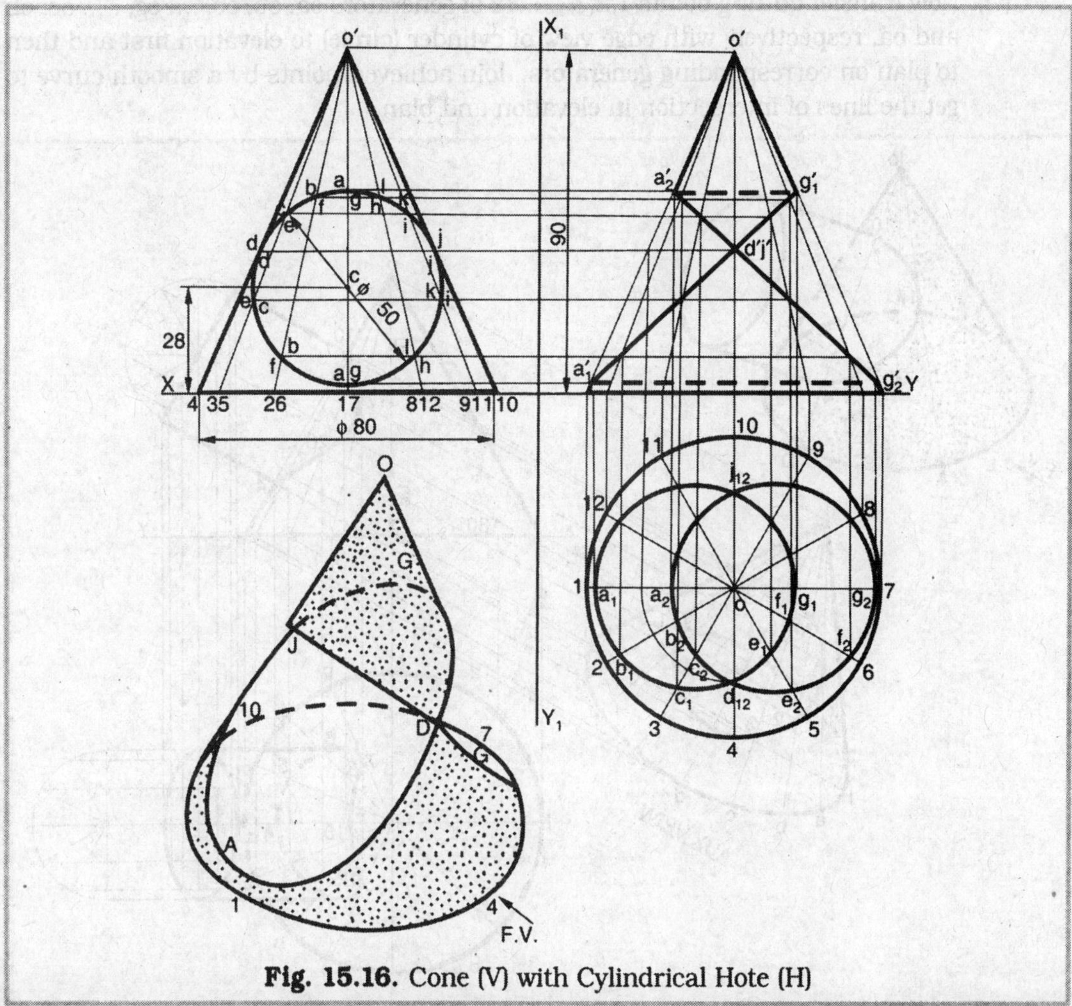

Fig. 15.16. Cone (V) with Cylindrical Hote (H)

Problem 17 : A cone, diameter of base 80 mm and height 80 mm, is resting on H.P. on its base. A cylinder, of 40 mm diameter, is intersecting the cone. The axis of the cylinder intersects the axis of cone at a point 20 mm above the base. Axis of cylinder is inclined to H.P. by 30° and it is parallel to V.P. Draw the plan, elevation and auxiliary plan on A.I.P. parallel to the base of the cylinder.

For solution see Fig. 15.17 and follow the procedure, as given below :

(i) First draw plan, elevation and auxiliary plan of the cone.

(ii) In elevation draw a line, at an angle of 30° to horizontal, from a point on axis 20 mm above the base.

(iii) Draw a circle, in auxiliary elevation, of 40 mm diameter with the centre as a point of intersection of cone axis with 30° line drawn from elevation.

(iv) In auxiliary plan, draw two generators o a and o g tangent to the circle, drawn as per (3) above. Transfer two generators in elevation and in plan.

(v) Divide arc ag of plan into six equal parts and mark generators oa, ob, oc, od, oe, of and og and transfer them in elevation and in auxiliary plan.

(vi) Now transfer cutting points 1,2,3,......12 of generators oa, ob, oc, ... og, ... oc, ob and oa, respectively with edge view of cylinder (circle) to elevation first and then to plan on corresponding generators. Join achieved points by a smooth curve to get the lines of intersection in elevation and plan.

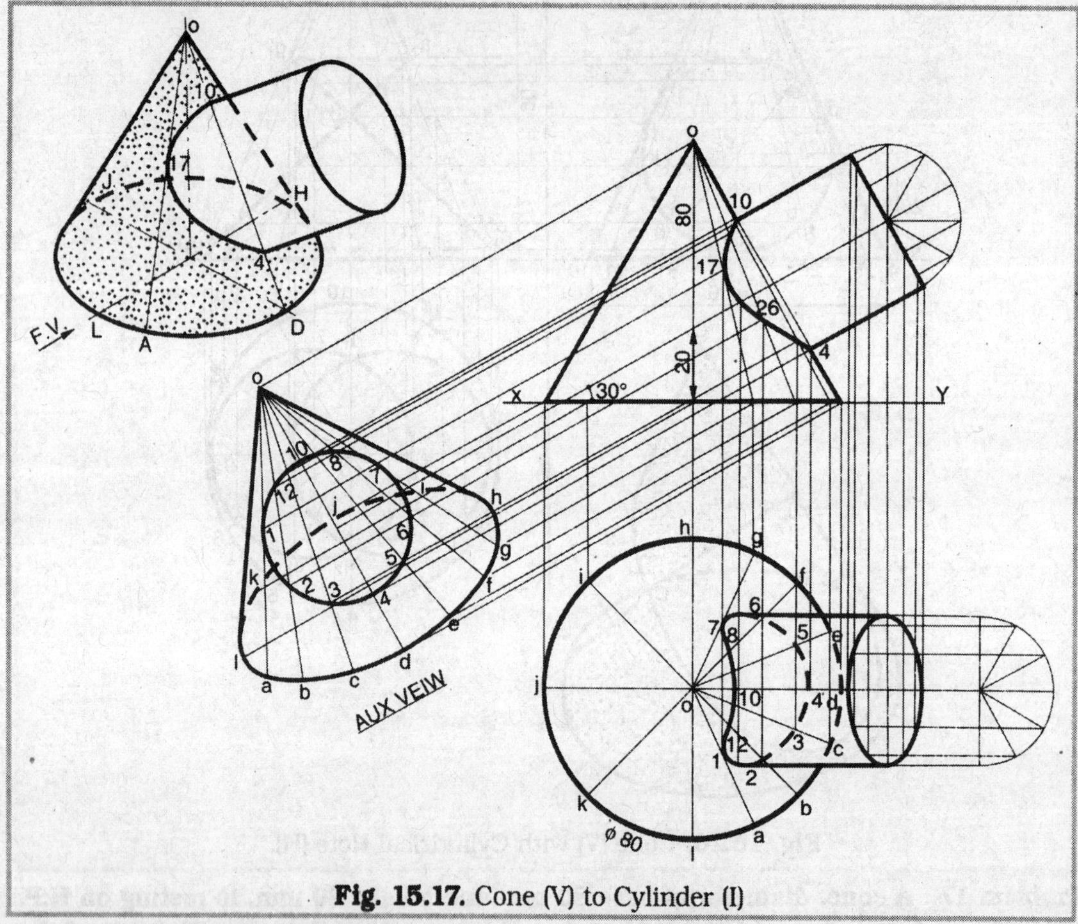

Fig. 15.17. Cone (V) to Cylinder (I)

Three demensional drawing is given to visualise the joint.

Problem 18 : A cone, diameter of base 70 mm and height 80 mm, is resting on H. P. on its base. A cylinder, of 40 mm diameter, is penetrating the cone from the top side. The axis of cylinder is parallel to the axis of the cone and distance between two axes is 10 mm. The plane containing both the axes is parallel to V. P. Draw the projections showing the curves of penetration. Solve it by the generator method.

For solution see Fig. 15.18 and follow the procedure, as given below :

(i) First draw plan and elevation of the cone and mark on them 8 generators.

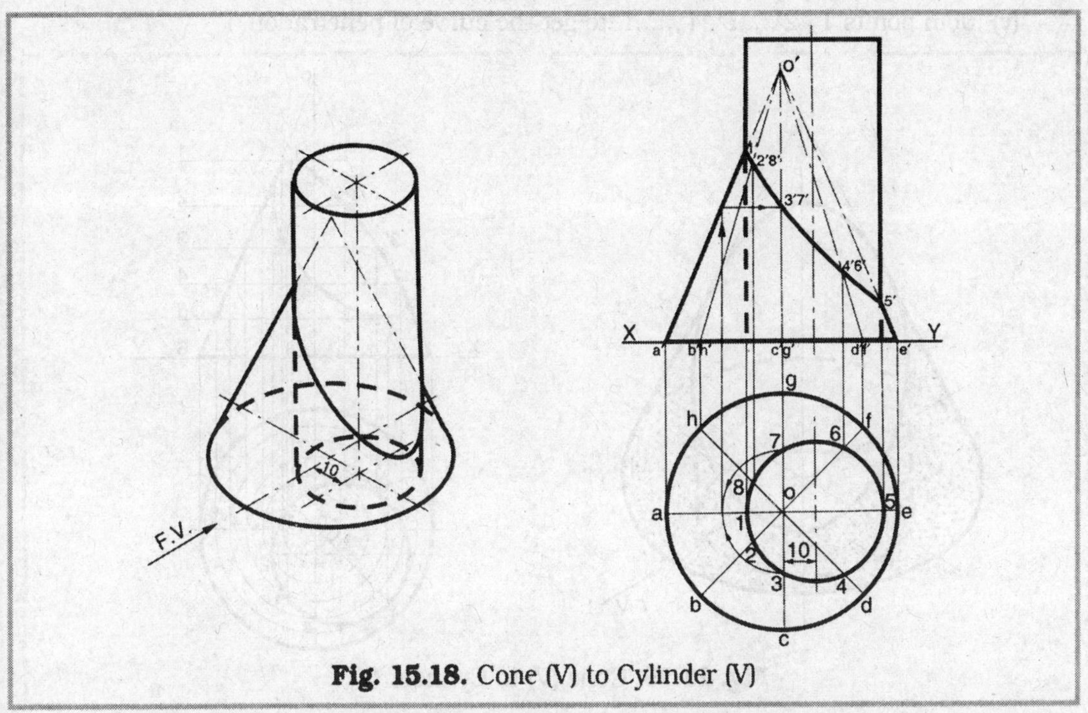

Fig. 15.18. Cone (V) to Cylinder (V)

(ii) Now mark the centre for circle of cylinder in plan, by taking 10 mm distance from o on a line parallel to xy from o. Draw circle of 40 mm diameter with the centre already marked. This circle represents plan of the cylinder. It is in edge view form.

(iii) Now transfer cutting points 1, 2, 3,......8 of generators oa, ob, oc,.....oh with edge view of the cylinder from plan to elevation on corresponding generators to get the points 1', 2', 3',......8' respectively. Do special construction to transfer points 3 and 7, as shown in the figure.

(iv) Join points 1', 2', 3',.......8' and 1'by a smooth curve to get the curve of penetration.

Problem 19 : In Problem 18, (1) Take plane containing both the axes inclined to V.P. by 45° instead of parallel. (2) Instead of cylinder take cylindrical hole. Solve the problem by the cutting plane method, or circle method.

For solution see Fig. 15.19 and follow the procedure, as given below :

(i) First draw plan and elevation of the cone.

(ii) Now mark the centre of cylindrical hole in plan by taking 10 mm distance from o on a line inclined to xy by 45° from o. Draw the circle of 40 mm diameter with the centre already marked.

(iii) In plan draw circles 1, 2, 3, 4 and 5 at nearly equal intervals with o as the centre. Circles 1 and 5 should be the smallest and largest circles respectively just touching the plan (circle) of cylindrical hole. Draw the corresponding cutting plane lines 1,2,3,4 and 5 in elevation parallel to xy.

(iv) Now transfer cutting points 1, (2, 2), (3, 3), (4, 4) and 5 of the circles, 1, 2, 3, 4 and 5 respectively with edge view (circle) of the cylindrical hole from plan to elevation on corresponding cutting plane lines 1,2,3,4 and 5 respectively to get points 1', (2', 2'), (3', 3') (4',4') and 5'.

(v) Join points 1', 2',....5', 4',.....1' to get the curve of penetration.

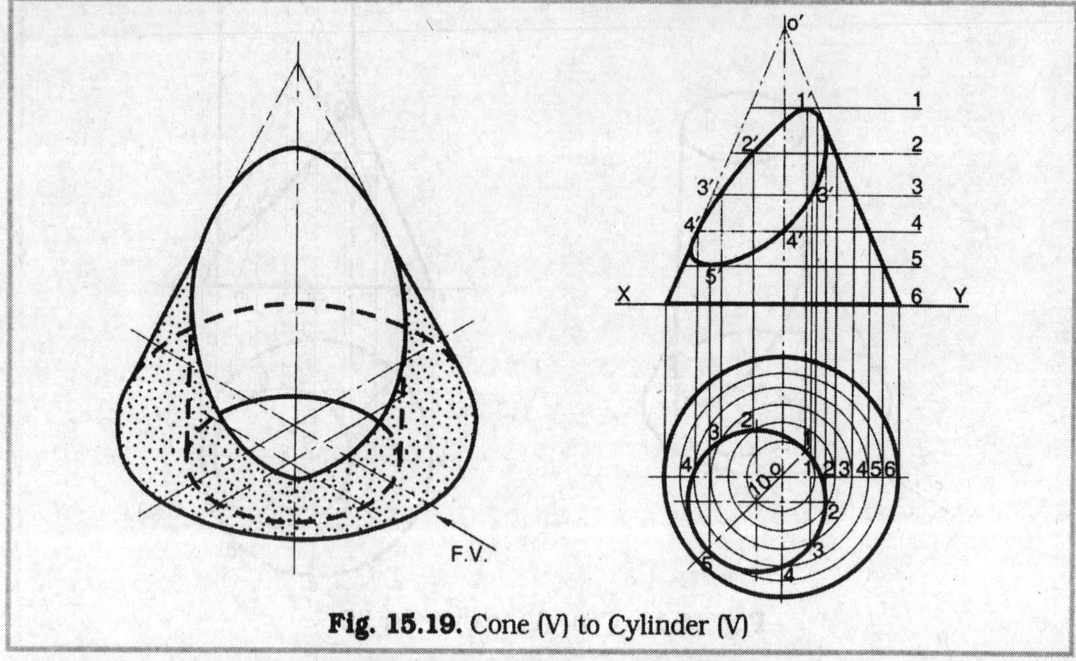

Fig. 15.19. Cone (V) to Cylinder (V)

Problem 20 : A cone, diameter of base 83 mm and height 90 mm, is resting on H.P. on its base. A square prism, side of base 50 mm, is penetrating the cone from the top side centrally. Two rectangular faces of prism are parallel to V.P. Draw the projections showing lines of penetration. Solve it by the cutting plane method.

For solution see Fig. 15.20 and follow the procedure, as adopted for the Problem 19. Three dimensional drawing given will help in understanding the problem.

Problem 21 : A cone, diameter of base 80 mm and height 90 mm, is resting on H.P. on its base. A vertical square prismatic hole, of 41 mm side, is cut through it from the top side. The vertical faces of square hole are equally inclined to V.P. and the axis of square hole is 10 mm nearer to the observer. The plane containing two axes is perpendicular to V.P. Draw the projections of cone with hole, showing the curves of intersection. Solve it by the generator method.

For solution see Fig. 15.21 and follow the procedure, as given below :

(i) First draw plan and elevation of cone and mark on them 12 generators.

(ii) Mark the centre of square hole in plan 10 mm from o on a line perpendicular to xy from o and nearer to observer. Draw the square of 41 mm side around that centre marked with faces equally inclined to xy.

(iii) Take 2 extra generators in plan passing through corners a and c of the square and 4 extra generators perpendicular to faces ab, bc, cd and da of the square. Draw corresponding 6 generators in elevation. This is required to get the points, where the curves of intersection will change their mode.

(iv) Now transfer 18 cutting points of 18 generators with square, edge view of square hole, from plan to elevation, on corresponding generators. Join them to get the curve of intersection, as shown in the figure.

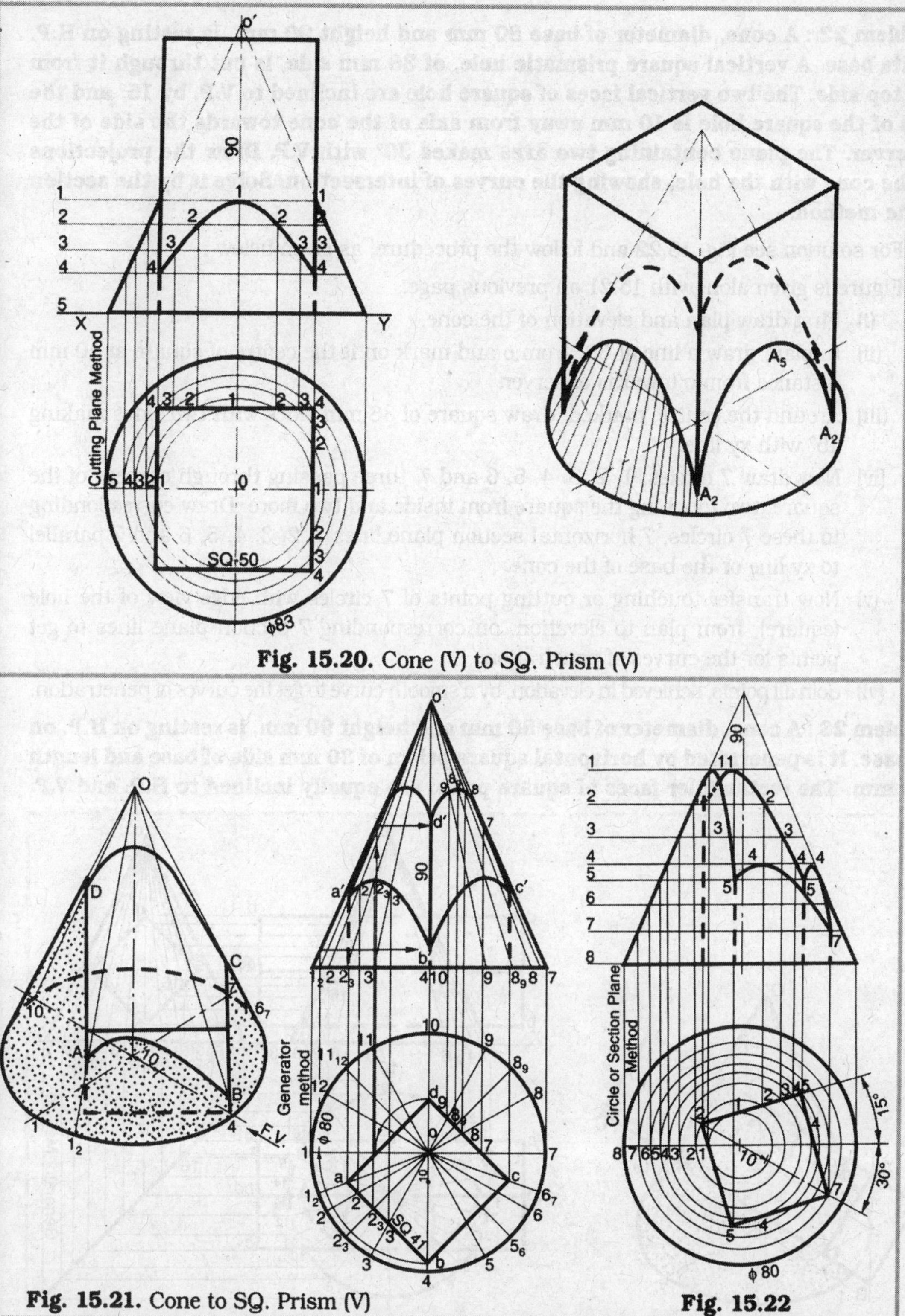

Fig. 15.20. Cone (V) to SQ. Prism (V)

Fig. 15.21. Cone to SQ. Prism (V)

Fig. 15.22

Understand and observe the nature of the curve from the three dimensional figure given.

Problem 22 : A cone, diameter of base 80 mm and height 90 mm, is resting on H.P. on its base. A vertical square prismatic hole, of 36 mm side, is cut through it from the top side. The two vertical faces of square hole are inclined to V.P. by 15° and the axis of the square hole is 10 mm away from axis of the cone towards the side of the observer. The plane containing two axes makes 30° with V.P. Draw the projections of the cone with the hole, showing the curves of intersection. Solve it by the section plane method.

For solution see Fig. 15.22 and follow the procedure, as given below :

Figure is given alongwith 15.21 on previous page.

(i) First draw plan and elevation of the cone.

(ii) In plan, draw a line at 30° from o and mark on it the centre of square at 10 mm distance from o towards observer.

(iii) Around the centre, marked, draw square of 36 mm sides with two sides making 15° with xy line.

(iv) Now draw 7 circles, 1, 2, 3, 4, 5, 6 and 7, three passing through corners of the square, two touching the square from inside and two more. Draw corresponding to these 7 circles, 7 horizontal section plane lines 1, 2, 3, 4, 5, 6 and 7 parallel to xy line or the base of the cone.

(v) Now transfer touching or cutting points of 7 circles with edge view of the hole (square), from plan to elevation, on corresponding 7 section plane lines to get points for the curves of penetration.

(vi) Join all points, achieved in elevation, by a smooth curve to get the curves of penetration.

Problem 23 : A cone, diameter of base 80 mm and height 90 mm, is resting on H.P. on its base. It is penetrated by horizontal square prism of 30 mm side of base and length 105 mm. The rectanguler faces of square prism are equally inclined to H.P. and V.P.

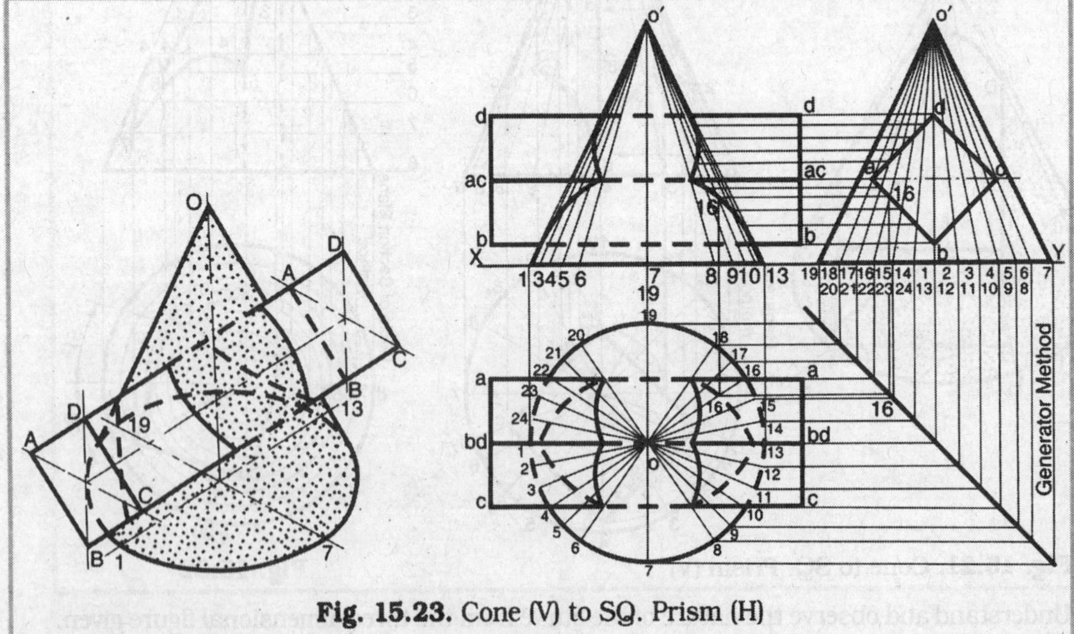

Fig. 15.23. Cone (V) to SQ. Prism (H)

both. The axis of prism intersects the axis of cone at right angles and at a point 28 mm above the base of cone. Draw the projections showing the curves of penetration. Solve the problem by the generator method.

For solution see Fig. 15.23 and follow the procedure, as given below :

Solution given in Fig. 15.23 is self explanatory. Here 24 generators are taken on the surface of the cone first selected and drawn in side view. Note the pattern of generators drawn in side view. Transfer them in other views and then solve the problem as usual.

Problem 24 : Draw the curve of intersection of two vertical cones, diameters of bases 105 mm and 85 mm and heights 80 mm and 70 mm respectively. The axes of two cones are parallel and 30 mm apart and the plane containing two axes is inclined to V.P. by 60°. Assume smaller cone nearer to the observer or away from V.P.

For solution see Fig. 15.24 and follow the procedure, as given below :

(i) First draw plan and elevation of the bigger cone.

(ii) From the centre of the plan of the bigger cone (circle), draw a line at an angle of 60° to xy line and mark on it, the centre of plan of smaller cone, at a distance 30 mm from the centre of the bigger cone.

(iii) Draw circle (plan of smaller cone) of 85 mm diameter, around the centre already marked and draw corresponding elevation of the smaller cone.

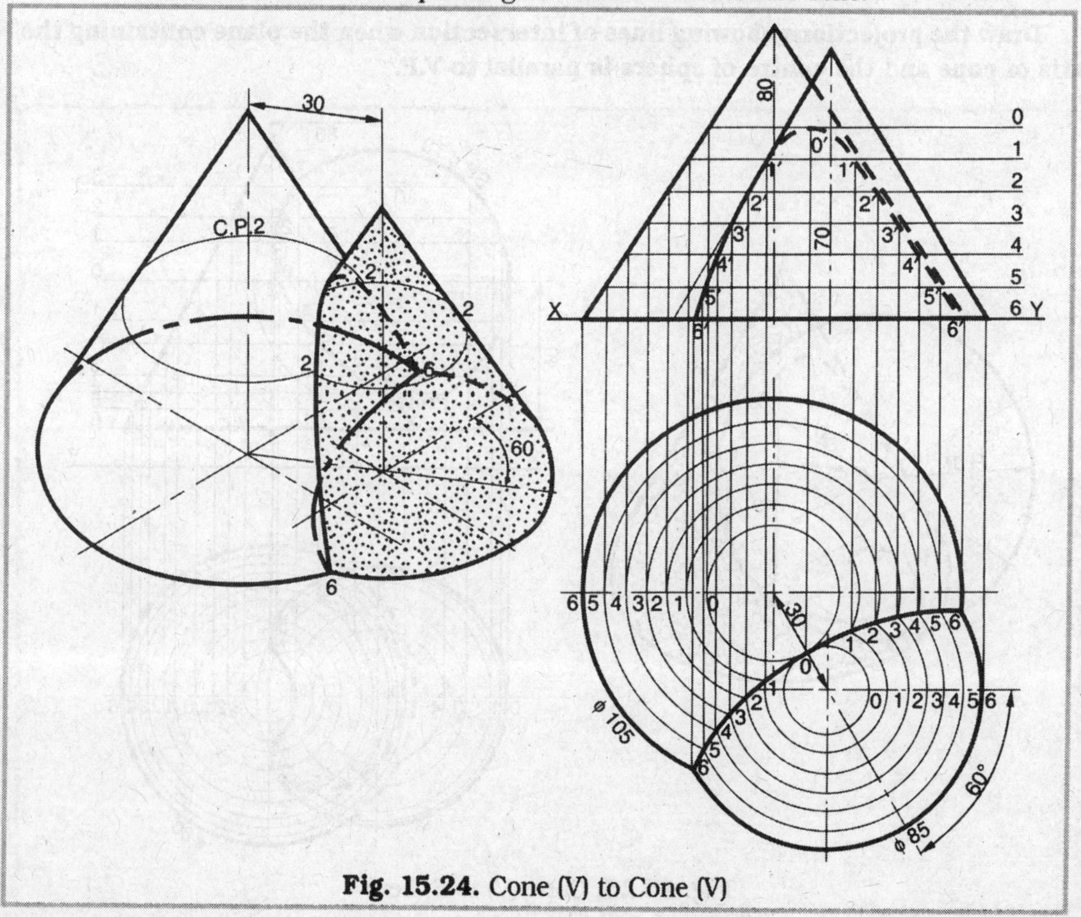

Fig. 15.24. Cone (V) to Cone (V)

(iv) Now take horizontal cutting planes 0, 1, 2, 3, 4, 5 and 6 at nearly equal distance in elevation.

We know that when a cone is cut by a section plane parallel to the base, its true section will be a circle.

So accordingly, draw (7 × 2) circles in plans of both the cones, as shown in the figure.

(v) Now transfer cutting points of the circles 0 with 0, 1 with 1, 2 with 2......and 6 with 6, from plan to elevation on the corresponding cutting plane lines 0, 1, 2,....6 respectively to get points for the curves of intersection.

(vi) Join cutting points in plan and elevation, as shown in the figure by a smooth curve to get the curves of intersection.

Three dimensional figure showing circle No.2 is given to understand the problem. Further this example resembles with the practical example of two mountains meeting each other. Normally a road route from one side of the two mountains to other side is constructed along the line of intersection of two mountains. Mountains are having conical shapes.

Problem 25 : A cone, diameter of base 80 mm and height 80 mm, is having its base parallel to H.P. A sphere of 70 mm diameter is attached to the cone such that the centre of the sphere remains 30 mm from the axis of the cone and 40 mm above the base of the cone.

Draw the projections showing lines of intersection when the plane containing the axis of cone and the centre of sphere is parallel to V.P.

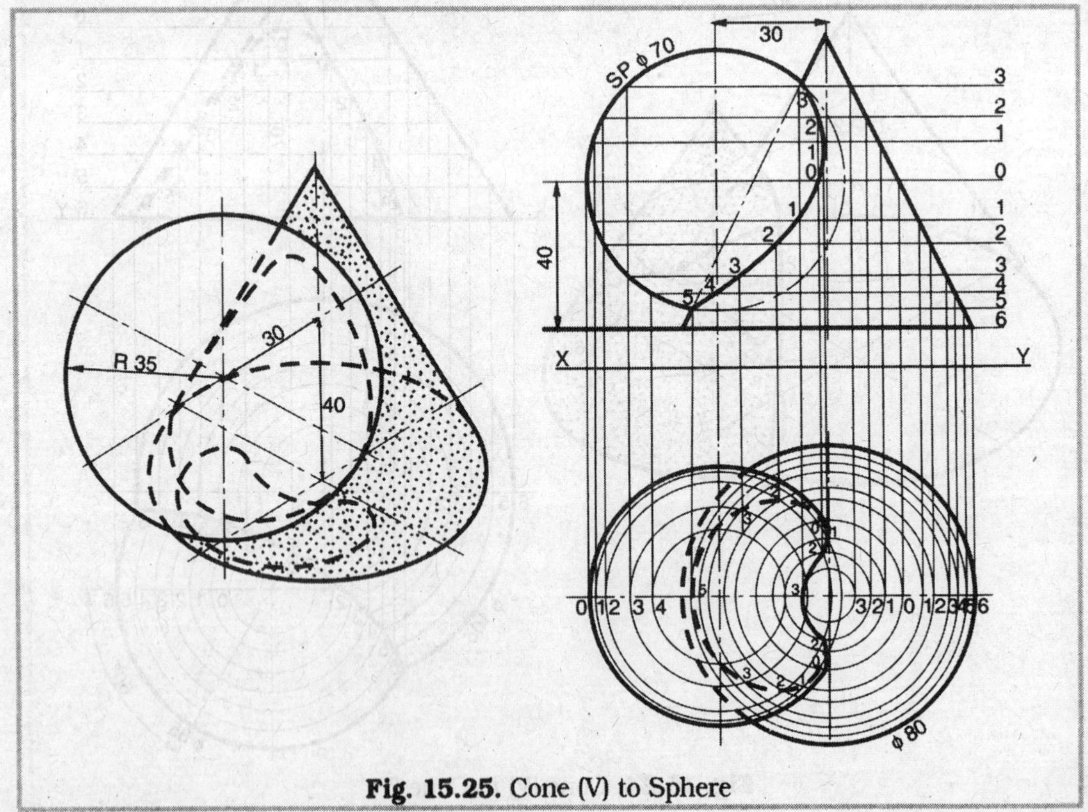

Fig. 15.25. Cone (V) to Sphere

For solution see Fig. 15.25 and follow the procedure, as given below :

(i) First draw plan and elevation of the cone.

(ii) Mark the centre of sphere in plan and in elevation, taking 30 mm on horizontal line from the centre of the cone in plan and taking 40 mm above the base of the cone in elevation. Draw a circle in plan and a circle in elevation of 70 mm diameter with their centres already marked.

(iii) Now take horizontal cutting planes 3, 2, 1, 0, 1, 2, 3, 4 and 5, as shown in elevation, to cut the cone and the sphere simultaneously in the region of intersection.

Sections of the cone and sphere both will be circles and hence draw corresponding circles in plan of sphere and cone.

(iv) Mark in plan cutting points of circles 3 of cone with 3 of sphere, 2 with 2, 1 with 1 and so on and transfer them on corresponding cutting plane lines in elevation.

(v) Join cutting points in plan and corresponding cutting points in elevation by a smooth curve to get the lines of intersection.

Three dimensional figure is given to visualise the joint.

Problem 26 : A cone, diameter of base 92 mm and height 64 mm, is resting on H.P. on its base. It is penetrated completely by another cone, diameter of base 63 mm and height of axis 96 mm. The axes of both the cones intersect at right angles at a point 38 mm above the base of a vertical cone and 56 mm from the base of the horizontal cone.

Draw the projections of two cones showing lines of intersection assuming the plane containing both the axes parallel to V.P.

For solution see Fig.15.26. Method of solution is clear from the figure and hence follow it.

Three dimensional drawing is given to understand the problem clearly.

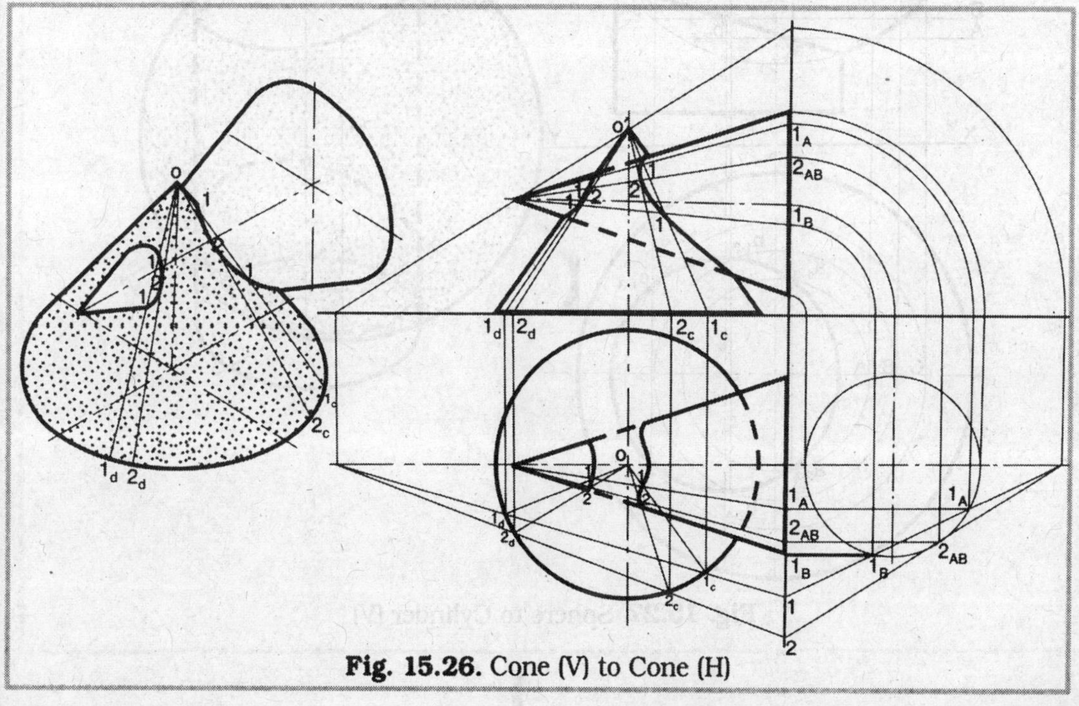

Fig. 15.26. Cone (V) to Cone (H)

10. SPHERE WITH CYLINDER AND SPHERE WITH PRISM

Problem 27 : A vertical cylinder, of 62 mm diameter, is penetrating a sphere of 110 mm diameter. The axis of the cylinder is 14 mm away from the centre of the sphere and the plane containing the axis of the cylinder and the centre of the sphere is parallel to V.P.

Draw the projections in 1st Angle System showing the curves of penetration.

For solution see Fig. 15.27 and follow the procedure, as given below :

(i) Draw first plan and elevation of the sphere. Both will be circles of 110 mm diameter.

(ii) In plan, mark the centre of cylinder, 14 mm away from the centre of the sphere, on a line parallel to xy. Draw a circle of 62 mm diameter around the centre already marked to represent the plan of the cylinder. Draw (rough) elevation of cylinder.

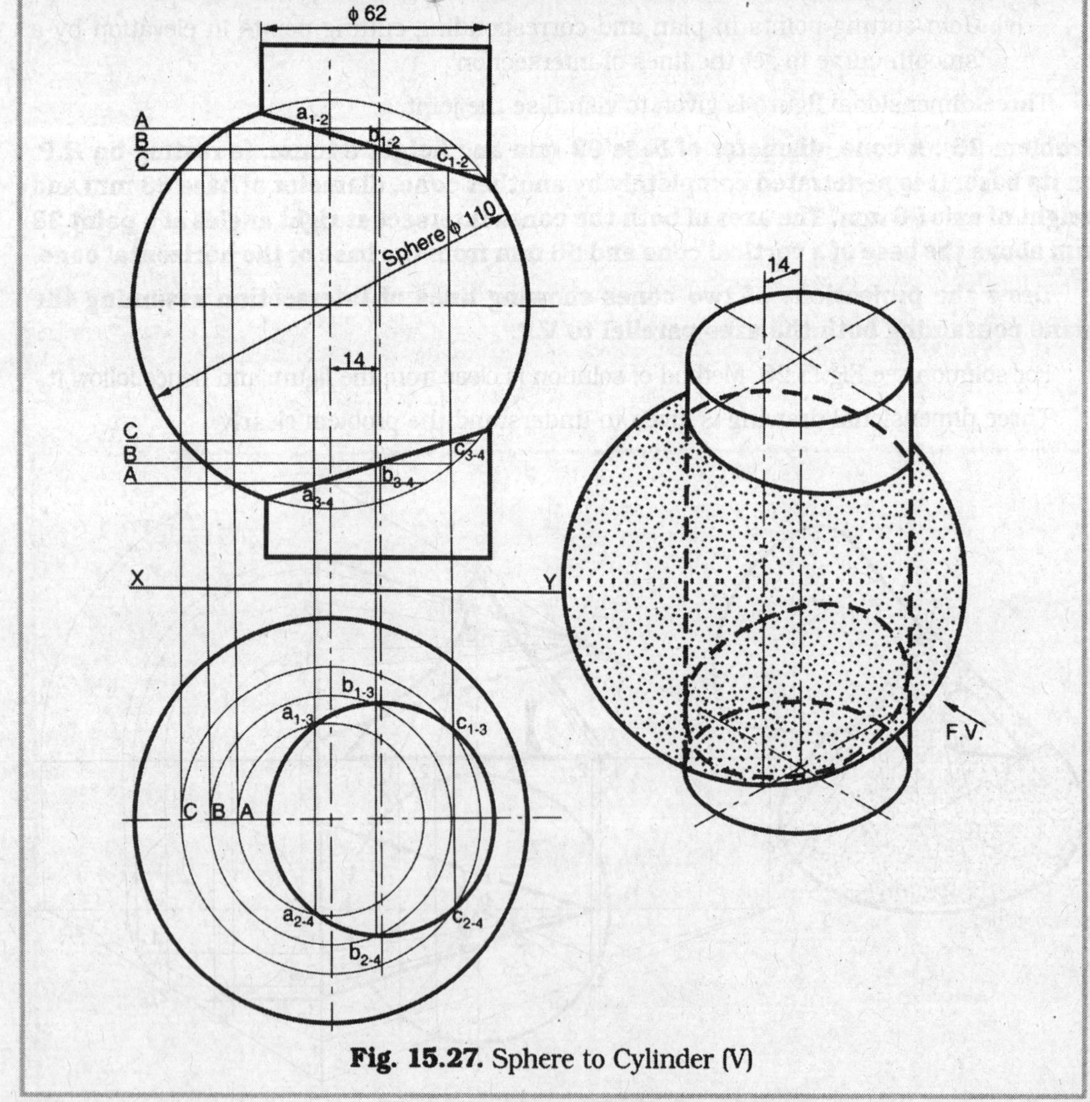

Fig. 15.27. Sphere to Cylinder (V)

(iii) Take horizontal cutting planes A, B, C and C, B, A in the region of intersection of two solids on elevation, as shown in the figure. Draw corresponding sections of the sphere (circles A, B and C) in plan.

(iv) Now transfer cutting points of circles A, B and C with circle, representing plan of cylinder, from plan to elevation on corresponding cutting plane lines. Join the points thus obtained to get the curves of intersection in elevation.

Problem 28 : A vertical cylinder, of 50 mm diameter, is penetrating a sphere of 80 mm diameter. The axis of the cylinder is 10 mm away from the centre of the sphere and the plane containing the axis of the cylinder and the centre of the sphere is inclined to V.P. by 45°. Cylinder is nearer to the observer than sphere.

Draw the projections showing the curves of penetration.

For solution see Fig.15.28 and follow the method followed in the previous problem.

Here circles 1, 2, 3, 4 and 5 are taken in plan, smallest No.1 is touching from inside and biggest No.5 is touching from outside and corresponding to them horizontal cutting planes 1, 2, 3, 4 and 5 are drawn on upper and lower side in elevation. Rest is clear from the figure.

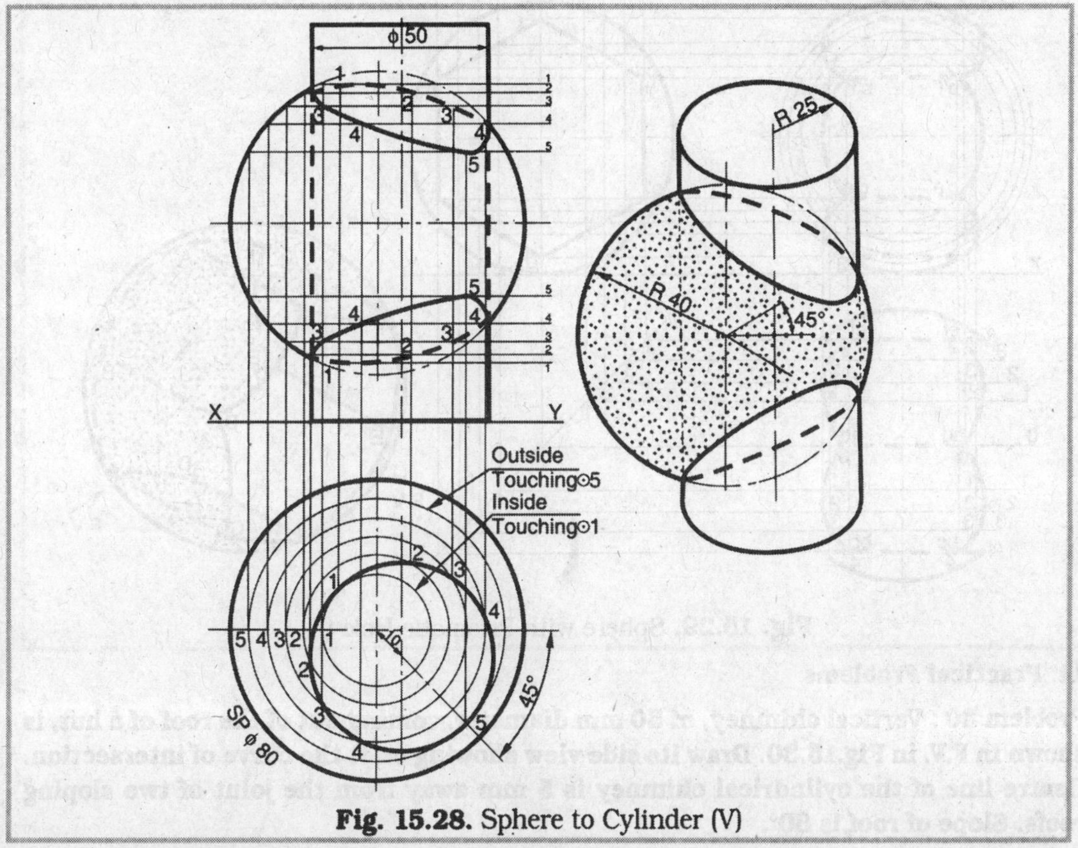

Fig. 15.28. Sphere to Cylinder (V)

Problem 29 : A sphere, of 80 mm diameter, has a hexagonal prismatic hole. The axis of the hole passes through the centre of the sphere. Two flat faces of the prismatic hole are perpendicular to H.P. and the axis of the hole is perpendicular to P.P. The sphere is resting on H.P. The side of the hexagon is 35 mm long.

Draw the projections of the sphere showing the curves of intersection of the hole.

For solution see Fig.15.29 and follow the procedure, as given below :

(i) First draw three projections of sphere (3 circles).

(ii) Draw regular hexagon of 35 mm side, in side view, around the centre of the sphere, keeping two edges of the hexagon perpendicular to the x - y line.

(iii) Now take cutting planes (parallel to V.P.), 4,3,2,1,0, 1, 2, 3 and 4 in plan and corresponding cutting planes, in side view, as shown in the figure. Cutting planes 4 and 4, 3 and 3, 2 and 2, 1 and 1 generate the same size circles in sections of the sphere. And hence, draw corresponding 5 circles 0, 1, 2, 3 and 4 in elevation.

(iv) Now transfer cutting points of section planes 0, 1, 2, 3 and 4 with edge view hexagon of the hole, from side view to elevation, on corresponding circles of elevation to get points of intersection. Transfer them from elevation to plan, on corresponding cutting plane lines 4, 3, 2, 1, 0, 1, 2, 3 and 4 to get points of intersection in plan.

(v) Join points in elevation and also in plan to get the curves of intersection of the hole with the sphere.

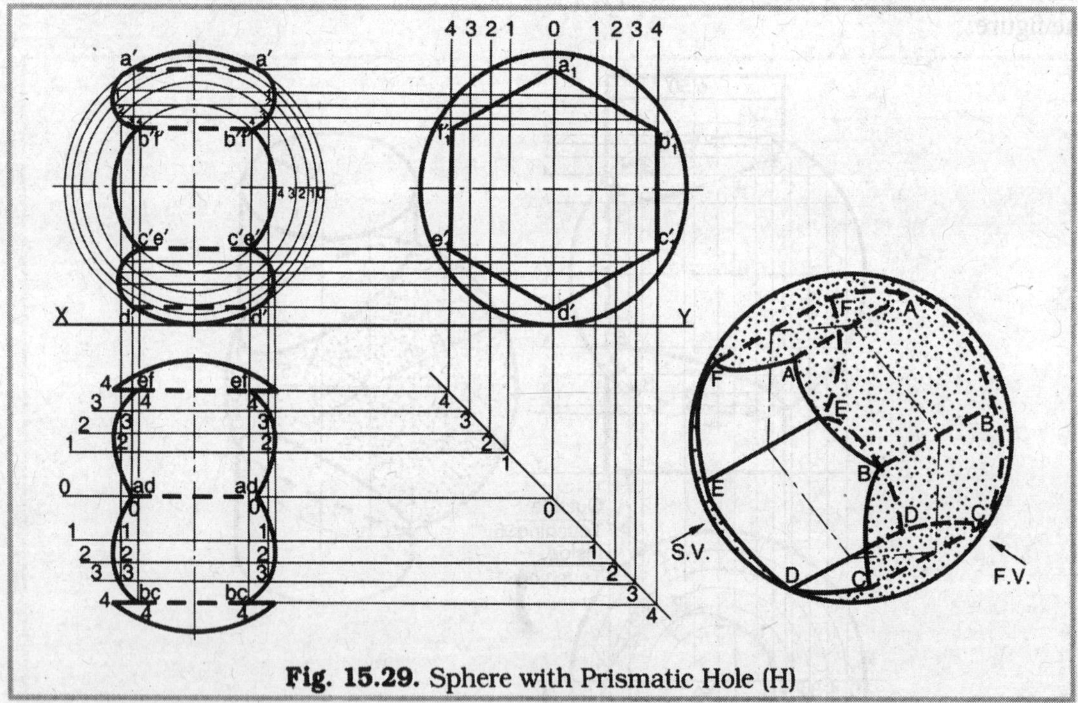

Fig. 15.29. Sphere with Prismatic Hole (H)

11. Practical Problems

Problem 30 : Vertical chimney, of 50 mm diameter, coming out of the roof of a hut, is shown in F.V. in Fig.15.30. Draw its side view showing on it the curve of intersection. Centre line of the cylindrical chimney is 5 mm away from the joint of two sloping roofs. Slope of roof is 30°.

For solution see Fig. 15.30 and follow the procedure, as given below :

(i) First draw the given F.V. Mark 12 generators on the curved surface of the cylindrical chimney in F.V.

(ii) Draw side view of hut, roof and chimney, as shown in the figure and mark 12 generators also on cylindrical chimney.

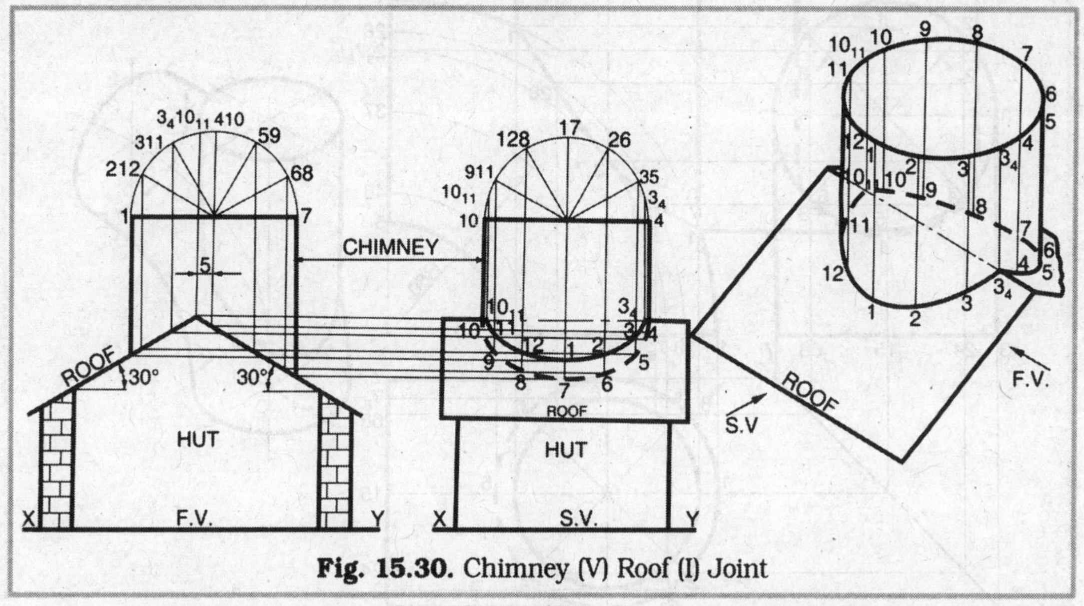

Fig. 15.30. Chimney (V) Roof (I) Joint

(iii) In elevation edge view of roof surfaces is seen. Utilise the edge view for the construction of curves of intersection.

(iv) Transfer cutting points of generators 1, 2, 3,.....12 with edge view of roof, from elevation to side view, on the corresponding generators to get the points of curves of intersection.

Do not forget to take two extra 3_4 and 10_{11} generators passing through the top edge of the roof.

Left side of roof is seen and hence points obtained due to left side roof are joined by full curve. Right side of roof is not seen from left side and hence the points due to right side are joined by dotted curve.

Problem 31 : A 90° pipe bend, of 60 mm diameter, has a branch of 50 mm diameter pipe. The radius of curvature of the bend is 60 mm. The axis of branch pipe passes through the centre of one of the end section of the bend.

Draw plan, elevation and side view of bend with branch pipe, showing on them the curve of penetration, when plane, containing the centre line of bend and axis of branch pipe, is kept parallel to V.P.

For solution see Fig. 15.31 and follow the procedure, as given below :

(i) First draw plan, elevation and side view of pipe bend.

(ii) Draw, in plan, a circle of 50 mm diameter concentric to circle of one end section of pipe bend, as shown in the figure. Draw elevation and side view of branch pipe roughly i.e. without penetration curve.

(iii) On the surface of the bend pipe draw 8 generators by dividing a circle of bend pipe in plan and the circle of the bend pipe in side view into 8 equal parts, as shown in the figure. Project and mark those 8 generators in plan and also in side view.

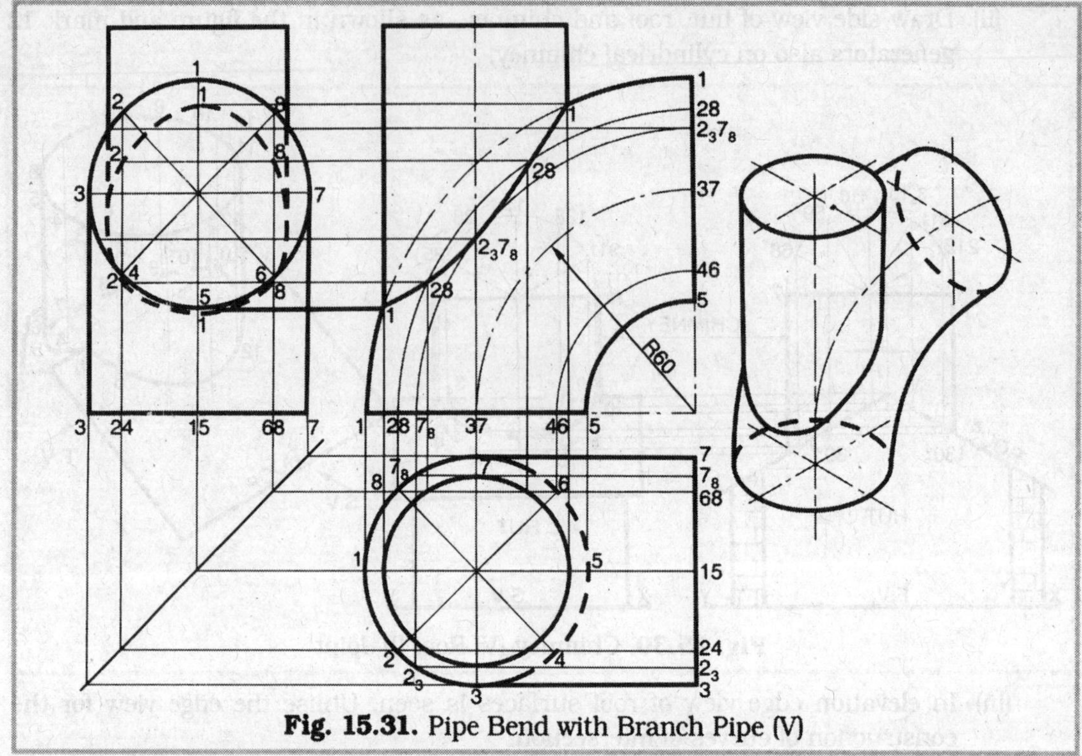

Fig. 15.31. Pipe Bend with Branch Pipe (V)

(iv) Now transfer cutting points of 8 generators with edge view of branch pipe (i.e. circle of 50 mm diameter) from plan to elevation and then to side view on the corresponding generators to get points of lines of intersection.

Do not forget to take 2 extra generators 2_3, and 7_8 on the surface of the bend pipe since these generators are just touching the edge view of the branch pipe.

(v) Join the points of elevation and side view, thus obtained, to get the curves of penetration.

Problem 32 : A connecting rod, 50 mm diameter, has a forged rectangular block at the end 50 mm wide and 17 mm thick. The smooth joint of rod with the block is having a fillet of radius 30 mm. Draw the projections of the rod with block end showing curve of intersection.

For solution see Fig. 15.32 and follow the procedure, as given below :

Discussion : As the rod approaches towards the forged rectangular block the rod diameter increases progressively. The change of the section occurs at a fillet radius of 30 mm. And hence the rod diameter increases from 50 mm to 110 mm [(25 + 30) × 2] as it reaches the block end. As the block width (50 mm) is less than the biggest diameter (110 mm) a curve of interpenetration is generated.

(i) First draw plan and elevation of the block end.

(ii) Draw plan of the rod and elevation of the rod. Connect, in elevation, the rod to the block by a fillet radius of 30 mm.

(iii) Now draw circles in plan 1, 2, 3, 4, 5 and 6 between 50 mm diameter and 110 mm diameter size and draw corresponding horizontal cutting plane lines 1, 2, 3, 4, 5 and 6 in the elevation, perpendicular to axis of the rod, as shown in the figure.

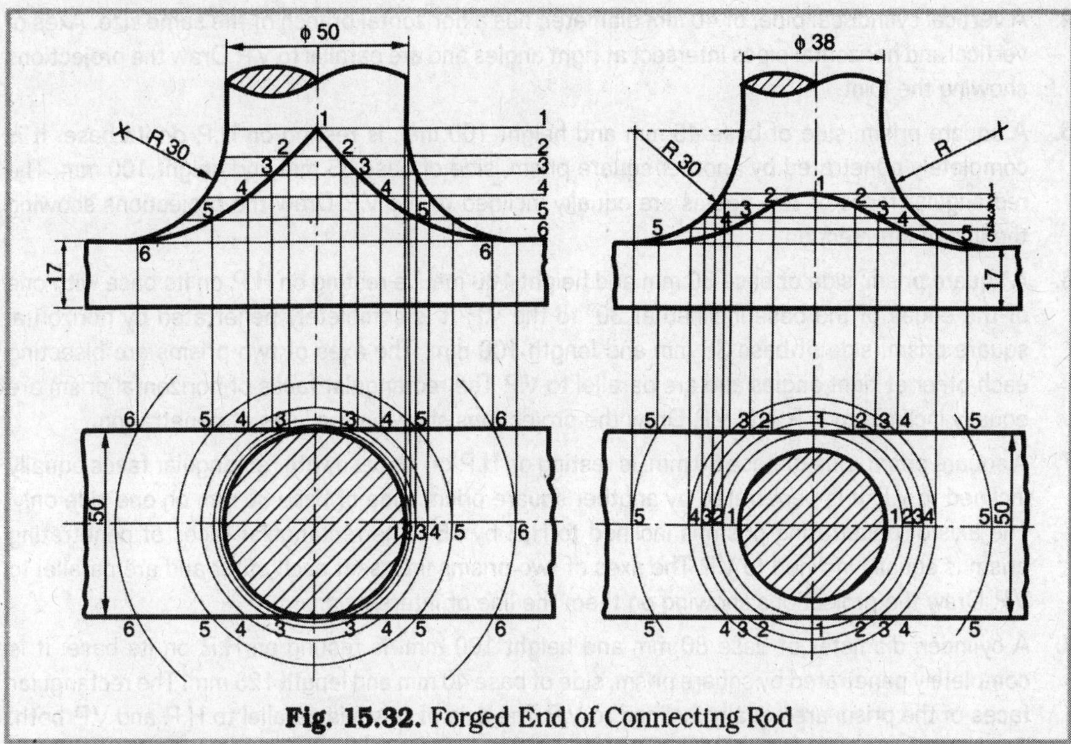

Fig. 15.32. Forged End of Connecting Rod

(iv) Now transfer sets of 4 cutting points 1, 2, 3, 4, 5 and 6 of circles 1, 2, 3, 4, 5 and 6 with edge view of rectangular block (rectangle), from the plan to elevation, on corresponding cutting plane lines 1, 2, 3, 4, 5 and 6 respectively to get the points of the curve of intersection. Join them by a smooth curve and get the curve of intersection.

A similar problem with the rod size 38 mm, lesser than the width of block, is solved in Fig. 15.32. The method of solution is the same. Note the nature of the curve of intersection.

EXERCISE

1. A cylinder, of 54 mm diameter, is resting on H.P. on its base. Another cylinder, of 40 mm diameter, penetrates the vertical cylinder. The axes of two cylinders intersect at right angle and they are parallel to V.P. Draw the projections of both the cylinders showing the line of intersection.

2. A cylinder, of 55 mm diameter, is resting on H.P. on its base. It is completely penetrated by another cylinder of 40 mm diameter. The axis of penetrating cylinder is parallel to V.P. and inclined to H.P. by 30°. The axis of penetrating cylinder is nearer to the observer by 10 mm as compared to vertical cylinder.

Draw projections showing the curves of penetration.

3. A cylinder, of 75 mm diameter, is resting on H.P. on its base. It is penetrated by another cylinder of the same size. The axis of penetrating cylinder is parallel to H.P. and V.P. and is 10 mm away from the axis of the vertical cylinder and is further away from V.P. Draw the projections showing the curves of penetration.

4. A vertical cylindrical pipe, of 40 mm diameter, has a horizontal branch of the same size. Axes of vertical and horizontal pipes intersect at right angles and are parallel to V.P. Draw the projections showing the joint.
5. A square prism, side of base 45 mm and height 100 mm, is resting on H.P. on its base. It is completely penetrated by another square prism, side of base 35 mm and height 100 mm. The rectangular faces of two prisms are equally inclined to the V.P. Draw the projections showing the line of intersection.
6. A square prism, side of base 50 mm and height 100 mm, is resting on H.P. on its base with one of the edges of the base inclined at 30° to the V.P. It is completely penetrated by horizontal square prism, side of base 35 mm and length 100 mm. The axes of two prisms are bisecting each other at right angles and are parallel to V.P. The rectangular faces of horizontal prism are equally inclined to H.P. and V.P. Draw the projections showing the lines of penetration.
7. A square prism, side of base 50 mm, is resting on H.P. on its base with rectangular faces equally inclined to V.P. It is penetrated by another square prism, side of base 30 mm on one side only. The axis of penetrating prism is inclined to H.P. by 45°. The rectangular faces of penetrating prism is equally inclined to V.P. The axes of two prisms intersect each other and are parallel to V.P. Draw the projections showing on them the line of intersection.
8. A cylinder, diameter of base 80 mm and height 100 mm, is resting on H.P. on its base. It is completely penetrated by square prism, side of base 40 mm and length 125 mm. The rectangular faces of the prism are equally inclined to V.P. The axis of prism is parallel to H.P. and V.P. both. The axis of prism is nearer to the observer by 20 mm as compare to the axis of cylinder. Draw the projections showing the curves of penetration.
9. A hexagonal prism, side of base 35 mm and height 150 mm, is resting on H.P. on its base with two edges of base perpendicular to V.P. It is completely penetrated by a square prism, side of base 40 mm and length 150 mm. The axis of the square prism is parallel to V.P. and inclined to H.P. by 30° and is away from V.P. by 15 mm as compare to the axis of hexagonal prism, The rectangular faces of square prism are equally inclined to V.P. Draw the projections showing the lines of intersection or penetration.
10. A cone, diameter of base 75 mm and height 100 mm, is resting on H.P. on its base. It is completely penetrated by a horizontal cylinder of 45 mm diameter. The axis of the cylinder is parallel to V.P. and intersects the axis of the cone at a point 30 mm above the base of the cone. Draw the projections of two solids showing the curves of penetration.
11. A cone, diameter of base 80 mm and height 100 mm, is resting on H.P. on its base. It is completely penetrated by a horizontal cylinder of 40 mm diameter in such a way that the cone and the cylinder both envelop a common sphere. The axes of both the solids intersect at right angles. A plane containing the axes of two solids is parallel to V.P. Draw the projections showing lines of penetration.
12. A cone, diameter of base 90 mm and height 100 mm, is resting on H.P. on its base. It is completely penetrated by a horizontal cylinder of 50 mm diameter. The axis of cylinder is parallel to V.P., away from V.P. by 10 mm and 40 mm above H.P. Draw the projections showing curves of penetration.
13. A vertical inverted conical funnel has a horizontal cylindrical outlet. The mouth of the conical funnel at the inlet is 100 mm in diameter and at the cylindrical outlet of the funnel is 40 mm in diameter. The cylindrical outlet pipe and conical inlet funnel both envelope a common sphere

and axes of both intersect at right angles. Draw the projections showing curves of joint when the axes of both the pieces are parallel to V.P.

14. A cone, diameter of base 60 mm and height 60 mm, is resting on H.P. on its base. It is penetrated by a vertical cylinder of 35 mm diameter from the top. The axes of two solids are parallel and 10 mm apart. Draw the projections showing the curves of penetration when,

(a) plane containing two axes is parallel to V.P. and

(b) plane containing two axes is inclined to V.P. by 60°

Assume cylinder nearer to the observer.

15. A cone, diameter of base 75 mm and height 100 mm, is resting on H.P. on its base. It is penetrated by a vertical square prism, side of base 30 mm and height 125 mm. The rectangular faces of the square prism are equally inclined to V.P. The axes of two solids are parallel and 12 mm apart. Draw the projections of the solids showing the curves of intersection when,

(a) plane containing two axes of solids is parallel to V.P. and

(b) plane containing two axes of solids is inclined to V.P. by 30°.

Assume square prism nearer to the observer.

16. Draw the curve of intersection of two vertical cones, diameters of bases 100 mm and 120 mm and heights 80 mm and 100 mm respectively. The axes of two cones are 30 mm apart when,

(a) plane containing two axes is parallel to V.P. and

(b) plane containing two axes is inclined to V.P. by 30°. Assume smaller cone nearer to the observer.

17. The vertical and horizontal cones, of base diameters 120 mm and 100 mm and heights 120 mm and 130 mm respectively, intersect each other. The plane containing two axes is parallel to V.P. The point of intersection of the two axes is at 50 mm from the base of the vertical cone and 60 mm from the base of the horizontal cone. Draw the projections showing on them the curves of penetration.

18. A vertical cylinder, of 75 mm diameter and height 100 mm, is penetrated by a horizontal cone, diameter of base 75 mm and length 120 mm. The axes of both the solids bisect each other and are parallel to V.P. Draw the projections showing the curves of penetration.

19. The big end of the connecting rod, 75 mm in diameter, has a forged rectangular block, 75 mm wide and 30 mm thick. The radius of curvature at the joint between forged block and rod is 25 mm. Draw the projections showing the curves of intersection.

20. A round vertical hole, of 500 mm diameter, is cut in a roof such that the centre line of the hole remains at the joint of two sloping roofs. Slopes of two sides of roofs are 30° and 60° with ground and the line, along which they meet is perpendicular to V.P. Draw plan, elevation and end view showing edge of the hole in a roof.

21. A square pyramid, side of base 50 mm and height 100 mm, is resting on H.P. on its base with edges of base equally inclined to V.P. It is penetrated by a horizontal cylinder of 50 mm diameter. The axes of both the solids bisect each other at right angles and are parallel to V.P. Draw the projections showing curves of penetration.

22. A circular pipe, of 30 mm diameter, is connected to the bottom of a semi cylindrical tank of diameter 100 mm, having rectangular section 100 mm × 60 mm length open on the top side.

Draw the projections showing the curves of penetration. Assume the axis of semi cylindrical tank perpendicular to V.P. and the axis of the circular pipe perpendicular to P.P. and 15 mm above the bottom most point of the tank.

23. A cone, diameter of base 60 mm and height 80 mm, is resting on H.P. on its base. The centre of the sphere intersecting with the cone, is 15 mm behind the axis of cone and is 30 mm above H.P. Diameter of sphere is 50 mm. Draw the projections showing curves of intersection.

24. A square pyramid, side of base 40 mm and height 55 mm, rests on H.P. on its base, with edges of base equally inclined to V.P. It is penetrated by a triangular prism of 25 mm side of base such that their axes intersect each other at right angles. The lower rectangular face of the prism is 8 mm above the H.P. and parallel to it. Draw the projections showing the curves of penetration.

25. A sphere, of 100 mm diameter, is penetrated by a vertical triangular prism, side of base 40 mm. The axis of the prism passes through the centre of the sphere. Two rectangular faces of the prism are equally inclined to V.P. and nearer to observer. Draw the projections showing the curves of penetration.

26. Fig.15.33 shows the elevation and plan of intersecting cylinders. Draw projections with the curves of penetration.

27. Fig.15.34 shows elevation and plan of intersecting cylinders. Draw projections with the curves of penetration.

28. Fig.15.35 shows the elevation, plan and side view of a square pyramid penetrated by a triangular prism. Draw the projections with lines of penetration.

29. Fig. 15.36 shows the elevation and plan of a cone penetrated by a cylinder. Draw the projections with the curves of penetration.

30. Fig. 15.37 shows the elevation and plan of a vertical cone penetrated by a vertical hexagonal prism. Draw the projections showing curves of penetration.

31. Fig.15.38 shows the elevation and plan of a cylindrical-conical hopper with offset cylindrical connection. Draw the projection with curves of intersection.

32. Fig.15.39 shows the elevation and plan of cylinder having a slot cut across it. Draw three projections showing the curves of intersection.

33. Fig.15.40 shows the top view of a cylinder having an equilateral triangular hole. Draw top view and elevation showing curves of intersection on them.

34. Fig.15.41 shows the top view of a sphere having a through slot. Draw three projections showing curves of intersection.

35. Fig.15.42 shows the top view and front view of a sphere with hexagonal hole. Draw three projections showing curves of intersection on them.

36. A hexagonal pyramid, of base edge 50 mm and height 110 mm, rests on its base in H. P. such that one pair of opposite base edges is parallel to V. P.

A hexagonal prism, of base edge 35 mm, is penetrating the pyramid vertically, such that their axes coincide. One pair of opposite base edges of the prism is perpendicular to V. P. Draw three views of the combination, showing the lines of intersection. [11] See Fig.15.43

[Pune University, May 1999]

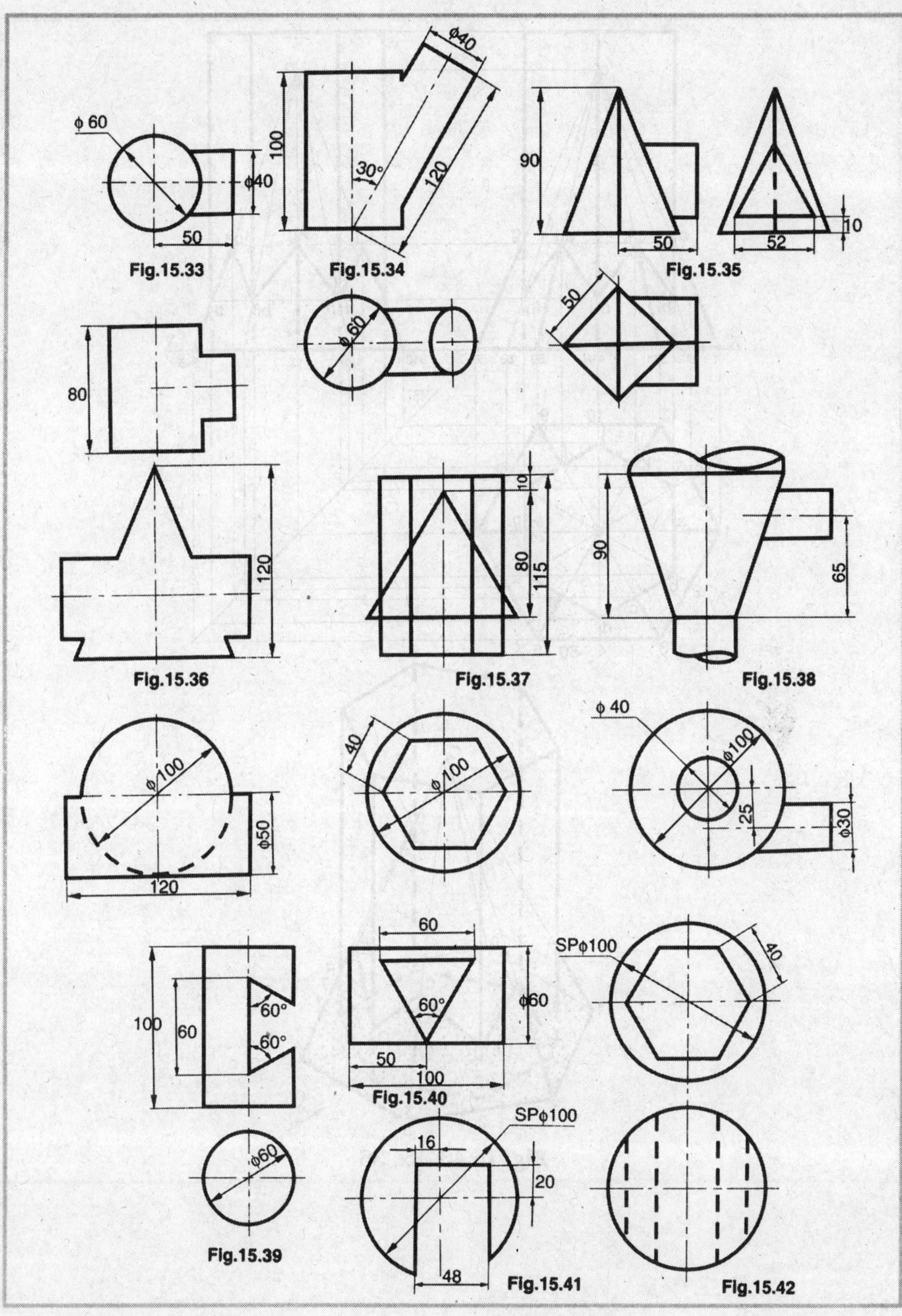

Fig.15.33 Fig.15.34 Fig.15.35 Fig.15.36 Fig.15.37 Fig.15.38 Fig.15.39 Fig.15.40 Fig.15.41 Fig.15.42

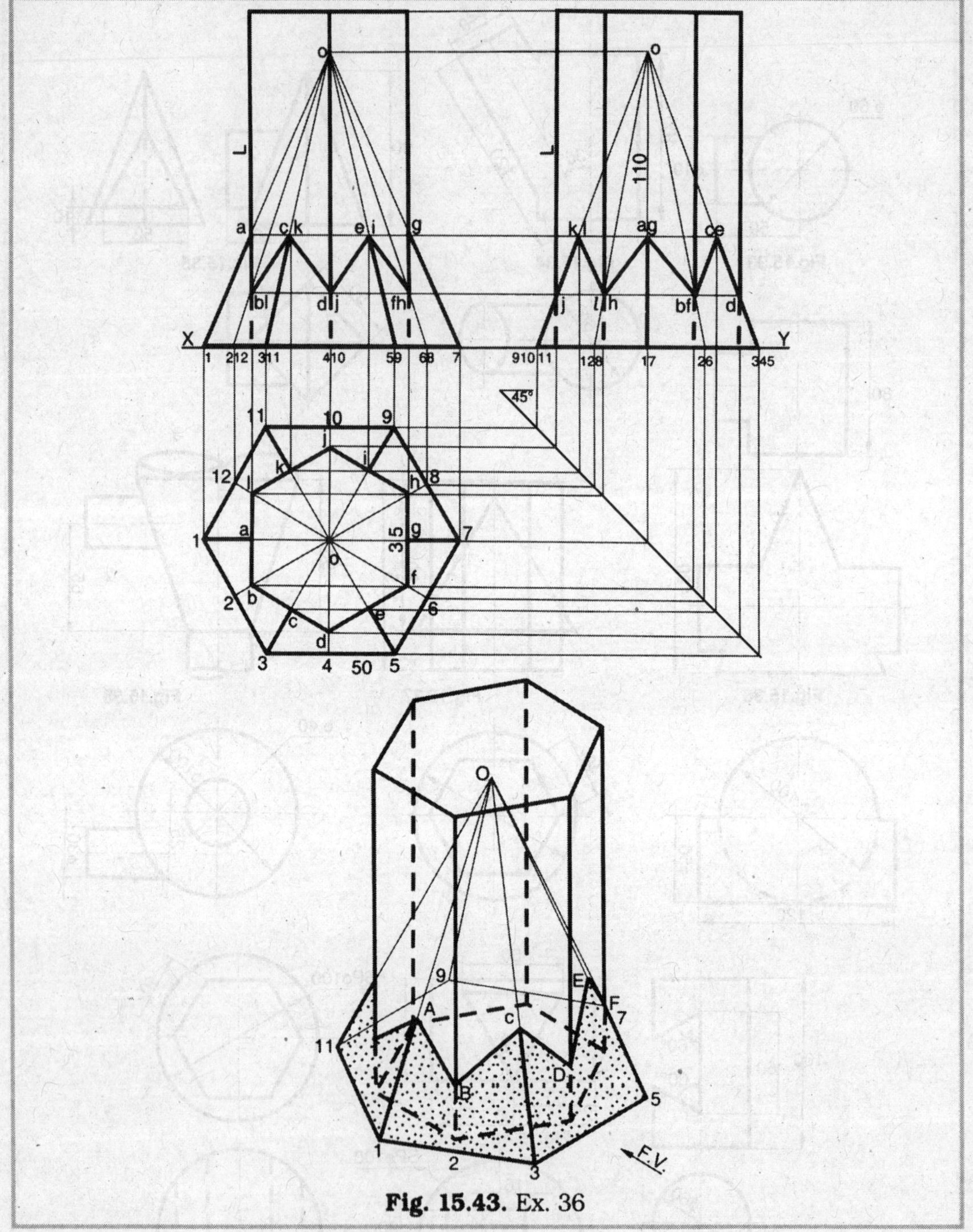

Fig. 15.43. Ex. 36

CHAPTER - 16

Orthographic Projections

1. Introduction :

Three kinds of projections are used by engineers to describe the objects.

1. Orthographic ; 2. Oblique ; 3. Perspective.

They differ from each other in the relationship of the point of sight and the rays of projections to the plane of projection. [See Fig. 16.1 (a) (b) (c)].

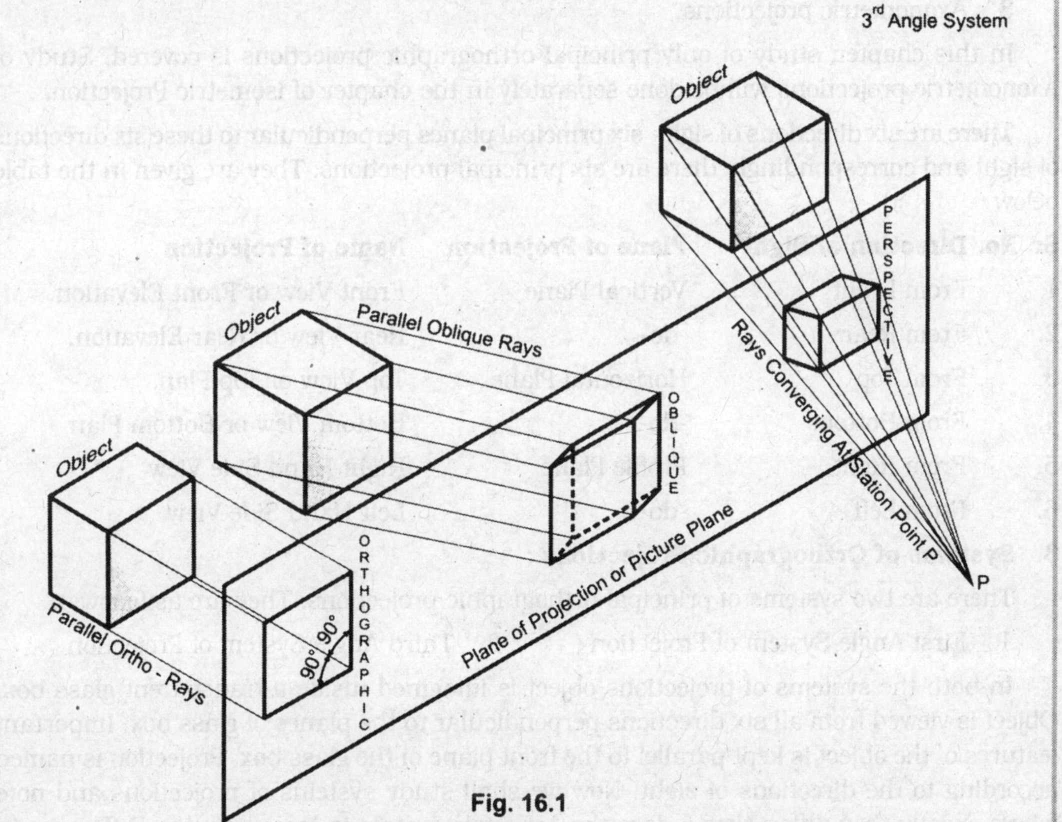

Fig. 16.1

Principles of (a) Orthographic (b) Oblique (c) Perspective Projections

In orthographic projection, it is assumed that observer is at infinite distance and rays or projection lines are parallel and perpendicular (i.e. ortho.) to the plane of projection. It is extensively used by engineers for the purpose of communication because these projections reveal exact shape and size of the object.

In oblique projection, it is assumed that observer is at infinite distance and rays or projection lines are parallel and not perpendicular (i.e. oblique) to the plane of projection. Oblique projection is a three dimensional projection. Principal face of the object is kept parallel to the plane of projection. Oblique projection is not discussed in this book.

In perspective projection, it is assumed that observer is at finite distance and rays or projection lines are converging at observer's eye known as 'station point'. Civil engineers and architects use perspective projection to show the apperance of the buildings and large complexes. It does not reveal exact size and shape of the object and hence is not used as a working drawing.

2. Orthographic Projection :

Orthographic projections are named according to the direction of sight or according to the picture plane on which projections are taken and drawn. They are as follows :

1. Principal orthographic projections;
2. Auxiliary orthographic projections;
3. Axonometric projections.

In this chapter, study of only principal orthographic projections is covered. Study of Axonometric projections will be done separately in the chapter of Isometric Projection.

There are six directions of sight, six principal planes perpendicular to these six directions of sight and correspondingly there are six principal projections. They are given in the table below :-

Sr. No.	Direction of Sight	Plane of Projection	Name of Projection
1.	From Front	Vertical Plane	Front View or Front Elevation.
2.	From Rear	-do-	Rear View or Rear Elevation.
3.	From Top	Horizontal Plane	Top View or Top Plan.
4.	From Bottom	-do-	Bottom View or Bottom Plan
5.	From Right	Profile Plane	Right Hand Side View
6.	From Left	-do-	Left Hand Side View

3. Systems of Orthographic Projections :

There are two systems of principal orthographic projections. They are as follows :

1. First Angle System of Projection ; 2. Third Angle System of Projection

In both the systems of projections object is imagined inside a transparent glass box. Object is viewed from all six directions perpendicular to the planes of glass box. Important features of the object is kept parallel to the front plane of the glass box. Projection is named according to the directions of sight. Now we shall study systems of projections and note where exactly they differ. Now a days the 1st angle system is more popular. Difference in the two systems is underlined.

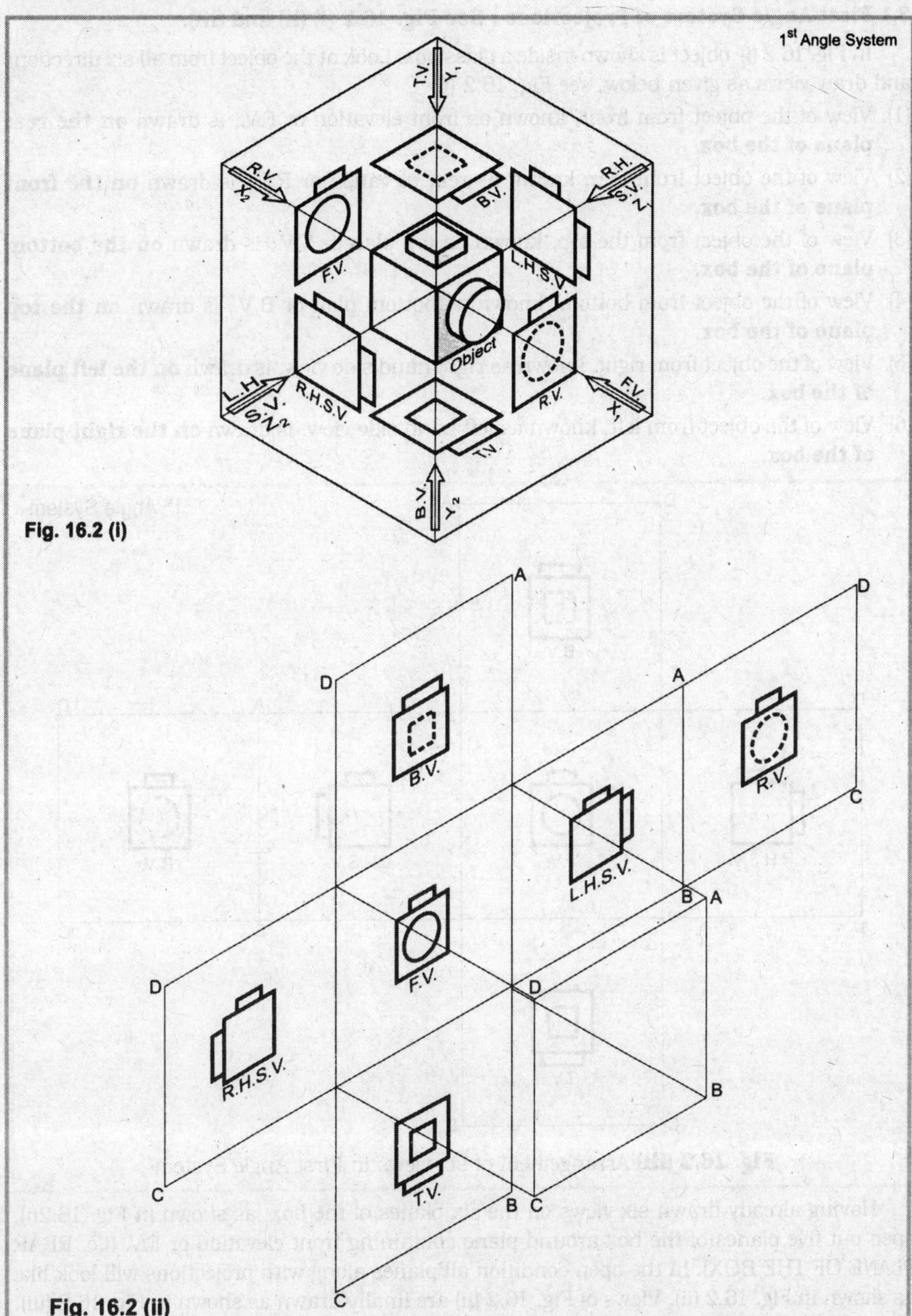

Fig. 16.2 (i)

Fig. 16.2 (ii)

3.1 First Angle System of Projections : See Fig. 16.2 (i) (ii) and (iii).

In Fig. 16.2 (i) object is shown inside a glass box. Look at the object from all six directions and draw views as given below, see Fig. 16.2 (i).

(1) View of the object from front, known as front elevation or F.V., is drawn **on the rear plane of the box.**

(2) View of the object from rear, known as rear elevation or R.V., is drawn **on the front plane of the box.**

(3) View of the object from the top, known as top plan or T.V., is drawn **on the bottom plane of the box.**

(4) View of the object from bottom, known as bottom plan or B.V., is drawn **on the top plane of the box.**

(5) View of the object from right, known as right hand side view, is drawn **on the left plane of the box.**

(6) View of the object from left, known as left hand side view, is drawn **on the right plane of the box.**

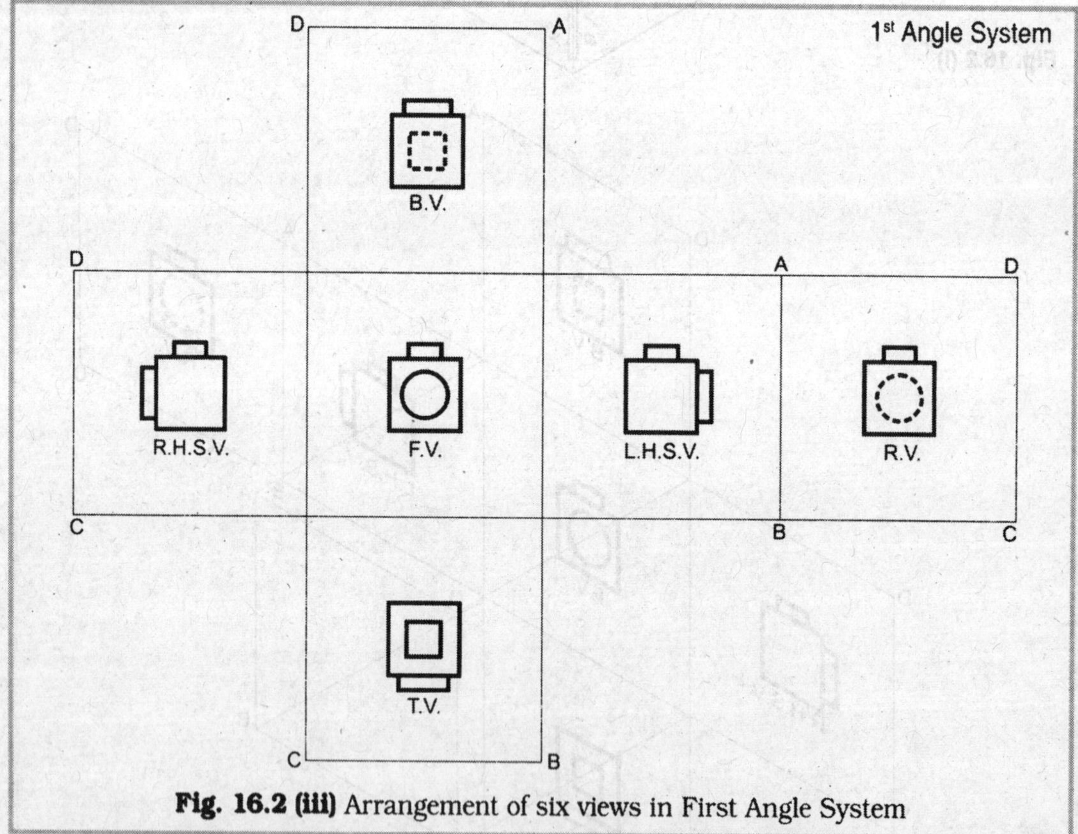

Fig. 16.2 (iii) Arrangement of six views in First Angle System

Having already drawn six views on the six planes of the box, as shown in Fig. 16.2(i), open out five planes of the box around plane containing front elevation or F.V. (i.e. REAR PLANE OF THE BOX). In the open condition all planes along with projections will look like as shown in Fig. 16.2 (ii). Views of Fig. 16.2 (ii) are finally drawn as shown in Fig. 16.2 (iii).

Observe these views to understand following conclusions for **1st Angle System of Projection.**

(1) Front Elevation or F.V. remains in the centre of all views.

(2) Top Plan or T.V. is vertically **below** F.V.

(3) Bottom Plan or B.V. is vertically **above** F.V.

(4) R.H.E.V. or R.H.S.V. is horizontally on **left** of F.V.

(5) L.H.E.V. or L.H.S.V. is horizontally on **right** of F.V.

(6) Rear Elevation or Rear View is horizontally on extreme right of F.V.

(7) Front portion of the object, in all views surrounding F.V., **faces away** from F.V.

(8) Rear portion of the object, in all views surrounding F.V., **faces** F.V.

(9) Length of F.V. and T.V. is same.

(10) Height of the F.V. and S.V. is same.

(11) The depth of the top view is the same as the depth of the side view.

3.2. Third Angle System of Projections : See Fig. 16.3(i), (ii) & (iii)

In Fig. 16.3(i) object is shown inside a glass box. Look at the object from all six directions and draw six views, as given below, see Fig. 16.3 (i).

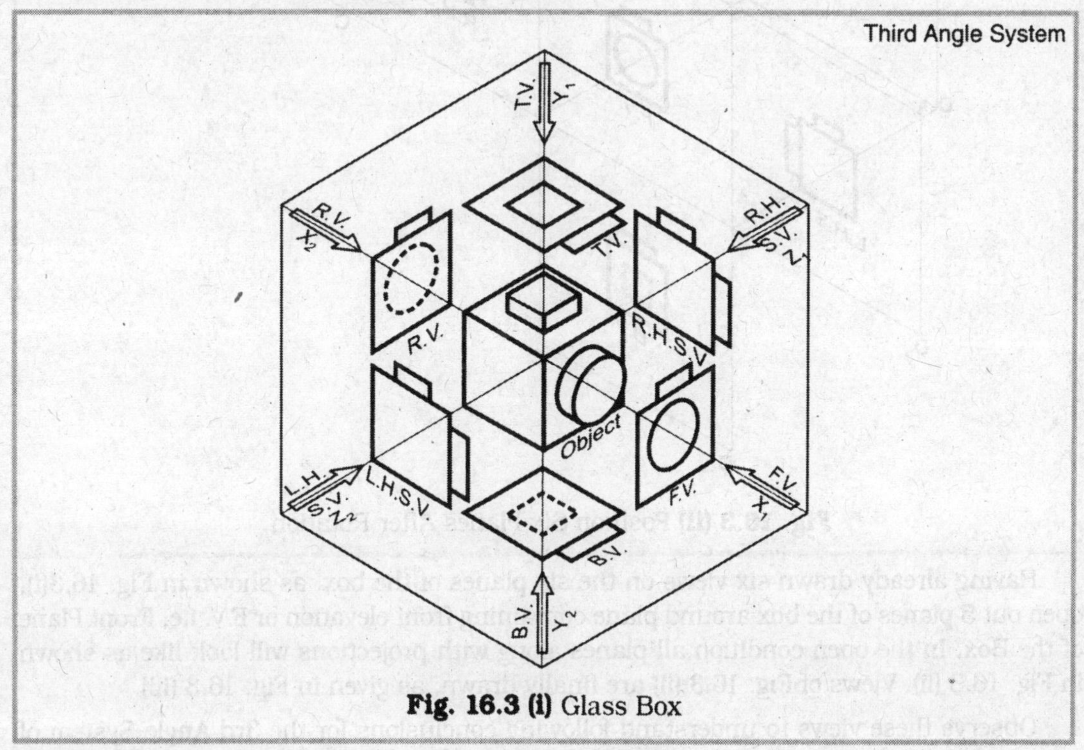

Fig. 16.3 (i) Glass Box

(1) View of the object from front, known as front elevation or F.V., is drawn **on the Front Plane of the Box.**

(2) View of the object from rear, known as rear elevation or R.V., is drawn **on the Rear Plane of the Box.**

(3) View of the object from top, known as top plan or T.V., is drawn **on the Top Plane of the Box.**

(4) View of the object from bottom, known as bottom plan or B.V., is drawn **on the Bottom Plane of the Box.**

(5) View of the object from right, known as right hand side view, is drawn **on the Right Plane of the Box.**

(6) View of the object from left, known as left hand side view, is drawn **on the Left Plane of the Box.**

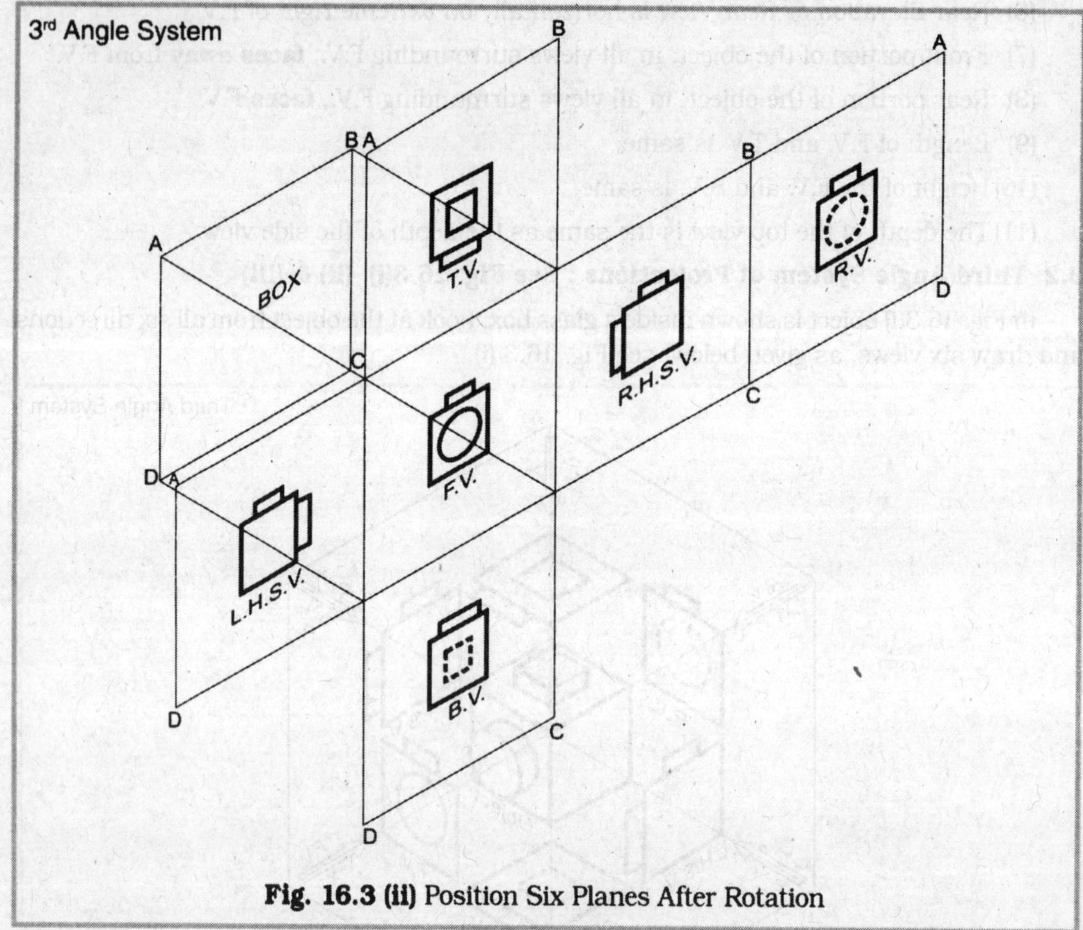

Fig. 16.3 (ii) Position Six Planes After Rotation

Having already drawn six views on the six planes of the box, as shown in Fig. 16.3(i), open out 5 planes of the box around plane containing front elevation or F.V. i.e. Front Plane of the Box. In the open condition all planes along with projections will look like as shown in Fig. 16.3 (ii). Views of Fig. 16.3 (ii) are finally drawn, as given in Fig. 16.3 (iii).

Observe these views to understand following conclusions for the 3rd Angle System of Projection.

(1) Front Elevation or F.V. remains in the centre of all views.

(2) Top Plan or T.V. is vertically **above** F.V.

(3) Bottom Plan or B.V. is vertically **below** F.V.

(4) R.H.E. V. or R.H.S.V. is horizontally on **right** of F.V.

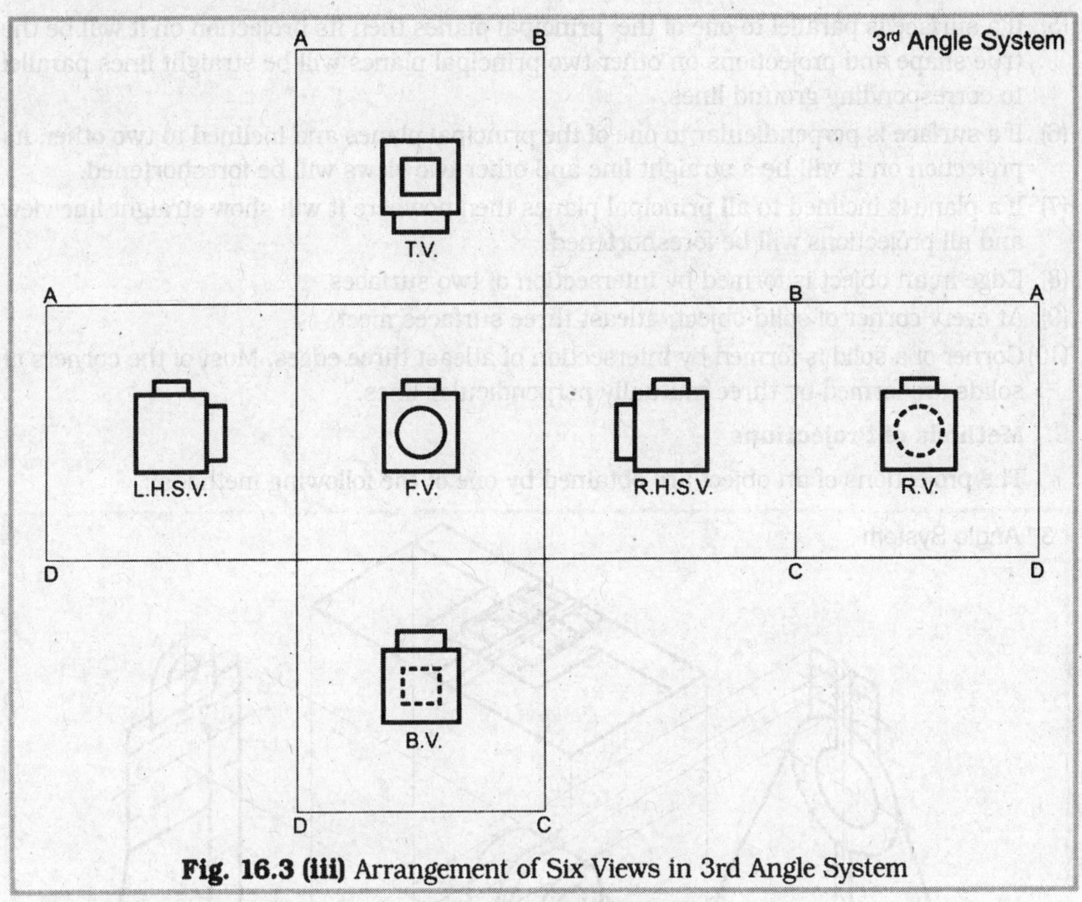

Fig. 16.3 (iii) Arrangement of Six Views in 3rd Angle System

(5) L.H.E.V. or L.H.S.V. is horizontally on **left** of F.V.

(6) Rear Elevation or Rear view is horizontally on extreme right of F.V.

(7) Front portion of the object, in all views, surrounding F.V. **faces** the front view.

(8) Rear portion of the object in all views surrounding F.V. **faces away** from the F.V. Conclusions (9) (10) and (11) of the 1st angle system remain same for the 3rd angle system.

4. Important Conclusions :

Now it is advisable to go through some important conclusions regarding projections of straight lines and projections of planes.

(1) If a line is perpendicular to one of the principal planes then it will be parallel to two other principal planes.

(2) If a line is perpendicular to one of the principal planes then its projection on that principal plane will be point view and on other two principal planes it will show the true length.

(3) If a line is parallel to one of the principal planes and inclined to two other P.Ps. then its projection on parallel P.P. will be the true length showing true inclinations and on other P.Ps it will show shorter lines parallel to ground lines (Vertical or Horizontal).

(4) If a line is inclined to all P.Ps. then its all projections will be shorter and inclined to ground lines.

(5) If a surface is parallel to one of the principal planes then its projection on it will be the true shape and projections on other two principal planes will be straight lines parallel to corresponding ground lines.

(6) If a surface is perpendicular to one of the principal planes and inclined to two other, its projection on it will be a straight line and other two views will be foreshortened.

(7) If a plane is inclined to all principal planes then nowhere it will show straight line view and all projections will be foreshortened.

(8) Edge in an object is formed by intersection of two surfaces.

(9) At every corner of solid object, atleast three surfaces meet.

(10) Corner of a solid is formed by intersection of atleast three edges. Most of the corners of solids are formed by three mutually perpendicular lines.

5. Methods of Projections

The projections of an object are obtained by one of the following methods :

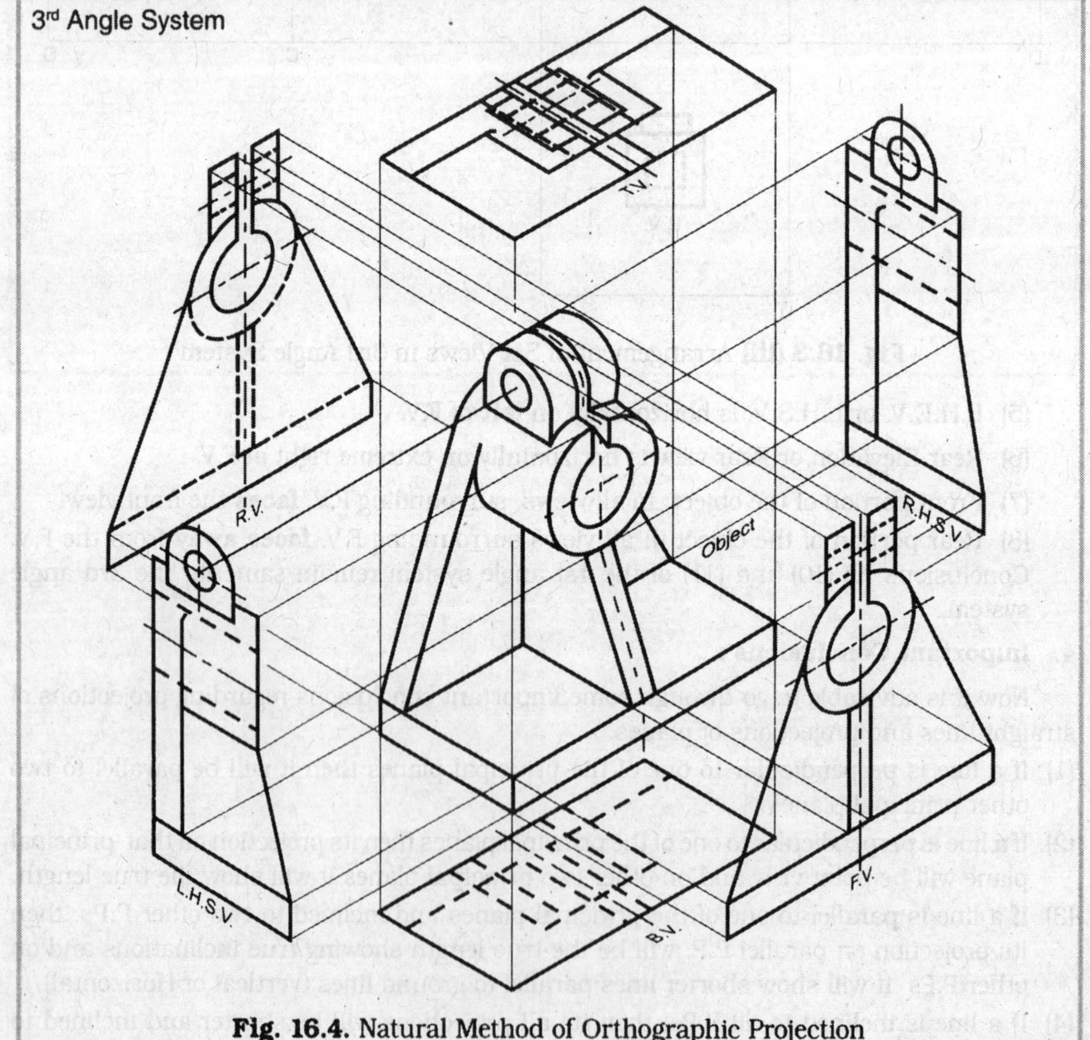

Fig. 16.4. Natural Method of Orthographic Projection

(1) The glass box method, see Fig. 16.2 (i) (ii) (iii) or 16.3 (i) (ii) (iii)

(2) The natural method, see Fig. 16.4

Results, i,e. projections, will be same whether you adopt the glass box method or you adopt the natural method.

(1) Glass Box Method : See Fig. 16.2 and Fig. 16.3

We have already studied glass box method while studing the systems of projections.

(2) Natural Method : See Fig. 16.4

Object is kept in a position such that its important features face front side. Now look at the object from front, top, bottom, right, left and rear to get front view, top view, bottom view, right hand side view, left hand side view and rear view respectively.

In Fig. 16.4, six principal projections according to 3rd Angle System are shown. In the application of this method either observer changes his positions and observes the object from six different positions or observer remains at one position and turn the object to observe 6 faces of the object to get six different projections.

6. Selection of Views

Decision about number of views to be drawn and selection of views, should be such that no misinterpretation occures about the shape of object to the reader. Whenever dispute arises between two firms the drawing on which the contract is made, becomes the important document. Number of views and selection of views should be such that the reader of drawing can imagine one and only one object. Views should clearly and completely describe the object. Superfluous views should be avoided. Views should be selected in such a way that dotted lines will be minimum in views.In case of more complex objects, sectional views, auxiliary views, etc. are drawn to make the object more clear. Projections/views showing contours, profiles, etc. of the object must be selected. Rear and bottom views are rarely used. Mostly important features of the object are taken on the front view, known as the principal view.

7. Analysis of Points, Lines and Surfaces in Principal Views or Projections

Exercise - 1 : In Fig. 16.5(i) three dimensional view of the object is given. Draw (i) F.V./Front Elevation (ii) L.H. S.V. and (iii) T. V. Use 1st Angle Projection System.

To get the front elevation look at it in the direction of arrow X. Elevation is the king of all views and hence selection of direction X should be done in such a way as to get maximum details in elevation.

(1) Draw rectangle mpqn of the same size as rectangle MPQN of the object, since rectangle MPQN is parallel to the plane of projection (V.P.). Its elevation mpqn will be T. S. (See conclusion : 5).

(2) Surface BFHJKCB is also parallel to V.P. or perpendicular to the line of sight and hence it will be seen in true shape in elevation. Draw it (According to the same conclusion 5)

(3) All lines OP, LM, IJ, GH, EF, AB, DC, KN, etc. are perpendicular to V.P. and hence they will be point views in elevation. (According to conclusion: 2).

(4) Now to draw left hand side view look at the object from left i.e. from Y. Surface L M P O, G H J I and A B F E will be seen in true shapes since those surfaces are parallel to P.P. (See conclusion 5)

(5) All lines MN, PQ, JK, HF, BC, AD, GE, LI, etc. are perpendicular to P.P. and hence they will be point views in side view (according to conclusion 2).

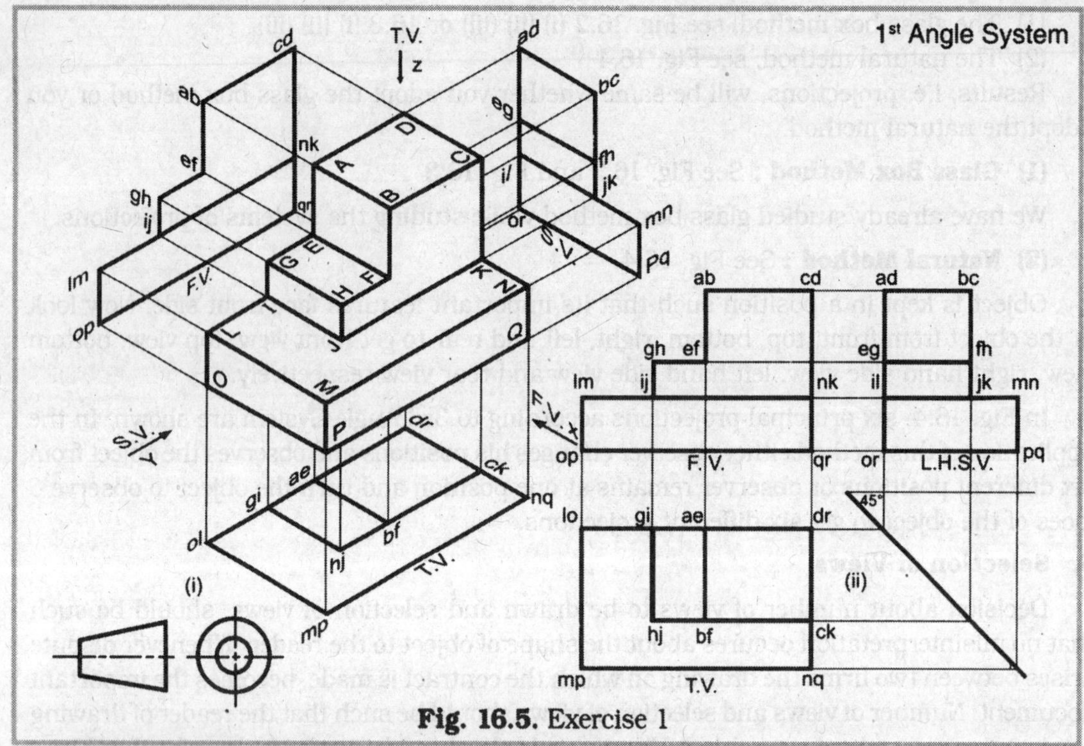

Fig. 16.5. Exercise 1

(6) Now to get plan look at the object from top or from Z direction. In plan surfaces ABCD, EFHG and KNMLIJK will be seen in true shapes since those surfaces are parallel to H.P. or are perpendicular to line of sight Z.

(7) Lines LO, MP, NQ, GI, HJ, AE, BF, CK, etc. are perpendicular to H.P. and hence they will be point views in plan (According to conclusion 2)

In Fig. 16.5 (ii) elevation, end view and plan are shown.

Students should draw the same views separately with left hand side view horizontally on right of elevation and top plan vertically below the elevation, as shown in Fig. 16.5(ii).

Exercise 2 : In Fig. 16.6 three dimensional view of the object is given. Draw elevation, top plan and left hand side view. Use 3rd Angle Projection System.

Follow the procedure, as given below :

(1) Surface ABCD is perpendicular to H.P. and inclined to V.P. and P.P. and hence its plan will be inclined straight line and its elevation and side view will be foreshortened. Draw height AD and BC of same length and AB and DC of projected length in elevation to get rectangle ABCD. (see conclusion 6)

(2) Surface PQRS is perpendicular to P.P. and inclined to V.P. and H.P. and hence its side view will be inclined straight line and its elevation and pla2n will be foreshortened.

Line PQ and line SR are parallel to V.P. and hence will be seen of T.Ls. in elevation.

Draw PS and QR in elevation of projected length and PQ and SR of T.L. to get foreshortened rectangle PQRS in elevation.

Δ23B is parallel to V.P. and hence will be seen in true shape.

With the similar concept, we can draw left hand end view and top view. In Fig. 16.6 (ii) elevation, left hand end view and top plan are given.

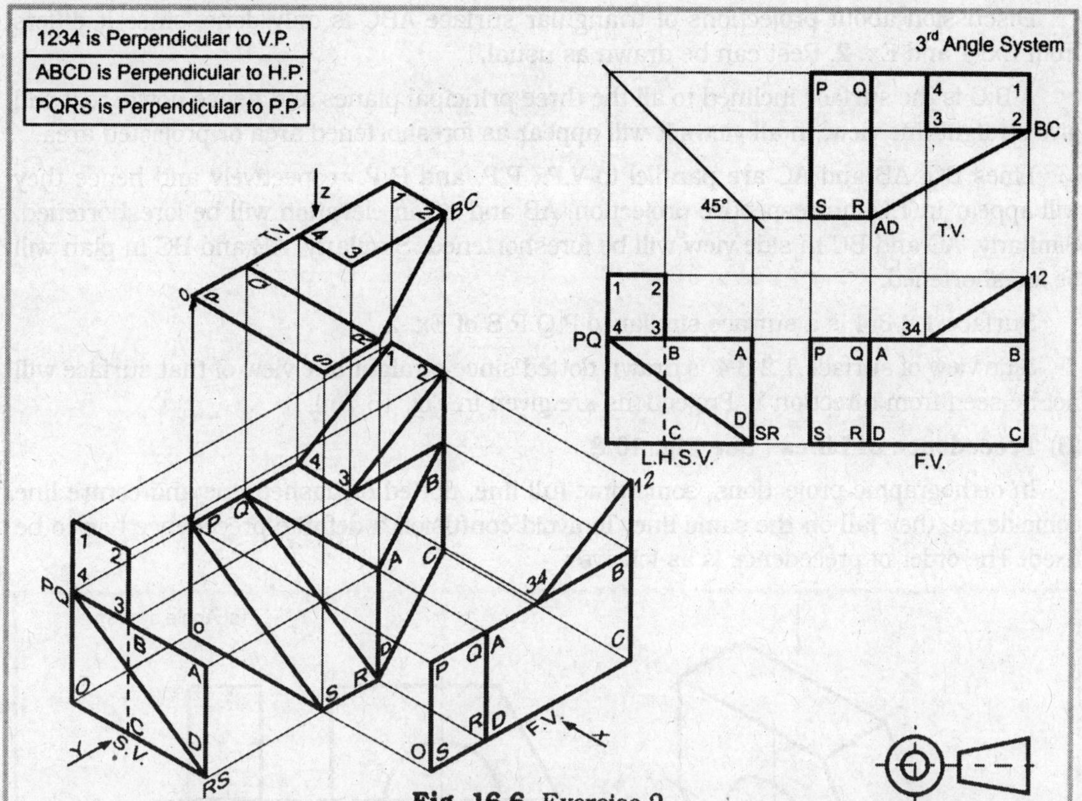

Fig. 16.6. Exercise 2

Exercise 3 : In Fig. 16.7 three dimensional view of the object is given. Draw the elevation, left hand side view and top plan. Use 1st Angle Projection System.

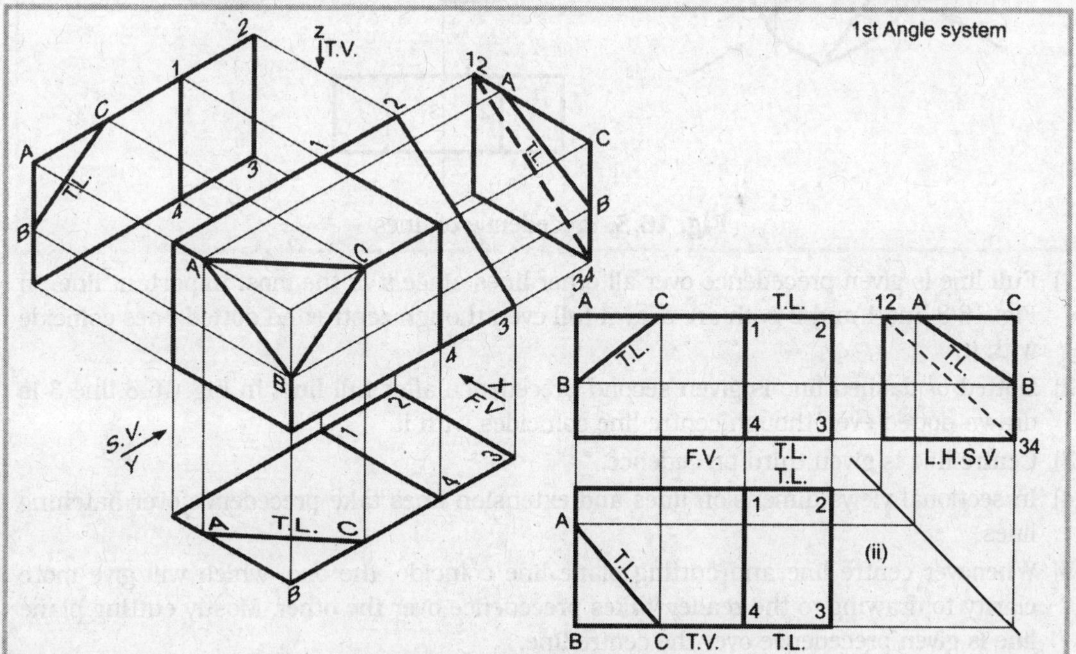

Fig. 16.7. Exercise 3

Discussion about projections of triangular surface ABC is only done, since it differs from Ex. 1 and Ex. 2. Rest can be drawn as usual.

A B C is the surface inclined to all the three principal planes and hence nowhere it will give straight line view. In all views it will appear as foreshortened area or projected area.

Lines BC, AB and AC are parallel to V.P., P.P. and H.P. respectively and hence they will appear in T.Ls. in respective projection. AB and AC in elevation will be foreshortened. Similarly, AC and BC in side view will be foreshortened. Similarly, AB and BC in plan will be foreshortened.

Surface 1 2 3 4 is a surface similar to P Q R S of Ex. 2.

Side view of surface 1 2 3 4 is drawn dotted since straight line view of that surface will not be seen from direction Y. Projections are given in Fig. 16.7(ii)

(8) Precedence of Lines : See Fig. 16.8.

In orthographic projections, some time full line, dotted or dashed line and centre line coincide i.e. they fall on the same line. To avoid confusion a definite precedence has to be fixed. The order of precedence is as follows :

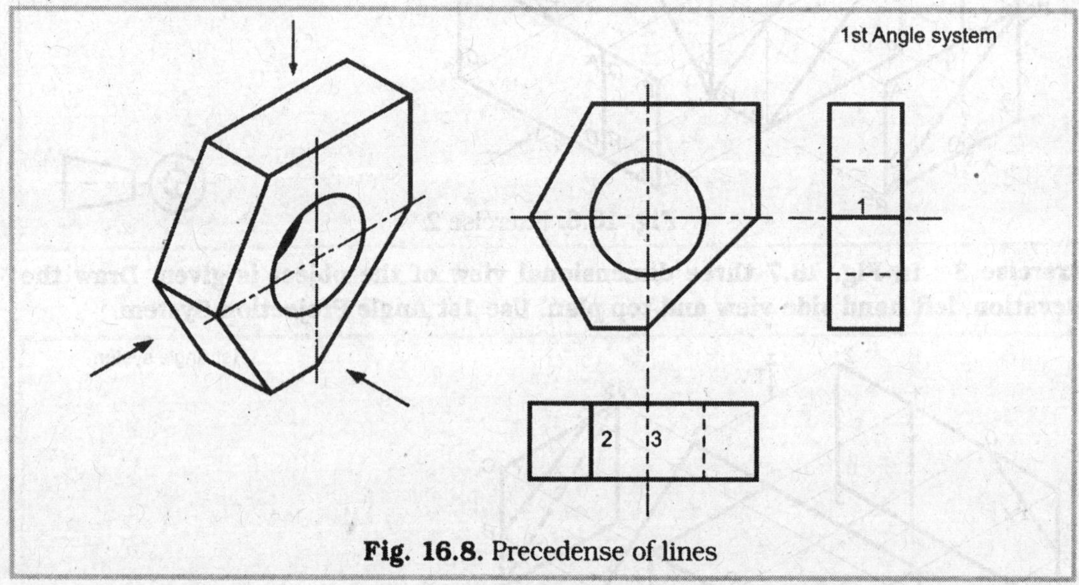

Fig. 16.8. Precedense of lines

(1) Full line is given precedence over all other lines since it is the most important line. In Fig. 16.8 line 1 and 2 both are drawn full even though centre and dotted lines coincide with it.

(2) Dotted or dashed line is given second precedence after full line. In Fig. 16.8 line 3 is drawn dotted even though centre line coincides with it.

(3) Centre line is given third precedence.

(4) In sectional views dimension lines and extension lines take precedence over hatching lines.

(5) Whenever centre line and cutting plane line coincide, the one, which will give more clarity to drawing to the reader, takes precedence over the other. Mostly cutting plane line is given precedence over the centre line.

(9) Procedure for Preparing Orthographic Projections to Scale

Follow the procedure step by step as given below :

(1) First of all decide the direction of principal view i.e. F.V. for the given pictorial view or a real object. Decide also whether pictorial drawing given is oblique drawing or isometric drawing.

(2) Do selection of views and decide about number of views to be drawn.

(3) Decide whether sectional and/or auxiliary views are required or not

(4) Prepare free hand sketches of required projections by soft pencil.

(5) Knowing overall dimensions of the views and knowing the size of the paper decide a suitable scale.

(6) Prepare border line and title block.

(7) Prepare rectangles for the views allowing sufficient space between two views and between views and the border line of the sheet. Keep in mind space required for dimensioning.

(8) Draw centre lines, if any, in all the views.

(9) Draw details simultaneously in all the views. It is a bad practice to prepare one view completely and then to start another view.

Draw details in the following order :

(i) Full, half and quarter circle.

(ii) Lines for the major shape of the object.

(iii) Lines for the minor shape of the object.

(10) After completing all the views, erase extra lines and fair required lines with 2H, 3H or 4H pencil.

(11) Give dimensions in all the views. As far as possible do not give dimensions inside the object, inside the hatched zone, etc. Give dimensions of features in appropriate views.

(12) Draw hatching lines in the sectional views.

(13) Draw cutting plane line and show arrows for direction of sight.

(14) Name the views.

(15) If views for many objects are drawn in one sheet then give numbers.

(16) Check carefully for missing dimensions and try to read the object from the views.

10. Types of Pictorial Projections

Having studied and grasped the system of projections, methods of projections, important conclusions and analysis of points, lines and planes in projections, we shall now take some most simple practice problems. Complicated problems are nothing but combination of many simple problems.

Along with the knowledge, imagination of object plays very important role. Pictorial projections given are either Axonometric Projections (i. e. Isometric, diametric & trimetric) or Oblique Projections. Isometric is very widely used projection, see Fig. 16.9. In pictorial projections mostly hidden parts are not shown even by dotted lines and hence we have to imagine the shape of hidden parts on the basis of symmetry.

To get principal orthographic projections we look at the object in the directions of axes of pictorial projection. Axes of projections for oblique as well as for isometric are shown in Fig. 16.9. along with both the type of projections. When we look at the object in the direction of any one axis, lines and dimensions of the object in that particular direction will not be seen. Dimensions and lines in the directions of two other axes will be seen in the true length, e.g. when you are on a railway track you cannot see the length of the train but you can see height and width of the guard's compartment or of the engine.

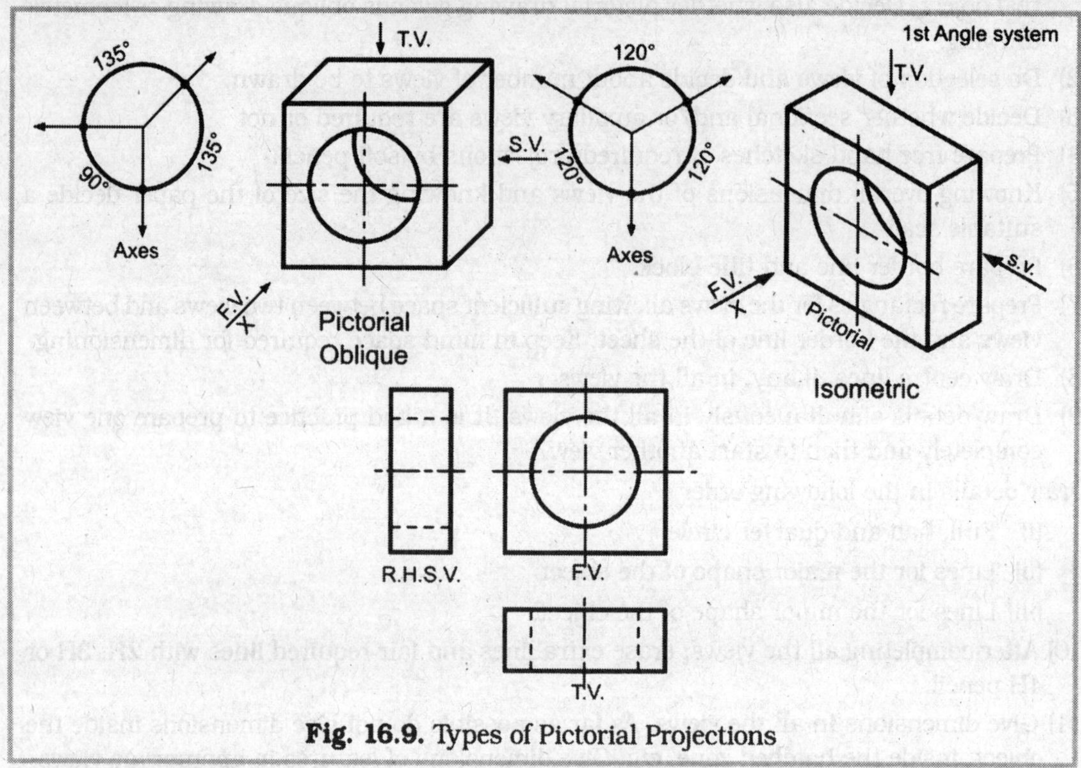

Fig. 16.9. Types of Pictorial Projections

11. Practice Problems

Problem 1 : In Fig. 16.10 (1) pictorial view of a rectangular plate is given, draw its (i) F.V. (ii) T. V. and (iii) R.H.S.V. Use 1st Angle System.

From direction X, dimension W will not be seen and dimensions L and H will be seen true and hence we get rectangle (L x H) as F. V. Similarly, R.H.S.V. will be rectangle (W x H) and plan will be rectangle (L x W). All three views are shown in Fig. 16.10(1)

Problem 2 : In Fig. 16.10 (2) pictorial projection of a plate is given, draw its (i) F.V. (ii) T.V. and (iii) R.H.S.V. Use 1st Angle System.

From direction X, (dimension W) thickness will not be seen and shape of front plane will be seen in true shape. And hence F.V. will be as shown in the figure. R.H.S.V. and T.V. will be same as the views of problem 1.

Problem 3 : In Fig. 16.10(3) pictorial projection of a plate is given, draw its (i) F.V. (ii) T.V. and (iii) R.H.S.V. Use 1st Angle System.

Projections are shown in the figure. Note that R.H.S.V. and T.V. are same as for problems 1 and 2.

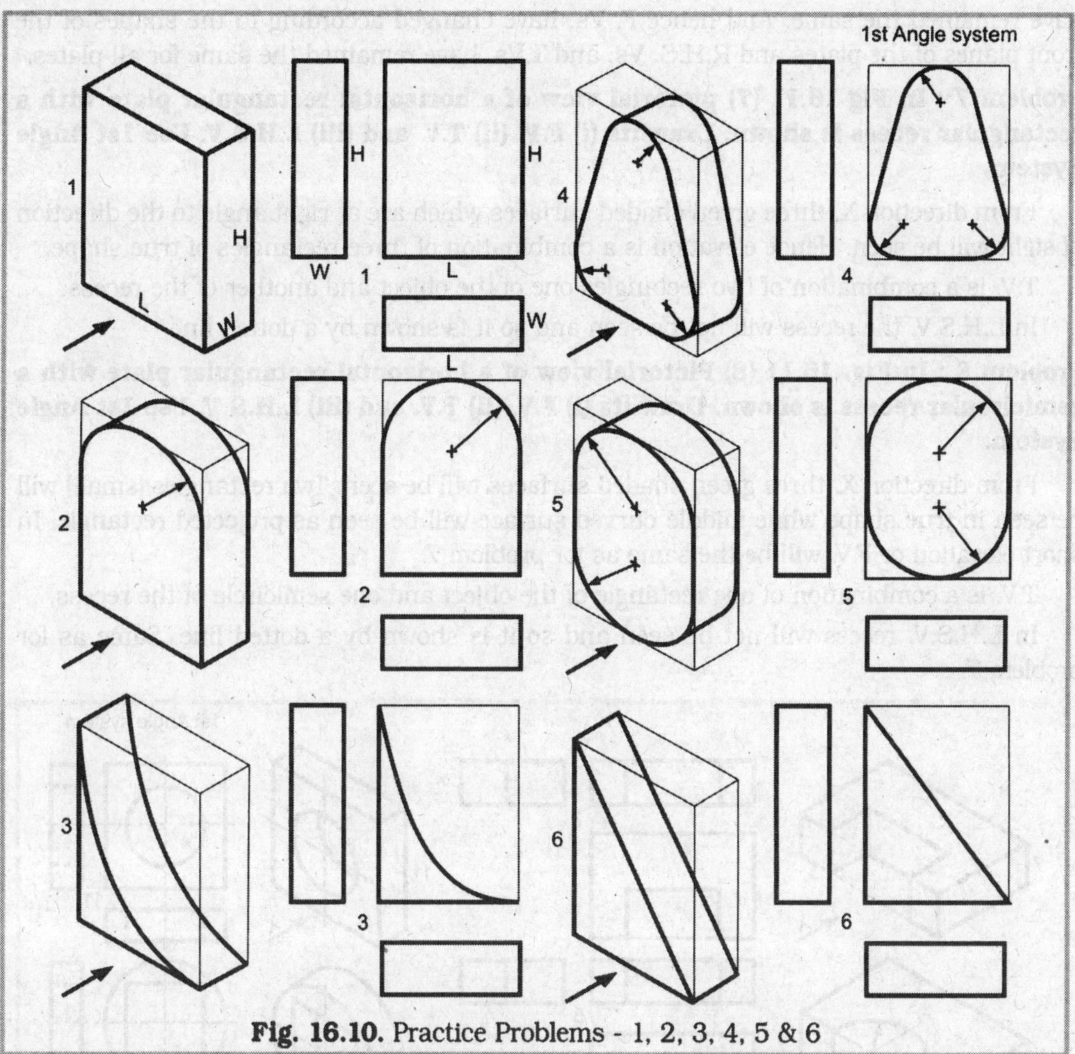

Fig. 16.10. Practice Problems - 1, 2, 3, 4, 5 & 6

Problem 4 : In Fig. 16.10(4) pictorial projection of a plate is given. Draw its (i) F.V. (ii) T.V. and (ii) R.H.S.V. Use 1st Angle System.

Projections are shown in the figure. Note that R.H.S.V. and T.V. are the same as for problems 1, 2, and 3.

Problem 5 : In Fig. 16.10(5) pictorial projection of a plate is given. Draw its (i) F.V. (ii) T.V. and (iii) R.H.S.V. Use 1st Angle System.

Projections are shown in the figure. Note that its R.H.S.V. and T.V. are the same as for problems 1, 2, 3 and 4.

Problem 6 : In Fig. 16.10(6) pictorial projection of a plate is given. Draw its (i) F.V. (ii) T.V. and (iii) R.H.S.V. Use 1st Angle System.

Projections are shown in the figure. Note that its R.H.S.V. and T.V. are the same as for problems 1, 2, 3, 4 and 5.

Conclusion : Plates 2, 3, 4, 5 and 6 can be assumed to be cut out of plate 1 without formation of any edge on the periphery of the plate. Overall dimensions L, H and W also have remained the same. And hence F. Vs. have changed according to the shapes of the front planes of the plates and R.H.S. Vs. and T.Vs. have remained the same for all plates.

Problem 7 : In Fig 16.11 (7) pictorial view of a horizontal rectangular plate with a rectangular recess is shown. Draw its (i) F.V. (ii) T.V. and (iii) L.H.S.V. Use 1st Angle System.

From direction X, three green shaded surfaces which are at right angle to the direction of sight will be seen. Hence elevation is a combination of three rectangles of true shape.

T.V. is a combination of two rectangles one of the object and another of the recess.

In L.H.S.V. the recess will not be seen and so it is shown by a dotted line.

Problem 8 : In Fig. 16.11 (8) Pictorial view of a horizontal rectangular plate with a semicircular recess is shown. Draw its (i) F.V. (ii) T.V. and (iii) L.H.S.V. Use 1st Angle System.

From direction X, three green shaded surfaces will be seen. Two rectangles (small) will be seen in true shape while middle curved surface will be seen as projected rectangle. In short elevation or F.V. will be the same as for problem 7.

T.V. is a combination of one rectangle of the object and one semicircle of the recess.

In L.H.S.V. recess will not be seen and so it is shown by a dotted line. Same as for problem 7.

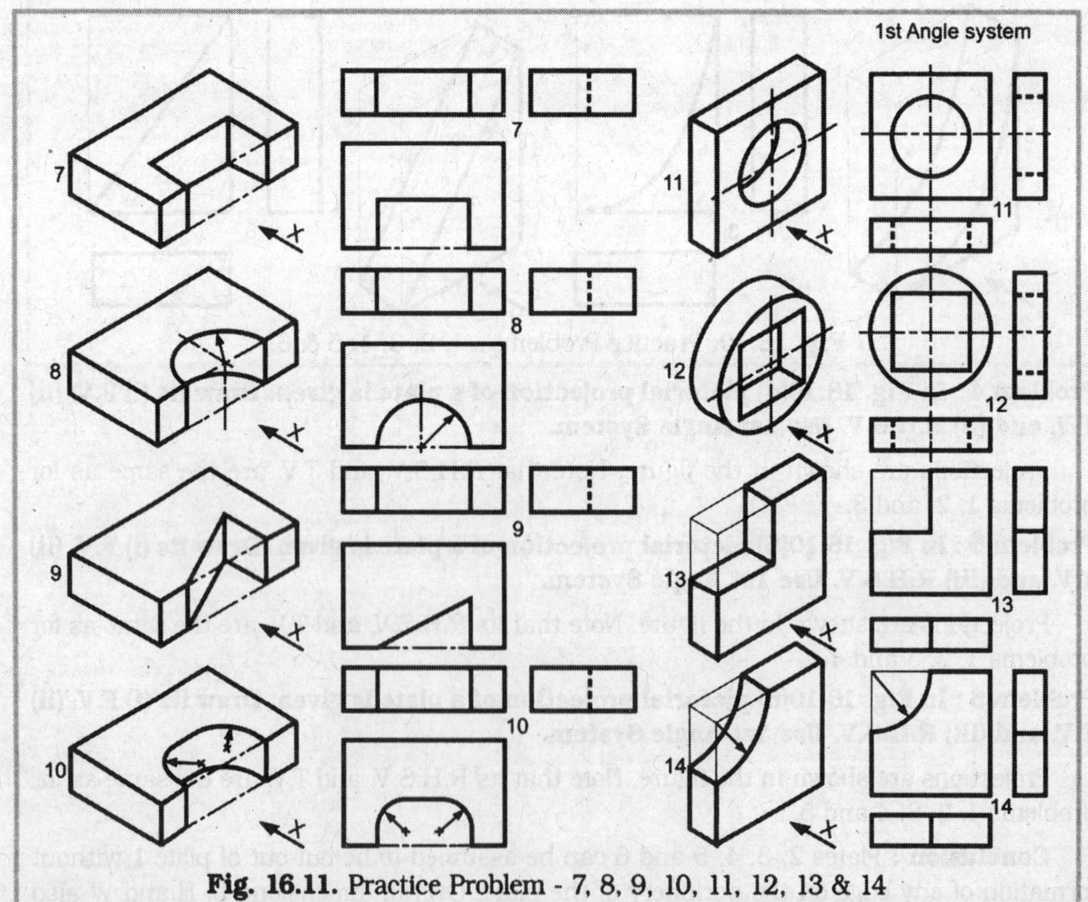

Fig. 16.11. Practice Problem - 7, 8, 9, 10, 11, 12, 13 & 14

Problem 9 : In Fig. 16.11(9) pictorial view of a horizontal rectangular plate with a triangular recess is shown. Draw its (i) F.V. (ii) T.V. and (iii) L.H.S.V. Use 1st Angle System.

F.V. and L.H.S.V. will be same as of problem 7 and 8. T.V. is a combination of one rectangle of the object and one triangle of the recess.

Problem 10 : In Fig. 16.11(10) pictorial view of a horizontal rectangular plate with recess is shown. Draw its (i) F.V. (ii) T.V. and (iii) L.H.S.V. Use 1st Angle System.

Three projections are given in the figure.

Conclusion : Recesses in plates of problems 8, 9 and 10 can be assumed to be cut out of rectangular recess of plate 7 without formation of any edge on the periphery of the recess. Overall dimensions of recess have also remained the same and hence T. Vs. have changed according to shapes of top planes of the plates and F. Vs. and L.H.S.Vs. have remained same for all plates.

Problem 11 : In Fig. 16.11(11) pictorial projection of a vertical square plate with a round hole is given. Draw its (i) F.V. (ii) T.V. and (iii) L.H. S.V. Use 1st Angle System.

From direction X, surface having square boundary and a circular hole will be seen. In T.V. and L.H.S.V. a rectangle will be seen and the hole will give pair of dotted lines at a distance of diameter of the hole.

Problem 12 : In Fig. 16.11(12) pictorial projection of a vertical round plate with square hole is given. Draw its (i) F.V. (ii) T.V. and (iii) L.H.S.V. Use 1st Angle System.

From direction X, surface having round boundary and square hole will be seen. T.V. and L.H.S.V. will be the same as for problem 11.

Problem 13/14 : In Fig. 16.11(13) (14) pictorial projection of a vertical square plate with a recess is shown. Draw its (i) F.V. (ii) T.V. and (iii) L.H.S.V. Use 1st Angle System.

In F.V. front plane of the object is seen in true shape in both the problems.

Irrespective of shape of recess T.Vs. and S. Vs. are the same for both the problems.

In short shapes of recesses are seen only in one view. In other views overall dimensions of recesses decide views irrespective of shapes of recesses.

Problem 15 : In Fig. 16.12 (i) pictorial projection of an object (complicated one) is given. Draw (i) F.V. (ii) T.V. and (iii) L.H.S.V. Use 1st Angle System of Projections.

As the problem is complicated one, to make it easier and simpler, let us divide the object into 5 simple parts, as shown in the Fig. 16.12(B) and draw first the F.V., T.V. and L.H.S.V. of each part separately. Views as they will appear are shown in Fig. 16.12(B).

Now, if we combine together F. Vs. of all parts taking care about their positions and locations with little changes where they join, we can easily draw the F.V. of the complete object. F.V. of full object is shown in Fig. 16.12 (A).

Similarly, if we combine together T.Vs. of all parts taking care about their positions and locations with little changes where they join, we can easily draw the T.V. of the complicated object. T.V. of full object is shown in Fig. 16.12(A).

Similarly, we can get L.H.S.V. also. It is shown in Fig. 16.12(A).

This above method is adopted in the beginning but after having sufficient practice of these types of problems you may solve the problem as a whole instead of dividing it into parts.

Shaded area [light black colour] indicates the area (zone) where the joint was existing before separate pieces were made.

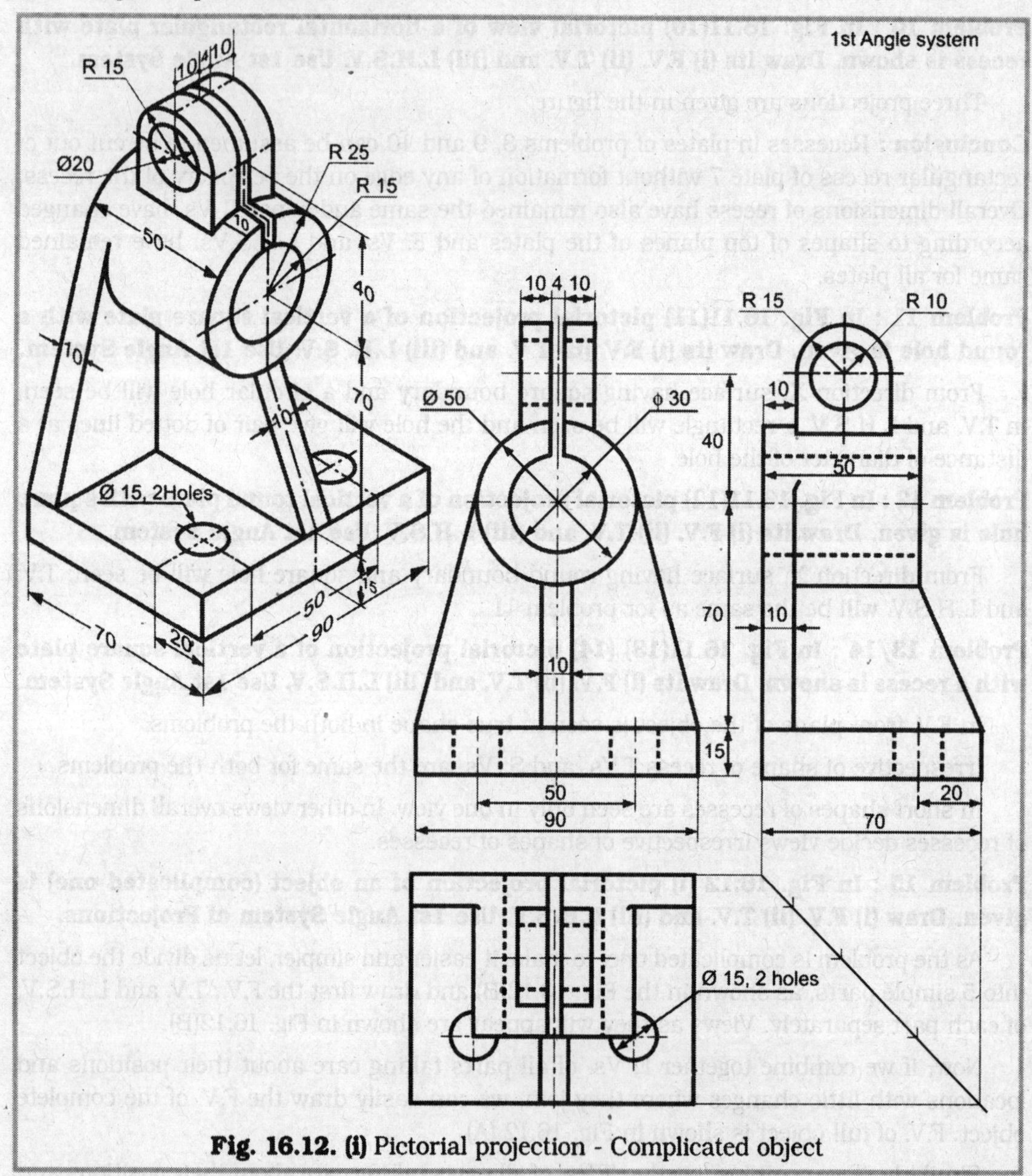

Fig. 16.12. (i) Pictorial projection - Complicated object

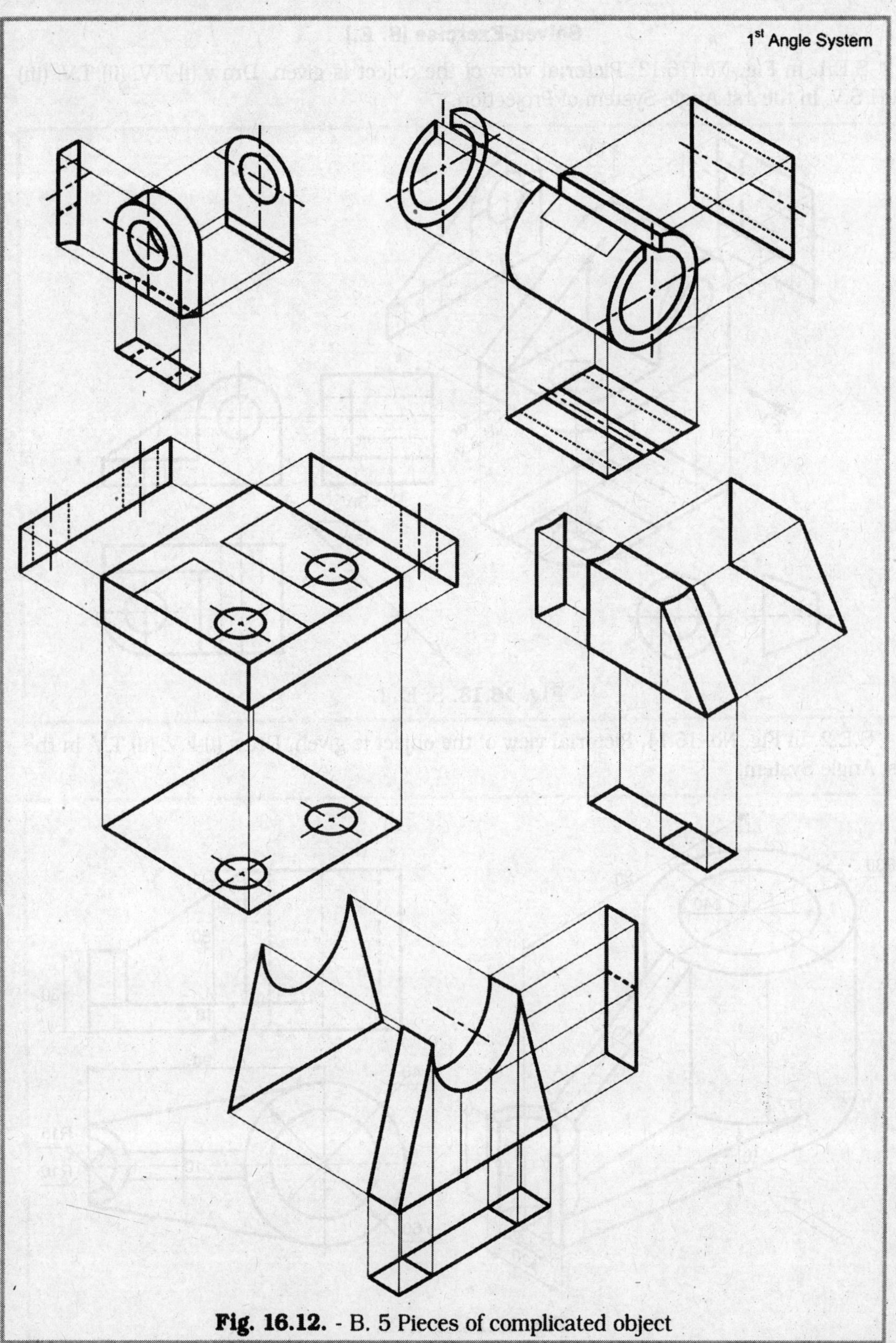

Fig. 16.12. - B. 5 Pieces of complicated object

Solved-Exercise [S. E.]

S.E.1. In Fig. No. 16.13, Pictorial view of the object is given. Draw (i) F.V. (ii) T.V. (iii) R.H.S.V. in the 1st Angle System of Projection.

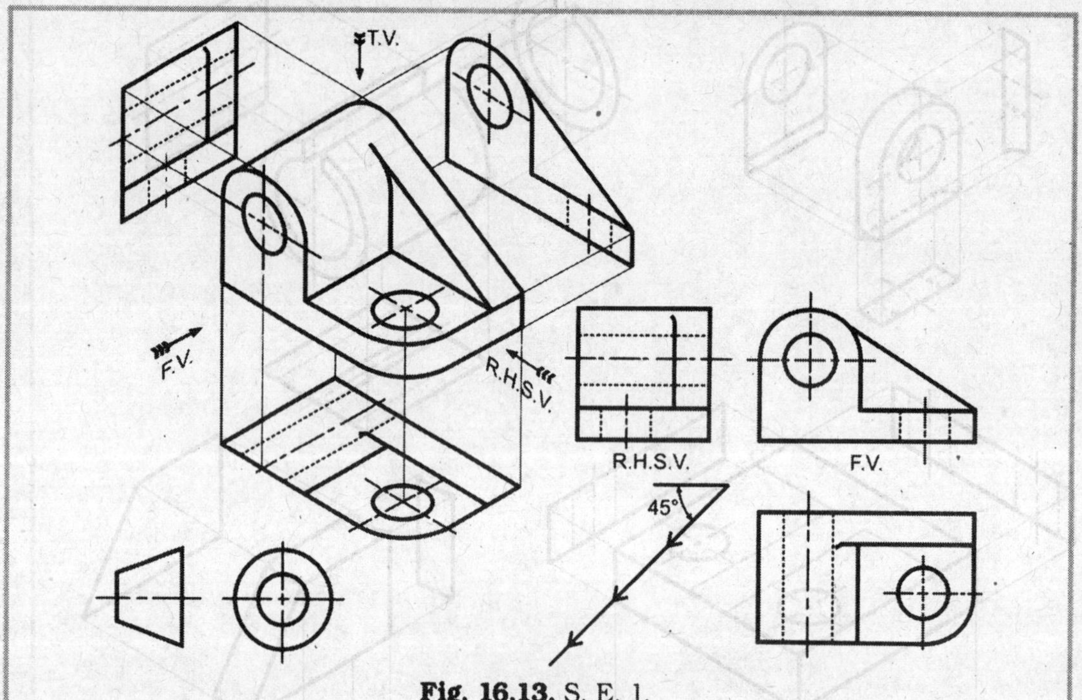

Fig. 16.13. S. E. 1.

S.E.2. In Fig. No. 16.14, Pictorial view of the object is given. Draw (i) F.V. (ii) T.V. in the 1st Angle System.

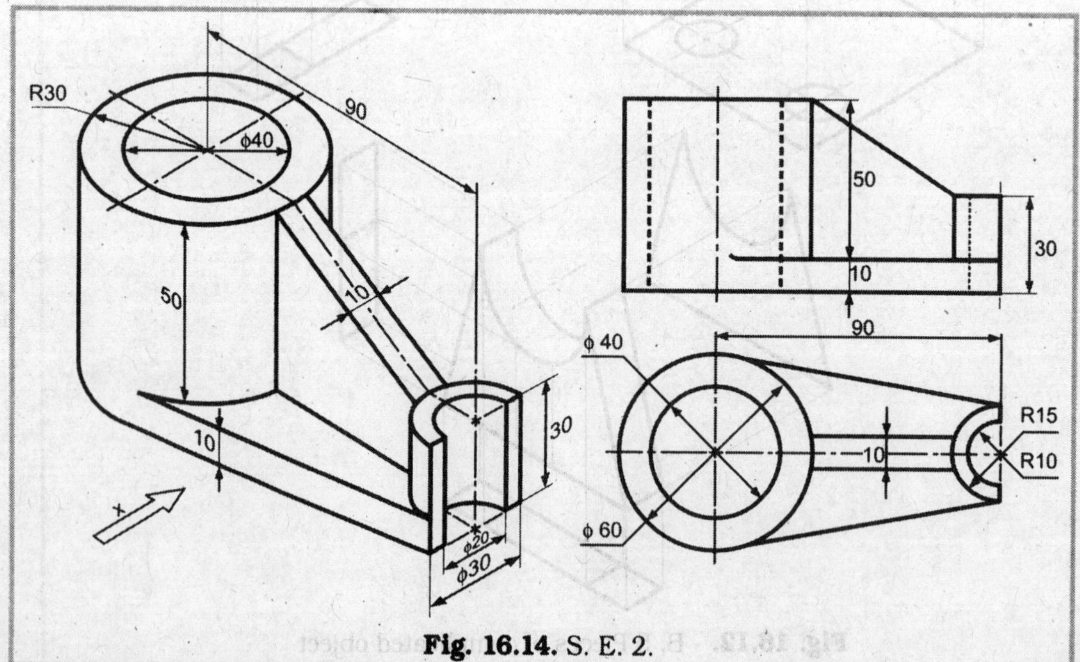

Fig. 16.14. S. E. 2.

S.E.3. In Fig. No. 16.15, Pictorial view of the object is given. Draw (i) F.V. (ii) T.V. (iii) L.H.S.V. in the 1st Angle System. Insert necessary dimensions.

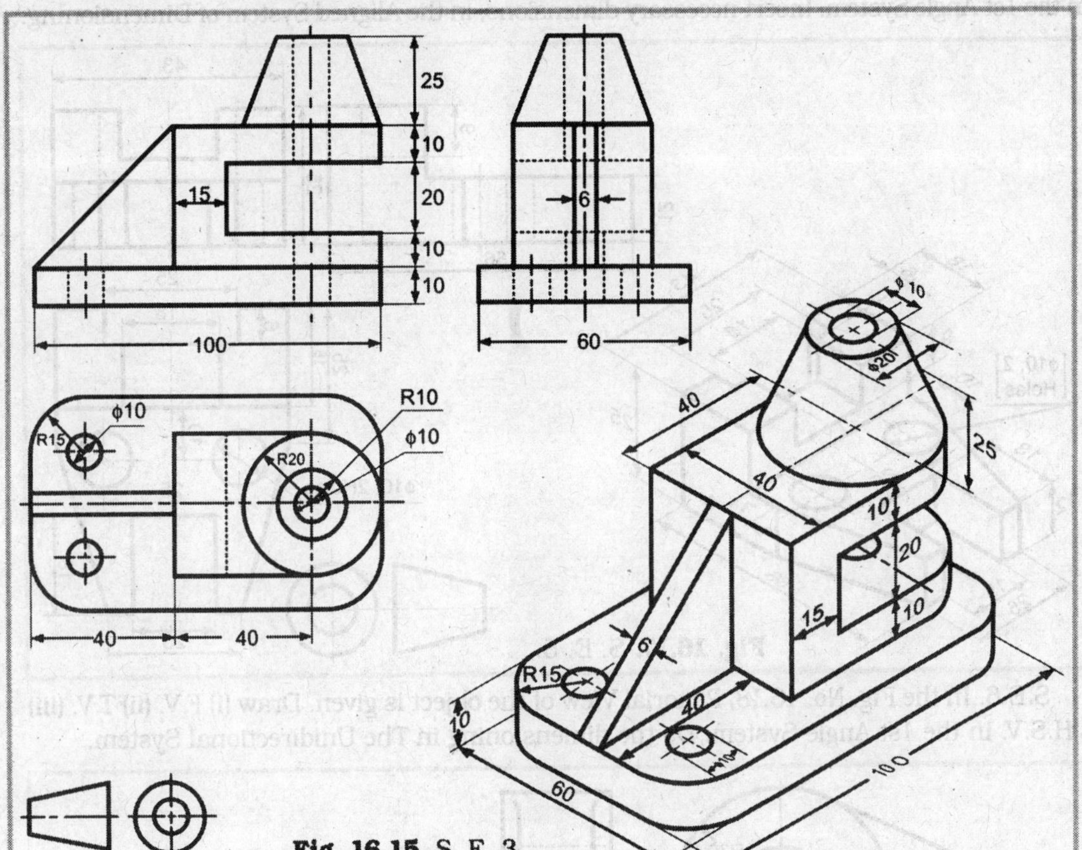

Fig. 16.15. S. E. 3.

S.E.4. In Fig. No. 16.16, Pictorial view of the object is given. Draw (i) F.V. (ii) L.H.S.V. and (iii) T.V. in the 3rd Angle System. Insert necessary dimensions.

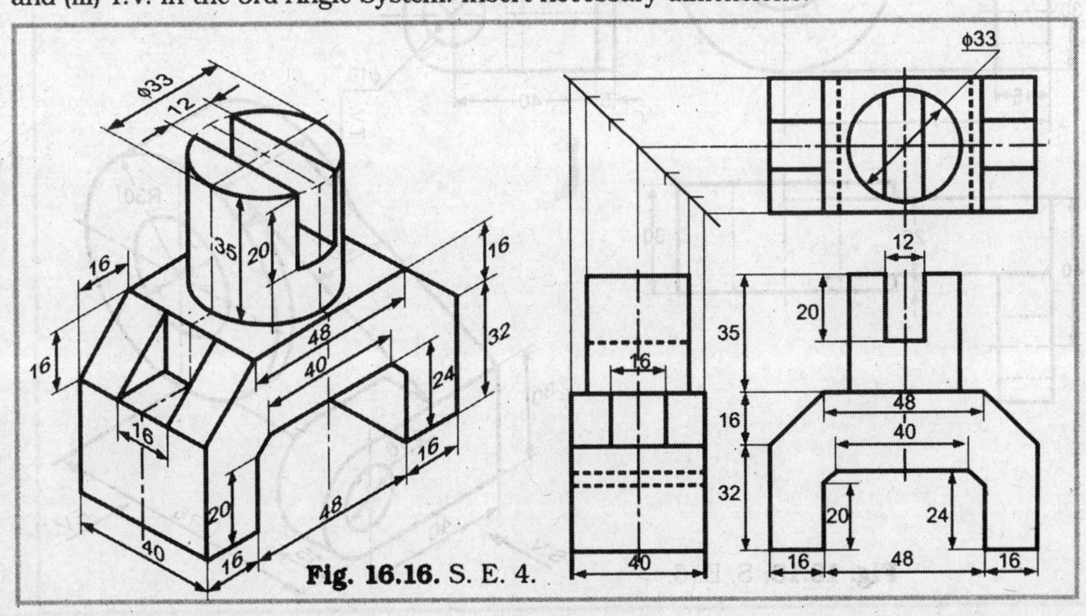

Fig. 16.16. S. E. 4.

S.E.5. In the Fig. No. 16.17, Pictorial view of the object is given. Draw (i) F.V. (ii) T.V. (iii) R.H.S.V. in the 1st Angle System. Insert necessary dimensions, in the Aligned System of Dimensioning.

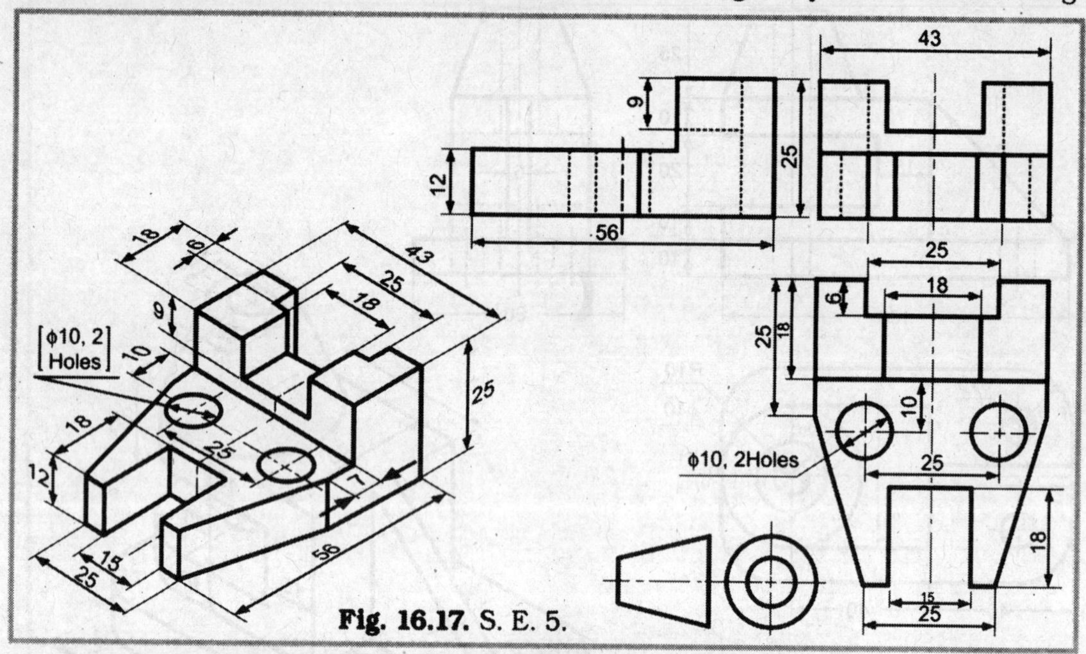

Fig. 16.17. S. E. 5.

S.E.6. In the Fig. No. 16.18, Pictorial View of the object is given. Draw (i) F.V. (ii) T.V. (iii) L.H.S.V. in the 1st Angle System. Do the dimensioning in The Unidirectional System.

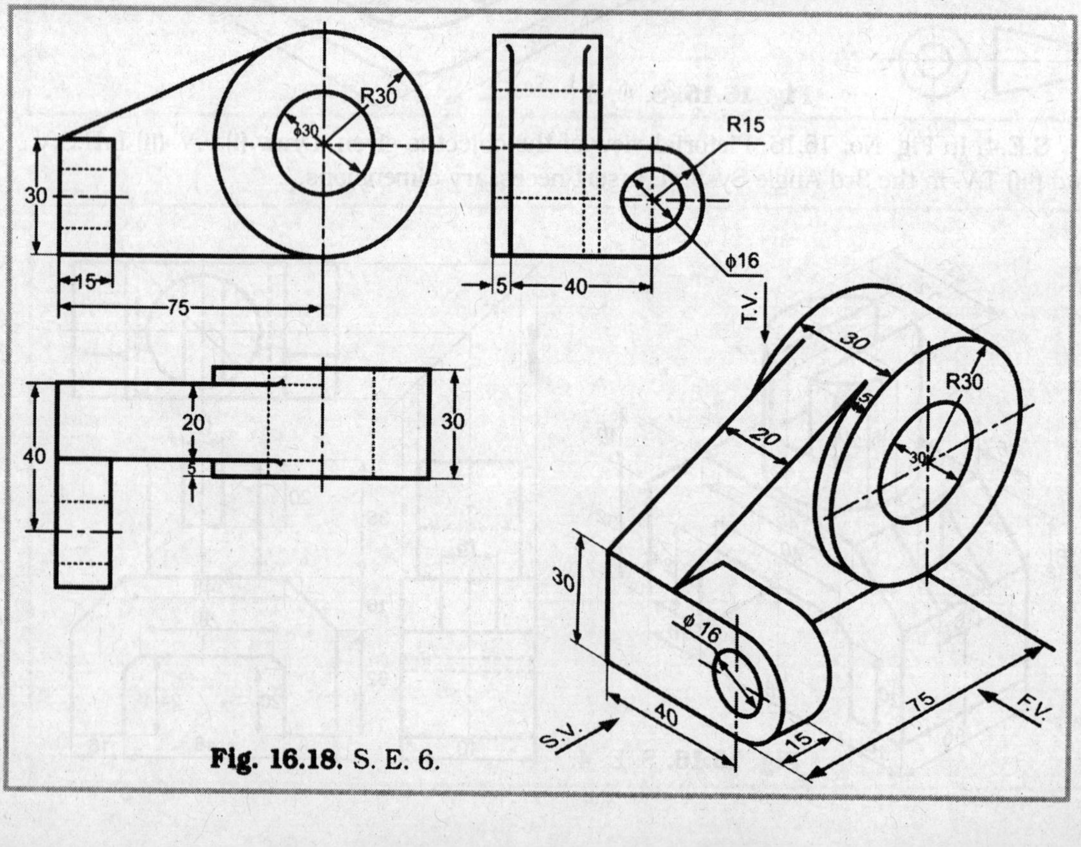

Fig. 16.18. S. E. 6.

S.E.7. In the Fig. NO. 16.19, Pictorial View of the object is given. Draw (i) F.V. (ii) L.H.S.V. in the 1st Angle System. Insert necessary dimensions in The Aligned System of Dimensioning.

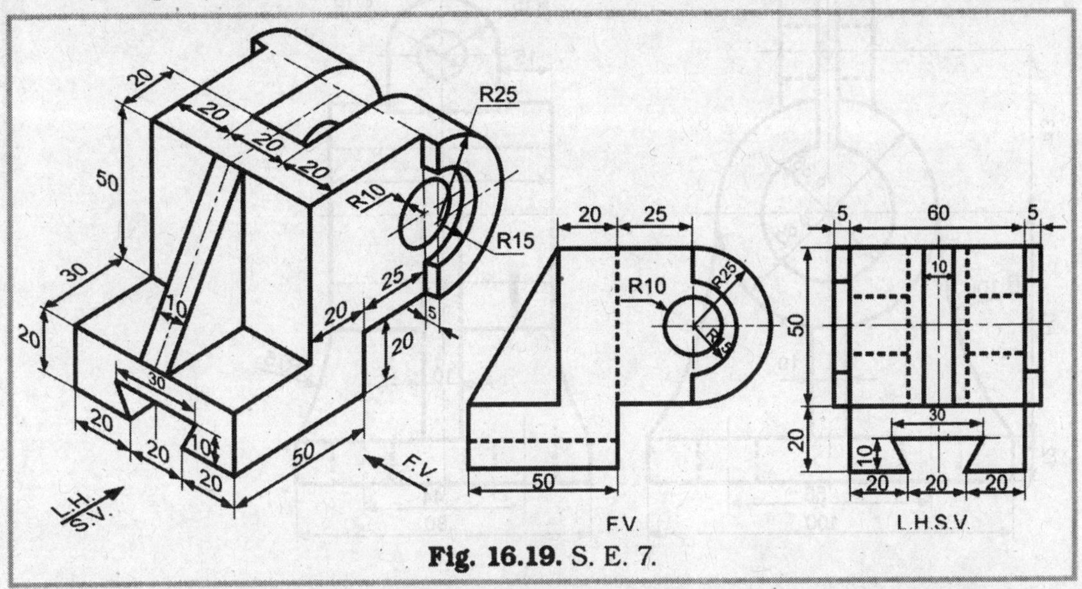

Fig. 16.19. S. E. 7.

S.E.8. In the Fig. No. 16.20, Pictorial view of the object is given. Draw (i) F.V. (ii) T.V. in The 3rd Angle System. Insert necessary dimensions in The Aligned System of Dimensioning.

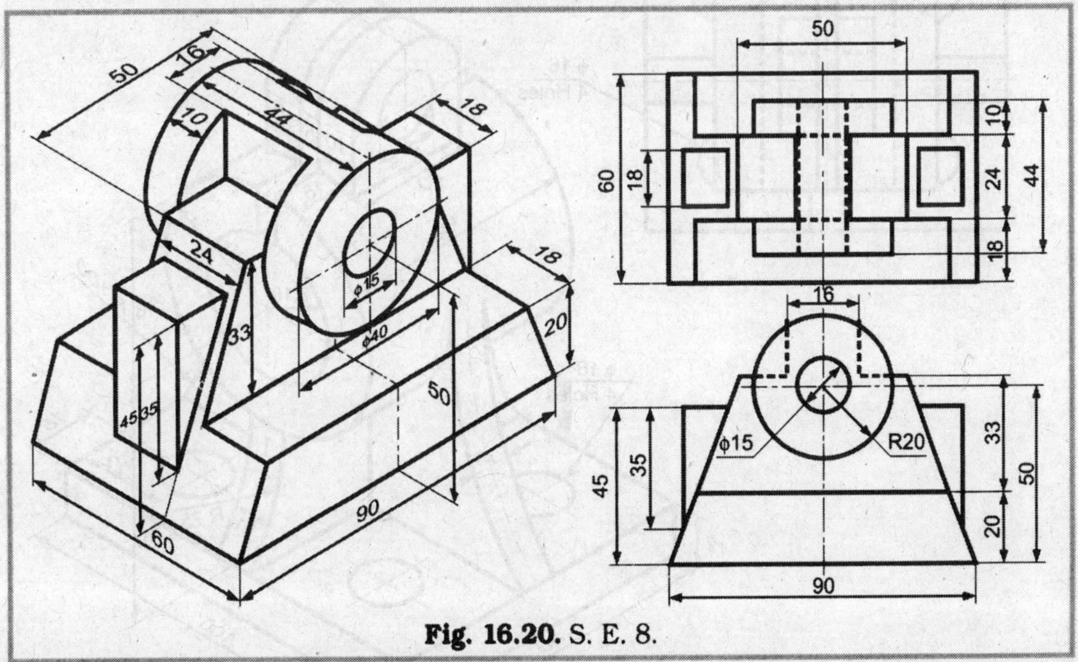

Fig. 16.20. S. E. 8.

S.E.9. In the Fig. NO. 16.21, The pictorial view of the object is given. Draw (i) F.V. (ii) T.V. (iii) L.H.S.V. in The 1st Angle System. Insert necessary dimensions in The Aligned System of Dimensioning.

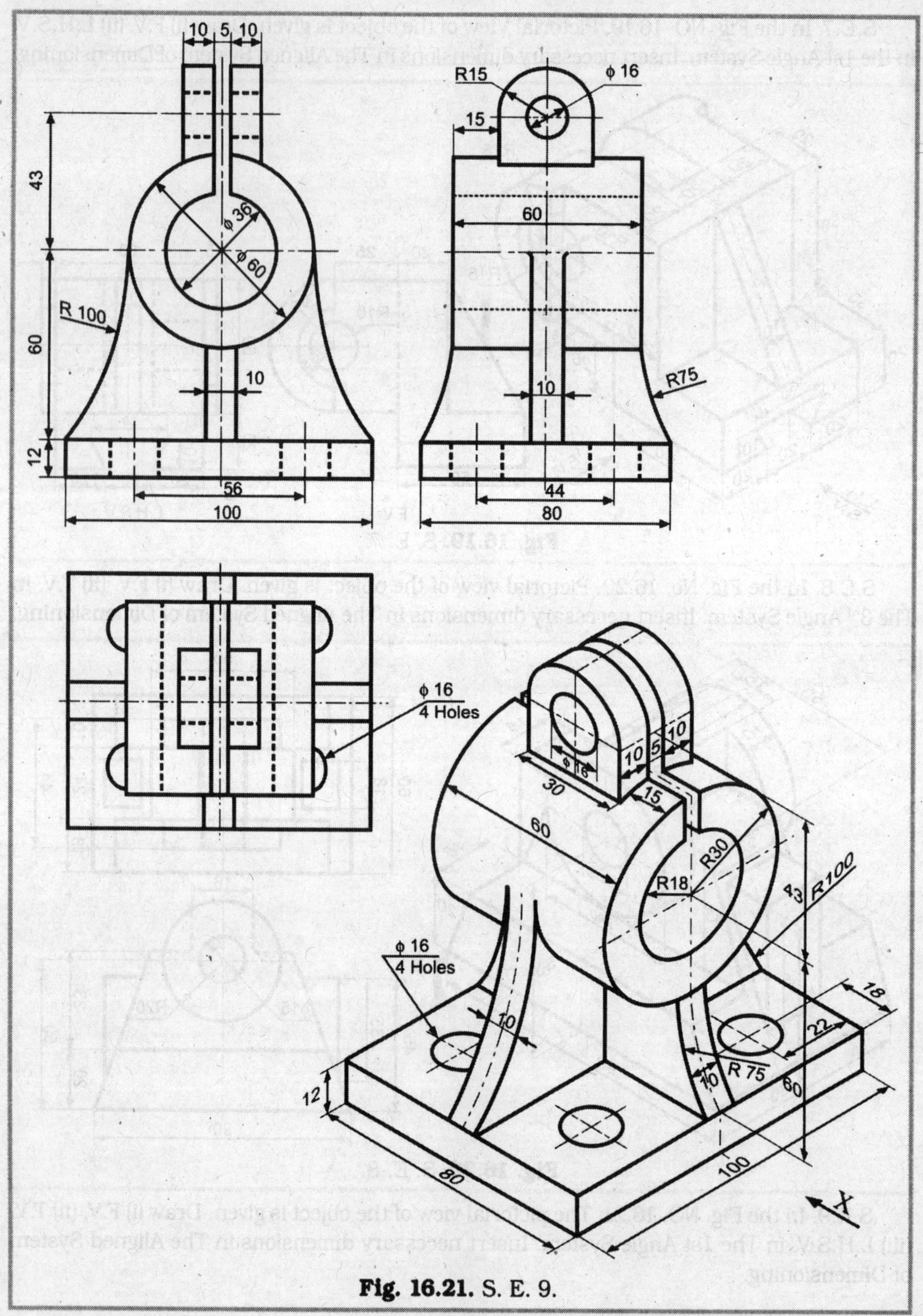
10
5
10
43
60
12
R 100
φ 36
φ 60
10
56
100
R15
φ 16
15
60
10
R75
44
80
φ 16
4 Holes
φ 16
4 Holes
10
5
10
φ 16
30
15
60
R30
R18
43
R100
18
22
10
R 75
10
12
60
80
100
X

Fig. 16.21. S. E. 9.

EXERCISE

Note : Pictorial views of the objects are given in Fig. 16.22 to 16.35. Draw orthographic views as stated in each exercise. Insert necessary dimensions on the projections. Elevation or front view should be drawn as seen from the direction of arrow X. All dimensions are in mm. Assume dimensions, if not given. (Solutions of problems are given in last pages of this book. In solution dimensions are not given.)

1. Fig. 16.22. Draw (i) Front view (ii) Left hand side view (iii) Top plan. Use the 3rd angle system.
2. Fig. 16.23. Draw (i) Front elevation (ii) Right hand side view (iii) Top plan. Use the 1st angle projection system.

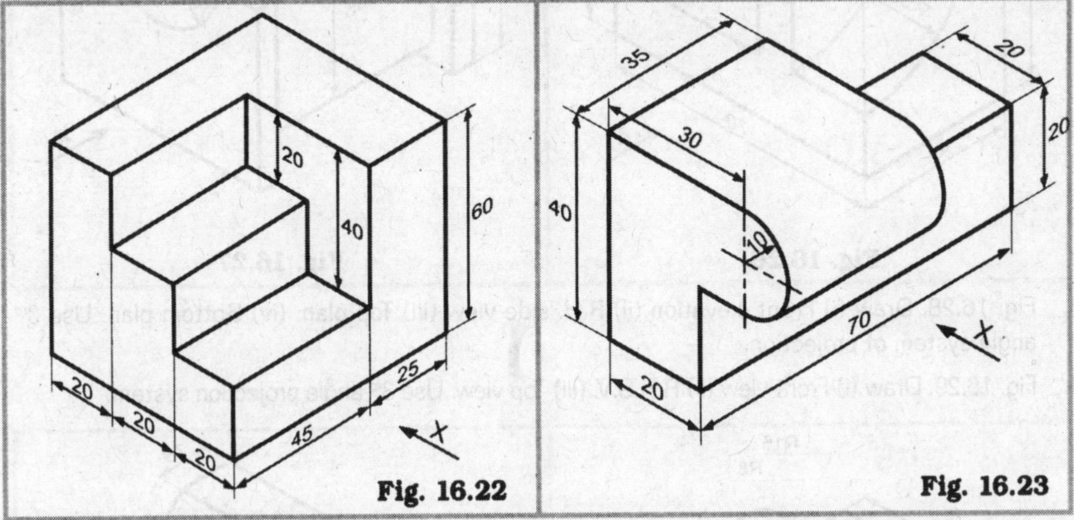

Fig. 16.22 Fig. 16.23

3: Fig. 16.24 Draw (i) Front view (ii) R.H. side view (iii) Top view. Use 3rd angle system.

4. Fig. 16.25 Draw (i) Elevation (ii) L.H. side view (iii) Top plan. Use 3rd angle system.

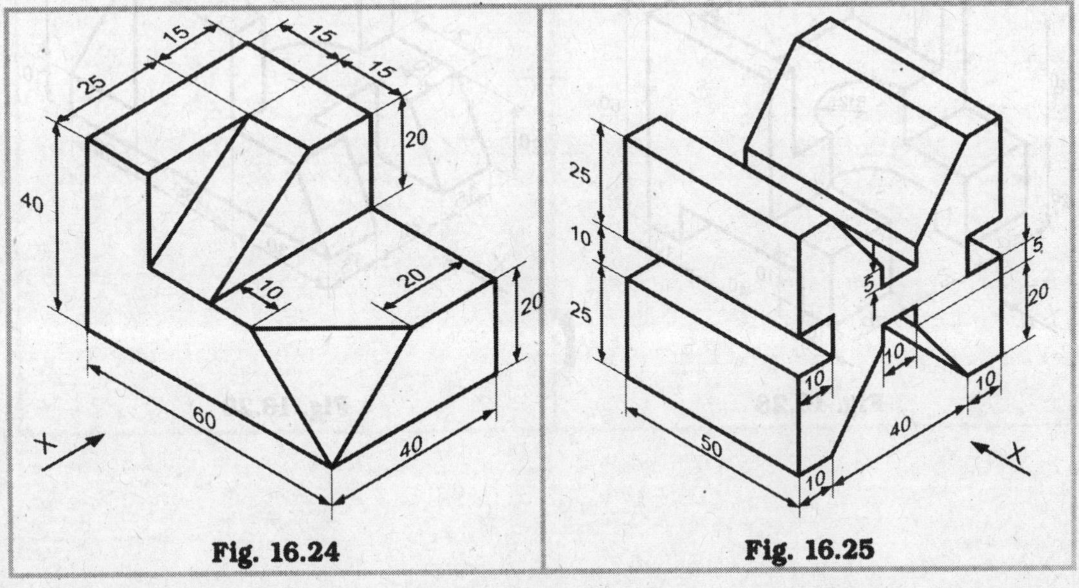

Fig. 16.24 Fig. 16.25

5. Fig. 16.26 Draw (i) Front view (ii) R.H. side view (iii) Top plan (iv) L.H. side view. Use the 1st angle projection system.

6. Fig. 16.27. Draw (i) Front view (ii) Top plan (iii) L.H. side view. Use the 3rd angle projection system.

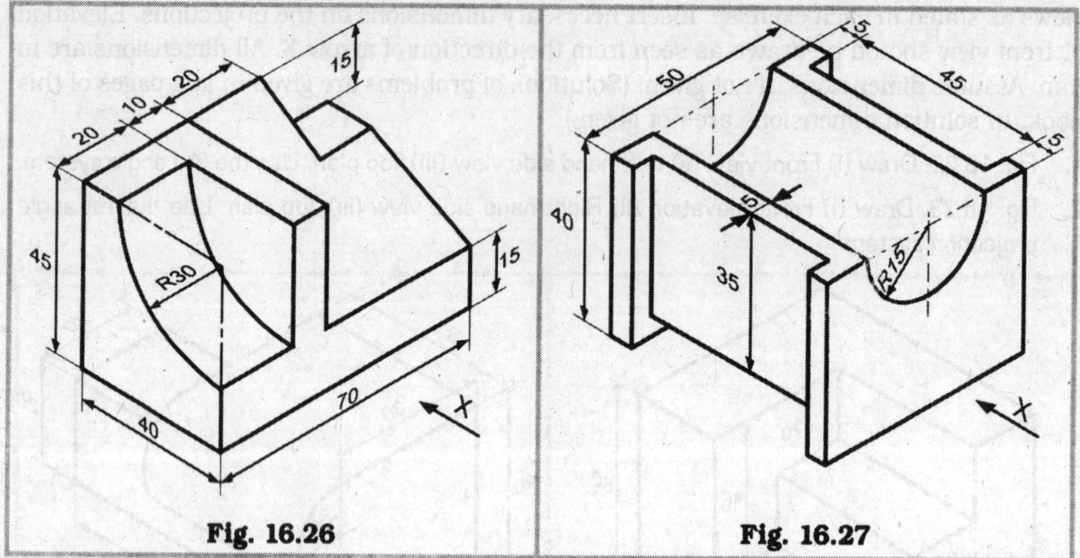

Fig. 16.26

Fig. 16.27

7. Fig. 16.28. Draw (i) Front elevation (ii) R.H. side view (iii) Top plan. (iv) Bottom plan. Use 3rd angle system of projection.

8. Fig. 16.29. Draw (i) Front view (ii) R.H.S.V. (iii) Top view. Use 3rd angle projection system.

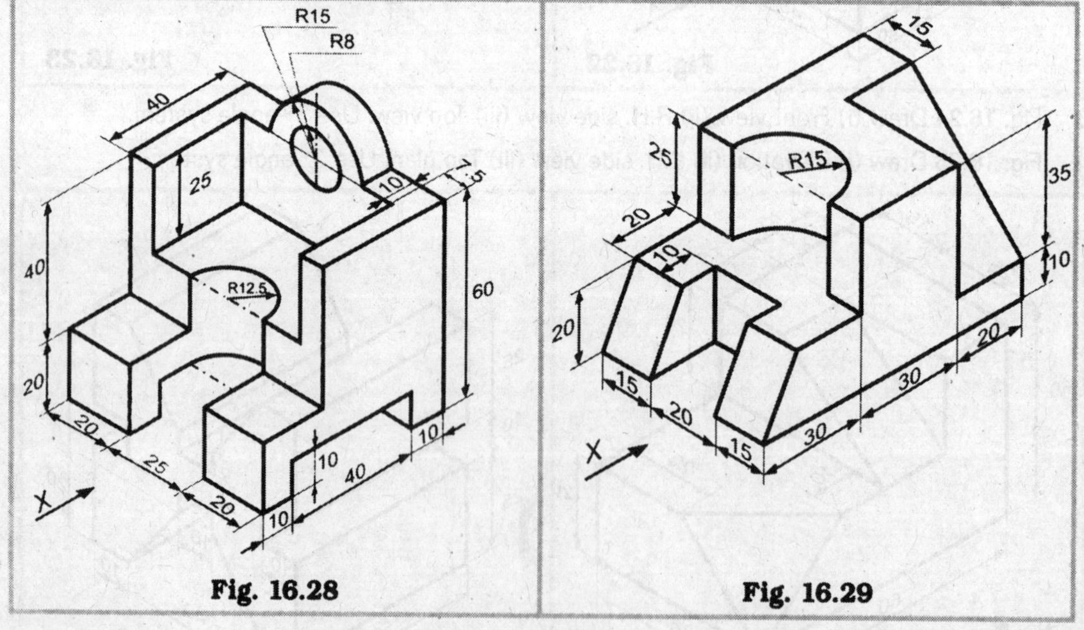

Fig. 16.28

Fig. 16.29

9. Draw (i) Front elevation (ii) Top plan (iii) R.H.S.V. Use the Ist angle projection system Fig. No. 16.30.

10. Fig 16.31. Draw (i) Front view (ii) R.H.S.V. side view (iii) Top plan. Use the 3rd angle projection system.

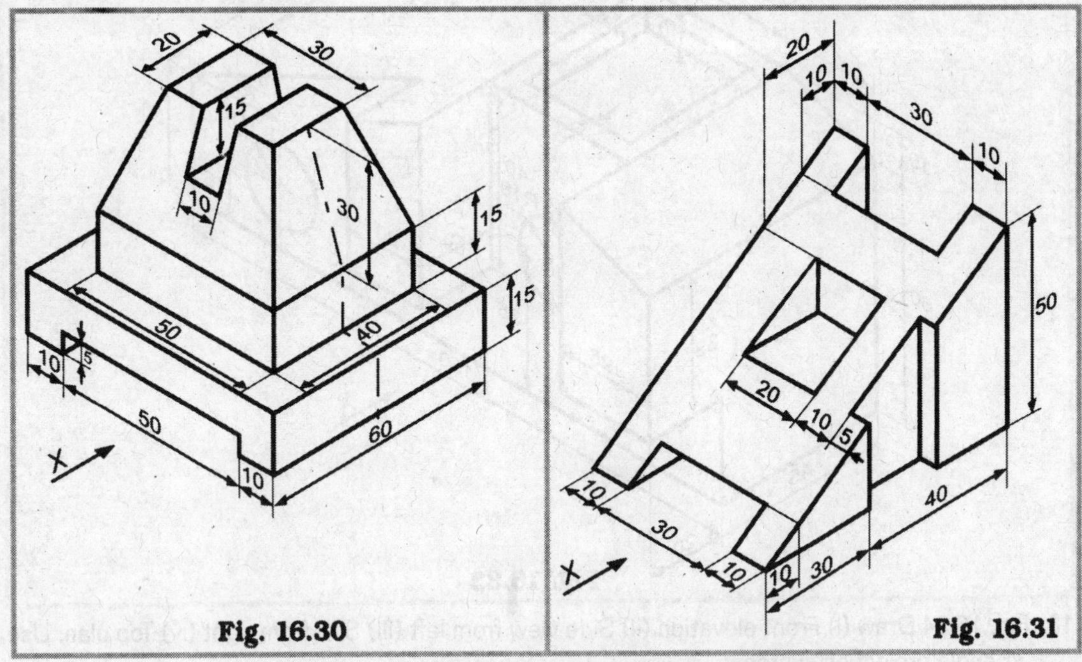

Fig. 16.30 **Fig. 16.31**

11. Fig. 16.32. Draw (i) Elevation (ii) R.H.S. V. (iii) Top Plan. Use 3rd angle system.

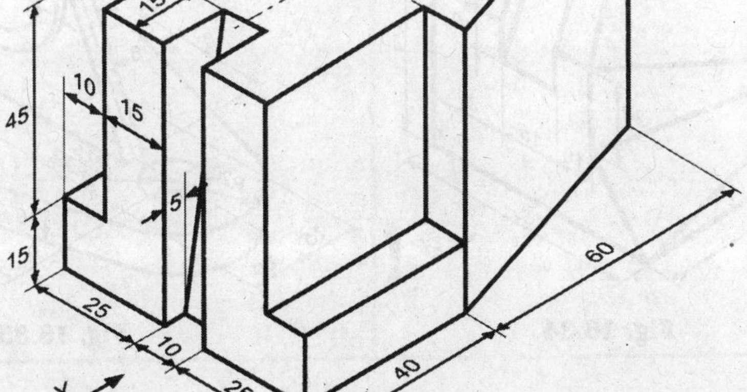

Fig. 16.32

12. Fig. 16.33. Draw (i) Front view (ii) Top Plan (iii) R. H. S. V. (iv) L H. S. V. Use the 1st angle projection method.

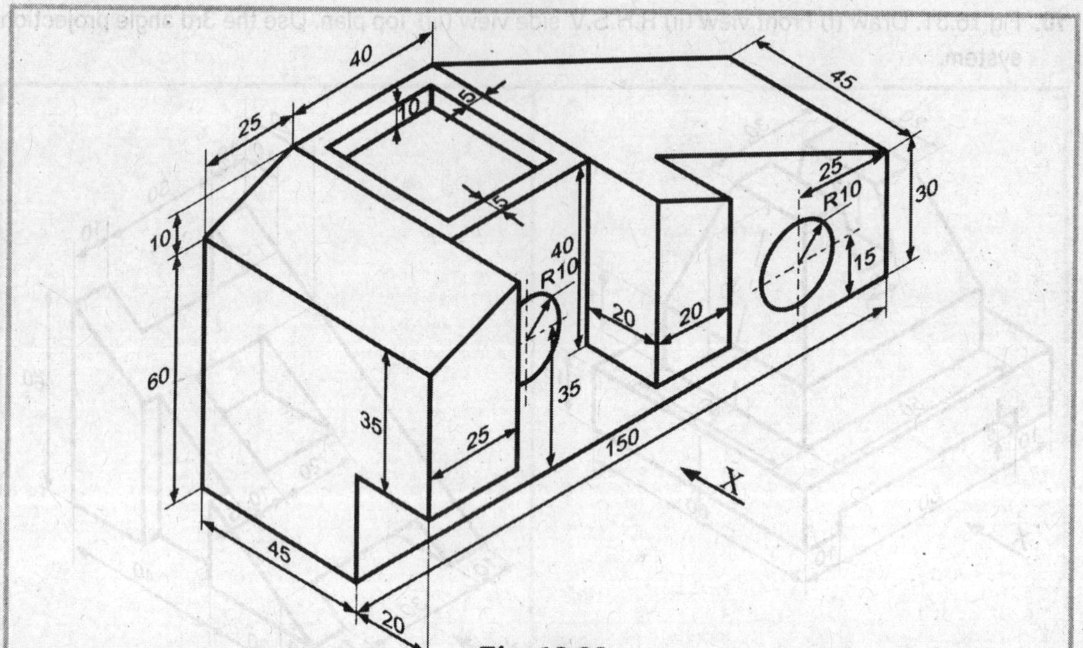

Fig. 16.33

13. Fig. 16.34 Draw (i) Front elevation (ii) Side view from left (iii) S.V. from right (iv) Top plan. Use 3rd angle projection system.

14. Fig. 16.35 Draw (i) Front view (ii) Top view (iii) Left hand side view. Use the 3rd angle projection system.

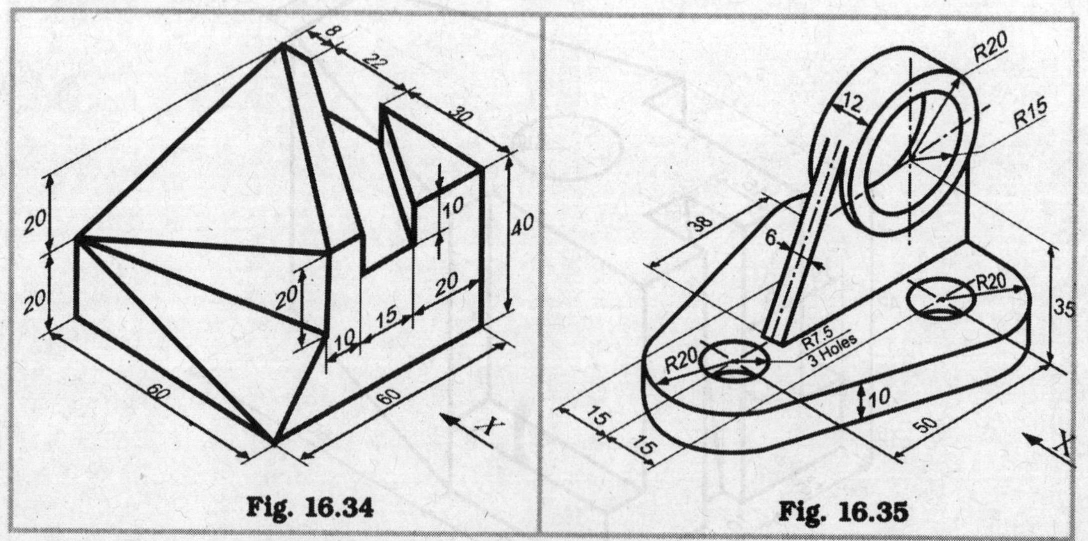

Fig. 16.34 **Fig. 16.35**

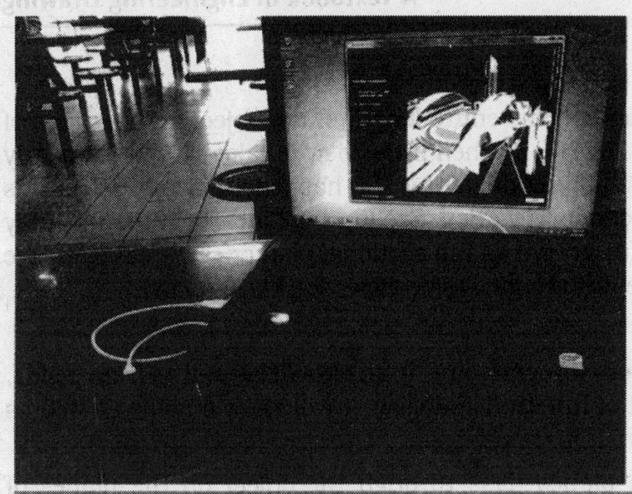

CHAPTER - 17

Sectional Orthographic Projections

1. Introduction

In orthographic projections, details of the object which are not visible from outside are shown by means of dotted or dashed lines. When the object is complex from inside, there will be many dotted lines in the projections. It will be difficult to interpret or read the projections of such object when there are too many dotted lines inside projections.

Orthographic projections of such a complex object is drawn after imagining it as an object cut by an imaginary cutting plane/planes parallel to principal planes passing through the details of the object. An imaginary plane by which the object is cut is known as cutting plane or section plane. See Fig. 17.1. Portion of the object between cutting plane and an observer is imagined to be removed and then the object is viewed by the observer and projection is drawn. Projection of cut object is known as sectional view or sectional projection. See sectional T.V. in Fig. 17.1 (ii) a. Generally dotted or dashed lines are omitted in the sectional projection. In the sectional projection along with the shape of the section, visible edges and the contours of the object behind the section plane are shown.

When one view is drawn sectional, other views are drawn as if the object is not cut and the entire object exists as a whole. It means other views are not affected by taking a section in one or many projections. Only cutting plane line is shown to locate the section and arrows are drawn to show the direction of sight. See cutting plane line A-A in Fig. 17.1 other than sectional projection. The cutting plane line known as line of section or trace of the cutting plane is drawn with one long dash and one short dash alternately and evenly spaced. Ends of the cutting plane line is drawn thicker and arrows to indicate direction of sight is shown on these ends. See C.P. line A-A in Fig. 17.1. (ii) (b) Change in the direction of cutting plane is also shown by a thicker line. See cutting plane line A-B, made thicker near centre, in Fig. 17.3 and Fig. 17.4

The elements of the object cut by cutting plane is made evident by drawing hatching lines in them. Hatching or section lines are drawn thinner than boundary lines and in normal circumstances they are drawn at 45°. Hatching lines are evenly spaced at about 2 to 3 mm apart.

The cutting plane need not be continuous but may be offset or turned through an angle thus giving rise to various types of sectional projections.

2. Full Section

When the section plane passes through the entire length of the object on the plane of symmetry, as shown in Figs. 17.1 (i), 17.2 (i) and then if the view is taken of half or nearly half object then that view is known as full sectional view. If it happens to be elevation, it is known as full sectional elevation or sectional elevation. See Fig. 17.3 (b). The word 'full' may be omitted. If it happens to be plan, it is known as full sectional plan or sectional plan, see Fig. 17.1(ii) (a). Similarly, there can be full sectional side view, see Fig. 17.2 (iv).

3. Half Section

When the object is symmetrical about its centre line, from outside as well as from inside, half sectional view is preferred to that of full sectional view. It will show outside as well as inside details of the object in one view.

When two perpendicular cutting planes pass through the object on the planes of symmetry, as shown in Figs. 17.3 (iii), 17.4 (i), (iii) to remove one-quarter of the object and then if the view of three quarters of the object is taken then that view is known as half sectional view. See Fig. 17.3 (iv) (c) and 17.4 (iv) (b). If it is elevation, it is called half sectional elevation. Similarly, we can have half sectional plan and half sectional end view or side view.

In half sectional view, the line, separating the sectional view and full view coincides with the centre line and as recommended by I.S. it is shown by dark continuous line. See Fig. 17.4 (iv)(b).

4. Off-set Section

It is likely that one cutting plane need not pass through all axes of symmetry of different portions of the object. In such circumstances instead of cutting the object by one cutting plane the direction of cutting plane may be changed (i.e. off setted) in order to cut the object through maximum possible axes of symmetry and thereby show maximum details of the object in one offset sectional view. Here it is absolutely necessary to show the cutting plane line in other adjoining view or views. For illustration see Fig. 17.5 and Fig. 17.6.

5. Illustrative Examples

Now to understand and grasp the above three types of sectional views let us study a few illustrative examples.

Example 1 : In Fig. 17.1.(i), object is shown in pictorial projection. Draw (i) Elevation (F.V.) and (ii) Full Sectional Plan (T.V.). Use 1st Angle System.

In the figure, horizontal cutting plane is shown cutting the object into two equal halves. Upper half part is assumed to be removed and top view is taken from the top. Front view of complete object is taken from the front. Both front view and sectional top view are shown in Fig. 17.1(ii). Section lines or hatching lines are drawn where the material of the object is cut by cutting plane.

Example 2 : In Fig. 17.2 (i), object is shown in pictorial projection. Draw (i) Elevation (F.V.) (ii) Sectional Plan (T.V.) along B-B and (iii) Sectional Side View along A-A. Use 1st Angle System.

Draw first the elevation, as shown in Fig. 17.2(iv). Now imagine the object after cutting it along B-B to get full sectional plan. The object after cutting will look like Fig. 17.2(ii). Draw sectional plan. Similarly, imagine the object after cutting it along A-A to get full sectional side view. The object after cutting along A-A will look like Fig. 17.2 (iii). Take end view from left and draw it. All views are shown in Fig. 17.2(iv).

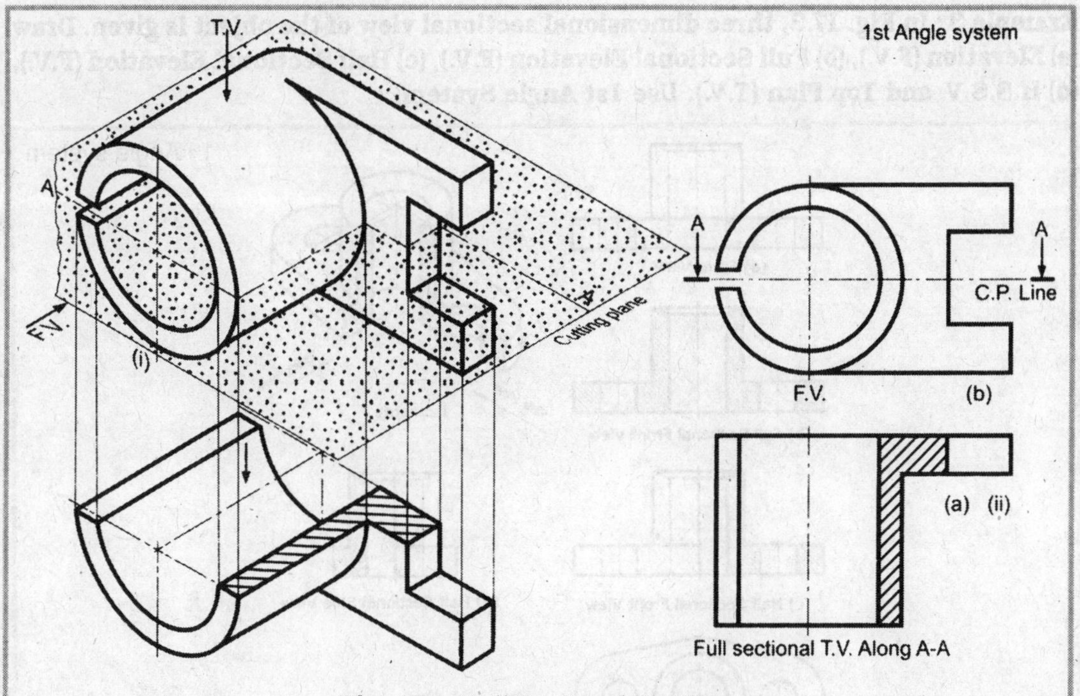

1st Angle system

Fig. 17.1. Full Sectional Top View

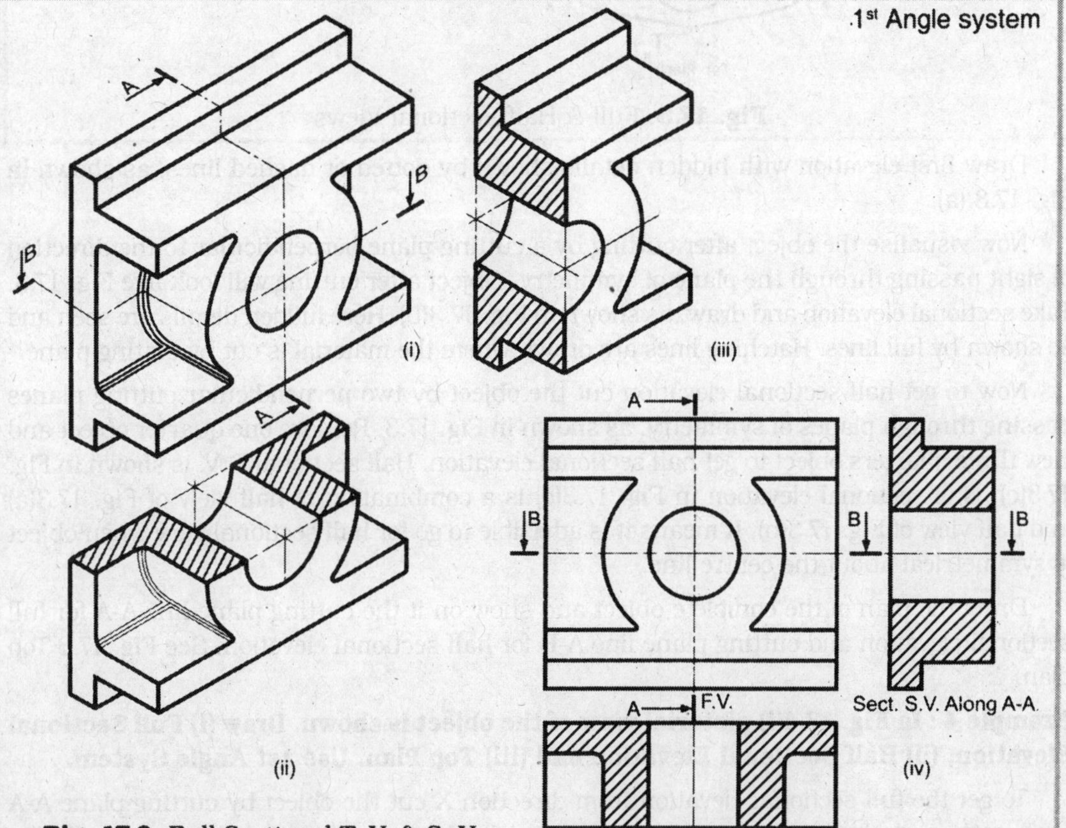

1st Angle system

Fig. 17.2. Full Sectional T. V. & S. V.

Example 3 : In Fig. 17.3, three dimensional sectional view of the object is given. Draw (a) Elevation (F.V.), (b) Full Sectional Elevation (F.V.), (c) Half Sectional Elevation (F.V.), (d) H.S.S.V. and Top Plan (T.V.). Use 1st Angle System.

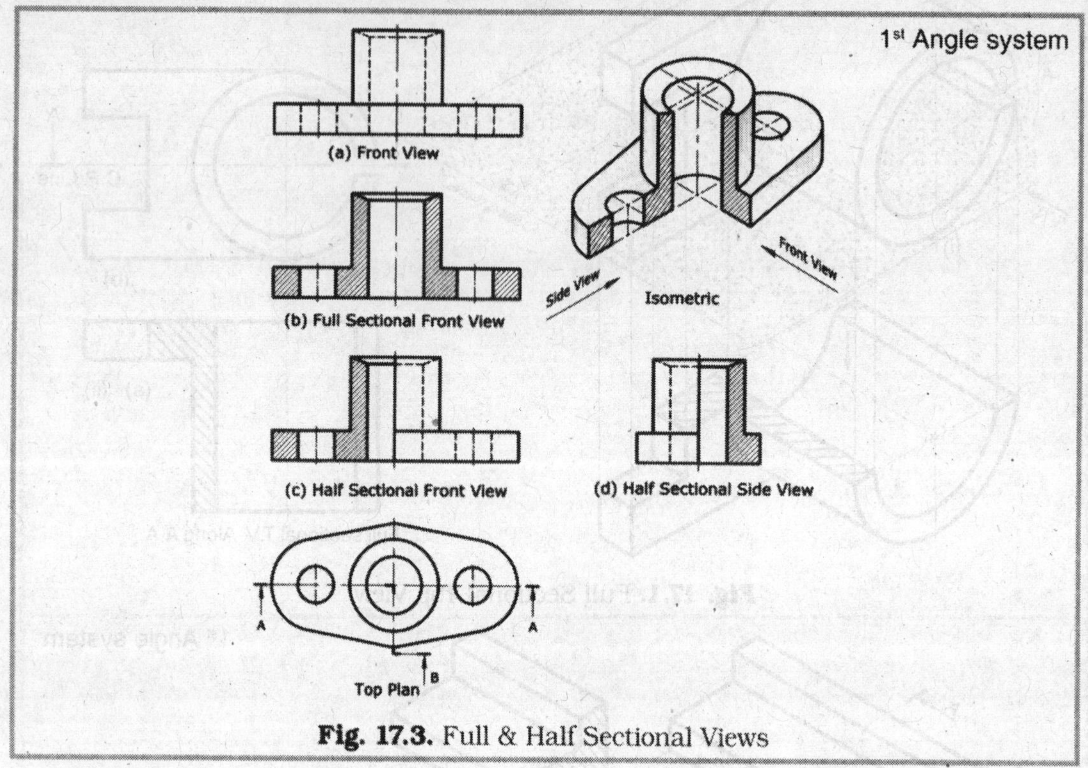

Fig. 17.3. Full & Half Sectional Views

Draw first elevation with hidden details shown by dotted or dashed lines, as shown in Fig. 17.3 (a).

Now visualise the object after cutting by a cutting plane perpendicular to the direction of sight passing through the plane of symmetry. Object after cutting will look like Fig. 17.3. Take sectional elevation and draw, as shown in Fig. 17.3(b). Here hidden details are seen and so shown by full lines. Hatching lines are drawn where the material is cut by cutting plane.

Now to get half sectional elevation cut the object by two perpendicular cutting planes passing through planes of symmetry, as shown in Fig. 17.3. Remove one quarter object and view three quarters object to get half sectional elevation. Half sectional F. V. is shown in Fig. 17.3(c). Half sectional elevation in Fig. 17.3(c) is a combination of half view of Fig. 17.3(a) and half view of Fig. 17.3(b). It means it is advisible to go for half sectional view when object is symmetrical about the centre line.

Draw the plan of the complete object and show on it the cutting plane line A-A for full sectional elevation and cutting plane line A-B for half sectional elevation. See Fig. 17.3 Top plan.

Example 4 : In Fig. 17.4(i) pictorial view of the object is shown. Draw (i) Full Sectional Elevation, (ii) Half Sectional Elevation and (iii) Top Plan. Use 1st Angle System.

To get the full sectional elevation from direction X cut the object by cutting plane A-A passing through the axis of symmetry and perpendicular to the direction X. After cutting, the half object will look like as shown in Fig. 17.4(ii). Take full sectional elevation from direction X and draw, as shown in Fig. 17.4(iv)(a).

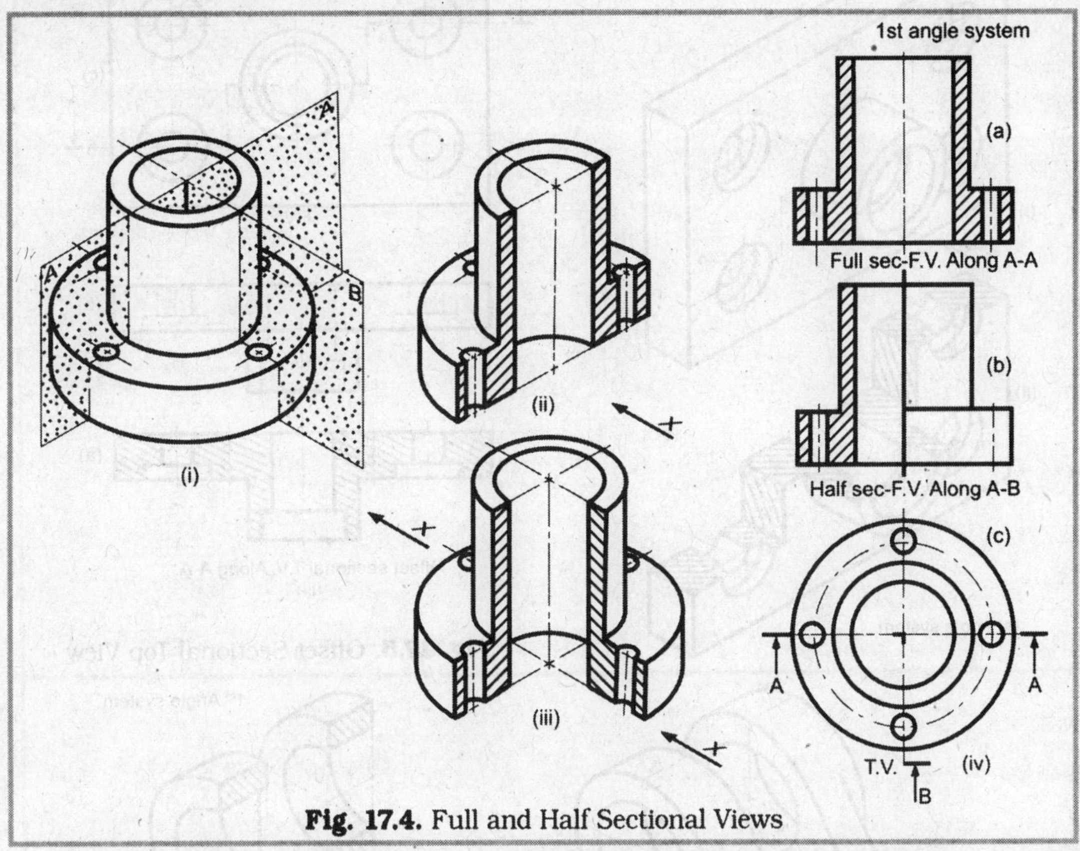

Fig. 17.4. Full and Half Sectional Views

To get the half sectional elevation from direction X cut the object by two mutually perpendicular planes A and B passing through axis of symmetry, as shown in Fig. 17.4(i). After cutting, the three quarter object will look like as shown in Fig. 17.4(iii). Take half sectional elevation from direction X and draw, as shown in Fig. 17.4(iv)(b).

Draw plan and show on it section plane line A-A for full sectional elevation and A-B for half sectional elevation. Draw hatching lines in areas where material is cut by section plane.

Example 5 : Isometric projection of the object is given in Fig. 17.5(i). Draw (i) Elevation/F.V., (ii) Top Plan (T.V.) and (iii) Offset Sectional Plan (T.V.) along A-A. Use 1st Angle Projection System.

Here if section is not taken by offset plane then holes will not be cut and details about holes will not appear in sectional plan. So instead of taking one plane section, offset section is taken. After cutting the object along A-A and removing the upper portion, which is between the observer and the cutting plane, the object will look like as shown in Fig. 17.5(ii).

Draw first F.V. and T.V. of complete object, as shown in Fig. 17.5(iii) (c) and Fig. 17.5(iii) (b) respectively. Now draw offset sectional plan looking from top of the cut object shown in Fig. 17.5(ii). Offset sectional T.V. is shown in Fig. 17.5(iii) (a).

Do not forget to show cutting plane line A-A in elevation. See Fig. 17.5(iii)(c).

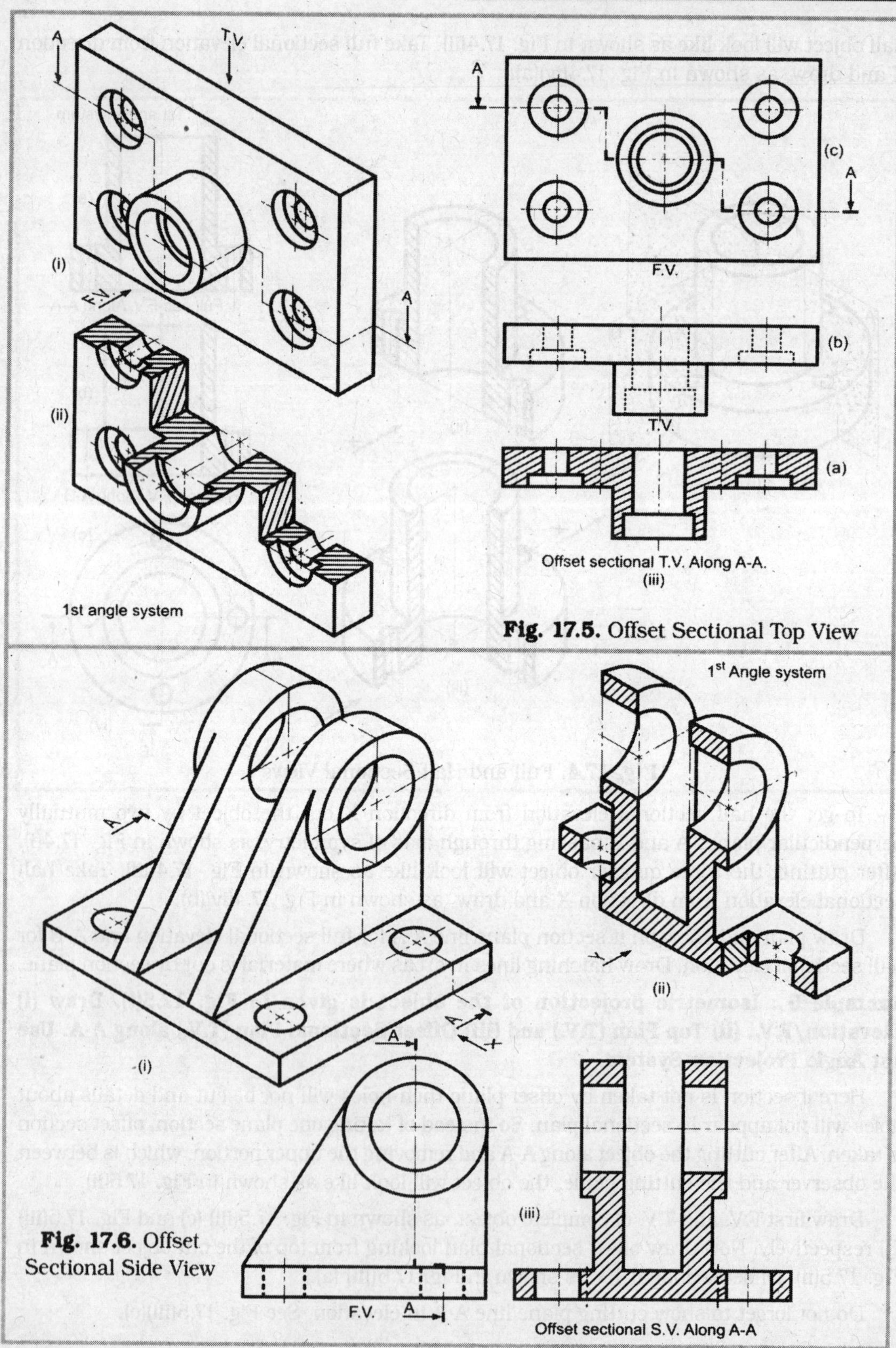

Fig. 17.5. Offset Sectional Top View

Fig. 17.6. Offset Sectional Side View

Example 6 : Pictorial view of an object is given in Fig. 17.6 (i). Draw (i) F.V. from direction X (ii) Offset Sectional S.V. from direction Y. Take section along A-A. Use 1st angle system.

After cutting the object along offset section A-A, it will look like Fig. 17.6(ii). Draw required views and do not forget to show section plane or cutting plane line A-A in F.V.

After solving the above six problems, students are requested to solve problems of solved exercises and unsolved exercises to grasp the subject and imbibe self confidence.

6. Partial or Local or Broken Out Section

If it is required to show only isolated inside details or one of the important features of a component or of an assembly, this type of section is used instead of full or half section. The imaginary cutting plane passes through particular important features, exposing its interior to view and leaving an irregular break line surrounding important feature as boundary. See Fig. 17.7 (a)(b) (c) and (d)

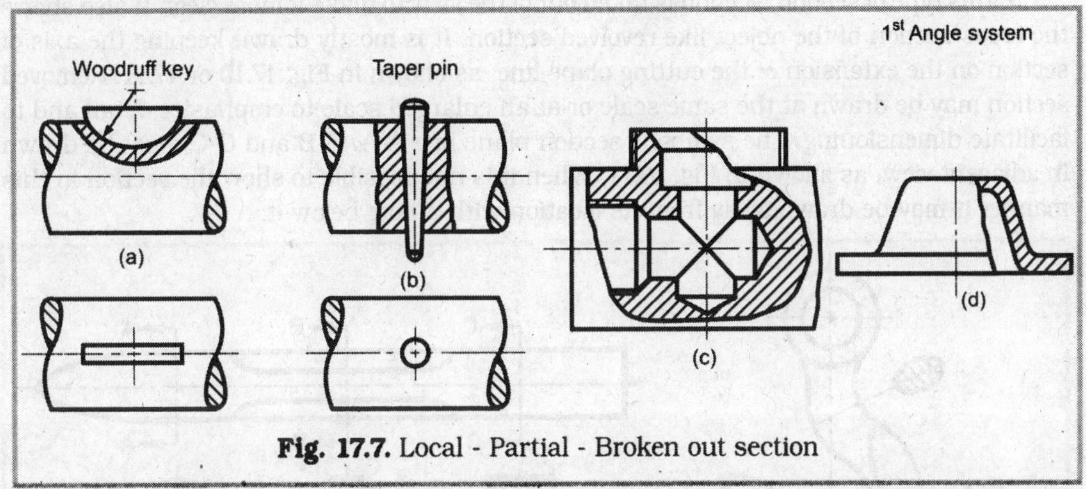

Fig. 17.7. Local - Partial - Broken out section

In Fig. 17.7(a) position of key and in Fig. 17.7(b) position of pin is shown by taking this type of section. Further in Fig. 17.7(c) inside details of the valve are also shown by this type of section. In Fig. 17.7(d) thickness of the vessel is shown.

7. Revolved Section

Revolved section is used to show the cross-section of objects of uniform section like arms, spokes, structural sections, bars, etc. Imaginary cutting plane is passed through the object perpendicular to the profile or longitudinal axis and then revolved on the centre line of the section by 90° to bring it in the plane of the paper. By this type of view we can avoid other view. See Fig. 17.8 for round section, I section and elliptical section.

Sometimes the view may be broken on both the sides of a revolved section, as shown in Fig. 17.8 and 17.9 to make the section more clear. I-Section of Fig. 17.8 and all sections of Fig. 17.9 are shown as broken revolved section.

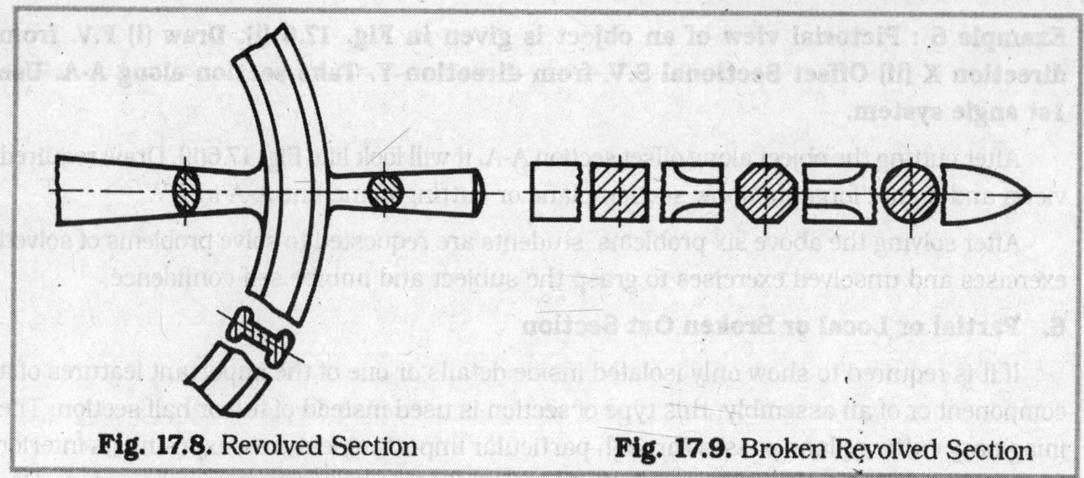

Fig. 17.8. Revolved Section **Fig. 17.9.** Broken Revolved Section

8. Details or Removed Section

In this type of section, section is taken out of the view to make it more clear. It also shows the cross-section of the object like revolved section. It is mostly drawn keeping the axis of section on the extension of the cutting plane line, as shown in Fig. 17.10 or 17.11. Removed section may be drawn at the same scale or at an enlarged scale to emphasize detail and to facilitate dimensioning. The symbolic section plane line A-A, B-B and C-C must be drawn in adjacent view, as shown in Fig. 17.11. When it is not possible to show the section in this manner it may be drawn away from its location with a note below it.

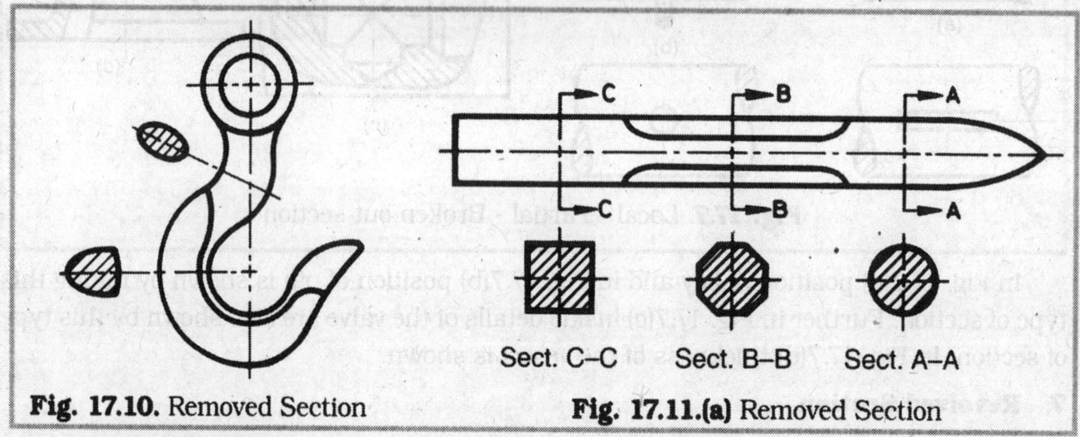

Fig. 17.10. Removed Section **Fig. 17.11.(a)** Removed Section

9. Auxiliary Section

A sectional view taken on auxiliary plane is called auxiliary section. See Fig. 17.12. An auxiliary section may be of any type studied so far i. e. full, half, removed, revolved, etc.

10. Hatching

Following rules are adopted for hatching :

(1) Area of a metal or a material, cut by a section plane is shown by hatching lines in projection. It is executed by thin lines normally at an angle of 45° to the axis or to the main outline or boundary of the sectional area. See Fig. 17.13, part A and C.

(2) If the axis or the boundary of section is at 45° then the horizontal or vertical hatching lines are drawn. See Fig. 17.13, part B and D.

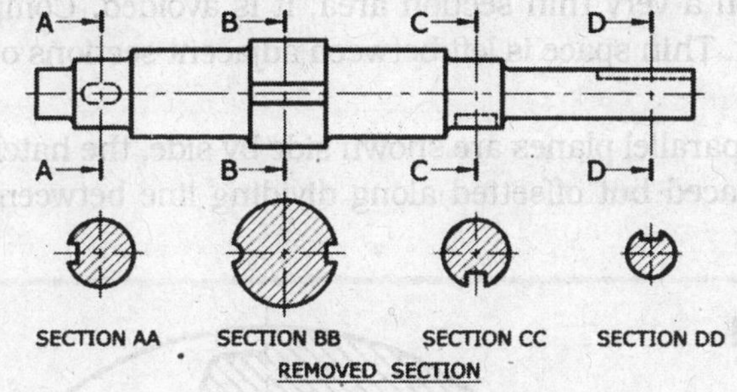

Fig. 17.11.(b) Removed Section

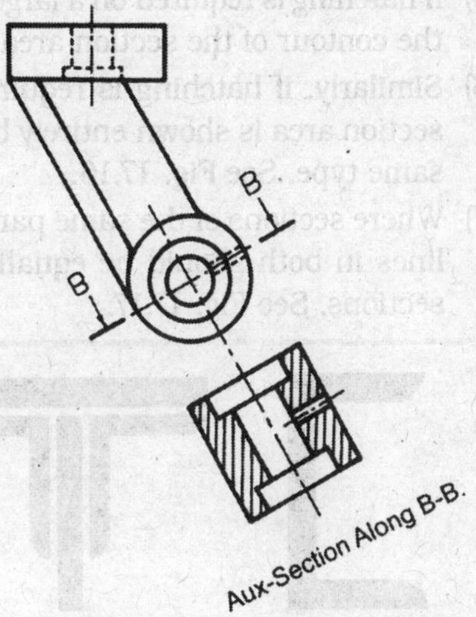

Fig. 17.12. Auxiliary Section

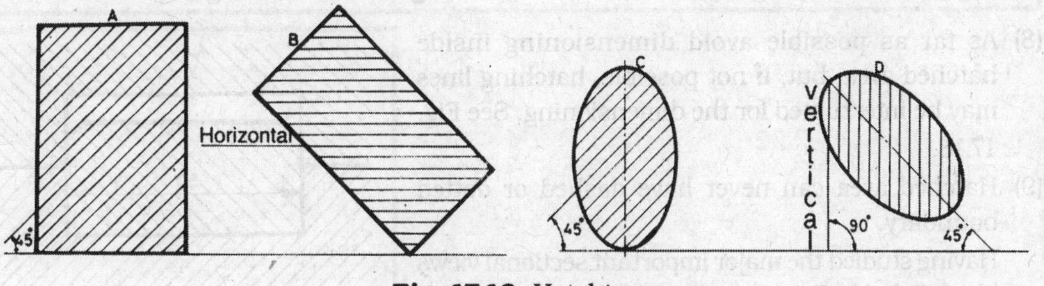

Fig. 17.13. Hatching

(3) Separated areas of a section of a single component are hatched in the same manner. Hatching lines in different projections for the same component, are drawn in the same manner. See Fig. 17.2(iv).

(4) Hatching lines on a second part, adjacent to the first are drawn at an angle of 45° but in the opposite direction. See Fig. 17.14, part A and B. Hatching lines on a third part adjacent to first two are drawn at an angle of 30° or 60°. See Fig. 17.14, part C adjacent to A and B. Sometimes spacing or pitch of the hatching lines is varied to separate it from the adjacent parts. See Fig. 17.14, part D, adjacent to parts A, B and C.

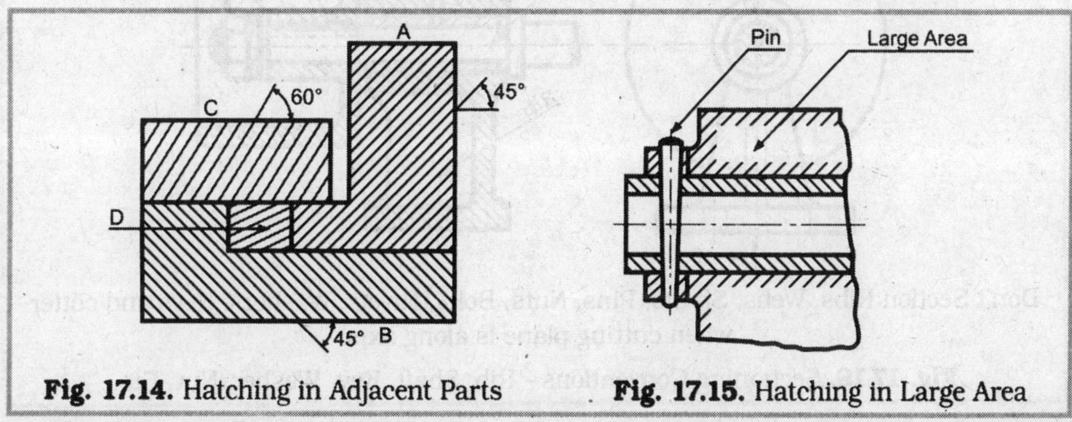

Fig. 17.14. Hatching in Adjacent Parts

Fig. 17.15. Hatching in Large Area

(5) If hatching is required on a large section area, it is avoided. It is limited to a zone following the contour of the section area required to be hatched. See Fig. 17.15.

(6) Similarly, if hatching is required on a very thin section area, it is avoided. Complete section area is shown entirely black. Thin space is left between adjacent sections of the same type. See Fig. 17.16.

(7) Where sections of the same part in parallel planes are shown side by side, the hatching lines in both should be equally spaced but offsetted along dividing line between the sections. See Fig. 17.17.

Fig. 17.16. Hatching in Thin Area

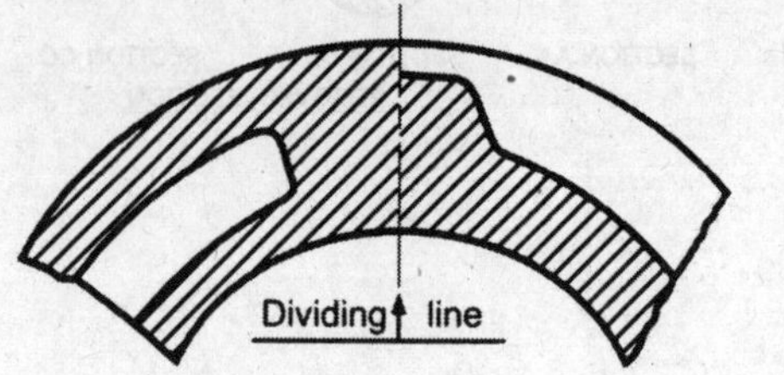

Fig. 17.17. Hatching in Parallel Planes

(8) As far as possible avoid dimensioning inside hatched area, but, if not possible, hatching lines may be interrupted for the dimensioning. See Fig. 17.18.

(9) Hatched area can never have dashed or dotted boundary.

Having studied the major important sectional views like full, half, offset, partial, revolved, removed, etc. now we shall study various conventional practices required to be followed as per Indian Standards.

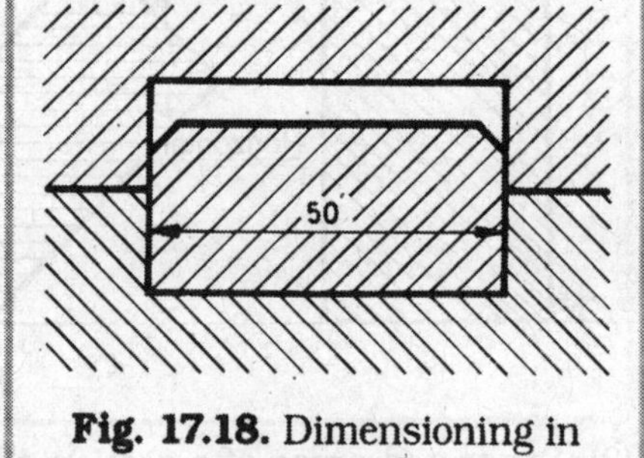

Fig. 17.18. Dimensioning in Hatched Area

11. Sectioning Conventions for - Ribs Webs, Shafts, Pins, Nuts, Bolts, Rivets, Washers, Keys, Cotters, Rods, Pulley Arms, etc.

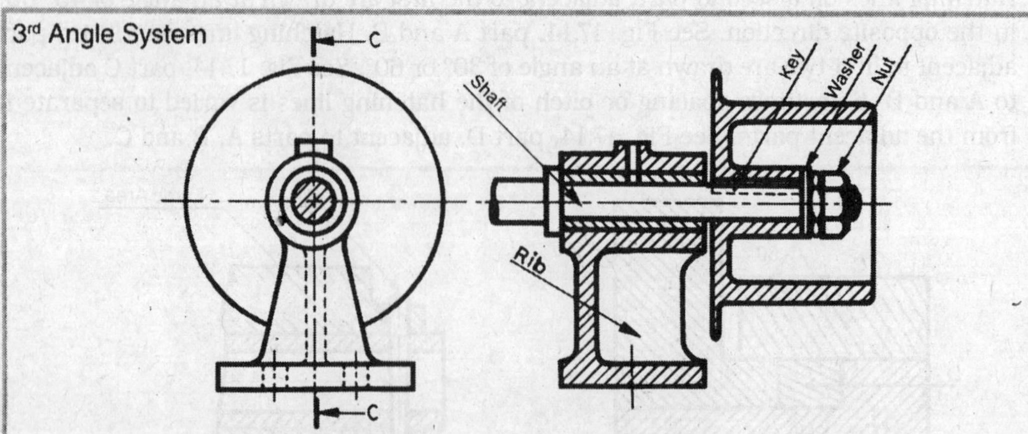

Don't Section Ribs, Webs, Shafts, Pins, Nuts, Bolts, Rivets, Washers, Keys and cotter when cutting plane is along axis.

Fig. 17.19. Sectioning Conventions - Rib, Shaft, Key, Washer, Nut, Etc.

Whenever, the cutting plane is passing through above machine elements and if C.P. contains the axis of above machine elements then those machine elements are not sectioned. See Figs. 17.19, 17.20, 17.21, 17.7 (a) & (b) and 17.22. However, when the cutting plane is cutting, elements perpendicular to the axis of the above machine elements, section is taken and in projection hatching lines are drawn, see section lines in rib in plan of Fig. 17.21 and section lines in shaft in S.V. of Fig. 17.19.

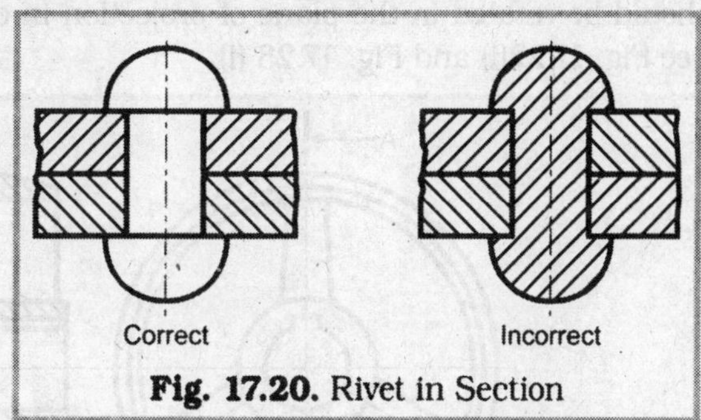

Fig. 17.20. Rivet in Section

These machine elements are simple, they do not have inner details and they are more easily recognised by their simple projection and hence they are not sectioned.

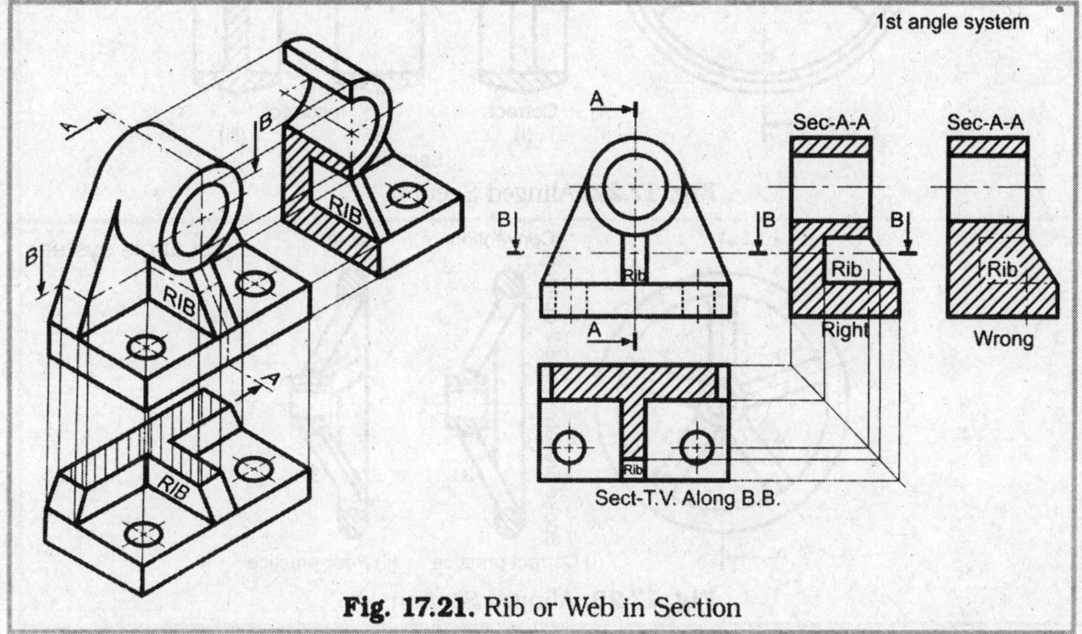

Fig. 17.21. Rib or Web in Section

Rib or web is a flat plate like part of the object used only to strengthen the object. Rib or web is not sectioned when cutting plane passes longitudinally through the centre of the rib or web. If section is taken in rib and if hatching lines are drawn then it will create wrong impression about the thickness or solidity of the component. See Fig. 17.21 side view with section lines in rib. However, if the cutting plane cuts the rib or web at right angle to the longitudinal surface of rib then section is taken and hatching lines are drawn in projection, see rib in section along B-B in plan in Fig. 17.21.

Spokes or arms of wheels, pulleys or flywheels are never sectioned when cutting plane passes through the axis of the arms, see Fig. 17.22(i). Here cutting plane A-A is passing through the axis of the arms and hence not sectioned in correct side view. Incorrect end views are also shown in Fig. 17.22(ii) and (iii).

12. Aligned Sections

Any part having odd number of spokes, ribs or holes will give an unsymmetrical and misleading section if the rules of projection are strictly followed. See Fig. 17.22(ii) and Fig. 17.23(ii). In such cases, to avoid confusion, the inclined or unsymmetrically placed elements should be rotated in the plane of projection in one view to bring it into the cutting plane. See Fig. 17.22(i) and Fig. 17.23 (i).

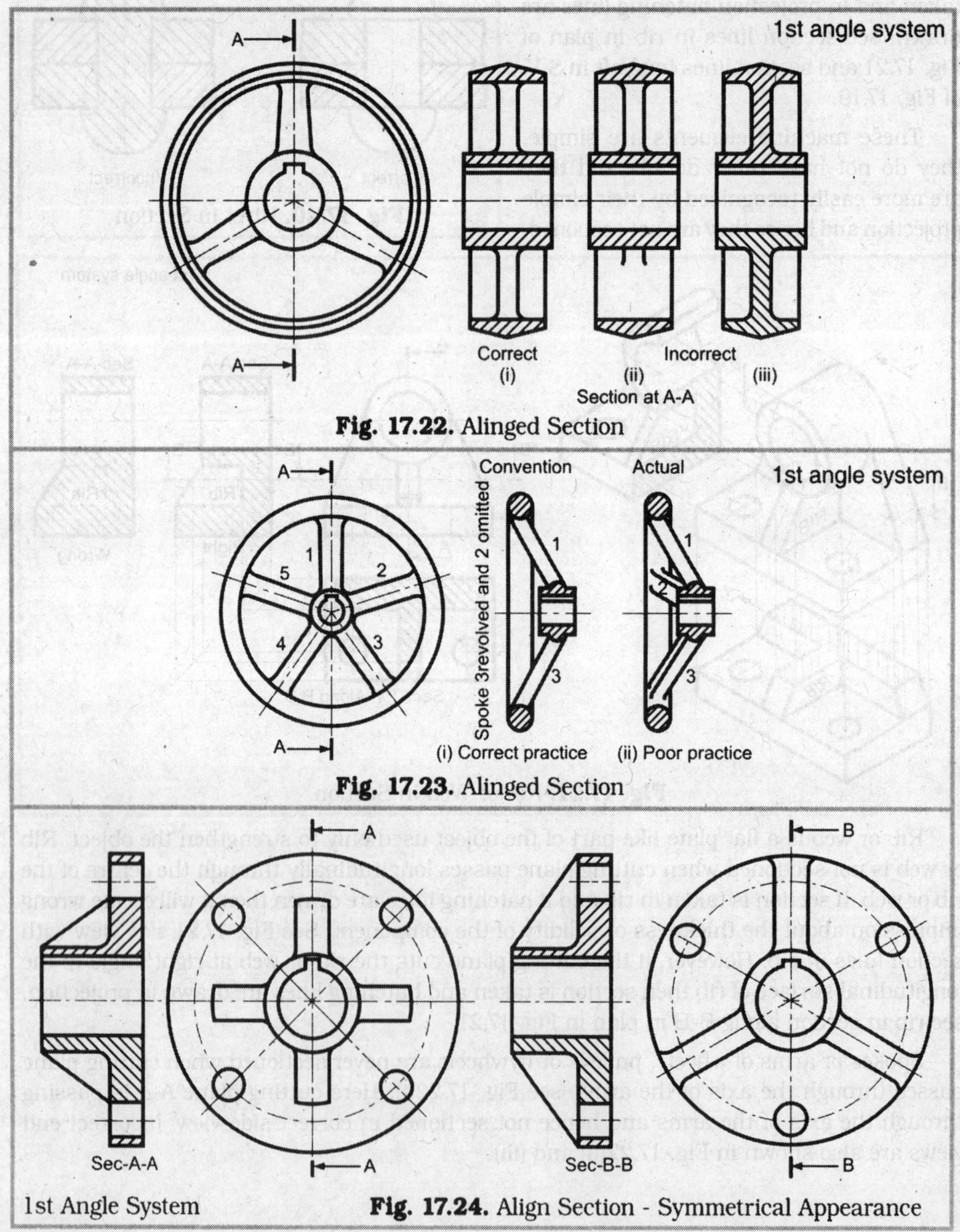

Fig. 17.22. Alinged Section

Fig. 17.23. Alinged Section

Fig. 17.24. Align Section - Symmetrical Appearance

In Fig. 17.24 two flanges are shown one with even number of ribs and holes and another with odd number of ribs and holes. In such cases whether the cutting plane passes through the hole or not they are rotated in the plane of projection to bring it into the cutting plane and shown in sectional view. Similarly, ribs are also rotated and shown in sectional view. This will create symmetrical appearance in sectional view and true positions of the ribs and the holes will be seen in adjacent view. See Fig. 17.24.

13. Conventional Breaks

Long members of uniform cross section are not drawn of full lengthwise since it occupies more space unnecessarily. Conventional break is given along the length, as shown in Fig. 17.25. Eventhough the drawing is drawn of lesser length, actual dimension is given on the drawing of the member. Earlier, we have seen such a break is given in revolved section.

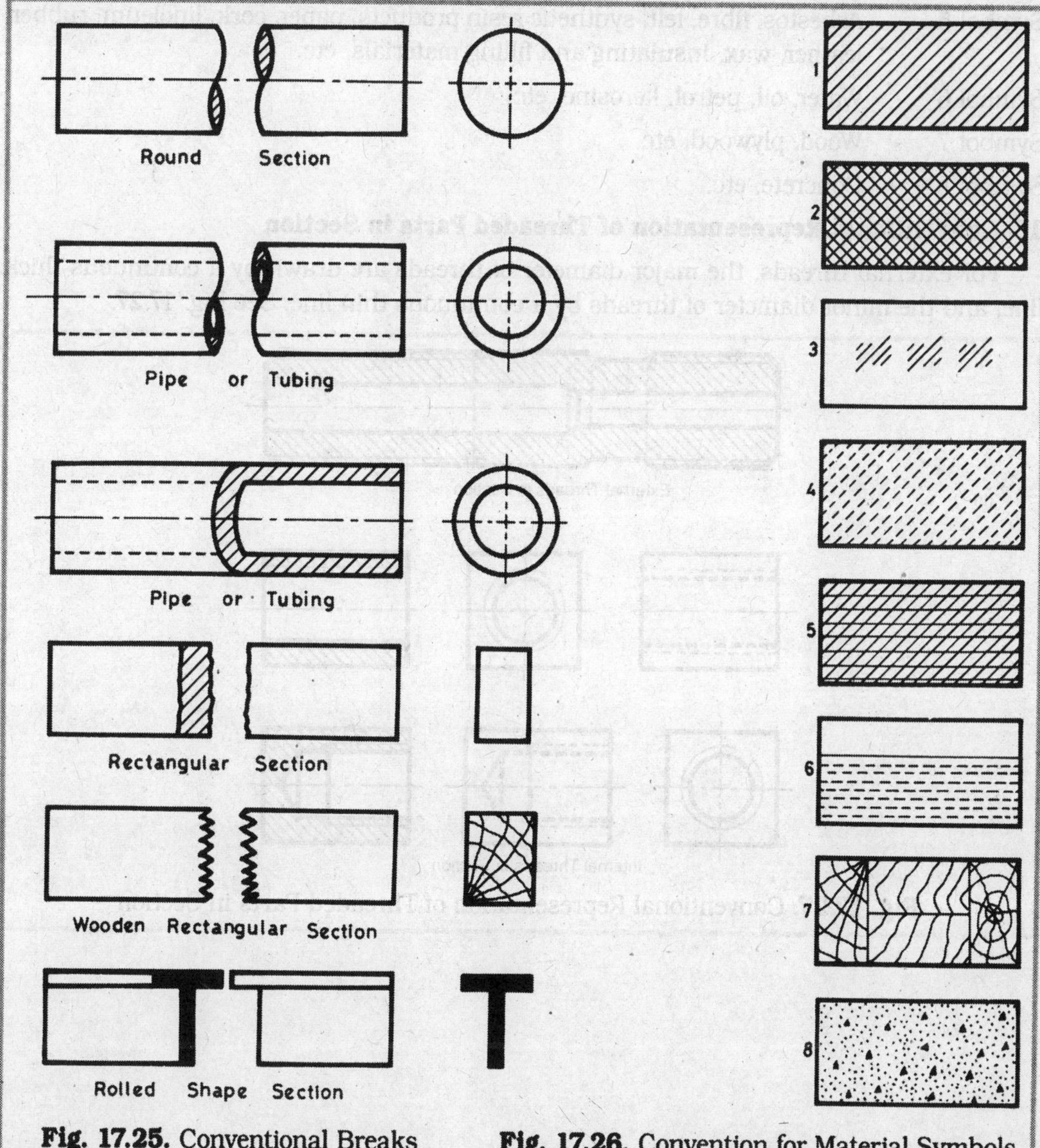

Fig. 17.25. Conventional Breaks

Fig. 17.26. Convention for Material Symbols

In Fig. 17.25 conventional methods, adopted to show broken ends of (i) round bars or shafts, (ii) pipes or tubes (iii) square or rectangular bars and (iv) wood scantlings, (v) Rolled shape section are given.

14. Convention for Material Symbols

Different metals and materials are shown symbolically in different manner in section. There are 8 different symbols adopted, as per I. S. as shown in Fig. 17.26 :

Symbol 1 - Used for steel, C. I, copper and its alloys, aluminium and its alloys, etc.

Symbol 2 - Used for lead, zinc, tin, white metal, etc.

Symbol 3 - Used for glass.

Symbol 4 - Glass, porcelain, stone ware, marble, slate, etc.

Symbol 5 - Asbestos, fibre, felt, synthetic resin products, paper, cork, linoleum, rubber, leather, wax, insulating and filling materials, etc.

Symbol 6 - Water, oil, petrol, kerosine, etc.

Symbol 7 - Wood, plywood, etc.

Symbol 8 - Concrete, etc.

15. Conventional Representation of Threaded Parts in Section

For external threads, the major diameter of threads are drawn by a continuous thick line, and the minor diameter of threads by a continuous thin line. See Fig. 17.27.

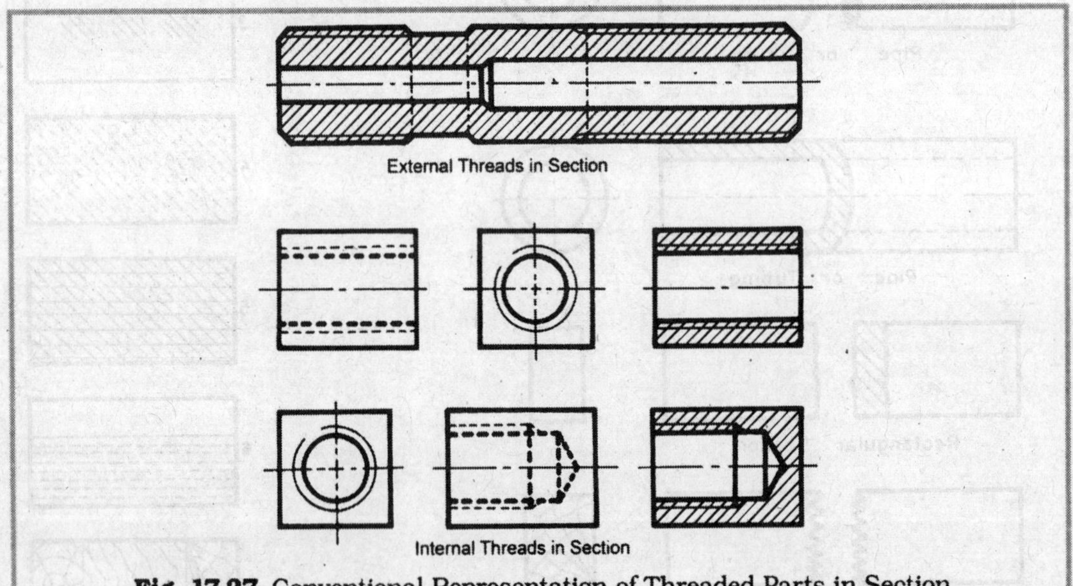

Fig. 17.27. Conventional Representation of Threaded Parts in Section

For internal threads, the major diameter of threads are drawn by a continuous thin line and the minor diameter of threads by a continuous thick line.

In section of threaded parts, hatching lines should be extended to the line defining the minor or major diameter of the threads for internal or external threads respectively. See Fig. 17.27.

16. Cutting Planes / Section Planes.

The following three types of cutting planes are used for the purpose of section in drawings :

(i) One plane section, Fig. 17.28.

(ii) Section in two parallel planes, Fig. 17.29.

(iii) Section in three continuous planes, Fig. 17.30.

(i) If important features of the object are in one plane then one plane section (like A-A) is used to get sectional view. See Fig. 17.28, section plane line A-A.

(ii) If important features of the object are on two parallel planes then two parallel planes are used to get sectional view. See Fig. 17.29, section plane line B-B.

(iii) If important features of the object are on three continuous planes then three continuous plane section is used to get sectional view. See Fig. 17.30, section plane line C-C.

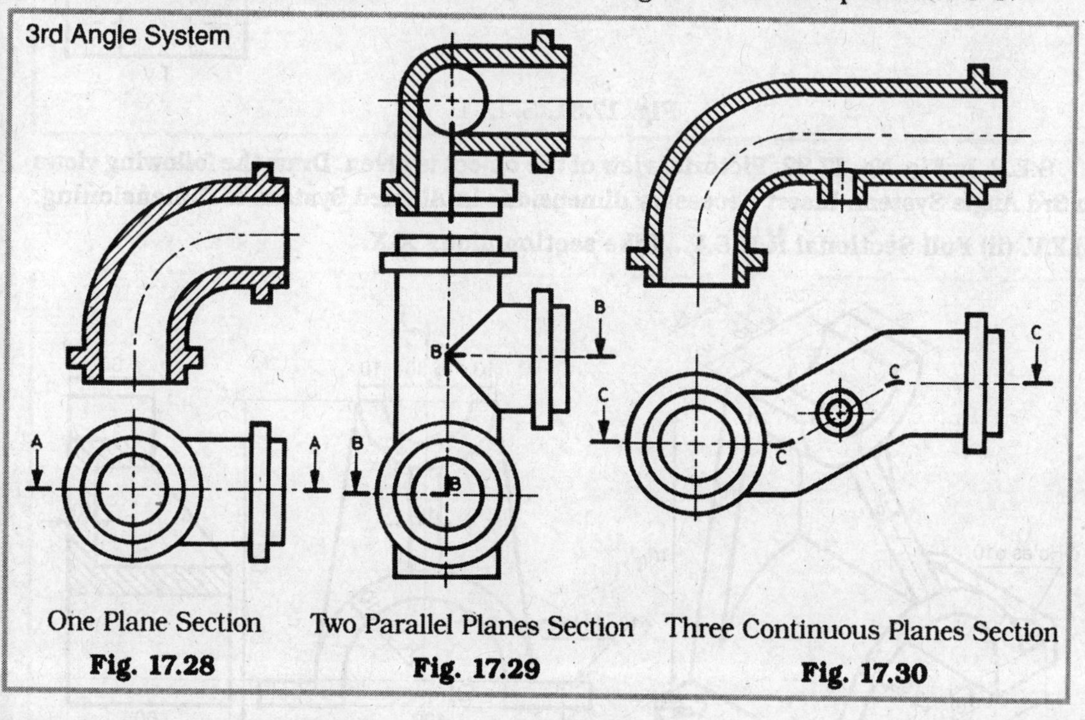

One Plane Section
Fig. 17.28

Two Parallel Planes Section
Fig. 17.29

Three Continuous Planes Section
Fig. 17.30

Solved Exercise

S.E.1. In Fig. No. 17.31, Pictorial View of the object is given. Draw (i) F.V. (ii) Full sectional Right Hand Side View along A-A. (iii) T.V. Use 1st Angle System.

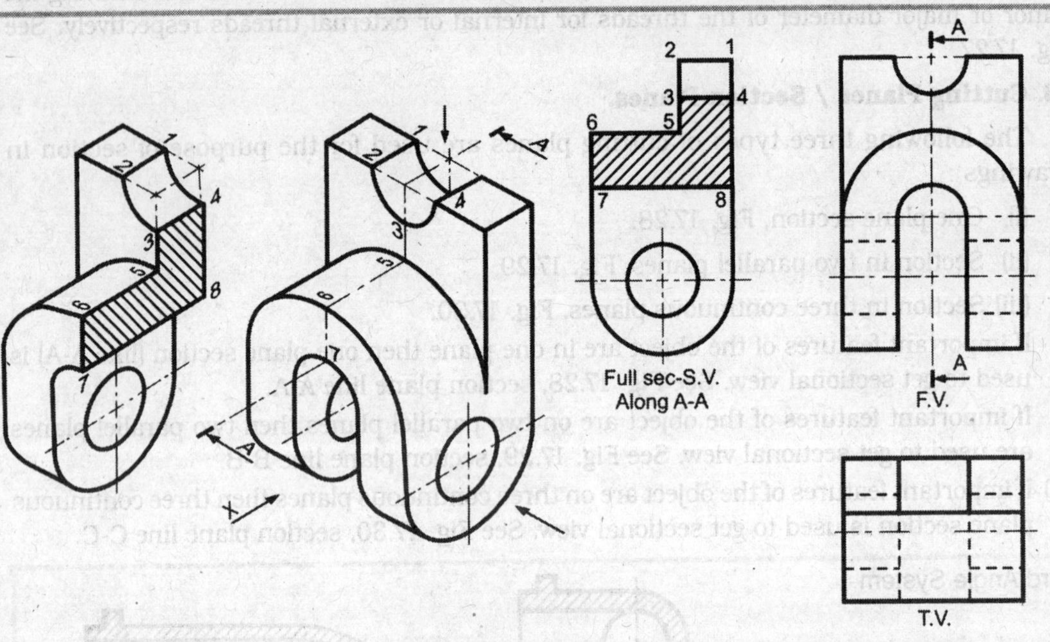

Full sec-S.V.
Along A-A

F.V.

T.V.

Fig. 17.31. S. E. 1

S.E.2. In Fig. No. 17.32, Pictorial view of the object is given. Draw the following views in 3rd Angle System. Insert necessary dimensions in Aligned System of Dimensioning. (i) F.V. (ii) Full Sectional R.H.S.V.. Take section along X-X.

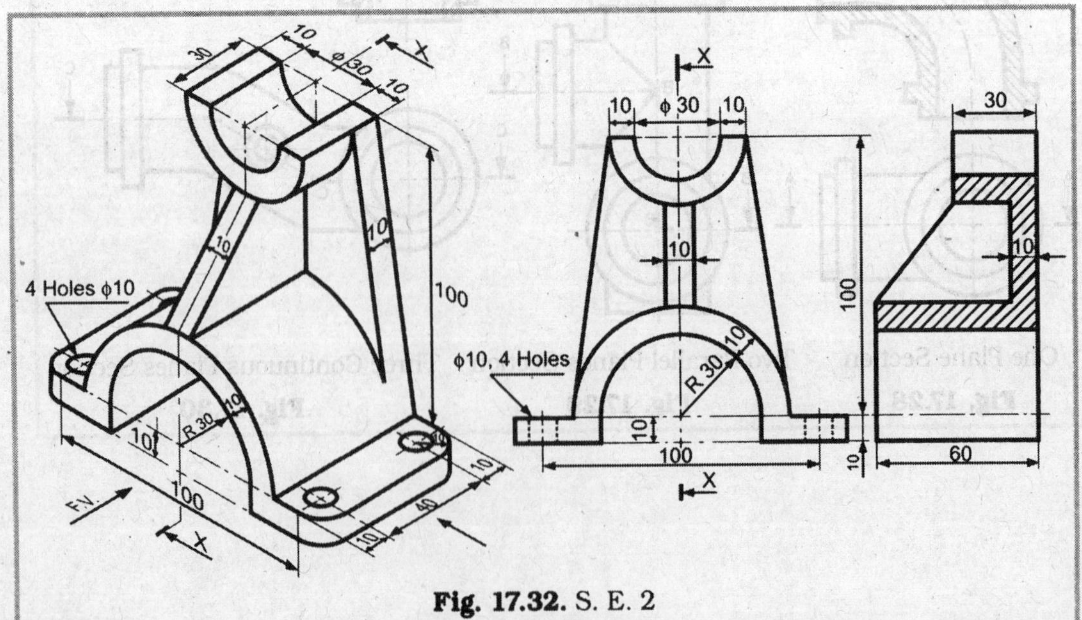

Fig. 17.32. S. E. 2

S.E.3. In Fig. No. 17.33, Pictorial View of the object is given. Draw (i) Full Sectional F.V. along B-B. (ii) T.V. (iii) L.H.S.V. Use 1st Angle System.

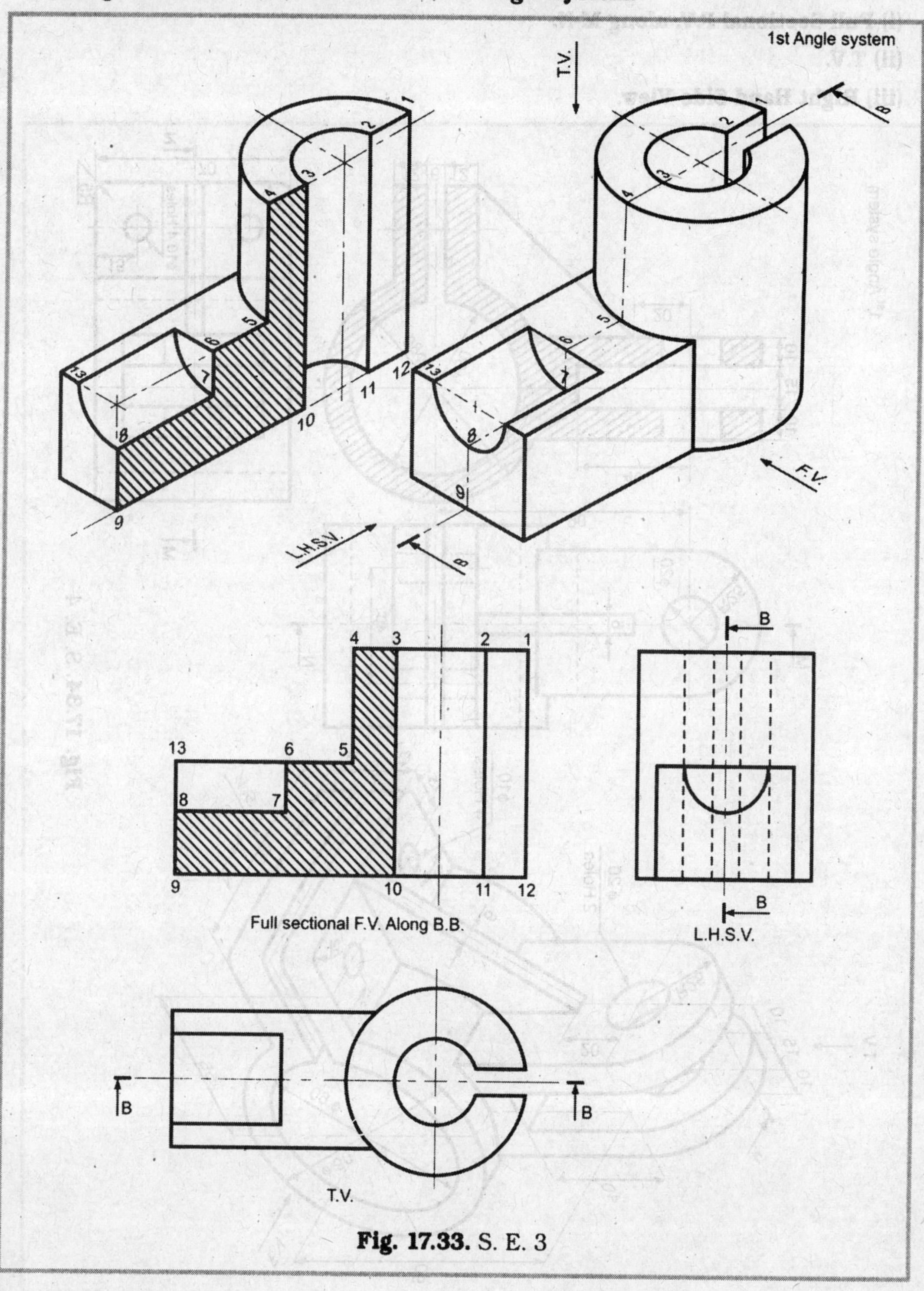

Fig. 17.33. S. E. 3

S.E.4. In Fig. No. 17.34, Pictorial view of the object is given. Draw following views in 1st Angle System and insert necessary dimensions in the projections.

(i) Full Sectional F.V. along M-N.

(ii) T.V.

(iii) Right Hand Side View.

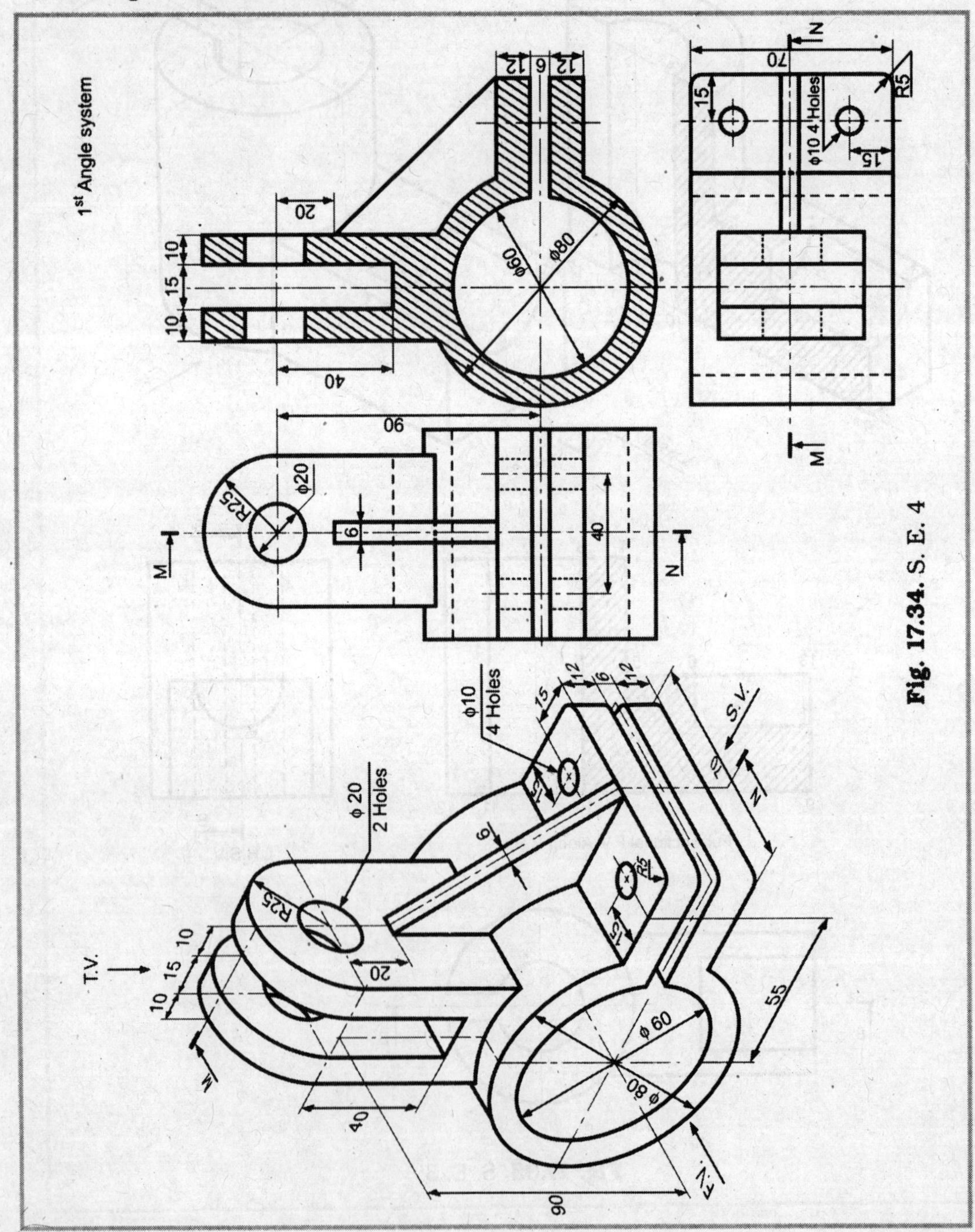

Fig. 17.34. S. E. 4

S.E.5. In Fig. No. 17.35 Pictorial view of the 3/4th object alongwith 1/4th cut portion of the object is given. Draw the following views in 1st Angle Projection System. Insert necessary dimensions (i) Half Sectional F.V. (ii) Half Sectional L.H.S.V. (iii) T.V. (Full)

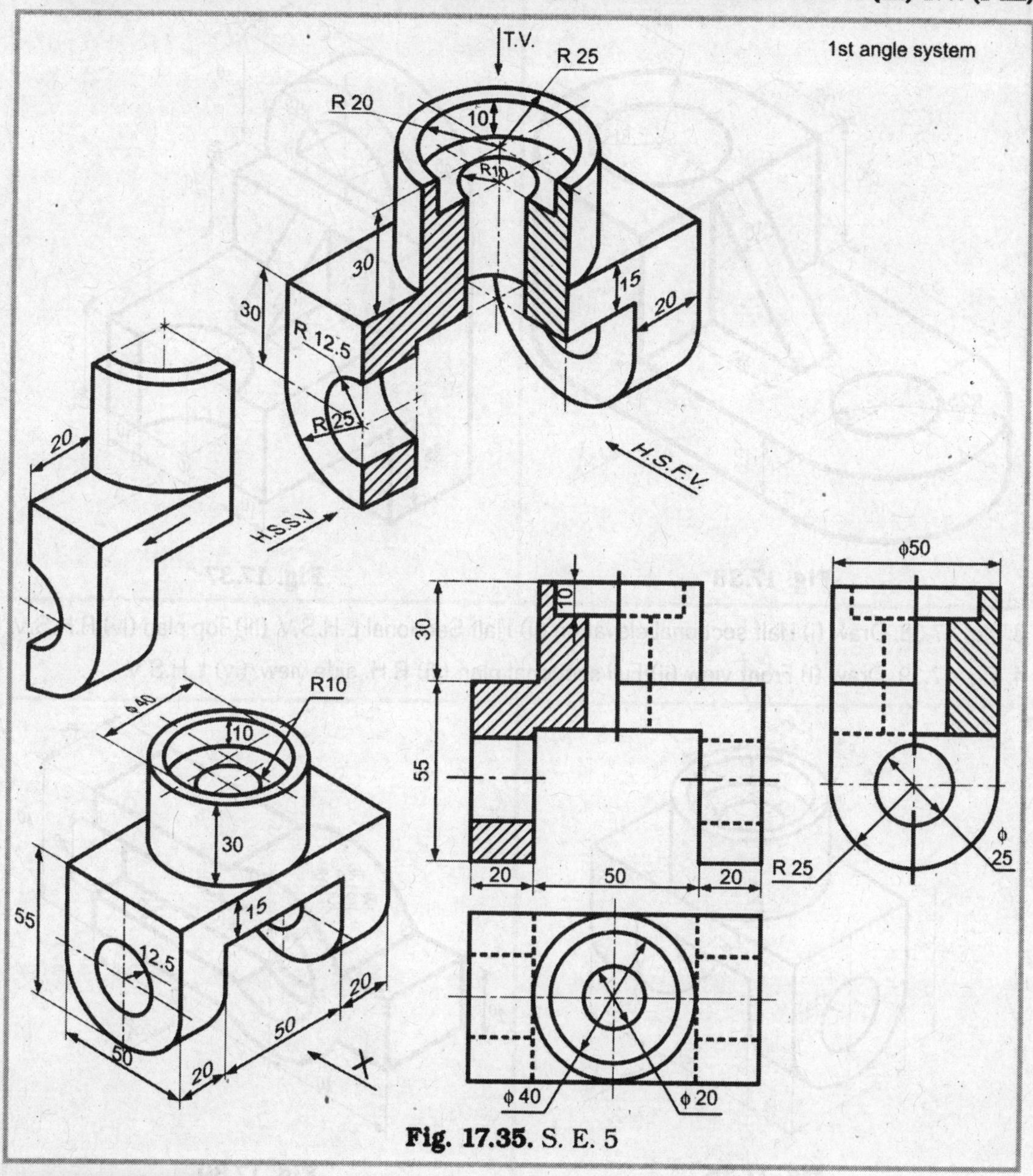

Fig. 17.35. S. E. 5

EXERCISE

Draw orthographic views, as stated in each exercise, from the pictorial views given in Figs. 17.36 to 17.61. Show cutting plane line/lines in respective view/views. Use conventional practices given in theory of this chapter as well as given in Indian Standards for engineering drawing. Insert dimensions in the views. Assume proportionate dimensions, if missing. Answers are given in last pages of this book.

1. Fig. 17.36 Draw (i) Full sectional elevation (ii) Top plan (iii) L.H.S.V.
2. Fig. 17.37. Draw (i) Front view (ii) Full sectional L.H. side view (iii) Top plan.

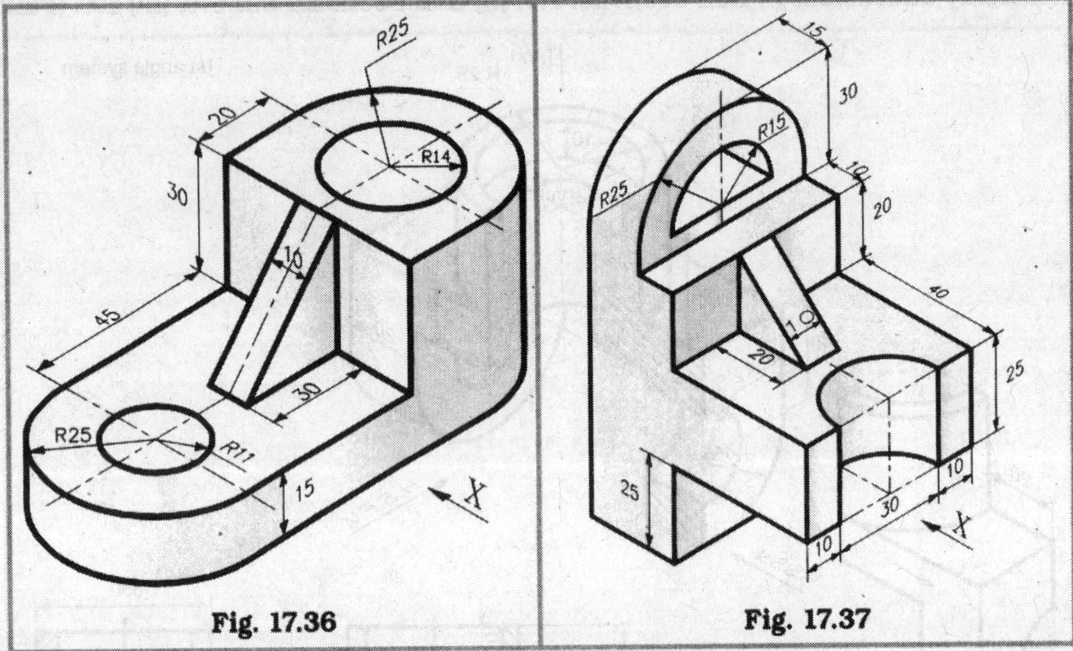

Fig. 17.36

Fig. 17.37

3. Fig. 17.38. Draw (i) Half sectional elevation (ii) Half Sectional L.H.S.V. (iii) Top plan (iv) R.H.S.V.
4. Fig. 17.39. Draw (i) Front view (ii) Full sectional plan (iii) R.H. side view. (iv) L.H.S.V.

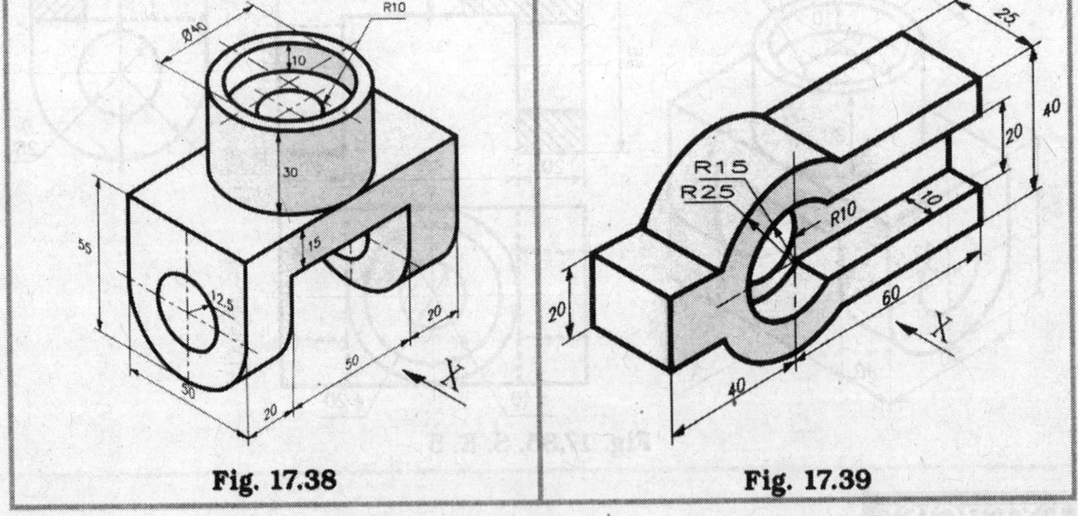

Fig. 17.38

Fig. 17.39

5. Fig. 17.40. Draw (i) Front elevation (ii) Full sectional L.H. side view (iii) Top Plan.

6. Fig. 17.41. Draw (i) Full sectional elevation (ii) L.H. side view (iii) R.H. side view (iv) Top plan.

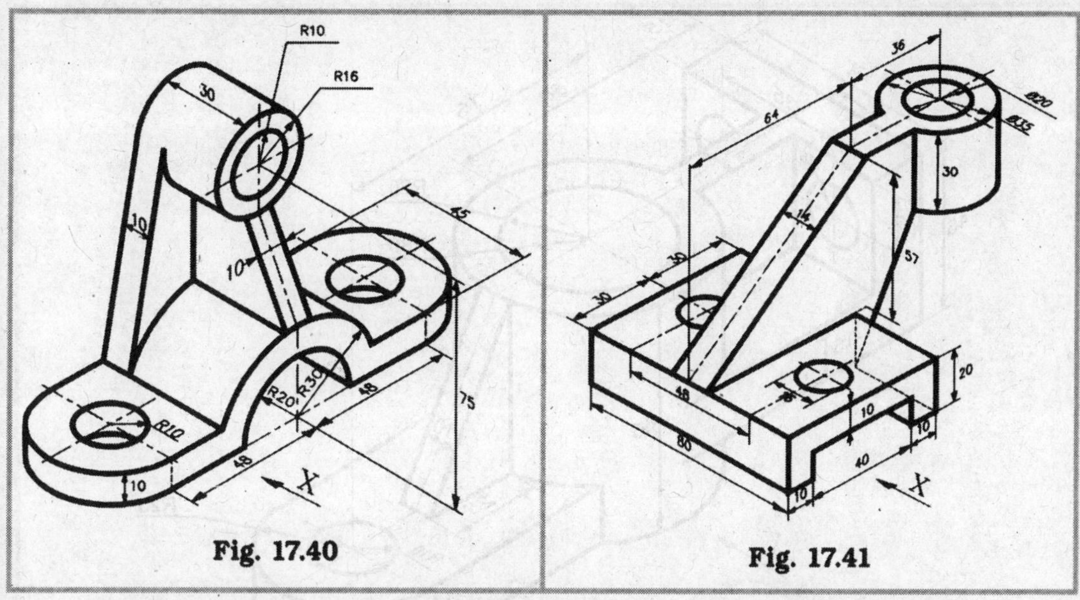

Fig. 17.40

Fig. 17.41

7. Fig. 17.42 Draw (i) Elevation (ii) R.H. sectional side view (iii) Rear view (iv) Top plan.

8. Fig. 17.43. Draw (i) Full sectional elevation (ii) L.H. side view (iii) Top plan.

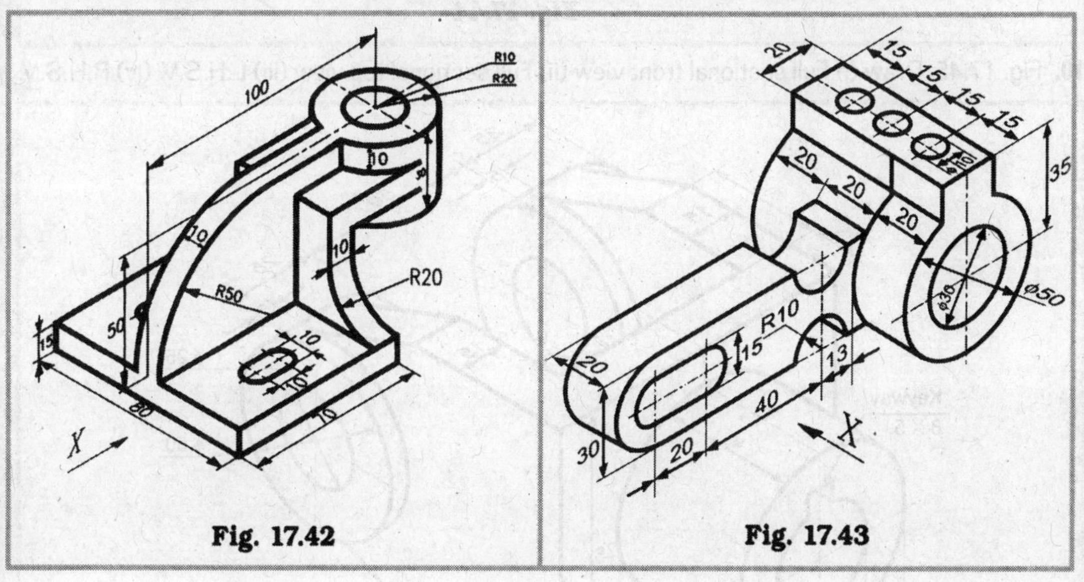

Fig. 17.42

Fig. 17.43

9. Fig. 17.44. Draw (i) Sectional elevation, take section along A-B (ii) R.H.S.V. (iii) Top plan. [Gujarat University November 1991]

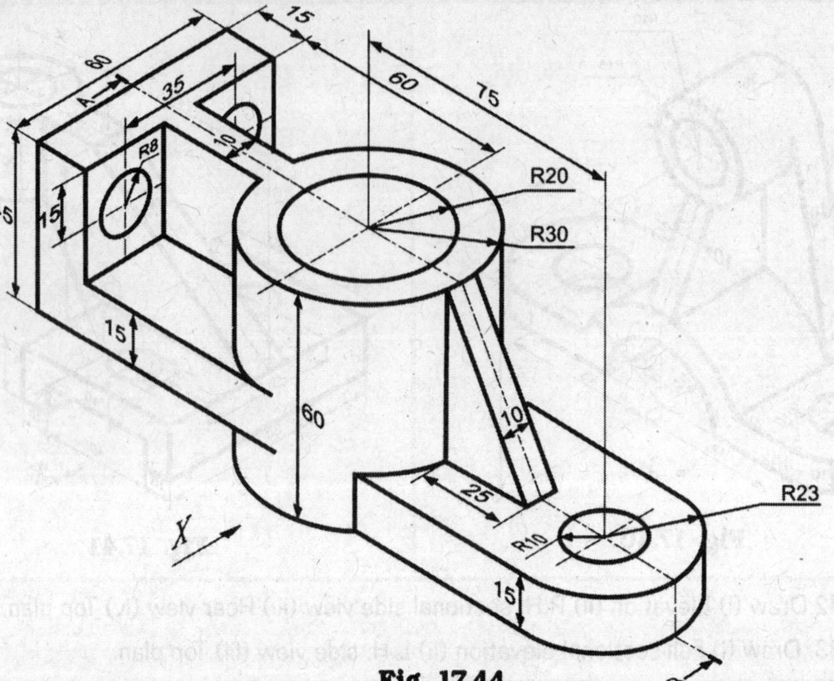

Fig. 17.44

10. Fig. 17.45. Draw (i) Full sectional front view (ii) Full sectional top view (iii) L.H.S.V. (iv) R.H.S.V.

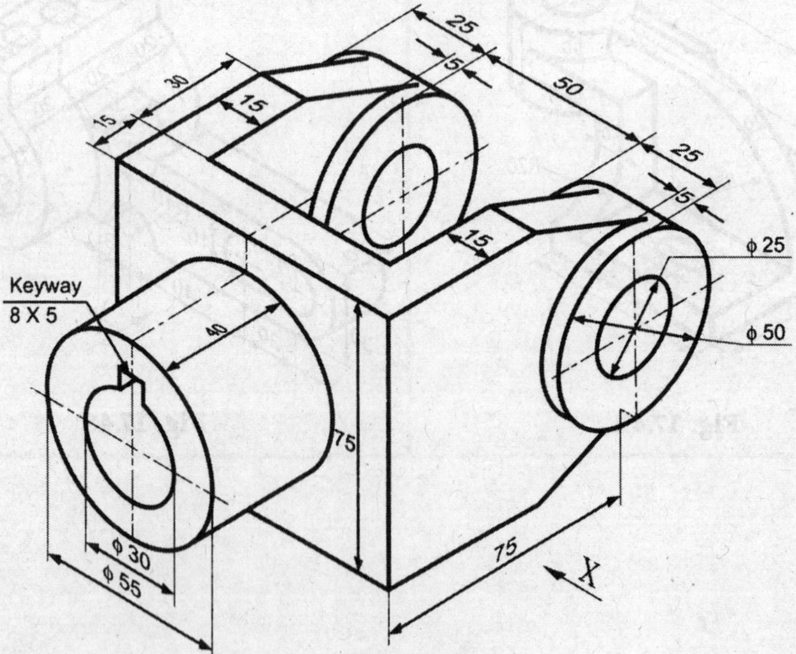

Fig. 17.45

11. Fig. 17.46. Draw (i) Elevation (ii) Full sectional L.H. side view (iii) Top plan

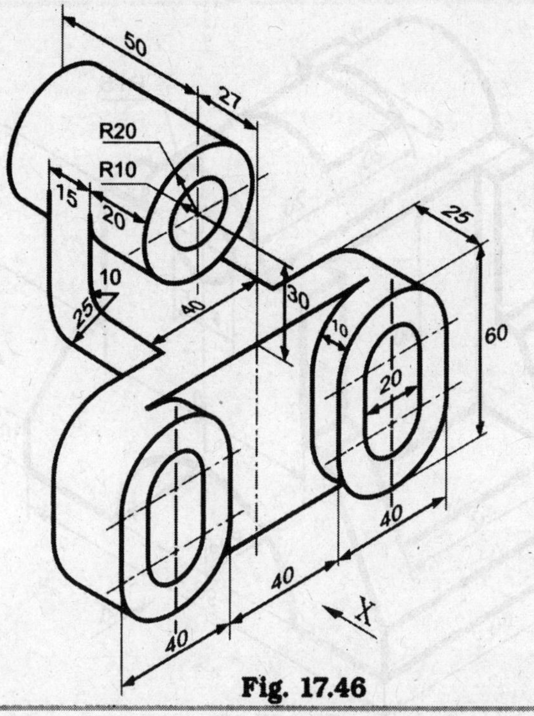

Fig. 17.46

12. Fig. 17.47. Draw (i) Offset sectional elevation, select suitable section. (ii) R.H. side view (iii) Top plan

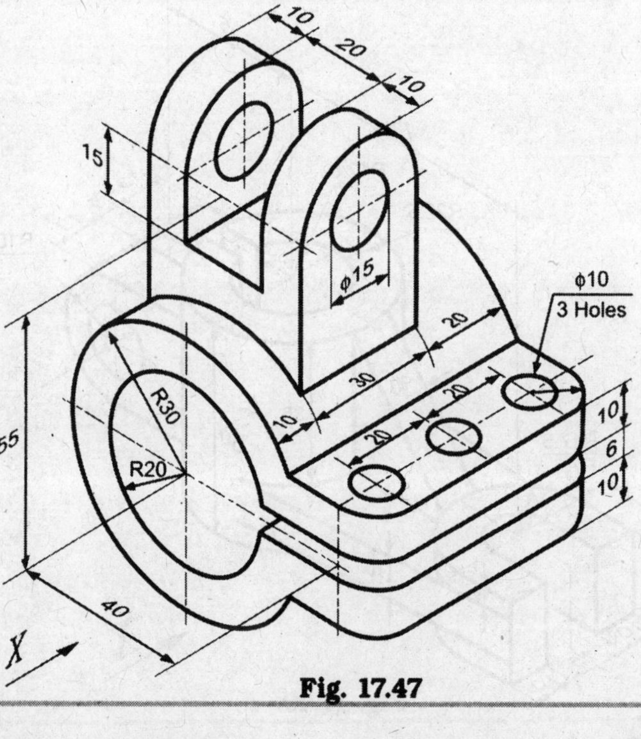

Fig. 17.47

13. Fig. 17.48. Draw (i) Elevation (ii) L.H. sectional side view (iii) Top plan

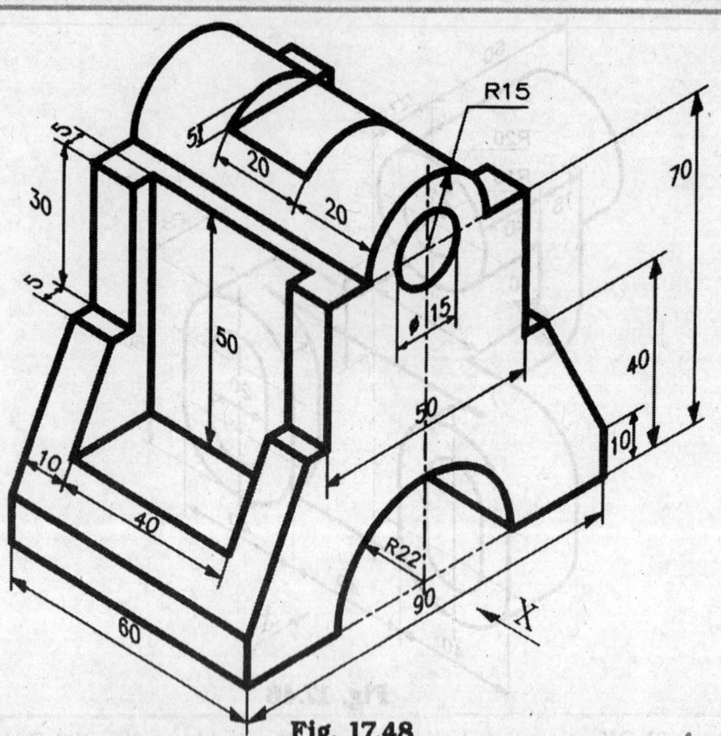

Fig. 17.48

14. Fig. 17.49. Draw (i) Full sectional elevation (ii) Top plan (iii) R.H. side view (iv) L.H. side view.

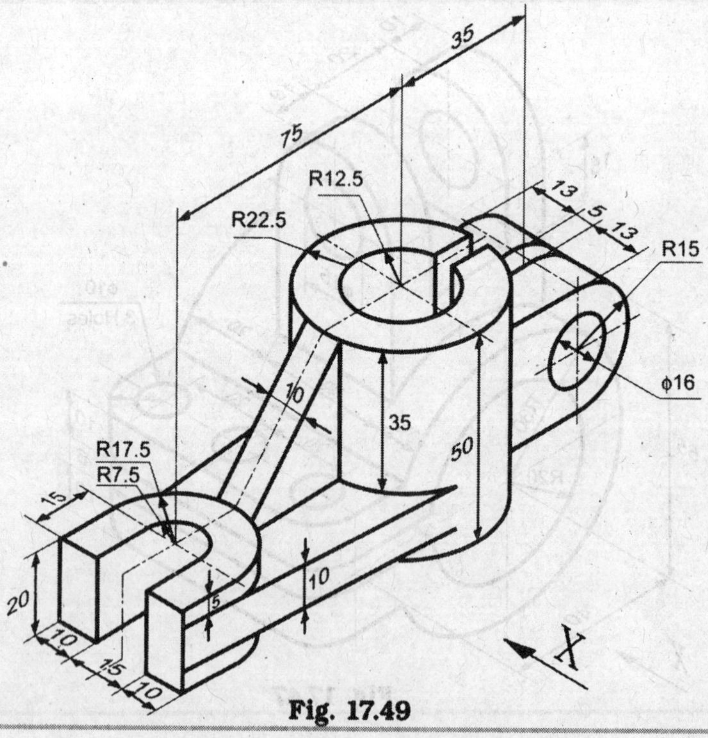

Fig. 17.49

15. Fig. 17.50. Draw (i) Half sectional elevation (ii) Half sectional L.H. side view (iii) R.H.S.V. (iv) Top plan (complete).

16. Fig. 17.51. Draw (i) Full sectional elevation (ii) R.H.S.V. (iii) L.H.S.V. (iv) Top plan.

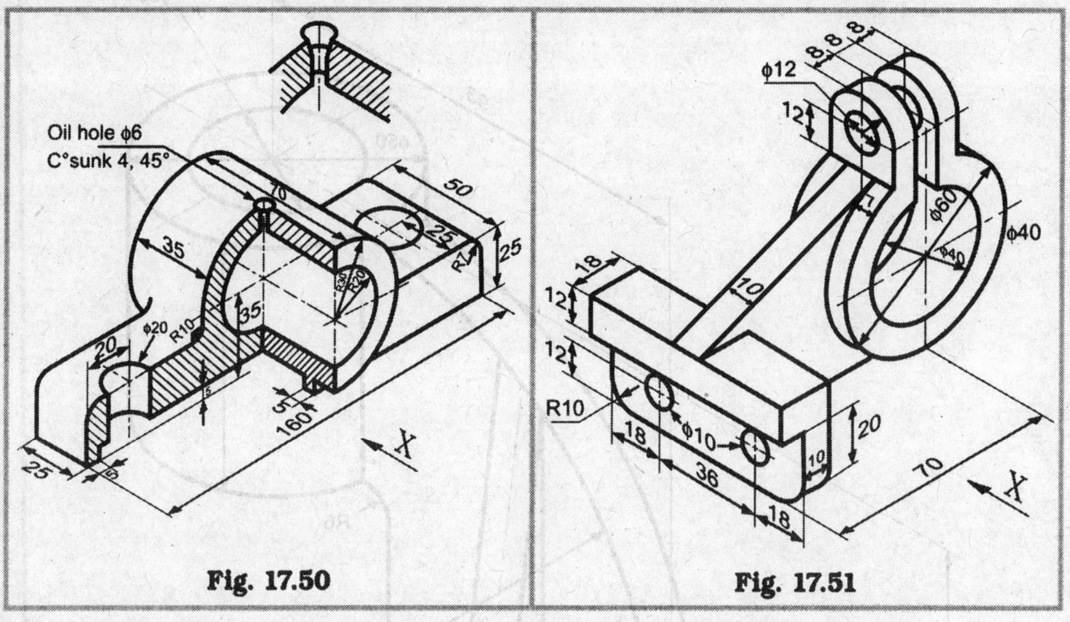

Fig. 17.50

Fig. 17.51

17. Fig. 17.52. Draw (i) Full sectional front view (ii) R.H.S.V. (iii) L.H.S.V. (iv) Top plan

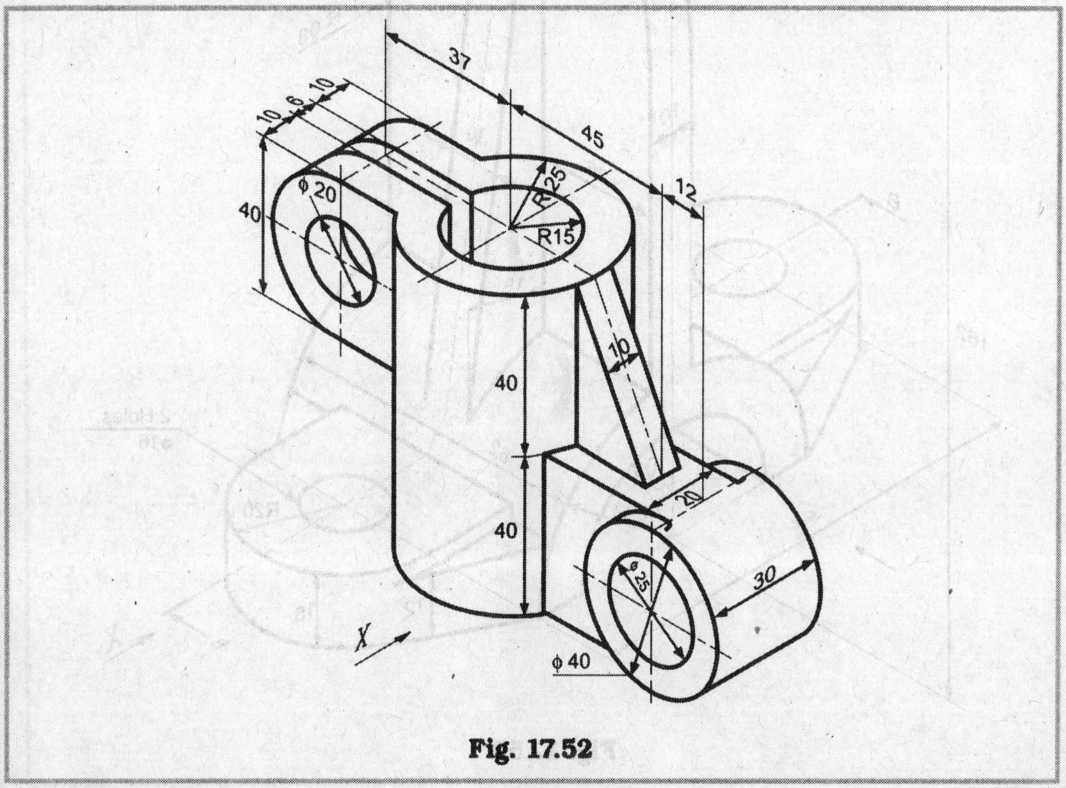

Fig. 17.52

18. Fig. 17.53. Draw (i) Elevation (ii) Sectional side view from left (Y) take section along B-B (iii) Top plan [Gujarat University June 1990]

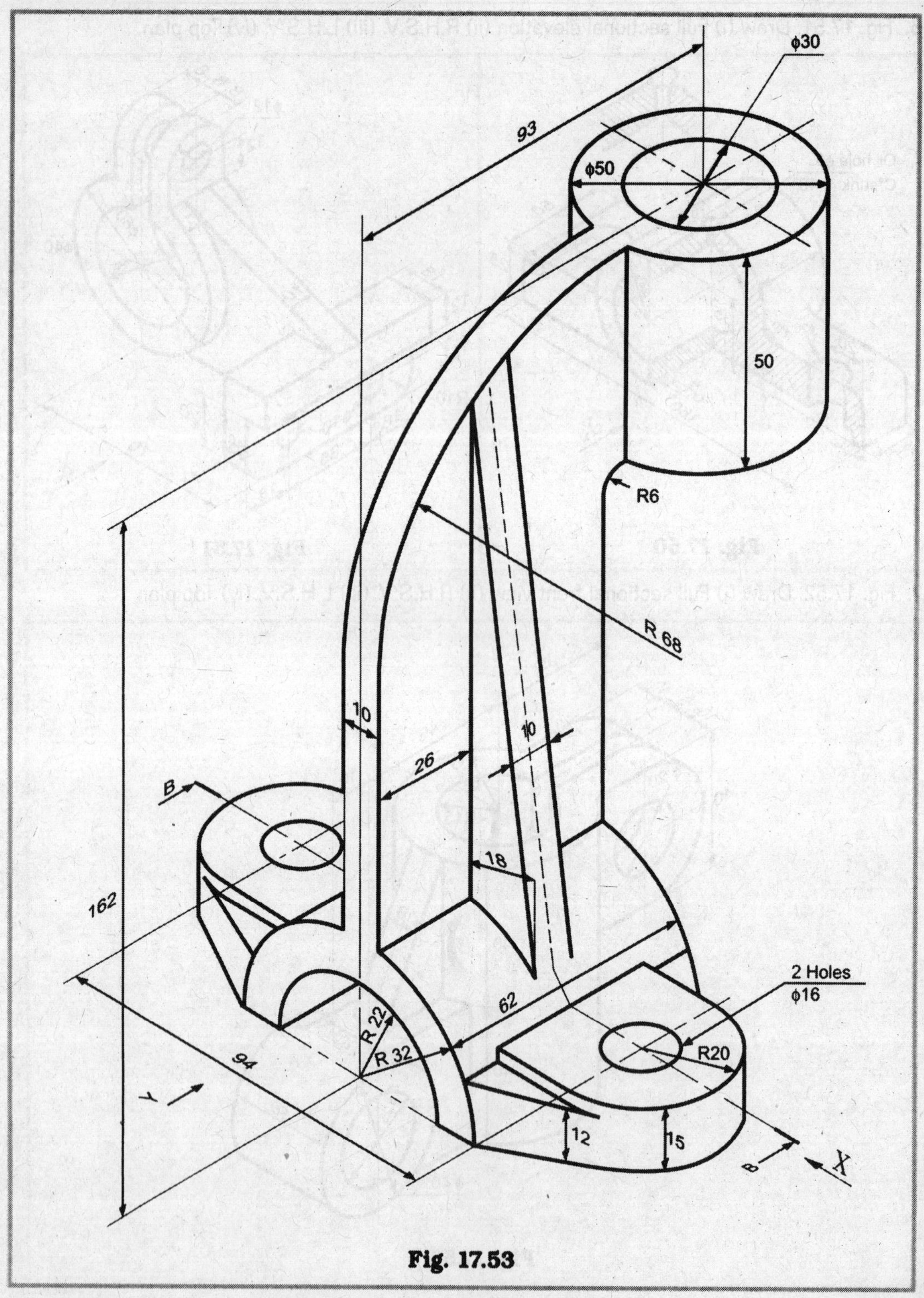

Fig. 17.53

19. Fig. 17.54. Draw (i) Front view (ii) Full sectional side view from right (iii) Top plan

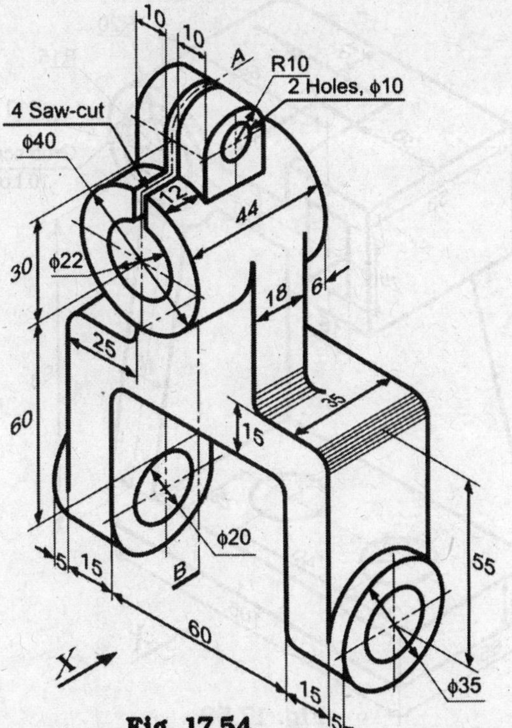

Fig. 17.54

20. Fig. 17.55. Draw (i) Front elevation (ii) Full sectional side view from left (iii) Top plan.

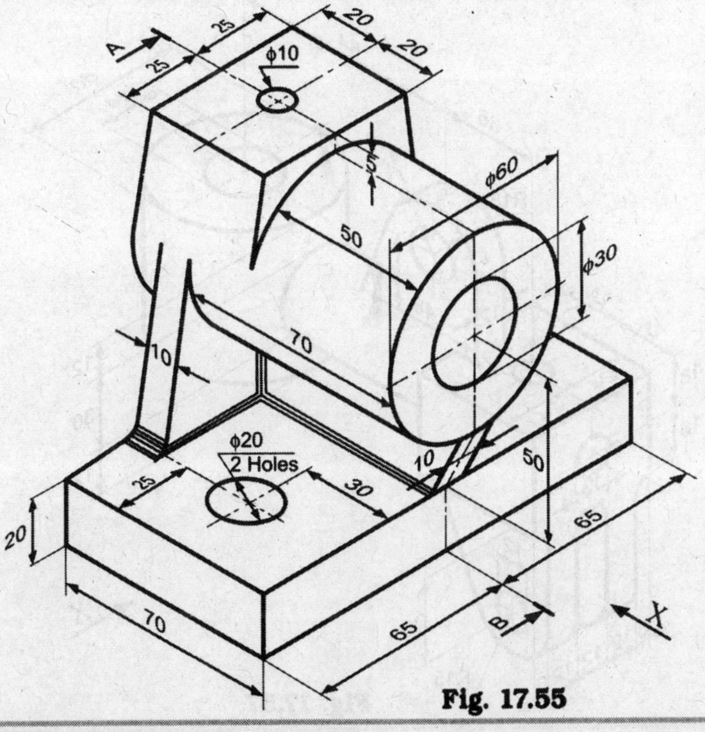

Fig. 17.55

21. Fig. 17.56. Draw (i) Front view (ii) Side view from left (iii) Top plan.

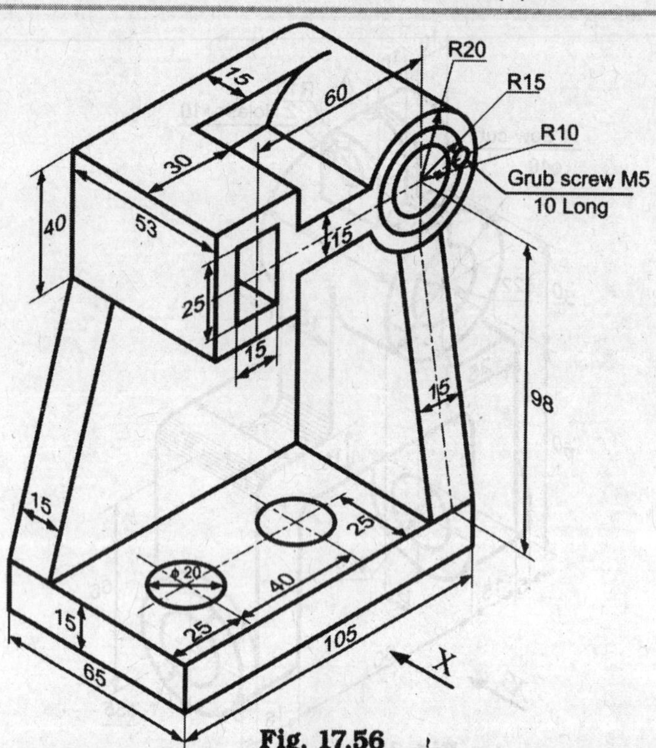

Fig. 17.56

22. Fig. 17.57. Draw (i) Full sectional F.V. (ii) L.H.S.V. (iii) Top plan (iv) Bottom plan

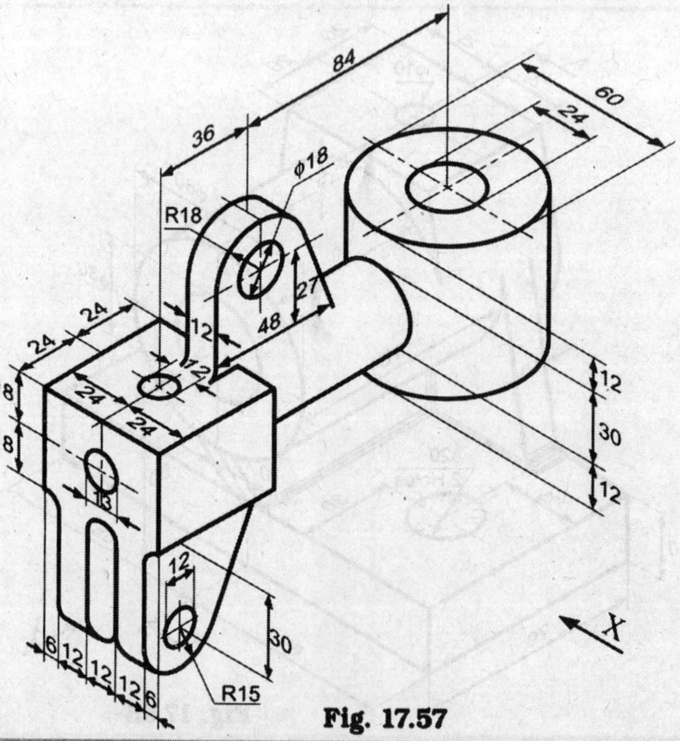

Fig. 17.57

23. Fig. 17.58. Draw (i) Front elevation (ii) Sectional side view from right (iii) Top plan (iv) L.H.S.V.

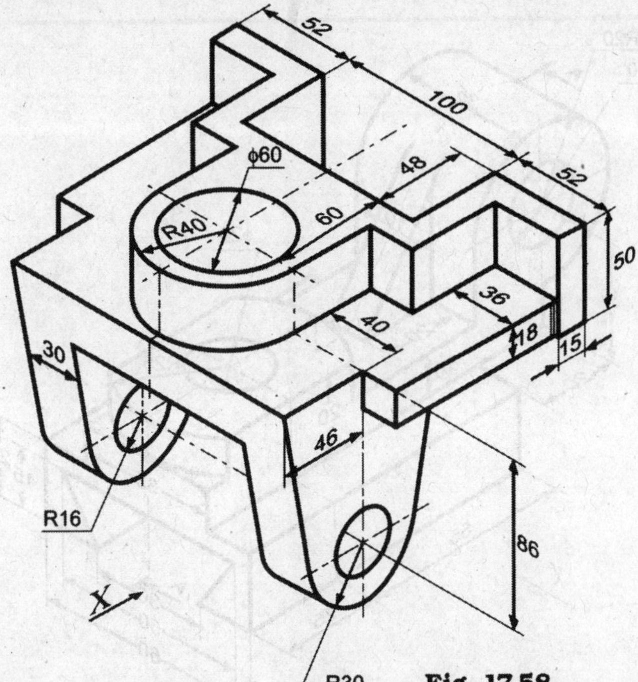

Fig. 17.58

24. Fig. 17.59. Draw (i) Sectional front view, take section along C-D (ii) Sectional side view from left, take section along A-B (iii) Top plan

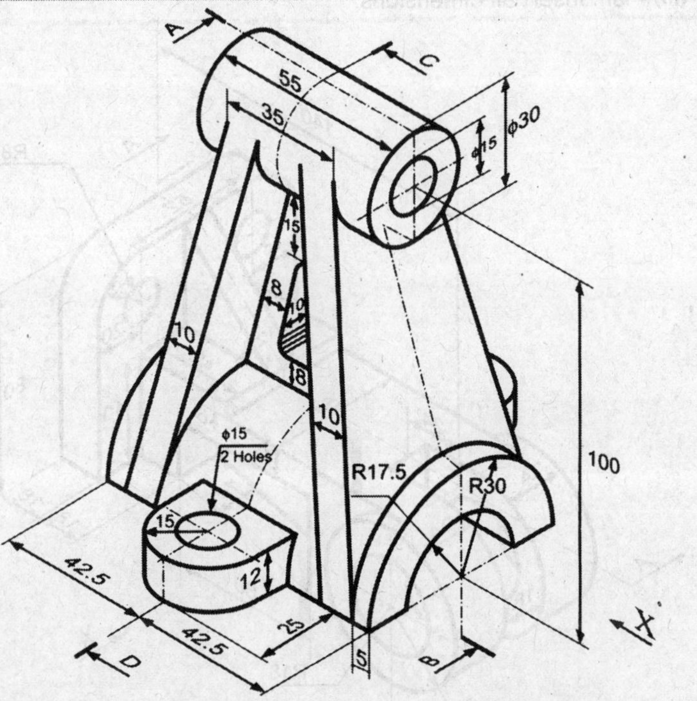

Fig. 17.59

25. Fig. 17.60. Draw (i) Full sectional front view along A-A. (ii) Right hand side view (iii) Top plan. Show all dimensions. [Gujarat University August 1993]

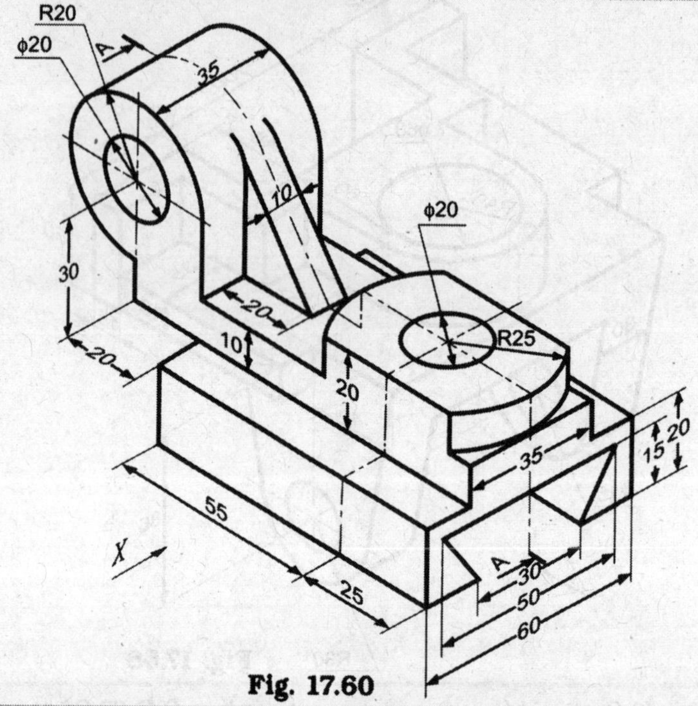

Fig. 17.60

26. Fig. 17.61. Draw the following views in 1st Angle System (i) Sectional Elevation-Section A-A. (ii) L.H.S.V. (iii) Plan. Insert all dimensions.

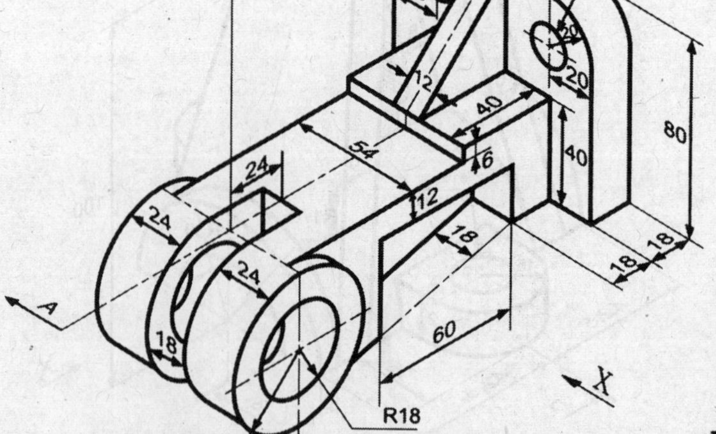

Fig. 17.61

27. Draw 6 Principal Orthographic Projections for the two pictorial views given.

(1) In First Angle Projection System and (2) In Third Angle Projection System.

Draw also symbol for the system of projection.

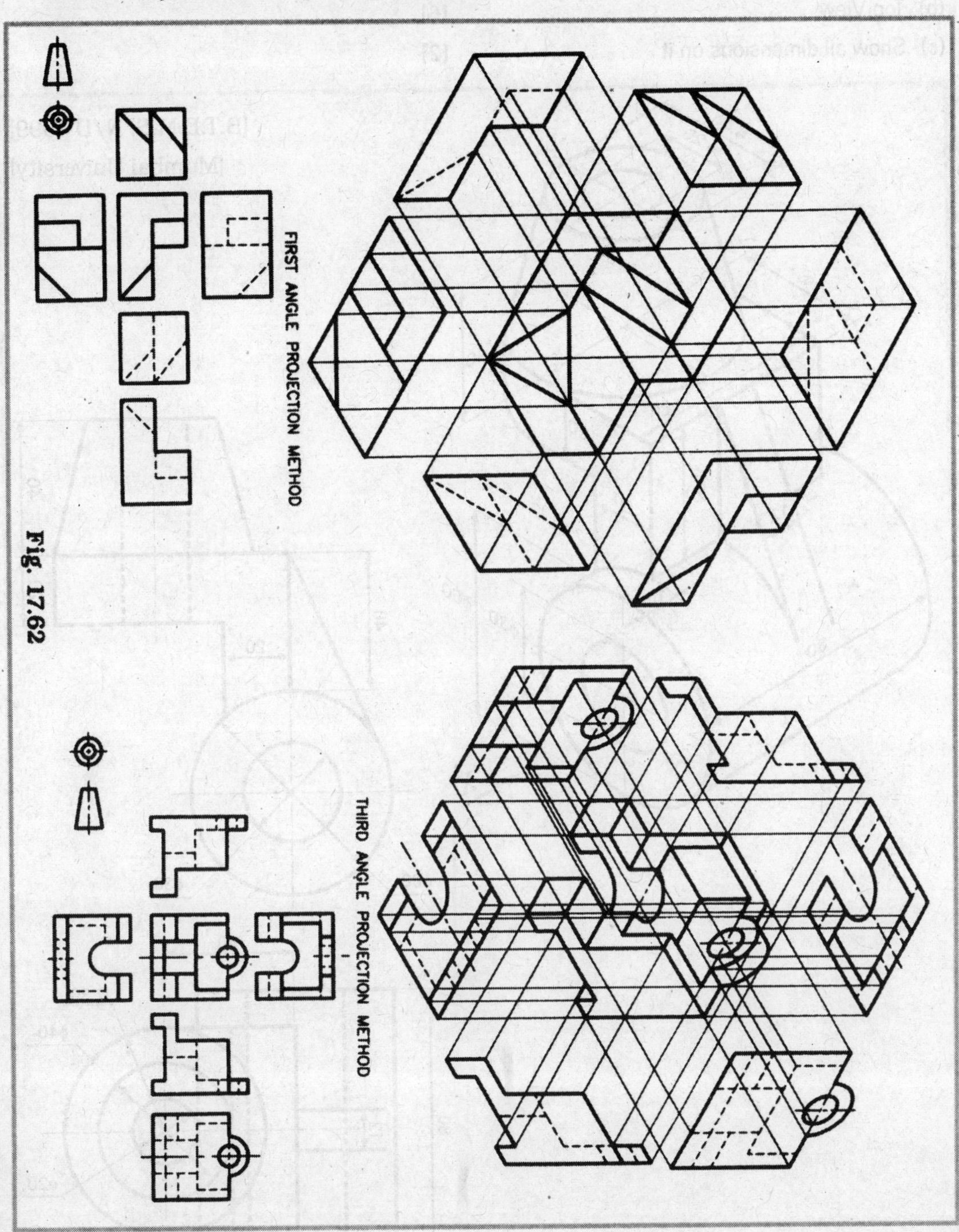

Fig. 17.62

28. Fig.17.63 shows a pictorial view of A.C.I. BLOCK. Draw to scale full size, the following views by using first angle method of projection.

(a) F. V. looking in the direction of arrow X [8]

(b) Top View [6]

(c) Show all dimensions on it [2]

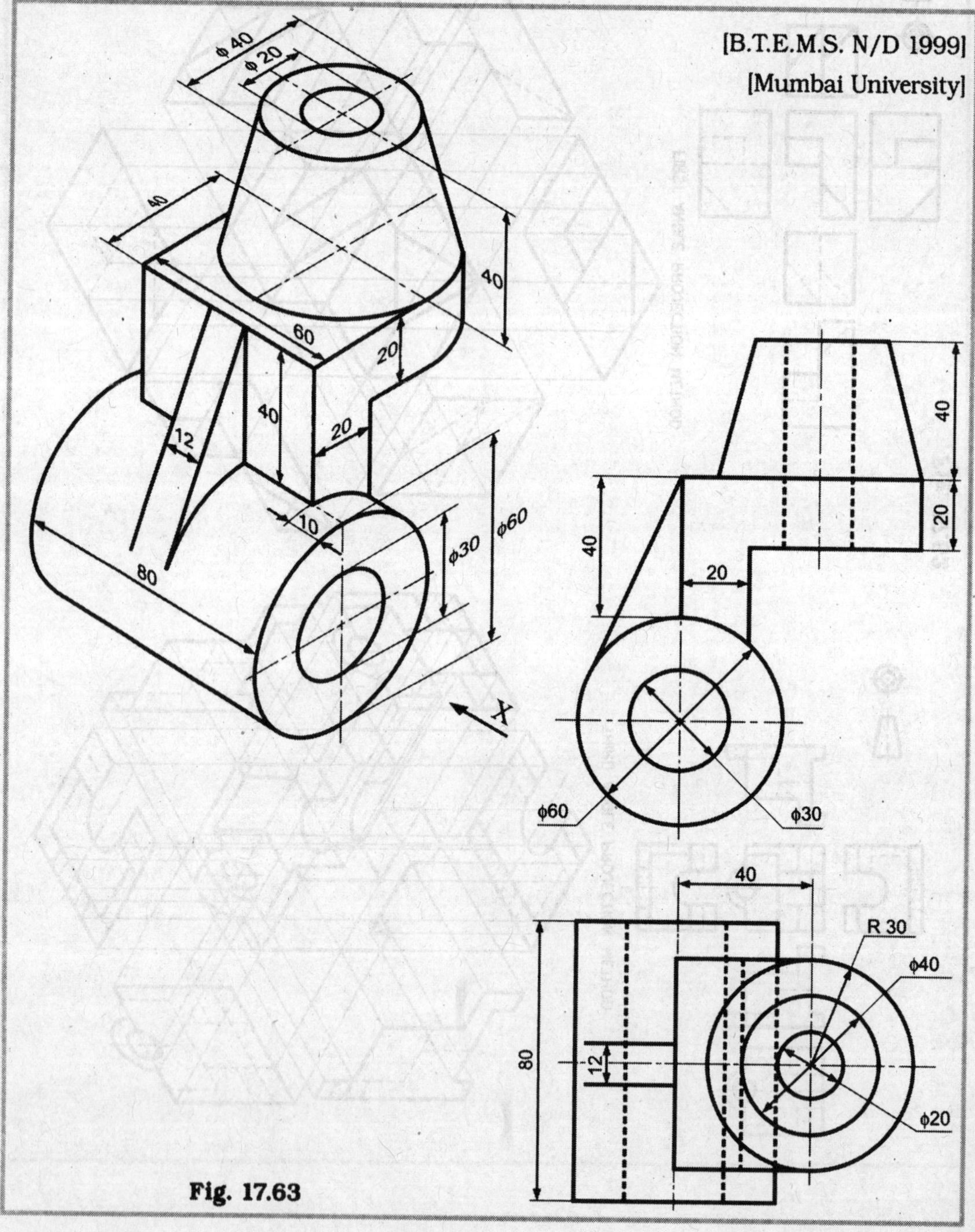

Fig. 17.63

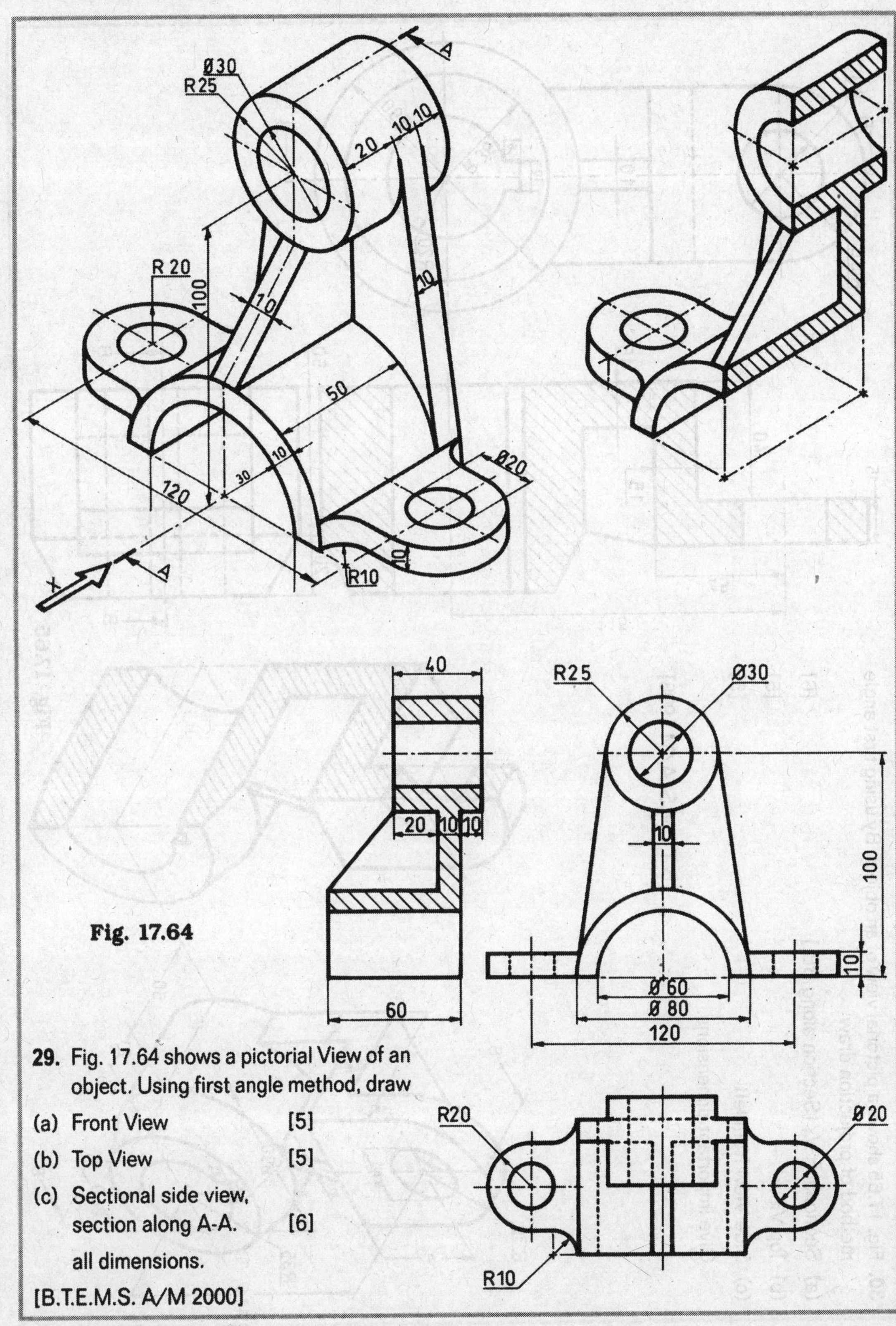

Fig. 17.64

29. Fig. 17.64 shows a pictorial View of an object. Using first angle method, draw

(a) Front View [5]

(b) Top View [5]

(c) Sectional side view, section along A-A. [6]

all dimensions.

[B.T.E.M.S. A/M 2000]

30. Fig. 17.65 shows a pictorial view of an object. By using first angle method of projection draw

(a) Sectional F. V., [Section along BB.] [6]

(b) Top View [5]

(c) Side View from left [5]

Give important dimensions.

[B.T.E.M.S. A/M 1998]

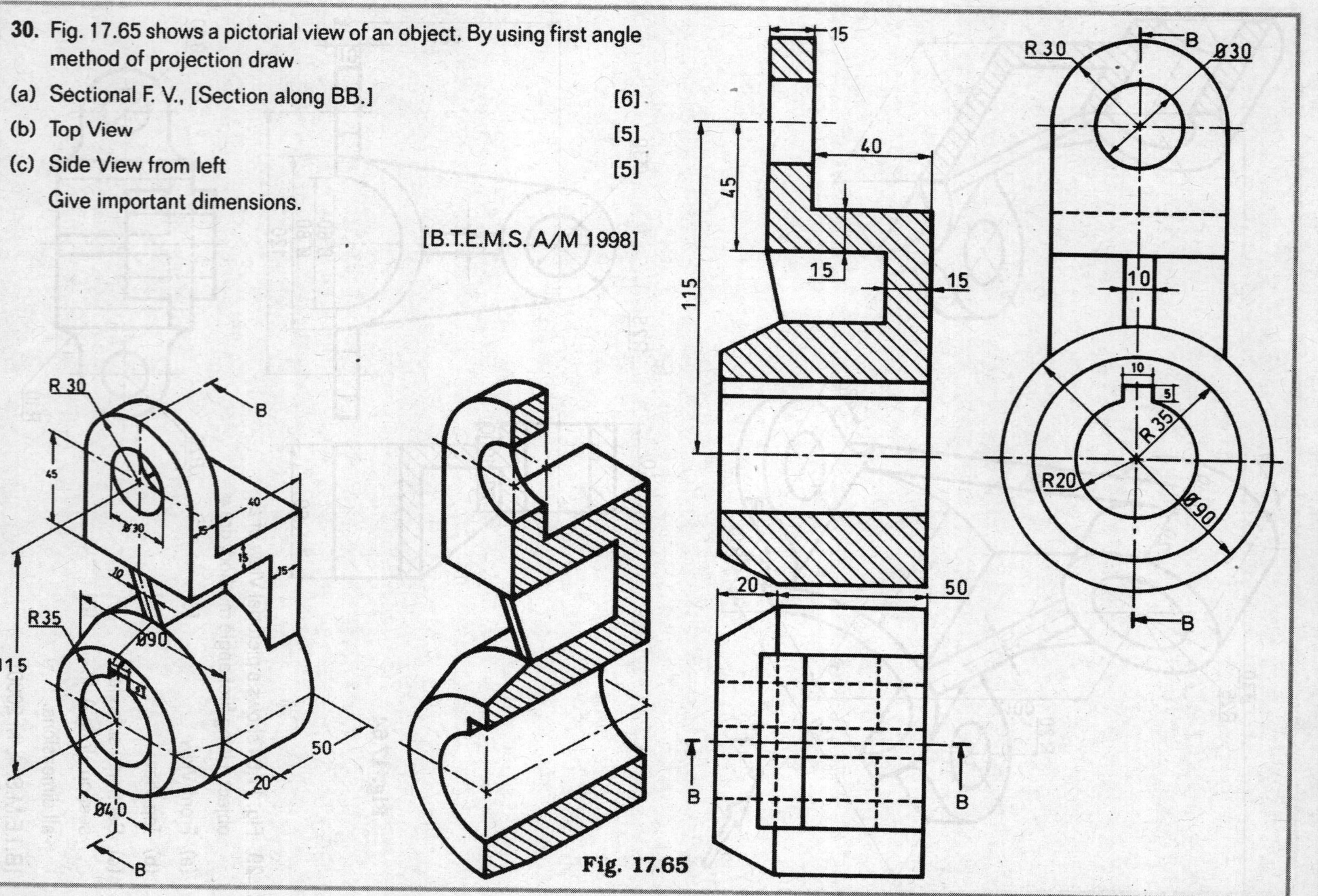

Fig. 17.65

31. Fig. 17.66 shows a pictorial view of A MACHINE PART. Draw to scale full size, the following views by using the first angle method of projection :

(a) Sectional F. V. in the direction of arrow X and section along A-A.

(b) Sectional L.H.S.V. taking section along B-B.

[Mumbai University, December, 1993]

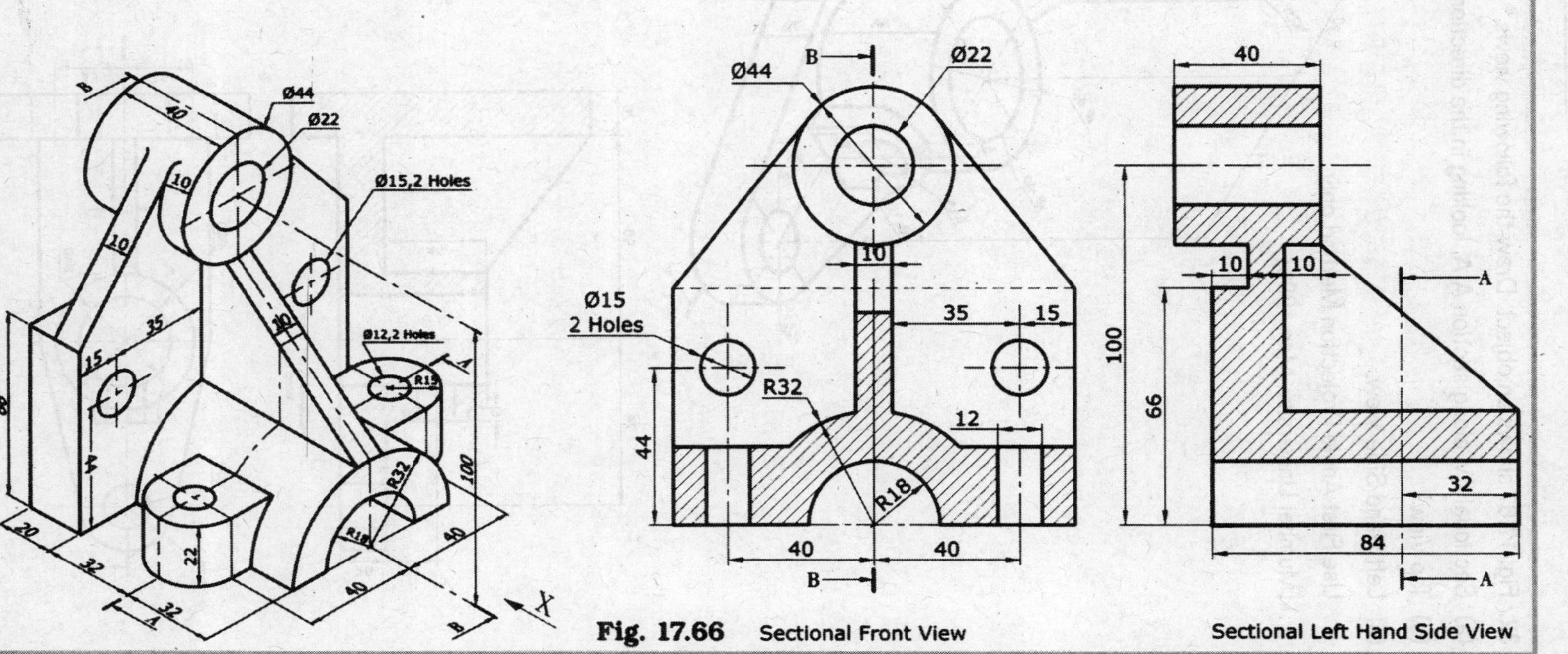

Fig. 17.66

32. Fig. 17.67 shows an object. Draw the following views

(a) Sectional F.V. along section A-A, looking in the direction of arrow X.

(b) Top View

(c) Left Hand Side View.

Use First Angle Projection Method only.

[Mumbai University, May 1996]

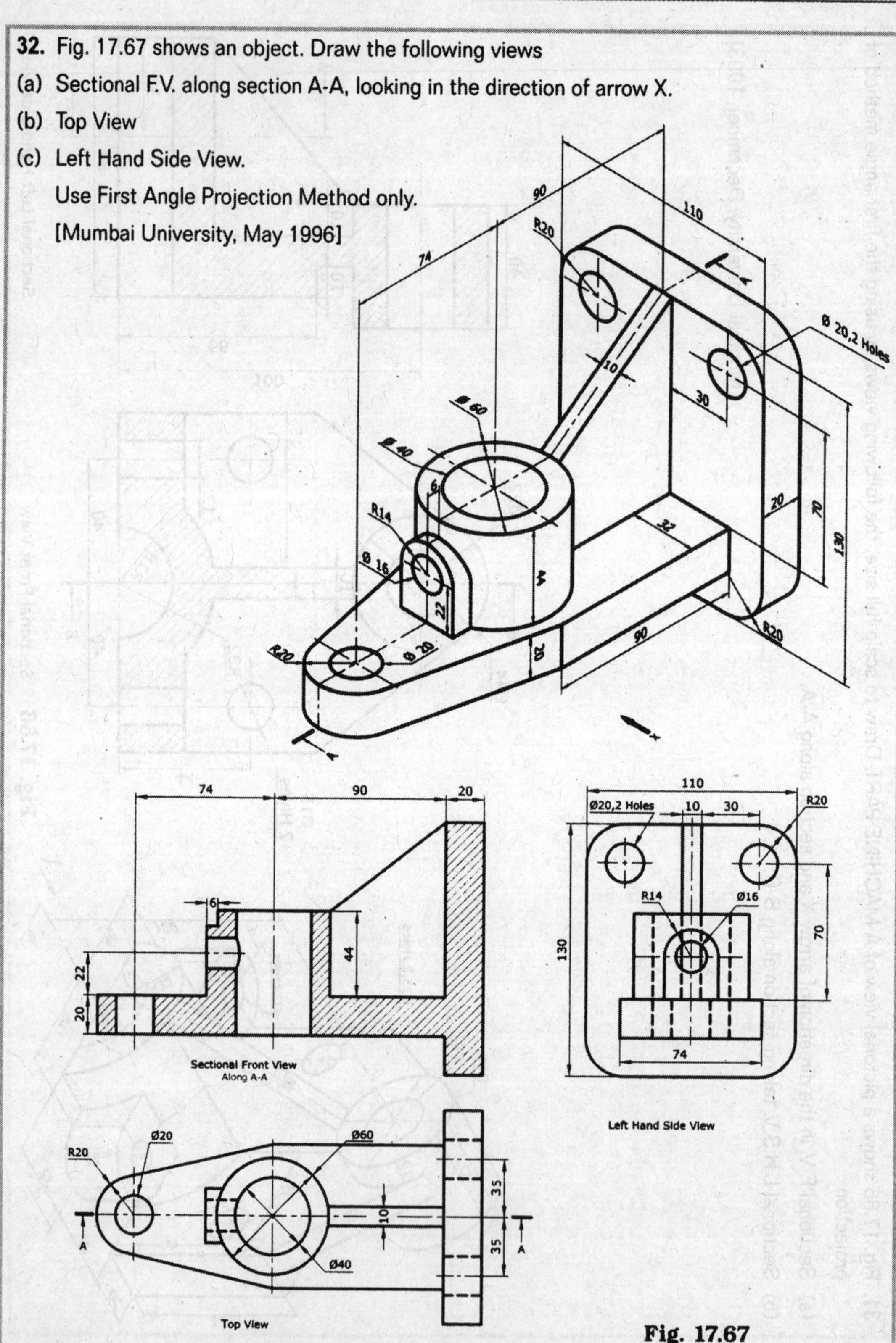

Fig. 17.67

33. Fig. 17.68 shows a pictorial view of A GUIDE-BRACKET. Draw to scale full size the following views.

(i) Sectional elevation on the section plane A-A in the direction of arrow-X

(ii) End view in the direction of arrow Y.

(iii) Plan.

[Mumbai University]

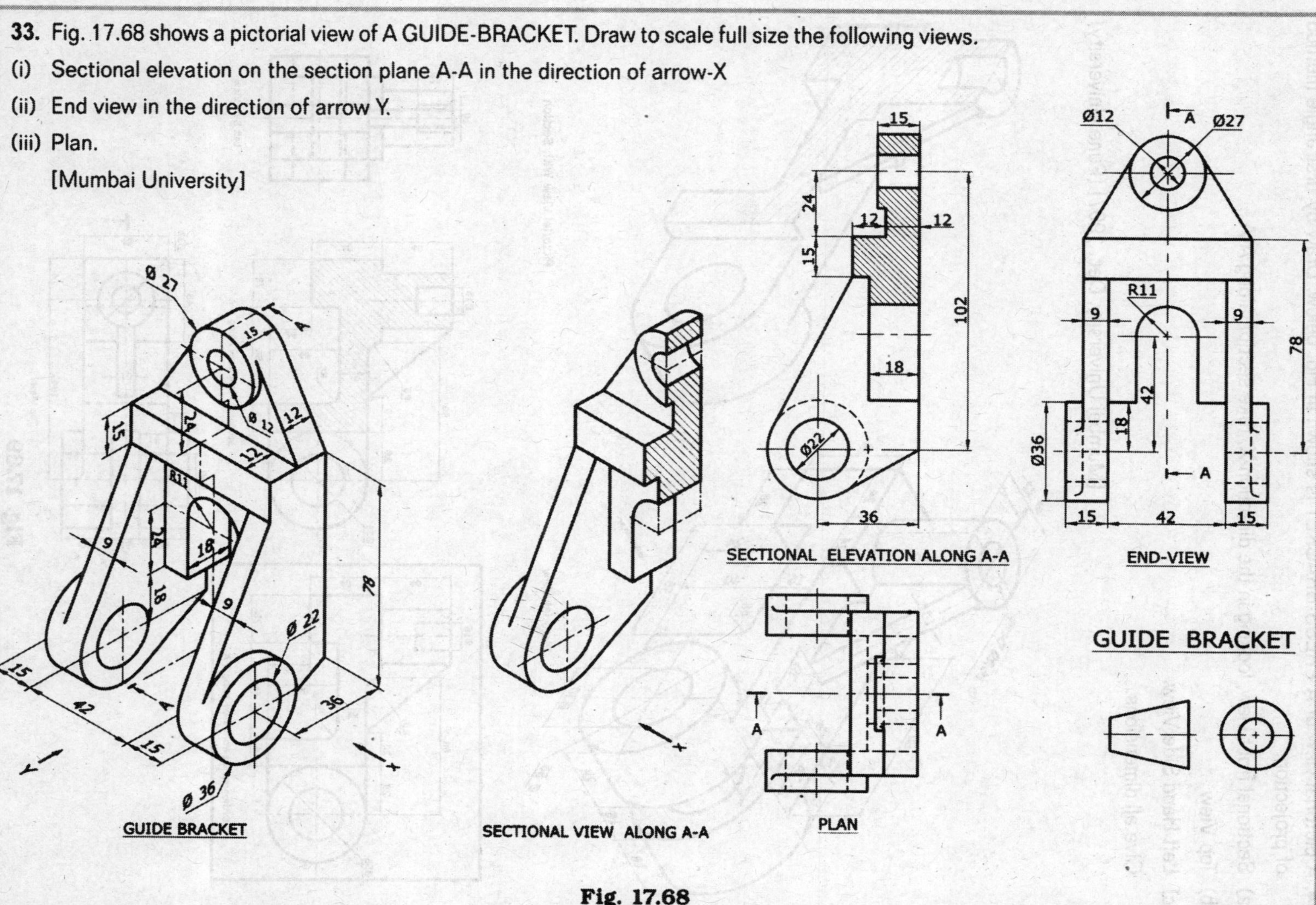

Fig. 17.68

34. A pictorial view of A PRESS-BRACKET is shown in Fig. below. Draw by first angle method of projection.

(a) Sectional front view, looking in the direction-X, take section along A-A.

(b) Top View

(c) Left Hand Side View.

Give all dimensions.

[Mumbai University, Dec. 1997] [Pune University]

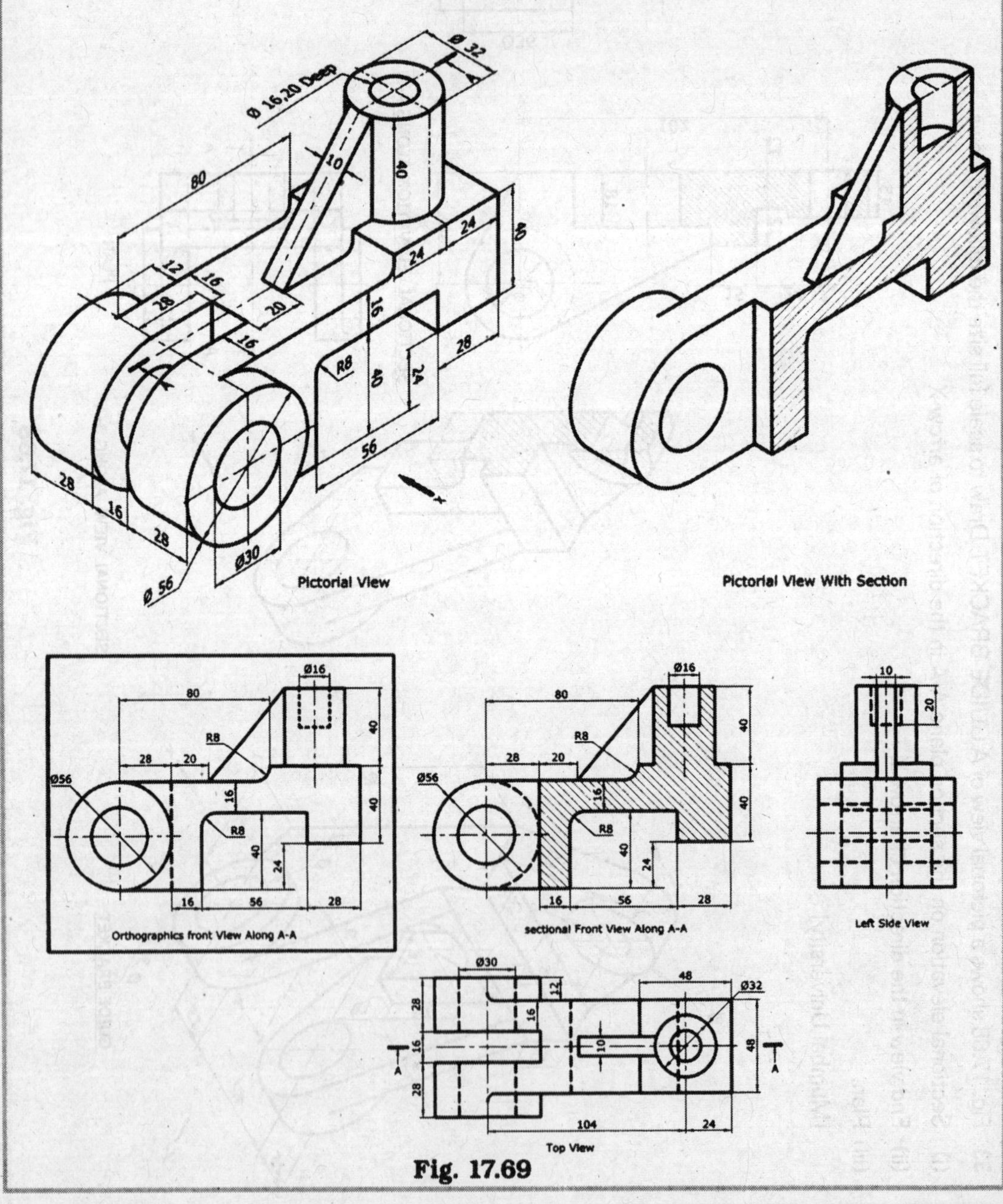

Fig. 17.69

35. Fig. shows a pictorial view of A SPINDLE BEARING. Draw to scale full size, the following views by using First Angle Method of Projection.

(a) Sectional F.V along Section A-A.

(b) Left Hand Side View.

(c) Top View.

Dimension the views.

[Mumbai University, Nov. 1987] [B.T.E.M.S.]

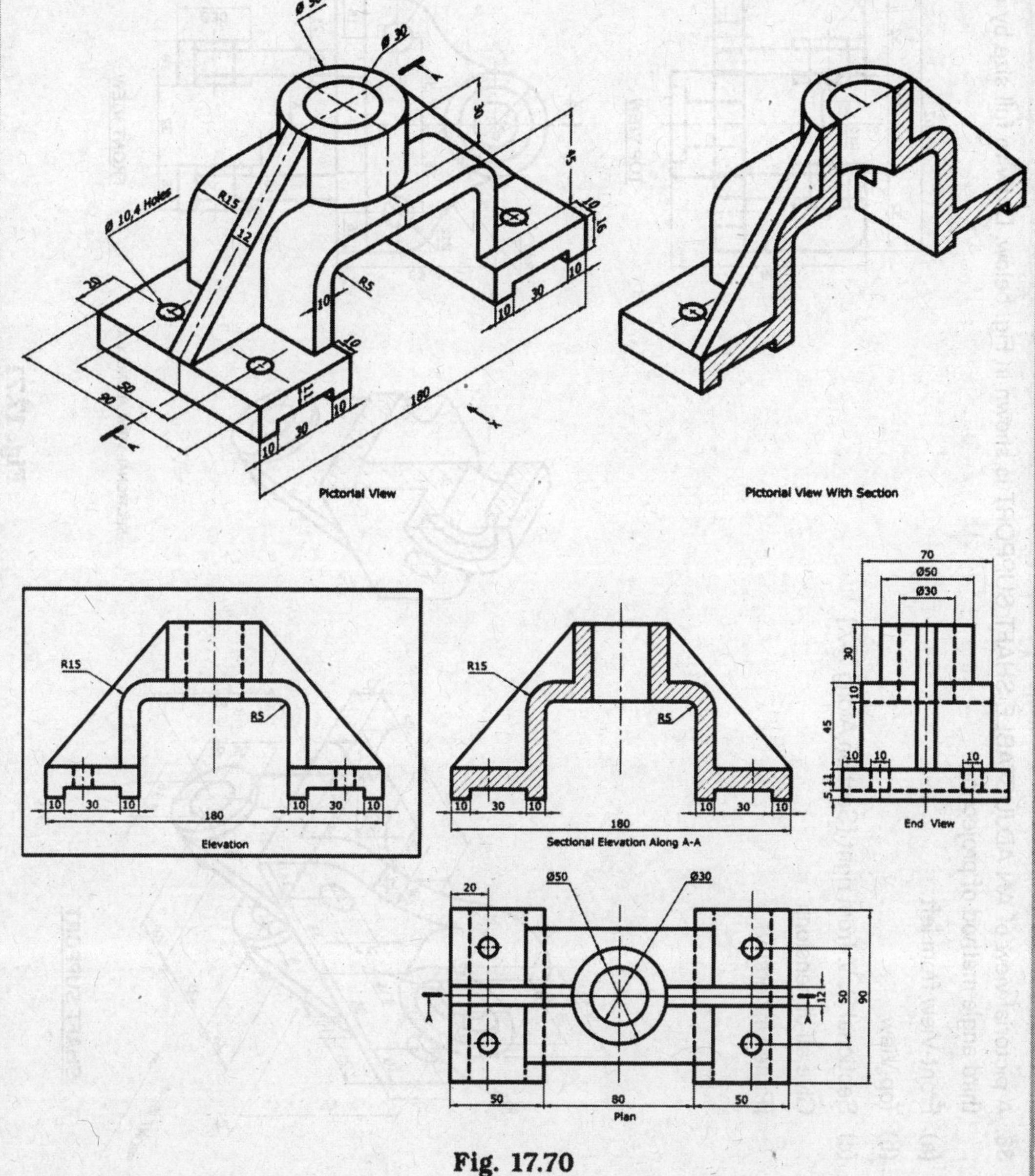

Fig. 17.70

36. A pictorial view of AN ADJUSTABLE SHAFT SUPPORT is shown in Fig. below. Draw to full size by using third angle method of projection.

(a) Front-View from left.

(b) Top-View.

(c) Sectional S. V. from right [Section Along X-X]

Give all dimensions.

[Pune University]

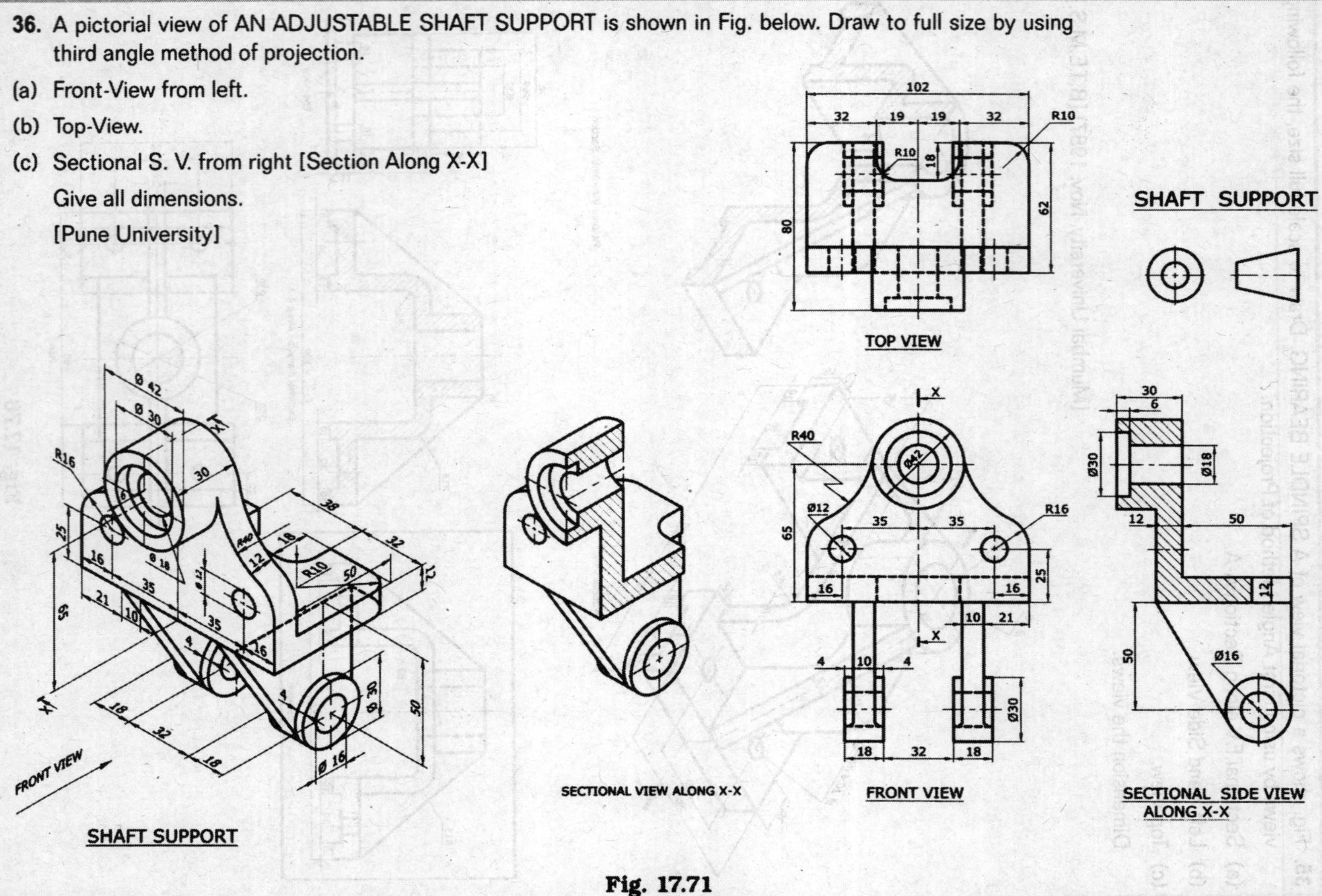

Fig. 17.71

37. Fig.17.72 shows a pictorial view of A BRACKET. Using third angle method of projection, draw the following views.

(a) Sectional front view, looking along direction X.

(b) Top View.

(c) Left Hand Side View.

Give all necessary dimensions.

[Pune University]

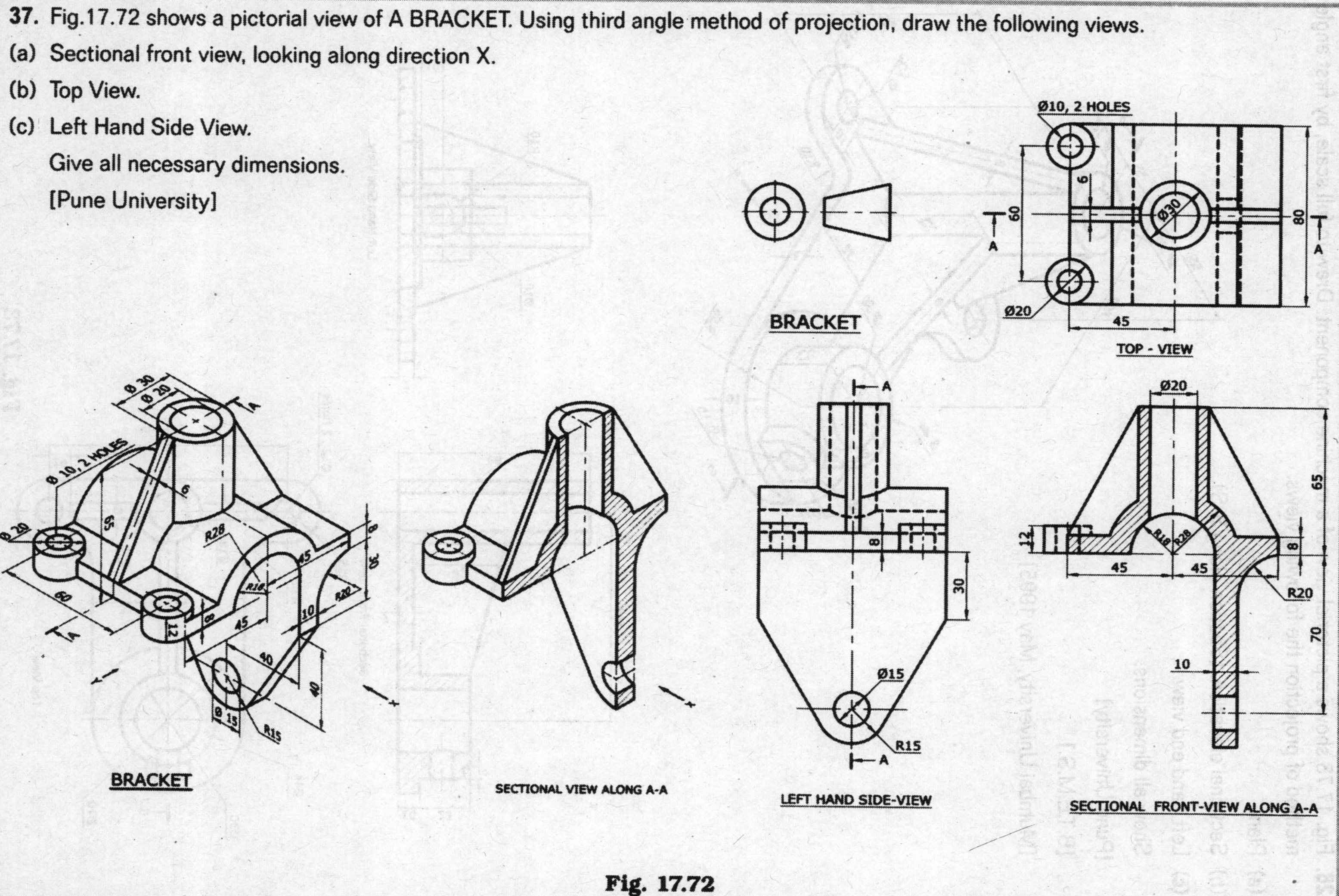

Fig. 17.72

28. Fig. 17.73 shows a pictorial view of a machine component. Draw to full scale, by first angle method of projection the following views.

(a) Plan

(b) Sectional elevation. (Section along S-S)

(c) Left hand end view.

Show all dimensions.

[Pune University]

[B.T.E.M.S.]

[Mumbai University, May 1995]

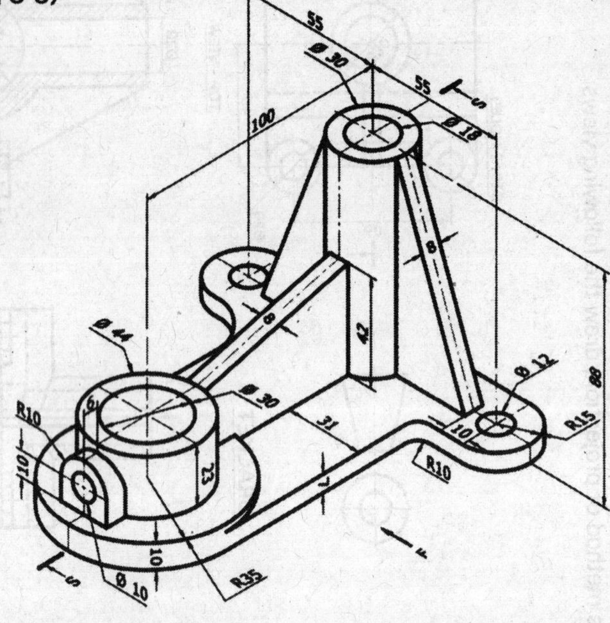

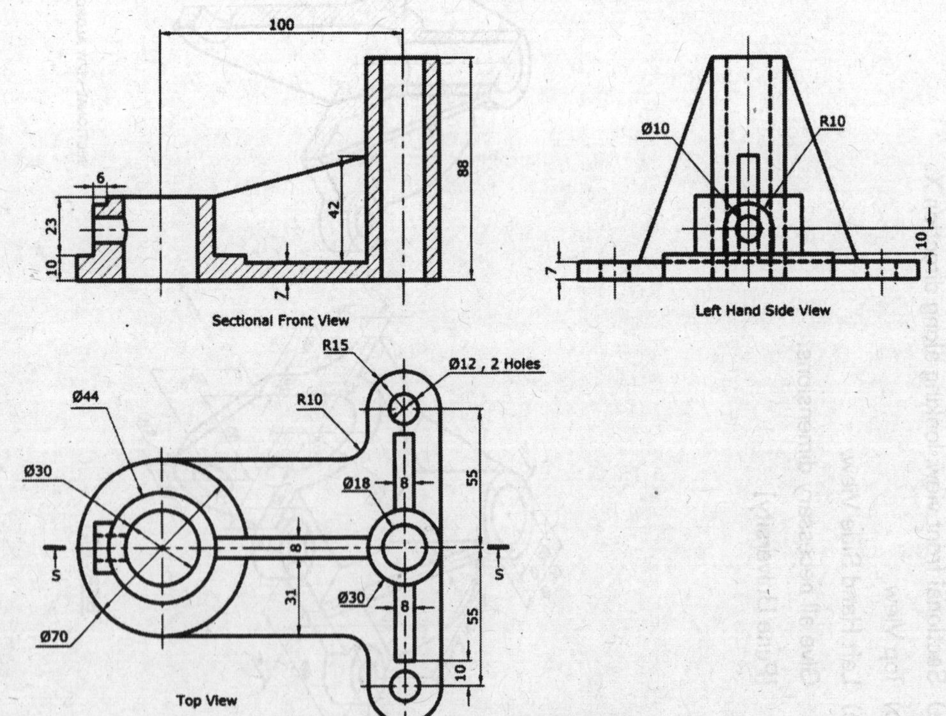

Fig. 17.73

39. Fig. 17.74 shows a pictorial view of A C. I. BRACKET. Draw to full size the following views, by using first angle method of projection.

(a) An elevation looking in the direction of arrow X.

(b) Sectional end view along A-A, looking in the direction of arrow Y.

(c) Plan

Show important dimensions. [Note : Solution not given] [B.T.E.M.S.]

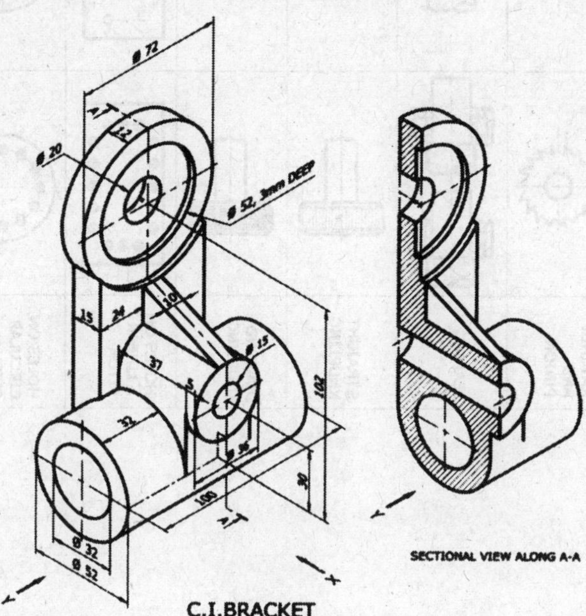

Fig. 17.74

40. A pictorial view of an object is shown in Fig. 17.75 using first angle method of projection, draw the following views.

(a) Sectional elevation in the direction of arrow X.

(b) Plan

(c) End view in the direction of arrow Y.

Give all dimensions. [Note : Solution not given] [Pune University]

Fig. 17.75 SECTIONAL VIEW ALONG A-A

Some common features, which have complicated and / or difficult to draw geometry, are presented in drawing by conventional representation instead of their actual drawing. They are given in 2-Tables below.

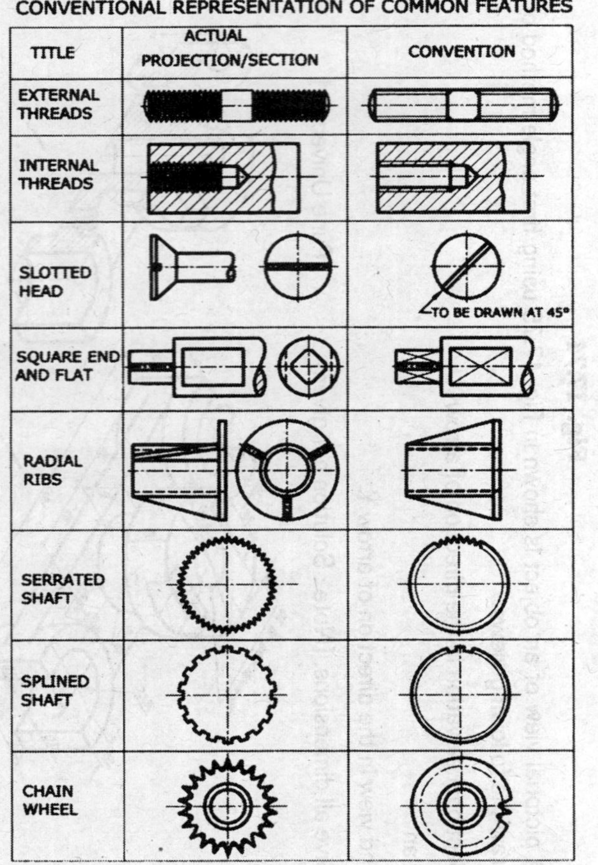

CONVENTIONAL REPRESENTATION OF COMMON FEATURES

TITLE	ACTUAL PROJECTION/SECTION	CONVENTION
EXTERNAL THREADS		
INTERNAL THREADS		
SLOTTED HEAD		TO BE DRAWN AT 45°
SQUARE END AND FLAT		
RADIAL RIBS		
SERRATED SHAFT		
SPLINED SHAFT		
CHAIN WHEEL		

CONVENTIONAL REPRESENTATION OF COMMON FEATURES

TITLE	ACTUAL PROJECTION/SECTION	CONVENTION
RATCHET AND PINION		
BEARING		
STRAIGHT KNURLING		
DIAMOND KNURLING		
HOLES ON A LINEAR PITCH		
HOLES ON CIRCULAR PITCH		
REPEATED PARTS		

CHAPTER - 18

Orthographic Reading

1. Introduction :

Engineering drawing is a language of all persons connected with technical activities. Language has writing and reading practices. To draw orthographic projections from pictorial view is a writing practice. Orthographic reading means to understand or to interpret or to visualise the shape and size of the object from the orthographic projections. In fact it is a reverse process to that of making orthographic drawing.

Reading orthographic projections is a bit difficult task to that of preparing orthographic projections from pictorial views. One can obtain expertise in reading practice by having thorough knowledge of the rules and conventional practices followed in preparing orthographic projections and by having sufficient amount of practice in preparing orthographic drawings.

Reading here does not mean speaking by the use of tongue but it is a mental work of correlating different lines, different shapes and different projections and thereby evolving in mind the three dimensional picture of the object with correct size. Most of the engineering objects are combination of simple objects like cylinder, cone, prism, pyramid, sphere, flat plates and their frustum or truncated frustum joined together in different positions and locations. One should not try to visualise the whole object at a time or at a glance but should try to visualise piece by piece and should keep in memory the shape, size and location of each piece separately, To help reading, the reader can prepare three dimensional pictorial sketch of all pieces, After visualising all pieces (simple objects) combine them together in respective positions and locations to get a picture of the whole complicated object in mind. Reader should do this systematically and patiently. Correlating respective lines of two views, three views or many views is of utmost importance in reading.

2. Important Findings

The following findings will be useful in reading orthographic projections.

(1) If one view, of any edge of an object, is seen as a point view and two other views of the same edge are seen parallel to ground lines (i. e. vertical or horizontal) then that edge must be perpendicular to the principal plane on which it shows point view and parallel to two other principal planes.

(2) If one view of any edge of an object is seen as inclined line and two other views of the same edge seen parallel to ground lines (i.e. horizontal or vertical) then that edge must be parallel to the principal plane on which it shows inclined line.

(3) If all views of any edge of an object are inclined lines then that edge must be inclined to all principal planes. In no view it will show the true length.

(4) If any flat surface of the object is seen as straight line views in two projections then that plane must give the true shape in the third projection. It will be parallel to the principal plane on which it shows projection in the true shape.

(5) If any flat surface of the object is seen as an inclined straight line in one view then that flat surface must be perpendicular to the principal plane on which it gives inclined straight line projection. Other two views will show the area less than the true shape.

(6) If any flat surface of the object is not giving in any projection straight line view means that plane will be inclined to all three principal planes and hence in all projections the area will be seen, which will not be the true shape.

(7) Area enclosed by one boundary in projection represents one surface either flat or curved.

(8) Intersection of two surfaces forms an edge of an object.

(9) Corner of a solid is formed by minimum three edges. Three surfaces of solid meet at a corner of a solid.

(10) Surface limit of the curved boundary is seen as a line in a drawing.

(11) A point in projection may represent a corner or an edge of the object.

(12) A line in projection may represent a line or surface of the object.

Moreover, students are requested to practise by imagining different possible objects for a given plan, for a given elevation and for given two views. This is necessary to train the mind in the direction of orthographic reading. Method of projecting other views from the given views must be known to the reader.

Now, we shall practise reading of orthographic projections by some illustrative problems :

3. Illustrative Problems

Problem 1 : In Fig. 18.1(a) front view is given. Imagine different possible objects for the same F.V. and draw corresponding top views.

In Fig. 18.1(b) five possible objects are shown whose front view will be the same as the given front view but whose top views will be different. Students are requested to imagine many more objects for the same front view and draw their corresponding top views. Three dimensional drawings of the objects shown in Fig. 18.1(b) are for the purpose of easy understanding. You are not required to draw these three dimensional drawings. You have to draw simply elevation and their corresponding top views as shown in Fig. 18.1(c). Pictorial drawings of the objects are kept in memory. Drawings are given in the 3rd angle projection system.

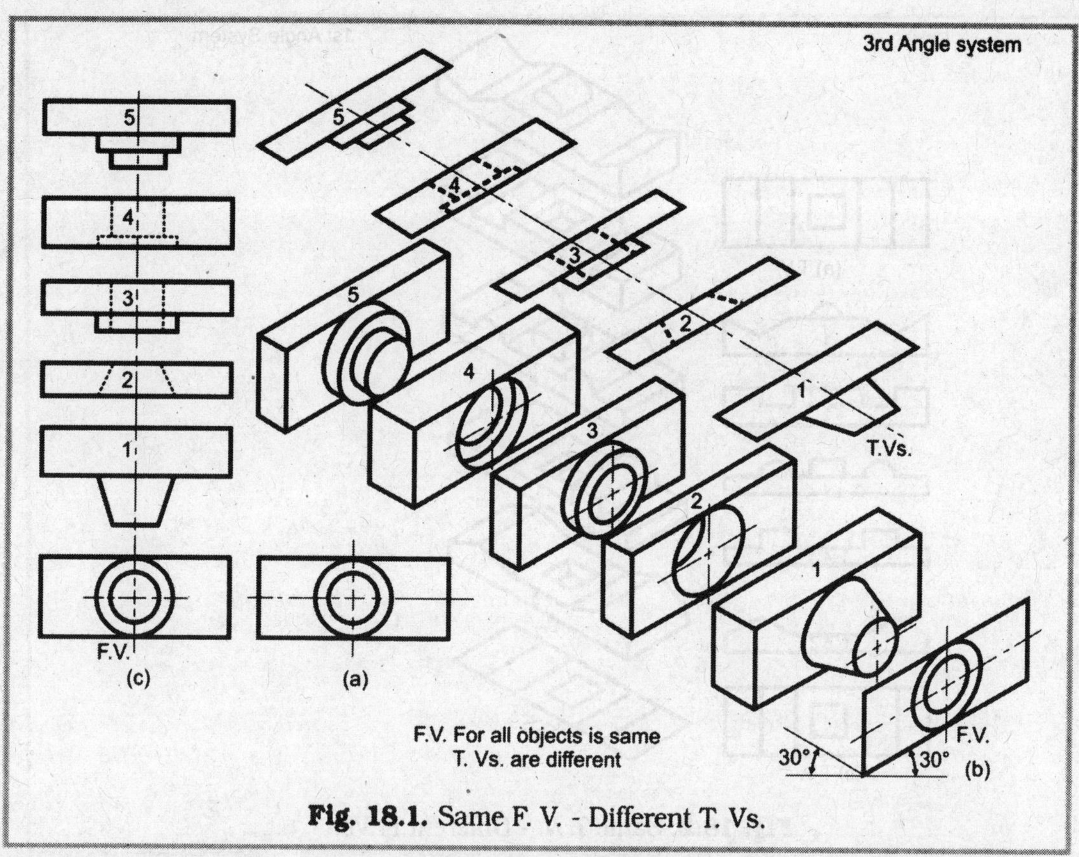

Fig. 18.1. Same F. V. - Different T. Vs.

Problem 2 : In Fig. 18.2(a) top view is given. Imagine different possible objects for the same T.V. and draw corresponding front views.

In Fig. 18.2(b) five possible objects are shown whose top view will be the same as the given T.V. but whose front views will be different. Imagine many more objects and draw corresponding front views. Solution is given in Fig. 18.2(c) in the 1st angle projection system.

Problem 3 : In Fig. 18.3(a) front view and top view are given. Imagine different possible objects for the same F.V. and T.V. and draw corresponding different side views.

In Fig. 18.3(b) six possible objects are shown whose F.V. and T.V. will be the same as the given but whose side views will be different. Imagine many more. Side views are shown in Fig. 18.3(c). Solution is in 3rd angle projection system.

Students are requested to solve many more such exercises to train the mind in mastering orthographic reading.

Now, we shall study the projection method to get other missing views when two or three views are given by taking same exercise for both the systems of projections i.e. 1st angle and 3rd angle systems.

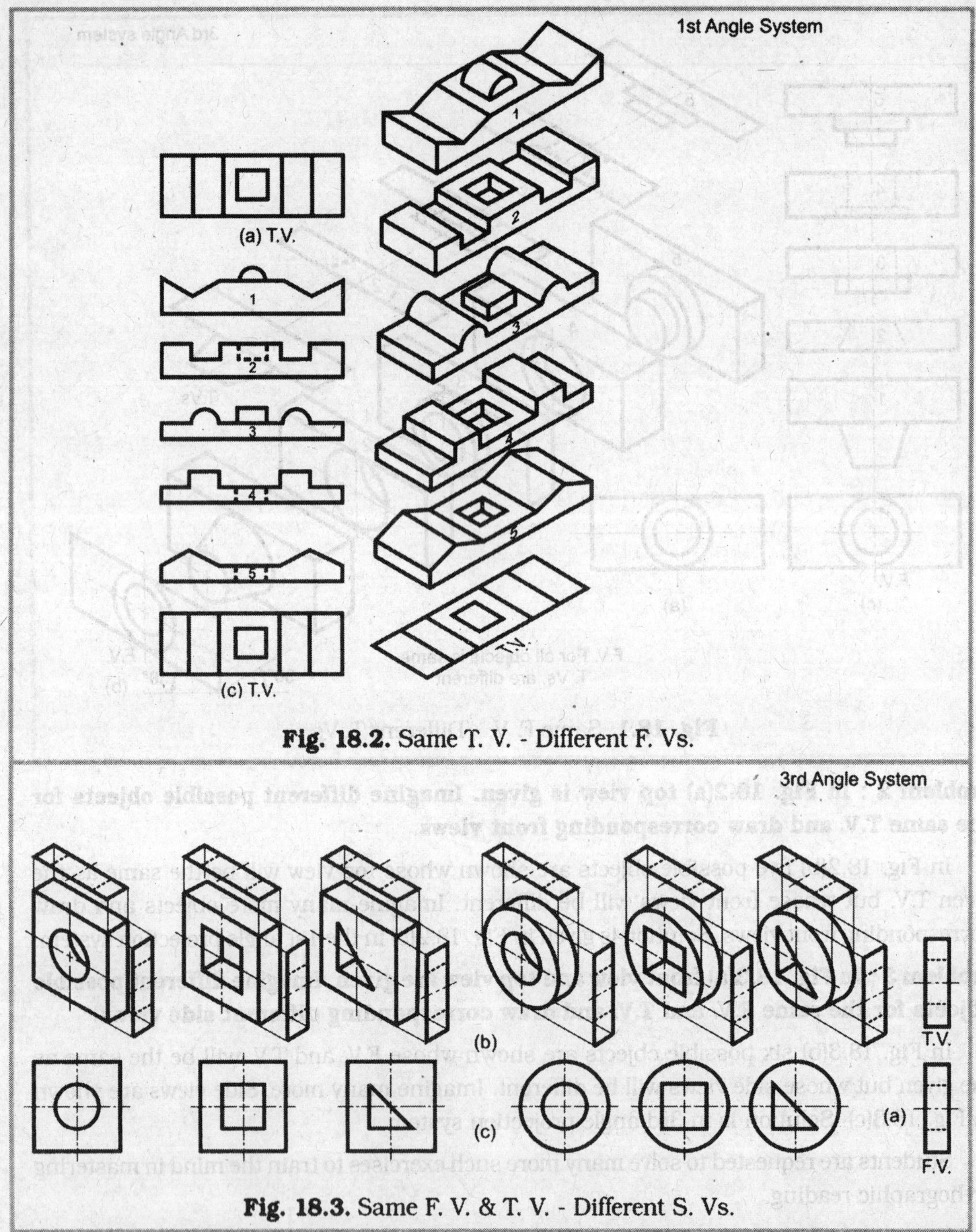

Fig. 18.2. Same T. V. - Different F. Vs.

Fig. 18.3. Same F. V. & T. V. - Different S. Vs.

Problem 4 : In Fig. 18.4 front view and left hand side view of an object are given. Draw four other missing views i.e. top view, bottom view, R.H. side view and rear view. Use 1st angle projection system.

Before proceeding for the solution of the problem, let us study some of the important aspects of first angle projection system.

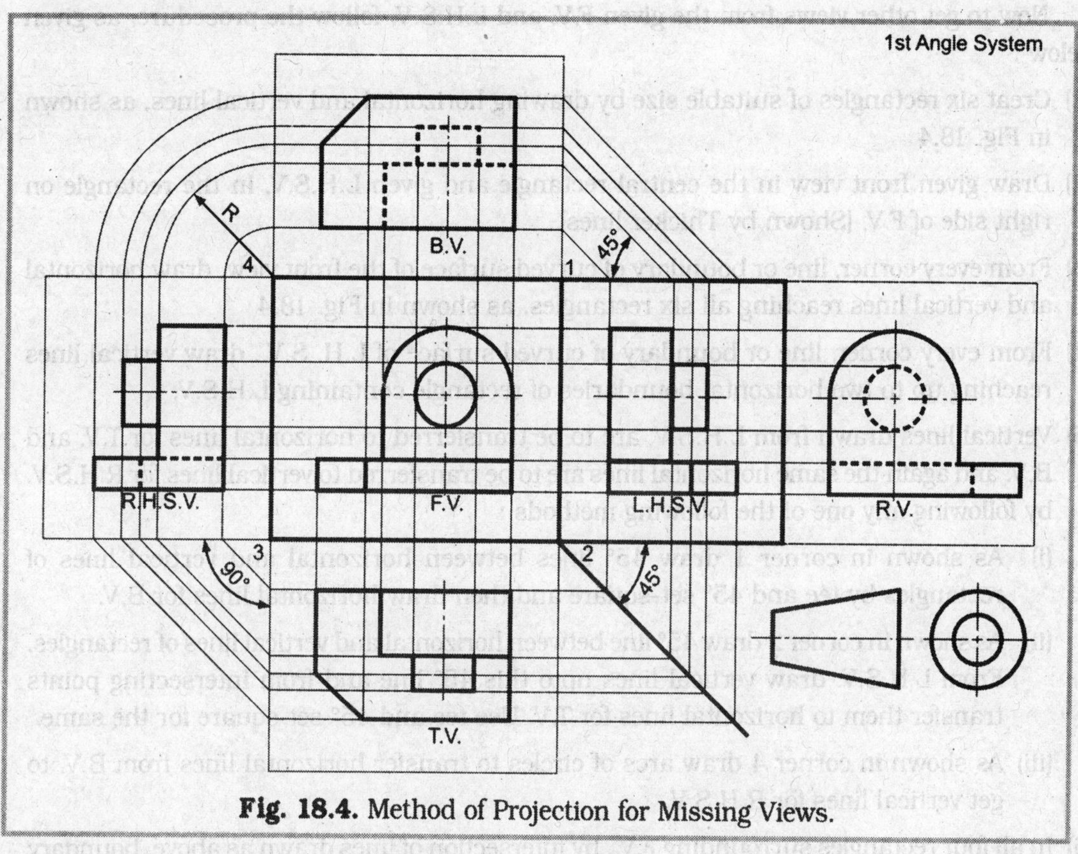

Fig. 18.4. Method of Projection for Missing Views.

(1) King view, front view is always situated at the centre.

(2) T.V. is situated below F.V.

(3) B.V. is situated above F.V.

(4) R.H.S.V. is situated on left side of F.V.

(5) L.H.S.V. is situated on right side of F.V.

(6) Rear view is situated on extreme right beside L.H.S.V.

(7) Portion of the object which is located on the front side of the object is seen away from front view, i.e. not facing front view, in all four views surrounding F.V. e.g. in Fig. 18.4 cylindrical portion, which is on front side of the object, is seen away from front view.

(8) Portion of the object, which is located on the rear side of the object, is seen nearer to front view, i.e. facing F.V. in all four views, surrounding F.V. e.g. in Fig. 18.4, plate (rectangular at the bottom side and semicylindrical on the upper side) situated on the rear side of the object is seen nearer (facing) to the front view in all four views surrounding the F.V.

In fact (7) and (8) aspects are the tests to be carried out to find out whether the given projections are in the 1st angle system or in the 3rd angle system. As the front view of cylindrical portion is full/continuous circle it can be concluded that it is on the front side of the object. In all four views surrounding F.V. it is seen not facing F.V. or away from F.V.

Now to get other views from the given F.V. and L.H.S.V. follow the procedure, as given below :

(1) Creat six rectangles of suitable size by drawing horizontal and vertical lines, as shown in Fig. 18.4

(2) Draw given front view in the central rectangle and given L.H.S.V. in the rectangle on right side of F.V. [Shown by Thicker lines]

(3) From every corner, line or boundary of curved surface of the front view, draw horizontal and vertical lines reaching all six rectangles, as shown in Fig. 18.4

(4) From every corner, line or boundary of curved surface of L.H. S.V., draw vertical lines reaching up to two horizontal boundaries of rectangle containing L.H.S.V.

(5) Vertical lines drawn from L.H.S.V. are to be transferred to horizontal lines for T.V. and B.V. and again the same horizontal lines are to be transferred to vertical lines for R.H.S.V. by following any one of the following methods :

   (i) As shown in corner 1 draw 45° lines between horizontal and vertical lines of rectangles by tee and 45° set-square and then draw horizontal lines for B.V.

   (ii) As shown in corner 2 draw 45° line between horizontal and vertical lines of rectangles. From L.H.S.V. draw vertical lines upto this 45° line and from intersecting points transfer them to horizontal lines for T.V. Use tee and 45° set-square for the same.

   (iii) As shown in corner 4 draw arcs of circles to transfer horizontal lines from B.V. to get vertical lines for R.H.S.V.

(6) In all four rectangles surrounding F.V., by intersection of lines drawn as above, boundary of projection is created. Imagine shapes of various parts of the object and accordingly make necessary lines thicker and darker to evolve projections.

(7) Rear view is drawn by taking dimensions on horizontal lines from F.V. In rear view, portion of the object of left side will be seen on right side and portion of the object of right side will be seen on left side.

(8) For drawing full and dotted lines one will have to imagine the object, Prepare freehand pictorial view of the object if you can.

Normally front portion of the object is seen full in front view and rear portion as dotted in front view. Rear portion is seen full in R.V. and front portion as dotted in R.V. Left side portion is seen full in L.H.S.V. and right side portion is seen dotted in L.H.S.V. Similarly, right side portion is seen full in R.H.S.V. and left side portion is seen dotted in R.H.S.V. Upper portion of the object is seen full in T.V. and lower portion of the object is seen dotted in T.V. Similarly, lower portion of the object is seen full in B.V. and upper portion of the object is seen dotted in B.V. These are guide lines only. They are to be applied with the imagination. Imagination plays important role in obtaining correct solution. Develop your mind for imagination.

Problem 5 : In Fig. 18.5 front view and top view of an object are given. Draw (4) four missing views i.e. B.V.; R.H.S.V.; L.H.S.V. and Rear view. Use 3rd angle projection system.

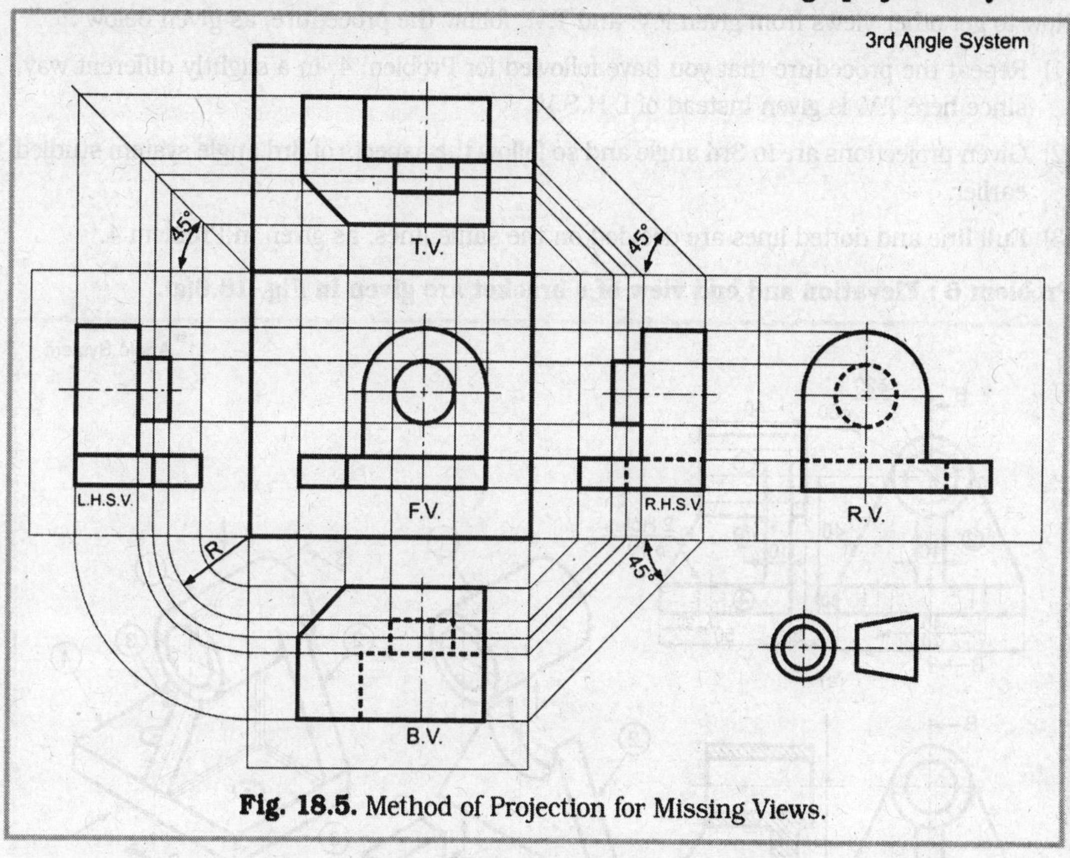

Fig. 18.5. Method of Projection for Missing Views.

Let us study first some of the aspects of 3rd angle projection system.

(1) King view, F.V. is always situated at the centre.

(2) T.V. is situated above F.V.

(3) B.V. is situated below F.V.

(4) R.H.S.V. is situated on right of F.V.

(5) L.H.S.V. is situated on left of F.V.

(6) Rear view is situated on the extreme right beside R.H.S.V.

(7) Portion of the object which is located on the front side of the object is seen nearer to F.V. in all four views surrounding F.V. e.g. in Fig. 18.5 cylindrical portion which is on front side of the object is seen nearer to F.V. in all four views.

(8) Portion of the object which is on rear side is seen away from the front view in all four views surrounding F.V. e.g. in Fig. 18.5 plate (rectangular at the bottom side and semicylindrical on the upper side), situated on rear side of the object, is seen away from front view in all four views surrounding the F.V.

In fact the above aspects at No.7 and 8 are the real tests to be carried out to decide whether the given projections are in 1st angle or 3rd angle system.

Now to get other views from given F.V. and T.V., follow the procedure, as given below :

(1) Repeat the procedure that you have followed for Problem 4, in a slightly different way, since here T.V. is given instead of L.H.S.V.

(2) Given projections are in 3rd angle and so follow the aspects of 3rd angle system studied earlier.

(3) Full line and dotted lines are decided on the same lines, as given in Problem 4.

Problem 6 : Elevation and end view of a bracket are given in Fig. 18.6(a).

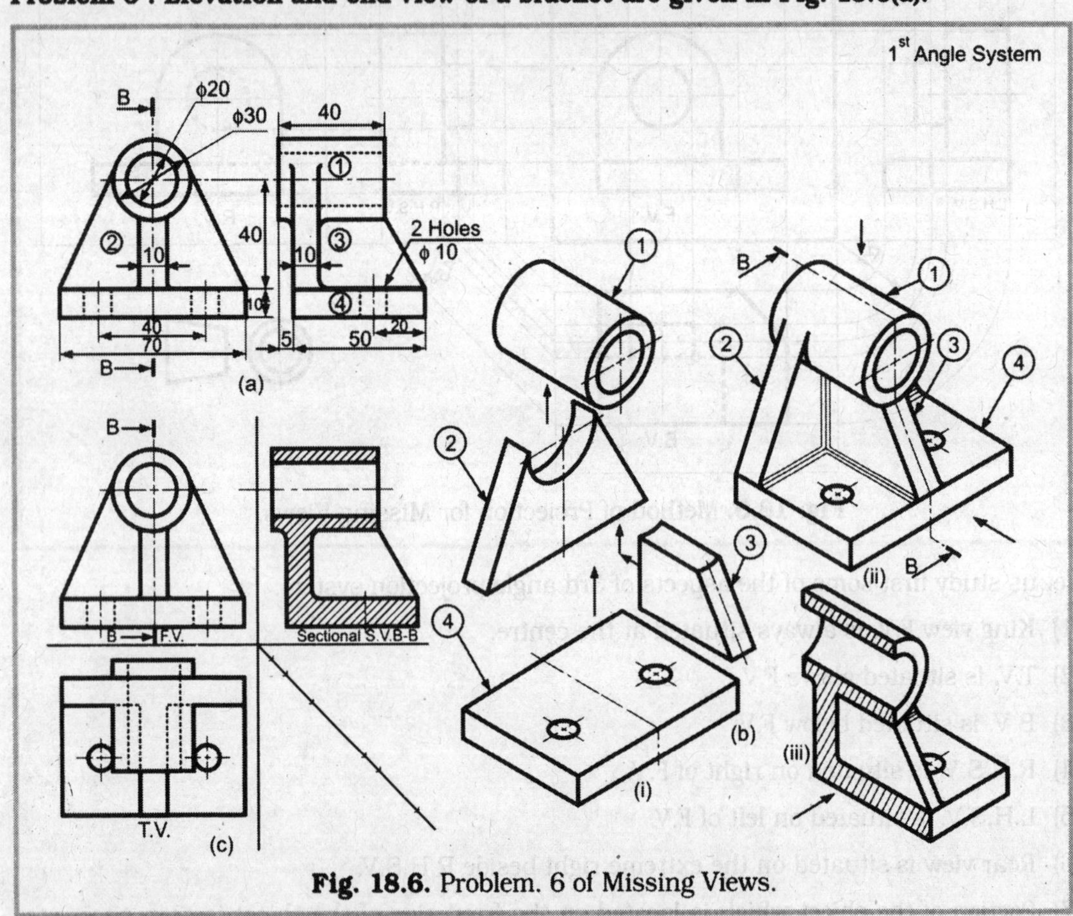

Fig. 18.6. Problem. 6 of Missing Views.

(i) Find in which system the projections are given ?

(ii) Draw front elevation, left hand sectional side view, take section at B-B and top plan in the same system of projections.

(i) Part No.3 i.e. rib must be in front of the bracket since in the front view it is shown by full lines. In side view rib is seen not nearer to F.V. but away from it and hence we can conclude that the projections given are in the 1st angle system.

(ii) To solve part (ii) first we have to imagine the object from the given two views.

As said earlier let us imagine the object piece by piece. Object can be conveniently divided into four pieces.

Correlating two concentric circles of F.V. with full and dotted rectangles of the side view we get hollow cylinder as piece No. 1. Similarly, pieces No. 2, 3 and 4 are imagined by correlating respective shapes of elevation and end view.

Having imagined four pieces No. 1, 2, 3 and 4, they are drawn in pictorial views as shown in Fig. 18.6(b)(i). Later on these pieces are joined by moving them in the direction shown in Fig. 18.6(b)(i), to get the pictorial drawing of complete bracket, as shown in Fig. 18.6(b)(ii).

Imagine now the cut object after cutting it along central plane B-B. Cut object is also shown in the Fig. 18.6(b)(iii).

Do not worry if you cannot draw pictorial views as shown in Fig. 18.6(b). Students are required to imagine such pictorial views in their minds.

Now having imagined the full object and the half cut object we can draw (i) Front Elevation (same as given), (ii) Full Sectional Side View and (iii) Top Plan. Solution is shown/given in Fig. 18.6(c) in 1st angle system.

Problem 7 : Elevation and side view are given in Fig. 18.7(i). (i) Find system of projection, (ii) Draw (a) Sectional Elevation along A-A, (b) Offset Sectional Side View along B-B and (c) Sectional Top Plan along C-C. Draw projections in the same system of projection.

(i) Given projections are in the 3rd angle system.

(ii) Shape of the component may be visualised by imagining the different pieces of the component separately and then by combining them into one piece. After visualising the complete component imagine its cutting along section A-A, along offset section B-B and along section C-C.

For your understanding three cut components after cutting it along A-A, B-B and C-C are shown in Fig. 18.7 (ii) in pictorial views.

Having imagined three different cut components, as shown in Fig. 18.7 (ii) draw (i) Sectional Elevation along A-A, (ii) Offset Sectional Side View along B-B and (iii) Sectional Plan along C-C, as shown in Fig. 18.7(iii).

Material that is cut by cutting plane is shown by means of section (hatching) lines. In normal circumstances dotted lines, to show hidden features, are avoided in the sectional views.

Having practised seven illustrative problems, students are required to practise many more problems of different varieties given as solved & unsolved exercise problems to grasp orthographic reading and develop self confidence in reading orthographic projection.

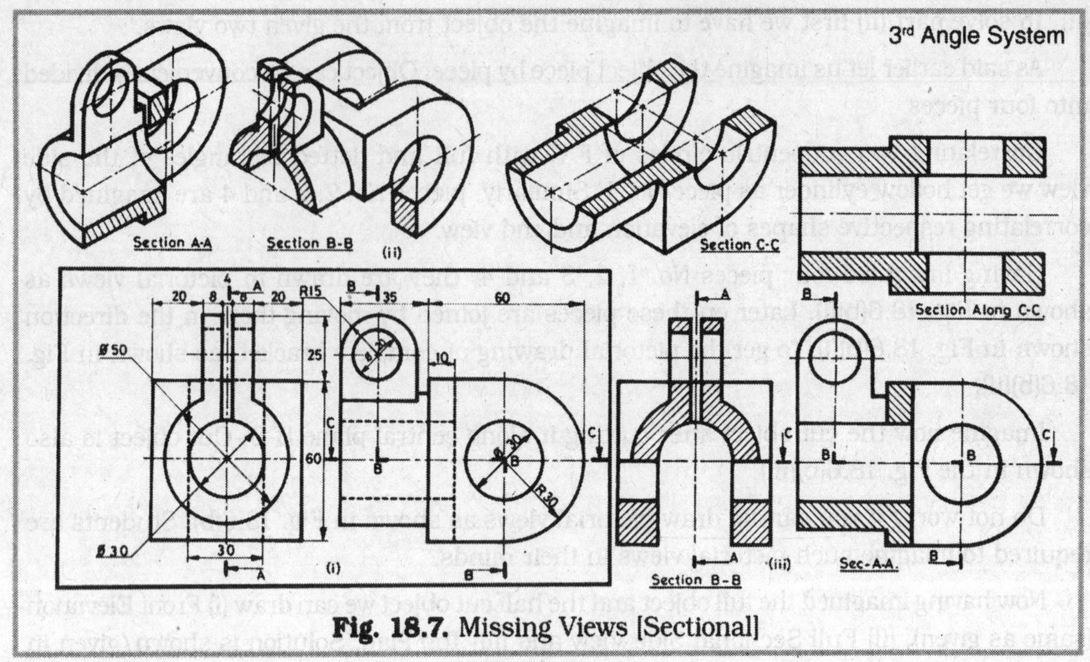

Fig. 18.7. Missing Views [Sectional]

Solved Exercise

S.E.1 The projections of a **'Bearing Cap'** are given in Fig. (A). Draw to full scale using 1st Angle Method of Projection the following views.

(i) H.S. Elevation (ii) H.S.L.H.S.V. Solution is given in Fig. (B)

S.E.2 Two views of **C.I. Bracket** are given in Fig. (C). Using the same method of projection draw the following views.

(i) Sectional F.V. (ii) Sectional T.V. (iii) S.V. from right

Solution is given in Fig. (D)

S.E.3 T.V. and F.V. of **A Block** are given in Fig. (E). Draw, in the same method of projection, the following views.

(i) Sectional F.V. (ii) R.H.S.V. (iii) L.H.S.V. Solution is given in Fig. (F).

S.E.4 **A C.I. machine part** is shown in Fig. (G). Using the same method of projection, draw the following views.

(i) Sectional F.V. (ii) Sectional T.V. (iii) S.V. Solution is given in Fig. (H)

S.E.5 Two views of **C.I. bracket** are given in Fig. (J). Draw to Full Size Scale and according to First Angle Projection Method the following views.

(i) Sectional F.V. (Sect. A-A)

(ii) T.V. Solution is given in Fig. (K)

(iii) Both Side Views

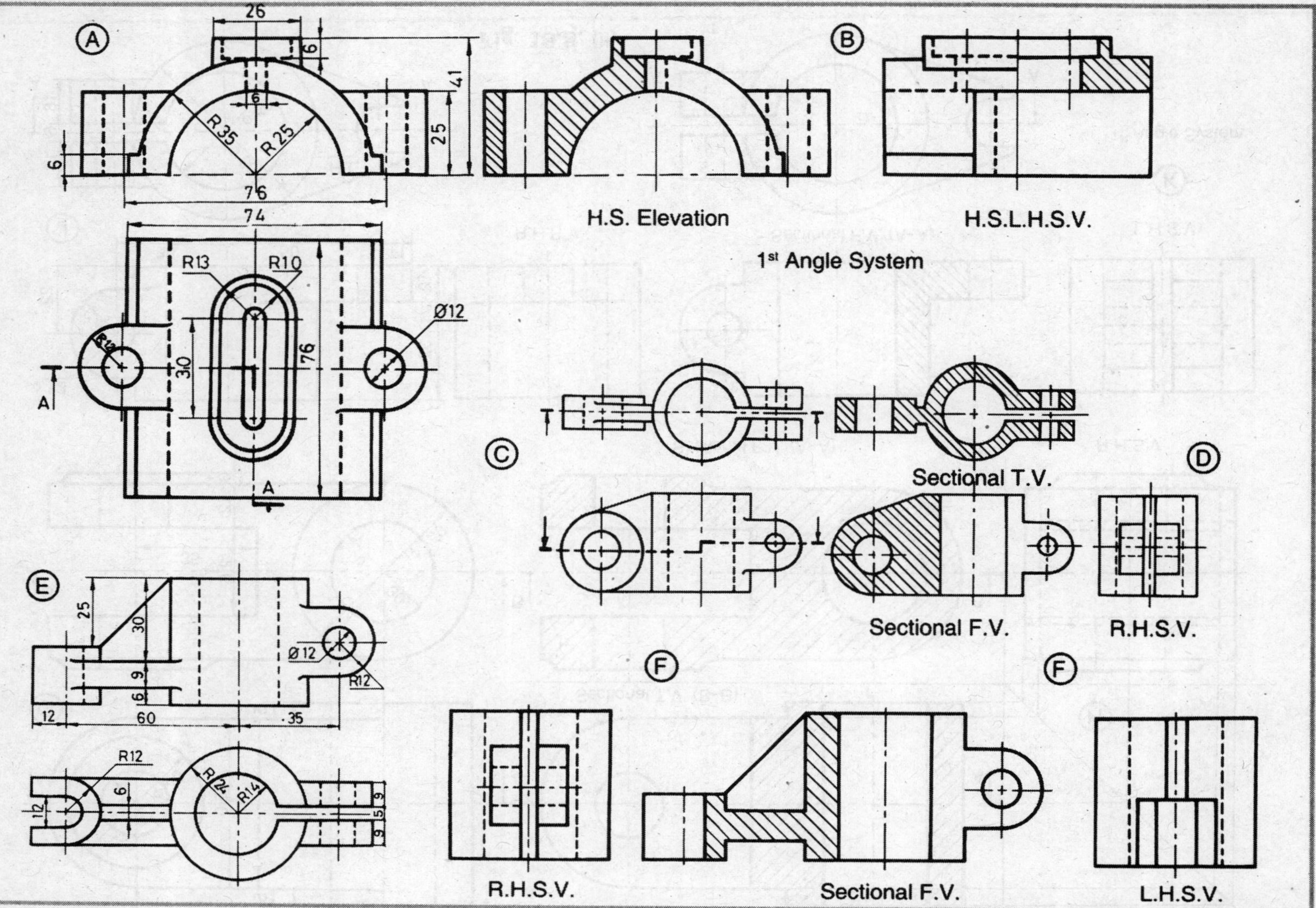
A
26
6
6
R35
R25
41
25
6
76
74
R13
R10
Ø12
R12
30
76
A
A
B
H.S. Elevation
H.S.L.H.S.V.
1st Angle System
C
Sectional T.V.
D
Sectional F.V.
R.H.S.V.
E
25
30
Ø12
R12
9
6
12
60
35
R12
6
R24
R14
12
9
5
9
F
F
R.H.S.V.
Sectional F.V.
L.H.S.V.

Fig. 18.8. (i)

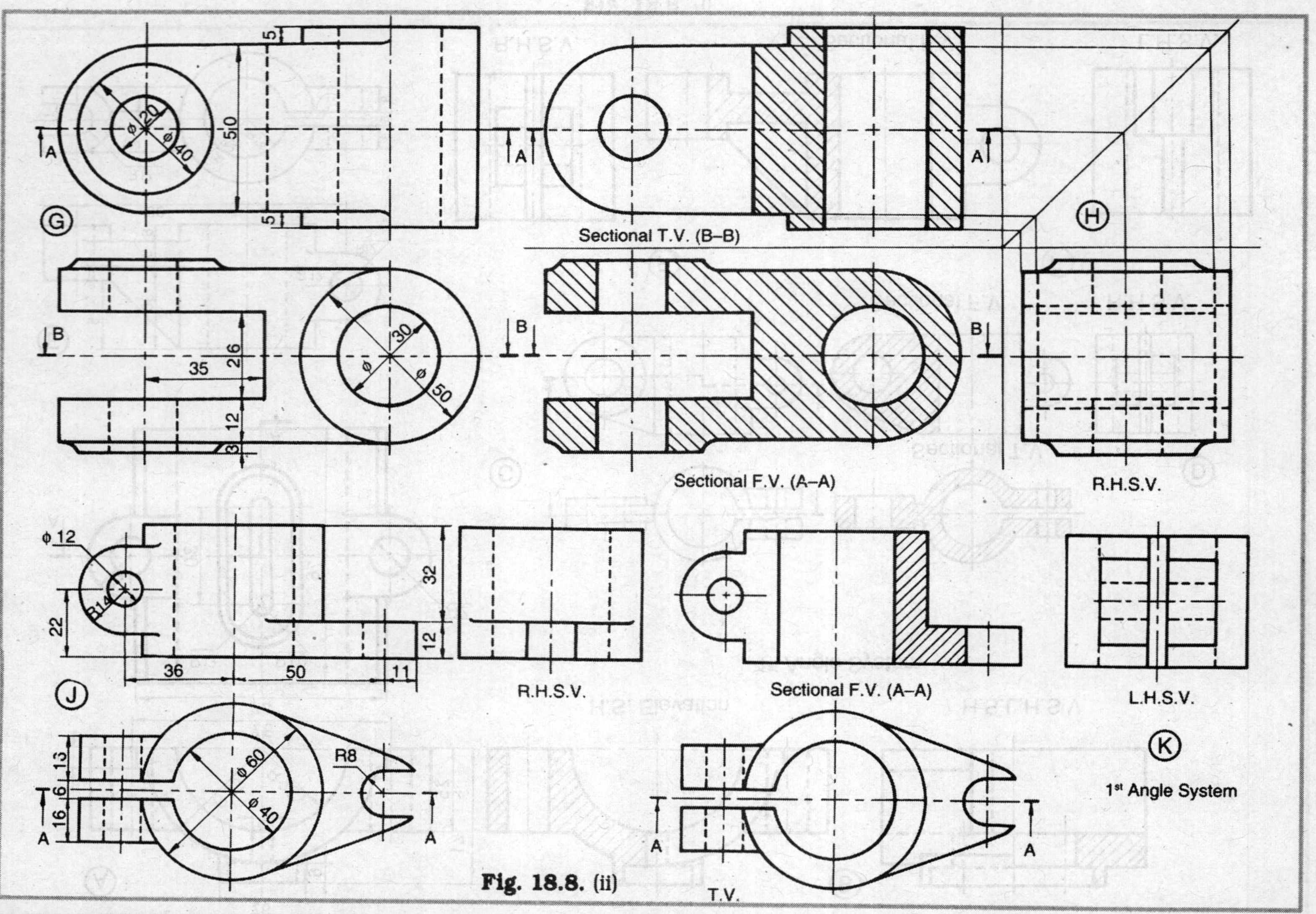
G
Sectional T.V. (B–B)
H
Sectional F.V. (A–A)
R.H.S.V.
J
R.H.S.V.
Sectional F.V. (A–A)
L.H.S.V.
K
1st Angle System
T.V.

Fig. 18.8. (ii)

EXERCISE

1. In Fig 18.9 T. V. is given. Draw according to the 1st angle projection system elevation of different objects whose Top View will be same as the given T. V.
2. In Fig. 18.10 T. V is given. Imagine different possible objects whose top view will be the same as given and then draw their elevations in the 3rd angle system of projection.
3. In Fig. 18.11 F. V. is given. Draw different plans and side views which satisfy the given F. V., in 1st angle system.

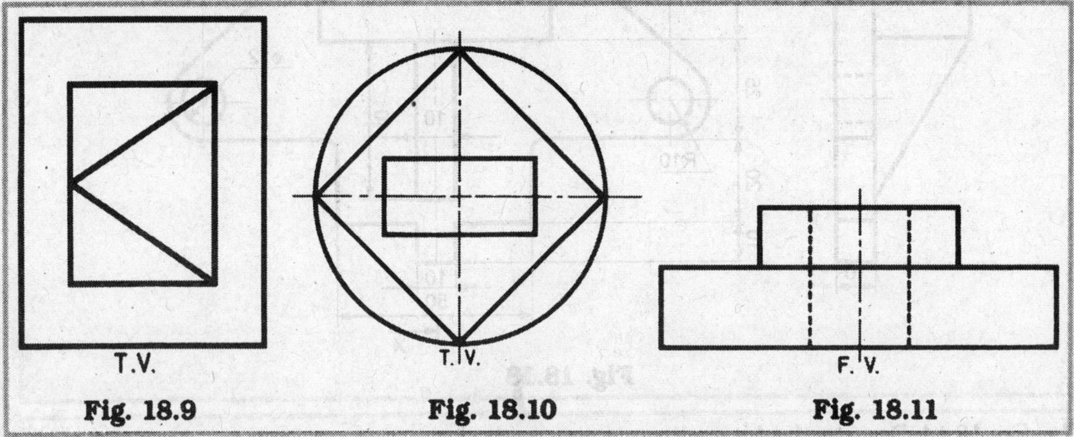

Fig. 18.9 Fig. 18.10 Fig. 18.11

4. In Fig. 18.12 (a) F. V. is given. Draw free hand pictorial sketches of the different objects whose front view will be the same as given. Draw also possible T. Vs. and S. Vs. in any system of projection.
5. In Fig. 18.12 (b) F. V. and T. V. are given. Imagine different objects whose F. V. and T. V. remain the same but whose side views will be different and then draw different possible side views.

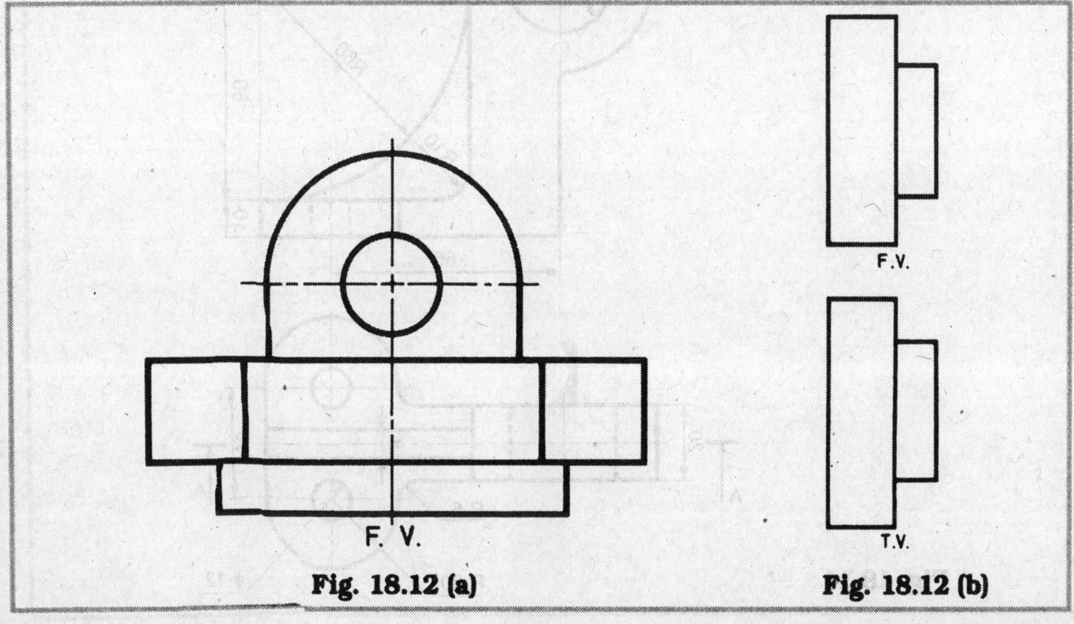

Fig. 18.12 (a) Fig. 18.12 (b)

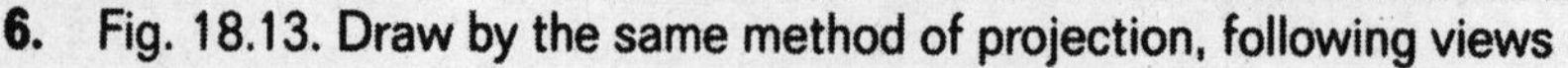

6. Fig. 18.13. Draw by the same method of projection, following views :

(i) Front View

(ii) Sectional Side View, take section along X-X

(iii) Top Plan/T.V.

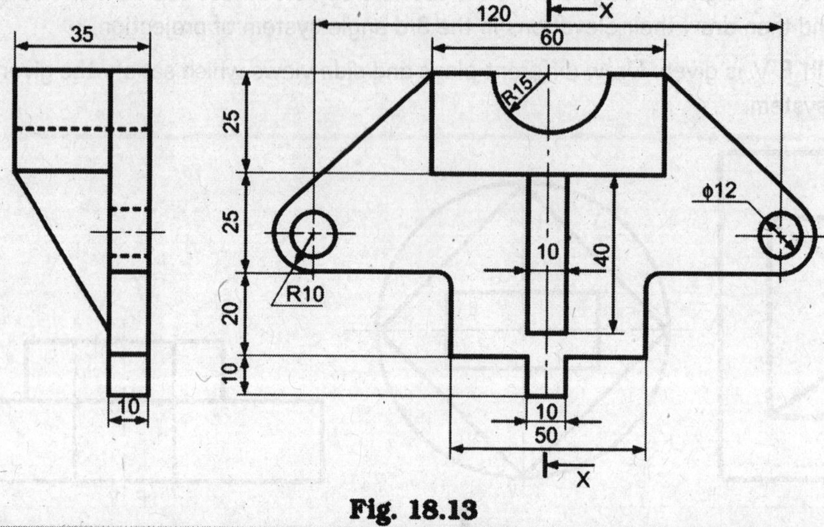

Fig. 18.13

7. Fig. 18.14. Draw by the 1st angle projection method :

(i) Sectional Front View, take section along A - A

(ii) Top Plan/T.V.

(iii) R. H. S. V. and L. H. S. V.

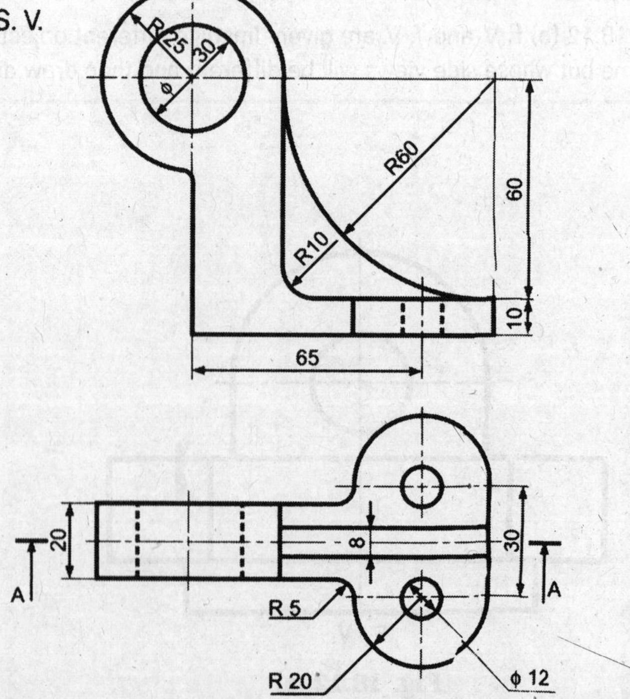

Fig. 18.14

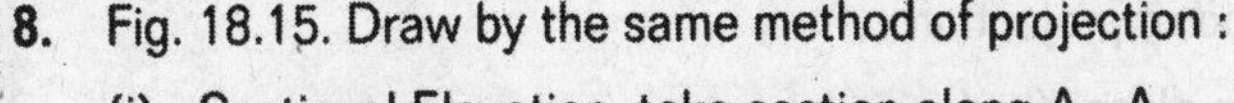

8. Fig. 18.15. Draw by the same method of projection :

(i) Sectional Elevation, take section along A - A

(ii) Bottom Plan/B.V.

(iii) L. H. Side View

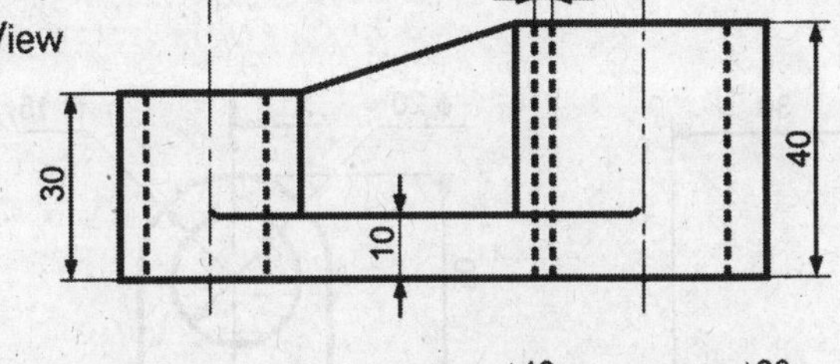

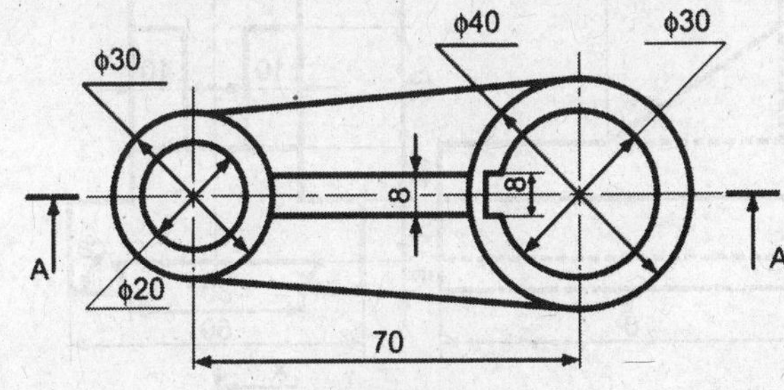

Fig. 18.15

9. Fig. 18.16. Draw by the same method of projection :

(i) Sectional elevation, take section along X - X

(ii) R. H. Side View

(iii) Top Plan/T.V.

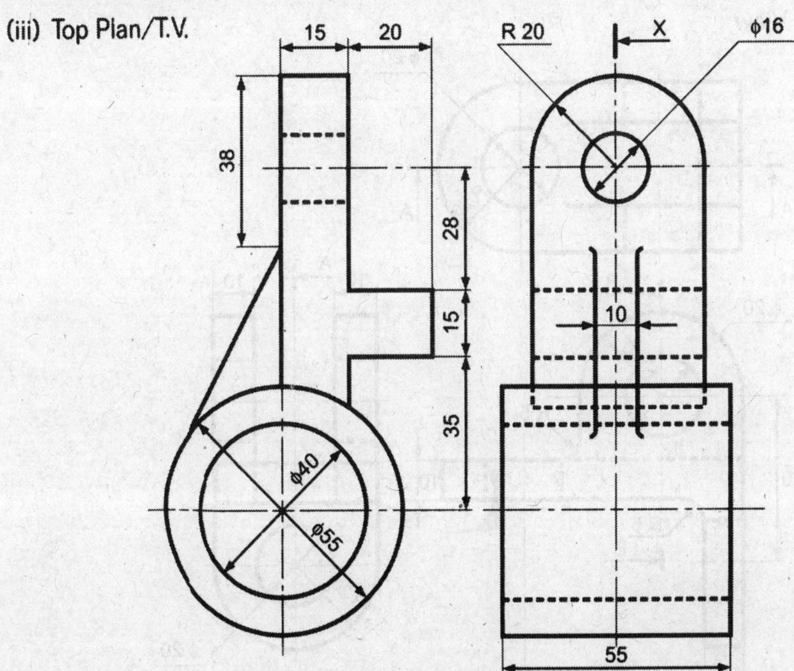

Fig. 18.16

10. Fig. 18.17. Draw by the same method of projection :

(i) Sectional Elevation, take section along X - X

(ii) Left Hand Side View

(iii) Top Plan/T.V.

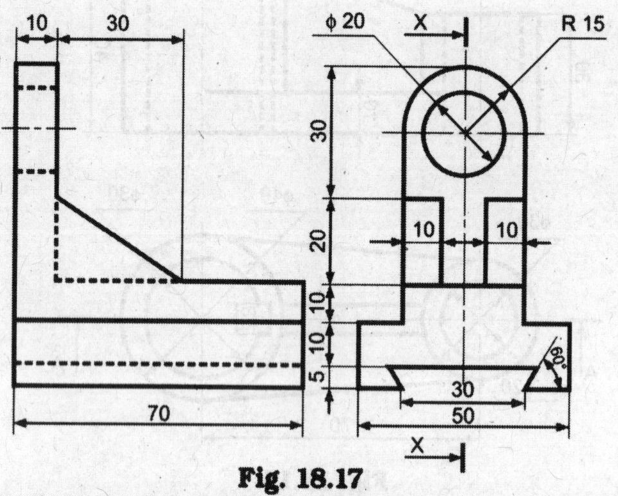

Fig. 18.17

11. Fig. 18.18. Draw by the same method of projection :

(i) Sectional Elevation along A - A

(ii) Sectional right hand side view along B - B

(iii) L. H. Side View

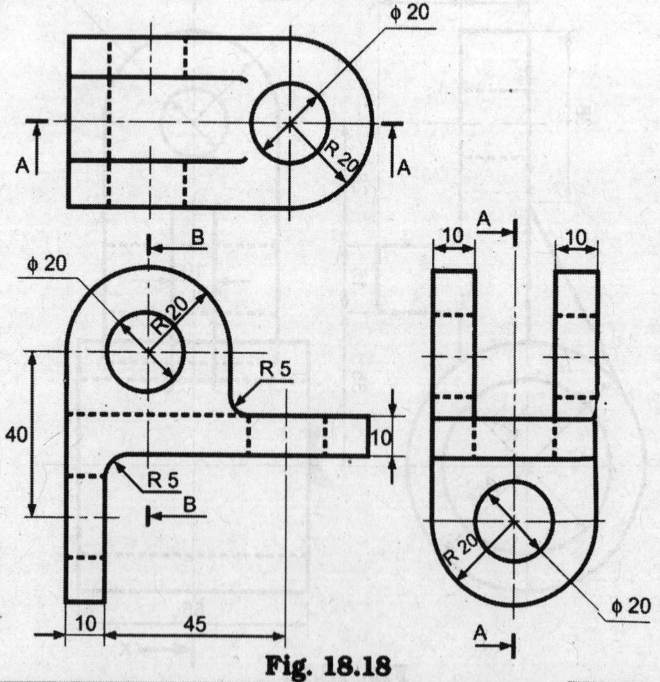

Fig. 18.18

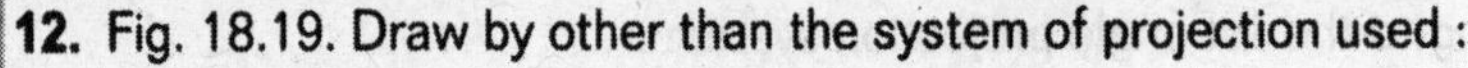

12. Fig. 18.19. Draw by other than the system of projection used :

(i) Sectional Elevation (Section X - X)

(ii) Bottom Plan

(iii) R. H. Side View

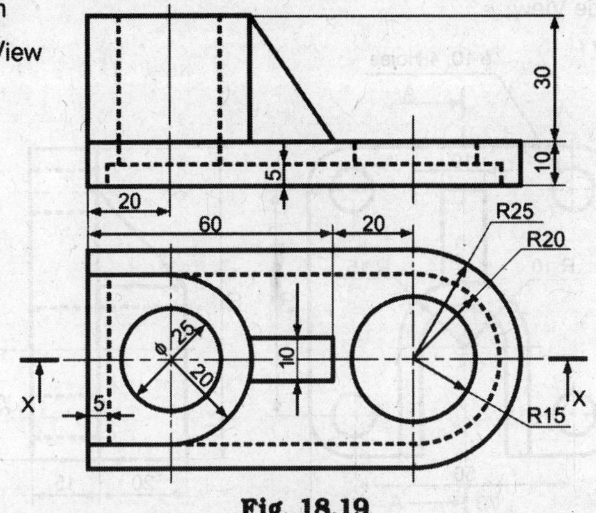

Fig. 18.19

13. Fig. 18.20. Draw by the 3rd angle projection method :

(i) Sectional Front View - Section along A - A

(ii) Sectional Side View - Section along B - B

(iii) Left Hand Side View

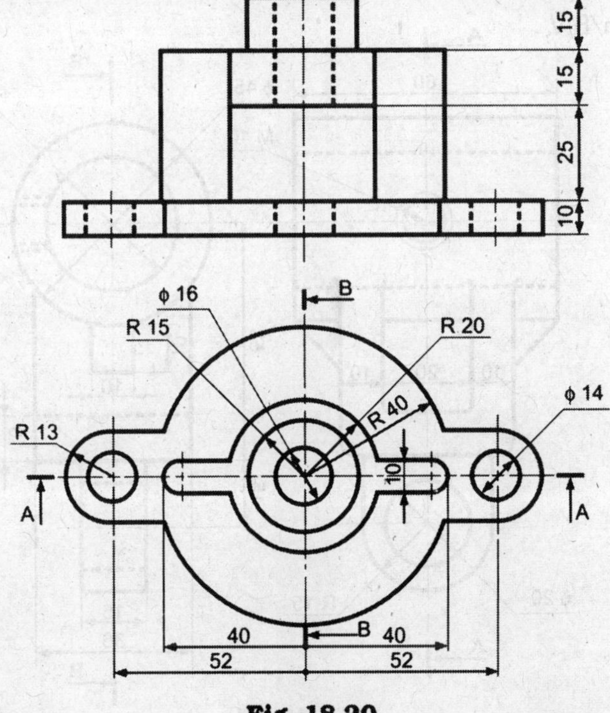

Fig. 18.20

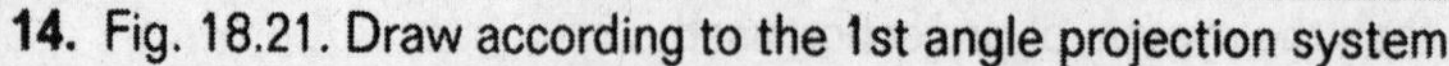

14. Fig. 18.21. Draw according to the 1st angle projection system :

(i) Front View

(ii) Sectional Side View

(iii) Top Plan/T.V.

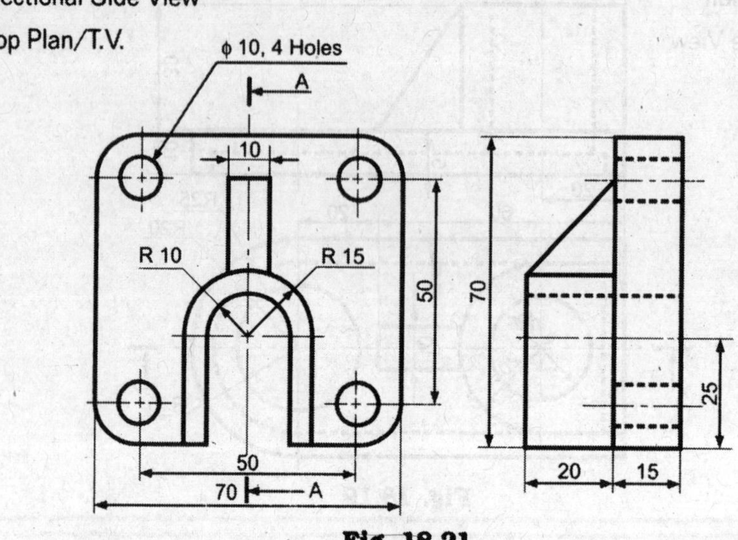

Fig. 18.21

15. Fig. 18.22. Draw by the same system of projection :

(i) Sectional Elevation - Section along B - B

(ii) Sectional Side View - Section along A - A

(iii) Bottom Plan/B.V.

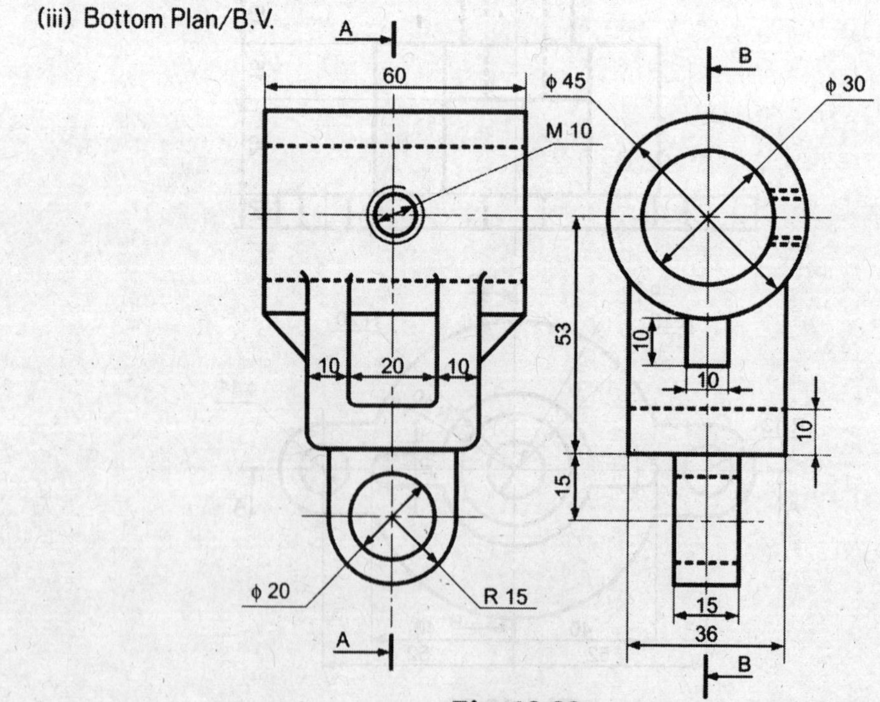

Fig. 18.22

16. Fig. 18.23. Draw by the 1st angle system of projection :

(i) Sectional Elevation Section along A - A

(ii) Sectional Side View Section along B - B

(iii) Top Plan/T.V.

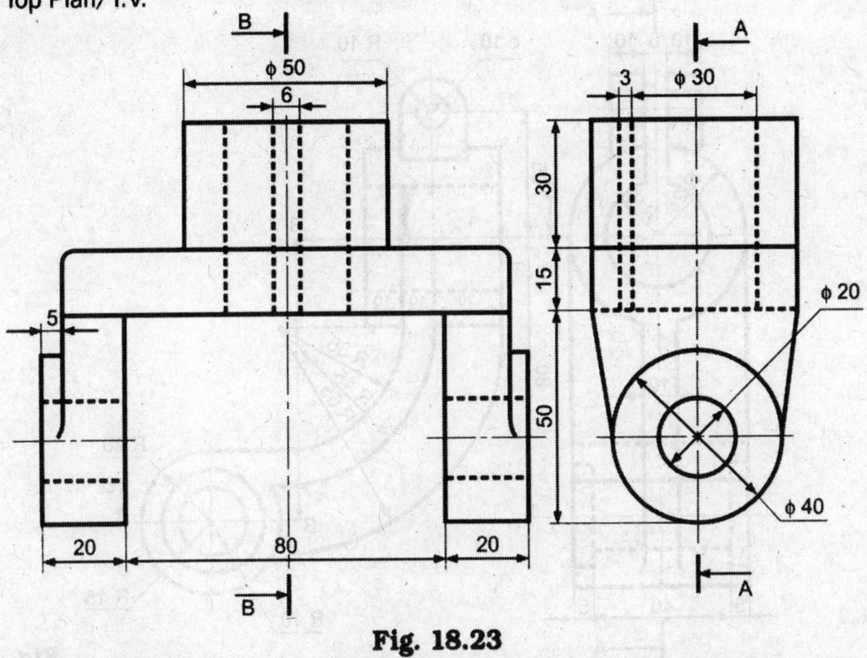

Fig. 18.23

17. Fig. 18.24. Draw by the 1st angle system of projection :

(i) Sectional Elevation - Section along A - A

(ii) Sectional Side View - Section along B - B

(iii) Left Hand Side View

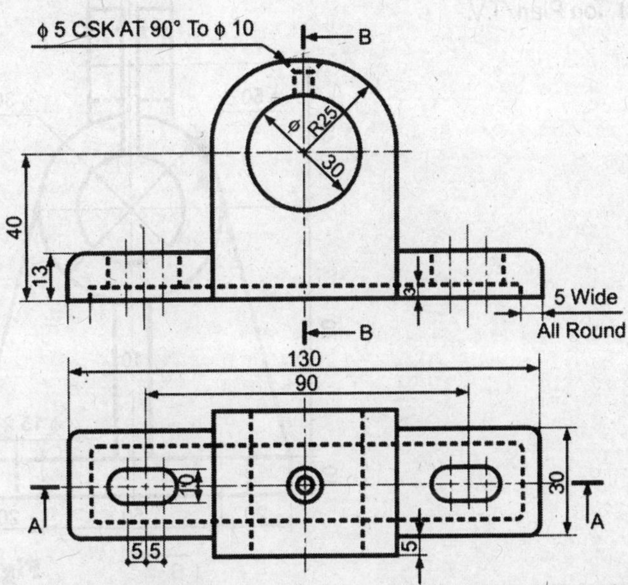

Fig. 18.24

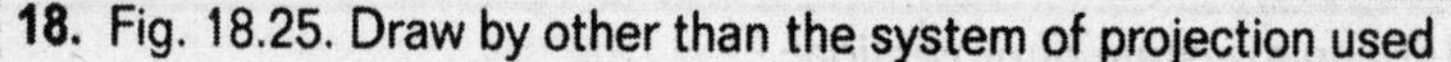

18. Fig. 18.25. Draw by other than the system of projection used :

(i) Sectional Elevation - Section along A - A

(ii) Sectional Top Plan - Section along B - B

(iii) R. H. Side View

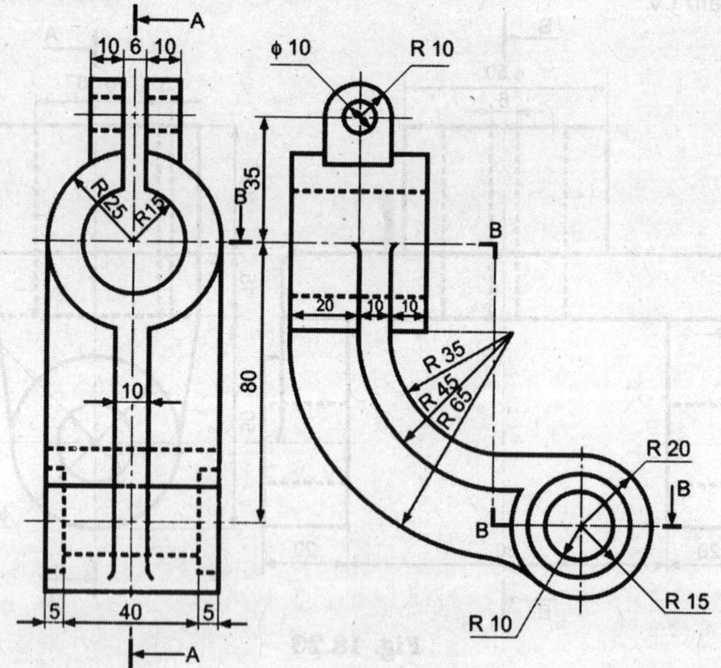

Fig. 18.25

19. Fig. 18.26. Draw according to the system of projection used :

(i) Sectional Elevation along A - A

(ii) Sectional Side View along B - B

(iii) Top Plan/T.V.

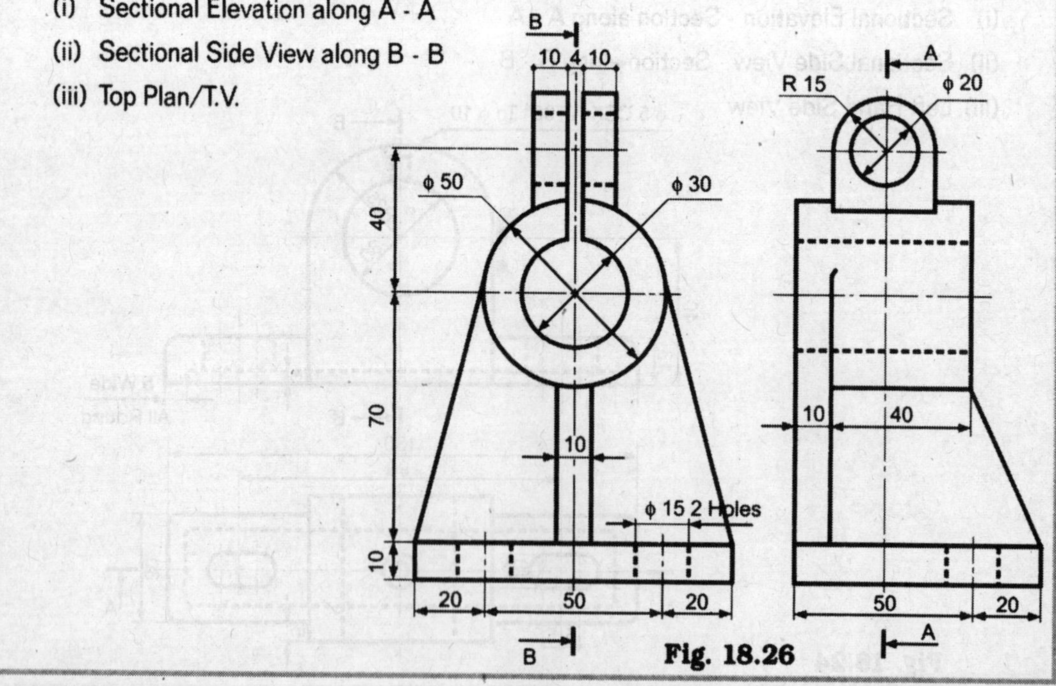

Fig. 18.26

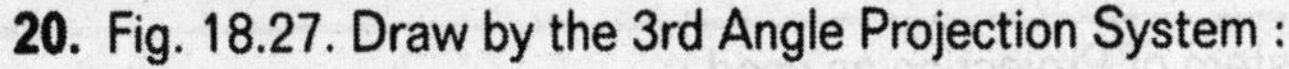

20. Fig. 18.27. Draw by the 3rd Angle Projection System :

(i) Sectional F. V. take section along X - X

(ii) R. H. Side View

(iii) L. H. Side View

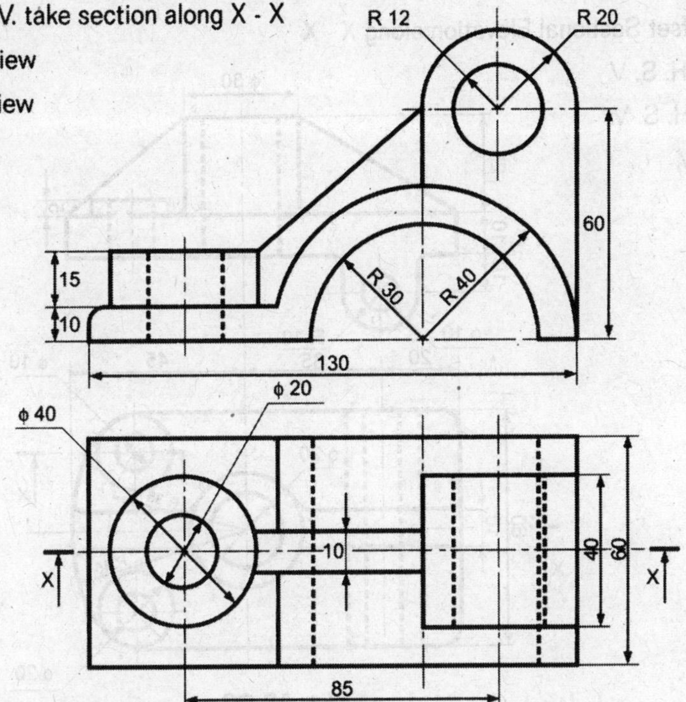

Fig. 18.27

21. Fig. 18.28. Draw by the 1st Angle System of Projection :

(i) Sectional Elevation along A - A

(ii) R. H. S. V.

(iii) L. H. S. V.

(iv) Sectional Plan along B - B

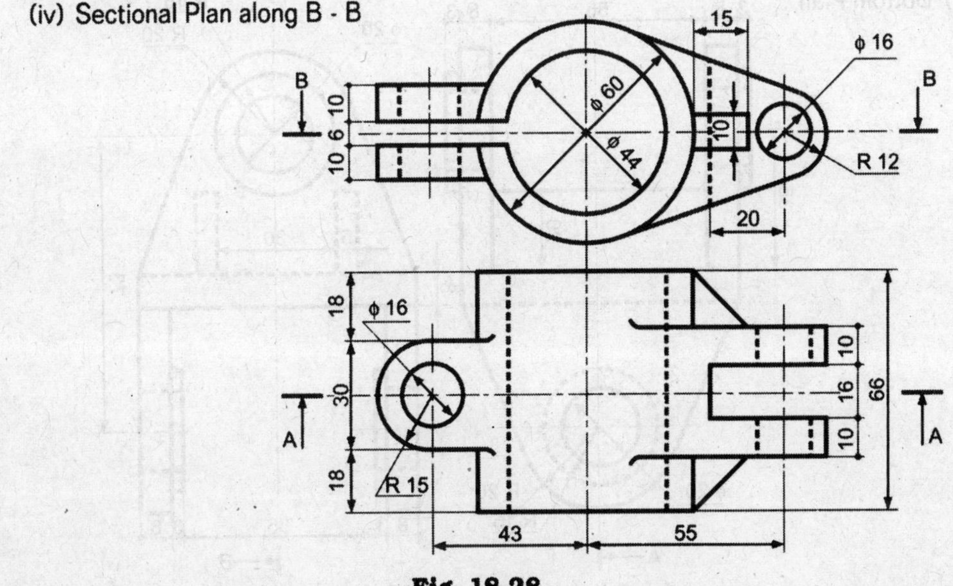

Fig. 18.28

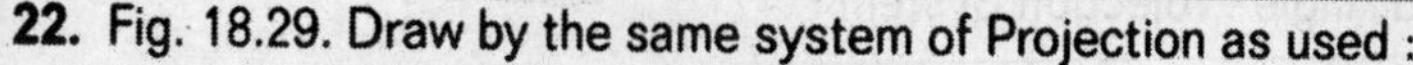

22. Fig. 18.29. Draw by the same system of Projection as used :

(i) Offset Sectional Elevation along X - X

(ii) R. H. S. V.

(iii) L. H. S. V.

(iv) T. V.

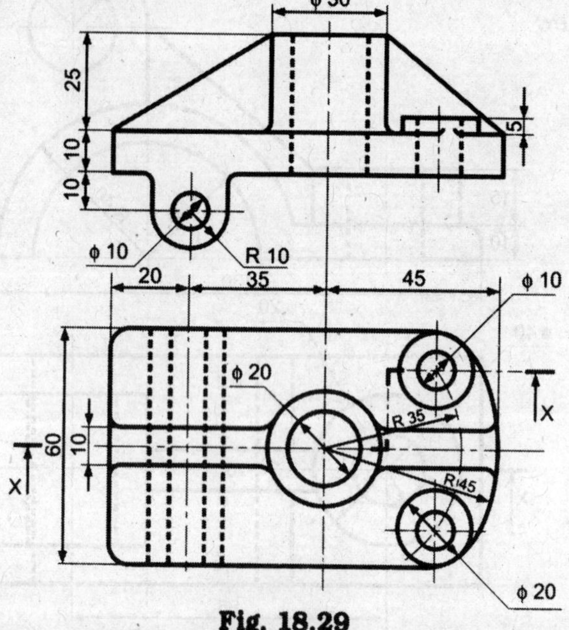

Fig. 18.29

23. Fig. 18.30. Draw by the 1st Angle Projection System :

(i) Sectional Elevation - Section along A - A

(ii) Sectional Side View - Section along B - B

(iii) Top Plan

(iv) Bottom Plan

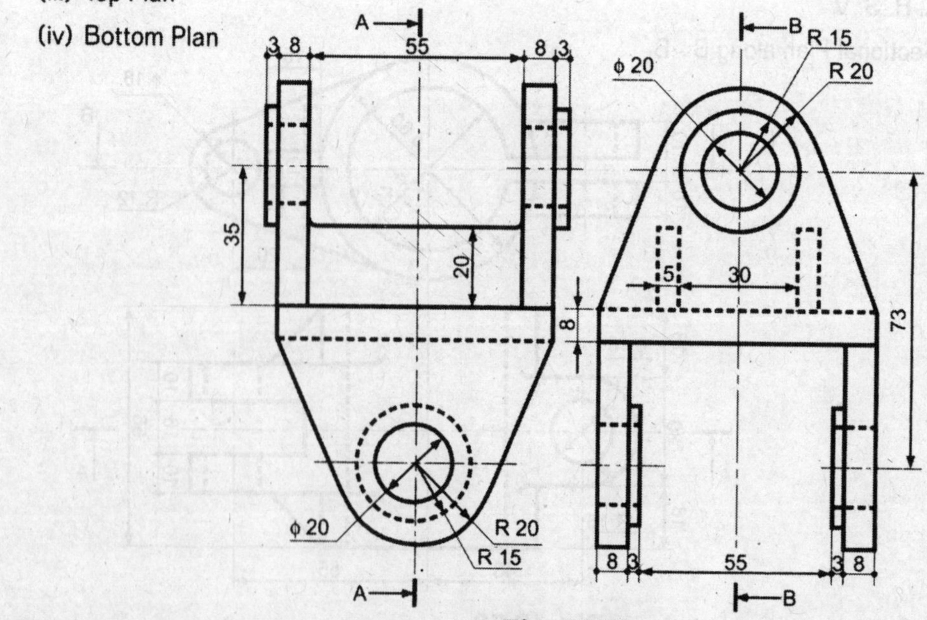

Fig. 18.30

24. Fig. 18.31. Draw by the same System of Projection :

(i) Full Sectional Elevation along A - A

(ii) Top Plan

(iii) Left Hand Side View

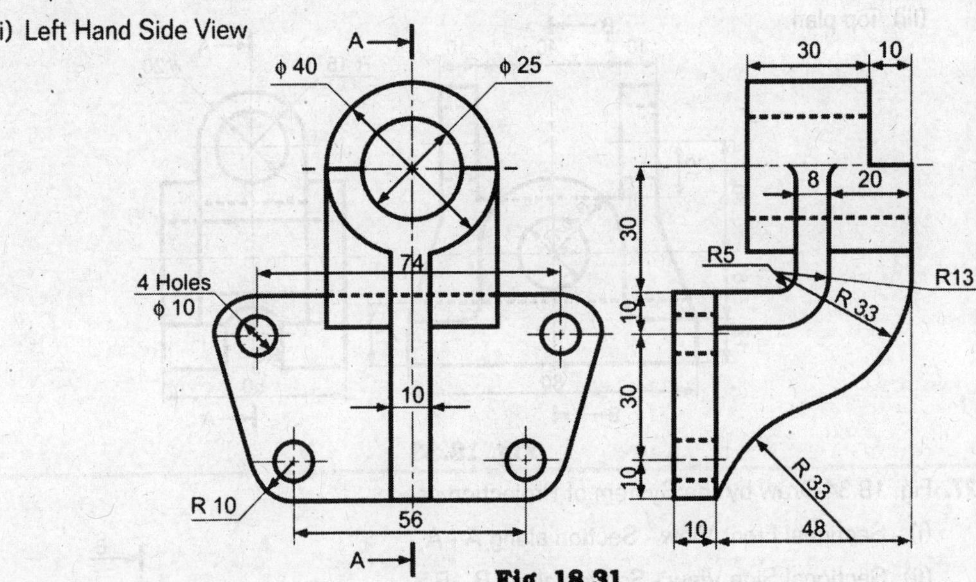

Fig. 18.31

25. Elevation and Top Plan of an object are given in Fig. 18.32 Draw in the same system of projection :

(i) Half Sectional Elevation, take section along A - B

(ii) Half Sectional R. H. Side View, take section along A - B

(iii) Top Plan

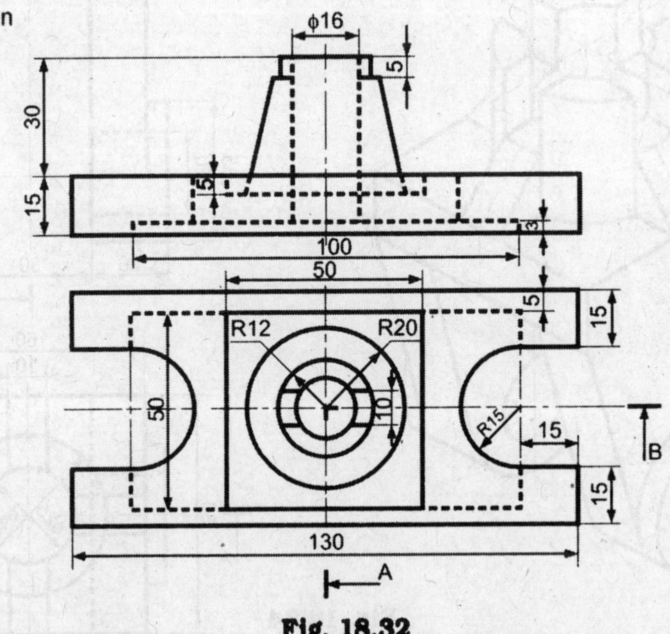

Fig. 18.32

26. Fig. 18.33. Draw in the 1st Angle Projection System :

(i) Sectional Elevation along A - A

(ii) Sectional Side View along B - B

(iii) Top plan

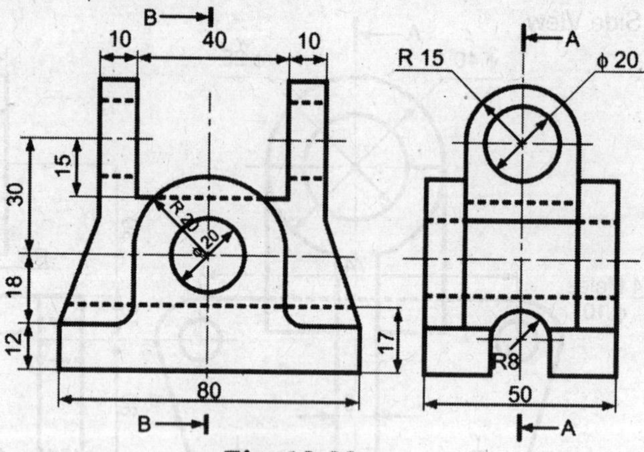

Fig. 18.33

27. Fig. 18.34 Draw by the System of Projection used :

(i) Sectional Front View - Section along A - A

(ii) Sectional Side View - Section along B - B

(iii) L. H. S. V.

Hint : As the component is difficult to visualise, pictorial drawing is given for understanding.

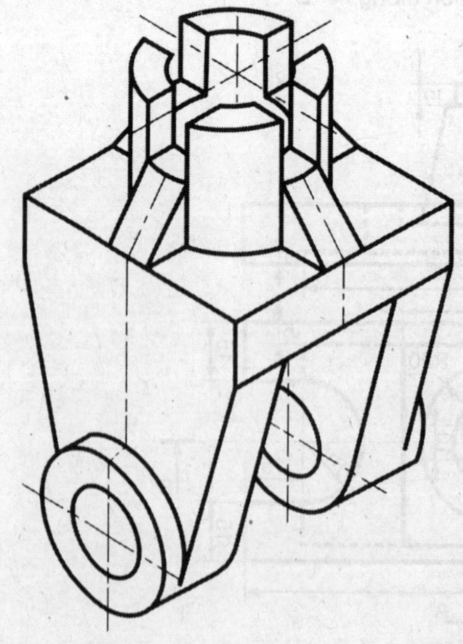

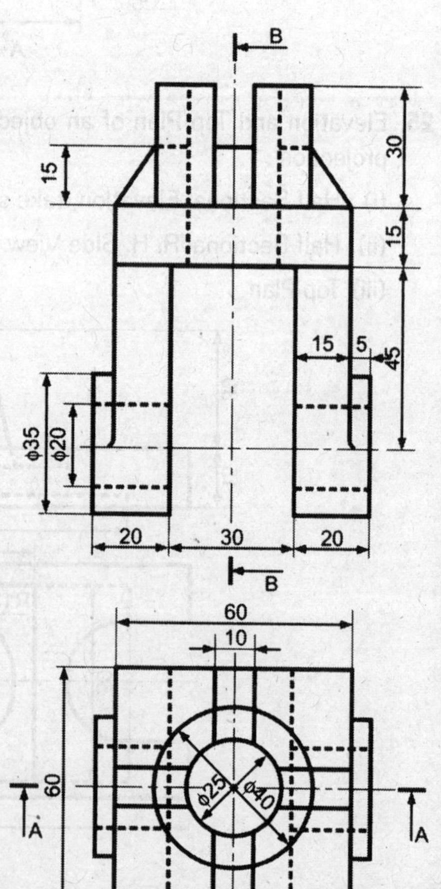

Fig. 18.34

28. Fig. 18.35. Draw by the same System of Projection :

(i) Sectional Elevation along A - A

(ii) Sectional Plan along B - B

(iii) Sectional Side View along C - C

(iv) Side View from left

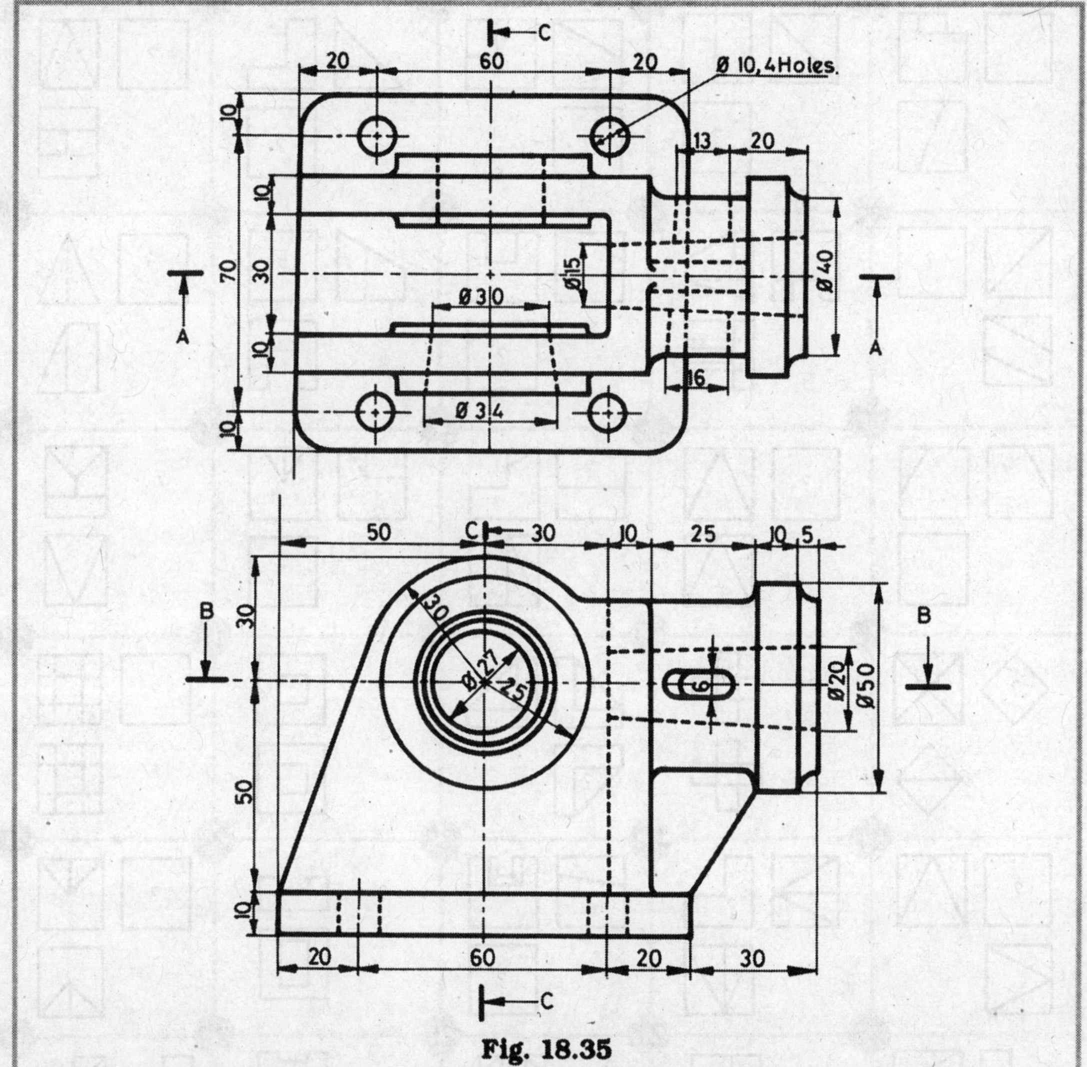

Fig. 18.35

29. Two or three orthographic views of 35 different objects, with either (i) a line or (ii) lines or (iii) a view missing in view marked by sign (?), are given in Fig. 18.36.

Imagine three dimensional view of each object and prepare a sketch of the same. Complete the view in which line or lines missing. Draw third missing view wherever not given.

(Answers are given in Fig. 18.37 in the form of three dimensional drawings. One can easily find the missing line or lines or view from the pictorial drawings given. The given Fig. 18.37 is not to scale.)

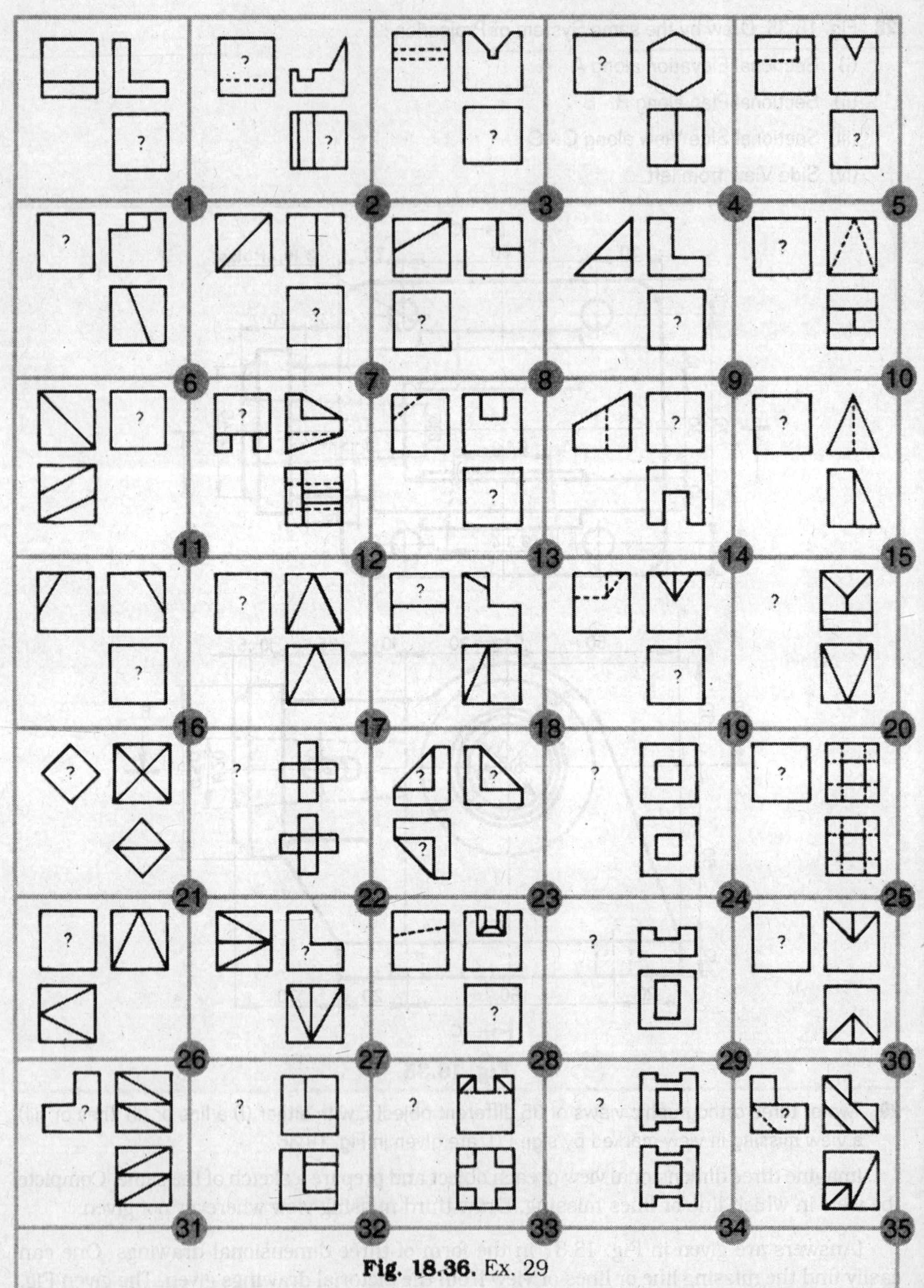

Fig. 18.36. Ex. 29

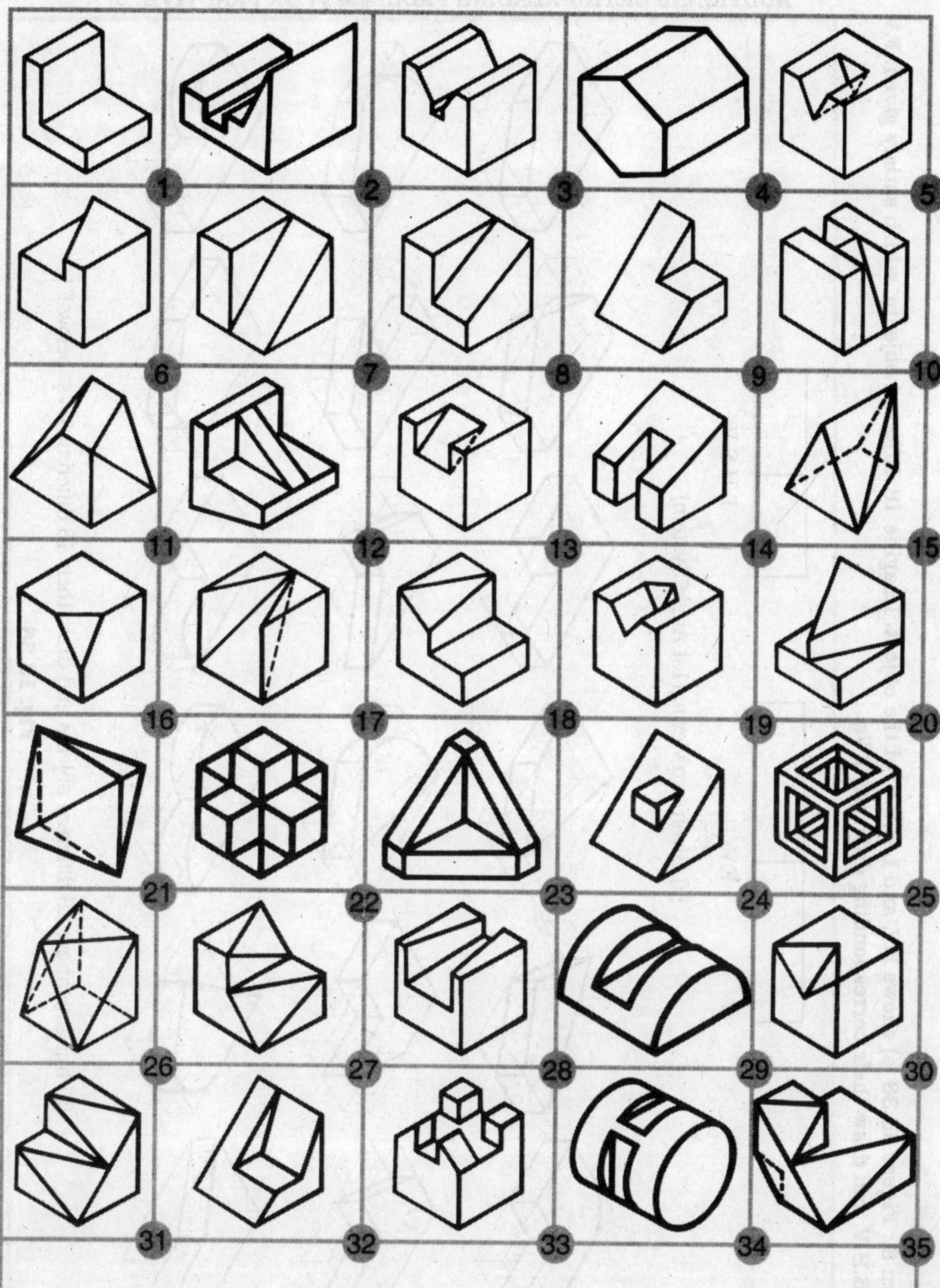

Fig. 18.37. Answers of Ex. 29

ADDTIONAL ORTHO-READING PROBLEMS FOR PRACTIVE

Problem 8 : Figure 18.38 (a) shows F.V. and L.H.S.V. of the object. Imagine 18 different objects, which satisfy given. F.V. and L.S.H.V. and draw their corresponding different T.Vs.

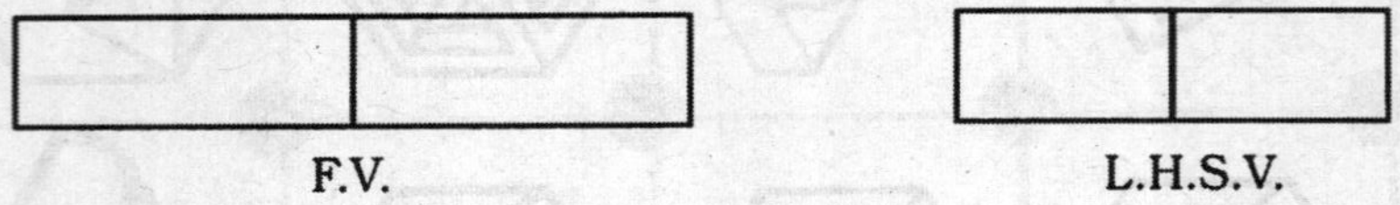

F.V.

L.H.S.V.

(a) Given two views. (1st Angle System)

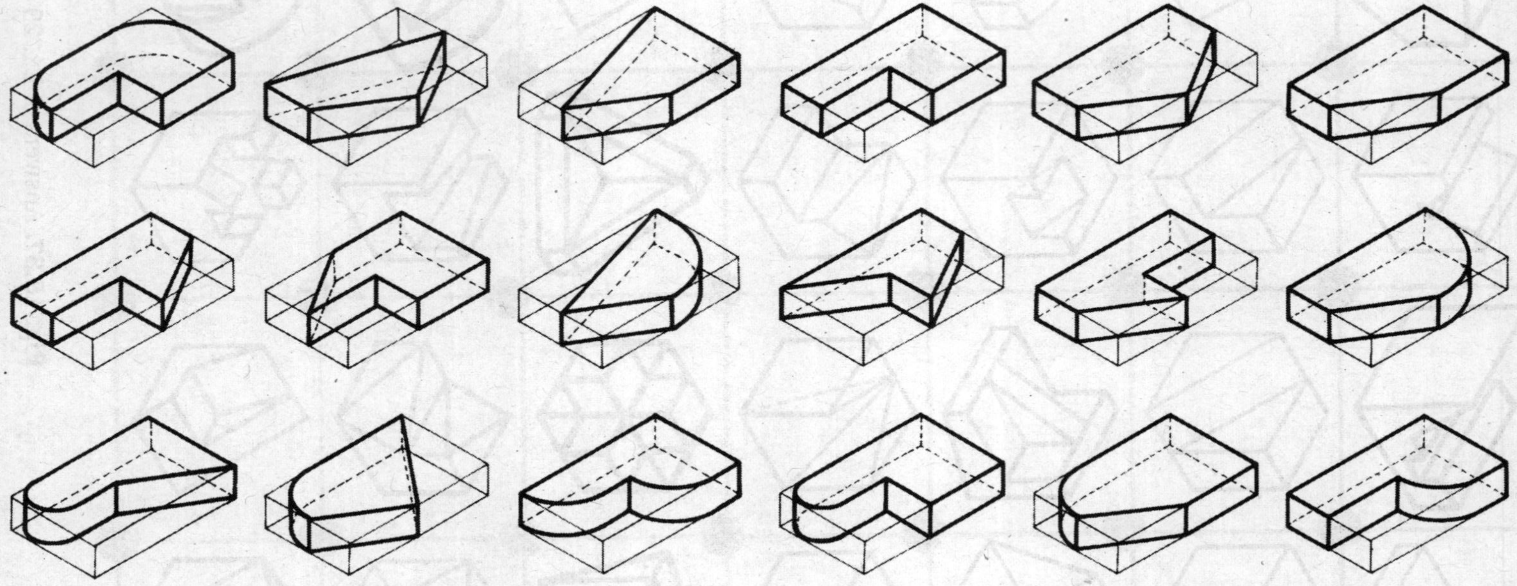

(b) Imagine above 18 different objects and draw their 18 different T.Vs. below F.V.

Fig. 18.38

Problem 9 : Figure 18.39 (a) shows two views of an object. Draw the following views.

(i) Elevation (Sectional)	**OR**	**(i) Sectional F. V.**	
(ii) Plan (T. V.)		**(ii) T. V.**	
(iii) R. H. S. V.		**(iii) R.H.S.V.**	

[Pune University, May 1994]

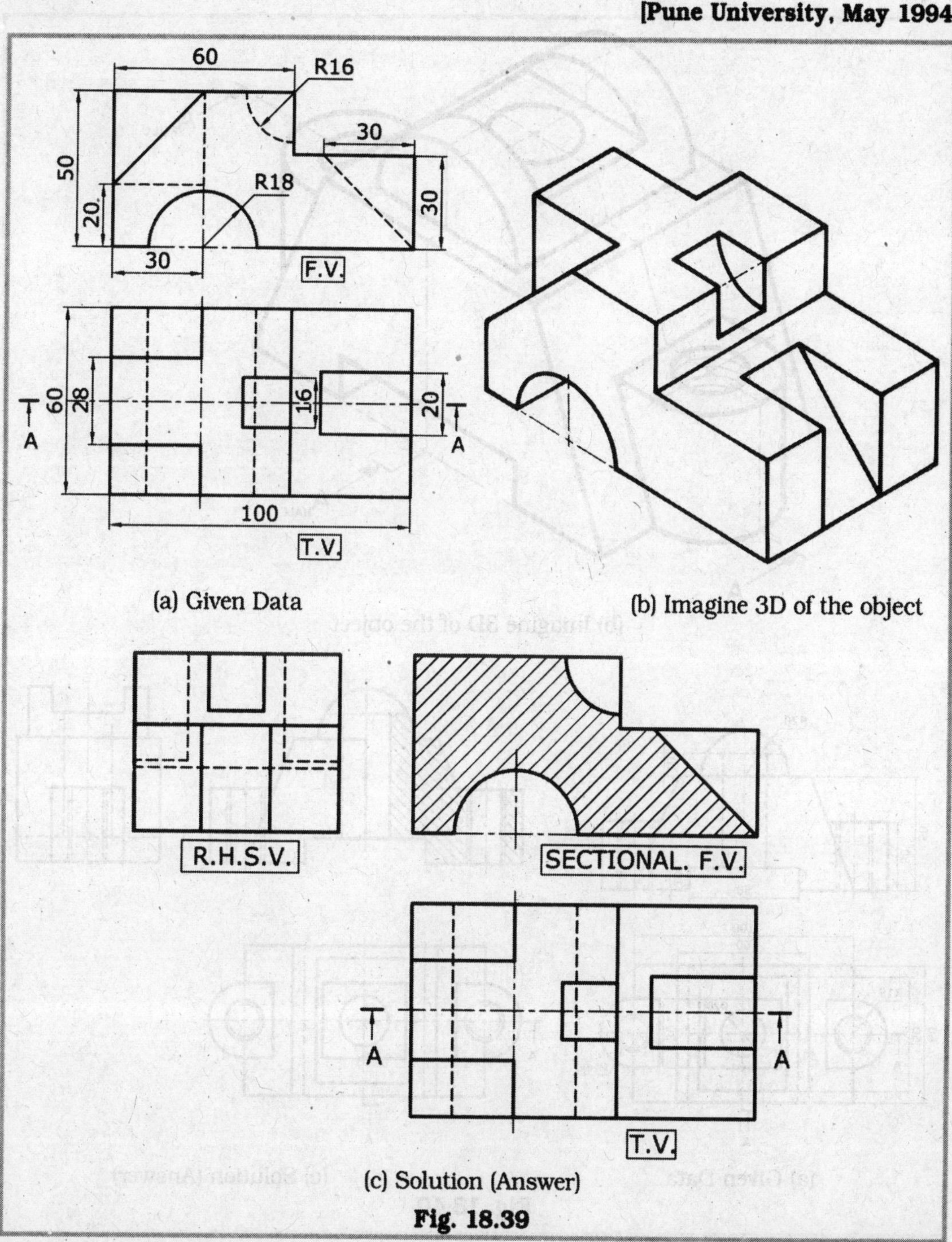

(a) Given Data

(b) Imagine 3D of the object

(c) Solution (Answer)

Fig. 18.39

Problem 10 : Figure 18.40 (a) shows the elevation and plan of an object. Draw the following views by 1st angle method of projections

(i) Half Sectional F. V. [Section along $A\text{-}A_{Half}$]

(ii) T. V.

(iii) L. H. S. V. **[Pune University]**

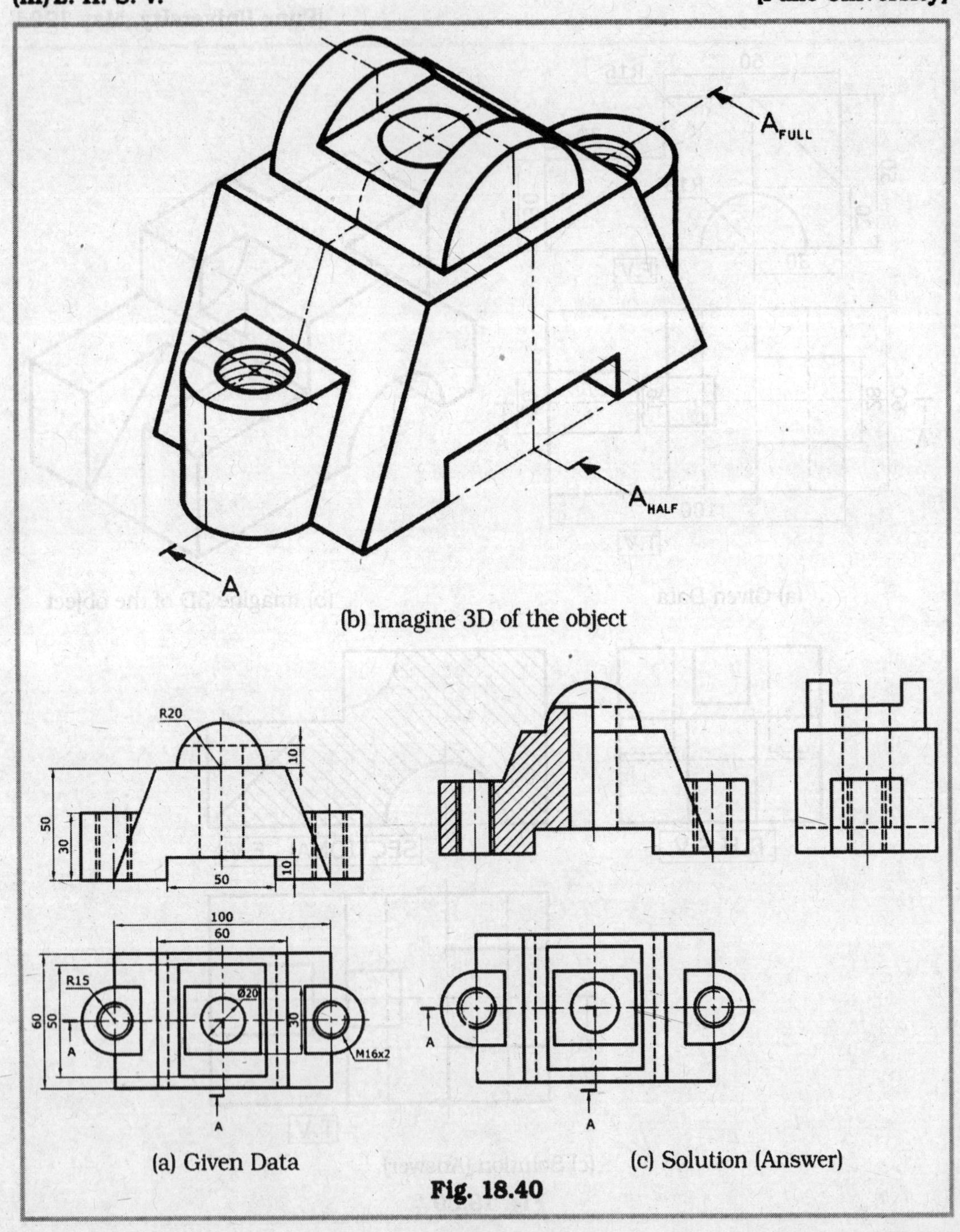

(b) Imagine 3D of the object

(a) Given Data

(c) Solution (Answer)

Fig. 18.40

Problem 11 : Figure 18.41 (a) shows F. V. and T. V. of an object. Using the same method of projections, draw.

(i) Sectional F. V. [Section along A-A]

(ii) T. V.

(iii) L. H. S. V. **[Pune University, May 1996]**

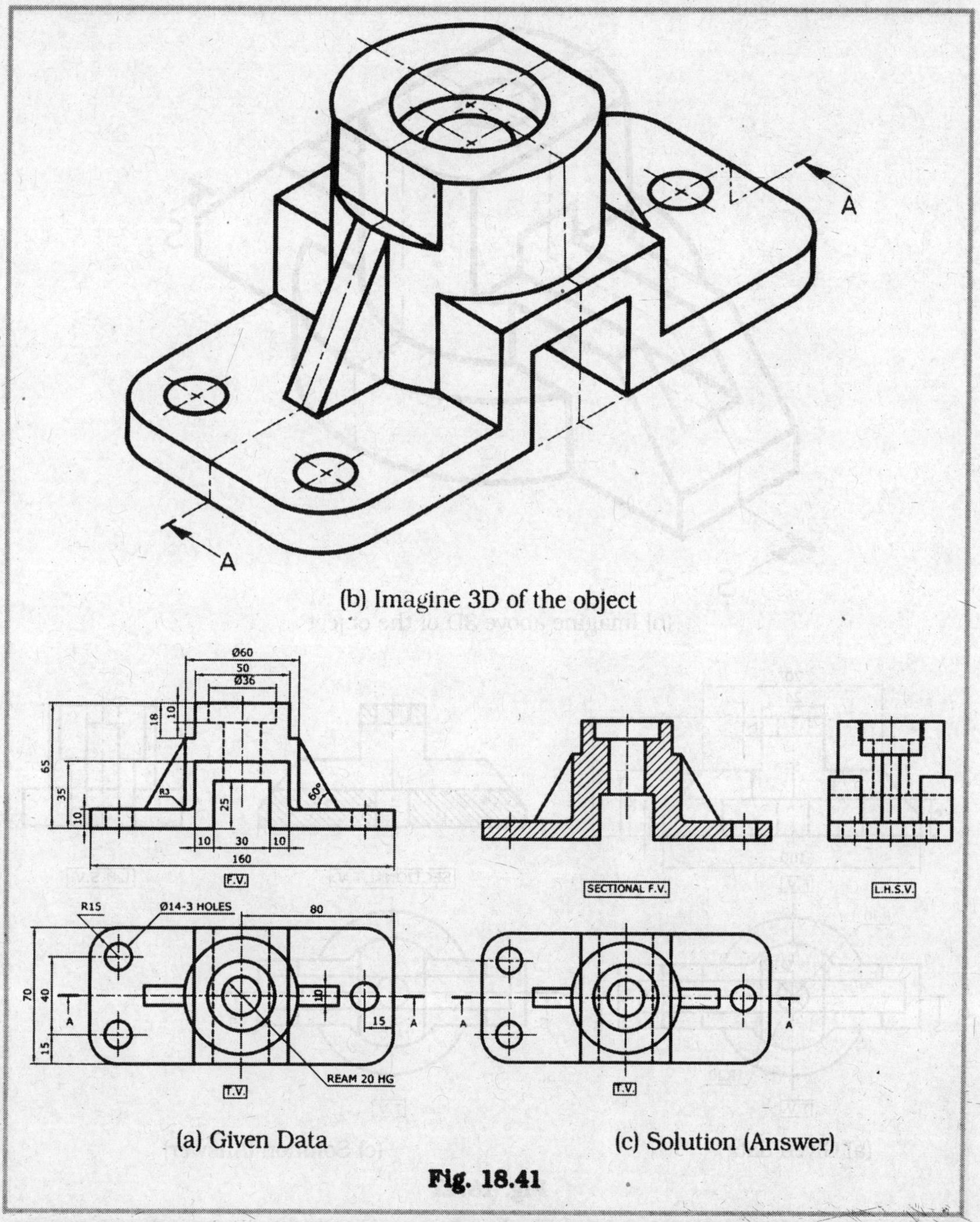

(b) Imagine 3D of the object

(a) Given Data

(c) Solution (Answer)

Fig. 18.41

Problem 12 : Figure 18.42 (a) shows F. V. and T. V. of an object. Draw

(i) Sectional F. V. [Section along S-S]

(ii) T. V.

(iii) L. H. S. V.

Using same method of projections **[Pune University]**

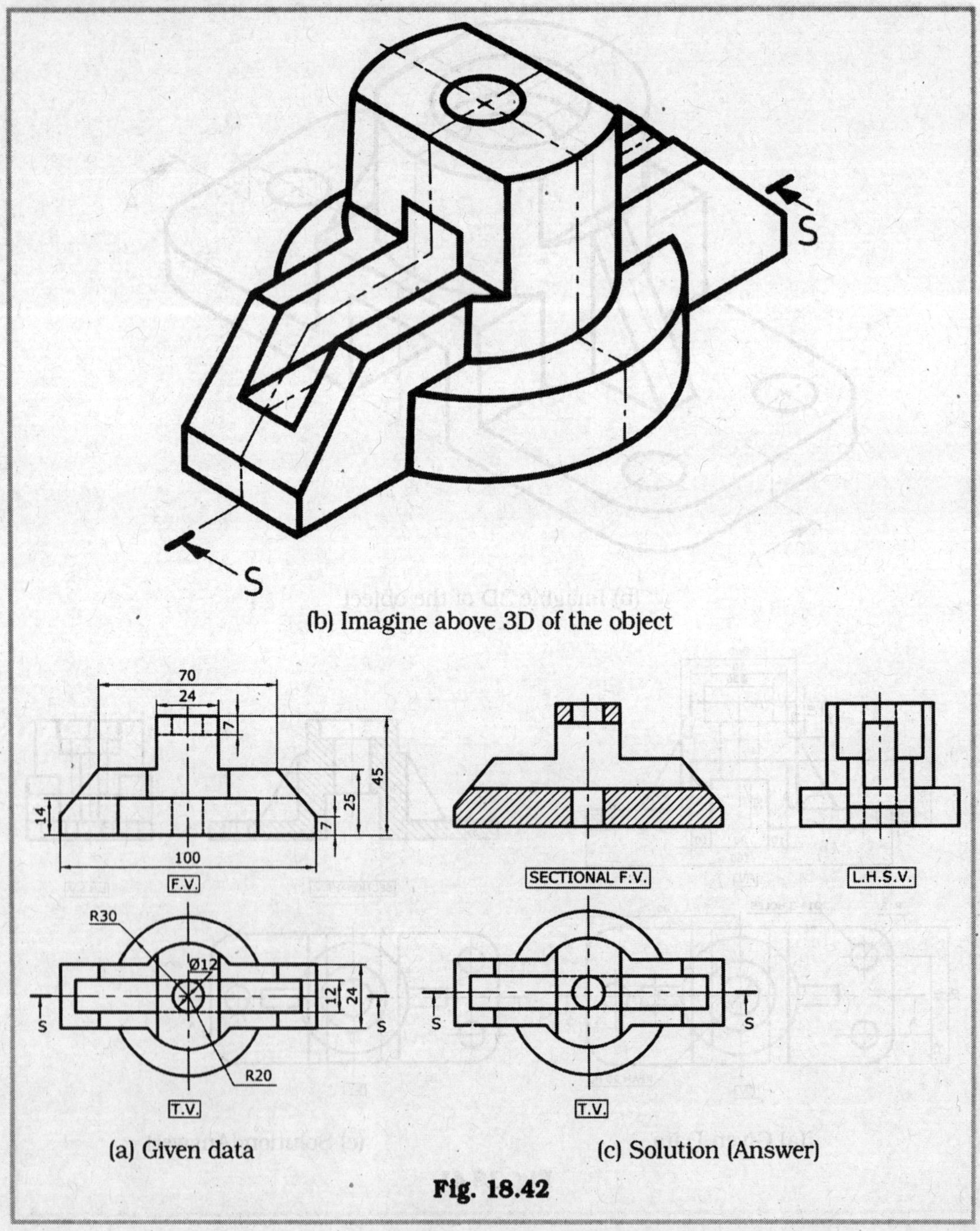

(b) Imagine above 3D of the object

(a) Given data

(c) Solution (Answer)

Fig. 18.42

Problem 13 : Figure 18.43 (a) shows F. V. and L. H. S. V. of A MACHINE PART. Draw the following views using 1:1 scale

(i) Sectional F. V. across section plane A-A.

(ii) Sectional L. H. S. V. across section plane B-B.

(iii) Missing T. V. showing all hidden details.

(iv) Insert atleast 10 important dimensions [Dimensioning not done]

[Do dimensioning as done in data] **[Mumbai University, November 1997]**

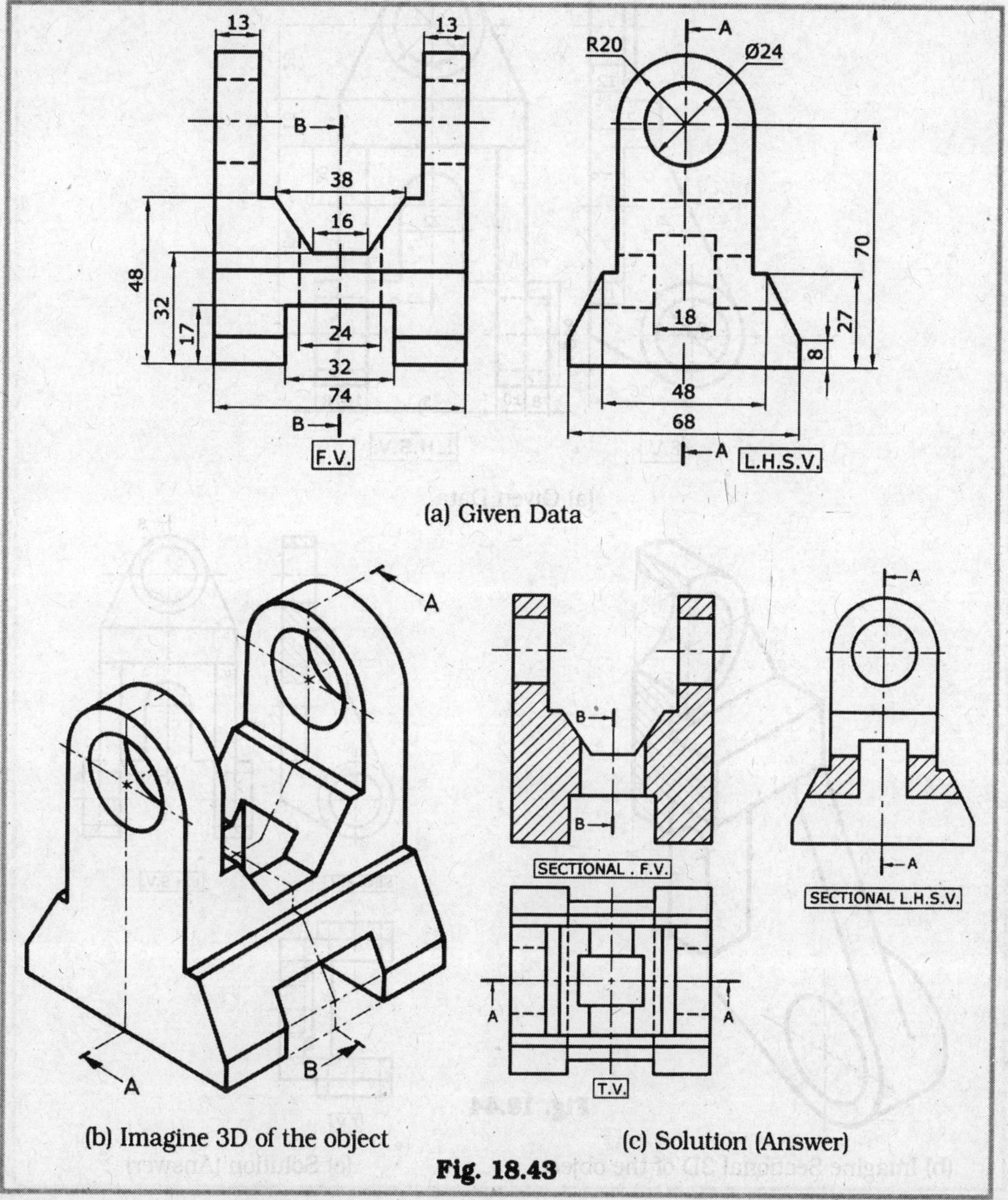

(a) Given Data

(b) Imagine 3D of the object

(c) Solution (Answer)

Fig. 18.43

Problem 14 : Figure 18.44 (a) shows F. V. and L. H. S. V. of A Bracket. Draw

(i) Sectional F. V. along S-S

(ii) L. H. S. V.

(iii) T. V.

Using same method of projections **[Pune University, November 1997]**

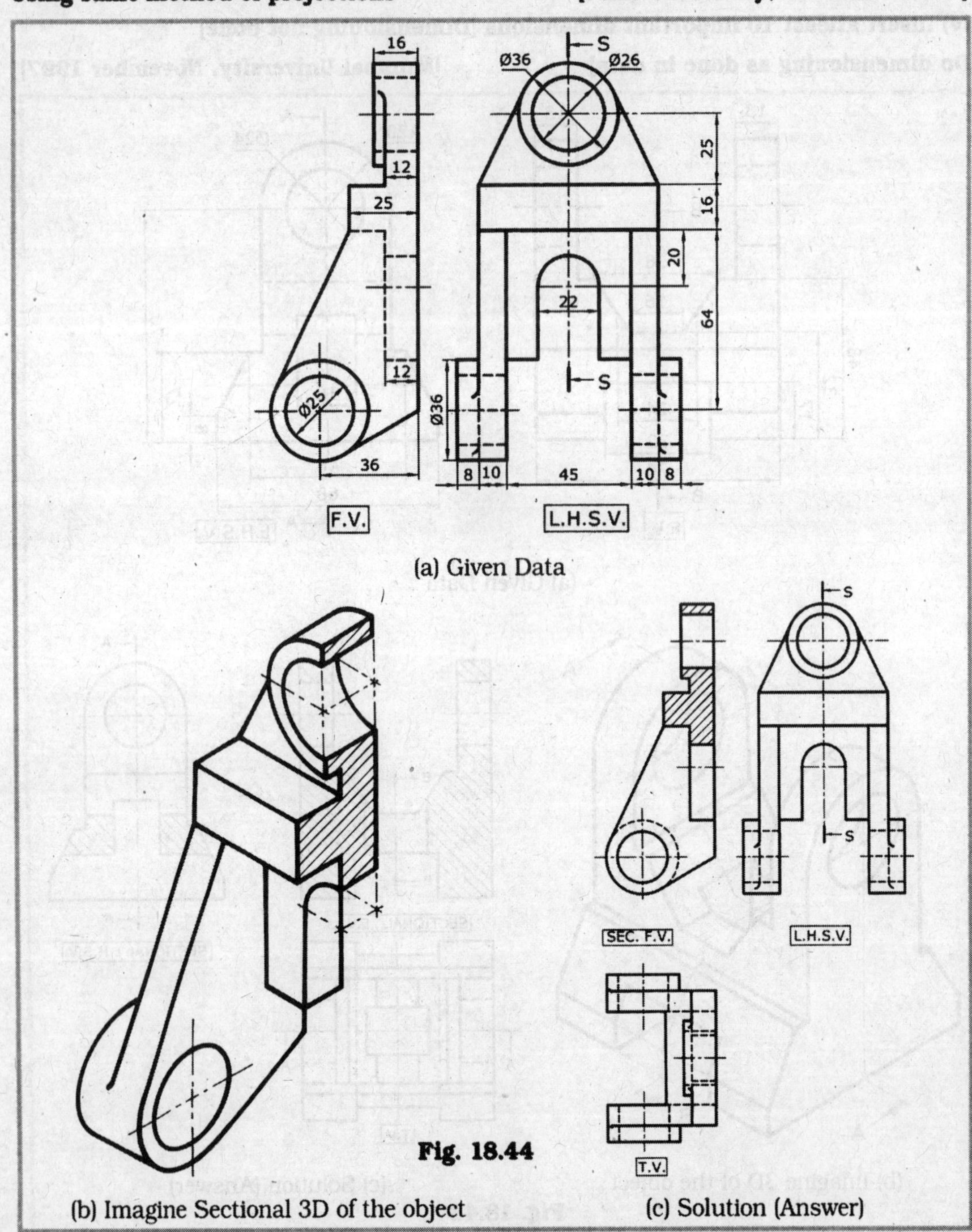

(a) Given Data

Fig. 18.44

(b) Imagine Sectional 3D of the object

(c) Solution (Answer)

Problem 15 : Figure 18.45 (a) shows two views of an object. Draw the following views by 1st angle-method.

(i) Sectional Elevation. [Section along A-A]

(ii) Plan (T. V.)

(iii) L. H. S. V. **[Pune University, November 1993]**

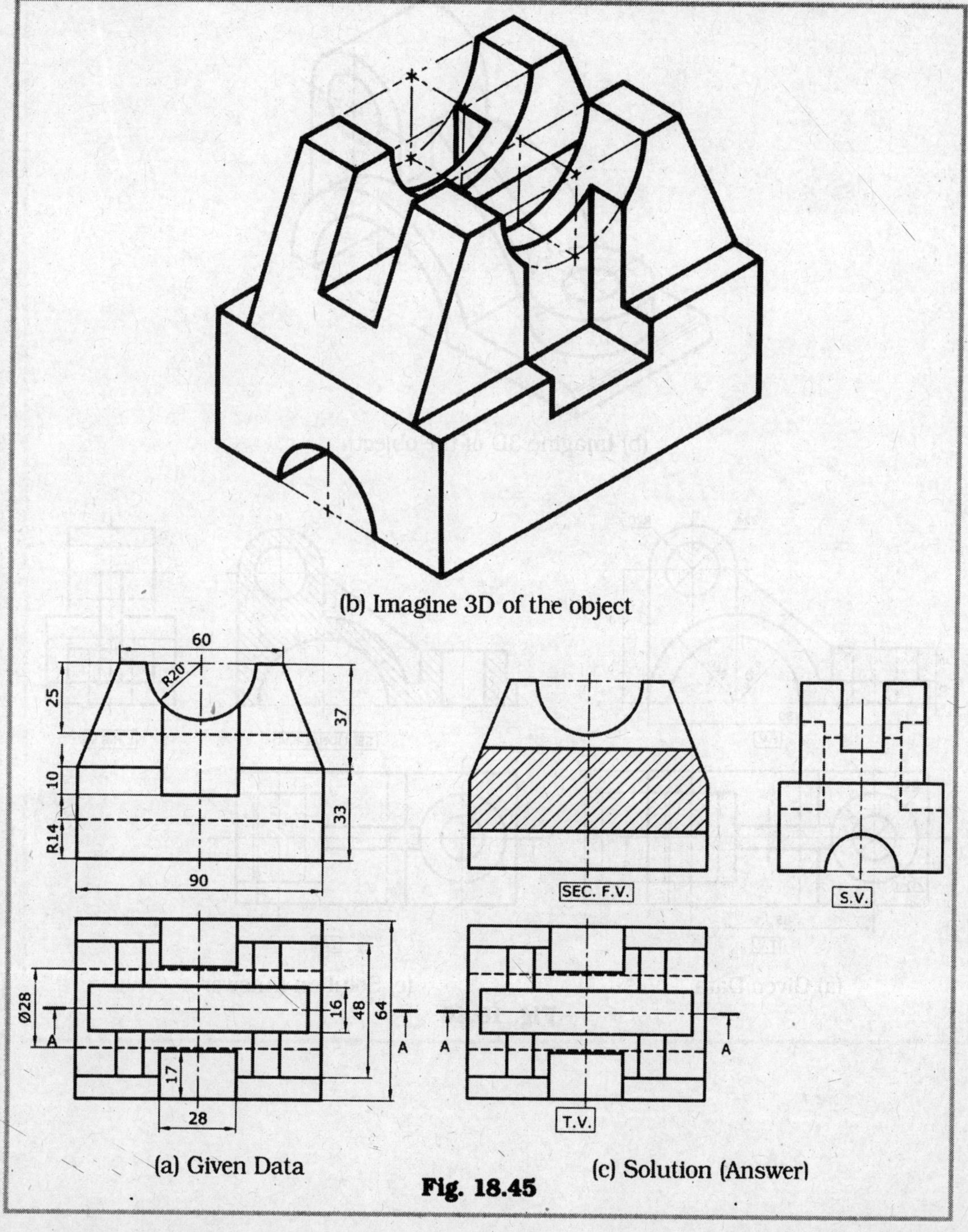

Fig. 18.45

Problem 16 : Figure 18.46 (a) shows two views of an object. Using the same method of projection, draw

(i) Sectional F. V. [Section along S-S]

(ii) Given T. V.

(iii) Left Hand Side View. **[Pune University, May 1996]**

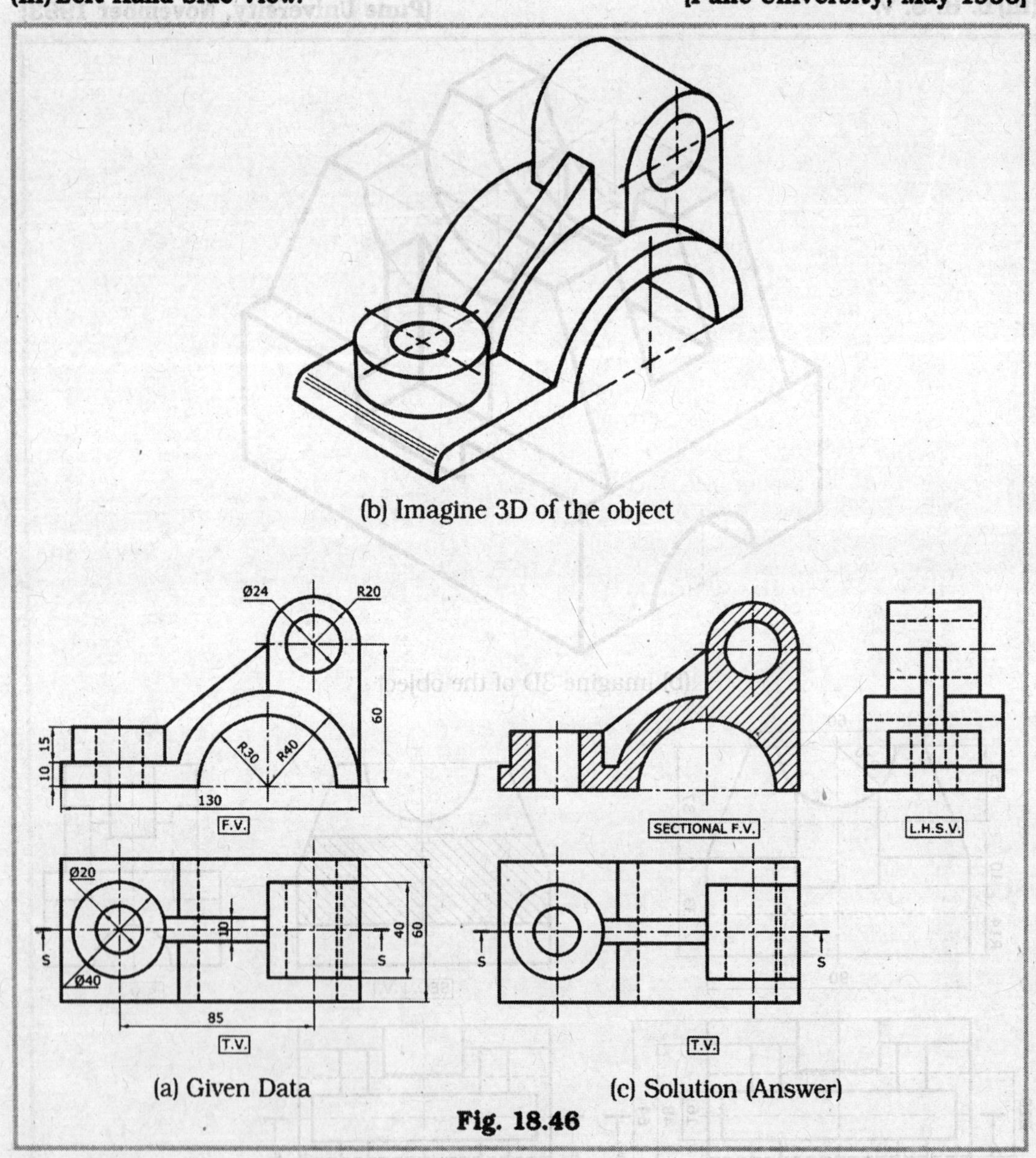

(b) Imagine 3D of the object

(a) Given Data (c) Solution (Answer)

Fig. 18.46

Problem 17 : Two views of an object are shown in Figure 18.47 (a).

Using 1st angle projection method, draw the following views.

(i) Sectional Elevation [Section along A-A]

(ii) Sectional Plan [Section along B-B]

(iii) Right Hand Side View. **[Mumbai University, November 1994]**

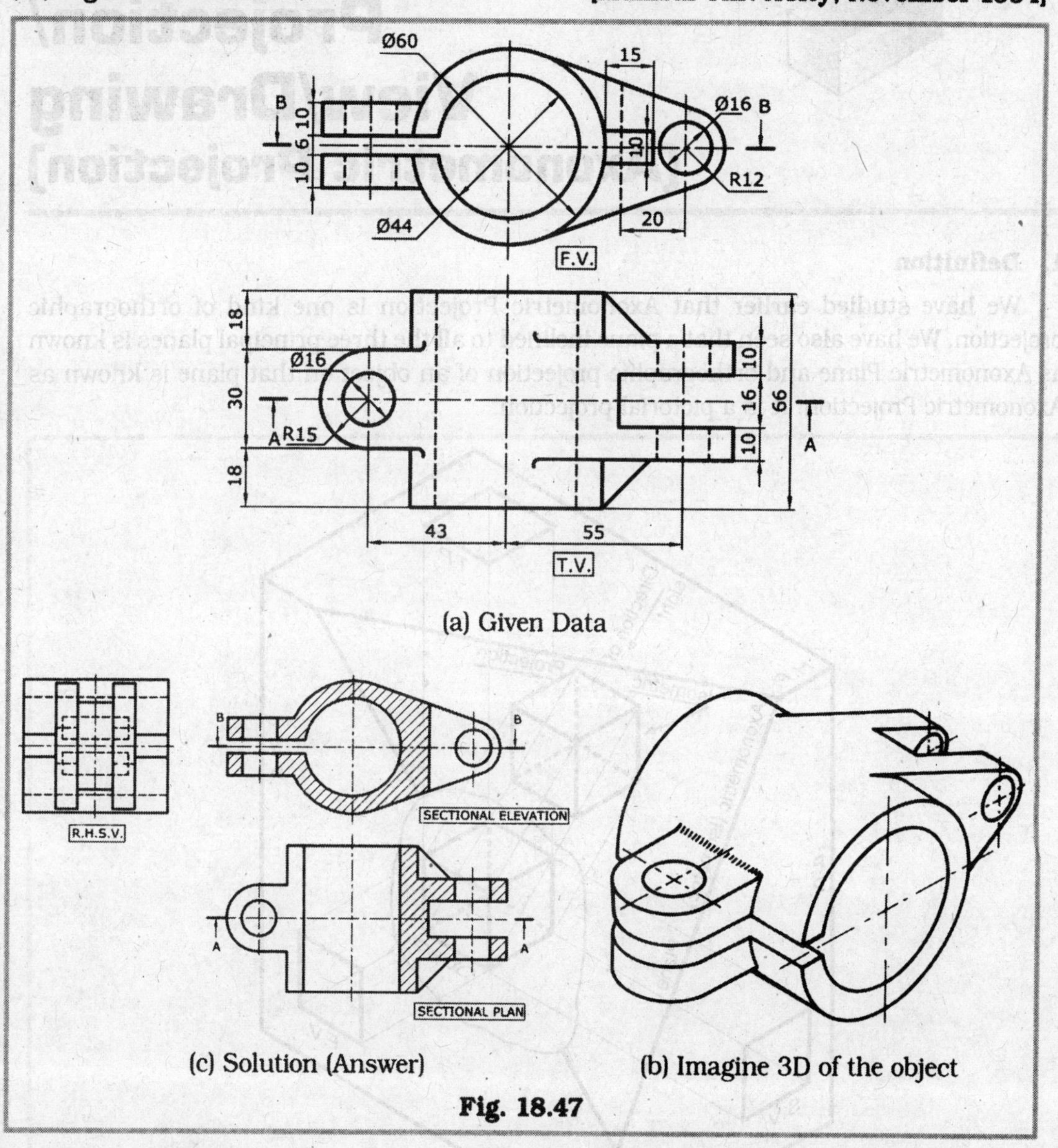

(a) Given Data

(c) Solution (Answer)

(b) Imagine 3D of the object

Fig. 18.47

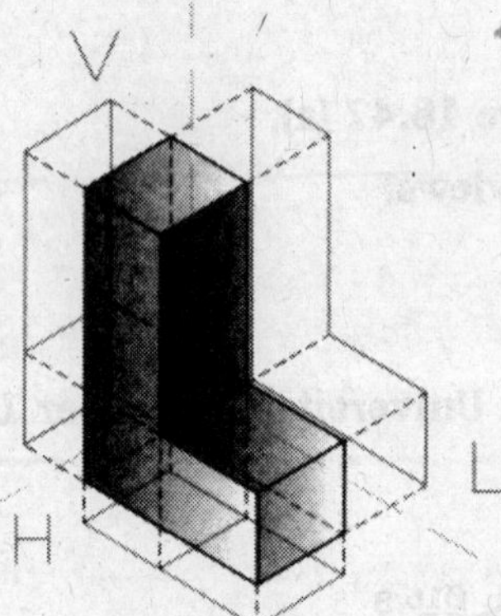

CHAPTER - 19

Isometric Projection/ View/Drawing (Axonometric Projection)

1. Definition

We have studied earlier that Axonometric Projection is one kind of orthographic projection. We have also seen that a plane inclined to all the three principal planes is known as Axonometric Plane and orthographic projection of an object on that plane is known as Axonometric Projection. It is a pictorial projection.

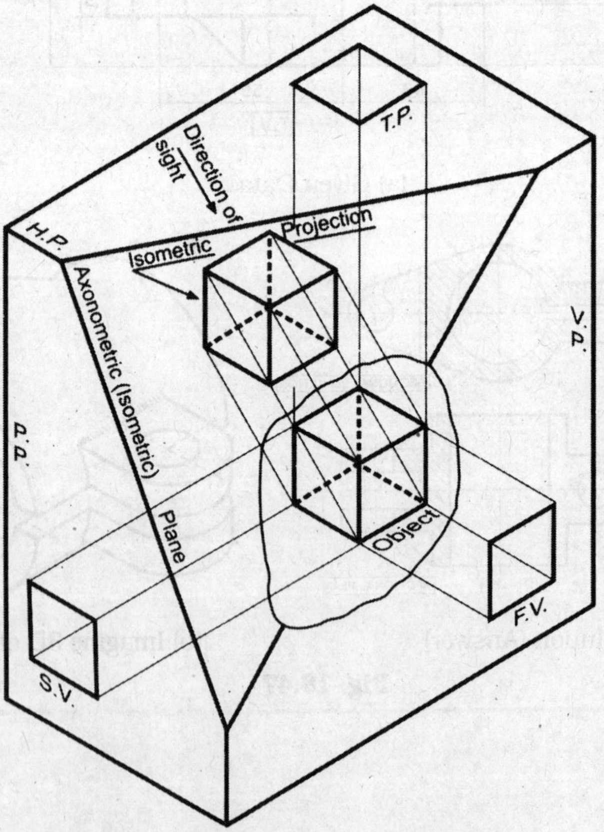

Fig. 19.1. Theory of Isometric Projection

If Axonometric Plane is equally inclined to all the three principal planes then that plane is known as the isometric plane and orthographic projection on it is known as isometric projection. Similarly, if Axonometric Plane is equally inclined to two of the three principal planes then that is known as the diametric plane and orthographic projection on it is known as diametric projection. Similarly, if Axonometric Plane is having different inclination with all the three principal planes, then that plane is known as the trimetric plane and orthographic projection on it is known as trimetric projection.

In Fig. 19.1, object (Cube) is kept inside the box with its faces parallel to principal planes i.e. (V.P., H.P. and P.P.). Principal orthographic projections of an object, that is F.V., T.V. and S.V., are shown on V.P., H.P. and P.P. respectively. In Fig. 19.1, Axonometric plane is shown and projection on it of the object is also shown. Here projection is isometric projection. Direction of sight is perpendicular to the plane of projection. i.e. Axonometric plane.

2. Isometric

As the picture plane, in the case of isometric projection, is equally inclined to all the three principal planes, dimensions of the object in all the three directions are equally foreshortened. And hence same scale can be used on all the three axes. Isometric means equal measurement. 4 edges of the cube are perpendicular to V.P., 4 edges are perpendicular to H.P. and 4 edges are perpendicular to P.P. These edges in a set of four are seen along parallel lines in three directions 120° apart, out of which one is vertical.

3. Diametric

As the picture plane, in the case of diametric projection, is equally inclined to two of the three principal planes dimensions of the object in two directions are equally foreshortened and in the third direction it is foreshortened by some other amount.

4. Trimetric

As the picture plane, in the case of trimetric projection, is unequally inclined to all the three principal planes, dimensions of the object in all the three directions are unequally foreshortened.

Diametric and trimetric projections are not discussed in this chapter. Only isometric projection as well as isometric view/drawing are discussed in full details.

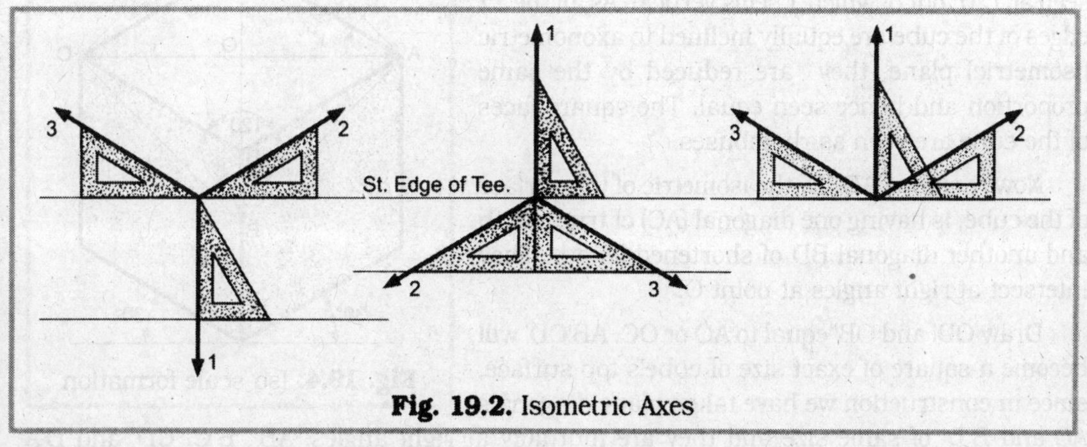

Fig. 19.2. Isometric Axes

5. Isometric Lines/Axes

Lines in space, perpendicular to principal planes, are known as Isometric Lines and these lines in Isometric projection are equally fore shortened and are seen along Isometric Axes. (Three lines 120° apart with one of them vertical). Isometric axes in three different formats are shown in the Fig. 19.2. All other lines are known as non-isometric lines and by what proportion they will be foreshortened is not known and hence dimensions of non-isometric lines are never used.

6. Isometric Planes

Any plane in space parallel to one of the principal planes and obviously perpendicular to two other P. Ps. is known as the Isometric Plane and its isometric projection can be drawn by the use of corresponding two axes as shown in Fig. 19.3 All the three isometric planes are shown in Fig. 19.3.

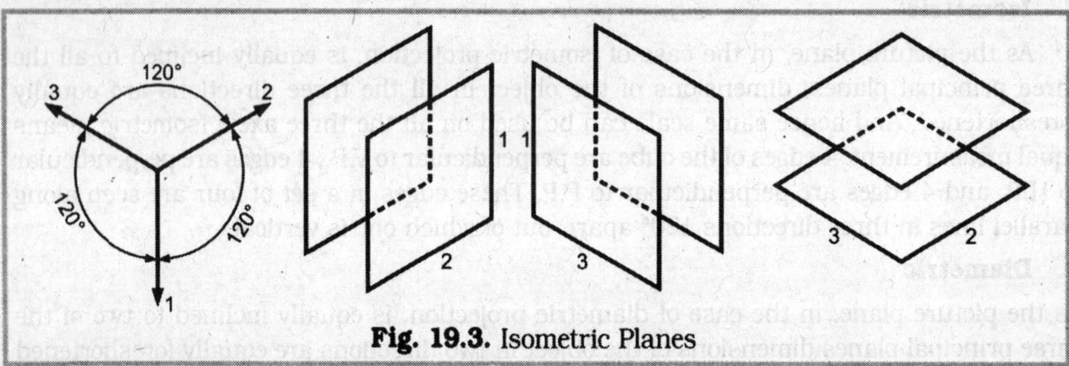

Fig. 19.3. Isometric Planes

7. Isometric Scale

Axonometric plane, which is equally inclined to all the three principal planes, is parallel to one of the diagonals of a cube's top square surface. Try to visualise this from Fig. 19.1

In the Fig. 19.4 diagonal AC, since it happens to be parallel to axonometric (isometric) plane, is shown in its true length while diagonal BD is shown of shortened length. Three sets of 4 edges of cube are seen at 120° out of which 1 set is vertical. As all the 12 edges of the cube are equally inclined to axonometric (isometric) plane, they are reduced by the same proportion and hence seen equal. The square faces of the cube are seen as rhombuses.

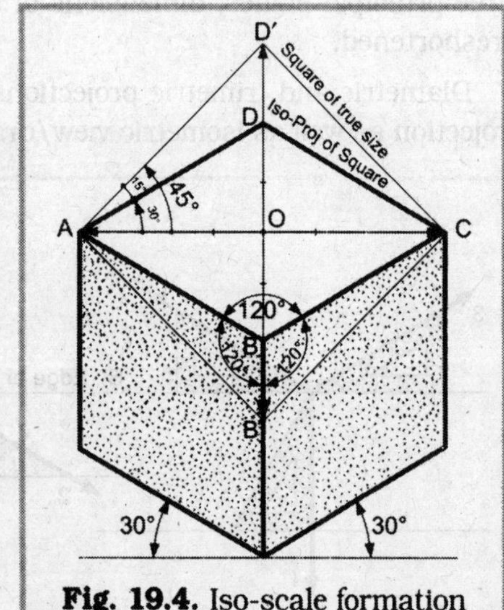

Fig. 19.4. Iso-scale formation

Now rhombus ABCD, the isometric of top surface of the cube, is having one diagonal (AC) of true length and another diagonal BD of shortened length. They intersect at right angles at point O.

Draw OD' and OB' equal to AO or OC. AB'CD' will become a square of exact size of cube's top surface, since in construction we have taken plane diagonals AC and B'D' of same size and they are mutually at right angles. AB', B'C, CD' and D'A represent true lengths. AB, BC, CD and DA represent lengths in isometric projection.

To understand the scale formation, we take a set of lines AD and AD'.

AD represents the isometric length. It is at an angle of 30° to horizontal. AD' represents true or actual or normal length. It is at an angle of 45° to horizontal. Angle between AD and AD' will obviously become 15°.

From the geometry of the Fig. 19.4, isometric scale of Fig. 19.5 is evolved.

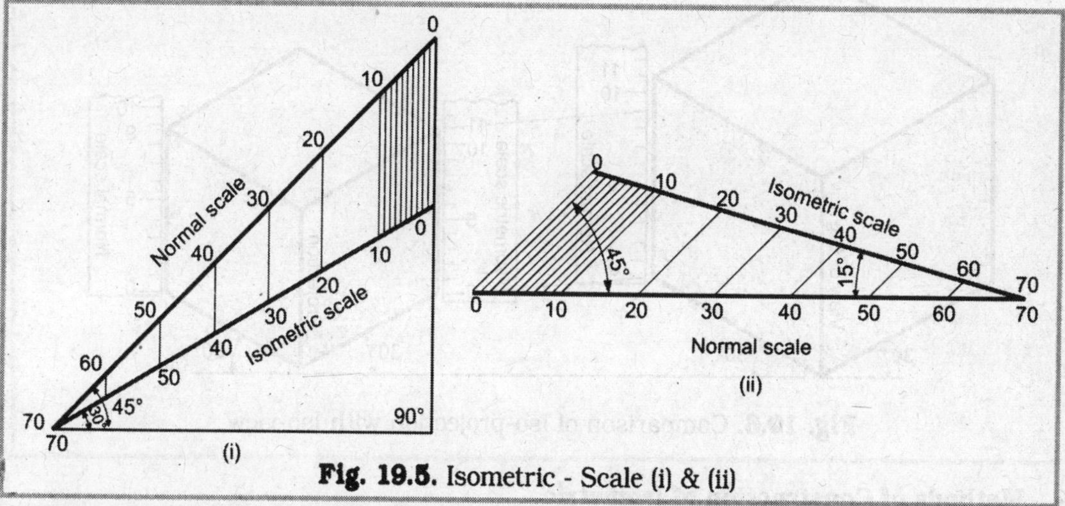

Fig. 19.5. Isometric - Scale (i) & (ii)

There are two different styles of preparing isometric scale.

(1) In the style first, normal, true or actual scale is drawn on a line at 45° to the horizontal line. By drawing vertical lines from normal scale line up to a line at 30° to the horizontal, we get isometric scale on a 30° line as shown in the Fig. 19.5(i)

(2) In the style second, normal, true or actual scale is drawn on a horizontal line. By drawing parallel lines at 45° to the horizontal line from normal scale line upto a line at 15° to the horizontal, we get isometric scale on a 15° line as shown in the Fig. 19.5 (ii).

Reduction of isometric line in isometric projection is

$$\frac{\text{Isometric length}}{\text{Normal, true or actual length}} = \frac{\sqrt{2}}{\sqrt{3}} = 0.815\,\Omega\,\frac{9}{11}$$

In short, figure in isometric projection is reduced lengthwise by approximately 81.5%. Volumewise it is reduced by (0.815 × 0.815 × 0.815 = 0.5413) or 54.13%.

8. Isometric View or Drawing

We are more interested in the shape of isometric of an object rather than its size and hence normally isometric view or drawing is drawn without reducing dimensions by isometric scale. When projection is drawn dimensions are reduced by the scale and then used. When iso-view or iso-drawing is drawn, then the actual dimensions are used i.e. without reducing them by the scale.

In Fig. 19.6, comparison of isometric view or drawing with isometric projection is shown. View or drawing of a cube is shown drawn by the normal scale while projection of a cube is shown drawn by the isometric scale. Method of drawing is absolutely same, there is only difference in the size. View or drawing is bigger in size (of actual size) while projection is smaller in size (of shortened size.)

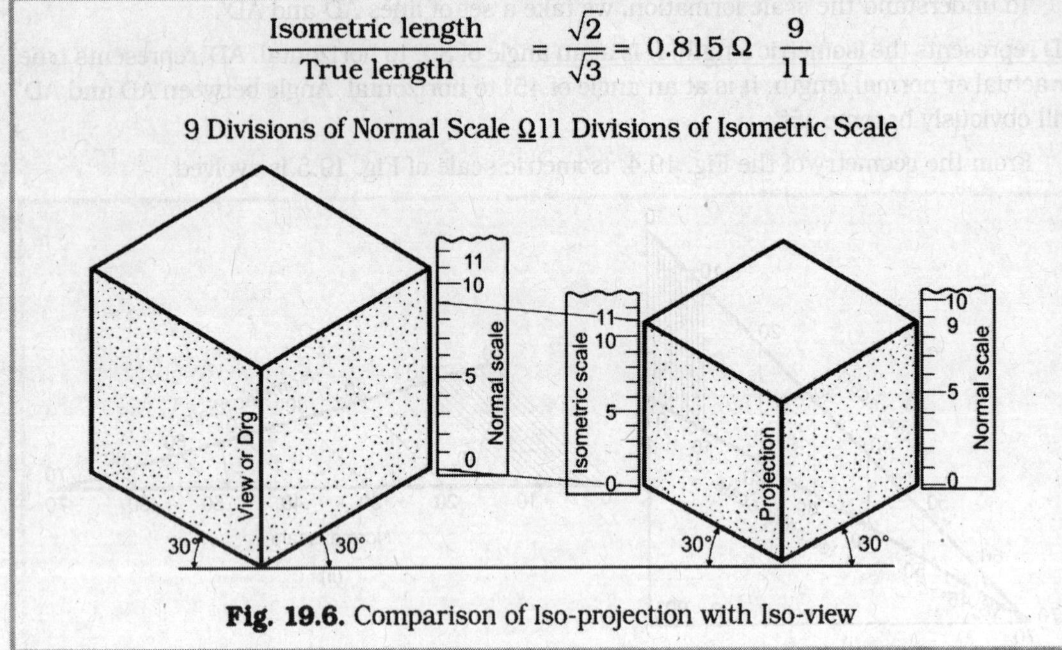

Fig. 19.6. Comparison of Iso-projection with Iso-view

9. Methods of Construction of Isometric

There are two methods of construction :

(i) Conventional method

(ii) Exact method

(i) Conventional Method :

This is a practical method. In this method for getting projection we have to use one scale for isometric, two scales for diametric and three scales for trimetric projections. Usually isometric drawing is drawn instead of projection by making measurements with a normal scale. While constructing isometric, measurements should be made only on or parallel to the three isometric axes.

(ii) Exact Method :

This method is theoretically correct and exact. It is a simple and superior method. In this method scales are not required to be used. Scales are taken care of automatically and exactly. Diametrics and trimetrics can be drawn very easily with perfect flexibility as to choice of angle of axes and position of object.

First of all we shall study the conventional method and solve problems also by the same method. Later on in the chapter we shall study the exact method and solve few problems also by the same method.

10. Isometric Projection of Plane Figures

We shall study this by taking different problems.

Problem 1 : The (i) Plan or (ii) Elevation or (iii) S. V. of a 40 mm square plate is given in Fig. 19.7. Draw its isometric view.

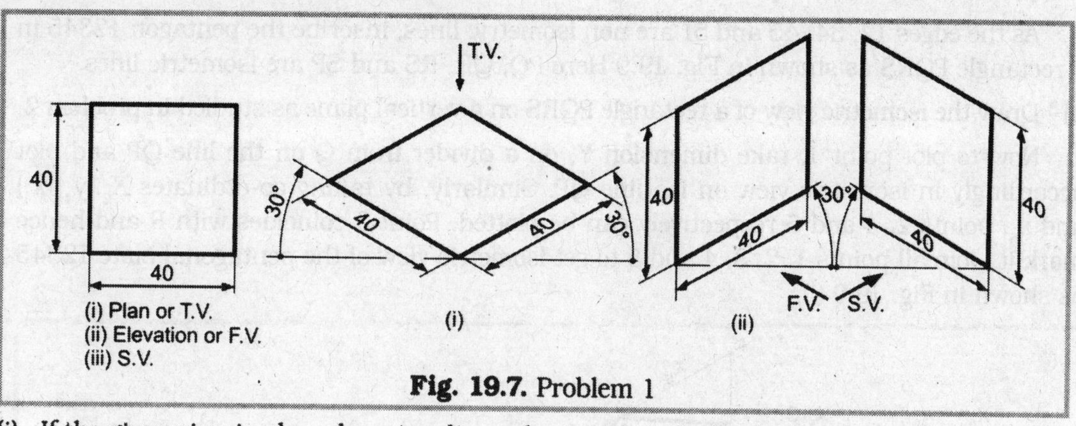

Fig. 19.7. Problem 1

(i) If the given view is plan, draw two lines along axes to form a horizontal plane. Here use two axes other than vertical. Vertical axis will be the direction of sight. Take 40 mm on two lines already drawn and complete the rhombus by drawing parallel lines to axes as shown in Fig. 19.7 (i).

(ii) If the given view is elevation, draw two lines along axes to form a vertical plane. Here use vertical axis and one suitable axis, remaining third axis will be the direction of sight. Take 40 mm on two lines already drawn. Complete the rhombus by drawing parallel lines to the axes as shown in Fig. 19.7 (ii).

(iii) If the given view is a side view, draw two lines along axes to form a plane parallel to the profile plane. Use vertical axis and one suitable axis. Direction of sight for S. V. will be the third remaining axis. On both the axes, already drawn, take 40 mm and then draw parallel lines to get rhombus as shown in Fig. 19.7 (iii).

Problem 2 : The (i) F. V. or (ii) S. V. or (iii) T. V. of a rectangular plate having its surface parallel to (i) V. P. or (ii) P. P. or (iii) H. P. respectively is given in the Fig. 19.8 Draw its isometric projection.

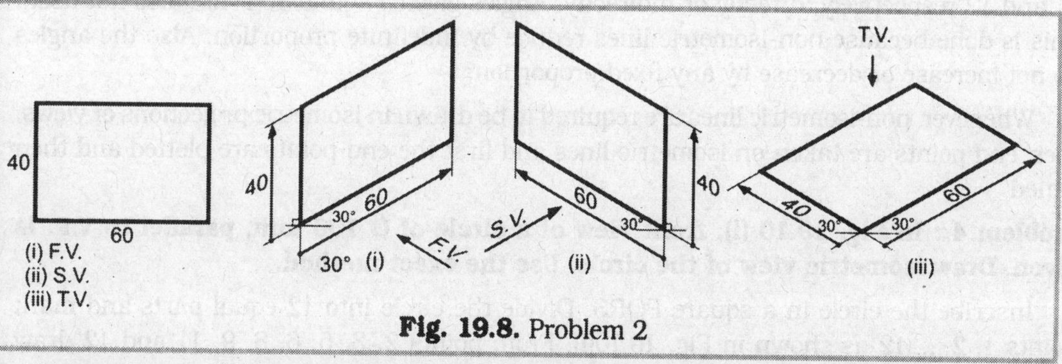

Fig. 19.8. Problem 2

Here take a set of two suitable axes according to the given view and take 60 mm and 40 mm (Iso-lengths) on the two axes and draw parallelograms to get isometric projections. See Fig. 19.8 (i), (ii) & (iii).

Here as the projection is required, reduce the dimensions 60 mm and 40 mm by the isometric scale.

Problem 3 : The (i) Elevation or (ii) Plan of a pentagonal plate having its surface parallel to (i) V. P. or (ii) H.P. respectively is given. Draw its isometric view.

As the edges 12, 34, 45 and 51 are non isometric lines, inscribe the pentagon 12345 in a rectangle PQRS as shown in Fig. 19.9 Here PQ, QR, RS and SP are isometric lines.

(i) Draw the isometric view of a rectangle PQRS on a vertical plane as studied in problem 2.

Now to plot point 1, take dimension Y_1 on a divider from Q on the line QP and plot accordingly in isometric view on iso-line QP. Similarly, by taking co-ordinates X_2,(y_4, x_4) and x_5, points 2, 4 and 5 respectively can be plotted. Point 3 coincides with R and hence mark it. Join all points 1, 2, 3, 4 and 5 to get isometric view of the pentagonal plate 12345 as shown in Fig. 19.9 (i)

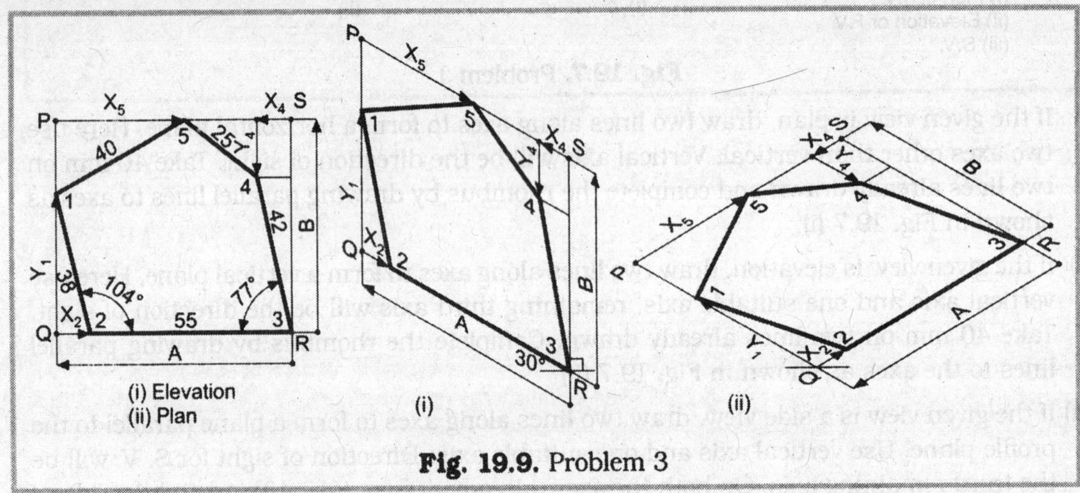

Fig. 19.9. Problem 3

(ii) Here isometric view of the rectangle PQRS should be drawn on a horizontal plane instead of vertical as shown in Fig. 19.9 (ii). Other procedure remains the same.

Students should note that during construction of the above isometric views (i) and (ii) we have not used anywhere given dimensions 42, 25, 40 and 38 of non isometric lines 34, 45 51 and 12 respectively, directly or indirectly. Angles given 77° and 104° are also not used. This is done because non-isometric lines reduce by indefinite proportion. Also the angles do not increase or decrease by any fixed proportion.

Whenever, non-isometric lines are required to be drawn in isometric projections or views, their end points are taken on isometric lines and first the end points are plotted and then joined.

Problem 4 : In Fig. 19.10 (i), front view of a circle of Ø 100 mm, parallel to V.P. is given. Draw isometric view of the circle. Use the exact method.

Inscribe the circle in a square PQRS. Divide the circle into 12 equal parts and mark points 1, 2,.....12 as shown in Fig. 19.10(i). From points 2, 3, 5, 6, 8, 9, 11 and 12 draw horizontal and vertical lines and find out co-ordinates (x_2, y_2), (x_3, y_3), (x_5, y_5), (x_6, y_6), (x_8, y_8), (x_9, y_9), (x_{11}, y_{11}) and (x_{12}, y_{12}) as shown in figure. The points 1, 4, 7 and 10 will become the mid points of PQ, QR, RS and SP respectively.

Now draw rhombus PQRS, the isometric view of the square PQRS as shown in Fig. 19.10(ii). Now with the known co-ordinates ($x_2 y_2$), for the point 2, plot the point 2 taking distances x_2 and y_2 along isometric axes from Q as shown in the Fig. 19.10(ii). Similarly, find points 3, 5, 6, 8, 9, 11 and 12. Find also the points 1, 4, 7 and 10 by marking the mid points of PQ, QR, RS and SP respectively.

Join all points 1, 2, 3,.....12 and 1 by a smooth curve to get the isometric view of the circle. It is an ellipse.

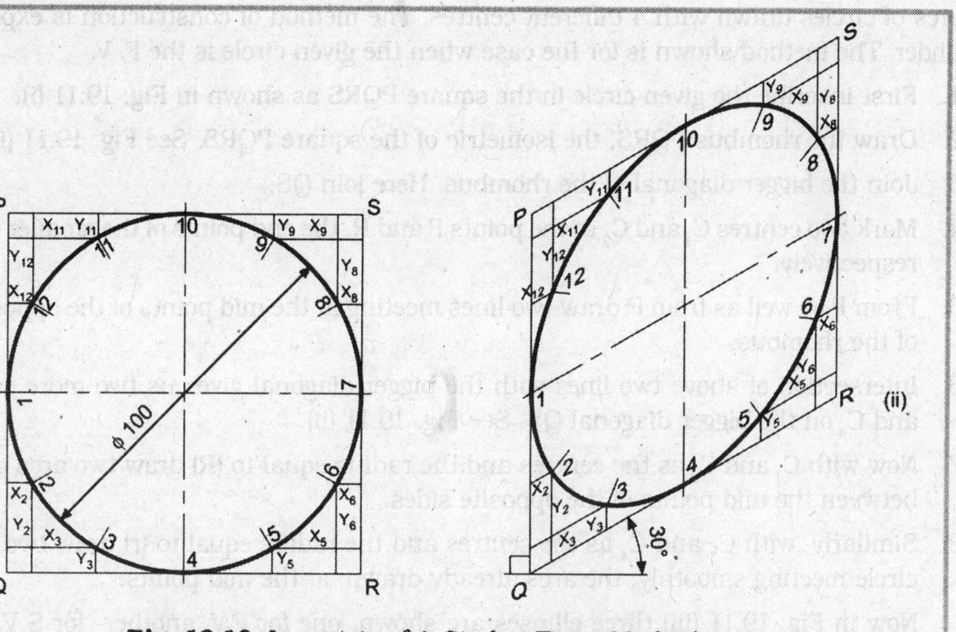

Fig. 19.10. Isometric of A Circle - Exact Method

Problem 5 : In Fig. 19.11 (i), front view of a circle, parallel to V.P. is given. Draw isometric view of the circle. Use four centre approximate method.

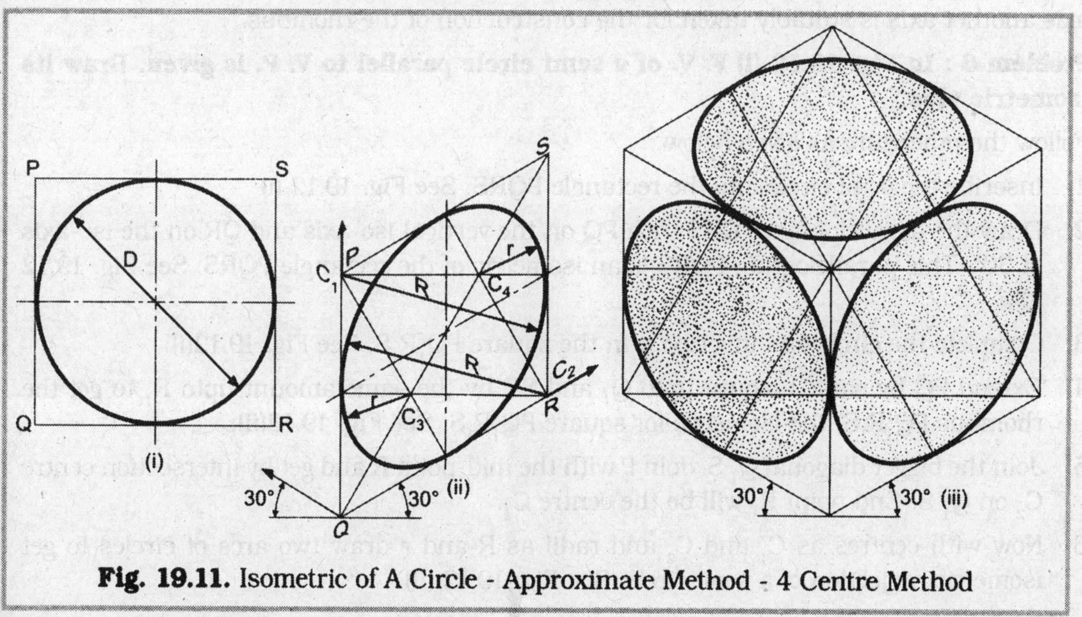

Fig. 19.11. Isometric of A Circle - Approximate Method - 4 Centre Method

In the problem 4 we have studied the exact method of drawing isometric of a circle. In the exact method, student is required to join 12 points by a smooth free hand curve. To join 12 points by a perfectly smooth curve is a difficult task. Eventhough it is an exact method, it is rarely used.

Ellipse, the isometric projection/view of a circle, is constructed by a combination of 4 arcs of circles drawn with 4 different centres. The method of construction is explained as under. The method shown is for the case when the given circle is the F. V.

1. First inscribe the given circle in the square PQRS as shown in Fig. 19.11 (i).
2. Draw the rhombus PQRS, the isometric of the square PQRS. See Fig. 19.11 (ii).
3. Join the bigger diagonal of the rhombus. Here join QS.
4. Mark two centres C_1 and C_2 at the points P and R, the end points of the smaller diagonal, respectively.
5. From P as well as from R draw two lines meeting at the mid points of the opposite sides of the rhombus.
6. Intersection of above two lines with the bigger diagonal gives us two more centres C_3 and C_4 on the bigger diagonal QS. See Fig. 19.11 (ii).
7. Now with C_1 and C_2 as the centres and the radius equal to (R) draw two arcs of a circle between the mid points of the opposite sides.
8. Similarly, with C_3 and C_4 as the centres and the radius equal to (r) draw two arcs of a circle meeting smoothly, the arcs already drawn, at the mid points.

Now in Fig. 19.11 (iii) three ellipses are shown, one for F.V. another for S.V. and the third for T.V.

To get the isometric of a circle, given in the plan, rhombus is constructed by two axes which are at 30° to the horizontal. For the given circle in F. V. and S. V. along with vertical axis another axis is suitably taken for the construction of the rhombus.

Problem 6 : In Fig. 19.12 (i) F. V. of a semi circle parallel to V. P. is given. Draw its isometric view.

Follow the procedure as given below :

1. Inscribe the semi circle into the rectangle PQRS. See Fig. 19.12 (i)
2. Draw the parallelogram PQRS with PQ on the vertical iso-axis and QR on the iso-axis at 30°. This parallelogram PQRS is an isometric of the rectangle PQRS. See Fig. 19.12 (ii).
3. Complete the circle and inscribe it in the square PQ_1R_1S. See Fig. 19.12(i).
4. Extend PQ by same amount upto Q_1 and SR by the same amount upto R_1 to get the rhombus PQ_1R_1S, the isometric for square PQ_1R_1S. See Fig. 19.12(ii).
5. Join the bigger diagonal Q_1 S. Join P with the mid-point R and get by intersection centre C_2 on Q_1 S. End point R_1 will be the centre C_1.
6. Now with centres as C_1 and C_2 and radii as R and r draw two arcs of circles to get isometric projection of a semi circle. See Fig. 19.12 (ii).

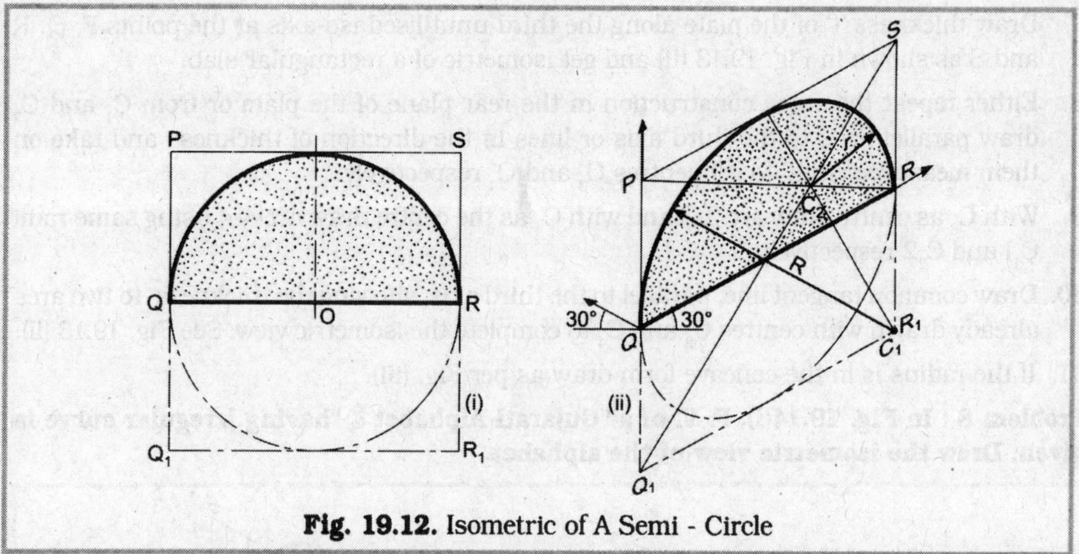

Fig. 19.12. Isometric of A Semi - Circle

Problem 7 : In Fig. 19.13 (i) Elevation and L. H. S. V., of a rectangular plate with top two corners curved are given. Draw the isometric view of a plate.

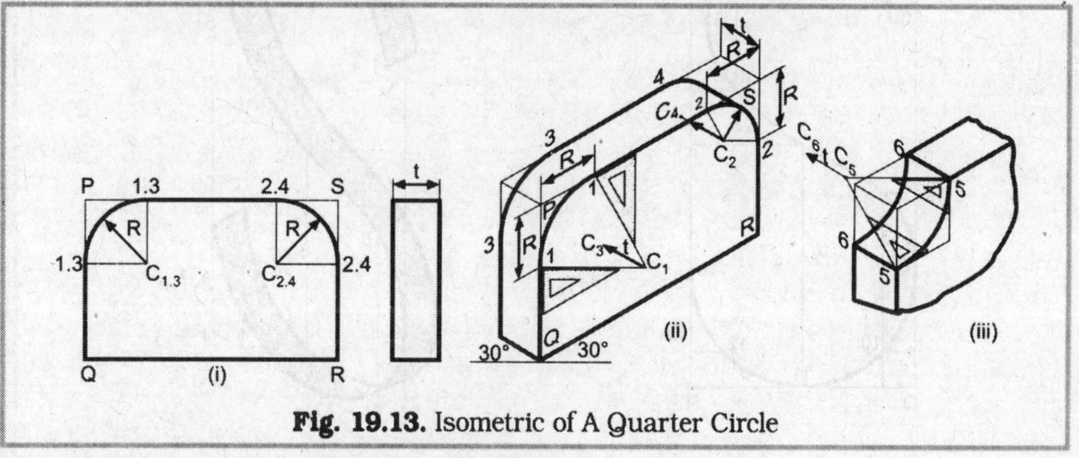

Fig. 19.13. Isometric of A Quarter Circle

For solution follow the procedure as given below :

1. Inscribe the elevation in a rectangle PQRS. See Fig. 19.13 (i).
2. Draw parallelogram PQRS with PQ along vertical isometric axis and QR along isometric axis at 30° to the horizontal. See Fig. 19.13 (ii) It represents the isometric of rectangle PQRS.
3. Along PS and PQ take distances R (the radius of curvature R of the plate) and mark two points 1 as shown in Fig. 19.13 (ii).
4. At two points 1 draw perpendicular lines to PS and PQ intersecting at the point C_1. Two setsquares indicate perpendicularity in the drawing. See Fig. 19.13 (ii).
5. Now with C_1 as the centre and radius equal to $C_1$1 draw an arc touching PQ and PS smoothly at points 1.
6. At the corner S repeat the same construction and get the centre C_2. Now with C_2 as the centre and radius equal to $C_2$2 draw an arc touching SR and SP smoothly at points 2.

7. Draw thickness 't' of the plate along the third unutilised iso-axis at the points P, Q, R and S as shown in Fig. 19.13 (ii) and get isometric of a rectangular slab.
8. Either repeat the same construction in the rear plane of the plate or from C_1 and C_2 draw parallel lines to the third axis or lines in the direction of thickness and take on them measurement 't' to get centres C_3 and C_4 respectively.
9. With C_3 as centre draw arc 3-3 and with C_4 as the centre draw arc 4-4 using same radii $C_1$1 and $C_2$2 respectively.
10. Draw common tangent line, parallel to the third axis, showing the thickness, to two arcs already drawn with centres C_2 and C_4 to complete the isometric view. See Fig. 19.13 (ii).
11. If the radius is in the concave form draw as per Fig. (iii).

Problem 8 : In Fig. 19.14(i), F. V. of a "Gujarati Alphabet ટ "having irregular curve is given. Draw the isometric view of the alphabet.

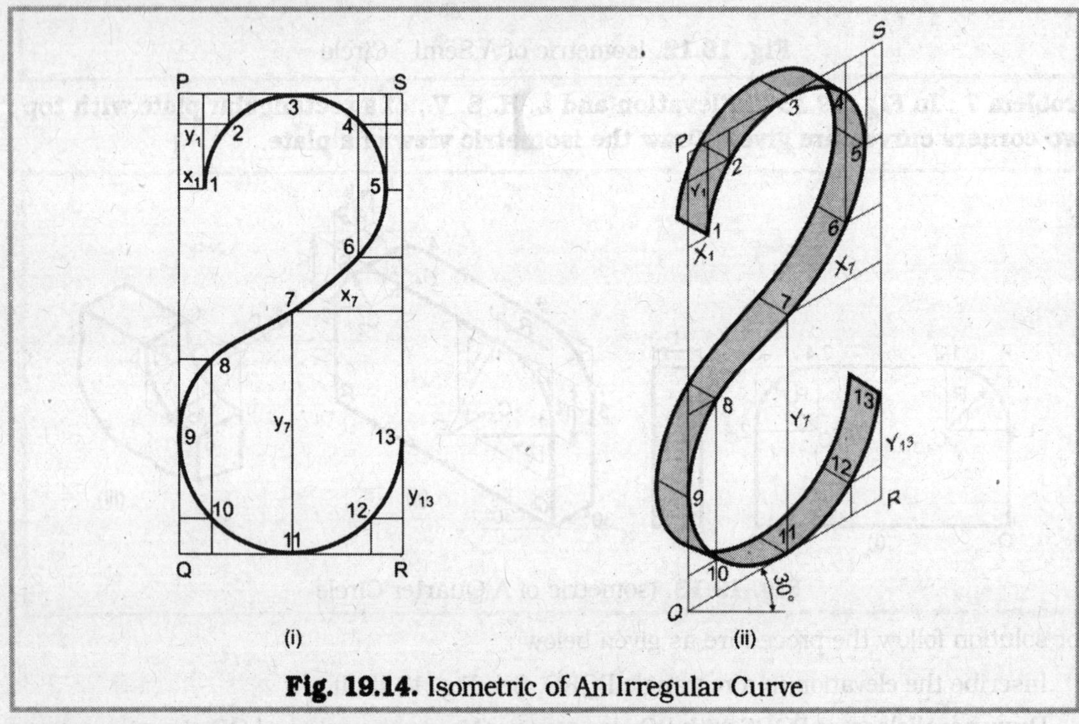

Fig. 19.14. Isometric of An Irregular Curve

For solution, follow the procedure as given below :

1. Inscribe the F. V. of alphabet in a rectangle PQRS.
2. Take points 1, 2,...13 on the irregular curve of alphabet at nearly equal distances. From points. 1, 2,.... 13 draw horizontal and vertical lines to get co-ordinates (x_1, y_1), (x_2, y_2),...... (x_{13} y_{13}) respectively w. r. t. the corner points of the rectangle.
3. Now draw PQRS parallelogram with PQ on vertical iso-axis and QR on iso-axis at 30° to the horizontal.
4. Take x_1 and y_1 distances from reference point P along the direction of iso-axes to get point 1 as shown in Fig. 19.14 (ii). Similarly, find points 2, 3... 13 by taking (x_2, y_2), (x_3, Y_3)....($X_{13}$$Y_{13}$) from corresponding reference corner points. Join points 1, 2, 3,... 13 by a smooth curve to get the isometric projection/view of the alphabet.

5. Draw equal thickness lines in the direction of 3rd axis to provide thickness to the alphabet as shown.

11. Isometric of Cylinders, Prisms, Pyramids, Cones, their Frustums and Truncated Frustums

Isometric projections/views of above solids are drawn mostly by box method. We shall study first the box method and then some important aspects relevant to the Box Method. In this method of isometric projection/view, solid object is assumed to be inside the box. The shape of the box, in which object is enclosed, is mostly square or rectangular prism. First, isometric of the square or rectangular prism is drawn and then by using isometric lines and iso-planes of the box, isometric projection/view of the object is constructed. By joing suitably the points, lines and planes thus obtained we can get isometric of the object.

This box-method can also be used for isometric projections/views of many complicated objects. Sometimes separate boxes for different portions of an object can be constructed at appropriate locations and in positions and thereby isometric of complicated object can be drawn.

Important Aspects :

While constructing isometric projections/views of solid objects, students should keep in mind the following important aspects :

1. Iso-axes for given F. Vs. and for given S. Vs. should be so selected that the given views will be totally visible from drawn isometric projections/views of the objects, e.g. whenever, F. V. and R. H. S. V. are given as shown in problems 29, 32, Figs. 19.36 and 19.39, the selection of iso-axes for F. V. and R. H. S. V. are so done that both views are totally visible.

   If side view is not given, selection of the iso-axes should be done in such a way that maximum details of the object will be visible in iso-projection or iso-view.

2. At every corner of a solid there must be at least three edges. Every corner of a plane has two edges. Out of the three edges of a visible corner of a solid at least two must be visible, if not all three. (in isometric) Check every corner for visible edges out of three edges.

3. Hidden details of the objects are generally not shown in isometric projections or views even with dashed or dotted lines. If hidden details of the objects are of such a nature that one cannot imagine them on the basis of visible portion then only those hidden details are shown by dotted lines.

4. In isometric, two edges (visible-full-continous) will never cross each other i.e. out of the two crossing edges one will be full and another will be dotted.

Now having studied the box method and important aspects for isometric projections of solids we shall solve few problems of cylinders, cones, prisms, pyramids, etc.

Problem 9 : In Fig. 19.15 (i) elevation and R. H. S. V. of an object are given. Draw the isometric view.

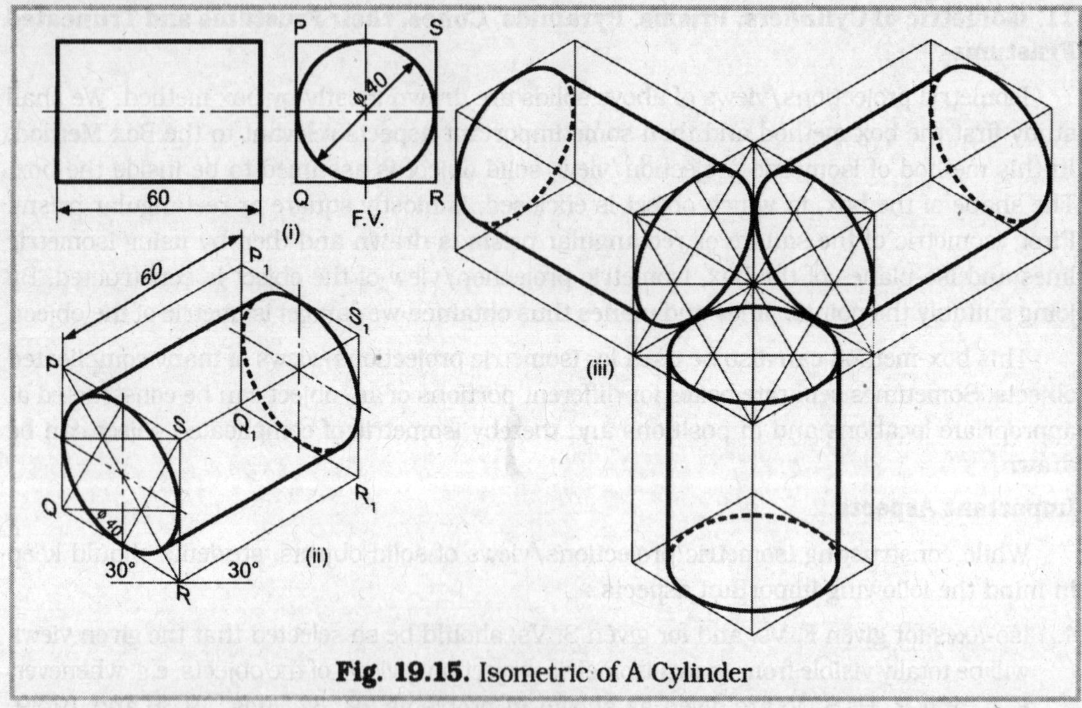

Fig. 19.15. Isometric of A Cylinder

For solution, follow the procedure as given below :

1. Inscribe the circle of elevation in a square PQRS. In fact by doing so we have enclosed the object (cylinder) in a square prism.
2. For elevation select vertical axis and left hand side 30° axis so that both F. V. and R. H.S.V. will be totally visible.
3. Draw rhombus PQRS with SR on vertical axis and QR on left hand side 30° axis. By four centre method construct the ellipse, the isometric of a circle.
4. On the third, right hand side 30°, axis take length available from side view and get P_1, Q_1, R_1 and S_1. Join them to get rhombus $P_1Q_1R_1S_1$ as shown in Fig. 19.15 (ii).
5. By four centre method construct the ellipse in rhombus $P_1Q_1R_1S_1$.
6. Join both the ellipses by two common tangent lines parallel to the third, right hand side 30°, axis.
7. Make visible lines of the object thicker and darker and retain all other construction lines thinner and lighter.
8. In Fig. 19.15 (iii) isometric drawings of cylinders in three different positions are shown. Depending upon the given views iso-axes are selected and correspondingly we can get isometric drawings of cylinders in three different positions as shown in Fig. 19.15 (iii).

Problem 10 : In Fig. 19.16 (i) F. V. and L. H. S. V. of an object are given. Draw the isometric drawing.

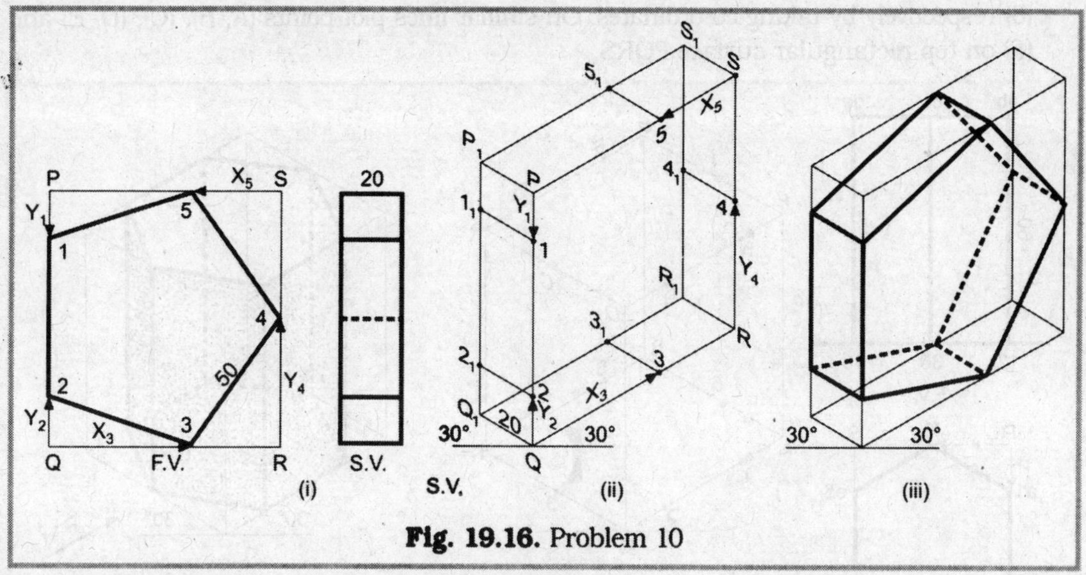

Fig. 19.16. Problem 10

For solution follow the procedure as given below :

1. Inscribe the pentagon 12345 in a rectangle PQRS. See Fig. 19.16 (i) In fact by doing so we have enclosed the object in a rectangular slab.
2. For F. V. select vertical axis and right hand side 30° axis so that both given F. V. and L. H. S. V. will be totally visible.
3. Draw parallelogram PQRS with PQ on vertical axis and QR on R.H. side 30° axis. By taking y_2 distance from F. V. and plotting it from Q on QP line in isometric drawing we get point 2 marked. Similarly, mark points 3, 4, 5, and 1 on lines QR, RS, SP and PQ by taking dimensions x_3, y_4, x_5 and y_1 respectively. Join them to get isometric drawing of a pentagon 12345 see Fig. (ii).
4. By taking thickness of rectangular slab from S. V. and plotting it along third iso-axis we get parallelogram $P_1Q_1R_1S_1$ as shown in Fig. 19.16 (ii).
5. Plot points 1_1, 2_1, 3_1, 4_1 and 5_1 by the same method or by drawing parallel lines to third axis equal to thickness from 1, 2, 3, 4, and 5 respectively.
6. Complete the figure as shown in Fig. 19.16 (iii).

Problem 11 : In Fig. 19.17(i) F. V. and T. V. of an object are given. Draw the isometric drawing

OR

Hexagonal prism is resting on H. P. on its base with two edges of the base perpendicular to V. P. Draw its isometric drawing.

For solution, follow the procedure as given below :

1. Inscribe the hexagon of plan in a rectangle PQRS. See Fig. 19.17 (i). In fact by doing so we have enclosed the hexagonal prism in a vertical rectangular prism.

2. First draw the iso-drawing of rectangular prism as shown in Fig. 19.17 (ii) and then on lines PQ, QR, RS and SP of bottom rectangular surface plot points (1, 2), (3), (4, 5) and (6) respectively by taking co-ordinates. On similar lines plot points (A, B), (C), (D, E) and (F) on top rectangular surface PQRS.

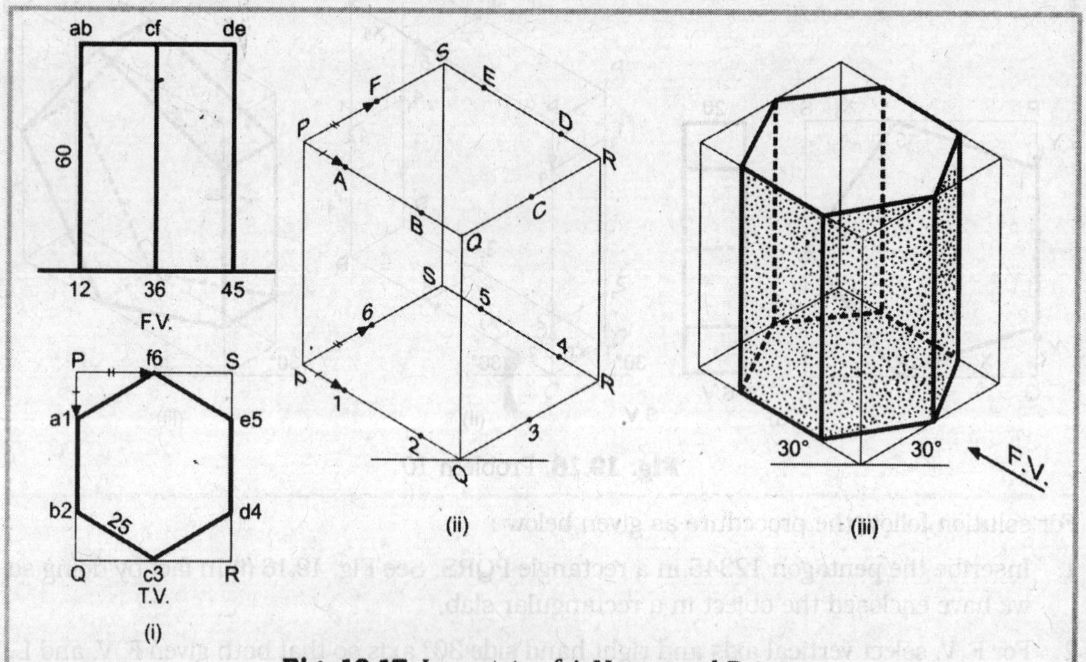

Fig. 19.17. Isometric of A Hexagonal Prism

3. Join these points. Visible lines should be drawn darker and thicker. Hidden lines may be avoided or drawn by medium thick dashed lines. Retain construction lines thinner and lighter. Final isometric is given in Fig. 19.17 (iii).

Problem 12 : In Fig. 19.18 (i) F. V. and T. V. of a frustum of a square pyramid are given in 3rd angle system. Draw the isometric view. (ii) In Fig. 19.18 (ii) F. V. and T. V. of a frustum of a cone are given in 3rd angle system. Draw the isometric view.

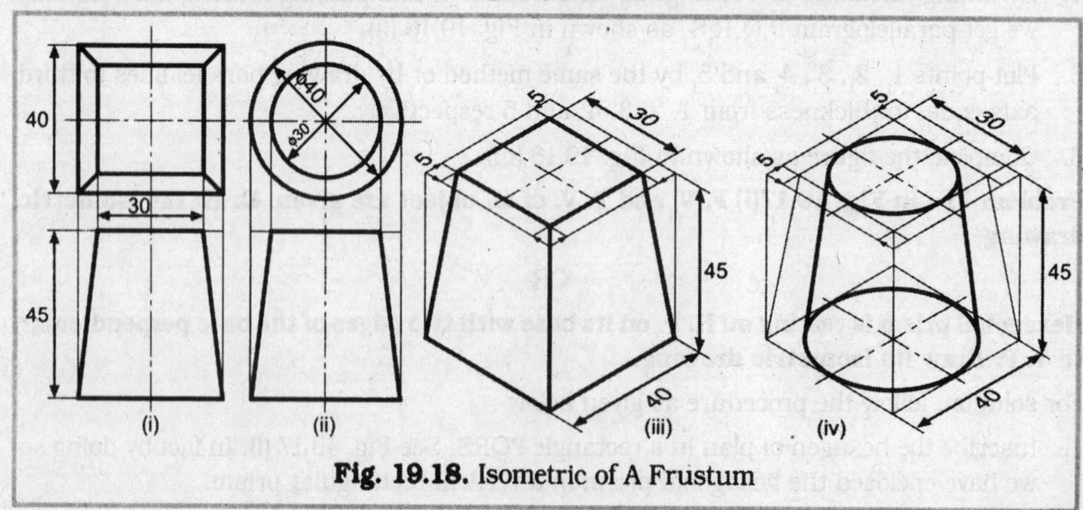

Fig. 19.18. Isometric of A Frustum

For solution follow the procedure as given below :

(i) 1. Inscribe F. V. in a rectangle. In fact by doing so, frustum of a sq-pyramid is enclosed in a sq-prism. See Fig. 19.18 (i).

2. First draw isometric view of a sq-prism with 40 mm edge of base and 45 mm height. See Fig. 19.18 (iii).

3. On the top sq-surface (40 × 40) of prism cut down 5 mm on all sides at corners to get square of 30 mm. See Fig. 19.18 (iii).

4. Join upper square (30 × 30) with lower square (40 × 40) to get frustum of a sq pyramid.

(ii) After constructing the frustum of a square-pyramid, proceed in the following manner to get the frustum of a cone.

1. Draw ellipses, one in the bottom rhombus (40 × 40) and another in the top rhombus (30 × 30) by four centre method.

2. Draw two common tangents to the two ellipses already drawn to get isometric view of a frustum of a cone. See Fig. 19.18 (iv).

Problem 13 : In Fig. 19.19 (i) F. V. and T. V. of a truncated frustum of a hexagonal pyramid are given. Draw its isometric drawing.

OR

A hexagonal pyramid is resting on H.P. on its base with two edges of the base parallel to V. P. Take edge of base 40 mm and height 80 mm. It is cut by a cutting plane perpendicular to V. P. and inclined to H.P. by 45° and passing through a point 32 mm from the apex on the axis. Draw the isometric drawing of the lower portion of the cut pyramid.

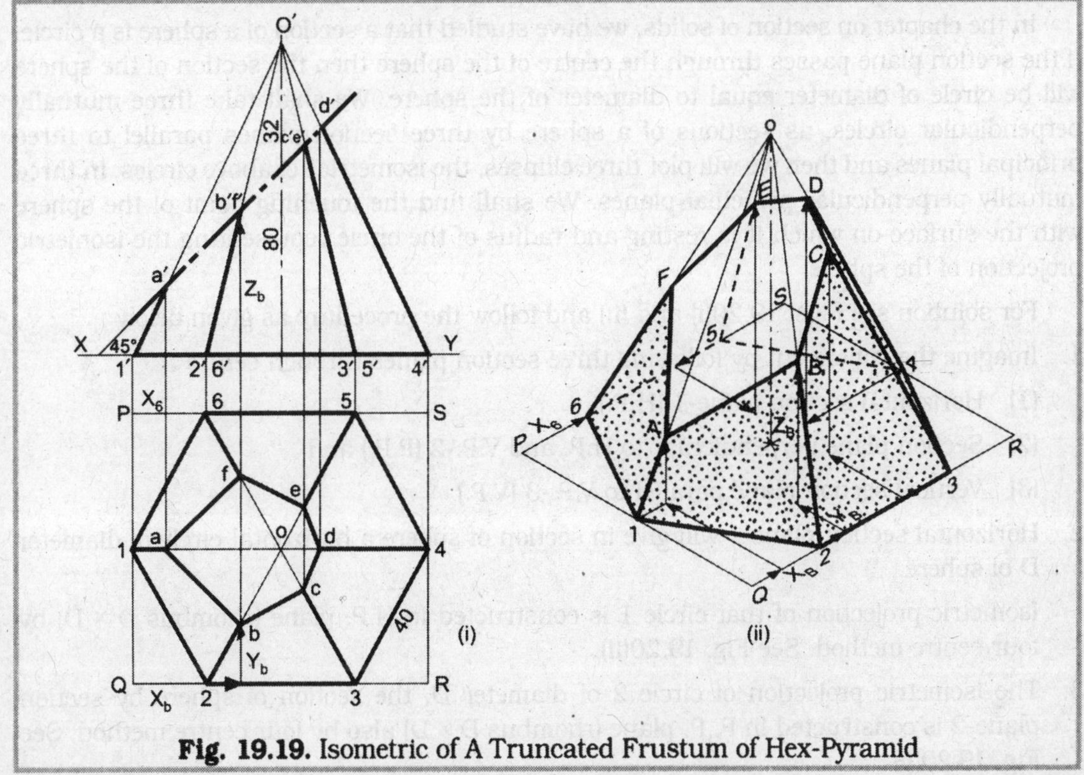

Fig. 19.19. Isometric of A Truncated Frustum of Hex-Pyramid

For solution see Fig. 19.19(ii) and follow the procedure as given below :

1. Inscribe hexagon 123456 in a rectangle PQRS.
2. Construct rhombus PQRS by taking two iso-axes other than vertical. Plot points 1, (2, 3), 4, (5 and 6) on lines PQ, QR, RS and SP respectively by taking co-ordinates. Similar to x_6 shown for point 6.
3. Now to plot points A, B, C, D, E and F, which are on an inclined section plane, take 3 co-ordinates for each point w.r.t. the known reference point.
4. For point B take reference point Q and find co-ordinates x_b, y_b (from plan) and x_b, z_b (from elevation) and plot them in isometric from the reference point Q in the direction of corresponding iso-axes as shown in Fig. 19.19(ii). Similarly, plot points C, D, E, F and A.
5. Join all points and get iso-drawing of the object.
6. If lines 1A, 2B,...6F are extended they will meet at O, the apex of the pyramid.
7. The other method of construction of truncated frustum is to first construct the full pyramid by plotting the point O by the use of co-ordinates (x_o, y_o, z_o) and then get point A, B,...F on lines 10, 20, 30,......60 by taking only two co-ordinates (x_a, y_a), (x_b, y_b)..... (x_f, y_f). Here heights of the vertical co-ordinates (z) are not required.

12. Isometric Projections/Views of A Sphere and Spherical Object

All types of orthographic projections of a sphere are circles of the true radius, i.e. F.V., T.V., S.V. Aux-Plan, Aux-Elev. Iso-projection, diametric projection and trimetric projection all are circles of the true radius.

In the chapter on section of solids, we have studied that a section of a sphere is a circle. If the section plane passes through the centre of the sphere then the section of the sphere will be circle of diameter equal to diameter of the sphere. We shall take three mutually perpendicular circles, as sections of a sphere by three section planes parallel to three principal planes and then we will plot three ellipses, the isometrics of above circles, in three mutually perpendicular principal planes. We shall find the touching point of the sphere with the surface on which it is resting and radius of the circle representing the isometric projection of the sphere.

For solution see Figs. 19.20(i) and (ii) and follow the procedure as given below :

1. Imagine the sphere cut by following three section planes through centre :

   (1) Horizontal section plane-1 (H.P.)

   (2) Section plane perpendicular to H.P. and V.P.-2 (P.P.) and

   (3) Vertical section plane parallel to V.P.-3 (V.P.)

2. Horizontal section plane 1 will give in section of sphere a horizontal circle of diameter D of sphere.

   Isometric projection of that circle 1 is constructed in H.P. plane (rhombus D × D) by four centre method. See Fig. 19.20(ii).

3. The isometric projection of circle 2 of diameter D, the section of sphere by section plane-2 is constructed in P. P. plane (rhombus D × D) also by four centre method. See Fig. 19.20(ii).

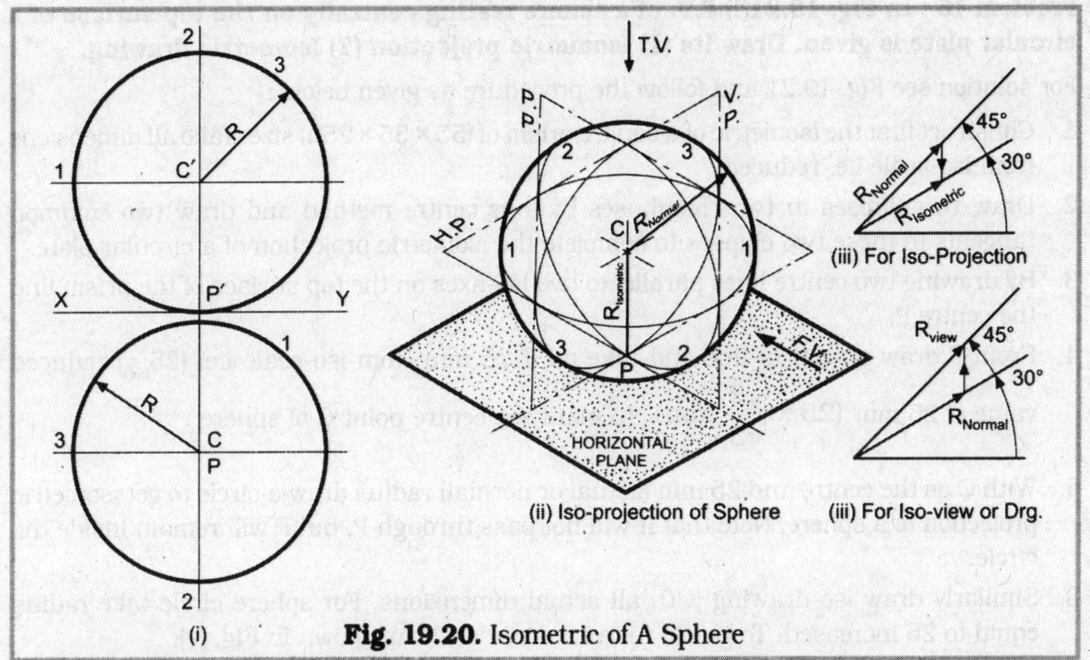

(i) **Fig. 19.20.** Isometric of A Sphere

4. Similarly, the isometric projection of circle-3 is constructed in V.P. plane (rhombus D × D), as shown in Fig. 19.20(ii).
5. Keep centres of all the rhombuses H.P; V.P. and P.P. at one point C which will eventally become the centre of the sphere in isometric.
6. Point of intersection P of two ellipses at the bottom most side will give touching point of sphere with the ground. Measured length CP will be equal to isometric length of the actual radius.
7. With centre C, draw a circle touching and enclosing all the three ellipses 1-2-3 drawn in mutually perpendicular planes. Note that radius of this circle will be equal to the actual radius of the sphere i.e. R_{normal}. In case of isometric view/drawing of sphere this radius will be $\frac{\sqrt{3}}{\sqrt{2}}$ times actual/normal radius.
8. $R_{isometric}$ and R_{view} can be achieved by scale shown in Fig 19.20 (iii)from known R_{normal}.

Important Note :

In construction of Isometric Projection of object containing sphere or spherical portion all dimensions are taken after reducing it by iso-scale except the radius of the circle, representing isometric of sphere or spherical portion. This radius of the circle is taken true or normal.

In construction of Isometric Drawing or View of object containing sphere or spherical portion all dimensions are taken true except the radius of the circle, representing isometric of sphere or spherical portion. This radius of the circle is taken after increasing it by iso-scale. Radius is increased by scale by factor $\frac{\sqrt{3}}{\sqrt{2}}$. R_{normal} is increased to R_{view} by scale as shown in Fig. 19.20(iii).

Problem 15 : In Fig. 19.21(i) F.V. of a sphere resting centrally on the top surface of a circular plate is given. Draw its (1) Isometric projection (2) Isometric drawing.

For solution see Fig. 19.21 and follow the procedure as given below :

1. Construct first the isometric of a square prism of (55 × 55 × 25h) size. Take all dimensions from iso-scale i.e. reduced.
2. Draw two ellipses in two rhombuses by four centre method and draw two common tangents to these two ellipses to complete the isometric projection of a circular plate.
3. By drawing two centre lines parallel to two iso-axes on the top surface of the prism find the centre P.
4. From P draw a vertical line and take on it 25 mm from iso-scale i.e. (25_{ISO}) reduced value of 25 mm ($25 \times \frac{\sqrt{2}}{\sqrt{3}}$ mm), to mark the centre point C of sphere.
5. With C as the centre and 25 mm (actual or normal) radius draw a circle to get isometric projection of a sphere. Note that it will not pass through P, but P will remain inside the circle.
6. Similarly draw iso-drawing with all actual dimensions. For sphere circle take radius equal to 25 increased. To get 25 increased use scale as shown in Fig. (v).

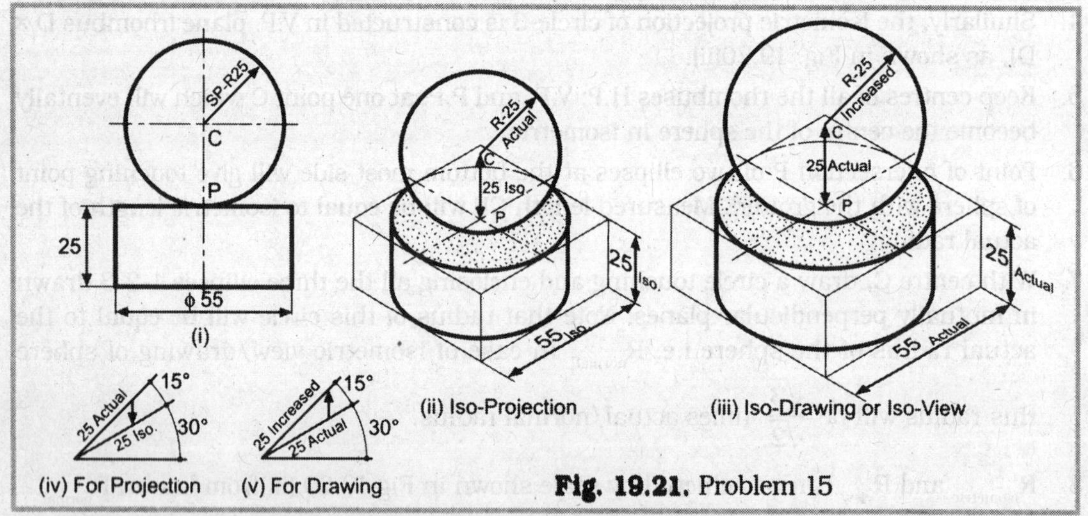

Fig. 19.21. Problem 15

13. Construction of Isometric by Centre Line Method

The box method discussed earlier can be used for all types of objects. However, when the object consists of a number of circular sections, square sections, regular sections, etc. lying in the same iso-plane or parallel iso-planes this centre line method is most convenient and rapid as well.

We shall solve some problems by this method and simultaneously study the important aspects of this method.

Problem 16 : Draw the isometric of a flower pot of 110 mm height, having circular cross-sections of 40, 25, 10, 20, 35, 45, 50, 45, 40, 30, 20 and 25 mm diameters, from bottom most plane to top most plane at an interval of 10 mm height. F.V. of the flower pot is given in Fig. No. 19.22(i)

For solution see Fig. 19.22(ii) and follow the procedure as given below :

1. Draw vertical line 1-12 of 110 mm height and mark on it points 2, 3,....11 at equal intervals of 10 mm each.
2. Draw two centre lines parallel to two iso-axes passing through point 1. Take (D/2) 20 mm on both the sides of the point 1 on two axes and mark points A_1, B_1, C_1 and D_1. See Fig. 19.22(ii).
3. Through these points A_1, B_1, C_1 and D_1 draw parallel lines to iso-axes and construct rhombus. In this rhombus construct ellipse, representing isometric of a circle of 40 mm diameter. Rhombus is not shown in the figure.
4. Similarly, construct ellipses to represent isometrics of circles around points 2, 3,....12 of 25, 10,....25 mm diameters respectively.
5. Now draw two smooth curves thicker and darker tangent to all 12 ellipses already drawn. Upper most ellipse should be made thicker and darker. Visible portion of the lower most ellipse should be made thicker and darker.
6. Handles of problems 39 and 43 are drawn by this method.

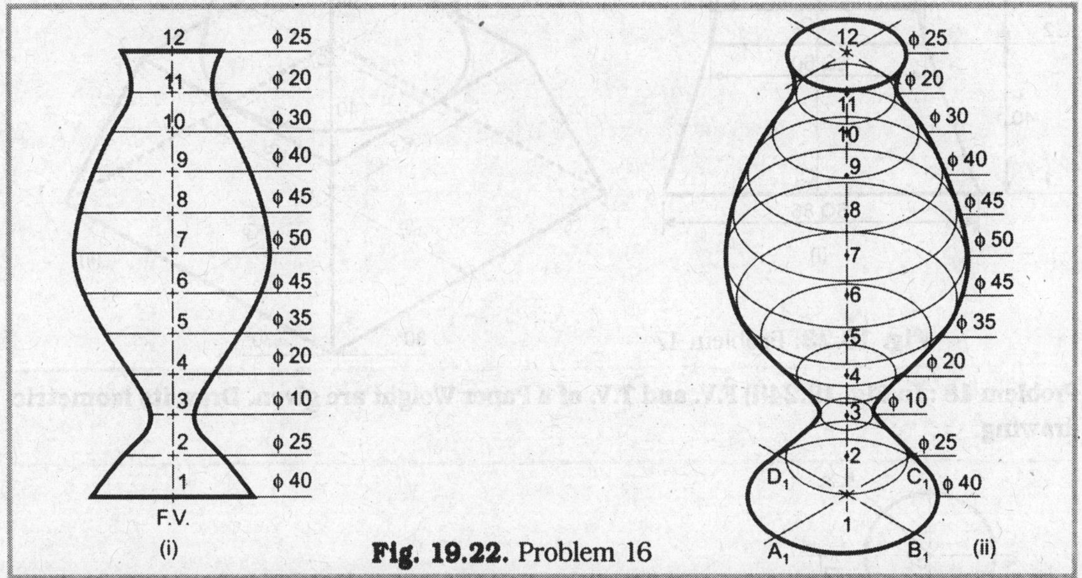

Fig. 19.22. Problem 16

Problem 17 : In Fig. 19.23(i) F.V. of three solids placed one above the other keeping their axes along one common line is shown. Draw the isometric view of the same.

For solution see Fig. 19.23(ii) and follow the procedure as given below :

1. Looking to the F.V., we can divide the object into 5 sections 1, 2, 3, 4 and 5 as shown in Fig. 19.23(i).
   (i) Along section 1, Sections is a square of 86 mm side.
   (ii) Along section 2, Sections are (1) square of 60 mm side and (2) circle of 60 mm diameter.
   (iii) Along section 3, Section is a circle of 40 mm diameter.
   (iv) At section 4, Centre of sphere is situated
   (v) Along section 5, Section is a circle of 35 mm diameter.
2. Draw vertical line and mark on it centres C_1, C_2, C_3, C_4 and C_5 at the intervals of heights 40, 50, 20 and 10 mm taken successively from bottom towards top side.

3. At section 1 i.e. around centre C_1 prepare a rhombus of 86 mm side. At section 2 i.e. around centre C_2 prepare a rhombus of 60 mm side and isometric of circle of 60 mm diameter. At section 3 prepare isometric of circle of 40 mm diameter and at section 5 prepare isometric of circle of 35 mm diameter. With centre C_4 draw a circle with $R_{increased}$ radius, (i.e. 20x11/9 mm radius) which will be touching smoothly to the ellipse of circle of 35 mm diameter. Join various sections by lines or tangent lines, depending upon object, to complete the isometric view.

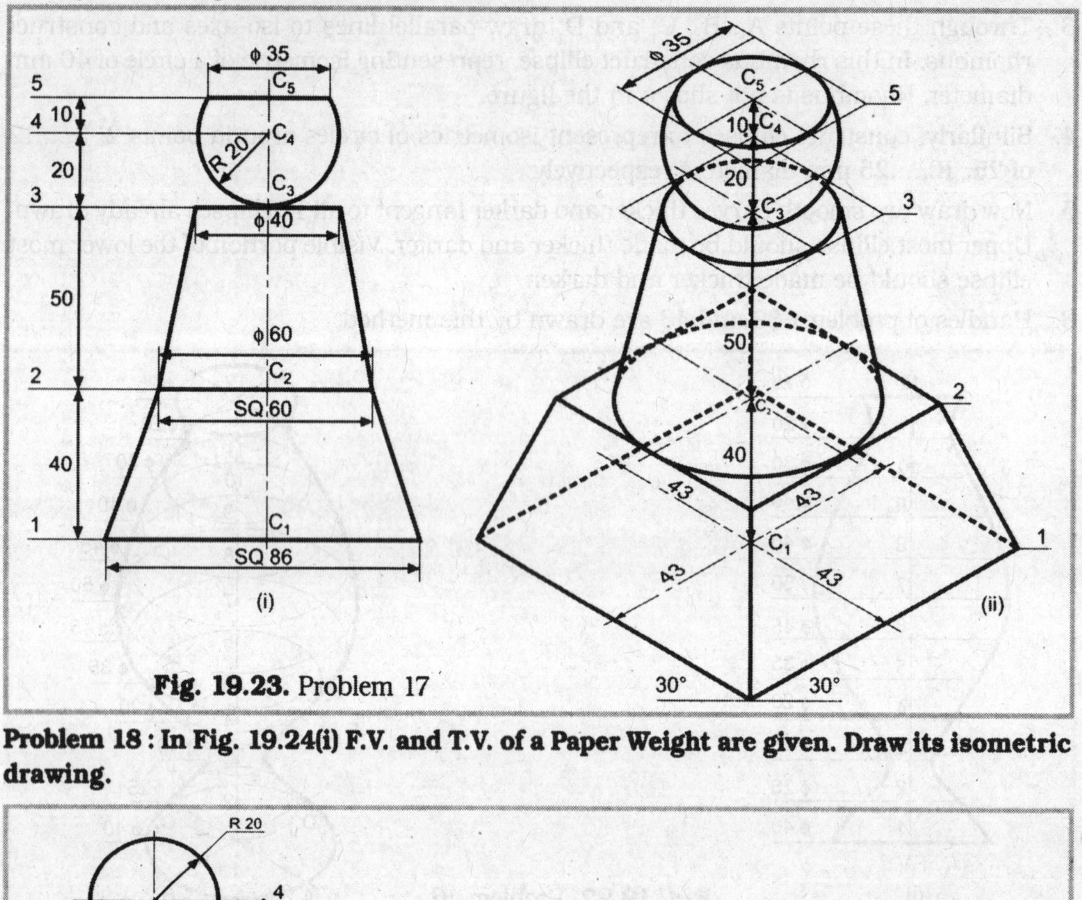

Fig. 19.23. Problem 17

Problem 18 : In Fig. 19.24(i) F.V. and T.V. of a Paper Weight are given. Draw its isometric drawing.

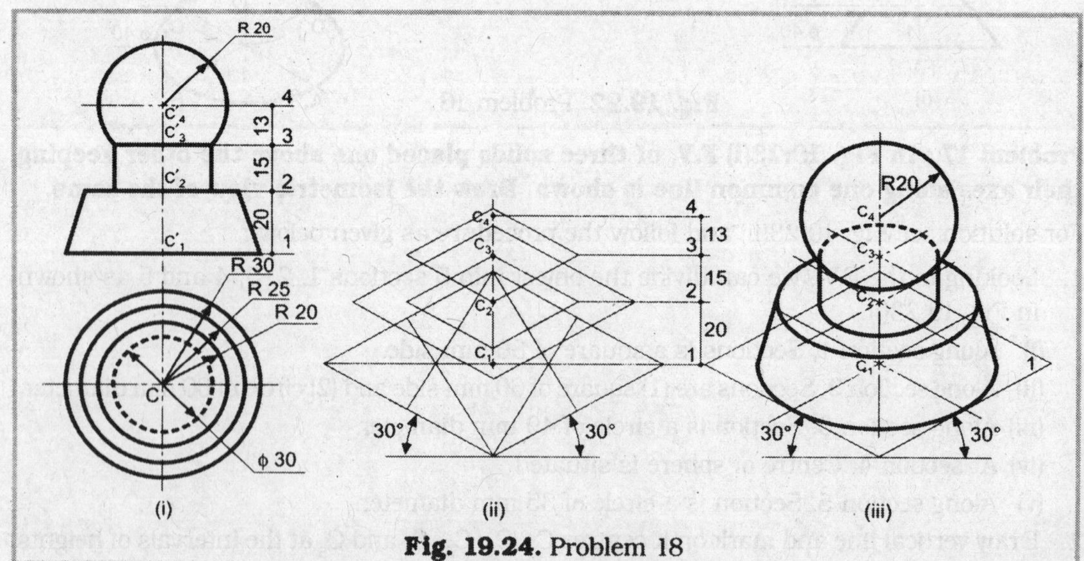

Fig. 19.24. Problem 18

Solve this problem in the same manner as problem 17. Fig. 19.24(ii) is a skeleton of construction leading to solution shown in Fig. 19.24(iii).

Problem 19 : In Fig. 19.25(i) F.V. and T.V. of a Square Headed Chamfered Bolt are given. Draw its isometric drawing.

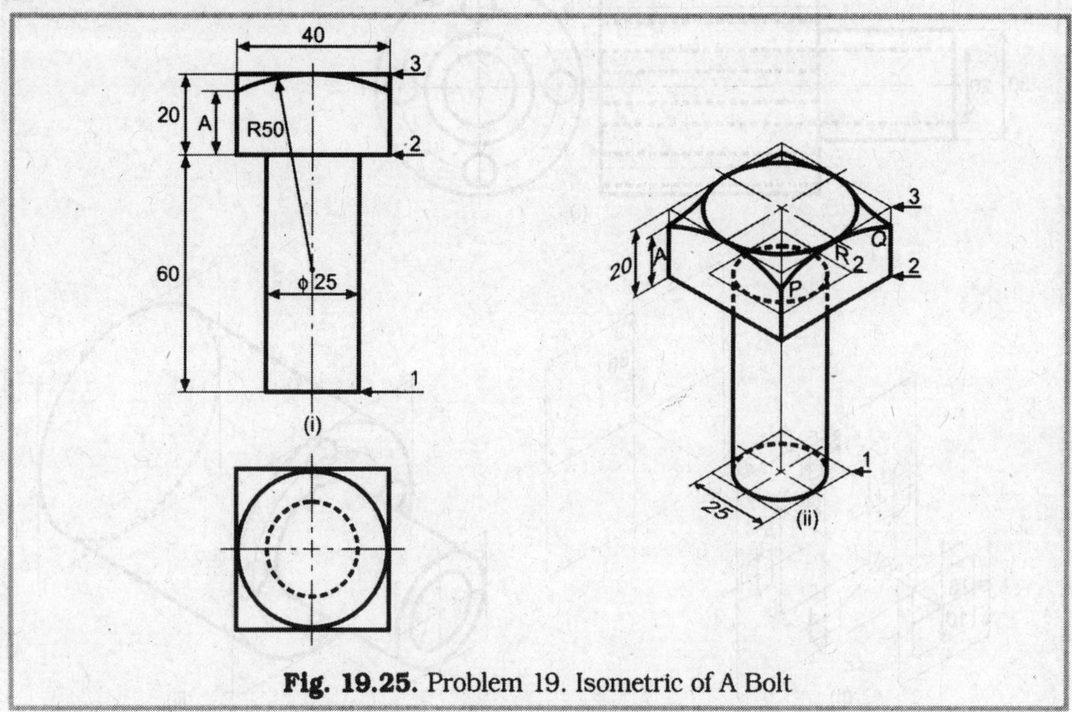

Fig. 19.25. Problem 19. Isometric of A Bolt

For solution see Fig. 19.25(ii) and follow the procedure as given below :

1. Divide the bolt in three sections and complete the isometric drawing of bolt without chamfer in the same manner as the previous problem.
2. To draw chamfer on the top surface of the bolt-head, take dimension A from F.V. and plot 4 points on 4 vertical lines of sq-prism for head e.g. point P and Q. Find also the centres of 4 top edges of head e.g. R.
3. Join P, R, Q by a smooth curve. Draw isometric of circle at centre 3.

   Join similar points on other faces also.

Problem 20 : In Fig. 19.26(i) F.V. and S.V. of a cylindrical flange are given. Draw its isometric drawing.

Skeleton work leading to solution is shown in Fig. 19.26 (ii)

In Fig. 19.26(ii) isometric layout of the principal centre lines is drawn. In the rhombuses shown construct ellipses by four centre method. Final solution is given in Fig. 19.26(iii).

Problem 21 : In Fig. 19.27(i) F.V. or T.V. or S.V. of a three way pipe flange is given. Draw its isometric view.

For solution see Figs. 19.27(i) and (ii) and follow the procedure as given below :

1. Take any point O and draw three lines along three iso-axes. Find points (x_1, x_2), (y_1, y_2) and (z_1, z_2) along three different axes at distances of (40, 15) respectively and successively from O.

2. Construct isometrics of 60 diameter circles around centres X_1, Y_1 and Z_1 in planes perpendicular to corresponding axis.

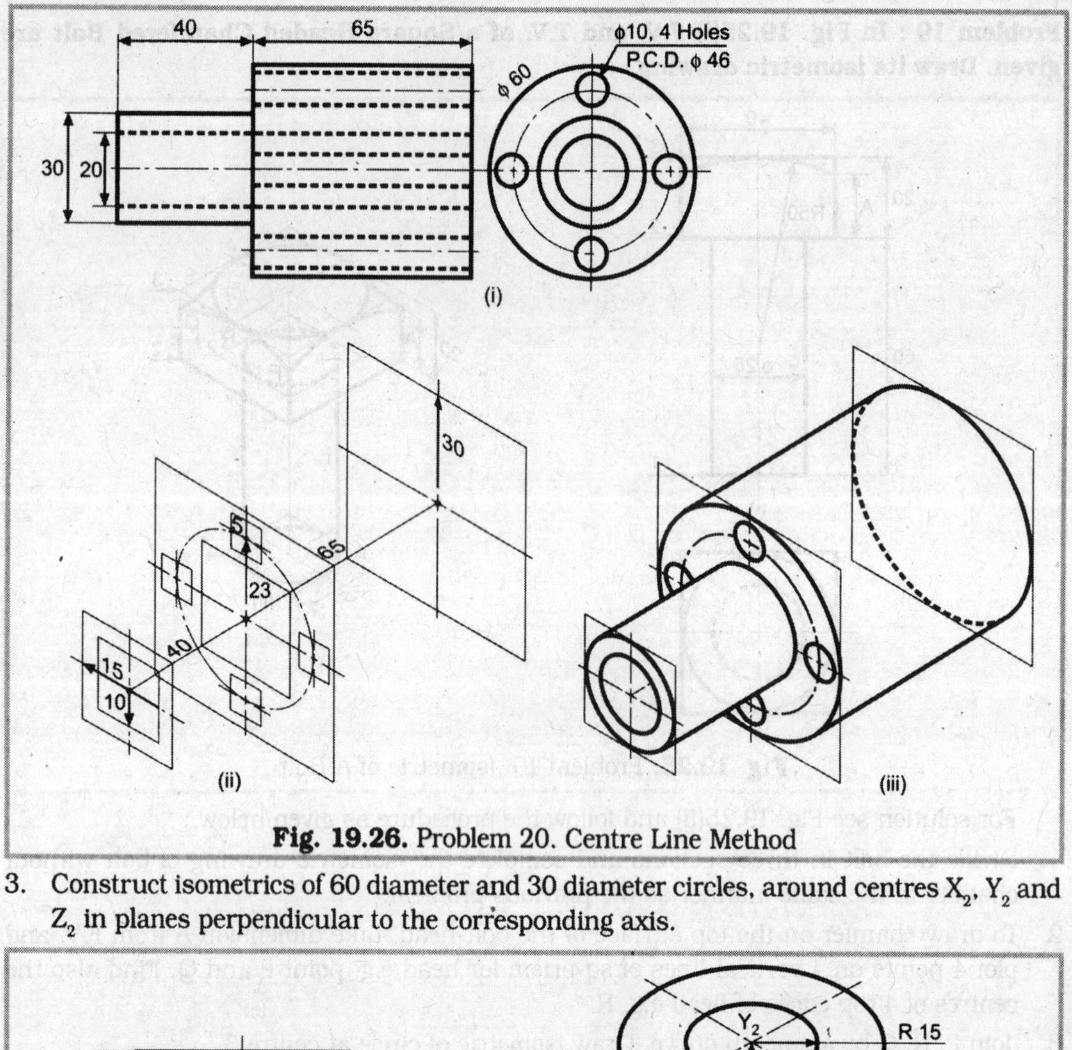

Fig. 19.26. Problem 20. Centre Line Method

3. Construct isometrics of 60 diameter and 30 diameter circles, around centres X_2, Y_2 and Z_2 in planes perpendicular to the corresponding axis.

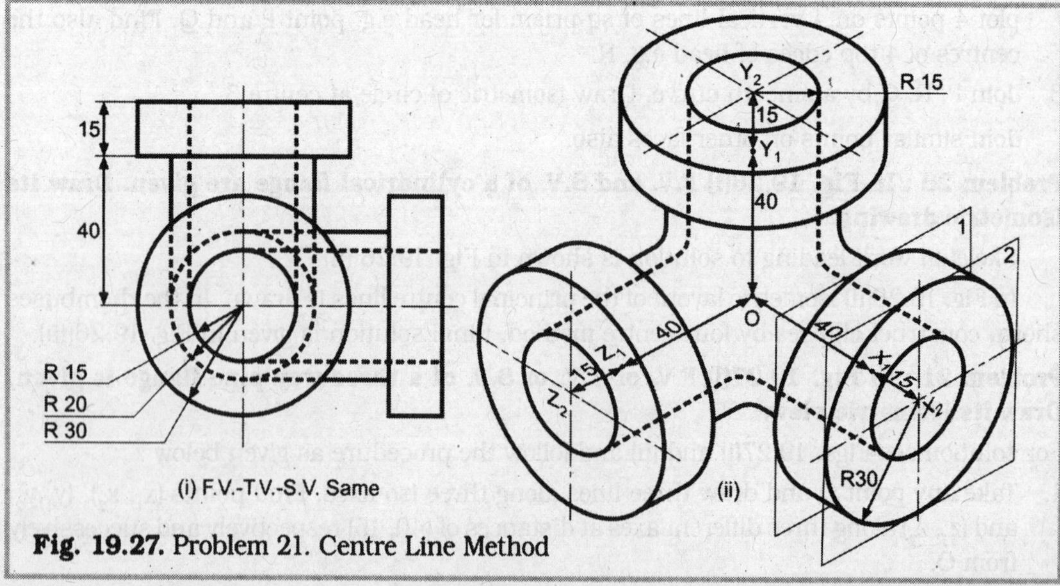

Fig. 19.27. Problem 21. Centre Line Method

4. Draw dotted lines parallel to iso-axes tangent to corresponding ellipses shown in the figure. Give curvature at the 3 points of intersections near the centre.
5. As the thickness of pipe is 5 mm draw parallel lines to dotted lines outside it and 5 mm away from it in the region where it is visible. This way isometric view is completed. See Fig. 19.27(ii).

14. Isometric of A Double Curved Surface

Some objects such as pipe bend, tube, ring, etc. have double curved surface. This curve does not lie in a single plane and hence cannot be constructed by plotting points by the usual methods studied so far.

Method of solution of this type of problem is discussed in the problem 22.

Problem 22 : In Fig. 19.28(i) T.V. of a circular ring (tube) having circular cross-section is shown. Draw its isometric.

For solution see Fig. 19.28(i) and (ii) and follow the procedure as given below :

1. First draw ellipse, to represent isometric of the centre line circle of a ring.
2. Imagine 8 or 12 spheres enclosed in the circular ring(tube) just touching from inside to ring at nearly equal distances.
3. Since a sphere projects as a circle in isometric, draw circles with 8 to 12 points at nearly equal distances taken on ellipse as centres and radius equal to 0.5 of the diameter of the cross section. See Fig. 19.28(ii).
4. Draw one curve, tangent to all circles from the inside and another curve tangent to all circles from the outside to complete the isometric projection of a ring.

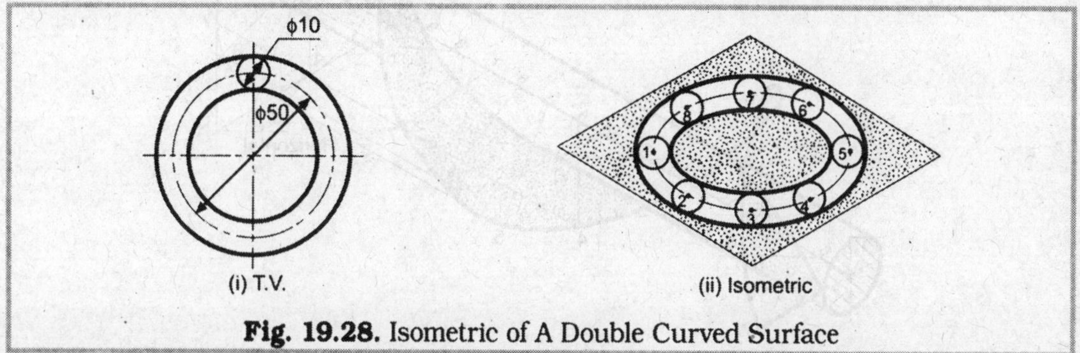

Fig. 19.28. Isometric of A Double Curved Surface

15. Construction of A Helicoid in Isometric

Problem 23 : In Fig. 19.29(i) F.V. and S.V. of a helical blade attached to a shaft on its circumference are given. Draw its isometric drawing.

For solution see Fig. 19.29(i) and (ii) and follow the procedure as given below. In isometric 1.5 times the given length is shown.

1. Draw isometric of the shaft (i.e. of cylinder)
2. Along centre line of the shaft mark points 1, 12, 11,...1, 12, 11,....7 as shown in the Fig. 19.29(ii) by transferring distances from F.V.
3. At the extreme right hand side i.e. around point 1 construct rhombus PQRS and 2 ellipses inside it to represent the isometrics of two circles of side view.

4. For dividing the two ellipses into 12 parts along circumferences mark 12 points on sides of rhombus, first by taking dimensions from S.V. along PQ, QR, RS and SP and then join them with centre point. 1. Mark points on both the ellipses. For keeping clarity of drawing only points 4, 5,....10 are marked.
5. Now from points 12, 11, 10,....1, 12,....7 of centre line draw parallel and equal length lines to 1-12, 1-11, 1-10,....1-1, 1-12,....1-7 lines obtained by construction at the extreme right. See Fig. 19.29(ii) and note how two points are marked on line 8-8.
6. Join obtained points 1, 12, 11,....1, 12,....7 by a smooth curve to get set of two curves one on the circumference of the shaft and another of the edge of helical blade.

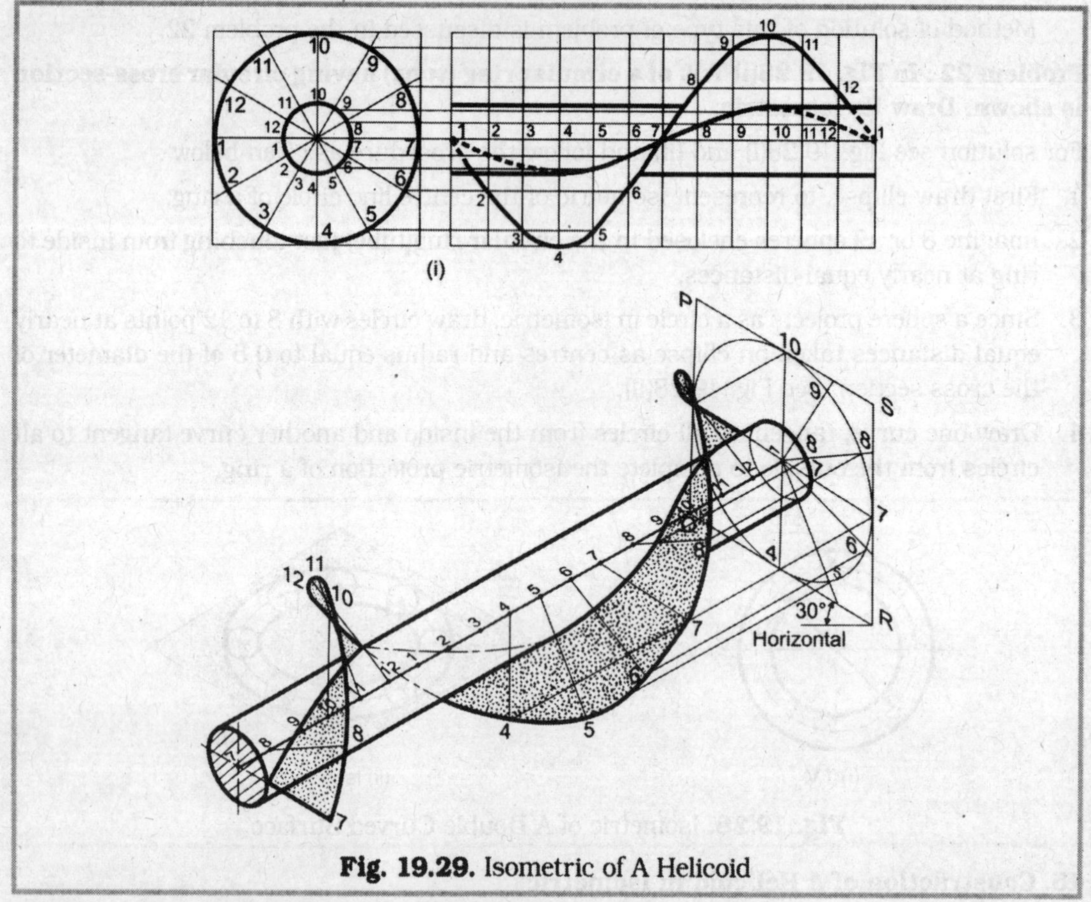

Fig. 19.29. Isometric of A Helicoid

16. Exact Method of Construction of Isometric

This is a simple method of drawing isometric, diametric and trimetric projections without the use of scale. Professor L. Eckhart of Vienna devised this method in the year 1937. This method cannot be used for isometric view or drawing.

Without going in details about the basic concepts of this method we shall study the application part of it and solve three problems.

Problem 24 : Elevation, Plan and S.V. of a lamp shade, hexagonal at the top end and square at the bottom end, are given. Draw its isometric projection by the exact method.

For solution see Fig 19.30 and follow the procedure as given below :

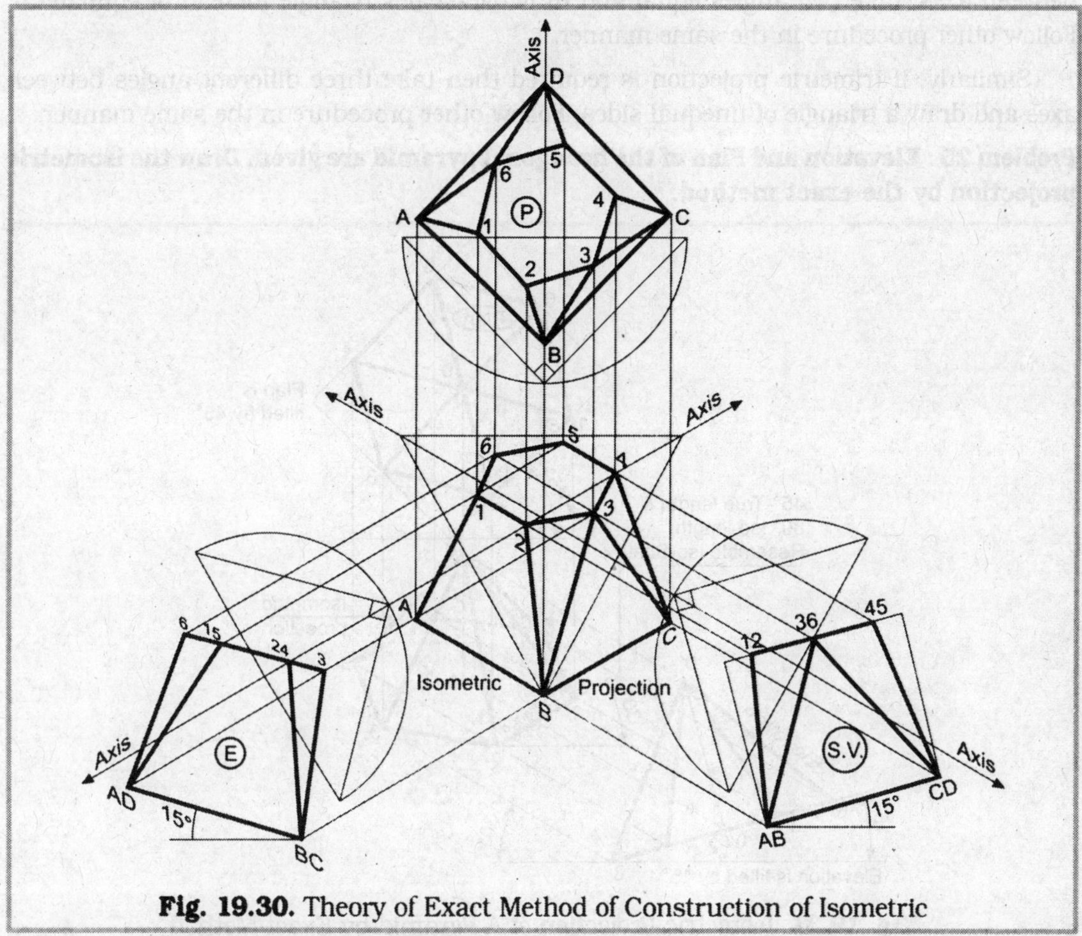

Fig. 19.30. Theory of Exact Method of Construction of Isometric

1. First draw three iso-axes 120° apart.
2. Construct equilateral triangle of suitable size keeping three corners of the triangle on three axes already drawn.
3. Parallel and equal to three sides of triangle draw three lines at suitable distance, as shown in the figure.
4. With these lines as diameters draw three semicircles towards the sides of traingle.
5. Mark 3 points where 3 axes intersect these three semicircles and join points of intersections with ends of corresponding diameter. By this construction we will get 3 sets of right angle lines.
6. Locate given elevation, plan and side view parallel to these 3 sets of right angle lines. For correct orientation see the figure.
7. Note that the base of elevation and side view will become inclined by 15° with the horizontal and plan will become equally inclined with the horizontal in the orientation.
8. From every corners of three views draw lines parallel to respective axis to get by intersection isometric projection in the centre, see Fig. 19.30.

By the same method, if diametric projection is required then out of the three angles between axes, take two angles equal and draw iso-scales triangle instead of equilateral. Follow other procedure in the same manner.

Similarly, if trimetric projection is required then take three different angles between axes and draw a triangle of unequal sides. Follow other procedure in the same manner.

Problem 25 : Elevation and Plan of the hexagonal pyramid are given. Draw the isometric projection by the exact method.

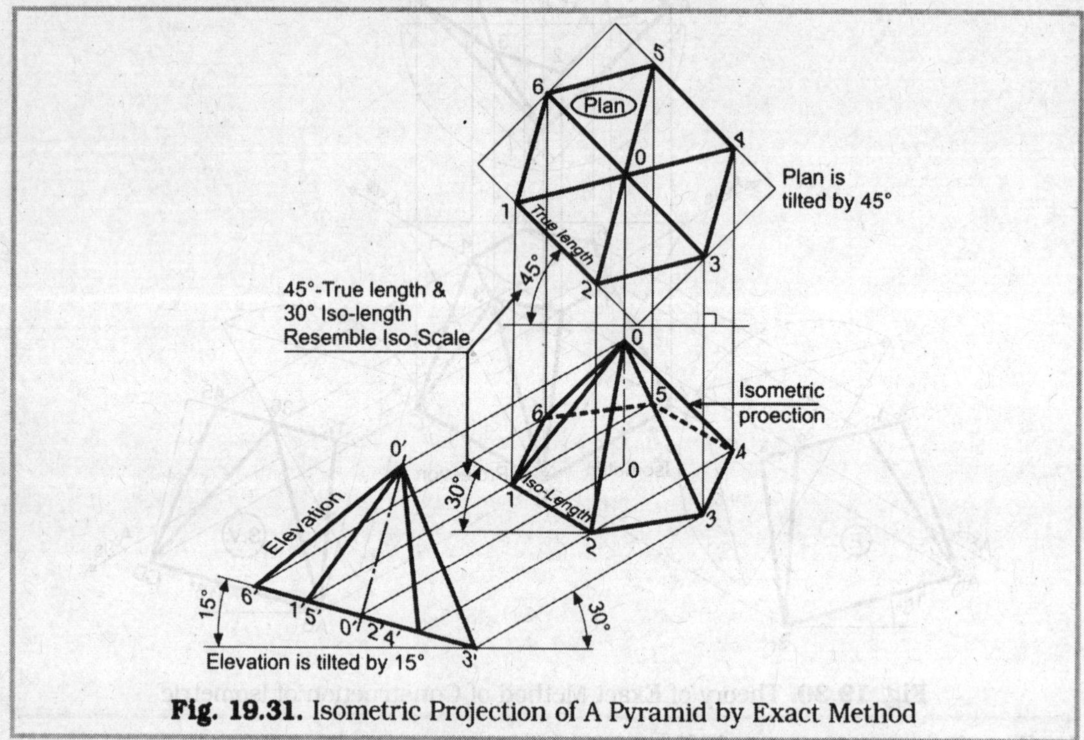

Fig. 19.31. Isometric Projection of A Pyramid by Exact Method

For solution see Fig. 19.31 and follow the procedure as given below :

1. Draw elevation tilted by 15° to the horizontal, as shown in the figure.
2. Draw plan tilted by 45° to the horizontal as shown in the figure.
3. From plan points, draw vertical lines and from elevation points draw lines at 30° to the horizontal. Intersection of the corresponding lines will generate the corresponding points in isometric projection. Join them and complete isometric projection.

Observe orientation of lines or edges and note that iso-scale is taken care of automatically.

Problem 26 : Front view and Top view of an object are given. Draw the isometric projection of an object by the exact method.

For solution of problem see Fig. 19.32 and follow the procedure as given below :

1. Draw front view tilted by 15° to the horizontal as shown.
2. Draw top view tilted by 45° to the horizontal as shown.
3. For circular or curved portion give same numbers to corresponding points in front view and top view.

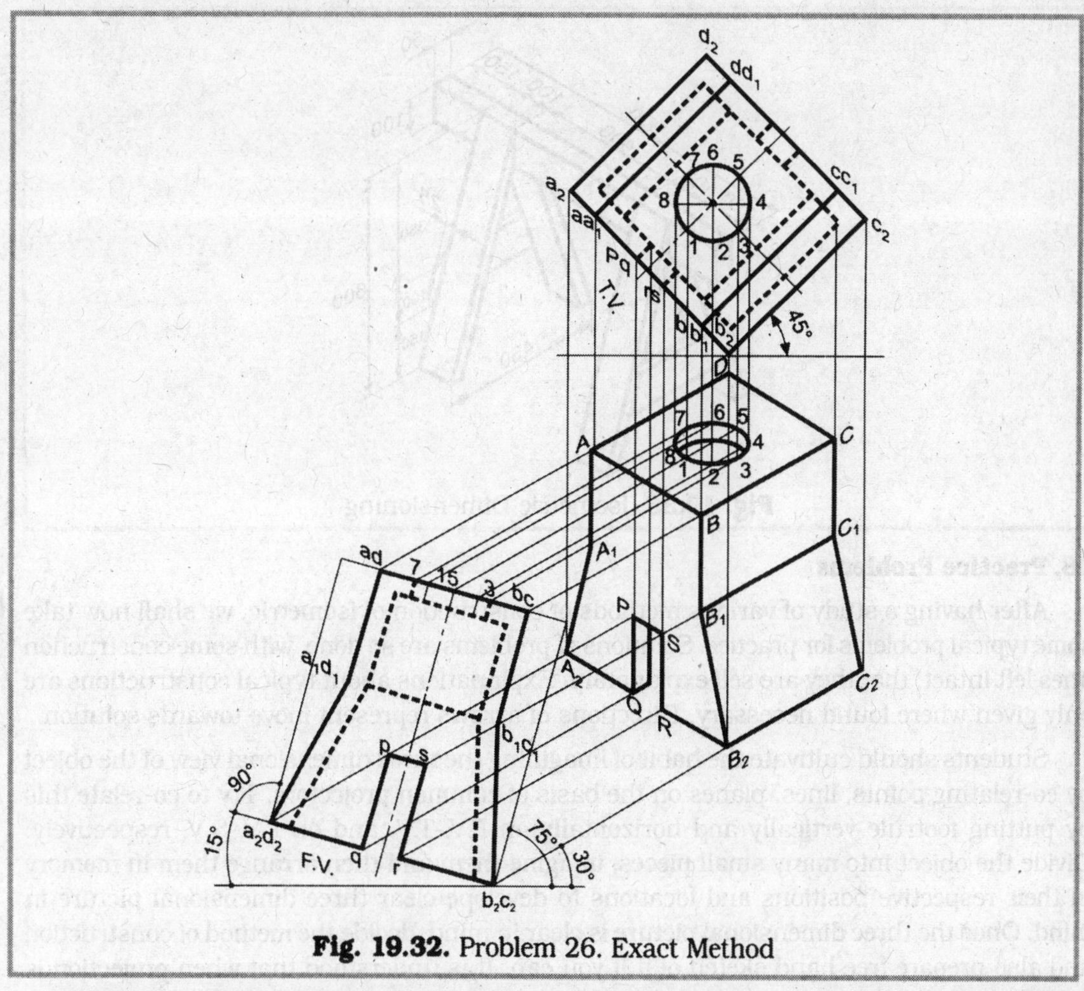

Fig. 19.32. Problem 26. Exact Method

4. As usual draw vertical lines from all plan points and draw 30° lines from all front view points. By intersection we will get isometric projection of an object.

This method is extremely simple. However, imagination of the object is necessary while joining points of intersections.

17. Dimensioning in Isometric

In dimensioning isometric usual rules, followed for dimensioning orthographic projections, should be followed. Two extension lines and one dimension line used for showing one particular dimension should be in the same isometric plane. Extension lines and dimension lines are drawn parallel to the isometric axes. Dimension figures are all written sloping parallel to the extension lines and dimension lines i.e. parallel to two of the isometric axes. As per usual rules dimensions should be placed to read horizontally or from the right.

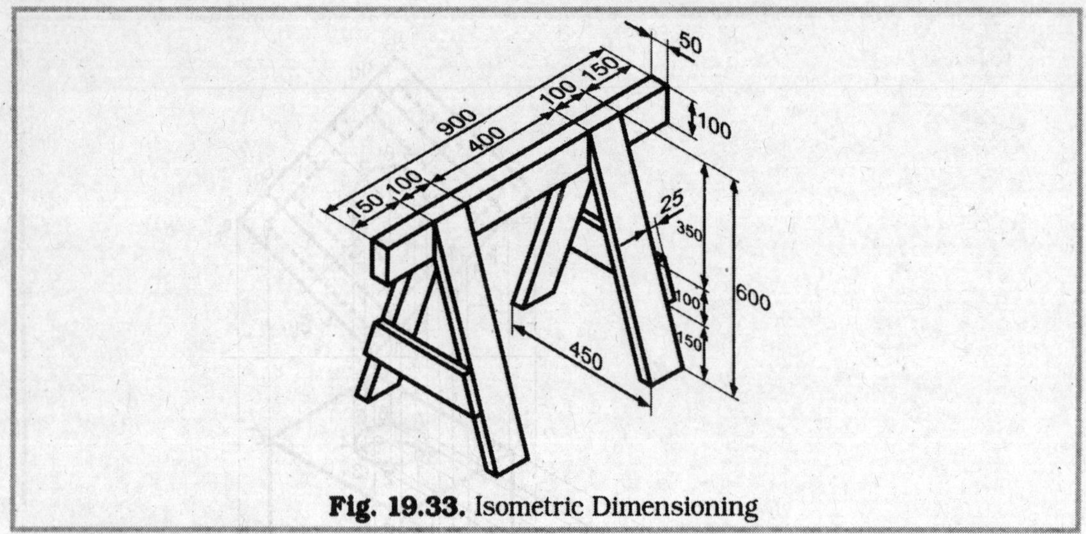

Fig. 19.33. Isometric Dimensioning

18. Practice Problems

After having a study of various methods of construction of isometric, we shall now take some typical problems for practice. Solutions of problems are so done, with some construction lines left intact, that they are self explanatory. Explanations about typical constructions are only given where found necessary. Directions of arrows represent move towards solution.

Students should cultivate the habit of imagining the three dimensional view of the object by co-relating points, lines, planes on the basis of common projectors. Try to co-relate this by putting footrule vertically and horizontally on F.V.-T.V. and on F.V.-S.V. respectively. Divide the object into many small pieces, imagine them and then arrange them in memory in their respective positions and locations to develope clear three dimensional picture in mind. Once the three dimensional picture is clear in mind, decide the method of construction and also prepare free hand sketch of it if you can. It is understood that when projection is asked reduced dimensions are taken from iso-scale except the radius for a sphere. Similarly, when view or drawing is asked dimensions are taken true or normal except the radius for a sphere. Here the radius of a sphere is increased by using iso-scale in reverse manner.

Always check every corner for visible lines out of three lines.

Problem 27 : In Fig. 19.34(i) F.V. and T.V. of a plate are given. Draw the isometric view.

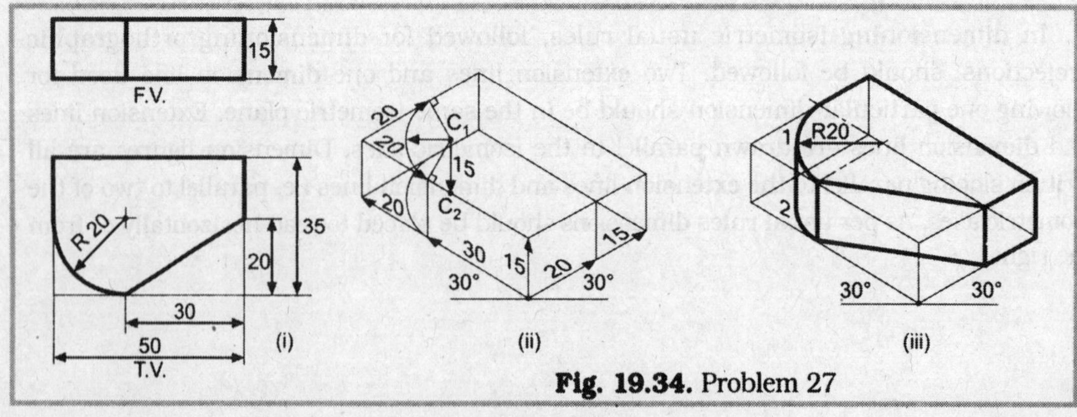

Fig. 19.34. Problem 27

For solution see Fig. 19.34 (ii) and (iii)

Problem 28 : In Fig. 19.35(i) one view of an object is given. This view represents F.V., T.V. as well as S.V. Draw the isometric projection.

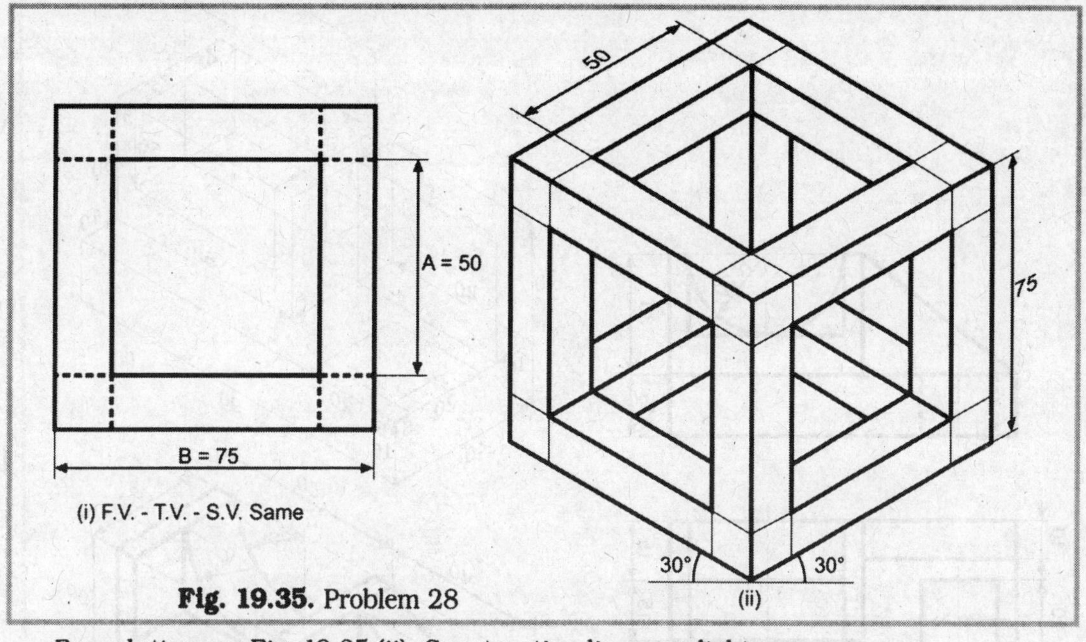

Fig. 19.35. Problem 28

For solution see Fig. 19.35 (ii). Construction lines are light.

Problem 29 : Draw the isometric drawing of the Alphabet 'B' from the given elevation and R.H.S.V. in Fig. 19.36.

For solution see Fig. 19.36(ii) and 19.36(iii).

After getting the centres in the front plane transfer them to rear plane by lines of thickness lengths along the 3rd axis.

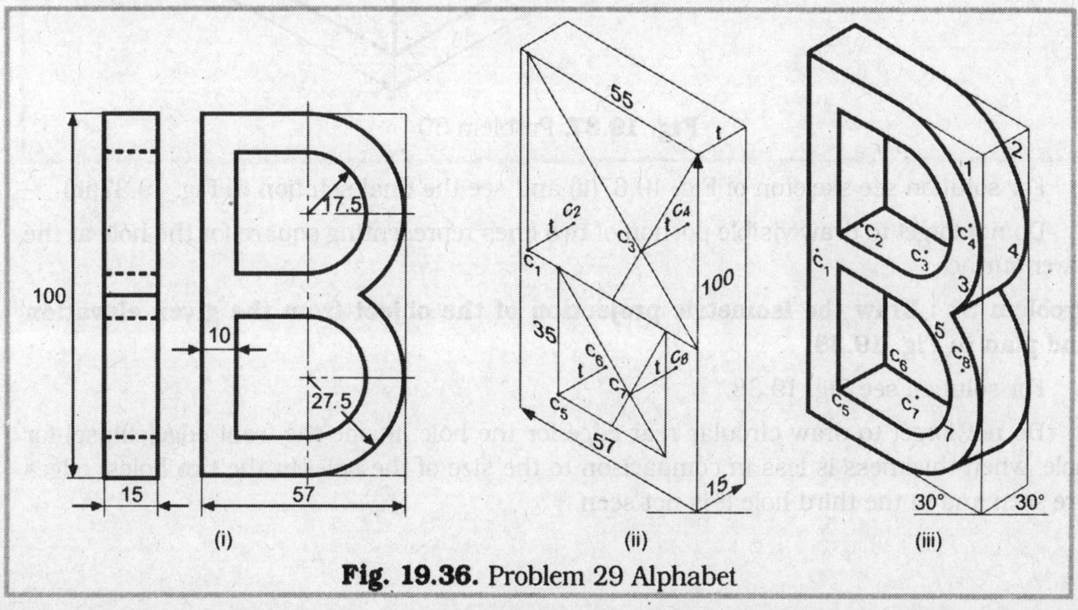

Fig. 19.36. Problem 29 Alphabet

Problem 30 : Draw the isometric of machined component from the given F.V. and T.V. in Fig. 19.37(i).

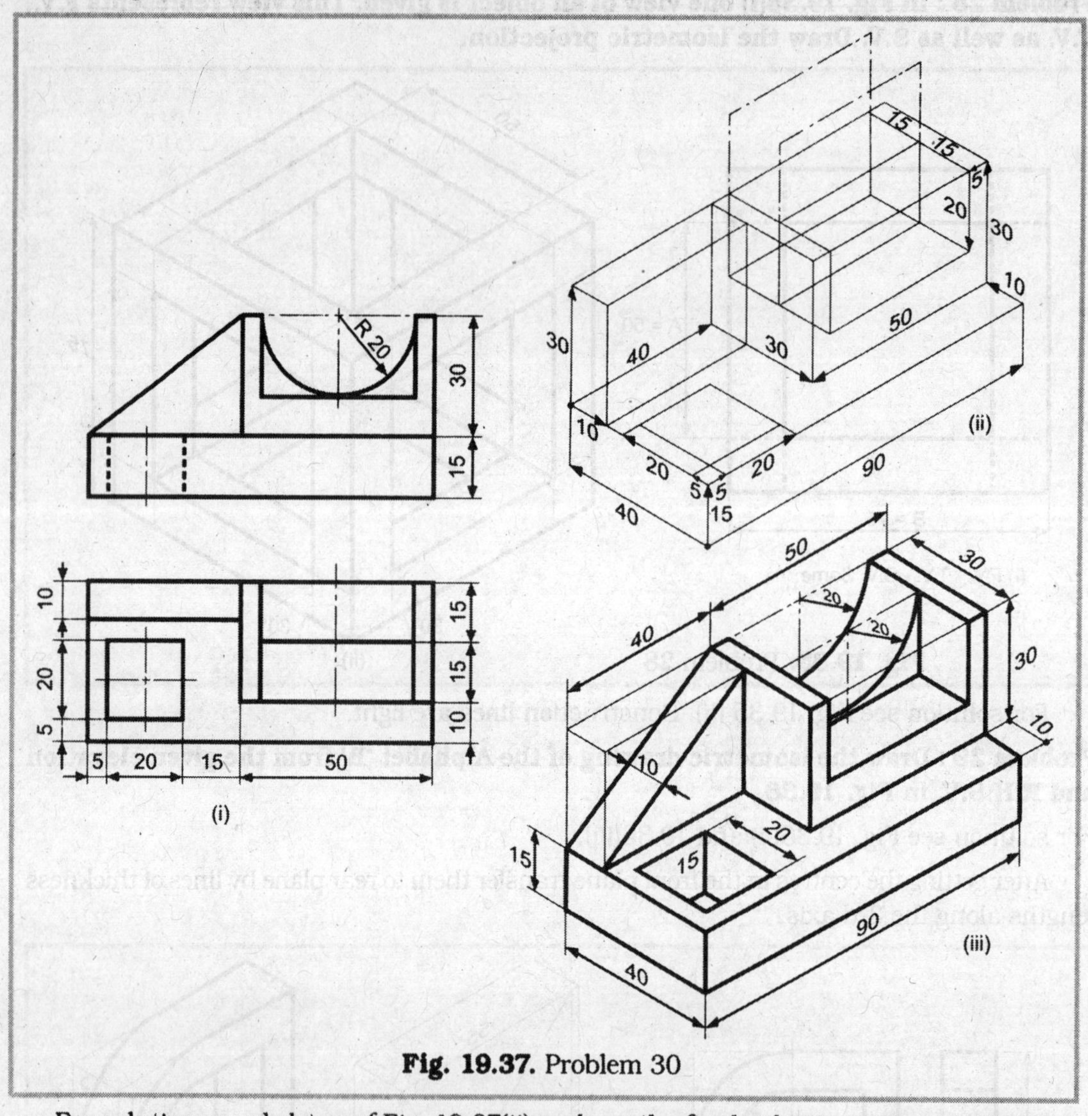

Fig. 19.37. Problem 30

For solution see skeleton of Fig. 19.37(ii) and see the final solution in Fig. 19.37(iii).

Do not forget to draw visible portion of two lines representing square for the hole at the lower surface.

Problem 31 : Draw the isometric projection of the object from the given elevation and plan in Fig. 19.38.

For solution see Fig. 19.38.

Do not forget to draw circular rear edge for the hole, inside the front edge(ellipse) for hole, when thickness is less in comparison to the size of the hole. In the two holes, edges are seen and in the third hole it is not seen.

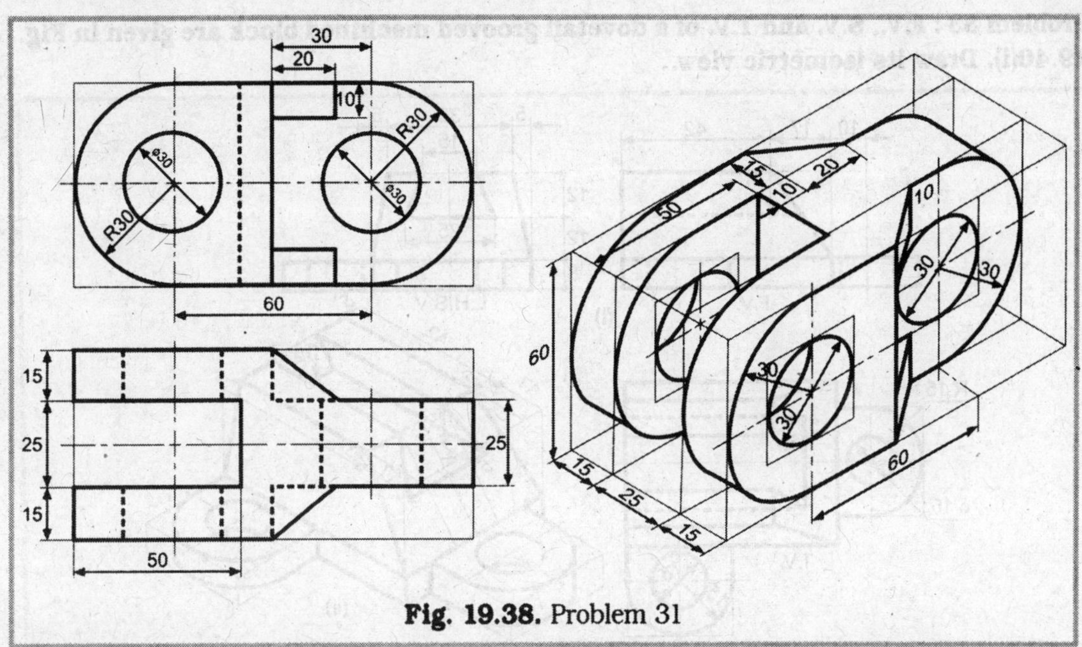

Fig. 19.38. Problem 31

Problem 32 : F.V. and R.H.S.V. of a machined C.I. block are given in Fig. 19.39(i). Draw its isometric drawing.

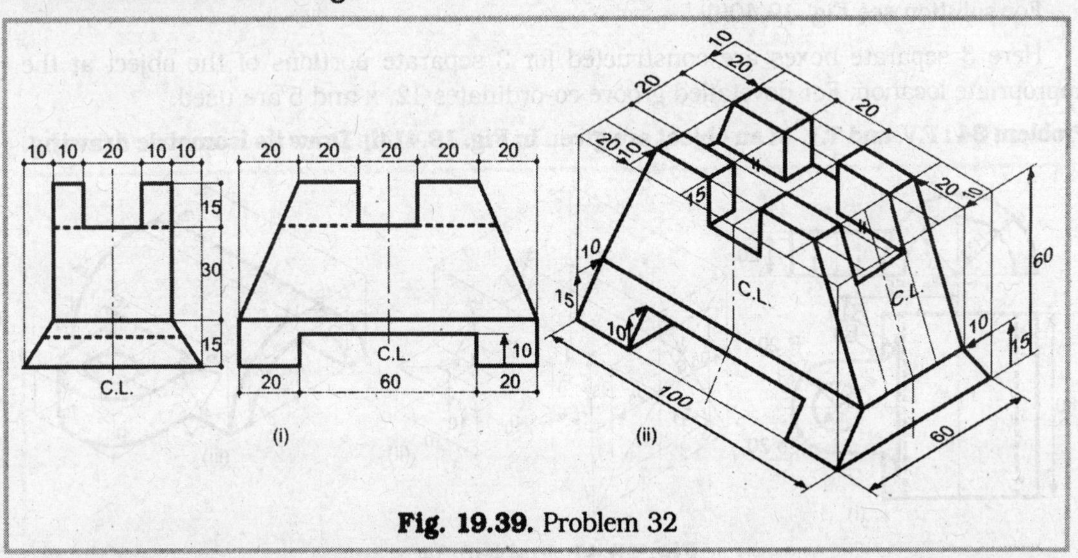

Fig. 19.39. Problem 32

For solution see Fig. 19.39(ii).

The lines for grooves on vertical surfaces are vertical and on sloping surfaces they are parallel to the centre line (C.L.). The height for groove is taken only along the vertical line.

Cutting of dimensions is shown by one sided arrow lines with dimensions on them.

Problem 33 : F.V., S.V. and T.V. of a dovetail grooved machined block are given in Fig. 19.40(i). Draw its isometric view.

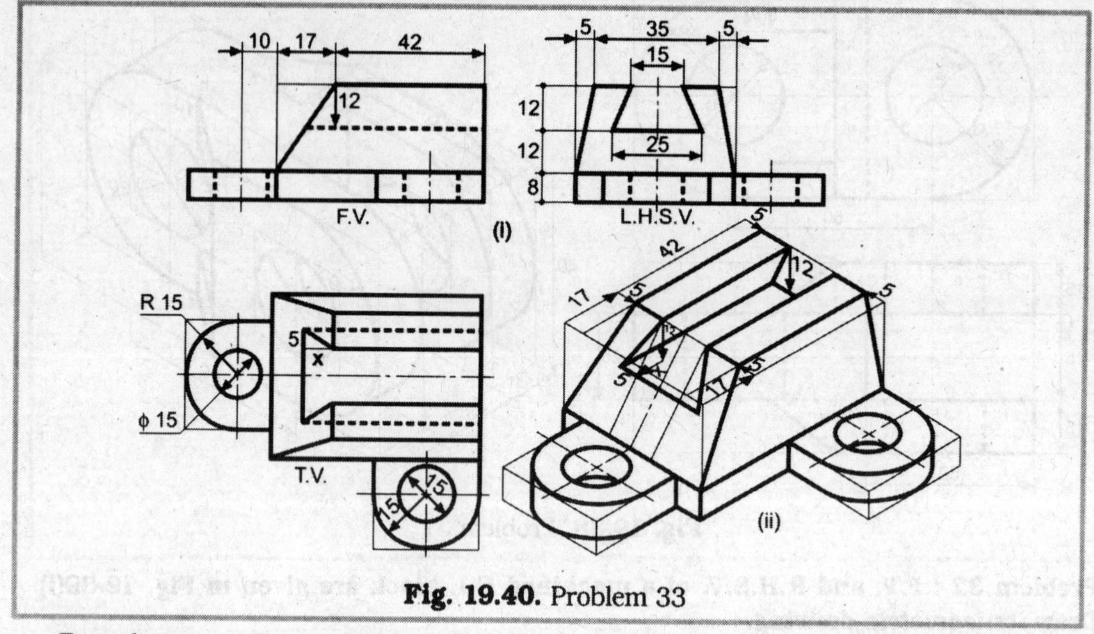

Fig. 19.40. Problem 33

For solution see Fig. 19.40(ii).

Here 3 separate boxes are constructed for 3 separate portions of the object at the appropriate location. For dovetailed groove co-ordinates 12, x and 5 are used.

Problem 34 : F.V. and T.V. of an object are given in Fig. 19.41 (i). Draw its isometric drawing.

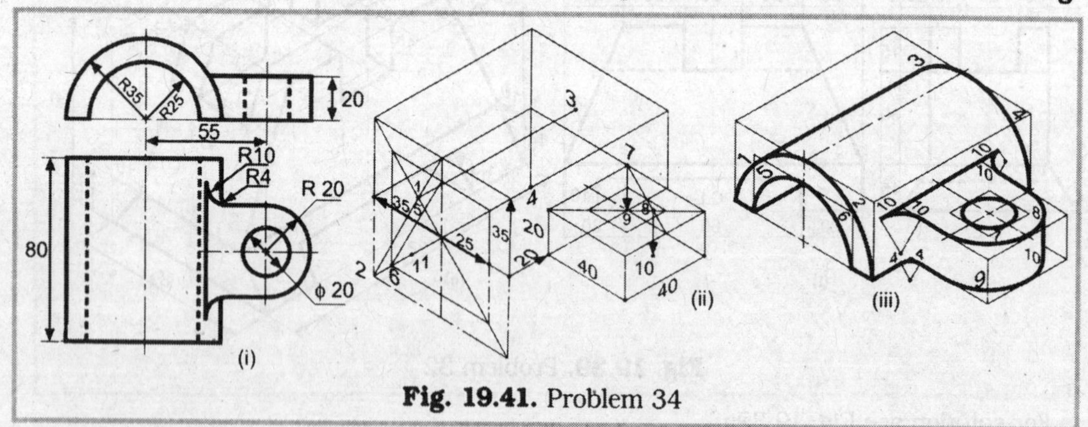

Fig. 19.41. Problem 34

For solution see Fig. 19.41(ii) and Fig 19.41(iii).

In this problem radius of the upper curve along which the upper surface is meeting the curved surface is given as 10 mm. Radius of the meeting curve at the lower suface is obtained by the construction as shown in Fig. 19.41 (iii). Fig. 19.41(ii) is a skeleton leading to solution.

Problem 35 : Draw the isometric view of a chamfered hexagonal nut for a bolt of diameter D. Avoid isometric of threads.

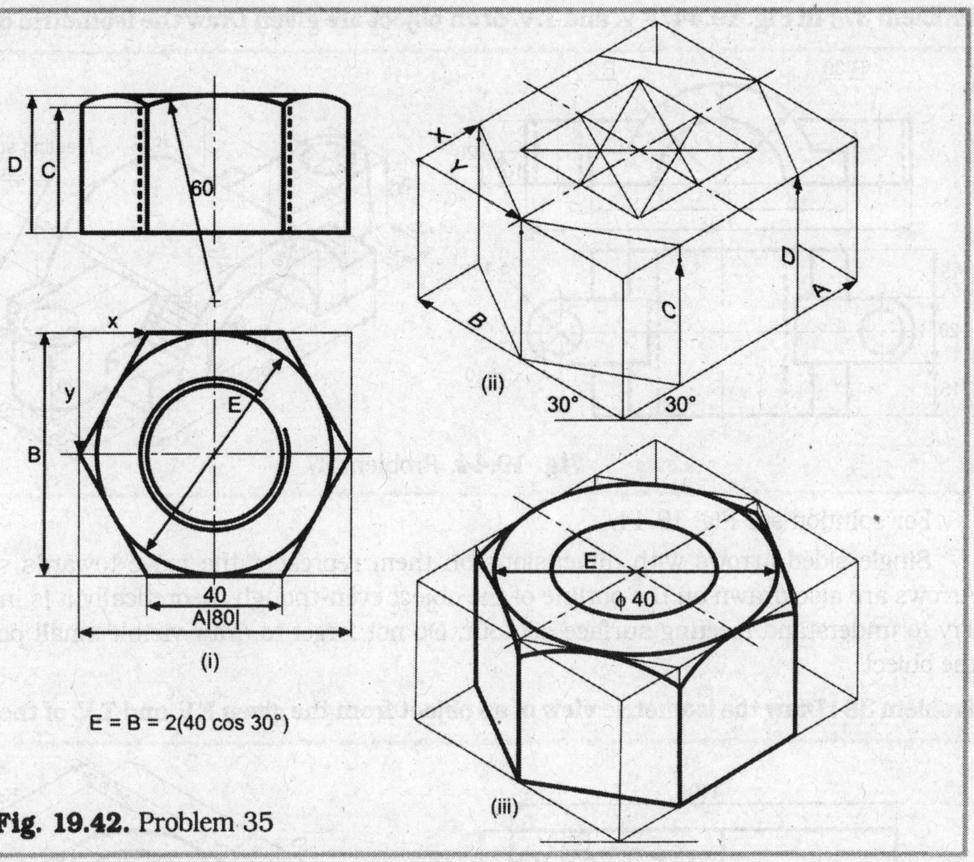

$E = B = 2(40 \cos 30°)$

Fig. 19.42. Problem 35

For solution see Fig. 19.42 (ii) and Fig. 19.42(iii)

First prepare F.V. and T.V. of a nut with standard proportions as shown in the figure. For isometric of chamfering use the dimension C and the midpoints of six upper edges of hexagonal prism. Fig. 19.42(ii) is a skeleton leading to the solution as given in Fig. 19.42(iii).

Problem 36 : In Fig. 19.43, F.V. and L.H.S.V. of a slotted truncated block are given. Draw the isometric drawing.

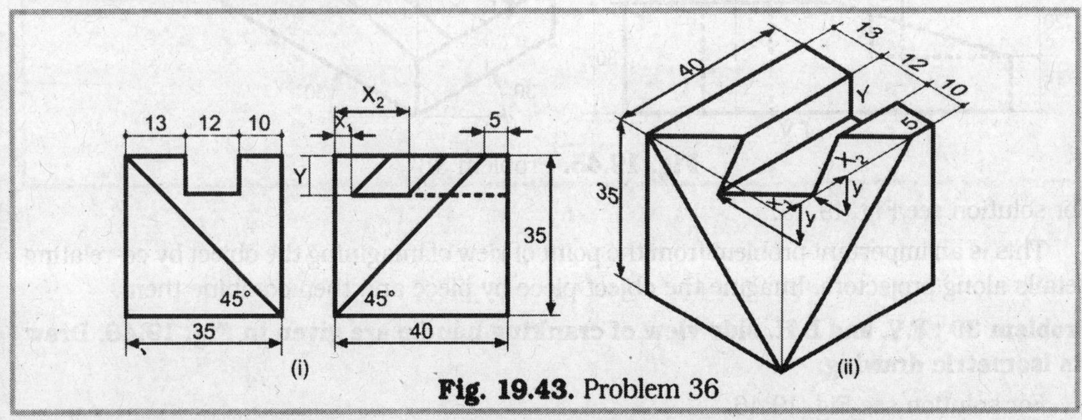

Fig. 19.43. Problem 36

For solution see Fig. 19.43. Use co-ordinate method to locate and construct slot on an inclined plane.

Problem 37 : In Fig. 19.44, F.V. and T.V. of an object are given Draw the isometric drawing.

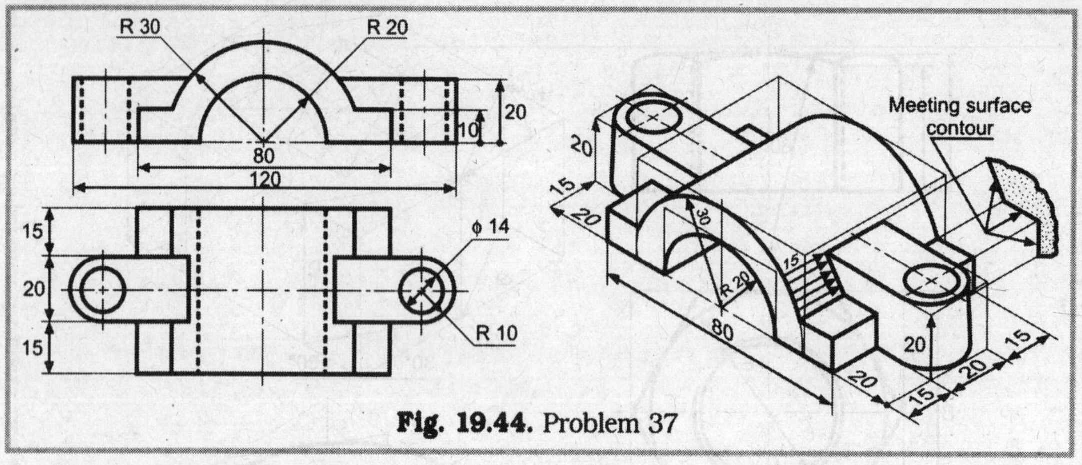

Fig. 19.44. Problem 37

For solution see Fig. 19.44.

Single sided arrows with dimensions on them represent the move towards solution. Arrows are also drawn on the outline of the object even-though theoretically it is incorrect. Try to understand meeting surface contour. Do not forget to draw visible small portion of the object.

Problem 38 : Draw the isometric view of an object from the given F.V. and T.V. of the object.

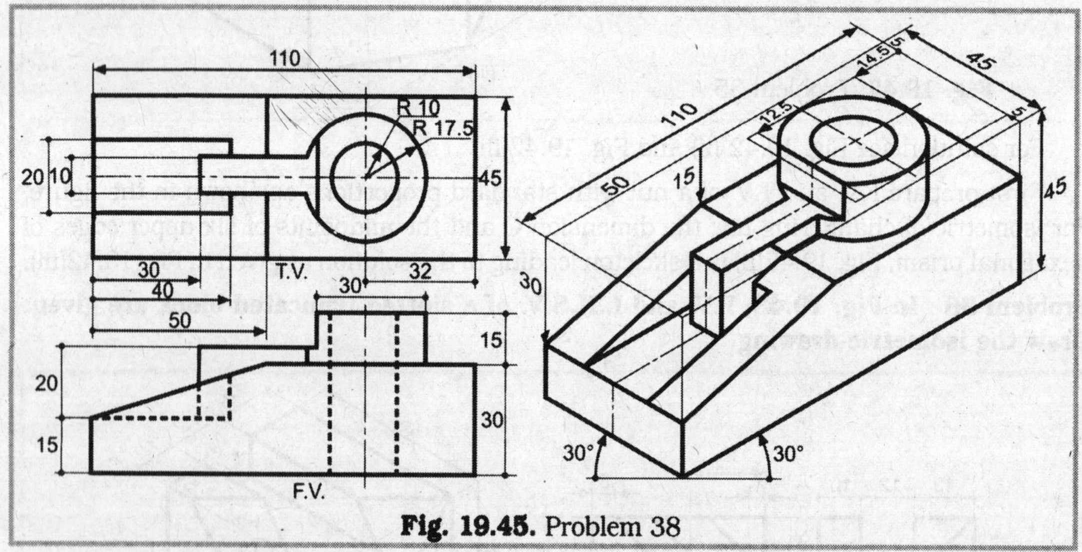

Fig. 19.45. Problem 38

For solution see Fig. 19.45.

This is an important problem from the point of view of imagining the object by co-relating details along projectors. Imagine the object piece by piece and then combine them.

Problem 39 : F.V. and L.H. side view of cranking handle are given in Fig. 19.46. Draw its isometric drawing.

For solution see Fig. 19.46.

Isometric of handle is constructed on the basis of centre line method. Dotted lines are drawn to make drawing more clear.

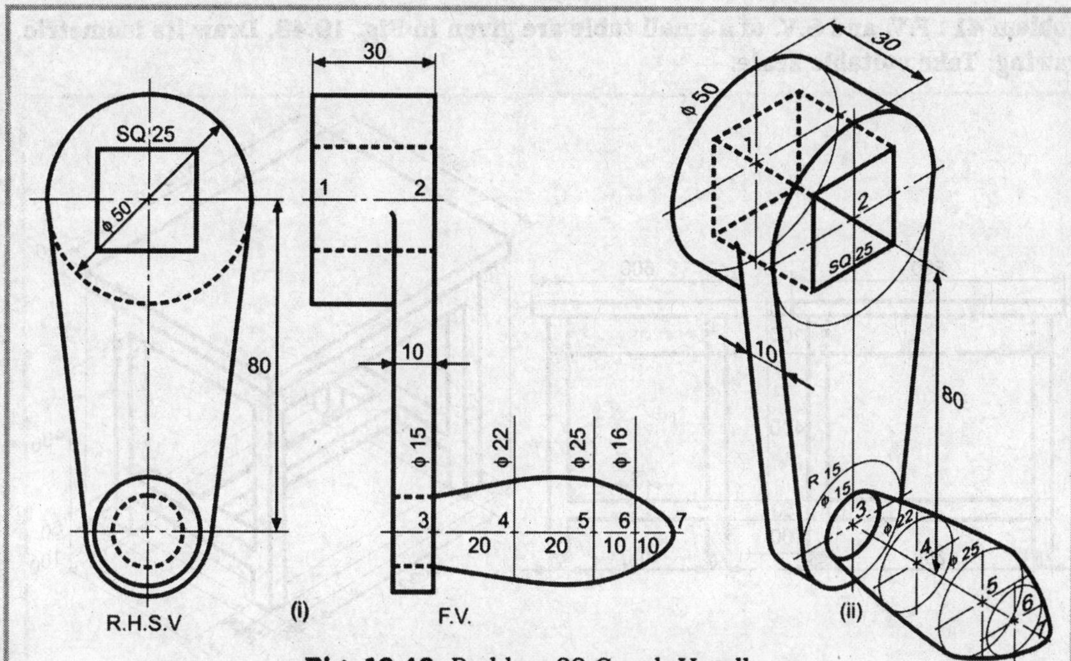

Fig. 19.46. Problem 39 Crank Handle

Problem 40 : F.V. and T.V. of two pieces of wooden joint are given in Fig. 19.47. Draw the isometric drawing of both the pieces ready in position for fitting.

Fig. 19.47. Carpentary Joint - Problem 40

For solution see Fig. 19.47.

Problem 41 : F.V. and S.V. of a small table are given in Fig. 19.48. Draw its isometric drawing. Take suitable scale.

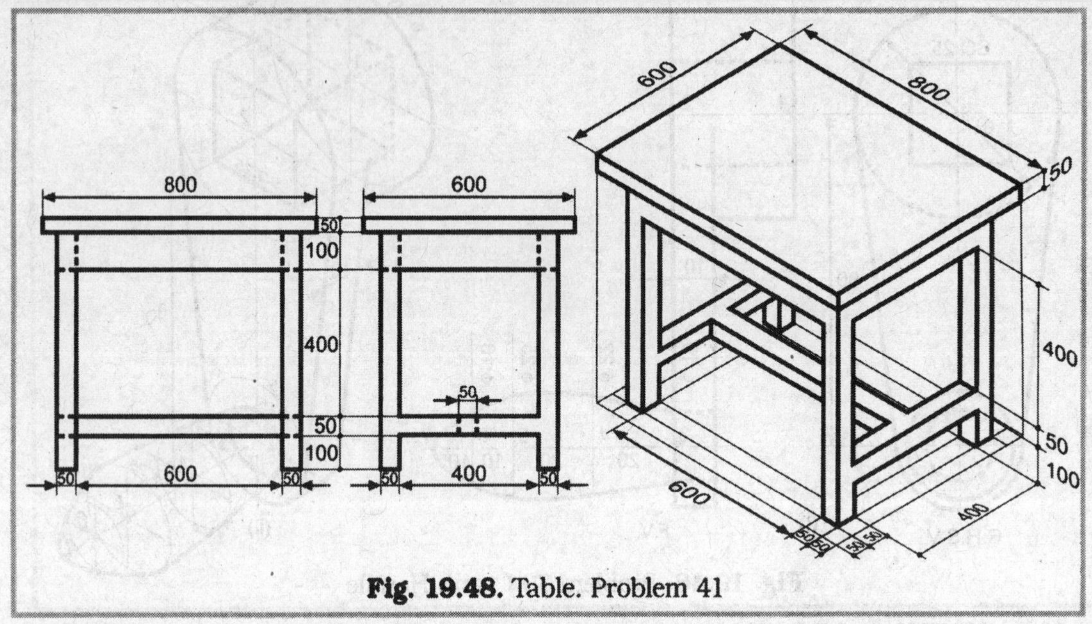

Fig. 19.48. Table. Problem 41

For solution see Fig. 19.48.

Problem 42 : F.V., T.V. and L.H.S.V. of a circular block with different cuttings are given in Fig. 19.49. Draw its isometric view.

In this problem it is desirable to draw 5 ellipses for five circular sections as shown in the figure. Dimensions are required to be taken on edges of rhombuses as shown in the figure for cutting cylinder at various places.

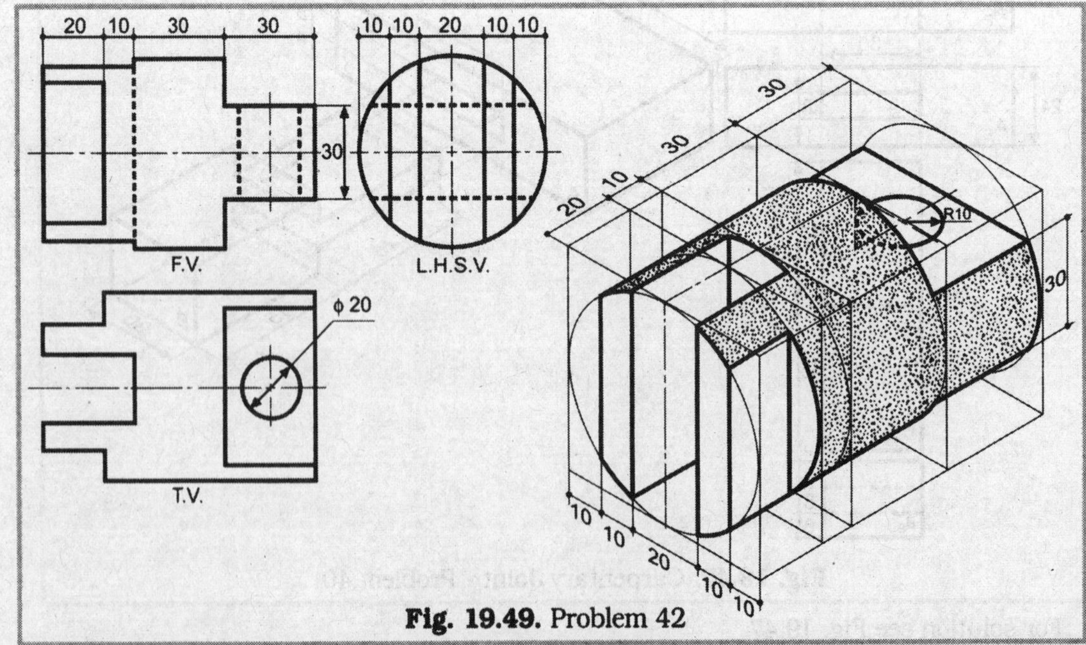

Fig. 19.49. Problem 42

For solution of problem see Fig. 19.49. To make drawing clear, curved surface of the cylinder is shaded.

Problem 43 : Draw the isometric drawing of the handle for the hand press, F.V. of it is given in Fig. 19.50.

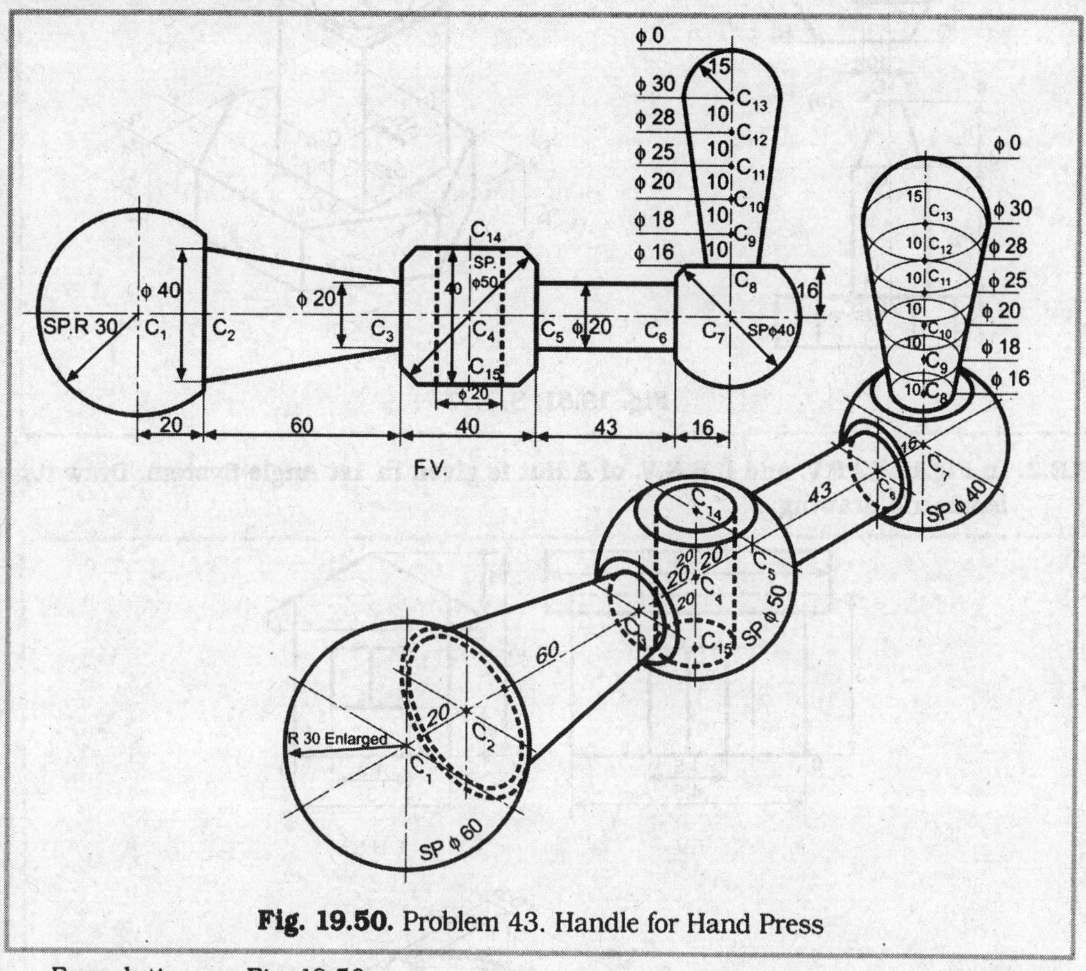

Fig. 19.50. Problem 43. Handle for Hand Press

For solution see Fig. 19.50.

This problem is solved by centre line layout method. First locate centres C_1, C_2,....C_{15} as marked in F.V. and then construct around each centre the section which exist there. For spherical portions with centres draw circles with enlarged radii. For the handle draw a smooth curve touching and enclosing all ellipses drawn along the centre line of the handle.

Solved Exercise

[Drawings of isometrics are drawn self explanatory and hence explanation is not given]

S.E.1. In Fig.19.51 F.V. and T.V. of the object are given in the 3rd Angle System. Draw its isometric view.

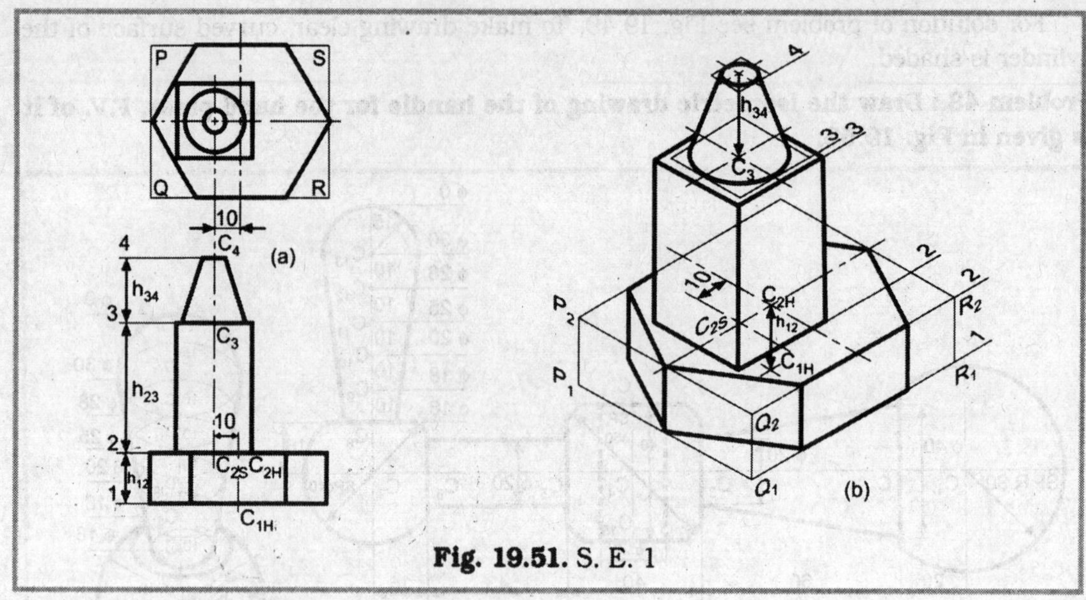

Fig. 19.51. S. E. 1

S.E.2. In Fig.19.52 F.V. and L.H.S.V. of A Hut is given in 1st Angle System. Draw its isometric drawing.

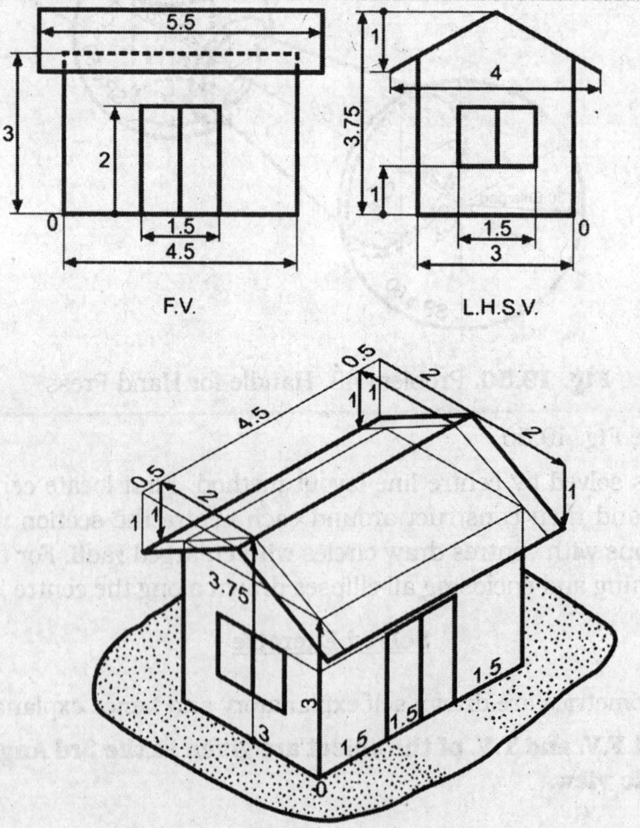

Fig. 19.52. S. E. 2

S.E.3. **In Fig.19.53 F.V. and T.V. of the object are given in 1st Angle System. Draw its isometric drawing [Gujarat University]**

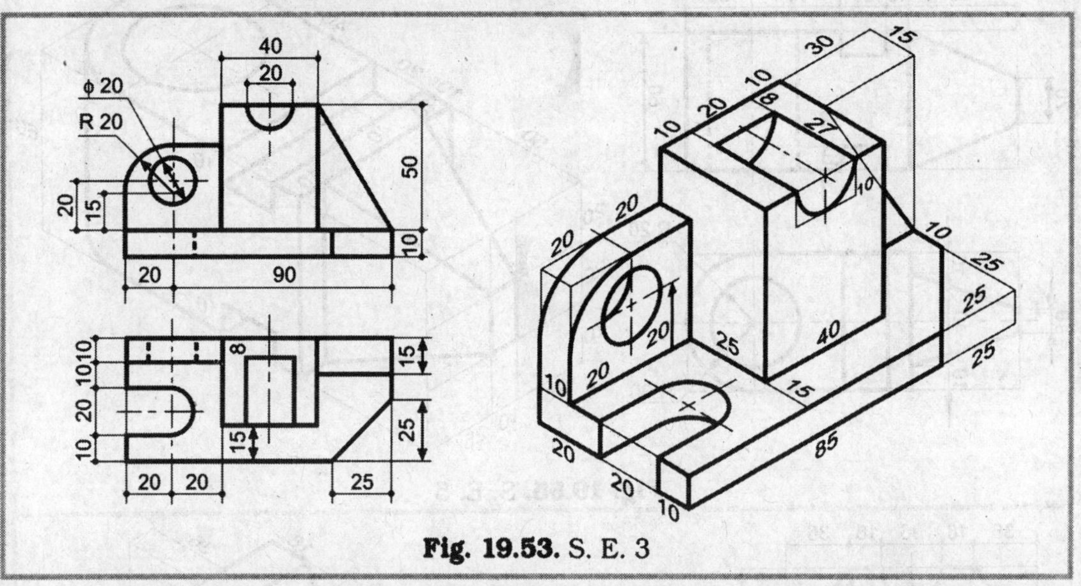

Fig. 19.53. S. E. 3

S.E.4. **In Fig.19.54 F.V. and T.V. of the object are given in 1st Angle System. Draw its isometric view. [Gujarat University]**

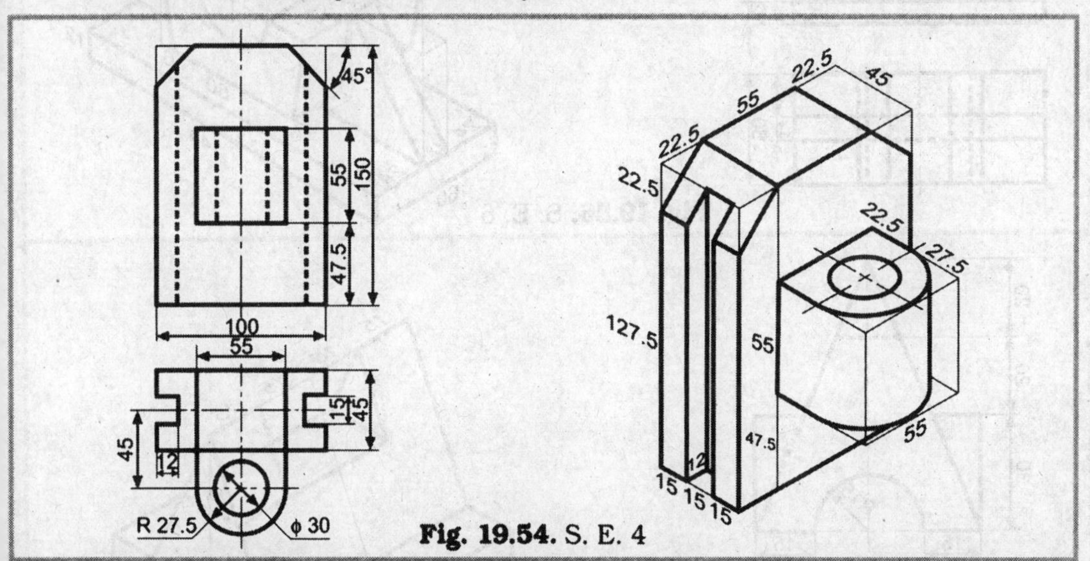

Fig. 19.54. S. E. 4

S.E.5. **In Fig.19.55 F.V. and T.V. of the object are given in 1st Angle System. Draw isometric drawing [Gujarat University]**

S.E.6. **In Fig.19.56 F.V. and T.V. of the object are given in 1st Angle System. Draw isometric view. [Gujarat University]**

S.E.7. **In Fig.19.57 F.V. and T.V. of the object are given in 1st Angle System. Draw isometric view.**

S.E.8. **In Fig.19.58 F.V. and T.V. of the object are given in 1st Angle System. Draw isometric drawing.**

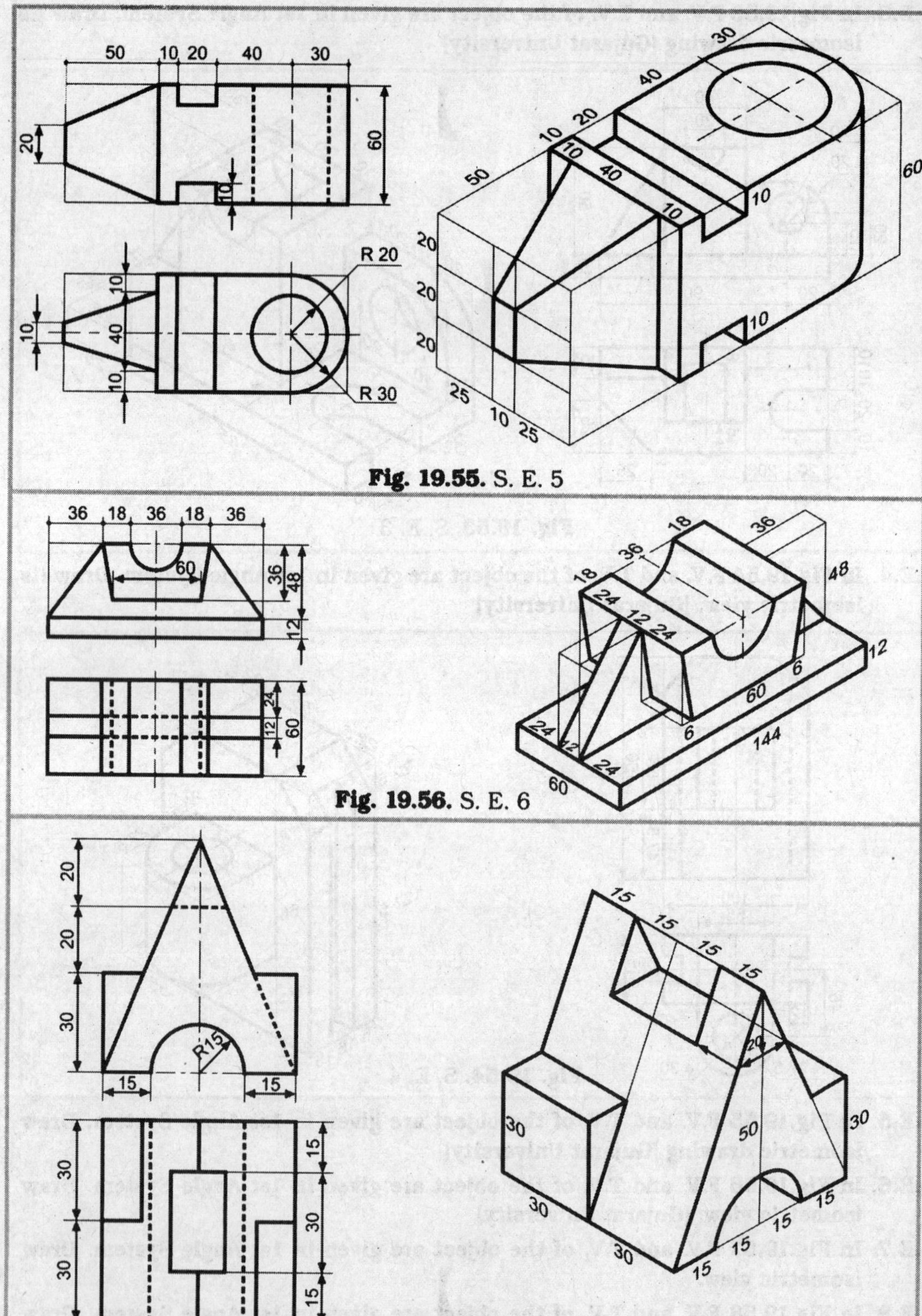

Fig. 19.55. S. E. 5

Fig. 19.56. S. E. 6

Fig. 19.57. S. E. 7

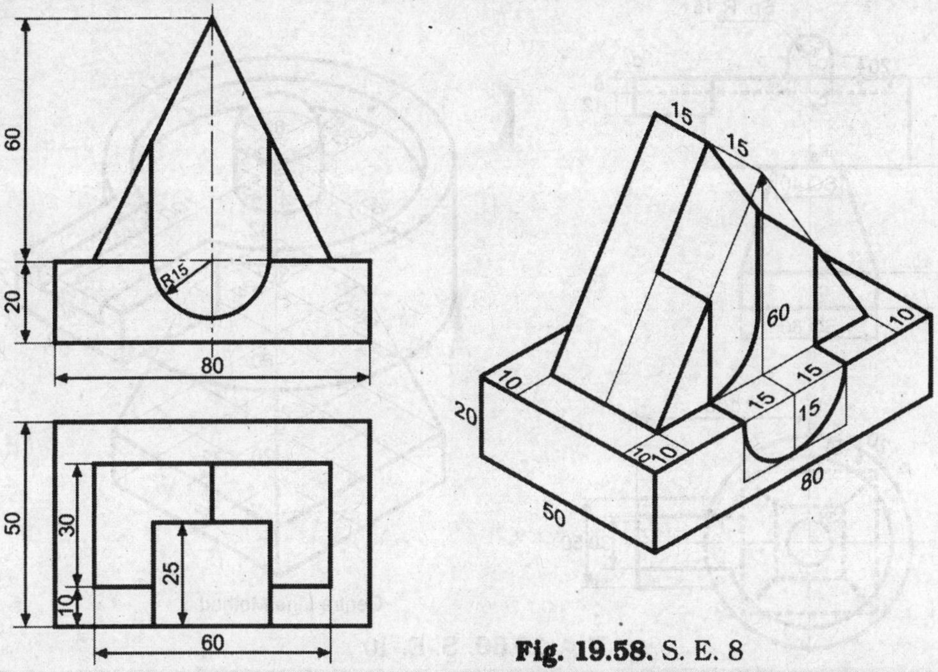

Fig. 19.58. S. E. 8

S.E.9. In Fig.19.59 F.V. and L.H.S.V. of the object are is given in 1st Angle System. Draw isometric view. Use Centre Line Method.

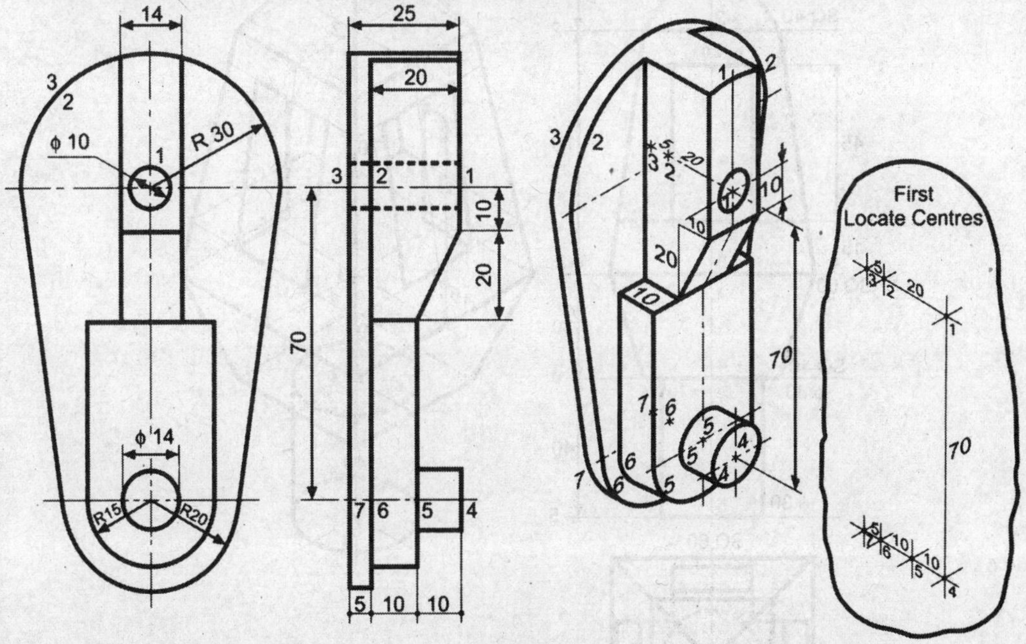

Fig. 19.59. S. E. 9

S.E.10. In Fig.19.60 Orthographic Projections of the object are given. Draw its isometric view. [Pune University]

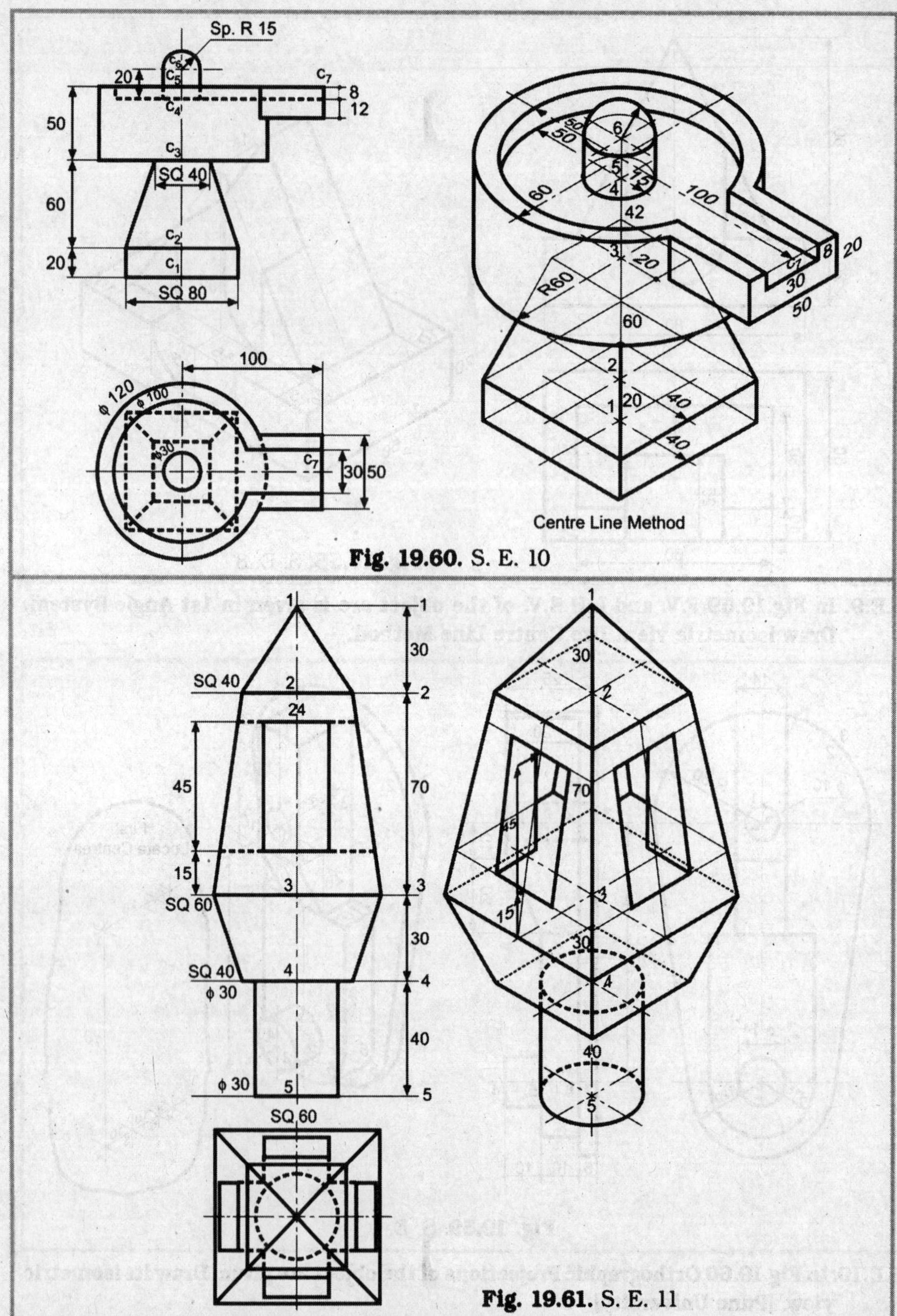

Fig. 19.60. S. E. 10

Fig. 19.61. S. E. 11

S.E.11. In Fig.19.61 Orthographic Views of the object are given. Draw isometric view. [Pune University]

S.E.12. In Fig.19.62 F.V. and R.H.S.V. are given. Draw isometric drawing.

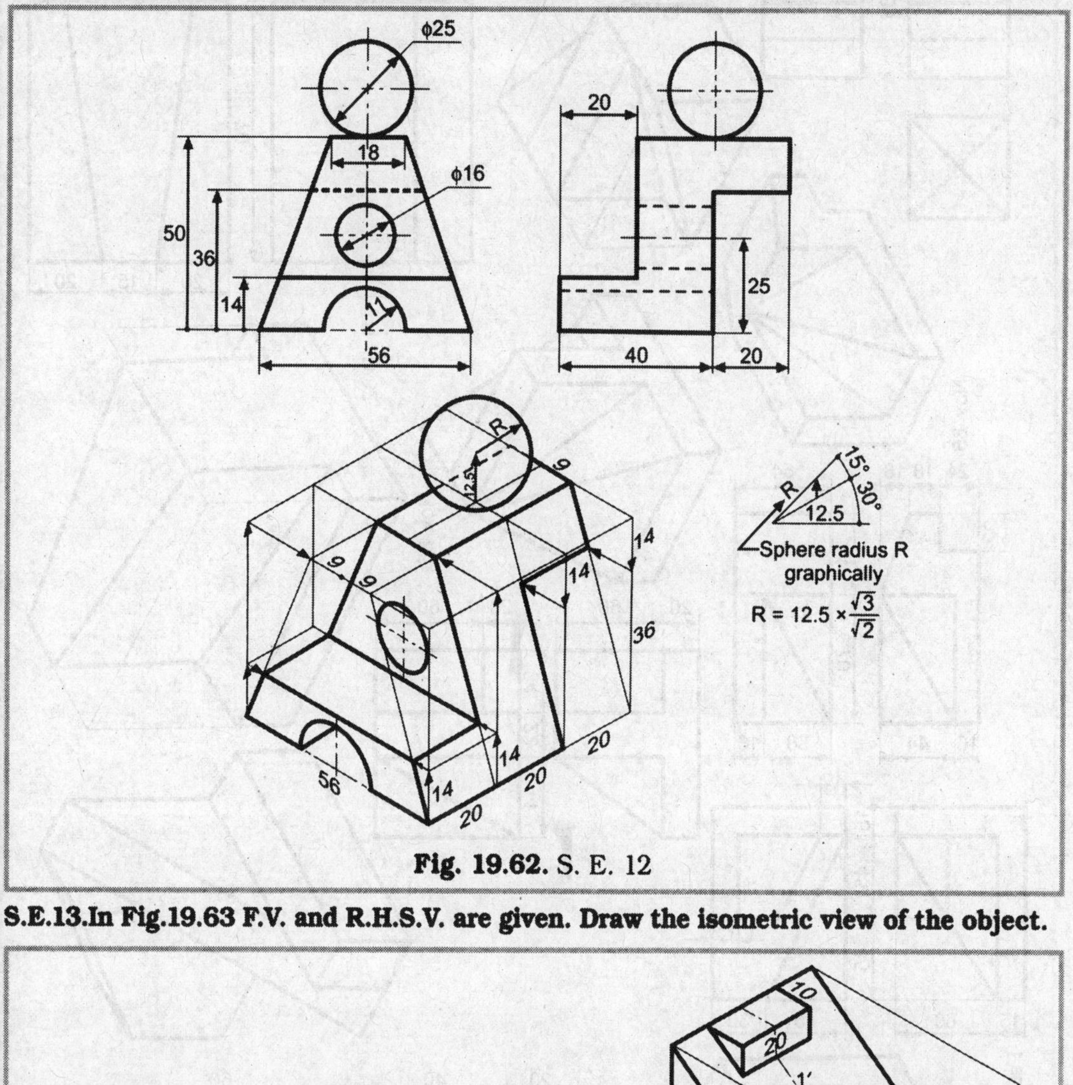

Sphere radius R graphically

$R = 12.5 \times \frac{\sqrt{3}}{\sqrt{2}}$

Fig. 19.62. S. E. 12

S.E.13. In Fig.19.63 F.V. and R.H.S.V. are given. Draw the isometric view of the object.

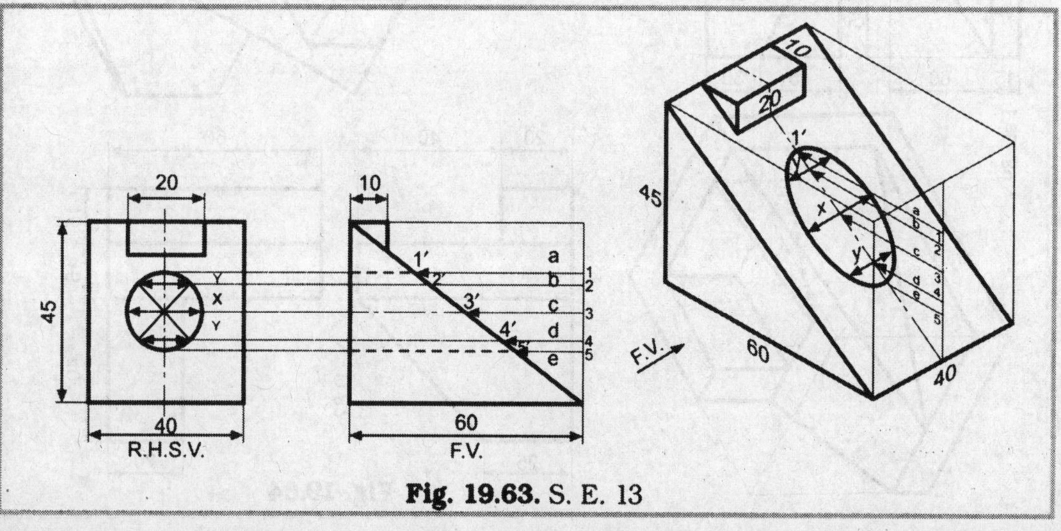

Fig. 19.63. S. E. 13

Few practice problems with solutions are given on next 4 full pages.

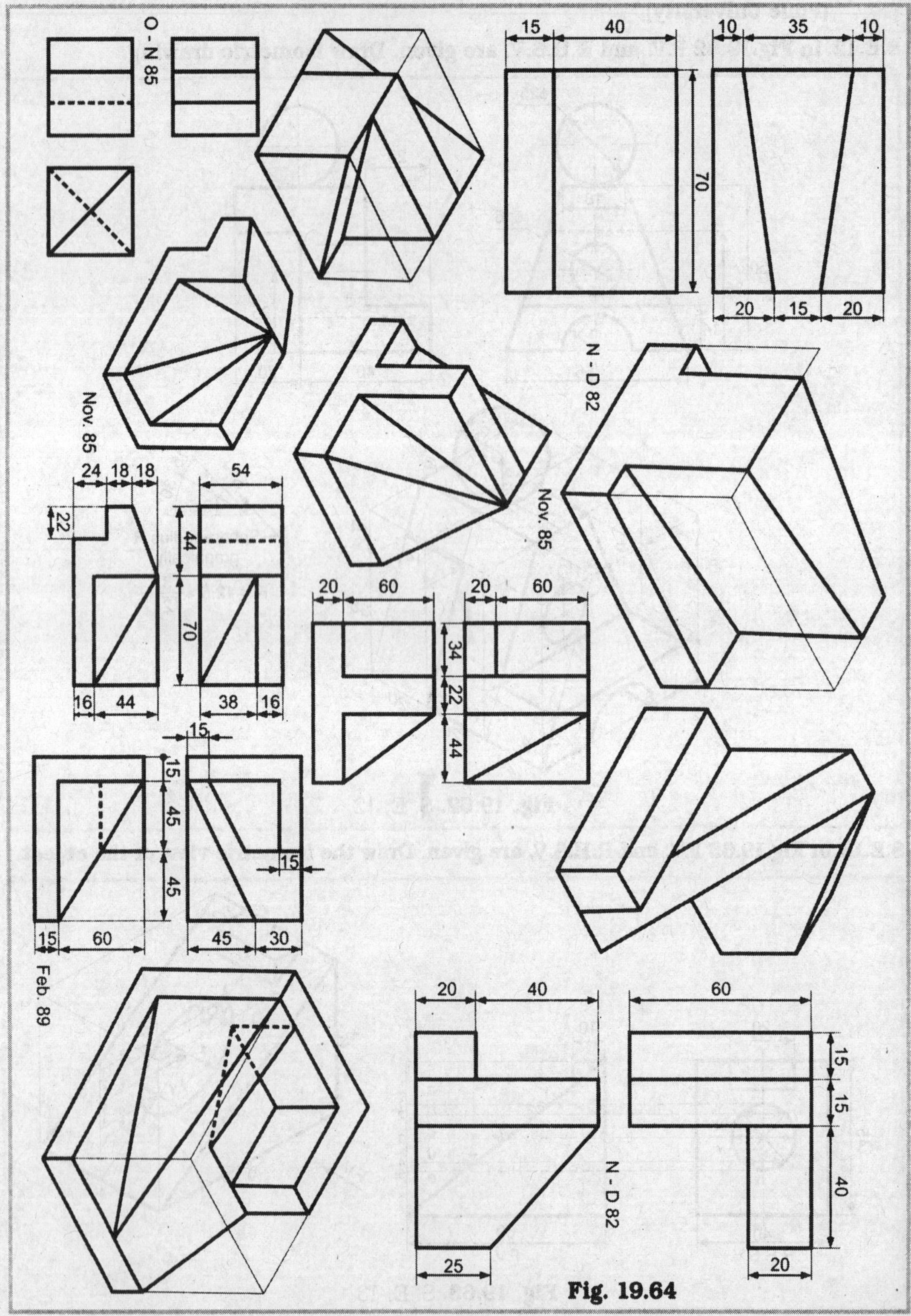

Fig. 19.64

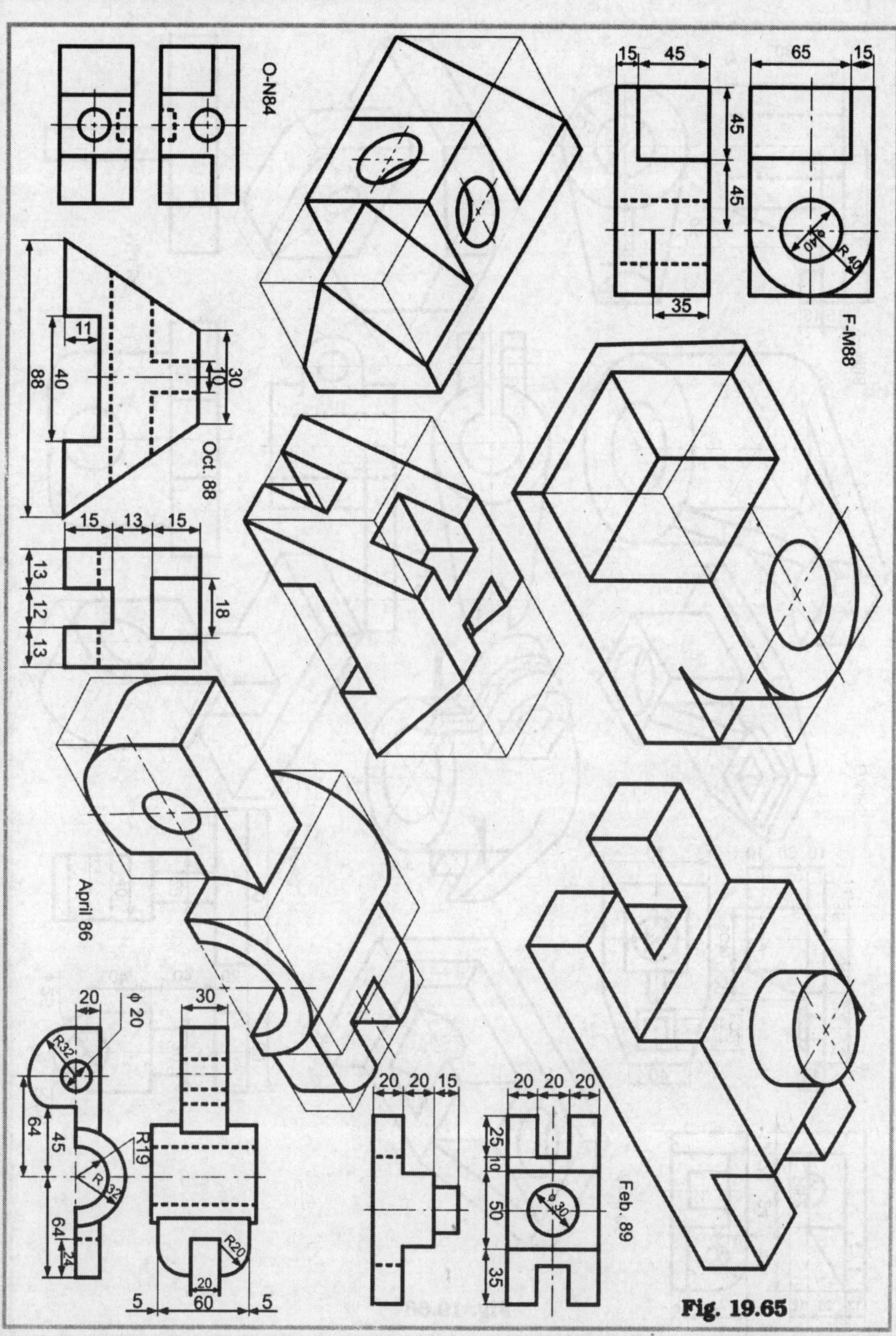
O-N84
Oct. 88
F-M88
April 86
Feb. 89
15
45
65
45
35
φ 40
R 40
11
40
88
10
30
13
12
18
20
φ 20
R32
64
R19
R
32
24
R20
60
5
25
50

Fig. 19.65

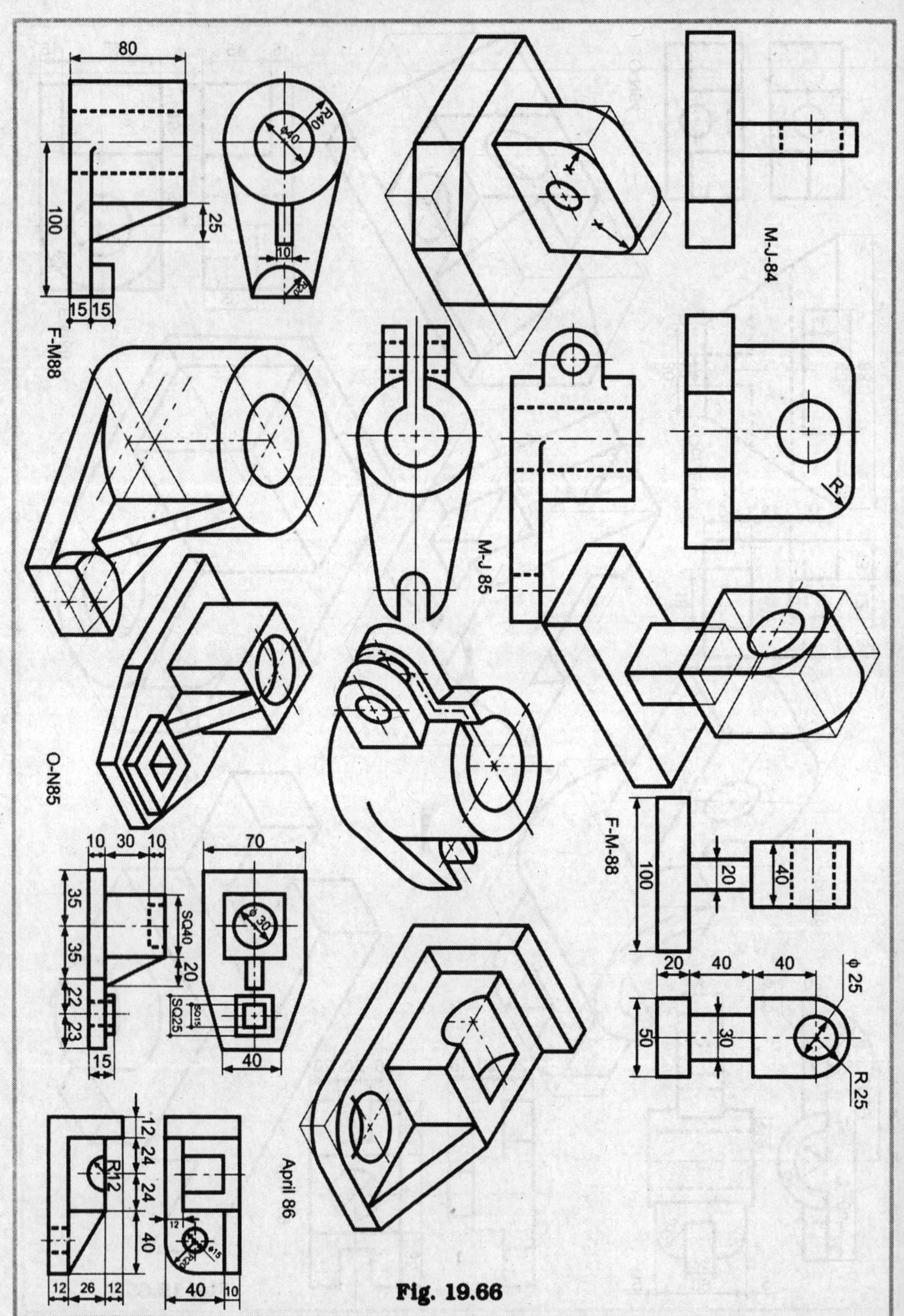
80
100
25
15 15
R40
φ40
10
R20
F-M88
M-J-84
R
M-J 85
O-N85
F-M-88
100
20
40
20 40 40
φ 25
50
30
R 25
10 30 10
35
35
22
23
15
SQ40
20
70
φ 30
SQ25
40
12
24
24
40
R12
12 26 12
40 10
φ15
April 86

Fig. 19.66

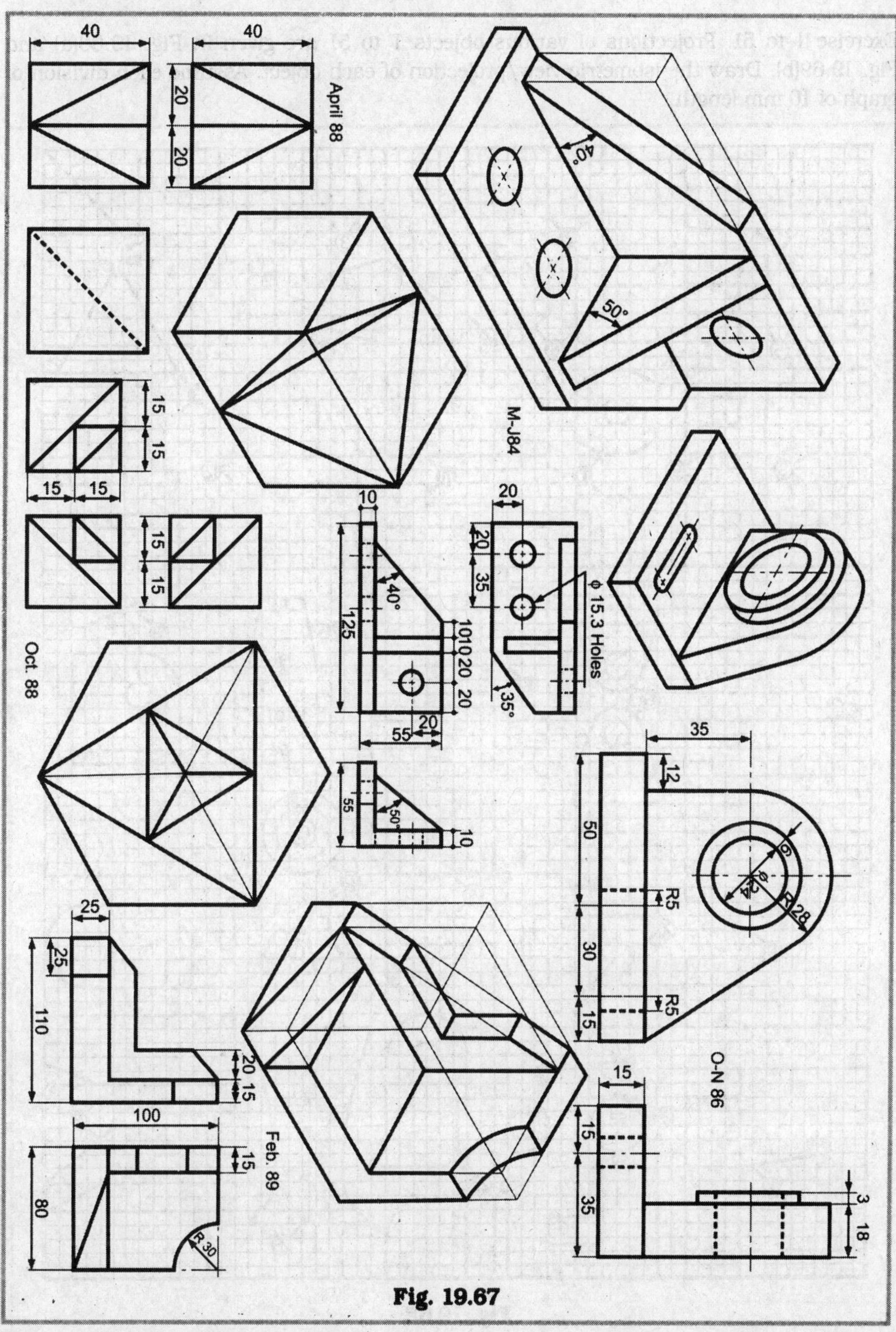

40
40
20
20
April 88
15
15
15
15
15
15
Oct. 88
M-J84
10
40°
125
10
10
20
20
20
55
20
20
35
ϕ 15.3 Holes
35°
55
50°
10
40°
50°
35
12
50
R5
30
R5
15
ϕ24
6
R28
O-N 86
25
25
110
20
15
100
15
80
R 30
Feb. 89
15
15
35
3
18

Fig. 19.67

Exercise 1 to 51. Projections of various objects 1 to 51 are given in Fig. 19.68(a) and Fig. 19.69(b). Draw the isometric view/projection of each object. Assume each division of graph of 10 mm length.

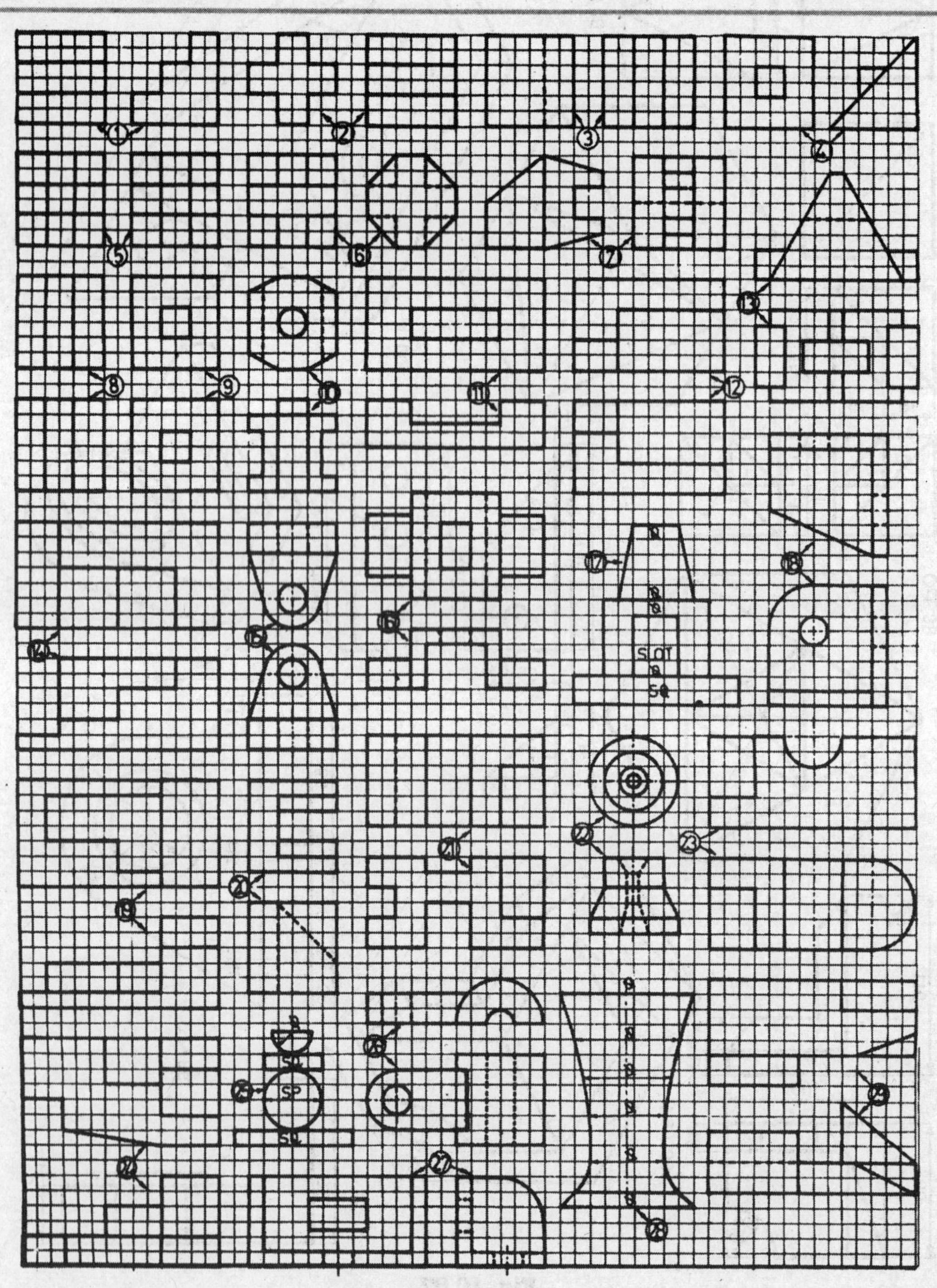

Fig. 19.68

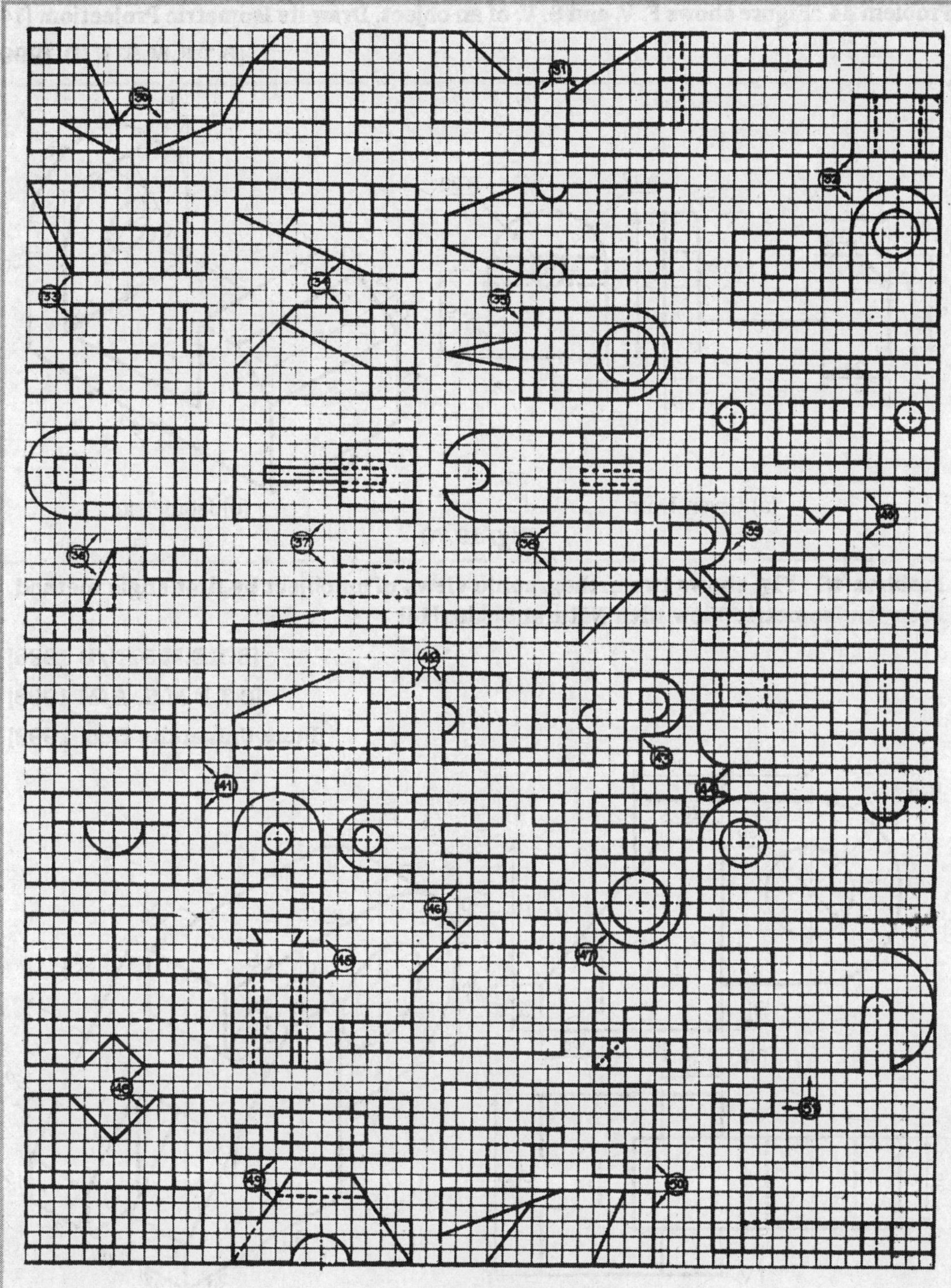

Fig. 19.69

Problem 44 : Figure shows F. V. and S. V. of an object. Draw its Isometric Projection. [14]

[B.T.E.M.S. N/D 1999]

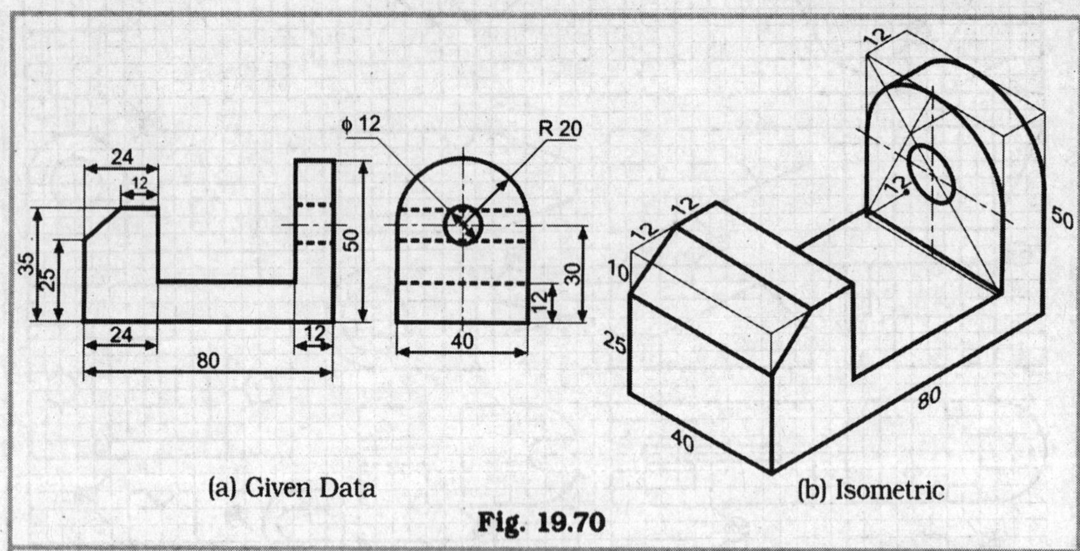

(a) Given Data (b) Isometric

Fig. 19.70

Problem 45 : Fig. shows two orthographic views of an object by first angle method. Draw An Isometric View with Natural Scale. [14]

[B.T.E.M.S. N/D 1996]

[B.T.E.M.S. A/M 1998]

[Pune University Dec. 1992]

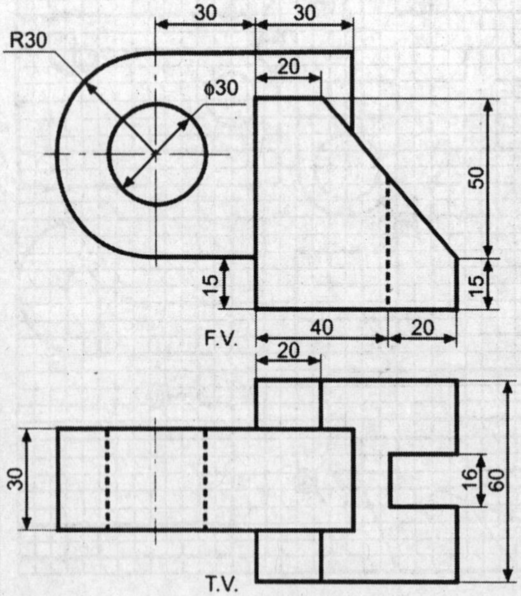

(a) Given Data 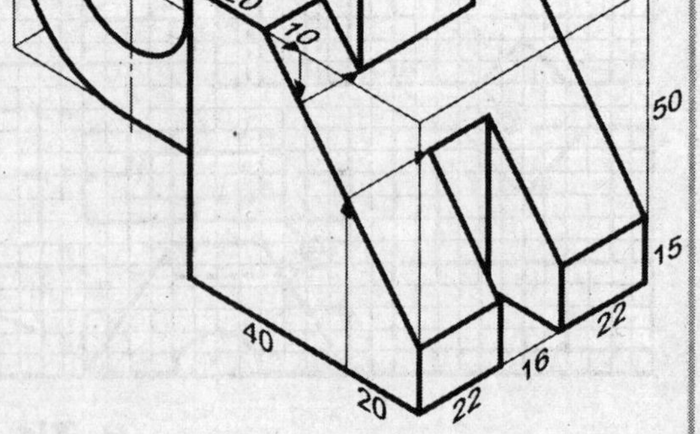 (b) Isometric

Fig. 19.71

Problem 46 : Two orthographic views are given in Figure by third angle method. Draw An Isometric View, taking O as origin. Use Natural Scale. [14]

[B.T.E.M.S. N/D 1998]

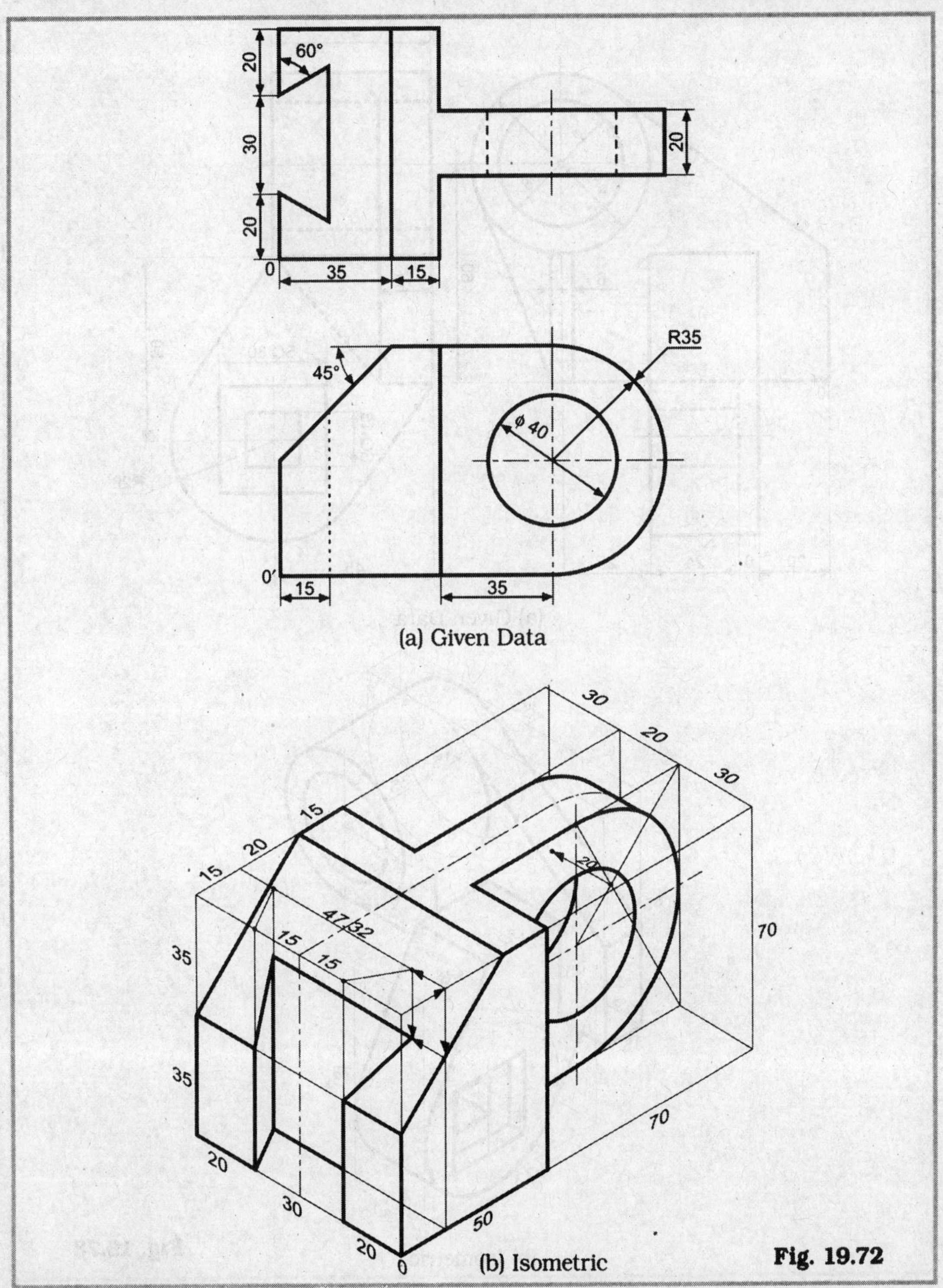

(a) Given Data

(b) Isometric

Fig. 19.72

Problem 47 : Orthographic views of an object are shown in Figure. Using Natural Scale, draw its Isometric View.

[Pune University]

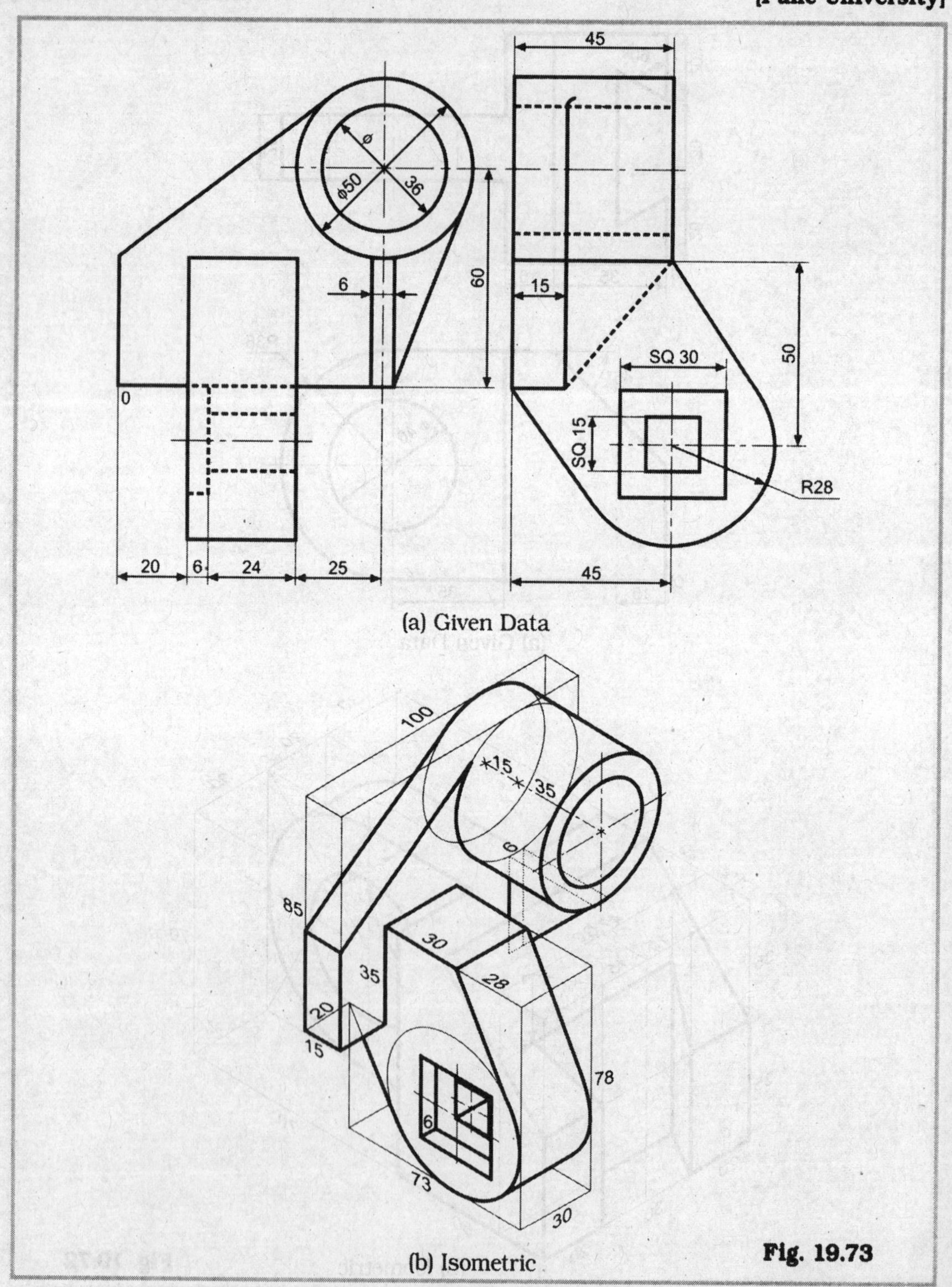

(a) Given Data

(b) Isometric

Fig. 19.73

Problem 48 : Two orthographic views of an object are given in Fig. by third angle method. Draw An Isometric View, taking O as origin. Use Natural Scale. [14]

[B.T.E.M.S. A/M 1998]

[B.T.E.M.S. A/M 2000]

[Pune University]

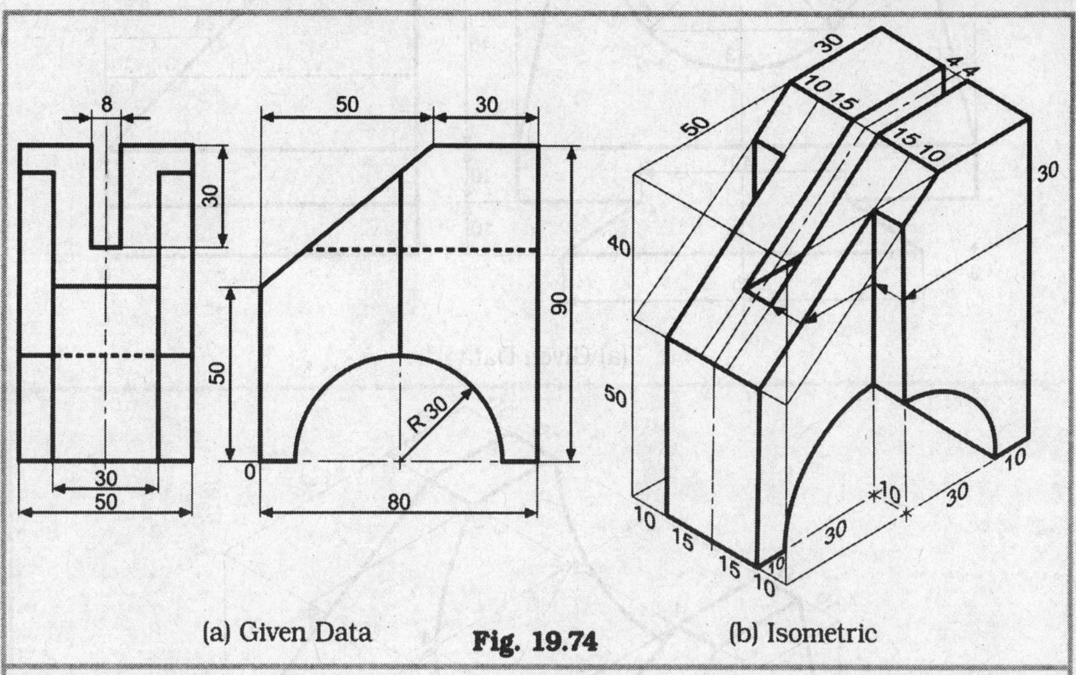

(a) Given Data **Fig. 19.74** (b) Isometric

Problem 49 : Fig. shows F. V. and S. V. of an object. Draw its Isometric Projection about an origin O. [14]

[B.T.E.M.S. A/M 1997]

(a) Given Data (b) Isometric

Fig. 19.75

Problem 50 : Refer figure and draw The Isometric View of the given object using Natural Scale. [Mumbai University, Dec. 2000]

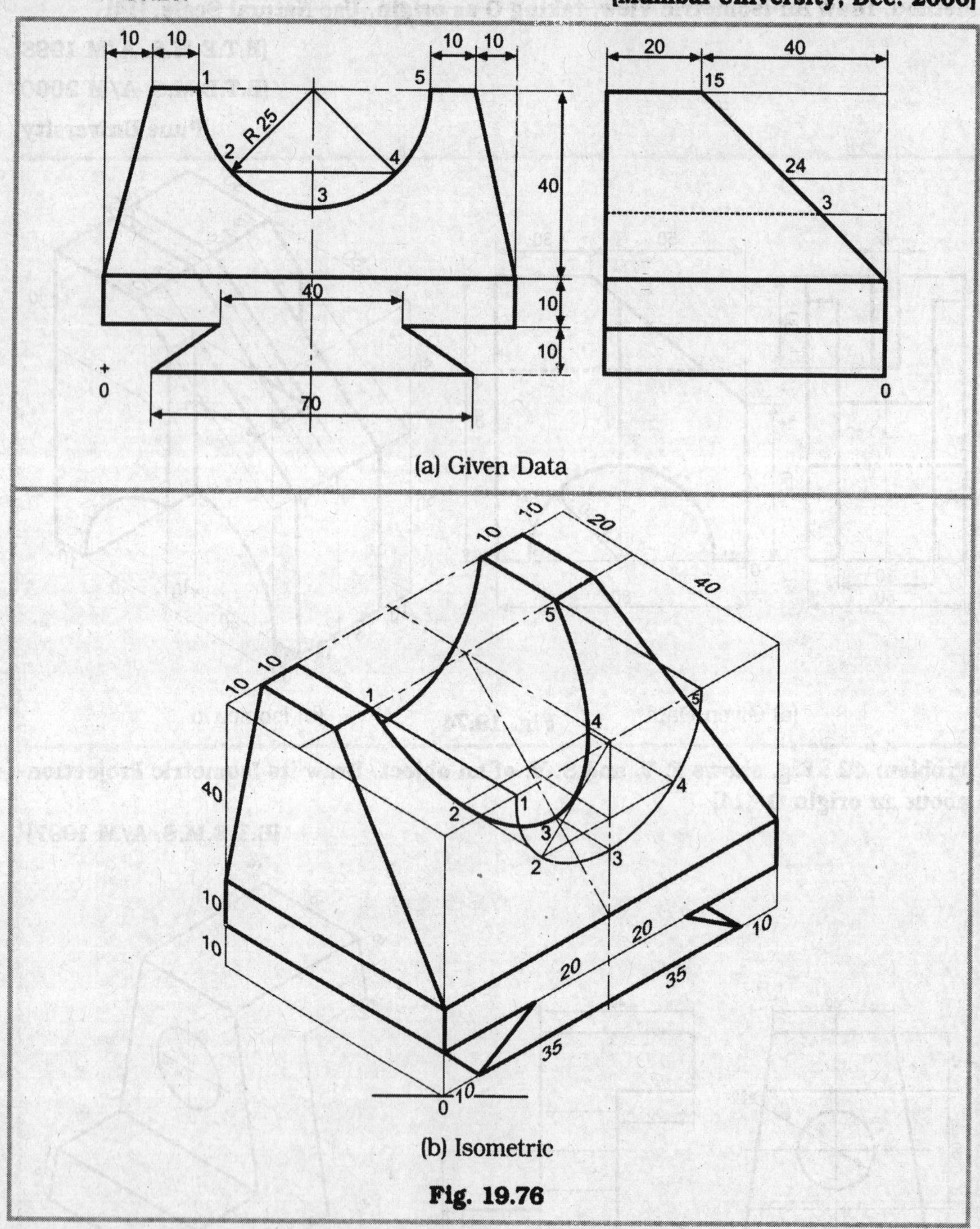

(a) Given Data

(b) Isometric

Fig. 19.76

Problem 51 : From the given view, draw An Isometric View. Co-ordinate method is used in drawing Isometric.

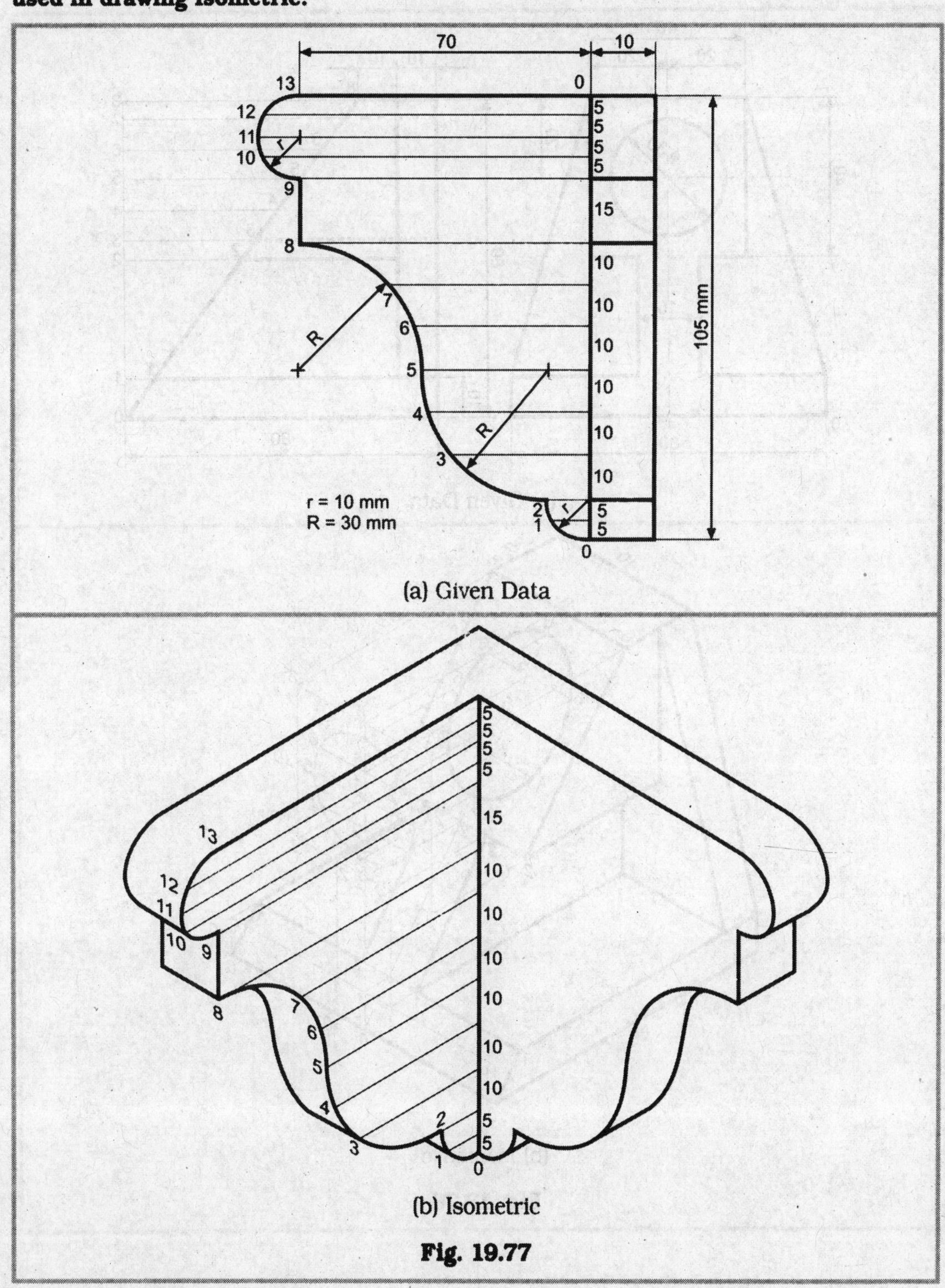

(a) Given Data

(b) Isometric

Fig. 19.77

Problem 52 : Fig. shows F.V. and L.H.S.V. of an object. Draw the Isometric view of the object using Natural Scale. **[Mumbai University, May 1997]**

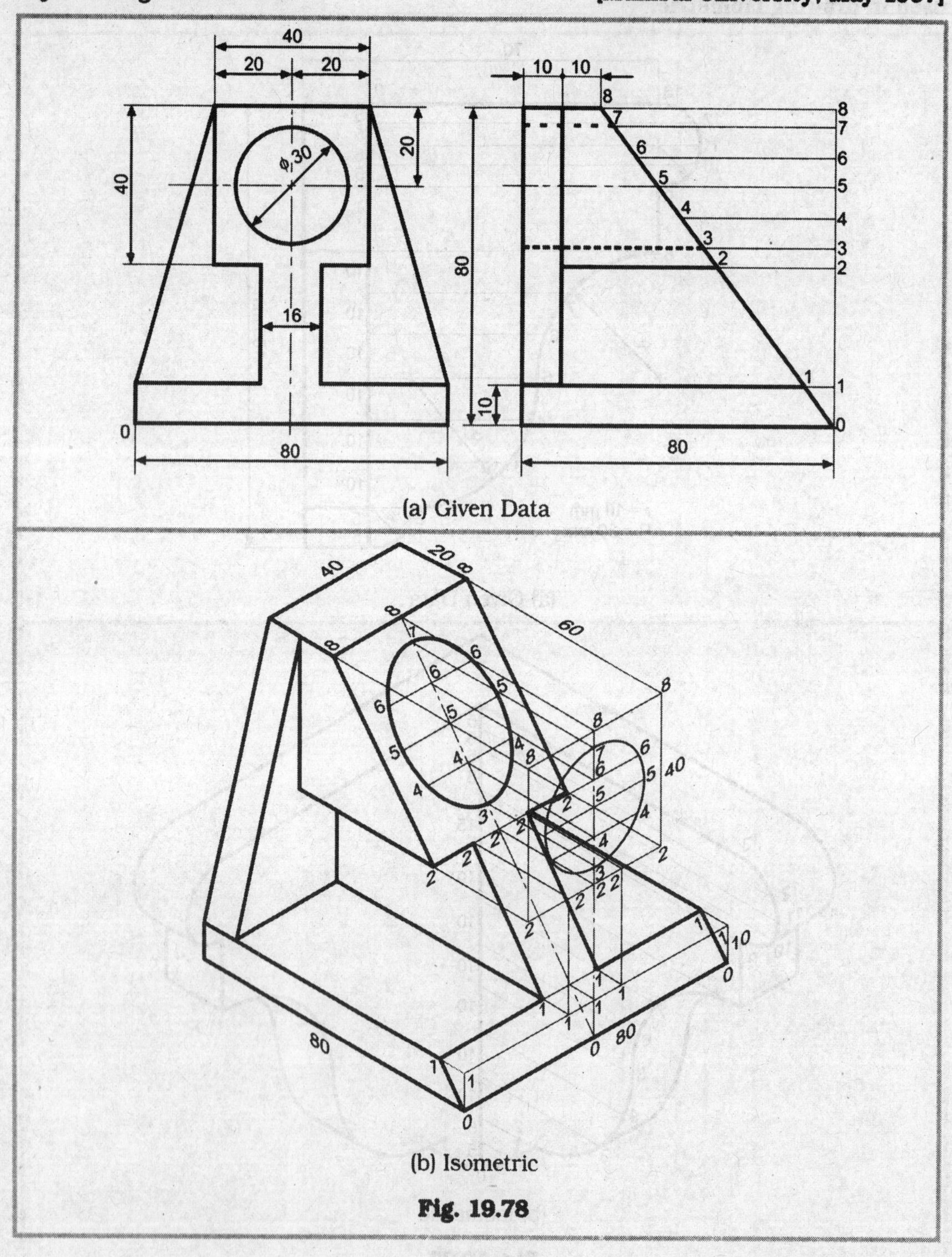

(a) Given Data

(b) Isometric

Fig. 19.78

Problem 53 : Fig. shows F. V. and T. V. of an object. Draw its Isometric View of the object using Natural Scale.

[Mumbai University Nov./Dec. 1996]

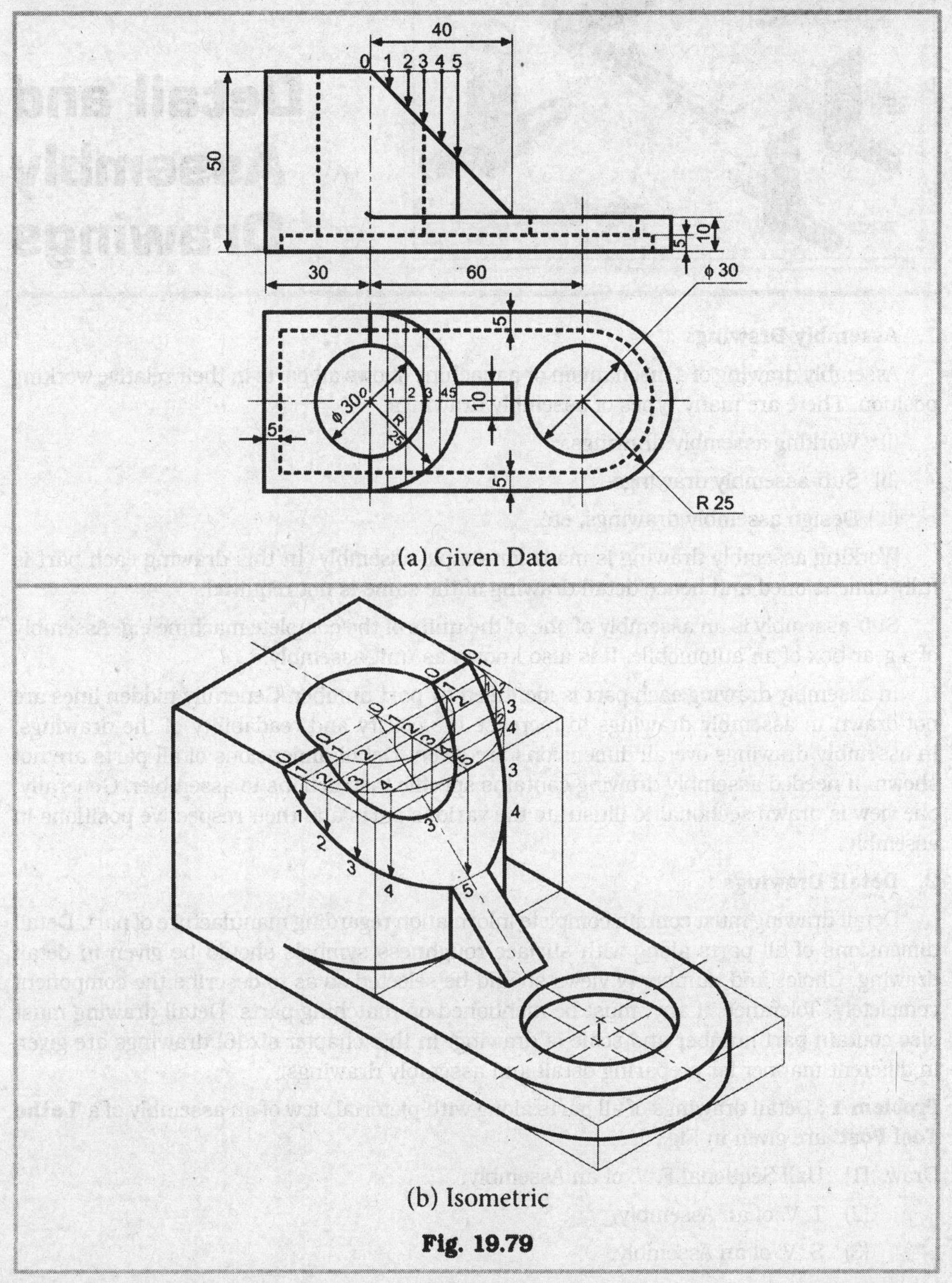

(a) Given Data

(b) Isometric

Fig. 19.79

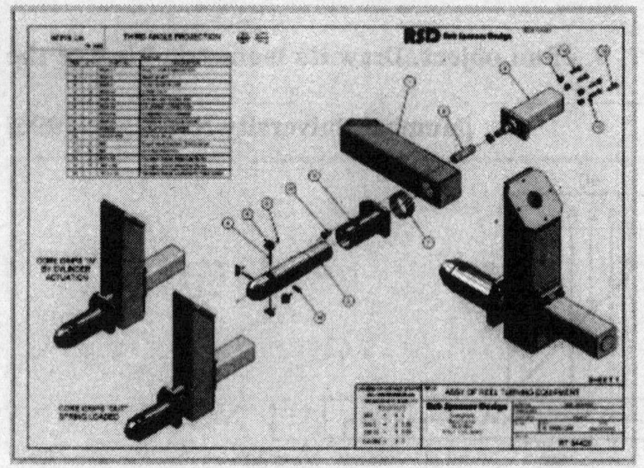

CHAPTER - 20

Detail and Assembly Drawings

1. Assembly Drawings :

Assembly drawing of a mechanism or a machine shows all parts in their relative working position. There are many types of assembly drawings.

(i) Working assembly drawings,

(ii) Sub-assembly drawings,

(iii) Design assembly drawings, etc.

Working assembly drawing is made for simple assembly. In this drawing each part is fully dimensioned and hence detail drawing of the same is not required.

Sub-assembly is an assembly of one of the units of the complete machine e.g. Assembly of a gear box of an automobile. It is also known as unit assembly.

In assembly drawing each part is identified by part number. Generally hidden lines are not drawn in assembly drawings to increase the clarity and readability of the drawings. In assembly drawings overall dimensions are given. Detail dimensions of all parts are not shown. If needed assembly drawing contains specific instructions to assembler. Generally, one view is drawn sectional to illustrate the various parts and their respective positions in assembly.

2. Detail Drawings :

Detail drawing must contain complete information regarding manufacture of part. Detail dimensions of all parts along with surface roughness symbols should be given in detail drawing. Choice and number of views should be selected so as to describe the component completely. Tolerance, if any, must be mentioned on matching parts. Detail drawing must also contain part number and scale of drawing. In this chapter six (6) drawings are given in different manner for preparing detail and assembly drawings.

Problem 1 : Detail drawings of all parts along with pictorial view of an assembly of a **'Lathe Tool Post'** are given in Fig.

Draw (1) Half Sectional F. V. of an Assembly.

(2) T. V. of an Assembly.

(3) S. V. of an Assembly.

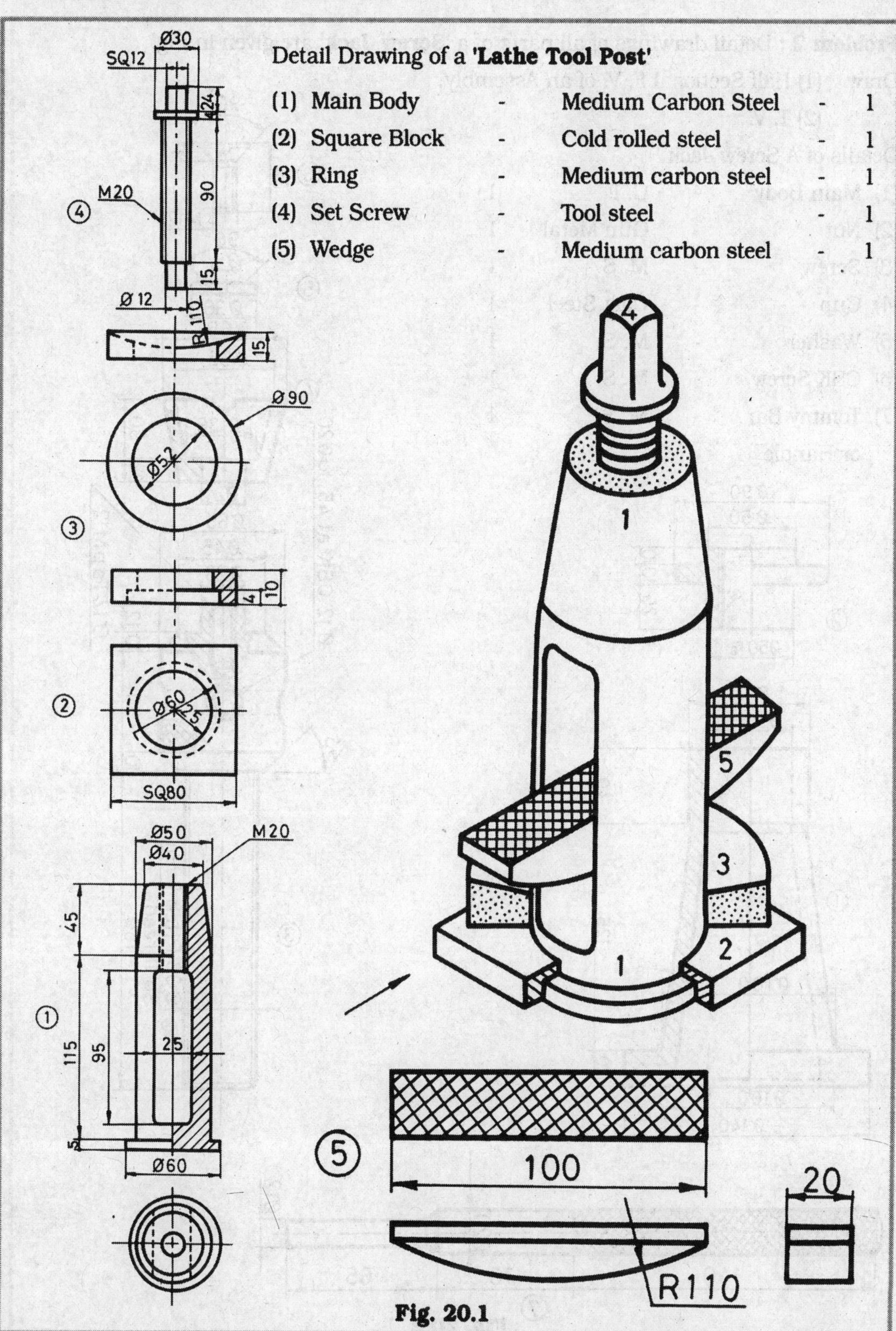
Detail Drawing of a 'Lathe Tool Post'
(1) Main Body - Medium Carbon Steel - 1
(2) Square Block - Cold rolled steel - 1
(3) Ring - Medium carbon steel - 1
(4) Set Screw - Tool steel - 1
(5) Wedge - Medium carbon steel - 1
Ø30
SQ12
24
90
M20
15
Ø 12
R110
15
Ø90
Ø52
10
4
Ø60
25
SQ80
Ø50
Ø40
M20
45
115
95
25
5
Ø60
100
20
R110

Fig. 20.1

Problem 2 : Detail drawings of all parts of a '**Screw Jack**' are given in Fig.

Draw (1) Half Sectional F. V. of an Assembly.

(2) T. V.

Details of A Screw Jack

(1)	Main Body	-	C. I.	1
(2)	Nut	-	Gun Metal	1
(3)	Screw	-	M. S.	1
(4)	Cup	-	Cast Steel	1
(5)	Washer	-	M. S.	1
(6)	CSK Screw	-	M. S.	1
(7)	Tommy Bar or Handle	-	M. S.	1

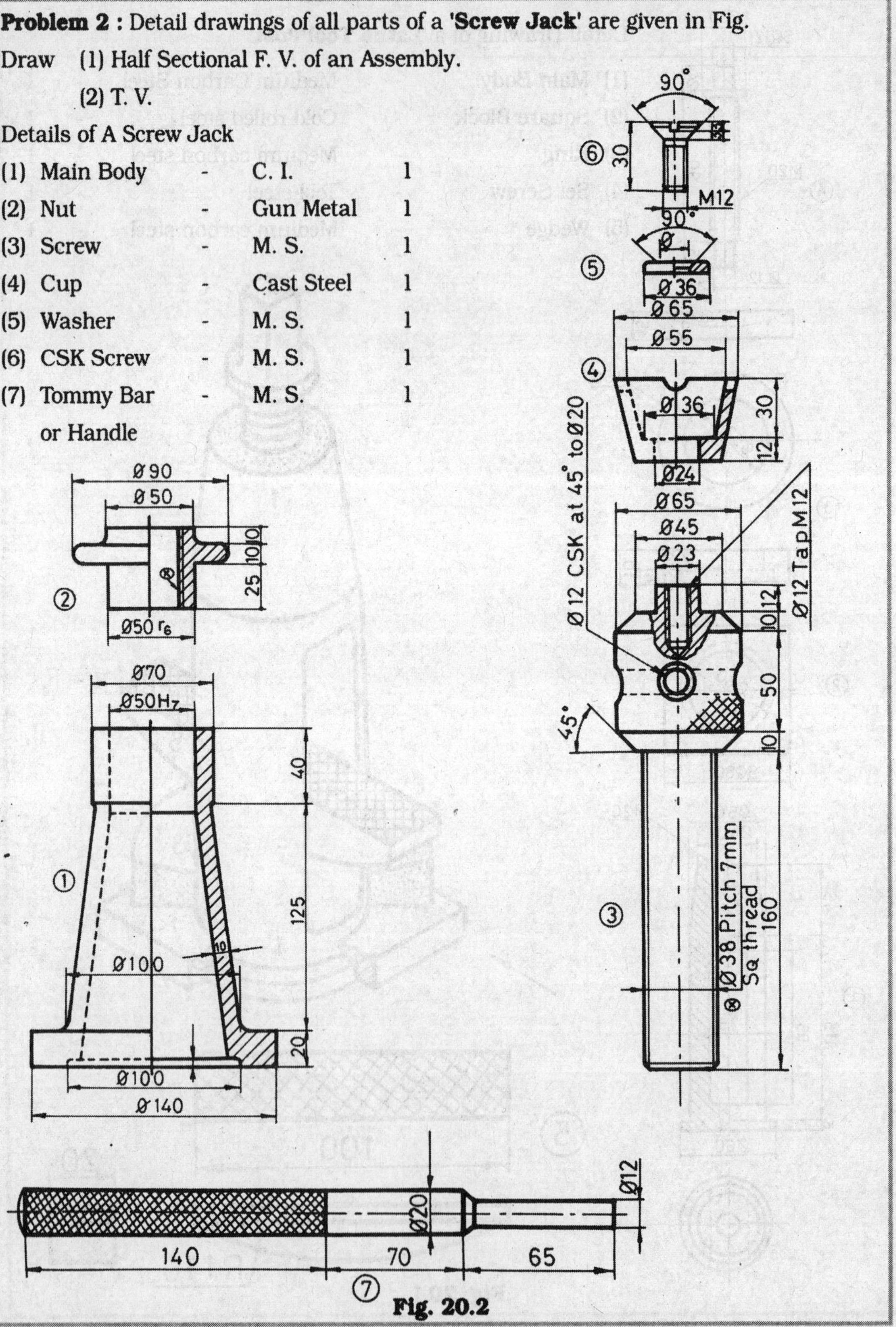

Fig. 20.2

Problem 3 : Pictorial views of all parts of a **'Big End of A Connecting Rod, with jib and cotter'** are given in Fig.20.3.

Draw (1) Detail drawings of all parts.

(2) Assembly drawings. (2 views)

Big end of a connecting rod with jib and cotter Part List.

(1) Strap	-	Forged Steel	-	1
(2) Connecting Rod End	-	Forged Steel	-	1
(3) Brass (Round)	-	Gun Metal	-	1
(4) Brass (Rectangular)	-	Gun Metal	-	1
(5) Jib	-	Steel	-	1
(6) Cotter	-	Steel	-	1
(7) Set Screw	-	M. S.	-	1

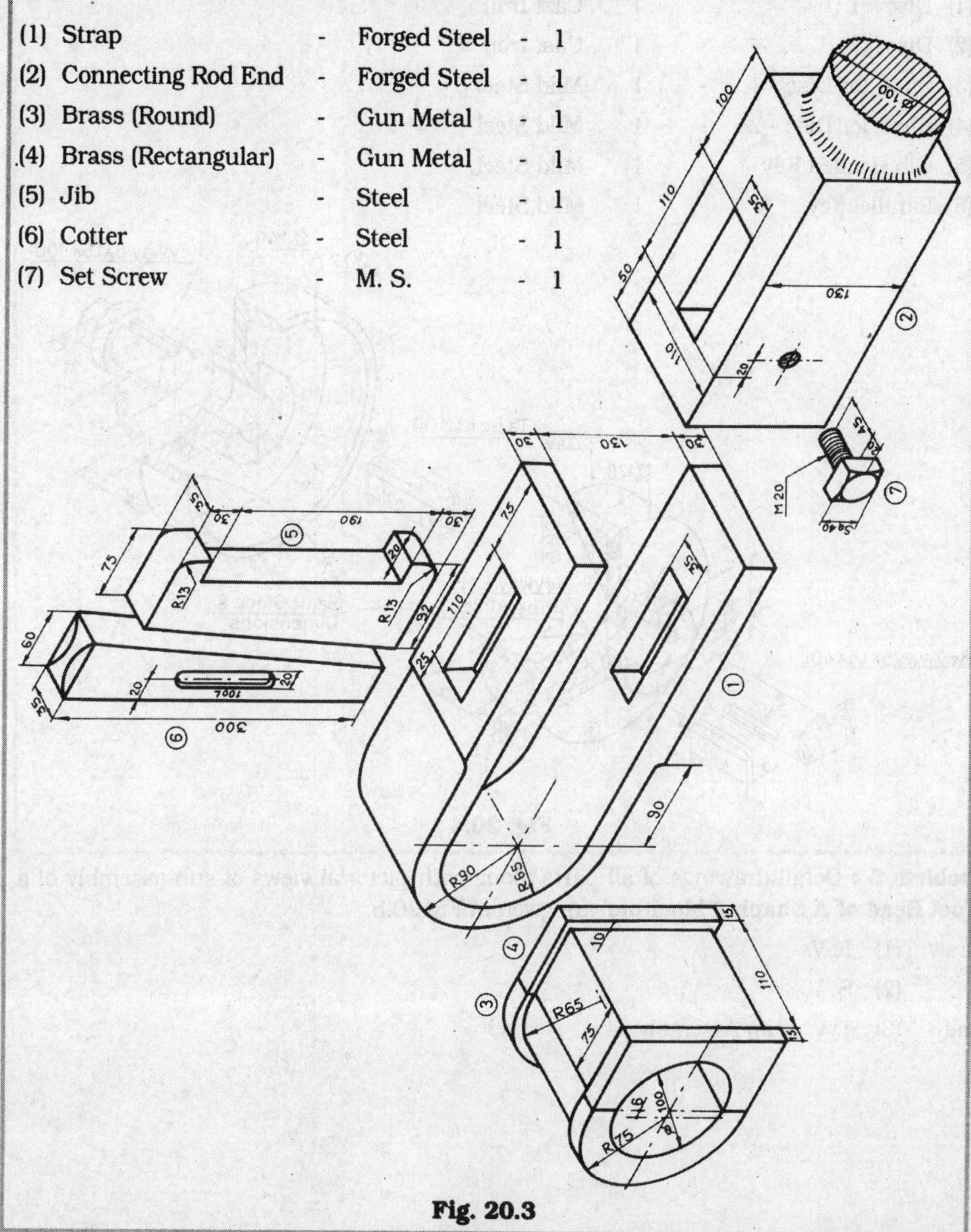

Fig. 20.3

Problem 4 : Pictorial views of all parts of a **'Friction Coupling'** are given in Fig.20.4.

Draw (1) Detail drawings.

(2) Assembly drawings. (2 views)

Friction Coupling Part List

(1) Disc - 1 - 1 Cast Iron

(2) Disc - 2 - 1 Cast Iron

(3) Shaft for Disc - 1 - 1 Mild Steel

(4) Shaft for Disc - 2 - 1 Mild Steel

(5) Gib Headed Key - 1 Mild Steel

(6) Parallel Key - 1 Mild Steel

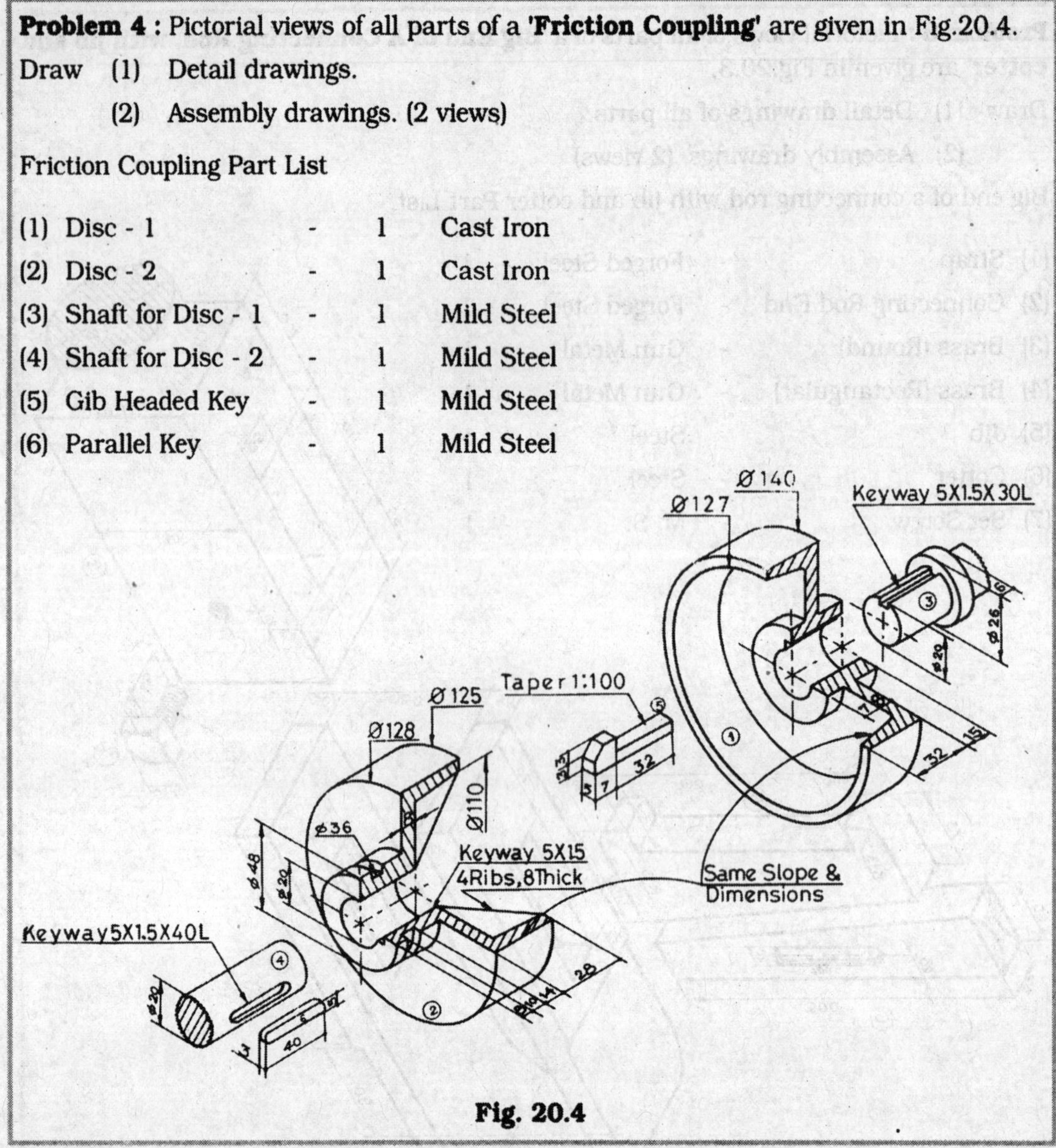

Fig. 20.4

Problem 5 : Detail drawings of all parts along with pictorial views of sub-assembly of a **'Tool Head of A Shaping Machine'** are given in Fig.20.5.

Draw (1) F. V.

(2) S. V.

and (3) T. V. of an Assembly.

Details of A Sub Assembly of 'A Tool Head of A Shaping Machine'

Sr. No.	Name of Part	Material	Nos.
(1)	Drag Release Plate	Cast Iron	1
(2)	Swivel Plate	Cast Iron	1
(3)	Swivel Plate Champing Screw	Case Hardened M. S.	1
(4)	Swivel Pin	M. S.	1
(5)	Tool Holder	M. S.	1
(6)	Washer for clamping Screw	Hardened Steel	1
(7)	Washer forTool Holder	Hardened Steel	1
(8)	Tool Holder clamping Screw	Case Hardened M. S.	1
(9)	Taper Pin	Case Hardened M. S.	1

Problem 6 : Detail drawings of all parts of a **'Heavy Duty Trolley Wheel'** are given in Fig.20.6.

Draw (1) F. V. and S. V. of an Assembly.

Details of Heavy Duty Trolly Wheel

Sr. No.	Name of Part	Material	Nos.
(1)	Main Body	C. I.	1 No.
(2)	Wheel	C. I.	1 No.
(3)	Rubber Tyre	Rubber	1 No.
(4)	Bush	Gun Metal	1 No.
(5)	Hex - Bolt	M. S.	1 No.
(6)	Plain Washer	M. S.	1 No.
(7)	Hex - slotted Nut	M. S.	1 No.
(8)	Split Pin	M. S.	1 No.
(9)	Plain Stud	M. S.	4 Nos.

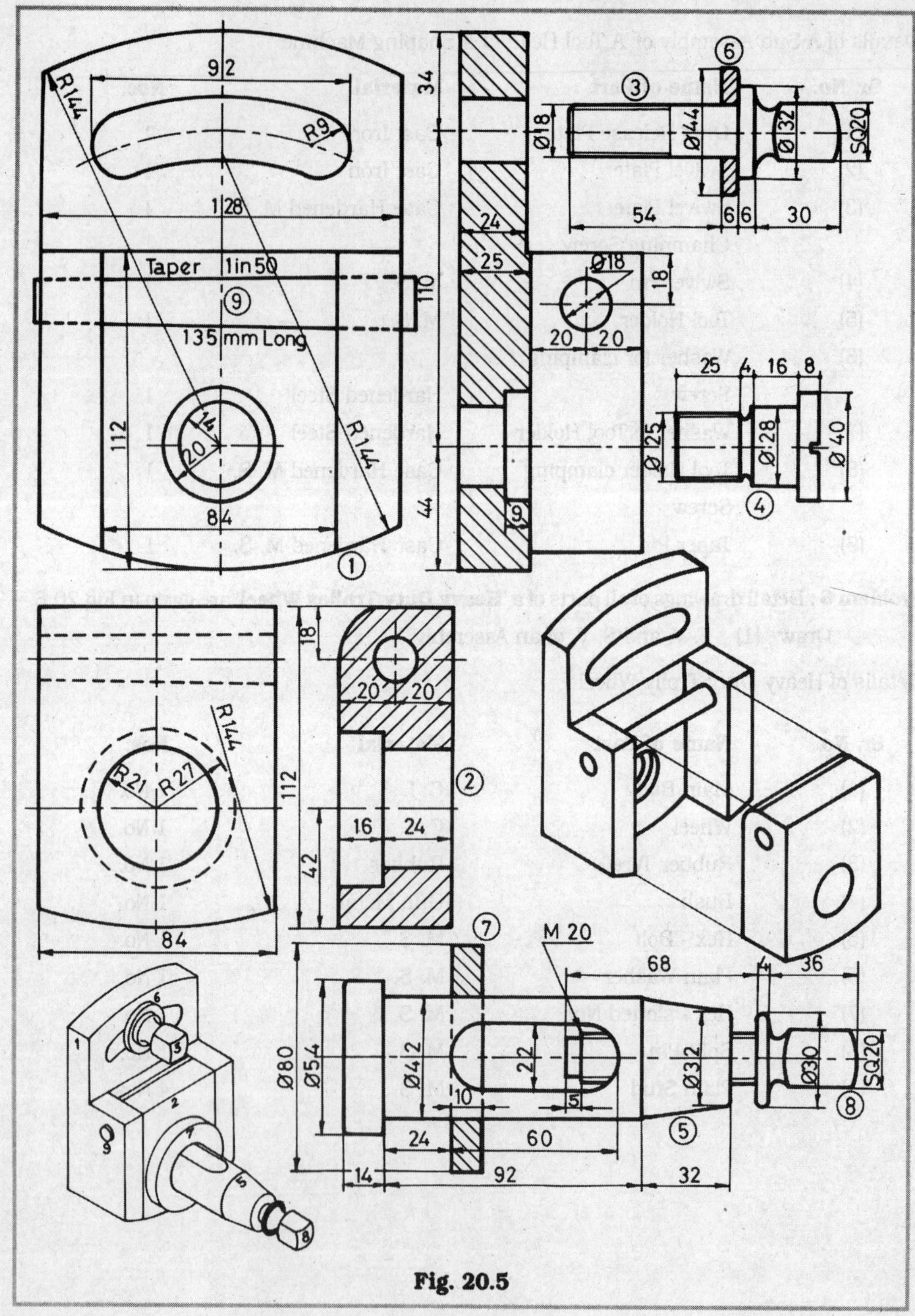
92
R144
R9
128
Taper 1in50
9
135 mm Long
110
112
14
20
R144
84
1
34
24
25
Ø18
18
20 20
44
9
6
3
Ø18
Ø44
Ø32
SQ20
54
6 6
30
25 4 16 8
Ø25
Ø28
Ø40
4
18
20 20
R144
R21
R27
112
16 24
42
2
84
7
M 20
68
4
36
Ø80
Ø54
Ø42
22
Ø32
Ø30
SQ20
8
10
5
24
60
5
14
92
32

Fig. 20.5

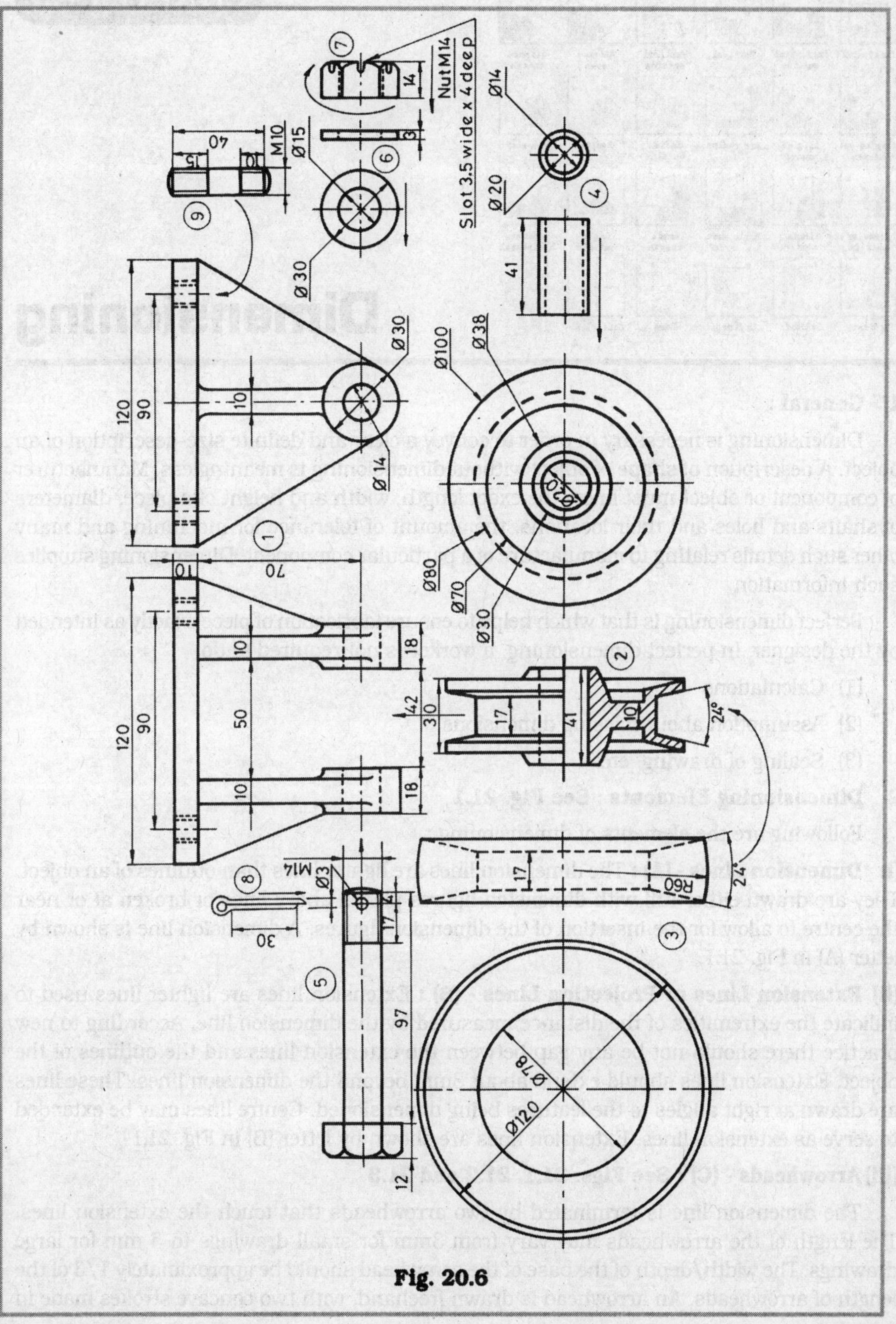

Nut M14
Slot 3.5 wide x 4 deep
M10
Ø15
Ø14
Ø20
Ø30
Ø100
Ø38
Ø80
Ø70
Ø120
Ø78
R60
24°

Fig. 20.6

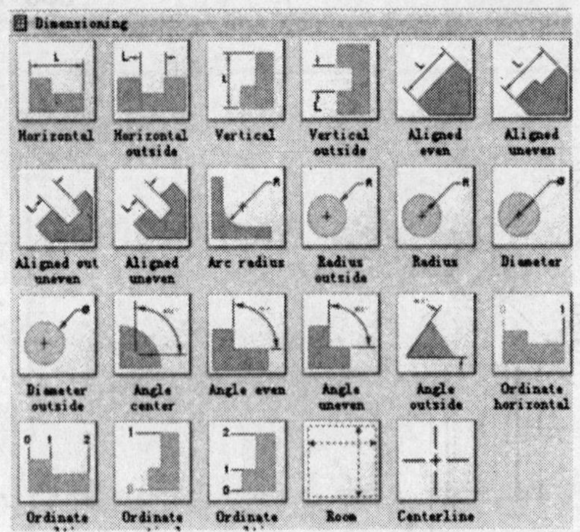

CHAPTER - 21

Dimensioning

1. General :

Dimensioning is necessary in order to convey a clear and definite size-description of an object. A description of shape of object without dimensioning is meaningless. Manufacturer of component or object must know the exact length, width and height of a piece, diameters of shafts and holes and their locations, the amount of tolerance for machining and many other such details relating to manufacture of a particular component. Dimensioning supplies such information.

Perfect dimensioning is that which helps to ensure fabrication of piece exactly as intended by the designer. In perfect dimensioning, a worker is not required to do

(1) Calculations

(2) Assumption about missing dimensions

(3) Scaling of drawing, etc.

2. Dimensioning Elements : See Fig. 21.1

Following are the elements of dimensioning :

(i) Dimension Lines - (A) : The dimension lines are lighter lines than outlines of an object. They are drawn either full with dimension figures placed above lines or broken at or near the centre to allow for the insertion of the dimension figures. A dimension line is shown by letter (A) in Fig. 21.1

(ii) Extension Lines or Projection Lines - (B) : Extension lines are lighter lines used to indicate the extremities of the distance measured by the dimension line. According to new practice there should not be any gap between the extension lines and the outlines of the object. Extension lines should extend about 3mm beyond the dimension lines. These lines are drawn at right angles to the features being dimensioned. Centre lines may be extended to serve as extension lines. Extension lines are shown by letter (B) in Fig. 21.1

(iii) Arrowheads - (C) : See Figs. 21.1, 21.2 and 21.3

The dimension line is terminated by two arrowheads that touch the extension lines. The length of the arrowheads may vary from 3mm for small drawings to 5 mm for large drawings. The width/depth of the base of the arrowhead should be approximately 1/3 of the length of arrowheads. An arrowhead is drawn freehand, with two concave strokes made in the direction of the point or may be made triangular in shape. (See Fig. 21.2) Arrowheads should be drawn within the limits of the dimensioned feature. When enough space is not available, they should be placed outside. Adjacent arrowheads may be replaced by a dot. A dot may also replace a single arrowhead to indicate the common origin of successive dimensions. (See Fig. 21.3) Some of the wrong practices are shown in Fig. 21.3A.

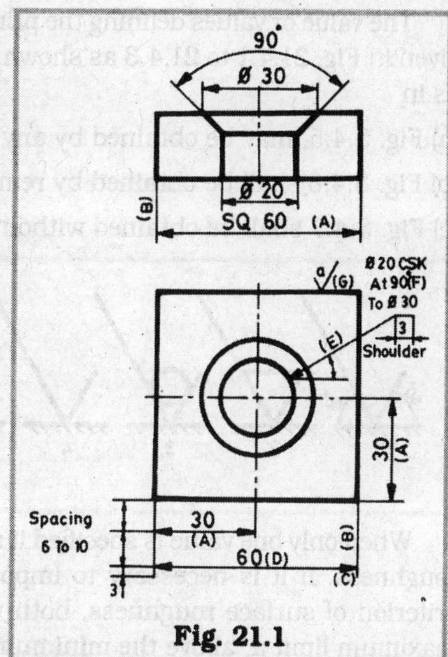

Fig. 21.1

(iv) Dimension Figures - (D) : The dimension figure may be placed in the centre above dimension line or in the break of the dimension line. All dimensions are mostly given in mm. The dimension figure is shown by letter (D) in Fig. 21.1

(v) Leaders or Pointer Lines - (E) :
See Fig. 21.1

The leader is a thin straight line that extends from a dimension figure or note towards the form to which it applies. One end of the leader ends in an arrowhead that touches the outline of the object and other end of it terminates with shoulder of about 3 mm length. Leaders are never drawn horizontal or vertical to avoid confusion with lines of the object but they are drawn at 30°, 45° or 60°. A leader is shown in Fig. 21.1 by letter (E).

Fig. 21.2

(vi) Notes - (F) : Instruction and basic informations relating to the manufacture of a part, such as "5 mm drilled & tapped", "6 mm drill and ream" or "chamfer 3 x 45° are placed adjacent to the features to which they apply and be connected to them by leaders. In Fig. 21.1 Note " Ø 20 CSK AT 90° to Ø 30" is shown.

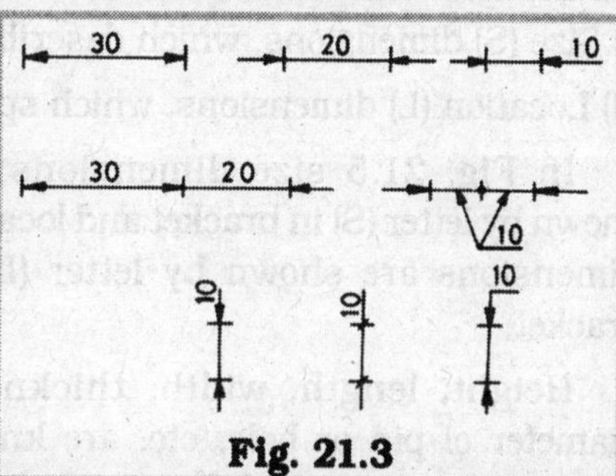

Fig. 21.3

(vii) Surface Roughness Symbols - (G) : See Fig. 21.4.

The basic symbol consists of two legs of unequal length inclined at approximately 60° to the line representing the surface under consideration as shown in Fig. 21.4.1

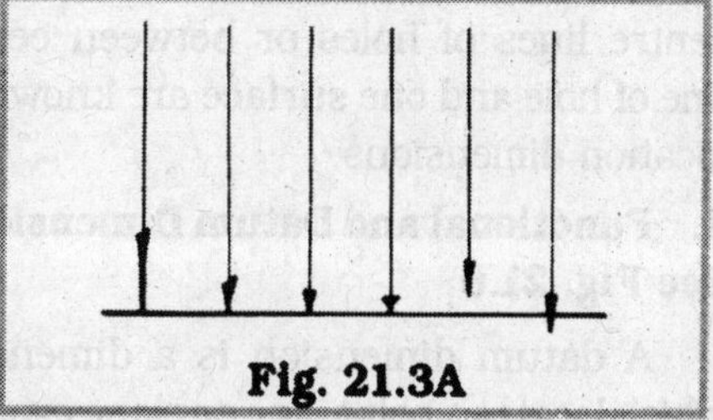
Fig. 21.3A

If the removal of material by machining is required, a bar is added to the basic symbol, as shown in Fig. 21.4.2.

If the removal of material is not permitted a circle is added to the basic symbol as shown in Fig. 21.4.3

When special surface characteristics have to be indicated a line is added to the longer leg of any of the above symbols as shown in Fig. 21.4.4

The value or values defining the principal criterion of roughness are added to the symbols given in Fig. 21.4.1 to 21.4.3 as shown in Fig. 21.4.5 to 21.4.7. Surface roughness specified as in

(a) Fig. 3.4.5, may be obtained by any production method

(b) Fig. 3.4.6 shall be obtained by removal of material by machining

(c) Fig. 3.4.7 shall be obtained without removal of material

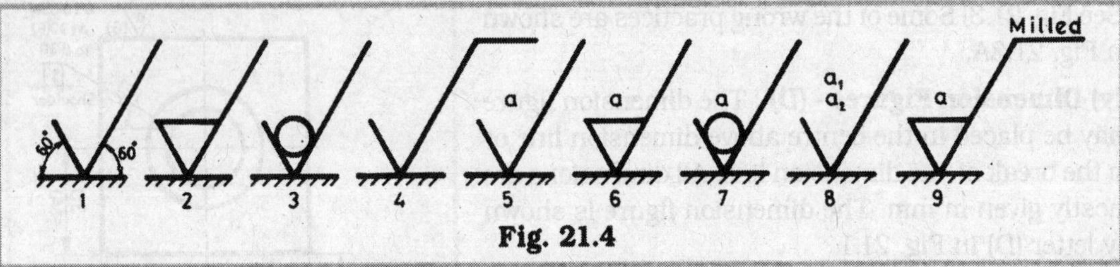

Fig. 21.4

When only one value is specified it represents the maximum permissible value of surface roughness. If it is necessary to impose maximum and minimum limits of the principal criterion of surface roughness, both values should be shown, as in Fig. 21.4.8 with the maximum limit a_1 above the minimum limit a_2.

If it is required that the final surface roughness be produced by one particular production method, this method should be indicated in plain language on an extension of the longer leg of the symbol given in Fig. 21.4.4 as shown in Fig. 21.4.9

3 Size and Location Dimensions : See Fig. 21.5

There are two kinds of dimensions -

(i) Size (S) dimensions, which describe the size of a feature

(ii) Location (L) dimensions, which specify the location of a feature.

In Fig. 21.5 size dimensions are shown by letter (S) in bracket and location dimensions are shown by letter (L) in bracket.

Height, length, width, thickness, diameter of pin or hole, etc. are known as size dimensions and distance between centre lines of holes or between centre line of hole and one surface are known as location dimensions.

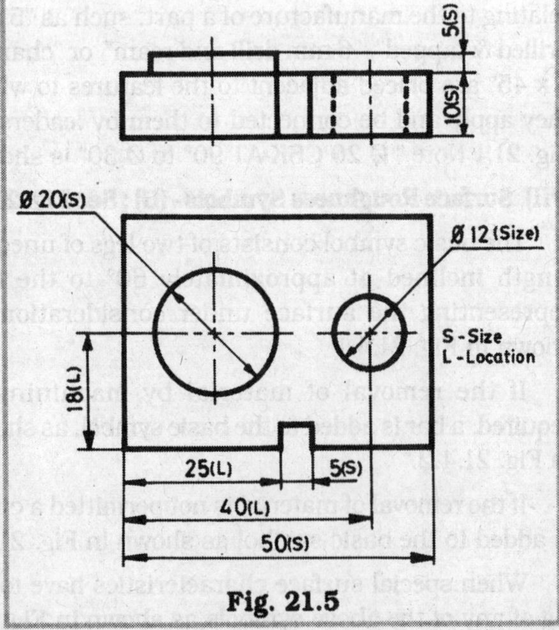

Fig. 21.5

4. Functional and Datum Dimensions: See Fig. 21.6

A datum dimension is a dimension which locates a point, line or plane exactly. It may also be used to specify the position of an ordinate or a point which establishes a true profile.

5. Placing of Dimensions :

As far as possible dimensions, notes, etc. should be placed outside the outlines of the view. However, sometimes it becomes necessary to place the dimension on or near the feature for clarity of presentation. If the feature is well away from the outline, it is impossible to place the dimension outside the outline of the view and hence it may be placed inside (See Fig. 21.7 and Fig. 21.8)

Dimensions should be placed on the view which shows the relevant features most clearly. (See Fig. 21.9) Dimensions, placed between two views, should apply to one view only. Dimensions should not be placed too close to each other or to the parts being dimensioned.

Dimension lines should be placed 6 mm away from the outlines of the object. Smaller dimension should be placed nearer to the outline and larger dimension should be placed away from the outline. Gap between two dimension lines should be nearly 6 mm. This method minimizes the possibility of crossed extension lines. (See Figs. 21.10 and 21.11)

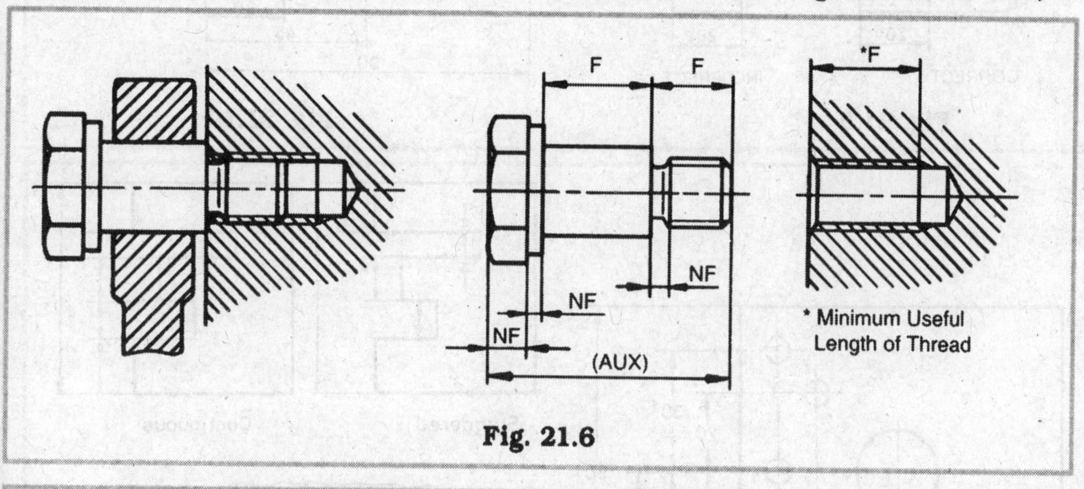

Fig. 21.6

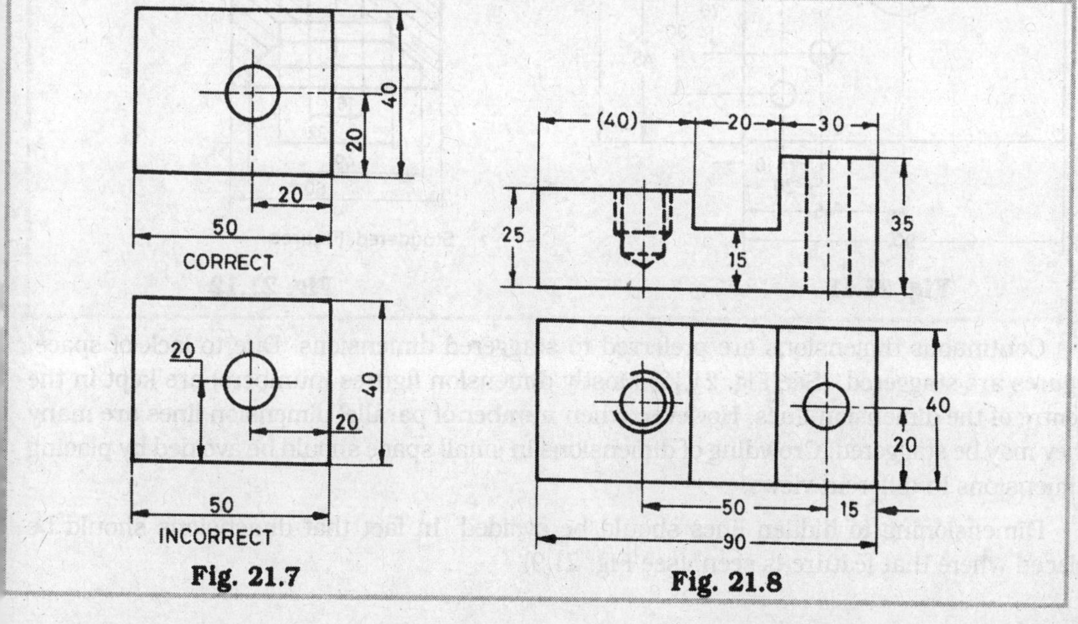

Fig. 21.7

Fig. 21.8

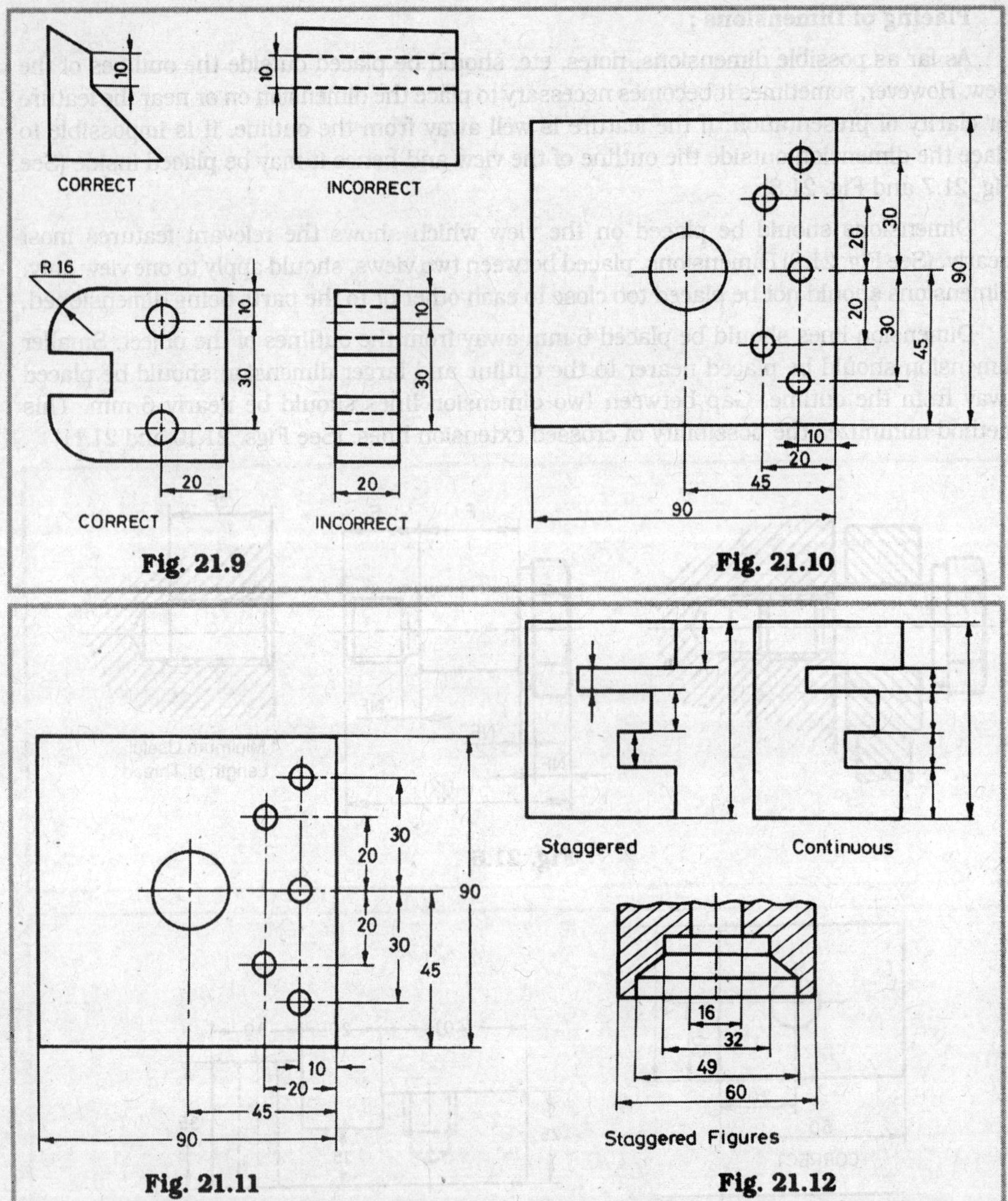

Fig. 21.9

Fig. 21.10

Fig. 21.11

Fig. 21.12

Continuous dimensions are preferred to staggered dimensions. Due to lack of space, figures are staggered. (See Fig. 21.12) Mostly dimension figures (numbers) are kept in the centre of the dimension lines. However, when number of parallel dimension lines are many they may be staggered. Crowding of dimensions in small space should be avoided by placing dimensions in different views.

Dimensioning to hidden lines should be avoided. In fact that dimensions should be placed where that feature is seen. (see Fig. 21.9)

6. **Reading Direction of Dimensions :**

The two recommended directions of placing dimensions are :

(i) Aligned system, See Figs. 21.13 and 21.10.

(ii) Unidirectional system, see Figs. 21.14 and 21.11.

(i) **Aligned System :**

In this system all dimension figures are so placed that they may be read from the bottom or from the right hand side of the drawing sheet and all dimension figures should be placed at right angles to and above the dimension lines.

In aligned system of dimensioning as far as possible placing of dimensions in the shaded area shown in the Fig. 21.15 should be avoided. Because dimensions in this area violate the principle of readability from bottom or from right hand side.

(ii) **Unidirectional System :**

In this system all dimensions are so placed that they may be read from the bottom edge of the drawing sheet. In this system there is no restriction about the direction of dimension lines. This system is advantageous on larger drawings, where it is inconvenient to read dimensions from the right hand side. It is extensively used in preparing drawings of aircraft and automobile industries.

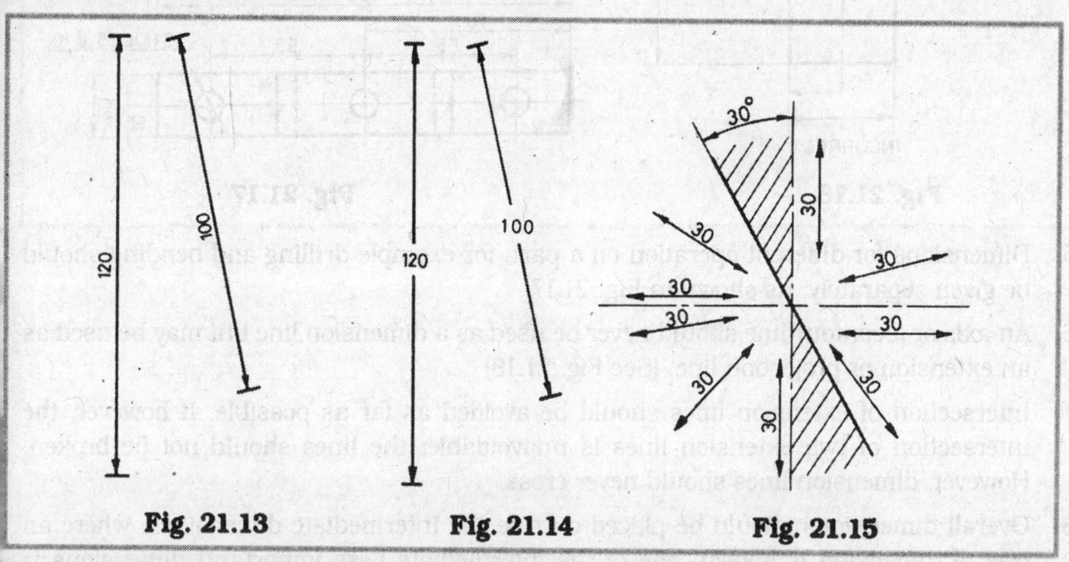

Fig. 21.13 **Fig. 21.14** **Fig. 21.15**

7. **General Rules, Methods and Guidelines for Dimensioning :**

Dimensioning is an art. One must study these rules and guidelines to master the art of dimensioning. Over and above what we have studied in dimensioning elements, placing of dimensions and reading direction of dimensioning, following are the rules and/or guide lines for dimensioning.

1. As far as possible, all the dimensions for one particular operation shall be specified in one view only, such as diameter and depth of a drilled hole or size and depth of a threaded hole.
2. Dimensions are to be given from visible outlines, rather than from hidden out lines. (See Fig. 21.9)

3. Dimensions are to be given from a base line, a centre line of a hole or cylindrical part, an important hole or a finished surface which may be readily established, based on design requirements and the relationship of other parts, (See Fig. 21.16)
4. Dimensioning to a centre line should be avoided except when the centre line passes through the centre of a hole, or a cylindrical part, as shown in Fig. 21.16

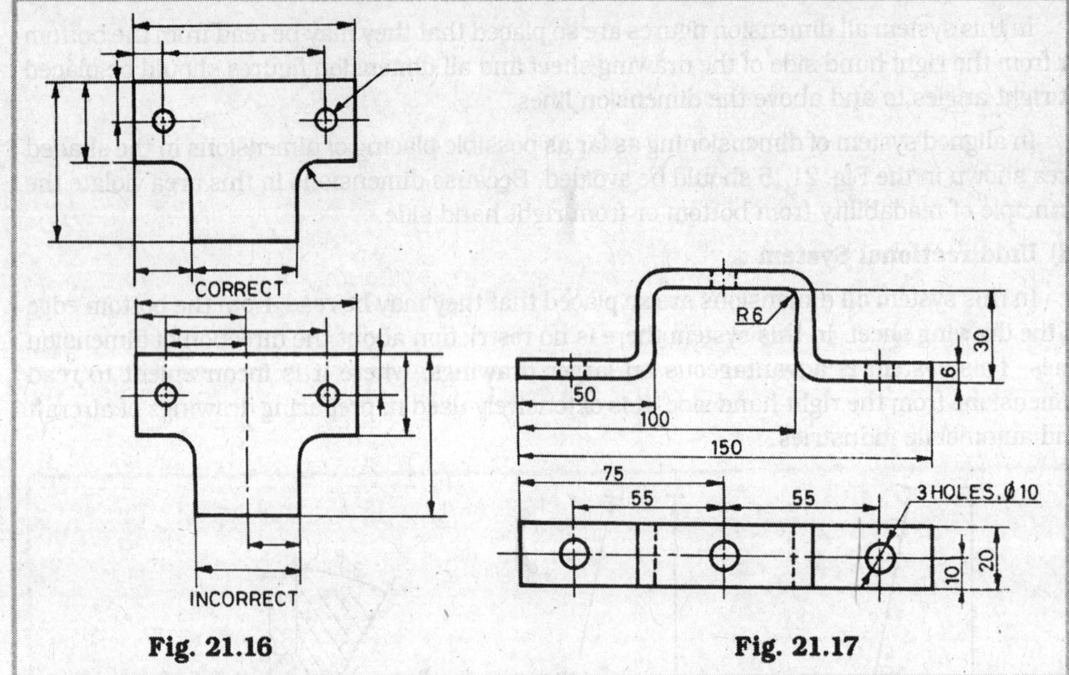

Fig. 21.16 Fig. 21.17

5. Dimensions for different operation on a part, for example drilling and bending should be given separately, as shown in Fig. 21.17
6. An axis or a contour line should never be used as a dimension line but may be used as an extension or projection line. (See Fig. 21.18)
7. Intersection of extension lines should be avoided as far as possible. If however, the intersection of two extension lines is unavoidable, the lines should not be broken. However, dimension lines should never cross.
8. Overall dimensions should be placed outside the intermediate dimensions, where an overall dimension is shown, one of the intermediate (less important) dimensions is redundant and should not be dimensioned. (See Fig. 21.19)
9. Dimensions should be quoted once only. It should not be repeated in other views.
10. As far as possible, all dimensions in one particular drawing be expressed in one unit only. The recommended unit is millimeter. There is no necessity to write mm with the number.
11. The decimal point in a dimension should be bold and should be in line with the bottom line of the figures. When the dimension is less than unity, a zero should precede the decimal point, such as 0.45.

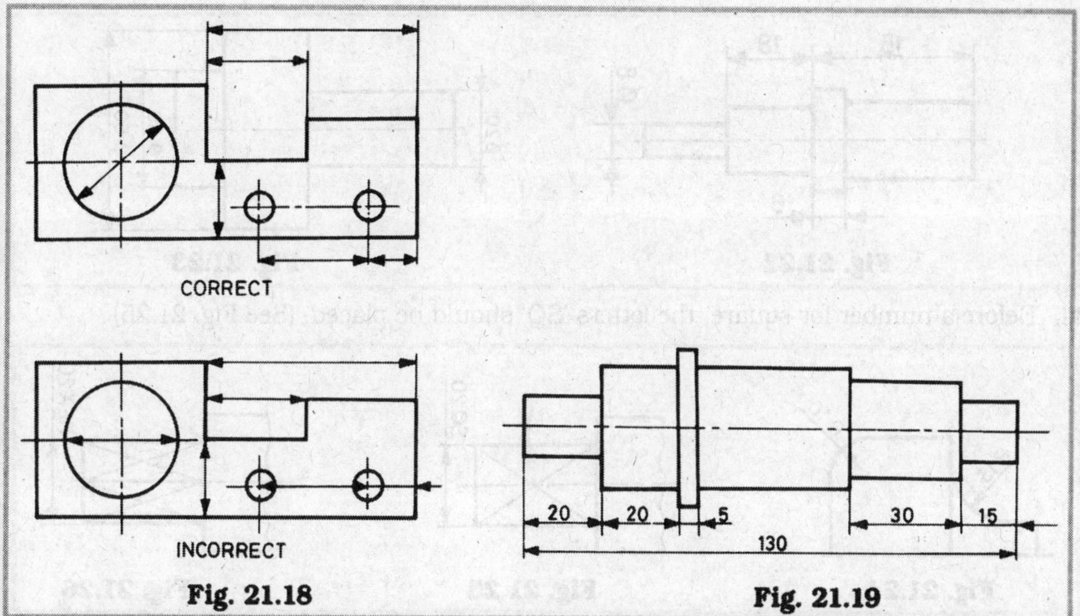

Fig. 21.18 **Fig. 21.19**

12. Depending upon requirements three arrangements of dimensions are practised.

(i) Chain dimensioning, see Fig. 21.20.

(ii) Parallel dimensioning, see Fig. 21.21.

(iii) Combined dimensioning.

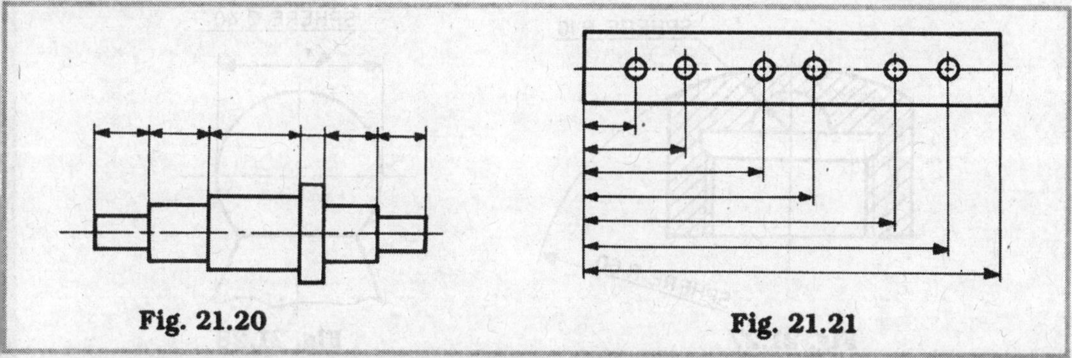

Fig. 21.20 **Fig. 21.21**

Chain dimensioning is used only where the possibility of accumulation of tolerances does not endanger the functional requirements of the part. Parallel dimensioning is used where number of dimensions of the same direction have a common datum feature.

Combine dimensioning is nothing but combination of chain and parallel dimensioning.

13. Numerals and letters should be large enough to ensure easy reading.

14. Dimensions of parts which are not drawn to scale should be underlined. (See Fig. 21.22) However, when the whole drawing is not to scale, the abbreviation 'NTS' may be used.

15. The following letters and symbols should be placed along with number to identify shape of a feature :

i. Before a number for diameter, the symbol 'Ø' should be placed. (See Fig. 21.23)

ii. Before a number for radius the letter 'R' should be placed. (See Fig. 21.24)

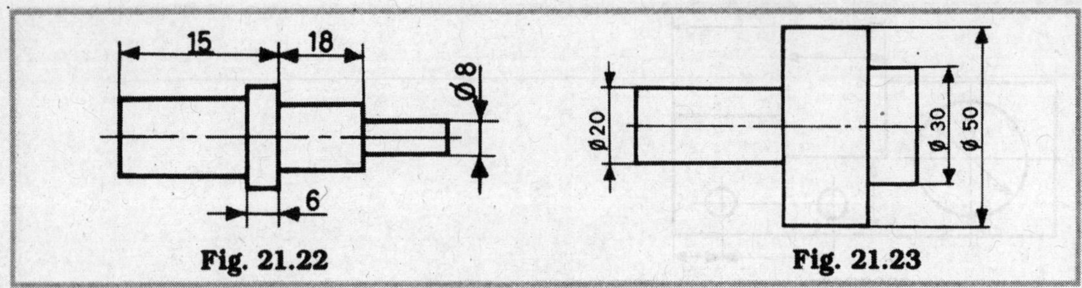

Fig. 21.22 Fig. 21.23

iii. Before a number for square, the letters 'SQ' should be placed. (See Fig. 21.25)

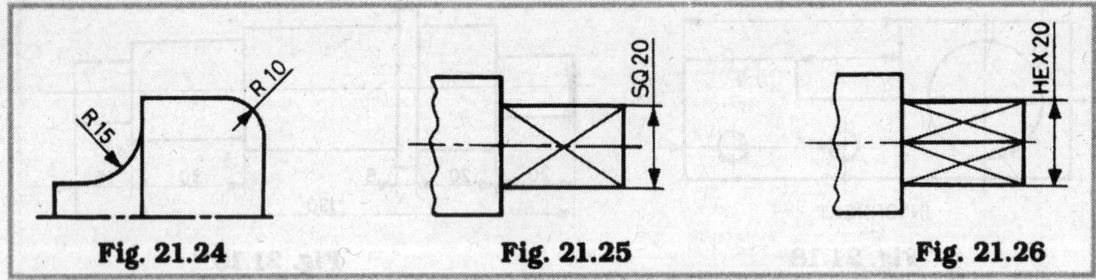

Fig. 21.24 Fig. 21.25 Fig. 21.26

iv. Before across dimension for a hexagon, the letter 'HEX' should be placed. (See Fig. 21.26)

If it clearly appears from the adjacent view that the dimension applies to a square or an hexagonal section, the letters 'SQ' or 'HEX' may be omitted. The radius or diameter dimension of a spherical surface should be preceded by the word 'SPHERE'. (See Figs. 21.27 and 21.28)

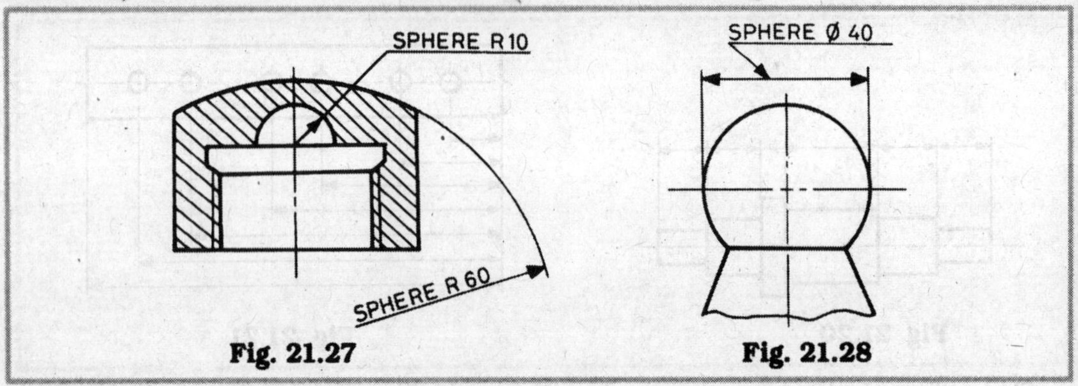

Fig. 21.27 Fig. 21.28

16. Where the space is limited, method shown in Fig. 21.29 is adopted.

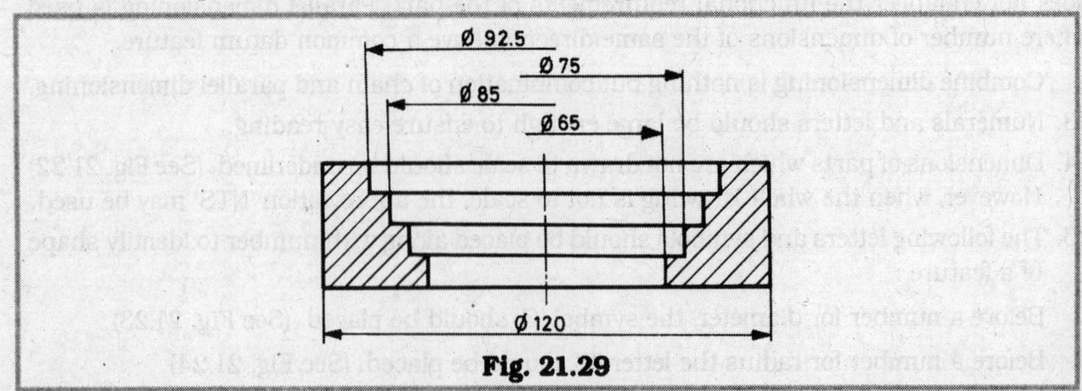

Fig. 21.29

17. Circles should be dimensioned by any one method shown in Fig. 21.30. Selection of method depends upon size of a circle.

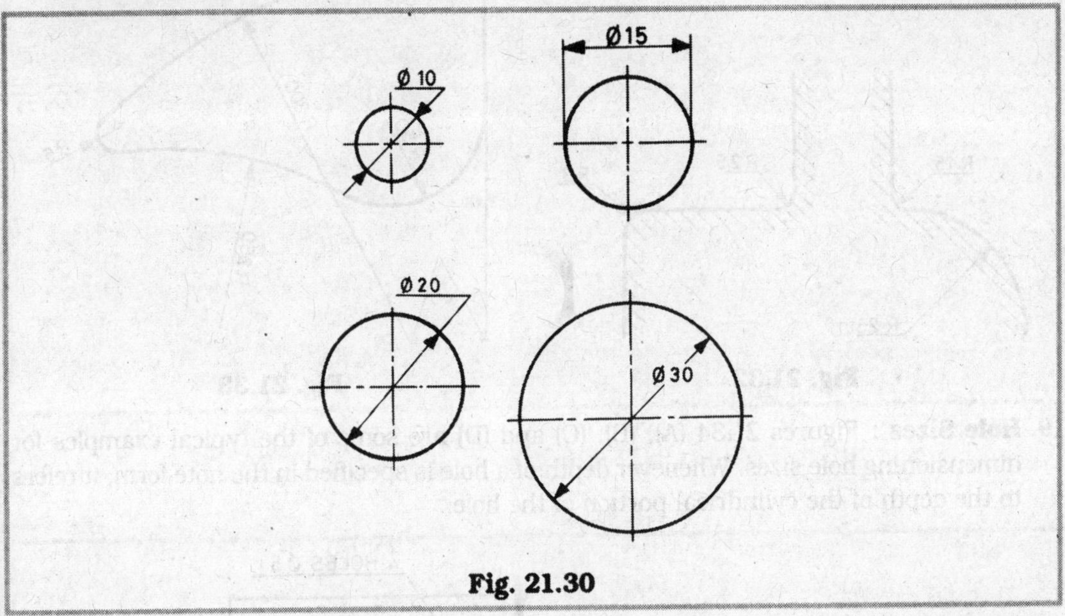

Fig. 21.30

18. Radii :

(a) As far as possible the dimension line of a quarter circle radius should pass through the centre of arc. Where the centre is required for its location, it should be marked by a clear dot. While dimensioning small radii, the arrow should be reversed, as shown in Fig. 21.31.

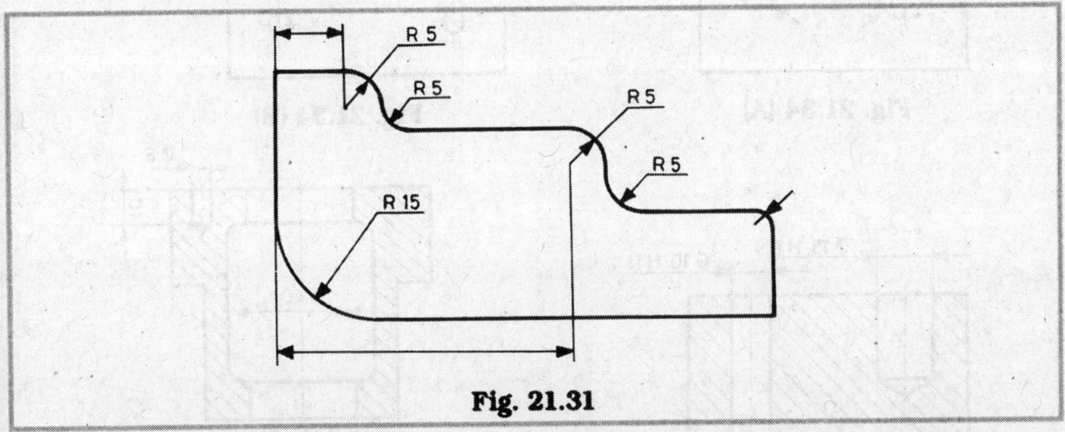

Fig. 21.31

(b) Radii of arcs which need not have their centres located should be dimensioned as shown in Fig. 21.32 and Fig. 21.33.

(c) When the centre of an arc falls outside the limits of the space available, the dimension line of the radius should be broken or interrupted according to whether or not it is necessary to locate the centre. (See Fig. 21.33)

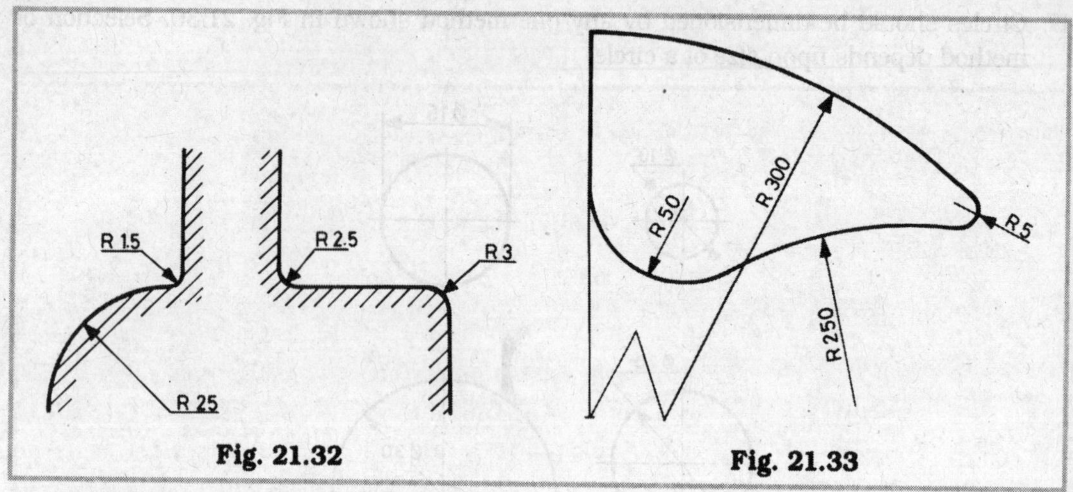

Fig. 21.32

Fig. 21.33

19. Hole Sizes : Figures 21.34 (A), (B), (C) and (D) are some of the typical examples for dimensioning hole sizes. Whenever depth of a hole is specified in the note form, it refers to the depth of the cylindrical portion of the hole.

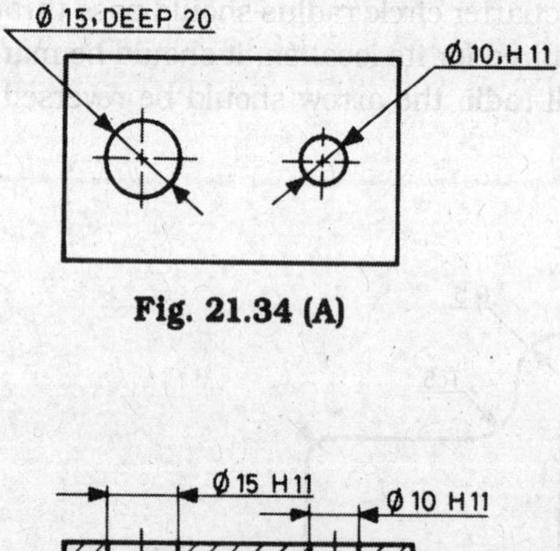

Fig. 21.34 (A)

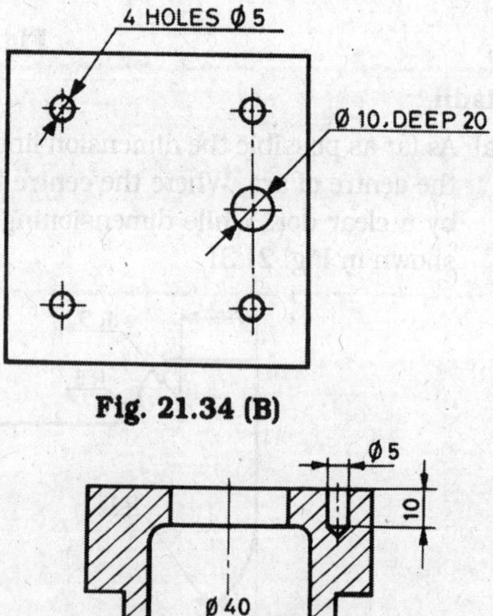

Fig. 21.34 (B)

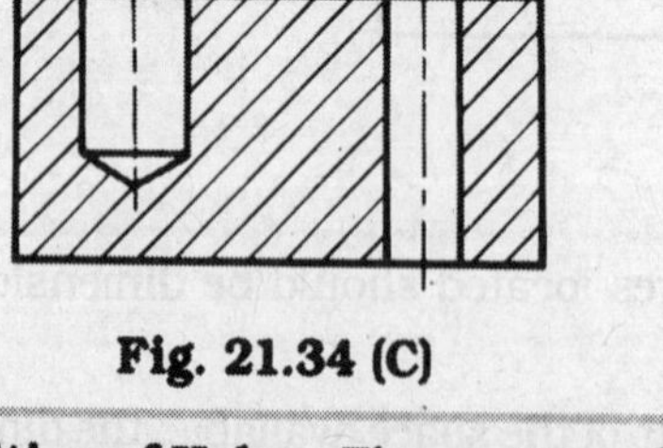

Fig. 21.34 (C)

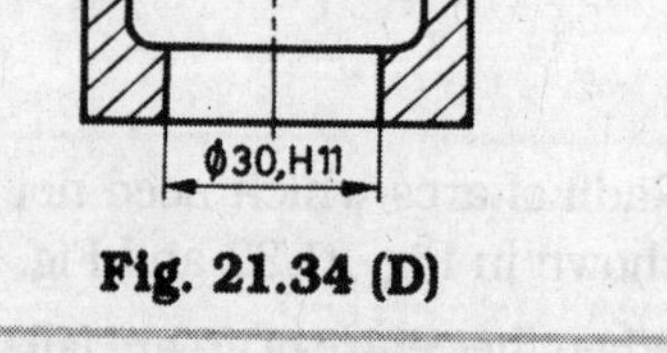

Fig. 21.34 (D)

20. Position of Holes : The position of holes and other features should be defined by either spacing them on pitch circles, as shown in Fig. 21.35 or by giving the rectangular co-ordinate, as shown in Fig. 21.36 (A) or by giving the centre distances, as shown in Figs.

21.36(B), (C) and (D). Where holes symmetrical to edges are to be dimensioned, the convention shown in Fig. 21.36 (E) is to be followed. Holes are dimensioned by using coordinates, even though they are lying on a pitch circle, see Fig. 21.36 (A).

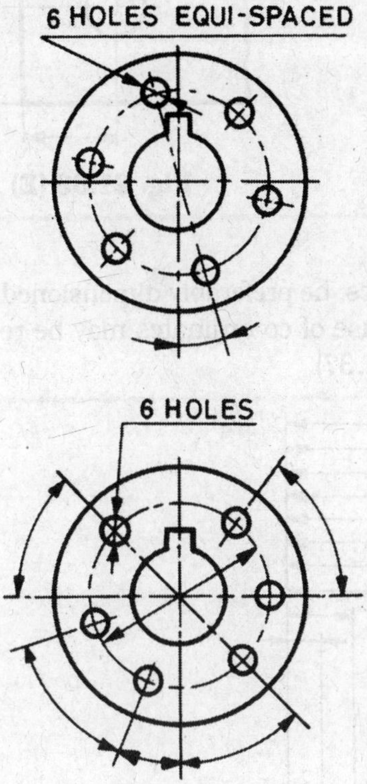

6 HOLES EQUI-SPACED

6 HOLES

Fig. 21.35

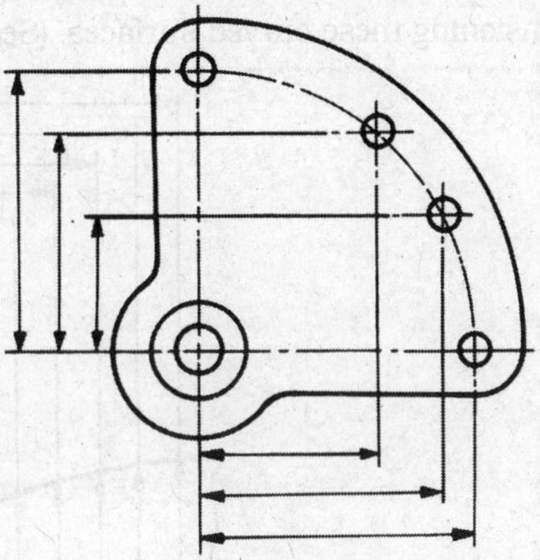

Fig. 21.36 (A)

Holes are dimensioned by using co-ordinates from the centre of a hole. [See Fig. 21.36 (B)]

Holes are positioned relative to one edge only. [See Fig. 21.36 (C)].

Symmetry of holes defined. [See Fig. 21.36 (D)]

Symmetry of hole with edges of plate is implied. [See Fig. 21.36 (E)]

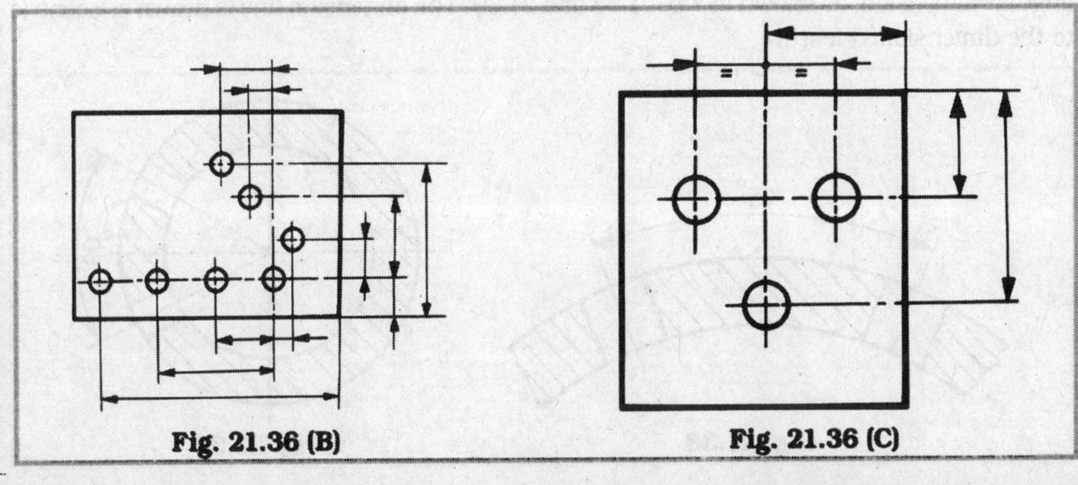

Fig. 21.36 (B)

Fig. 21.36 (C)

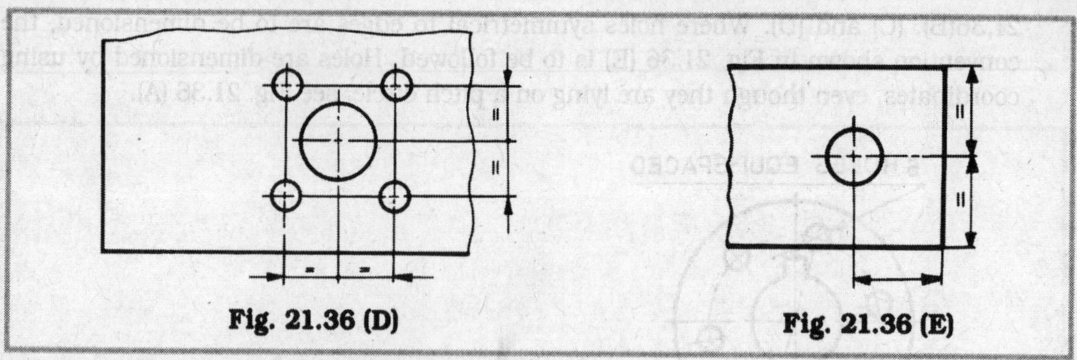
Fig. 21.36 (D)

Fig. 21.36 (E)

21. Dimensioning Curved Surfaces :

A curved line or surface composed of circular arcs, be preferably dimensioned by using radii. However, if this method is impracticable the use of co-ordinates may be resorted to for dimensioning these curved surfaces. (See Fig. 21.37)

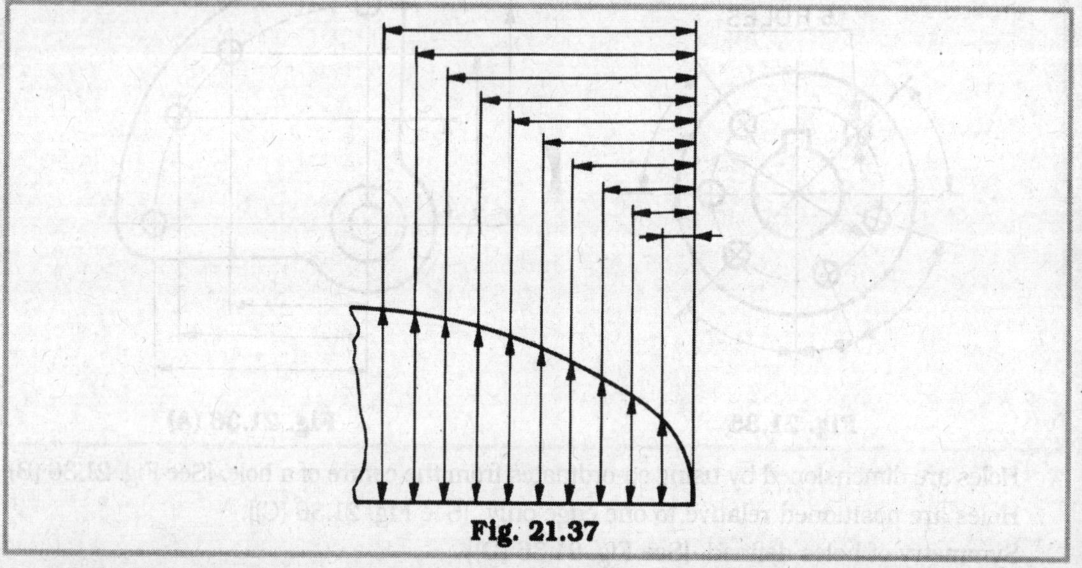
Fig. 21.37

While dimensioning the positions of holes on curved surfaces, it is preferable to use angular dimension, as shown in Fig. 21.38 and 21.39. The dimension line is drawn concentric to the dimensioned feature.

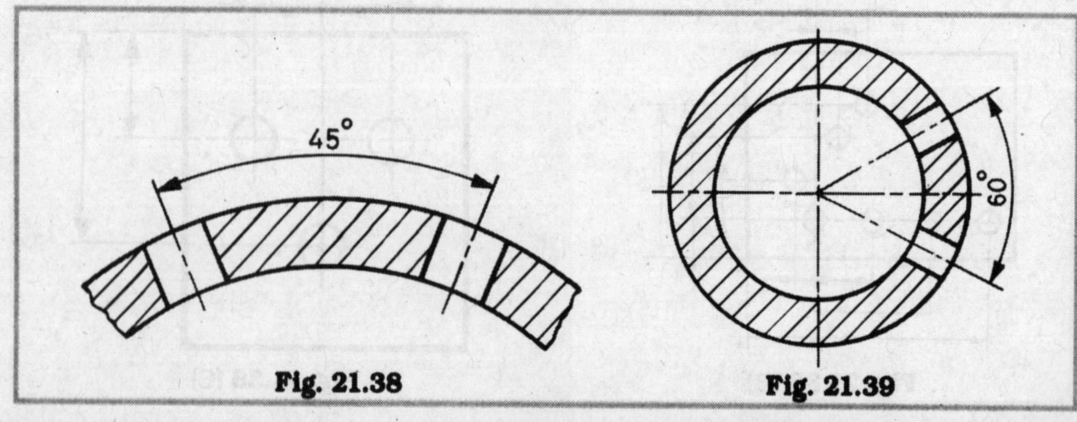

Fig. 21.38

Fig. 21.39

22. Angular Dimensioning : Angular dimensions and their deviations are dimensioned in the same manner as that of dimensioning linear dimensions. However, dimension figures for angular dimensions should be placed, as shown in Fig. 21.40 (A).

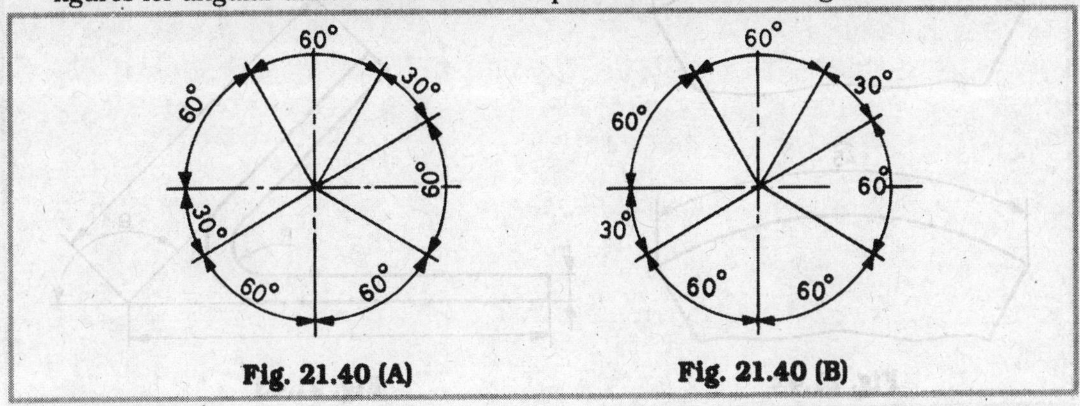

Fig. 21.40 (A) **Fig. 21.40 (B)**

In certain cases, they may be writtten horizontally if this improves clarity, as shown in Fig. 21.40 (B)

Some typical examples are given in Fig. 21.41 to give an idea of the methods of indicating angular dimensions.

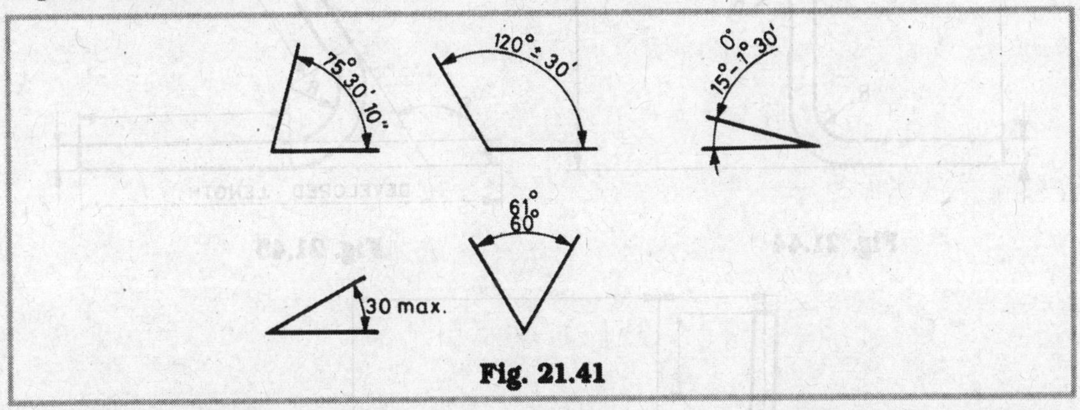

Fig. 21.41

23. Dimensioning of Chords and Arcs : Dimension method is shown in Fig. 21.42 while a chordal dimension refers to the straight distance, arcual distance is the exact arc length, both being equal to the length of the corresponding dimension line. In the case of arcual dimensions, the dimension shall be identified by (⌒).

24. Dimensioning Bent Parts : The method of dimensioning bent parts is shown in Fig. 21.43, Fig. 21.44 and Fig. 21.45. Bends in tubes and pipes should preferably be dimensioned to the centre line. (See Fig. 21.46)

25. Dimensioning of Countersinks, Counter bores and Spot Faces : Some typical examples for the method of dimensioning these features are shown in Fig. 21.47 A, B, C, D, E and F.

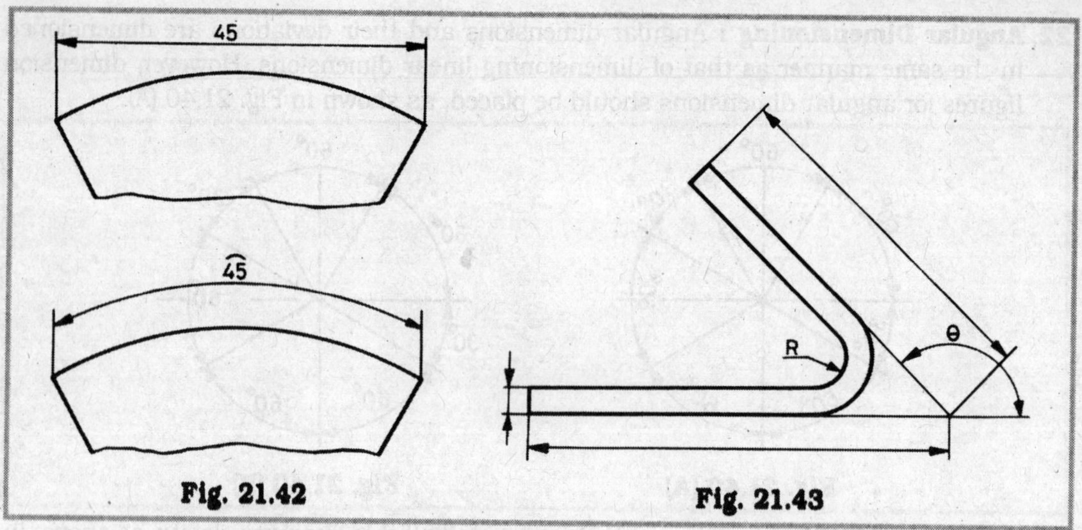

Fig. 21.42

Fig. 21.43

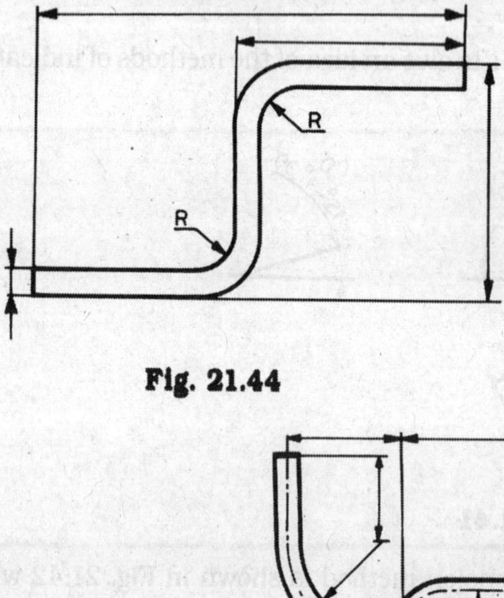

Fig. 21.44

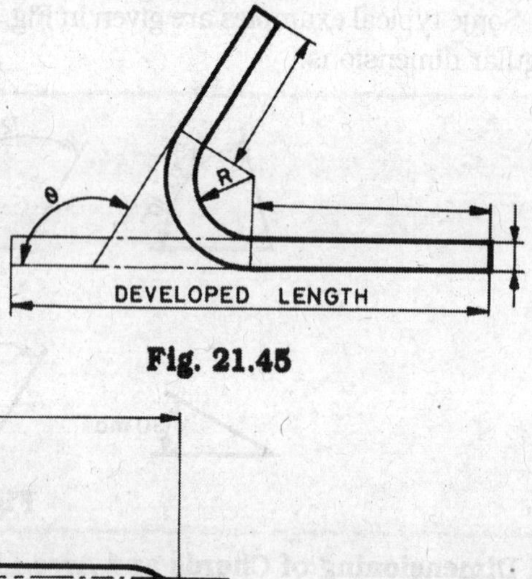

DEVELOPED LENGTH

Fig. 21.45

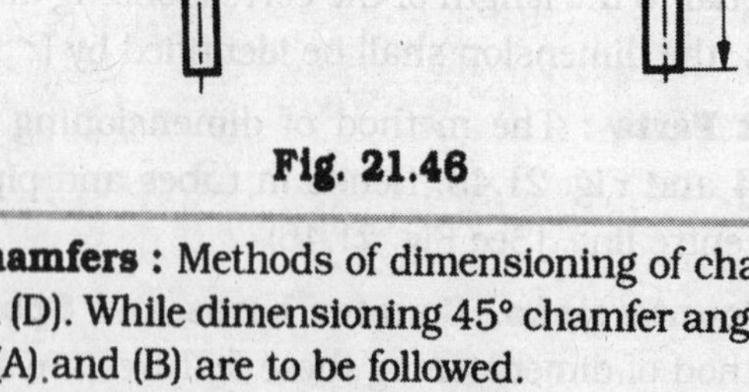

Fig. 21.46

26. Dimensioning of Chamfers : Methods of dimensioning of chamfers are shown in Fig. 21.48 (A), (B), (C) and (D). While dimensioning 45° chamfer angles, the (usual) methods, shown in Fig. 21.48 (A) and (B) are to be followed.

While dimensioning chamfer angle other than 45°, methods, shown in Fig. 21.48 (C) and (D) are to be followed.

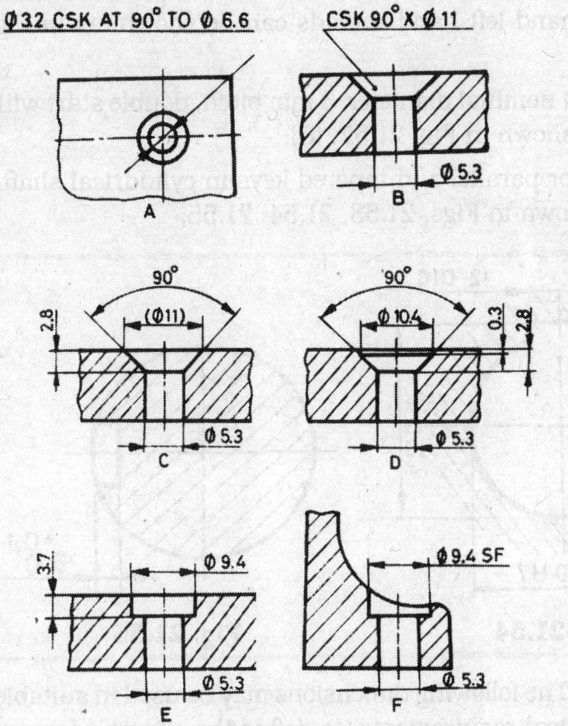

Fig. 21.47

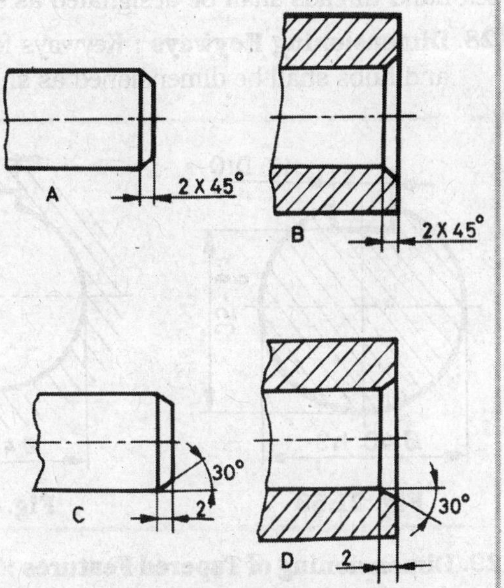

Fig. 21.48

27. Methods of Dimensioning Screw Threads : Screw threads are dimensioned by indicating the corresponding designation for the size of the threads. Fig. 21.49 shows the method of dimensioning external threads. Length of threads are to be dimensioned only to the useful length (run out not to be included). (See Fig. 21.50)

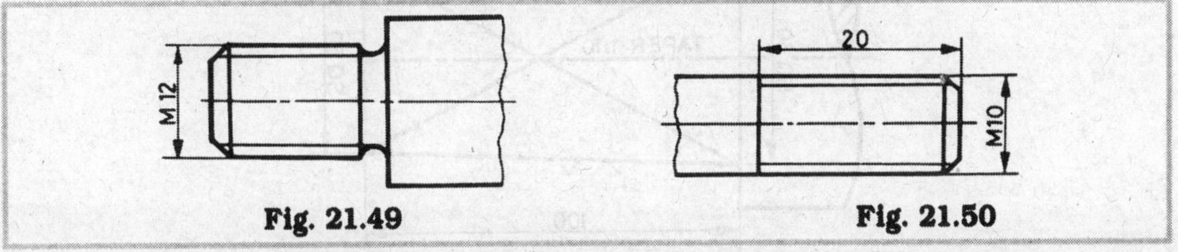

Fig. 21.49

Fig. 21.50

While dimensioning internal threads, it is also necessary to dimension the length of drilled hole before tapping. (See Fig. 21.51)

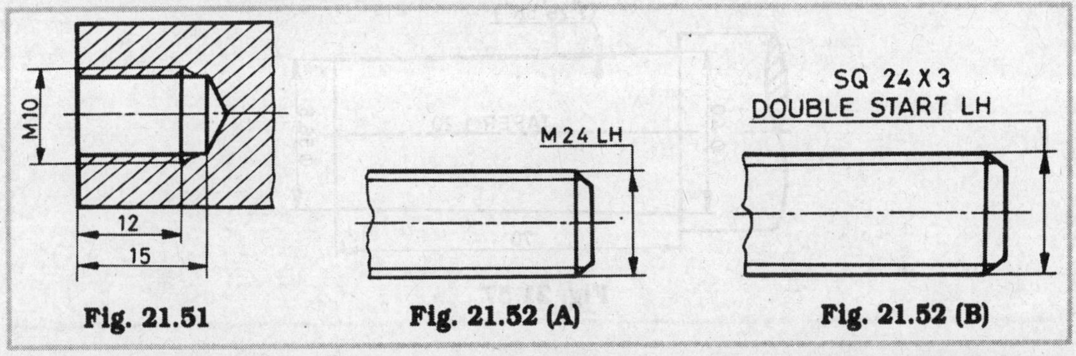

Fig. 21.51

Fig. 21.52 (A)

Fig. 21.52 (B)

Number of start of threads or right hand-left hand threads can be shown, as in Fig. 21.52.

For example, square threads of 24 mm nominal diameter, 3 mm pitch, double start with left hand threads shall be designated as shown in Fig. 21.52. (B).

28. Dimensioning Keyways : Keyways for parallel and tapered keys in cylindrical shafts and hubs shall be dimensioned as shown in Figs. 21.53, 21.54, 21.55.

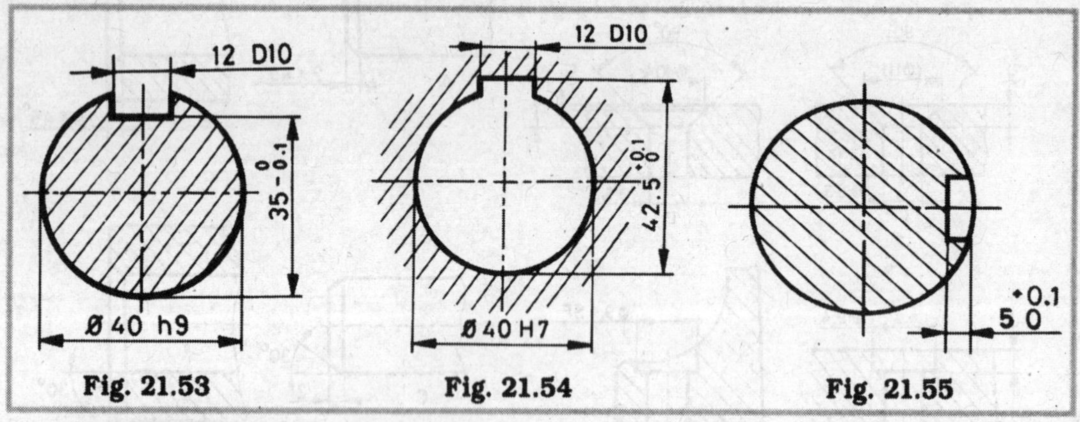

Fig. 21.53 **Fig. 21.54** **Fig. 21.55**

29. Dimensioning of Tapered Features : The following dimensions may be used in suitable combinations depending on technological requirements, to define the size and form of a tapered feature.

(a) Diameter (or width) at each end of the tapered feature, see Fig. 21.56 or 21.57.

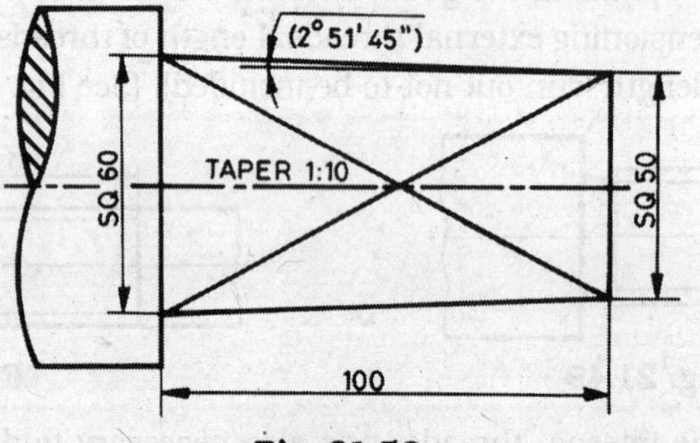

Fig. 21.56

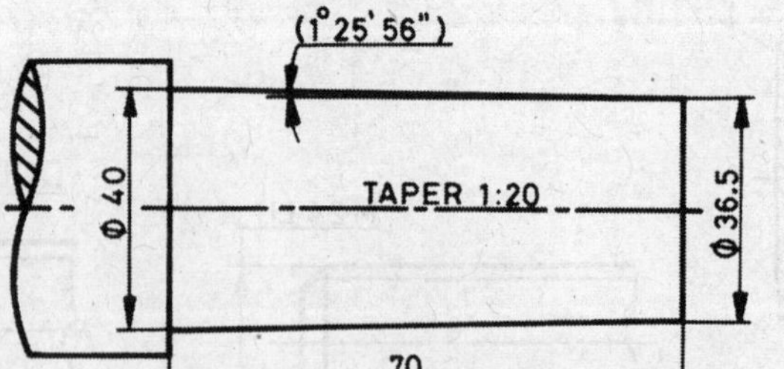

Fig. 21.57

(b) Length of tapered feature.

(c) The half angle of taper is indicated in bracket.

(d) Taper is placed parallel to centre line.

EXERCISE

1. Why is dimensioning necessary? What is perfect dimensioning?
2. Enlist dimensioning elements and show them on a drawing.
3. Discuss in details various aspects of dimensioning elements.
4. Show clearly the difference between size and location dimensions.
5. Explain clearly with illustrations two systems of placing of dimensions.
6. Write an essay on rules and guidelines for dimensioning, giving illustrations.
7. What do you understand by (1) Chain Dimensioning (2) Parallel Dimensioning (3) Combine Dimensioning? Explain with sketches.
8. Write down letters and symbols used for identifying shape, features of elements.
9. How are hole size and radii dimensioned in drawing from smallest to be biggest sizes?
10. How are curved surfaces dimensioned?
11. What is angular dimensioning? How is it done?
12. How are the following elements dimensioned? Explain with sketches : (a) Chords and Arcs, (b) Bent Parts, (c) Countersinks, Counter bores and Spot faces, (d) Chamfers, (e) Screw threads, (f) Key ways, (g) Taper Features.

CHAPTER - 22

Limits, Fits and Tolerances

1. Introduction:

It is impossible to manufacture any article to the exact size. Hence the article is manufactured not to the exact size but close to the design, nominal or basic size. For example if the design size of the shaft diameter is 50 mm, then shaft is manufactured with diameter of shaft between limits say 50.05 mm and 49.95 mm. The diameter 50.05 mm is known as **The Upper Limit** and the diameter 49.95 mm is known as **The Lower Limit.** See Fig. 22.1.

If the limits are closer to the basic size and the difference between two limits is small then the cost of the manufacturing of the article will be higher. By widening the two limits cost of manufacturing can be reduced, at the cost of quality.

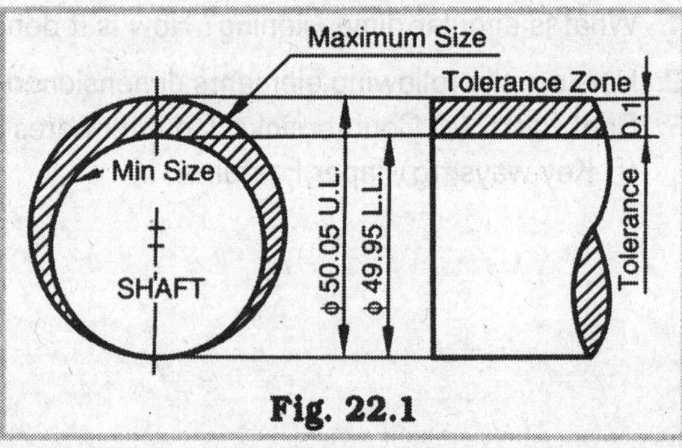

Fig. 22.1

By manufacturing the articles between two well defined limits interchangeability can be achieved. In the modern age, mass production interchangeability is achieved by manufacturing fitting parts within well defined limits.

The difference between upper and lower limits of the size of the parts is known as **Tolerance.**

Tolerance is not desirable but it is envitable inaccuracy in manufacturing. No manufacturing activity can continue without tolerance. In short it is a margin allowed for the variations in the workmanship.

2. Tolerance: See Fig. 22.1.

Tolerance is defined as the total permissible variation in size of a part. In the example of shaft manufactured within limits 50.05 mm and 49.95 mm, the tolerance is 0.1 mm [50.05 – 49.95].

In fact it is the algebraic difference between maximum size permissible and minimum size permissible. Students should note the manner of presentation of tolerance and tolerance zone.

3. Types of Fits : See Fig 22.2.

When two parts are to be assembled the relation resulting from the difference between their sizes before assembly is called **A Fit.**

There are three main types of fits.

1. Clearance Fit
2. Interference Fit
3. Transition Fit

3.1. Clearance Fit: See Fig. 22.2A

When limits of size of male and female parts are so prescribed that when they are assembled together clearance always exists between them. Clearance is the positive difference between female (Hole) and male (Shaft) components.

Varying amount of clearance generates various types of clearance fits. In short in this fit hole is always bigger than the shaft. In the example of the figure clearance varies from minimum 0.005 mm to maximum 0.03 mm.

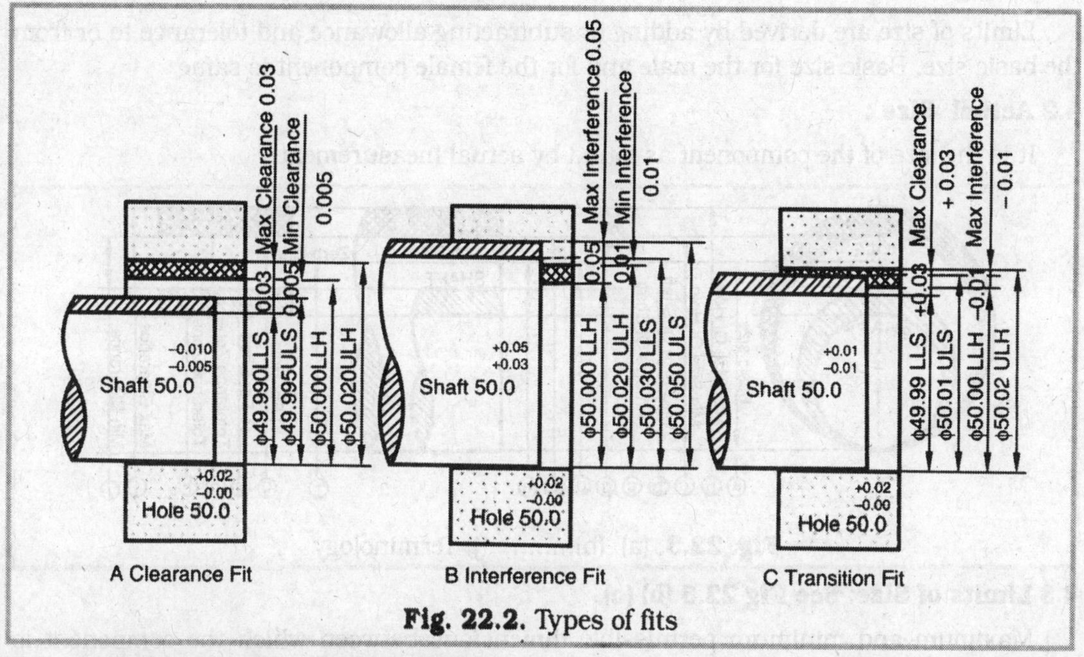

Fig. 22.2. Types of fits

3.2. Interference Fit: See Fig. 22.2.B

When limits of size of male and female parts are so prescribed that when they are assembled together interference [Negative clearance] always exists between them. Interference is the negative difference between female (Hole) and male (Shaft) components. Varying amount of interference generates various types of interference fits.

In short in this fit hole is always smaller than the shaft. In the example of figure interference varies from 0.01 to 0.05 mm.

3.3. Transition Fit: see Fig. 22.2C

When limits of size of male and female parts are so prescribed that when they are assembled together either clearance or interference occure between them.

In short in this type of fit, hole may be bigger than the shaft or may be smaller than the shaft.

For example if two limits of the shaft size prescribed are 49.99 mm and 50.01 mm and two limits of hole size prescribed are 50.00 mm and 50.02 mm, then it is possible that maximum possible clearance will be[50.02 - 49.99] + 0.03 mm and minimum possible clearance will be [50.00 – 50.01] – 0.01 mm. Negative sign indicates that it is not clearance but it is an interference. In this fit depending upon actual size of the shaft and the actual size of the hole clearance fit or interference fit may occure.

4. Terminology:

In order to study the topics of limits, fits and tolerances one must study and understand the following terminology. For convenience, out of two mating parts male part is represented by shaft and female part is represented by hole.

4.1. Basic Size [Nominal Size]: See Fig 22.3(a)

It is the exact size of the component decided by the designer. It is also called nominal size. It is generally full number such as 25.0, 60.0, 75.0 etc.

Limits of size are derived by adding or subtracting allowance and tolerance to or from the basic size. Basic size for the male and for the female component is same.

4.2 Actual Size :

It is the size of the component as found by actual measurement.

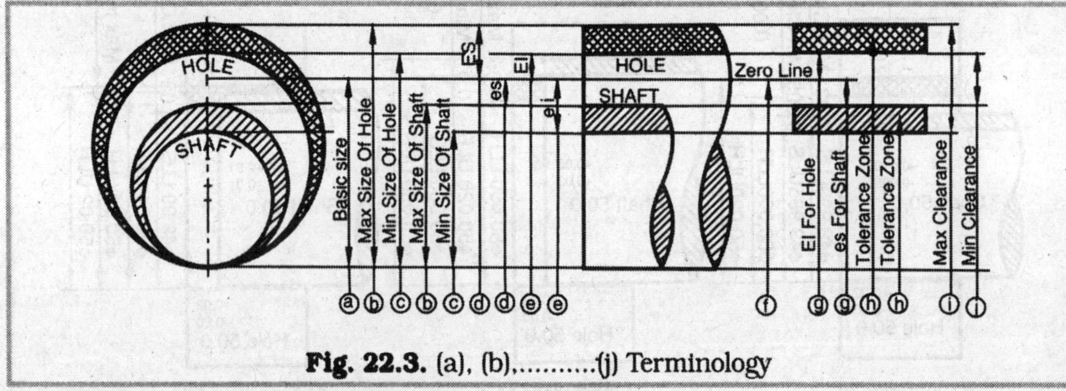

Fig. 22.3. (a), (b),..........(j) Terminology

4.3 Limits of Size: See Fig 22.3 (b) (c)

Maximum and minimum permissible dimensions between which the component is acceptable are known as **Limits** of the size of the component.

4.4. Maximum Limit of Size: See Fig 22.3 (b)

It is the permissible greater (upper) size of the component.

4.5 Minimum Limit of Size: See Fig. 22.3 (c)

It is the permissible smaller (Lower) size of the component.

4.6 Deviation:

It is defined as the algebraic difference between actual size and the corresponding basic size.

4.6.1 Upper Deviation: See Fig. 22.3 (d)

It is the algebraic difference between the maximum limit of size [For male or female] and the corresponding basic size. It is denoted by letters 'ES' for hole and 'es' for shaft.

['ES' or 'es' means Ecart Superior]

4.6.2 Lower Deviation: See Fig. 22.3 (e)

It is the algebraic difference between minimum limit of size [For male or female] and the corresponding basic size. It is denoted by letters 'EI' for hole and 'ei' for shaft.

4.7. Zero Line:See Fig. 22.3 (f)

It is a line representing the basic size. Deviation is zero. Positive deviations are shown above the zero line and negative deviations are shown below the zero line.

4.8 Fundamental Deviations: See Fig. 22.3 (g)

It is one out of the two deviations conveniently chosen to locate the position of the tolerance zone w. r. t. zero line. It may be upper or lower deviation closer to the zero line. In the figure 'EI' for hole and 'es' for shaft are fundamental deviations.

4.9. Tolerance Zone : See Fig. 22.3(h)

It is a graphical presentation of tolerance by a rectangle bounded by two limits [maximum and minimum] of a size of component. Thickness of zone represents the magnitude of tolerance and the position of zone w.r.t. zero line represents the fundamental deviation.

4.10.1. Maximum Clearance: See Fig. 22.3 (i)

It is the difference between the maximum size of the female component and the minimum size of the male component in case of clearance and transition fit.

4.10.2. Minimum Clearance: See Fig. 22.3 (j)

It is the difference between the minimum size of the female component and maximum size of the male component in case of clearance fit.

4.11.1. Maximum Interference: See Fig. 22.4 (a)

It is the negative difference between the minimum size of the female component and maximum size of the male component before assembly in case of transition and interference fit.

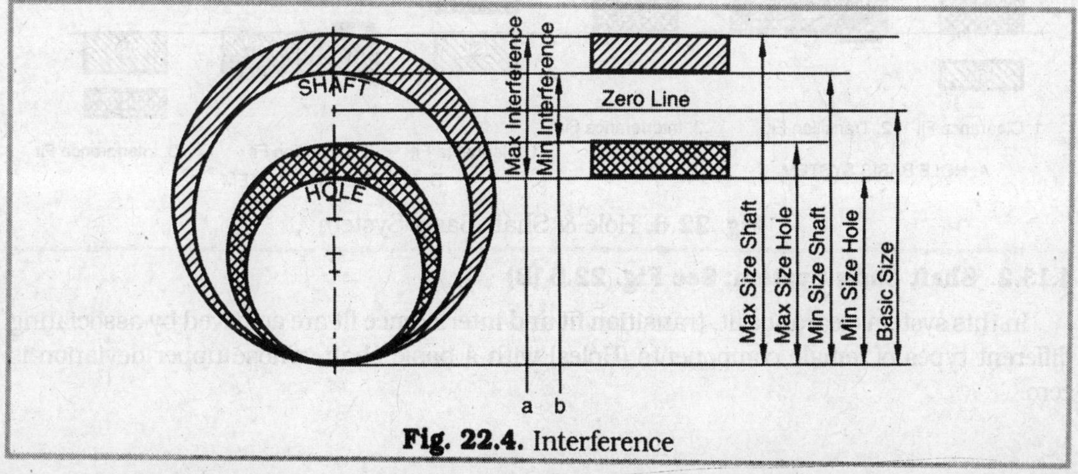

Fig. 22.4. Interference

4.11.2. Minimum Interference: See Fig. 22.4 (b)

It is the negative difference between the maximum size of the female component and the minimum size of the male component before assembly in the case of the interference fit.

4.12.1. Basic Hole: See Fig. 22.5A.

It is a female (Hole) component whose lower deviation is zero i.e., basic size is the lower limit of the female component 'H' hole is a basic hole.

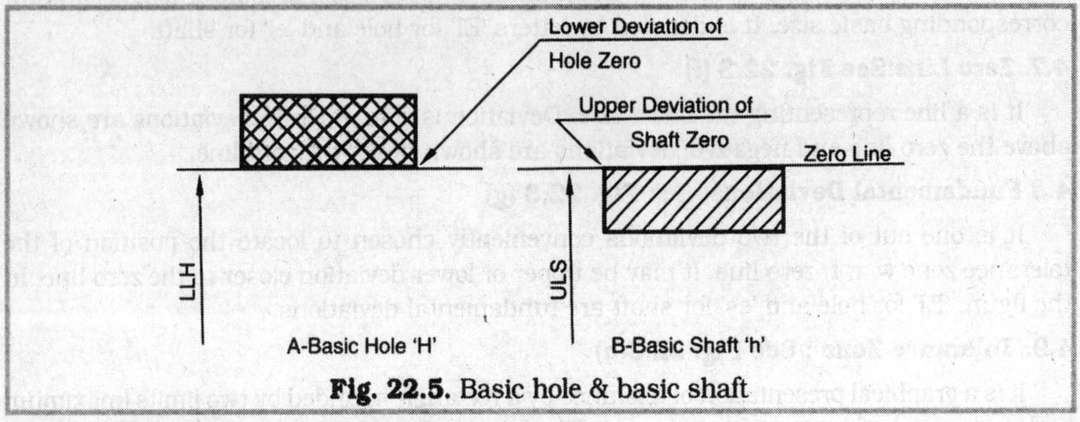

Fig. 22.5. Basic hole & basic shaft

4.12.2. Basic Shaft: See Fig. 22.5(B)

It is a male (shaft) component whose upper deviation is zero i.e., basic size is the upper limit of the male component. 'h' shaft is a basic shaft.

4.13.1. Hole Basis System: See Fig. 22.6(A)

In this system clearance fit, transition fit and interference fit are achieved by associating different types of male components (shafts) with a basic hole, whose lower deviation is zero.

Holes are produced by drills and reamers, while shafts are produced by turning and grinding. Drills and reamers are available in standard sizes and hence variations in size of hole is difficult to achieve. Variations in size of shafts can be easily achieved. Hence it is a standard practice to use **"Hole Basis System"** in comparision to **"Shaft Basis System"**.

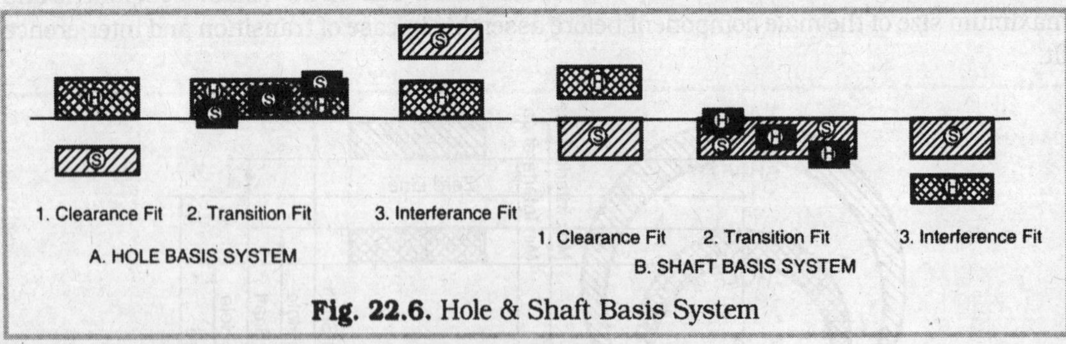

Fig. 22.6. Hole & Shaft Basis System

4.13.2. Shaft Basis System: See Fig. 22.6 (B)

In this system clearance fit, transition fit and interference fit are achieved by associating different types of female components (Holes) with a basic shaft, whose upper deviation is zero.

4.14. I.S. Specification for Grades of Tolerances: See Fig. 22.7

According to Indian Standards there are 18 grades of tolerances. They are designated as IT01, IT0, IT1, IT2, ... IT16. IT01 is used for parts manufactured with accuracy of the highest order. i.e., for parts manufactured with least tolerance e.g. slip gauges, while IT16 is used for parts manufactured with accuracy of the lowest order i.e. for parts manufactured with maximum tolerance. Say sand cast parts.

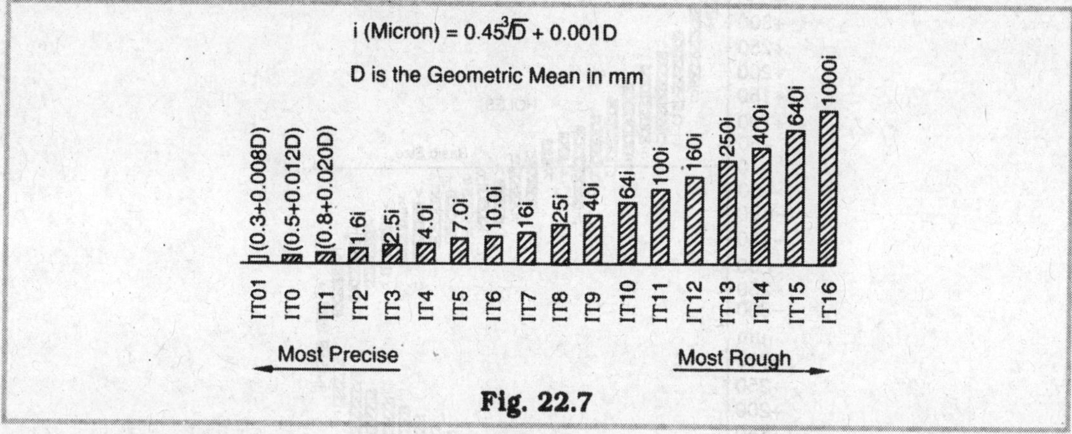

Fig. 22.7

Use of grades of tolerances along with the class of work or accuracy associated with them can be understood perfectly from the table given below. As you go down in the table class of accuracy decreases.

Sr. No.	Tolerance Grade	Class of Work/Manufacturing Process/Applications etc.
1.	IT01, IT0, IT1	Slip Gauge Blocks
2.	IT2	High Quality Gauges, Plug Gauges.
3.	IT3	Good Quality Gauges, Gap Gauges.
4.	IT4	Gauges, Precision Fits produced by lapping.
5.	IT5	Machine Lapping, Fine Boring and Grinding.
6.	IT6	Grinding, Fine Honing.
7.	IT7	High Quality Turning, Broaching, Honing.
8.	IT8	Centre Lathe Turning, Boring, Reaming.
9.	IT9	Work of Semi automatic & Automatic Lathes.
10.	IT10	Milling, Slotting, Planing, Rolling, Extrusion.
11.	IT11	Drilling, Boring, Rough Turning
12.	IT12	Light Press-work, Tube Drawing
13.	IT13	Press-work, Tube Rolling.
14.	IT14	Die-casting
15.	IT15	Stamping
16.	IT16	Sand casting, Flame Cutting.

4.15. I.S. Specifications For Fundamental Deviations: See Fig. 22.8

The value of size of tolerance is decided by grades of tolerances as studied in the previous article 4.14. It is just not sufficient to define the upper and the lower limits of the size of the component. Another value required is the position of the tolerance zone relative to basic size or zero line.

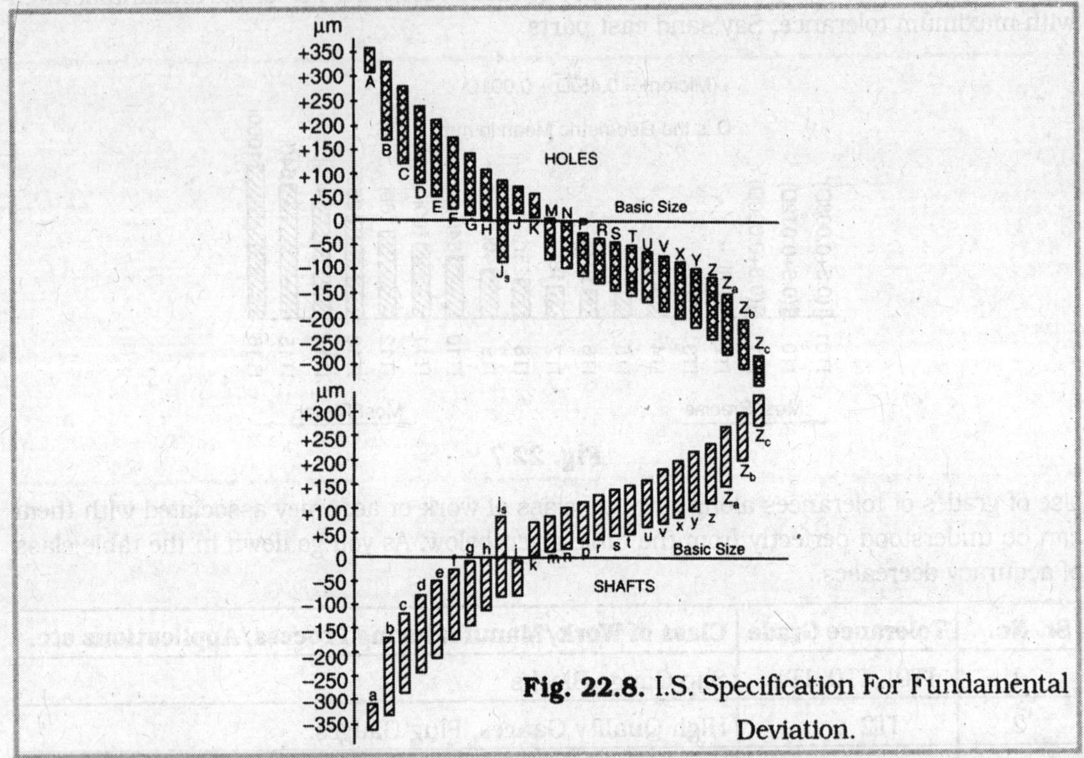

Fig. 22.8. I.S. Specification For Fundamental Deviation.

According to I.S. Specification, there are 25 Fundamental Deviations and they are designated by alphabets. Capital Alphabets for holes and small Alphabets for shafts.

25 Fundamental Deviations for holes are designated by A, B, C, D, E, F, G, H, J_s, J, K, M, N, P, R, S, T, U, V, X, Y, Z, Z_a, Z_b and Z_c. Similarly Fundamental Deviations for shafts are designated by a, b, c, d, e, f, g, h, j_s, j, k, m, n, p, r, s, t, u, v, x, y, z, z_a, z_b and z_c.

For position of tolerance zone relative to basic size or zero line see Fig. 22.8 for holes and shafts both. This figure indicates only the position of the tolerance zone and not the size, of the tolerance zone. Size of the tolerance zone is fixed by grades of tolerance as studied in article 4.14.

4.16. Designation of Hole and Shaft:

Hole is designated by capital alphabet having numeral suffix, indicating grade of tolerance e.g. H_8, F_{12} etc. Here alphabet H and F indicate the position of the tolerance zones w.r.t. zero line and numbers 8 and 12 indicate the sizes of tolerances or tolerance zones for holes.

Similarly shaft is designated by small (OR) lower case alphabets having numeral suffix indicating grades of tolerance e.g., e_7, g_5 etc. Here alphabets e and g indicate the positions of the tolerances w.r.t. zero line and numbers 7 and 5 indicate the size of tolerances or tolerance zones for shafts.

For designating hole and shaft completely basic size is always written prior to alphabet with suffix. For example $50H_8$, $50F_{12}$, $50e_7$, $50g_5$ etc.

Here 50 represents the basic size or nominal size of hole or shaft in mm.

4.17. Designation of Fits:

So far we have studied 25 deviations and 18 grades of tolerances for hole and shaft both. Therefore theoretically there exits an enormous number of possible combinations of hole and shaft representing different fits. However in practice small percentage of hole range is only used.

In brief a fit of mating pair is designated as $50H_8g_7$. In designation of fit basic size (50) common to both male and female components is followed by alphabet (H_8) representing fundamental deviation alongwith numeral suffix representing tolerance grade, first for hole and then similarly for the shaft (g_7).

4.18. Selection of Fits:

Hole is produced generally by drill (OR) reamer and hence it is produced of a fixed size and not of the odd size. Therfore **"Hole Basis System"** is extensively used with 'H' hole having fundamental deviation zero. Some of the commonly used fits are given in the table below.

Table. Commonly Used Fits.

Category Of Fit	Sub Category Of Fit	Designation Of Shaft	Designation Of Hole H_6	H_7	H_8	Application
(A) Clearance Fit	(i) Loose Running Fit	d_8		✓	✓	(1) Plummer Block Bearings (2) Loose Pulleys.
	(ii) Easy Running Fit	e_7	✓			(1) Properly Lubricated Bearings.
		e_8		✓	✓	(2) Heavily loaded Bearings of Turbogenerators.
		e_9			✓	(3) Electric Motor Bearings.
	(iii) Normal Running Fit	f_6	✓			(1) Greece or oil lubricated Bearings.
		f_7		✓		(2) Electric Motor Bearings.
		f_8			✓	(3) Gear Box Bearings.
	(iv) Close Running Fit (OR) Sliding Fit	g_5	✓			(1) Spigot or Location Fit
		g_6		✓		
		g_7			✓	
	(v) Precision Sliding Fit	h_5	✓			(1) Fine Spigot (OR) Location Fit.
		h_6		✓		
		h_7			✓	
		h_8			✓	

Category Of Fit	Sub Category Of Fit	Designation Of Shaft	Designation Of Hole			Application
			H_6	H_7	H_8	
(B) Interference Fit	(i) Light Press Fit	p_5	✓			(1) Assembly of Ferrous and Nonferrous Parts.
		p_6		✓		
	(ii) Light Drive Fit	r_5	✓			(1) -do-
		r_6		✓		(2) Light Drive Fit for Nonferrous Parts.
	(iii) Heavy Drive Fit	s_5	✓			(1) Permanent and Semipermanent assembly of ferrous parts.
		s_6		✓		(2) Presssed collars on shafts.
		s_7			✓	(3) Valve Seatings.
	(iv) Force Fit (OR) Permanent Fit	t_5	✓			(1) Permanent Assembly of Ferrous Parts
		t_6		✓		
		t_7			✓	

EXERCISE

1. What is tolerance? Why it is provided?
2. Explain (i) Unilateral Tolerance (ii) Bilateral Tolerance (iii) Tolerance Zone (iv) Maximum Tolerance (v) Minimum Tolerance.
3. What is fit? Explain types of fits with examples.
4. What is interference? When and where it is kept?
5. Explain terminology of limits, fits and tolerances, with diagrams.
6. Explain (i) Upper Deviation (ii) Lower Deviation (iii) Zero Line (iv) Fundamental Deviation
7. Explain (i) Maximum Clearance (ii) Minimum Clearance (iii) Maximum Interference (iv) Minimum Interference.
8. Explain with diagrams
   (i) Hole Basis System
   (ii) Shaft Basis System, with diagrams
9. Which system is preferred and why?
10. Explain in detail with diagram I.S. Specifications for grades of tolerances.
11. Explain in detail I.S. Specification for fundamental deviations.
12. How fit is designated? How fit is selected?

CHAPTER - 23

Fasteners

1. INTRODUCTION

Fasteners are used to hold together or to assemble together seperate parts. Fasteners are of two types :

1. **Permanent Fasteners**
2. **Temporary Fasteners**

Seperate parts are permanently fastened together by welding or riveting and are temporarily fastened by threaded fasteners. In case of threaded fasteners,seperate parts can be assembled and disassambled as and when needed. Other temporary fastening devices are pins,cotters,keys, etc.

2. SCREW THREAD TERMINOLOGY

2. (a) Screw Thread :

A ridge of uniform section in the form of a helix on the outside or inside surface of a cylinder or a cone is known as screw thread.

2. (b) Straight Screw Thread :

The thread formed on the surface of a cylinder is called straight thread.

2. (c) Taper Screw Thread : (Not shown)

The thread formed on the surface of a cone is called taper thread.

2. (d) External Thread : (See Fig. 23.1)

The thread formed on the external surface of the cylinder or a cone is known as external thread.

2. (e) Internal Thread : (See Fig. 23.1)

The thread formed on the internal surface of the cylinder or a cone is known as internal thread.

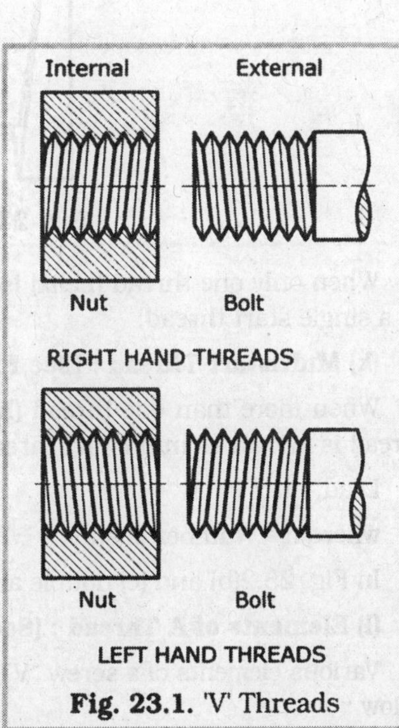

Fig. 23.1. 'V' Threads

2. (f) Right Hand Thread : (See Fig. 23.1)

A nut screws on a bolt,when turned clockwise,then the thread is called right hand thread. Unless otherwise stated,threads are always right hand.

2. (g) Left Hand Thread : (See Fig. 23.1)

A nut screws off a bolt,when turned clockwise,then the thread is called as left hand thread.

2. (h) V-Thread or Triangular Thread : (See Fig. 23.1)

When a ridge of uniform section of the thread is triangular in shape then the thread is named as V-thread or triangular thread.

2. (i) Square Thread : (See Fig. 23.2)

When a ridge of uniform section of thread is square in shape then the thread is named as square thread.

2. (j) Single Start Thread : [See Fig. 23.2(a)]

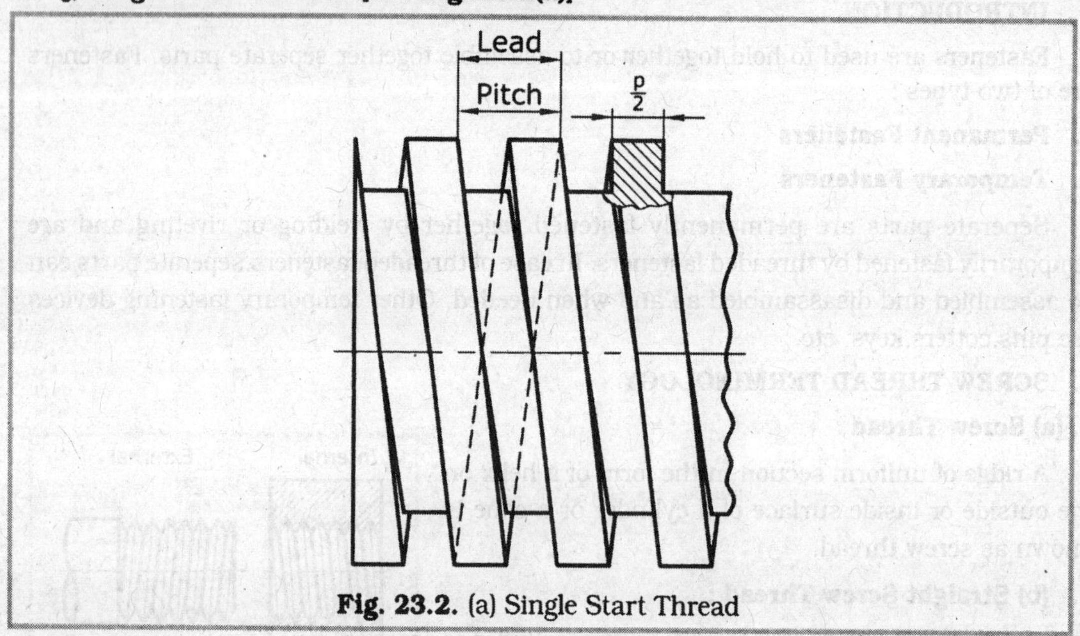

Fig. 23.2. (a) Single Start Thread

When only one thread (helix) is wrapped round the cylinder then the thread is known as a single start thread.

2. (k) Multistart Thread : [See Fig. 23.2(b) and (c)].

When more than one thread (helix) are wrapped parallel round the cylinder then the thread is known as multistart thread. In case of multistart thread-

Lead = n x Pitch

where n = Number of start.

In Fig. 23.2(b) and (c) double and triple start threads are shown.

2. (l) Elements of A Thread : (See Fig. 23.3)

Various elements of a screw "V" thread are shown in Fig. 23.3. Each element is defined below :

2. (I) (i) Crest : It is the outer-most edge in the external thread and inner-most edge in the internal thread.

2. (I) (ii) Root : It is the inner-most portion of an external thread and outer-most portion of an internal thread.

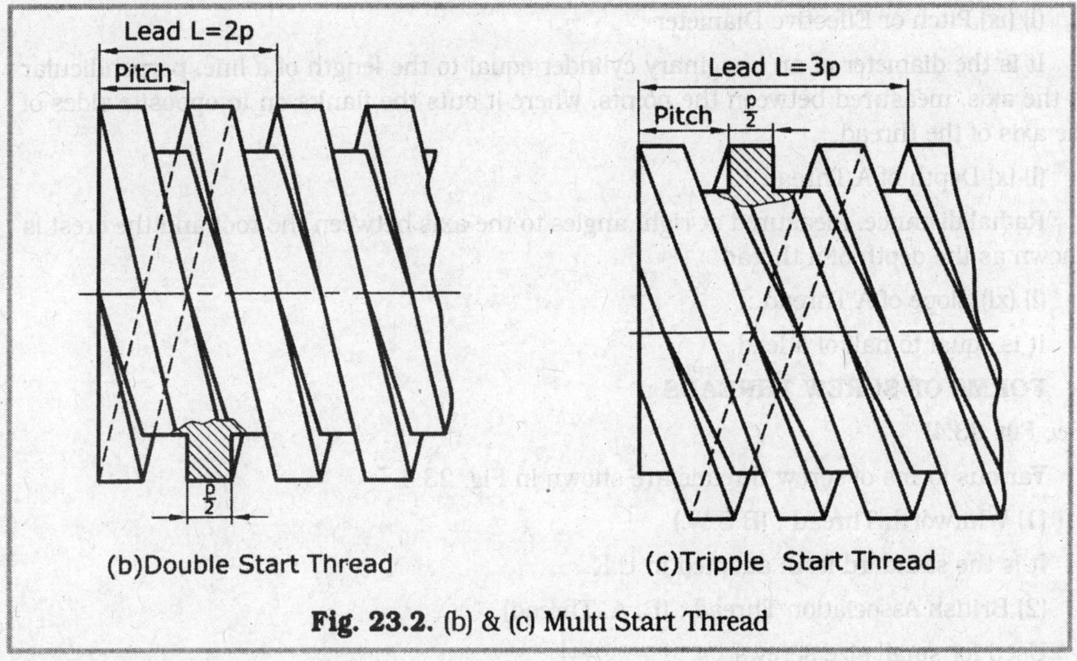

Fig. 23.2. (b) & (c) Multi Start Thread

2. (I) (iii) Flank : The surface of the thread between crest and root is called flank of the thread.

2. (I) (iv) Thread Angle : It is the angle between two adjacent flanks measured along a plane passing through the axis.

2. (I) (v) Pitch : The distance, between corresponding points on adjacent thread forms, measured parallel to the axis, is called pitch. The pitch can also be described as number of thread forms per unit length of thread.

2. (I) (vi) Lead : The distance between corresponding points on the same (helix) thread form, measured parallel to the axis, is called lead.

In other words, it is the distance by which threaded part move axially with respect to a fixed mating part when given one complete revolution.

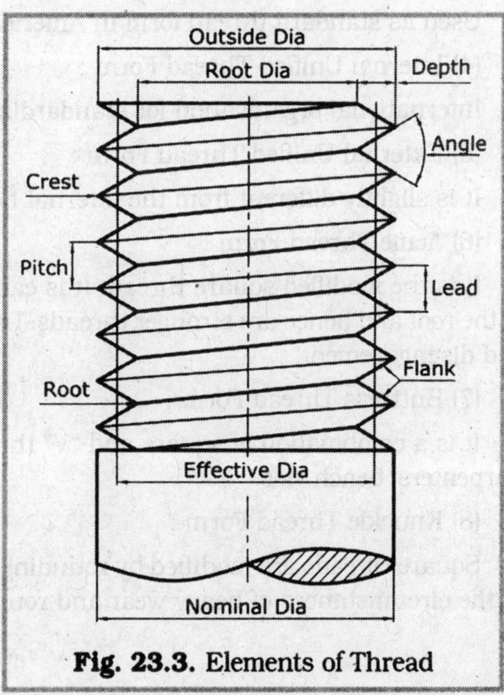

Fig. 23.3. Elements of Thread

2. (l) (vii) Major or Nominal Diameter :

It is the largest diameter of screw thread.

2. (l) (viii) Minor or Root or Core Diameter :

It is the smallest diameter of a screw thread.

2. (l) (ix) Pitch or Effective Diameter :

It is the diameter of an imaginary cylinder equal to the length of a line, perpendicular to the axis, measured between the points, where it cuts the flanks on to opposite sides of the axis of the thread.

2. (l) (x) Depth of A Thread :

Radial distance, measured at right angles to the axis between the root and the crest is known as the depth of a thread.

2. (l) (xi) Slope of A Thread :

It is equal to half of a lead.

3. FORMS OF SCREW THREADS :

(See Fig. 23.4)

Various forms of screw threads are shown in Fig. 23.4

3. (1) Whitworth Thread : (B.S.W.)

It is the standard form adopted in U.K.

3. (2) British Association Thread : (B. A. Thread)

Used for small size screws.

3. (3) Sellers or American Standard Thread :

Used as standard thread form in America.

3. (4) Internal Unified Thread Form :

International organisation for standardization adopt this thread as the basic profile.

3. (5) External Unified Thread Form :

It is slightly different from the internal type.

3. (6) Acme Thread Form :

It is the modified square thread. It is easier to manufacture. Acme threads are thicker at the root and hence are stronger threads. Used in half nuts, used for frequent engagement and disengagement.

3. (7) Buttress Thread Form :

It is a combination of square and "V" threads. Vertical flanks take axial force. Used in Carpenters' bench vice.

3. (8) Knuckle Thread Form :

Square threads are modified by rounding off the corners of the square thread. It is used in the circumstances of heavy wear and rough usage.

3. (9) Square Thread Form : Used in power transmission screws.

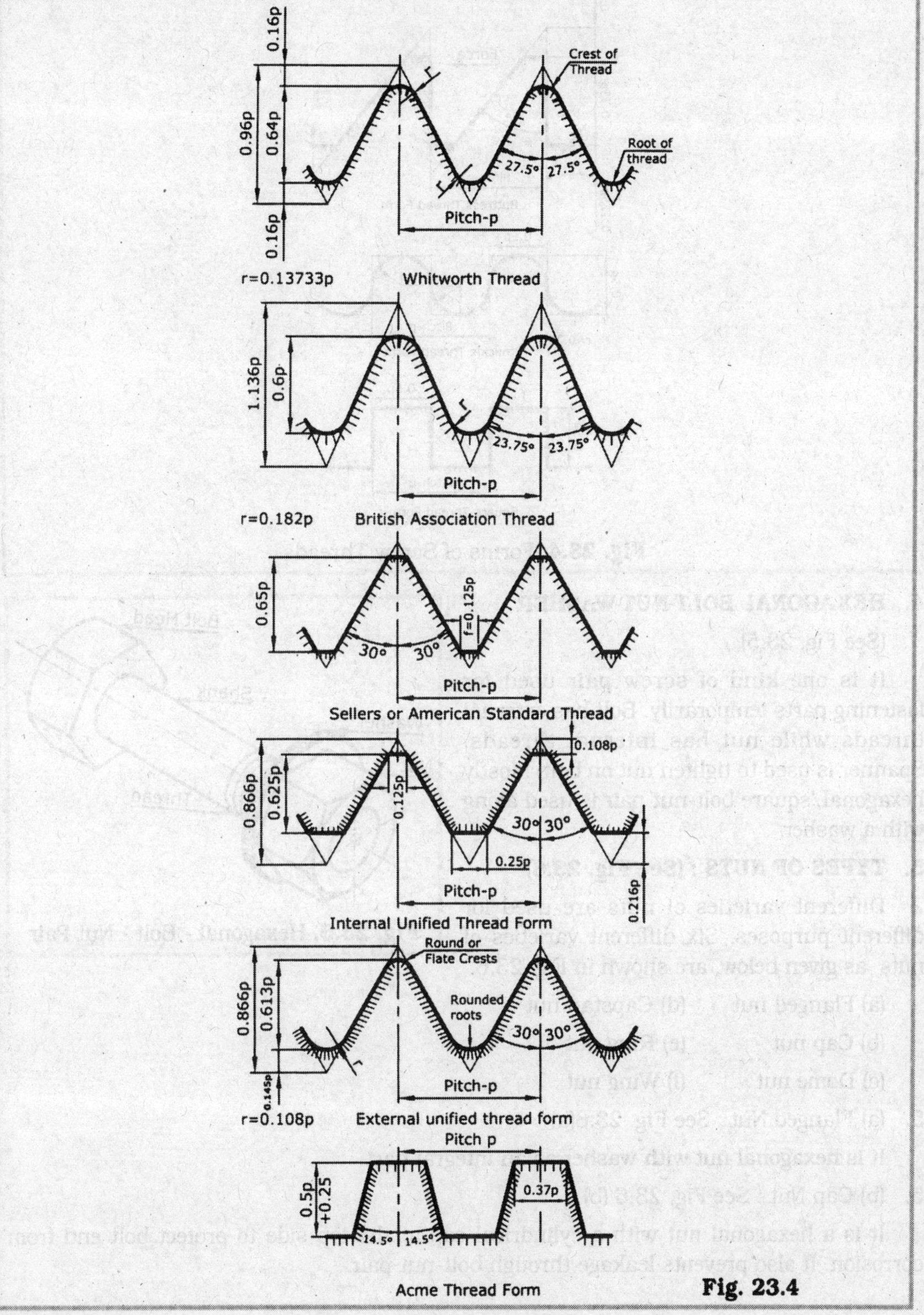

Fig. 23.4

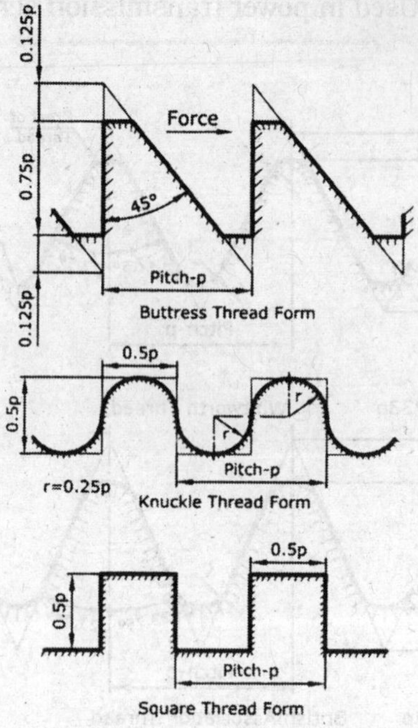

Fig. 23.4. Forms of Screw Threads

4. HEXAGONAL BOLT-NUT-WASHER

(See Fig. 23.5)

It is one kind of screw pair used for fastening parts temporarily. Bolt has external threads while nut has internal threads. Spanner is used to tighten nut on bolt. Mostly hexagonal/square bolt-nut pair is used along with a washer.

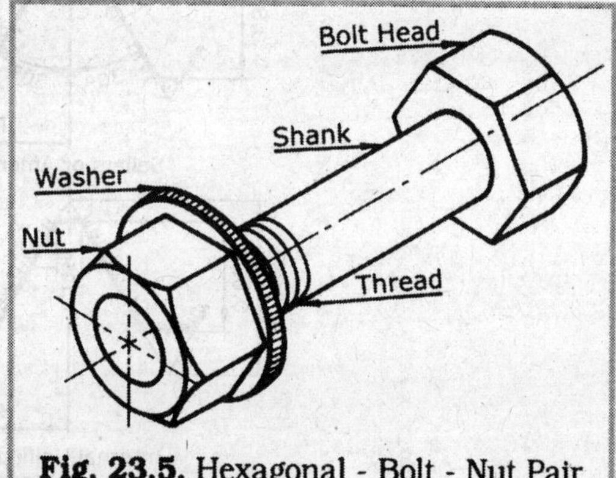

Fig. 23.5. Hexagonal - Bolt - Nut Pair

5. TYPES OF NUTS : (See Fig. 23.6)

Different varieties of nuts are used for different purposes. Six different varieties of nuts as given below, are shown in Fig. 23.6.

(a) Flanged nut	(d) Capstan nut
(b) Cap nut	(e) Ring nut
(c) Dome nut	(f) Wing nut

5. (a) Flanged Nut : See Fig. 23.6(a)

It is hexagonal nut with washer as an integral part.

5. (b) Cap Nut : See Fig. 23.6 (b)

It is a hexagonal nut with a cylindrical cap at the top side to protect bolt end from corrosion. It also prevents leakage through bolt-nut pair.

5. (c) Dome Nut : See Fig. 23.6(c)

It is a hexagonal nut with a spherical dome at the top side. By providing smooth spherical dome harmful sharp edges are avoided.

5. (d) Capstan Nut : See Fig. 23.6(d)

It is a cylindrical nut with equispaced six holes drilled on the curved surface as shown. Tommy bar tool is used for turning the nut on the bolt.

5. (e) Ring Nut :See Fig. 23.6(e)

It is a cylindrical ring nut with equispaced slots provided on the circumference as shown. C-spanner tool is used for turning the nut.

5. (f) Wing Nut : See Fig. 23.6(f)

This type of nut is used where it is required to be operated frequently and that too by hand i.e. by thumb and fingers. No special tool is required.

Drawings in Fig. 23.6(d) (e) and (f) are given in third angle projection system.

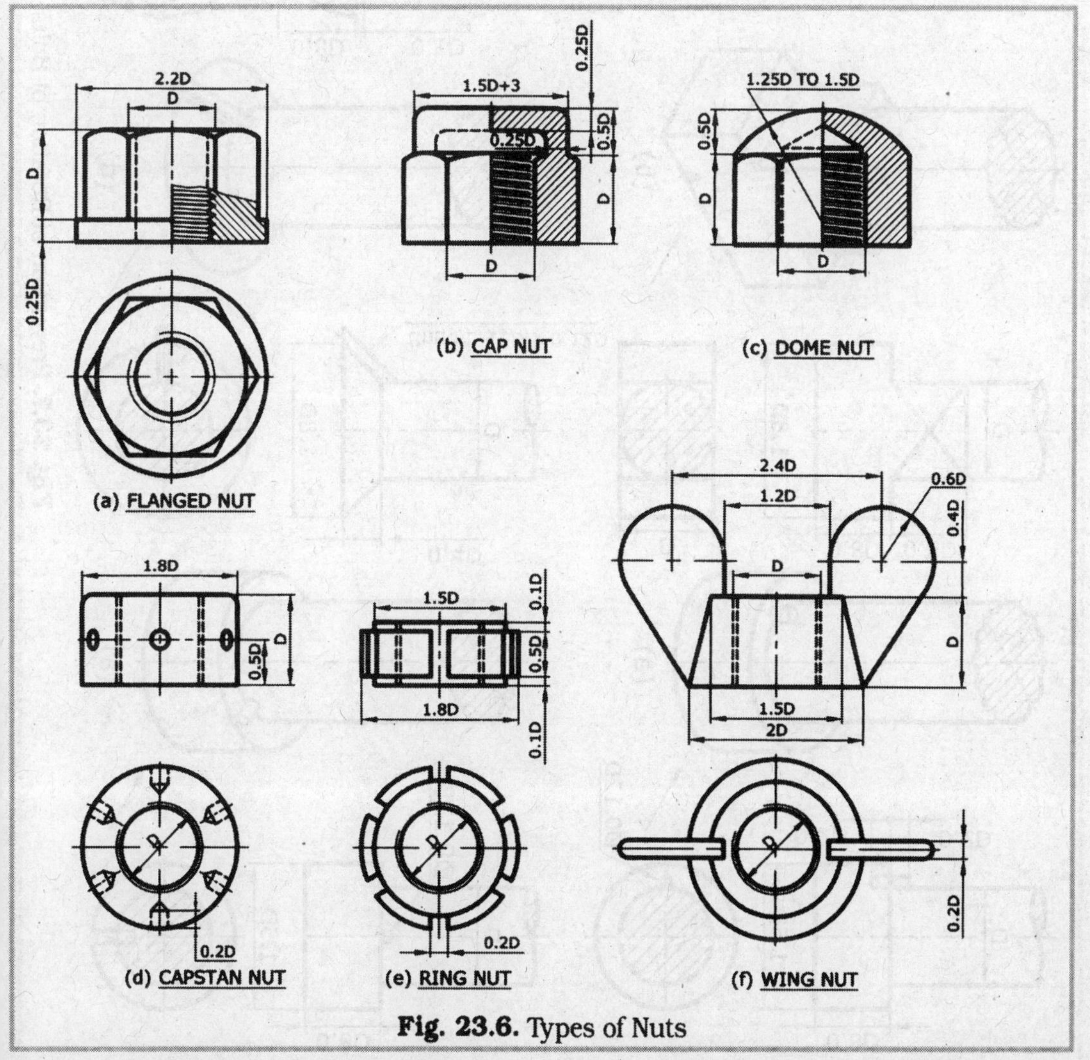

Fig. 23.6. Types of Nuts

6. PREVENTION OF ROTATION OF A BOLT (See Fig. 23.7)

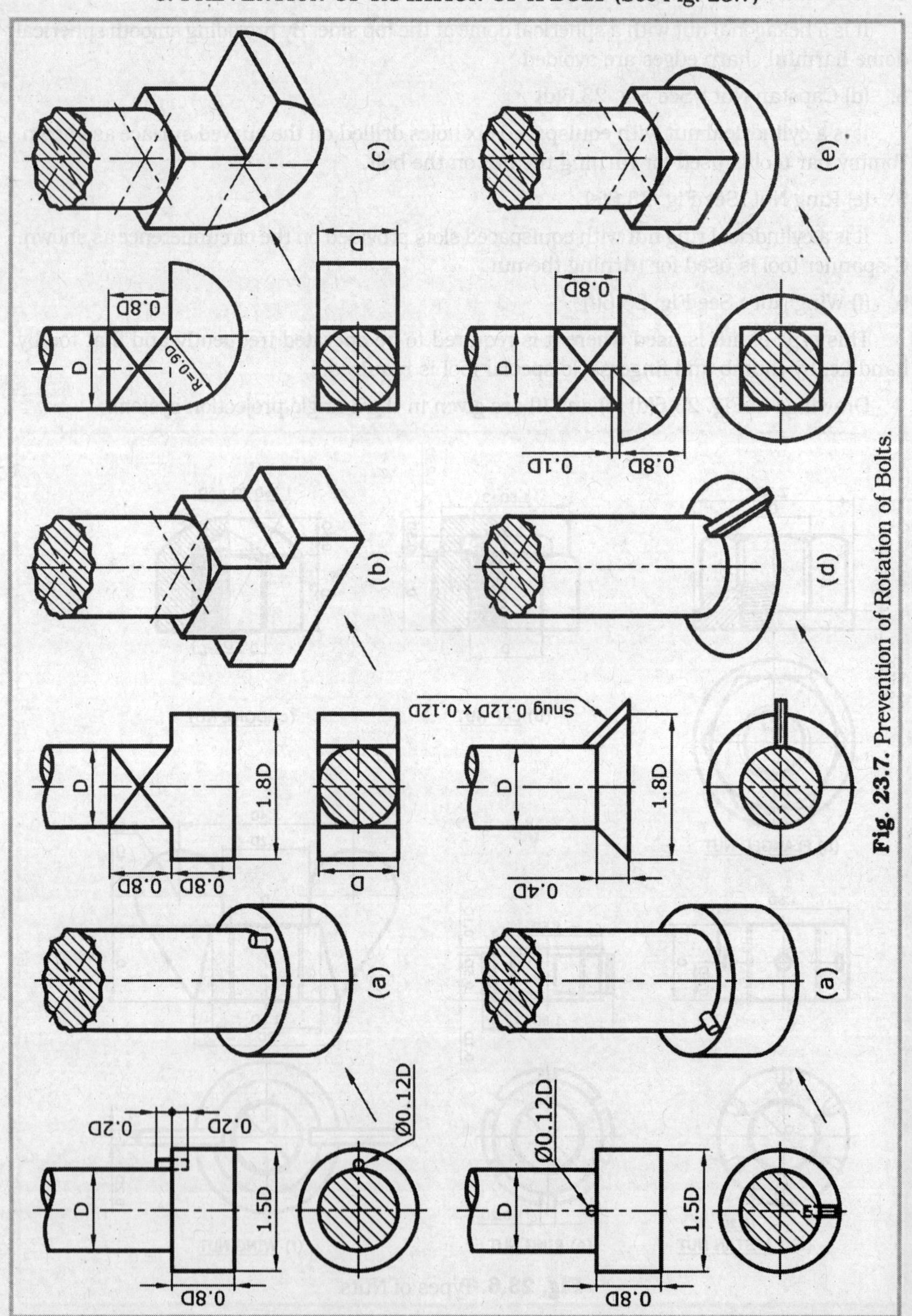

Fig. 23.7. Prevention of Rotation of Bolts.

While tightening a nut on a bolt, the bolt rotates along with the nut and hence the nut cannot be tightened.

There are two ways of preventing bolt rotation along with rotation of nut.

1. By holding the bolt head by means of a spanner.
2. By providing some rotation prevention technique.

When it is not convenient or not possible to hold a bolt head by means of a spanner, rotation prevention technique, using pin or square neck or snug, is used, as shown in Fig. 23.7.

Six different types of bolts, as given below, are used to prevent the rotation of the bolt itself.

(a) Cheese headed bolt - 2 types -Fig. 23.7(a)

(b) T-headed bolt - Fig. 23.7 (b)

(c) Hook bolt - Fig. 23.7 (c)

(d) Countersunk headed bolt with snug - Fig. 23.7(d)

(e) Cup headed bolt with square neck - Fig. 23.7(e)

Figures of different types of bolts are self explanatory and hence detail discussion is not given. A pin, a square neck or a snug prevent a bolt from rotation.

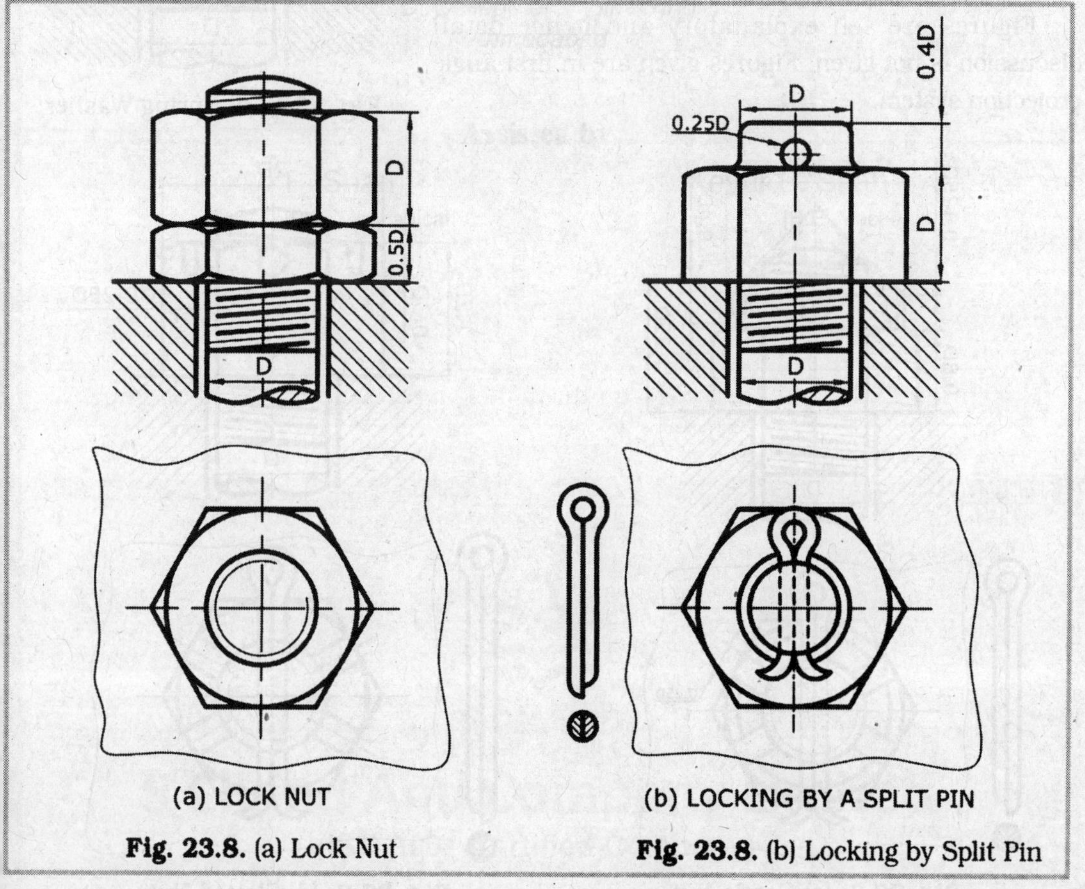

Fig. 23.8. (a) Lock Nut **Fig. 23.8.** (b) Locking by Split Pin

7. LOCKING ARRANGEMENTS OF NUTS : See Fig. 23.8

Due to the vibration, nut has tendancy to screw off the bolt. This tendancy sometimes may cause serious accident, perticularly in vehicles. To prevent nut from getting loose, some methods are employed.

Some of the important methods, known as locking devices, are listed below and shown in Fig. 23.8.

(a) Lock nut or Check nut - see Fig. 23.8(a)

(b) Locking by split pin - See Fig. 23.8(b)

(c) Castle nut - See Fig. 23.8 (c)

(d) Slotted nut - see Fig. 23.8 (d)

(e) Sawn nut - See Fig. 23.8 (e)

(f) Locking by a pin - See Fig. 23.8 (f)

(g) Penn, ring or grooved nut - See Fig. 23.8 (g)

(h) Nut with a locking plate - See Fig. 23.8 (h)

(i) Nut with a spring washer - see Fig. 23.8 (i)

Figures are self explanatory and hence detail discussion is not given. Figures given are in first angle projection system.

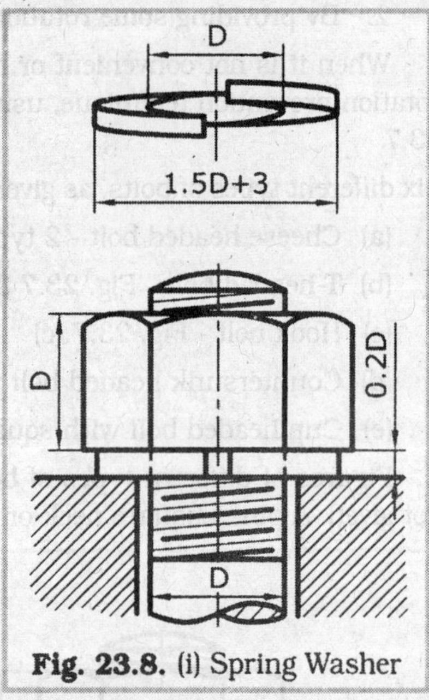

Fig. 23.8. (i) Spring Washer

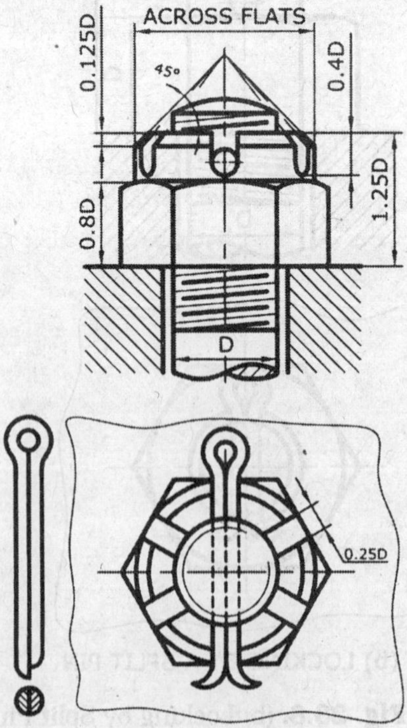

ACROSS FLATS
0.125D
45°
0.4D
1.25D
0.8D
D
0.25D

Fig. 23.8. (c) Castle Nut

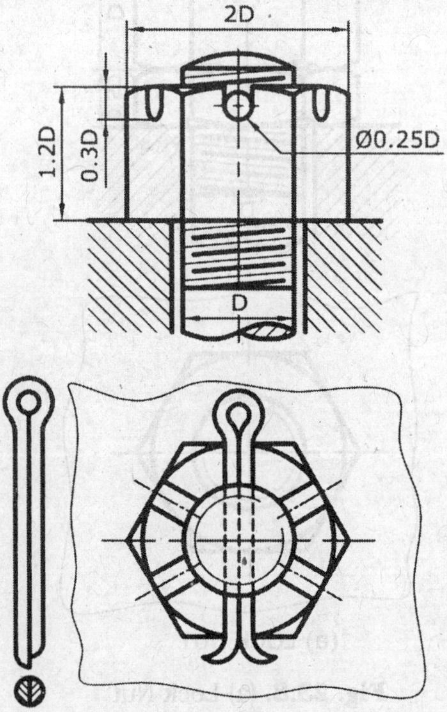

2D
1.2D
0.3D
Ø0.25D
D

Fig. 23.8. (d) Slotted Nut

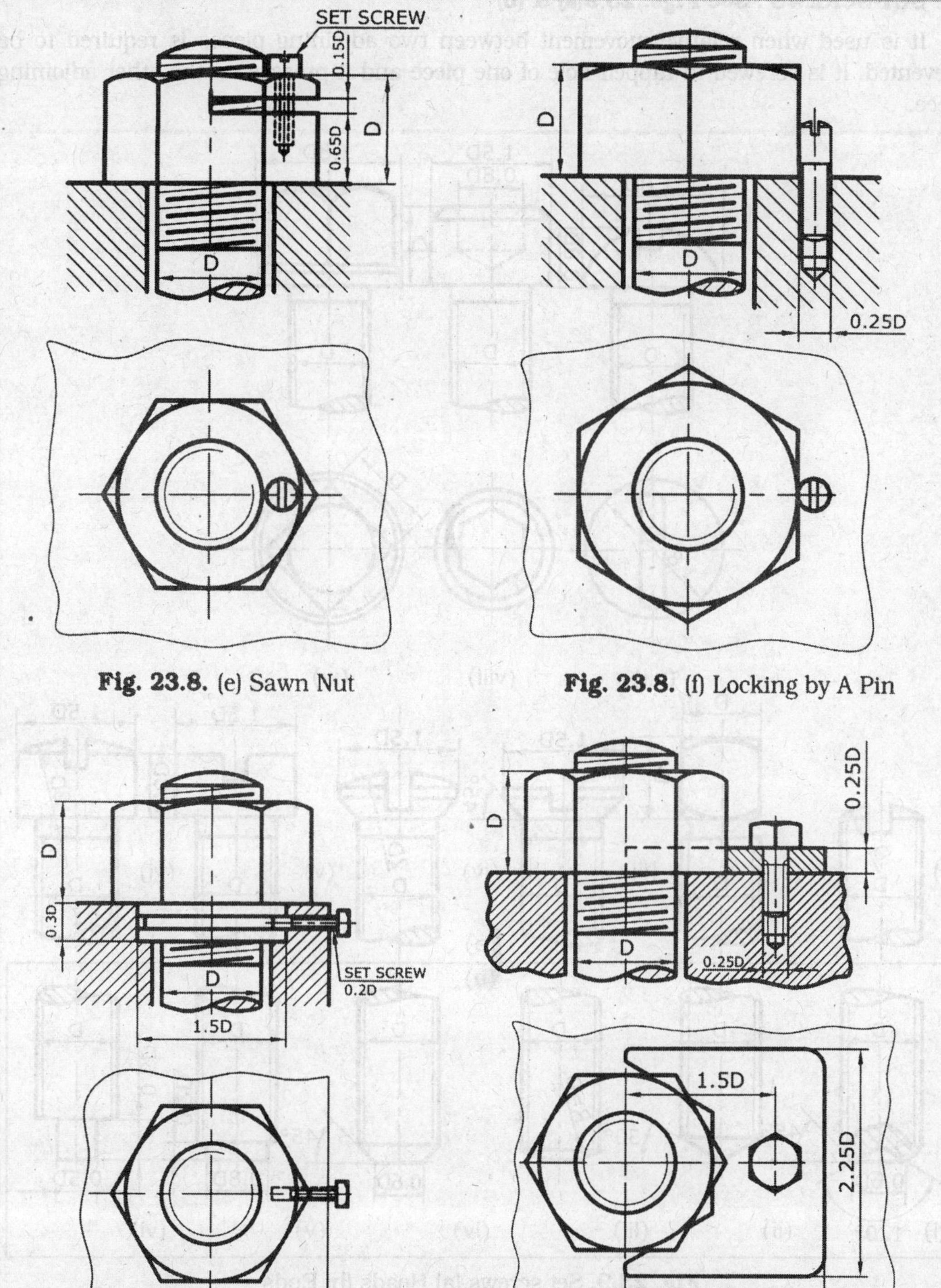

Fig. 23.8. (e) Sawn Nut

Fig. 23.8. (f) Locking by A Pin

Fig. 23.8. (g) Penn, Ring or Grooved Nut

Fig. 23.8. (h) Locking Plate

8. SET-SCREWS : See Figs. 23.9(a) & (b)

It is used when relative movement between two adjoining pieces is required to be prevented. It is screwed in tapped hole of one piece and it presses on the other adjoining piece.

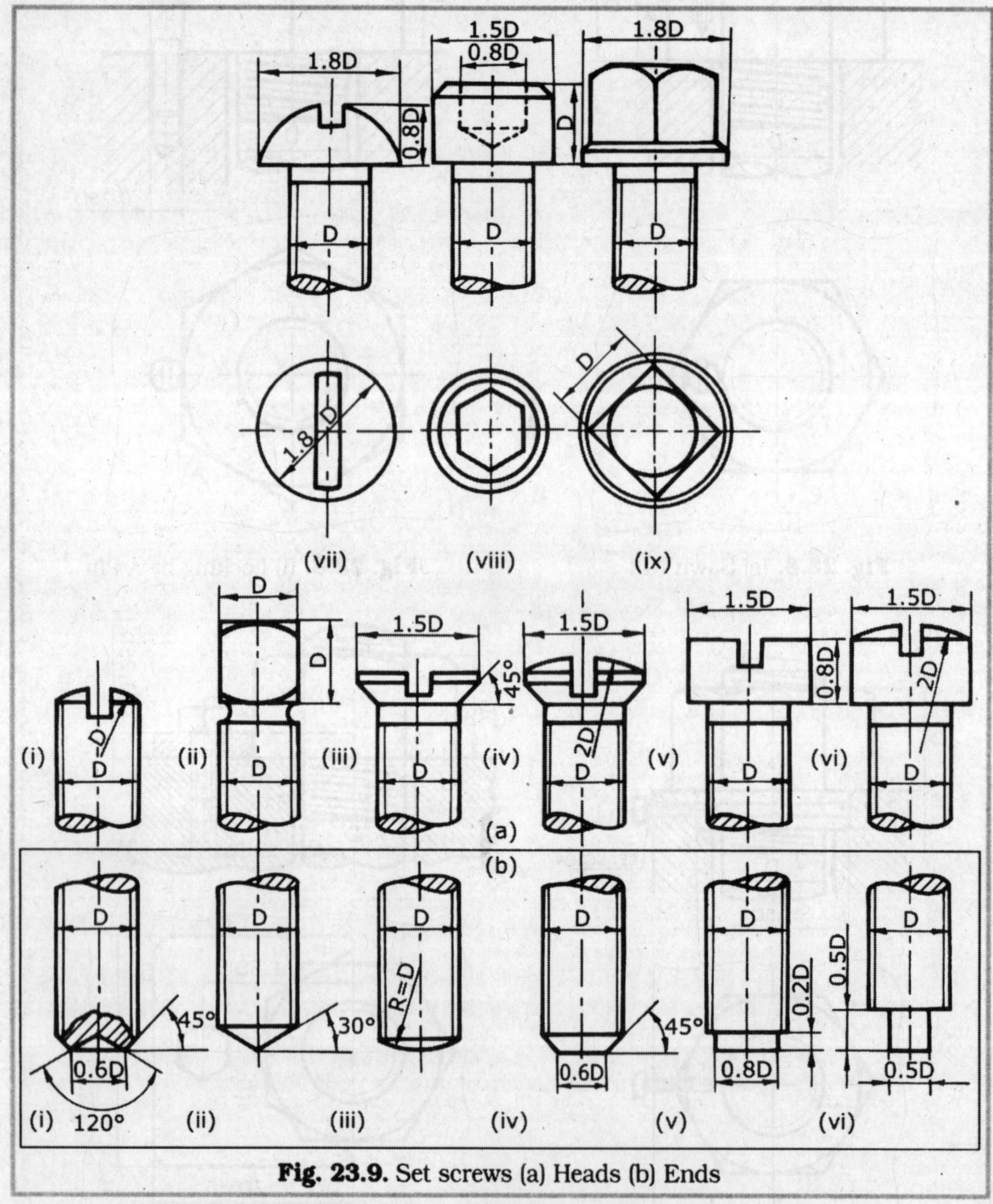

Fig. 23.9. Set screws (a) Heads (b) Ends

Set screws have heads of different varieties as given below: See Fig. 23.9(a).

(i) Headless grub screw - Fig 23.9(a) (i)

(ii) Square head - Fig. 23.9(a) (ii)

(iii) Countersunk head - Fig. 23.9(a) (iii)

(iv) Rounded counter sunk head - Fig. 23.9(a) (iv)

(v) Cylindrical or cheese head - Fig. 23.9(a) (v)

(vi) Fillister - Fig. 23.9(a) (vi)

(vii) Round or cup head - Fig. 23.9(a) (vii)

(viii)Socket head - Fig. 23.9 (a) (viii)

(ix) Collar screw with square head - Fig. 23.9 (a) (ix)

Over and above these heads, square and hexagonal heads are also used. Similarly, set screws have ends of different varieties as given below. See Fig. 23.9(b)

(i) Cup end - Fig. 23.9(b) (i)

(ii) Conical end - Fig. 23.9 (b) (ii)

(iii) Oval end - Fig 23.9 (b) (iii)

(iv) Flat end - Fig 23:9(b) (iv)

(v) Half dog end - Fig. 23.9(b) (v)

(vi) Full dog end - Fig. 23.9(b) (vi)

Fig. 23.9(a) and Fig. 23.9(b) are self explanatory and hence detail discussion is not given here. Screw drivers,spanners, wrenches, etc. are used to operate heads of set screws.

9. STUD OR STUD BOLT : See Fig. 23.10

It is a cylindrical piece threaded at both ends. see Fig.23.10(a). One end of the stud is for insertion in a tapped hole and the other threaded end of the stud receives nut or nuts.

Fig. 23.10(b) illustrates the use of a stud. It is used for holding cylinder heads of automobile engines. The stud replaces bolt when it is not convenient to accomodate a bolt head. Stud with square neck, as shown in Fig. 23.10(c),can be tightened by engaging square neck in a spanner.

A stud with collar as shown in Fig. 23.10 (A) is used in stuffing box assembly .

10. (A) FOUNDATION BOLTS & (B) LIFTING EYE BOLTS : See Fig. 23.11 (a), (b), (c), (d), (e) & (f).

Foundation bolts are used for putting machines to their foundations i.e. fixing them on floor.

Foundation bolts are suspended from holes provided in the base plate of the machine, inside the holes digged earlier in the floor. After levelling the machines cavities of floor surrounding foundation bolts are filled up by a mixture of water, cement, sand, grit, etc. as shown in the figures. Materials like lead, sulphur, etc. can also be used in place of cement.

There are many varieties of foundation bolts. Some of them are as under:

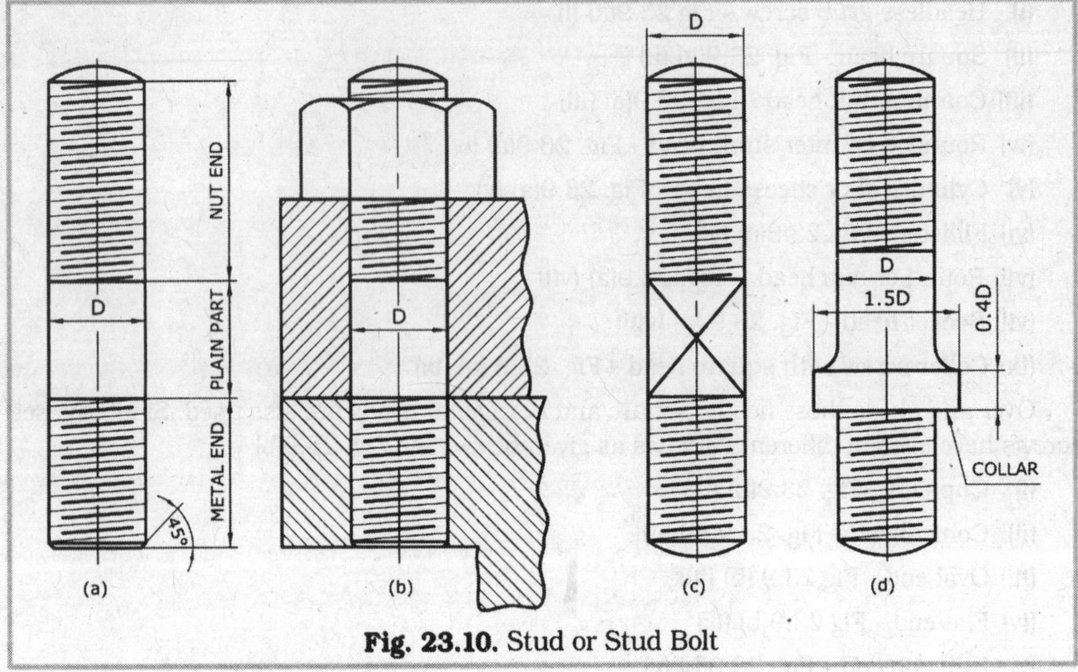

Fig. 23.10. Stud or Stud Bolt

(a) Bent or twisted foundation bolt - Fig. 23.11(a)

(b) Eye foundation bolt - Fig. 23.11(b)

(c) Rag foundation bolt - Fig. 23.11(c)

(d) Lewis foundation bolt Fig. 23.11(d)

(e) Cotter foundation bolt - Fig. 23.11(e)

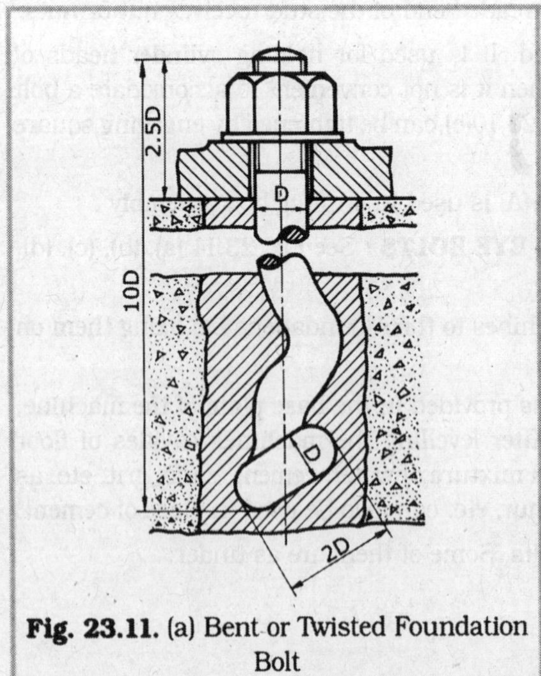

Fig. 23.11. (a) Bent or Twisted Foundation Bolt

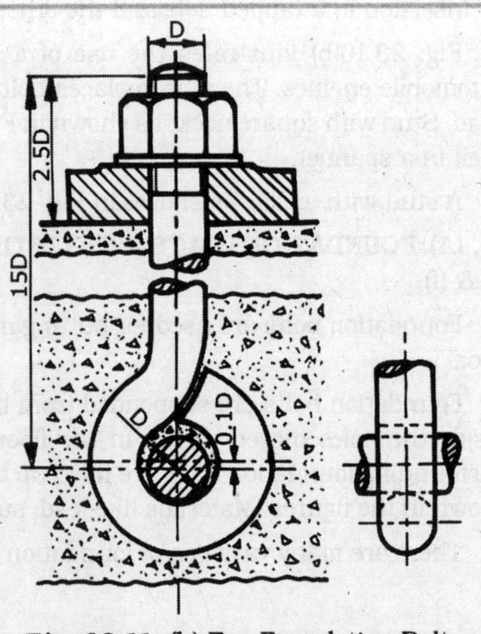

Fig. 23.11. (b) Eye Foundation Bolt

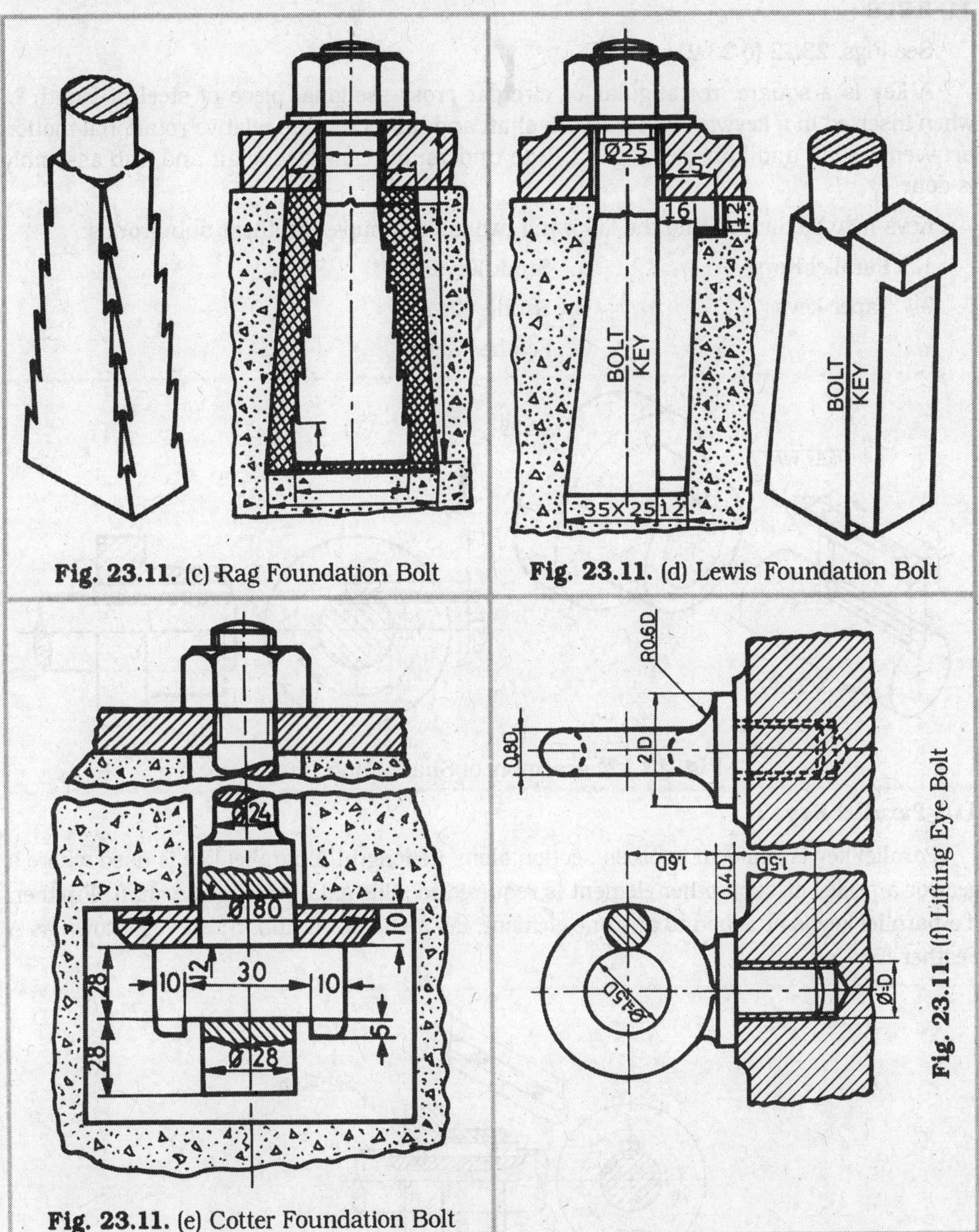

Fig. 23.11. (c) Rag Foundation Bolt

Fig. 23.11. (d) Lewis Foundation Bolt

Fig. 23.11. (e) Cotter Foundation Bolt

Fig. 23.11. (f) Lifting Eye Bolt

Figures of foundation bolts are self explanatory and hence a detailed discussion is not given.

10 (B) Lifting Eye Bolt : This bolt is fitted in the tapped hole, provided on the top side approximately in the centre of a heavy machine. Normally the location of this lifting eye bolt is vertically above the C. G. of the machine. Machine is lifted by inserting crane hook in the eye of this bolt. See Fig. 23.11

11. KEYS :

See Figs. 23.12 to 23.21

A key is a square, rectangular or circular cross-sectional piece of steel of length L, when inserted in a keyway kept between shaft and hub, prevents relative rotational motion between a shaft and hub. See Fig. 23.12 to understand how key, shaft and hub assembly is done.

Keys may be divided into the following two main groups and three sub-groups:

1. Parallel Keys
2. Taper keys

A. Sunk Keys
B. Saddle Keys
C. Feather Keys

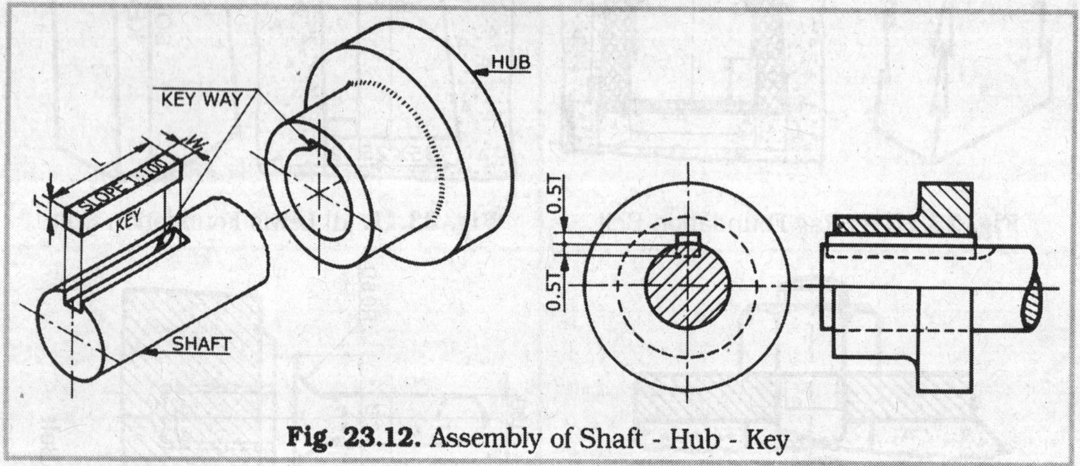

Fig. 23.12. Assembly of Shaft - Hub - Key

11.1 Parallel Keys :

Parallel key is uniform in cross-section along its length. A parallel key is used where a gear or a pulley or such other element is required to slide axially along the shaft. Further, if a parallel key is attached to any one element, i.e. a shaft or a hub, the key is known as A Feather Key.

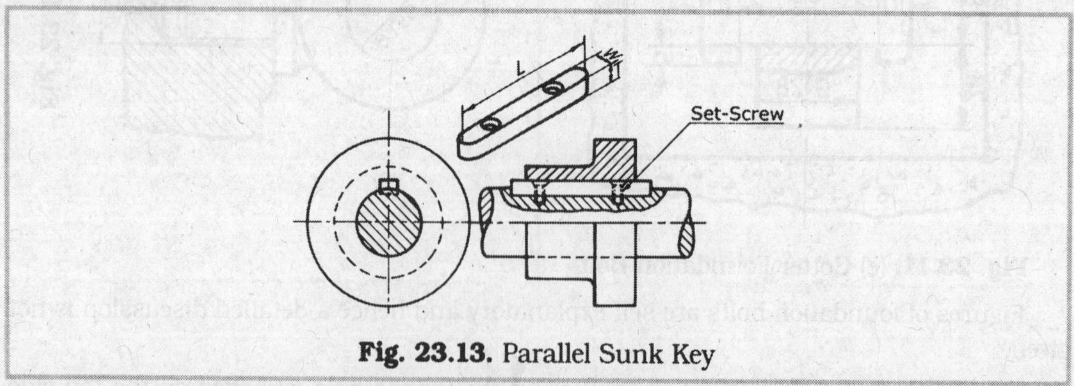

Fig. 23.13. Parallel Sunk Key

Feather sunk keys, attached to the shaft by two counter sunk headed screws, is shown in Fig. 23.13.Feather sunk keys, attached to the hubs by peg, screw or two heads are shown in Figs. 23.14, 23.15 and 23.16 respectively.

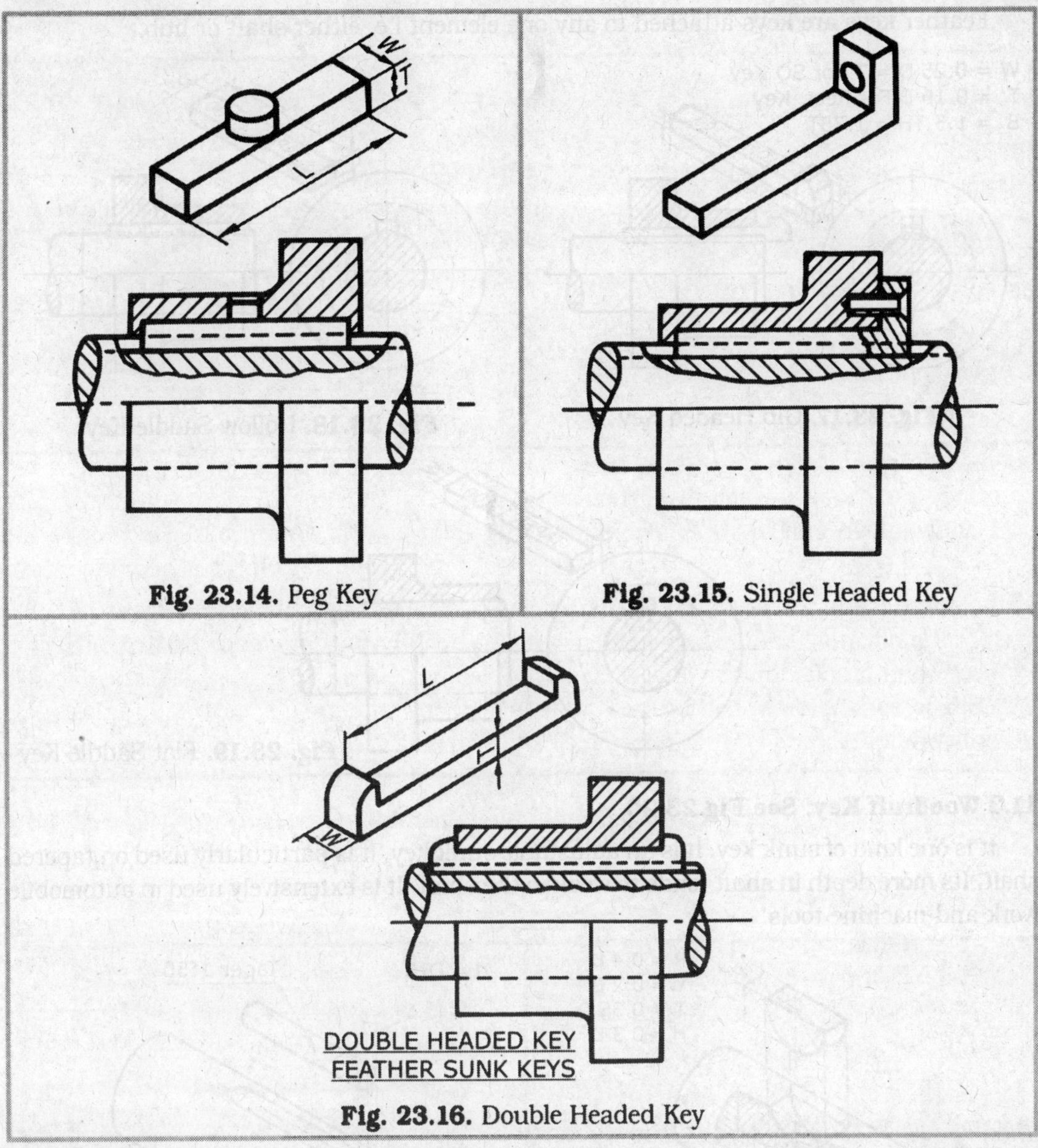

Fig. 23.14. Peg Key

Fig. 23.15. Single Headed Key

Fig. 23.16. Double Headed Key

Taper Keys :

Taper keys are uniform in width and tapered along thickness. Taper keys are used when along with relative rotational motion relative axial movement is also to be prevented. Taper keys are of two main varieties.

1. Taper Sunk Keys
2. Taper Saddle Keys

Gib headed key shown in Fig. 23.17 is a tapered sunk key. Hollow saddle key shown in Fig. 23.18 and flat saddle key shown in Fig. 23.19 are taper saddle keys.

A key that engages a key slot, in both the shaft and the hub, is known as a sunk key.

In case of saddle keys, key way is not provided in the shaft and key does not engage in the shaft.

Feather keys are keys attached to any one element i.e. either shaft or hub.

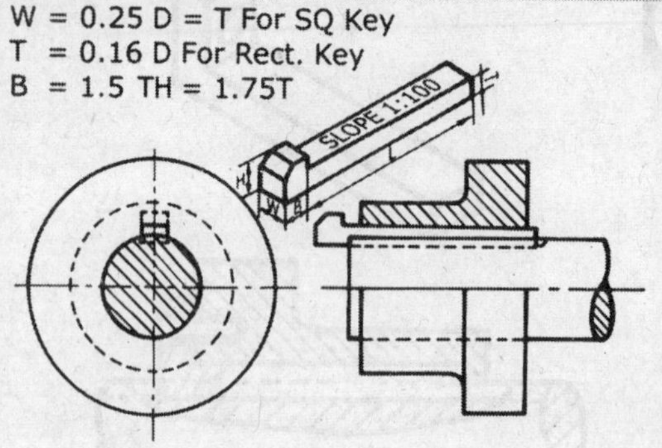

W = 0.25 D = T For SQ Key
T = 0.16 D For Rect. Key
B = 1.5 TH = 1.75T
SLOPE 1:100
L
H
W
B

Fig. 23.17. Gib Headed Key

W
L
TAPER 1:100
T
D

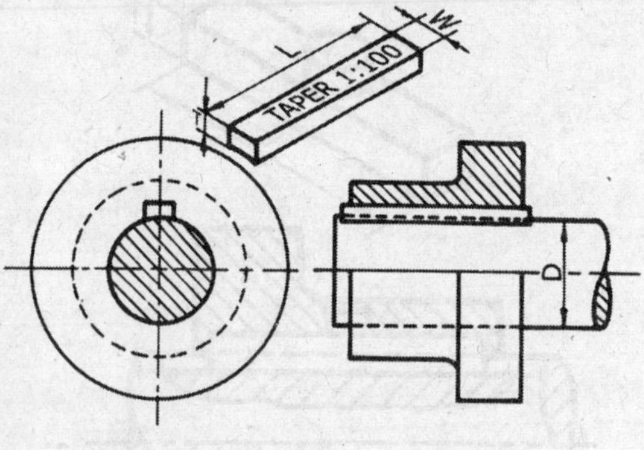

Fig. 23.18. Hollow Saddle Key

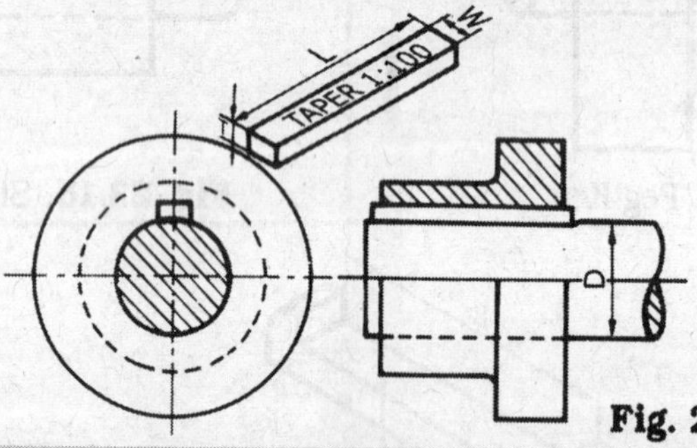

W
L
TAPER 1:100
T
D

Fig. 23.19. Flat Saddle Key

11.3 Woodruff Key: See Fig.23.20

It is one kind of sunk key. It is an adjustable sunk key. It is particularly used on tapered shaft. Its more depth in shaft tends to weaken the shaft. It is extensively used in automobile work and machine tools.

R = 0.4 D
W = 0.2 D
T = 0.35 D
H = 0.3 D
W
T
H

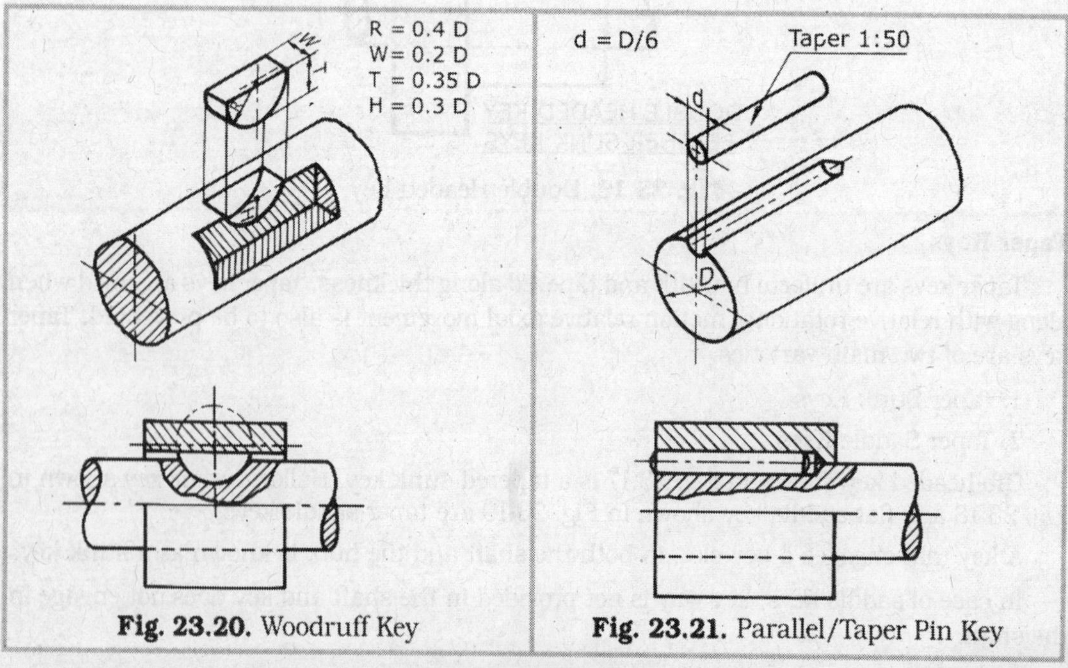

Fig. 23.20. Woodruff Key

d = D/6
Taper 1:50
d
D

Fig. 23.21. Parallel/Taper Pin Key

11.4 Pin Key/Round Key: See Fig. 23.21

It is a simple key with circular cross section. It can be made with taper or without taper. This type of key is suitable for low power transmission.

Constructional features along with proportional dimensions of various types of keys are shown in Fig.23.12 to Fig. 23.21. Figures are self explanatory and hence a detailed discussion is not given.

12. COTTER JOINT : See Fig. 23.22

A cotter is used to connect two rods rigidly by inserting cotter at right angles to the axes of two rods. Cotter joint is capable of transmitting tensile as well as compressive forces. Cotter is uniform in thickness and taper in width. Taper is generally given on one side only.

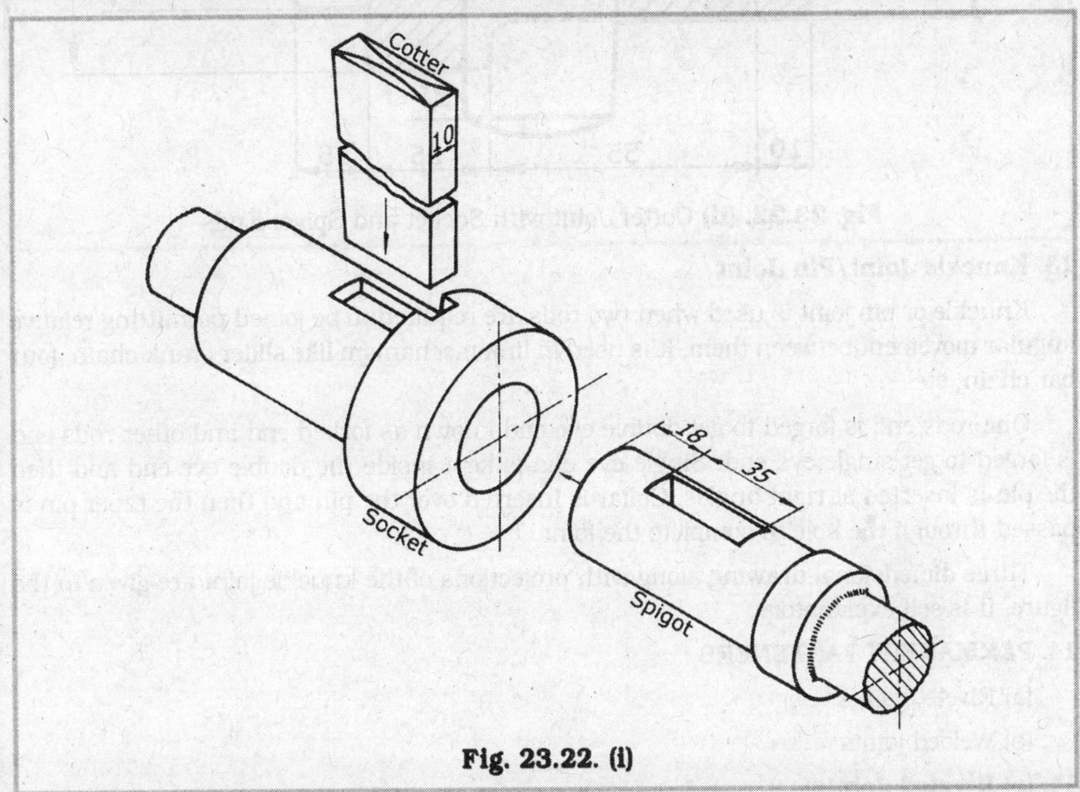

Fig. 23.22. (i)

Socket and spigot cotter joint is shown in Fig. 23.22. One of the rods end is forged to form socket end and other rod's end is forged to get spigot end. Spigot end is inserted inside the socket end and then the cotter is inserted at right angles, as shown in figure to get a joint. Clearances shown in the figure are absolutely essential for proper functioning of the joint.

Constructional features and proportionate dimensions are clearly shown in Fig. 23.22 and hence further detail discussion is not necessary.

Sleeve cotter joint can be made by inserting two rods from opposite sides of the sleeve and then inserting two cotters at right angles.

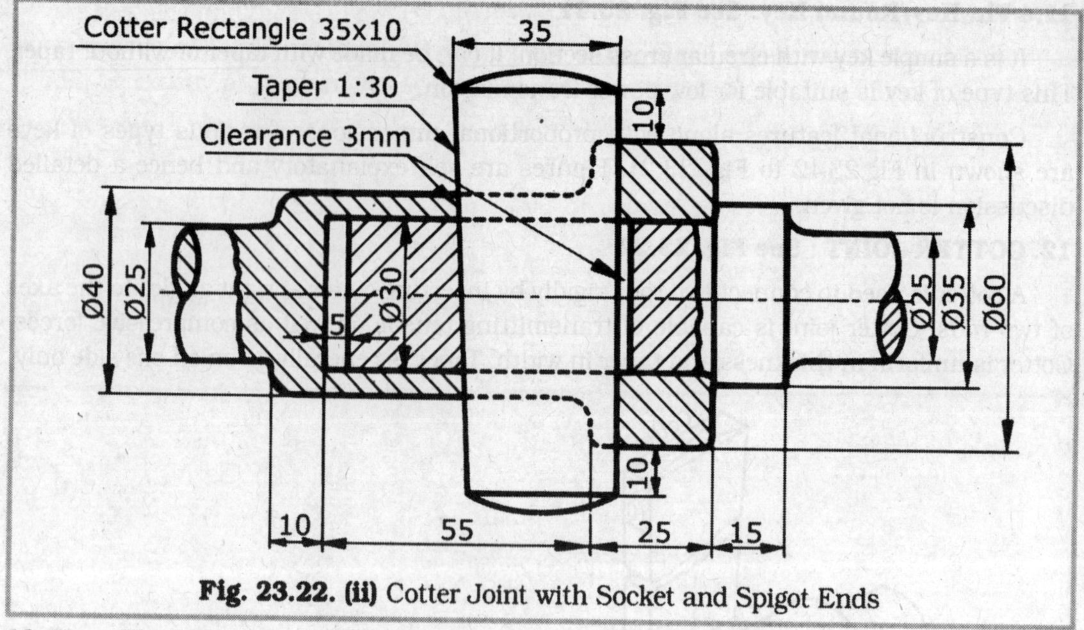

Fig. 23.22. (ii) Cotter Joint with Socket and Spigot Ends

13. Knuckle Joint/Pin Joint

Knuckle or pin joint is used when two rods are required to be joined permitting relative angular movement between them. It is used in link mechanism like slider crank chain, four bar chain, etc.

One rods end is forged to get double eye end known as forked end and other rods end is forged to get single eye end. Single eye end is kept inside the double eye end and then the pin is inserted at right angles. Collar is inserted over the pin and then the taper pin is passed through the hole to complete the joint.

Three dimensional drawing along with projections of the knuckle joint are given in the figure. It is self explanatory.

14. PERMANENT FASTENERS

(a) Riveted joints

(b) Welded joints

14. (a) Riveted Joints:

Riveted joints are used in body building of automobiles,ships,etc. They are used in pressure vessels, tanks, boilers, etc. They are also used in steel structures, bridges, etc.

Joints created by rivets are of permanent nature and cannot be disassembled without chipping off heads of rivets.

Rivets are made of ductile material like mild steel, wrought iron, brass, aluminium, etc.

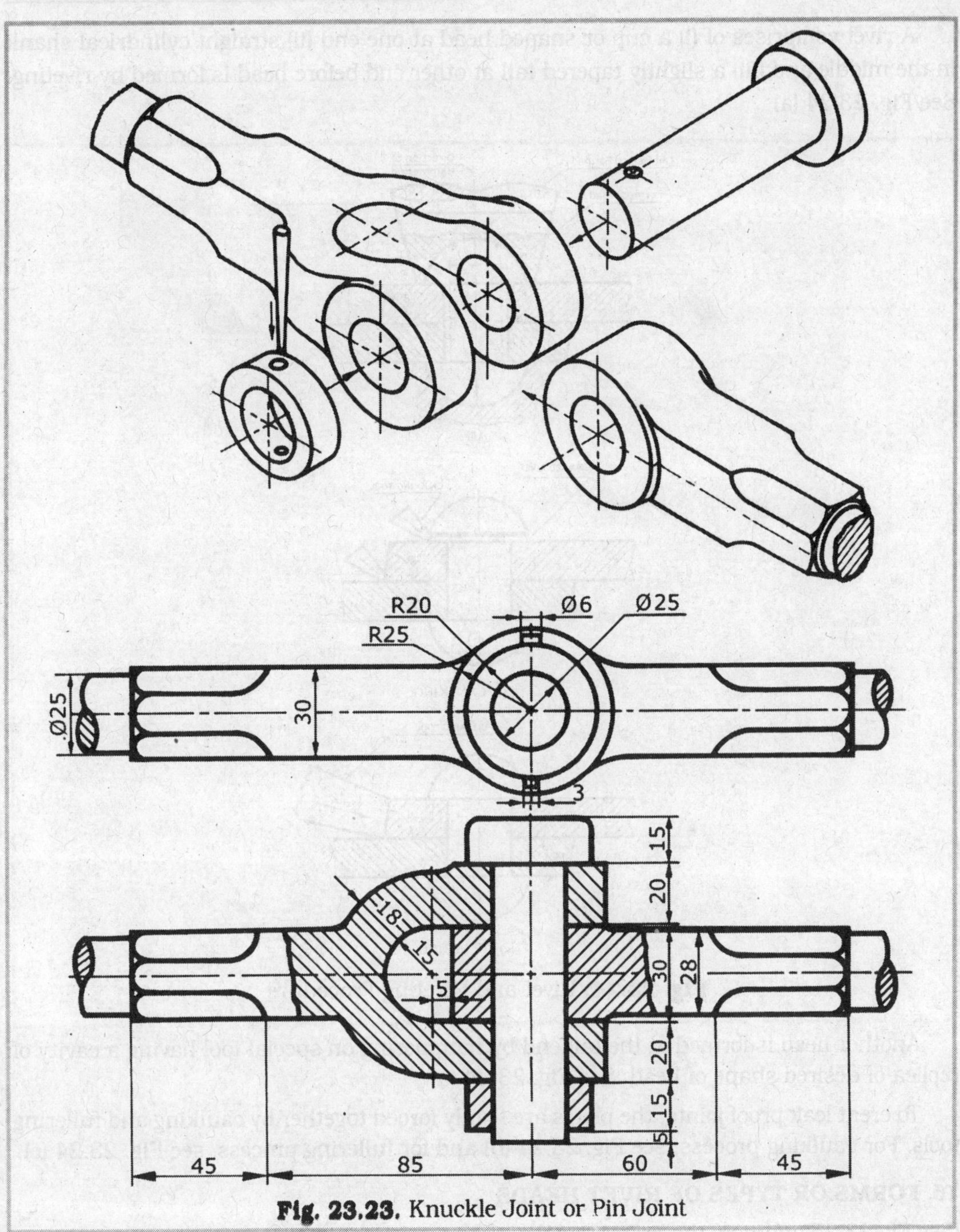

Fig. 23.23. Knuckle Joint or Pin Joint

A rivet comprises of (i) a cup or snaped head at one end (ii) straight cylindrical shank in the middle and (iii) a slightly tapered tail at other end before head is formed by riveting. See Fig. 23.24 (a)

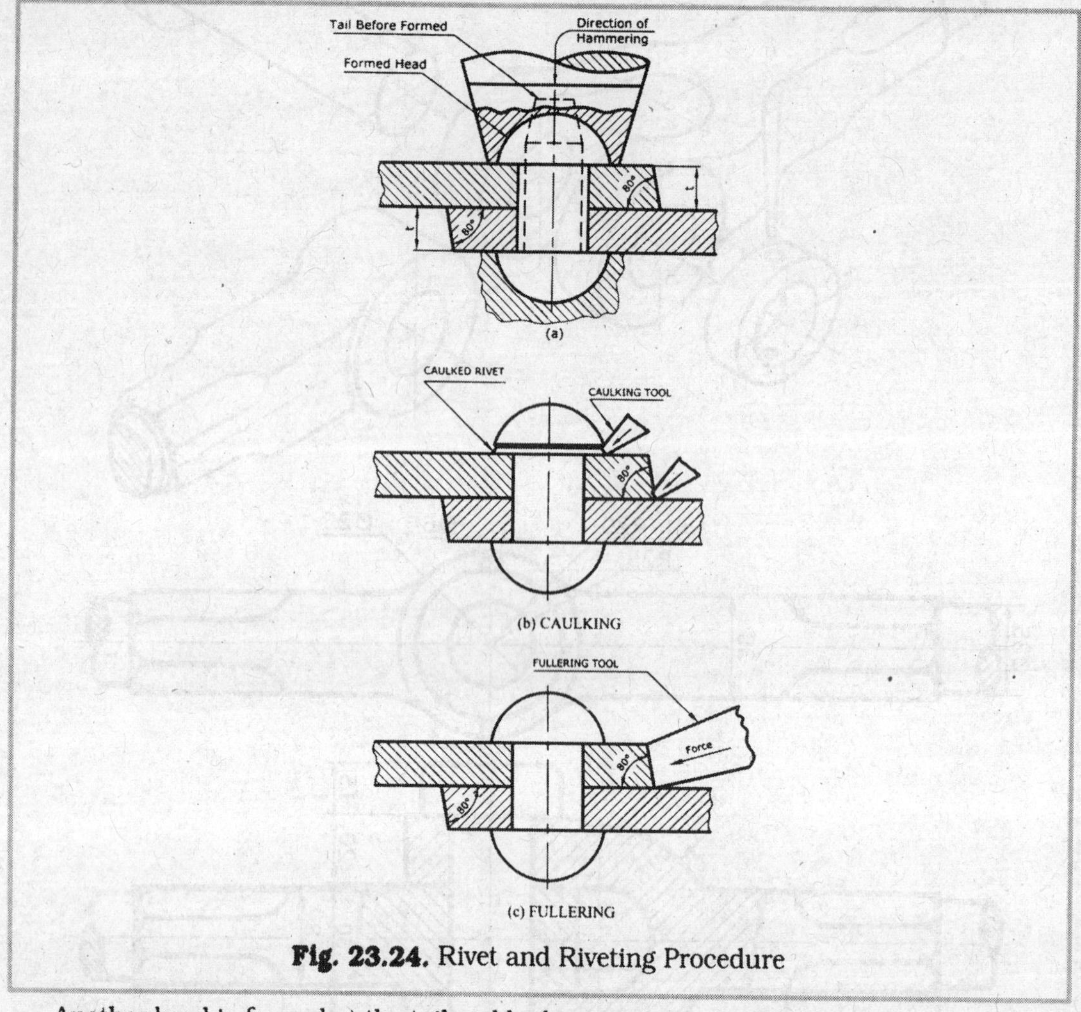

Fig. 23.24. Rivet and Riveting Procedure

Another head is formed at the tail end by hammering on special tool having a cavity of replica of desired shape of head. See Fig. 23.24 (a)

To creat leak proof joints, the plates are firmly forced together by caulking and fullering tools. For caulking process, see Fig. 23.24 (b) and for fullering process, see Fig. 23.24 (c).

15. FORMS OR TYPES OF RIVET HEADS

(See Fig.23.25 and note the proportions)

1. Snap or cup head - Fig. 23.25(i)
2. Pan head - Fig. 23.25(ii)
3. Pan head with tapered neck - Fig. 23.25 (iii)

4. Rounded countersunk head - Fig. 23.25(iv)
5. Countersunk head - Fig. 23.25 (v)
6. Snap head 2 - Fig. 23.25(vi)
7. Ellipsoidal head - Fig. 23.25(vii)
8. Pan head type 1 - Fig. 23.25(viii)
9. Conical head - Fig. 23.25(ix)
10. Pan head type 2 - Fig. 23.25(x)

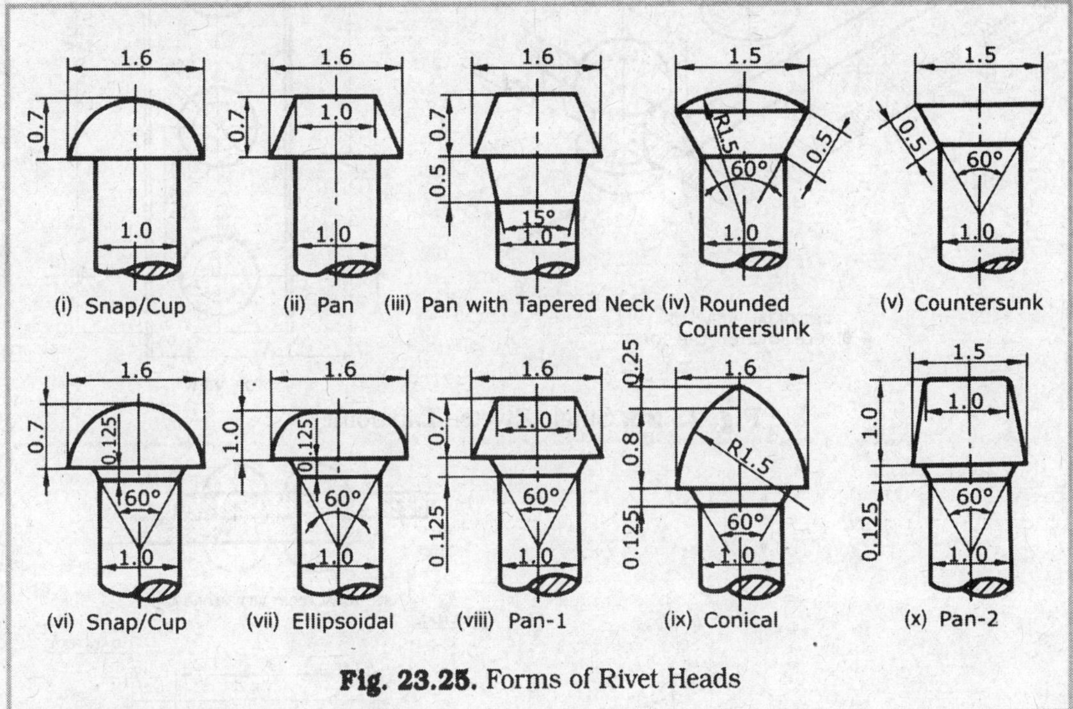

Fig. 23.25. Forms of Rivet Heads

16. TYPES OF RIVETED JOINTS

There are two main types of riveted joints :

(a) Lap joints

(b) Butt joints.

16. (a) Lap joints:

When the two plates, to be connected are overlapped in position in the joint then the joint is called the lap joint. Depending upon the number of rows of rivets, the joints are further classified as single riveted lap joint, double riveted lap joint,triple riveted lap joint, etc. Further depending upon the arrangement of rivets in adjoining rows of rivets they are classified as chain riveting and zigzag riveting.

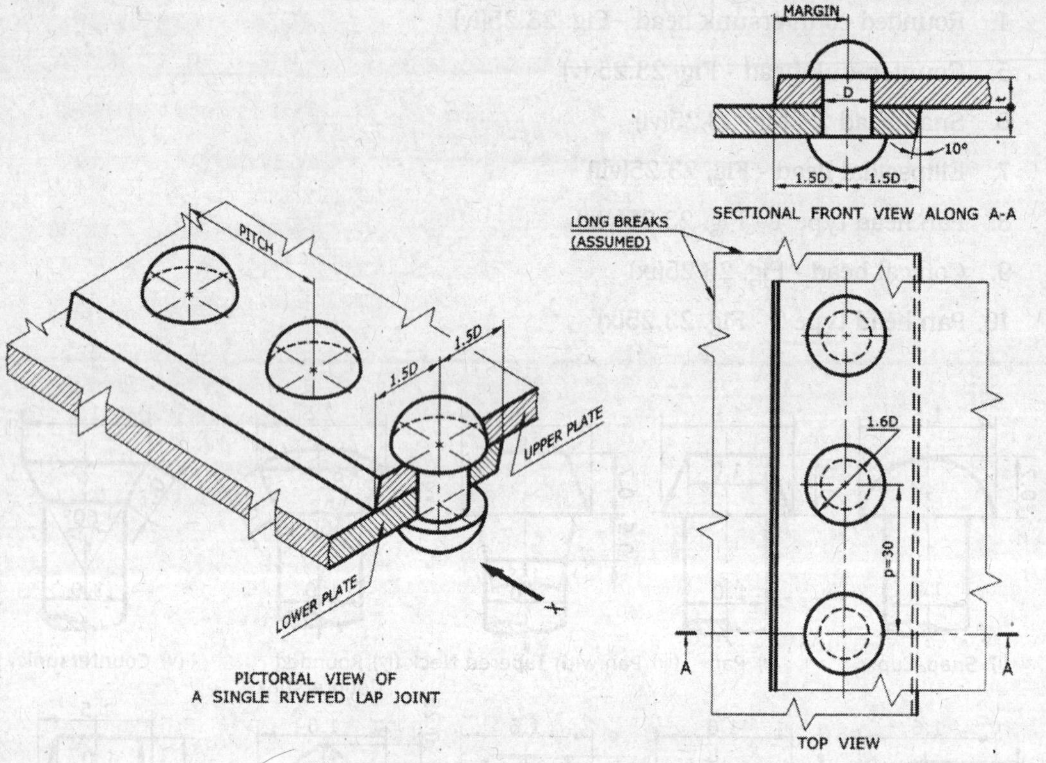

Fig. 23.26. Single Riveted Lap Joint

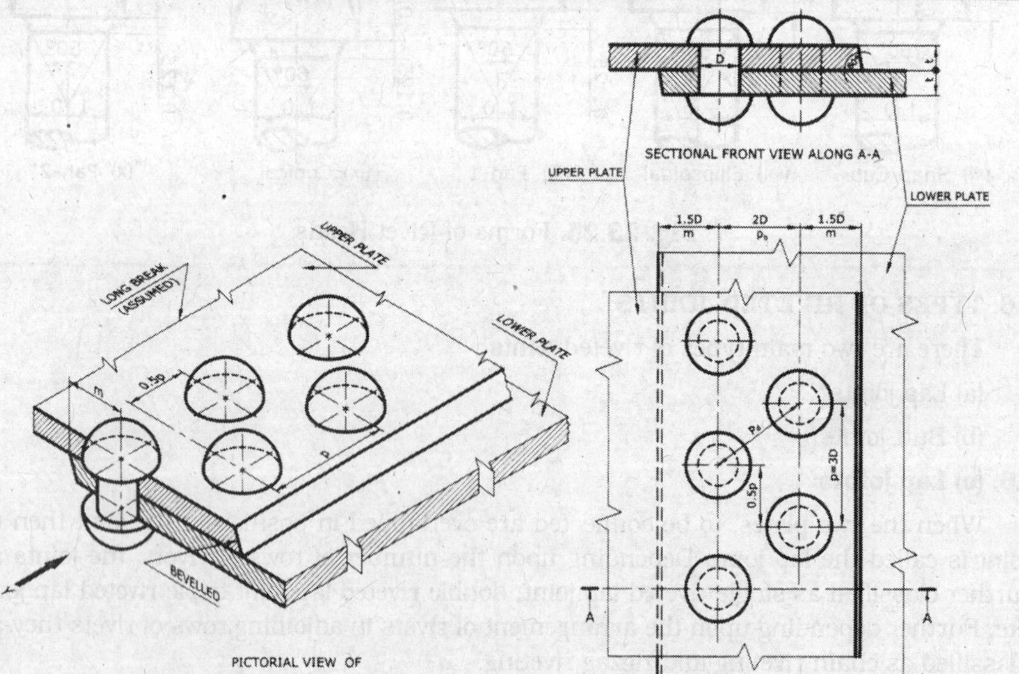

Fig. 23.27. Double Riveted Zigzag Lap Joint

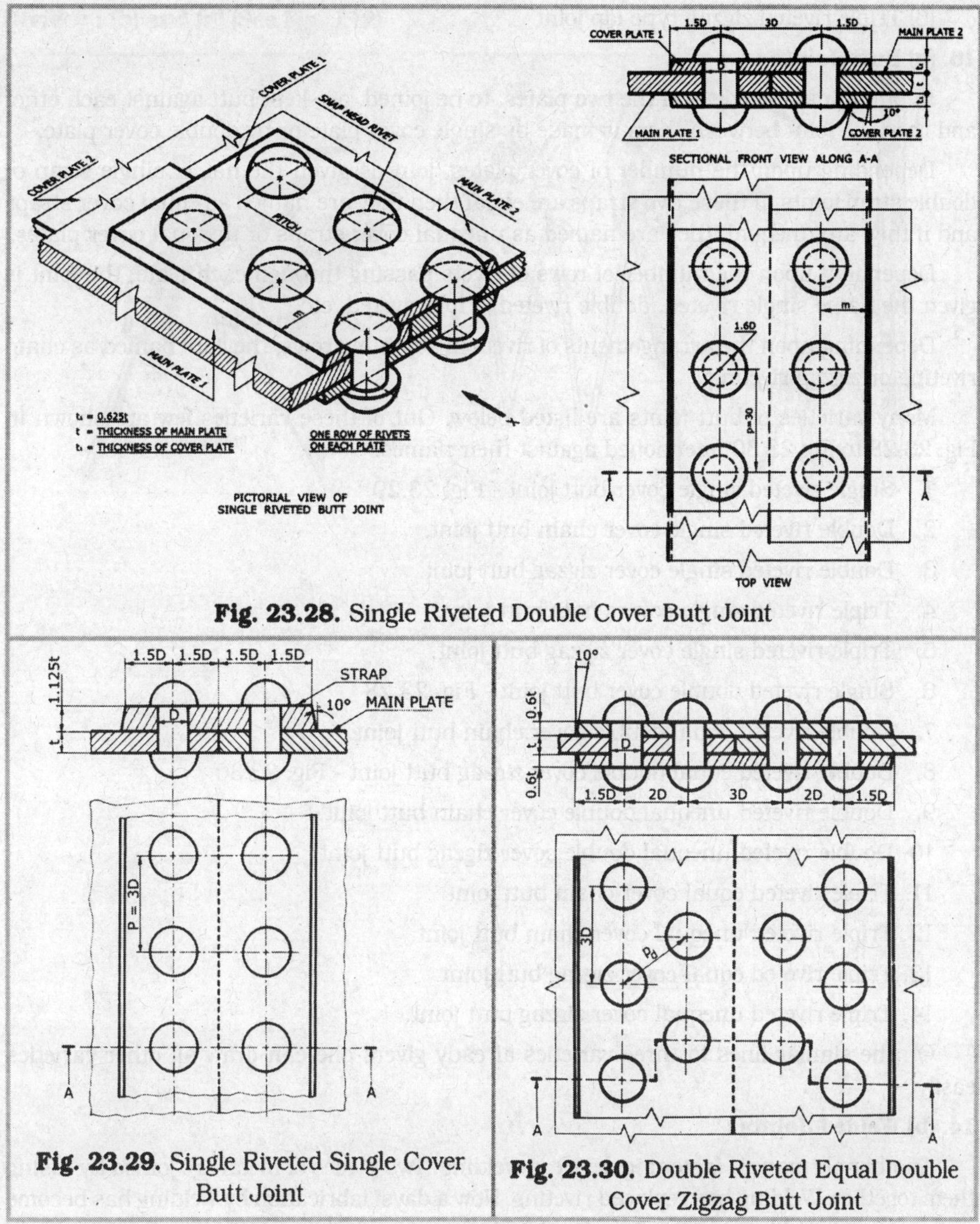

Fig. 23.28. Single Riveted Double Cover Butt Joint

Fig. 23.29. Single Riveted Single Cover Butt Joint

Fig. 23.30. Double Riveted Equal Double Cover Zigzag Butt Joint

Out of many varieties of lap joints two varieties are shown in Fig. 23.26 and Fig. 23.27.

In Fig. 23.26, a single riveted lap joint is shown and in Fig. 23.27, a double riveted zigzag lap joint is shown.

Other types of lap joints, given below, can be drawn on the similar lines.

(1) Double riveted chain type lap joint

(2) Triple riveted chain type lap joint

(3) Triple riveted zigzag type lap joint

16. (b) Butt Joints :

In butt joints the edges of the two plates, to be joined, are kept butt against each other and then the joint between them is made by single cover plate or by double cover plate.

Depending upon the number of cover plates, joint is given the name, single strap or double strap joints. If these two straps are equal then they are named as equal cover straps and if they are unequal, they are named as unequal cover straps or unequal cover plates.

Depending upon the number of rows of rivets passing through each plate, the joint is given the name single riveted, double riveted, triple riveted, etc.

Depending upon the arrangements of rivets in adjoining rows, they are named as chain riveting or zigzag riveting.

Many varieties of butt joints are listed below. Out of these varieties few are shown in Fig. 23.28 to Fig.23.30, mentioned against their names.

1. Single riveted single cover butt joint - Fig. 23.29
2. Double riveted single cover chain butt joint.
3. Double riveted single cover zigzag butt joint.
4. Triple riveted single cover chain butt joint.
5. Triple riveted single cover zigzag butt joint.
6. Single riveted double cover butt joint - Fig. 23.28
7. Double riveted equal double cover chain butt joint.
8. Double riveted equal double cover zigzag butt joint - Fig. 23.30
9. Double riveted unequal double cover chain butt joint.
10. Double riveted unequal double cover zigzag butt joint.
11. Triple riveted equal cover chain butt joint.
12. Triple riveted unequal cover chain butt joint.
13. Triple riveted equal cover zigzag butt joint.
14. Triple riveted unequal cover zigzag butt joint.

On the similar lines to three varieties already given, one can draw all other varieties easily.

14. (b) Welded Joints :

Another permanent fastening device is welding. Two pieces of metal are joined by fusing them together. Welding has replaced riveting. Now a days, fabrication by welding has become very common. Welding can be done by electric arc, gas, electric resistance, etc.

Some of the forms of the welded joints as per IS:696-1960 with their sectional representations and appropriate symbols are shown in Fig. 23.31. Students should study these symbols critically and should remember them. In drawing and design office one is required to use them frequently.

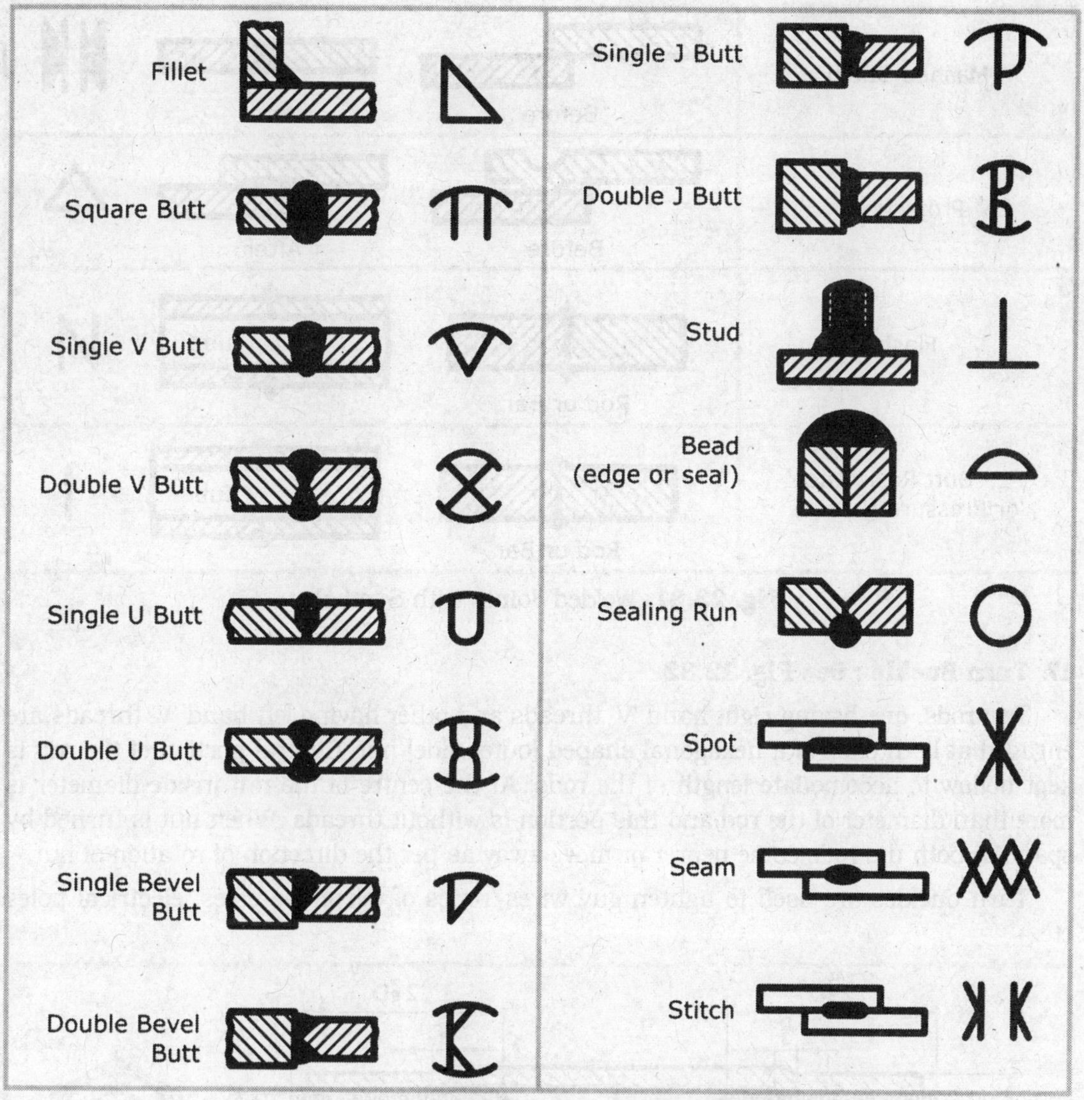

Fillet
Square Butt
Single V Butt
Double V Butt
Single U Butt
Double U Butt
Single Bevel Butt
Double Bevel Butt
Single J Butt
Double J Butt
Stud
Bead (edge or seal)
Sealing Run
Spot
Seam
Stitch

Fig. 23.31. Welded Joints with Symbols

Plug or Slot
Backing Strip
Mashed Seam
Before
After

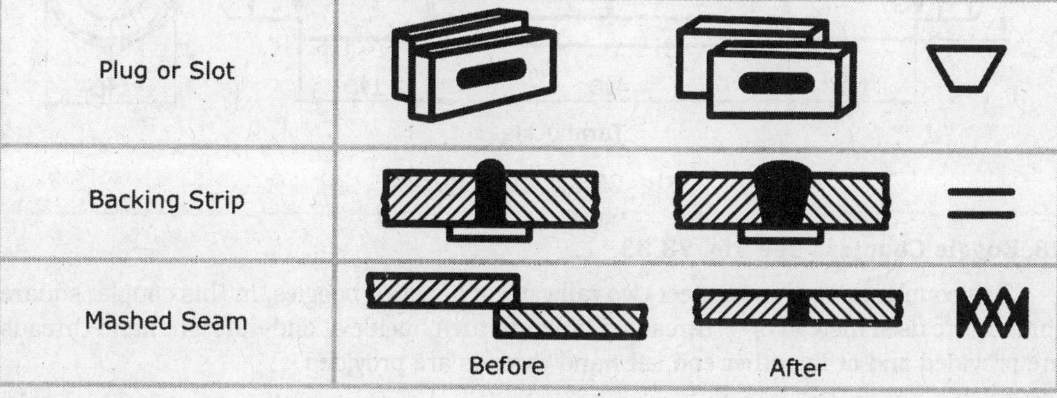

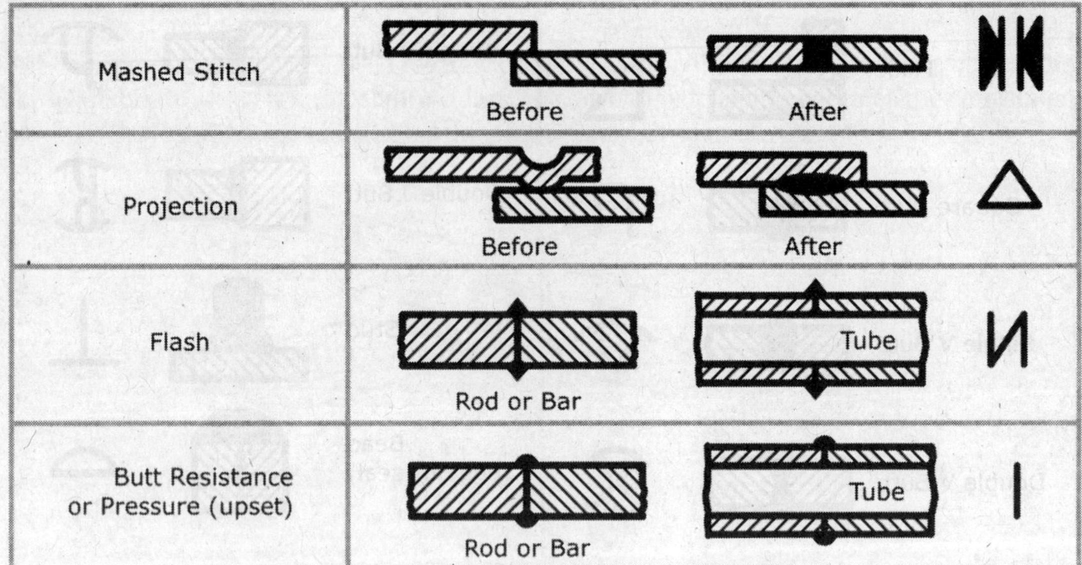

Fig. 23.31. Welded Joints with Symbols

17. Turn Buckle : See Fig. 12.32

Two rods, one having right hand 'V' threads and other having left hand 'V' threads are engaged at both ends of a hexagonal shaped [outer side] nut. Central portion of the nut is kept hollow to accomodate length of the rods. At the centre of the nut inside diameter is more than diameter of the rod and this portion is without threads. When nut is turned by spanner, both the rods come nearer or move away as per the direction of rotation of nut.

Turn buckles are used to tighten guy wires/ropes of telephone poles, electrical poles etc.

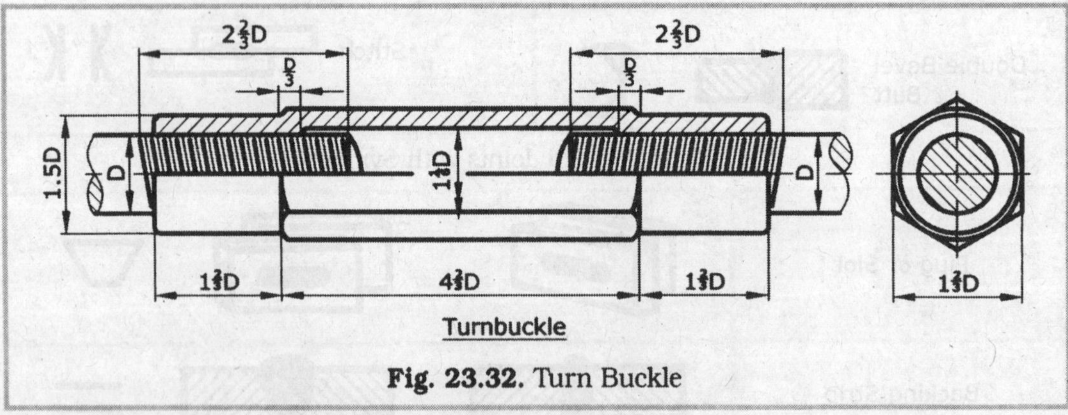

Fig. 23.32. Turn Buckle

18. Boggie Coupler : See Fig. 23.33

This coupler is used to connect two railway carriages or boggies. In this coupler square threads are used instead of 'V' threads. Similar to turn buckle at one end right hand threads are provided and at the other end left hand threads are provided.

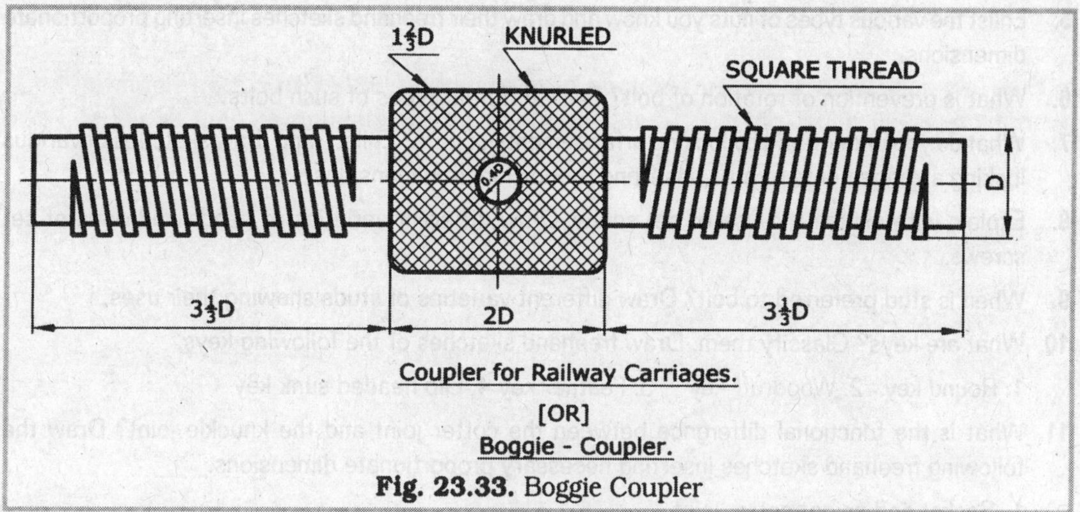

Coupler for Railway Carriages.

[OR]

Boggie - Coupler.

Fig. 23.33. Boggie Coupler

19. Cotter Joint With Sleeve : (OR) Double Cotter Joint ; See Fig. 23.34

In this joint two ends of the two rods are inserted from the opposite sides in a hollow cylindrical sleeve as shown. Two cotters are driven tightly in the slots provided in the sleeve as well as in the rods as shown.

Outer edges of the cotter are perpendicular to the axis of the joint, while inner edges are tapered. Bars subjected to axial forces are joined with this joint.

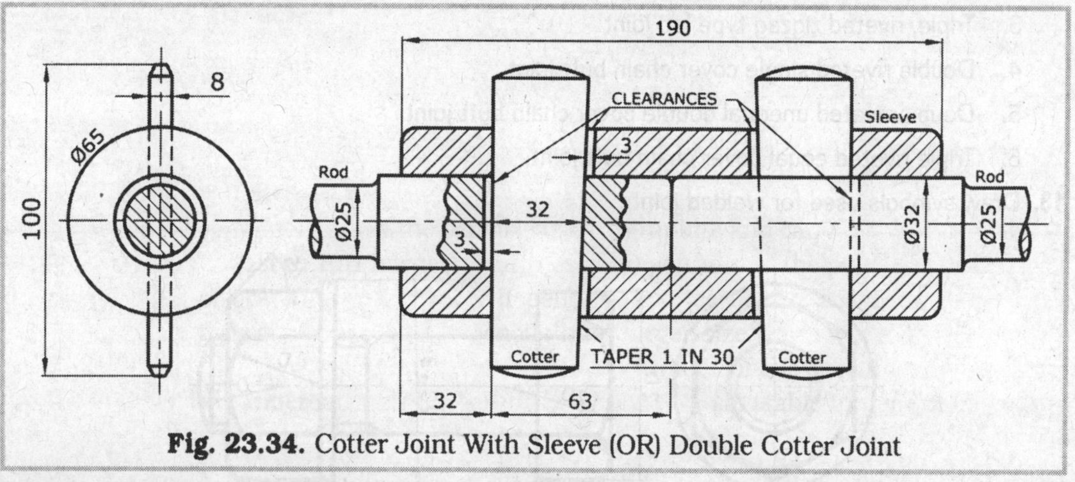

Fig. 23.34. Cotter Joint With Sleeve (OR) Double Cotter Joint

EXERCISE

1. What are fasteners? Explain the difference between permanent and temporary fasteners.
2. Show on sketch the different terms used for screw thread and define each term.
3. Explain the difference between single start and multistart threads with the help of a sketch. Explain lead and pitch of thread.
4. Draw freehand proportionate sketches of the following forms of screw threads.

   1. Acme thread 2. Buttress thread 3. Whitworth thread 4. Unified internal and external threads 5. Sellers thread.

5. Enlist the various types of nuts you know and draw their freehand sketches inserting proportionate dimensions.

6. What is prevention of rotation of bolt? Draw different types of such bolts.

7. What do you understand by locking arrangements of nuts? Enlist and draw sketches of various locking arrangements of nuts, inserting proportionate dimensions.

8. Explain with sketch the use of set-screw. Draw different varieties of heads and ends of set screws.

9. When is stud preferred to bolt? Draw different varieties of studs showing their uses.

10. What are keys? Classify them. Draw freehand sketches of the following keys.

1. Round key 2. Woodruff key 3. Feather key 4. Gib headed sunk key.

11. What is the functional difference between the cotter joint and the knuckle joint? Draw the following freehand sketches inserting necessary proportionate dimensions.

1. Socket and spigot cotter joint
2. Sleeve cotter joint.
3. Knuckle joint.

12. Prepare freehand sketches of the following, inserting proportionate dimensions.

1. Rivet head forms
2. Double riveted chain type lap joint
3. Triple, riveted zigzag type lap joint
4. Double riveted single cover chain butt joint
5. Double riveted unequal double cover chain butt joint
6. Triple riveted equal cover chain butt joint.

13. Draw symbols used for welded joints.

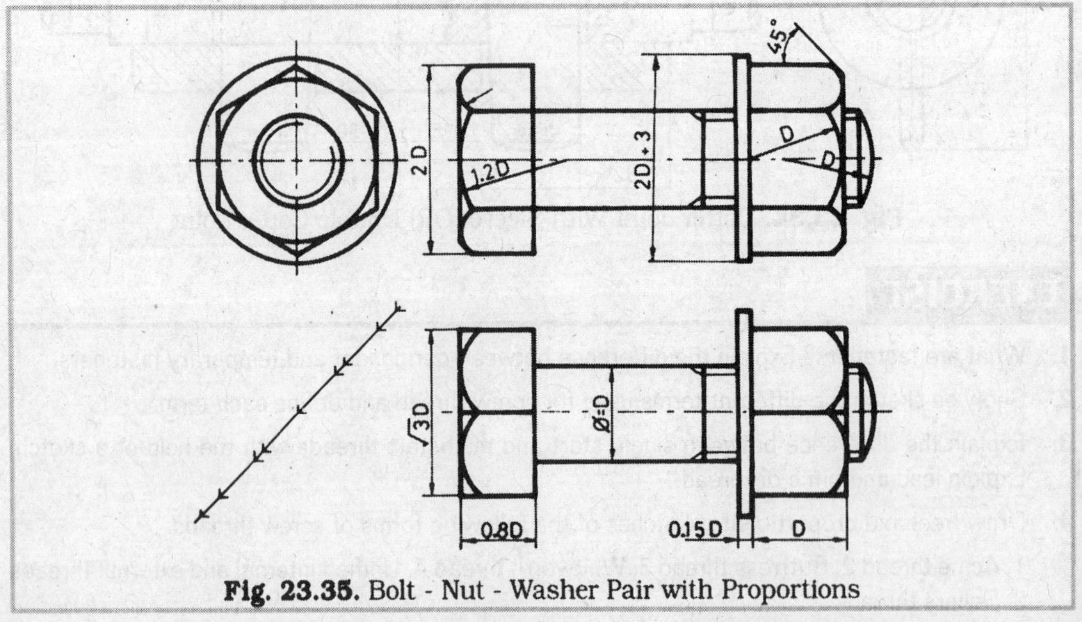

Fig. 23.35. Bolt - Nut - Washer Pair with Proportions

CHAPTER - 24

Shaft Couplings

1. INTRODUCTION

Shaft couplings are used to connect two shafts for the purpose of transmission of motion and power. e.g. an electric motor is connected to a centrifugal pump by coupling so as to drive it. Couplings are classified as : (i) Rigid or Fast Couplings. (ii) Flexible Couplings (iii) Loose or Disengaging Couplings. (iv) Non-aligned Couplings.

2. RIGID OR FAST COUPLINGS : (Fig. 24.1 to 24.6)

Rigid couplings are used to connect only co-axial shafts. This coupling provides rigid connection between two shafts and does not allow any relative motion between shafts.

Rigid couplings are further sub-classified as under:

(a) Muff Couplings - (i) Box or Muff Couplings : Fig. 24.1

(ii) Half Lap Muff Couplings : Fig. 24.2 (iii) Split Muff Couplings : Fig. 24.3

(b) Flanged Couplings - (i) Flanged Couplings : Fig. 24.4

(ii) Protected Type Flanged Couplings : Fig. 24.5 (iii) Solid or Forged Flanged Couplings : Fig. 24.6

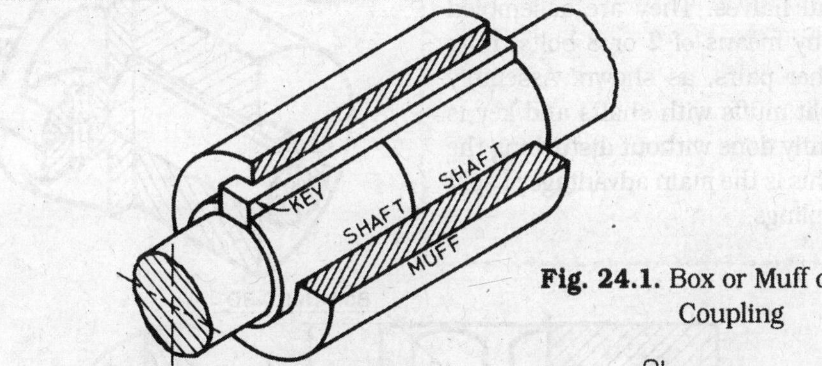

Fig. 24.1. Box or Muff or Sleeve Coupling

2. (a) (i) BOX OR MUFF COUPLINGS: Fig 24.1

It consists of cast iron cylindrical sleeve called "Muff". Two shafts are butt against each other in the middle of the muff, as shown. A key is driven in the keyway between the muff and the shaft. A sunk key is inserted on both the shafts.

2. (a) (ii) HALF LAP MUFF COUPLINGS : (Fig. 24.2)

Two shafts are in overlapped position inside the muff instead of butt position. The taper provided in the overlap prevents the axial separation of shafts. Hollow saddle key is used. Overlap portion helps in power transmission.

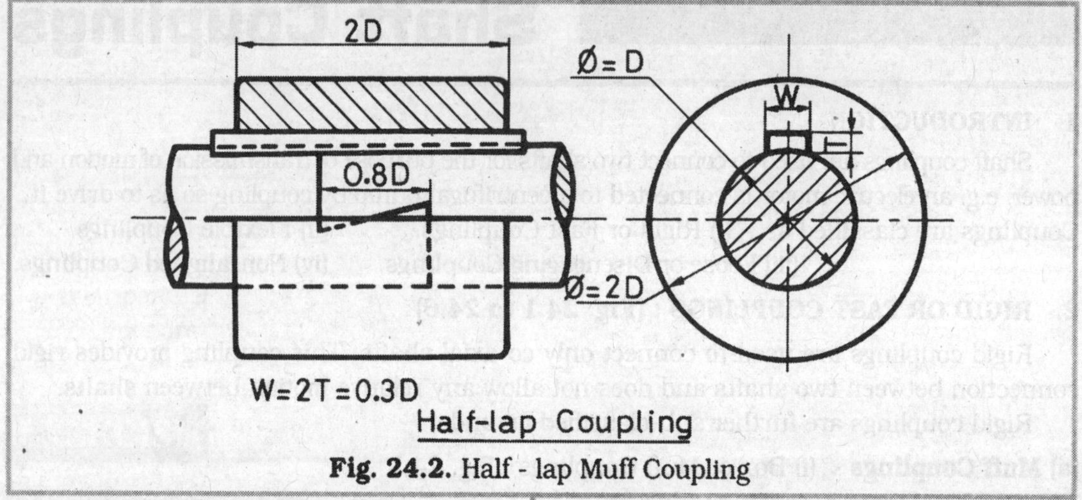

Fig. 24.2. Half - lap Muff Coupling

2. (a) (iii) SPLIT MUFF COUPLINGS : (Fig. 24.3)

Muff is made in two equal semi-cylindrical halves. They are assembled together by means of 2 or 3 bolts, nuts and washer pairs, as shown. Assembly of two split muffs with shafts and key is conveniently done without disturbing the set-up. This is the main advantage of split muff couplings.

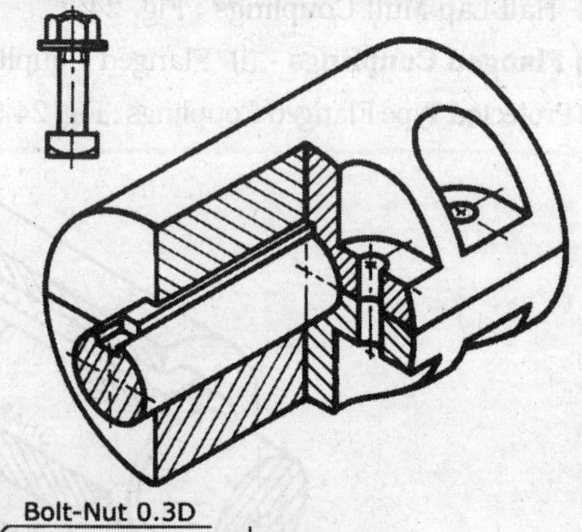

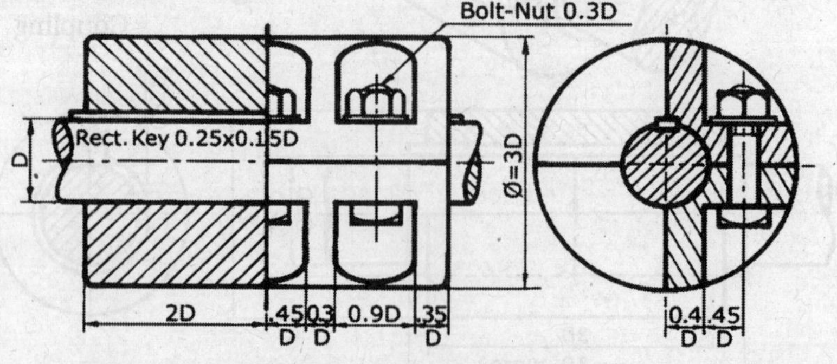

Fig. 24.3. Split Muff Coupling

2. (b)(i) FLANGED COUPLINGS (UNPROTECTED) : (Fig. 24.4)

This is very widely used coupling. Two similar cast iron flanges are keyed to the two ends of shafts. Taper sunk keys are commonly used. For the purpose of alignment, cylindrical portion of one flange is fitted into a corresponding recess of another flange. Bolt-Nut pairs, 4 to 6 in numbers, are used to connect two flanges.

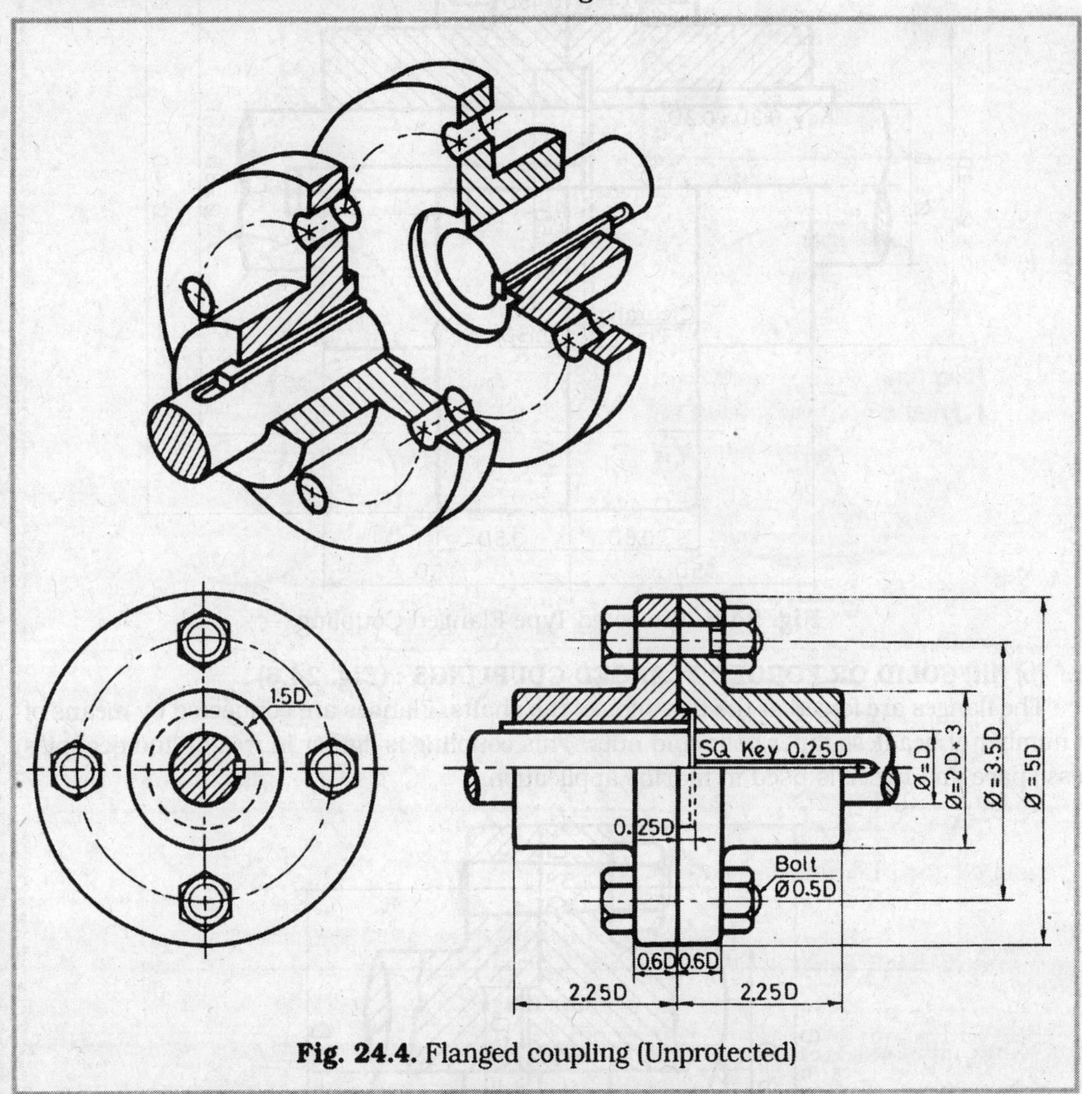

Fig. 24.4. Flanged coupling (Unprotected)

2. (b) (ii) PROTECTED TYPE FLANGED COUPLINGS : (Fig. 24.5)

In this coupling both the flanges are provided with annular projection just to cover bolt-head, on one side and nut on the other side. This arrangement prevents bolts and nuts from catching the clothes of workmen and thus ensures safety. All other things are similar to the flanged coupling.

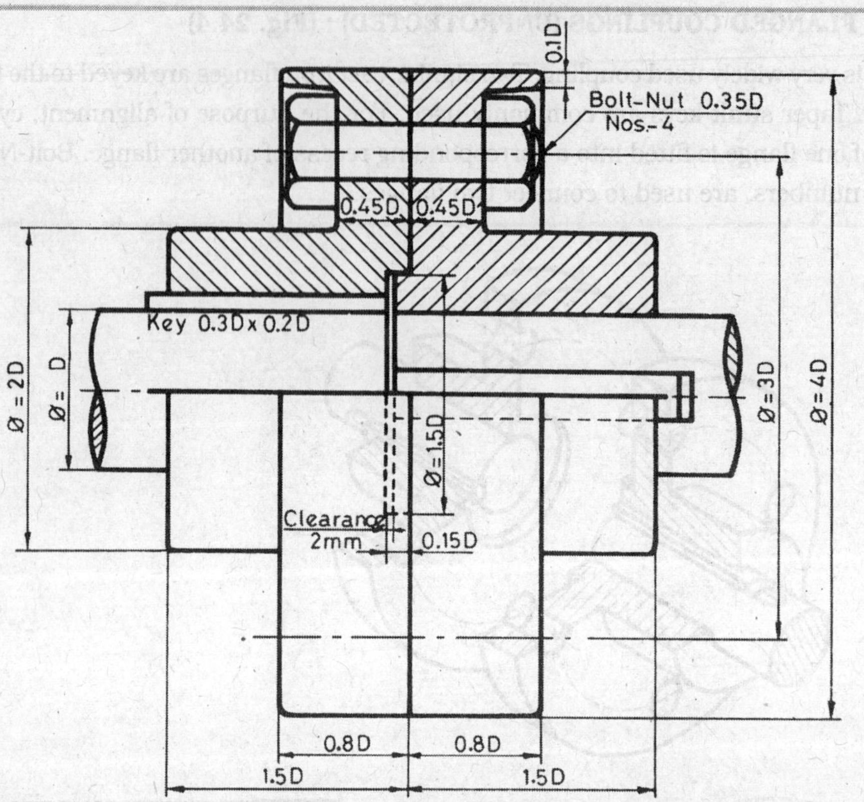

Fig. 24.5. Protected Type Flanged Coupling

2. (b) (iii) SOLID OR FORGED FLANGED COUPLINGS : (Fig. 24.6)

The flanges are forged at the two ends of the shafts. Flanges are connected by means of a number of headless taper bolts and nuts. This coupling is lighter in weight and occupies less space and hence is used in marine application.

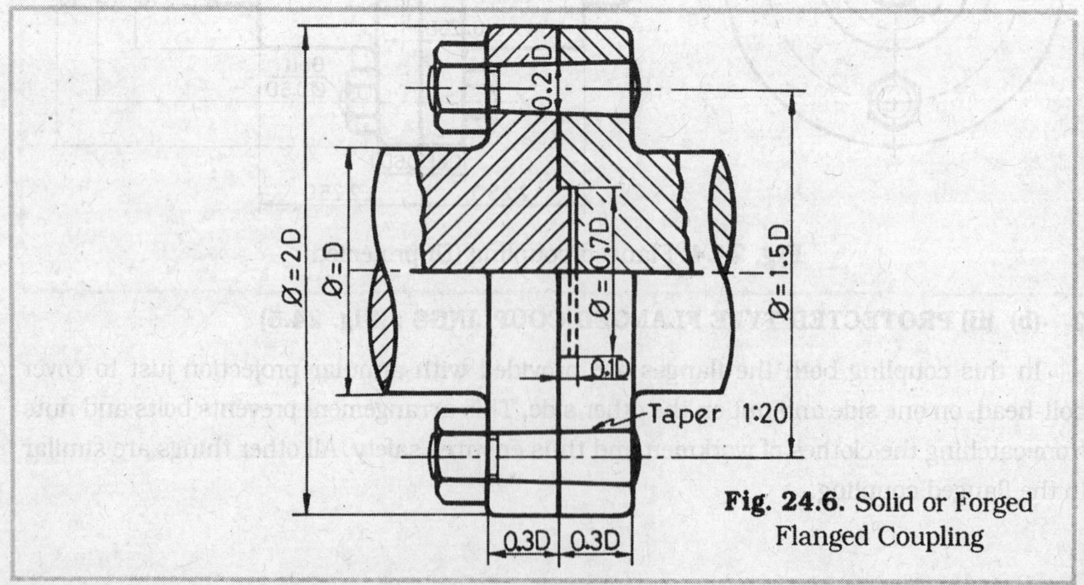

Fig. 24.6. Solid or Forged Flanged Coupling

3. PIN-TYPE FLEXIBLE FLANGED COUPLINGS :(Fig. 24.7)

A flexible coupling allows variation in the alignment of shafts and permits limited amount of relative motion. Rubber or leather washers, as shown in the figure, are used to absorb shocks. These washers also act as insulators.

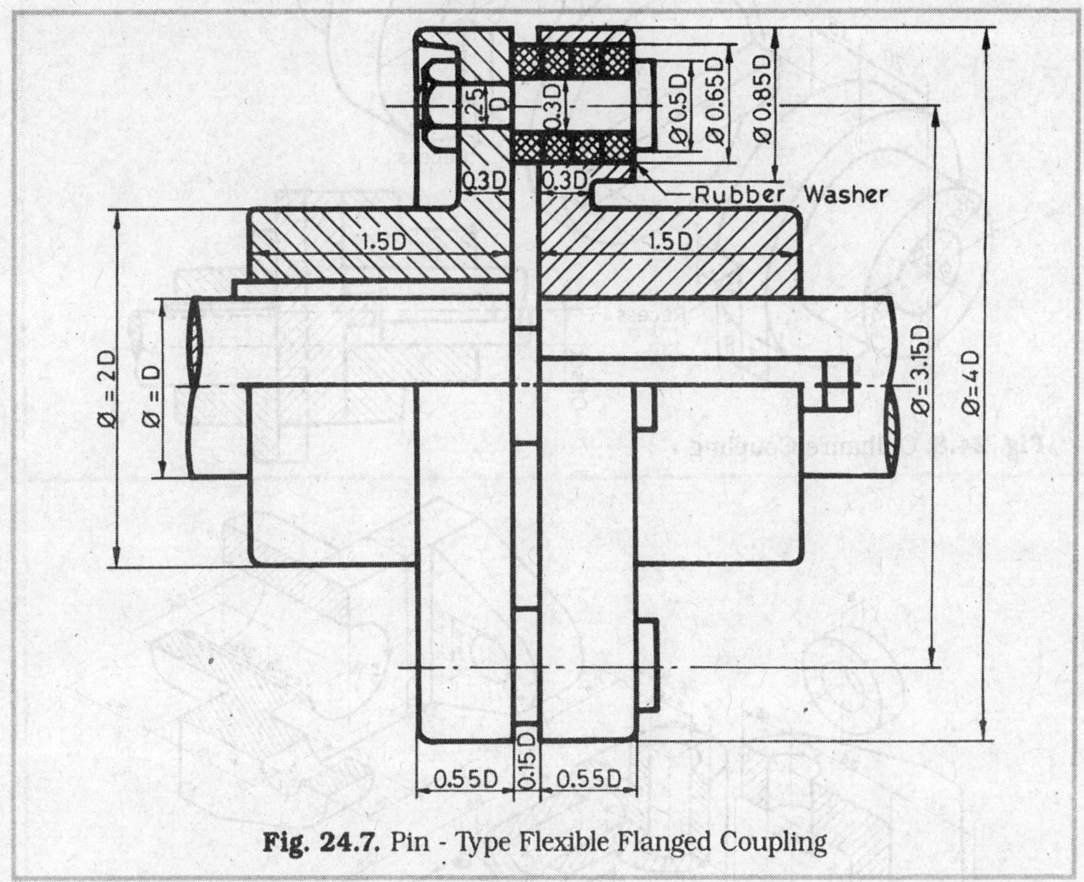

Fig. 24.7. Pin - Type Flexible Flanged Coupling

4. LOOSE OR DISENGAGING COUPLINGS :

In this coupling shafts can be engaged and disengaged at will. Following are the couplings of this type :

(i) Claw Couplings

(ii) Friction Clutches

(iii) Dog Clutches, etc.

Illustrations of the above couplings are not given.

5. NON-ALIGNED COUPLINGS (OLDHAM'S COUPLINGS) : (Fig. 24.8)

This type of coupling is used for connecting two shafts whose axes are not co-axial but are parallel. Oldham's coupling is such a coupling. The two mutually right-angled tongues, provided on two opposite sides of central circular disc, slide inside corresponding recesses provided on two flanges, keyed on two shafts. The figure is self explanatory.

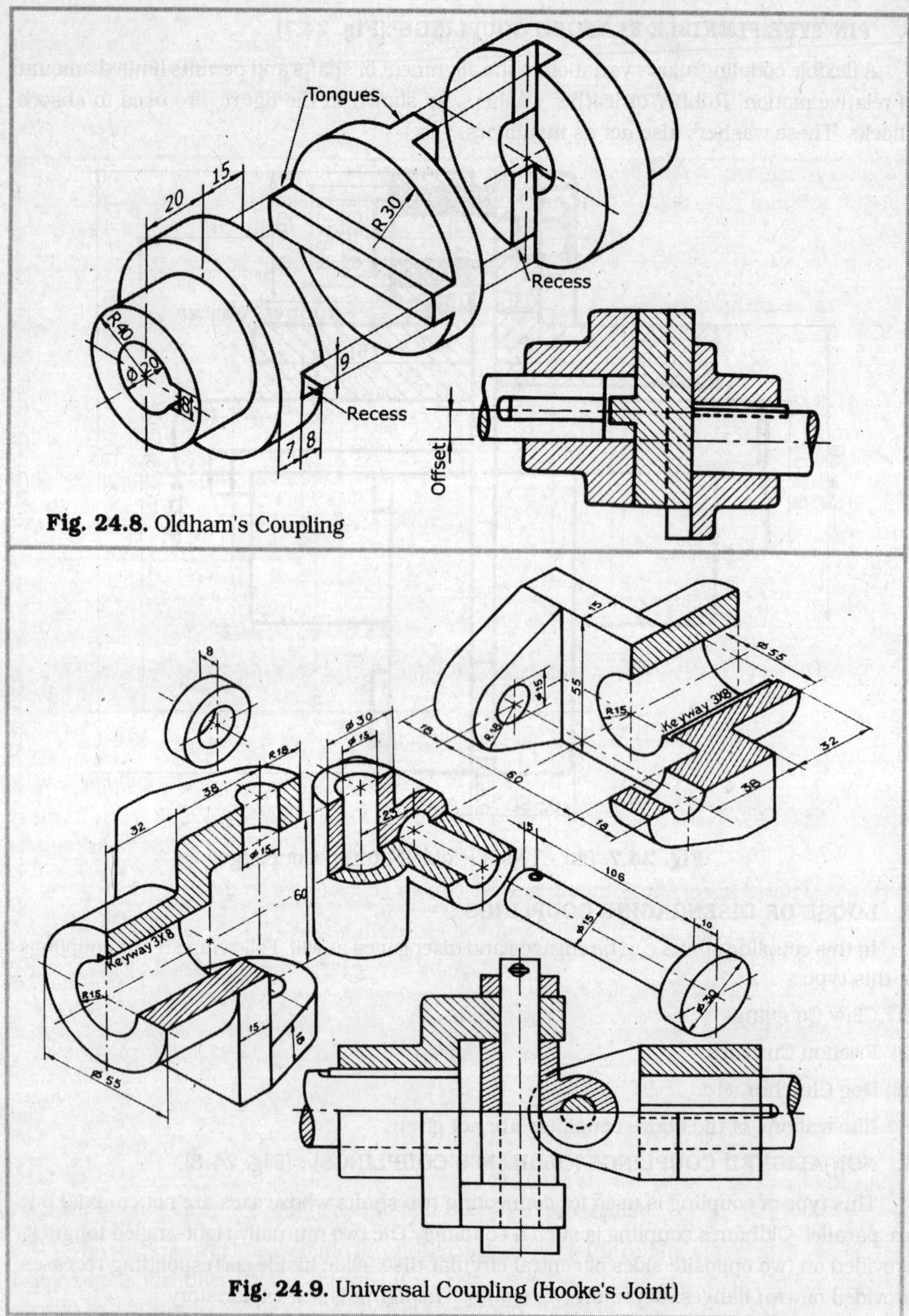

Fig. 24.8. Oldham's Coupling

Fig. 24.9. Universal Coupling (Hooke's Joint)

6. UNIVERSAL COUPLINGS (HOOKE'S JOINT) : (Fig. 24.9)

This coupling is used to connect two shafts whose axes intersect at an angle. Variation in angle between shafts is permitted even during rotation. This type of couplings are used in automobiles and machine tools. See figure on page 632.

EXERCISE

1. Classify couplings?
2. List different types of couplings you know, stating special features and use of each one.
3. What is the difference between rigid and flexible couplings? Where are they used?
4. Name and sketch the coupling used for coupling parallel shafts, having smaller distance between them. Give proportionate dimensions.
5. Name and sketch the coupling used for joining two shafts at some angle. Give proportionate dimensions.
6. Name and draw sketches of all types of muff couplings showing clearly the difference between them. Give proportionate dimensions.
7. Name and draw sketches of all types of flanged couplings showing clearly the difference between them. Give proportionate dimensions.

CHAPTER - 25

Bearings

1. INTRODUCTION :

The function of bearings is to support rotating shafts. A shaft rotates inside the bearing practically with zero friction.

The bearings are classified in two different manners.

(1)	(i)	A Journal Bearing	[Fig. 25.1(a)]
	(ii)	A Pivot Bearing	[Fig. 25.1(b)]
	(iii)	A Collar or Thrust Bearing	[Fig. 25.1 (c)]
(2)	(i)	Sliding Bearings	[Fig. 25.2 to 25.6]
	(ii)	Rolling or Antifriction Bearings	[Fig. 25.7]

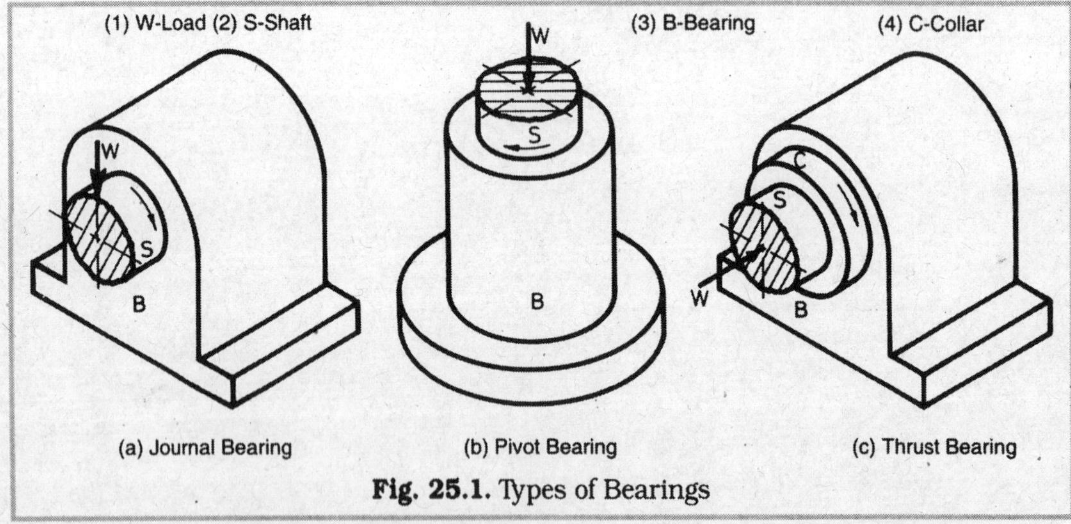

Fig. 25.1. Types of Bearings

2. A JOURNAL BEARING : [Fig.25.1 (a)]

A bearing, in which the load on the shaft is perpendicular to the axis, is known as journal bearing. See Fig. 25.1(a). The portion of the shaft rotating within the bearing is known as journal.

Some of the journal bearings are :

(i) Solid Bearing - Fig. 25.2

(ii) Bushed Bearing - Fig. 25.3

(iii) Pedestal Bearing or Plummer Block Bearing - Fig. 25.4

2. (i) SOLID BEARINGS : [Fig. 25.2]

A solid bearing is the simplest kind of a journal bearing. It consists of a cast iron block with a hole to accomodate the journal (shaft). Oil hole is drilled at the top for the purpose of lubrication. A rectangular base plate or the sole has two holes for bolting down the bearing in the position. This type of bearing is used only for light load and for slower speeds, since there is no provision for adjustment on account of wear.

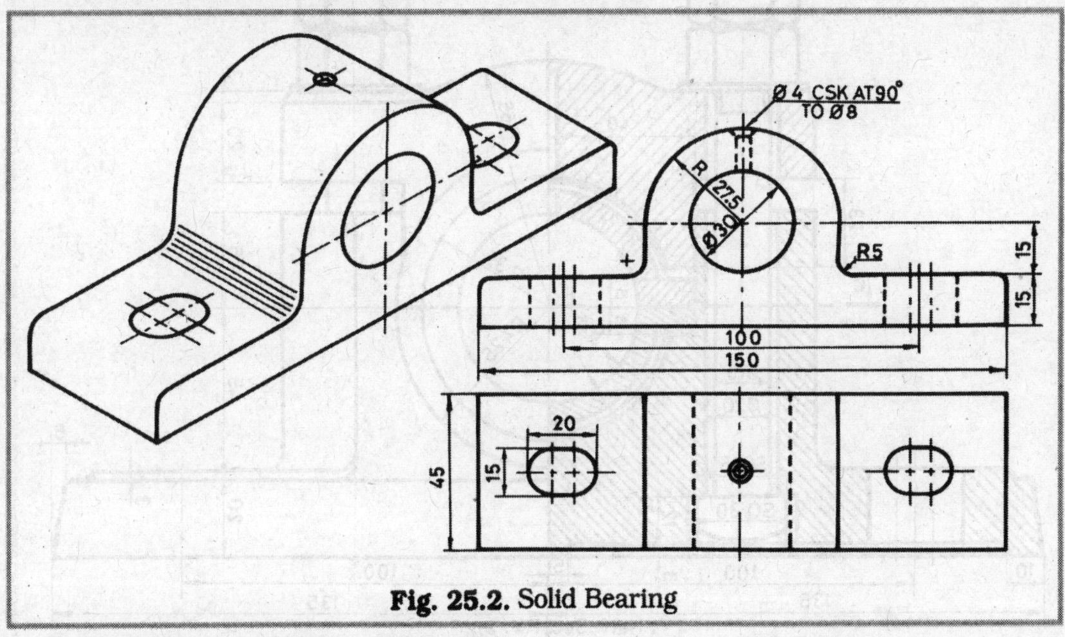

Fig. 25.2. Solid Bearing

2. (ii) BUSHED BEARINGS : [Fig. 25.3]

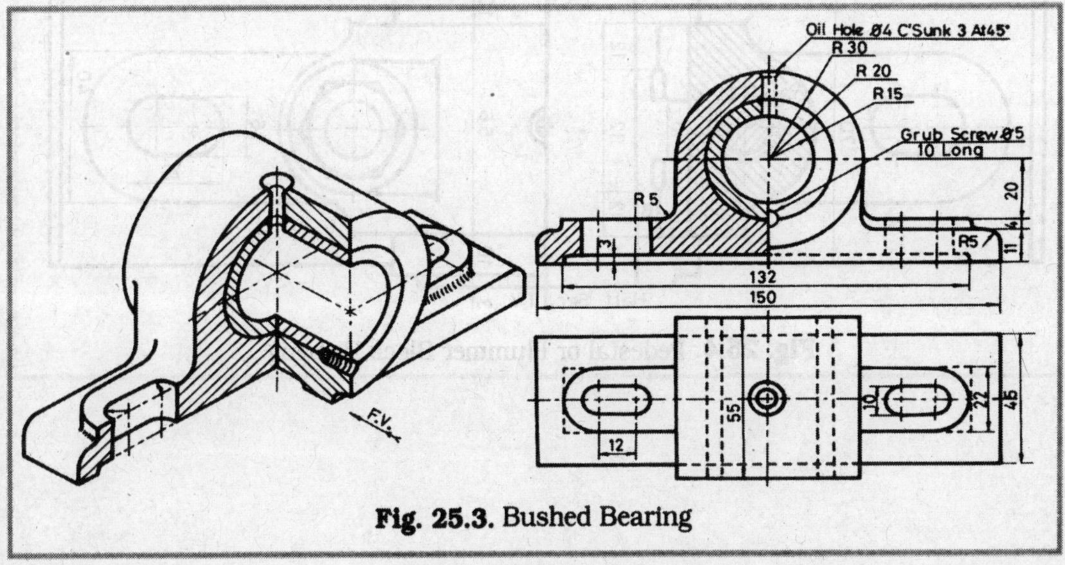

Fig. 25.3. Bushed Bearing

It is a modified form of solid bearing. Brass, phospher bronze or gun metal bush is press-fitted inside the bearing. Bush is hollow cylindrical piece forced or press-fitted inside the solid bearing. The journal or the shafts rotates in the bush. Wear and tear takes place only on bushes and hence periodically they are replaced. Grub-screw or a dowel-pin prevents rotation of bush along with rotation of shaft.

In the above two bearings, insertion of the shaft in the bearing is endwise only.

2. (iii) PEDESTAL BEARINGS OR PLUMMER BLOCK BEARINGS : [Fig. 25.4]

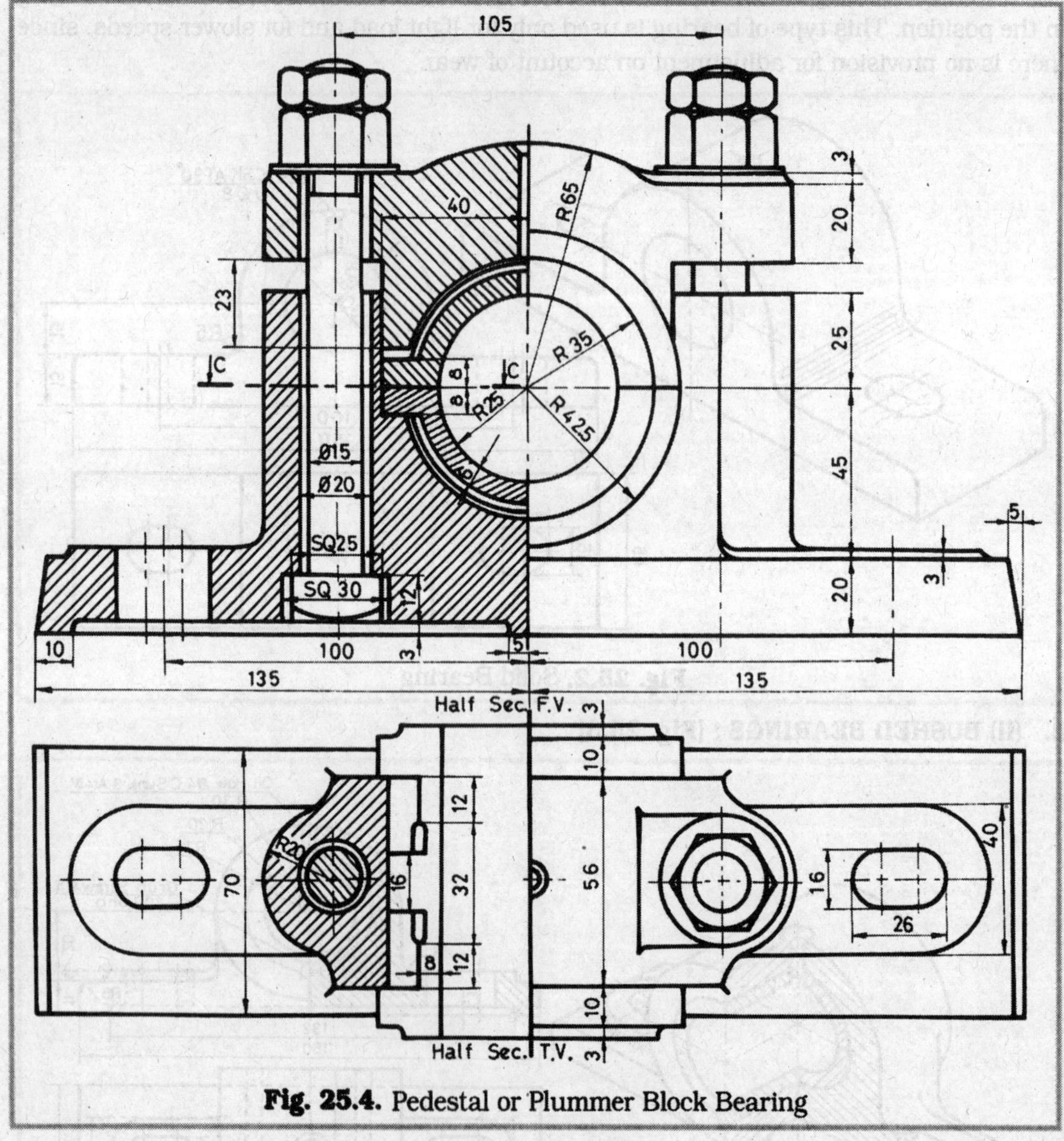

Fig. 25.4. Pedestal or Plummer Block Bearing

It consists of mainly,

(1) A cast iron pedestal with a sole,

(2) Gunmetal or phospher bronze brasses in two halves.

(3) A cast iron cap,

(4) Two pairs of square headed bolt, nut and washer made of mild steel.

Provision of separate cap and brasses in two halves facilitates easy placing and removal of the shaft in and from the bearing. Adjustment for wear can also be easily done in this bearing.

Brasses are prevented from rotating along with shaft by any one of the methods discussed in the later part of this chapter. Collars at either end on brasses prevent axial movement of brasses inside the bearing.

Elliptical bolt holes in the sole permit minor adjustment in the position of bearing. Square headed bolts seat in square recesses at the bottom of the pedestal and hence it does not rotate when the nut is being tightned. A gap purposely kept between the cap and the main body, permits adjustment for wear and tear.

These types of bearings are widely used for heavy loads, higher speeds and larger sizes.

3. METHODS OF PREVENTION OF ROTATION OF BRASSES IN A BEARING : [Fig.25.5]

Brasses are prevented from rotating along with the rotating shaft by adopting one of the methods as shown in the Fig. 25.3 and the Fig. 25.5

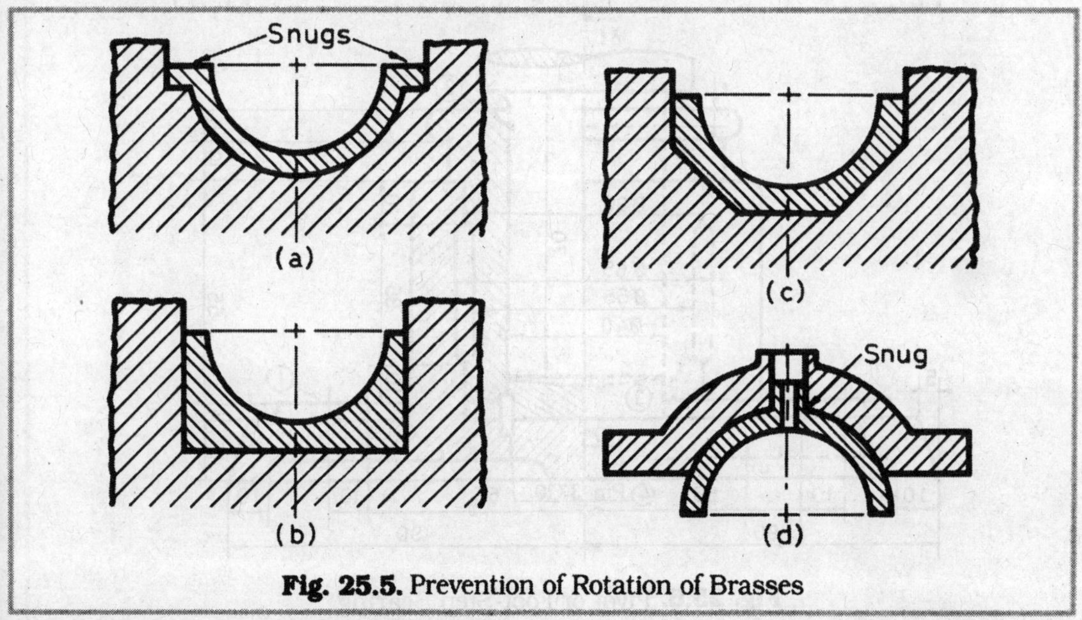

Fig. 25.5. Prevention of Rotation of Brasses

(1) By a grub-screw as shown in the Fig. 25.3 - see Fig. 25.3

(2) By snugs at the sides - see Fig. 25.5(a)

(3) By a snug at the top - see Fig. 25.5(d).

(4) By making outer shape of one of the brasses rectangular - see Fig. 25.5(b)

(5) By making outer shape of one of the brasses octagonal - see Fig. 25.5(c)

Outer rectangular or octagonal shape of the brasses may be kept for entire length or may be kept at the ends only.

Brasses are made thicker where they are likely to wear more. Thicker brasses, at the place of wear, take care of wear and tear and hence they are used for longer duration.

4. PIVOT BEARINGS : [Fig. 25.1(b) and Fig. 25.6)

These bearings are used for vertical shafts with axial loads. The bottom surface of the shaft end rests on the surface of the bearing. See Fig. 25.1(b) and Fig 25.6.

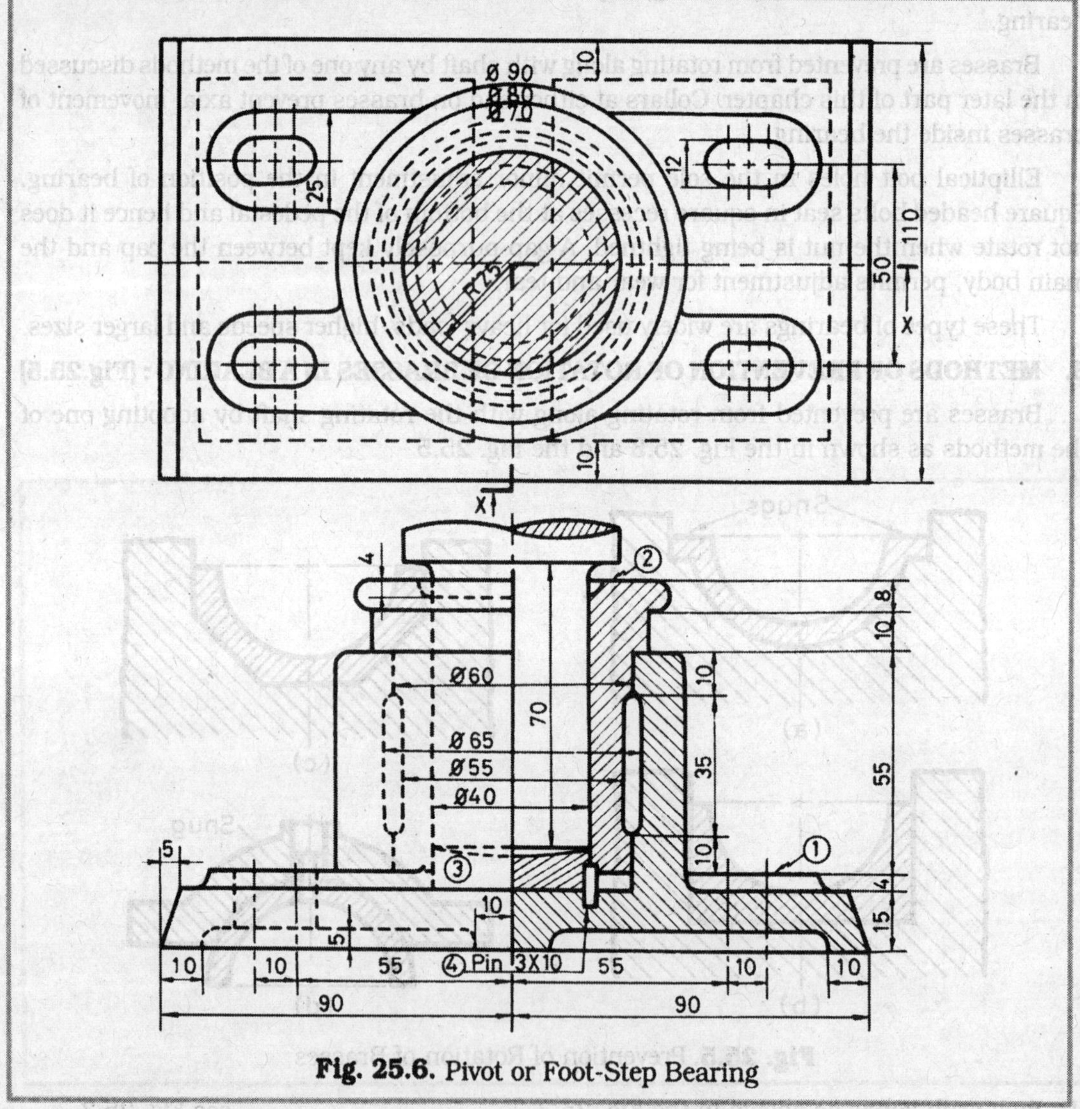

Fig. 25.6. Pivot or Foot-Step Bearing

It consists of mainly,

(1) A cast iron block with sole

(2) A gun metal bush with a collar at the top.

(3) Concave steel disc placed at the bottom in the bush.

(4) M.S. pin half in the disc, half in the bush and half in the housing located off centre.

Concave steel disc takes axial thrust and the bush takes the radial load. M.S. pin prevents, the disc and the bush both from rotating along with the rotating shaft. The collar of the bush serves as an oil cup. Oil is used for the purpose of lubrication. Soles in most of the bearings are recessed at the bottom to reduce machining.

5. ANTI-FRICTION BEARINGS : (Fig. 25.7)

So far we have studied sliding bearings. Friction in sliding bearings is much more than rolling bearings. Anti-friction bearings have rolling friction. Ball and roller bearings are Anti-friction bearings. Depending upon the direction of load acting on the shaft, anti-friction bearings are classified as journal bearings and thrust bearings.

Antifriction bearing consists of mainly,

(1) An inner race (ring)

(2) An outer race (ring)

(3) Balls or rollers (Rolling elements of various shapes)

(4) Cage or separator.

The outer race remains stationary in the housing of a machine body and inner race rotates along with the shaft. Balls and rollers roll inside the groove between inner race and outer race. The fit between the shaft and the inner race is force fit.

Radial bearings and thrust bearings are further classified according to the shape of the elements, such as :

(1) Ball bearings,

(2) Cylindrical roller bearings,

(3) Spherical roller bearings,

(4) Taper roller bearings, etc.

Most of the radial bearings can carry a thrust load and some of the thrust bearings can carry a radial load and, therefore, from the point of view of the acting forces no clear division is done.

Deep groove ball bearings are suitable for greater radial loads and smaller axial loads. Roller bearings have greater load carrying capacity than ball bearings since rollers have line contact with rings(races). Cylindrical roller bearings are not suitable for the axial loads. Roller bearings are cheaper than ball bearings. Taper roller bearings are suitable for radial as well as axial load. Taper roller bearings are widely used for automobiles, machine tools, etc. Thrust bearings have loose races or rings grooved to accomodate the balls or rollers. Thrust ball bearings are used for the axial loads. If the thrust is likely to occur in both the directions double direction ball thrust bearing is used.

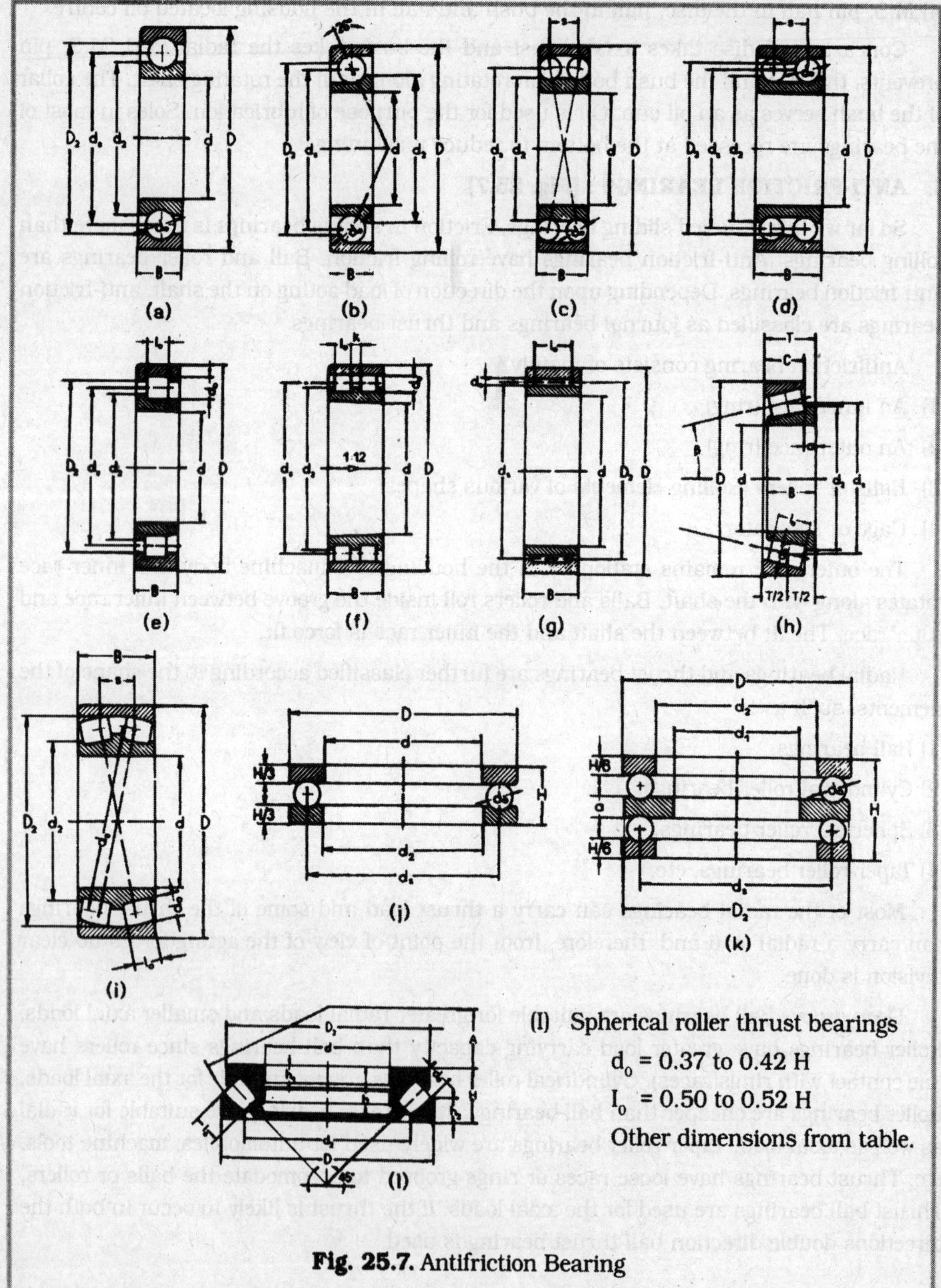

Fig. 25.7. Antifriction Bearing

(a) Single row deep groove ball bearings

$d_s = 0.5 (D + d)$

$d_0 = 0.3 (D - d)$

$d_2 = d_s - 0.6 d_0$

$D_2 = d_s + 0.6 d_0$

(b) Single row angular contact ball bearings

$d_s = 0.5 (D + d)$

$d_0 = 0.3 (D - d)$

$d_2 = d_s - 0.6 d_0$

$D_2 = d_s + 0.6 d_0$

(c) Double row self-aligning ball bearings

$d_s = 0.5 (D + d)$

$d_0 = 0.25 (D - d)$

$d_2 = d_s - 0.8 d_0$

$S = B - d_0$

(d) Double row angular contact ball bearings

$d_s = 0.5 (D + d)$

$d_0 = 0.28 (D - d)$

$d_2 = d_s - 0.6 d_0$

$D_2 = d_s + 0.6 d_0$

$S = 0.5 B$

(e) Single row cylindrical roller bearings

$d_s = 0.5 (D + d)$

$d_0 = 0.25 (D - d)$

$I_0 = 1 \text{ to } 1.5 d_0$

$d_2 = d_s - 0.6 d_0$

$D_2 = d_s + 0.6 d_0$

(f) Double row cylindrical roller bearings with tapered bore

$d_s = 0.5 (D + d)$

$d_0 = 0.23 (D - d)$

$I_0 = d_0$

$d_2 = d_s - 0.6 d_0$

$K = 0.3 d_0$

(g) Single row needle roller bearings

$d_s = 0.5 (D + d)$

$d_0 = 0.1 (D - d)$

$I_0 = 0.7B$

$D_2 = d_s - 0.6 d_0$

(h) Single row tapered roller bearings

$d_s = 0.5 (D + d)$

$d_0 = 0.25 (D - d)$

$I_0 = 0.8 \text{ to } 1 C$

$d_2 = 0.5 (D + d)$

$\alpha = 0.2 C$

$\beta = 12°$

(i) Double row spherical roller bearings

$d_s = D_2 - d_0$

$d_0 = 0.22 \text{ to } 0.25 (D - d)$

$I_0 = 0.38B$

$D_2 = 0.5 (D + d) + d_0$

$d_2 = 0.45 (D + d)$

$S = 0.53 B$

(j) Single direction ball thrust bearings

$d_s = 0.5 (D + d)$

$d_0 = 0.36 (D - d)$

d_2 = indicated in respective Bearing dimension tables

(k) Double direction ball thrust bearings

$d_s = 0.5 (D + d_2)$

$d_0 = 0.35 (D - d_2)$

α indicated in respective bearing dimension

d_1 tables

d_2

Some of the rolling bearings are as under :

1. Single Row Deep Grooved Ball Bearings - Fig. 25.7(a)
2. Single Row Angular Contact Ball Bearings - Fig. 25.7(b)
3. Double Row Self-aligning Ball Bearings - Fig. 25.7(c)
4. Double Row Angular Contact Ball Bearings - Fig. 25.7(d)
5. Single Row Cylindrical Roller Bearings - Fig. 25.7(e)
6. Double Row Cylindrical Roller Bearings With - Tapered Bore - Fig. 25.7(f)
7. Single Row Needle Roller Bearings - Fig. 25.7(g)
8. Single Row Tapered Roller Bearings - Fig. 25.7(h)
9. Double Row Spherical Roller Bearings - Fig. 25.7(i)
10. Single Direction Ball Thrust Bearings - Fig. 25.7(j)
11. Double Direction Ball Thrust Bearings - Fig. 25.7(k)
12. Spherical Roller Thrust Bearings - Fig. 25.7(l)

EXERCISE

1. Classify bearings? Draw sketches of each type.
2. What is a bush? Why it is used in bearing?
3. What are brasses ? Why are they used in bearings?
4. Explain the method of prevention of rotation of brasses in bearings with illustration.
5. Draw or sketch following bearings for a shaft of 50mm diameter.
   (i) Solid bearings
   (ii) Bushed bearings
   (iii) Plummer block bearings
   (iv) Pivot bearings
6. What are antifriction bearings? How do they differ from the sliding bearings?
7. List different types of sliding and rolling bearings stating special characteristics of each type.

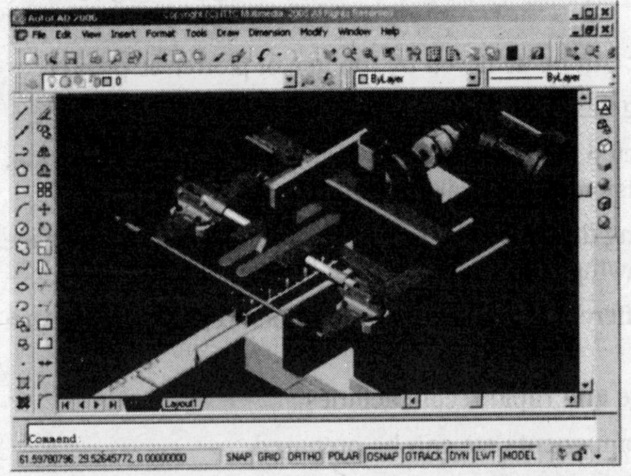

CHAPTER - 26

AutoCAD

Introduction :

The AutoCAD design package is a powerful tool to prepare desired drawings. According to the instructions given by mouse/key board etc it prepares drawings to the scale. AutoCAD prepares exact, precise, neat and clean drawings. Drawings prepared by AutoCAD looks better than manually prepared drawings. AutoCAD drawings are accurate as compared to manually prepared drawings. AutoCAD provides facility, to correct errors, to increase or decrease the size, to store the drawing in disc, to send the drawings to other places by storing inside disc, to repeat drawings in as many numbers as you want, to rotate the drawings as you desire etc.

Fig. 26.1 Computer and Laptop

AutoCAD provides various commands like Line, Circle, Polygon, Ellipse, Arc, Zoom, Erase, Redraw, Copy, Chamfer, Mirror, Rotate, Hatch etc. Prepared drawings, on AutoCAD, can be plotted on paper with a pen plotter, dotmatrix printer, laser printer or inkjet printer. CAD is used in Industry, Business, Education, Research project etc.

Advantages of AutoCAD:

1. With the help of AutoCAD highly accurate, uniform in thickness quality, precise drawings can be prepared in comparatively smaller time than manually prepared drawings.
2. Drawing skill and drawing instruments are not essential. Hand skill is not required.
3. Prepared drawings by AutoCAD can be saved (OR) preserved for long time in the internal memory of hard disc in computer. These drawings can be opened as and when required. It can be transferred in floppy and moved to any place as and when necessary.

4. Prints, of these drawings, on paper can be achieved by the use of plotter or printer.
5. AutoCAD has in built dimensioning facilities.
6. Units of measurements can be selected in AutoCAD. e.g. Decimal system, Engineering, Architectural etc. Standard of precision in the drawing can be also varied.
7. Storage space required for preserving drawings is comparatively very small than space required for manually prepared drawings.
8. Drawing in different colours and different line thicknesses can be prepared by the use of layer command of AutoCAD.
9. The computer has unlimited storage and rapid recall facilities.
10. Visual models of various objects, components etc can be prepared.
11. Colour graphics help us in highlighting important features.
12. No scrap drawings. No paper wastage.
13. It helps in increasing the productivity.
14. Multicolour display makes visualization more effective.
15. Hatch command facilitates faster uniform hatching in the hatched area of the object.
16. Template can be manufactured by AutoCAD drawing.
17. AutoCAD helps in finding out area and perimeter of 2D drawing and volume of 3D drawing.
18. Alterations and improvements in the drawing can be easily done on screen.
19. Objects can be viewed from various angles. For example from the car body 3D drawing, how it will look from front, rear, side, top and bottom can be visualised.
20. Sectional drawings can be prepared from full body 3D drawings very easily.
21. Mirror command can give the mirror image of the prepared drawing easily. e.g. In flat scheme if 4 similar flats are on the same floor then it's drawing can be prepared by preparing drawing of only one flat. This is impossible in manual drawings.
22. Scale command can increase or decrease the size of drawing to definite scale very easily. Which is impossible in manual drawings.
23. Copy command facilitates to copy same drawing at various locations. e.g. 100 windows and 50 doors of same size and same details are drawn in the drawing of multistorey building by copy command very easily, which is impossible in manual drawings.
24. Divide command divides the length and circumference into required number of equal parts. This is useful in preparing the cutting size of marble tiles to be fitted in a particular size room. Entire room will have same size marble tiles. Equal divisions along circumference will help in circular design. e.g. Cricket stadium seating arrangements, marble fittings in circular designs etc.
25. Array command facilitates copying of selected object in rectangular and circular format.
26. Computer aided manufacturing is also possible.

Computer System:-

Computer system has main two elements.

(1) Computer Hardware

(2) Computer Software

These two elements work together to make up a computer system.

(1) Computer Hardware has

(i) Input units

(A) Key Board

(B) Mouse

(C) Joystick

(D) Digitizer

(E) An Image Scanner

(ii) Central Processing and Memory Units

(A) Arithmatic & Logic Unit (ALU)

(B) Control Unit

(C) Registers

(iii) Output Devices

(A) Monitor

(B) Printer and

(C) Plotter

There are two types of memories used in computers.

(i) Primary storage (or) Main memory (or) Core memory

(ii) Secondary storage (or) Auxiliary memory.

Read Only Memory (ROM) and Random Access Memory are the two types of primary storage. Floppy disc, Hard disc and magnetic tape are secondary storage devices.

Floppy disc is a thin disc of plastic, coated with magnetic oxide. Hard disc is a non removable, non flexible storage device having a large storage capacity.

A Keyboard:-

It is a text input device.

Keyboard is needed to type required words, commands etc. Function keys has different functions to perform as given below.

Key Board contains:

Fig. 26.2. Key Board

(i) Alphanumeric keys : Alphabetic, numeric and symbol keys are arranged in a pattern similar to the normal typewriter. Shift key is used for capital & lower case letters. A separate numeric keyboard is provided on the right side of keyboard.

(ii) Function keys has different functions to perform. F - I to F - II.

(iii) Cursor control keys : There are four cursor control keys. They are useful in editing programs, controlling cursor in graphic applications etc.

(iv) Special keys are

(1) Control (2) Alt (3) Number lock (4) Page up (5) Page down (6) Print etc.

Functional Keys:-

Functions of various functional keys, mounted on key board, are as under.

Key	Function
F1	It provides online help
F2	Command window on and off
F3	Osnap on and off
F4	Tablet on and off
F5	Isoplane Toggle
F6	Co-ordinates on and off
F7	Grid on and off
F8	Ortho mode on and off
F9	Snap mode on and off
F10	Polar tracking on and off
F11	Objects snap tracking on and off
ESC	To cancel any command
↵	Enter.

A Mouse:

Mouse is a input device attached to the computer. It is a small box having three buttons. It is moved on a horizontal flat surface, named as mouse pad. Instructions are passed on to the processor to quickly positioned the cursor on the screen of monitor. Mouse is a very widely used input device to-day.

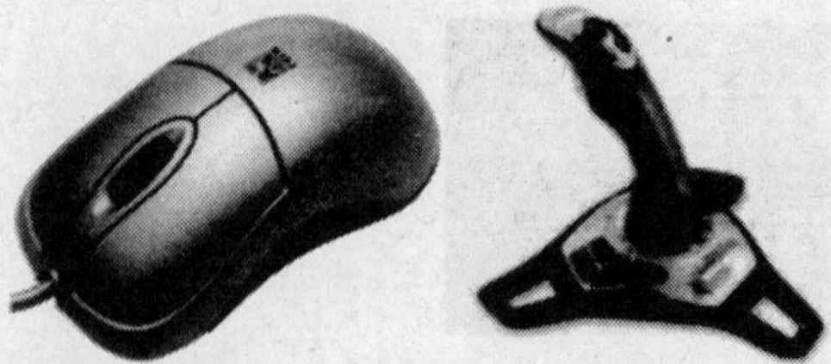

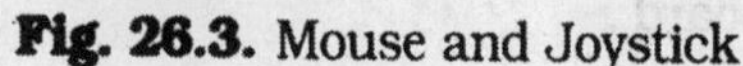

Fig. 26.3. Mouse and Joystick

Fig. 26.4. Digitizer

A Joystick :

Joystick consists of a box with toggle stick in the centre. Joystick feed signals to the computer. Operator should be able to co-ordinate cursor movement on screen and hand movement on the Joystick. Side movement produces change in x-co-ordinate and front-back movement produces change in y-co-ordinate. The cursor appears as a bright spot on the screen.

A Digitizer:

It's a electromagnetic graphic input device. It is used to copy existing drawing. Operator moves the stylus on the flat surface of the table. This motion is converted into digital signal. Ultimately graphic is generated corresponding to movement of the stylus.

An Image Scanner:

A scanner convert a picture or a text on a paper into a similar graphic in the computer.

Graphical quantities are inserted in the computer by digitizer, scanner etc.

A Monitor:

A screen similar to television screen may be black and white or colour is used as output device. The resolution of the display is important. Finer resolution is better. Cost depends on resolution.

Fig. 26.5. Flatbed Scanner

Fig. 26.6. Monitor

Output Devices:

CAD system is complete only when at the end it gives us hard copies of the designs. One has to select equipment most suitable for the application.

Hard copies are sometimes used for the purpose of reviewing the designs also.

Selection of the equipment depends upon quality of hard copy, cost of hard copy, speed at which it can be achieved etc.

Hard Copy Units:

(1) Printers

(a) Dot Matrix Printer (b) Laser Printer (c) Inject Printer

1. (a) Dot Matrix Printer:

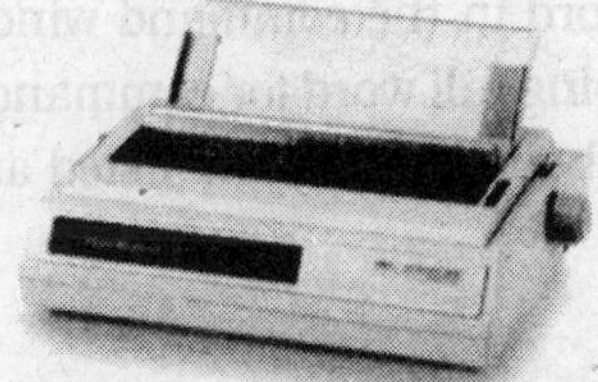

Fig. 26.7. Dot Matrix Printer

Fig. 26.8. Laser Printer

In this characters are made up of small dots. Small pins strike ribbon to form small dots. Multiple ribbons or single ribbon with different colour bands are used. Quality of copy is poor. It is used for high speed, low cost work.

1. (b) Laser Printer:

Laser beam is used for printing. It is one of the latest output device. Quality of printing is very good.

Printer parts do not touch the paper during printing. This laser printers are available with different dpi (Dot per inch) capacity.

1. (c) Inkject Printer:

Fig. 26.9. Inkjet Printer

Fig. 26.10. Plotter

Images are generated by propelling fine droplets of ink on to the printing medium, Droplets fixed on to the printing surface produces graphics.

(2) Plotters:

(a) Drum Plotters (b) Flat Bed Plotters

2. (a) Drum Plotters:

Paper is attached to the drum and pen is mounted on the slide. Drum rotates and the pen slides over it. Relative motion of the two generates graphics.

It is fast and less expensive. It makes drawing of limited width and unlimited length.

2. (b) Flat Bed Plotters:

Paper is mounted on a flat horizontal or vertical surface. Servo-controlled pens or stylus moves in the direction of two axes. Pen plotters are slow and requires more maintenance.

x-y directional movement of the pens generates required graphics.

Commands Aliases

To operate command one has to type command's word in the command window. This takes lot of time and it is inconvenient also. Instead of typing full word for command you can type it's short form named as commands aliases. Alphabetically they are listed as under.

1. A - Arc
2. AA - Area
3. AL - Align
4. AR - Array
5. B - Block
6. BH - Bhatch
7. BR - Break
8. C - Circle
9. CHA - Chamfer
10. COL - Colour
11. CO - Copy
12. D - Dimstyle
13. DAL - Dim aligned
14. DAN - Dim angular
15. DBA - Dim baseline
16. DCE - Dim centre
17. DCO - Dim continue
18. DDA - Dim disassociate
19. DDI - Dim diameter
20. DED - Dimedit
21. DI - Distance
22. DIV - Divide
23. DLI - Dim linear
24. DO - Donut
25. DOR - Dim ordinate
26. DRA - Dim radius
27. DST - Dimstyle
28. DT - D Text
29. E - Erase
30. EL - Ellipse
31. EX - Extend
32. EXT - Extrude
33. F - Fillet
34. H - Bhatch
35. HI - Hide
36. I - Insert
37. IN - Intersect
38. L - Line
39. LA - Layer
40. LI - List
41. LO - Layout
42. LT - Linetype
43. LTS - LTSCALE
44. LW - Lweight
45. M - Move
46. MI - Mirror
47. ML - Mlinc
48. O - Offset
49. OP - Options
50. OS - Osnap
51. P - Pan
52. PE - Pedit
53. PL - Pline
54. PO - Point
55. POL - Polygon
56. PS - Pspace
57. R - Redraw
58. RE - Regen

59. REC - Rectangle
60. REV - Revolve
61. RO - Rotate
62. SC - Scale
63. SHA - Shade
64. SL - Slice
65. SN - Snap
66. ST - Style
67. SU - Subtract
68. TR - Trim
69. UN - UNITS
70. UNI - Union
71. X - Explode
72. Z - Zoom

Different Methods of Selecting Object in AutoCAD: See Figures below.

For the purpose of editing, object/objects is/are selected in AutoCAD. Selected object will be seen as dotted. Methods are as under.

(1) By clicking the object with mouse, object is selected. Selected object will be seen dotted.

(2) Simple Window:

Simple window is of continuous line. First lower left corner is clicked and then upper right corner is clicked by means of mouse. Rectangular window is generated.

Objects, which are 100% in this window, are selected. If the slightest portion of the object is outside the window, the object is not selected.

(3) Crossing Dotted Window:

Crossing window is of dotted line. First upper right corner is clicked an then lower left corner is clicked by means of mouse. Rectangular dotted window is generated.

Objects, which are 100% inside the window as well as objects, which have touched the window are selected. 100% outside objects are not selected.

(4) By giving all option all objects are selected.

(5) If you want to select objects which were previously selected earlier give previous option by typing alphabet P.

(6) Remove option can remove unwanted selected object. Option is given by typing alphabet R.

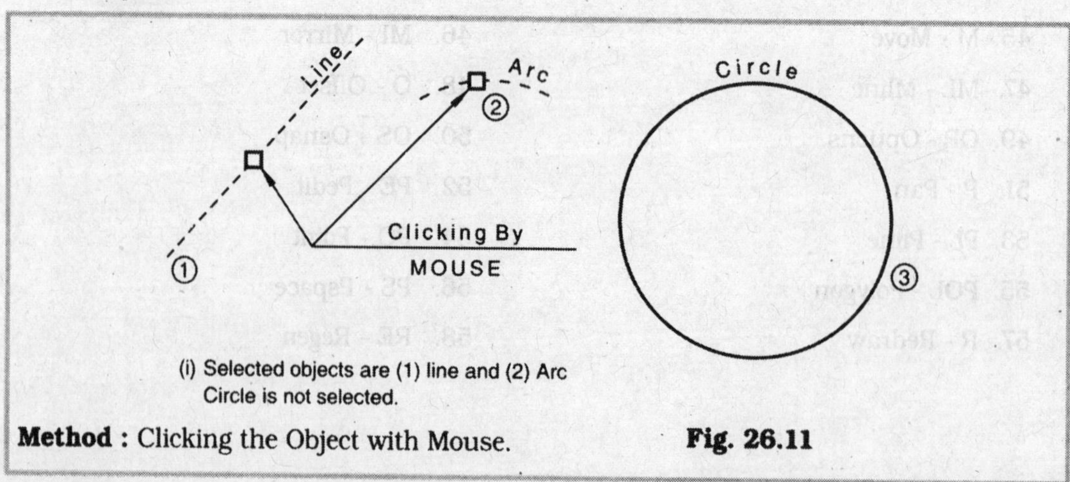

Method : Clicking the Object with Mouse. **Fig. 26.11**

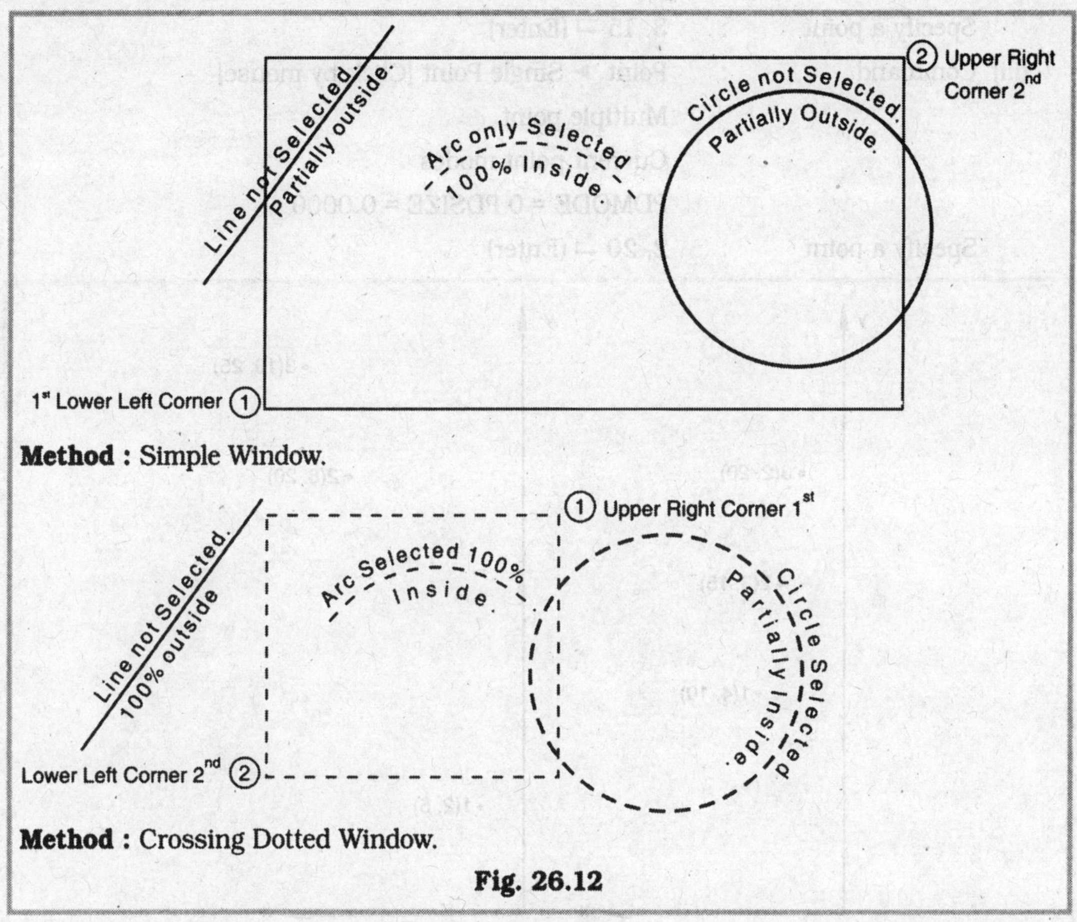

Fig. 26.12

1. Point Command / Multiple Point Command

The point command plots a point in the drawing. Point is used for location and for snap point. Point is dimensionless but it represents position.

Point command can be given in three different ways as under.

(1) Click by mouse on 'Point' toolbar icon.

(2) Type by key board word 'Point' in command window.

(3) Select point option from the draw menu.

Problem 1 : Locate points having co-ordinates [4, 10], [3, 15] and [2, 20].

(i)	Command	:	Point [Click on point icon]
			Current point modes
			PDMODE = 0 PDSIZE = 0.0000
	Specify a point	:	4, 10 ↵ (Enter)
(ii)	Command	:	Point [Type] ↵ (Enter)
			Current point modes
			PDMODE = 0 PDSIZE = 0.0000

Specify a point : 3, 15 ↵ (Enter)

(iii) Command : Point ▶ Single Point [Click by mouse]

Multiple point

Current point modes

PDMODE = 0 PDSIZE = 0.0000

Specify a point : 2, 20 ↵ (Enter)

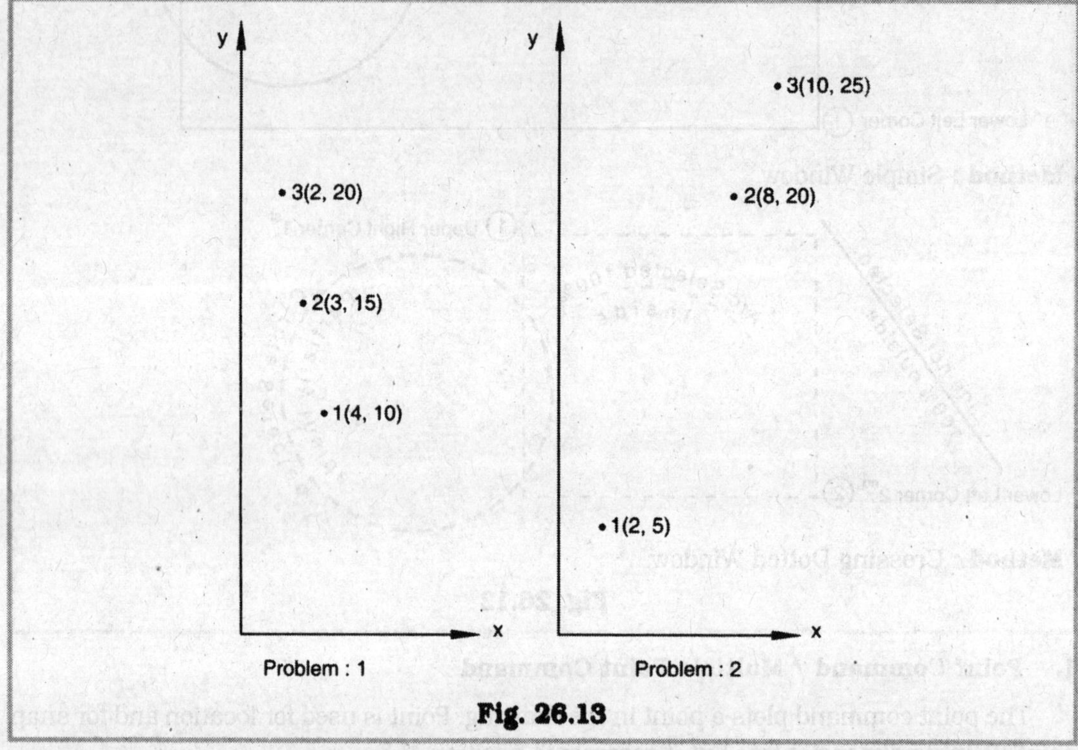

Fig. 26.13

Problem 2 : Locate multiple points having co-ordinates [2, 5], [8, 20] and [10, 25].

Take command from draw menu.

Command : Point ▶ Single Point

Multiple Point [Click by mouse]

Current point modes

PDMODE = 0 PDSIZE = 0.0000

Specify a point : 2, 5 ↵ (Enter)

Specify a point : 8, 20 ↵ (Enter)

Specify a point : 10, 25 ↵ (Enter)

2. Line Command

To draw a line, line command is used. Line command can be given by following ways.

1. Click on 'Line' icon.
2. Type, by keyboard, 'Line' in command window.
3. By selecting Line option from DRAW menu.

Having given Line command, supply co-ordinates of starting point and end point. Starting point and end point can be clicked also.

Problem 3 : Draw a horizontal line (→) of 150 units, having starting point-1 (50, 50).

(i) Absolute Co-ordinate Method:

Solution :

Command		:	Line ↵ (Enter)
From	Point-1	:	(50, 50) ↵ (Enter)
To	Point-2	:	(200, 50) ↵ (Enter)

Line will be ready on screen.

OR

(ii) Relative Co-ordinate Method:

Solution:

Command		:	Line ↵ (Enter)
From	Point-1	:	(50, 50) ↵ (Enter)
To	Point-2	:	@ (150, 0)

OR

(iii) Polar Co-ordinate Method:

Solution:

Command		:	Line ↵ (Enter)
From	Point-1	:	(50, 50) ↵ (Enter)
To	Point-2	:	@ 150 < 0°.

Problem 4 : Draw a vertical line (↓) of 150 units, having starting point-1 (50, 160).

(i) Solution:

Command		:	Line ↵ (Enter)
(Absolute Co-ordinate Method)			
From	Point -1	:	(50, 160) ↵ (Enter)
To	Point-2	:	(50, 10) ↵ (Enter)

Vertical Line (↓) will be ready on screen.

OR

(ii) Solution:

Command	:	Line ↵ (Enter)
(Relative Co-ordinate Method)		
From Point-1	:	(50, 160) ↵ (Enter)
To Point-2	:	@ (0, -150) ↵ (Enter)

OR

(iii) Solution:

Command	:	Line ↵ (Enter)
(Polar Co-ordinate Method)		
From Point-1	:	(50, 160) ↵ (Enter)
To Point-2	:	@ (150 < -90°).

Problem 5 : Draw inclined line 12 (↖) of 100 units, having starting point-1 (100, 30).

(i) Absolute Co-ordinate Method:

Solution:

Command	:	Line ↵ (Enter)
From Point-1	:	(100, 30) ↵ (Enter)
To Point-2	:	(20, 90) ↵ (Enter)

Line will be ready on screen.

(ii) Relative Co-ordinate Method:

Solution:

Command	:	Line ↵ (Enter)
From Point-1	:	(100, 30) ↵ (Enter)
To Point-2	:	@ (-80, 60) ↵ (Enter)

Same line will be ready on screen.

(iii) Polar Co-ordinate Method:

Solution:

Command	:	Line ↵ (Enter)
From Point-1	:	(100, 30) ↵ (Enter)
To Point-2	:	@ 100 < 143.13°

Same line will be ready on screen.

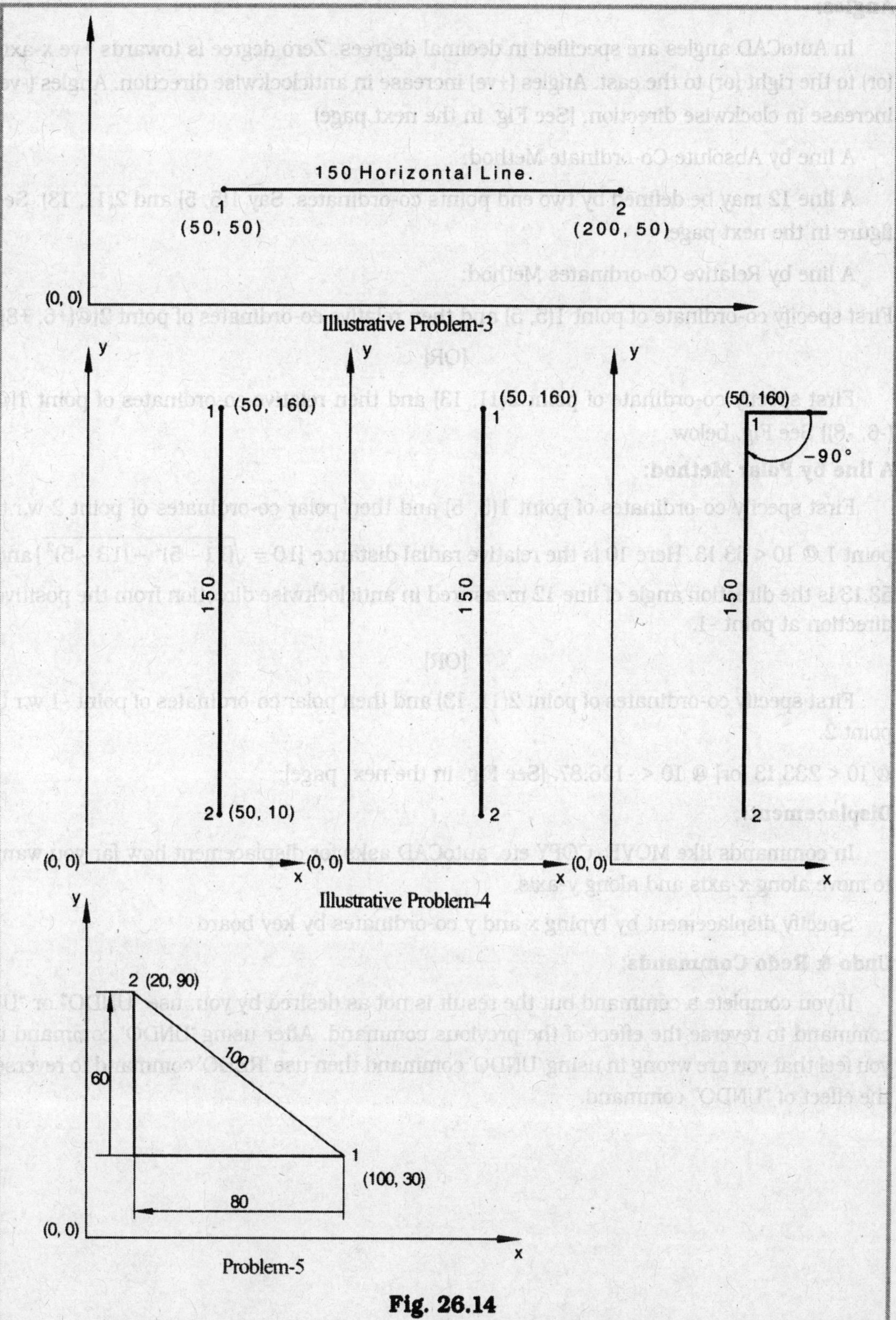
150 Horizontal Line.
1
(50, 50)
2
(200, 50)
(0, 0)
Illustrative Problem-3
y
1 (50, 160)
150
2 (50, 10)
(0, 0)
x
y
(50, 160)
1
150
2
(0, 0)
x
y
(50, 160)
1
−90°
150
2
(0, 0)
x
Illustrative Problem-4
y
2 (20, 90)
100
60
1
(100, 30)
80
(0, 0)
x
Problem-5

Fig. 26.14

Angles:

In AutoCAD angles are specified in decimal degrees. Zero degree is towards +ve x-axis (or) to the right (or) to the east. Angles (+ve) increase in anticlockwise direction. Angles (-ve) increase in clockwise direction. [See Fig. in the next page)

A line by Absolute Co-ordinate Method:

A line 12 may be defined by two end points co-ordinates. Say 1(5, 5) and 2(11, 13). See figure in the next page.

A line by Relative Co-ordinates Method:

First specify co-ordinate of point 1(5, 5) and then relative co-ordinates of point 2[@(+6, +8)]

[OR]

First specify co-ordinate of point 2(11, 13) and then relative co-ordinates of point 1[@ (-6, -8)] See Fig. below.

A line by Polar Method:

First specify co-ordinates of point 1(5, 5) and then polar co-ordinates of point 2 w.r.t. point 1 @ 10 < 53.13. Here 10 is the relative radial distance $[10 = \sqrt{(11-5)^2 + (13-5)^2}]$ and 53.13 is the direction angle of line 12 measured in anticlockwise direction from the positive direction at point -1.

[OR]

First specify co-ordinates of point 2(11, 13) and then polar co-ordinates of point -1 w.r.t. point 2.

@ 10 < 233.13 [or] @ 10 < -126.87. [See Fig. in the next page].

Displacement:

In commands like MOVE, COPY etc. autoCAD asks for displacement how far you want to move along x-axis and along y-axis.

Specify displacement by typing x and y co-ordinates by key board.

Undo & Redo Commands:

If you complete a command but the result is not as desired by you, use "UNDO" or "U" command to reverse the effect of the previous command. After using 'UNDO' command if you feel that you are wrong in using 'UNDO' command then use 'REDO' command to reverse the effect of "UNDO" command.

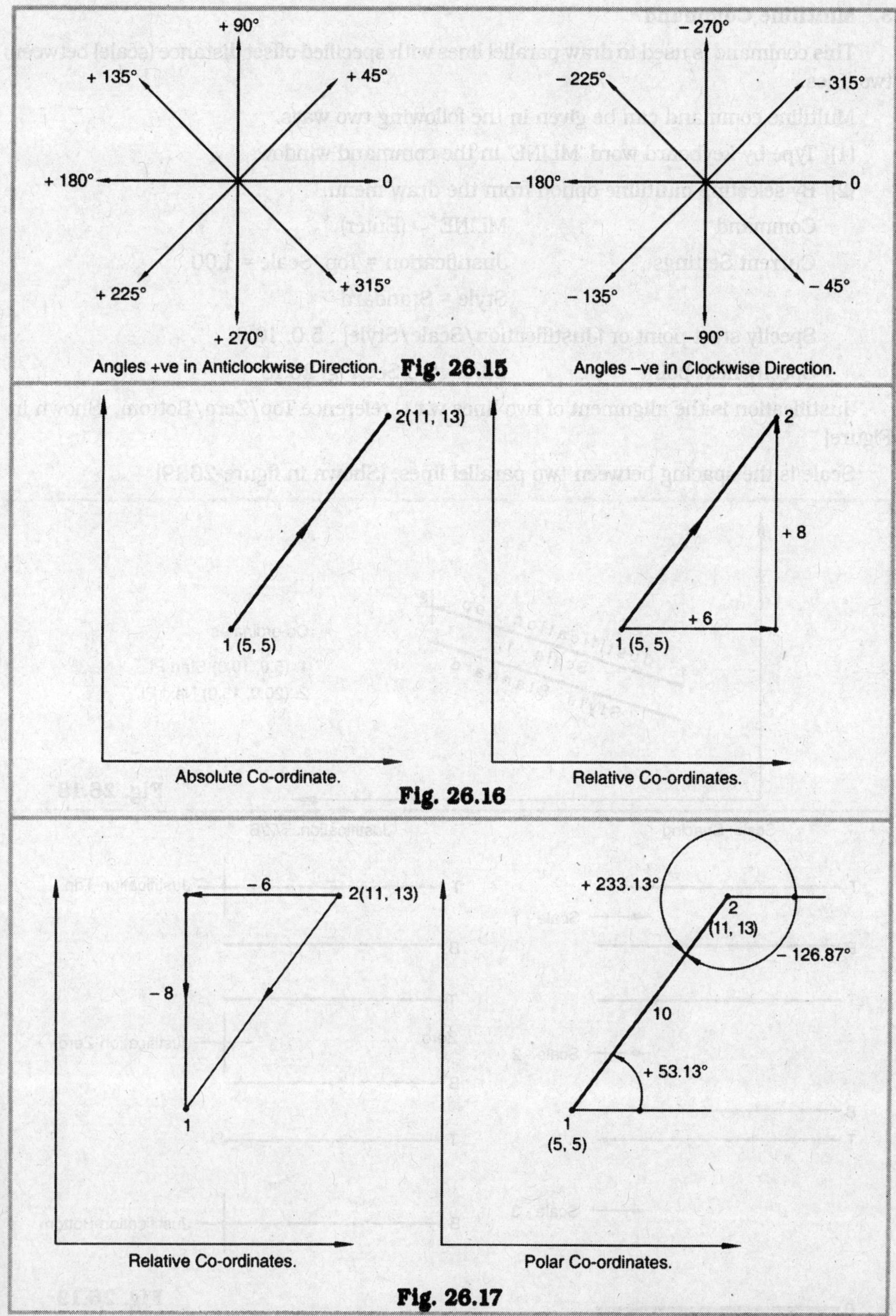

Fig. 26.15

Fig. 26.16

Fig. 26.17

3. Multiline Command

This command is used to draw parallel lines with specified offset distance (scale) between two lines.

Multiline command can be given in the following two ways.

(1) Type by keyboard word 'MLINE' in the command window.

(2) By selecting multiline option from the draw menu.

Command : MLINE ↵ (Enter)

Current Settings : Justification = Top, Scale = 1.00

Style = Standard

Specify start point or [Justification/Scale/Style] : 5.0, 10.0

Specify next point : 20.0, 15.0 See Fig. 26.18

Justification is the alignment of two lines w.r.t. reference Top/Zero/Bottom. [Shown in Figure]

Scale is the spacing between two parallel lines. [Shown in figure-26.19]

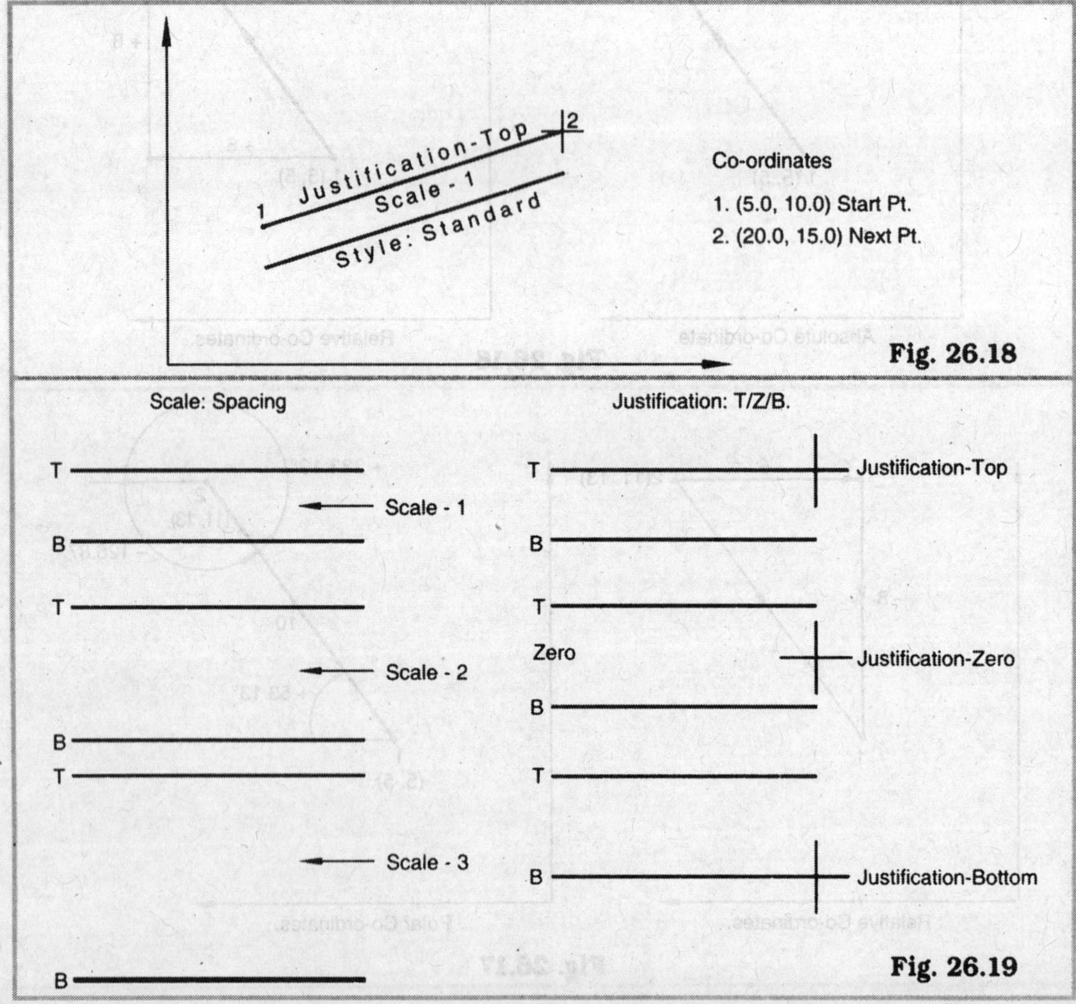

Fig. 26.18

Fig. 26.19

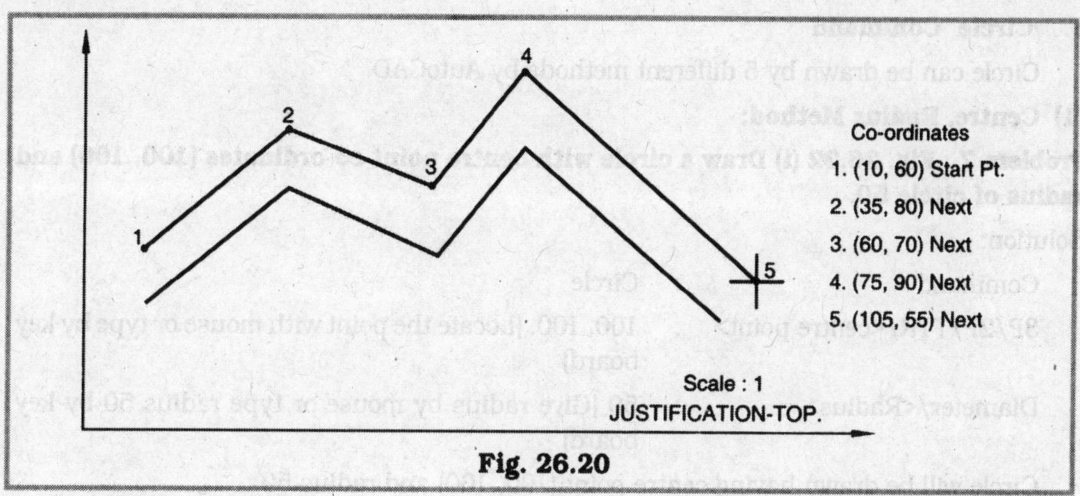

Fig. 26.20

4. Rectangle Command

Rectangle is a quadrilateral having two pairs of opposite sides are equal and 4 angles are each of 90°.

Rectangle command can be given in three different ways as under.

(1) Click by mouse on 'Rectangle' tool bar icon.

(2) Type by keyboard word 'Rectangle' in command window.

(3) Select 'Rectangle' option from draw menu.

Problem 6 : Draw a rectangle having diagonally opposite points 1, (40, 30) and 2 (170, 110).

Solution:

Command	:	Rectangle
Specify first corner point	:	40, 30 ↵ (Enter)
or [Chamfer/Elevation/Fillet/Thickness/Width]		
Specify other corner point	:	170, 110 ↵ (Enter)
or [Dimensions]		

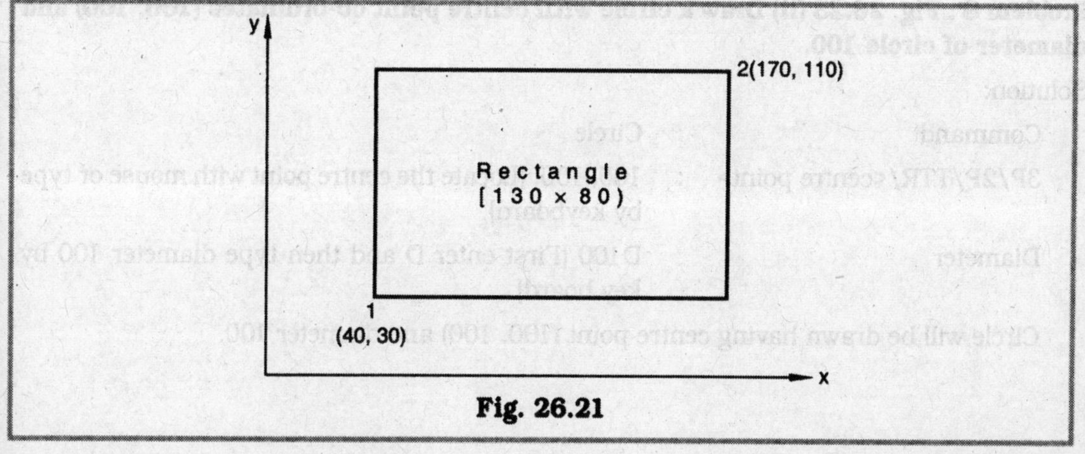

Fig. 26.21

5. Circle Command

Circle can be drawn by 5 different methods by AutoCAD.

(1) Centre, Radius Method:

Problem 7 : Fig. 26.22 (i) Draw a circle with centre point co-ordinates (100, 100) and radius of circle 50.

Solution:

Command	:	Circle
3P/2P/TTR/<centre point>	:	100, 100. [Locate the point with mouse or type by key board]
Diameter/<Radius>	:	50 [Give radius by mouse or type radius 50 by key board]

Circle will be drawn having centre point (100, 100) and radius 50.

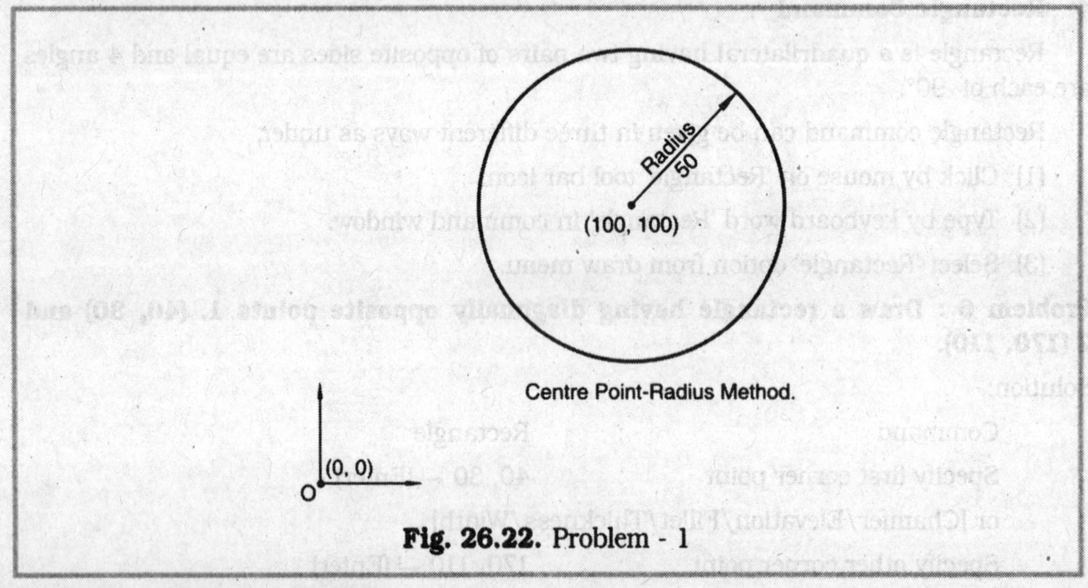

Fig. 26.22. Problem - 1

(2) Centre, Diameter Method:

Problem 8 : Fig. 26.23 (ii) Draw a circle with centre point co-ordinates (100, 100) and diameter of circle 100.

Solution:

Command	:	Circle
3P/2P/TTR/<centre point>	:	100, 100. [Locate the centre point with mouse or type by keyboard].
Diameter	:	D100 [First enter D and then type diameter 100 by key board]

Circle will be drawn having centre point (100, 100) and diameter 100.

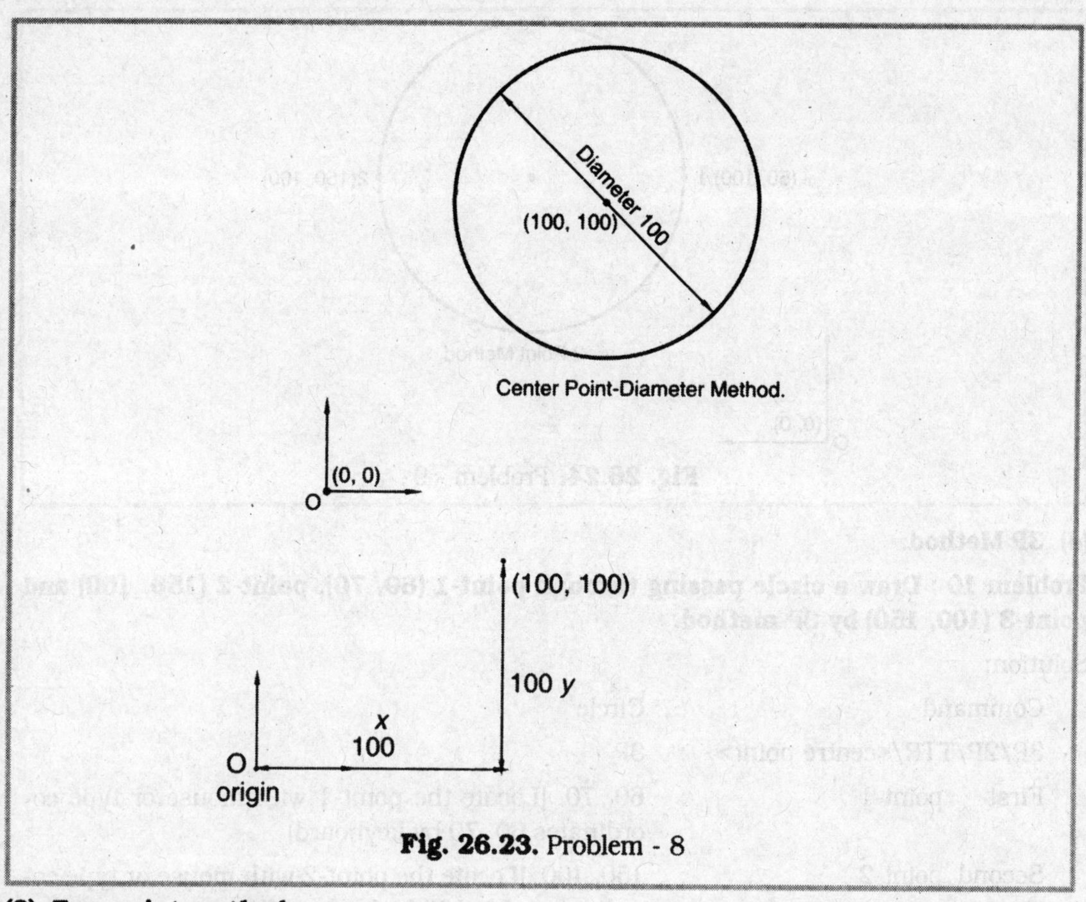

Fig. 26.23. Problem - 8

(3) Two point method:

Problem 9 : Fig. 26.24 (iii) Draw a circle with diametrically opposite points (1) 50, 100 and (2) 150, 100.

Solution:

Command	:	Circle
3P/2P/TTR/<centre point>	:	2P
First point 1	:	50, 100 [Locate the point-1 with mouse or type co-ordinates 50, 100 by key board]
Second point 2	:	150, 100. [Locate the point-2 with mouse or type co-ordinates 150, 100 by keyboard]

Circle will be drawn having diametrically opposite 2 points (50, 100) and (150, 100).

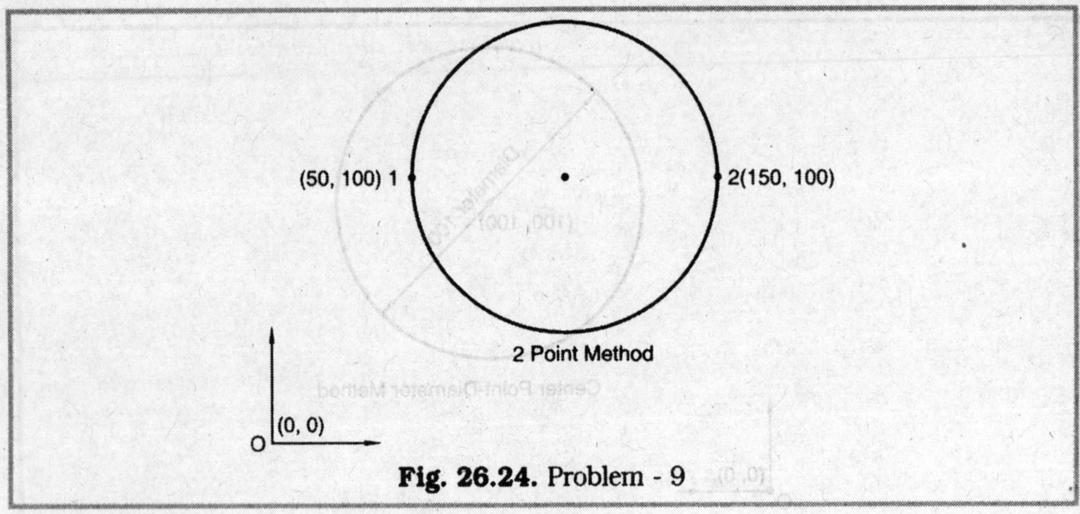

Fig. 26.24. Problem - 9

(4) 3P Method:

Problem 10 : Draw a circle passing through point-1 (60, 70), point-2 (150, 100) and point-3 (100, 150) by 3P method.

Solution:

Command	:	Circle
3P/2P/TTR/<centre point>	:	3P
First point-1	:	60, 70. [Locate the point-1 with mouse or type co-ordinates 60, 70 by keyboard]
Second point-2	:	150, 100. [Locate the point-2 with mouse or type co-ordinates 150, 100 by keyboard]
Third point-3	:	100, 150. [Locate the point-3 with mouse or type co-ordinates 100, 150 by keyboard.]

Circle will be drawn passing through three points having co-ordinates (60, 70), (150, 100) and (100, 150).

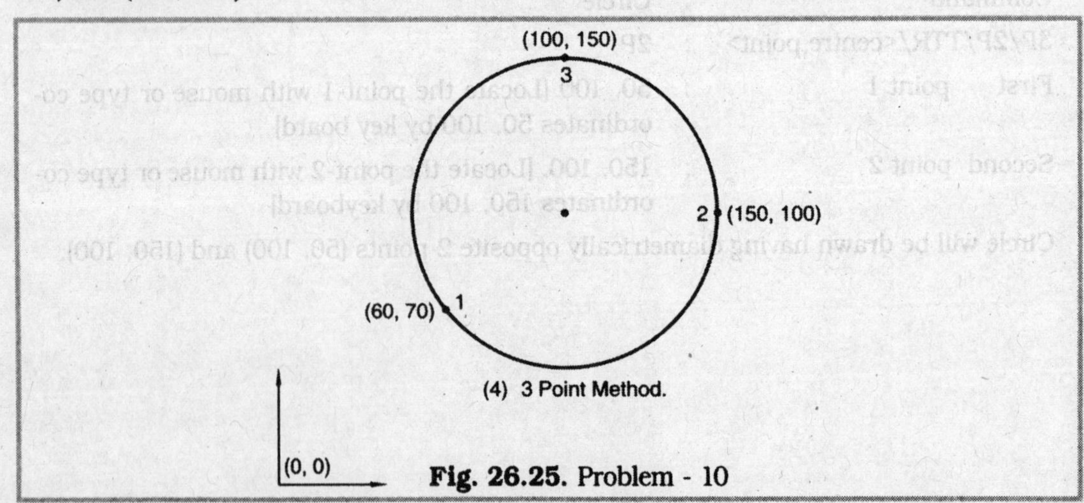

Fig. 26.25. Problem - 10

(5) Tangent-Tangent-Radius Method:

Problem 11 : Draw a circle with given two tangent lines (1) Line-1 and Line-2, as shown, and radius of the circle 50.

Solution:

Command	:	Circle		
3P/2P/TTR/<centre point>	:	TTR (Tangent-Tangent-Radius)		
Line-1	:	Command	:	Line
		From Point	:	10, 30
		To Point	:	10, 160.
Line-2	:	Command	:	Line
		From Point	:	80, 30
		To Point	:	170, 120

Line-1 and Line-2 can be drawn using mouse.

Radius : 50.

Circle will be drawn having radius 50 and Line-1 and Line-2 as tangent to the circle.

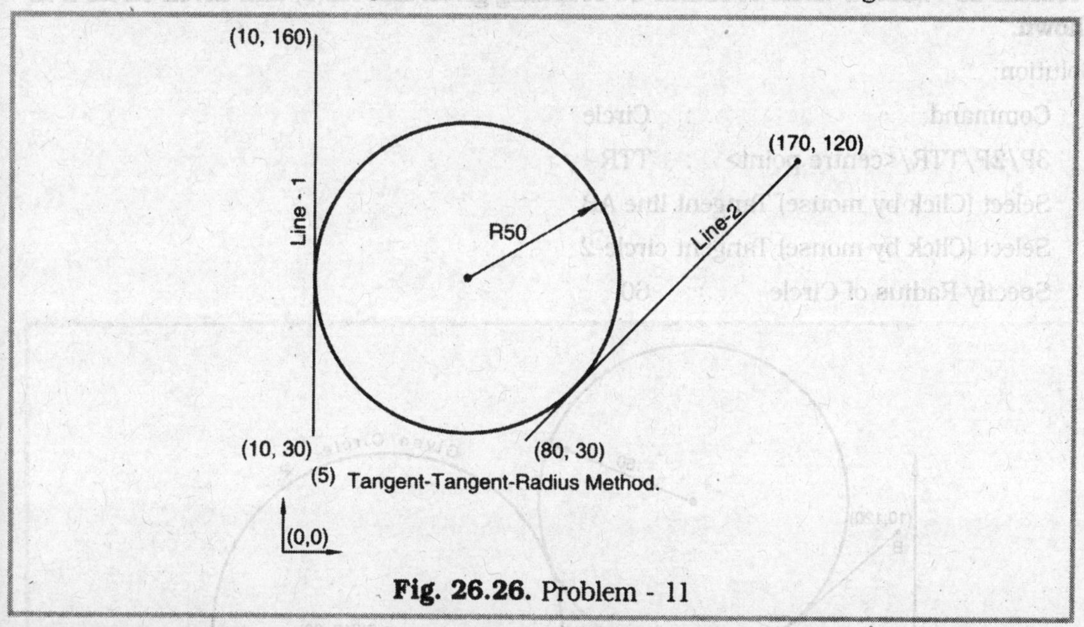

Fig. 26.26. Problem - 11

Problem 12 : Draw a circle of radius 40 touching given two circles 1 & 2 as shown.

Solution:

Command : Circle

3P/2P/TTR/<centre point> : TTR

(Select) Click by mouse tangent Circle-1

(Select) Click by mouse tangent Circle-2

Specify Radius of Circle : 40.

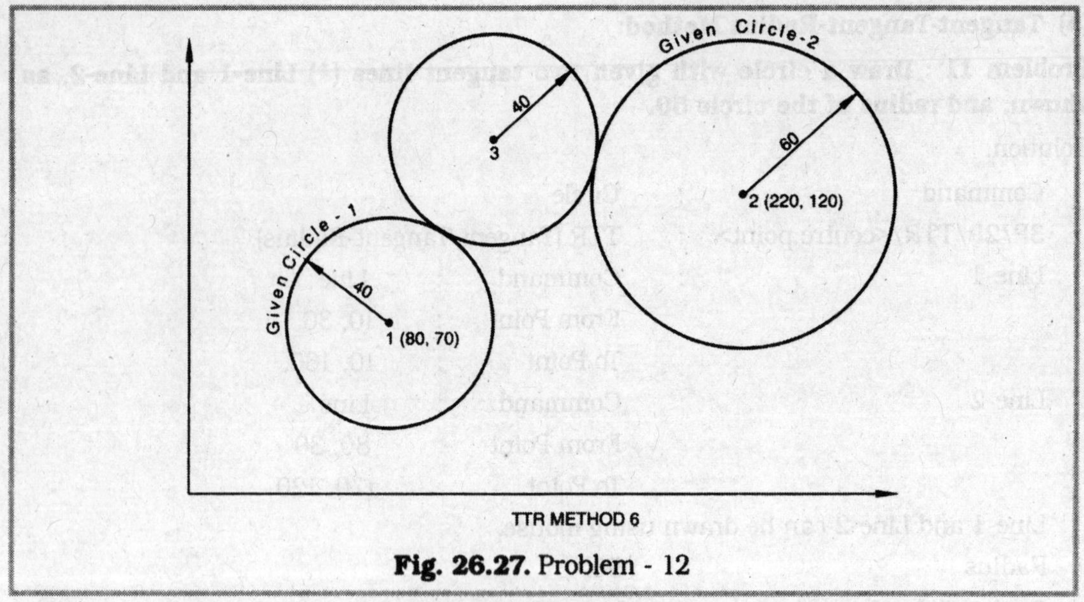

Fig. 26.27. Problem - 12

Problem 13 : Draw a circle of radius 60 touching given line AB(1) and Given circle-2 as shown.

Solution:

Command	:	Circle
3P/2P/TTR/<centre point>	:	TTR

Select (Click by mouse) Tangent line AB

Select (Click by mouse) Tangent circle-2

Specify Radius of Circle : 60.

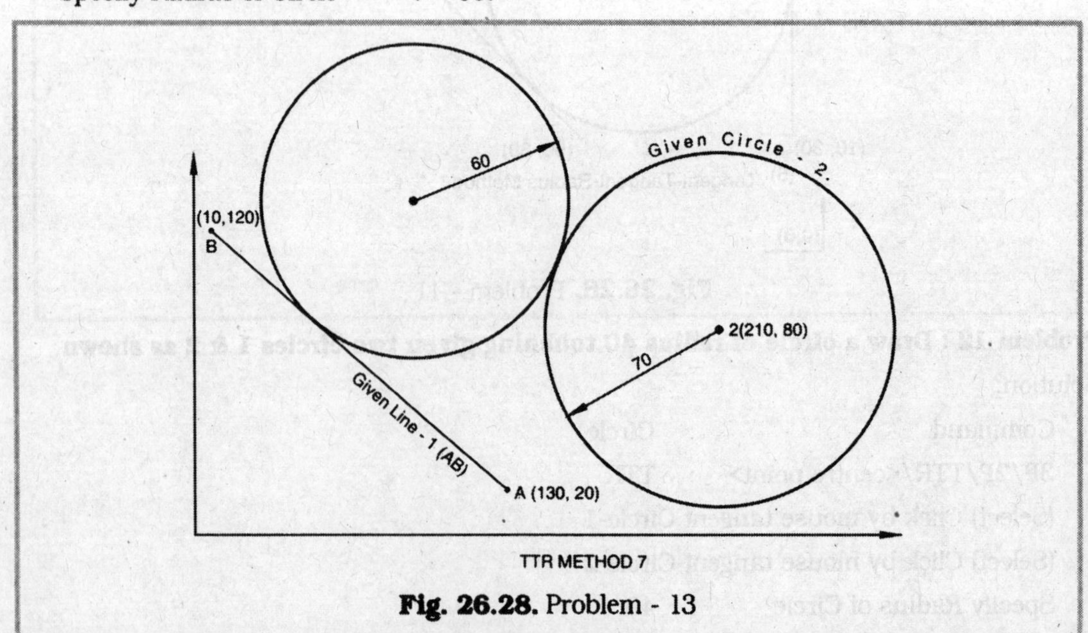

Fig. 26.28. Problem - 13

6. Arc command

Arc command is used to draw arc of a circle.

Arc command can be given in three different ways as under.

(1) Click by mouse on 'Arc' toolbar icon.

(2) Type by keyboard word 'Arc' in command window.

(3) Select 'Arc' option from draw menu.

Method: 1 See Fig. 26.29.1

Problem 14 : Draw an arc of a circle passing through two end points 1(120, 70) and 3(40, 30) and also passing through point 2 (70, 120).

Solution:

Command	:	Arc ↵ (Enter)
1st start point-1 of arc or [Centre]	:	120, 70 ↵ (Enter)
2nd point-2 of arc or [Centre/End]	:	70, 120 ↵ (Enter)
3rd end point-3 of arc	:	40, 30 ↵ (Enter)

Method : 2 See Fig. 26.29.2

Problem 15 : Draw an arc of a circle having centre (70, 70), 1st start point of arc 120, 70 and end point of arc 40, 30.

Solution:

Command	:	Arc ↵ (Enter)
Specify start point of arc or [C Enter]	:	CE ↵ (Enter)
Centre point of arc	:	(70, 70) ↵ (Enter)
1st start point of arc	:	120, 70 ↵ (Enter)
End point of arc	:	40, 30 ↵ (Enter)

Method : 3 See Fig. 26.29.3

Problem 16 : Draw an arc of a circle having centre point 70, 70, starting end point 120, 70 and angle of arc 233.13°.

Solution:

Command	:	Arc ↵ (Enter)
Start point of arc or [C Enter]	:	CE ↵ (Enter)
Centre point of arc	:	70, 70 ↵ (Enter)
Start point of arc	:	120, 70 ↵ (Enter)
End point of arc or [Angle/chord Length]	:	a ↵ (Enter)
Included angle	:	233.13° ↵ (Enter)

Method : 4 See Fig. 26.29.4

Problem 17 : Draw an arc of a circle having (1) centre point 70, 70, (2) Start point 120, 70 and length of chord 89.44.

Solution:

Command	:	Arc ↵ (Enter)
Start point of arc or [C Enter]	:	C ↵ (Enter)

Centre point of arc	:	70, 70 ↵ (Enter)
Start point of arc	:	120, 70 ↵ (Enter)
End point of arc or [Angle/Chord length]	:	L ↵ (Enter)
Chord length	:	89.44 ↵ (Enter)

Datas for drawing arc of a circle are as under. There are 4 methods by which arc can be drawn. They are discussed by 4 problems as under.

(1) Centre point

(2) Start point

(3) End point

(4) Included angle

(5) Length of chord (Length of line connecting start point to end point)

(6) Middle point

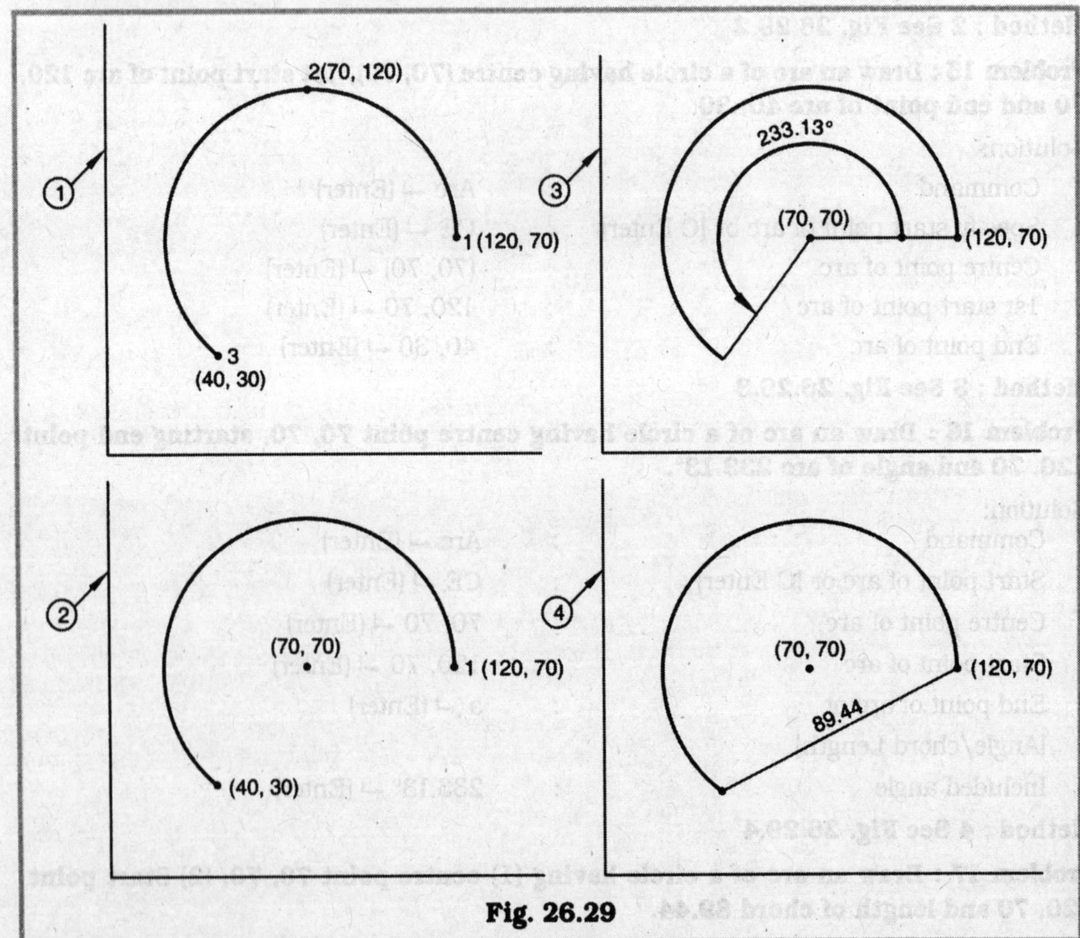

Fig. 26.29

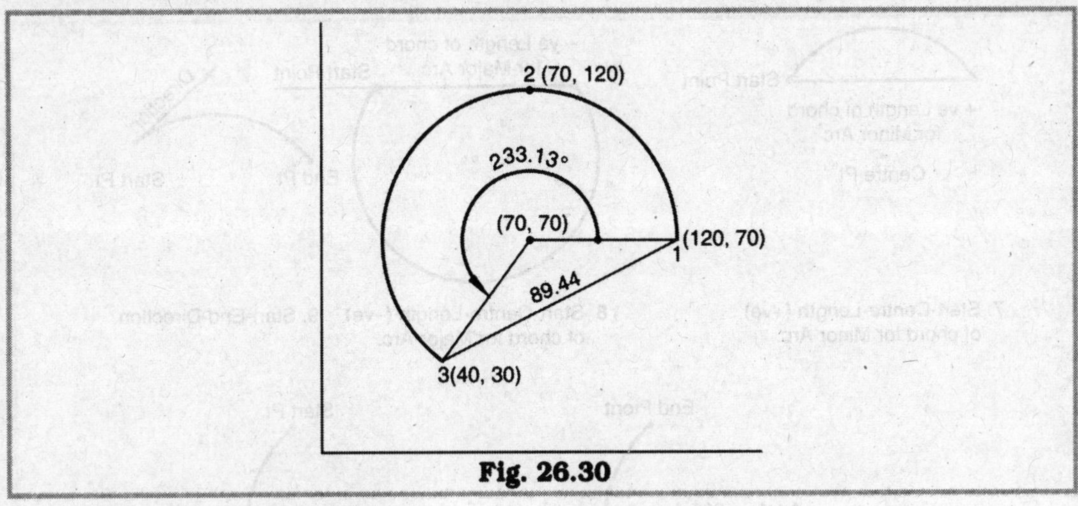

Fig. 26.30

Arc Command from Draw Menu

Arc→	3P	Fig. 26.31.1
	Start - Centre - End	Fig. 26.31.2
	Start - Centre - Angle	Fig. 26.31 (2/5)
	Start - Centre - Length	Fig. 26.31 (7/8)
	Start - End - Angle	Fig. 26.31 (10/11)
	Start - End - Direction	Fig. 26.31.9
	Start - End - Radius	Fig. 26.31 (4/6)
	Centre - Start - End	Fig. 26.31.3
	Centre - Start - Angle	Fig. 26.31 (2/5)
	Centre - Start - Length	Fig. 26.31 (7/8)

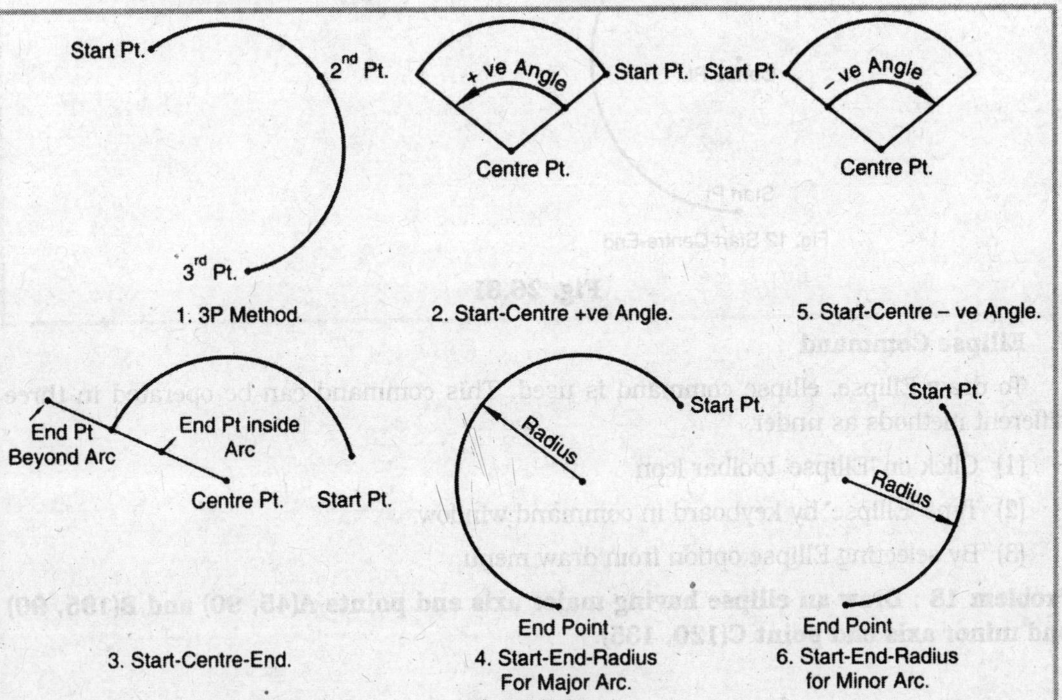

1. 3P Method.

2. Start-Centre +ve Angle.

5. Start-Centre – ve Angle.

3. Start-Centre-End.

4. Start-End-Radius For Major Arc.

6. Start-End-Radius for Minor Arc.

Fig. 26.31

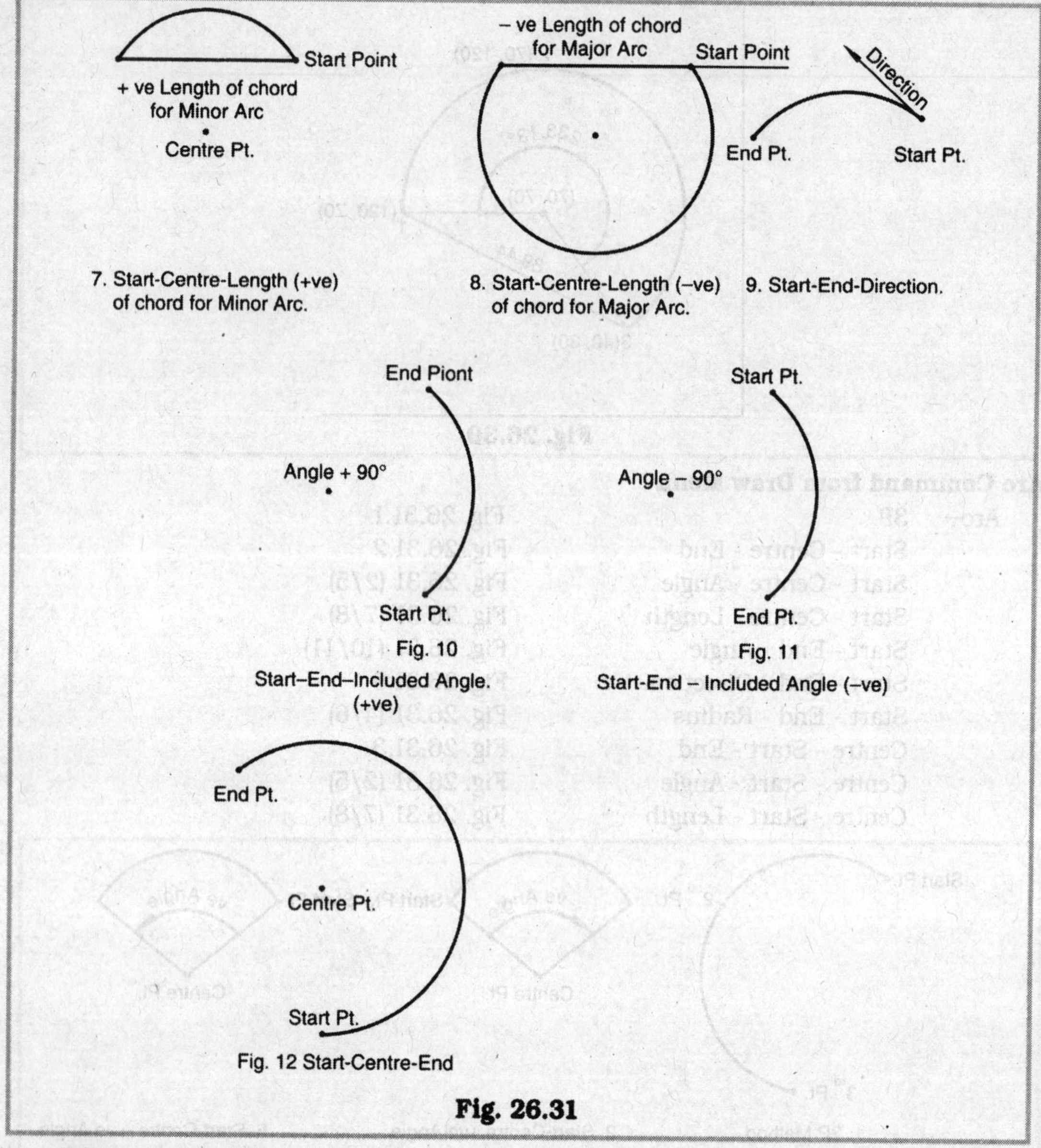

Fig. 26.31

7. Ellipse Command

To draw Ellipse, ellipse command is used. This command can be operated in three different methods as under.

(1) Click on 'Ellipse' toolbar icon.

(2) Type 'Ellipse' by keyboard in command window.

(3) By selecting Ellipse option from draw menu.

Problem 18 : Draw an ellipse having major axis end points A(45, 90) and B(195, 90) and minor axis end point C(120, 135).

Solution:

	Command	:	Ellipse ↵ (Enter)
1.	Specify axis end point of ellipse or [Arc/Centre]	:	45, 90 ↵ (Enter)
2.	Specify other end point of axis	:	195, 90 ↵ (Enter)
3.	Specify distance to other axis or [Rotation]	:	120, 135 ↵ (Enter)

Ellipse will be drawn having

1. Major Axis Length = (195 - 45) = 150.
2. Minor Axis Length = 2(135-90) = 90.

One can select the point by clicking the mouse at axis end point, centre etc. instead of giving co-ordinates of the said points.

Problem 19 : Draw an ellipse having centre 0(90, 60), major axis end point B(140, 60) and minor axis end point C(90, 90).

Solution:

	Command	:	Ellipse ↵ (Enter)
1.	Specify centre of ellipse	:	90, 60 ↵ (Enter)
2.	Specify endpoint of axis	:	140, 60 ↵ (Enter)
3.	Specify distance to other axis or (Rotation)	:	90, 90 ↵ (Enter)

Ellipse will be drawn having

1. Major Axis = 2(140 - 90) = 100
2. Minor Axis = 2(90 - 60) = 60

Problem 20 : Draw an ellipse having major axis end points [42, 50], [158, 50] and 55° degree rotation around major axis.

Command	:	Ellipse ↵ (Enter)
Specify axis end point of ellipse or [Arc/Centre]	:	42, 50 ↵ (Enter)
Specify other end point of axis	:	158, 50 ↵(Enter)
Specify distance to other axis or [Rotation]	:	r ↵ (Enter)
Specify rotation around major axis	:	55 ↵ (Enter)

Problem 21 : Draw an ellipse having major axis end point [135, 50], Centre point (90, 50) and rotation angle 55°.

Command	:	Ellipse ↵ (Enter)
Specify Centre of Ellipse	:	90, 50 ↵ (Enter)
Specify endpoint of axis	:	135, 50 ↵ (Enter)
Specify distance to other axis or [Rotation]	:	r ↵ (Enter)
Rotation around major axis	:	55° ↵ (Enter)

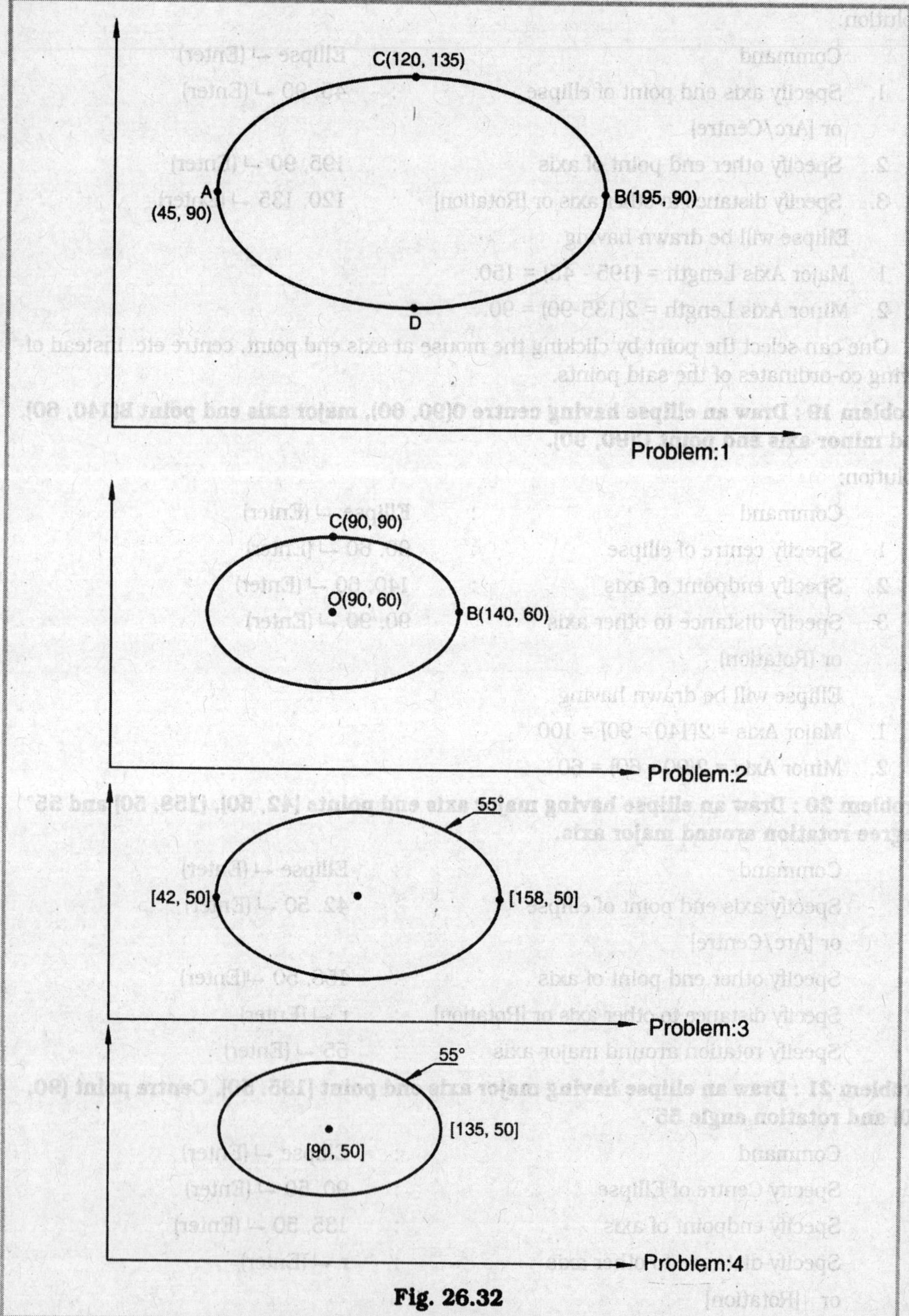
C(120, 135)
A
(45, 90)
B(195, 90)
D
Problem:1
C(90, 90)
O(90, 60)
B(140, 60)
Problem:2
55°
[42, 50]
[158, 50]
Problem:3
55°
[135, 50]
[90, 50]
Problem:4

Fig. 26.32

8. Polygon command

Polygon command is used to draw regular polygonal shapes. Although one can draw polygon also with line command but it takes more time. Polygon command draws polygon faster with high accuracy.

3 sides, i.e. triangle, to 1024 sides regular polygons can be drawn with polygon command. Polygon command can be given in 3 different methods as under.

(1) Click on Polygon toolbar icon.

(2) Type Polygon by key board in command window.

(3) Select Polygon option from draw menu.

There are three methods of drawing polygons.

Method : 1

Problem 22 : Draw a regular polygon (Hexagon n = 6) inscribed in a circle of radius 50 units with centre (100, 100)

Solution:

Command	:	Polygon ↵ (Enter)
Number of Sides	:	6 ↵ (Enter)
Centre of Polygon	:	(100, 100) ↵ (Enter)
Inscribed in circle/Circumscribed about circle (I/C)	:	I ↵ (Enter)
Radius of Circle	:	50 ↵ (Enter)

Method : 2

Problem 23 : Draw a regular polygon (Pentagon n = 5) circumscribed about circle of radius 50 units with centre (100, 100).

Solution:

Command	:	Polygon ↵ (Enter)
Number of Sides	:	5 ↵ (Enter)
Centre of Polygon	:	(100, 100) ↵ (Enter)
Inscribed in Circle/Circumscribed about Circle (I/C)	:	C ↵ (Enter)
Radius of Circle	:	50 ↵ (Enter)

Method : 3

Problem 24 : Draw a polygon (Octagon n = 8). End points co-ordinates of the edge are 1. (60, 20) and 2. (100, 20).

Solution:

Command	:	Polygon ↵ (Enter)
Number of Sides	:	8 ↵ (Enter)
First end point of edge	:	(60, 20) ↵ (Enter)
Second end point of edge	:	(100, 20) ↵ (Enter)

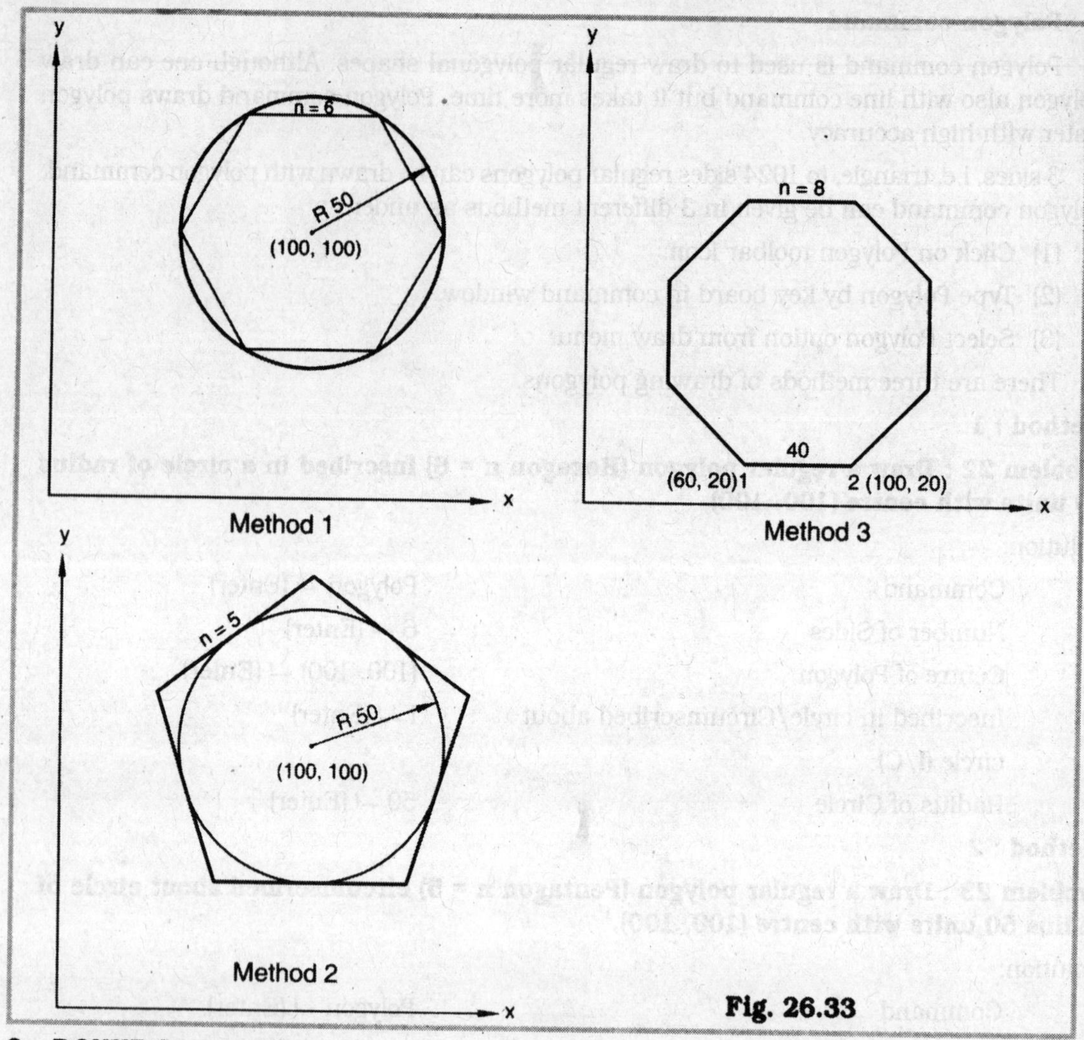

Fig. 26.33

9. DONUT Command

The 'DONUT' [or DOUGHNUT] command enables us to draw filled rings and solid circles. Required datas are (1) Centre point (2) Outside diameter (3) Inside diameter.

'DONUT' command can be given in two different ways as under.

(1) Type by keyboard word 'DONUT' in command window.

(2) Select 'DONUT' option from draw menu.

It is not available on toolbar icon. This command enables us to draw washers.

Problem 25 : Draw filled rings having (1) Inside diameter 80 (2) Outside diameter 120 and (3) Centre point 80, 80.

Solution:

Command	:	'DONUT' ↵ (Enter)
Inside diameter of donut	:	80 ↵ (Enter)
Outside diameter of donut	:	120 ↵ (Enter)
Centre of donut	:	80, 80 ↵ (Enter)

Problem 26 : Draw solid filled circle having (1) Inside diameter 0(zero) (2) Outside diameter 60 and (3) Centre point 90, 90.

Solution:

Command	:	'DONUT' ↵ (Enter)
Inside diameter of donut	:	0 ↵ (Enter)
Outside diameter of donut	:	60 ↵ (Enter)
Centre point of donut	:	90, 90 ↵ (Enter)

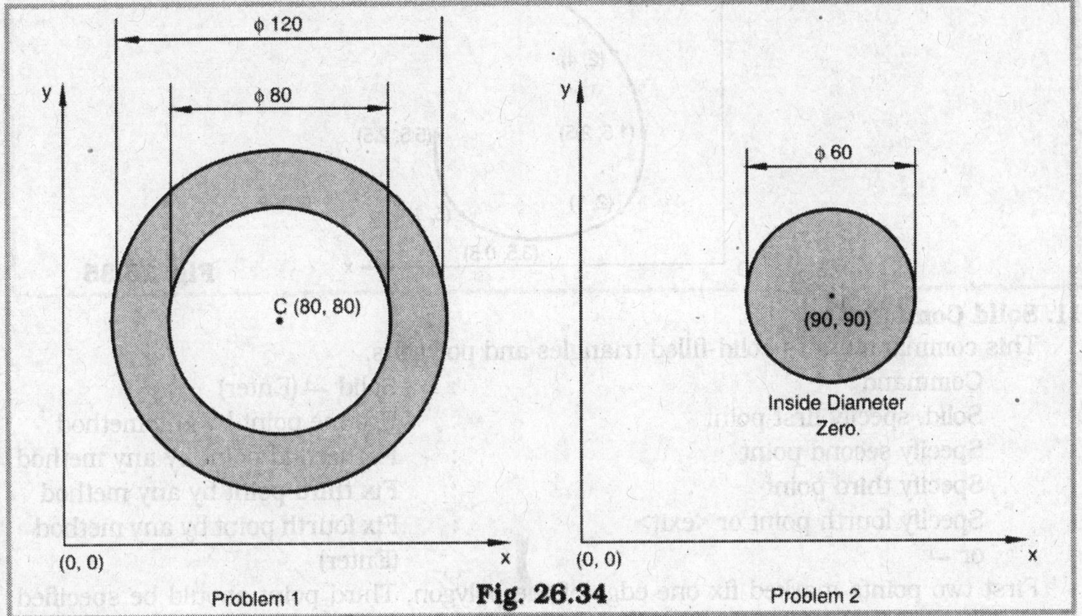

Fig. 26.34

10. Spline command

Spline command is used to draw a smooth curve passing through number of given points.

Spline command can be given in three different ways as under.

(1) Click by mouse on 'SPLINE' toolbar icon.

(2) Type by keyboard word 'SPLINE' in the command window.

(3) Select 'SPLINE' option from draw menu.

Problem 27 : Draw a curve (ᘔ ┘) passing through given points (1.5, 6.5), (3, 7.5), (4.5, 6.5), (4, 5.5) (2, 4), (1.5, 2.5), (2, 1), (3.5, 0.5), (5, 1) and (5.5, 2.5).

Command	:	Spline ↵ (Enter)
Specify first point or [object]	:	1.5, 6.5 ↵ (Enter)
Specify next point	:	3, 7.5 ↵ (Enter)
Specify next point or [close, fit tolerance]	:	4.5, 6.5 ↵ (Enter)
(Start Tangent)		
— do —	:	4, 5.5 ↵ (Enter)
— do —	:	2, 4 ↵ (Enter)
— do —	:	1.5, 2.5 ↵ (Enter)
— do —	:	2, 1 ↵ (Enter)
— do —	:	3.5, 0.5 ↵ (Enter)

— do — : 5, 1 ↵ (Enter)

— do — : 5.5, 2.5 ↵ (Enter)

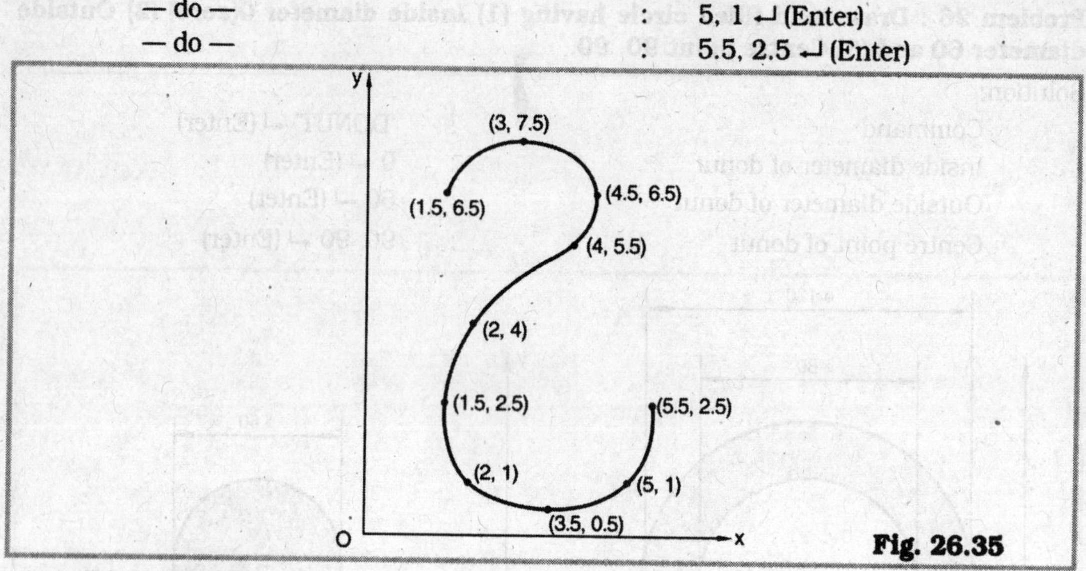

Fig. 26.35

11. Solid Command

This command draws solid-filled triangles and polygons.

Command	:	Solid ↵ (Enter)
Solid, specify first point	:	Fix one point by any method
Specify second point	:	Fix second point by any method
Specify third point	:	Fix third point by any method
Specify fourth point or <exit> or ↵	:	Fix fourth point by any method (Enter)

First two points marked fix one edge of the polygon. Third point should be specified diagonally opposite to the second point. Specify fourth point.

AutoCAD 2004 takes the last two points of the first polygon as the first two points for the next polygon.

Once the command solid is on it will continue unless (entered). i.e. exit.

How different solid shapes are generated is shown in the figure below. Note the sequence of points marked and corresponding shapes generated.

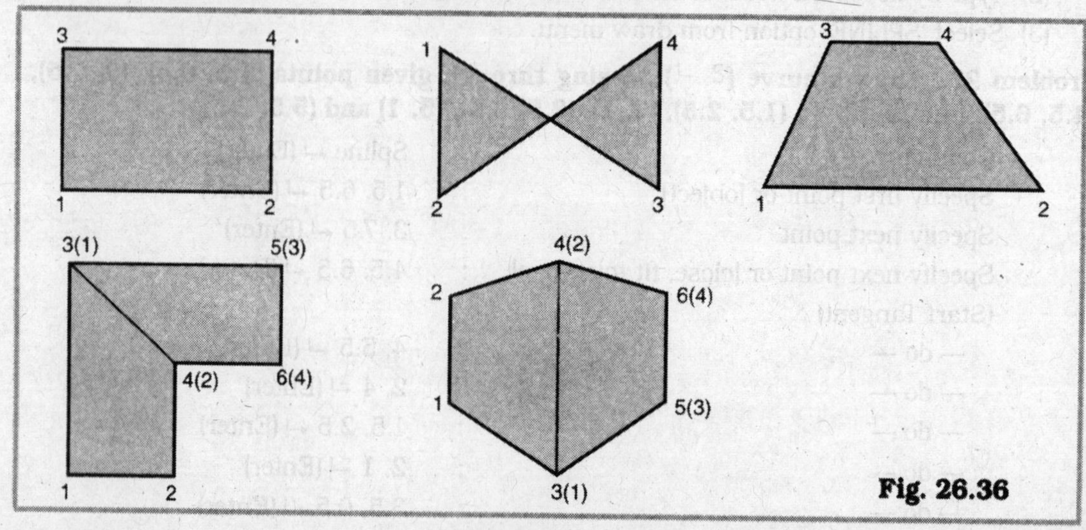

Fig. 26.36

12. Pan command

Pan command moves the drawing on the screen. Display does not change in shape and size only it's location changes. In fact it moves the display window.

Cursor changes to a hand shape cursor. By holding down the pick button you are locking the cursor to it's current location. Now cursor is moved in the desired direction and along with cursor graphics also move. When pick button is released panning stops. Only hand cursor move.

At the end to stop panning either press (Enter) ↵ or escape ESC.

Figure below shows has original display is moved upward as well as rightward by 'PAN' command.

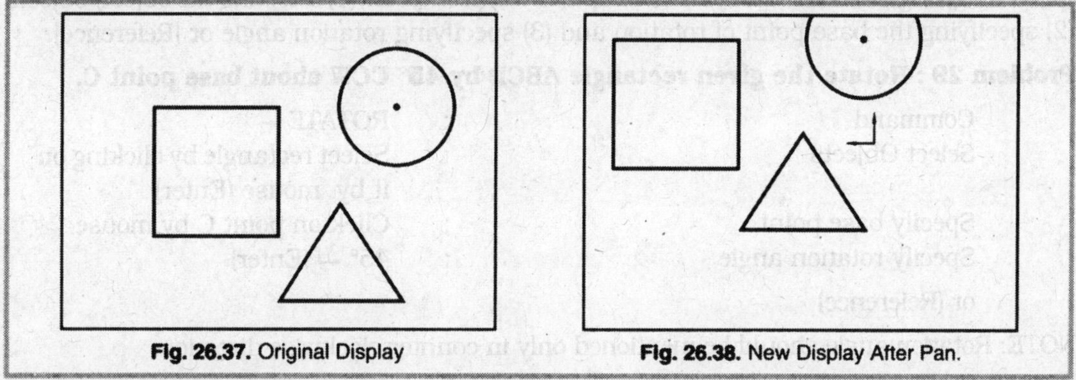

Fig. 26.37. Original Display **Fig. 26.38.** New Display After Pan.

13. Erase command

Erase command is used to remove object/objects from the drawing. It's an editing tool. It is nothing but eraser/rubber as we use for manual drawing. This eraser leave zero impression of object/objects removed.

Eraser command can be given in three different ways as under.

(1) Click by mouse on 'ERASE' toolbar icon.

(2) Type by keyboard word 'ERASE' in the command window.

(3) Select 'ERASE' option from modify menu.

Problem 28 : Remove circle (O) from the drawing.

Command	:	erase
Select objects	:	Select by simple window totally or by crossing window totally or partially circle will become dotted.
Select objects	:	↵(Enter) object circle will disappear.

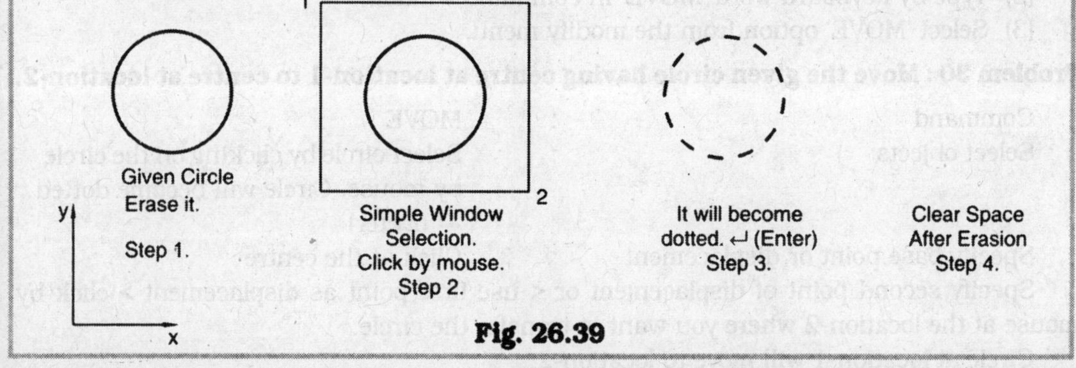

Fig. 26.39

14. Rotate Command

Rotate command facilitates to rotate the object/objects at defined angle. This command is used to rotate object/objects about a base point maximum by 360°. Whereas move command facilitates to move the object anywhere in the drawing

Rotate Command can be given in three different ways.

(1) Click by mouse on 'ROTATE' toolbar icon.

(2) Type by keyboard word 'ROTATE' in the command window.

(3) By selecting 'ROTATE' option from modify menu.

After giving 'ROTATE' command, computer will prompt for (1) selecting the object/objects (2) specifying the base point of rotation and (3) specifying rotation angle or [Reference].

Problem 29 : Rotate the given rectangle ABCD by 45° CCW about base point C.

Command	:	ROTATE
Select Objects	:	Select rectangle by clicking on it by mouse (Enter)
Specify base point	:	Click on point C by mouse
Specify rotation angle or (Reference)	:	45° ↵ (Enter)

NOTE: Rotation angle should be mentioned only in counter clockwise direction.

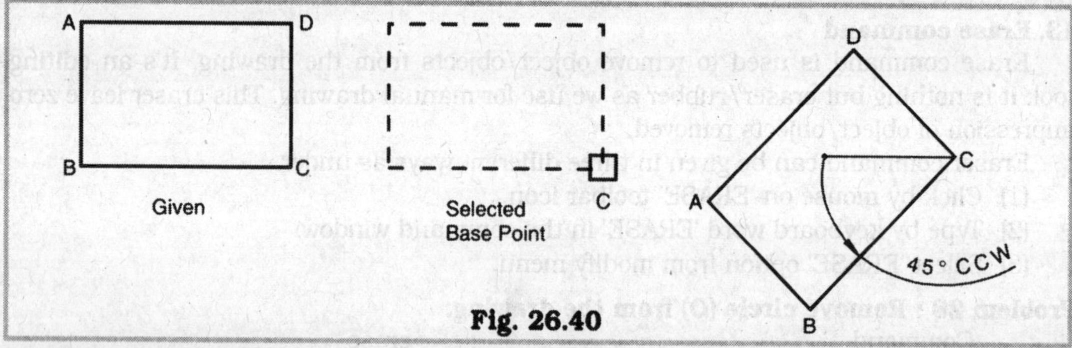

Fig. 26.40

15. Move Command

Move command is used when we want to displace object/objects to a specified location. Move command can be given in three different ways as under.

(1) Click by mouse on 'MOVE' tool bar icon.

(2) Type by keyboard word 'MOVE' in command window.

(3) Select 'MOVE' option from the modify menu.

Problem 30 : Move the given circle having centre at location-1 to centre at location-2.

Command	:	MOVE
Select objects	:	Select circle by clicking on the circle by mouse. Circle will become dotted ↵ (Enter)
Specify base point or displacement	:	Click at the centre

Specify second point of displacement or < use first point as displacement > click by mouse at the location-2 where you want to transfer the circle.

Circle at location-1 will move to location-2.

16. Copy Command

Copy command is used to copy or copies of given object/objects.

Copy command can be given in three different ways as under.

(1) Click by mouse on 'COPY' tool bar icon.

(2) Type by keyboard word 'COPY' in command window.

(3) Select 'COPY' option from the modify menu.

Problem 31 : Copy (OR) copies the given circle having centre at location-1 to centre at location-2 (OR) centres at locations 3, 4, 5, ... n etc.

(1) Command : COPY

(2) Select objects : Select circle by clicking on the circle by mouse. Circle-1 will become dotted. ↵ (Enter)

1. (3) Specify base point of displacement or (Multiple) : Click at the centre-1.

   (4) Specify second point of displacement : Click at the location-2 or <use first point as displacement> where you want to move the circle.

2. (3) Specify base point of displacement or (Multiple) : M (Enter)

   (4) Specify base point : Click at the centre-1.

   (5) Specify base point : Specify second point of displacement (or) <use first point as displacement < click at 3, 4, 5, ... n ↵ (Enter)

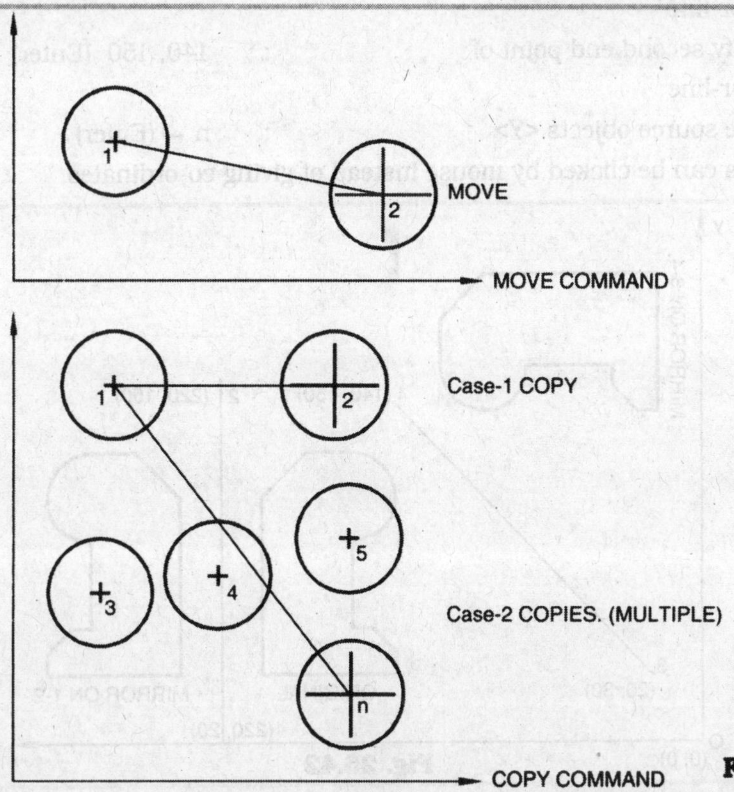

Fig. 26.41

17. Mirror command

Mirror copy of original drawing/object is achieved by operating mirror command. Mirror copy is exactly similar to which we can see in mirror.

Mirror command can be given in three different ways as under.

(1) Click by mouse on 'MIRROR' toolbar icon.

(2) Type by keyboard word 'MIRROR' in command window.

(3) Select MIRROR option from modify menu.

Problem 32 : Prepare 2 mirror images of the given original object on line 1-2 and line 3-4. Points 1(220, 20), 2(220, 150), 3(20, 30) and 4(140, 150).

Solution 1:

Command	:	MIRROR ↵ (Enter)
Select objects	:	Select original object ↵ (Enter)
Specify 1st end point of mirror line	:	220, 20 ↵ (Enter)
Specify 2nd end point of mirror line	:	220, 150 ↵ (Enter)
Delete source objects (N)	:	n (or) press ↵

Solution 2:

Command	:	MIRROR ↵ (Enter)
Select objects	:	Select original object ↵ (Enter)
Specify one end point of Mirror-line	:	(20, 30) ↵ (Enter)
Specify second end point of Mirror-line	:	140, 150 (Enter)
Delete source objects <Y>	:	n ↵ (Enter)

Points can be clicked by mouse instead of giving co-ordinates.

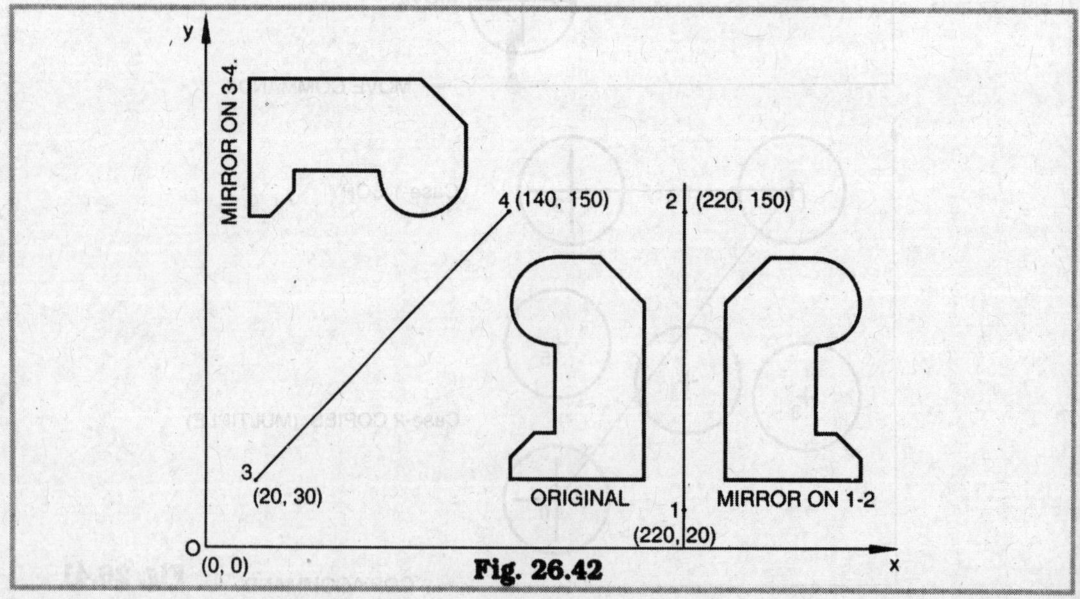

Fig. 26.42

18. Zoom Command

Drawing display can be enlarged or reduced by zoom command. When working on drawing it becomes necessary to bring some portion of drawing to focus on the screen, for editing.

Command	:	Zoom ↵ (Enter)
		[Type word zoom by keyboard]
[All/Centre/Dynamic/Extends/Previous/Scale/Window]		
<real time>	:	Type here first letter of the option by which you want to zoom.

Explanation regarding different options are given below. Some are illustrated in figure below.

Real time	:	It provides convex lens tool.
All	:	Full drawing is visible in the current view port.
Centre	:	First it will ask for centrepoint. Prompt for scale.
Dynamic	:	Zooms to displace the prepared portion of drawing.
Extends	:	Full drawing will be visible on the screen.
Previous	:	Zooms to display the previous view.
Scale	:	Zooms according to specified scale factor.
Window	:	Zooms to display portion specified by a rectangular window.

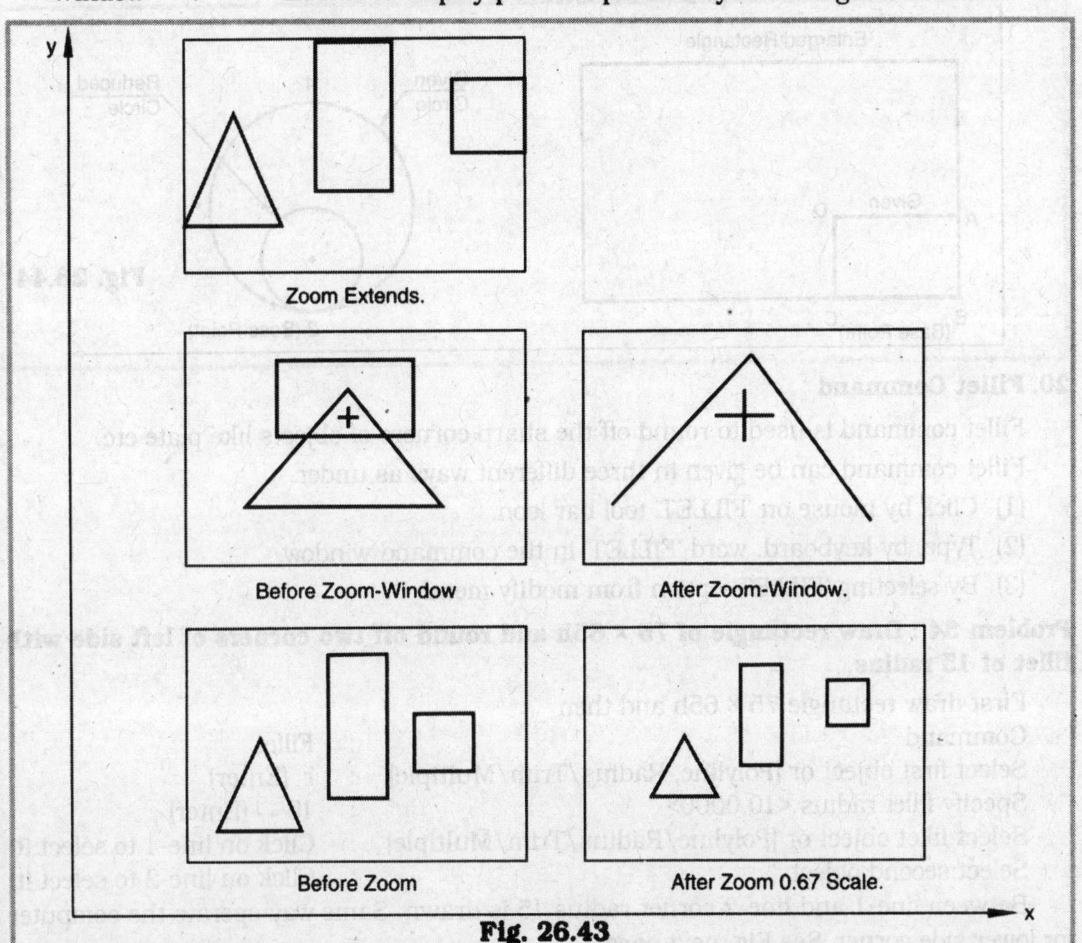

Fig. 26.43

19. Scale Command

Scale command enlarges or reduces the selected shape/object equally in two/three directions. This enlargement or reduction occurs around selected base point. Location of the base point remains same i.e. it does not move. Only the image enlarges or reduces.

Scale command can be given in three different ways, as under.

(1) Click by mouse on 'SCALE' toolbar icon.

(2) Type by keyboard word 'SCALE' in the command window.

(3) By selecting 'SCALE' option from modify menu.

Problem 33 : By using 'SCALE' command

(1) Enlarge the given rectangle ABCD three times around base point B.

(2) Reduce the given circle 1234 0.5 times around base point 2.

Command	:	SCALE
Select objects	:	Click by mouse on (1) Rectangle (2) Circle. (Enter)
Specify base point	:	Click by mouse on (1) Point B of rectangle (2) Point 2 of circle.
Specify scale factor or (Reference)	:	(1) 3 (2) 0.5 ↵ (Enter)

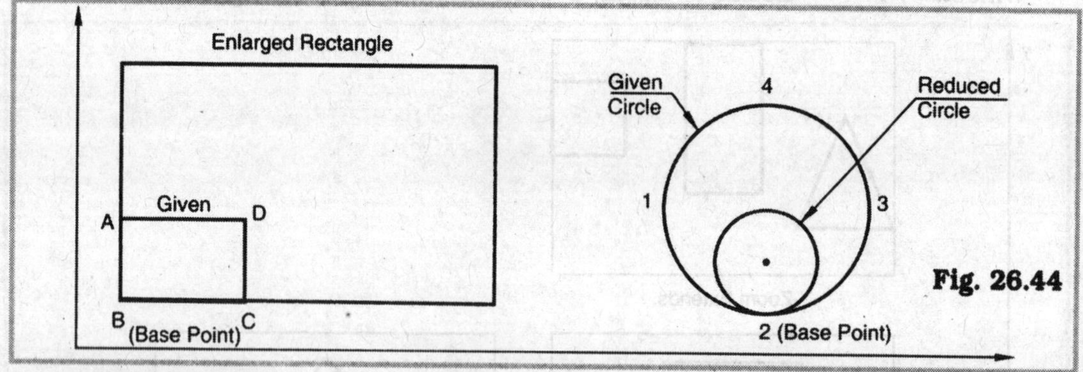

Fig. 26.44

20. Fillet Command

Fillet command is used to round off the sharp corners of objects like plate etc.

Fillet command can be given in three different ways as under.

(1) Click by mouse on 'FILLET' tool bar icon.

(2) Type, by keyboard, word 'FILLET' in the command window.

(3) By selecting 'FILLET' option from modify menu.

Problem 34 : Draw rectangle of 75 × 65h and round off two corners of left side with fillet of 15 radius.

First draw rectangle 75 × 65h and then

Command	:	Fillet
Select first object or [Polyline, Radius/Trim/Multiple]	:	r (Enter)
Specify fillet radius <10.0000>	:	15 ↵ (Enter)
Select fillet object or [Polyline/Radius/Trim/Multiple]	:	Click on line-1 to select it.
Select second object	:	Click on line-2 to select it.

Between line-1 and line-2 corner radius 15 is drawn. Same way operate the computer for lower side corner. See Fig. next page.

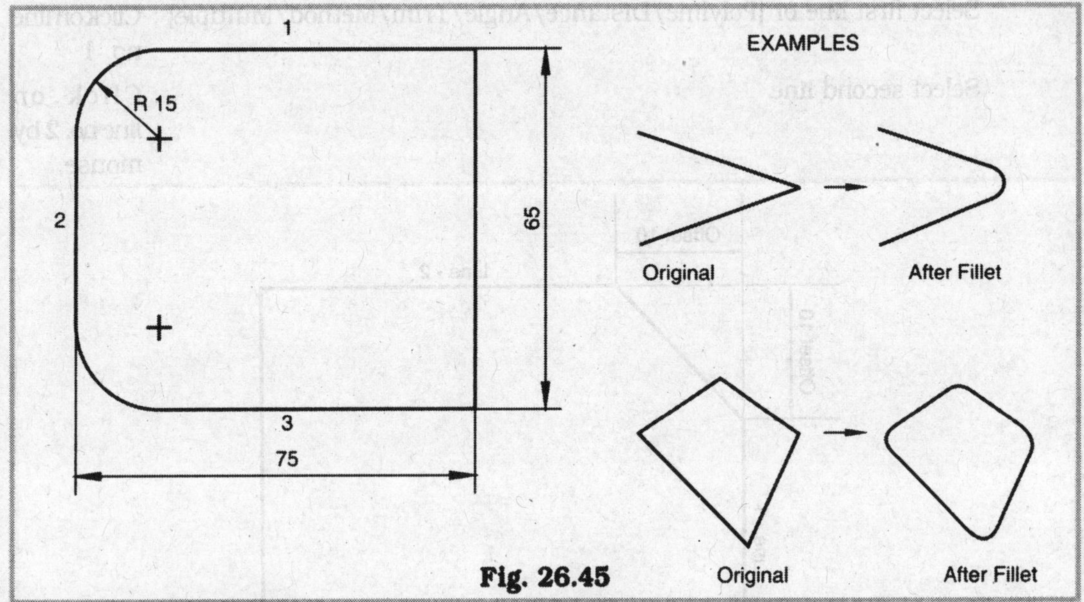

Fig. 26.45

21. Chamfer Command

Chamfer command draws a line at the corner of the object. The chamfer line is drawn by the offset distances along two selected lines of the corner or by one offset distance and the angle.

Chamfer command can be given in the following three ways.

(1) Click by mouse on 'CHAMFER' toolbar icon.

(2) Type by key board word 'CHAMFER' in command window.

(3) By selecting 'CHAMFER' option from modify menu.

Method 1. Two offset distances.

Command	: Chamfer
(Trim mode) current chamfer Dist. 1 = 0.0000, Dist. 2 = 0.0000.	
Select first line or [Polyline/Distance/Angle/Trim/Method/Multiple	: d ↵ (Enter)
Specify first chamfer distance <0.0000>	: 10 ↵ (Enter)
Specify second chamfer distance <0.0000>	: 10 ↵ (Enter)
Select first line or [Polyline/Distance/Angle/Trim/Method/Multiple]	: Select line-1 by clicking over it.

Select second line : Select line-2 by clicking over it by mouse.

Method 2. One offset distance and angle.

Command	: Chamfer
(Trim mode) current chamfer distance 1 = 0.0000 Dist. 2 = 0.0000	
Select first line or [Polyline/Distance/Angle/Trim/Method/Multiple]	: a ↵ (Enter)
Specify chamfer length on the first line < 0.0000 >	: 10 (Enter)
Specify chamfer angle from the first line < 0 >	: 45 ↵ (Enter)

Select first line or [Polyline/Distance/Angle/Trim/Method/Multiple] : Click on line no. 1

Select second line : Click on line no. 2 by mouse.

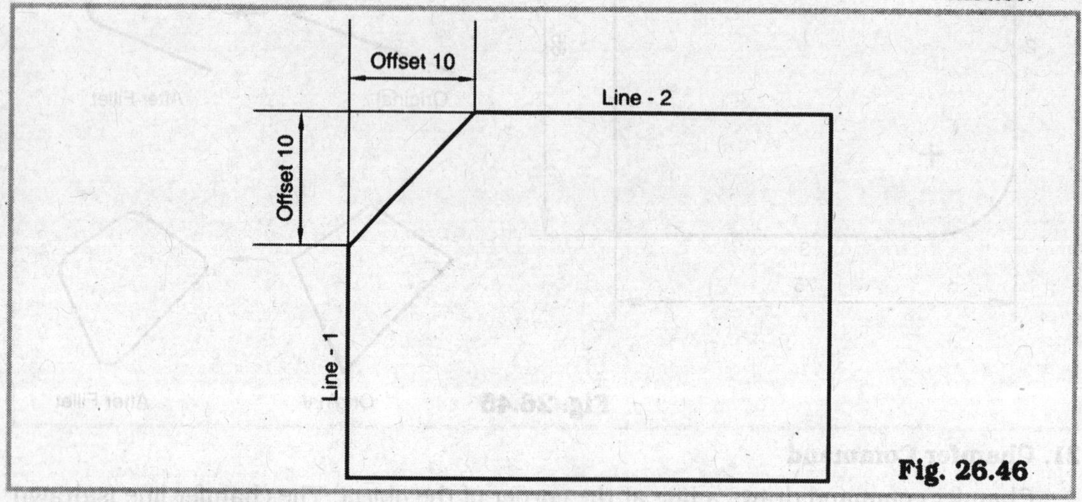

Fig. 26.46

22. Offset Command

Offset command generates a new shape similar to the given shape at a specified distance inside and/or outside the given shape. It can generate new shape also through a specified point. Given shapes such as Circle, Rectangle and Polygon might be get scaled up or down without change in similarity. This command is also used to draw parallel lines, arcs, curves etc.

Offset command can be given in three different ways as under.

(1) Click by mouse on 'OFFSET' toolbar icon.

(2) Type by keyboard word 'OFFSET' in the command window.

(3) By selecting 'OFFSET' option from modify menu.

Problem 35 : Draw

(1) Concentric circle parallel to given circle outside it at a given offset distance.

(2) Rectangle offset to given rectangle inside it at a given offset distance.

(3) Hexagon offset to given hexagon outside it at a given offset distance.

Command : Offset

Specify offset distance or [Through] <5.0000>

↵ (Enter)

Select object to offset or <exit> : Click by mouse on circle or use any method of selection

Specify point on side to offset : Click by mouse any outside point of circle.

Similar procedure is required to draw inside rectangle parallel to given rectangle and outside hexagon parallel to given hexagon.

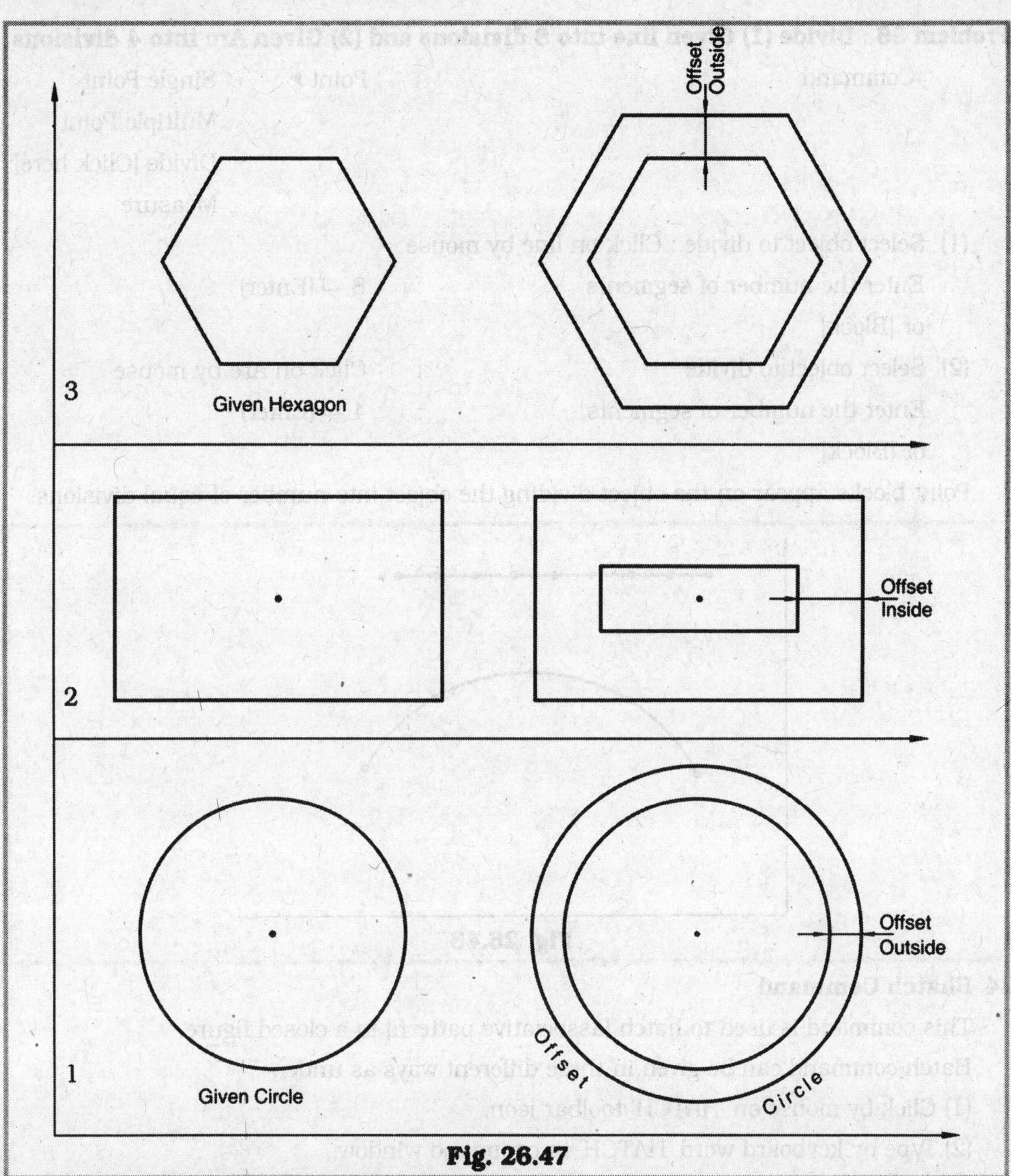

Fig. 26.47

23. Divide Command {Line & Arc Both}

(1) Line and (2) Arc of a circle can be divided into number of equal divisions by 'Divide' command.

This command can be given by draw menu. In the draw menu different commands are given. On the right side, of Point Command, we will have 4 commands as below.

Point ▸ Single Point

Multiple Point

Divide [click the mouse here for divide command]

Measure

Problem 36 : Divide (1) Given line into 8 divisions and (2) Given Arc into 4 divisions.

Command	:	Point ▸	Single Point
			Multiple Point
			Divide [Click here]
			Measure

(1) Select object to divide : Click on line by mouse

Enter the number of segments or [Block] : 8 ↵ (Enter)

(2) Select object to divide : Click on Arc by mouse

Enter the number of segments or [Block] : 4 ↵ (Enter)

Point blocks appear on the object dividing the object into number of equal divisions.

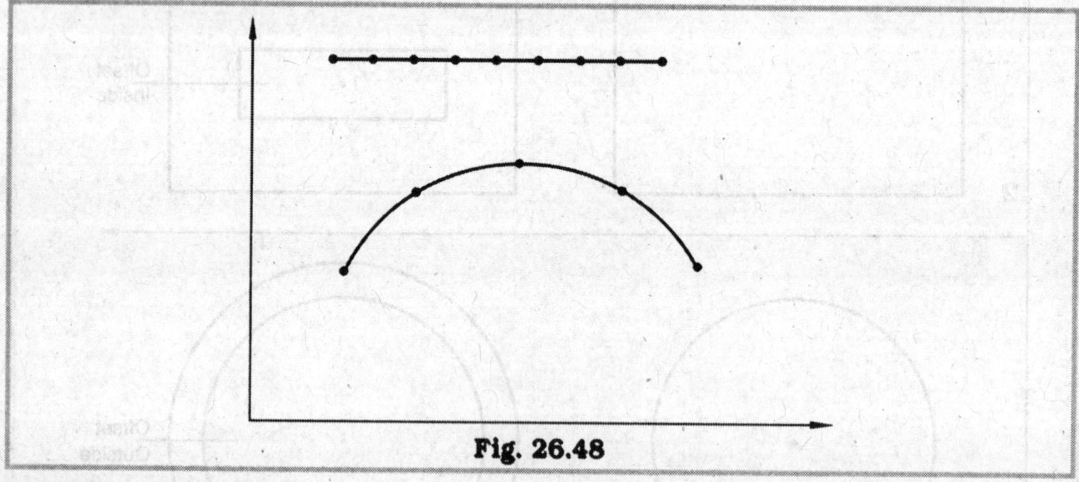

Fig. 26.48

24. Bhatch Command

This command is used to hatch [associative pattern] in a closed figure.

Hatch command can be given in three different ways as under.

(1) Click by mouse on 'HATCH' toolbar icon.

(2) Type by keyboard word 'HATCH' in command window.

(3) By selecting 'HATCH' option from draw menu.

Moment you select 'HATCH' command a dialog box will appear with various options.

Left side:

Type	:	Decide predefined or user defined pattern
Pattern	:	One can select any one predefined standard hatch pattern out of many available. If you click the side button you will see different patterns with their names. Whichever pattern you select, sample of it will be visible below.
Angle	:	One can decide angle of pattern. This is the angle of inclination of hatch lines.
Scale	:	Write value for scale. This defines the distance between hatching lines.

Right Side:

Pick Points : Pick by mouse an internal point in the space to be hatched.

Click now o.k. button. You will see on screen area already hatched by the pattern, you have selected.

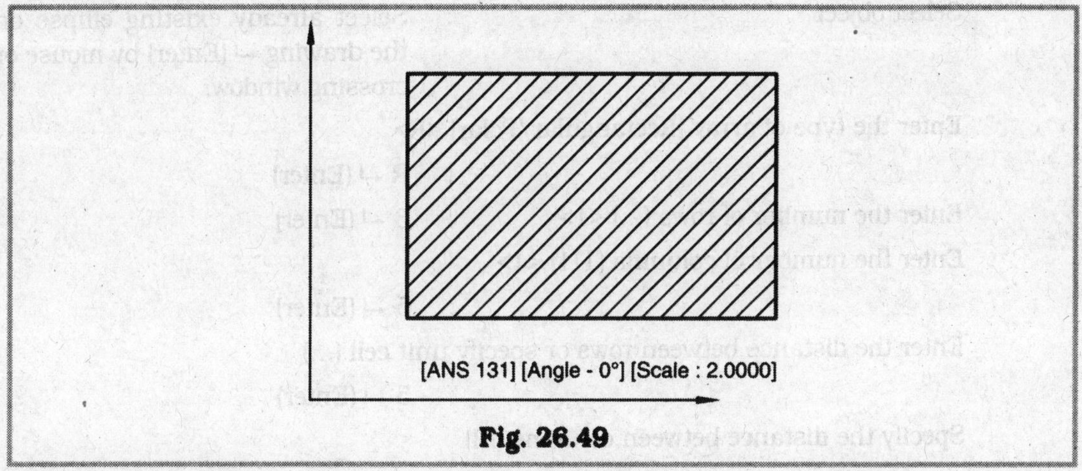

Fig. 26.49

25. Array Command

Array command facilitates us to create multiple copies of selected object in a rectangular or a polar/circular pattern. Rectangular array is used for seating arrangements in auditorium, tiles fitting in hall etc. Polar array is used for gear teeth, sprocket teeth etc.

Array command can be given in three different ways as under

(1) Click by mouse on 'Array' toolbar icon.

(2) Type by keyboard word 'ARRAY' in command window.

(3) Select 'ARRAY' option from the modify menu.

After selecting ARRAY command computer will ask for the followings.

(1) Select the object to be copied.

(2) Rectangular or circular/polar pattern.

(3) Number of rows for rectangular pattern

(4) Number of columns for rectangular pattern

(5) Unit cell or distance between rows

(6) Distance between columns

OR

(3) Centre point of ARRAY for circular/polar pattern

(4) Number of items in the array

(5) Angle to fill

(6) Rotate arrayed objects.

Two illustrative problems are given here to practice and grasp the 'ARRAY' command. First problem is for rectangular pattern and second problem is for polar/circular pattern.

Problem 37 :

Command : ARRAY ↵ (Enter)

Select object : Select already existing ellipse on the drawing ↵ (Enter) by mouse or crossing window.

Enter the type of array [Rectangular/Polar] <R>

: R ↵ (Enter)

Enter the number of rows (...) <1> : 3 ↵ (Enter)

Enter the number of columns (111) <1>

: 5 ↵ (Enter)

Enter the distance between rows or specify unit cell (...)

: 5 ↵ (Enter)

Specify the distance between columns (III)

: 3 ↵ (Enter)

Problem 38 :

Command : ARRAY ↵ (Enter)

Select Object : Select already existing ellipse on the drawing ↵ (Enter) by mouse or crossing window

Enter the type of array [Rectangular/Polar] <R>

: P ↵ (Enter)

Locate centre point of array : 130, 140 ↵ (Enter)

Specify number of items in the array

: 6 ↵ (Enter)

Specify angle to fill (+ = CCW, – = CW) <360> 90 (Enter)

Rotate arrayed objects: [Yes/No] <y> ↵ (Enter)

Here CCW = Counter Clock-wise [+]

CW = Clock-wise [–]

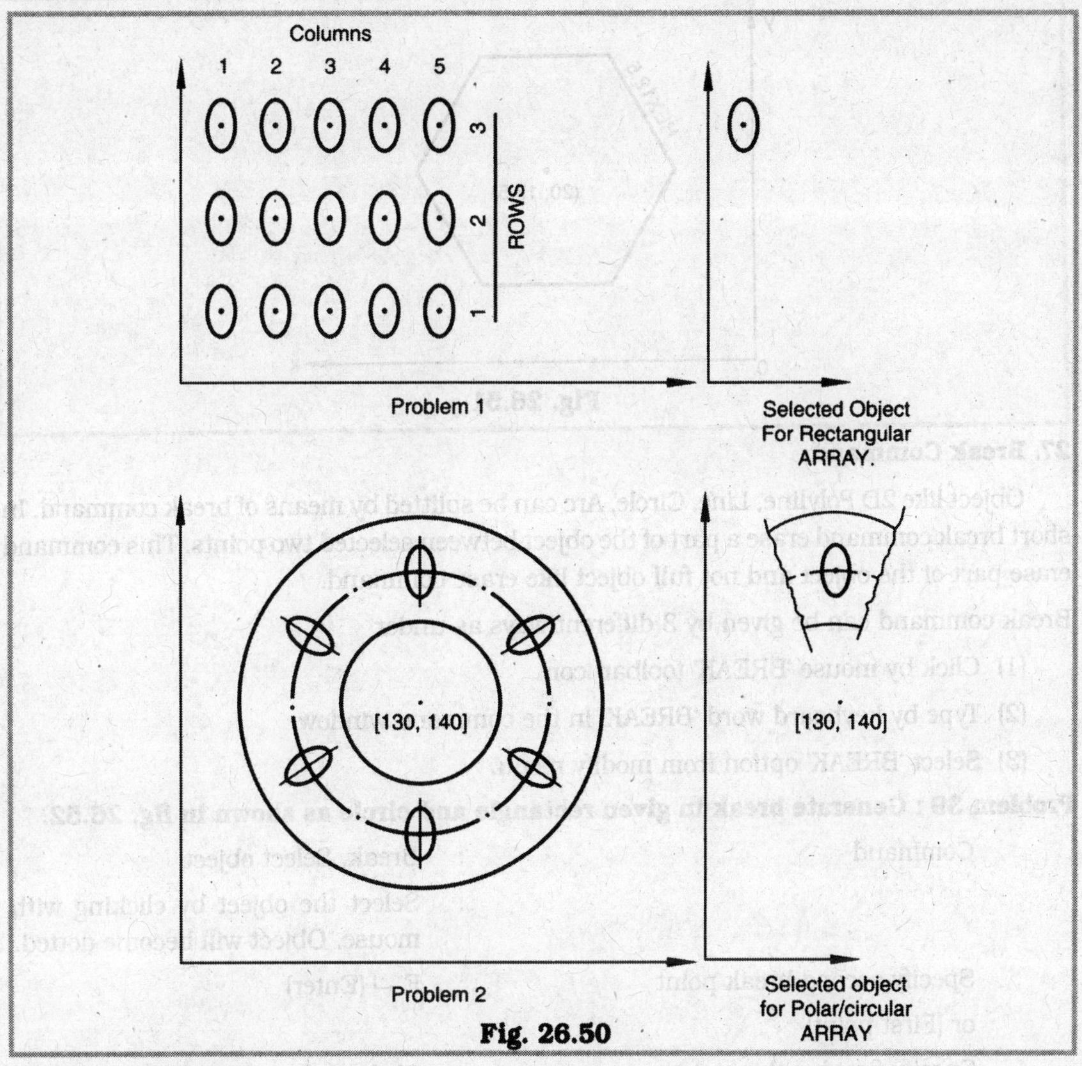

Fig. 26.50

26. Text Command

Text command creates a single line text in the drawing.

Text command can be given in three different ways as under.

(1) Click by mouse on 'TEXT' toolbar icon.

(2) Type by keyboard word 'TEXT' in the command window.

(3) By selecting 'TEXT' option from draw menu.

Command	:	TEXT ↵ (Enter)
Current text style "standard" text height	:	0.5000
Specify start point of text or [Justify / Style]	:	7, 19 (Enter)
Specify height <0.5000>	:	2.0000 ↵ (Enter)
Specify rotation angle of text <0>	:	60
Enter text	:	HEX 12.5

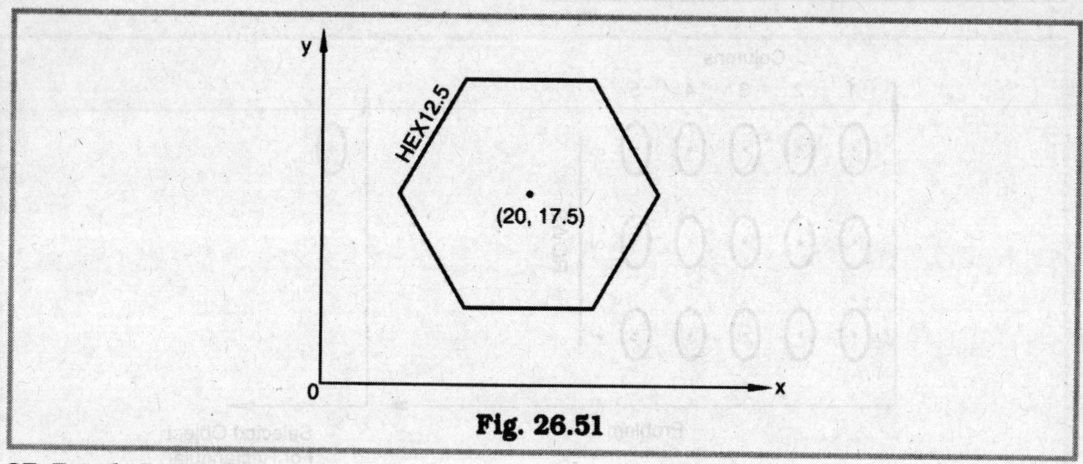

Fig. 26.51

27. Break Command

Object like 2D Polyline, Line, Circle, Arc can be splitted by means of break command. In short break command erase a part of the object between selected two points. This command erase part of the object and not full object like erase command.

Break command can be given by 3 different ways as under.

(1) Click by mouse 'BREAK' toolbar icon.

(2) Type by keyboard word 'BREAK' in the command window.

(3) Select 'BREAK' option from modify menu.

Problem 39 : Generate break in given rectangle and circle as shown in fig. 26.52.

Command	:	Break, Select object
	:	Select the object by clicking with mouse. Object will become dotted.
Specify second break point or [First point]	:	F ↵ (Enter)
Specify first break point	:	Click on first break point
Specify second break point	:	Click on second break point

In a closed object, such as circle, portion between first point and second point is removed in a counter clockwise direction.

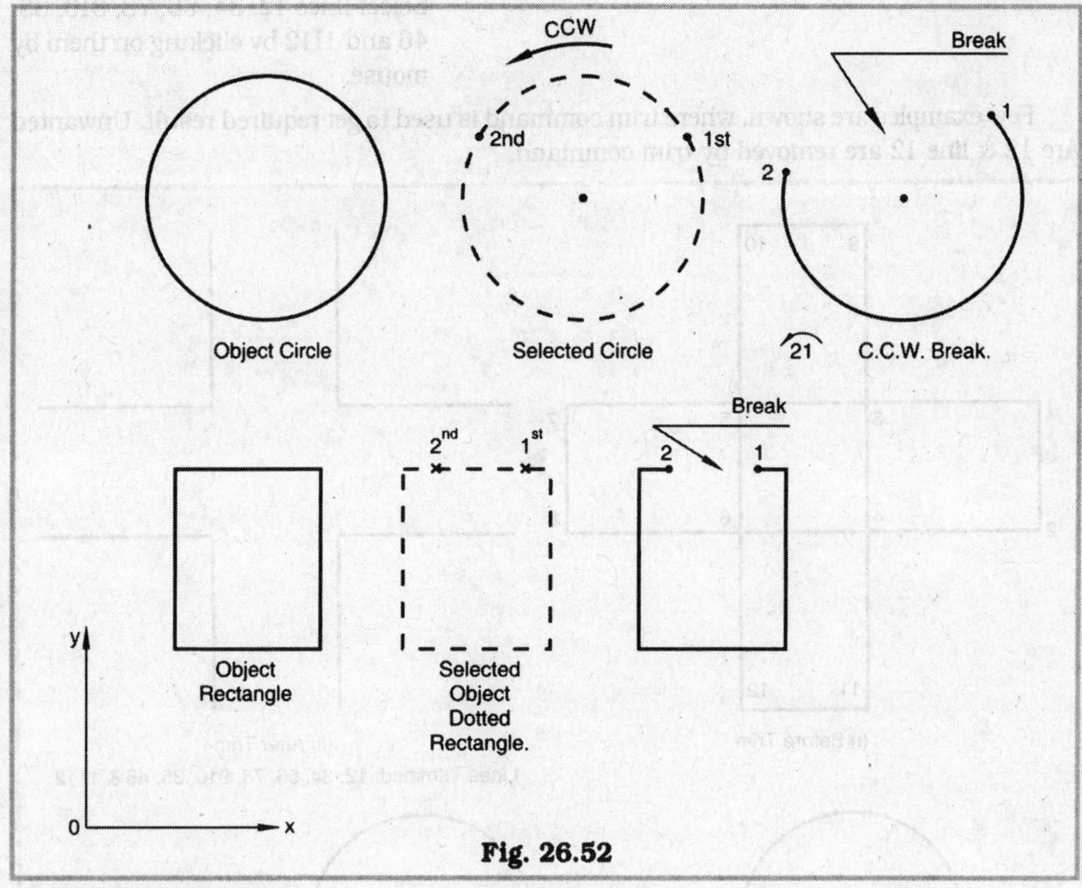

Fig. 26.52

28. TRIM Command

TRIM command erases the portions of selected entities that cross a specified boundary. In fact unwanted entities are removed.

Trim command can be given in the following three ways.

(1) Click by mouse on 'TRIM' toolbar icon.

(2) Type by keyboard word 'TRIM' in command window.

(3) By selecting TRIM option from modify menu.

After selecting 'TRIM' command, computer will ask for selecting objects and selecting objects to trim.

Problem 40 : Two rectangles 135786421 and 910561211439 are crossing as shown. Using trim command trim lines 12, 34, 56, 78, 910, 35, 46 and 11 - 12. Example of two roads crossing at right angle.

Command	:	TRIM
Select objects	:	Click by mouse both rectangles. (Enter)
Select objects to trim or shift		
Select to extend or [Project/Edge/Undo]		

: Select lines 12, 34, 56, 78, 910, 35, 46 and 1112 by clicking on them by mouse.

Few examples are shown, where trim command is used to get required result. Unwanted Arc 12 & line 12 are removed by trim command.

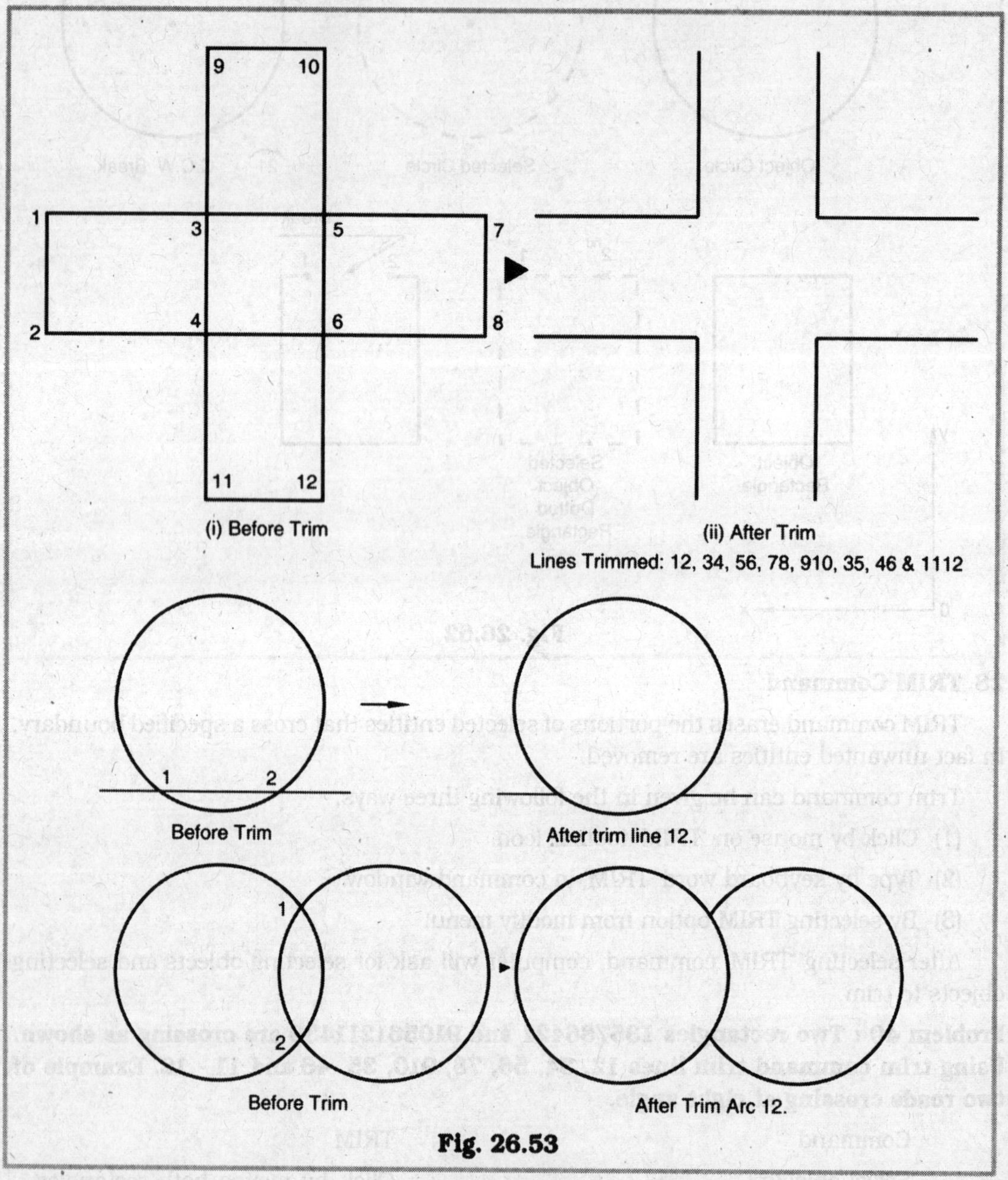

Fig. 26.53

Problems of Computer Aided Drawings:

Having studied various commands on AutoCAD, let us draw various shapes/drawings on AutoCAD. To understand, grasp and master different commands one must do sufficient practice of solving problems.

Problem 41 : Draw 'KITE' shown in fig. below, using CADD.

(1)	Command	:	Line [click toolbar icon
	Line specify first point	:	110, 10 ↵ (Enter) [Point-1]
	Specify next point or (Undo)	:	@ 30 < 0 ↵ (Enter) [line 1-2]
	Specify next point or (Undo)	:	@ – 100, 100 ↵ (Enter) [Line 2-3]
	Specify next point or [Close/Undo]	:	@ 70, 70 ↵ (Enter) [line 3-4]
	Specify next point or [Close/Undo]	:	C ↵ (Enter) [line 4-1]
(2)	Command	:	Mirror ↵ (Enter)
	Select objects	:	Select lines 1-2, 2-3 and 3-4 by click ing over these lines ↵ (Enter)
	Specify first point of Mirror line	:	110, 10 ↵ (Enter)
	Specify second point to Mirror line	:	@ 0, 130 ↵ (Enter)
	Delete source objects [Yes/No]	:	n ↵ (Enter)
(3)	Command	:	Arc [start, end, direction]
	Specify start point of arc or [centre]	:	40, 110 ↵ (Enter)
	Specify end point of arc	:	@ 140, 0 ↵ (Enter)
	Specify tangent direction for the start point of arc	:	45 ↵ (Enter)

'KITE' will be ready on the screen.

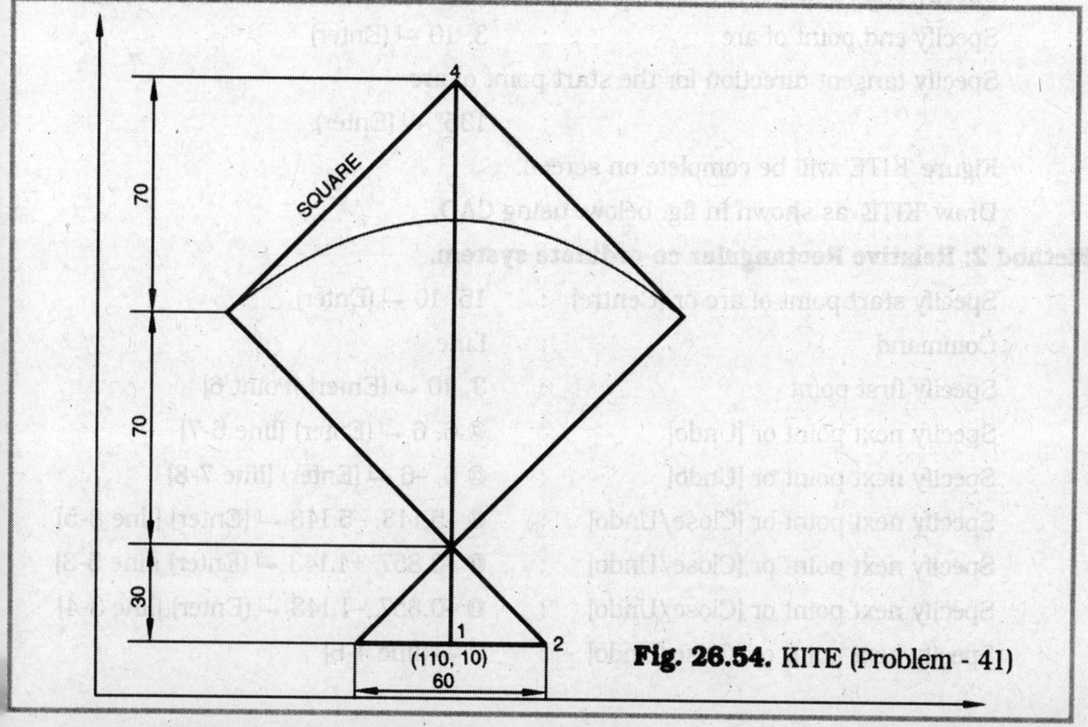

Fig. 26.54. KITE (Problem - 41)

Problem 42 : Draw 'KITE' as shown in fig. below, using CAD.

Method 1. Absolute coordinate system.

Command	:	Line
Specify first point	:	3, 10 ↵ (Enter)
Specify next point or (Undo)	:	9, 16 ↵ (Enter)
Specify next point or [Undo]	:	15, 10 ↵ (Enter)
Specify next point or [Close/Undo]	:	9.857, 4.857 ↵ (Enter)
Specify next point or [Close/Undo]	:	9, 6 ↵ (Enter)
Specify next point or [Close/Undo]	:	8.143, 4.857 ↵ (Enter)
Specify next point or [Close/Undo]	:	C ↵ (Enter)
Command	:	Line
Specify first point	:	8.143, 4.857 ↵ (Enter)
Specify next point or (Undo)	:	6, 2 ↵ (Enter)
Specify next point or (Undo)	:	12, 2 ↵ (Enter)
Specify next point or [Close/Undo]	:	9.857, 4.857 ↵ (Enter)
Command	:	Line
Specify first point	:	9, 2 ↵ (Enter)
Specify next point or [Undo]	:	9, 16 ↵ (Enter)
		↵ (Enter)
Command	:	Arc [Start, end, direction]
Specify start point of arc or [Centre]	:	15, 10 ↵ (Enter)
Specify end point of arc	:	3, 10 ↵ (Enter)
Specify tangent direction for the start point of arc		
	:	135° ↵ (Enter)

Figure 'KITE' will be complete on screen.

Draw 'KITE' as shown in fig. below, using CAD.

Method 2: Relative Rectangular co-ordinate system.

Specify start point of arc or [Centre]	:	15, 10 ↵ (Enter)
Command	:	Line
Specify first point	:	3, 10 ↵ (Enter) [Point 6]
Specify next point or [Undo]	:	@ 6, 6 ↵ (Enter) [line 6-7]
Specify next point or [Undo]	:	@ 6, –6 ↵ (Enter) [line 7-8]
Specify next point or [Close/Undo]	:	@ –5.143, –5.143 ↵ (Enter) [line 8-5]
Specify next point or [Close/Undo]	:	@ –0.857, +1.143 ↵ (Enter) [line 5-3]
Specify next point or [Close/Undo]	:	@ –0.857, –1.143 ↵ (Enter) [line 3-4]
Specify next point or [Close/Undo]	:	C ↵ [line 4-6]

Command	:	Line
Specify first point	:	8.143, 4.857 ↵ (Enter) [Point 4]
Specify next point or [Undo]	:	@ –2.143, –2.857 ↵ (Enter) [line 4-1]
Specify next point or [Undo]	:	@ 6, 0 ↵ (Enter) [line 1-2]
Specify next point or [Undo]	:	@ –2.143, 2.857 ↵ (Enter) [line 2-5]
	:	↵ (Enter) Command completed.
Command	:	line
Specify first point	:	9, 2 ↵ (Enter) [Point 9]
Specify next point or [Undo]	:	@ 0, 14 ↵ (Enter) [line 9-7]
		↵ (Enter) Command completed.
Command	:	Arc [Start, End, Direction]
Specify starting point of arc or (Centre)	:	15, 10 ↵ (Enter)
Specify end point of arc	:	@ –12, 0 ↵ (Enter)
Specify tangent direction for the start point of arc	:	135° ↵ (Enter)

Fig 'KITE' will be complete on screen.

save it.

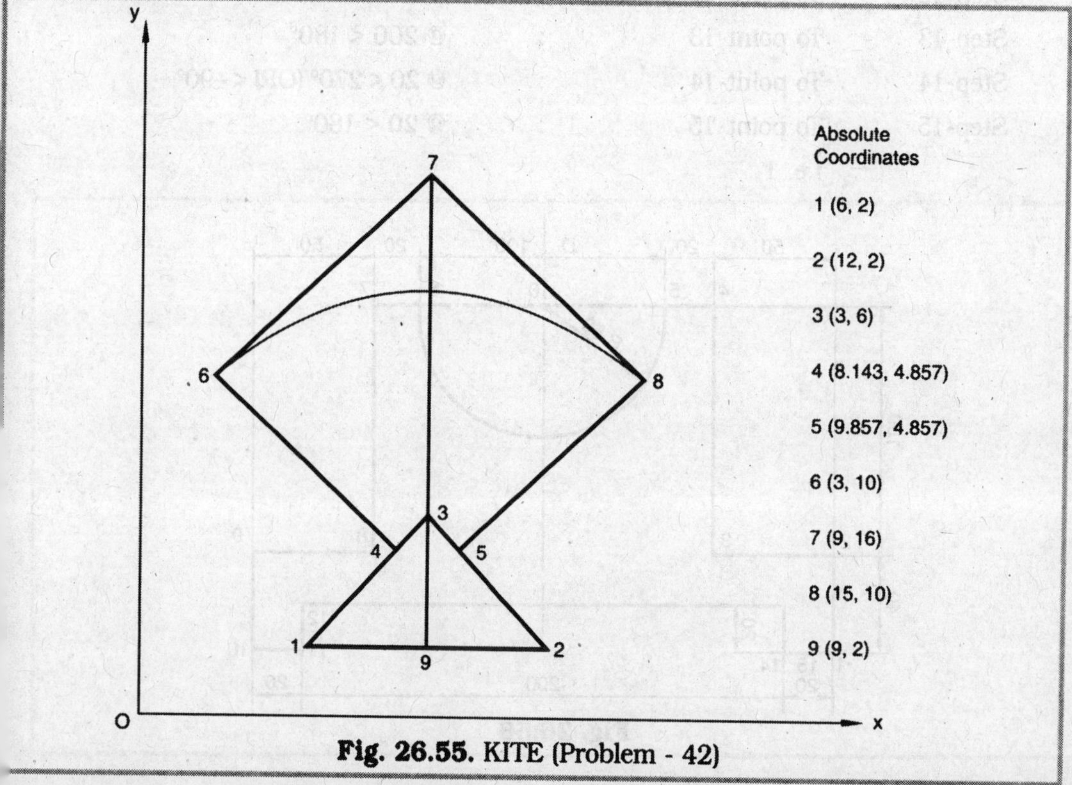

Fig. 26.55. KITE (Problem - 42)

Problem 43 : Prepare given drawing by CAD to scale and write steps to prepare it.

Step-1	Pick up any point-1 and start.		
Step-2	Command	:	Line
	To point-2	:	@ 40 < 90°
Step-3	To point-3	:	@ 50 < 0°
Step-4	To point-4	:	@ 100 < 90°
Step-5	To point-5	:	@ 20 < 0°
Step-6	Command	:	ARC
	Starting Point	:	5
	End Point	:	6 @ 100 < 0°
	Radius	:	50
	Centre-0	:	@ 50 < 0°
Step-7	Command	:	Line
	To point-7	:	@ 20 < 0°
Step-8	To point-8	:	@ 100 < 270° (OR) < –90°
Step-9	To point-9	:	@ 50 < 0°
Step-10	To point-10	:	@ 40 < 270° (OR) < – 90°
Step-11	To point-11	:	@ 20 < 180°
Step-12	To point-12	:	@ 20 < 90°
Step-13	To point-13	:	@ 200 < 180°
Step-14	To point-14	:	@ 20 < 270° (OR) < –90°
Step-15	To point-15	:	@ 20 < 180°
	i.e. 1		

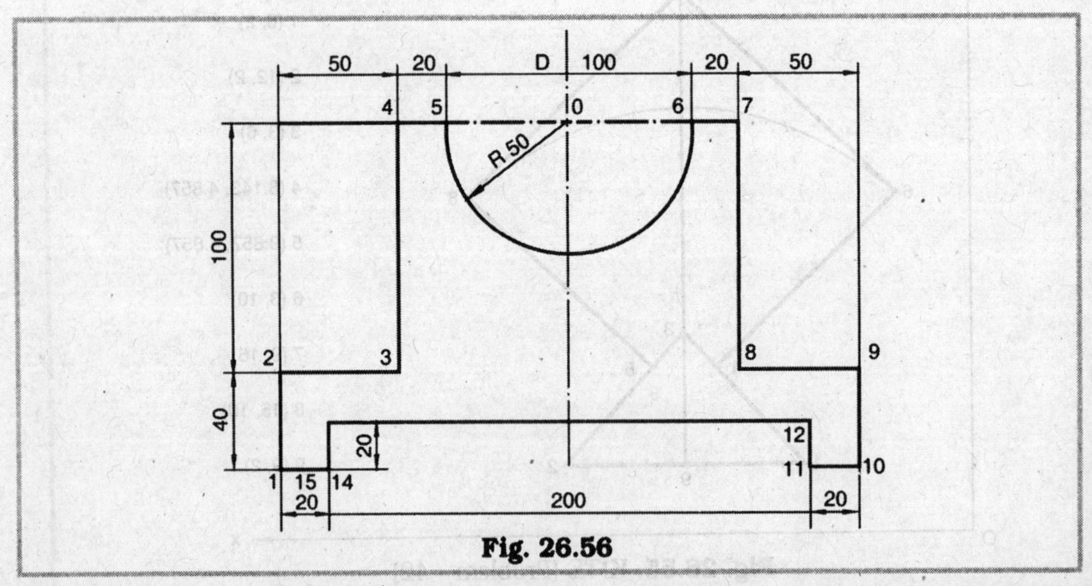

Fig. 26.56

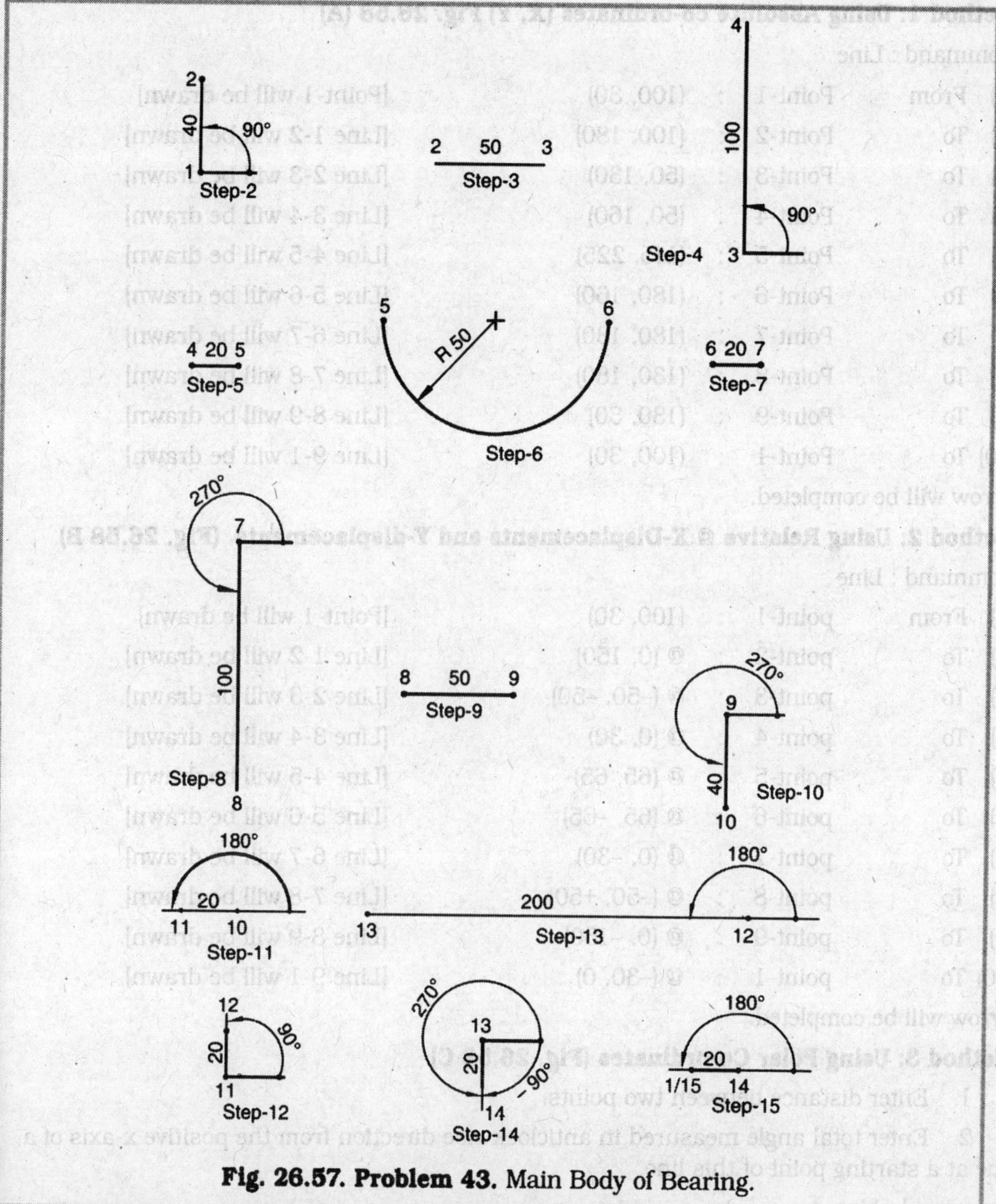

Fig. 26.57. Problem 43. Main Body of Bearing.

Problem 44 : Draw Arrow as shown in figure using line segments. Use line command and three methods as under.

(1) Using Absolute Co-ordinates Fig. 26.58 (A)

(2) Using Relative Co-ordinates Fig. 26.58 (B)

(3) Using Polar Co-ordinates Fig. 26.58 (C)

Method 1: Using Absolute co-ordinates (X, Y) Fig. 26.58 (A)

Command : Line

(1) From Point-1 : (100, 30) [Point-1 will be drawn]

(2) To Point-2 : (100, 180) [Line 1-2 will be drawn]

(3) To Point-3 : (50, 130) [Line 2-3 will be drawn]

(4) To Point-4 : (50, 160) [Line 3-4 will be drawn]

(5) To Point-5 : (115, 225) [Line 4-5 will be drawn]

(6) To Point-6 : (180, 160) [Line 5-6 will be drawn]

(7) To Point-7 : (180, 130) [Line 6-7 will be drawn]

(8) To Point-8 : (130, 180) [Line 7-8 will be drawn]

(9) To Point-9 : (130, 30) [Line 8-9 will be drawn]

(10) To Point-1 : (100, 30) [Line 9-1 will be drawn]

Arrow will be completed.

Method 2: Using Relative @ X-Displacements and Y-displacements. (Fig. 26.58 B)

Command : Line

(1) From point-1 : (100, 30) [Point-1 will be drawn]

(2) To point-2 : @ (0, 150) [Line 1-2 will be drawn]

(3) To point-3 : @ (–50, –50) [Line 2-3 will be drawn]

(4) To point-4 : @ (0, 30) [Line 3-4 will be drawn]

(5) To point-5 : @ (65, 65) [Line 4-5 will be drawn]

(6) To point-6 : @ (65, –65) [Line 5-6 will be drawn]

(7) To point-7 : @ (0, –30) [Line 6-7 will be drawn]

(8) To point-8 : @ (–50, +50) [Line 7-8 will be drawn]

(9) To point-9 : @ (0, –150) [Line 8-9 will be drawn]

(10) To point-1 : @ (–30, 0) [Line 9-1 will be drawn]

Arrow will be completed.

Method 3: Using Polar Co-ordinates [Fig. 26.58 C]

1. Enter distance between two points.

2. Enter total angle measured in anticlockwise direction from the positive x-axis of a line at a starting point of this line.

Command : Line

(1) From Point-1 : (100, 30) [Point-1 will be drawn]

(2) To Point-2 : @ 150 < 90° [Line 1-2 will be drawn]

(3) To Point-3 : @ 70.71<225° [Line 2-3 will be drawn]

(4) To Point-4 : @ 30 < 90° [Line 3-4 will be drawn]

(5) To Point-5 : @ 91.92<45° [Line 4-5 will be drawn]

(6) To Point-6 : @ 91.92<315°OR–45° [Line 5-6 will be drawn]

(7) To Point-7 : @ 30 < 270° OR –90° [Line 6-7 will be drawn]

(8) To Point-8 : @ 70.71 < 135° [Line 7-8 will be drawn]

(9) To Point-9 : @ 150 < 270° [Line 8-9 will be drawn]

(10) To Point-1 : @ 30 < 180° [Line 9-1 will be drawn]

Arrow will be completed.

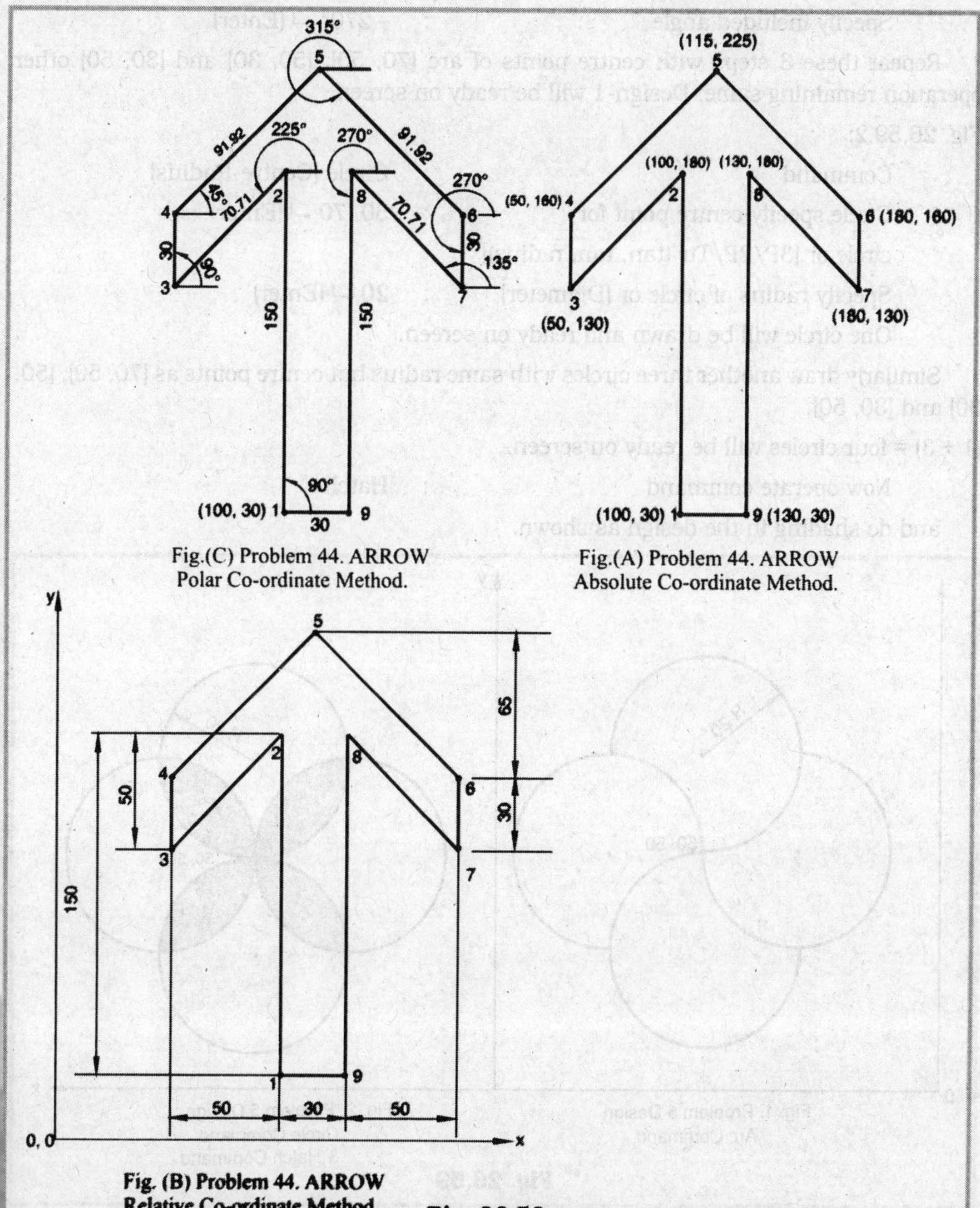

Fig.(C) Problem 44. ARROW
Polar Co-ordinate Method.

Fig.(A) Problem 44. ARROW
Absolute Co-ordinate Method.

Fig. (B) Problem 44. ARROW
Relative Co-ordinate Method.

Fig. 26.58

Problem 45 : Draw given two simple designs using CAD.

Command	:	Arc [Start – Centre – Angle]
Arc Specify start point of arc or [Centre]	:	50, 50 ↵ (Enter)
Specify Centre point of arc	:	50, 70 ↵ (Enter)
Specify included angle	:	– 270° ↵ (Enter)

Repeat these 3 steps with centre points of arc [70, 50], [50, 30] and [30, 50] other operation remaining same. Design-1 will be ready on screen.

Fig. 26.59.2:

Command	:	Circle [Centre-Radius]
Circle specify centre point for circle or [3P/2P/Ttr (tan, tan, radius)]	:	50, 70 ↵ (Enter)
Specify radius of circle or [Diameter]	:	20 ↵ (Enter)

One circle will be drawn and ready on screen.

Similarly draw another three circles with same radius but centre points as [70, 50], [50, 30] and [30, 50].

(1 + 3) = four circles will be ready on screen.

Now operate command : Hatch

and do shading in the design as shown.

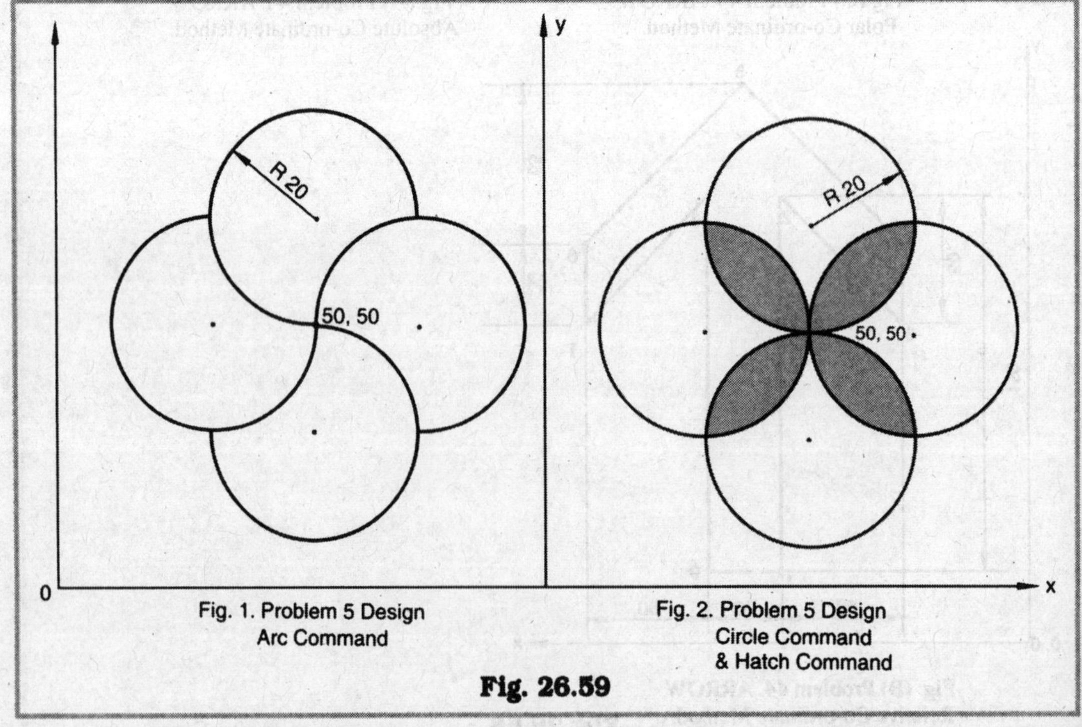

Fig. 1. Problem 5 Design Arc Command

Fig. 2. Problem 5 Design Circle Command & Hatch Command

Fig. 26.59

Problem 46 : Draw 'JET PLANE' by using CAD.

(1)	Command	:	Line [click by mouse on toolbar icon	
	line specify first point	:	5, 15 ↵ (Enter)	[Point-1]
	Specify next point or [Undo]	:	@ 15, 10 ↵ (Enter)	[line 1-2]
	Specify next point or [Undo]	:	@ 105, 0 ↵ (Enter)	[line 2-3]
	Specify next point or [Close/Undo]	:	@ 40, –10 ↵ (Enter)	[line 3-4]
			↵ (Enter)	
(2)	Command	:	Line	
	line, specify first point	:	45, 25 ↵ (Enter)	[Point-5]
	Specify next point or [Undo]	:	@ 25, 50 ↵ (Enter)	[line 5-6]
	Specify next point or [Undo]	:	@ 10, 0 ↵ (Enter)	[line 6-7]
	Specify next point or [Close/Undo]	:	@ –10, –30 ↵ (Enter)	[line 7-8]
	Specify next point or [Close/Undo]	:	@ 10, –20 ↵ (Enter)	[line 8-9]
			↵ (Enter)	
(3)	Command	:	Line	
	line, specify first point	:	125, 25 ↵ (Enter)	[Point-3]
	Specify next point or [Undo]	:	@ 15, 30 ↵ (Enter)	[line 3-10]
	Specify next point or [Undo]	:	@ 15, 0 ↵ (Enter)	[line 10-11]
	Specify next point or [Close/Undo]	:	@ 0, –37.5 ↵ (Enter)	[line 11-12]
			↵ (Enter)	
(4)	Command	:	Mirror [click by mouse on toolbar icon	
	Select objects	:	Click on lines 1-2, 2-3, 3-4, 5-6, 6-7, 7-8, 8-9, 3-10, 10-11, 11-12.	
			[These lines will be dotted]	
			↵ (Enter)	
	Specify first point of mirror line	:	5, 15 ↵ (Enter)	
	Specify second point of mirror line	:	@ 165, 0 ↵ (Enter)	
	Delete source objects [Yes/No]	:	n ↵ (Enter)	
(5)	Command	:	line	
	line specify first point	:	20, 25 ↵ (Enter)	[point-2]
	Specify second point or [Undo]	:	@ 0, –20 ↵ (Enter)	[line 2-13]
			↵ (Enter)	
(6)	Command	:	line	
	line, specify first point	:	125, 25 ↵ (Enter)	[point-3]
	Specify second point or [Undo]	:	@ 0, –20 ↵ (Enter)	[line 3-14]
			↵ (Enter)	

Fig. 26.60 'JET PLANE' will be ready on screen. Save it.

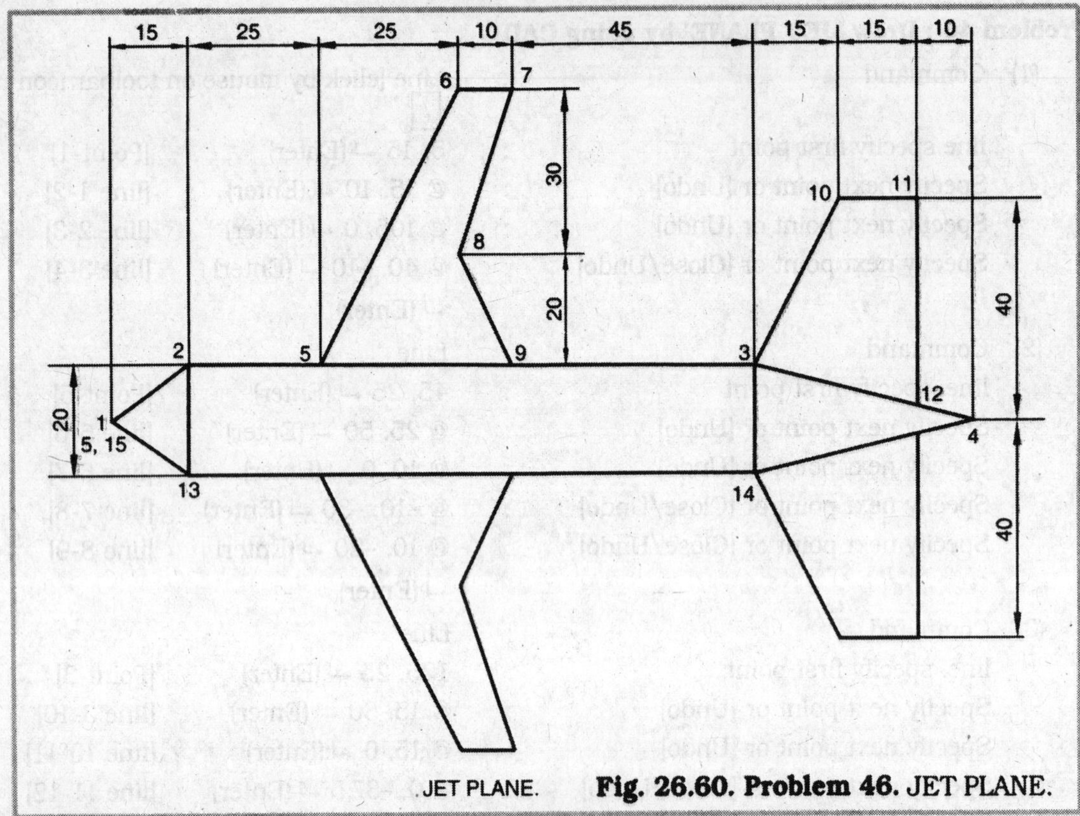

Fig. 26.60. Problem 46. JET PLANE.

Problem 47 : Draw given design using following commands of the CAD.

(1) Line Command

(2) Arc Command

(3) Copy Command

(4) Mirror Command

(5) Erase Command, if required.

(1)	Command	:	Line	
	line specify first point	:	10, 10 ↵ (Enter)	Point-1
	Specify next point or [Undo]	:	90, 10 ↵ (Enter)	Line (1-2)
	Specify next point or [Undo]	:	90, 30 ↵ (Enter)	Line (2-3)
	Specify next point or [Close/Undo]	:	10, 30 ↵ (Enter)	Line (3-4)
	Specify next point or [Close/Undo]	:	C ↵ (Enter)	Line (4-1)
	Rectangle 1-2-3-4 is drawn.			
(2)	Command	:	line	
	line specify first point	:	10, 20 ↵ (Enter)	Point-5
	Specify next point or [Undo]	:	@ 10, 10 ↵ (Enter)	line 5-6
	Specify next point or [Undo]	:	@ 5, –5 ↵ (Enter)	line 6-7
			↵ (Enter)	

(3)	Command	: Arc [Start, End, Direction]
	arc specify start point of arc or [Centre]	: 10, 30 ↵ (Enter)
	Specify end point of arc	: @ 10, 0 ↵ (Enter)
	Specify tangent direction for the start point of arc	: 90° ↵ (Enter)
	Semi Circular Arc 4-6 is drawn.	
(4)	Command	: Copy (Multiple Copy)
	Select Object	: Select semi circle 4-6, line 5-6 and line 6-7 by clicking mouse over arc 4-6, line 5-6 and line 6-7.↵ (Enter)
	Specify base point or displacements or [Multiple]	: M ↵ (Enter)
	Specify base point	: 20, 30 ↵ (Enter)
	Specify second point of displacement or <use first point as displacement>	: 30, 30 ↵ (Enter) 40, 30 ↵ (Enter) 50, 30 ↵ (Enter) 60, 30 ↵ (Enter) 70, 30 ↵ (Enter) 80, 30 ↵ (Enter) 90, 30 ↵ (Enter) ↵ (Enter)
(5)	Command	: Erase
	Select object	: line similar to 6-7 only outside the rectangle 1234 after copying click over it. ↵ (Enter)
(6)	Command	: Mirror
	Select objects	: Semicircles 4-6 and similar. line 5-6, 6-7 and similar line 8-3. ↵ (Enter)
	Specify first point of Mirror line	: 10, 20 ↵ (Enter)
	Specify second point of Mirror line	: @ 80, 0 ↵ (Enter)
	Delete source objects [Yes/No]	: n ↵ (Enter)
(7)	Command	: ARC [Start/End/Direction]
	arc specify start point of arc or [Centre]	: 10, 30 ↵ (Enter)
	Specify end point of arc	: @ 0, – 20 ↵ (Enter)
	Specify tangent direction for the start point of arc	: 180 ↵ (Enter)

(8) Similarly draw arc on rightside of rectangle on line 23.

Command	:	arc [Start, end, direction]
arc specify start point of arc or [Centre]	:	90, 10 ↵ (Enter)
Specify end point of arc	:	90, 30 ↵ (Enter)
Specify tangent direction for the start point of arc	:	0 ↵ (Enter)

Figure will be ready on screen. Save it.

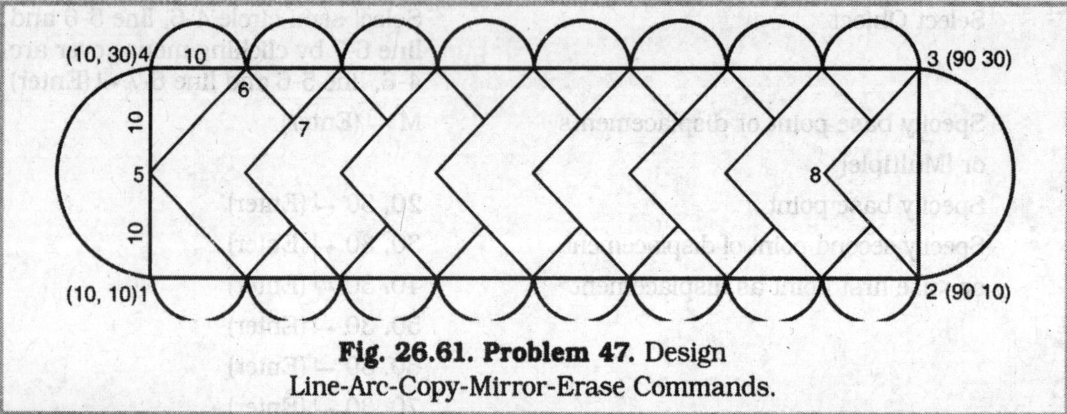

Fig. 26.61. Problem 47. Design Line-Arc-Copy-Mirror-Erase Commands.

EXERCISE

1. What is AutoCAD? Explain?
2. Enlist and explain in details advantages of AutoCAD. Compare with mannual drawing.
3. Explain different elements of computer.
   (i) Key Board (ii) Mouse (iii) Joy Stick
   (iv) Digitizer (v) Emage Scanner (vi) Moniter
   (vii) Dot Matrix Printer (viii) Laser Printer (ix) Inject Printer
   (x) Plotter.
4. Make a list of Commands Aliase.
5. Explain with illustrations, different methods of selecting objects for editing purpose.
6. Explain with illustration following commands.
   (a) Point/Multipoint
   (b) Line
   (c) Multiline
   (d) Rectangle
   (e) Circle
   (f) Arc
   (g) Ellipse
   (h) Polygon
   (i) Donut
   (j) Spline
   (k) Solid
   (l) Pan
   (m) Erase
   (n) Rotate
   (o) Move
   (p) Copy
   (q) Mirror
   (r) Zoom
   (s) Scale
   (t) Fillet
   (u) Chamfer
   (v) Offset
   (w) Divide
   (x) Bhatch
   (y) Array
   (z) Break & Trim

SOLUTIONS TO EXERCISES

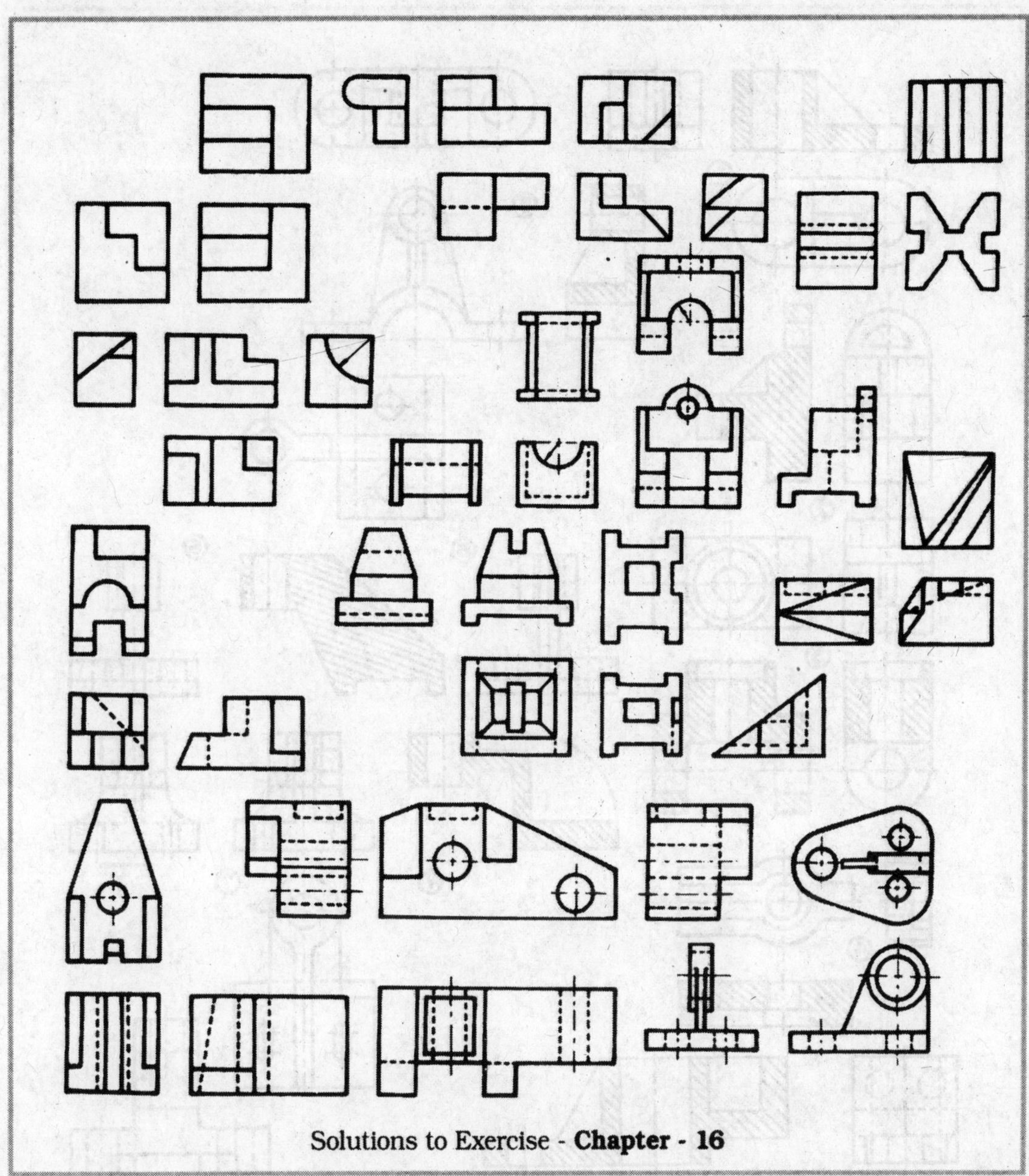

Solutions to Exercise - **Chapter - 16**

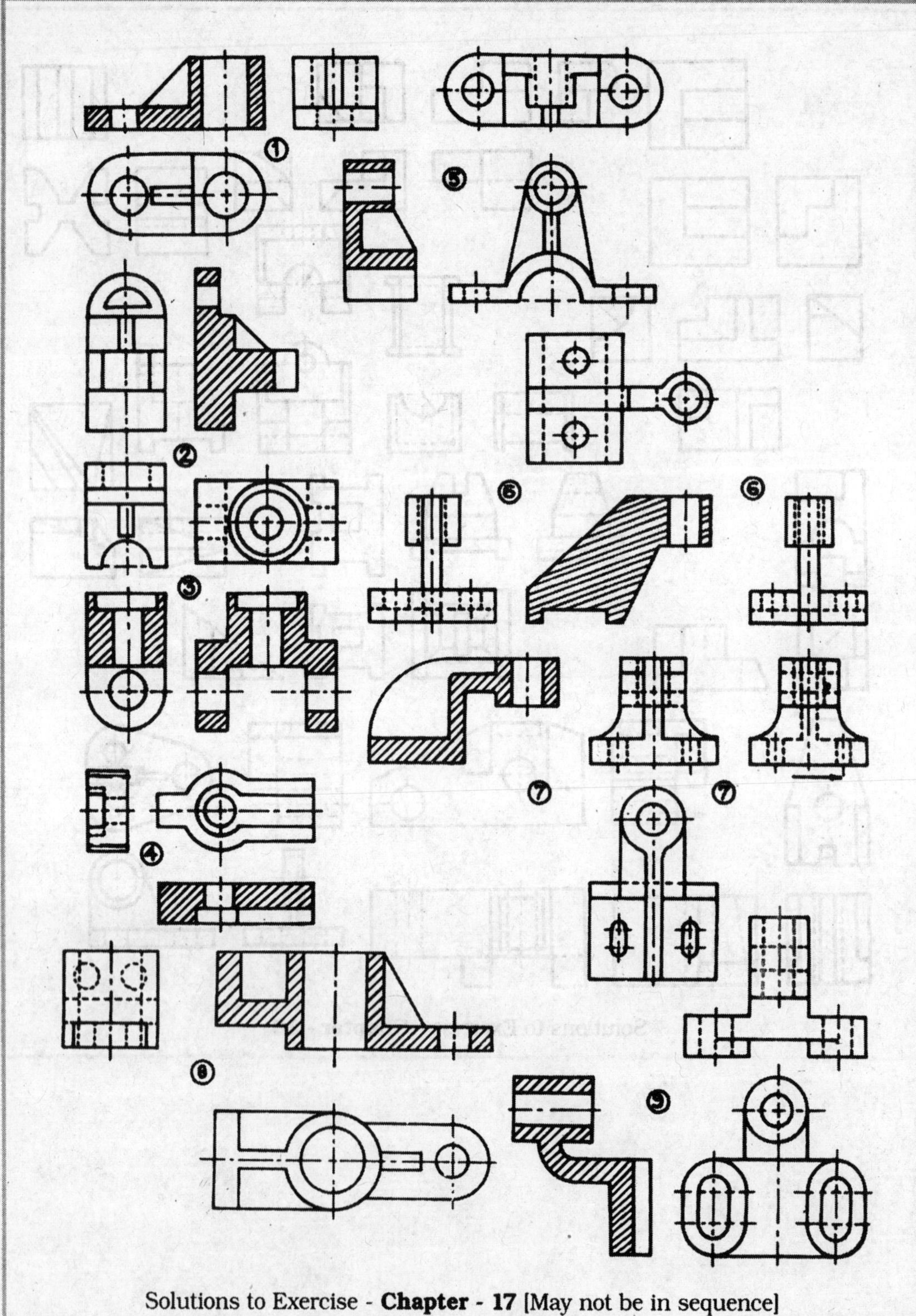

Solutions to Exercise - **Chapter - 17** [May not be in sequence]

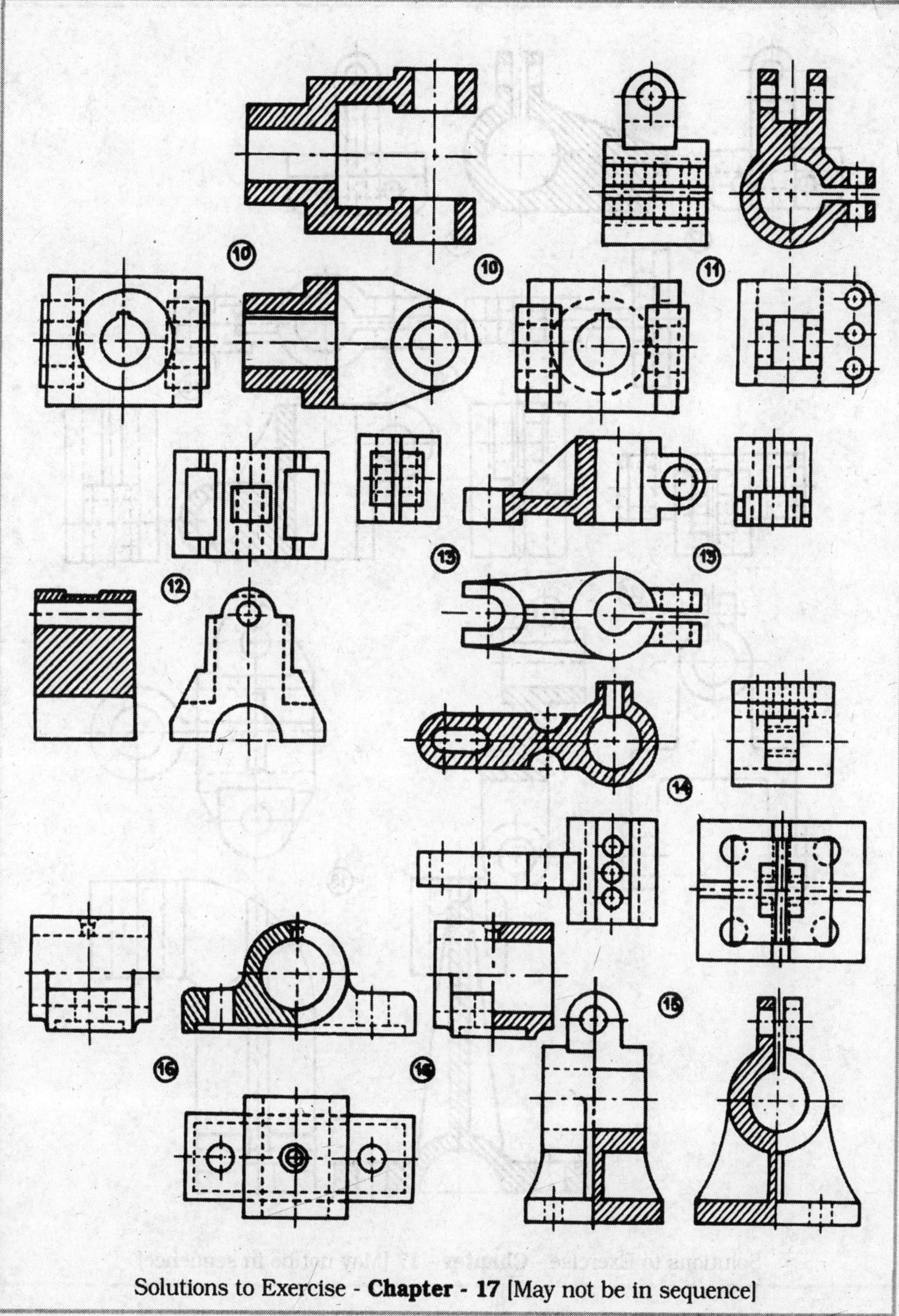

Solutions to Exercise - **Chapter - 17** [May not be in sequence]

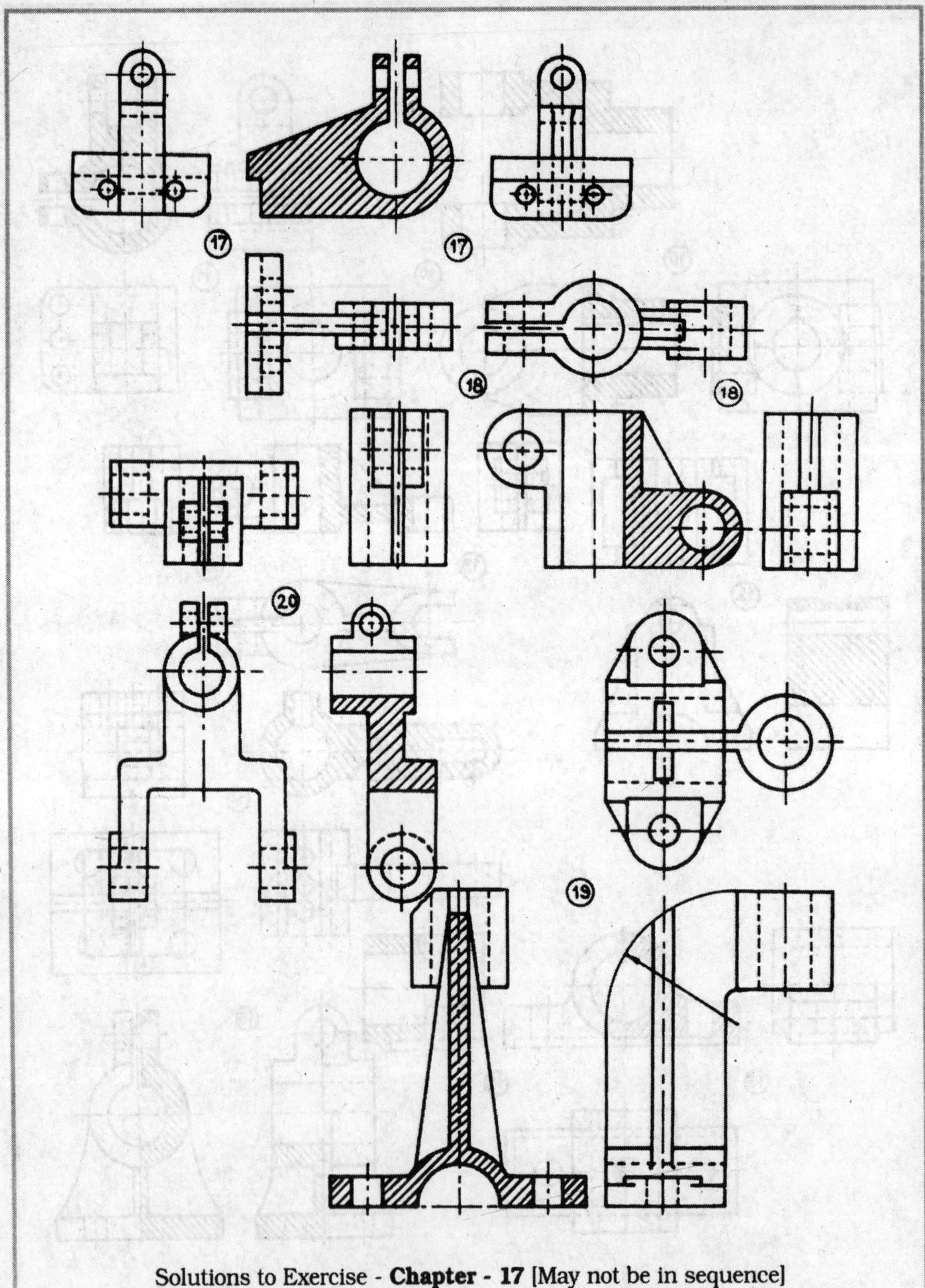

Solutions to Exercise - **Chapter - 17** [May not be in sequence]

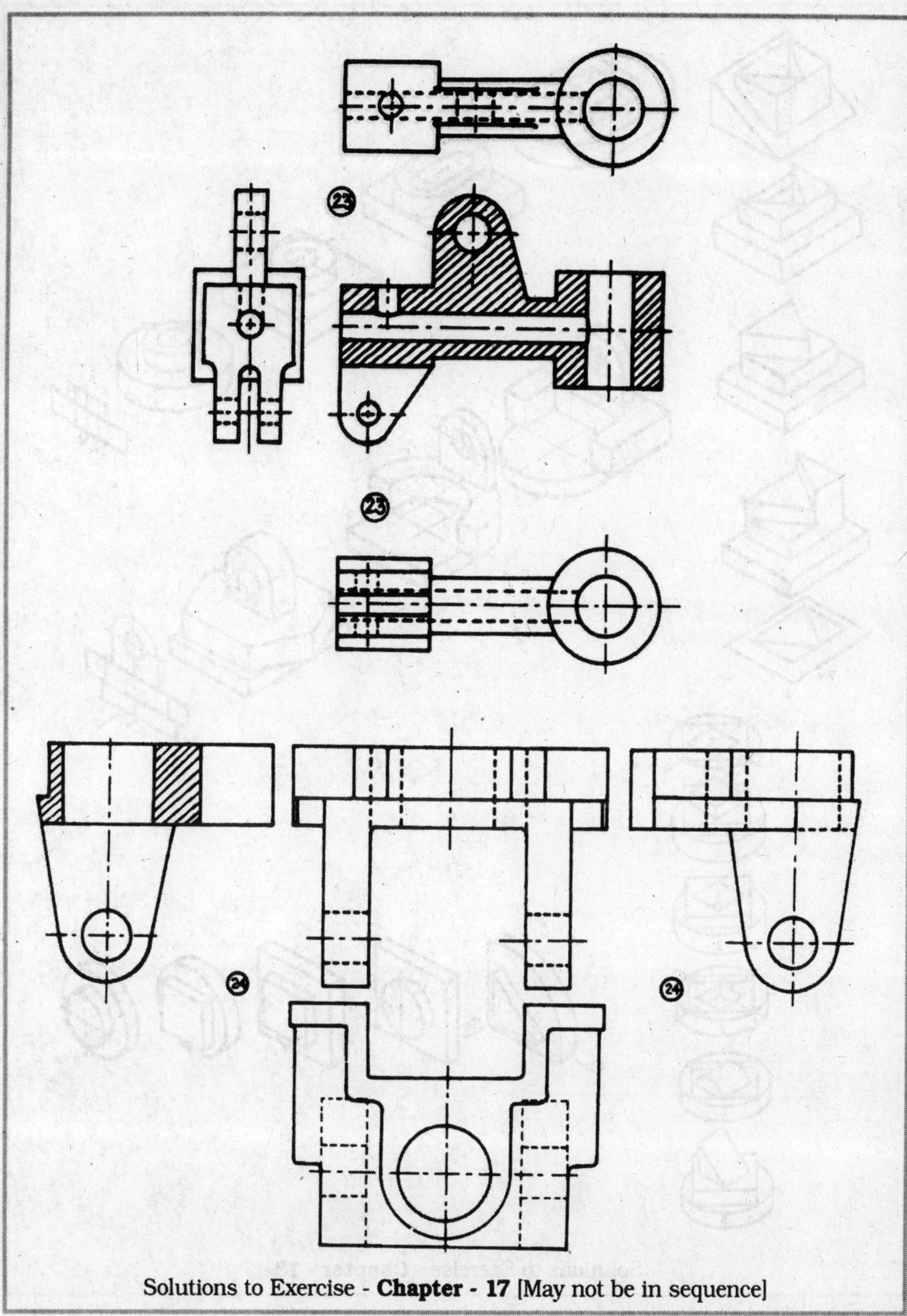

Solutions to Exercise - **Chapter - 17** [May not be in sequence]

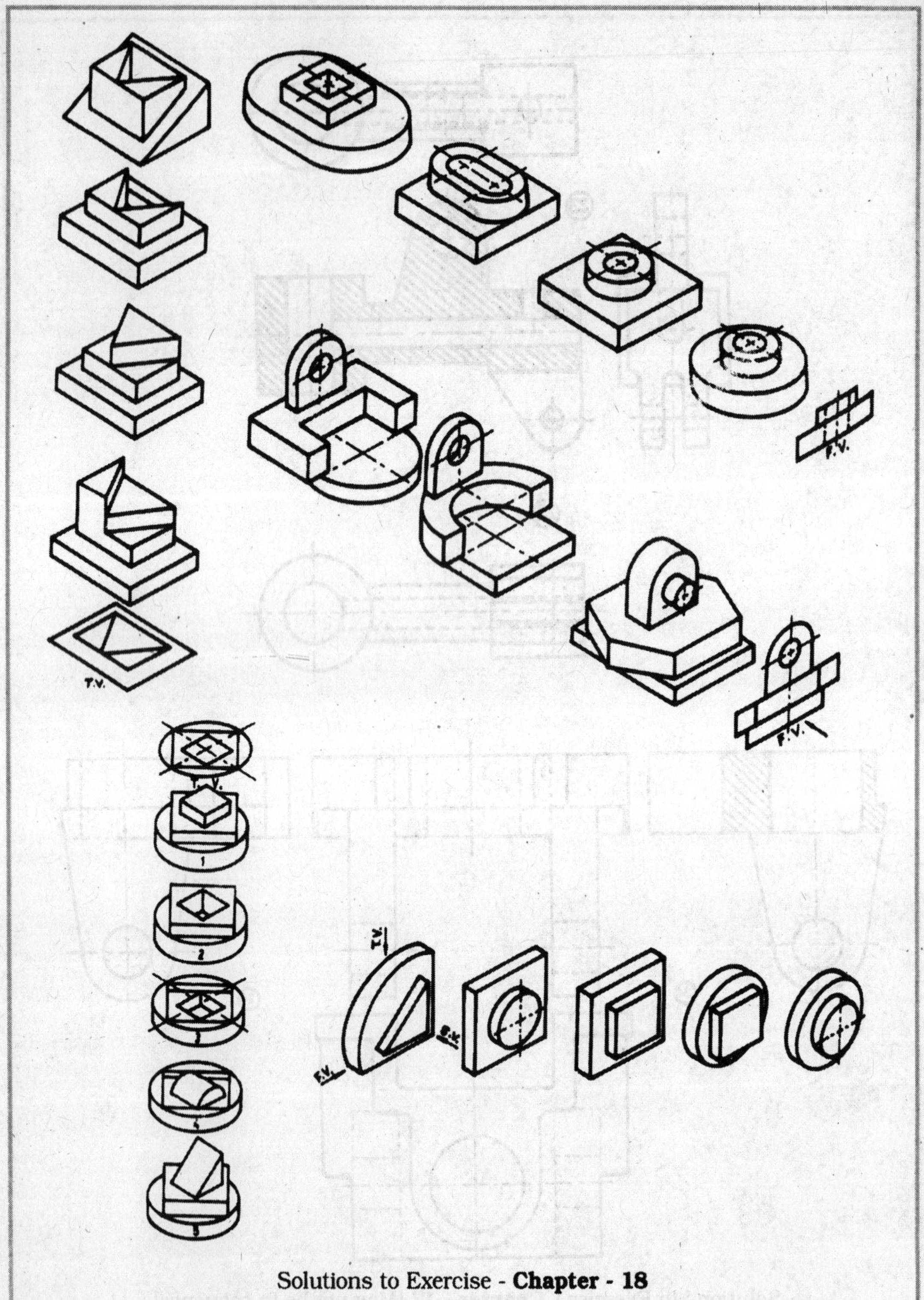

Solutions to Exercise - **Chapter - 18**

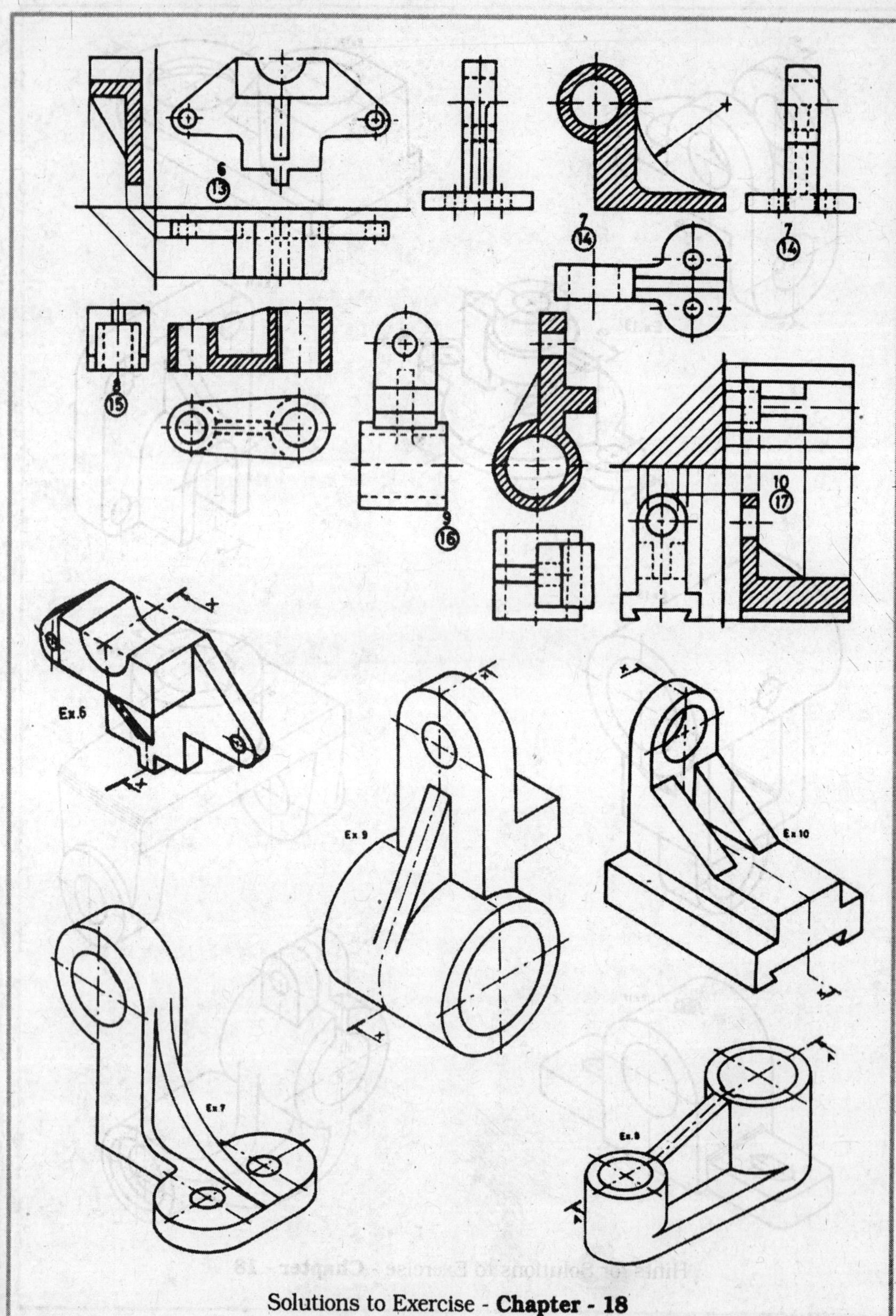

Solutions to Exercise - **Chapter - 18**

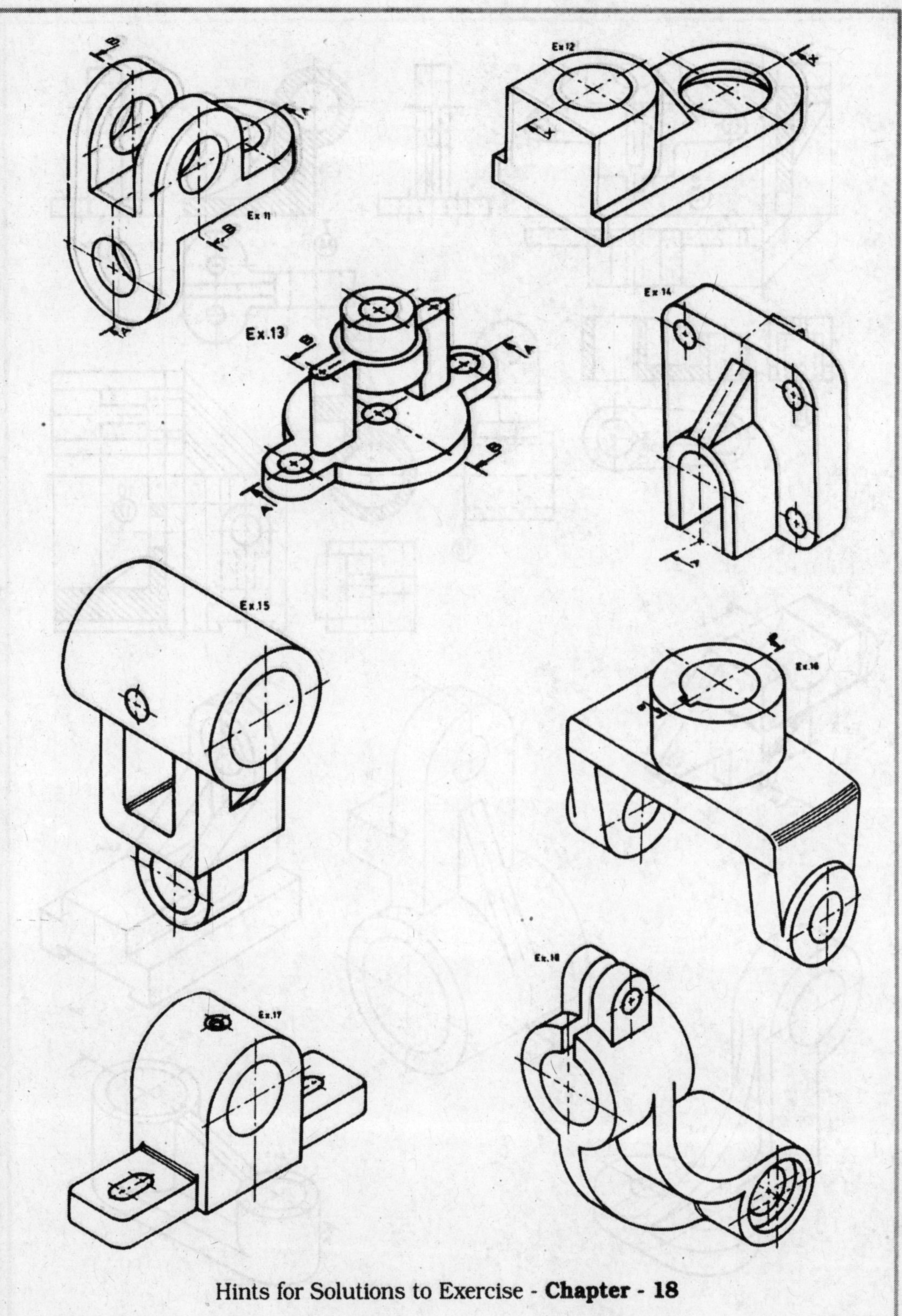

Hints for Solutions to Exercise - **Chapter - 18**

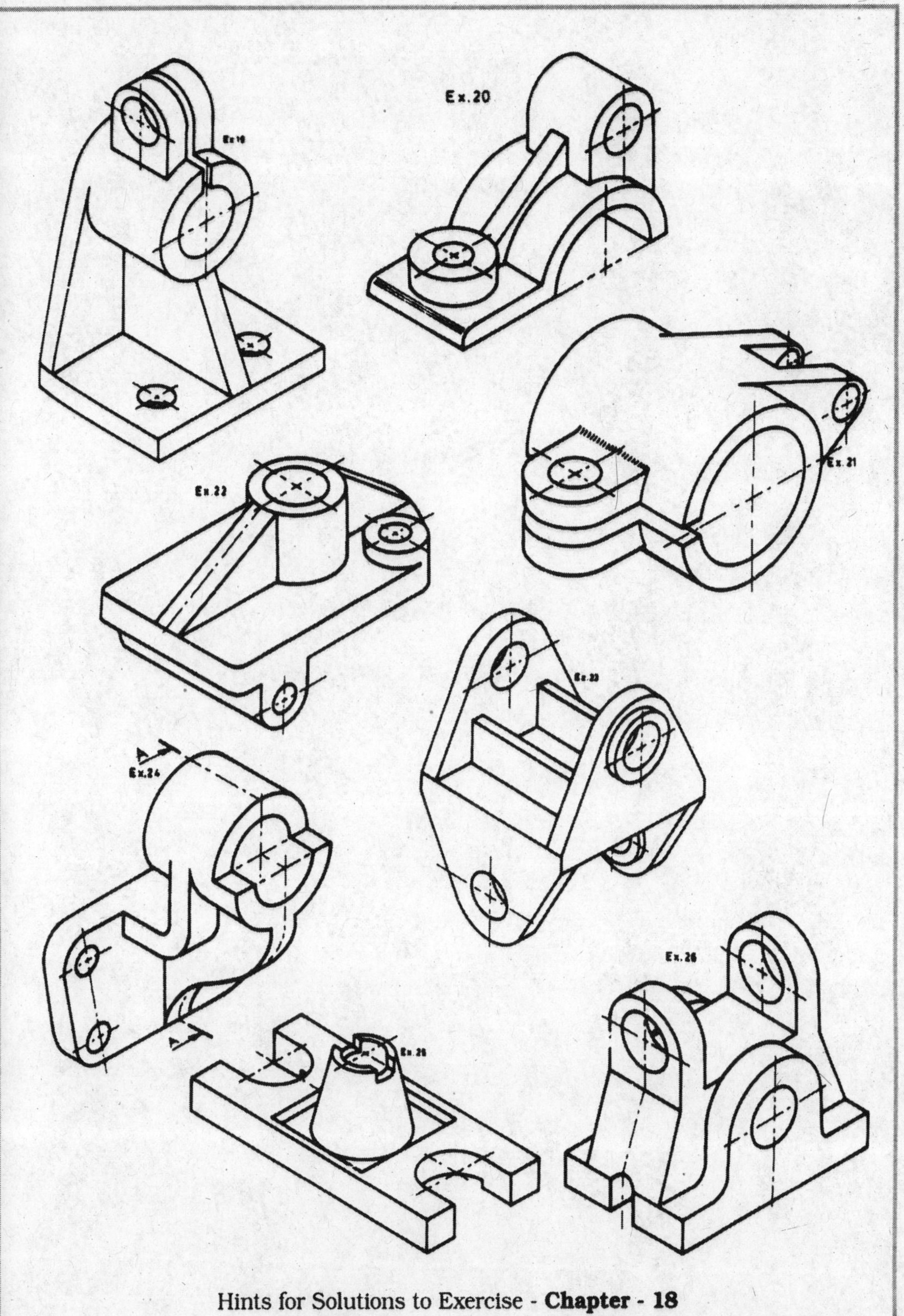

Hints for Solutions to Exercise - **Chapter - 18**